P9-CCC-505

*Hans-Georg Elias*
**Macromolecules**

*Macromolecules.* Hans-Georg Elias.

ISBN: 978-3-527-31173-6

## 1807–2007 Knowledge for Generations

Each generation has its unique needs and aspirations. When Charles Wiley first opened his small printing shop in lower Manhattan in 1807, it was a generation of boundless potential searching for an identity. And we were there, helping to define a new American literary tradition. Over half a century later, in the midst of the Second Industrial Revolution, it was a generation focused on building the future. Once again, we were there, supplying the critical scientific, technical, and engineering knowledge that helped frame the world. Throughout the 20th Century, and into the new millennium, nations began to reach out beyond their own borders and a new international community was born. Wiley was there, expanding its operations around the world to enable a global exchange of ideas, opinions, and know-how.

For 200 years, Wiley has been an integral part of each generation's journey, enabling the flow of information and understanding necessary to meet their needs and fulfill their aspirations. Today, bold new technologies are changing the way we live and learn. Wiley will be there, providing you the must-have knowledge you need to imagine new worlds, new possibilities, and new opportunities.

Generations come and go, but you can always count on Wiley to provide you the knowledge you need, when and where you need it!

*William J. Pesce*
President and Chief Executive Officer

*Peter Booth Wiley*
Chairman of the Board

*Hans-Georg Elias*

# Macromolecules

Volume 2: Industrial Polymers and Syntheses

WILEY-VCH Verlag GmbH & Co. KGaA

**The Author**

*Prof. Dr. Hans-Georg Elias*

Michigan Molecular Institute
1910 West St. Andrews Road
Midland, Michigan 48540
USA

| | |
|---|---|
| Volume 1 | Chemical Structures and Syntheses (Published in 2005) |
| Volume 2 | Industrial Polymers and Syntheses |
| Volume 3 | Physical Structures and Properties |
| Volume 4 | Processing and Application of Polymers |



**Library of Congress Card No.:**
applied for

**British Library Cataloguing-in-Publication Data**
A catalogue record for this book is available from the British Library.

**Bibliographic information published by the Deutsche Nationalbibliothek**
The Deutsche Nationalbibliothek lists this publication in the Deutsche Nationalbibliografie; detailed bibliographic data are available in the Internet at http://dnb.d-nb.de.



Printed in the Federal Republic of Germany
Printed on acid-free paper

**Printing** Strauss GmbH, Mörlenbach
**Binding** Litges & Dopf Buchbinderei, Heppenheim
**Cover Design** Gunther Schulz, Fußgönheim

**ISBN:** 978-3-527-31173-6

The polymers, those giant molecules,
like starch and polyoxymethylene,
flesh out, as protein serfs and plastic fools,
the Kingdom with life's stuff.

John Updike
The Dance of the Solids
("Midpoint and Other Poems"
A. Knopf, New York 1968)

# Preface

The preceding Volume I of the 4-volume work "Macromolecules" discusses chemical structures and syntheses of natural and synthetic macromolecules. It is mainly concerned with the principles of structures and polymerizations which by necessity involves ideal and idealized structures and processes. Industrial polymerizations are rarely ideal, however. Monomers are usually not pure by laboratory standards and reactions are often neither isothermal nor homogeneous. Variations in the type of initiator or catalyst, in the nature of the process (homogeneous versus heterogeneous, continuous versus discontinuous), deliberately introduced side reactions, or small proportions of comonomers produce very different *grades* of polymers from the same monomer.

The present Volume II is thus concerned with what can loosely be called the *chemical technology* of polymers. Like Volume I, it is based on a volume of the 6th German edition of this work, i.e., Volume III (2001). It is not a cover-to-cover translation, though, but an updated and expanded version as attested to by the size (635 versus 566 pages) and the fact that no less than 43 % of all of the figures are new.

Commodity polymers such as poly(ethylene), poly(propylene), poly(styrene), poly(vinyl chloride), poly(hexamethylene adipamide), etc., result from a few types of monomers that are especially simple to obtain from intermediates and raw materials. On the other hand, very many different types of monomers are used not only for specialty plastics, fibers, and rubbers but also for other polymer applications such as thickeners, in optoelectronics, for surface treatments, and the like. Information about such monomers and polymers can rarely be found in standard textbooks of polymer science. Monographs about specific polymers, on the other hand, usually concentrate on applications but not on syntheses, structures, and fundamental properties. The bulk of the chapters of this book thus discusses the *chemical technology* of industrially produced and/or used polymers, both synthetic ones and those based on natural sources.

Chemical technology is concerned with the delicate and complex interplay between the availability and chemistry of raw materials, base chemicals, intermediates, monomers, and polymers, the chemistry and technology of polymerization processes and plants, the properties and use of polymers, and the influence of economics, politics, and environmental concerns.

Chapter 1 of this book introduces the field. Chapter 2 reviews very briefly some of the chemical and physical polymeric terms that are used in later chapters of this volume. Although these aspects are discussed in great detail in the other three volumes of this series, it was deemed necessary to include such a chapter in order to make the book self-contained as much as possible.

Chapter 3 discusses raw materials for polymers (coal, natural gas, crude oil, wood, etc.) as well as their competing uses as energy sources, including economic and environmental aspects. Chapter 4 briefly outlines the sources of intermediates and monomers for polymers without detailing the chemistry of their syntheses, which is beyond the scope of this book.

Chapter 5 reviews basic polymerization mechanisms, processes, and techniques as well as polymerization reactors and their influence on polymer properties. The book does not venture into chemical engineering proper: hardware is treated in only a cursory manner and software such as process modeling and scale-up not at all.

Chapters 6–12 constitute the bulk of the book since they survey industrial syntheses and basic properties of the industrially and scientifically most important polymers. Each chapter comprises polymers of the same basic chain structure which usually coincides with an organization according to their syntheses and/or their intermediates and raw materials. Chapter 6 is therefore concerned with carbon-chain polymers, Chapter 7 with carbon-oxygen chains, Chapter 8 with polysaccharides, Chapter 9 with carbon-sulfur chains, Chapter 10 with carbon-nitrogen chains, Chapter 11 with peptides and proteins, and Chapter 12 with inorganic and semi-inorganic polymers.

Even a book of this size cannot discuss *all* polymers. It concentrates therefore on major industrial commodity and specialty polymers, including abandoned ones if they teach a lesson. Also included are some polymers that are not yet industrial but are interesting because their structures, syntheses, and/or properties are often not easy to generalize. "Polymer" is used in the widest meaning: it includes many biopolymers as well as several inorganic polymers that are of great economic importance but are usually not treated in polymer textbooks.

The book concentrates on industrial syntheses; physical properties and applications of resulting polymers are discussed in a qualitative manner but not in great detail (see Volume III for principles of general physical properties and structure–property relationships). This book does describe average properties of unmodified standard grades before processing, however; these data were taken from handbooks or company literature.

Compounding and processing of polymers to plastics, elastomers, fibers, coatings, etc., is not discussed here but in Volume IV of this work. Volume IV also contains data on filled and reinforced polymers as well as adjuvants for polymers.

Volume II is not an encyclopedia. Space requirements and the impossibility of reporting all available information restricts the treatment of individual polymers and processes to exemplary data. A complete list of all polymer properties, process variations, and applications is neither possible nor attempted. The volume should however provide a glimpse at the many possible syntheses of base chemicals, intermediates, monomers, and polymers and the almost infinite variations in the structures and properties of the resulting polymers. The content of Volume II is therefore much more descriptive and less quantitative than that of Volumes I (fundamental chemistry) and III (physical chemistry and physics).

The whole "Macromolecules" series can probably be best described as an expanded textbook: the four volumes are not monographs and therefore also not reference books. More in-depth information can be found in the extensive lists of monographs and review articles which includes some older books that I used as sources of information for historic data. Primary publications are usually too specialized for the scope of this work; they are therefore only cited as sources of graphs and tables.

I am again indebted to my good friends and former colleagues at Michigan Molecular Institute, Professors Petar R. Dvornic and Steven E. Keinath, who read and checked the final draft of all chapters and made many helpful suggestions. My profound thanks go also to Dr. Friedrich Schierbaum, Potsdam, for the fast and detailed help with the section "Starch."

Midland, Michigan (Summer 2006)

Hans-Georg Elias

# List of Symbols

Symbols for physical quantities follow the recommendations of the International Union of Pure and Applied Chemistry (IUPAC), symbols for physical units those of the International Standardization Organization (ISO). Exceptions are indicated.

I.Mills, T.Cvitas, K.Homann, N.Kallay, K.Kuchitsu, Eds., (International Union of Pure and Applied Chemistry, Division of Physical Chemistry), "Quantities, Units and Symbols in Physical Chemistry", Blackwell Scientific Publications, Oxford 1988.

### Symbols for Languages

D = German (deutsch), F = French, G = (classic) Greek, L = (classic) Latin.
The Greek letter υ (upsilon) was transliterated as "y" (instead of the customary phonetic "u") in order to make an easier connection to written English (example: πολυς = polys (many)). For the same reason, $\chi$ was transliterated as "ch" and not as "kh."

### Symbols for Chemical Structures

A: symbol for a monomer or a leaving group (polycondensations)
B: symbol for a monomer or a leaving group (polycondensations)
L: symbol for a leaving molecule, for example, $H_2O$ from the reaction of –COOH + HO–
R: symbol for a monovalent substituent, for example, $CH_3$– or $C_6H_5$–
Z: symbol for a divalent unit, for example, –$CH_2$– or –*p*-$C_6H_4$–
Y: symbol for a trivalent unit, for example, –C(R)< or –N<
X: symbol for a tetravalent unit, for example, >C< or >Si<
*: symbol for an active site: radical ($^{\bullet}$), anion ($^{\ominus}$), cation ($^{\oplus}$)
*p*Ph *para*-phenylene (in text)
*p*-$C_6H_4$ *para*-phenylene (in line formulas)

### Prefixes of Words (in systematic polymer names in *italics*)

alt alternating
at atactic
blend polymer blend
block block (large constitutionally uniform segment)
br branched. IUPAC recommends sh-branch = short chain branch, l-branch = long chain branch, f-branch = branched with a branching point of functionality *f*
co joint (unspecified)
comb comb
compl polymer–polymer complex
cyclo cyclic
ct cis-tactic
eit erythrodiisotactic
g graft
ht heterotactic
ipn interpenetrating network

it isotactic
net network; μ-net = micro network
per periodic
r random (Bernoulli distribution)
sipn semi-interpenetrating network
star star-like; f-star, if the functionality *f* is known; f is then a number
st syndiotactic
stat statistical (unspecified distribution)
tit threodiisotactic
tt trans-tactic

**Quantity Symbols** (unit symbols, see Chapter 13, Appendix)

Quantity symbols follow in general the recommendations of IUPAC.

[C] amount-of-substance concentration of substance C = amount of substance C per total volume = "mole concentration of C"
$c$ concentration = mass concentration (= mass-of-substance per total volume) = "weight concentration." IUPAC calls this quantity "mass density" (quantity symbol $\rho$). The quantity symbol $c$ has, however, traditionally been used for a special case of mass concentration, i.e., mass-of-substance per volume of solution and the quantity symbol $\rho$ for another special case, the mass density ("density") = mass-of-substance per volume of substance
$E$ energy
$f$ fraction (unspecified), see also $x$, $w$, $\phi$, etc.
$f$ functionality
$G$ Gibbs energy ($G = H - TS$); formerly: free enthalpy
$H$ enthalpy; $\Delta H_{mix}$ = enthalpy of mixing, $\Delta H_{mix,m}$ = molar enthalpy of mixing
$i$ variable ($i$th component, etc.)
$K$ general constant; $K_n$ = equilibrium constant
$k$ rate constant (always with index); $k_i$ = rate constant of initiation; $k_p$ = rate constant of propagation, $k_t$ = rate constant of termination, $k_{tr}$ = rate constant of transfer
$k_B$ Boltzmann constant ($k_B = R/N_A = 1.380\ 658 \cdot 10^{-23}$ J K$^{-1}$)
$L$ length (always geometric)
$M$ molecular weight. In this volume used instead of the correct symbol $M_r$ (= relative molar mass with physical unit 1). Note that most physical methods measure molar masses (physical unit: mass per amount-of-substance) and not molecular weights. Molar mass and molecular weight are numerically identical if the molar mass is given in g/mol. $\overline{M}_n$ = number-average molecular weight; $\overline{M}_w$ = weight-average molecular weight
$m$ mass
$N$ number of entities
$N_A$ Avogadro constant ($N_A = 6.022\ 136\ 7 \cdot 10^{23}$ mol$^{-1}$)
$n$ amount of substance (in mol); formerly: mole number
$p$ extent of reaction (fractional conversion); $p_A$ = extent of reaction of A groups
$Q$ polymolecularity index (= "polydispersity index"), e.g., $Q = \overline{M}_w/\overline{M}_n$

| | |
|---|---|
| $R$ | molar gas constant ($R \approx 8.314\ 510$ J K$^{-1}$ mol$^{-1}$) |
| $R$ | rate of reaction |
| $r$ | copolymerization parameter |
| $r_0$ | initial ratio of amounts of substances in copolymerizations |
| $S$ | entropy |
| $T$ | temperature (always with units). In physical equations always as thermodynamic temperature with unit kelvin; in descriptions, either as thermodynamic temperature (unit: kelvin) or as Celsius temperature (unit: degree Celsius). Mix-ups can be ruled out because the physical unit is always given. IUPAC recommends for the Celsius temperature either $t$ as a quantity symbol (which can be confused with $t$ for time) or $\theta$ (which can be confused with $\Theta$ for the theta temperature). $T_c$ = ceiling temperature, $T_G$ = glass temperature, $T_M$ = melting temperature |
| $t$ | time |
| $u$ | fractional conversion of monomer molecules ($p$ = fractional conversion of groups; $y$ = yield of substance) |
| $V$ | volume |
| $w$ | mass fraction = weight fraction |
| $X$ | degree of polymerization of a molecule with respect to monomeric units (not to repeating units); $\overline{X}_n$ = number-average degree of polymerization of a substance; $\overline{X}_w$ = mass-average degree of polymerization of a substance |
| $x$ | mole fraction (amount-of-substance fraction); $x_u$ = mole fractions of units, $x_i$ = mole fraction of isotactic diads, $x_{ii}$ = mole fraction of isotactic triads, etc. |
| $x_{br}$ | degree of branching |
| $y$ | yield of substance |
| $z$ | coordination number, number of neighbors |
| $\varepsilon_r$ | relative permittivity (formerly: dielectric constant) |
| $\eta$ | dynamic viscosity, e.g., $\eta_0$ = viscosity at rest (Newtonian viscosity), $\eta_1$ = viscosity of solvent |

| | | | |
|---|---|---|---|
| $\eta_r$ | $= \eta/\eta_1$ | = relative viscosity, | |
| $\eta_i$ | $= (\eta - \eta_1)/\eta_1$ | = relative visc. increment | (= specific viscosity, $\eta_{sp}$), |
| $\eta_{inh}$ | $= (\ln \eta_r)/c$ | = inherent viscosity | (= logarithmic visc. number), |
| $\eta_{red}$ | $= (\eta - \eta_1)/(\eta_1 c)$ | = reduced viscosity | (= viscosity number, $\eta_{sp}/c$), |
| $[\eta]$ | $= \lim \eta_{red,c \to 0}$ | = limiting visc. number | (= intrinsic viscosity) |

| | |
|---|---|
| $\lambda$ | wavelength ($\lambda_0$ = wavelength of incident light) |
| $\nu$ | kinetic chain length |
| $\nu$ | frequency |
| $\pi$ | mathematical constant pi |
| $\rho$ | density (= mass/volume of the same matter) |
| $\phi$ | volume fraction; $\phi_f$ = free volume fraction |
| $\phi$ | angle |

# Table of Contents

# 1 Introduction

Polymer technology comprises industrial syntheses of polymers, i.e., the chemical technology (this volume), and their commercial and military applications, i.e., mainly physical technology (Volume IV). Polymers may be natural products, semisynthetics from natural products, or fully synthetic substances. They are used as working materials, for clothing, as information carriers, and for a great number of other applications. Polymer technology in a wider sense comprises other applications of polymers, for example, in the food industry and in medical technology. By definition, functions of biological macromolecules (nucleic acids, proteins, polysaccharides, etc.) in living beings are excluded except those that are prepared and applied by humans for specific therapeutic purposes (blood plasma expanders, etc.).

Polymers are *substances* composed of polymer molecules (macromolecules) of high molecular weight (for the terminology, see Chapter 2). Macromolecules are composed of thousands (and sometimes millions) of covalent connected atoms; electron-deficient and coordination bonds are relatively rare. No sharp dividing line exists between high-molecular weight and low-molecular weight compounds.

Most industrially synthesized and used macromolecules are organic molecules, i.e., molecules that consist exclusively or predominantly of carbon and hydrogen atoms and, in some cases, also of some oxygen, nitrogen, and sulfur atoms, and, in rare cases, also of phosphorus atoms. Examples are poly(ethylene)s with the idealized chemical constitution $+CH_2-CH_2+_n$, poly(oxymethylene), $+O-CH_2+_n$, nylon 6, $+NH(CH_2)_5CO+_n$, and poly(phenylene sulfide), $+S-(1,4-C_6H_4)+_n$. Semiinorganic polymers are less often used; the most prominent example is poly(dimethylsiloxane), $+O-Si(CH_3)_2+_n$. These macromolecules are "linear" (i.e., unbranched) or slightly branched (Chapter 2); some of them are crosslinked during processing to insoluble three-dimensional networks. Most completely inorganic macromolecules, on the other hand, exist only as highly ordered three-dimensional networks. They are used as such (stones, etc.) or as ceramic masses.

Syntheses and chemical reactions of macromolecules are investigated by macromolecular chemistry (Volume I), a relatively young branch of science that owes its existence to the *finding* that many organic colloids such as natural rubber, cellulose, proteins, and "metastyrene" are composed of macromolecules and not of aggregates of small molecules. Empirical applications of macromolecular substances date back many thousands of years, however. They were based on the *knowledge* of the existence and behavior of naturally occurring macromolecules that could be exploited by mankind using empirical technologies. Examples are the use of timber as construction material, of wool from animals for clothing, of leather from animal skins for clothing and other purposes, of starch, collagen, and asphalt as glues, etc. Semisynthetic polymers based on natural polymers have been produced since ca. 1850, completely synthetic polymers since ca. 1930 (Table 1-1; see also see Volume I, Table 1-3).

The many atoms of a macromolecule, their interconnection by chemical bonds, and the many possible spatial arrangements of atoms (microconformations) in the resulting chains generate great numbers of different possible physical structures of single macromolecules (macroconformations) and assemblies of macromolecules (morphologies). The many physical structures from a single chemical structure allow the many different applications of a given polymer.

Table 1-1 Some early industrial polymers. * Not application oriented. ** No longer produced.

| Polymers | Discovery | Production | Main applications |
|---|---|---|---|
| Elastomers | | | |
| From natural products | | | |
| Natural rubber (vulcanization) | 1839 | 1851 | tires and other rubber articles |
| Chlorinated rubber | 1859* | 1928 | industrial rubber articles |
| Synthetic polymers | | | |
| Poly(butadiene), random | 1911 | 1929 | elastomers (number Buna rubbers) |
| Poly(2-chlorobutadiene) | 1925 | 1929 | industrial rubber articles |
| Polysulfide rubbers | 1926 | 1929 | elastomers, sealants |
| Poly(isobutylene) | 1934 | 1937 | elastomers (inner tubes of tires) |
| Styrene-butadiene rubbers | 1926 | 1937 | elastomers (letter Buna rubbers) |
| Polysulfide rubbers | 1926 | 1939 | elastomers (Thiokol®) |
| Poly(isoprene) | 1879* | 1955 | tires |
| Poly(butadiene), *cis*-1,4 | | 1956 | elastomers ** |
| Poly(isoprene), *cis*-1,4 | 1909 | 1958 | elastomers ** |
| SBS triblock copolymers | | 1965 | thermoplastic elastomers |
| Fibers | | | |
| From natural products | | | |
| Cellulose (mercerization) | 1844 | | cotton fibers |
| Cellulose (nitration) | 1846* | 1889 | gun cotton |
| Cellulose (2 1/2) acetate | 1865* | 1927 | acetate fibers, cigarette filters |
| Cellulose (via $Cu^{2+}/NH_3$) | 1890 | 1900 | cuprammonium rayon (copper silk) |
| Cellulose (via xanthate) | 1892 | 1899 | rayon fibers (viscose silk) |
| Synthetic polymers | | | |
| Poly(vinyl chloride) | 1838* | 1934 | Pe-Ce fiber** (1938: theromoplastics) |
| Polyamide 66 | 1934 | 1938 | nylon fibers (1941: thermoplastics) |
| Polyamide 6 | 1938 | 1939 | Perlon® fibers, thermoplastics |
| Poly(acrylonitrile) | 1894* | 1942 | acrylic fibers |
| Poly(ethylene terephthalate) | 1941 | 1953 | polyester fibers (later: thermoplastics) |
| Polyamides, aromatic | | 1961 | high-modulus fibers |
| Thermoplastics | | | |
| From natural products | | | |
| Cellulose nitrate | 1846* | | gun cotton (1921: nitro lacquers) |
| Cellulose nitrate + camphor | 1865 | 1869 | spectacle frames (1883: cast films) |
| Cellulose acetate | 1865* | 1927 | photographic films |
| Synthetic polymers | | | |
| Poly(methyl methacrylate) | 1880* | 1928 | thermoplastics (organic glass) |
| Poly(styrene) | 1839* | 1930 | foams, small containers, etc. |
| Poly(vinyl chloride) | 1838* | 1938 | thermoplastics |
| Poly(vinylidene chloride) | 1838* | 1939 | packaging films |
| Vinylidene chloride copolymers | 1838 | 1939 | thermoplastics (packaging films) |
| Poly(ethylene), low-density | 1932 | 1939 | packaging films, bottles, etc. |
| Poly(tetrafluoroethylene) | 1939 | 1950 | plastics, fibers |
| ABS polymers | | 1954 | thermoplastics |
| Poly(ethylene), high-density | 1953 | 1955 | thermoplastics, plastic foams |
| Polycarbonate, bisphenol A- | 1953 | 1959 | films, sheets, etc. |
| Poly(oxymethylene) | 1859 | 1959 | thermoplastics |
| Poly(propylene), isotactic | 1954 | 1959 | thermoplastics, fibers |
| Thermosets | | | |
| From natural products | | | |
| Natural rubber + sulfur | 1851 | | ebonite (working material) |
| Casein + formaldehyde | 1897 | 1904 | Galalith® (haberdashery) |

Table 1-1 (continued)

| Polymers | Discovery | Production | Main applications |
|---|---|---|---|
| Thermosets | | | |
| Synthetic polymers | | | |
| Phenol + formaldehyde | 1906 | 1909 | electric insulation (1969: fibers) |
| Alkyd resins | 1847 | 1926 | coatings |
| Amino resins | 1904 | 1928 | thermosets |
| Unsaturated polyesters | 1930 | 1936 | thermosets |
| Polyurethanes | 1935 | 1940 | fibers**, foamed products, etc. |
| Silicones | 1901* | 1942 | fluids, elastomers, resins |
| Epoxy resins | 1939 | 1946 | adhesives, lacquers |
| Polyimides | 1945 | 1960 | high-temperature plastics |
| Thickeners, adhesives, etc. | | | |
| Poly(vinyl acetate) | 1912 | 1930 | adhesives, poly(vinyl alcohol) |
| Poly(ethylene oxide) | 1859 | 1931 | thickeners, sizes |
| Poly(vinyl ether)s | 1928 | 1936 | adhesives, plasticizers |
| Poly(*N*-vinyl pyrrolidone) | 1939 | 1939 | thickeners, flocculants, hair setting |
| Specialty polymers | | | |
| Poly(acetylene) | 1976 | | semiconductors |
| Dendrimers | 1982 | 1991 | diagnostics, additives for inks |

The great early time intervals between the discovery of early polymers on one hand and their industrial synthesis and application on the other hand were caused by several factors. The synthesis of many early polymers was more or less accidental (e.g., cellulose nitrate) and not systematic (e.g., poly(isoprene)). The usefulness of many of these polymers was not immediately recognized and it took 50–100 years until a market evolved.

Technically and/or economically directed syntheses need much shorter time spans of 1–15 years from the laboratory to industrial products. The times required for the introduction of new polymers is still great if industrial syntheses of monomers are not available or are too expensive.

An example is the synthesis of poly(ethylene terephthalate) from ethylene glycol and dimethyl terephthalate which was a laboratory curiosity at the time of the discovery of the polymer. It took 12 years from laboratory synthesis to industrial production of polyester fibers but many additional years to work out a synthesis of sufficiently pure terephthalic acid as a replacement for dimethyl terephthalate and again many years to produce so-called bottle-grade poly(ethylene terephthalate)s.

The conversion of laboratory polymerizations to industrially useful products is also delayed if no industrial process is available for the processing of the polymer or if energy requirements are too high. For example, poly(acrylonitrile) fibers could only be produced after *N*,*N*-dimethylformamide (DMF) was found and became industrially available as a solvent for fiber spinning. Syndiotactic poly(styrene) (sPS) failed commercially despite excellent material properties because it needs considerably more energy for processing than its atactic cousin (PS) (processing of sPS is at $T > T_M = 270°C$ whereas that of PS is at $T > T_G = 100°C$).

Some polymers have such extraordinary properties that price is not an issue. Examples are special polymers for certain medical, military, or space applications. In most cases, however, new polymers compete with either established polymers or with other materials. They will succeed only if they can be offered at a reasonable price per property and density (not per weight since they are applied by volume) and can be processed economically with existing equipment.

Production costs are mainly determined by the costs of capital, monomers, initiators, and solvents and the energy requirements for the polymerization and work-up. They decrease with the size of the production plant (see Section 5.5). The last decades saw additional environmental costs for the production of raw materials, energy, monomers, and polymers, the processing of polymers, and the disposal and recycling of old polymeric products. For example, increased environmental costs (especially the work-up of water) led to a decline in the production of rayon from wood in Western countries which was not compensated by new rayon plants in Asia.

The last 50 years saw a strong increase in the production of synthetic polymers (Table 1-2). It was mainly caused by four factors: growth of the world population from ca. 2.5 billion in 1950 to 6.5 billion in 2005, increases in living standards in Western countries and more recently also in some Asian countries, availability of relatively inexpensive raw materials such as crude oil and natural gas, and subsequent partial replacement of older materials by newer polymer-based ones. In the past, growth was mainly determined by technology and economics. In the future, it will also be determined by ecology and attempts to redistribute technologies and wealth (see Section 3.2.5).

Between 1950 and 2000, higher living standards caused an increase of the per capita consumption of fibers by a factor of 2.4, that of rubbers by a factor of 3.1, and that of plastics by a factor of 108. Great differences exist between different parts of the world. In highly developed countries such as the United States, annual plastics consumption per capita is ca. 120 kg, in sub-Saharan Africa only ca. 1 kg (world: 25 kg (2005)).

The world production of polymers can therefore be safely predicted to increase further from the present level (plastics and synthetic fibers in 2004: $250 \cdot 10^6$ t/a). There will be a shift in production sites, however. In the past, production of both bulk and specialty polymers took place predominantly in Western countries. Since patents have expired and know-how became more freely available, bulk polymers are now increasingly produced by Asian countries. Petroleum-exporting Arab countries are in the process of establishing a petrochemical industry that will produce not only intermediates but also hydrocarbon-based polymers. There may be also shifts in the raw material basis: from crude oil to natural gas, from petrochemicals to coal, and from fossil raw materials to a green technology based on bioethanol from sugar cane, grasses, and wood.

Table 1-2 World production of polymers for plastics, fibers, and rubbers in million tons per year.

| Type | | 1940 | 1950 | 1960 | 1970 | 1980 | 1990 | 2000 |
|---|---|---|---|---|---|---|---|---|
| Plastics | | 0.36 | 1.6 | 6.7 | 31 | 59 | 100 | 172 |
| Fibers, | synthetic | 0.005 | 0.069 | 0.70 | 5.0 | 11.5 | 15.7 | 31.7 |
| | regenerated natural (rayon) | 1.1 | 1.6 | 2.6 | 3.4 | 3.3 | 3.2 | 2.8 |
| | natural textile fibers | 8.7 | 8.0 | 12.8 | 14.0 | 17.7 | 21.0 | 21.1 |
| Rubbers, | synthetic | 0.043 | 0.54 | 1.94 | 5.9 | 8.7 | 9.9 | 10.8 |
| | natural | 1.44 | 1.89 | 2.02 | 3.1 | 3.9 | 5.2 | 7.3 |

## Literature to Chapter 1

ENCYCLOPEDIAS AND HANDBOOKS

Houben-Weyl, Methoden der organischen Chemie, 4th ed. (E.Müller, Ed.), vols. XIV/1 (1962), Makromolekulare Stoffe, and XIV/2 (1963), Makromolekulare Stoffe; 5th ed., vols. XX-A, XX-B, and XX-C (1987); G.Thieme, Stuttgart

H.Mark, N.G.Gaylord, N.M.Bikales, Eds., Encyclopedia of Polymer Science and Technology, Wiley, New York, 1966-1976 (16 vols. and 2 supplemental vols.)

H.Mark, C.Overberger, G.Menges, N.M.Bikales, Eds., Encyclopedia of Polymer Science and Engineering, 2nd ed., Wiley, New York 1985-1990 (19 vols. + 2 supplemental vols.). Short version: J.I.Kroschwitz, Ed., Concise Encyclopedia of Polymer Science and Engineering, Wiley, NewYork 1990

W.A.Kargin, Ed., Enciclopedia Polimerov, Sovietskaya Enciklopedia Publ., Moscow 1972 (3 vols.)

Ullmann's Encyclopedia of Industrial Chemistry, VCH, Weinheim, 5th ed., vols. A1-A28 (1985-1996), vols. B1-B8 (1990-1995). -, Wiley-VCH, 7th ed. 2005

Kirk-Othmer, Encyclopedia of Chemical Technology, Wiley, New York, 4th ed., 27 vols. + 1 supplemental volume + 1 index volume (1991-1998). Short version: Concise Encyclopedia of Chemical Technology, Wiley, New York 1999. –, –, Wiley, Hoboken (NJ), 5th ed. 2004 ff.

G.Allen, J.C.Bevington, Eds., Comprehensive Polymer Science, Pergamon, Oxford, 7 vols. (1989), First Supplement 1992, Second Supplement (1996)

J.E.Mark, Ed., Physical Properties of Polymers Handbook, AIP Press, Williston (VT) 1996 (contains data collections)

J.C.Salamone, Ed., Polymeric Materials Encyclopedia, CRC Press, Boca Raton (FL) 1996, 12 vols. or 1 CD-ROM. Short version: Concise Polymeric Materials Encyclopedia, CRC Press, Boca Raton (FL) 1999

E.S.Wilks, Ed., Industrial Polymers Handbook. Products, Processes, Applications, 4 vols., Wiley-VCH, Weinheim 2000

J.L.Atwood, J.W.Steed, Eds., Encyclopedia of Supramolecular Chemistry, Dekker, New York 2004

PROGRESS REPORTS

Modern Plastics Encyclopedia, McGraw-Hill, New York (annually in October)

H.-G.Elias, New Commercial Polymers 1969-1975, Gordon and Breach, New York 1977

H.-G.Elias, F.Vohwinkel, New Commercial Polymers 2, Gordon and Breach, New York 1986

THESAURI

The International Plastics Selector, Commercial Names and Sources, Cordura Publ., San Diego (CA) 1978

Parat (Fachinformationszentrum Chemie, Berlin), Index of Polymer Trade Names, VCH, Weinheim 1987

Gardner's Chemical Synonyms and Trade Names, Gower Publ., Brookfield (VT), 9. Aufl. 1987

DATA COLLECTIONS (see also COMPUTER READABLE DATABASES)

J.Brandrup, E.H.Immergut, Eds., Polymer Handbook, Wiley, New York, 1st ed. (1966), 2nd ed. (1975), 3rd ed. (1989); J.Brandrup, E.H.Immergut, E.A.Grulke, Eds., 4th ed. (1999)

O.G.Lewis, Physical Constants of Linear Homopolymers, Springer, Berlin 1968

W.J.Roff, J.R.Scott, Handbook of Common Polymers, Butterworths, London 1971

R.E.Schramm, A.F.Clark, R.P.Reeds, Eds., A Compilation and Evaluation of Mechanical, Thermal, and Electrical Properties of Selected Polymers, U.S. National Bureau of Standards, Washington (DC) 1973

P.A.Schweitzer, Corrosion Resistance Tables (Metals, Plastics, Nonmetallics, Rubbers), Dekker and Hanser, Munich, 2nd ed. 1987

D.W. van Krevelen, Properties of Polymers - Correlation with Chemical Structure, Elsevier, Amsterdam, 2nd ed. 1976, 3rd ed. 1989

I.S.Grigoriev, E.Meilikhov, Handbook of Physical Quantities, CRC Press, Boca Raton (FL) 1997

N.A.Waterman, M.F.Ashby, Eds., The Materials Selector, Chapman & Hall, Boca Raton (FL) 1999

J.E.Mark, Ed., Polymer Data Handbook, Oxford University Press, New York 1999

-, The International Plastics Selector, Internat.Plast.Sel., San Diego (CA) 1977

Fachinformationszentrum Chemie, Berlin, Ed., Parat-Index of Polymer Trade Names, VCH, Weinheim, 2nd ed. 1992

H.Domininghaus, Plastics for Engineers, Hanser, Munich 1993
H.-J.Saechtling, International Plastics Handbook, Hanser, Munich, 3rd ed. 1995

COMPUTER READABLE DATABASES

J.E.Williams, Ed., Computer-Readable Databases, American Library Association, Chicago 1985
Deutsches Kunststoff-Institut, Darmstadt, and Fachinformationszentrum Chemie, Berlin:
POLYMAT: Database with 30-50 properties each of 13 000 types of plastics from 130 producers);
POLYTRADE: 4000 trade names of plastics from more than 950 producers;
POLYVOC: electronic dictionary with ca. 60 000 terms.
CAMPUS® (= Computer Aided Material Preselection by Uniform Standards)
(free discs with properties of plastics produced by 30 companies belonging to the CAMPUS system; about 50 properties per type, including stress-strain curves, etc.
Fachinformationszentrum Chemie, Berlin: PLASPEC (Database)
–, Plastics Databases, ASM International, Materials Park (OH)
J.E.Mark, Ed., Polymer Data Handbook, Oxford University Press, New York 1999
–, Polymers: A Property Database, CRC Press, Boca Raton (FL) 2006 (CD-ROM)

DICTIONARIES

PARAT (Fachinformationszentrum Chemie, Berlin), Index of Polymer Trade Names, VCH, Weinheim 1987
M.Ash, I.Ash, Encyclopedia of Plastics, Polymers and Resins, Chem.Publ.Co., New York 1982-1988 (4 vols.)
A.M.Wittfoht, Plastics Technical Dictionary, Hanser, Munich 1992
M.B.Ash, I.A.Ash, Handbook of Plastics Compounds, Elastomers, and Resins, An International Guide by Category, Tradename, Composition and Supplier, VCH, Weinheim 1992
R.J.Heath, A.W.Birley, Dictionary of Plastics Technology, Blackie and Son, Glasgow 1992
D.V.Rosato, Rosato's Plastics Encyclopedia and Dictionary, Hanser, Munich, 2nd ed. 1993
T.Whelan, Polymer Technology Dictionary, Chapman & Hall, London 1994
M.S.M.Alger, Polymer Science Dictionary, Chapman & Hall, London, 2nd ed. 1997

BIBLIOGRAPHIES

E.R.Yescombe, Sources of Information on the Rubber, Plastics and Allied Industries, Pergamon, Oxford 1968; Plastics and Rubbers: World Sources of Information, Appl.Sci.Publ., Barking (Essex) 2nd ed. 1976
G.J.Patterson, Plastics Book List, Technomic Publ., Westport (CT) 1975
O.A.Battista, The Polymer Index, McGraw-Hill, New York 1976
S.M.Kaback, Literature of Polymers, Encyclopedia of Polym.Sci.Technol., 1st ed., vol. **8** (1968) 273
J.T.Lee, Literature of Polymers, Encyclopedia of Polym.Sci.Eng., 2nd ed., vol. **9** (1987) 62
R.T.Adkins, Ed., Information Sources in Polymers and Plastics, K.G.Saur, New York 1989

NOMENCLATURE

International Union of Pure and Applied Chemistry, Macromolecular Division, Commission on Macromolecular Nomenclature, Compendium of Macromolecular Nomenclature, Blackwell Scientific, Oxford 1991; Glossary of Basic Terms in Polymer Science (Provisional Recommendations 1994), Pure Appl.Chem. **66** (1994) 2483

# 2 Structure and Properties of Polymers

## 2.1 Chemical Structure

### 2.1.1 Fundamental Terms

A **macromolecular substance** is composed of many **macromolecules**, i.e., molecules with large molecular masses (G: *macros* = large; L: *molecula* = small mass (diminutive of *moles*)). If all molecules of a macromolecular substance possess exactly the same chemical structure and thus the same molecular mass, then the substance has a well-defined **molar mass** $M$ with the physical unit mass per amount of substance, e.g., g/mol. Such a substance also has a well-defined **relative molar mass** with the physical unit of unity. This "dimensionless" relative mass is also called **molecular weight**.

The simplest macromolecules consist of long chains which, with respect to *chemical constitution*, are composed of **chain units, monomeric units, repeating units**, and **endgroups**. With respect to chemical synthesis, such macromolecules consist of **monomeric units (mers)** and endgroups, albeit of a different type than those of the constitutional formula. Endgroups of macromolecules are often unknown. They are therefore usually not specified.

An example is poly(hexamethylene adipamide), known as polyamide 66 (nylon 66), $H{+}NH(CH_2)_6NHCO(CH_2)_4CO{+}_nOH$, that can be obtained by the polycondensation of hexamethylenediamine, $H_2N(CH_2)_6NH_2$, and adipic acid, $HOOC(CH_2)_4COOH$, and also by the ring-opening polymerization of the cyclic compound $C_{12}H_{22}O_2N_2$ according to

(2-1)

$$\begin{matrix} N\,H_2N(CH_2)_6NH_2 \\ + \\ N\,HOOC(CH_2)_4COOH \end{matrix} \xrightarrow{-(N-1)\,H_2O} H[NH(CH_2)_6NHCO(CH_2)_4CO]_NOH \xleftarrow{+\,H_2O} N \begin{matrix} HN(CH_2)_6NH \\ | \qquad\quad | \\ OC(CH_2)_4CO \end{matrix}$$

The resulting polyamide 6.6 molecule consists of $N$ hexamethylene adipamide repeating units, $-NH(CH_2)_6NHCO(CH_2)_4CO-$, which in turn are composed of $N$ monomeric units $-NH(CH_2)_6NH-$ and $N$ monomeric units $-CO(CH_2)_4CO-$ (if from polycondensation) or $N$ monomeric units $-NH(CH_2)_6NHCO(CH_2)_4CO-$ (if from ring-opening polymerization). In the latter case, repeating units and monomeric units are identical. The **chain units** of this molecule are always $-NH-$, $-CH_2-$, and $-CO-$.

The **endgroups** are H– and –OH with respect to constitution but $-NH_2$ and $-COOH$ with respect to chemical reactions. In systematic names (IUPAC names), the molecule consists of the endgroups $H_2N(CH_2)_6NH-$ and $-CO(CH_2)_4COOH$ as well as $(N-2)$ repeating units $-NH(CH_2)_6NHCO(CH_2)_4CO-$ (Volume I).

The number of *repeating units* per molecule is its **degree of polymerization**, $X$. This term is process-related and not structure-related. The polymer molecule of Eq.(2-1) has thus a degree of polymerization $X = 2\,N$ if from polycondensation, but $X = N$ if from ring-opening polymerization. Note that the degree of polymerization as well as the molar mass (or molecular weight) are *average quantities* if the molecules of a *substance* have different degrees of polymerization (see p. 13).

### 2.1.2 Names of Polymers

Macromolecules are the smallest molecular units of a macromolecular compound. According to IUPAC, "macromolecule" is synonymous with "**polymer molecule**" and "macromolecular compound" is synonymous with "**polymer**." The distinction between molecules and substances as collections of like molecules is much more important in polymer science than in low-molecular weight chemistry since a macromolecular substance is usually *not* composed of macromolecules of the same molecular weight and macroconfiguration and sometimes not even of the same monomeric units. In small-molecule chemistry, on the other hand, a substance such as "benzene" consists only of identical benzene molecules. Whether "benzene" denotes the molecule or the substance is usually clear from the context.

Common usage does not consider "macromolecule" synonymous with "polymer molecule", however. A macromolecule is simply a very large molecule whereas "polymer molecule" refers not only to a molecule consisting of many smaller units (G: *polys* = many; *meros* = part) but, in the original meaning of the word, to molecules of a substance consisting of many *equal* units (see Volume I, Section 1.3). Molecules of poly(styrene) with the constitution $R\text{-}[CH_2\text{-}CH(C_6H_5)]_N\text{-}R'$ are thus usually considered to be polymer molecules whereas enzyme molecules with the constitution ala-gly-lys...val-gly are commonly not counted as *polymer* molecules since they consist of many *different types* of monomeric units. Both poly(styrene) molecules and enzyme molecules are of course macromolecules.

The *naming* of macromolecules and polymers has changed considerably during the last years (see the example in Section 10.2.2). The first polymeric compounds were usually given **trivial names** that referred to the origin (e.g., cellulose for the substance from the cells of plants) or the action (e.g., catalase for a catalytically acting enzyme). However, there are many different celluloses (Section 8.3.1) which are usually described by a common *idealized* chemical structure.

Idealized constitutions of *molecules* (and not *substances*) are the basis of the so-called **systematic names** of the *Chemical Abstracts Service* (CAS) which replaced the previously proposed names of macromolecules by the *International Union of Pure and Applied Chemistry* (IUPAC). Different rules exist for the systematic names of organic and inorganic macromolecules (see Volume I). For example, names of organic macromolecules are composed of the names of the smallest repeating units which are considered to be biradicals and named according to the substitution principle of organic chemistry. These names are then added according to the additivity principle of inorganic chemistry, using arbitrary rules for seniorities. An example is the repeating unit of poly(hexamethylene adipamide) which is not written as common for chemical reactions (see Eq.(2-1)) but as $-NHCO(CH_2)_4CONH(CH_2)_6-$. This polymer has the systematic name poly(imino(1,6-dioxohexamethylene)iminohexamethylene).

Systematic names give no hint of the origin of repeating units. They are also too cumbersome for daily use. This book therefore uses mainly so-called **generic names** that are composed of the monomer name(s) and the prefix "poly." These names are therefore **poly(monomer)** names. An example is poly(styrene), the generic name of the macromolecules $-[CH_2-CH(C_6H_5)]_N-$ of the polymers of styrene, $CH_2{=}CH(C_6H_5)$. The systematic name of poly(styrene) is poly(1-phenylethylene).

In some cases, systematic names are used because they are identical with the old generic names. An example is the systematic name poly(ethylene) for the ideal structure $\text{+}CH_2\text{–}CH_2\text{+}_n$ of *polymers* of eth(yl)ene (indices: $n$ (amount) for a substance, $N$ (number) for a molecule). The new generic name of these polymers would be poly(ethene).

In generic polymer names, this book also writes *all* names of monomers in parentheses, regardless of whether the names consist of many words or just one word. Examples are poly(styrene) instead of the common polystyrene and poly(vinyl chloride) instead of polyvinyl chloride. Names of chemical groups are not written in parentheses, however, because they are not generic names and do not refer to monomers. An example is "polyamide" for polymers containing an amide group –NH–CO– in the main chain.

Industry also uses trivial names which are either short forms of generic names or now freely usable former trade names. Examples are "vinyl plastics" for plastics from poly(vinyl chloride) and "nylon" for aliphatic polyamides. Some other trivial names are still protected trade names such as Bakelite®.

Also widely used are abbreviations and acronyms of polymer names which are however often specific for a certain industry. Poly(ethylene terephthalate)s, the aromatic polyesters from ethylene glycol and terephthalic acid, carry the acronym PET in the plastics industry but the acronym PETE for recycling purposes and the acronym PES for polyester fibers. Differences also exist between different organizations and countries; American codes are usually different from everybody else's. The PE-HD of ISO (*International Standardization Organization*) and DIN (German industrial standards) is the HDPE of ASTM (*American Society for Testing and Materials*), i.e., poly(ethylene) of high density. The Appendix contains a table of common abbreviations and acronyms.

### 2.1.3 Constitution

The constitution of a polymer is described by structure-related or process-related terms. The description of the idealized chemical structure is then supplemented by that of constitutional errors. Note that the reported properties of most industrial polymers are also affected by impurities such as residual monomers, catalyst residues, added stabilizers or nucleating agents, etc.

The smallest structure-related unit is the **constitutional repeating unit** (CRU), the smallest process-related unit the **monomeric unit** or **mer** (erroneously called "the monomer" in physics). A CRU may consist of one monomeric unit or several of them. Each monomeric unit contains at least one **chain unit** which in turn is comprised of at least one **chain atom** and its substituents. For example, the constitutional repeating unit of nylon 66 by polycondensation (Eq.(2-1), left) is composed of two monomeric units that contain a total of 14 chain units: 2 –NH– units, 10 $\text{–}CH_2\text{–}$ units, and 2 –CO– units. The chain atoms are C, and N, the "substituents" H and O.

#### Structure-related Terms

The chains of **homochain polymers** consist of a single type of chain atom; an example is poly(ethylene) $\text{+}CH_2\text{–}CH_2\text{+}_n$ with carbon atoms as chain atoms. **Heterochain polymers** have chain atoms of two or more types, for example, C and N in nylon 66.

Polymers are called **regular** if their molecules consist of a single type of constitutional repeating unit in identical arrangements. **Irregular polymers** contain more than one type of constitutional repeating unit and/or more than one type of arrangement of these units. An example of polymers with more than one type of CRU are copolymers from two or more different types of monomer. Irregular polymers may also result from one type of monomer if the monomer can form different types of monomeric units. An example is butadiene, $CH_2{=}CH{-}CH{=}CH_2$, which may lead to monomeric units $-CH_2-CH{=}CH-CH_2-$ and $-CH_2-CH(CH{=}CH_2)-$. Irregularity may also arise through different relative arrangements of monomeric units, for example, **head-to-head** and **tail-to-tail** connections, $-CHR-CH_2-CH_2-CHR-$ and $-CH_2-CHR-CHR-CH_2-$, instead of the usual **head-to-tail** arrangements, $-CH_2-CHR-CH_2-CHR-$.

The simplest polymer molecules consist of unbranched chains –M–M–M–...–M–M– of monomeric units M that are either **open chains** with two endgroups each (so-called **linear chains**) or **cyclic chains** with no endgroups (ring polymers, cyclic polymers, but not "cyclopolymers"!). "Unbranched" does not mean the absence of substituents since branching is not defined topologically in macromolecular chemistry but via the process. Neither the repeating units $-CH_2-CH(C_2H_5)-$ of poly(1-butene) nor the polymer itself are thus called "branched." The free-radical polymerization of ethene, on the other hand, leads not only to simple ethylene units $-CH_2-CH_2-$ but also to small proportions of $-CH_2-CH(CH_2-CH_3)-$ units (see Section 6.2.3); these poly(ethylene)s are thus said to be branched.

**Branched macromolecules** containing one branching point per molecule are called **stars** (star(-like) molecules) (Fig. 2-1). They possess at least three arms which may or may not be of equal length. Constitutionally different arms of the same molecule are said to be **mikto arms** (G: *miktos* = mixed). Molecules with more or less regularly spaced branches are **comb molecules**.

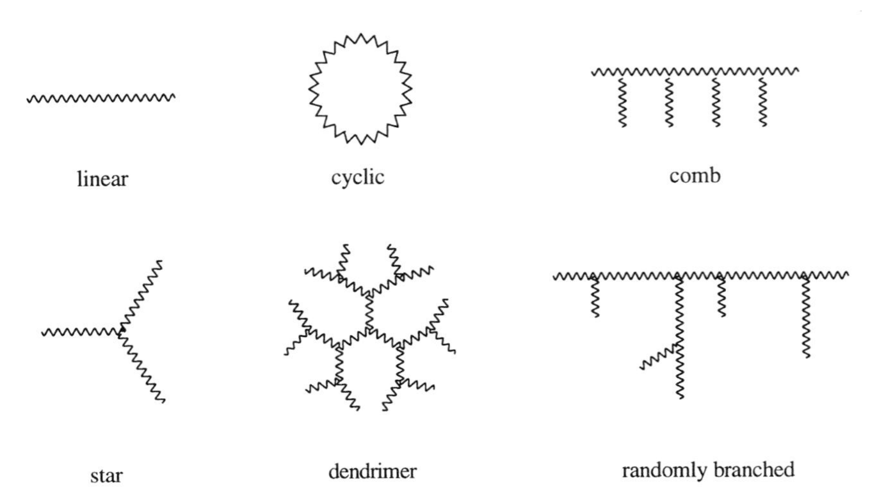


Fig. 2-1 Two-dimensional projection of two unbranched chains (linear and cyclic) and four differently branched polymer molecules on a plane. These molecules may exist in different three-dimensional shapes (macroconformations), depending on their constitution, architecture, and surroundings (solutions, melts, glass, crystalline states, etc.). A linear macromolecule may form random coils in the melt or glassy state but folded and/or helical chains in the crystalline state.

Star molecules with regular branch-upon-branch structure are known as **dendrimers** (G: *dendron* = tree). They consist of a **core** to which several **dendrons** are bound (3 in the dendrimer of Fig. 2-1).

**Hyperbranched** molecules are highly branched polymer molecules. They are usually defined via their polymerization processes because their constitution is very difficult to elucidate. A trifunctional molecule of the type $AYB_2$ with a trifunctional central atom −Y< and two different types A and B of functional groups can lead to 4 different classes of monomeric units (end units, and linear, branched, and dendritic chain units) (Fig. 2-2) if A can only react with B and neither A with A nor B with B.

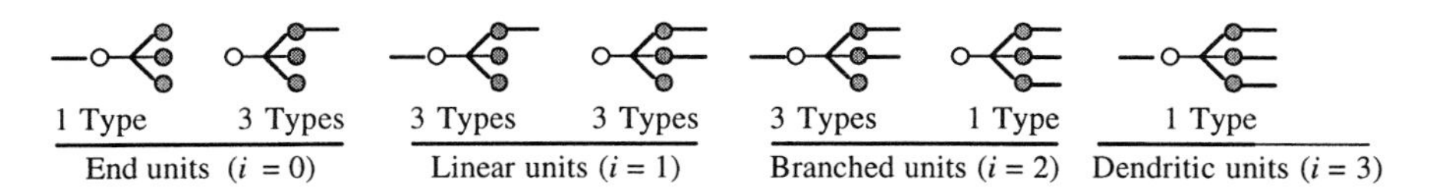


Fig. 2-2 Classes and types of monomeric units in hyperbranched macromolecules.

**Randomly branched macromolecules** are defined as molecules that contain two or more randomly distributed branch points. Depending on the length of branches, short-chain branches are distinguished from long-chain ones.

Chains can be furthermore regularly interconnected with other chains to form "one-dimensional" **spiro chains** and **double-strand chains** (= **ladder chains**), "two-dimensional" **layer polymers**, and "three-dimensional" **lattice polymers** (Fig. 2-3). One-, two- and three-dimensional polymers are also known as **catena** (L: *catena* = chain), **phyllo** (G = *phyllon* = leaf), and **tecto** polymers (G: *tecton* = carpenter, builder). In silicate chemistry, one-dimensional polymers are called **ino** chains (G: *ino* = genitive of *is* = fiber). Irregularly interconnected chains are said to be **crosslinked**.

Polymer molecules may also by electrically neutral or charged. **Macroions** carry one electric charge per molecule; **polyions** many of them. They may be present as **macroanions** or **macrocations** on one hand and as **polyanions, polycations, polyacids, polybases,** and **polysalts** on the other. Correspondingly, **macroradicals** carry one unpaired electron and **polyradicals** many of them.

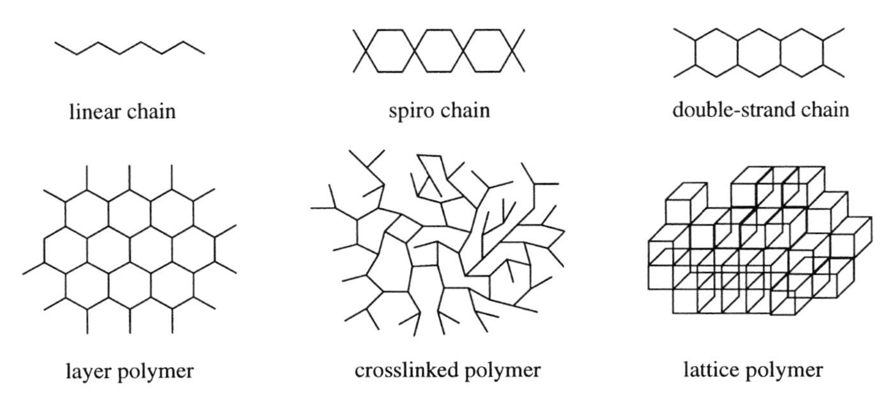


Fig. 2-3 Two-dimensional representations of sections of linear, multi-strand, and crosslinked polymer molecules (schematic).

**Process-related Terms**

According to IUPAC, **homopolymers** are macromolecular substances that were generated from only one type of monomer. **Copolymers** are correspondingly polymers from two or more types of monomers; in the old literature, they were also called **interpolymers** or **heteropolymers** (not to be confused with "heterochain polymer").

"Copolymer" is an umbrella term which comprises **bipolymers** (from 2 types of monomers), **terpolymers** (3 types of monomers), **quaterpolymers** (4 types of monomers), **quinterpolymers** (5 types of monomers), etc. However, "copolymer" is often used with the same meaning as "bipolymer."

The term "copolymer" is a process-related and not a structure-related term since it implies the *generation* of a polymer from two or more types of monomer and not the composition of two or more units. Polymers consisting of two or more types of mers that were not generated by direct polymerization are therefore called **pseudocopolymers**. An example is the polymer from the partial saponification of poly(vinyl acetate). This polymer contains both remaining vinyl acetate units, $–CH_2–CH(OOCCH_3)–$ and newly generated vinyl alcohol units, $–CH_2–CHOH–$.

In **random copolymers**, the sequence of monomeric units follows Bernoullian statistics (zeroth order Markov statistics). **Gradient copolymers** have a systematic changing average composition along the chain whereas **periodic copolymers** have an orderly succession of two or more types of monomeric units such as ...abababab..., ...abbabbabb..., ...abcabcabcabc..., etc. The special case ...abababab... is known as **alternating copolymer**, or, more precisely, **alternating bipolymer**.

IUPAC distinguishes **block *co*polymers** from the copolymerization of two or more different types of monomers and **block polymers** that are obtained by other types of syntheses, for example, the coupling of two homopolymer chains. Such block (co)polymers may be **diblock (co)polymers** $a_m b_n$, **triblock (co)polymers** $a_m b_n a_m$ or $a_m b_n c_p$, etc. Multiblock copolymers with many short blocks are called **segment(ed) copolymers**.

**Molecular Weights**

**Polymer molecules** contain many atoms, monomeric units, repeating units, etc., per macromolecule and **oligomer molecules** only few of them (G: *oligos* = few). No sharp dividing line exists between oligomers and polymers.

In industry, oligomers are known as **prepolymers**, especially, if further processing of this so-called **B-stage** leads to crosslinked polymers (**C-stage**). Such prepolymers are also often called **resins** but this term often comprises all types of polymers before further processing to molded parts, etc. (see also Section 3.9).

Macromolecular substances are **uniform** if all of their macromolecules are identical with respect to constitution and molecular mass. However, practically all synthetic polymers are **non-uniform** with respect to the number, the arrangement, and the type of their repeating units and/or monomeric units. Depending on the type of deviation from the ideal situation, one thus distinguishes **constitutionally non-uniform** polymers (type, arrangement of units) and **molecularly non-uniform** polymers that are non-uniform with respect to the molecular mass of their molecules (**polymolecular polymers**).

Literature calls these different types of polymeric substances **monodisperse** and **polydisperse**. However, "disperse" refers to multi*phase* systems. Furthermore, something cannot be *mono*disperse since "dispersity" always involves two or more types of phases. See also Volume I, p. 44.

A macromolecule has a **molecular mass** (e.g., in gram or kilogram), $m_{mol}$, and correspondingly a **relative molecular mass**, $M_r = m_{mol}/m_u$, where $m_u$ is the **atomic mass constant**, $m_u = m_a(^{12}C)/12$, and $m_a(^{12}C) = 1.660\,540\,2 \cdot 10^{-27}$ kg the atomic mass of a $^{12}C$ atom. The **relative molecular mass** is often called **molecular weight**. This "dimensionless" quantity is not directly available from experiments but only indirectly with certain assumptions about the molecule structure, for example, the number of charges per molecule in mass spectroscopy or the number of endgroups per molecule in endgroup determinations. Most physical methods do not deliver molecular weights but molar masses.

The **molar mass**, $M = m/n$, is given by the ratio of the mass $m$ to the amount-of-substance $n$ of the molecules (e.g., in g/mol). Molar masses are obtained without any assumptions about the chemical structure of the molecules, for example, by membrane osmometry, static light scattering, or ultracentrifugation (see Volumes I and III).

It has recently become customary to report molecular weights and/or molar masses in "daltons" (symbol: da or Da), using the physical unit "kilogram." However, the "dalton" is by definition identical with the unified atomic mass (1 Da ≡ 1 $m_u$ ≡ 1 u), a very small quantity (see above); molecular weights of *molecules* are certainly not in the kilogram range. Neither "dalton" nor its symbol "Da" are approved as physical units by the international *Conférence Générale des Poids et Mesures.*

Synthetic polymers are usually polymolecular. The measured molar masses are therefore average quantities that depend on the experimental methods for the determination. Membrane osmometry delivers the **number-average molar mass** of all $i$ molecules present in the polymer:

$$\text{(2-2)} \qquad \overline{M}_n \equiv \frac{\sum_i n_i M_i}{\sum_i n_i} = \frac{\sum_i N_i M_i}{\sum_i N_i} = \sum_i x_i M_i = \frac{\sum_i m_i}{\sum_i (m_i / M_i)} = \frac{\sum_i c_i}{\sum_i (c_i / M_i)}$$

where $n_i$ = amount-of-substance ("moles") of molecules of type $i$, $N_i = n_i N_A$ = number of molecules, $N_A \approx 6.022 \cdot 10^{23}$ mol$^{-1}$ = Avogadro constant, $M_i$ = molar mass of molecules, $m_i = n_i M_i$ = mass of molecules, $x_i$ = amount-of-substance fraction ("mole fraction"), and $c_i = m_i/V$ = mass concentration. $M_i$ has to be replaced by $\overline{M}_{n,i}$ if $i$ refers to polymolecular fractions instead of molecules. Number-average molar masses, $\overline{M}_n$, become numerically identical to **number-average molecular weights**, $\overline{M}_{r,n}$, if the molar mass is given in gram per mol.

Static light scattering delivers **mass-average molar masses**, $\overline{M}_w$, that are defined via

$$\text{(2-3)} \qquad \overline{M}_w \equiv \frac{\sum_i m_i \overline{M}_i}{\sum_i m_i} = \sum_i w_i \overline{M}_i = \frac{\sum_i n_i \overline{M}_i^2}{\sum_i n_i \overline{M}_i}$$

where $w_i$ = mass-fraction of molecules of type $i$. If these fractions are not molecularly uniform but polymolecular, then $\overline{M}_i$ has to be replaced by $\overline{M}_{n,i}$ and $\overline{M}_i^2$ by $\overline{M}_{n,i}\overline{M}_{w,i}$. Again, mass-average molar-masses $\overline{M}_w$ are identical to **weight-average molecular-weights** $\overline{M}_{r,w}$ if the molar mass is reported in g/mol.

Polymerization processes deliver polymers with different types and widths of molar mass distributions of their molecules. The width of the molar mass distribution is often described by a **polymolecularity index** ("polydispersity index") that is given by the ratio of two molar-mass averages such as $\overline{M}_w / \overline{M}_n$. These indices range from 1 (molecularly homogeneous), 1.05 (certain living polymers), and 2.0 (polymers from linear equilibrium polycondensations) to values of 60 or more for hyperbranched polymers.

The molecular weights reported for industrial polymers are often neither number-averages nor mass-averages but either viscosity-average molar masses (or molecular weights), some ill-defined molar masses from size-exclusion chromatography, or some melt-viscosity related data.

Dilute solution viscosities $\eta$ can be easily measured as a function of mass concentration $c$. They are reported as **relative viscosities** $\eta_r = \eta/\eta_0$, **specific viscosities** $\eta_{sp} = \eta_r - 1$, or **intrinsic viscosities** $[\eta] = \lim (\eta_{sp}/c)_{c\to 0}$ where $\eta_0$ is the viscosity of the solvent. Intrinsic viscosities are related to molar masses by the empirical equation $[\eta] = K_v \overline{M}_v^{\alpha}$ where $K_v$ and $\alpha$ are system-specific constants and $\overline{M}_v$ is the so-called viscosity-average molar mass. The exponent $\alpha$ adopts values of 0 for hard spheres, 0.5 for unperturbed polymer coils, 0.764 for completely disturbed polymer coils (see p. 21), and 2 for infinitely long and stiff rods (see Volume III, Chapter 4)

Some older industrial polymers such as poly(styrene) and poly(vinyl chloride) are occasionally still characterized by so-called Fikentscher numbers K that are measured at a defined concentration $c$ (in g/dL) and calculated from

$$\text{(2-4)} \qquad \lg(\eta/\eta_o) = \left[\frac{75\cdot 10^{-6}\,\mathrm{K}^2}{1+1.5\cdot 10^{-3}\,\mathrm{K}c} + 10^{-3}\,\mathrm{K}\right]c$$

K was introduced as a measure of the molar mass because it seemed to be practically independent of concentrations. **K values** depend less strongly on molecular weights than intrinsic viscosities or specific viscosities (Fig. 2-4).

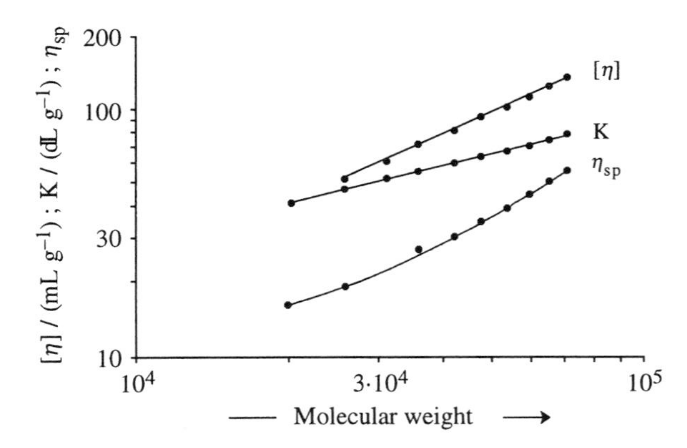


Fig. 2-4 Dependence of intrinsic viscosities $[\eta]$, Fikentscher K values, and specific viscosities $\eta_{sp}$ on the molecular weights of poly(vinyl chloride)s [1]. $[\eta]$ was measured in cyclohexane at 30°C, K values at 1 g/mL in cyclohexanone at 25°C, and $\eta_{sp}$ at $c$ = 0.4 g/mL in nitrobenzene at 30°C. The average of the molecular weight was not specified.

Molar masses of industrial thermoplastics are very often characterized by their **melt flow indices** (**MFI**) because MFIs are also indicators for processing requirements. MFIs are shear-dependent measures of apparent dynamic fluidities (= inverse apparent dynamic viscosities) (see Volumes III and IV). The melt flow index indicates how many grams of polymer are extruded from a standard plastometer by a standard load.

Multiplication of MFI by the density of the melt leads to the **melt volume index** (**MVI**), a measure of the apparent kinematic fluidity (= apparent inverse kinematic viscosity). Both MVI and MFI are measures of molar masses since dynamic viscosities $\eta_0$ at zero shear gradients depend on molar masses $M$ according to $\eta_0 = K_\eta M^\varepsilon$. The proportionality constant $K_\eta$ is specific for the type of polymer whereas the exponent $\varepsilon$ is universal with a value of 1 below and values of ca. 3.4 above the critical molar mass for the onset of polymer entanglement (see Volume III).

Molar masses of rubbers are usually described by **Mooney viscosities**. These data are obtained by deforming the rubbers in a cone-plate viscometer for a certain time at constant rotational speed and temperature. The resulting retraction force is a measure of viscosity.

### 2.1.4 Configuration

The tetrahedral orientation of valences of carbon atoms leads to stereoisomerisms in carbon chains (G: *stereos* = solid, hard, rigid) (Volume I, Chapter 4). By definition, **configurational isomers** have a "high" energy barrier for the conversion of one configurational isomer into another and **conformational isomers** a "low" one. Configurational isomers are therefore stable entities whereas conformational isomers interconvert rapidly.

"Configuration" originally referred only to the arrangement of atoms in space and not to the problem whether chemical compounds can be obtained as pure isomers. The term "configuration" is still used in this sense by physicists, who do not distinguish between configurational and conformational isomers as do chemists.

Stereoisomers are also divided according to their symmetry properties. Two **enantiomers** relate to each other like image and mirror-image (G: *enantios*, from *en* = in, *antios* = opposite) but two **diastereomers** do not (G: *dios* = two, *a*- not) (Volume I, Chapter 4).

Molecules of **stereoregular polymers** consist of only one type of species of **stereorepeating units**. All symmetry centers of these units are defined. **Tactic polymers**, on the other hand, consist of **configurational repeating units** where centers of stereoisomerism are not necessarily defined (G: *tattein* = to arrange). Hence, a stereoregular molecule is always tactic but a tactic molecule is not necessarily stereoregular. An example is poly(2-pentene, $\text{-[-CHR–CHR'-]-}_n$ with R = $C_2H_5$ and R' = $CH_3$ where all repeating units have the same tacticity (Fig. 2-5).

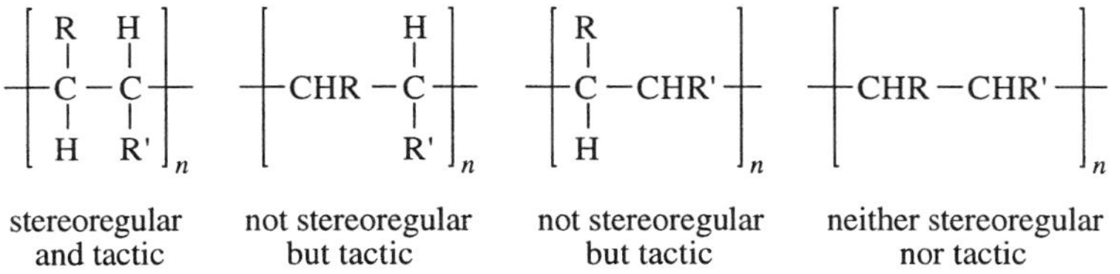

| $\left[\begin{array}{ccc} R & & H \\ \vert & & \vert \\ C & - & C \\ \vert & & \vert \\ H & & R' \end{array}\right]_n$ | $\left[\begin{array}{ccc} & & H \\ & & \vert \\ CHR & - & C \\ & & \vert \\ & & R' \end{array}\right]_n$ | $\left[\begin{array}{ccc} R & & \\ \vert & & \\ C & - & CHR' \\ \vert & & \\ H & & \end{array}\right]_n$ | $\left[\begin{array}{ccc} CHR & - & CHR' \end{array}\right]_n$ |
|---|---|---|---|
| stereoregular and tactic | not stereoregular but tactic | not stereoregular but tactic | neither stereoregular nor tactic |

Fig. 2-5 Stereoregular and/or tactic polymer molecules $\text{-[-CHR–CHR'-]-}_n$. Notations –CHR– and –CHR'– indicate unknown spatial arrangements (in Fischer projections: above or below the plane).

"Tacticity" always relates to the *relative* configurations about stereogenic centers if one follows the chain from, e.g., left to right. An **isotactic repeating unit** (IT) consists of a single constitutional repeating unit. An isotactic polymer molecule thus consists solely of isotactic repeating units (G: *isos* = equal). **Syndiotactic repeating units** (ST) are com-

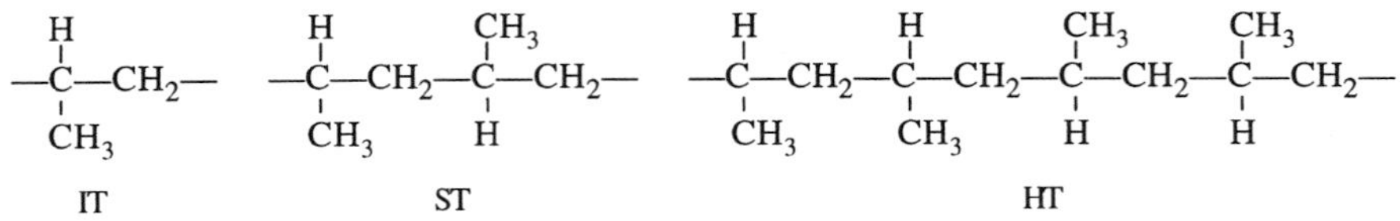

IT: $-\overset{H}{\underset{CH_3}{C}}-CH_2-$ ST: $-\overset{H}{\underset{CH_3}{C}}-CH_2-\overset{CH_3}{\underset{H}{C}}-CH_2-$ HT: $-\overset{H}{\underset{CH_3}{C}}-CH_2-\overset{H}{\underset{CH_3}{C}}-CH_2-\overset{CH_3}{\underset{H}{C}}-CH_2-\overset{CH_3}{\underset{H}{C}}-CH_2-$

Fig. 2-6 Fischer projection of the three simplest repeating units of poly(propylene).

posed of two enantiomeric configurational repeating units (G: *syn* = together; *dios* = two) and **heterotactic repeating units** (HT) of four enantiomeric configurational units (Fig. 2-6). The repetition of HT generates a heterotactic chain, composed of alternating isotactic and syndiotactic diads (it-st)$_n$. As the following picture shows, repetition of a heterotactic triad consisting of three monomeric units does not lead to a heterotactic polymer chain but to a chain (it-st-st)$_n$.

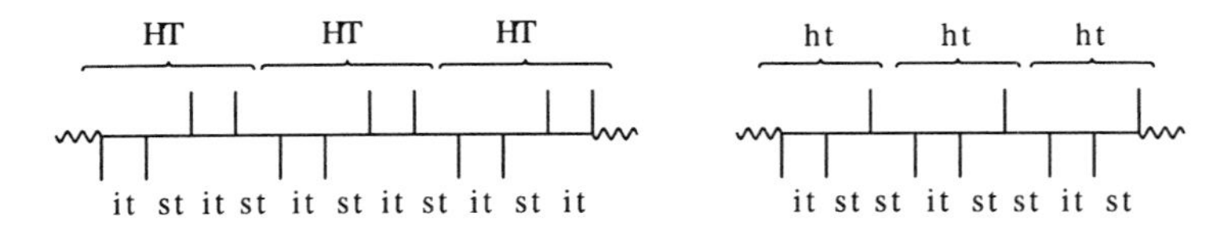


The poly(propylene)s of Fig. 2-6 thus contain only one center of stereoisomerism per configurational repeating unit: they are **monotactic**. The polymer molecules of Fig. 2-5 comprise two stereoisomeric centers per configurational monomeric unit: they are thus **ditactic**. Truly **atactic polymers**, on the other hand, have equal proportions of all possible configurational units with a Bernoullian distribution from molecule to molecule. In literature, however, "atactic" often means " not predominantly tactic."

Textbooks often claim that isotactic macromolecules have all substituents "on the same side" above or below the chain itself. This statement is only true for Fischer projections on a plane but not for stereo formulas. Since tacticities are defined as *relative* configurations, substituents of isotactic macromolecules can be only on the same *spatial* side if the constitutional repeating units consists of an even number of chain units; an example is poly(propylene) (Fig. 2-7).

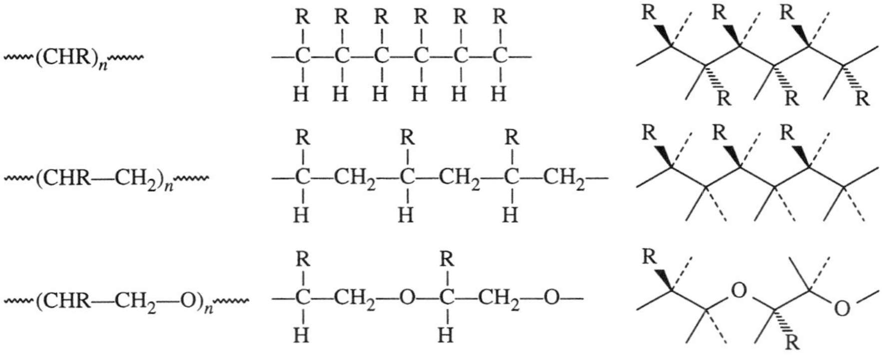

~(CHR)$_n$~ $-\overset{R}{\underset{H}{C}}-\overset{R}{\underset{H}{C}}-\overset{R}{\underset{H}{C}}-\overset{R}{\underset{H}{C}}-\overset{R}{\underset{H}{C}}-\overset{R}{\underset{H}{C}}-$

~(CHR—CH$_2$)$_n$~ $-\overset{R}{\underset{H}{C}}-CH_2-\overset{R}{\underset{H}{C}}-CH_2-\overset{R}{\underset{H}{C}}-CH_2-$

~(CHR—CH$_2$—O)$_n$~ $-\overset{R}{\underset{H}{C}}-CH_2-O-\overset{R}{\underset{H}{C}}-CH_2-O-$

Fig. 2-7 Fischer projections = projections of cis conformations on a plane (center) and stereo formulas (right) of isotactic polymer molecules with odd numbers of chain units per monomeric units (top: poly(methyl methylene), bottom: poly(oxypropylene)) and with an even number of chain units per monomeric unit (center: poly(propylene)). R = $CH_3$.

Substituents of isotactic macromolecules are on opposite spatial sides if the constitutional repeating unit consists of an odd number of chain units; an example is poly(oxypropylene). It is just the opposite for syndiotactic polymers: even numbers of chain units per monomeric unit lead to opposite spatial sides and odd numbers to positions on the same side of the chain (Fig. 2-8).

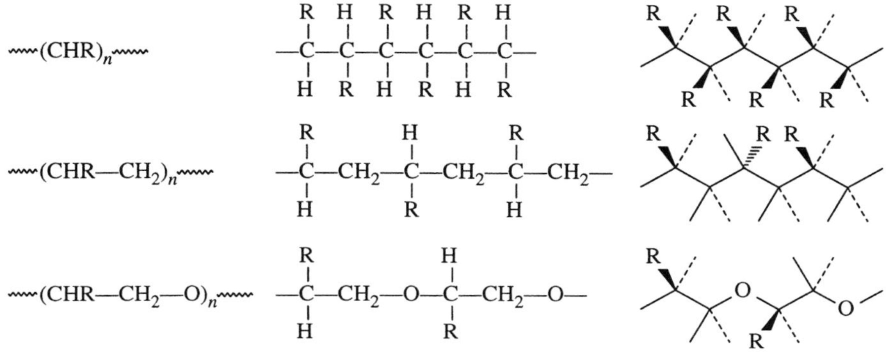

~~(CHR)$_n$~~

~~(CHR—CH$_2$)$_n$~~

~~(CHR—CH$_2$—O)$_n$~~

Fig. 2-8 Fischer projections (center) and stereo representations (right) of syndiotactic poly(methylene), poly(propylene), and poly(oxypropylene) (left, top to bottom).

Polymer molecules with **torsional stereoisomerism (geometric isomerism)** may have tactic arrangements around multiple bonds. For example, polymer molecules with the constitutional repeating unit $–CH_2–C(CH_3)=CH–CH_2–$ may have 1,4-cis-tactic (ct; E form) or 1,4-trans-tactic (tt; Z form) units in addition to 1,2 or 3,4 isotactic of syndiotactic units (Fig. 2-9).

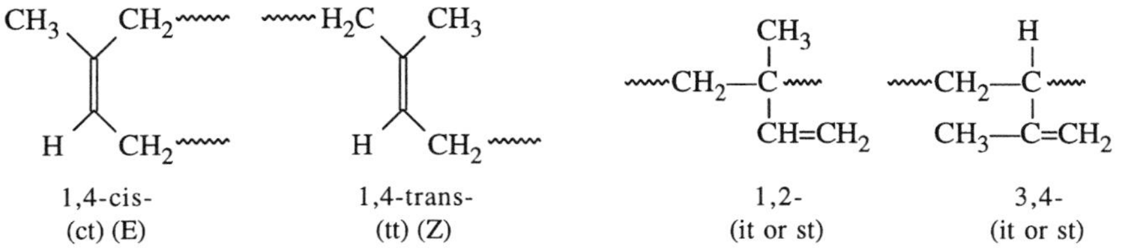

1,4-cis- (ct) (E) | 1,4-trans- (tt) (Z) | 1,2- (it or st) | 3,4- (it or st)

Fig. 2-9 Constitution and configuration of monomeric units from $CH_2=C(CH_3)–CH=CH_2$.

## 2.2 Physical Structure

### 2.2.1 Microconformations

The rotation of atoms or groups around single bonds creates spatial arrangements of these ligands that are called conformations in organic chemistry, **microconformations** in macromolecular chemistry, and configurations in statistical mechanics. The sequence of microconformations around chain atoms determines the **macroconformation** of the macromolecule, i.e., its shape and size.

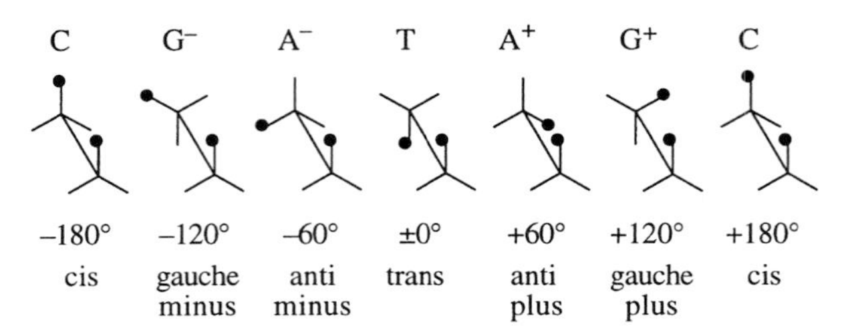


Fig. 2-10 Energetically distinguished microconformations of molecules $RCH_2–CH_2R$ (● = R). Numbers indicate the macromolecular designations of conformational angles between indicated groups R. Organic chemistry sets cis as 0° and trans as 180°.

In principle, an infinite number of microconformations is possible around each chain bond. However, because of attractive and repulsive forces between adjacent and nearby chain units, only a few types of positions are preferred (Fig. 2-10). Only these are designated as **microconformations**.

Macromolecules prefer trans and gauche positions. Anti and cis microconformations seem to be rare. They are present in some crystallized polymers, albeit alternating with either gauche or trans microconformations.

### 2.2.2 Macroconformations

In *ideal crystalline* states, constitutionally and configurationally regular molecules possess conformationally regular structures. In poly(ethylene) segments, $\text{+}CH_2–CH_2\text{+}_n$, for example, successive methylene groups are always in trans microconformations and the chains adopt the macroconformation $(T)_n$ of a **zigzag chain** (Fig. 2-11).

The subsequent methyl substituents of isotactic poly(propylene), $\text{+}CH_2–CH(CH_3)\text{+}_n$, repel each other, which forces every second microconformation into a gauche position. The two possible gauche positions $G^+$ and $G^-$ are energetically equivalent but sequences ...$TG^+TG^-TG^+TG^-$... are sterically impossible. An it-PP segment is therefore composed of either $(TG^+)_n$ or $(TG^-)_n$ sequences, both in equal amounts. The alternate succession of trans and identical gauche positions forces the isotactic poly(propylene) *molecules* into the macroconformation of a **helix** with three monomeric units per complete turn ($3_1$ helix). The *substance* it-PP consists of equal amounts of $(TG^+)_n$ and $(TG^-)_n$, since these two macroconformations are energetically equal.

Because of the rather small size of methyl substituents, helices of it-PP are rather compact. Depending on the size of substituents and their intermolecular distances, many other types of helices are possible (see Volume III), for example, $4_1$ helices with 4 monomeric units per 1 turn or $7_2$ helices with seven monomeric units per 2 turns (i.e., $3.5_1$ helices). The so-called α-helix of polymers of $\alpha$-amino acids is a right-turning helix (P helix) with 3.6 amino acid units per complete turn.

In crystalline polymers, only chain sections and not whole chains can adopt their ideal macroconformations because of constitutional and configurational errors in molecules, non-equilibrium conditions during crystallization, and so on (Fig. 2-11). Crystallization from dilute solution may lead to **chain folding** and crystallization from the melt to so-called **fringed micelles** in addition to **folded micelles**.

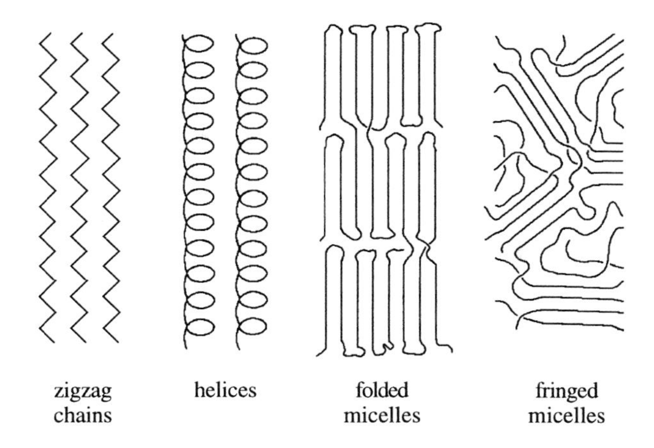


Fig. 2-11 Macroconformations of polymer chains. From left to right: chains in all-trans conformations (zigzag chains); chains in helical conformations; lamellae of chain segments in either zigzag or helical macroconformations (both symbolized as straight lines) with either ordered or disordered loops; fringed micelles with ordered and disordered regions.

### 2.2.3 Morphology of Solid Polymers

The physical structure of (semi)crystalline polymers is affected not only by intrinsic molecular properties such as constitution and configuration but also by external factors such as polymer concentration and cooling rates during crystallization. These factors lead not only to folded and fringed micelles but also to other morphologies. The cooling of polymer *melts* often starts crystallization from nuclei which may not be only external materials such as particles of nucleation agents but also internal ones such as crystalline regions that have not completely melted in the past. Folded micelles may grow from these nuclei into the three spatial directions. Because of the cooling of the melt, the mobility of the remaining chain segments is severely impeded and the space between ordered regions is filled with less-ordered chain segments. The overall structure of the resulting entities is spherical; they are thus called **spherulites** (Fig. 2-12).

The flow of crystallizing *dilute polymer solutions* forces chain segments to orient in the flow direction; they crystallize in extended chain conformations. The resulting bundles of chain segments serve as nucleation agents for other segments which crystallize subsequently with chain folding. The resulting structures resemble **shish-kebabs**.

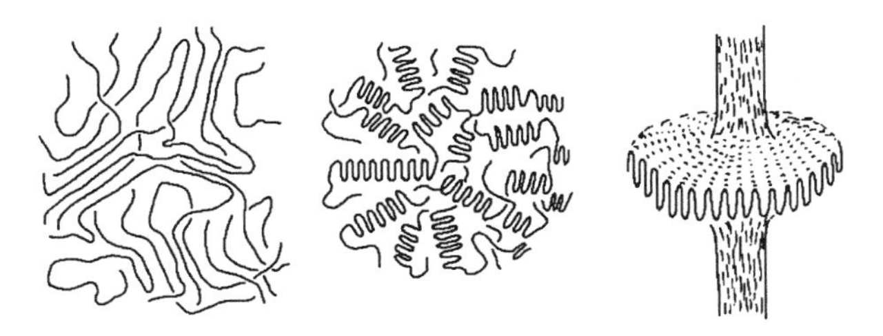

Fig. 2-12 Schematic representation of polymer morphologies. From left to right: folded micelles, cross-section through a spherulite, and shish-kebab structure.

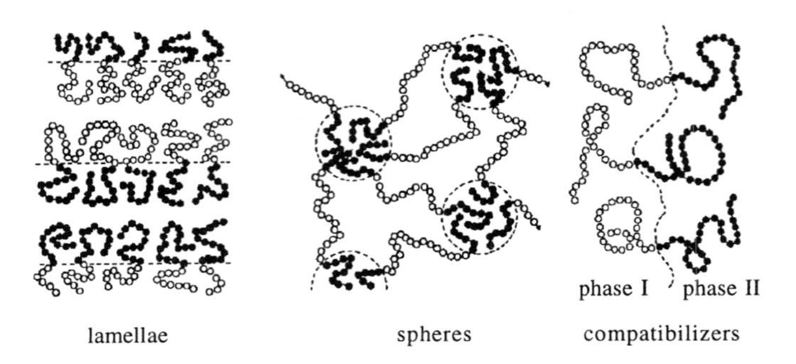


Fig. 2-13 Schematic representation of the morphology of diblock polymers (lamellae and spheres) and their action as compatibilizers in biphasic polymeric materials. ● A units, o B units.

Special morphologies are obtained from block polymers with incompatible blocks (Fig. 2-13). Blocks $A_m$ and $B_n$ of diblock polymer molecules $A_m$-*block*-$B_n$ that are of equal size ($m = n$ with respect to space requirements and not with respect to numbers of monomeric units), incompatible, and non-crystallizing will phase-separate so that the $A_m$ blocks face other $A_m$ blocks and $B_n$ blocks face other $B_n$ blocks. Therefore, the diblock polymer will form alternating lamellae of $A_m$ and $B_n$ blocks. Alternating lamellae will also form from triblock polymers of the type $A_m$-*block*–$B_n$–*block*-$A_m$ if $n \approx 2m$.

Diblock copolymers with very different space requirements of the two types of blocks will lead to spheres composed of the minor blocks embedded in an amorphous matrix of the larger blocks. If the space requirements of the two types of blocks are neither the same nor widely different, rods of the minor blocks are formed in a matrix of the major blocks since rods can be viewed as stretched spheres or shrunk lamellae.

In triblock polymers with widely different space requirements for the blocks, major blocks act as crosslinking agents for the spheres of minor blocks. The resulting structures act as **thermoplastic elastomers** if the glass temperature $T_{G,sph}$ of the spherical "phase" is far above the testing temperature $T$ and the glass temperature $T_{G,mat}$ of the matrix is below $T$. An example is the triblock polymer $S_m$–$B_n$–$S_m$ from styrene S and butadiene B with $T_{G,S} \approx 95°C$ and $T_{G,B} \approx -30°C$.

Diblock polymers act as **compatibilizers** for two non-miscible phases I and II if one block is compatible with phase I and the other block is compatible with phase II. It is not necessary that the monomeric units of blocks and phases be constitutionally identical.

### 2.2.4 Structure in Solutions, Melts, and Glasses

On *melting* of crystalline polymers, thermal energy is usually sufficient to surmount the potential energy barriers that lead to long regular sequences of microconformations. The various types of microconformations now follow each other at random and the chains convert to the macroconformations of **random coils** with sequences of monomeric units without long-range order (Fig. 2-14, left). Short sections of chains however may be in ordered sequences of conformations, for example, helices (Fig. 2-14. center). So-called stiff macromolecules in all-helix macroconformations may also form random coils in solution (Fig. 2-14, right). Certain macromolecules may also generate lyotropic **mesophases** in solution.

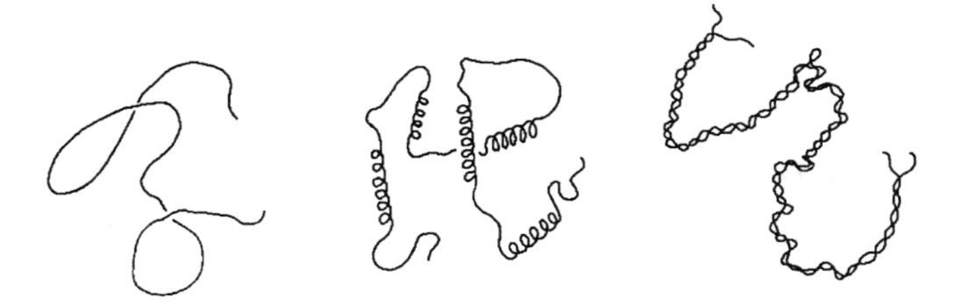

Fig. 2-14 Schematic representation of coil structures of chain molecules.

The density of isolated random coils (i.e., in solution) is very small, usually in the range of volume fractions of $\phi \approx 0.01$. Since the density of molten polymers is much higher (ca. 0.7-1.3 g/mL), a coil in the melt must be filled with segments of other coils. Sufficiently long chains may become **entangled** and the melt behaves as a temporary physical network with a correspondingly high melt viscosity.

In the melt, each chain segment is surrounded by other chain segments of the same type. Attraction and repulsion forces thus compensate each other and the polymer chain appears to be infinitely thin. Such phantom chains can cross each other and the polymer molecule adopts the shape of an **unperturbed random coil**.

In solution, chain segments are surrounded not only by other chain segments but also by solvent molecules. Strong solvent–solute interactions exist in thermodynamically good solvents. The chain segments are stiffened and the polymer coils are **perturbed** and expanded. In dilute polymer solutions of thermodynamically bad solvents, segment–segment and segment–solvent interactions just compensate each other and the polymer coil appears as unperturbed. Such solvents are called **theta solvents** (see Volume III).

With increasing concentration, individual coils start to overlap and interpenetrate each other. At a certain critical concentration, sufficiently long chains become entangled and form temporary networks which in turn cause a strong increase of viscosity with concentration.

Cooling of a polymer melt either freezes the structure of the melt (i.e., the entangled coils) at the so-called **glass** (transition) **temperature** $T_G$ or leads to a crystallization of chain segments. The first case generates a glassy material without any long-range order but with coiled and often entangled chains. In the second case, crystallization is never complete because the transport of segments to their ideal lattice positions is impeded by coiled and entangled chains. So-called crystalline polymers are thus only **semi-crystalline**. Their degree of crystallinity varies with the experimental method and so does their **melting temperature** $T_M$.

## 2.3 Physical Properties

Most polymers are used industrially because of their solid-state properties and here especially because of their mechanical ones. Examples are plastics, fibers, elastomers, coatings, and adhesives. The same monomeric units can lead to very different properties because of differences in molecular weights and molecular weight distributions, constitutional errors, differences in tacticity, variations in morphologies, presence of additives, catalyst residues, residual solvent, etc.

Product **p**roperties are not only affected by **p**olymer structures but very much by **p**rocess conditions (**3 P rule**). Processing can lead to vast differences in segmental orientation, crystalline structure, degree of crystallinity, lattice disturbances, structural gradients, etc. The property data reported in Chapters 6–12 are therefore only typical and not general values; values of polymer grades may vary widely.

### 2.3.1 Thermal Properties

Polymers are divided according to their *mechanical* and *thermal properties* into thermoplastics, elastomers, thermoplastic elastomers, and thermosets and according to their *electrical properties* into insulators and semiconductors.

**Thermoplastics** are uncrosslinked linear or branched polymers with physical transformation temperatures that are above their application temperatures. For amorphous polymers, this is the glass temperature $T_G$ at which segmental movements stop and the melt solidifies to an amorphous glass on cooling. The solidification is not a thermodynamic second order process and therefore not a glass *transition* temperature.

The highest physical transition temperature of a semicrystalline polymer is the melting temperature $T_M$. Since semicrystalline polymers contain not only crystalline regions but also amorphous ones, they also exhibit glass temperatures $T_G < T_M$ in addition to melting temperatures. Thermoplastics can be repeatedly molten and solidified.

**Thermosets**, on the other hand, are highly crosslinked polymers with only short chain segments between crosslinks. These segments cannot crystallize (no $T_M$) and are usually too short for concerted segmental movements (no $T_G$). Thermosets result from low-molecular weight thermosetting **resins** (A-stage) that are polymerized to branched but still soluble oligomeric B-stages. The subsequent reaction and simultaneous shaping leads to the insoluble C-stages. In contrast to thermoplastics, thermosets can neither be remolten nor dissolved.

**Elastomers** result from the *chemical* crosslinking of **rubbers** which are high-molecular weight chain molecules with glass temperatures below the application temperature $T$. Elastomers are only lightly crosslinked. Because $T_G < T$, their long chain segments between crosslinking sites are mobile and can respond to deformation of the part. Segments return to their original position after the load is removed.

**Thermoplastic elastomers** (TPEs) are *physically* crosslinked polymers consisting of domains with $T_G > T$ in matrices with $T_G < T$. The domains "melt" at $T > T_G$ and the TPEs can be processed into the desired shape. On cooling, the elastomeric properties return.

**Fibers** are usually oriented "one-dimensional" thermoplastics but may also be elastomers or thermosets.

Thermal properties of industrial polymers are characterized not only by melting and glass temperatures but more often by so-called Vicat (VT), Martens (MT), and heat distortion temperatures (HDT). VTs employ depressions, MTs bendings, and HDTs deflections under constant loads or forces (see Volume IV). All methods depend not only on thermal transitions and relaxations but also on elasticities of specimens. An example is a nylon 66 with $T_M$ = 265°C, HDT-B = 246°C, VT-B = 230°C, HDT-A = 75°C, MT = 55°C, and $T_G$ = 50°C. Neither VT, MT, HDT nor $T_G$ are indicators of the continuous service temperature (Volume IV), which is ca. 100°C for nylon 66.

### 2.3.2 Mechanical Properties

Industry measures mechanical properties of polymers using standardized test specimens (rectangular bars or dumbbells for thermoplastics, fibers bundles for fibers, etc.) and standardized testing methods (temperature, load, speed, etc.) (see Volume III). Testing conditions vary with the regulating agency; they are different for ISO (International Standardization Organization) and ASTM (American Society for Testing and Materials). Unfortunately, these conditions are often not reported in the literature.

Most mechanical data of polymers are obtained from tensile testing in which a standardized specimen with a length $L_0$ is drawn to a length $L$ with constant, standardized speed. The initial slope of tensile stress $\sigma$ as a function of elongation, $\varepsilon = (L - L_0)/L_0$, is the **modulus of elasticity (tensile modulus, Young's modulus)** (Fig. 2-15). **Rigid polymers** ("hard polymers", but hardness is a surface property) have large moduli of elasticity, but **soft polymers** have small ones. **Brittle polymers** break at small elongations and **rigid-brittle** polymers at small tensile stresses *and* small elongations.

Many polymers yield on further extension. Those plastics with an upper yield point and extensions at break of less than 20 % are called **strong**, those with larger ultimate extensions, **ductile** (in the USA, they are often said to be **tough**, but toughness is an impact property and not a tensile property). For example, poly(styrene) is a rigid-brittle polymer whereas low-density poly(ethylene) is soft-ductile.

The tensile stress sometimes passes through a lower yield point before the polymer finally breaks at the **elongation at break (fracture elongation, $\varepsilon_B$)** with a **tensile strength at break (fracture strength, $\sigma_B$)**.

**Impact strengths** are determined by letting a pendulum fall against a test specimen with or without a notch. In **Izod** tests, the pendulum falls against the free end of a specimen that is clamped on one side only. In **Charpy tests**, the pendulum strikes the center of a specimen that is clamped on both ends.

In the United States, impact strengths are usually reported in foot-pounds per square inch (often erroneously as pounds per square inch), instead of Joule per meter, which corresponds to the initiation of fracture in infinitely thin specimens. In Europe, data are reported in $J/m^2$ which refers to the propagation of fracture in infinitely thick specimens.

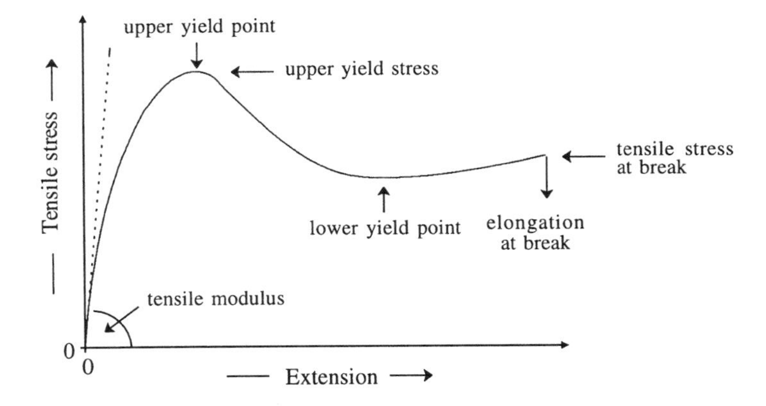


Fig. 2-15 Mechanical property data from tensile testing. - - - Initial slope.

### 2.3.3 Electrical Properties

Most polymers are electrical insulators; many of them are used because of their dielectric properties (see Volume IV). The **dielectric constant** $\varepsilon_r$ of these materials is now called **relative permittivity** according to ISO, IUPAP, and IUPAC. It is the ratio of capacitances of a condensor in the specimen and in vacuo.

The **resistivity** (old name: specific volume resistivity) is the product of electric resistance (measured in ohm) and length (thus the unit of Ω cm) whereas the **resistance** (old name: surface resistivity) depends on the distance between electrodes and is thus measured in Ω. The resistivity of dielectric polymers decreases with increasing relative permittivity and finally becomes constant at $\varepsilon_r > 8$.

The **dissipation factor** (formerly: loss tangent, loss factor), tan $\delta$, is the ratio of power loss to available power. Polymers with high **dielectric loss indexes** $\varepsilon_r$ tan $\delta$ can be heated (and thus welded) by high-frequency fields. Polymers with low loss indexes are excellent insulators for high-frequency conductors.

The relative permittivity is composed of a real and an imaginary part, both of which are frequency-dependent. The imaginary part is caused by the dissociation of polar groups which in turn leads to the development of heat. The low thermal conductivities of dielectric polymers do not allow the heat to dissipate. The temperature of the specimen rises and so does the electric conductivity until a **breakdown (arc through)** occurs. The resistance against such a breakdown is measured by the **electric strength (breakdown field strength)** of the plastic. A breakdown can also occur through **tracking** on the surface (see Volume IV).

### 2.3.4 Other Properties

**Limiting oxygen indices (LOI)** indicate the minimum volume percentage of oxygen that sustains the burning of materials after ignition (usually from above (upper LOI)).

**Water absorption** of a polymer is usually measured after 24 hours; in most cases, it is not an equilibrium value.

## Literature to Chapter 2

For detailed discussions of chemical structures and physical properties including lists of books and reviews, see the other volumes of this work: Volume I (2005): Chemical Structure and Properties; Volume III (planned for 2007): Physical Structures and Properties; Volume IV (planned for 2008): Applications of Polymers (fibers, elastomers, plastics, paints, thickeners, etc.).

## Reference to Chapter 2

[1] Data of A.W.Cooker, R.W.Wypert, in A.J.Wickson, Ed., Handbook of PVC Formulating, Wiley, New York 1993

# 3 Raw Materials and Energy

## 3.1 Feedstocks

Polymers for human use (plastics, fibers, elastomers, coatings, thickeners, catalysts, fuels, food stuffs, etc.) are obtained *directly* or *indirectly* (i.e., after modification or regeneration) from fossil materials, plants, animals, and minerals (this chapter) or *completely synthetically* (Chapter 6 ff.) from monomers via intermediates (Chapter 4) that are gained from some of the fossil materials that also serve for the generation of energy.

### 3.1.1 Natural Raw Materials

Natural raw materials for polymers are very often polymers themselves. Examples are natural rubber, wool, cotton, wood, hemicelluloses, casein, collagen, coal, asphalt, bitumen, starch, pectins, and certain silicates.

*Mineral raw materials* are often polymeric. They serve mainly as building materials, mostly directly but also after chemical transformations. Examples are cement (Section 12.3.9) and ceramic masses (Section 12.3.10).

*Fossil raw materials* are rarely employed themselves as working materials (examples: asphalt and bitumen (Section 3.5.3)) but mainly as energy carriers (Section 3.2) and as feedstocks for the synthesis of aliphatic and aromatic monomers (natural gas, crude oil, coal, lignite; Sections 3.4-3.7 and Chapter 4).

*Animal raw materials* are usually proteins (Section 11 and Volume I) such as wool from the hairs of sheep, goats, llamas, etc., silk from the cocoons of silk worms, and casein from milk. The skins of animals are used directly as pelts or, after physical and/or chemical modification, as leathers (from hides of large animals, from skins of smaller ones). The collagen of skins and bones is also converted to gelatin. The mucopolysaccharide chitin is obtained from the shells of crustaceans such as lobsters.

Animals can never be a very significant source of raw materials since they feed on other animals or plants and the latter are much more accessible. Only fur, leather, wool, and silk are commercially significant "direct" animal materials. All other commercial products of animal origin result from byproducts of meat production such as skins, bones, shells, etc.

*Plant raw materials* are mostly obtained from specially grown plants. This group comprises all plants that deliver cellulosic fibers such as wood (Section 3.7), lignins (Section 3.8), and fibers such as cotton, jute, etc. (Volume IV). Grains and potatoes are the main sources of starch, a mixture of polysaccharides amylose and amylopectin (Section 8.2 and Volume I). Gums are water-soluble polysaccharides (Section 3.9; see p. 414) while pectins, a group of acidic polysacharides, are obtained from fruits (Section 8.7.5).

Directly used are also polyprenes such as natural rubber, chicle, balata, and gutta-percha, all of which are obtained from the latices of certain trees and bushes (Section 6.3).

Wood is a natural composite of celluloses, lignins, hemicelluloses, water, and air (Section 3.7) that is used as such for buildings and furniture or as raw material for cellulosic fibers and thickeners. In less-industrialized countries, wood and other plant materials are predominant sources for heating and cooking (and $CO_2$ and soot!).

Many other natural products serve as raw materials, intermediates, or monomers for polymers. This group includes natural resins (Section 3.9), fats and fatty oils (Section 3.10), various sugars (Section 7.1), and so-called biomass (Section 3.10), which sometimes also includes wood, especially waste wood.

The land area of Earth produces annually ca. $170 \cdot 10^9$ tons of biomass (Section 3.10). Humans consume annually for food, materials, and energy about $2 \cdot 10^9$ tons of wood, $2 \cdot 10^9$ tons of grain, and $2 \cdot 10^9$ tons of other plants (fruits, vegetables, oil-producing plants, beets, cane, etc.). Hence, the total use of plants and plant materials of $6 \cdot 10^9$ tons per year corresponds approximately to the total annual use of crude oil, natural gas, and coal, which is $(7–8) \cdot 10^9$ tons of oil equivalents.

In principle, man may thus replace the ecologically questionable consumption of **fossil fuels** (crude oil (petroleum), natural gas, natural gas liquids, coal, lignite, peat, oil shales, and tar sands) by using the hitherto untapped $164 \cdot 10^9$ tons of biomass. However, this is not possible for agricultural, political, chemical, and economic reasons.

For example, the replacement of synthetic fibers by natural fibers would require considerable acreage (example: all of German agrarian lands to satisfy the annual German fiber consumption). Such a use of land would leave no land for the production of food and certainly none for the production of alcohols or biodiesel as fuel. It is true that plants "grow by themselves" thanks to free energy from the sun. But the cultivation of plants requires additional energy, for example, for the manufacture of synthetic fertilizers to maintain high yields of crops and for the transport of crops over long distances to processing plants (cf. the effect of transportation costs, p. 465). The supply of raw materials and intermediates from plants is furthermore somewhat unreliable because quality and quantity of these materials depend on variations of the climate. Political decisions may furthermore lead to a complete stop of imports.

Materials from renewable resources also have an additional disadvantage compared to materials from crude oil and natural gas: they have high ratios of oxygen to carbon and small ratios of carbmmen soll.difficult and costly to use renewable resources to produce all or even some of the base chemicals that are presently demanded by industry and consumers (see also Chapter 4).

Renewable resources are advantageous if they deliver directly or in very few steps products that can be used directly. However, the properties of such products can only be varied within a relatively narrow range which limits their use. Chemical modification of polymers from renewable resources does lead to different properties but these properties are often not very much different from those of starting materials since these modifications leave the backbones of polymer chains intact. Such modifications are also very costly. This is the reason why the production of regenerated cellulosic fibers has stagnated or decreased in Western countries (Volume IV).

### 3.1.2 Fossil Feedstocks

Fully synthetic polymers can be "tailor-made" for their intended applications. Their synthesis from scratch requires suitable raw materials and readily available energy, both of which are presently provided mainly by crude oil. Smaller amounts of intermediates are obtained from natural gas, coal, and wood and even smaller ones from natural oils, fats, fatty oils, lignins, and biomass.

This situation is not going to change dramatically within the next decades. Fossil raw materials will still provide the bulk of intermediates and monomers. They will also be the main source of energy since "renewable energies" from the sun, wind, geothermal sources, and tides will not add substantially to energy production for various reasons. The production of hydroelectricity has geological limits, wind power is unreliable, costly, and needs back-up, and nuclear power is not always welcome.

Polymers are usually synthesized in many steps. The main **raw materials** crude oil, natural gas, coal, and wood deliver **feedstocks** (syngas, naphtha, etc.), then **base chemicals** (ethene, benzene, etc.), **intermediates** (cyclohexane, etc.), **monomers**, and finally **polymers**. An example is the synthesis of polyamide 6: crude oil → naphtha → benzene → cyclohexane → cyclohexanone → ε-caprolactam → polyamide 6 → nylon fiber. The resulting polymers are rarely applied as such but are usually provided with adjuvants (fillers, dyestuffs, plasticizers, etc.) before, during, and/or after processing depending on their intended use as plastics, fibers, elastomers, films, coatings, thickeners, etc.

The classification of a chemical substance or material as raw material, feedstock, intermediate, end-product. etc., is not an absolute one for a particular substance. It depends on the intended use since the end-product of a processing stage is often the starting material for the next step. An example is ethene which is an end-product of oil refining and also a monomer for polymerizations. Polymers are the resultant products for the polymer industry but the base materials for the plastics compounder, etc.

Early chemical industry was based on coal. However, coal was not used directly as raw material. The chemical industry depended first on coal tar, a byproduct of the production of coke for the steel industry. Coal was later utilized directly, first for the synthesis of acetylene as a base product (via calcium carbide) and later for the production of synthesis gas (a mixture of $H_2$ and CO). About 50–60 years ago, coal as a raw material was pushed aside by crude oil, which led to ethene as a base product. Some decades later, crude oil was partially replaced by natural gas which furnishes methane as a base product. Ethanol from biomass is presently a niche base product and a hydrogen chemistry based on water and nuclear energy is at this time more a dream (and for some a nightmare) than reality. The historical sequence is therefore (see also Sections 3.3–3.6):

| | | |
|---|---|---|
| Coal tar | → | aromatic compounds |
| Coal | → | acetylene |
| Coal (later also oil and natural gas) | → | synthesis gas |
| Crude oil | → | ethene, propene, etc. |
| Natural gas | → | methane |
| Biomass | → | ethanol |
| Water (+ nuclear energy) | → | hydrogen (?) |

All major raw materials for the chemical industry (coal, crude oil, natural gas, biomass) are also the main sources for mechanical, thermal, and electrical energy. There is therefore a delicate interplay between production capacity, production, and demand on one hand and price on the other, not only for the raw material (e.g., crude oil) but between these production and market forces and the subsequent refinery products (e.g., naphtha), base chemicals (e.g., benzene and ethene), intermediates (e.g., ethyl benzene), monomers (e.g., styrene), and polymers (e.g., poly(styrene)) (Figs. 3-1 and 3-2).

For example, the years 1992–1998 saw a slight annual increase in the production of crude oil and oil (Fig. 3-1). The price of crude oil fluctuated somewhat but, on an annual average, stayed about constant as did the price of naphtha, the oil fraction that is used as raw material for chemical syntheses. The 1997 financial crisis in Southeast Asia led to a drop in the price of oil and a subsequent reduction in its annually produced volume by ca. 7 % in 1999 and 2000. This, in turn, resulted in lower prices for naphtha (Fig. 3-2), the intermediates benzene and ethene, the monomer styrene (from benzene and ethene), and poly(styrene). Much more dramatic, however, was the effect of a reduced output of benzene and then of ethene in 1994–1995 which led to price increases by factors of ca. 1.7 (ethene), ca. 1.5 (benzene), and ca. 2.5 (styrene). The price for poly(styrene) was thus raised by a factor of ca. 1.6 which did not recover all costs and led to a drop in profits.

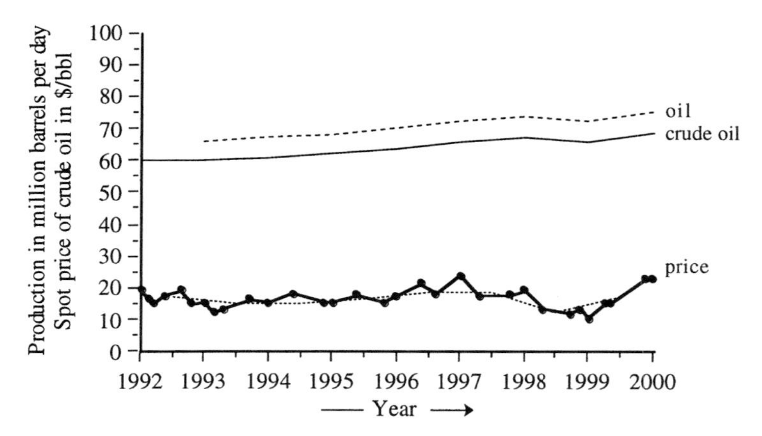


Fig. 3-1 Average daily production of crude oil [1–3] and oil [4a], annual average price of crude oil (····) [4b], and spot price of crude oil on selected days (●) [5] during 1992–2000. "Oil" includes crude oil, shale oil, oil from oil sands, and natural gas liquids (the separately recovered liquid content of natural gas). Spot prices refer to crude oil from Dubai. 1 barrel oil = 1 bbl = 42 US gallons = 159 L.

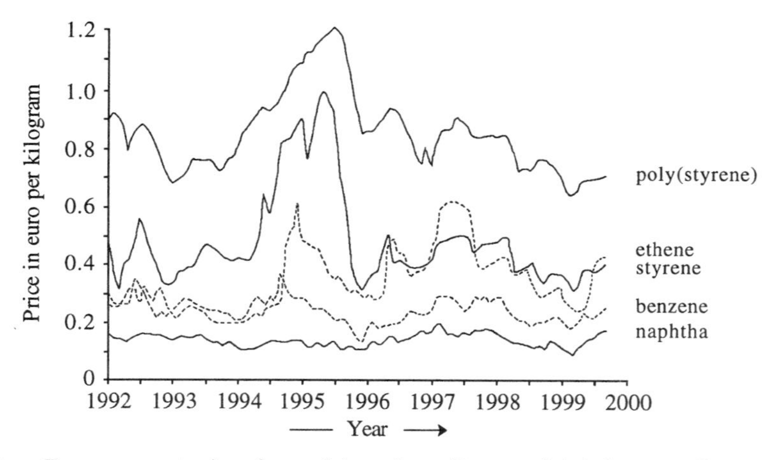


Fig. 3-2 Western European spot prices for naphtha, ethene (from naphtha), benzene (from coal or naphtha), styrene (from ethene and benzene), and poly(styrene) (from styrene) during the 8-year period 1992–1999 (redrawn from [5]). Year marks indicate the beginning of the year.

## 3.2 Energy

The synthesis of monomers and polymers requires not only base materials and intermediates but also auxiliary chemical compounds (hydrogen, chlorine, alkali, water, air, etc.) and especially energy. The consumed energy is usually reported as **primary energy**, both for energy carriers and **primary electricity** (nuclear, hydro, wind power, sun).

### 3.2.1 Physical Units for Energy

Based on tradition and various degrees of scientific inclination, energy production and consumption is reported in various physical units. The scientific unit "Joule" is used by the International Energy Agency (IEA) but industry and commerce often prefer reports in older energy units or units referring to matter such as mass or volume. The British Petroleum Company LTD (BP) publishes world statistical data in "Tons of Oil Equivalent" whereas the World Coal Institute naturally prefers "Tons of Coal Equivalent." In the United States, the United Kingdom, and Germany one can still find "Coal Units":

| | | |
|---|---|---|
| German ton Steinkohleeinheit (hard coal) | 1 t SKE | ≙ 29.31 GJ |
| U.S. Coal unit (hard coal) | 1 CU | ≙ 27.92 GJ |
| Short ton bituminous coal | 1 T | ≙ 26.58 GJ |
| U.K. pit coal unit (hard coal) | | ≙ 24.61 GJ |

The United States Census Bureau continues to report energy data in British thermal units, oil production in U.S. barrel petroleum (1 bbl = 42 US gallons = 159 L), natural gas production in cubic feet (1 cu.ft = 1 CF = 0.028 317 $m^3$), and coal production in short tons (1 sh t ≈ 907.185 kg).

The **British thermal unit** (BTU, Btu) is defined as the quantity of heat that is needed to raise the temperature of 1 pound (≈ 453.59 g) of water by 1 degree Fahrenheit. There are several BTUs, depending on the selected temperature range. The BTU(mean) is defined as the 180th part of the quantity of heat to bring one pound of water from 32°F to 212°F; it equals 1055.79 J. The international BTU ($Btu_{IT}$) equals 1055.06 J.

Quantities of non-oil fuels such as natural gas, coal, or wood are often expressed in **tons of oil equivalents** (TOE, toe). Factors for the conversion of TOEs into other units vary each year since oils, coals, etc., from various countries and fields have various energy contents (see Sections 3.3-3.6) and the contribution of fuels from the different fields to the total world energy production and consumption varies annually. In 2004, energy equivalents were 1 TOE = $4.1868 \cdot 10^{10}$ J = $1 \cdot 10^{10}$ cal = $3.968 \cdot 10^{7}$ BTU = $1.163 \cdot 10^{7}$ W h. Up to the 1970s, conversion factors of 1 TOE = $1.05 \cdot 10^{10}$ cal were used widely. Matter-related equivalents are 1 t crude oil = 1.165 kiloliters = 7.33 barrels = 307.86 US gallons. "Thermie" and "therm" are older units: 1 thermie = 999.7 cal = 3967 Btu, 1 therm = $25.2 \cdot 10^{6}$ cal = $10^5$ Btu = 29.30 kW h.

Quantities of natural gas (NG) and liquefied natural gas (LNG) are also often converted from volumetric units into energy units and vice versa. The 2004 conversions were $10^3$ $m^3$ NG = 0.73 t LNG = 0.9 TOE. The gas industry also uses special symbols for prefixes: M for $10^3$ (L: *mille*), MM or $\overline{M}$ for $10^6$ (L = *mille* × *mille*), B for $10^9$ (billion), T for $10^{12}$ (trillion), and Q for $10^{15}$ (quadrillion, often simply as "quad").

### 3.2.2 Energy Production

In commerce, one distinguishes between primary and secondary energy. **Primary energy** is produced directly from fossil energy carriers (oil, natural gas, coal, combustible renewables, and waste) as thermal, mechanical, or electrical energy and as electricity directly from hydroelectric and nuclear power plants as well as from the wind, sun, tides, and geothermal production. **Secondary energy** is obtained from the conversion of primary energy to other energy forms. These energy conversions work with considerable differences in efficiency (Table 3-1).

Table 3-1 Efficiencies of energy conversions.

| Primary energy | Conversion by | Conversion to | Efficiency in % |
|---|---|---|---|
| Heat (combustion) | fuel cell | electrical energy | 60 |
| | hard coal power plant | electrical energy | 45 |
| | gas and steam turbines | electrical energy | 58 |
| | gas and steam turbines | hot water | 89 |
| | Diesel engine | mechanical energy | 38 |
| | Otto engine | mechanical energy | 25 |
| | oil heater | thermal energy | 65 |
| | fire place (wood) | thermal energy | 15 |
| Heat (geothermal) | gas and steam turbines | electrical energy | 10 |
| Electricity | electrical engine | mechanical energy | 92 |
| | electrical heating | thermal energy | 100 |
| | neon lamp | radiation (light) | 20 |
| | electric light bulb | radiation (light) | 3 |
| Radiation (radioactive) | nuclear power plant | electrical energy | 33 |
| Radiation (sun) | solar cell | electrical energy | 10 |

During the last thirty years, the world supply of primary energy from oil, coal, natural gas, nuclear, hydro, combustible renewables and waste, and other (see explanations to Table 3-2) has increased faster than the increase in world population: on average, each earthling now consumes ca. 1.63 TOE per year.

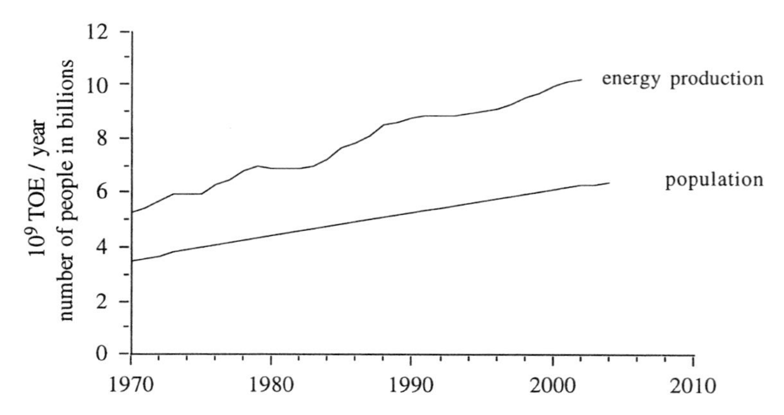


Fig. 3-3 World production of primary energy in tons of oil equivalents per year and world population in billions [6]. For the list of energy carriers included in "energy" see Table 3-2.

Table 3-2 World supply of total primary energy in 1973 and 2003 [6]. For definitions, see below.

| Year | Unit | Oil | Coal | Gas | Waste | Nuclear | Hydro | Other | Total |
|---|---|---|---|---|---|---|---|---|---|
| 1973 | $10^6$ TOE | 2715 | 1496 | 998 | 676 | 54 | 109 | 6 | 6 034 |
| 2003 | $10^6$ TOE | 3639 | 2581 | 2242 | 1143 | 687 | 233 | 53 | 10 578 |
| 1973 | % of total | 45.0 | 24.8 | 16.2 | 11.2 | 0.9 | 1.8 | 0.1 | 100 |
| 2003 | % of total | 34.4 | 24.4 | 21.2 | 10.8 | 6.5 | 2.2 | 0.5 | 100 |
| 2003/1973 | TOE/TOE | 1.34 | 1.73 | 2.25 | 1.69 | 12.7 | 2.14 | 8.83 | 1.75 |

The increase in energy supply from 1970 to 2002 (Fig. 3-3) was accompanied by a shift of relative proportions of contributions by the various energy sources (Table 3-2). Oil is still the major provider of energy but its relative contribution is now lower. Coal maintained its second place. The strong increase of supply of gas was mainly caused by progress in transportation (pipe lines, tankers). Output of energy by nuclear power plants increased dramatically whereas hydroelectric power and energy from waste (and renewables) doubled. The contributions of "other" (solar, wind, etc.) is still very minor. The supply of energy by region is shown in Table 3-3.

The meaning of "oil", "gas", etc., in statistics differs by government agencies, organizations, and companies. In Table 3-2, the various terms are meant to include the following sources:

Oil — crude oil (Section 3.4), natural gas liquids (Section 3.6), refinery feedstocks (Section 3.4), and hydrocarbons from sources such as oil shales and oil sands (Section 3.5);

Coal — all types of hard and brown coal plus peat and coke (Section 3.3);

Gas — natural gas (Section 3.6) without natural gas liquids and gas from gas works;

Waste — renewables and waste therefrom that is burned in *central plants* for heat and/or power by either industry or local authorities. "Waste" includes wood and other vegetable waste, animal materials and wastes, ethanol, sulfide lyes (from production of regenerated celluloses), and municipal wastes (residential, commercial, and public service sectors);

Nuclear — heat equivalents of electricity assuming a thermal efficiency of 33 % (Section 3.2.6);

Hydro — energy content of electricity from hydro power plants, excl. gas from pumped storage plants;

Other — geothermal heat and electricity (assuming an average thermal efficiency of 10 %), solar (heat and photovoltaic), wind, and tides.

Note that all reports of energy supply, production, etc. refer only to *commercially traded* fuels. The reports do not include the use of wood energy (ca. 4.5 % of the primary world energy). Wood energy includes energy from woodfuel (= fuelwood + charcoal) and black liquor (see p. 73). Woodfuel contributes 22 % of the total energy in Africa, 14.5 % in Central and South America, but only 5.5 % in Asia, 2% in North America, and 1.5 % in Europe. Wood has a calorific value of ca. 15 000 kJ/kg.

Table 3-3 Primary energy supply in 2003 [6]. * Including nuclear and hydro but not woodfuel. For member countries of OECD (*O*rganization for *E*conomic *C*ooperation and *D*evelopment), see Fig. 3-4.

| Source | $10^6$ TOE per year | Supply of primary energy in percent of TOE | | | | | | | |
|---|---|---|---|---|---|---|---|---|---|
| | | OECD states | China | Asia w/o China | Former USSR | Africa | Latin America | Middle East | Non-OECD Europe |
| Total * | 10 579 | 50.9 | 13.5 | 11.6 | 9.1 | 5.3 | 4.4 | 4.2 | 1.0 |
| Crude oil | 3 712 | 27.1 | 4.4 | 4.8 | 13.8 | 10.8 | 9.1 | 29.7 | 0.3 |
| Natural gas | 2 447 | 41.5 | 1.4 | 8.9 | 28.2 | 5.5 | 4.4 | 9.4 | 0.7 |
| Hard coal | 2 692 | 34.6 | 37.3 | 12.5 | 7.9 | 6.1 | 1.5 | 0 | 0.1 |

### 3.2.3 Energy Consumption

Energy is needed for the production of goods, transport of goods and people, agriculture, commercial and public service, and residential needs. The energy requirements of industry are thus competing with those of commerce and the general population.

Statistics often divides the consumption of energy equivalents among four groups:

Industry: energy consumed in production of metals, petrochemicals, minerals, etc., but without transportation;

Transport: energy needed for the movement of goods and people by road, railway, air, shipping, and in pipelines;

Other sectors: energy used in agriculture and fishing, for residences, commercial applications, public services, and non-specified uses;

Non-energy: use of energy equivalents of fossil fuels as plastics, fibers, elastomers, etc.

The world energy consumption by these sectors is shown in Table 3-4 for the years 1973 and 2002. The percent of consumption of oil for transport, of coal for industry, and of gas and electricity for Other increased. The percent consumption of oil equivalents for non-energy purposes remained constant whereas that of coal increased. In 2002, industry used a smaller share of consumed energy but transportation used more.

The same trend is observed for the United States, except that commerce and residences enlarged their shares (Table 3-4). More than half of the energy consumed by this sector is for space heating, another 15 % for cooling. People used about 60 % of the energy consumed by the transport sector, most of it in cities.

The biggest US industrial user of energy is the steel industry followed by the petrochemical and chemical industry (ca. 8 % of the total). The paper industry consumes ca. 1.5 % of the total energy. About 30 % of the industrially used energy is for direct process heat, 30 % for process steam, and 40 % for other purposes.

Table 3-4 Annual world [6] and U.S. [7] consumption of primary energy and energy equivalents by sectors. * "Other" plus "Non-energy." US oil consumption: $938 \cdot 10^6$ t/a ≈ 25 % of the world (2004).

| Source | Source | Year | Consumption in percent of total TOE/year | | | |
|---|---|---|---|---|---|---|
| | | | Industry | Transport | Other | Non-energy |
| World | Oil | 1973 | 26.7 | 42.3 | 24.6 | 6.4 |
| | | 2003 | 19.9 | 57.8 | 15.7 | 6.6 |
| World | Coal | 1973 | 57.6 | 5.3 | 36.3 | 0.8 |
| | | 2003 | 76.1 | 1.0 | 20.5 | 2.4 |
| World | Gas | 1973 | 56.7 | 2.6 | 40.7 | ~ 0 |
| | | 2003 | 45.3 | 5.2 | 49.5 | ~ 0 |
| World | Electricity | 1973 | 51.3 | 2.4 | 46.3 | ~ 0 |
| | | 2003 | 42.2 | 1.8 | 56.0 | ~ 0 |
| World | Total | 1973 | 39.7 | 24.9 | 31.7 | 3.7 |
| | | 2003 | 31.9 | 26.0 | 39.1 | 3.0 |
| USA | Total | 1970 | 43.7 | 23.7 | 32.6 * | |
| | | 2002 | 33.4 | 27.2 | 39.4 * | |

It must be noted that only certain percentages of the consumed energies and energy equivalents are used efficiently: 55 % by industry, 45 % by trade and households, and just 17 % by transportation. The remainder is lost by interconversion of energy carriers or energy into other energy forms (for example, coal into electricity), transport of energy (for example, overhead power lines), and waste heat (see Table 3-1).

Energy consumption increases with the standard of living and levels off once a certain standard is reached, as shown by a plot of energy consumption per capita as a function of the annual gross domestic product (GDP) per capita (Fig. 3-4). The GDP counts the market value of all produced goods and delivered services *within* the country whereas the gross national product (GNP) is the total market value of all goods and services produced by its natural and corporate *citizens* within the country *and* abroad.

The consumption of energy by OECD countries increases steeply with increasing wealth of a nation and then seems to become constant at GDPs of ca. 50 000 US-$/capita. The consumption spreads over a certain band with no clear indication of the effect of industrialization (cf. New Zealand and Switzerland, both without heavy industry, or Belgium and the United Kingdom, both with heavy industry). Four countries fall outside the band (Canada, Iceland, Norway, and the United States), all of which are characterized by conditions (d) and (e) mentioned below.

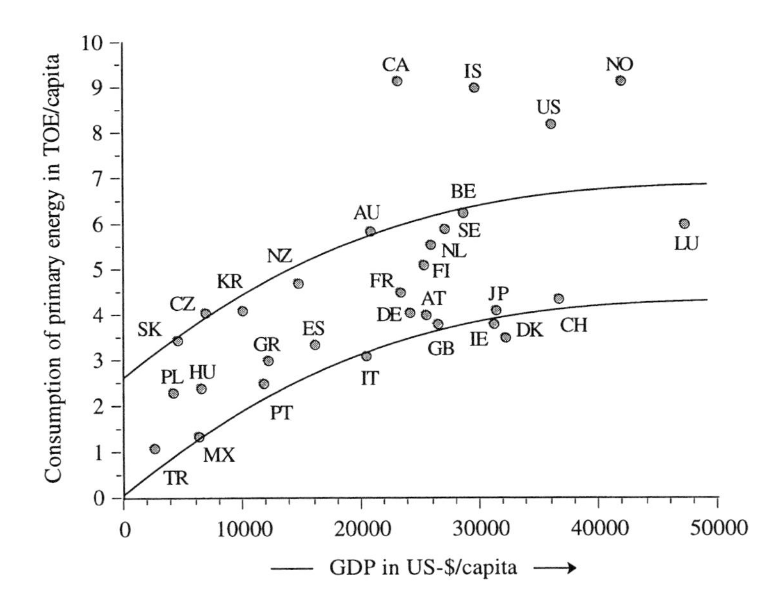


Fig. 3-4 Consumption of primary energy of OECD countries in TOE per capita as a function of the GDP per capita in 2001 [4, 7]. Data for developing countries were not reported; GDP/capita for Second World countries is ca. 1000–5000 US-$ and that of Third World countries ca. 100–500 US-$ if differences in purchasing power of local currencies are neglected.

ISO two-letter country code: AT = Austria, AU = Australia, BE = Belgium, CA = Canada, CH = Switzerland (Confoederatio Helvetica), CZ = Czech Republic, DE = Germany (Deutschland), DK = Denmark, ES = Spain (España), FI = Finland, FR = France, GB = United Kingdom (Great Britain), GR = Greece, HU = Hungary, IE = Ireland, IS = Iceland, IT = Italy, JP = Japan, KR = Republic of Korea (South Korea), LU = Luxembourg, MX = Mexico, NL = Netherlands, NZ = New Zealand, NO = Norway, PL = Poland, PT = Portugal, SE = Sweden, SK = Slovakia, TR = Turkey, US = United States of America.

The GDP is often taken as a measure of wealth of a country but should be used with caution for conclusions about thrifty and wasteful countries: (a) GDPs are based on official and often artificial exchange rates; (b) GDPs include services (banking, tourism, salaries of government employees, government credits, etc.) that consume little energy or none at all; (c) hidden energy in the import of goods is not considered; (d) the more or less compulsory use of energy for heating in cold climates and air-conditioning in hot ones is neglected as are (e) transportation needs in sparsely populated areas.

The last 20 years saw a strong shift in the pattern of energy consumption by world regions (Table 3-5). The proportions of energy consumption by OECD countries, European non-OECD countries, and the countries of the former USSR fell, whereas that of Africa, China, other Asian countries and of the Middle East increased dramatically. Responsible for this development are three factors: improved living standards (except in some African countries), increased production (especially in China and India), and most of all, a strong increase in world population (it almost doubled during the last 40 years).

Table 3-5 Annual consumption of fuel by world region in 1973 and 2002 [6] and population data for 1980 and 2003 [7, 8]; population data for 1973 are not available.

| Year | $10^6$ TOE per year | Consumption of fuel in percent | | | | | | | |
|---|---|---|---|---|---|---|---|---|---|
| | | OECD | China | Asia w/o China | Former USSR | Africa | Latin America | Middle East | Non-OECD Europe |
| 1973 | 4549 | 62.5 | 5.8 | 4.9 | 14.1 | 2.8 | 5.0 | 2.1 | 2.8 |
| 2002 | 7095 | 52.1 | 11.6 | 12.0 | 8.6 | 5.7 | 5.0 | 4.1 | 0.9 |
| TOE | 2003/1973 | 0.83 | 2.00 | 2.37 | 0.61 | 2.94 | 1.00 | 1.95 | 0.32 |
| Population | 2003/1980 | 1.21 | 1.31 | 1.63 | 1.08 | 1.87 | 1.51 | 1.93 | 1.02 |

## 3.2.4 Range of Fossil Fuels

The increasing demand for oil and gas as the main sources for energy and chemicals have always fueled fears about the reach of reserves and resources of fossil fuels. **Reserves** are proved (demonstrated) deposits that can be extracted economically by present technologies (see also Sections 3.3 ff.). **Resources** are speculative (inferred) deposits that may be found and exploited with a certain degree of probability (here 50 %). World **stocks** (= reserves and resources) of oil and natural gas have increased during 1984–2003 whereas coal resources decreased (Table 3-6). These changes are caused by the discovery of new deposits, improved production methods, and changes in judgment.

Table 3-6 Annual world energy consumption, world reserves, world resources, and world stocks (= reserves and resources) of fossil fuels in 1984 and 2003 in million tons of oil equivalents (TOE) [9].

| Energy carrier | Consumption | Reserves | | Resources | | Stocks | |
|---|---|---|---|---|---|---|---|
| | 2003 | 1984 | 2003 | 1984 | 2003 | 1984 | 2003 |
| Oil (petroleum) | 3 697 | 95 000 | 117 800 | 300 000 | 426 000 | 395 000 | 543 800 |
| Oil (sands and shale) | ? | 0 | 140 000 | 420 000 | 927 000 | 420 000 | 1 067 000 |
| Natural gas | 1 833 | 91 000 | 120 500 | 19 000 | 315 000 | 110 000 | 435 500 |
| Coal | 2 519 | 460 000 | 750 000 | 7 000 000 | 5 250 000 | 7 460 000 | 6 000 000 |

Reported reserves may be higher or lower than in reality. Oil reserves sometimes include only conventional low-viscosity ("light") oils but sometimes also the high-viscosity ("heavy") oils of tar sands and shale oil. OPEC countries (p. 49) are interested in boosting their reserves: the higher the reserves, the greater are OPEC production quotas. Oil companies may also overestimate their reserves in order to push their stock prices higher, or they may choose to underestimate reserves in order to lower taxes on assets, etc.

Calculations of the number of years that reserves etc. will last can thus be only rough estimates that depend on the assumptions made. In the simplest case, both annual consumption $C$ and proven reserves $R$ are assumed to be independent of time. The time $t_{lin}$ until the total exhaustion is simply $t_{lin} = R/C$. This *linear* approach estimates the range of world reserves of oil as 32 years for oil, 66 years for natural gas, and 298 years for coal, using the year 2003 as the base.

More realistic is the assumption of a consumption that increases exponentially with time, $C = C_o \exp(kt)$, until all resources are exhausted at a time $t_{ex}$. The decrease of resources with time is then given by

$$-\mathrm{d}R/\mathrm{d}t = C_o \exp(kt) \tag{3-1}$$

where $C_o$ = energy consumption in the base year (here: 2003) and $k$ the rate constant of the exhaustion of the reserves. Integration from $R = R_o$ to $R = 0$ for the resources and from $t = 0$ to $t_{ex}$ for the time results in a time $t_{ex}$ of exhaustion of reserves of

$$t_{ex} = (1/k) \ln (1 + (kR_o/C_o)) \tag{3-2}$$

In the 20 years from 1984 to 2003 ($\Delta t$ = 20 a), the annual world oil consumption climbed to $C = 3.18 \cdot 10^9$ t from $C_o = 2.86 \cdot 10^9$ t. With $C = C_o \exp(kt)$, this leads to a rate constant of $k = 5.3 \cdot 10^{-3}$ a$^{-1}$, i.e., an average of 0.53 % per year. For the year 2003, one obtains with Eq.(3-2) for the exhaustion of oil reserves a time of $t_{ex}$ = 29.4 years and for that of oil stocks (reserves + resources) a time of $t_{ex}$ = 109 years.

However, the rate of consumption certainly does not increase exponentially until all reserves, resources, or stocks are exhausted. According to historical experience, rates will rather decrease because it becomes less and less economical to work marginal fields. One can, for example, assume that consumptions increase *exponentially* with time until a fraction $f$ of the range has been consumed. After that point, consumption decreases *linearly* with time until everything is consumed:

$$kR_o/C_o = 1 - [1 + (1/2)\, kt_{ex,lin}(1 - f)] \exp (fk_{ex,lin}t_{ex,lin}) \tag{3-3}$$

Alternatively, one can assume a *linearly increasing* consumption that is followed by an equally *linearly decreasing* consumption per year:

$$t_{lin,lin} = \left(\frac{1+f}{2\,fk}\right)\left[\left(1 + \frac{8\,fkR_o}{(1+f)^2 C_o}\right)^{1/2} - 1\right] \tag{3-4}$$

An 0.5 % increase/decrease with $k$ = 0.0053 a$^{-1}$ will lead to a range of 40 years for the oil reserves with 2003 as the base year.

### 3.2.5 Ecology

The increasing use of fossil fuels combined with the anticipated harmful effects of global warming have recently led to efforts to reduce the anthropogenic output of so-called greenhouse gases (carbon dioxide, carbon monoxide, methane, higher hydrocarbons, ozone, nitrogen oxides, ammonia, chlorofluorocarbons, etc.). There is no question about the existence of a global warming but there are doubts about the impact of anthropogenic causes and especially of the role of industrialization.

On Earth, Ice Ages alternate with brief interglacial periods because of regular oscillations in Earth's orbit. Some 25–45 million years ago, atmospheric $CO_2$ dropped from ca. 1000–1500 ppm to ca. 300 ppm. The regular cycles were interrupted about 10 000 years ago according to the analysis of atmospheric gases trapped in ice in Greenland and Siberia. Instead of entering a new Ice Age, Earth warmed because of the actions of man. Approximately 8000 years ago, agriculture was developed in the Mediterranean, which led to deforestation and an increase of *anthropogenic* $CO_2$ from ca. 0 ppm in 6000 B.C. to ca. 45 ppm in 1800 A.D. The beginning of rice farming in East Asia caused anthropogenic methane emissions to rise from ca. 0 ppb in 3000 B.C. to 250 ppb in 1800 A.D. (as greenhouse gas, $CH_4$ is about 40 times more effective than $CO_2$).

Since ca. 1800, anthropogenic emissions of $CO_2$ climbed from 45 ppm to 365 ppm and those of $CH_4$ from 250 ppb to 1750 ppb. The main cause seems to be the increase in world population: ca. $10^6$ in 6000 B.C., ca. $10^9$ in 1800 A.D., and $6.6 \cdot 10^9$ in 2004. The same basic energy needs per person in 1800 and 2004, would lead to $6.6 \cdot 45$ ppm = 293 ppm $CO_2$ which amounts to 80 % of the total increase to 365 ppm: only 20 % can be caused by industrialization *and* higher living standards.

Furthermore, not only ca. 5–7 billion tons of carbon from *traded* fossil fuels are burned each year but also 3–5 billion tons of biomass. About 60 % of the $CO_2$ from biomass is caused by the slash-and-burn of tropical forests, another 30 % from the man-made burning-off of savannas. It is also not well known how much biomass is used as fuel for heating and cooking; for the majority of people, this is the only fuel they have.

The amounts of stored and circulated $CO_2$ are decades greater than anthropogenic emissions. About $190\,000 \cdot 10^9$ t $CO_2$ are bound in nature, of which 79.3 % is in the seas, 18.4 % in the carbon equivalents of fossil fuels, 0.6 % each in humus and plants, and 1.1 % in the atmosphere (inorganically bound $CO_2$ ($CaCO_3$, etc.) adds practically nothing to the $CO_2$ cycle). Each year, $760 \cdot 10^9$ t $CO_2$ are exchanged (0.4 % of the total) of which $200 \cdot 10^9$ t occurs between the oceans and the atmosphere. Ocean plants assimilate ca. $460 \cdot 10^9$ t/a and land-based plants ca. $60 \cdot 10^9$ t/a.

Despite all uncertainties, the majority of nations has decided that global warming is (1) not beneficial (e.g., by preventing a new Ice Age) but rather harmful (e.g., by melting glaciers and rising oceans); (2) essentially man-made (no non-anthropogenic causes) by burning of fossil fuels, wastes, and savannas (but not of tropical forests, fire wood, etc.); (3) predominantly caused by industrialized countries (see Table 3-7).

According to the **Kyoto Protocol** (1992, ratified in 2004), industrialized countries should reduce within 10 years the output of greenhouse gases from fuel, transport, industrial processes (metal, chemicals), agriculture, waste incineration, etc. The envisioned target is generally 92 % of the 1990 $CO_2$ emissions (but not that of the other greenhouse gases: $CH_4$, $N_2O$, HFCs (hydrofluorocarbons), PFCs (perfluorocarbons, $SF_6$)) of the targeted countries (European Community, etc.) but the United States is allowed 93 %, the Russian Federation 100 %, Australia 108 %, and Iceland 110 %. The Protocol was not accepted by the United States, Australia, Laos, North Korea, Singapore, and Taiwan.

Table 3-7 Annual carbon dioxide emissions from fossil fuels (petroleum, natural gas, coal, flaring of natural gas) by region and selected countries [10]. Regions: North America (Canada, Mexico, United States); Western Europe (Europe without Former Eastern Block), Industrialized Asia (Japan, Australia). Former Eastern Block (former Soviet Union, Eastern Europe) plus former Yugoslavia).
[1] Mexico: no limit. [2] Except Norway (101) and Iceland (110). NA = not applicable.

| Region | $CO_2$ emission in $10^6$ t/a | | | Tons of $CO_2$/capita | | | Kyoto goal |
|---|---|---|---|---|---|---|---|
| | 1982 | 1992 | 2002 | 1982 | 1992 | 2002 | % of 1990 |
| World | 18 264 | 21 430 | 24 532 | 4.0 | 3.9 | 3.9 | NA |
| Industrialized regions | 9 671 | 10 638 | 12 058 | 11.0 | 11.2 | 11.7 | NA |
| All other regions | 8 593 | 10 792 | 12 474 | 2.3 | 2.4 | 2.4 | NA |
| *Industrialized regions* | | | | | | | |
| North America | 5 083 | 5 862 | 6 705 | 15.5 | 15.8 | 15.9 | 93-94 [1] |
| Western Europe | 3 496 | 3 454 | 3 763 | 8.4 | 7.9 | 8.1 | 92 [2] |
| Industrialized Asia | 1 092 | 1 322 | 1 590 | 8.2 | 9.3 | 10.8 | 94-108 |
| *Other regions* | | | | | | | |
| Developing Asia | 2 565 | 4 310 | 6 254 | 1.1 | 1.5 | 1.9 | no limit |
| Former Eastern block | 4 420 | 4 142 | 3 114 | 11.4 | 10.0 | 7.6 | 92-100 |
| Middle East | 503 | 813 | 1 183 | 5.1 | 5.8 | 6.7 | no limit |
| Latin America | 624 | 771 | 1 006 | 2.1 | 2.1 | 2.3 | no limit |
| Africa | 574 | 756 | 918 | 1.2 | 1.2 | 1.1 | no limit |
| *Selected countries* | | | | | | | |
| United States | 4 390 | 5 067 | 5 749 | 19.0 | 19.9 | 20.0 | 93 |
| China | 1 500 | 2 449 | 3 322 | 1.5 | 2.1 | 2.6 | no limit |
| Russia | NA | 1 887 | 1 522 | NA | 12.7 | 10.6 | 100 |
| Japan | 884 | 1 046 | 1 179 | 7.5 | 8.4 | 9.2 | 94 |
| India | 345 | 661 | 1 026 | 0.5 | 0.8 | 1.0 | no limit |
| Germany (West) | 701 | 886 | 838 | 11.4 | 11.0 | 10.1 | 92 |
| Germany (East) | 306 | | | 18.3 | | | |
| Canada | 423 | 477 | 592 | 17.2 | 16.8 | 18.9 | 94 |
| United Kingdom | 574 | 574 | 552 | 10.2 | 10.0 | 9.3 | 92 |
| South Korea | 142 | 284 | 451 | 3.6 | 6.5 | 9.5 | no limit |
| Italy | 371 | 416 | 449 | 6.6 | 7.3 | 7.8 | 92 |
| Australia | 208 | 276 | 410 | 13.7 | 15.9 | 21.0 | 108 |
| France | 432 | 381 | 407 | 7.9 | 6.6 | 6.8 | 92 |
| Ukraine | NA | 570 | 388 | NA | 11.0 | 7.9 | 100 |
| South Africa | 272 | 317 | 378 | 9.2 | 8.2 | 8.4 | no limit |
| Saudi Arabia | 166 | 236 | 329 | 16.1 | 13.4 | 14.0 | no limit |
| Poland | 387 | 326 | 268 | 10.7 | 8.5 | 6.9 | 94 |
| Switzerland | 38 | 46 | 44 | 6.0 | 6.6 | 6.2 | 92 |
| Norway | 28 | 35 | 46 | 6.9 | 8.2 | 10.2 | 101 |

It is clear from the discussion that such a reduction reduces only marginally the concentration of greenhouse gases. Since industrialized countries emit ca. 50 % of the $CO_2$ from fossil fuels (see Table 3-7), it is in effect a decrease to 96 % which means a reduction of the output of anthropogenic $CO_2$ from 345 ppm to 331 ppm, a value that is still considerably higher than the pre-1800 value of 45 ppm. The true goal of the Kyoto Protocol is rather to "promote, facilitate and finance ... the transfer of ... technologies ... to developing countries" (Article 10) by providing "new and additional financial resources to meet the full costs ..." "for the transfer of technology, needed by the developing country" (Article 11). In other words, industrialized countries are to pay for the transfer of their technologies so that they become less competitive.

### 3.2.6 Non-Fossil Energy Sources

The emission of greenhouse gases could be considerably reduced if fossil fuels are replaced by alternative energy sources. Primary energies from these sources are electricity (hydroelectricity, nuclear power, wind energy, tides), electricity or heat (solar power), heat (heat pumps for space heating, geothermal energy for electricity, both expensive), or heat or mechanical energy (biomass). Hydrogen is *not* a primary energy source.

*Hydroelectricity* is the oldest of these energies. Its production has climbed steadily but seems to have reached a plateau (Fig. 3-5). Hydroelectric power plants require sufficient water supplies, suitable geologies, and large investments in dams and power lines. For these reasons, hydroelectricity provides Norway with 99 % of its electricity, Brazil 83 %, and Canada 58 % (world: 16.6 %). The largest producers of hydroelectricity were Canada (13.1 % of the world total), China (10.8 %), and Brazil (10.7 %) (2002). Disadvantages are the destruction of the countryside and problems with fisheries.

*Nuclear power* has also reached a plateau (Fig. 3-2). Most of the electricity from nuclear reactions is produced by fission of uranium or thorium atoms in light-water reactors (boiling or pressurized water) that work at temperatures below 300°C. High-temperature reactors require cooling liquids that can withstand 1000°C. Breeder reactors are only used in France and in Japan. Nuclear power supplies 78 % of the electricity in France, 46 % in Sweden, and 45 % in the Ukraine (2002 data). The largest producers of nuclear electricity are the United States (30.3 % of the world total of 2660 TW h), France (16.4 %), and Japan (11.1 %). Because of the (mainly emotional) problems with the storage and recycling of spent nuclear fuel and the fear of a Chernobyl-type catastrophe, Sweden, Germany, Austria, and the Netherlands plan to abandon nuclear power. But China, India, Japan, South Korea, Taiwan, and Pakistan plan 76 new reactors (2005).

Nuclear fusion, if possible and feasible, would probably provide an "endless energy." The process would fuse hydrogen atoms in a plasma of more than 100 million kelvins. The reactor must be enclosed in a very strong magnetic field so that it can withstand these extremely high temperatures without melting. At present, nuclear fusion can only be maintained for very short time periods because of the inability to reach the "ignition point", i.e., the temperature at which more energy is delivered than is taken up for heating to the operation temperature.

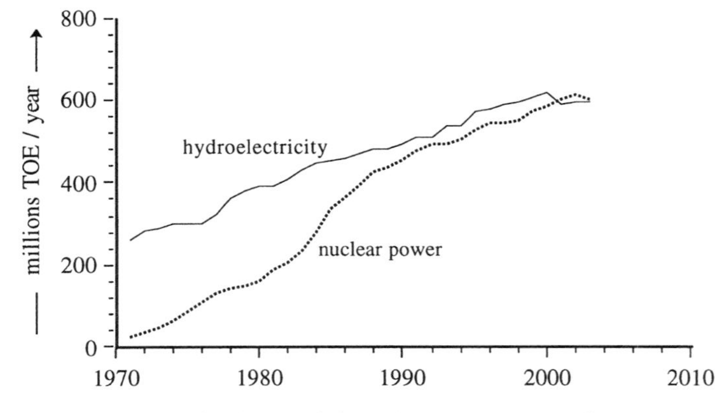


Fig. 3-5 World consumption of hydroelectricity and nuclear power [4, 6].

Table 3-8 Generation of electricity (exclusive of pumped storage) [6, 9]. "Other" includes geothermal, solar, wind, and combustible renewables and waste.

| | | (TW h)/a | Generation of electricity in percent by Coal | Gas | Nuclear | Hydro | Oil | Other |
|---|---|---|---|---|---|---|---|---|
| 1973 | World | 6 111 | 32.8 | 12.1 | 3.4 | 21.0 | 24.7 | 6.0 |
| 2003 | World | 16 661 | 40.1 | 19.4 | 15.8 | 15.9 | 6.9 | 1.9 |
| 2003 | USA | 3 883 | 51 | 17 | 20 | 7 | 3 | 2 |

The reach of nuclear power depends on the type of nuclear reactor and the processing of spent fuel. With the present technology, all uranium will be exhausted in ca. 40 years. A new type of fast reactor would extend the reach to the year 2080.

Nuclear energy and hydroelectricity now provide a considerable proportion of the electricity that is produced and consumed world-wide (Table 3-8). About the same proportion is contributed by the burning of natural gas, whereas the electricity production by burning of oil has become less important. Coal-based power plants are still the main providers of electricity, followed by gas-powered plants (Table 3-8).

The contribution of other energy sources is relatively small, only 1.9 % world-wide (2002) and ca. 3 % in the United States (2003). *Wind power* is too unreliable for industrial purposes; even for community use it needs back-up by conventional power plants. 27 000 rotors of 600 MW each would be needed in order to replace a single nuclear power plant. The transport of wind electricity would also be very costly.

*Solar heating* is a useful (but still expensive) process for the passive production of limited amounts of hot water. Electricity by solar energy is possible (*photovoltaics*) but very costly (Table 3-9) and not available during nights and cloudy days. For commercial purposes, back-up power plants of the same capacity must be installed, e.g., thermal or nuclear power plants. *Geothermal energy* has not been exploited, except in Iceland.

This leaves the production of *hydrogen* as energy carrier and the dream of a $CO_2$-free energy production. At the present time, ca. $500 \cdot 10^9$ $m^3$ $H_2$ are produced annually worldwide, 96 % from fossil fuels and only 4 % by hydroelectricity from water. For every kilogram of liquid hydrogen from natural gas, 20 kg of $CO_2$ are generated! Hydrogen production by solar power is at least 5 times as expensive as that from oil at 50 $/barrel, even in the tropics. Fuel cells are as yet not ripe for mass production.

Table 3-9 Data for the life cycle of power plants (actual annual data + annualized construction data) that annually produce 1 GW electricity [11]. Aluminum is an example of the many construction materials required. * Modern type with gas and steam turbine. 1 acre = 4046.7 $m^2$. 1) Reservoir, 2) on land (0 for existing roofs), 3) production of devices, none for the power production itself.

| | Physical unit | Hydro-electricity | Nuclear power | Gas power plant* | Photovoltaic 3 kW h |
|---|---|---|---|---|---|
| Required area | acre | 10.6 1) | 1.2 | 1.2 | 2760 2) |
| Aluminum | ton | 22 | 140 | 88 | 18 780 |
| Emissions ($CO_2$ equivalents) | ton | 35 000 | 157 000 | 3 436 000 | 1 016 000 3) |
| Residuals | ton | 245 | 7 540 | 12 030 | 44 819 |
| Highly active wastes | ton | 0.06 | 98.6 | 0.04 | 6.5 |

# 3.3 Coals

## 3.3 Chemical Structure of Coals

Coals are the oldest fossil fuels that were utilized by industry for energy and by the chemical industry for chemical base compounds. Coals are inhomogeneous, porous, black to brown, solid mixtures of highly aromatic macromolecular compounds that contain inorganic components, moisture, and often also methane.

Coals result from the coalification of plants over geological time periods, first to humic acids, and then to peat, brown coal, lignite, bituminous coal, and anthracite. The average composition of ash-free, dry coals vary from ca. $C_{75}H_{140}O_{56}N_2S$ for peat to ca. $C_{240}H_{90}O_4NS$ for anthracite (Table 3-10).

Designations of coals vary from country to country. In energy statistics, such as in Tables 3-2 and 3-6, "coals" include humic coals and the fuels derived from humic coals (coke, coke oven gas, furnace gas, etc.). "Humic coals" are coals derived from humus, i.e., decayed plant matter. The degree of coalification is called the **rank** which increases from peat, brown coal, lignite, subbituminous coal, and bituminous coal to anthracite. "Hard coal" is the name given to anthracites in the United States but to anthracites *and* bituminous coals in Europe, except in Wales where the term is unknown. "Hard coals" are called "pit coals" in the United Kingdom. "Soft coals" range from near lignite to near anthracite in the United States. In Europe, "brown coals" comprise both the fibrous "hard brown coals" (= lignites) and the "soft brown coals" (= "brown coals" in the U.S.). The coal industry subdivides coals according to their main uses (for steam, coking, etc.); again, these designations vary from country to country.

Table 3-10 Compositions of ash-free humic coals.

| | Composition of coals in wt% | | | | | | |
|---|---|---|---|---|---|---|---|
| | Moisture as found | Volatiles in dry coals | Elementary composition of solids | | | | |
| | | | C | H | O | N | S |
| Anthracites | 1.5–3.5 | 4–12 | 92–96 | 3–4 | 2–3 | 0.5–2 | ≈ 1 |
| Bituminous coals | 1–20 | 12–45 | 75–92 | 4–6 | 3–20 | 1.3–2 | ≈ 1 |
| Lignites, brown coals | 30–50 | 45–60 | 60–75 | 5–8 | 17–35 | 0.7–2 | 0.5–3.0 |
| Peat | 70–90 | 45–75 | 45–60 | 4–7 | 20–45 | 0.5–3 | 0.1–0.5 |

Coals have irregular, highly aromatic structures (Fig. 3-6) which were exploited by the early chemical industry for the synthesis of chemical compounds that were based on aromatic compounds, e.g., dyestuffs. Oxygen in coal is found mainly in carbonyl, carboxylic acid, hydroxyl, and methoxy groups. Nitrogen is mainly present in aromatic structures such as aromatic nitriles, carbazoles, pyrroles, pyridines, and quinolines. Sulfur-containing groups comprise thiols, dialkyl and alkyl-aryl thioethers, disulfides, and thiophenes. Hard coals also contain ca. 1 carbon radical per 5000 carbon atoms.

Coals contain crosslinked (insoluble) and branched (soluble) molecules, the latter with molecular weights between 3000 and 500 000. Segments between crosslinking points have molecular weights of ca. 1000 according to swelling experiments.

Coal has a lower energy content than oil and gas: oil 42 000 kJ/kg, natural gas 40 000 kJ/kg, hard coal 34 000 kJ/kg, brown coal 8000-14 000 kJ/kg, and wood 15 000 kJ/kg. The energy content varies with the type of coal and also from country to country. Caloric values (in TOE/t) for steam coal are 0.659 (USA), 0.611 (Australia), 0.572 (Germany), 0.564 (South Africa), 0.541 (China), 0.516 (Ukraine), and 0.441 (India).

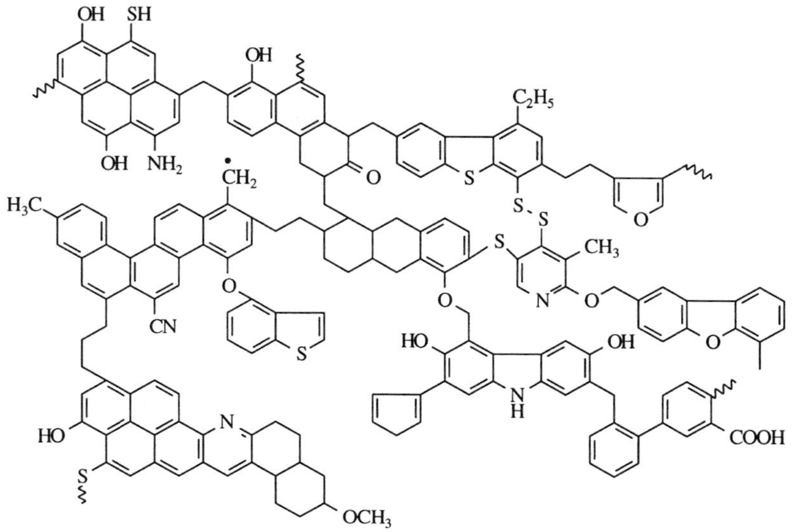

Fig. 3-6 Schematic representation of a crosslinked section of a hard coal. Pyrolysis severs $CH_2$–$CH_2$, $CH_2$–O, and S–S bonds. The resulting fragments are "soluble coal".

### 3.3.2 Coal Production and Reserves

The world production of coal climbs steadily (Fig. 3-7) albeit less fast than the world population. Coal is competitive with oil for the generation of energy (see Table 3-8), a major source for aromatic compounds, and irreplaceable for the production of graphite, tars, and carbon black (Section 3.3.3).

The increase of annual coal production is not smooth, but is rather characterized by "bumps" that are caused by overproduction on one hand and periodic economic downturns on the other. These swings lead to great price fluctuations (Fig. 3-8).

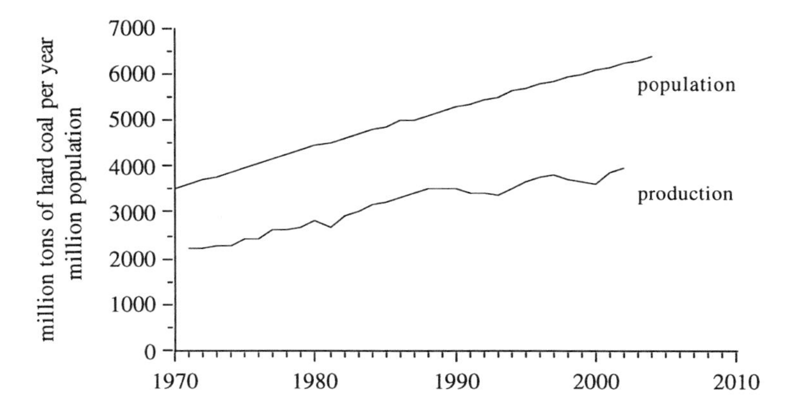


Fig. 3-7 Annual world production of hard coal (anthracite and bituminous) and change of world population in 1970–2002 [6, 9].

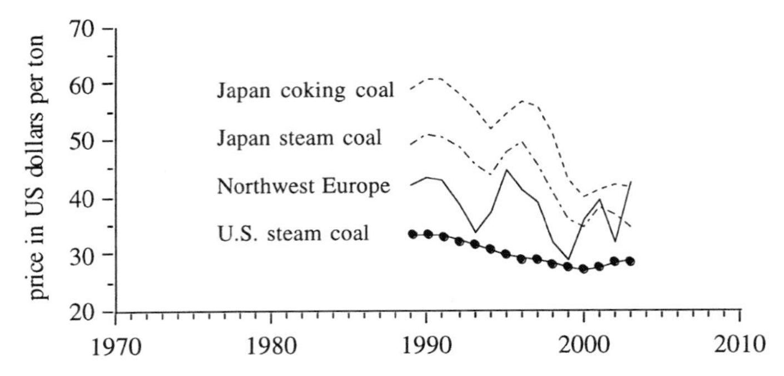


Fig. 3-8 Price of US coal receipts at steam-electric utility plants, market prices in Northwest Europe, and Japan import cif prices for coking and steam coal [4]. cif = cost + insurance + freight.

Price fluctuations are often dramatic for countries and regions that have to import most of their coal such as Japan and Northwest Europe (Fig. 3-8), but fairly smooth for supplies from domestic productions, for example, in the United States.

Over the last 50 years, several shifts in coal production occurred as the example of U.S. coal mining shows (Fig. 3-9). Like the world production, the total U.S. production of coals also climbed steadily, albeit with occasional bumps. During the last 33 years, the annual production of bituminous coal hovered between $560 \cdot 10^6$ and $660 \cdot 10^6$ short tons, whereas that of subbituminous coal climbed steadily from $16 \cdot 10^6$ short tons in 1970 to ca. $430 \cdot 10^6$ short tons in 2003. The production of anthracite decreased from ca. $43 \cdot 10^6$ short tons in 1949 to ca. $1.3 \cdot 10^6$ short tons in 2003.

One major reason for the decrease of coal prices in the United States (Fig. 3-8) is the increasing importance of surface mining relative to underground mining (Fig. 3-9) and the resulting decrease of production costs. The same effect is observed worldwide: the annual coal production in Germany, the United Kingdom, and Poland dropped drastically because deep mining of shallow coal bands cannot compete with surface mining of large coal layers as, for example, in Australia (Table 3-11).

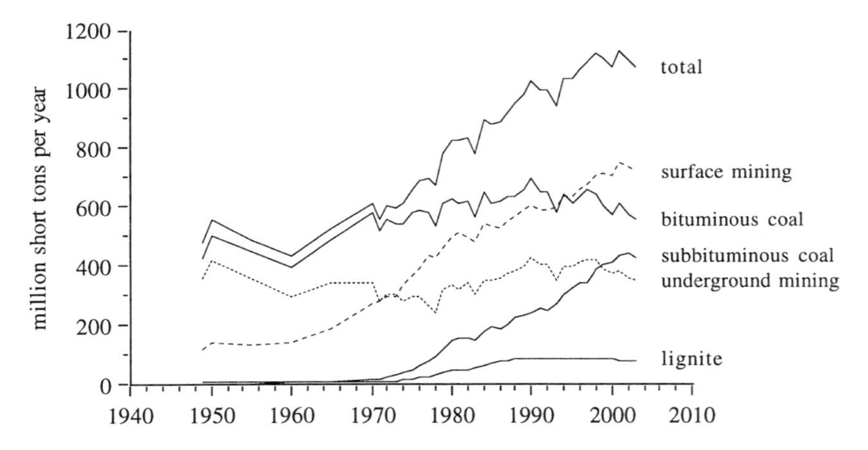


Fig. 3-9 U.S. coal production: total; bituminous, subbituminous, and lignite; and by underground or surface mining [12]. The annual production of anthracite decreased from $42.7 \cdot 10^6$ short tons in 1949 to $1.3 \cdot 10^6$ short tons in 2003. 1 short ton ≈ 907.2 kg.

Table 3-11 Annual production by the world and the 15 largest coal-producing countries (in 1985 and 2002, respectively) [12] and proved (demonstrated) reserves of coal in the world and in countries with reserves of more than $5000 \cdot 10^6$ t [4]. Coal = anthracite, bituminous, subbituminous, lignites, and brown coals. *) Except brown coal. NA = not available.

| | Production in $10^6$ t/a | | | Proved reserves in $10^6$ t (*) | | |
|---|---|---|---|---|---|---|
| | All coal 1985 | All coal 2002 | Lignite 2003 | Total 2002 | Anthracite and bituminous 2002 | Subbituminous and lignite 2002 |
| World | 5408 | 5734 | | 984 453 | 519 062 | 465 391 |
| China | 1060 | 1677 | | 114 500 | 62 200 | 52 300 |
| United States | 974 | 1206 | 84 | 249 994 | 115 891 | 134 103 |
| India | 211 | 433 | | 84 396 | 82 396 | 2 000 |
| Australia | 190 | 417 | | 82 090 | 42 550 | 39 540 |
| South Africa | 211 | 270 | | 49 520 | - | 49 520 |
| Russian Federation | NA | 266 | 110 | 157 010 | 49 088 | 107 922 |
| Germany | 637 | 255 | 193 | 66 000 | 23 000 | 43 000 |
| Poland | 303 | 196 | | 22 160 | 20 300 | 1 860 |
| Indonesia | 2 | 123 | | 5 370 | 790 | 4 580 |
| Ukraine | NA | 103 | | 34 153 | 16 274 | 17 879 |
| United Kingdom | 115 | 36 | | 1 500 | 1 000 | 500 |
| Kazakhstan | NA | 89 | | 34 000 | 31 000 | 3 000 |
| Canada | 73 | 80 | | 6 578 | 3 471 | 3 107 |
| Colombia | 11 | 53 | | 6 648 | 6 267 | 381 |
| Czech Republic | NA | 77 | | 5 678 | 2 114 | 3 564 |

China is the largest producer of coal (Table 3-11) which provided in 2000 ca. 75 % of its energy requirements. China's coal mining industry suffers from underground smoldering fires of its methane-rich coal. These fires consume ca. $20 \cdot 10^6$ t of coal per year and are difficult to extinguish; one of the fires has burned for the last 100 years!

About 30 % of the proved coal reserves (without brown coal) are in Asia and Oceania, 25 % in North America, 25 % in the former Soviet Union, 10 % in Europe, and ca. 5 % each in Central/South America and in Africa. As always, these numbers are to be treated with caution. There are reports that the newly discovered coal fields in Inner Mongolia would add another 1 000 000 million tons, thus doubling the world coal reserves!

### 3.3.3 Coals as Raw Materials

Coals serve mainly for direct heating and the production of process steam. Large proportions are also used for the various coal gasification processes that produce town gas, generator gas, synthesis gas, and water gas (see Section 3.6). 50–100 years ago, bituminous coal was also *the* raw material of the chemical industry since it provided almost all necessary feedstocks for chemical syntheses. However, the production of base chemicals by coal conversion as the main process cannot compete with processes based on oil and natural gas. Nevertheless, some aromatic base chemicals are byproducts of coking and other coal refining processes for the production of graphites, tars, and carbon blacks. Coal gasification also delivers ca. 16 % of all hydrogen (80 % of $H_2$ stems from cracking of crude oil or natural gas; only 4 % from electrolysis of water!).

Table 3-12 Reaction conditions and yields of coal refining processes.

| Process | Reaction conditions | | Yield per ton of coal | |
|---|---|---|---|---|
| | $T$/°C | $p$/bar | L of liquids | $m^3$ of gas |
| Pyrolysis | 1000–1400 | 1–70 | 160–240 | 100–140 |
| Hydrogenation (Bergius process) | 400–480 | 150–700 | 400–560 | 60–85 |
| Extraction | 370–430 | 60–70 | 320–480 | 100–130 |
| Gasification (Fischer–Tropsch process) | 170–200 | < 30 | 240–320 | 230–280 |

The conversion of complex coal structures to simple base chemicals, intermediates, or monomers requires rather drastic processes: pyrolysis, hydrogenation, extraction, and gasification (Table 3-12).

**Pyrolysis** is the thermal degradation of coal in the absence of air. This *dry distillation* is also called **carbonization** or **coking** since coke is the desired main product for heating and steel-making (metallurgical coke). In Europe, coke was also the byproduct from the gasification of coal for town gas (Section 3.6). A small proportion of coke is still reacted with calcium oxide for the synthesis of calcium carbide, the hydrolysis of which delivers acetylene according to $CaC_2 + 2\ H_2O \rightarrow Ca(OH)_2 + C_2H_2$.

Coking of bituminous coal at 1000°C produces about 75 % coke, 20 % light gas (mainly $CH_4$ and $H_2$; used as fuel for the coking process), 4 % coal tars, and ca. 0.3 % $NH_3$ (converted to $(NH_4)_2SO_4$ (fertilizer)).

Until ca. 1950, coal tar was the main source of aromatic chemicals for the chemical industry. Most of these compounds are now produced by the petrochemical industry, but coking processes still deliver ca. 11 % of all benzene and ca. 95 % of higher aromatics. Coal tar is now mainly used for manufacturing carbon black, wood preservatives, binders for road surfaces, and anticorrosion coatings.

The distillation of coal tar delivers first light oils (bp 150°C), followed by wash oils (bp 150–300°C), heavy oils (bp 300–390°C), and pitch (bp > 390°C). Light oil consists of the BTX fraction (benzene, toluene, xylenes), also called *naphtha*. Wash oils are obtained as individual fractions: carbolic oils (phenols) with bp of 150–190°C, naphthalene oils (naphthalene, pyridine) with bp of 190–230°C, and creosote oils (cresols) with bp of 230–300°C. Heavy oils consist of anthracene oils (anthracene, phenanthrene, carbazoles; bp 300–350°C) and residual oils (bp 350–390°C). Note that "naphtha", "light oil", and "heavy oil" have different meanings in the petroleum industry (Section 3.4).

Yields are 15–20 % chemical oil ("naphthenic oil"), 30 % creosote, and 50–55 % pitch. Pitch, the semi-solid to solid component of tar, is used for electrodes in the aluminum industry. Creosote serves to impregnate telephone poles and railroad ties. Anthracene oil and residual oil are feedstocks for carbon black. Centrifuging of naphthalene oil, creosote oil, and anthracene oil delivers, naphthalene, anthracene, and other polynuclear aromatic compounds.

**Hydrogenation** is performed at far lower temperatures than pyrolysis. It delivers liquid hydrocarbons and is therefore also called **hydroliquefaction**. The original **Bergius process** was a one-stage process in which a slurry of 50–60 % solids in heavy oils was reacted at high pressures with hydrogen (from electrolysis) by tungsten or molybdenum sulfide catalysts (F.Bergius, Nobel prize 1931, together with C.Bosch). In 1927–1944, the Bergius process was used in Germany for the production of synthetic gasoline.

Since 1980, two-stage hydrogenation processes have been developed that deliver less gas and more distillation products of higher quality. These developmental processes furnish oils that boil at lower temperatures and are lower in hydrogen and higher in oxygen and nitrogen than typical crude oils.

**Extraction** uses 1–2 mm large particles of hard or soft coals that are dried and converted to a paste by addition of tetralene, naphthalene, and acidic oils such as cresols or phenols. Heating the paste under pressure (Table 3-12) results in an ash-free, pitch-like mass that melts at ca. 200°C (hard coal extracts) or 100°C (brown coal extracts). Hydrogenation delivers gasoline. The process is outdated and unlikely to be revived.

**Gasification** is a 2-stage process in which coal or coke is first reacted with a mixture of oxygen and water vapor to **producer gas** (historically called **water gas**) that contains ca. 40 % CO, 50 % $H_2$, and 6 % $CO_2$. In the original **Fischer–Tropsch** low-pressure synthesis (1925), purified (sulfur-free) synthesis gas was converted to hydrocarbons by a cobalt catalyst in an exothermic reaction, $n\ CO + 2\ n\ H_2 \rightarrow (CH_2)_n + n\ H_2O$. The product consisted of ca. 10–15 % liquid gas ($C_3$–$C_4$), 50 % gasoline ($C_5$–$C_{10}$), 15 % Kogasin I ($C_{10}$–$C_{14}$), 12 % Kogasin II ($C_{14}$–$C_{18}$), 8 % slack wax ($C_{18}$–$C_{28}$), and 3 % paraffins (> $C_{28}$) (Kogasin = from the German *Ko*ks-*Gas*-Benz*in* = coke-gas-gasoline).

The process was used in Germany to produce motor fuels and fatty acids, albeit on a smaller scale than the Bergius process ($0.6 \cdot 10^6$ t/a versus $4 \cdot 10^6$ t/a); the last plant was closed in 1962. The presently operating coal-based **Sasol** process of South Africa and Malaysia delivers primarily 1-olefins, including odd-numbered ones such as 1-pentene, which are not obtainable from other sources. 15 new natural gas-based Fischer-Tropsch plants (GTL = gas-to-liquid) will go on stream in 2006–2010, many of them in Qatar.

**Coal liquefaction by grafting** is an interesting, though not industrial, process. Coals become soluble in aliphatic hydrocarbons at 140°C and atmospheric pressure when aliphatic monomers are grafted and in aromatic solvents, when aromatic monomers are used. Curiously, grafting also removes a considerable proportion of sulfur. Liquid coal can be transported in pipelines and used as liquid fuel.

Experimental processes try to preserve the cyclic structures in coals. Examples are the hydrogenation of coals that are suspended in heavy oils and the extraction of coals by solvents (e.g., tetralene) or with supercritical gases (e.g., toluene).

## 3.4 Oil

### 3.4.1 Crude Oil

#### Occurrence

**Crude oils** (**petroleums**) are light yellow, brown, or black viscous liquids (G: *petros* = stone, *petra* = rock; L: *oleum* = oil) that occur in porous sedimentary rocks at depths ranging from just below the surface to 20 km deep (Gulf coast region of the United States). Oil is commonly combined with methane and other gaseous hydrocarbons.

Prevailing theory holds that crude oil and the accompanying methane were formed in geological time spans from the biomass of plankton and from proteins, fats, and carbohydrates of aquatic microorganisms in oxygen-free standing waters. Most oil seems to have been generated by anaerobic bacteria at temperatures of 100–120°C. However, new experiments show that methane forms from $CaCO_3$ and $H_2O$ at temperatures of 500–1500°C and pressures of 5–11 GPa that prevail in the Earth's mantle at 350 km and below. Since crude oil can be made from methane, it is possible that oil was (and maybe still is!) formed inorganically. This would provide an additional, or even alternative, explanation for the fact that "exhausted" oil fields start to deliver again after some decades (the standard explanation is a recurrent migration from side reservoirs to the main pool).

Production of oil or gas from pools requires porous reservoir rocks (most often sandstone or limestone) and capping impermeable covers, the traps. There are two main types of trap: stratigraphic layers and structural traps such as domes of impermeable rocks or salt that dip away in all directions. Sedimentary reservoir rocks were originally filled with water which was subsequently replaced by the less-dense oil.

Oil was first discovered in 1840 in Baku (Russia; now: Azerbaijan, in 1855 in Persia (now: Iran), 1857 in Romania, 1859 in the United States (Oil Creek, PA), 1862 in Canada, 1927 in Iraq, 1936 in Saudi Arabia, and 1971 in Norway (see Table 3-13).

**Production and Reserves**

In some reservoirs, gas pressure is so large that drilling through the impermeable trap produces a gusher. However, most oil is obtained by pumping. Conventional pumping removes ca. 15–30 % of the oil in a reservoir (**primary oil recovery**); a higher yield is often obtained by reinjecting the produced gas into the reservoir in order to maintain gas pressure, especially if the gas has little commercial value (see Section 3.6). Some gases are also burned by so-called **flaring**. Reinjections are called **cycling** or **recycling** if a gas field is operated only for recovery of the (liquid) **condensates** (= $C_5$ and higher).

Yields can be pushed to 30–40 % if water or $CO_2$ is injected into the drilling holes (**secondary oil recovery**). An even higher (and more costly) yield can be obtained by pumping in water vapor or by chemical flooding with aqueous solutions of surfactants, alkali, or polymers such as poly(acrylamide) or xanthan (**tertiary oil recovery**).

The pumped oil often contains sand and water which are removed by settling. The density of the resulting crude oil is usually 0.82–0.94 g/mL but can also be as low as 0.65 g/mL or as high as 1.02 g/mL. Crude oils fluorescence yellowish to green-bluish because of their contents of higher aromatic compounds. The color of crude oil darkens on exposure to light because of a polymerization of unsaturated compounds to asphalt-like masses.

Crude oil contains 95–98 % hydrocarbons and 2–5 % of oxygen-, nitrogen-, and/or sul–fur-containing compounds (see also the structure of coal, Fig. 3-6). The hydrocarbons are mainly paraffins, in part naphthenes (= alkylcyclopentanes and (alkyl)-cyclohexanes), and, to a small extent, also aromatic compounds (mainly benzene, toluene, and xylenes)). Crude oils are called **sour** if they contain hydrogen sulfide and mercaptans and **sweet** if they do not. **Light oil crudes** have API gravities of more than 30°, **heavy oil crudes** API gravities of 7–22° (API = American Petroleum Institute).

The API gravity is a dimensionless quantity that is calculated from the specific gravity by API gravity = [141.5/(specific gravity at 60°F)] – 131.5 where the specific gravity is defined as the ratio of the mass of a body to the mass of an equal volume of water at the specified temperature (here 60°F).

On average, an oil well produces daily ca. 2 t in the United States and in Germany but ca. 1250 t in Saudi Arabia. The cost of producing oil varies accordingly: in 1995, it was only ca. 1–2 $ per barrel in Iraq (oil near the surface!) but 2–3 $ per barrel in Saudi

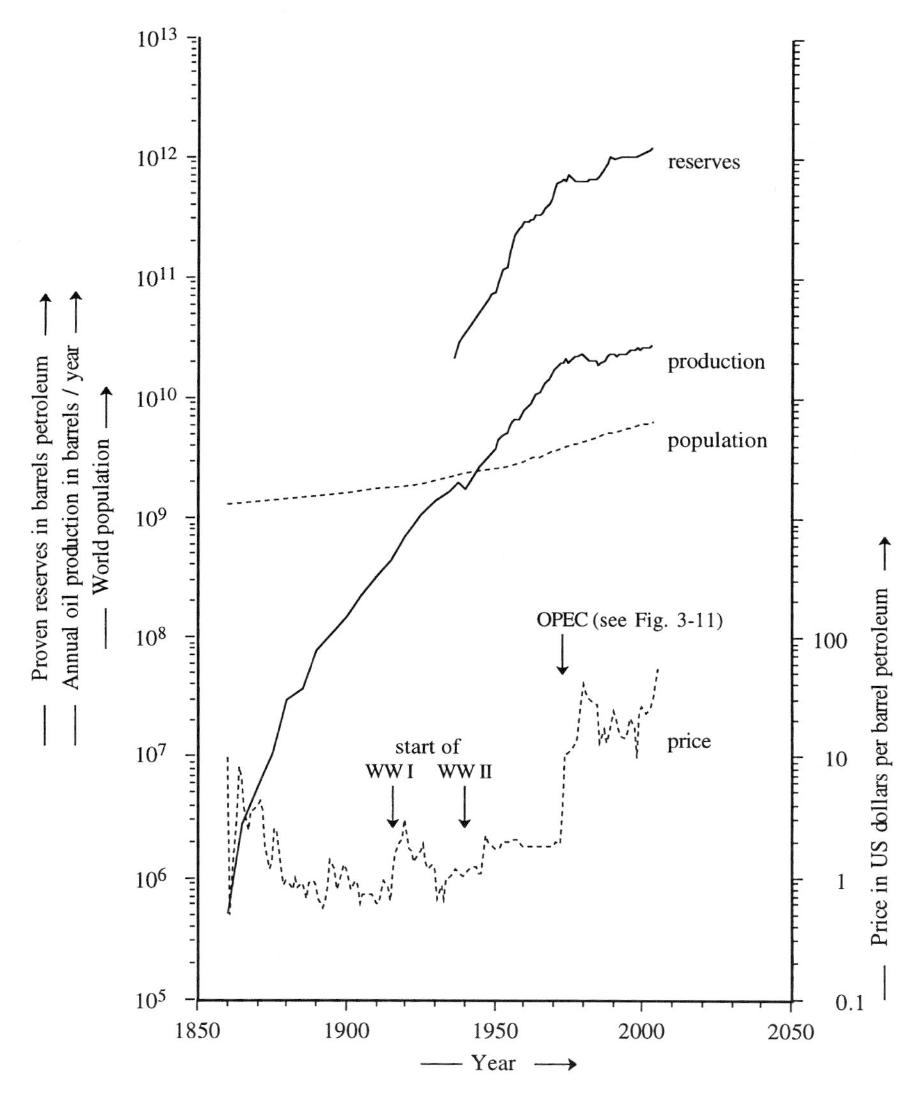


Fig. 3-10 Proved world oil reserves, annual world oil production, world population, and end-of-year price of oil in historic dollars (1860–1899: Pennsylvania; 1900–1944: U.S. average; 1945–1975: Arabian Light posted at Ras Tanura; 1976 ff.: Dubai) [4, 7, 13]. Note the logarithmic scales.

Arabia, ca. 10 $ per barrel in Texas, and ca. 15 $ per barrel in the North Sea. Secondary oil is about twice as expensive as primary oil, tertiary oil about three times.

The world production of crude oil has climbed steadily since 1860 (Fig. 3-10) although the *rate* of increase (logarithmic scale in Fig. 3-10!) has decreased since 1980. The world proved reserves have kept pace with the annual world production of oil (Fig. 3-10, Table 3-13) so that the reach has not decreased: it rather climbed from 22 years in 1925 to 43 years in 1995 for the annual productions and proved reserves known at the time. Oil reserves usually do not contain less-valuable oils such as tars and heavy crudes. Resources are speculative and often estimated as 5–6 times proved reserves.

Some of the speculative resources will turn into proved reserves by new explorations. But new explorations and investments are expensive and differ vastly in cost from region to region. Investment costs are about 10 times higher in the United States and about 15 times higher in the North Sea than in Saudi Arabia. Explorations may not only be costly but even total losses. The Mukluk concession (Northwest of Prudhoe Bay, Alaska) with suspected 1.5 billion barrels was auctioned off in 1982 to 11 partners for 1.7 billion dollars. The construction of an artificial island cost another 100 million dollars. All that the drillings produced was water.

Table 3-13 Annual production of crude oil, shale oil, oil sands, and natural gas liquids (where recovered separately) per year in 1925, 1950, 1975, and 2000 and proved reserves at the end of 1970 and 2003, all in million barrels (bbl), of important oil-producing countries [1, 4, 13-15]. Saudi Arabia and Kuwait: countries plus one-half of the neutral zone between Saudi Arabia and Kuwait. World data in tons include crude oil, shale oil, oil sands, and natural gas liquids. * OPEC countries. NA = not available. 1 bbl = 42 US gallons petroleum = 0.158 987 $m^3$ ≙ 0.0931 TOE (tons oil equivalent).

| Country | Production since | Production in $10^6$ bbl/a | | | | Reserves in $10^6$ bbl | |
|---|---|---|---|---|---|---|---|
| | | 1925 | 1950 | 1975 | 2000 | 1970 | 2003 |
| Saudi Arabia * | 1936 | 0 | 200 | 2635 | 3393 | 141 400 | 262 700 |
| USA | 1859 | 764 | 1974 | 3057 | 2830 | 45 400 | 30 700 |
| Russia | 1863 | NA | NA | NA | 2422 | | 69 100 |
| Iran * | 1913 | 34 | 242 | 1565 | 1394 | 70 000 | 130 700 |
| Mexico | 1901 | 115 | 72 | 288 | 1259 | 4 500 | 16 000 |
| Norway | 1971 | 0 | 0 | 69 | 1220 | | 10 100 |
| Venezuela * | 1917 | 20 | 547 | 885 | 1212 | 14 000 | 77 200 |
| China | 1939 | 0 | 1 | 544 | 1187 | | 23 700 |
| Canada | 1862 | ≈ 0 | 29 | 633 | 993 | 10 200 | 16 900 |
| United Kingdom | 1919 | ≈ 0 | ≈ 0 | 11 | 970 | | 4 500 |
| Iraq * | 1927 | 0 | 50 | 825 | 943 | 32 000 | 115 000 |
| United Arab Emirates * | 1962 | 0 | 0 | 619 | 912 | 11 800 | 97 800 |
| Kuwait * | 1946 | 0 | 126 | 779 | 768 | 80 000 | 96 500 |
| Nigeria * | 1958 | 0 | 0 | 652 | 768 | 9 300 | 34 300 |
| Algeria * | 1914 | ≈ 0 | ≈ 0 | 372 | 576 | | 11 300 |
| Libya * | 1961 | 0 | 0 | 540 | 538 | 29 200 | 36 000 |
| Indonesia * | 1893 | 22 | 49 | 476 | 531 | 10 000 | 4 400 |
| Brazil | 1940 | 0 | < 1 | 64 | 413 | | 10 600 |
| Oman | 1967 | 0 | 0 | 108 | 350 | | 5 600 |
| Qatar * | 1949 | 0 | 12 | 189 | 312 | 4 300 | 15 200 |
| Argentina | 1907 | 0 | 23 | 64 | 299 | | 3 200 |
| Australia | 1964 | 0 | 0 | 138 | 295 | | 4 400 |
| Malaysia | 1910 | 0 | < 1 | 37 | 289 | | 4 000 |
| India | 1889 | 0 | 2 | 54 | 285 | | 5 600 |
| Egypt | 1911 | 0 | 1 | 16 | 285 | | 3 600 |
| Kazakhstan | ? | NA | NA | NA | 272 | | 9 000 |
| Angola | 1956 | 0 | 0 | 7 | 271 | | 8 900 |
| Syria | ? | 0 | NA | 46 | 201 | | 2 300 |
| Ecuador | 1917 | ≈ 0 | 3 | 59 | 149 | | 4 600 |
| Denmark | 1972 | 0 | 0 | 1 | 133 | | 700 |
| Gabon | 1957 | 0 | 0 | 68 | 119 | | 2 400 |
| Azerbaijan | ? | NA | NA | NA | 103 | | 7 000 |
| Germany | 1880 | ≈ 0 | 8 | 41 | 49 | | |
| Total World ($10^6$ bbl) | | 367 | 3 801 | 20 331 | 27 254 | 611 200 | 1 147 700 |
| Total World ($10^6$ t) | | 149 | 525 | 2 740 | 3 604 | | |

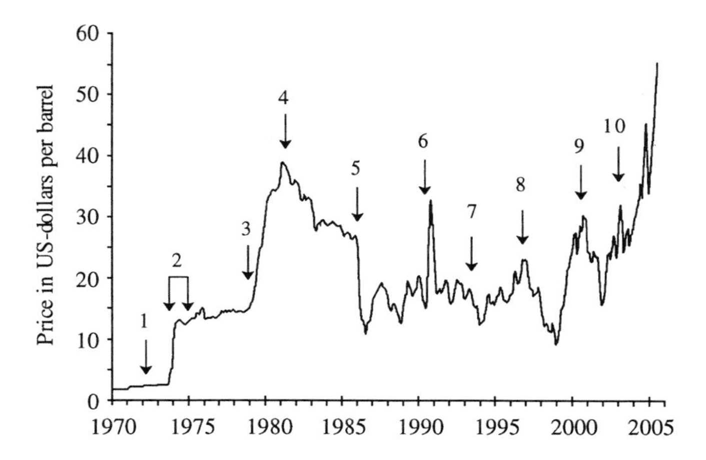


Fig. 3-11 Refiner acquisition cost of imported crude oil (historical dollars, not adjusted for inflation) [9]. Numbers indicate major events:

1. OPEC begins nationalizations. 2. Period of oil embargo, ostensibly to help in the fight against Israel. 3. OPEC starts raising prices in three steps of 14.5–15 % each. 4. First major fighting in Iran–Iraq war. 5. OPEC output reaches 18 million bbl per day. 6. Iraq invades Kuwait. 7. OPEC production reaches 23 million bbl per day. 8. U.S. launches missiles into southern Iraq following Iraqi attacks on Kurdish safe havens. 9. Strong demand and OPEC cutbacks. 10. Anticipation of U.S. invasion of Iraq.

The largest oil reserves and resources are in the Middle East (Saudi Arabia, Iran, Iraq, United Arab Emirates, Kuwait, etc.), in Venezuela, in Russia, and in the United States (Table 3-13). There may be rather large oil fields in other, not-so-well explored regions of the world, for example, in several successor states of the former Soviet Union (Kazakhstan, Azerbaijan, etc.). It is speculated that oil fields near the Falkland Islands contain $50{\cdot}10^9$ barrels and those near the coast of West Africa ca. $100{\cdot}10^9$ bbl.

Most industrialized nations do not have large oil fields and therefore haveto import oil from other regions. The oil price was fairly constant (considering inflation) between 1930 and 1970 as long as the United States was a major producer and the oil companies either owned oil fields or had long-range contracts. The situation changed after several major oil-exporting nations banded together in 1960 to form **OPEC**, the *O*rganization of *P*etroleum *E*xporting *C*ountries, ostensibly to ensure stable supplies of oil and fair prices. Present members of OPEC include Algeria, Indonesia, Iran, Iraq, Kuwait, Libya, Nigeria, Qatar, Saudi Arabia, the United Arab Emirates (UAE), and Venezuela.

Twelve years later, OPEC nations began to nationalize oil and transfer Western assets. Since this time, oil prices have been subject to many political and speculative influences (Fig. 3-11): there is no relationship between the cost of production and the price of oil.

### 3.4.2 Petroleum Processing

Crude oil (petroleum) is a mixture of about 100 000 different chemical compounds with 1-85 carbon atoms per molecule. Petroleum refining (petroleum processing) recovers these components or generates usable products from oil by distillation, extraction, and cracking which is the degrading of higher molecular weight compounds.

**Refineries**

A refinery is actually a group of various plants (Fig. 3-12) that varies with the type of crude oil and the desired refinery products. Light sweet crudes (West Texas, Alaska) need less-sophisticated layouts than heavy sour crudes (Canada, Russia, most OPEC countries). Refineries always deliver a broad range of different products; a refinery that produces petrochemicals but not gasoline is not profitable.

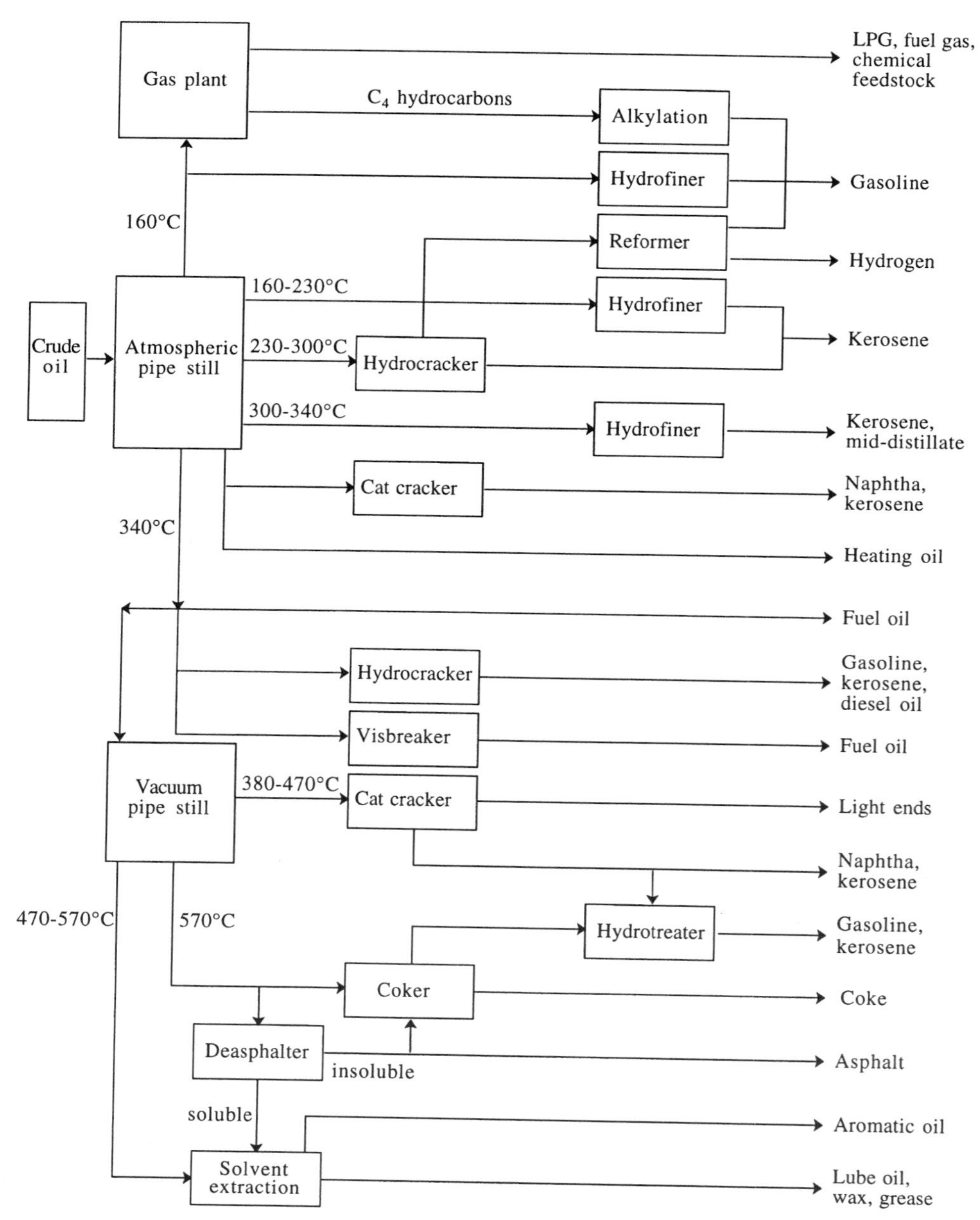


Fig. 3-12 Schematic diagram of refinery operations [16].

Table 3-14 Composition of world refinery products [6].

| Year | Production in $10^9$ tons | Refinery products in percent | | | | | |
|---|---|---|---|---|---|---|---|
| | | Ethane, LPG, naphtha | Motor gasoline | Aviation fuels | Middle distillates | Heavy fuel oil | Other products |
| 1973 | 2.719 | 5.8 | 21.0 | 4.2 | 26.0 | 33.8 | 9.2 |
| 2003 | 3.517 | 8.1 | 24.7 | 6.0 | 32.0 | 17.0 | 12.2 |

During the last 30 years, the product spectrum has changed considerably (Table 3-14). The percentage of heavy fuel oils was cut in half whereas the demand for transportation fuels (motor gasoline, aviation fuels) and chemical feedstocks increased. Changes in types of crudes and product demands led to larger and more complex refineries (cost per plant: ca. $600 \cdot 10^6$ $). In the United States, the throughput in barrels per day increased from ca. $15.0 \cdot 10^6$ in 1980 to $16.8 \cdot 10^6$ in 2004, whereas the number of refineries declined from ca. 300 to 149. This causes occasional regional gasoline shortages since climates and different environmental standards by states demand different gasoline formulations. Further U.S. refinery closings are pending because of new regulations for reformulated gasoline and ultralow-sulfur diesel oils (15 instead of 500 ppm sulfur). The United States is now an importer not only of crude oil but also of gasoline. Import is however hampered because of a tanker shortage which in turn leads to price increases.

**Distillation**

Refining begins with atmospheric distillation (Fig. 3-12) by letting oil flow through pipes in a large furnace. Vapors are processed in the gas plant and less-volatile fractions are atmospherically distilled at various temperatures. The highest boiling fractions are then distilled in vacuum. The resulting primary fractions are equilibrium mixtures that are stripped of their lower-boiling constituents before storage or further processing.

The initial atmospheric distillation of crude oil at ca. 250°C under light pressure of 2–3 bar is called **fractionation**. The resulting **fractionates** consist of **gases** and various **distillates (light, middle)** (Table 3-15) that are heavier than naphtha and range to light lubricating oils. They are known by different names, depending on industry and country. All of them practically consist only of saturated hydrocarbons.

**Light hydrocarbons** comprise ethane, propane, and butanes. **Heavy hydrocarbons** are most often $C_5$ and higher hydrocarbons; "**heavies**" include **naphtha** (bp 40–200°C) and **gas oils** (bp 200–590°C). Crude oil fractions consisting of a large number of various types of hydrocarbons are called **refined products** (e.g., gasoline, heating oils, waxes, etc.). Products of high purity with only one or two types of hydrocarbons are classified as **petrochemicals**; examples are ethene, propene, benzene, toluene, etc.

**Gases** boil below 15°C (Table 3-15). They consist of saturated compounds with $N$ = 1–4 carbon atoms that are used as **refinery gases** for further refining operations or as **liquid gases** for fuel. Methane is usually considered separately from the light hydrocarbons with $C_2$–$C_4$. The mixture of propane and butanes under pressure at room temperature, i.e., in liquid state, is known as **liquefied petroleum gas (LPG)**. Its major non-chemical use is as fuel. A mixture of ethane and propane, when used as feedstock for steam cracking, is also called LPG.

Table 3-15 Boiling ranges $\Delta T_s$/°C, fractions, and intermediates from the distillation of crude oil. LPG = liquefied petroleum gases. Note that "naphtha" has a different meaning in coal processing.

| $\Delta T_s$/°C | Crude oil fractions | Intermediates | Products from intermediates |
|---|---|---|---|
| *Atmospheric distillation (fractionation)* | | | |
| < 15 | gases | $CH_4$, $C_2H_6$, $C_3H_8$, $C_4H_{10}$ | gases in cylinders, LPG |
| 15–175 | light distillates | primary flash distillates (PFD) | |
| | | straight run gasolines (SRG) | gasolines |
| | | straight run benzine (SRB) | |
| | | light distillate feedstock (LDF) | |
| | | cat reformer feedstocks | |
| 175–350 | middle distillates | naphtha | derivatives (chem. feedstock) |
| | | kerosine | jet fuels (kerosine) |
| | | gas oils | domestic heating oils, fuel oils, diesel oils |
| > 350 | atmospheric distillates | | fuel oils |
| *Vacuum distillation* | | | |
| 350–550 | vacuum distillates | spindle oil | lubricting oils, paraffin waxes |
| | | lubricating oils | |
| | | cat cracker feed | |
| | | hydrocracker feed | |
| > 550 | vacuum residues | bitumens | bitumen or asphalt |
| > 550 | deasphalted residues | bright stock | lubricating oils |
| | | cylinder stock | micro wax |
| - | residues | asphaltenes | fuel oils, petroleum cokes |

**Light distillates** boil between 15°C and 175°C. Primary flash distillates comprise, for example, **petroleum ether** (petroleum spirits, white spirits), a $C_5$-$C_6$ fraction that boils at 40-70°C, smells like ether, and is used as a solvent. Straw-colored fractionates distilling off from petroleum between 70°C and 90°C are sometimes called **naphtha** (Babylonian: *neptu* = petroleum). Light naphtha approaches **petroleum ether (ligroine)**, heavy naphtha is more like gasoline. The term "naphtha" is also used for heavier distillates from petroleum, for fractions from cracking operations, for various light oils in the distillation of coal tar, and for crude oils that are rich in cyclohexanes. Light naphtha (chemical gasoline) is used as feedstock for ethene, propene, butadiene, isoprene, and ammonia. The price of olefins is therefore always linked to the price of naphtha.

**Benzines** are mixtures of $C_5$–$C_{12}$ hydrocarbons that range from petrolethers and light benzine (bp 70–90°C, mainly $C_6$ and $C_7$), middle benzine (gasolines; bp 90–180°C), to heavy benzine (ligroins, bp 150–180°C). Again, designations and boiling ranges vary.

**Straight-run gasolines** (SRGs) boil at 95–175°C. They have low octane values and are thus prone to engine knock. For **motor gasolines** (US: gasoline, gas; UK: petrol), SR gasolines are therefore subjected to **catalytic reforming** (cat reforming) at 400–500°C in the presence of hydrogen and platinum which leads to a higher proportion of branched hydrocarbons. Gasolines vary widely in composition. They are always blended.

**Kerosine** (ASTM, ACS) (= kerosene (general language)), also called "coal oil" is the general term for fractionates from the distillation of crude oils with boiling points ranging between those of benzines and diesel oils (G: *keros* = wax; in International English, kerosine is called "paraffin"). Kerosine is mainly used for jet fuels.

**Gas oil** serves for heating and Diesel oils, a small portion also as feedstock for ethene.

**Refining**

In general, distillation does not produce specific fractions. Distillation alone also does not generate the desired product mix, especially not the desired large quantities of gasoline. Distillation is therefore followed by a series of other processes (Fig. 3-12).

**Cracking** is the degradation of larger molecules to smaller ones by free-radical kinetic chain reactions, i.e., pyrolyses. At 800°C, the reaction is spontaneous and non-catalytic (**thermal cracking**). Cracking processes are mainly used to produce ethene and other olefins, especially by **steam cracking** (e.g., 1 s at 800°C or 0.5 s at 900°C), where water vapor is added. The water vapor reduces the partial pressure of hydrocarbons which, in turn, suppresses their polymerization to tars.

**Visbreaking** is a "mild" thermal cracking operation (450–510°C, 3–20 bar) that increases the proportion of light heating oil which is needed for fuel oil specifications.

Gasolines are produced by cracking at temperatures of 400–500°C using catalysts such as zeolites or magnesium, aluminum, or molybdenum silicates (**catalytic cracking, cat cracking**). Cat cracking in fluidized beds is the preferred method for gasolines since it delivers larger fractions of branched hydrocarbons (octane numbers (ON) are based on isooctane which has ON ≡ 100). Cat cracking delivers not only gasolines but also gases such as methane, ethane, ethene, propene, butenes, etc.

Higher hydrocarbons are degraded by cat cracking in the presence of hydrogen (**hydrocracking**). Distillation residues are cracked under mild conditions (1–5 s at 460–500°C) to lower-viscosity products that are added to heavy heating oils (viscosity breaking, **vis breaking**). **Coking** uses finely divided coke instead of catalysts. It delivers not only gasoline but also **petroleum coke** (**oil coke**) for the manufacturing of electrodes, specialty steels, etc.

**Reforming** also converts low-octane gasolines into higher ones, using as feedstock saturated hydrocarbons or naphtha with substantial proportions of naphthenes. Thermal reforming (0.2 s at 650°C) has been supplanted by **catalytic reforming** (490–540°C, 8–40 bar, partial $H_2$ pressure) in the presence of hydrogenation-dehydrogenation catalysts such as platinum and other precious metal catalysts. Solvent extraction of aromatics from the resulting **reformate** yields **raffinate**, a feedstock for steam cracking.

**Treating**

Crude oils and their fractionates from distillation contain sulfur compounds such as hydrogen sulfide, sulfides, disulfides, mercaptans, and thiophenes. Fractionates are therefore desulfurized (**sweetened**) by various procedures that depend on the type and proportions of the sulfur compounds. Immediately after distillation, fractionates are washed with aqueous solutions of lye (sodium hydroxide or caustic soda) which converts hydrogen sulfide to sodium sulfide and mercaptans, followed by oxidation to disulfides. This procedure avoids the formation of sulfur from the reaction of hydrogen sulfide and mercaptans with oxygen from air.

Higher boiling fractionates such as paraffinic kerosines and lubricating oils are desulfurized mainly by treatment with sulfuric acid; other acids are less often used.

Lubricating oils and waxes are still treated with clays which remove unwanted odors and colors as well as traces of asphaltic compounds.

Solvents such as benzene, ketones, alkyl chlorides, etc., are used to dewax fractionates and solvents as diverse as propane, phenol, furfural, etc., to extract certain compounds.

### 3.4.3 Petrochemistry

Direct distillation delivers only relatively small amounts of olefins and other intermediates for chemical syntheses. Instead, the majority of these compounds is obtained from petrochemical plants that are usually integrated with refinery operations. These plants are usually fairly inflexible: a refinery that was built for an optimal production of gasoline cannot be switched to an optimal production of heating oil and vice versa, and certainly not to a drastically enhanced production of base chemicals and intermediates.

Prices of intermediates and monomers are thus coupled with prices of transportation and heating fuels which are the main business of refineries. However, prices of gasoline and heating oils are often political prices that are held low to keep the population happy. Refineries therefore often compensate the reduced profits or losses from sales of gasoline and heating oil by raising prices of chemical feedstocks. This, in turn, may lead to rather large price swings for monomers, as Fig. 3-2 shows for the petrochemical products ethene and styrene as compared to the refinery products naphtha and benzene.

Base chemicals such as ethene, propene, butadiene, etc., are obtained by **thermal** or **catalytical cracking** of hydrocarbon gases (usually as so-called **liquid gases**) on one hand and naphtha or gas oils on the other (Table 3-16). Liquid gases comprise LPG = liquified petroleum gases (p. 51) and NGL = natural gas liquids (see Section 3.6). Direct cracking of crude oil, which would have made the chemical industry less dependent on the oil industry, is not competitive.

Cracking involves both chain scission (Eq.(3-5)) and dehydrogenation (Eq.(3-6)):

$$(3\text{-}5) \qquad C_{m+n}H_{2(m+n)+2} \rightarrow C_mH_{2m} + C_nH_{2n+2}$$

$$(3\text{-}6) \qquad C_nH_{2n+2} \rightarrow C_nH_{2n} + H_2$$

Table 3-16 Composition of products from cracking of gases, naphtha, and crude oil crackers and various feedstocks [17]. [1] Part of fuel gas. [2] Part of "Other gases" (XV). [3] Included in $C_4$ (X). [4] Wherefrom 68 % aromatics (BTX). [5] Wherefrom 52 % aromatics (BTX). AGO = atmospheric gas oil.

| | Product | Percent of products from various crackers and feedstocks | | | | |
|---|---|---|---|---|---|---|
| | Cracker → | Gas crackers | | Naphtha crackers | | Crude crackers |
| | Feedstock → | Ethane | Propane | Naphtha | AGO | Crude oil |
| I | Fuel gas: Hydrogen | 5.8 [1] | 1.9 [1] | | | [2] |
| II | Fuel gas: Methane | 7.5 [1] | 27.0 [1] | 16.7 | 11.6 | 12.4 |
| III | Fuel gas: Other components | 0.5 [1] | 0.2 [1] | | | 0 |
| IV | Ethane | | | | | 1.6 |
| V | Ethene | 80.0 | 44.2 | 32.0 | 25.3 | 24.4 |
| VI | Ethyne (= acetylene) | 0 | 0 | 0 | 0 | 2.5 |
| VII | Propene | 1.9 | 16.6 | 13.4 | 14.1 | 6.9 |
| VIII | Propyne (methyl acetylene), allene | | | | | 0.7 |
| IX | Butadiene | 2.4 | 3.1 | 4.6 | 4.7 | [3] |
| X | Other $C_4$ compounds | 0.7 | 1.4 | 3.4 | 5.7 | 5.1 |
| XI | $C_5$ fraction | | | | | 1.6 |
| XII | $C_6$-$C_7$ fraction | | | | | 14.3 |
| XIII | Fuel oil, including $C_8$ | | | 6.2 | 18.5 | 28.5 |
| XIV | Crack gasoline (pyrobenzine) | 1.2 | 5.6 | 23.7 [4] | 20.1 [5] | |
| XV | Other gases ($CO$, $CO_2$, $H_2$) | | | | | 1.7 |
| XVI | $H_2S$ | | | | | 0.3 |

The preferred feedstocks for cracking are liquid gases in the United States and naphtha or gas oil in Europe and Japan. The reason is the different demand for main refinery products: gasolines in the United States but heating oils in Europe and Japan. The production of the higher molecular weight heating oils requires less drastic refinery operations than that of the lower molecular weight gasoline. The former thus produces relatively more naphtha or gas oil as byproducts than that of gasoline, which leads to relatively more liquid gas. The byproducts are thus used as feedstocks for cracking.

In 1995, naphtha was the feedstock for 97 % of ethene in Japan, 71 % in Western Europe, and only 11 % in the United States. Liquid gas (LPG, NGL) delivered 76 % of ethene in the United States but only 17 % in Western Europe. In both the United States and Europe, 9 % of ethene was derived from gas oil and 3–4 % from refinery gases.

Cracking reactions are free-radical chain reactions that involve thermal α-scissions of C–C bonds (activation energy $\Delta E^{\ddagger} \approx 350$ kJ/mol), for example,

$$(3\text{-}7) \qquad CH_3(CH_2)_4\text{–}(CH_2)_3CH_3 \underset{\text{termination}}{\overset{\text{thermal scission}}{\rightleftharpoons}} CH_3(CH_2)_4^{\bullet} + {}^{\bullet}(CH_2)_3CH_3$$

β-scissions with activation energies of ca. 45 kJ/mol, for example,

$$(3\text{-}8) \qquad \begin{array}{lcll} CH_3CH_2CH_2CH_2CH_2^{\bullet} & \longrightarrow & CH_3CH_2CH_2^{\bullet} & + \; CH_2{=}CH_2 \\ CH_3CH_2CH_2CH_2^{\bullet} & \longrightarrow & CH_3CH_2^{\bullet} & + \; CH_2{=}CH_2 \\ CH_3CH_2CH_2^{\bullet} & \longrightarrow & CH_3^{\bullet} & + \; CH_2{=}CH_2 \end{array}$$

hydrogen transfer reactions, for example,

$$(3\text{-}9) \qquad CH_3CH_2^{\bullet} + CH_3(CH_2)_7CH_3 \longrightarrow CH_3CH_3 + CH_3(CH_2)_5C^{\bullet}(H)CH_2CH_3$$

and termination reactions, for example, (re)combinations with $\Delta E^{\ddagger} \approx 0$ kJ/mol,

$$(3\text{-}10) \qquad CH_3(CH_2)_5C^{\bullet}(H)CH_2CH_3 + {}^{\bullet}CH_2CH_3 \longrightarrow CH_3(CH_2)_5CH(C_2H_5)CH_2CH_3$$

or terminations by disproportionation ($\Delta E^{\ddagger} \leq 15$ kJ/mol), for example,

$$(3\text{-}11) \qquad CH_3CH_2CH_2^{\bullet} + {}^{\bullet}CH_2CH_2CH_3 \longrightarrow CH_3CH_2CH_3 + CH_2{=}CH(CH_3)$$

Because of the high activation energies, reactions (3-7) and (3-8) increase strongly with increasing temperature, which in turn delivers relatively more ethene and relatively less higher olefins. The upper reaction temperatures are limited by the thermal stability of the steels used for the reactors. They are now 850–900°C.

The yields of the various components can be varied not only by the cracking temperature but also by the residence time in the cracker. The yield of ethene increases with decreasing residence time and by dilution of the feed with water vapor. The reason for the latter effect is the suppression of bimolecular reactions (3-9)–(3-11) by lowering the partial pressure of their components, whereas the rate of the unimolecular reactions (3-7) and (3-8) is not affected. Injection of water vapor also reduces the formation of coke. It has the disadvantage that bigger reactors and more energy are required. The largest naphtha steam crackers now annually produce 920 000 tons of ethene and 550 000 tons of propene.

Table 3-17 $C_4$ fractions (in weight percent) from gas oil or naphtha [15]. $t_r$ = residence time.

| $C_4$ compound | Boiling temperature in °C | Composition of $C_4$ fraction from catcracking of gas oil | steam cracking of naphtha low severity < 800°C $t_r$ = 1 s | high severity < 900°C $t_r$ = 0.5 s |
|---|---|---|---|---|
| Butane | −1 | 13 | 4 | 3 |
| Isobutane | −12 | 37 | 2 | 1 |
| 1-Butene | −6 | 12 | 20 | 14 |
| 2-Butene, *trans* | 1 | 12 | 7 | 6 |
| 2-Butene, *cis* | 4 | 11 | 7 | 5 |
| Isobutene | −4 | 15 | 32 | 22 |
| 1,3-Butadiene | −4 | 0.5 | 26 | 47 |
| 1,2-Butadiene, ethyl and vinyl acetylene | | 0 | 2 | 2 |

Syntheses of petrochemicals from naphtha compete with those from synthesis gas, town gas, and synthetic natural gas (see Section 3.6). Naphtha as petrochemical feed is therefore increasingly replaced by higher fractionates. This, in turn leads to higher proportions of $C_4$–$C_6$ compounds and smaller proportions of ethene.

Ethene (99 % purity) and propene from cracking are either used directly as monomers (Chapter 6) or processed further to other intermediates and monomers (Chapter 4).

The components of the $C_4$ fraction cannot be separated by distillation because their boiling temperatures are in a very narrow range (Table 3-17). Therefore, 1,3-butadiene is first extracted by *N*-methyl pyrrolidone or *N*,*N*-dimethylformamide. The remaining gas is then reacted with water which generates almost exclusively *t*-butanol and no other butanols because isobutene reacts about 1000 times faster than the other butenes. The remaining butenes are subsequently extracted and distilled. The separated *t*-butanol is dehydrated back to isobutene.

The $C_5$ fraction contains $C_5$ hydrocarbons and some $C_4$ and $C_6$ compounds (Table 3-18); it is similarly worked up. **Crack gasoline = pyrolysis gasoline = pyrobenzine** (Table 3-16) is a source of aromatics or used as an octane enhancer for gasolines.

Table 3-18 Composition of a $C_5$ fraction from the steam cracking of light ends. na = Not applicable.

| Component | bp/°C | wt% | Component | bp/°C | wt/% |
|---|---|---|---|---|---|
| Pentane | 36.0 | 22.1 | Cyclopentane | 49.3 | 0.9 |
| *i*-Pentane | 27.8 | 15.0 | Cyclopentene | 44.2 | 1.8 |
| | | | Cyclopentadiene | 41.7 | 14.5 |
| 1-Pentene (α-amylene) | 29.9 | 3.4 | + dicyclopentadiene | 171 | |
| 2-Pentene, *trans* | 36.3 | 3.3 | | | |
| 2-Pentene, *cis* | 36.9 | ≈ 0 | Hexane | 68.7 | 2.1 |
| 2-Methyl-1-butene | 38.6 | 4.7 | 2-Methylpentane | 60.2 | 0.6 |
| 3-Methyl-1-butene | 20.0 | 0.6 | 3-Methylpentane | 63.2 | 0.3 |
| 2-Methyl-2-butene | 38.0 | 2.6 | 2,2-Dimethylbutane | 40.7 | 0.1 |
| | | | 2,3-Dimethylbutane | 57.9 | 0.1 |
| 1,4-Pentadiene | 26.0 | 1.6 | | | |
| 1,3-Pentadiene, *trans* | 42.0 | 5.5 | $C_4$ compounds | na | 0.5 |
| 1,3-Pentadiene, *cis* | 44.1 | 3.2 | | | |
| Isoprene | 34.0 | 15.0 | Other compounds | na | 3.1 |

## 3.5 Kerogens and Bituminous Substances

### 3.5.1 Oil Shale

Oil shales are kerogen-containing porous rocks of low permeability that are found on all continents. **Kerogen** is a wax-like insoluble organic material (G: *keros* = wax, *genes* = born) that seems to have originated from plants or animal organisms. Chemically, it is a highly crosslinked polymer that decomposes thermally to alkanes, olefins, isoprenoids, terpenoids, porphyrins, etc. The composition of the degradation products depends strongly on the origin of oil shale. For example, alkyl pyridines, pyrrols, and indoles total 40 % of the degradation products of kerogen from the Wyoming Green River deposit, whereas Estonian kerogen delivers 50 % of alkyl phenols, naphthols, and resorcinols.

The world-wide resources of **shale oil** extracted from oil shales are estimated as ca. $80 \cdot 10^{12}$ barrels petroleum (bbl), i.e., ca. 50 times the proved reserves of ca. $1.15 \cdot 10^{12}$ bbl of crude oil (Table 3-13). A shale oil deposit ca. 200 miles (≈ 320 km) west of Denver is said to contain 1 trillion bbl, i.e., ca. 4 times the proved reserves of Saudi Arabia.

Shale oil is difficult to extract because kerogen and rocks are intimately mixed. Retorting of oil shale at ca. 500°C, either in situ or above ground after mining and crushing, produces gas, liquid gas, and shale oil; ca. 20 % of the kerogen are converted to coke that remains in the rock. Large quantities of shale has to be moved for ground retorting since one ton of oil shale delivers only between 20 and 420 L shale oil, consisting on average of ca. 20 % naphtha, 50 % kerosine/gas oil, and 30 % fuel oil. The whole operation thus has to be performed at the source. It requires large amounts of water, ca. 1–3 liter per 1 liter of produced oil, which is an environmental headache. At crude oil prices of more than ca. 50 $/bbl, production of shale oil is economical (2005).

Crude shale oil has to be prerefined before it can be subjected to refining since it contains ca. 2 % nitrogen compounds, which are catalyst poisons for refining operations. Distillation of crude shale oil delivers naphtha, light oils, heavy oils, and residue. The subsequent hydrogenation leads to a **synthetic crude oil** (**syncrude**) which can be conventionally processed petrochemically.

Several refining procedures have been tested, for example, for gasolines (hydrotreating plus fluid catalytic cracking), for jet fuels (hydrotreating plus hydrocracking), and for diesel fuels (coking followed by hydrotreating).

### 3.5.2 Tar Sands

Tar sands (**oil sands, bituminous sands**) consist of 5–18 % highly viscous, bituminous oils in sands or water-swollen clays. The oils are rich in aromatic compounds and sulfur.

Tar sands are found in all regions of the world. The largest deposits are in the Athabasca region of the Province of Alberta (Canada) and in Venezuela. The Athabasca field comprises 34 000 square kilometers (≈ 13 000 square miles) with ca. $500 \cdot 10^{12}$ bbl proved reserves and $500 \cdot 10^{9}$ bbl resources of oil (2005 production: ca. $9 \cdot 10^{8}$ bbl/a). The Venezuelan Orinoco region holds $270 \cdot 10^{12}$ bbl reserves and $930 \cdot 10^{12}$ bbl resources. The United States, the former Soviet Union, Madagaskar, Peru, Trinidad, the Balkan states, Italy, and the Philippines have total reserves of ca. $2.5 \cdot 10^{12}$ bbl oil.

At present, only Canada and Venezuela produce oil from tar sands. In Canada, tar sands are mined if the deposit is shallow and the overburden not thicker than the deposit itself. **Oil mining** involves the drilling of mine shafts into the oil rock so that the oil drains into the shaft from where it can be pumped out. Alternatively, tar sands at the surface can be scraped off and transported to a plant where the oil is extracted.

Non-mining techniques are applied in Canada for tar sands with a thick overburden. They are essentially secondary oil recovery methods with injected hot water or water vapor, and maybe added detergents. This **steam-assisted gravity drainage** (SAGD; "**huff-and-puff process**") employs two underground wells. High-pressure steam is injected into the oil sands through the higher tap well where it melts the bitumen. Gravity causes the bitumen to collect in the lower well since bitumen is more dense than water at room temperature but less dense at temperatures above 80°C. The bitumen is then pumped to the surface for processing. Costs are estimated as 10–20 $/barrel (2005).

The Orinoco deposits are extracted by injection of water containing an emulsifier and 12 % NaCl. Heating of the resulting primary emulsion reduces the water to ca. 2 %. Addition of new water and new emulsifier results in an emulsion with 70 % bitumen and 30 % water that is stable up to a year and can be used directly in thermal power plants. This feed produces less $CO_2$ per produced energy than hard coal but the plant needs desulfurization devices.

### 3.5.3 Bitumens and Asphalts

**Bitumens** are semi-solid to solid mixtures of high-molecular weight hydrocarbons (L: *bitumen* = pitch; from *pix tumens* = tomb pitch because it was used in Egypt for embalming). The term "bitumen" is used for different matter from various sources. For example, the United Nations divides degassed petroleums in "crude oils" with in situ viscosities of less than 10 Pa s and "bitumens" with in situ viscosities of more than 10 Pa s. In industry, the term is now used for both naturally occurring bitumens (Fig. 3-13) and the viscous residues from crude oil distillation (Table 3-15).

The terminology is further confused by the interchangeable use of "bitumen" and "asphalt" in the United States. Internationally, "asphalt" is the name for a naturally occurring or synthetically produced mixture of bitumenous substances and inorganic minerals (G: *asphaltos* = bonding material for stones; from *a-* = non and *sphallein* = to cause a fall). An example is Trinidad asphalt, which is a naturally occurring composite of 47 % bitumen, 28 % clay, and 25 % water.

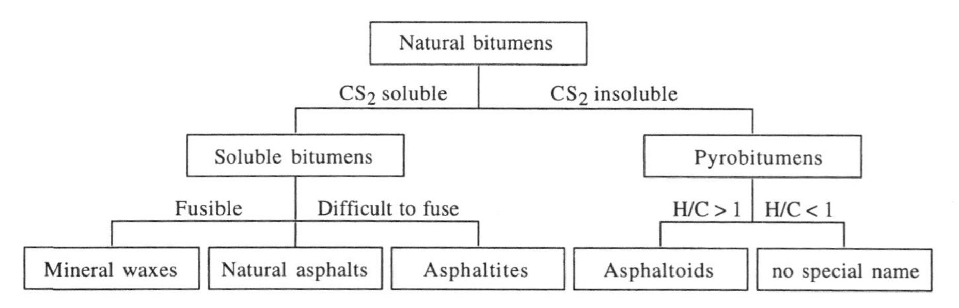


Fig. 3-13 Classification of natural bitumens (adopted from [18]).

Bitumens are mixtures of many different types of molecules: branched aliphatics, cycloaliphatics, aromatics, heterocyclics, etc.; molecular weights of soluble species vary between ca. 300 and ca. 2000. Bitumens also contain heteroelements, predominantly sulfur and nitrogen, but also oxygen, vanadium, nickel, and iron.

Depending on the solubility of natural bitumens in $CS_2$, one distinguishes **soluble bitumens** and **pyrobitumens** (Fig. 3-13). *Soluble bitumens* are furthermore subdivided according to their fusibility into **mineral waxes** (example: ozocerite = ozokerite), **natural asphalts** (examples: Trinidad asphalt and Athabasca tar sands), and **asphaltites** (example: gilsonite). *Pyrobitumens* include **asphaltoids** with high ratios of hydrogen to carbon ($H/C > 1$; example: elaterite) and pyrobitumens with a ratio of $H/C < 1$ (no special term).

**Mineral waxes**, also called **earth waxes**, such as **ozocerite** are light yellow to black, soft to hard-brittle mixtures of aliphatic, aromatic, and naphthenic hydrocarbons. High-molecular weight ozocerites do not smell, but low-molecular weight ones smell either aromatic or like crude oil (G: *ozein* = to smell, *keros* = wax). Purified ozocerites are white and non-smelling; they were traded as **ceresine**.

Natural ozocerites are found in the United States (Utah, Texas), in Poland, and in Russia near the Caspian Sea and the Baikal where they were mined at depths of 250–450 m. The presently traded ozocerites are either synthetic materials from crude oil refining or blends of synthetic and natural ozocerites.

**Natural asphalts** have widely varying contents of inorganics and water. The asphalts of Trinidad contain 47 % bitumen but Athabasca tar sands have less than 18 %

**Asphaltites** are hard natural asphalts with few minerals. The most important representative is **gilsonite**, a brown mineral that is only found in Utah. Gilsonite reserves are estimated at 5 million tons, resources at 10 million tons. This **pitch** is the preferred raw material for graphite that is used in nuclear reactors. The car industry uses it for solvent-free joints. It is also important for printing inks.

**Asphaltoids** have low contents of oxygen groups. An example is the mineral **elaterite** (see p. 215). Pyrobitumens with H/C ratios smaller than 1 are hard coals (Section 3.3); an example is anthracite with 90/240.

Residues from the distillation of petroleum are man-made bitumen; they are called **pyrogenous asphalts** in the United States. **Catback asphalts** are bitumens that are liquefied with petroleum distillates whereas bitumens mixed with rosin ester are **modified asphalts**, and those treated with $P_2O_5$ are known as **catalyzed asphalts**. An **emulsified asphalt** is an emulsion of asphalt in water.

Bitumen ("asphalt") is a thermoplastic material consisting of a micellar solution of 20–30 % asphaltenes in maltenes. **Maltenes (petrolenes)** have a high H/C ratio and molecular weights of $M \approx 300–400$ whereas **asphaltenes** have low H/C ratios and $M > 1000$.

Bitumen is a thermoplastic. Its production volume surpasses that of poly(ethylene) (USA 2002: $36 \cdot 10^6$ t/a versus $16 \cdot 10^6$ t/a). Because of its low cost (less than 10 US-cts/lb), it is mainly used in paving (USA: 75 %), for roofing (USA: 15 %), as adhesive, coating and casting material, for electrical insulation, etc.

The main disadvantages of bitumen are its black color, rutting (formation of furrows), and the strong temperature-dependence of viscosity. Properties can be improved by additives: flow and application by small proportions of diblock polymers (better dispersion of asphaltenes in maltene), increased plasticity by added polyolefins, and improved thermomechanical behavior by crosslinking with sulfur, maleic anhydride, or sulfur trioxide.

# 3.6 Natural Gas

## 3.6.1 Composition

Nature produced and is still producing large amount of organic gases. **Marsh gas**, a flammable mixture of methane and carbon dioxide, results from the anaerobic reduction of $CO_2$ and of organic compounds of plants and dead animals by bacteria on the oxygen-poor bottom of marshes and lakes. Enormous amounts of **methane** are generated by termites, in rice plantations by bacteria, and in the digestive tracts of ruminants. Ocean floors at 500–900 m depth and permafrost areas also contain **methane hydrate** in estimated amounts of $1700 \cdot 10^{15}$ $m^3$ $CH_4$ that probably surpass those of the combined reserves and resources of all other fossil fuels (cf., natural gas: $0.5 \cdot 10^{15}$ $m^3$).

**Natural gas** is the industry term for gases with high contents of aliphatic hydrocarbons. It is found in the Earth's crust, either dissolved in crude oil (**dissolved gas**), in immediate contact with crude oil but without dissolution (**associated gas**), or neither in contact with nor dissolved in crude oil (**non-associated gas**).

Natural gas (**NG**) contains methane and higher aliphatic hydrocarbons and also carbon dioxide, nitrogen, dihydrogen sulfide, mercaptans, water vapor, and traces of other chemical compounds. The composition of natural gas varies strongly with the region (Table 3-19). European natural gas is very rich in methane, whereas Saudi Arabian and Iranian natural gases contain proportionally more higher hydrocarbons. French natural gas from the Lacq region is so rich in dihydrogen sulfide that the latter can be used to produce elemental sulfur.

Some sulfur compounds are highly toxic and corroding. Natural gas with high concentrations of such sulfur compounds that cannot be produced technically and/or economically is called **sour gas**. A **sweet gas**, on the other hand, can be produced and processed without additional operations.

Higher hydrocarbons are easy to liquefy. Virgin or partially processed natural gases with such hydrocarbons are called **wet gases**. A **dry gas**, on the other hand, contains no or only a little economically obtainable liquefiable gases; a dry gas may also be a dehydrated natural gas. **Natural gas liquids** (**NGL**s) are mixtures of hydrocarbons (generally $C_2$–$C_5$) that exist as gases in the reservoirs because of the temperatures and pressures prevailing therein but can be obtained as liquids by condensation or absorption.

**Liquefied petroleum gases** (**LPG**s), on the other hand, are byproducts of the production of natural gas or crude oil. They usually consist of $C_3$–$C_4$ hydrocarbons. The concentration of LPGs in crude oils is usually too small for economic recovery; these LPGs are flared at the production well. An exception is American LPGs.

## 3.6.2 Production, Consumption, and Reserves

Natural gas was originally worthless because it could not be transported to consumers. Dissolved or associated gases were thus either reinjected or just flared (Table 3-20), only a small proportion was directly processed near production sites. There was no incentive to search for non-associated gas.

The situation changed after very expensive pipelines were constructed, for example, from Russia to Germany, and suitable rock formations were identified that could be uti-

Table 3-19 Composition of natural gas from various regions of the world.

| Region | Proportion in weight percent | | | | | | | |
|---|---|---|---|---|---|---|---|---|
| | $CH_4$ | $C_2H_6$ | $C_3H_8$ | $C_4H_{10}$ | $CO_2$ | $H_2S$ | $N_2$ | Other |
| USA (Rio Arriba, NM) | 93.5 | 2.4 | 0.5 | 0.2 | 2.2 | | 1.1 | 5.1 |
| Algeria | 86.9 | 9.0 | 2.6 | 1.2 | | | | 0.3 |
| North Sea | 85.5 | 8.1 | 2.7 | 0.0 | | | | 3.7 |
| Iran | 74.9 | 13.0 | 7.2 | 3.1 | | | | 1.8 |
| USA (Amarillo, TX) | 51.4 | 5.6 | 3.7 | 2.3 | | | 35.0 | 2.0 |
| France (Lacq) | 49.6 | 4.0 | 2.7 | 1.5 | 19.5 | 22.6 | 0 | 0.1 |
| Saudi Arabia | 48.1 | 18.6 | 11.7 | 4.6 | | | | 17.0 |

lized for the subterranean storage of natural gas. The situation changed again after special ocean tankers were built that allow the transport of **liquefied natural gas (LNG)** from gas and oil fields, for example, from Arabian oil fields to Japan.

For this purpose, natural gas is liquefied at –163°C at slightly elevated pressure. The liquefaction removes dissolved oxygen, carbon dioxide, water, and sulfur compounds and reduces the volume to ca. 1/600. The remaining LNG is nearly pure methane (bp –161.3°C at atmospheric pressure).

Tankers can transport 150 000–200 000 $m^3$ LNG, i.e., up to ca. 200 000 tons. They have well insulated double-walled tanks that keep the LNG in "boiling condition"; no external refrigeration is required. A small proportion of LNG vents or boils off, a part of of it into air but another part is collected and used as fuel.

Costs vary widely: $10^6$ BTUs sell for 0.85 US-$ in Qatar, but for 8.50 US-$ in the United States. The cost of a full LNG facility (drilling, pipelines, liquefaction, tankers, etc.) is estimated as $(4\text{-}5)\cdot 10^9$ US-$ (2005). Both LNG-producing and -importing countries therefore demand safeguards against sudden price swings. For this reason and because of the coupling of the production of dissolved and associated natural gas with that of crude oil, the price of NG and LNG is not free but coupled to the price of crude oil.

The world production of natural gas climbed steadily from $1513\cdot 10^9$ $m^3$/a in 1980 to $2620\cdot 10^9$ $m^3$/a in 2003. The same is true for most world regions except Europe/Eurasia where it dipped during 1992–1995 (Fig. 3-14) because of the dissolution of the Soviet Union and the replacement of the planned economy by a more free market one.

Table 3-20 Production, sales, reinjection, and flaring of natural gases in some OPEC countries in the year 1980 [13]. Data in million cubic meters. Most flaring has now been replaced by liquefaction.

| Country | Production volume | Sales | Use of produced volume Reinjection | Flaring | Other/unknown |
|---|---|---|---|---|---|
| Saudi Arabia | 53 265 | 11 431 | 270 | 38 368 | 3 196 |
| Algeria | 43 427 | 11 647 | 14 366 | 9 714 | 7 700 |
| Venezuela | 35 449 | 13 854 | 16 535 | 2 235 | 2 825 |
| Indonesia | 29 612 | 18 503 | 6 296 | 1 700 | 3 113 |
| Nigeria | 24 552 | 1 070 | 0 | 23 482 | 0 |
| Libya | 20 380 | 5 170 | 10 650 | 4 560 | 0 |
| Iran | 20 080 | 7 138 | 2 340 | 9470 | 1 132 |

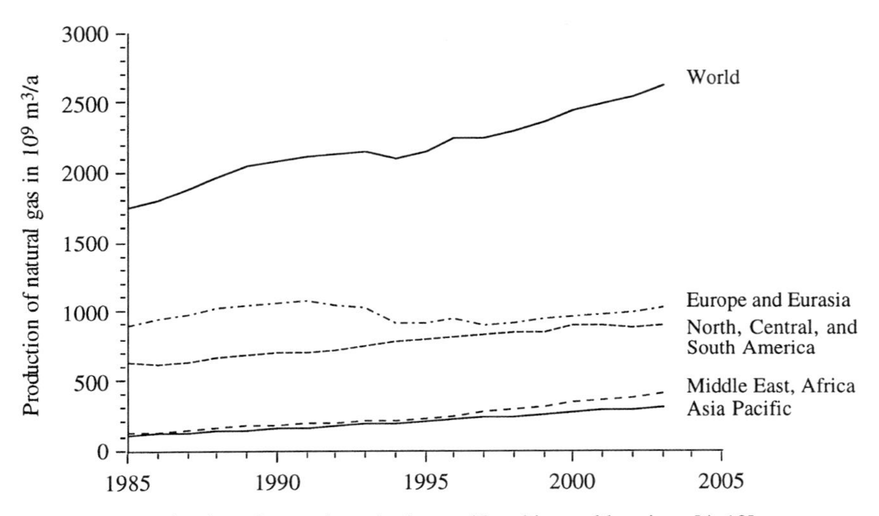


Fig. 3-14 Annual production of natural gas in the world and in world regions [4, 10].

Russia is the largest producer of natural gas, followed by the United States (Table 3-21). Russia is also the largest exporter, especially to Germany, which is the second largest importer after the United States. The world production of ca. $2.71 \cdot 10^{12}$ m$^3$ per year corresponds to $3.18 \cdot 10^9$ tons of coal units or $2.22 \cdot 10^9$ tons of oil equivalents (ca. 2/3 of the energy equivalent of oil). Note that production volumes do not correspond to energy contents which vary between 42 MJ/m$^3$ (Algeria) and 33 MJ/m$^3$ (Netherlands).

Table 3-21 The nine largest natural gas-producing, -exporting, or -importing countries in 2004 [6]. Data in $10^9$ m$^3$/a. A blank indicates that this nation was not one of the top 9 nations in this category.

| Country | Production | Export | Import |
|---|---|---|---|
| Russia | 620.1 | 194.8 | |
| United States | 532.0 | 24.2 | 120.6 |
| Canada | 182.6 | 103.1 | |
| United Kingdom | 101.2 | | |
| Algeria | 88.3 | 64.5 | |
| Indonesia | 79.5 | 38.6 | |
| Iran | 81.3 | | |
| Norway | 82.3 | 75.9 | |
| Netherlands | 86.0 | 53.6 | |
| Turkmenistan | | 44.1 | |
| Austria | | | |
| Malaysia | | 25.5 | |
| Germany | | | 90.1 |
| Japan | | | 81.2 |
| Ukraine | | | 54.4 |
| Italy | | | 67.9 |
| France | | | 44.0 |
| South Korea | | | 28.9 |
| Spain | | | 27.0 |
| Belarus | | | 19.6 |
| *World* | *2794.5* | *788.3* | *794.4* |

Table 3-22 Production and consumption of liquid gas (mainly from gas fields) and condensates in the United States [1b]. "Condensate" is also called "field condensate" or "natural gasoline."

| Feedstock | | Source | Million cubic meters 1980 | 1990 | 1997 |
|---|---|---|---|---|---|
| Liquid gas | $C_2$–$C_4$ | Gas fields | 70.1 | 72.3 | 84.1 |
| | | Oil refineries (production minus feed) | 19.2 | 11.9 | 24.5 |
| | | Imports minus exports | 11.1 | 8.6 | 8.9 |
| | | Stock | −18.4 | −15.6 | −18.3 |
| | | Total consumption of liquid gas | 82.0 | 77.2 | 99.2 |
| Condensate | $C_5$ and higher | Gas fields (production and consumption) | 20.0 | 17.8 | 18.0 |

Production and transport consume ca. 7 % of the natural gas. The bulk of natural gas is used as fuel. Only a small portion of dry natural gas (i.e., methane!) is converted to chemicals (see Chapter 4), either directly or after conversion to syngas (= synthetic gas) which is a mixture of CO and $H_2$ in various proportions (see Section 4.1.3)

Liquid gas ($C_2$–$C_4$) and condensates ($C_5$ and higher), on the other hand, serve exclusively as feedstocks for the chemical industry, especially in the United States (see Tables 3-17 and 3-18 and Chapter 4). They are both obtained from the wet natural gases of oil fields. Liquid gas from gas fields is usually supplemented from the net production of liquid gas by oil refineries and from net imports (Table 3-22). Condensates for the chemical industry are obtained exclusively from gas fields.

New explorations have increased the reserves of natural gas in the world and most of its regions between 1983 and 2003 (Table 3-23). They stayed about constant in Europe and Eurasia (former Soviet Union) between 1993 and 2003 but declined sharply in the United States since 1983. Resources, i.e., speculative reserves, are estimated as twice the amount of proven reserves. Europe/Eurasia still has the largest stocks (= reserves + resources), followed by the Middle East.

The 2003 production of natural gas of $2718.8 \cdot 10^9$ $m^3/a$ (Table 3-21) constitutes just ca. 1.5 % of the proven reserves of $175\,780 \cdot 10^9$ $m^3$ (Table 3-23). The reach would thus be 65 years if the consumption stays constant, production methods and yields per field remain unchanged, no resources are touched, and no new fields are found.

Table 3-23 Proved reserves of natural gas at the end of 1983, 1993, and 2003 [4] and resources in 1999. Data in $10^{12}$ $m^3$.

| Region | Reserves 1983 | 1993 | 2003 | Resources 1999 |
|---|---|---|---|---|
| North America | 10.40 | 8.75 | 7.31 | 30.2 |
| Central and South America | 3.18 | 5.54 | 7.19 | 10.4 |
| Europe and Eurasia | 40.48 | 63.62 | 62.30 | 176.1 |
| Middle East | 26.38 | 44.43 | 71.72 | 79.2 |
| Africa | 6.29 | 10.01 | 13.78 | 17.9 |
| Asia Pacific | 5.95 | 8.73 | 13.47 | 19.9 |
| *Total World* | *92.68* | *141.08* | *175.78* | *333.7* |

# 3.7 Wood

## 3.7.1 General Aspects

**Structure**

In *botany*, wood is the tissue that grows from the cambium of gymnosperms and angiosperms into the interior of the plant whereas the outward growth produces bark. Just below the bark is the cambium, a thin layer (L: *cambium* = to exchange) of gymnosperms (plants with naked seeds; G: *gymnos* = naked) and angiosperms (flowering plants with seeds in an ovary; G: *angeion* = vessel). Newly formed wood cells transport liquids; this layer is thus called **sapwood**. With time, inner sapwood cells become inactive and the wood becomes **heartwood**.

In *chemistry*, wood is a naturally occurring material consisting of cellulose, polyoses, lignin, water, and air. The composition of dry woods varies widely (Table 3-24). Celluloses are linear glucose polymers (Section 8.3); lignins are crosslinked polymers with various aromatic monomeric units (Section 3.8). Hemicelluloses are polysaccharides composed of different sugar units; they are also called polyoses (Section 8.9).

Cellulose is found in the cell walls of all higher plants and also in green algae and some fungi. It is also synthesized by tunicates and the bacterium *Acetobacter xylinum*. The textile industry uses celluloses from cotton, basts, leaves, and wood, the paper industry celluloses from wood and grasses.

In *technology*, wood behaves like a composite of crosslinked lignin that is "foamed" by air, filled with polyoses ("hemicelluloses"), plasticized by water, and reinforced by cellulose fibers. The air-filled hollow space (lumen) of wood comprises ca. 70 % of the total volume. Freshly cut ("green") wood contains ca. 40–60 wt% water, air-dried wood ca. 10–20 wt%. Densities of air-dried wood vary between ca. 0.17 g/cm$^3$ (balsa wood) and 1.25 g/cm$^3$ (quebracho, an exceedingly hard wood containing 24 % tannins).

According to the Food and Agricultural Organization (FAO) of the United Nations, wood is a *material* from trees. Wood is subdivided into **softwoods** (= coniferous = *gymnospermae*) and **hardwoods** (non-coniferous = broad-leaved = *angiospermae*).

Table 3-24 Composition of some dry woods and natural fibers. The extract consists of proteins, waxes, and resins.

| Material | | Content in wt% of | | | | |
|---|---|---|---|---|---|---|
| Type | Source | Celluloses | Polyoses | Lignins | Extract | Ash |
| Wood | beech tree | 38 | 35 | 25 | 1.6 | 0.4 |
| Wood | oak tree | 35 | 32 | 26 | 2.0 | 0.5 |
| Wood | willow tree | 43 | 30 | 23 | 3.6 | 0.4 |
| Wood | fir tree | 43 | 26 | 29 | 1.7 | 0.3 |
| Wood | poplar tree | 43 | 31 | 23 | 1.5 | 0.5 |
| Straw | wheat | 30 | 45 | 15 | 5 | 5 |
| Bagasse (dry pulp) | sugar cane | 44 | 29 | 24 | | 3 |
| Stem | bamboo | 60 | 14 | 24 | | 2 |
| Fiber | jute | 72 | 14 | 11 | 2 | 1 |
| Fiber | ramie | 76 | 16 | 1 | 6 | 1 |
| Fiber | cotton | 95 | 0.3 | 1 | 2.5 | 1.2 |

Table 3-25 Forested areas in 2000 [19]. 1 hectare ≈ 2.471 acres. *) 1990–2000.

| Region | Area in million hectares Total area | Total forest | Forest plantations | Forests as % of total land | Plantations as % of forests | Annual change of forest cover in percent *) |
|---|---|---|---|---|---|---|
| Africa | 2978 | 650 | 8 | 21.8 | 1.2 | –0.8 |
| Asia | 3084 | 548 | 116 | 17.8 | 21.1 | –0.1 |
| Europe w/o Russian Federation | 571 | 188 | 15 | 32.9 | 8.0 | +0.1 |
| Russian Federation | 1689 | 851 | 17 | 50.3 | 2.0 | ? |
| North America | 2029 | 526 | 16 | 25.9 | 3.0 | ? |
| Central America | 108 | 24 | 1.3 | 21.8 | 5.4 | ? |
| South America | 1755 | 886 | 10.5 | 50.5 | 1.2 | –0.4 |
| Oceania | 849 | 198 | 2.8 | 23.3 | 1.4 | –0.2 |
| *World* | *13063* | *3869* | *186* | *29.6* | *4.8* | *–0.2* |

**Growth, Production, and Consumption**

For thousands of years, mankind has used wood from trees and shrubs as construction material, fuel, and as a source of chemicals. The tremendous growth of the world population during the last two centuries has caused concerns that this natural source of materials is soon to be not only exhausted but also lost as a "green lung" for the recycling of carbon dioxide. Deforestation started 7000 years ago in the Mediterranean as mankind converted from hunting and collecting to agriculture. At present, about 30 % of the land is covered by forests from which only less than one-half (ca. 15 % of the total land) is accessible and about one-quarter is exploited (7.5 % of total land) (Table 3-25).

The percentage of forested areas is highest in South America and Russia (50 %) and lowest in Asia (18 %), Africa (22 %) and Central America (22 %). Forests of Europe are almost totally accessible (96 %), those of Australia (21 %) and Latin America (30 %) very little. Europe exploits and usually reforests 98 % of the accessible wooded areas, whereas in Latin America only 25 % are commercially exploited. The FAO estimated in 1980 that about 0.4 % of the total wooded area (including savannas and bush) are cut and never reforested. For 2000, FAO reported a +0.1 % annual increase of forested areas in Europe and a +0.2 % annual increase in the U.S., but strong annual decreases in South America (–0.4 %) and especially in Africa (–0.8 %).

The forest stand (US: timber stand) and the production and consumption of wood is measured in tons, cubic meters, "festmeters" (= 1 $m^3$ of solid timber, no space between logs), board feet (1000 bd ft = 2.36 $m^3$), etc., depending on country and industry.

U.S. commerce uses the following definitions (U.S. Forest Service). "Wood" is a material cut from a tree. A "tree" is a wooded plant with a single erect stem and a height of at least 12 feet that has a diameter of 3 inches or more at a height of 4 $^1/_2$ feet above ground. In the United States, "timber" refers to standing trees with diameters of at least 11 inches (hardwood) or at least 9 inches (softwood) at 4 $^1/_2$ feet above ground that provide at least one 12-foot sawlog or two noncontinuous 8-foot logs. In International English, "timber" refers to wood for construction.

Removal of branches converts cut timber ("trunks") into "logs" that are cut in the forest to 24-foot lengths. The resulting "roundwood" is sawed to various lengths and cross-sections to become "sawnwood" ("lumber" if used for construction). Roundwood yields sawnwood plus bark, chips, sawdust, etc.). Roughly squared logs are called "balks". Sawn and planed rectangular pieces of 10"-11" width and 2"-4" thickness are "planks" whereas "boards" are less than 2" thick but may have any width. The size of finished lumber refers to widths before planing: a 2×4 actually measures 1 $^1/_2$ in. by 3 $^1/_2$ in. Metric lengths are actual lengths, however. 1 board foot measures 1 ft × 1 in. × 1 in.

Table 3-26 Production of forest products in 2002 [19]: woodfuel (= fuelwood, charcoal), industrial roundwood, sawnwood, and wood-based panels plus chemical pulp for paper and paper + paperboard.

| Area | Production in $10^6$ $m^3$ | | | | Production in $10^6$ t | |
|---|---|---|---|---|---|---|
| | Woodfuel | Roundwood | Sawnwood | Panels | Chem. pulp | Paper |
| Africa | 545.9 | 66.8 | 7.8 | 2.3 | 2.6 | 3.3 |
| Asia | 782.2 | 222.6 | 61.1 | 58.8 | 40.3 | 97.8 |
| Europe | 106.9 | 480.1 | 127.8 | 63.3 | 47.0 | 102.0 |
| North and Central America | 159.0 | 615.1 | 152.3 | 57.5 | 78.9 | 106.5 |
| Oceania | 13.0 | 49.6 | 8.7 | 3.9 | 4.4 | 3.5 |
| South America | 190.0 | 153.5 | 33.2 | 9.6 | 11.5 | 11.5 |
| *World (total)* | *1797.0* | *1587.7* | *390.9* | *195.4* | *184.7* | *324.6* |

The world contains ca. 140 000·$10^6$ $m^3$ wood. About 3400·$10^6$ $m^3$ were removed in 2002 (3040·$10^6$ $m^3$ in 1983); practically the same volume regrows. About 53 % of the cut wood is used as woodfuel and about 47 % as industrial roundwood (Table 3-26). Of the total roundwood, Africa consumes 89 % as woodfuel, Asia 79 %, South and Central America 59 %, Europe 18 %, and North America 15 %.

About 60 % of the industrial roundwood is converted to sawnwood, which in turn serves for lumber (Section 3.7.2) and for the manufacture of panels, etc. (Section 3.7.4). The manufacture of sawnwood from roundwood generates large proportions of bark, woodchips, sawdust, etc., which have to be marketed, converted, burnt, etc.

Decorticated roundwood is also converted to mechanical pulp for cardboard and hard paper (Section 3.7.7) or chemical pulp for paper (Section 3.7.9). Chemical pulping produces not only fibrous cellulose but also lignin waste, which serves as an energy source.

In 2002, the largest producer of lumber was the United States (207·$10^6$ $m^3$) (Table 3-27), followed by Canada (69·$10^6$ $m^3$), Russia (19.$10^6$ $m^3$), Germany and Brazil (15·$10^6$ $m^3$ each), and Japan (14·$10^6$ $m^3$).

The consumption of lumber was greatest in the United States (286·$10^6$ $m^3$), Canada (26·$10^6$ $m^3$), Japan (22·$10^6$ $m^3$), Germany (16·$10^6$ $m^3$), Brazil (16·$10^6$ $m^3$), Russia (14·$10^6$ $m^3$), and China (13·$10^6$ $m^3$).

In the United States, the consumption of lumber stayed approximately constant from 1900–2004, although the population increased from 72 million to 290 million. Consumption of plywood and veneer increased first strongly but slackened off since 1985. The consumption of pulp climbed steadily to 181·$10^6$ $m^3$/a because of the additional demand for paper thanks to "paperless offices" with printers and copy machines.

Table 3-27 U.S. production, imports, exports, and consumption of timber products in 2002 [20].

| Product | Volume in million cubic meters | | | |
|---|---|---|---|---|
| | Production | Imports | Exports | Consumption |
| Lumber | 206.5 | 89.8 | 10.2 | 286.1 |
| Plywood and veneer | 30.4 | 5.8 | 0.9 | 35.3 |
| Pulp products | 161.6 | 41.7 | 22.2 | 181.1 |
| *Total* | *368.5* | *137.3* | *33.3* | *502.5* |

### 3.7.2 Natural Wood

Wood is easily accessible and easy to work with; it has been used for construction for thousands of years. Lumber owes its strength to cellulose fibers which are usually aligned in the major growth direction of trees. Wood is therefore an anisotropic material with different mechanical properties parallel to the grain (longitudinal) and perpendicular to the grain (vertical). An exception is the wood of the guaiac tree where the fibers crisscross each other. The wood of this tree is therefore used for chopping blocks. Heating guaiac wood delivers waxes and oils.

Densities of trees grown in the United States vary mostly between 0.3 and 0.8 g/cm$^3$ (Fig. 3-15) if measured at moisture contents of 12 %. Densities of green woods can be up to ca. 10 % lower because these woods contain relatively more vacuoles. Static flexural moduli increase with increasing density. At the same density, they are generally higher for hardwoods than for softwoods. Bur oak and live oak are exceptions.

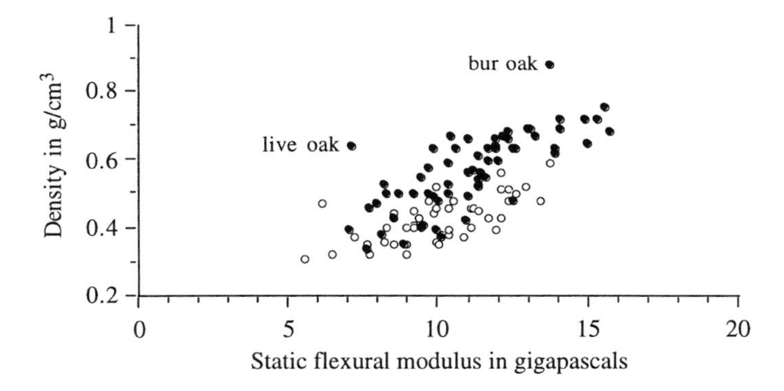


Fig. 3-15 Variation of static flexural modulus (bending modulus) with density for hardwoods (●) and softwoods (O) grown in the United States [21]. Measurements at a moisture content of 12 %. Flexural moduli are moduli of elasticity in bending. Examples of moduli of elasticity in tension for hardwoods are 12.3 GPa (white oak), 12.0 GPa (white ash), and 8.1 GPa (quaking aspen) and for softwoods are 13.4 GPa (Douglas fir) and 9.9 GPa (white spruce).

Woods have very high flexural moduli between 5 and 16 GPa (Fig. 3-15) which rival those of man-made, fiber-reinforced composites (Volume IV) and surpass those of semi-crystalline or amorphous polymers (Volume III). The tropical high-density woods kaneelhart (*Licaria* spp.; $\rho$ = 0.96 g/cm$^3$) and ipe (*Tabebuia* spp., $\rho$ = 0.92 g/cm$^3$) even have flexural moduli of 28 GPa and 21 GPa, respectively. Even the low-density tropical wood balsa (*Ochroma pyramidale*; $\rho$ = 0.16 g/cm$^3$) has a flexural modulus of 3.4 GPa (all data at a standard moisture content of 12 %).

Tensile and compressive strengths perpendicular and shear strengths parallel to the grain are by a factor of ca. 200 lower, however (Fig. 3-16). In each case, woods of coniferous trees have in general lower strengths than woods of broadleaved ones, hence the designation "softwood" of the former and "hardwood" of the latter (these designations have nothing to do with "hardness" which scientifically refers to *surface* hardness (see Volumes III and IV)). However, softwoods and hardwoods of comparable density have about the same compressive strength perpendicular to the grain (Fig. 3-16, top). Large compressive strengths are found for kaneelhart (120 MPa) and ipe (89 MPa).

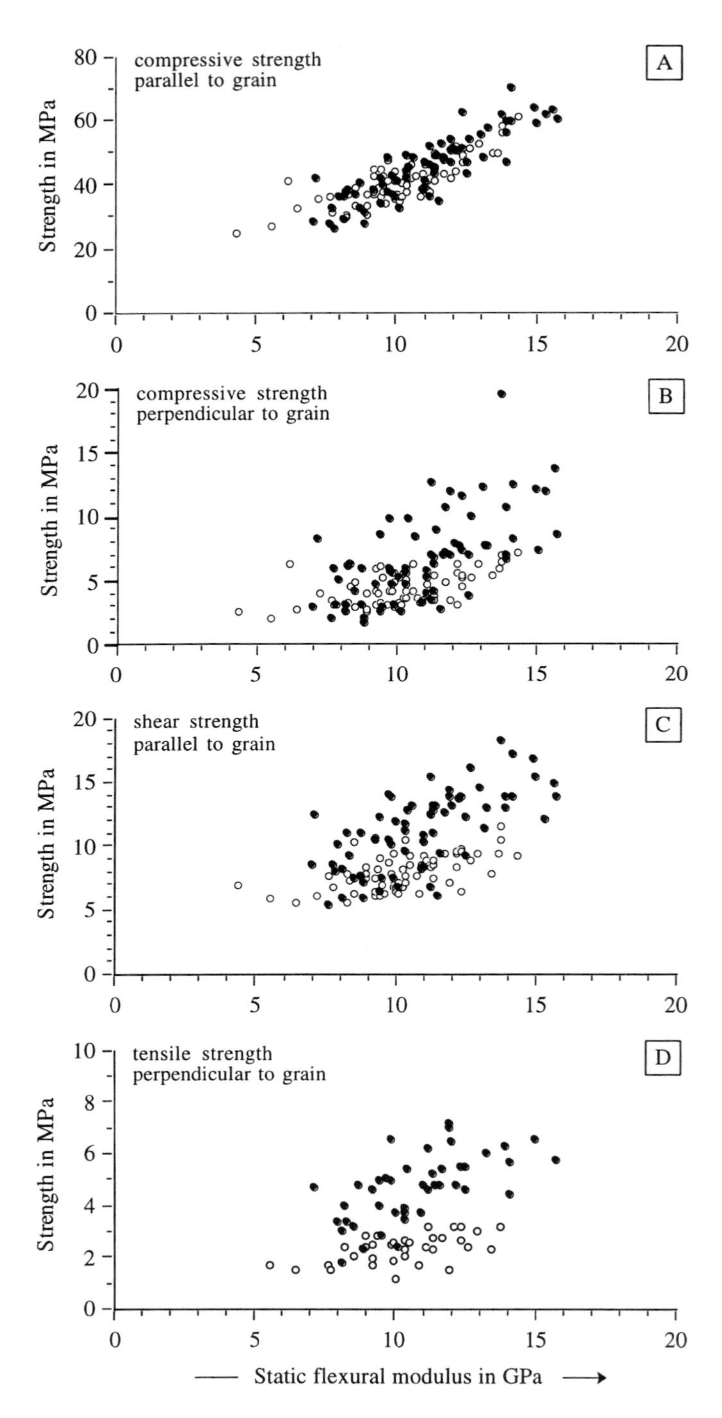


Fig. 3-16 Strength of US and Canadian hardwoods (●) and softwoods (O) as function of the static flexural modulus. Data of [21] for moisture contents of 12 %. Green strengths are usually one-half of the strengths measured at 12 % moisture.

Compressive strengths of woods are about 10 times higher parallel to the grain than perpendicular to it (Fig. 3-16, A and B). Flexural and tensile moduli, tensile and shear strengths, and other mechanical properties also differ in longitudinal and vertical direction by factors of 2–10. These differences led to the development of plywood and veneer (Section 3.7.4).

Wood has also other disadvantages such as swelling in water, combustibility, attack by fungi and termites, and low resistance against abrasion. Mankind has tried to overcome these disadvantages by charring or painting wood surfaces or treating the wood with tar, phosphates, chromates, or ammonium salts. Newer developments are pressurized wood (Section 3.7.3) and polymer wood (Section 3.7.5).

### 3.7.3 Pressurized Wood

For the manufacture of pressurized wood, beech wood is machine-dried and subsequently shaped by machining. The part is then *pressurized* (compressed in all three spatial directions) by pressures up to 300 bar (= 30 MPa) and temperatures up to 150°C. This causes pore volumes to decrease to practically zero and densities to increase to values up to ca. 1.44 g/cm$^3$ (crystalline cellulose has a density of ca. 1.5 g/cm$^3$). Pressurized wood is also known as densified wood, high-density wood, and high-duty wood.

The process preserves the fiber direction. Compression, impact, and flexural strength in the fiber direction increase strongly, however. Pressurized wood can only be worked by high-speed machining. It can no longer be nailed.

Pressurized wood has a high alternating flexural strength and is therefore used for springs at transport chutes. The textile industry uses pressurized wood for bearings and impact plates since it has a high splinter resistance and needs no lubrication. Dirt on fibers is furthermore pressed into the wood surface instead of into the woven goods. Mallets from pressurized wood prevents sparks.

### 3.7.4 Plywood and Veneer

**Plywood** consists of uneven numbers of thin sheets of wood that alternate in the direction of grain, usually 90°. The layers are glued together under pressure by thermosetting resins, usually starch pastes, casein, bone glue, or urea resins for interior applications and phenolic resins, alkyd resins, or polyurethanes for exterior ones.

**Fiber board** is made from wood chips that are converted to a paste by action of water vapor, heat, and pressure. After addition of thermosetting resins and additives (protection against fire, pests, etc.), boards are formed, cured (i.e., crosslinked), and dried.

**Chip boards** are **particle boards** from wood chips that are glued together by phenol, urea, or melamine resins using heat and pressure. **Hard boards (molded) fiber boards, wood waste panels**) are made from chips from waste wood; the name has nothing to do with hard wood or hard fibers. Hard boards are often laminated with plastic sheets.

A **veneer** (G: *furniren* = to furnish) consists of a thin layer of high-grade hardwood for decoration or economics that is glued on low-cost fiber board, chip board, or hard board by starch paste, casein glue, dispersion adhesives, or thermosetting resins.

### 3.7.5 Polymer Wood

Polymer wood is wood that is filled with synthetic resins. It is prepared by degassing wood and soaking it in a monomer, a monomer-initiator mixture, or a prepolymer, depending on the type of wood. Phenol-formaldehyde resins, for example, can enter the lumen of the wood and condense therein to a thermoset. The resulting polymer wood is more thermally resistant and more resistant against decay than wood. The toughness is reduced while all other mechanical properties are either improved or remain unaffected.

The hardening of phenolic resins produces water (Section 6.6.3) that acts as a plasticizer for wood. Many other thermosetting resins and condensation monomers are unsuited for the production of polymer wood since the released low-molecular weight compounds can permeate out only with difficulty or not at all.

Monomers for chain polymerization include cyclic compounds or those with carbon-carbon double bonds, provided they do not swell the wood. Not all such monomers can be used for the preparation of polymer wood, however. For example, poly(acrylonitrile) does not dissolve in its monomer; hence, it forms powdery deposits in wood instead of interconnected polymer regions. Its glass temperature ($T_G \approx 98°C$) is furthermore far higher than the allowable polymerization temperature. Vinyl chloride presents the same problem ($T_G \approx 81°C$) and, in addition, is gaseous at room temperature ($T_b = -14°C$). Vinyl acetate, on the other hand, is unsuited because the glass temperature of poly(vinyl acetate) is too low ($T_G = 32°C$).

Polymerization may be initiated free-radically by peroxides, redox systems, etc., or by $\gamma$-rays. The latter is the method of choice because it avoids the heating of wood. It requires monomers that have sufficiently high G values (see Volume I); those with too low G values would need too high radiation doses that will destroy the wood. The selection of monomers for $\gamma$-ray polymerizations is thus restricted. In the past, either methyl methacrylate, styrene/acrylonitrile mixtures, or unsaturated polyesters were used (note: "unsaturated polyester" is the industry term for the *mixture* of unsaturated polyester molecules with styrene or methyl methacrylate). Today, melamines functionalized with acrylic groups are preferred.

Polymerization of monomers in wood by $\gamma$-rays is probably a graft polymerization on cellulose and lignin molecules since electron spin resonance indicates the formation of radicals on both both of them. Furthermore, a part of the newly formed polymers cannot be extracted. Intermolecular crosslinking can be ruled out because all extractable polymer molecules are linear, whereas branched molecules are expected as precursors for intermolecular crosslinks.

Polymerizations by free-radical initiators are impeded by certain ingredients of wood, for example, quercetin, which reacts with oxygen to a quinone that inhibits the polymerization (see Volume I). This inhibition is unavoidable but can be suppressed, for example, by a mixture of fast- and slow-decomposing free-radical initiators.

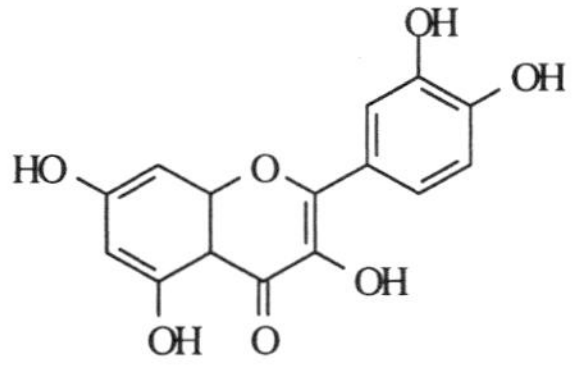

quercetin
= cyanidanol
= 3,3',4',5,7-pentahydroxyflavone

Because of its improved mechanical properties, polymer wood is used for window frames, sports equipment, musical instruments, and boats. Parquet from polymer wood does not need to be sealed. The US annual business is ca. 1 billion dollars.

### 3.7.6 Plasticized Wood

On benzylation with benzoyl chloride or benzyl alcohol or by treatment with the corresponding lauroyl compounds, wood flour (wood meal, saw dust) is converted to a pale yellow mass that can be hot-press molded to transparent sheets.

Wood flour and wood chips also dissolve in phenols, ketones, or alcohols after 30–180 min at temperatures of 200–250°C. The resulting paste-like solutions contain up to 70 % wood.

### 3.7.7 Mechanical Pulp

Two processes convert solid wood to wood fibers: mechanical pulping (known since 1844) and chemical pulping (invented in 1874) (Section 3.7.8). The resulting pulps are used to prepare fibers, paper, cardboard, and other cellulosic products (Volume IV).

In **mechanical pulping**, decorticated pieces of coniferous or non-coniferous wood are milled or pressed against rotating grinding wheels while being cooled by water. The process tears off fine fibers of ca. 0.2 mm length that have practically the same composition as the constituent wood and thus a very high lignin content (Table 3-24: ca. 29 %). The fibers are too short to produce paper; mechanical pulp therefore serves mainly for cardboard and hard paper. The high lignin content provides these products with a brownish color which can be lessened by treating the pulp with brightening agents; i.e., hydrogen peroxide or (sodium) hydrosulfite (= sodium hyposulfite, sodium dithionite; $Na_2S_2O_4$).

In **thermomechanical wood pulping**, wood is first preheated under pressure before being ground or milled. **Semichemical wood pulping** adds chemicals before preheating in a pressure vessel and mechanical pulping.

"Wood pulp" in Table 3-28 includes pulps by the three mechanical wood pulping operations mentioned above as well as chemical wood pulps (Section 3.7.8). "Pulp of other fibers" refers to pulps from straw, bagasse, bamboo, etc. In 2002, the main producers of non-wood pulps were Pakistan (425 000 t/a), United States (243 000 t/a), Mexico (186 000 t/a), Italy (165 000 t/a), and Colombia (156 000 t/a).

"Dissolving pulps" in Table 3-28 are chemical pulps by the sulfate, soda, and sulfite processes that are used for other purposes than making of paper, paperboards, carton, etc. These uses include regenerated cellulose fibers, cellulose plastics, explosives, lacquers, etc. Examples are rayon, cellulose acetates, cellulose nitrates, cellulose propionates, etc. (see Section 8.3 and Volume IV). This sector is in steady decline because of the decreasing demand for rayon in Western countries, which is mainly due to competition with other fibers and the enormous costs and ecological problems connected with the process water in the manufacture of rayon fibers (see Volume IV).

"Paper and paper board" in Table 3-28 refers to all kinds of paper products (ca. 12 % for newspapers, 30 % for other printing and writing papers, and 58 % for all other).

Table 3-28 Production and production capacities of various pulps as well as paper and paperboard by 39 reporting countries in 2002 [22]. All other countries did not supply the Food and Agricultural Organization (FAO) of the United Nations with data, including China, India, and Russia. It is estimated by FAO that the 39 nations, which did respond, produce ca. 85 % of all pulp and paper products.

| Country | Production of dry materials in 1000 t/a | | | |
|---|---|---|---|---|
| | Wood pulp | Pulp of other fibers | Dissolving pulp | Paper and paperboard |
| Production | | | | |
| United States | 52 553 | 243 | 947 | 82 770 |
| Canada | 25 522 | 0 | 0 | 20 076 |
| Japan | 10 588 | 4 | 77 | 30 671 |
| Sweden | 11 354 | 0 | 0 | 10 724 |
| Germany | 3 149 | 0 | 0 | 18 520 |
| Brazil | 7 915 | 96 | 137 | 7 661 |
| Austria | 1 556 | 1 | 196 | 4 419 |
| 32 other reporting countries | 16 089 | 1 495 | 423 | 53 708 |
| *Total production by 39 countries* | *128 726* | *1 839* | *1 780* | *228 549* |
| *Total world production (estimated)* | *151 442* | *2 164* | *2 094* | *263 000* |
| *Total world capacity* | *166 050* | *2 272* | *2 505* | *264 106* |
| *Estimated use of capacity in percent* | *91* | *95* | *84* | *99* |

### 3.7.8 Chemical Pulp

Chemical pulping treats cellulosic materials (wood, sawdust, etc.) with acids or alkali. The process removes large proportions of lignins and hemicelluloses of plants and reduces the chain lengths of cellulose molecules. The resulting chemical pulp is obtained as short, 1–3 mm long fibers that can be used for paper making but not as textile fibers since the resulting "celluloses" still contain too high percentages of polyoses ("hemicelluloses"), usually pentosans. The celluloses of wood pulp also always have carbonyl and carboxyl groups.

For the production of rayon fibers from chemical pulp, sheets are prepared that are steeped in caustic soda in order to remove more of the alkali-soluble hemicelluloses. After shredding and aging, celluloses are further processed by xanthation (also called "sulfidation"), etc. (Section 8.3.7).

Chemical pulp is obtained by three major processes: alkali, sulfate (ca. 75 %), and sulfite. **Alkali processes** can utilize all woods and also parts of short-lived plants such as bagasse, straw, etc.; the latter contribute 5–6 % to celluloses from chemical pulping. The main alkali process cooks the wood in ca. 8 % aqueous caustic soda solution (caustic lye) under pressure for several hours at 40–170°C. It converts lignin to lignin phenolates that diffuse out of the resulting soda pulp (soda paper, sodium paper, "cellulose") and form very dark aqueous solutions (black liquors, waste liquors, spent liquors) that are used as an energy source. Alkali processes are now used less and less.

**Acid processes** (sulfite and sulfate processes), on the other hand, are less universal since some types of trees contain tanning agents and other chemical compounds that react with lignin under acidic conditions. These reactions lead to additional crosslinking of lignins which in turn slows down their subsequent degradation.

In **sulfite processes**, non-coniferous wood is cooked with aqueous solutions of hydrogen sulfites for several hours at 140–150°C, either with calcium hydrogen sulfite (**sulfite process**) or now also with sodium, magnesium, or ammonium hydrogen sulfite, either with a surplus of sulfur dioxide (**acidic bisulfite process**) or without (**bisulfite process**). The reaction converts insoluble lignins to soluble lignin sulfonic acids and hydrolyzes hemicelluloses to mono- and oligosaccharides. The remaining cellulose is defibrillated in horizontal drums with spikes and the resulting **sulfite cellulose** (**sulfite pulp®**) subsequently bleached with chlorine, hypochlorous acid (HOCl), calcium chloride hypochlorite (approximate composition 3 $CaCl(OCl)\cdot Ca(OH)_2\cdot 5\ H_2O$; known as chloride of lime or bleach), chlorine dioxide ($ClO_2$), or hydrogen peroxide ($H_2O_2$).

The original sulfite process led to viscous "black liquors" that contained calcium salts of lignin sulfonates. The dry mass could not be burned and the black liquors were thus simply discharged into rivers with disastrous results for plants, fish, etc. The newer bisulfite processes produce sodium, ammonium, or magnesium salts of lignin sulfonates. The burning of these black liquors delivers process energy and the reusable compounds NaOH, $NH_3$, or $Mg(OH)_2$.

The **sulfate process** uses non-coniferous, coniferous, or highly resinous woods, saw dust, etc. The wood is cooked for several hours in an aqueous solution of NaOH, $Na_2S$, and $Na_2CO_3$ at 165–175°C until the lignin content is reduced to ca 5 %. Further pulping would reduce the strength of cellulose fibers. The process probably leads to a partial replacement of the OH groups of cellulose by SH groups. These mercaptan groups are unstable in alkaline solutions; they cleave the ether bridges of lignin and replace them by sulfide bridges. The resulting sulfide bridges are split hydrolytically: lignin is degraded and becomes soluble "alkali lignin".

The reaction mixture is concentrated by evaporation of water. Sodium sulfate is added (hence the name sulfate process) and the solution of alkali lignin (**black liquor**) separated by filter presses from cellulose fibers called **sulfate cellulose** or **sulfate pulp®**.

The black liquor is condensed to a dry mass that contains lignin sulfonates and all added inorganic compounds. Burning this residue delivers process energy. At temperatures of ca. 1150°C, carbon is formed which reduces the sulfate to sulfide. The also produced sodium carbonate is causticated with calcium hydroxide to sodium hydroxide. Both sodium sulfate and sodium hydroxide are reused. Waste gases from the processing of resin-rich pine trees are condensed to deliver sulfate turpentine and tall oil (p. 77-78).

Sulfate cellulose is more opaque and voluminous than sulfite cellulose and leads to stronger (i.e., more tear-resistant) paper and paperboard. The sulfate process is therefore also called **kraft process** (G: *Kraft* = strength, power, force). Like sulfite cellulose, sulfate cellulose must also be bleached.

Dissolution and steam explosion are newer, pilot-stage pulping processes. Boiling of wood chips in acidic or alkaline ethanol-water mixtures degrades and dissolves the lignin, hydrolyzes hemicelluloses to soluble sugars, and transforms the insoluble cellulose fibers to a highly valued pulp. A similar process works with mixtures of phenol and water.

In the steam-explosion process, wood chips are subjected for short times (seconds to minutes) to pressurized water vapor (35–70 bar) at 200–300°C. The sudden release of the vapor pressure "explodes" the bonds between lignin and cellulose. Lignin is then extracted by alkali.

### 3.7.9 Wood Gasification

Wood has served as source of low-molecular weight chemicals in the past and is now being rediscovered as renewable raw material, especially as source of ethanol (Section 3.7.11). The oldest wood conversion processes used brute force to convert components of wood to commercially useful products. These processes are known as wood gasifications although they deliver not only gases but also liquids.

Heating wood to 300–500°C anaerobically produces wood gas, water vapor, wood vinegar, and wood tar. Yields of these components vary strongly with the type and humidity content of wood as well as process conditions. For example, 100 kg of beech wood yields 19 kg combustible **wood gas** composed of 49 % $CO_2$, 34 % CO, 13 % $CH_4$, 2 % $CH_2{=}CH_2$, and 2 % $H_2$. During World War II, gasoline was reserved for military purposes in Continental Europe, and wood gas was therefore used to drive civilian cars that had to be outfitted for this purpose with a kind of large retort.

**Wood vinegar**, also called **pyroligneous acid**, from wood gasification is liquefied by cooling. It is a reddish-brown aqueous solution containing ca. 12 % acetic acid and its homologues, 2 % methanol, 1 % acetone plus methyl acetate, and 10 % dissolved tar. Treatment with lime milk, a suspension of $Ca(OH)_2$ in water, removes the acetic acid as calcium acetate. The remaining **wood spirit**, a mixture of methanol, acetone, and methyl acetate, was formerly an important source of so-called **wood alcohol** (= methanol!).

**Wood tar** is obtained by wood gasification in yields of 130 kg per ton of dry wood. It is used directly for various purposes, depending on the type of wood: as binder and sealing compound for wooden boats and roof paper (roofing felt) and for the smoking of meat products. Distillation of wood tar delivers **wood-pine oil** for lacquers and oil-based paints (wood-pine oil is also obtained from the extraction of roots of dead coniferous trees, p. 79), as **birch oil** from birches for the preparation of Russian leather (Russian calf, Russia), etc.

Oils are also obtained by cooking wood in the presence of carbon monoxide, water, and catalysts at temperatures of 350–400°C and pressures of up to 300 bar. Gasification of wood in the presence of hydrogen at 300–800°C and pressures of 30–100 bar delivers mainly methane. Gasification at temperatures of ca. 1000°C leads to synthesis gas and then to base chemicals albeit in low yields: 1000 tons of wood deliver 52 t ethene, 8 t acetylene, 5 t propane, 18 t benzene, and 3 t toluene. Wood gasification in (electric) arcs at temperatures of 2000–2500°C generates up to 15 % acetylene. At present, all wood gasification processes are not economical for base chemicals.

**Charcoal** is obtained by three processes. The traditional method covers large heaps of wood with earth and lets the wood burn slowly for about a month. Gaseous byproducts escape and are not recovered. The charcoal has a high caloric content and is thus the energy source of choice for cooking in many countries (see also p. 213).

Industrial processes enclose air-dried wood billets in a retort which is then heated anaerobically to 275°C. Without further heating, temperatures climb to 340–400°C. After 4–5 hours at this temperature and subsequent cooling, industrial charcoal is obtained in yields of ca. 35 %. It consists of ca. 95 % carbon, 3 % ash, and 2 % other materials. Industrial retorting allows the recovery of wood spirits.

Charcoal is also a byproduct in the distillation of wood. Ten cords of wood (ca. 36 $m^3$) yield 5.4 t of charcoal, 10 000 L of pyroligneous liquor, and 5000 kg of gas.

Charcoal is used industrially for many purposes, for example, making carbon steel, refining of copper, gun powder, absorbants, etc., and also for the syntheses of $CS_2$, NaCN, and ferrosilicon.

**Activated charcoal** is obtained from petroleum, coal (as a**ctivated carbon** or **filter carbon**), and from the retorting of hard wood, coconut shells, or peach pits. It has a large surface area and is mainly used for water purification and filters of gas masks.

### 3.7.10 Wood Saccharification

Wood has been used as source of sugars in the past. It is now being eyed as raw material for the production of ethanol (see next section).

More than 200 processes are known for the hydrolysis of celluloses of wood, straw, and other plant materials to glucose but none of these is presently economical. For example, treating coniferous wood at room temperature with 38.5 % hydrochloric acid delivers a polymeric "**dry sugar**" that can be used as fodder for animals.

Subsequent treatment of the dry sugar with 10 % acetic acid results in a mixture of low-molecular weight sugars. For example, 100 kg dry coniferous wood delivers 31 kg glucose, 17 kg mannose, 3 kg galactose, 1 kg fructose, and 5 kg xylose as well as 2 kg acetic acid, 3 kg resin, and 33 kg lignin. The yield of glucose can be increased to 55 kg if 3–6 % sulfuric acid or hydrochloric acid are used at 140–180°C and 6–9 bar. The lignin serves as a source of process energy.

Wood needs to be ground/milled to extraordinarily fine particles before lignocelluloses can be enzymatically attacked. For example, the hydrolysis of cellulose (ca. 100 % glucose units) by the enzyme cellulase of *Trichiderma viride* delivers glucose in yields of only ca. 50 %.

### 3.7.11 Bioethanol and Sundiesel

Sugars from wood and other plant materials are now increasingly used as base chemcals for the production of ethanol as partial or even complete replacement of motor gasoline. So-called **anhydrous ethanol** ("dry ethanol") can be added to gasoline in proportions up to 10 vol% (E 10) without modification of the engine. E 85 (85 vol% anhydrous alcohol) requires modified engines (flexible fuel vehicles). Cars can run also on 100 % **hydrous ethanol** (E 100), i.e., ethanol containing ca. 5 vol% water), albeit with specially designed engines, pipes, etc. Per volume, ethanol produces about 2/3 of the energy of gasoline but has a higher octane value (fuel of choice for race cars).

**Bioethanol** from wood competes with that from agricultural residue (straw, orange peels, downgraded potatoes, beet molasses, etc.), and cultivated grasses, grains, bulbs, sugar beets, and sugar cane. The simplest source of sugars is sugar cane which delivers 8–13 t saccharose per hectare (10 000 $m^2$) (Table 3-29) as well as considerable amounts of bagasse (p. 393) which is used as energy source and for the production of bagasse paper. Saccharose is then fermented to ethanol by yeasts or bacteria. Yeasts contain the enzyme invertase which catalyzes the reaction of saccharose to equal proportions of glucose and fructose (both $C_6H_{12}O_6$): $C_{12}H_{22}O_{11} + H_2O \rightarrow 2\ C_6H_{12}O_6$.

Table 3-29 Plants and plant products for the conversion to bioethanol. Data for sugar cane refer to Brazil, for orange peels to Florida, and for all others to Europe. Conversion of primary products is by acid hydrolysis (H) and fermentation (F), all others by fermentation; wheat also by malting (M). Yields of primary and secondary products are given in tons (t) per year (a) and hectare (ha; 1 ha ≈ 2.27 acres). Ethanol prices are based on 2002 British data [23] that were converted to US dollars using May 2006 exchange rates. Prices do not contain subsidies (USA) or taxes (US, Europe).

| Product | | Conversion | Culture in t/(ha a) | | Ethanol yield | | Ethanol price |
|---|---|---|---|---|---|---|---|
| Primary | secondary | | primary | secondary | L/ha | L/t | $/gal |
| Wood | - | H - F | | | 2500 | 200 | 1.52 |
| Straw | - | H - F | | | | 340 | 2.91 |
| Orange peels | - | H - F | | | | 250 | |
| Wheat | starch | - M F | 5.5 | | 2040 | 380 | 2.12 |
| Corn | starch | - - F | 8.3 | 4 | 3000 | 390 | 1.11 |
| Jerusalem artichoke | starch | - - F | 40 | | 3600 | 90 | |
| Sugar cane | saccharose | - - F | | 8-13 | 5500 | 72 | 0.89 |
| Sugar beet | saccharose | - - F | 60 | 5- 9 | 6000 | 110 | 2.41 |
| Sugar beet | molasses | - - F | - | (2.4) | (720) | 300 | |

Yeasts also contain the enzyme zymase that converts $C_6H_{12}O_6$ to ethanol according to $C_6H_{12}O_6 \rightarrow 2\ C_2H_5OH + 2\ CO_2$. Fermentation is at 250–300°C for 2–3 days.

Another direct source of saccharose is sugar beets that also produce 40 kg molasses per ton sugar beet. Molasses contains ca. 50 % sugar.

Less productive than sugar cane and sugar beet is the use of starches (glucose polymers, Section 8.2) from grain. One hectare yields about 4 tons of corn starch plus valuable gluten (a protein) and hull. The starch is then directly fermented to ethanol. The yield is up ca. 3000 L (≈ 2 t) ethanol per hectare (Table 3-29). Wheat delivers a similar yield per ton; however, it requires malting before fermentation, which boosts the price.

Wood as well as other biomass refuse (straw, peels, etc.) and grasses need more complicated treatment since they not only contain cellulose (another glucose polymer, see Section 8.3) but also hemicelluloses (xylose and mannose polymers, Section 8.9), pectins (various sugars, Section 8.7.5), and lignins (Section 3.8), depending on the plant.

At present (2006), biomass (including logging residues) with a 10 wt% water content is treated with 1.25 times its weight of 70–77 % sulfuric acid, then diluted with water to 20–30 % solids content, and heated 1 hour at 100°C. The resulting gel is pressed and the mixture of acid and sugars chromatographically separated. Hemicelluloses are hydrolyzed with 0.7 % sulfuric acid at 190°C followed by hydrolysis of celluloses with 0.4 % sulfuric acid at 215°C. After neutralization, resulting sugars are fermented to ethanol.

Much more efficient than wood-to-ethanol promises to be wood-to-diesel (BtL = biomass-to-liquid). The BtL process converts chopped wood by low-temperature carbonization at 400-500°C to a gas and finely divided charcoal. The tar-containing gas is mixed with the charcoal and burned at 1400°C to a raw synthesis gas (see p. 89) which is separated from dust, chlorine, and sulfur and finally converted in a Fischer-Tropsch reactor (p. 45) to so-called **sun diesel.** Yields of sun diesel are ca. 4000 L per hectare and year.

By the end of 2007, the United States will have a bioethanol capacity of $6.7 \cdot 10^9$ gal/a which corresponds to ca. 3.2 % of the energy content of the $140 \cdot 10^9$ gal gasoline that are burned each year. Since a $1.7 \cdot 10^8$ L/a ethanol plant costs $3 \cdot 10^8$ \$, a complete replacement of gasoline by ethanol would require an investment of $1.4 \cdot 10^{12}$ \$.

## 3.8 Lignins

In wood technology, "lignin" is the part of wood that cannot be dissolved by dilute acids or organic solvents (L: *lignum* = wood). Lignin is mainly concentrated in the lamellae of plants from where it successively grows into the primary and secondary cell walls. The lignin content of plants ranges from ca. 1 % in cotton to ca. 50 % in barks (see also Table 3-24).

In chemistry, "lignin" refers to a group of amorphous, high-molecular weight, aromatic compounds with high contents of methoxy groups that derive from the three trans-monolignols: *p*-coumaryl alcohol, coniferyl alcohol, and sinapyl alcohol:

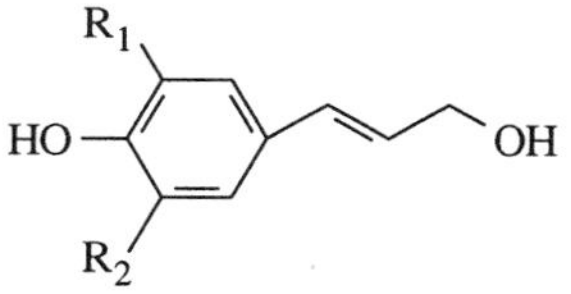

| $R_1$ | $R_2$ | |
|---|---|---|
| H | H | *p*-coumaryl alcohol |
| H | $OCH_3$ | coniferyl alcohol |
| $OCH_3$ | $OCH_3$ | sinapyl alcohol |

Lignin is synthesized in nature from these monolignols by dehydrogenating polymerizations that are catalyzed by cell wall-bound peroxidases (Volume I, p. 564). Due to the random nature of propagation steps, highly branched polymers result. It is unclear whether the insolubility of lignin is due to crosslinking (as most people think) or to inclusion in the matrix of cellulose fibers.

The composition and constitution of lignins differ from plant to plant and change with the age of the plant; an example is shown in Fig. 3-17. Lignins of coniferous trees contain practically only coniferyl units, those of non-coniferous trees coniferyl and sinapyl units, and those of grasses all three monolignol units. Depending on the plant within each group, monolignol units are interconnected in different ways. In addition, pulping attacks different bonds in different proportions. Lignins are therefore labeled according to their origin (fir lignin, birch lignin, etc.) and the type of pulping (fir Kraft lignin, spruce dioxane lignin, etc.).

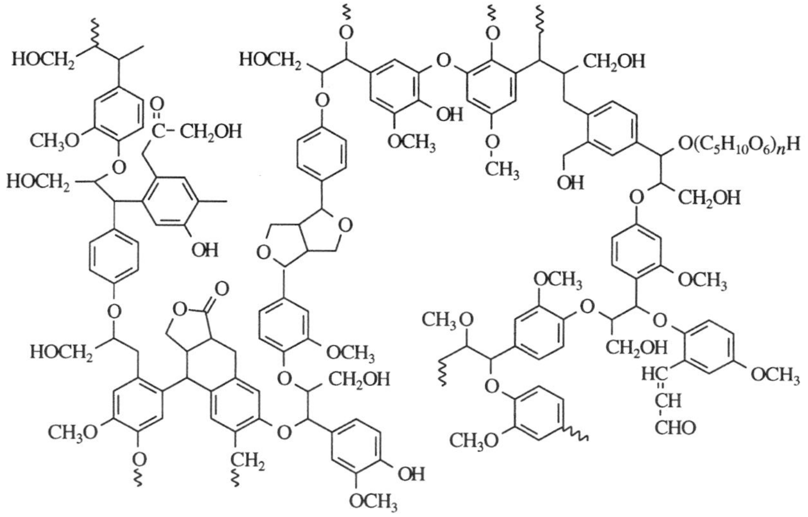

Fig. 3-17 Schematic representation of a lignin from coniferous wood.

Each year, chemical pulping delivers about 50 million tons of lignin, either as sulfate or kraft lignins or as lignin sulfonates from the sulfite process with molecular weights between 4000 and 100 000 (p. 73). Most lignins are burned in order to improve the energy balance. Small amounts of thickened waste sulfite liquor are used as binder in foundries, as flotation and drilling auxiliaries, ion exchangers, etc. Even smaller amounts serve as base compounds for chemicals such as syringa aldehyde, gallic acid, etc. In 1980, waste sulfite liquor was the main source for vanillin (4-hydroxy-3-methoxy benzaldehyde). Vanillin is still made that way in Norway and China but plants in the United States and Canada were shut down for environmental reasons. The bulk of vanillin is now obtained by direct synthesis.

In nature, lignin is decayed by the white fungus *Phanerochaete chrysosporium* which uses the enzyme ligninase. The end product of this degradation is carbon dioxide but the intermediates are not known. Some of the possible intermediates have skeletons that resemble those of known environmental poisons, e.g., benzopyranes.

## 3.9 Natural Resins

Natural resins (**resina**) are solid to semi-solid, mostly amorphous organic materials with relatively low molecular weights and relatively high glass temperatures that usually break conchoidally. Examples are **oleoresins** of tree saps, some exudations of insects, shellac, and some mineral hydrocarbons. Early synthetic prepolymers and polymers resembled natural resins; they were thus called **synthetic resins** or simply **resins**. The designation "resin" was later retained for synthetic polymers and oligomers that serve as raw material for plastics, lacquers, printing inks, etc., i.e., *before* formulating and/or cross-linking. Synthetic resins include thermoplastics such as poly(ethylene) and nylons as well as thermosetting resins such as amino and phenolic resins.

Several natural resins are also known as **gums** or **balsams**; the terms are historical and not well defined. The dividing line between resins, gums, and balsams is not sharp:

- **Hard resins** are hard and brittle at room temperature. Examples are dammar, rosin, and amber, a fossilized hard resin (original D: *glassa* (now: Bernstein); G: *elektron*).
- **Soft resins** are also called **balsams** (Hebrew: *basam* = pleasant). Balsams are either liquid resins or solutions of these resins in essential oils. Examples are styrax, turpentine, and balsam of Tolu. **Essential oils** are natural esters that dissolve easily in alcohols. These solutions are essences for perfumes, hence the name "essential oil."
- **Oleoresins** are natural resins containing essential oils.
- **Gums** are plant exudates that harden in air. Examples are gum arabic (*G. arabic*) and *G. tragacantha*. However, "gum" also designates the wood of the tree *liquidambar styraciflua*. **Gum resins** are mixtures of natural gums and resins.
- **Animal resins,** for example, shellac.

Natural resins are mixtures of various chemical compounds that are usually based on diterpenes and sometimes also on triterpenes and sesquiterpenes. Important ingredients of resina are resin acids (rosin acids), resin alcohols (resinols), esters of these acids with alcohols (resines), their unsaturated derivatives (resenes), tan-like derivatives of phenols (resino tannols), etc.

**Dammar** (dammer) is any of the hard resins that are obtained from slitting the Indo-Malayan trees of the genera *Shorea* and *Hopea* (Malay: damar = resin). Dammar is used as a binder for protective coatings, as glue for theater hairdos, etc.

**White** or **hard dammar** is the Malaysian name of **copal** (Mexican: *copalli* = incense), semi-fossil lumps of resins from long-gone tree-like *Papilionaceae* that are found in sandy shores of the tropics. Copals were used as binders for oil paints.

**Amber** is a fossil resin from coniferous trees of the Tertiary Period that is mainly found at the shores of the Baltic Sea. It is now mainly used for jewelry but served formerly also as a binder for lacquers.

**Naval stores** were originally resins and pitches that were used to caulk the seams of wooden ships during American colonial times. Today, naval stores is the name for all resins and oils from pines, for example, pine oil, rosin, tall oil, and turpentine.

**Turpentine** is a sticky mixture of resin and oil from pine logs; on distillation, it delivers **(gum) rosin (pine gum**, a hard resin) and **oil** (or **spirits**) of **turpentine** (also called turpentine). Rosin can also be obtained from the roots of pine stumps and from **tall oil**, the byproduct of sulfate pulp (Swedish: *tall* = pine). See also p. 267.

Removal of volatiles from crude turpentine by steam distillation and extraction of rosin from pine stumps yields **colophony**, a light yellow to black residue that is used as drier for paints, for gluing writing papers, pressure-sensitive adhesives, in solders, glazing of roasted coffee beans, etc. The name "colophony" stems from the antique city of Kolophon (near Ephesos ("Ephesus"), Asia Minor) where resins were distilled. The world production of colophony is now more than 1 million tons per year; 400 000 t/a are produced in China and the rest mainly in the United States, Russia, and Portugal.

Colophony is a complex mixture of monocarboxylic acids of alkylated $C_{20}$ hydroxyphenanthrenes, mainly abietic acid and pimaric acid (see top of the next page).

**Mastic** is the exudate from slitting pistacia trees (G: *mastiche* = resin from pistacia trees). It is used in the manufacture of adhesives and varnishes, glazing of coffee beans, coatings of foods, and resination of Greek wines (retsina).

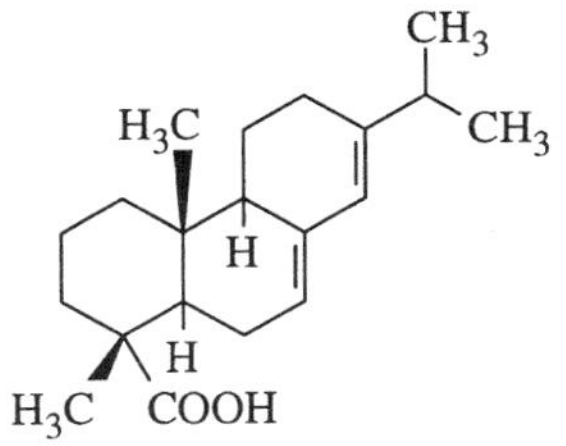

$CH_3$ $H_3C$ $CH_3$ H H $H_3C$ COOH

abietic acid

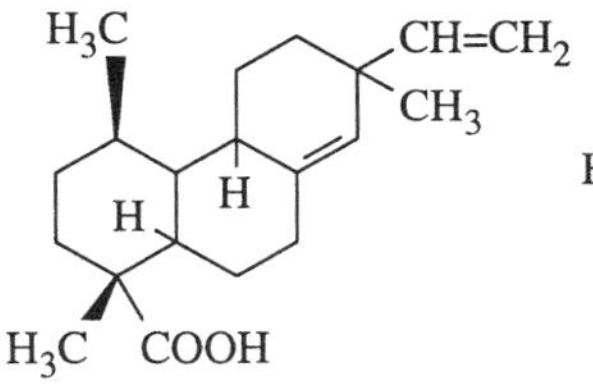

$H_3C$ $CH=CH_2$ $CH_3$ H H $H_3C$ COOH

pimaric acid

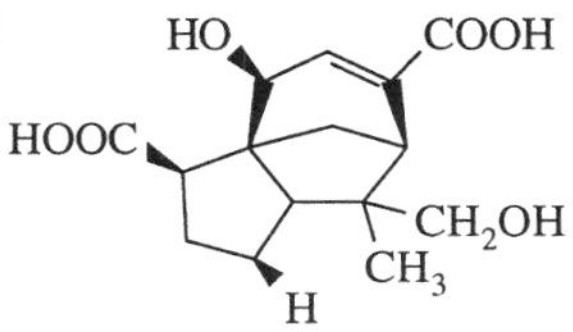

HO COOH HOOC $CH_2OH$ $CH_3$ H

shellolic acid

**Shellac** is purified lac resin which is produced by the female scale insect *Kerria lacca* (formerly *Tachardia lacca*) as protective coat (shell) for the eggs it deposits on leaves of Indian softwood trees. The 3–10 mm thick coat is scratched off (stick lac), milled, extracted with water or dilute soda solution, dried, and often also bleached (seed lac).

Shellac is the only known commercial natural resin from animals. Its name is derived from "shell" and the Sanskrit word *laksha* for hundred thousand because 100 000 insects are needed to produce 1 ounce (28.4 g) of shellac. Production is ca. 9000 kg/a.

Commercial shellac contains 94 % resin, 5 % waxes, and 1 % of a dye that is already in the lac resin; this dye provides shellac with its characteristic red color. The resin consists of derivatives of various polyhydroxy acids, mainly aleuritic acid (= 9,10,16-trihydroxy palmitic acid, $HO(CH_2)_6CH(OH)CH(OH)(CH_2)_7COOH$), and shellolic acid, $C_{13}H_{16}(OH)_2(COOH)_2$. The molecular weights of 1000-1500 are relatively high for a natural resin. On heating, shellac can be processed like a thermoplastical before cross-linking occurs. Shellac is an excellent varnish for wooden articles. The first phonograph records consisted of shellac that was reinforced with cotton fluff, carbon black, and mineral fillers.

In contrast to colophony and shellac, all other natural resins require only little work-up and purification. However, their collection is very labor-intensive. Dammar, copal, mastic, etc., are therefore increasingly replaced by synthetic resins. But many different gums are still used as thickeners for foodstuffs, protective colloids in cosmetics, and the like (see Chapter 8 and Volume IV).

## 3.10 Natural Oils and Biomass

Natural oils are solid to liquid plant and animal products. Most consist of mixed triglycerides of higher fatty acids with even numbers of carbon atoms, but some contain free fatty acids (tall oil) and some are not fatty acid (esters) at all (e.g., turpentine, a mixture of $\alpha$- and $\beta$-pinene). About 80 % of all produced fats and fatty oils are used as foodstuffs while 14 % serve for chemical and technical purposes. In Europe, especially in Germany, rape-seed oil is converted to rape diesel oil (see below).

Fatty acids may be saturated, mono-unsaturated, or multi-unsaturated. In industry, one usually distinguishes between drying oils with high contents of linoleic and linolenic acid (iodine number > 170), semi-drying acids with high contents of linolenic and oleic acid (iodine number 100–170), and non-drying oils with high contents of oleic acid (iodine number < 100). The **iodine number** is a measure of the degree of unsaturation; it is the grams of iodine that are taken up per 100 grams of oil by the addition reaction $-CH{=}CH- + I_2 \rightarrow -CHI{-}CHI-$.

**Drying oils** are rich in triple-unsaturated fatty acids (Table 3-29). They include linseed oil from the seeds of the flax plant (*Linum usitatissimum*), Tung oil from the nuts of the tree *Aleurites fordii*, hemp oil, poppyseed oil, nut oil from walnuts, and perilla oil from the seeds of the labiate *Perilla ocymoides*.

**Semi-drying oils** contain many mono- and double-unsaturated fatty acids. Examples are soy bean oil, cotton seed oil, corn oil, rape(-seed) oil, and dehydrogenated castor oil.

**Non-drying oils** are rich in saturated and mono-unsaturated fatty acids. They include castor oil, olive oil, and arach(id)ic oil from peanuts (*Arachis hypogea*).

Drying oils autoxidize in air, a property which is utilized in painting. The autoxidation is accelerated by metal salts such as cobalt naphthenate or manganese linoleate. Oxidation leads to trans isomerization of cis-double bonds, scission of carbon-carbon double bonds, formation of volatile products, and finally to crosslinks. Polymerization at elevated temperatures in the absence of oxygen results in products with increased stability against alkali. The "drying" of these oils is therefore not a physical process but a series of chemical reactions that convert a liquid to a solid.

Fatty oils are not only used as such but also as base chemicals for monomers. Examples are oils from castor beans, soy beans, rice bran, rapeseeds, and crambe.

Table 3-29 Weight percent of fatty acids $R(CH_2)_7COOH$ in some drying, semi-drying, and non-drying oils and world production of crops (seeds, nuts, beans) and oils therefrom in 2002 [22]. CA = castor oil, CN = coconut oil (incl. oil from dried kernels (copra)), CS = cotton seed oil, LS = linseed oil, SB = soy bean oil, T = Tung oil. Symbols for double bonds: = cis, $\hat{=}$ trans.

| Fatty acid Trivial name | Substituent R | Weight percent in oils: Drying LS | T | Semi-drying CS | SB | Non-drying CA | CN |
|---|---|---|---|---|---|---|---|
| Caprylic acid | H | | | | | | 6 |
| Capric acid | $H(CH_2)_2$ | | | | | | 6 |
| Lauric acid | $H(CH_2)_4$ | | | | | | 44 |
| Myristic acid | $H(CH_2)_6$ | | | 1 | | | 18 |
| Palmitic acid | $H(CH_2)_8$ | 6 | 4 | 29 | 11 | 2 | 11 |
| Stearic acid | $H(CH_2)_{10}$ | 4 | 1 | 4 | 4 | 1 | 6 |
| Palmito(lino)leic a. | $H(CH_2)_6CH{=}CH$ | | | 2 | | | |
| Oleic acid | $H(CH_2)_8CH{=}CH$ | 22 | 8 | 24 | 25 | 7 | 7 |
| Ricinoleic acid | $H(CH_2)_6CH(OH)CH_2CH{=}CH$ | | | | | 87 | |
| 9,11-Linoleic acid | $H(CH_2)_6CH{=}CHCH{=}CH$ | | | | | | |
| 9,12-Linoleic acid | $H(CH_2)_5CH{=}CHCH_2CH{=}CH$ | 16 | 4 | 40 | 51 | 3 | 2 |
| Linolenic acid, $\alpha$- | $H(CH_2)_2(CH{=}CHCH_2)_2CH{=}CH$ | 52 | 3 | | 9 | | |
| Eleostearic acid, $\beta$- | $H(CH_2)_4CH \hat{=} CHCH \hat{=} CHCH{=}CH$ | | 80 | | | | |
| Erucic acid | $H(CH_2)_8CH{=}CH(CH_2)_4$ | (from rapeseed oil = colza oil, canbra oil, canola oil) | | | | | |
| World production of crops in $10^6$ t/a in 2004 | | 1.9 | 0.5 | 44 | 204 | 1.3 | 55 |
| World production of oils in $10^6$ t/a in 2002 | | | | 3.8 | 30 | | 3.1 |
| World production of oils in $10^6$ t/a in 1980 | | 1.0 | | 3.5 | 15.0 | 0.4 | 3.0 |

The main component (80–85 %) of **castor oil** is the glyceride of ricinoleic acid. This ester can be cleaved by alkali or by heat. Alkali cleavage of ricinoleic acid (I) yields 2-octanol (II) and sebacic acid (III):

$$\underset{\text{I}}{H(CH_2)_6\underset{\displaystyle OH}{\underset{|}{C}}HCH_2CH{=}CH(CH_2)_7COOH} \longrightarrow \underset{\text{II}}{H(CH_2)_6\underset{\displaystyle OH}{\underset{|}{C}}HCH_3} + \underset{\text{III}}{HOOC(CH_2)_8COOH} \tag{3-12}$$

For thermal cleavage, castor oil is first subjected to methanolysis since methyl ricinoleate (IV) can be split thermally in much higher yields than the glyceride of ricinoleic acid. The pyrolysis proceeds at 550°C with short residence times and delivers heptanal (enanthol (V)) and methyl undecylenate (VI):

$$\underset{\text{IV}}{CH_3(CH_2)_5{-}C(H)(OH){-}CH_2{-}C(H){=}C(H){-}(CH_2)_7COOCH_3} \longrightarrow \underset{\text{V}}{CH_3(CH_2)_5{-}C(H){=}O} + \underset{\text{VI}}{H_2C{=}CH{-}CH_2{-}(CH_2)_7COOCH_3} \tag{3-13}$$

The methyl ester (VI) is saponified to undecylenic acid which is then reacted in an anti-Markovnikov reaction with HBr to 11-bromoundecanoic acid, $Br(CH_2)_{10}COOH$. Subsequent reaction with $NH_3$ leads to $NH_2(CH_2)_{10}COONH_4$ and, on acidification, the free 11-aminoundecanoic acid, $NH_2(CH_2)_{10}COOH$. Polycondensation of this $\omega$-amino acid delivers polyamide 11, $H{+}NH(CH_2)_{10}CO{+}_nOH$ (Chapter 10).

Methanolysis of **soy bean oil** leads to the methyl esters of the constituent fatty acids (see Table 3-29). The reducing ozonolysis of these esters delivers the corresponding $C_9$-aldehyde acid methyl ester which is subsequently converted to the amino ester and then to 9-aminopelargonic acid (9-aminononanoic acid), the monomer for polyamide 9, e.g.,

(3-14) $$H(CH_2)_5CH{=}CHCH_2CH{=}CH(CH_2)_7COOCH_3 \xrightarrow{+O_3} CHO(CH_2)_7COOCH_3$$

$$\xrightarrow{+NH_3,\ +H_2} NH_2(CH_2)_8COOCH_3 \xrightarrow[-CH_3OH]{+H_2O} NH_2(CH_2)_8COOH$$

**Rice (bran) oil** is rich in oleic acid which, on ozonolysis, delivers pelargonic acid and azelaic acid (monomer for polyamide 6.9):

(3-15) $$CH_3(CH_2)_7CH{=}CH(CH_2)_7COOH \longrightarrow CH_3(CH_2)_7COOH + HOOC(CH_2)_7COOH$$

The glycerol esters of the fatty acids of the plant *Crambe abyssinica* are rich in *cis*-erucic acid (*cis*-13-docosenic acid) (ca. 55 %). Ozonolysis of the methyl ester of erucic acid leads to the monomethyl ester of brassidic acid which can be converted by ammonium hydroxide + sulfur oxydichloride to the corresponding nitrile. Hydrogenation of the nitrile delivers the amino acid, the monomer for polyamide 13:

(3-16) $$H(CH_2)_8CH{=}CH(CH_2)_{11}COOCH_3 \longrightarrow CH_3OOC(CH_2)_{11}COOH \longrightarrow$$

$$CH_3OOC(CH_2)_{11}CN \longrightarrow HOOC(CH_2)_{12}NH_2$$

**Rape (colza)** is easy to grow and therefore has long been used as a source of animal fodder (ca. 22 % proteins) and rapeseed oil which is a mixture of glycerol esters of various fatty acids. Oil of older rape strains is not healthy for humans and animals because of its high content of erucic acid. Oils from newer strains (**canola oil**) are fit for human consumption since they contain far less erucic acid and considerably more oleic acid (in parentheses: percent in old rapeseed oil): 0.5 % (48 %) erucic acid, 63 % (15 %) oleic acid, 20 % (13.5 %) linoleic acid, 9 % (8.5 %) linolenic acid, 4 % (2.5 %) palmitic acid, 2% (2 %) hexadecenic acid, 1 % (5 %) icosenic acid, 0.5 % (1 %) docosadienic acid, and 2.5 % (5 %) other.

In Europe, rapeseed oil is transesterified with methanol which converts the glycerol esters to the corresponding methyl esters which are called **biodiesel**. The glycerol from this process is now replacing synthetic glycerol (Chapter 4).

One hectare produces ca. 1500 L/a biodiesel ($\rho$ = 0.684 g/mL) which has a similar efficiency with respect to crude oil-based diesel as sun diesel (91 % versus 93 %). In Germany, 10 % of agricultural acreage is now planted with rape. If all German agricultural acreage would grow rape, the total output of biodiesel of $19 \cdot 10^6$ t/a would still be only a small percentage of the total mineral oil consumption of Germany of $130 \cdot 10^6$ t/a.

Biodiesels are also obtained from other plant materials, especially straw. It is estimated that the whole Austrian need for diesel fuels could be satisfied by biodiesel from Austrian straw. As with all plant raw materials, high costs are encountered for the collection and transportation of plants to processing plants.

Another, widely untapped, source of intermediates and monomers are sugars from oligosaccharides and polysaccharides (Chapter 8). The hydrolysis of **hemicelluloses** (see Section 8.9) of corn cobs and other plant waste products leads to pentoses, for example, xylose (I). Dilute sulfuric acid converts xylose to furfurol (II), which, on pyrolysis at 400°C, delivers furan (III). Hydrogenation of III results in tetrahydrofuran (IV):

(3-17) HO OH HO HO CHO (I) $\xrightarrow{-3\,H_2O}$ O CHO (II) $\xrightarrow{-CO}$ O (III) $\xrightarrow{+H_2}$ O (IV)

Tetrahydrofuran (IV) is the monomer for poly(tetrahydrofuran) and the base material for a number of other monomers. The hydrolysis of IV delivers 1,4-butanol, one of the monomers for poly(butylene terephthalate). Reaction of tetrahydrofuran with HCl yields 1,4-dichlorobutane that is subsequently converted to the dinitrile, $NC(CH_2)_4CN$. The dinitrile is hydrogenated to 1,6-hexamethylene diamine, $H_2N(CH_2)_6NH_2$, one of the monomers for aliphatic AA/BB polyamides such as nylon 6.6.

## Literature to Chapter 3

3.0 SURVEYS

D.G.Altenpohl, Materials in World Perspective, Springer, Berlin 1980

P.Kennedy, The Rise and Fall of the Great Powers, Random House, New York 1988

3.2 ENERGY

J.T.McMullan, R.Morgan, R.B.Murray, Energy Resources and Supply, Wiley, London 1976

D.N.Lapedes, Ed., Encyclopedia of Energy, McGraw-Hill, New York 1976

D.A.Tillman, Wood as an Energy Source, Academic Press, New York 1978

D.K.Rider, Energy. Hydrocarbon Fuels and Chemical Resources, Wiley, New York 1981

3.3 COALS

M.E.Hawley, Coal, Academic Press, New York 1976 (2 volumes)

J.A.Cusumano, R.A.Dalla Betta, R.B.Levy, Catalysis in Coal Conversion, Academic Press, New York 1979

G.J.Pitt, G.R.Millward, Eds., Coal and Modern Coal Processing, Academic Press, London 1979

C.Y.Wen, E.S.Lee, Coal Conversion Technology, Addison-Wesley, Reading (MA) 1979

R.M.Davidson, Molecular Structure of Coal, IEA Coal Research, London 1980

D.D.Whitehurst, T.O.Mitchell, M.Farcasiu, Coal Liquefaction, Academic Press, New York 1980

C.H.Fuchsman, Peat. Industrial Chemistry and Technology, Academic Press, New York 1980

R.A.Meyers, Ed., Coal Structure, Academic Press, New York 1982

R.K.Hessley, J.W.Reasoner, J.T.Riley, Coal Science. An Introduction to Chemistry, Technology and Utilization, Wiley, New York 1986

D.L.Wise, Ed., Bioprocessing and Biotreatment of Coal, Dekker, New York 1990

N.Berkowitz, An Introduction to Coal Technology, Academic Press, Orlando (FL), 2nd ed. 1994

M.Hofrichter, A.Steinbüchel, Eds., Biopolymers, Vol. II: Lignin, Humic Substances and Coal, Wiley-VCH, Weinheim 2002

3.4 OIL

A.L.Waddams, Chemicals from Petroleum, Murray, London 1973

G.D.Hobson, W.Pohl, Eds., Modern Petroleum Technology, Wiley, New York, 4th ed. 1973

H.K.Abdel-Aal, R.Schmelzlee, Petroleum Economics and Engineering, Dekker, New York 1976

D.O.Shah, R.S.Schechter, Eds., Improved Oil Recovery by Surfactant and Polymer Flooding, Academic Press, New York 1977
G.Jenkins, Oil Economists' Handbook, Elsevier Applied Science, London, 4th ed. 1986
P.H.Spitz, Petrochemicals. The Rise of an Industry, Wiley, New York 1988
J.G.Speight, The Chemistry and Technology of Petroleum, Dekker, New York, 3rd ed. 1999

3.5.1 OIL SHALE

T.F.Yen, Ed., Science and Technology of Oil Shale, Ann Arbor Sci.Publ., Ann Arbor (MI) 1976
T.F.Yen, G.V.Chilingarian, Oil Shale, Elsevier, Amsterdam 1976
S.Lee, Oil Shale Technology, CRC Press, Boca Raton (FL) 1990

3.5.2 TAR SANDS and 3.5.3 BITUMENS AND ASPHALTS

H.Abraham, Asphalts and Allied Substances, Van Nostrand, Princeton. 6th ed. 1961-1963 (5 Vols.)
E.J.Barth, Asphalt: Science and Technology, Gordon and Breach, New York 1962
A.J.Hoiberg, Bituminous Materials. Asphalts, Tars and Pitches, Interscience, New York 1965 (2 Vols.)
P.Zakar, Asphalts, Chem.Publ.Co., New York 1971
G.V.Chilingarian, T.F.Yen, Bitumen, Asphalts and Tar Sands, Elsevier, Amsterdam 1978
T.F.Yen, G.V.Chilingarian, Eds., Asphaltenes and Asphalt, Elsevier, Amsterdam, Vol. 1 (1989)
R.F.Meyer, What good's all that gunk?, CHEMTECH (July 1991) 432
A.M.Usmani, Ed., Asphalt Science and Technology, Dekker, New York 1997
I.Kett, Asphalt Materials and Mix Design Manual, Noyes Publ., Park Ridge (NJ) 1999

3.6 NATURAL GAS

W.L.Lom, A.F.Williams, Substitute Natural Gas, Halsted, New York 1976
W.L.Lom, Liquefied Natural Gas, Appl.Sci.Publ., London 1977
M.T.Gillies, Ed., $C_1$-Based Chemicals from Hydrogen and Carbon Monoxide, Noyes Publ., Park Ridge (NJ) 1982
E.D.Shoan, Clathrate Hydrates of Natural Gases, Dekker, New York 1998
R.L.Kleinberg, P.G.Brewer, Probing Gas Hydrate Deposits, Amer.Scientist **89** (2001) 244

3.7.2 NATURAL WOOD

H.F.J.Wenzl, The Chemical Technology of Wood, Academic Press, New York 1970
F.A.Loewus, V.C.Runeckles, Eds., The Structure, Biosynthesis and Degradation of Wood (= Recent Advances in Phytochemistry, vol. 11), Plenum, New York 1977
J.F.Kennedy, G.O.Phillips, P.A.Williams, Eds., Wood and Cellulosics, Wiley, New York 1988
D.Fengel, G.Wegener, Wood Chemistry, Ultrastructure, Reactions, De Gruyter, Berlin 1989
M.Lewin, I.S.Goldstein, Eds., Wood Structure and Composition, Dekker, New York 1991
D.-N.S.Hon, N.Shiraishi, Eds., Wood and Cellulosic Chemistry, Dekker, New York, 2nd ed. 2000
E.Sjöström, Wood Chemistry: Fundamentals and Applications, Academic Press, San Diego (CA), 2nd ed. 1993
K.A.McDonald, D.E.Kretschmann, Eds., Wood Handbook: Wood as an Engineering Material, USDA Forest Service, Madison (WI) 1999 (out of print); see www.fpl.fs.fed.us/documents/fplgtr113

3.7.3 PRESSURIZED WOOD

F.P.Kollmann, E.W.Kuenzi, A.J.Stamm, Wood Based Materials, Springer, New York 1974

3.7.4 PLYWOOD AND VENEER

T.Sellers, Jr., Plywood and Adhesive Technology, Dekker, New York 1985

3.7.5 POLYMER WOOD

J.A.Meyer, Wood-Plastic Materials and Their Current Commercial Applications, Polym.-Plastics Technol.Eng. **9** (1977) 181

3.7.7 MECHANICAL PULP and 3.7.8 CHEMICAL PULP

J.P.Casey, Ed.,Pulp and Paper: Chemistry and Chemical Technology, Interscience, New York, 2nd ed. 1960 (3 Vols.)
A.J.Stamm, E.E.Harris, Chemical Processing of Wood, Chem.Publ., New York 1963
S.A.Rydholm, Pulping Processes, Krieger, Melbourne (FL) 1985 (reprint of 1965 edition)

3.8 LIGNINS
I.A.Pearl, The Chemistry of Lignin, Dekker, New York 1967
K.Freudenberg, A.C.Neish, Constitution and Biosynthesis of Lignin, Springer, Berlin 1968
K.V.Sarkanen, C.H.Ludwig, Eds., Lignins. Occurrence, Formation, Structure and Reactions, Wiley, New York 1971
J.F.Kennedy, G.O.Phillips, P.A.Williams, Lignocellulosics: Science, Technology, Development and Use, Prentice-Hall, Englewood Cliffs (NJ) 1992
D.N.-S.Hon, Ed., Chemical Modification of Lignocellulosic Materials, Dekker, New York 1995
M.Hofrichter, A.Steinbüchel, Eds., Biopolymers, Vol. II: Lignin, Humic Substances and Coal, Wiley-VCH, Weinheim 2002

3.9 NATURAL RESINS
E.Hicks, Shellac. Its Origins and Applications, Chem.Publ.Co., New York 1961
Angelo Bros., Shellac, Angelo Bros., Cossipore, Calcutta, 3rd ed. 1965
S.Maiti, S.S.Ray, A.K.Kundu, Rosin: A Renewable Resource for Polymers and Polymer Chemicals, Progr.Polym.Sci. **14** (1989) 297
D.A.Grimaldi, Amber: Window to the Past, H.N.Abrams, New York 1996

3.10 NATURAL OILS AND BIOMASS
D.Swern, Ed., Bailey's Industrial Oil and Fat Products, Wiley, New York, 4th ed. 1979 and 1982 (2 Vols.)
L.E.St.Pierre, G.R.Brown, Eds., Future Sources of Organic Raw Materials - ChemRawn I, Pergamon Press, Oxford 1980
E.Campos-López, Ed., Renewable Resources, Academic Press, New York 1980
C.E.Carraher, Jr., L.H.Sperling, Eds., Polymer Applications of Renewable-Resource Materials, Plenum, New York 1983
H.H.Szmant, Industrial Utilization of Renewable Resources: An Introduction, Technomic, Lancaster (PA) 1986
S.Manjula, C.K.S.Pillai, Naturally Occurring Organic Bio-Monomers as Possible Future Sources for Polymers, Polymer News **12** (1987) 359
W.A.Wood, S.T.Kellog, Eds., Biomass, (Methods of Enzymology, Vols. 160 and 161), Academic Press, New York 1988
F.W.Lichtenthaler, Ed., Carbohydrates as Organic Raw Materials, VCH, Weinheim 1991 (mono- and disaccharides but not starch, cellulose and other polysaccharides)
D.P.Mobley, Ed., Plastics from Microbes, Hanser, Munich 1995
G.A.Olah, A.Goeppert, G.K.Surya Prakash, Beyond Oil and Gas: The Methanol Economy, Wiley-VCH, Weinheim 2006

## References to Chapter 3

[1] U.S.Census Bureau, Statistical Abstract of the United States: 1999
[2] U.S.Census Bureau, Statistical Abstract of the United States: 2001, No. 880
[3] U.S.Census Bureau, Statistical Abstract of the United States: 2003, No. 880
[4] BP 2004, Statistical Review of World Energy, www.bp.com, (a) p. 6, (b) p. 14
[5] M.Balsam, C.Lach, R.-D.Maier, Nachr.Chem. **48**/3 (2000) 39, data of Figs. 1a and 1b
[6] International Energy Agency, Key World Energy Statistics, 2004 Edition
[7] U.S. Census Bureau, Statistical Abstracts of the United States, 2004-2005
[8] U.S. Census Bureau, Statistical Abstracts of the United States, 1998
[9] U.S. Energy Information Administration, March 2005 (www.eia.doe.gov)
[10] U.S. Energy Information Administration, March 2005, World Energy Use and Carbon Dioxide Emissions, 1980-2001 (www.eia.doe.gov/iea/), 25 October 2004
[11] W.Kröger, Neue Zürcher Zeitungs **220**/103 (6 May 1999) 17
[12] U.S. Energy Information Administration, Annual Energy Review 2003 (www.eia.doe.gov)
[13] G.Jenkins, Oil Economist's Handbook, Elsevier Applied Science, London, 4th ed. 1986

[14] R.Famighetti, Ed., The World Almanac and Book of Facts, World Almanac Books, Mahwah (NJ) 1997
[15] K. Weissermel, H.-J.Arpe, Industrielle organische Chemie, Verlag Chemie, Weinheim, 5th ed. (1998), p. 75
[16] J.Speight, in Kirk-Othmer, Concise Encyclopedia of Chemical Technology, Wiley-Interscience, New York, 4th ed. 1999, p. 1495
[17] –, Chemical & Engineering News (28 May 1979) p. 33-35
[18] R.F.Meyer, W. de Witt, Jr., CHEMTECH (July 1991) 432, Fig. 1
[19] United Nations, Forest and Agricultural Organization, ftp://ftp.fao.org/docrep/fao, 11.06.2005
[20] USDA, Agricultural Statistics 2004, Table 12-33
[21] K.A.McDonald, D.E.Kretschmann, Eds., Wood Handbook: Wood as an Engineering Material, USDA Forest Service, Madison (WI) 1999, Table 4-3a
[22] FAO, ftp://ftp.fao.org/decrep/fao, accessed 2005-07-10
[23] www.dft.gov.uk/stellent/groups/dft_roads/documents, accessed 2006-05-25

# 4 Base Chemicals, Intermediates, and Monomers

## 4.1 Base Chemicals

### 4.1.1 Survey

Until the 1950s, coal was the major source of both energy and chemicals in Europe whereas the United States already relied on crude oil as the major feedstock. The reason was the absence of significant crude oil fields in Europe on one hand and readily accessible oil fields in the United States on the other (see Table 3-13). In addition, there were different demands in the United States, Western Europe, and Japan. Europe had a highly developed extensive rail system for transporting people and goods that used coal-powered engines (except Switzerland which used electricity from its many hydroelectric plants). The United States, on the other hand, relied on gasoline powered cars for the transport of people and diesel-powered trucks and trains for the shipment of goods.

The coal-based economy of Europe provided plenty of coke. The European chemical industry was thus based on coke as a feedstock for aromatic compounds and synthesis gas (Section 4.1.3), also from coke, as base materials for aliphatics. The United States, in contrast, had a ready supply of aliphatics as byproducts of the refining of crude oil for gasoline and motor and diesel fuels.

The situation changed in the 1950s after oil fields were developed in the Middle East, and then again in the 1980s with the beginning of the exploitation of natural gas fields. Processing of crude oil and natural gas by petroleum refining and subsequent petrochemical operations delivered methane, ethene, propene, butadiene, benzene, toluene, xylenes, and methanol as base products. This group of chemicals and their downstream products thus became known as **petrochemicals**.

In the United States, crude oil delivers ca. 50 % of petrochemicals, natural gas liquids ca. 40 %, natural gas ca. 9 %, and coal just 0.5 %. In Europe and Japan, petrochemical naphtha (crude gasoline) is the major feedstock. The Middle East is presently developing a petrochemical industry based on ethane which was just flared in the past.

Natural gas delivered first butane and then butene and butadiene. A number of intermediates followed later, especially ethane and propane, and then ethene and higher hydrocarbons. Natural gas is presently the main feedstock for methanol.

Natural gas liquids are the main raw material sources for ethene and even-numbered olefins. Odd-numbered olefins (pentenes, etc.) are based on coal. Naphtha from crude oil is the major source for propene (Section 4.4.1) and aromatics (Section 4.7).

At present, there are three major feedstocks:

- **methane** (Section 4.2.1) from natural gas (Section 3.6);
- **naphtha** from petroleum processing (Section 3.4.2);
- **syngas** (Section 4.1.3), the mixture of carbon monoxide and hydrogen in various proportions, that is obtained from naphtha, natural gas, or coals.

Some specialty monomers are also derived from natural oils or biomass (see Section 3.12).

### 4.1.2 From Feedstocks and Base Chemicals to Polymers

Polymers can be roughly divided into bulk and specialty polymers. Bulk polymers are usually obtained downstream from feedstocks or readily available base chemicals in as few steps as possible. They are produced in huge quantities and serve for many different applications for which they often compete with each other. Because of that competition and an occasional oversupply of feedstocks and/or base chemicals, bulk polymers are often in search of additional applications.

Monomers for bulk polymers are either primary or base chemicals that are obtained directly from feedstocks (example: ethene from naphtha or syngas) or indirectly via very few intermediates such as formaldehyde from methanol (and methanol from syngas or, to a minor extent, from propane/butane).

Specialty polymers, on the other hand, are valued for very specific applications. Very often, their recognized special properties started a search for economical monomers, intermediates, and/or base chemicals. Sometimes a new method allowed a monomer to be isolated from a fraction or waste stream that then sought an application in the polymer field. An example is cyclopentene.

Ideally, syntheses of base chemicals, intermediates, and monomers should involve as few steps as possible. Furthermore, each step should involve only one **educt** E (= starting material, also called **substrate** in the biosciences (L: *educere* = to bring out, evolve (*ex* = out, *ducere* = to lead)) and one **product** P since separation processes are costly. The reaction of an educt E with another chemical compound A is usually avoided since A is an additional cost factor. Employment of A is, of course, required if the product results from the addition of A to E according to $E + A \rightarrow P$. In general, a reaction $E \rightarrow P$ is preferred, which often means high-temperature gaseous-state reactions that are catalyzed by metals or metal oxides and have short residence times (see also p. 130, Appendix A 4).

Also, E should react to P without producing a byproduct B. However, a valuable byproduct may improve the economy of the process, especially, if it is used in-house (external marketing requires additional efforts). Finally, the reaction of E to P should also not involve solvents because of the high costs of solvents and their recycling.

Since a feedstock may deliver many different base chemicals, a base chemical many different intermediates, and an intermediate many different monomers, very complex flow schemes result. The situation is furthermore complicated by the fact that feedstocks, base chemicals, and intermediates also serve for the syntheses of many products other than monomers and polymers. However, since this book is only concerned with the chemical technology of monomers and polymers, other products such as solvents, detergents, dyestuffs, pharmaceuticals, etc., are generally not depicted in flow schemes.

The purpose of this chapter is to show the interconnection between feedstocks and polymers. Hence, only educts and products are mentioned in flow schemes. In some cases, gross reactions are discussed but not mechanisms because mechanisms: a) are usually not known or not known in detail; and b) doing so would digress too far from the main theme of this book which is industrial polymers and their monomers

The reader will notice that not many new industrial polymers have been introduced during the last two decades and that many have disappeared from the market. The reason is economy. It takes 10–30 years for a new industrial chemical to become a big product and a turnover of ca. 40 million dollars to justify the new business.

The long and ardous path from a raw material to a finished polymeric good may be illustrated by the preparation of poly(ethylene) film from crude oil as raw material (Fig. 4-1). The primary distillation of 100 000 kg of crude oil leads to just 10 000 kg of chemical naphtha but to 90 000 kg of other products. The 10 000 kg of naphtha furnish 3000 kg of ethene plus propene, butene, and other chemicals. On polymerization of ethene to poly(ethylene), some polymer is lost as wall deposits; furthermore, some charges will fail totally. On average over many charges, 2900 kg of poly(ethylene) is obtained that is extruded to 56 wt% of sellable poly(ethylene) film. The remaining 44 wt% is lost in the start-up, through trimming, confectioning, and the like. The total yield of product (poly(ethylene) film) is thus only 1.6 wt% with respect to the raw material.

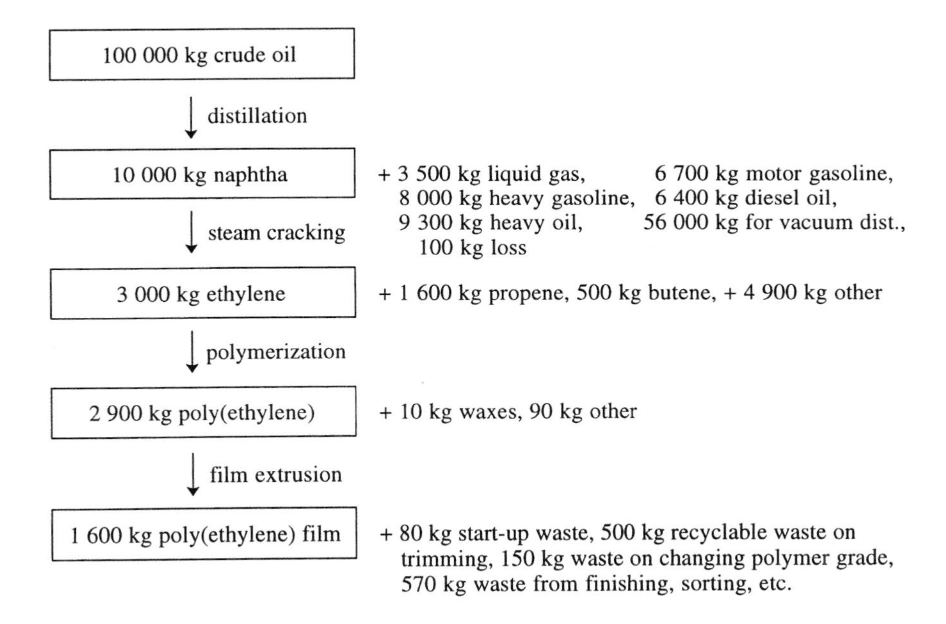


Fig. 4-1 Yield of feedstock, fractionates, monomers, polymer, polymer film, and polymer waste.

### 4.1.3 Synthesis Gas

**Definitions**

Synthesis gas (**syngas**) is the general name for mixtures of the gases $CO + H_2$ that are used as chemical feedstocks for the syntheses of methanol, formaldehyde, and other chemical base materials. However, the term "synthesis gas" is also used for **ammonia gas**, the mixture of $N_2 + 3\ H_2$ for the synthesis of ammonia.

Synthesis gas was first obtained from hard coal, brown coal, or coke as carbon sources by gasification with oxygen and water vapor. This synthesis gas is also called **water gas**. It has the same composition, $CO + H_2$, as the **oxo gas** from naphtha cracking (see below) that is used for hydroformylation which is the reaction of olefins (especially 1-olefins) with oxo gas to aldehydes according to $R{-}CH{=}CH_2 + CO + H_2 \rightleftarrows R{-}CH_2{-}CH_2{-}CHO$ (oxo reaction). **Methanol synthesis gas** consists of $CO + 2\ H_2$ (see Eq.(4-9)).

**Town gases** may be water gas, gases from coking, gases from low-temperature carbonification, etc. They consist mainly of hydrogen (40–67 %) and methane plus nitrogen, CO, $CO_2$, and higher hydrocarbons.

Town gases were formerly used to heat and illuminate houses, light the streets, etc. Town gas for street lights was replaced by electricity many decades ago. For cooking, city dwellers in Europe now use mainly natural gas or electricity and in the United States electricity, propane, or LPG. In some rural areas, firewood is still the main fuel for cooking and heating.

**Synthetic natural gas** (**substitute natural gas, SNG**) is a potential chemical feedstock that is not yet commercial. It results from the methanization of coal by hydrogen.

### Syngas from Coke

Synthesis gas (**syngas**) was originally obtained from carbon (as coke from hard coal or lignite), oxygen, and water.

The main reactions in the synthesis of synthesis gas from *coke*, oxygen, and water vapor are the exothermal partial combustion of carbon, Eq.(4-1), and the endothermal formation of water gas, Eq. (4-2):

(4-1) $2\,C + O_2 \rightleftarrows 2\,CO$ ($\Delta H = -\,246$ kJ/mol)

(4-2) $2\,C + 2\,H_2O \rightleftarrows 2\,CO + 2\,H_2$ ($\Delta H = +\,119$ kJ/mol)

These reactions are accompanied by a number of other reactions, mainly the endothermic formation of carbon monoxide (Boudouard equilibrium),

(4-3) $C + CO_2 \rightleftarrows 2\,CO$ ($\Delta H = +\,162$ kJ/mol)

the hydrogenation of carbon,

(4-4) $C + 2\,H_2 \rightleftarrows CH_4$ ($\Delta H = -\,87$ kJ/mol)

the methanization of carbon monoxide,

(4-5) $CO + 3\,H_2 \rightleftarrows CH_4 + H_2O$ ($\Delta H = -\,206$ kJ/mol)

and the water gas reaction,

(4-6) $CO + H_2O \rightleftarrows CO_2 + H_2$ ($\Delta H = -\,42$ kJ/mol)

The overall reaction, Eq.(4-7), can be approximated by Eqs.(4-1) and (4-2):

(4-7) $4\,C + O_2 + 2\,H_2O \rightleftarrows 4\,CO + 2\,H_2$ ; H/C = 1

Today, the main carbon sources for synthesis gases are *crude oil* and *natural gas* because they deliver a greater proportion of hydrogen. As Eqs.(4-7) and (4-8) show, the hydrogen/carbon H/C ratios in synthesis gas are only 1 for coke (Eq.(4-7)) but 4 for methane as feedstock (Eq.(4-8); they are ca. 2 for crude oil and ca. 2.4 for gasoline.

(4-8) $2\,CH_4 + O_2 \rightleftarrows 2\,CO + 4\,H_2$ ; H/C = 4

High temperatures of 900–1000°C are necessary for the gasification of coke in order to achieve high reaction rates and a shift of the Boudouard equilibrium, Eq.(4-3), to carbon monoxide (CO yields of 2 %, 57.7 %, 97.2 %, and 99.3 % are obtained at temperatures of 450°C, 700°C, 900°C, and 1000°C, respectively).

**Syngas from Natural Gas, Crude Oil, and Naphtha**

Natural gas and crude oil can be converted to syngas by two types of processes: steam reforming and partial oxidation. *Steam reforming* utilizes hydrocarbons with boiling temperatures of less than 200°C (naphtha) that are cracked by external heating using first $H_2$ and a desulfurization catalyst ($CoO\text{-}MoO_3/Al_2O_3$) at 350–450°C and then $H_2O$ and a cracking catalyst ($Ni\text{-}K_2O/Al_2O_3$) at 700–830°C and 15–40 bar.

*Partial oxidation* can employ all hydrocarbons from methane to heavy heating oil. It uses $H_2O$ and a less than stoichiometric amount of $O_2$ in an autothermal reaction at 1200–1500°C and 30–80 bar.

Syngases contain not only CO and $H_2$ but also other gaseous components ($H_2S$, COS, $CO_2$) that need to be removed before the gas can be used for the synthesis of intermediates. $H_2S$ and COS are catalyst poisons and $CO_2$ can either lead to side reactions or to the formation of gas cushions that impede the desired chemical reactions.

All three deleterious gases, $H_2S$, COS, and $CO_2$, can be removed from synthesis gas by absorption in methanol, *N*-methyl pyrrolidone, poly(ethylene glycol dimethyl ether)s, aqueous sulfolane–diisopropanol amine solutions, aqueous $K_2CO_3$, or by adsorption on molecular sieves. These absorbing or desorbing systems are then regenerated by degassing at elevated temperatures or reduced pressures. A part of the world-wide production of $H_2S$ is oxidized to $SO_2$ which is then reacted with $H_2S$ to elemental sulfur.

**Uses of Syngas**

There are three ways to use synthesis gas: directly, after separation of its components, or after reaction with other chemicals.

1. The main direct use is the synthesis of methanol according to

$$\text{(4-9)} \qquad CO + 2\,H_2 \rightleftarrows CH_3OH \qquad (\Delta H = -92 \text{ kJ/mol})$$

either by low-pressure processes with Cu-Al-Zn oxide or modified CuO-ZnO catalysts at 240–260°C (ICI, Lurgi) or a high-pressure process at 340 bar and 320–280°C with $ZnO\text{-}Cr_2O_3$ catalysts. Some plants have capacities up to $10^6$ t/a. Methanol is obtained with a purity of ca. 99.99 wt% and used for many syntheses (Section 4.2.2).

A second direct use of syngas is the methanization of carbon monoxide to **synthetic natural gas (SNG)** according to Eq.(4-5). SNG serves as an energy source.

The third direct use of syngas is the production of hydrocarbons by the Fischer-Tropsch process, based on either coal (South Africa) or natural gas (Malaysia), using Fe catalysts, temperatures of 170–350°C, and pressures of less than 28 bar (many variants):

$$\text{(4-10)} \qquad n\,CO + (1 + 2\,n)\,H_2 \rightleftarrows H(CH_2)_nH + n\,H_2O$$

Hydrocarbons obtained range from gasoline to waxes.

2. Synthesis gas is also a major source for the isolation of carbon monoxide (Section 4.1.4) and/or hydrogen (Section 4.1.5).

3. Hydroformylation is the reaction of syngas, either with olefins to aldehydes (Fig. 4-5, Route VII or Fig. 4-7, Route IX) or with oxirane (Fig 4-5, Route Xa).

### 4.1.4 Carbon Monoxide

Pure carbon monoxide is obtained from synthesis gas by either low-temperature fractionation or by CO absorption. Raw synthesis gas consists of $CO/H_2$ mixtures and unreacted $CH_4$ (Section 4.1.3). Cooling this mixture under pressure (40 bar) to –180°C liquefies the $CH_4/CO$ mixture whereas hydrogen remains gaseous and is removed. Subsequent reduction of the pressure to 2.5 bar yields a CO fraction at the top of the column that contains less than 0.1 vol% of $CH_4$.

Alternatively, CO can be obtained from $CO/H_2/CH_4$ mixtures by two types of absorption processes which both use copper(I) compounds as absorbent. In the high-pressure process, CO is absorbed at pressures up to 300 bar by either HCl-acidic aqueous solutions of CuCl or $NH_3$-basic aqueous solutions of Cu(I) carbonate or formate. Lowering the pressure and increasing the temperature to 40–50°C releases the CO.

The cosorb process, on the other hand, absorbs CO in a solution of CuCl and $AlCl_3$ in toluene at 25°C and pressures of less than 20 bar. The resulting Cu(I)-CO complex dissociates at 110–140°C and 1–4 bar and releases the CO.

The resulting carbon monoxide is employed in *carbonylation* of methanol to acetic acid ($CH_3OH + CO \rightarrow CH_3COOH$) or of methyl acetate to acetic anhydride ($CH_3COOCH_3 + CO \rightarrow (CH_3CO)_2CO$). Carbon monoxide is also used in carbonylation of nucleophilic partners, for example, of acetylene to acrylic acid ($CH{\equiv}CH + CO + H_2O \rightarrow CH_2{=}CHCOOH$) or of ethene to propionic acid ($CH_2{=}CH_2 + CO + H_2O \rightarrow CH_3CH_2COOH$). Replacement of $H_2O$ by alcohols leads to the corresponding esters.

### 4.1.5 Hydrogen

Approximately 80 % of the world production of pure hydrogen stems from petrochemical processes, ca. 16 % from coal gasification, and only 4 % from the electrolysis of water or other processes. The "green dream" of hydrogen as a clean motor fuel from the electrolysis of water requires large inexpensive sources of electricity which are nowhere in sight: hydroelectricity is mostly at its geological limits (p. 38) except in China, wind power is costly and unreliable, and nuclear power is decidedly "ungreen." This leaves the thermal generation of hydrogen: production of 1 kg of liquid hydrogen releases ca. 20 kg of $CO_2$ if from natural gas and ca. 30 kg of $CO_2$ if from coal!

Hydrogen is the desired main product of the *electrolysis* of water and a welcome byproduct of other electrolyses, such as $2\ HF \rightarrow H_2 + F_2$, $2\ HCl$ (to $H_2O$) $\rightarrow H_2 + Cl_2$, and $2\ NaCl + 2\ H_2O \rightarrow H_2 + Cl_2 + 2\ NaOH$. Hydrogen from electrolysis of water is very pure (> 99 vol%) whereas hydrogen from oil products needs considerable purification.

*Steam reforming* of lower hydrocarbons delivers synthesis gas (p. 91). According to the general equation for the formation of synthesis gas (reaction from right to left in

Eq.(4-10)), water is the source of 33 % of the hydrogen of the synthesis gas if methane is the feedstock but the source of ca. 47 % of the hydrogen if hexane is employed. From synthesis gas, $H_2S$ and COS have to be removed by absorption or desorption processes (see p. 91, top). The undesired CO is then reacted with $H_2O$ to $CO_2 + H_2$. $CO_2$ is removed by pressure washing. Remaining traces of CO are then subjected to a methanizing reaction at 300–400°C , $CO + 3\ H_2 \rightarrow CH_4 + H_2O$.

*Hydrocracking* of light crude oil distillates leads to the formation of $H_2$ from cyclization and aromatization processes. The resulting **refinery gas** is an important source of hydrogen for hydrofining and hydrotreating in refineries. Hydrogen is obtained from refinery gas by fractional condensation at low temperatures, by absorption in molecular sieves, or by flow through semipermeable membranes from polysulfones or polyimides.

In the *electrothermal cracking* of light hydrocarbons in an arc, the gaseous mixture attains temperatures of ca. 1800°C for seconds. The mixture is then rapidly cooled to 200°C by water. The process is used for the production of acetylene (see Section 4.3.3) but its greatest volume fraction is the "byproduct" hydrogen.

Some 67 % of the world production of hydrogen is consumed by the synthesis of ammonia, 20 % by petroleum refining, 9% for the synthesis of methanol, and 4 % for oxo processes, hydrogenations (including hardening of fats), and other chemical reactions.

## 4.2 $C_1$ Compounds

### 4.2.1 Methane

Methane is obtained as "dry natural gas" from natural gas (Section 3.7.2) as well as from wellhead and refinery gases. It is used primarily as fuel and secondarily as a base chemical, mainly after conversion to synthesis gas (Section 4.1.3). Methane is also directly converted to acetylene by dehydrogenation (pyrolysis), to hydrogen cyanide by reaction with ammonia, to chlorinated $C_1$ compounds by chlorination, and to dimethyldichlorosilane by reaction with silicon (Fig. 4-2).

Acetylene and hydrogen cyanide are important base chemicals. Their many further reactions, including the synthesis of monomers for polymers, are discussed in Sections 4.2.5 (HCN) and 4.3.3 ($C_2H_2$), respectively.

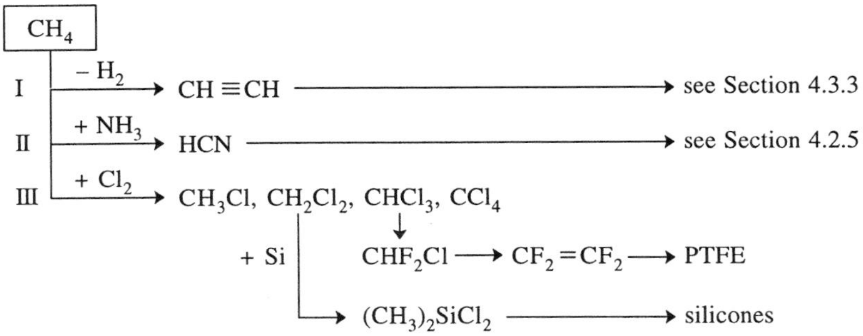

$CH_4$

I $\xrightarrow{-H_2}$ $CH{\equiv}CH$ $\longrightarrow$ see Section 4.3.3

II $\xrightarrow{+NH_3}$ HCN $\longrightarrow$ see Section 4.2.5

III $\xrightarrow{+Cl_2}$ $CH_3Cl$, $CH_2Cl_2$, $CHCl_3$, $CCl_4$

$CHCl_3 \downarrow$ $CHF_2Cl \longrightarrow CF_2{=}CF_2 \longrightarrow$ PTFE

+ Si $\longrightarrow (CH_3)_2SiCl_2 \longrightarrow$ silicones

Fig. 4-2 Base chemicals from methane. PTFE = poly(tetrafluoroethylene).

Reaction of dichloromethane, $CH_2Cl_2$, with silicon leads to $(CH_3)_2SiCl_2$ (plus $SiCl_4$, $CH_3SiCl_3$, and $(CH_3)_3SiCl$), the monomer for poly(dimethyl siloxane), $-[Si(CH_3)_2-O]_n-$.

Chloroform delivers tetrafluoroethylene, $CF_2{=}CF_2$, in two steps. Chloroform is reacted with hydrogen fluoride to difluorochloromethane which is then thermolyzed to tetrafluoroethene, $CF_2{=}CF_2$, the monomer for poly(tetrafluoroethylene), PTFE:

$$CHCl_3 \xrightarrow[-\,HCl]{+\,HF} CHFCl_2 \xrightarrow[-\,HCl]{+\,HF} CHF_2Cl \xrightarrow[-\,2\,HCl]{\Delta} CF_2{=}CF_2 \qquad (4\text{-}11)$$

### 4.2.2 Methanol

Methanol (formerly: carbinol, wood alcohol) is presently industrially obtained almost exclusively from methanol synthesis gas ($CO + 2\,H_2$, p. 91) and to a very small extent from oxidizing propane-butane mixtures. The total world production of methanol is ca. $30 \cdot 10^6$ t/a). Wood alcohol from wood spirits (Section 3.7.10) is industrially obsolete.

Ca. 50 % of $CH_3OH$ has polymer-related uses; the rest is used to synthesize methyl *t*-butylether (MTBE; antiknocking agent for gasoline (27 %)), methyl halogenides (7 %), methyl amines (4 %), as a solvent (4 %), etc. Polymer-related uses comprise (Fig. 4-3):

I. Formaldehyde is obtained by dehydrogenation (Ia) or oxidation (Ib) of $CH_3OH$. HCHO is mainly used for thermosets (UF, PF, MF) and poly(oxymethylene) (POM).

II. Carbonylation of methanol is one of the newer processes for acetic acid which is used for vinyl acetate (for poly(vinyl acetate), PVAc) and cellulose acetate (CA).

III. Various syntheses (p. 97, 101) employ methanol for the synthesis of methyl methacrylate (for poly(methyl methacrylate), PMMA).

IV. Esterification of terephthalic acid delivers dimethyl terephthalate (DMT), a monomer for the synthesis of poly(ethylene terephthalate), PET.

V. A new process (p. 107) converts methanol to propene for poly(propylene) .

VI. Oxidative carbonylation produces dimethyl oxalate which is then reacted with $H_2$ to give ethylene glycol (EG) + methanol by hydrogenation/hydrolysis.

Methanol is also fermented to single cell protein (SCP), an animal fodder.

| $CH_3OH$ | Reagent | Product | Use of methanol | Use for polymer |
|---|---|---|---|---|
| Ia | $-\,H_2$ | HCHO | 35 % | UF, PF, MF, POM |
| Ib | $+\,(1/2)\,O_2,\ -\,H_2O$ | HCHO | | |
| II | $+\,CO$ | $CH_3COOH$ | 7 % | PVAc, CA, PET *) |
| III | + various | $CH_2{=}C(CH_3)COOCH_3$ | 3 % | PMMA |
| IV | $+\,1,4\text{-}C_6H_4(COOH)_2$ | $C_6H_4(COOCH_3)_2$ | 2 % | PET |
| V | $-\,H_2O$ | $CH_2{=}CH(CH_3)$ | | PP |
| VI | $+\,CO + (1/2)\,O_2$ | $CH_3OC(O){-}C(O)OCH_3 \xrightarrow[-\,2\,CH_3OH]{+\,4\,H_2}$ EG | | PET, etc. |

Fig. 4-3 Polymer-related industrial reactions of methanol (use data: 1996). *) As solvent for TPA.

### 4.2.3 Formaldehyde

Formaldehyde, HCHO, is synthesized from $CH_3OH$ by oxydehydrogenation at 600-720°C with Ag as catalyst or by oxidation at 350-450°C with an $Fe_2O_3$-$MoO_3$ catalyst (Fig. 4-3). The first step of oxydehydrogenation is an exothermal dehydrogenation, $CH_3OH \rightleftarrows HCHO + H_2$. The released hydrogen is immediately oxidized to water by an added understoichiometric amount of oxygen; formic acid is not formed. The overall reaction is thus the same as in the direct oxidation, $CH_3OH + (1/2)\ O_2 \rightleftarrows HCHO + H_2O$.

Most of the world production of HCHO of ca. $5\cdot10^6$ t/a is converted directly or indirectly to polymers (Fig. 4-4). Ca. 3 % are used for poly(oxymethylene) (POM) by

I. direct polymerization of HCHO, or

II. polymerization of 1,3,5-trioxane, the cyclic trimer of HCHO.

III. HCHO is used directly as a comonomer for thermosetting resins (Route III): urea resins (UF; 33 %), phenolic resins (PF; 11 %), and melamine resins (MF; 4 %).

IV. The hydrating carbonylation of HCHO by sulfuric acid delivers glycolic acid, $HOCH_2COOH$ (ca. 50 000 t/a), which is used for many purposes, including the cyclodimerization to glycolide and further to poly(glycolide). Glycolic acid is no longer esterified by methanol and then hydrogenated to ethylene glycol, $HOCH_2CH_2OH$.

V. The condensation of HCHO with propanal (from the hydroformylation of ethene) at 160-210°C and 40-80 bar in the presence of a secondary amine and acetic acid to methacrolein with subsequent oxidation is one of the newer methods for the synthesis of methacrylic acid, which is either polymerized to poly(methacrylic acid), PMAA, or esterified to methyl methacrylate, the monomer for poly(methyl methacrylate), PMMA.

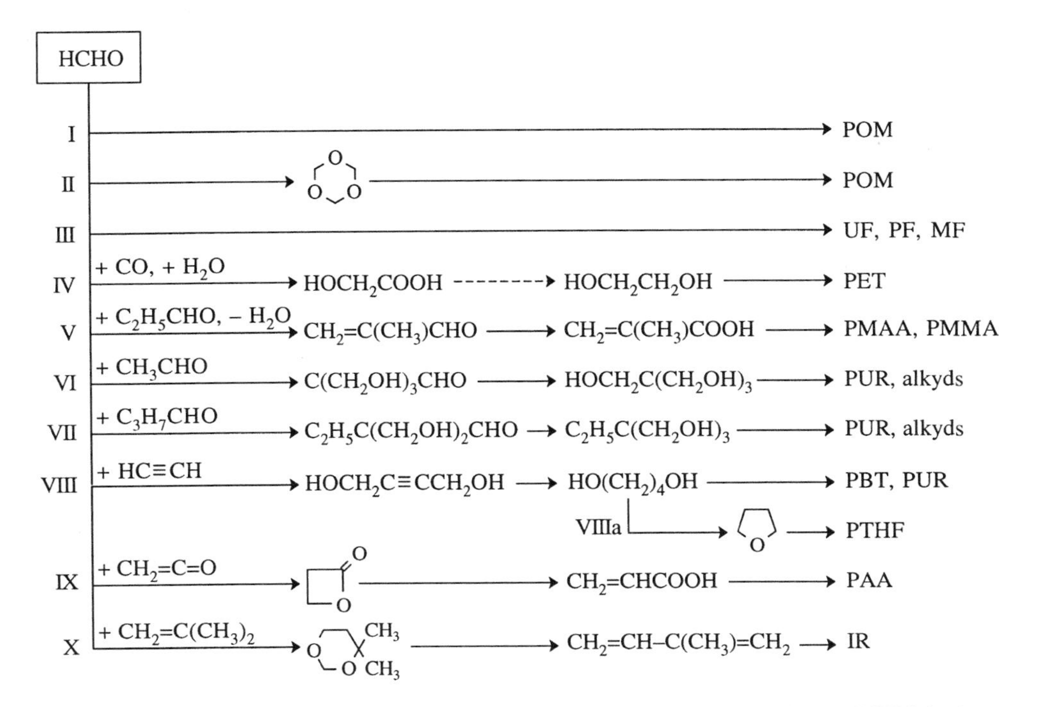


Fig. 4-4 Formaldehyde as a base chemical for polymers. Stoichiometry is not shown. HCHO is also used for the syntheses of 1,4-butanediol (for PBT), methylenediisocyanate (for PUR), etc.

VI + VII. Triple mixed aldolizations of HCHO with acetaldehyde (VI) or butyraldehyde (VII) in aqueous $Ca(OH)_2$ or NaOH (VI) with subsequent Cannizarro reactions deliver (VI) pentaerythritol, $C(CH_2OH)_4$, or (VII) trimethylol propane, $C_2H_5C(CH_2OH)_3$. Both polyols as well as 1,4-butanediol from Route VIII serve as components in polyurethanes (PUR) and alkyd resins.

HCHO is also used as an intermediate in one of the syntheses of methylene diisocyanate, a PUR monomer (Eq.(4-36)). These PUR-related uses of formaldehyde consume ca. 8 % of the HCHO production.

VIII. The Reppe process reacts an aqueous HCHO solution with acetylene at 100-110°C and 5–20 bar in the presence of modified copper acetylide (see also p. 105, VI). The resulting 2-butyne-1,4-diol is then hydrogenated to 1,4-butanediol which serves as a comonomer in poly(butylene terephthalate) (PBT), $-[O(CH_2)_4OOC(p\text{-}C_6H_4)CO]_n-$, and polyurethanes. The dehydration of 1,4-butanediol (Route VIIIa) is also one of the routes to tetrahydrofuran, $C_4H_8O$, which is the monomer for poly(tetrahydrofuran), PTHF, a component of elastic polyurethane fibers.

IX. The addition of ketene, $CH_2{=}C{=}O$, to formaldehyde leads to $\beta$-propiolactone which stabilizes itself to the polyester, $CH_2{=}CHCOO{-}[CH_2CH_2COO]_n{-}H$. Depolymerization of this polyester delivers acrylic acid which is polymerized to poly(acrylic acid) (PAA) or converted to alkyl acrylates and then polymerized to poly(alkyl acrylate)s.

X. Formaldehyde is also used in one of the routes to isoprene. The sulfuric acid-catalyzed Prins reaction of isobutene, $CH_2{=}C(CH_3)_2$ with 2 HCHO (aq) at ca. 80°C leads to 4,4-dimethyl-1,3-dioxane which is subsequently fragmented by phosphate catalysts at 240–400°C to isoprene, $CH_2{=}C(CH_3){-}CH{=}CH_2$, $H_2O$, and HCHO. Isoprene is the monomer for synthetic natural rubber (IR).

### 4.2.4 Formic Acid

Formic acid is exclusively produced from carbon monoxide, $CO + H_2O \rightleftarrows HCOOH$, at 115–150°C and 8–30 bar with NaOH or $Ca(OH)_2$ as catalysts by insertion of carbon monoxide into the HO–H bond of water. It is also a byproduct in the oxidation of butane or of light naphtha to acetic acid. Methyl formate, $HCOOCH_3$, is similarly obtained from CO and $CH_3OH$.

Both formic acid and methyl formate have many non-polymeric applications (fermentation of green fodder for animals, use in the textile and leather industries, intermediate in organic syntheses, etc.). The main polymer-related use of methyl formate is its reaction with dimethylamine to *N,N*-dimethylformamide, the solvent from which poly(acrylonitrile) is spun to acrylic fibers.

### 4.2.5 Hydrogen Cyanide

Hydrogen cyanide, $H{-}C{\equiv}N$, is an important base chemical that serves mainly for the extension of carbon chains. It is obtained from methane (p. 93) or higher hydrocarbons, from formamide, or as a byproduct in the synthesis of acrylonitrile from propene.

The formation of hydrogen cyanide from methane is a Pt-catalyzed exothermic ammonoxidation at 1000–1200°C with short residence times (Andrussov process):

(4-12) $CH_4 + NH_3 + (3/2)\ O_2 \longrightarrow HCN + 3\ H_2O$ $\Delta H = -473$ kJ/mol

Even higher temperatures are employed in oxygen-free processes such as the Pt-, Ru-, or Al-catalyzed endothermic reaction of ammonia with methane at 1200–1300°C ($CH_4 + NH_3 \rightarrow HCN + 3\ H_2$) or with higher hydrocarbons such as propane at 1300–1600°C ($C_3H_8 + 3\ NH_3 \rightarrow 3\ HCN + 7\ H_2$).

Hydrogen cyanide is also obtained from the endothermic dehydration of formamide at 380-430°C with Fe phosphate or Al phosphate as catalyst, $HCONH_2 \rightarrow HCN + H_2O$.

Hydrogen cyanide is used primarily for the syntheses of methacrylates and methacrylonitrile via acetone cyanohydrin. The alkali hydroxide- or carbonate-catalyzed addition of HCN to acetone at 40°C delivers acetone cyanohydrin I:

(4-13) $$(H_3C)_2C{=}O + HCN \xrightarrow{+[OH^{\ominus}]} (H_3C)_2C(CN)(OH)\quad \text{I}$$

Acetone cyanohydrin is subsequently saponified by concentrated sulfuric acid at 80–140°C to an amide sulfate II. The reaction of II with methanol delivers methyl methacrylate III (Eq.(4-14), top):

(4-14)
$$(H_3C)_2C(CN)(OH)\ \text{I} \xrightarrow[+ H_2SO_4]{+ H_2O} (H_3C)_2C(CONH_2{\cdot}H_2SO_4)(OH)\ \text{II} \xrightarrow[- H_2O,\ - NH_4[HSO_4]]{+ CH_3OH} CH_2{=}C(CH_3){-}COOCH_3\ \text{III}$$
$$(H_3C)_2C(CN)(OH)\ \text{I} \xrightarrow[\text{cat.}]{+ H_2O} (H_3C)_2C(CONH_2)(OH)\ \text{IV} \xrightarrow[- HCONH_2]{+ HCOOCH_3} (H_3C)_2C(COOCH_3)(OH)\ \text{V} \xrightarrow{- H_2O} \text{III}$$

This route delivers ammonium hydrogen sulfate, $NH_4[HSO_4]$ as an undesired by-product, a compound which has no good use (ammonium sulfate, $(NH_4)_2SO_4$, is the most important nitrogen-delivering fertilizer but ammonium hydrogen sulfate delivers only one $NH_3$ per molecule (for problems with the use of $(NH_4)_2SO_4$ as a byproduct of chemical syntheses, see Section 10.2.3)). The elimination of $NH_4[HSO_4]$ can be avoided if acetone cyanohydrin is partially hydrolyzed to $\alpha$-hydroxy isobutyramide (IV) which is subequently reacted with methyl formate to the $\alpha$-hydroxy isobutyric acid methyl ester (V) and then catalytically dehydrated to methyl methacrylate, MMA (III). The by-product formamide is recycled to HCN. MMA is polymerized to PMMA.

The second important use of HCN is direct hydrocyanation of 1,3-butadiene by $Ni^0$ phosphite catalysts which leads to a mixture of pentenenitriles and methylbutenenitriles that are then isomerized to predominantly 3- and 4-pentenenitriles (Fig. 4-15). The subsequent anti-Markovnikov addition of HCN delivers adiponitrile which is then hydrogenated to hexamethylenediamine, a monomer for many aliphatic polyamides.

(4-15) $$CH_2{=}CH{-}CH{=}CH_2 \xrightarrow[\text{[cat]}]{+ HCN} \left[ CH_2{=}CH{-}CH_2{-}CH_2{-}CN \rightleftarrows CH_3{-}CH{=}CH{-}CH_2{-}CN \right] \xrightarrow[\text{[cat]}]{+ HCN} NC(CH_2)_4CN$$

Hydrogen cyanide is also reacted with chlorine in aqueous solution at 20–40°C to gaseous cyanogen chloride, ClCN (VI, bp = 14°C). Most of VI is reacted above 300°C on charcoal to cyanuric chloride (= 2,4,6-trichloro-1,3,5-triazine, VII) with a world production of more than 100 000 t/a.

(4-16) $$HCN \xrightarrow[- HCl]{+ Cl_2} \underset{VI}{ClCN} \xrightarrow{cat.} \underset{VII}{C_3N_3Cl_3} \longrightarrow \underset{VIII}{C_3H_3N_3}; \quad VII \xrightarrow{+ NH_3} \underset{IX}{C_3N_3(NH_2)_3}$$

80 % of VII is used for triazine-based herbicides. Cyanuric chloride is also reacted with bisphenol A to the dicyanate $NCO(p\text{-}C_6H_4)\text{–}C(CH_3)_2\text{–}(p\text{-}C_6H_4)OCN$, which upon polymerization with bismaleiimides yields so-called BT resins (Section 10.7.2). Neither cyanuric chloride VII nor triazine VIII are monomers for so-called triazine resins which contain either polyisocyanurate or melamine structures (Section 10.8.4). The reaction of cyanuric chloride VII with ammonia does deliver melamine (IX) but this monomer for melamine resins (MF) is now industrially prepared exclusively from urea, $NH_2CONH_2$.

The Reppe process, $HC{\equiv}CH + CO + ROH \rightarrow CH_2{=}CHCOOR$ for the synthesis of acrylic esters (or acrylic acid if ROH is replaced by $H_2O$) with $Ni(CO)_4$ as catalyst is no longer used industrially. The ethylene cyanohydrin process, ethylene oxide + HCN $\rightarrow$ $HOCH_2CH_2CN$, is also obsolete. The resulting hydroxypropionitrile was previously used for the synthesis of acrylic esters according to $HOCH_2CH_2CN + ROH + H_2SO_4 \rightarrow CH_2{=}CHCOOR + NH_4[HSO_4]$ which delivered the unwanted byproduct $NH_4[HSO_4]$.

## 4.3 $C_2$ Compounds

### 4.3.1 Ethane

Ethane is presently not a direct base chemical for petrochemical syntheses, in part, because it is always obtained from refinery operations as a mixture with other hydrocarbons, and, in part, because it is not very reactive. As mentioned before (p. 87), an ethane-based petrochemistry is now being developed in the Middle East, probably similar to methane chemistry (Section 4.2.1).

The conversion of the rather inert ethane to intermediates and monomers requires drastic reaction conditions. The direct chlorination of ethane seems to have been tried, albeit as a pilot process. The Lummus-Armstrong Transcat process produces vinyl chloride by reacting ethane with $Cl_2$ at 450–500°C in an alkali chloride melt of $CuCl_2{\cdot}CuO$. The reaction consists of a series of coupled reactions: chlorination of ethane by $Cl_2$ to $CH_3CH_2Cl$ + HCl, subsequent dehydrochlorination of $CH_3CH_2Cl$ to $CH_2{=}CH_2$, and chlorination of $CH_2{=}CH_2$ to $ClCH_2CH_2Cl$ that is dehydrochlorinated to $CH_2{=}CHCl$.

Simultaneously, the copper salt is reduced to CuCl according to $CuO{\cdot}CuCl_2 + 2\ HCl + CH_2{=}CH_2 \rightarrow 2\ CuCl + H_2O + ClCH_2CH_2Cl$. Reaction of CuCl with $HCl/O_2$ regenerates the catalyst.

### 4.3.2 Ethene

Ethene and propene are the two most important base chemicals that are produced worldwide in amounts of ca. $80 \cdot 10^6$ t/a ($C_2H_4$) and $40 \cdot 10^6$ t/a ($C_3H_6$). In countries with highly developed petrochemical industries, both olefins are obtained from crude oil and/or natural gas. In regions without a strong petrochemistry base but with a production of fermentation alcohol (Brazil, India, Pakistan, Peru), ethene is still obtained by dehydration of ethanol.

However, crude oil and natural gas contain little or no ethene which therefore must be obtained by cracking feedstocks that contain compounds with two or more carbon atoms per molecule: naphtha, liquid petroleum gases (LPG), natural gas liquids (NGL), and gas oil. In the United States, steam cracking of LPG and NGL delivers most of the ethene (Table 4-1). In Western Europe and Japan, naphtha is more attractive because it is technically (and often also economically) more attractive than steam cracking of gas oils. Older processes such as isolation of ethene from blast furnace gas and partial hydrogenation of acetylene from calcium carbide are no longer important.

LPG, NGL, and refinery gases contain mostly $C_2$–$C_4$ compounds; their cracking therefore leads to high yields of ethene but only to small yields of propene. The increasing use of propene (Section 4.4) sometimes produces bottlenecks so that cracking processes for feedstocks with higher hydrocarbons, such as gas oil, have been developed.

Ethene is mainly used as a monomer for poly(ethylene)s and ethylene copolymers with propene, butene, vinyl acetate, etc.: 57 wt% in the United States, 54 wt% in Western Europe, and 47 wt% in Japan (1996). Approximately 15 wt% serve for the production of vinyl chloride and ca. 12 wt% for that of ethylene oxide (oxirane). Ethene is also the base chemical for many other monomers. Polymer-related syntheses of ethene are depicted schematically in Fig. 4-5.

I. *Polymerization*: Ethene polymers comprise the various homopolymers of ethene (LDPE, HDPE, etc.; Section 6.2.3), modified homopolymers (Section 6.2.4), and copolymers with various comonomers (E/P, EPDM, EVAC, etc.; Section 6.2.5).

II. The *Ziegler alfen-process* is an $Al(C_2H_5)_3$-mediated oligomerization of ethene to $\alpha$-olefins with even number of carbon atoms (alfen: from alpha (*alf*a) alk*ene*), either as a 2-step low-temperature process with stoichiometric amounts of $Al(C_2H_5)_3$ or as a 1-step high-temperature process with catalytic amounts. The mechanism involves an insertion of ethene into the Al–$C_2H_5$ bond (see Volume I).

Table 4-1 Contribution of various feedstocks (in wt%) to the production of ethene in various world regions and maximum obtainable wt% of ethene and propene in the resulting crack products [1a]. Data for 1994/1995. The trends are to use relatively more NGL and LPG in Europe (because of the availability of natural gas from the North Sea and from Russia) and more naphtha in the United States.

| Feedstock | Contribution of feedstock in wt% to total ethene production | | | | Wt% of crack products | |
|---|---|---|---|---|---|---|
| | World | USA | Western Europe | Japan | Ethene | Propene |
| Naphtha | 46 | 11 | 71 | 97 | 35 | 17 |
| LPG, NGL | 42 (LPG, NGL + refinery gases) | 76 | 17 | 3 (LPG, NGL + refinery gases) | 81 (LPG, NGL + refinery gases) | 3 (LPG, NGL + refinery gases) |
| Refinery gases | | 4 | 9 | | | |
| Gas oil | 12 | 9 | 3 | 0 | 25 | 12 |

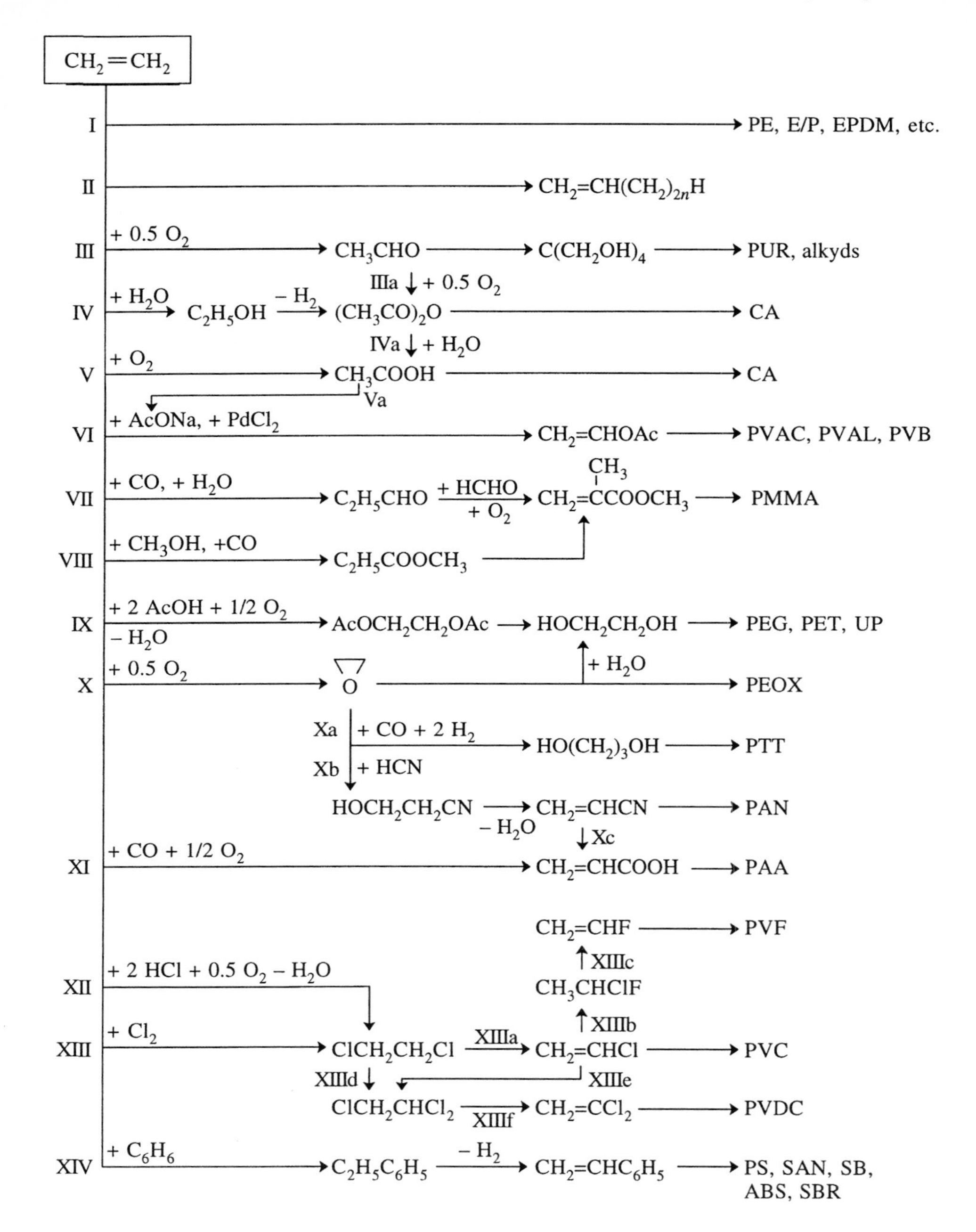


Fig. 4-5 Reactions of ethene to intermediates, monomers, and polymers. Ac = $CH_3CO$-.

III. The exothermal catalytic *partial oxidation* of ethene in aqueous, HCl-acidic solution with $PdCl_2$ as catalyst and $CuCl_2$ as oxidizing agent is the dominant process for the synthesis of acetaldehyde, $CH_3CHO$. The reaction proceeds via a charge-transfer complex (π-complex) of ethene with $PdCl_2/OH^{\ominus}$ to the complex $[Cl_2PdOH/CH_2=CH_2]^{\ominus}$.

This complex is transformed to the σ-complex $[Cl_2Pd{-}CHOH]^{\ominus}$, which dissociates to $CH_3CHO + Pd^0 + 2\ Cl^{\ominus} + H^{\oplus}$. Palladium is converted back to $PdCl_2$ according to $Pd^0 + 2\ CuCl_2 \rightleftarrows PdCl_2 + 2\ CuCl$; the copper(I) chloride is reoxidized to $CuCl_2$.

Small amounts of acetaldehyde are still obtained from the hydration of acetylene, the dehydrogenation of ethanol, the oxidation of $C_3/C_4$ alkanes, and the oxidation of ethene. In general, the oxidation of ethene to acetaldehyde is in decline because its most prominent downstream product (route IVa-Va) acetic acid is now obtained by carbonylation of methanol (Fig. 4-3). In polymer chemistry, acetic acid and acetic anhydride are used to esterify cellulose to cellulose acetate (CA). Sodium acetate also serves in the acetoxylation of ethene to vinyl acetate (route VI).

Another use of acetaldehyde is for the mixed aldolization with formaldehyde (route IV in Fig. 4-4) to pentaerythritol which serves in alkyd resins.

IV. The *hydration* of ethene is the most important industrial route to ethanol, either indirectly with concentrated sulfuric acid at ca. 80°C to the mono- and disulfates with subsequent hydrolysis or directly with $H_3PO_4/SiO_2$ catalysts at 300°C and 70 bar with short residence times. The first process has an ethene selectivity of 86 % whereas the second one has an ethanol yield of just 4 % which requires recycling of the ethene.

V. The *direct oxidation* of ethene to acetic acid (V) at 150–160°C with Pd/heteropolyacid catalysts is performed in Japan in a *relatively* small plant (100 000 t/a!).

VI. The *acetoxylation* of ethene (VI) is a variation of route IV with aqueous acetic acid/sodium acetate instead of aqueous HCl solutions. The original liquid-phase process delivers many byproducts such as formic acid and chloride ions which led to a strong corrosion of the reactors. This expensive process was thus replaced by a gas-phase reaction at 175–200°C and 5–10 bar. The only byproduct is carbon dioxide.

The resulting vinyl acetate, $CH_2{=}CHOOCCH_3$, is polymerized to poly(vinyl acetate), $\text{+}CH_2{-}CH(OOCCH_3)\text{+}_n$, (PVAc). A part of PVAC is transesterified with *n*-butanol to poly(vinyl alcohol), $\text{+}CH_2{-}CH(OH)\text{+}_n$, (PVAL), and butyl acetate, a valuable solvent. A part of PVAL is acetalized, either with butyric aldehyde to poly(vinyl butyrate), PVB, or with formaldehyde to poly(vinyl formal) (PVFM) (see Section 6.8.6).

VII. The *hydroformylation* (*oxo reaction*) of olefins to aldehydes,

$$R{-}CH{=}CH_2 + CO + H_2 \longrightarrow RCH_2CH_2CHO \text{ or } RCH(CHO)CH_3 \qquad (4\text{-}17)$$

with carbonyl–hydrogen complexes of cobalt or rhenium such as $HCo(CO)_3$ at 100–200°C and 20–300 bar yields propioaldehyde, $C_2H_5CHO$, from ethene (R = H). Propioaldehyde is then condensed with formaldehyde at 160–210°C and 40–80 bar in the presence of a secondary amine. The resulting methacrolein, $CH_2{=}C(CH_3)CHO$, is oxidized to methacrylic acid (MAA), $CH_2{=}C(CH_3)COOH$. Esterification of MAA delivers methyl methacrylate (MMA), $CH_2{=}C(CH_3)COOCH_3$, the monomer for poly(methyl methacrylate, $\text{+}CH_2{-}C(CH_3)COOCH_3\text{+}_n$, (PMMA).

VIII. A new route to MMA is the *carbonylation* of ethene with CO and $CH_3OH$ by a Pd-based catalyst to methyl propionate, $C_2H_5COOCH_3$, a reaction that proceeds with 99.9 % selectivity. Methyl propionate is then reacted with formaldehyde to MMA and water according to $CH_3CH_2COOCH_3 + HCHO \rightarrow CH_2{=}C(CH_3)COOCH_3 + H_2O$.

IX. The *acetoxylation* of ethene with $CH_3COOH$, $O_2$, and a $TeO_2/HBr$ catalyst to glycol diacetate in a 360 000 t/a plant was abandoned one year after it started because of corrosion. It would have been a very attractive path to ethylene glycol, $HOCH_2CH_2OH$.

X. Ethylene glycol is now obtained by *partial air oxidation* of ethene with silver catalysts to ethylene oxide (= EO = oxirane = *cyclo*-[$CH_2CH_2O$]) at 250–300°C and pressures of 1-2 MPa, followed by an acidic hydrolysis of EO to ethylene glycol by an ca. 10 times molar excess of water at 50-70°C. Also used is a catalyst-free vapor-phase air oxidation at 140–230°C and 20–40 bar. Both processes follow directly the ethene → ethylene oxide process.

The old two-step chlorohydrin process to EO is practically abandoned. It consisted of an intermediary formation of ethylene chlorohydrin according to $CH_2{=}CH_2 + Cl_2 + H_2O \rightarrow HOCH_2CH_2Cl + HCl$, followed by a subsequent cyclization with CaO according to 2 $HOCH_2CH_2Cl$ + CaO → 2 *cyclo*-[$CH_2CH_2O$] + $CaCl_2$ + $H_2O$. The process consumed expensive chlorine and delivered more than 300 kg of $CaCl_2$ per 100 kg of ethylene oxide; in addition, there were considerable problems with effluents.

Oxirane is directly polymerized to poly(ethylene oxide), PEOX. The hydroformylation of oxirane (route Xa) leads to trimethylene glycol, $HO(CH_2)_3OH$, which is polycondensed with terephthalic acid to poly(trimethylene terephthalate) (PTT).

Ethylene oxide is a relatively expensive intermediate which is the reason why route Xb to ethylene cyanohydrin and further to acrylonitrile, $CH_2{=}CHCN$, is no longer used. Acrylonitrile is rather synthesized by ammonoxidation of propene (Fig. 4-7, Route X).

Xa. A Japanese plant hydrolyzes acrylonitrile by $H_2O/H_2SO_4$ to acrylamide sulfate, $CH_2{=}CHCONH_2{\cdot}H_2SO_4$, which is subsequently esterified by alcohols ROH to acrylic acid esters or hydrolyzed by water to acrylic acid (AA), $CH_2{=}CHCOOH$. Polymerization of AA delivers poly(acrylic acid), PAA. The problem here is the stoichiometric production of the byproduct ammonium hydrogensulfate (see p. 97).

XI. The *oxycarbonylation* of ethene by CO + $^1/_2$ $O_2$ delivers directly acrylic acid. This process uses $PdCl_2{\cdot}CuCl_2$ as catalyst, acetic acid as solvent, and acetic anhydride for the removal of the byproduct water (to acetic acid). The major byproduct is $\beta$-acetoxypropionic acid, $CH_3COOCH_2CH_2COOH$. This process competes with that of acrylic acid from propene via acrolein (Section 4.4.1), i.e., the cost of ethene versus that of propene.

XII and XIII. 1,2-Dichloroethane, $ClCH_2CH_2Cl$, is the key product for the synthesis of vinyl halides. It is obtained by chlorination (route XIII) or oxychlorination (route XII). The uncatalyzed direct chlorination in the gas phase at 90–130°C is probably a free-radical reaction, whereas that in 1,2-dichloroethane as solvent with $CuCl_2$, $FeCl_3$, or $SbCl_3$ as catalysts at 40–70°C and slight pressure is probably ionic. The oxychlorination at 220–240°C and 2–4 bar works with oxygen or air and *dry* HCl (no corrosion!).

$ClCH_2CH_2Cl$ is dehydrochlorinated to vinyl chloride, $CH_2{=}CHCl$, in the gas phase at 400–500°C at 25–35 bar (XIIIa); dehydrochlorination by alkali in the liquid phase is no longer practiced. More than 95 % of vinyl chloride is then polymerized free-radically to poly(vinyl chloride) (PVC), commonly called "vinyl" or "polyvinyl."

At slightly elevated temperatures, vinyl chloride readily adds hydrogen fluoride, HF, to 1-chloro-1-fluoroethane (route XIIIb). The subsequent dehydrochlorination of this intermediate to vinyl fluoride (route XIIIc) proceeds at pyrolytic conditions of 500–600°C without catalyst. $CH_2{=}CHF$ can also be obtained from acetylene (Fig. 4-6, Route III). Vinyl fluoride is then polymerized to poly(vinyl fluoride), $+\!CH_2{-}CHF\!+_n$ (PVF).

Chlorination of vinyl chloride (route XIIIe) or 1,2-dichloroethane (route XIIId) delivers 1,1,2-trichloroethane, $Cl_2CH_2CH_2Cl$, which is dehydrochlorinated at 100°C by aqueous $Ca(OH)_2$ or NaOH to vinylidene chloride, $CH_2{=}CCl_2$, (Route XIIIf). All commercial poly(vinylidene chloride)s are not homopolymers, $\text{+}CH_2\text{–}CCl_2\text{+}_n$, but copolymers with several percent of other monomeric units (see Section 6.8.2).

XIV. Styrene, $CH_2\text{=}CHC_6H_5$, is obtained by *alkylation* of benzene with ethene to ethyl benzene, $CH_3CH_2C_6H_5$. The classic reaction in the liquid phase at ca. 90°C with, e.g., $AlCl_3$ as catalyst delivers a mixture of benzene, ethyl benzene, diethyl benzenes, etc. Subsequent separation of the mixture by distillation, recovery of the catalyst by washing and drying, and use of expensive corrosion-resistant reactors have led to alkylations in the gas phase at 300°C and 40–65 bar with $Al_2O_3/SiO_2$ or $H_3PO_4/SiO_2$ as catalysts. These plants now have nameplate capacities of up to 500 000 t/a.

Ethyl benzene is then dehydrogenated to styrene, $CH_2\text{=}CHC_6H_5$, at 550–620°C by iron oxide catalysts with, e.g., $Cr_2O_3/KOH$ as promoter. Addition of water vapor reduces the partial pressure of ethyl benzene and shifts the equilibrium therefore to styrene; it also removes carbon by the water gas reaction, $C + 2\ H_2O \rightarrow CO_2 + 2\ H_2$.

Styrene is polymerized to poly(styrene) (PS), $\text{+}CH_2\text{–}CHC_6H_5\text{+}_n$, and copolymerized with many other monomers to SAN (with acrylonitrile), SB or SBR (with butadiene), ABS (with acrylonitrile and butadiene), etc. The world-wide nameplate capacity for styrene is now ca. $20 \cdot 10^6$t/a.

### 4.3.3 Ethyne

Ethyne, commonly still called **acetylene**, lost its dominant role as a base chemical in the United States in the 1940s and in Western Europe in the 1950s. It was replaced by hydrocarbon feedstocks (crude oil, natural gas, natural gas liquids, refinery products, etc.) that are less expensive, safer to handle, and more easy to transport, e.g., by pipelines. These advantages outweigh by far the lesser reactivity of these hydrocarbons. Acetylene is therefore now used only if coal as the raw material for ethyne can be easily and inexpensively mined (Australia, India), electricity is inexpensive, older plants are amortized and still economical, special products are desired (e.g., higher vinyl esters, *N*-vinyl carbazole, other *N*-vinyls, etc.), and/or no other processes are available (e.g., for 1,4-butanediol, acetylene carbon black, etc.).

The world acetylene production of ca. 800 000 t/a is presently ca. 25 % of the 1960 production. Most acetylene is now obtained in special plants by pyrolysis of hydrocarbons. Some acetylene is produced as a byproduct of cracking: the high-severity cracking of naphtha delivers ca. 2 wt% $C_2H_2$ and the advanced cracking ca. 10 wt%.

The classic route to acetylene uses calcium carbide from the electrothermal reaction of coke or anthracite with calcium oxide at 2200–2300°C according to $CaO + 3\ C \rightarrow CaC_2 + CO$. Calcium carbide is subsequently hydrolyzed exothermally to acetylene, $CaC_2 + 2\ H_2O \rightarrow CH{\equiv}CH$.

The most important petrochemical process for ethyne is the autothermal Sachsse-Bartholomé process in which methane (sometimes also crack naphtha) and under-stoichiometric oxygen are separately heated to 500–600°C, mixed, and then burned. The overall reaction, $2\ CH_4 \rightarrow C_2H_2 + 3\ H_2$ is endothermal; it includes the endothermal dehydrogenation of $CH_4$ to $C_2H_2 + H_2$ and the exothermal oxidation of a part of $CH_4$. The process delivers ca. 8 wt% $C_2H_2$ and ca. 5 kg of carbon black per 100 kg of acetylene. The main product is crack gas (with 57 wt% $H_2$ and 26 wt% CO) that is used as synthesis gas.

Natural gas, liquefied petroleum gases, and refinery gases are also cracked in a 1 m-long electric arc at 1800°C (at the center: 20 000°C!), followed by rapid cooling to ca. 200°C by injection of water. This process delivers ca. 56 kg of acetylene and 23 kg of ethene per 100 kg hydrocarbon.

Fig. 4-6 shows the most important routes from acetylene to monomers and polymers.

I. The classic route to chloroprene, the monomer for poly(chloroprene) (CR), involved the dimerization of acetylene to vinyl acetylene at 80°C by an aqueous HCl-acidic solution of CuCl + $NH_4Cl$, followed by addition of HCl. Main byproducts were divinyl acetylene in the first step and methyl vinylketone and 1,3-dichloro-2-butene in the second. Newer chloroprene syntheses are all based on butadiene (Section 4.5.1).

II. The addition of HCl (byproduct of chlorinations) to obtain vinyl chloride at 140–200°C is catalyzed by $HgCl_2$ on activated charcoal. Despite low investment and operational costs, the process is no longer used by most Western countries because ethene as a base chemical (see Fig. 4-5, XIII) is much more economical than acetylene.

III. The addition of hydrogen fluoride is either catalyzed by mercury compounds at 40–150°C or by fluorides of Al, Sn, or Zn at 250–400°C. The byproduct is difluoroethane which is dehydrofluorinated to vinyl fluoride, the monomer for poly(vinyl fluoride), PVF. The process competes with routes XIIIb and XIIIc of Fig. 4-5.

$CH{\equiv}CH$

| Route | Reagent | Intermediate | Monomer | Polymer |
|---|---|---|---|---|
| [I] | | $CH{\equiv}C-CH{=}CH_2$ → (+ HCl) | $CH_2{=}C(Cl)-CH{=}CH_2$ | CR |
| (II) | + HCl | | $CH_2{=}CHCl$ | PVC |
| III | + HF | | $CH_2{=}CHF$ | PVF |
| [IV] | + HCN | | $CH_2{=}CHCN$ | PAN |
| (V) | + $CH_3COOH$ | | $CH_2{=}CHOAc$ | PVAC, etc. |
| VI | + 2 HCHO | $HOCH_2-C{\equiv}C-CH_2OH$ | $HO(CH_2)_4OH$ | PBT |
| (VII) | + $H_2O$ | $CH_3CHO$ | see Fig. 4-5 | |
| VIII | + ROH | | $CH_2{=}CHOR$ | P(vinyl ether)s |
| [IX] | + $(CH_3)_2CO$ | $CH_3-C(CH_3)(OH)-C{\equiv}CH$ | $CH_2{=}C(CH_3)-CH{=}CH_2$ | IR |
| X | + 2-pyrrolidone (N–H, =O) | | N-vinyl-2-pyrrolidone (N–CH=CH$_2$, =O) | PVP |

Fig. 4-6 Intermediates and monomers from acetylene. [ ] Route is now obsolete; ( ) route no longer generally used in the United States and Western Europe but still in use in countries without a large petrochemical industry. Ac = $CH_3CO$.

IV. Acrylonitrile is no longer obtained by the $CuCl/NH_4Cl$–catalyzed addition of HCN to acetylene in the liquid phase (Nieuwland process). Most acrylonitrile syntheses now use propene as a base chemical (see Fig. 4.7, Route X).

V. In some Eastern European and Asian countries, vinyl acetate is still produced by the addition of acetic acid to acetylene in the gas phase at 170–250°C using $Zn(OAc)_2$ on charcoal as catalyst. Most vinyl acetate is now obtained from ethene (Fig. 4-5).

VI. In the United States, most 1,4-butanediol is still obtained by ethynylation of acetylene by aqueous formaldehyde solutions at 100–110°C and 5–20 bar in the presence of copper acetylide. The resulting 2-butyne-1,4-diol is subsequently hydrogenated to 1,4-butanediol, for example, with Raney nickel at 70–100°C and 250–300 bar.

VII. The hydration of acetylene to acetaldehyde is economical only if inexpensive acetylene is available from electrothermal processes, i.e., inexpensive electricity (Italy, Switzerland). For acetaldehyde as a base chemical and intermediate, see Section 4.3.5.

VIII. Acetylene is the base chemical for most syntheses of vinyl ethers by the Reppe process, using potassium alcoholate as catalyst in the liquid-phase process at 120–180°C.

IX. Snamprogretti (Italy) used a 3-step process to obtain isoprene from acetylene. The KOH-catalyzed addition of acetone to acetylene in liquid ammonia to methylbutynol was followed by a selective hydrogenation to methylbutenol, and finally a dehydration of this compound to isoprene.

X. Acetylene is also used to vinylate 2-pyrrolidone to *N*-vinyl pyrrolidone which serves as a monomer and comonomer in special polymers, for example, as blood plasma expanders (p. 279).

### 4.3.4 Ethanol

The industrial production of ethanol by hydration of ethene provides only 17 % of the world demand. About 83 % of ethanol is obtained from agricultural resources, mainly by a 2-step enzymatic process. In the first step, starch is hydrolyzed by $\alpha$-amylase to a hemiacetal. The remaining glucosidic bonds are broken by glucoamylase. The resulting glucose is then fermented by yeast to ethanol. Besides starch, other raw materials for ethanol include molasses from sugar cane, wastes from the sulfite process (Section 3.7.8.), and products from wood saccharification (Section 3.7.9).

A new 1-step process binds $\alpha$-amylase from *Streptococcus bovis* and glucoamylase from *Rhizopus eryzae* via functional groups to the surface of yeast cells. After 70 h at 30–38°C, ethanol is obtained in 92 % yield with respect to glucose.

In countries with a strong petrochemistry, ethanol is no longer a major intermediate for monomers. In countries with a sizable production of fermentation alcohol (Brazil, Poland, Russia), $C_2H_5OH$ is converted to 1,3-butadiene by the Lebedev process, a simultaneous dimerization–dehydration–dehydrogenation at 380°C with $MgO–SiO_2$ or $Al_2O_3–SiO_2$ catalysts according to $2\ C_2H_5OH \rightarrow CH_2{=}CH{-}CH{=}CH_2 + 2\ H_2O + H_2$. Fermentation alcohol also serves as a source of ethene (Section 4.3.2).

China and India catalytically dehydrogenate ethanol to acetaldehyde, similarly to the synthesis of formaldehyde from methanol (Fig. 4-3). Acetaldehyde is then converted to butadiene in a 3-step process: aldolization to acetaldol followed by a reduction to 1,3-butandiol, and then a dehydration to 1,3-butadiene:

(4-18)

$$2\ CH_3CHO \longrightarrow CH_3\underset{\displaystyle OH}{CH}CH_2CHO \xrightarrow{+H_2} CH_3\underset{\displaystyle OH}{CH}CH_2CH_2OH \xrightarrow[-2\ H_2O]{} CH_2{=}CH-CH{=}CH_2$$

### 4.3.5 Acetaldehyde

Acetaldehyde is an important intermediate for many large-volume chemicals, especially acetic acid and compounds based on acetic acid (acetic anhydride, cellulose acetate, vinyl acetate, etc.). It is gradually losing its importance because acetic acid is now considerably less expensive to make by carbonylation of methanol (Fig. 4-3): its syntheses from ethene and acetylene suffer from the high prices of the base chemicals, and its synthesis from $C_3/C_4$ cuts is unspecific.

Monomers from acetaldehyde include butadiene (Eq.(4-18)) and pentaerythritol (Fig. 4-4, route VI) as well as various pyridines (from $CH_3CHO$, $HCHO$, and $NH_3$). In China and India, acetaldehyde and acetic anhydride are reacted to ethylidene diacetate, $CH_3CH(OOCCH_2)_2$, which is subsequently dehydrogenated to acetic acid, $CH_3COOH$, and vinyl acetate, $CH_2{=}CHOOCCH_3$.

### 4.3.6 Acetic Acid

Acetic acid, $CH_3COOH$, is either obtained from natural resources or petrochemical reactions. Biosyntheses of acetic acid include oxidative fermentation of ethanol-containing substrates, fermentation of molasses from sugar cane, and the coking of wood.

Petrochemistry synthesizes acetic acid by three routes:

I. $C_1 + C_1$ reactions: carbonylation of methanol by CO with rhodium carbonyl–iodine catalysts with the overall reaction $CH_3OH + CO \rightarrow CH_3COOH$. The largest plant now has a nameplate capacity of 330 000 t/a.

II. $C_2$ transformations: free-radical oxidation of acetaldehyde via peracetic acid, $CH_3COOOCH_3$, with the overall reaction $2\ CH_3CHO + O_2 \rightarrow 2\ CH_3COOH$.

III. $C_4$–$C_8$ degradation: oxidative degradation of butane, butenes, or light gasolines. These processes deliver several byproducts, depending on feed and reaction conditions.

The world production of synthetic acetic acid is ca. $6 \cdot 10^6$ t/a. Approximately 43 % is used for the synthesis of vinyl acetate (Fig. 4-5), 13 % for the esterification of cellulose to cellulose acetate, and 12 % as a solvent for the synthesis of dimethyl terephthalate or terephthalic acid, the monomers for poly(ethylene terephthalate).

## 4.4 $C_3$ Compounds

### 4.4.1 Propene

Fifty years ago, propene (historic name: **propylene**) from refinery operations was mainly converted to gasoline. It became a very important base chemical with a worldwide production of ca. $50 \cdot 10^6$ t/a (2000) only after petrochemistry was developed.

Approximately 69 % of propene is now obtained as a coproduct to ethene from the steam cracking of petrochemical naphtha, liquid petroleum gases (LPG), natural gas liquids (NGL), petroleum gases, and gas oil (Section 4.3.2). The production of gasoline by fluidized catalytic cracking and catalytic reforming delivers another 28 %.

Less than 3 % of the propene production stems from dehydrogenation of propane or from olefin metathesis. Both processes are stop-gap procedures for propene shortages.

The catalytic dehydrogenation of propane to propene at ca. 650°C requires the costly isolation of propane from LPGs, NGLs, or crude oil distillates. The process converts relatively expensive propane to propene that cracking delivers less expensively.

Propene by olefin metathesis is the reversal of the Phillips triolefin process (1 olefin as educt, 2 olefins as product) that was operated until 1972. At that time, the demand for propene increased and the reverse Phillips process became known as "olefin conversion technology". In this technology, ethene is first dimerized to 2-butene which is then reacted with ethene to propene:

$$(4\text{-}19) \qquad 2\ CH_2{=}CHCH_3 \underset{\text{olefin conversion technology}}{\overset{\text{Phillips triolefin process}}{\rightleftarrows}} CH_2{=}CH_2 + CH_3CH{=}CHCH_3$$

Bottlenecks in propene production will also be overcome by dehydration of methanol (from syngas, Section 4.2.2), $3\ CH_3OH \rightarrow CH_2{=}CHCH_3 + 3\ H_2O$, starting in 2007 with a production of $10^5$ t/a. This process will use a zeolite-based catalyst and will deliver propene in yields of ca. 70 % (with dimethyl ether as a byproduct).

Fig. 4-7 shows propene-based syntheses to intermediates, monomers, and polymers:

I. Almost 50 wt% of propene is directly polymerized to poly(propylene) (PP) or copolymerized with ethene and a diene monomer to EPDM (p. 229).

II. The liquid-phase dimerization of propene by $Na/K_2CO_3$ at 150°C and 40 bar delivers 4-methyl-1-pentene that serves as a monomer for its homopolymer P4MP and as a comonomer in some linear low-density poly(ethylene)s (LLDPE).

III. Dimerization of propene by $Al(i\text{-}C_3H_7)_3$ at 200°C and 200 bar yields 2-methyl-1-pentene that is isomerized to 2-methyl-2-pentene. This intermediate is then disproportionated to isoprene and methane by overheated water vapor at 650–800°C with HBr as catalyst. However, this route is no longer economical.

IV. Propene is oxidized to acrolein, $CH_2{=}CHCHO$, at temperatures of 350–400°C by $Cu_2O$ or Cu and $I_2$ as promoter. The acid-catalyzed hydration of acrolein delivers 3-hydroxypropionic aldehyde which is then catalytically hydrogenated by Ni to 1,3-propane diol, one of the two monomers for poly(trimethylene terephthalate) (PTT). This route to 1,3-propanediol now has to compete with a new process based on glucose.

V. Chlorine adds to propene at lower temperatures but substitutes the methyl group of propene at temperatures of 500–510°C (hot chlorination). The resulting allyl chloride is mainly used for the synthesis of epichlorohydrin. The older 3-step process (Route Va) from propene reacts allyl chloride in aqueous solution with HOCl to dichlorohydroxypropane (2 isomers: $ClCH_2\text{–}CHOH\text{–}CH_2Cl + HOCH_2\text{–}CHCl\text{–}CH_2Cl$) that is then converted at 50–90°C by lime milk (aqueous $Ca(OH)_2$) to epichlorohydrin (Eq.(4-20)). A new Solvay process (2007) reacts glycerol with HCl to $ClCH_2\text{–}CHOH\text{—}CH_2Cl$ (2007).

$$(4\text{-}20) \qquad 2\ \underset{\substack{|\\ Cl}}{CH_2}\text{—}\underset{\substack{|\\ OH}}{CH}\text{—}CH_2Cl \xrightarrow[-\ CaCl_2,\ -\ 2\ H_2O]{+\ Ca(OH)_2} 2\ H_2C\underset{O}{\overset{}{-}}CH\text{—}CH_2Cl$$

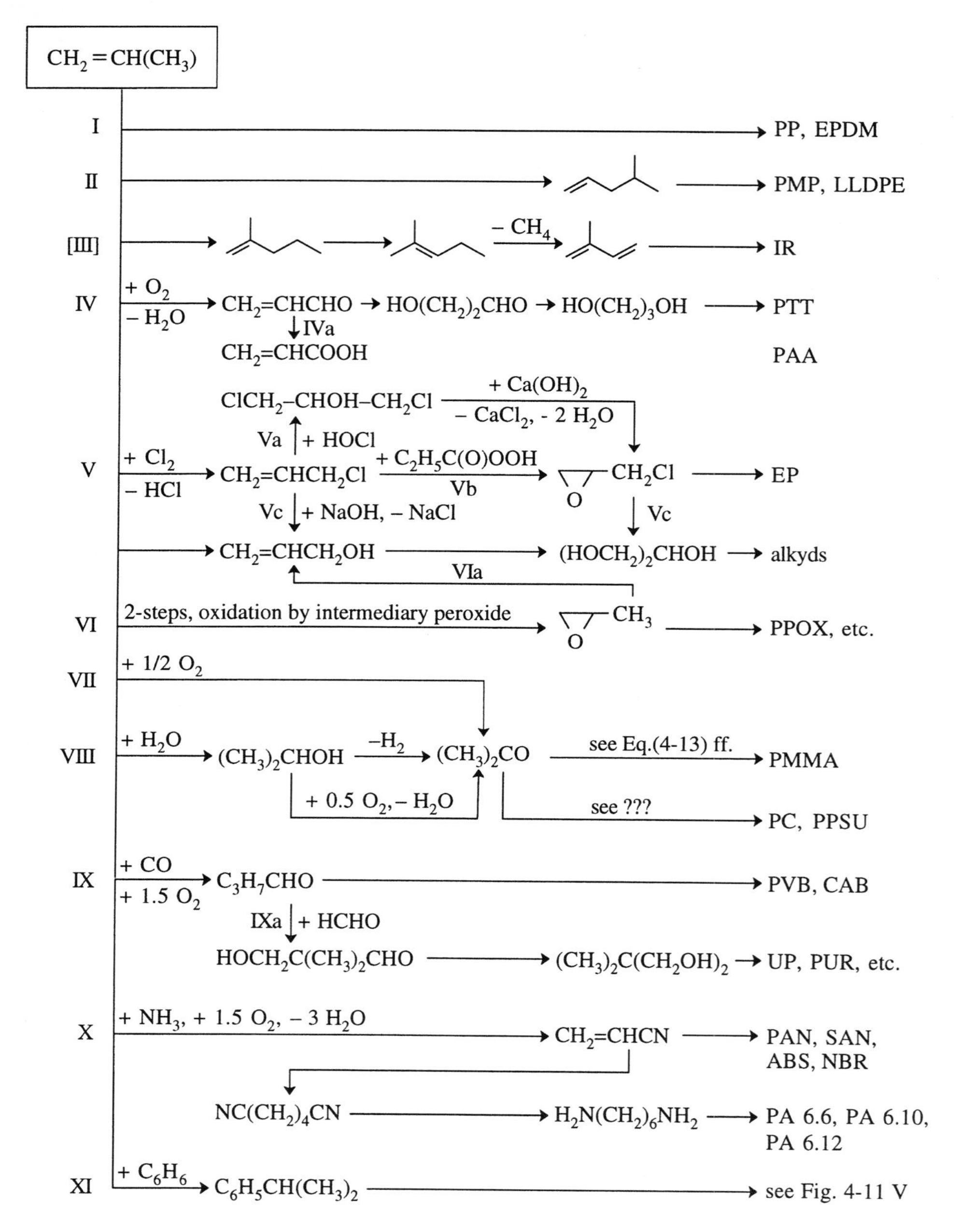


Fig. 4-7 Routes from propene to intermediates, monomers, and polymers. [ ] Route is now obsolete.

Vb. Route Va produces as a byproduct very inexpensive calcium chloride from expensive chlorine. A newer Route Vb therefore oxidizes allyl chloride directly by perpropionic acid (from propionic acid + $H_2O_2$) to epichlorohydrin.

VI. The successful oxidation of ethene to ethylene oxide (Fig. 4-5, X) has as yet not found an economical counterpart in the direct oxidation of propene to propylene oxide since either yields and/or selectivities are too small. The epoxidation of propene by hydrogen peroxide to propylene oxide and water is deemed too expensive because of the cost of $H_2O_2$ and has apparently never made it beyond the pilot stage.

Instead, industry uses a 2-step process with intermediary formation of hydroperoxides ROOH. In a first reactor, a hydrocarbon with a labile H atom is oxidized by air or oxygen in the presence of a free-radical initiator at ca. 130°C and 35 bar to the hydroperoxide. An example is the oxidation of *t*-butane, $(CH_3)_3CH + O_2 \rightarrow (CH_3)_3COOH$, which, in part, subsequently reacts to *t*-butanol, $(CH_3)_3COH$. In a second reactor, propene in *t*-butanol is then epoxidized by the hydroperoxide to propylene oxide at 90–130°C and 15–65 bar using a molybdenum-based catalyst. About 67 % of propylene oxide (PO) is used world-wide for the syntheses of poly(propylene oxide) (PPOX), propylene glycol, and polyols (multifunctional alcohols), etc., mainly for use in polyurethanes (PUR).

PO is also isomerized to allyl alcohol, $CH_2{=}CHCH_2OH$ by $Li_3PO_4$ or $Cr_2O_3$ (Route VIa). Oxidation of allyl alcohol by $H_2O_2$ with $WO_3$ as catalyst delivers glycerol (glycerin). However, most of the glycerol, $HOCH(CH_2OH)_2$, is obtained industrially from allyl chloride via epichlorohydrin (Route Vb–Vc) and increasingly as byproduct of the production of biodiesel from the reaction of animal fats or vegetable oils with methanol.

VII. Propene is directly oxidized to acetone by $PdCl_2/CuCl$ catalysts at ca. 115°C and 10–14 bar in analogy to the oxidation of ethene (p. 100, III). The process is attractive because it can directly use the propene/propane mixture of crack gas without isolation of propene; propane is inert under process conditions.

VIII. Another route to acetone is via isopropanol from the hydration of propene. Since isopropanol is a major intermediate (Section 4.4.2), many process variants are known. In the 2-step liquid-phase process, sulfuric acid reacts with propene to the ester $(CH_3)_2CH{-}OSO_3H$ which is then hydrolyzed to $(CH_3)_2CHOH + H_2SO_4$. Direct hydration of propene in the presence of a strongly acidic ion exchange resin works at far higher temperatures of up to 160°C and pressures of up to 100 bar. Even more drastic conditions of up to 270°C and 250 bar are used in gas-phase processes with $H_3PO_4/SiO_2$ as catalyst in the medium-pressure process and $WO_3{-}ZnO/SiO_2$ in the high pressure one.

Isopropanol can be (oxy)dehydrated to acetone which is an intermediate in the syntheses of methyl methacrylate (see Eq.(4-13) ff.) and bisphenol A Fig. 4-12, Route II).

IX. The hydroformylation of propene by oxo gas $CO + H_2$ to a mixture of butyraldehyde and isobutyraldehyde is catalyzed by $H[Co(CO)_4]$ which is formed almost instantaneously from CO and Co or $Co(OH)_2$ under the reaction conditions of 140–180°C and 250–300 bar. The spent catalyst is either thermolyzed to metallic Co, precipitated as $Co(OH)_2$ by alkali hydroxides, or extracted as $H[Co(CO)_4]$ by an aqeous $NaHCO_3$ solution. The mixture of aldehydes and other compounds is then distilled.

Butyraldehyde reacts with poly(vinyl alcohol) to poly(vinyl butyral) (PVB) and, in combination with acetaldehyde, with cellulose to cellulose acetobutyrate (CAB).

Mixed aldolization of isobutyraldehyde and formaldehyde results in hydroxypivaldehyde, $HOC(CH_3)_2CHO$, that is hydrogenated to neopentylglycol, $(CCH_3)_2C(CH_2OH)_2$ (Route IXa). This diol serves as a monomer for polyester resins, e.g., unsaturated polyesters (UP), and lacquers as well as a component in polyurethanes, PUR.

X. The ammonoxidation of propene by $NH_3 + {}^3/_2\ O_2$ in a fluidized-bed reactor at ca. 450°C and 1.5 bar leads directly to acrylonitrile, $CH_2{=}CHCN$, (Sohio process). The presently used catalyst is not known, but its forerunners were $Bi_2O_3{\cdot}MoO_3$, $UO_2{\cdot}Sb_2O_3$, and $Bi_2O_3 \cdot MoO_3$ + Fe compounds. The process proceeds probably via an adsorption of propene on the surface of the catalyst, followed by a step-wise elimination of two hydrogen atoms to the adsorbed entity $[CH_2{=}CH{-}CH]_{ads}$ which subsequently reacts with adsorbed dehydrogenated ammonia, $[NH]_{ads}$. The Sohio process delivers a fiber-grade acrylonitrile (AN) with a purity greater than 99 %. Byproducts per 1000 kg of AN are 160 kg of HCN (see Section 4.2.5), 35 kg of acetonitrile (usually burned), and acrolein.

Acrylonitrile is mainly polymerized to poly(acrylonitrile), PAN. It also serves as a comonomer in polymers with styrene (SAN), butadiene + styrene (ABS), and with butadiene in so-called nitrile rubbers (NBR) and for the synthesis of polyamide 4.2 (p. 456).

Approximately 10 % of the world production of acrylonitrile is converted electrochemically to adiponitrile by hydrogenating dimerization. Adiponitrile is then hydrogenated to hexamethylenediamine, a monomer for many aliphatic polyamides such as PA 6.6, PA 6.10, and PA 6.12.

XI. The alkylation of benzene by propene leads to cumene = isopropylbenzene, $C_6H_5CH(CH_3)_2$, the base chemical for phenol (see Fig. 4-11, Route V).

### 4.4.2 Isopropanol and Acetone

*Isopropanol*, $(CH_3)_2CHOH$, (correct name: 2-propanol since there is no "isopropane") is a major chemical (production ca. $2 \cdot 10^6$ t/a). It is produced by addition of water to propene (Fig. 4-7, VIII), either with intermediate formation of the half-ester of sulfuric acid or by one of the many direct catalytic hydration processes. In Western Europe, about 36 % of 2-propanol is used for the synthesis of acetone; in the United States, only ca. 6 %. In the past, isopropanol was obtained by hydrogenation of acetone from fermentation processes.

2-Propanol is mainly used as a solvent, antifreeze, etc. Its only direct polymer-related use is as isopropyl acetate in combination with ethyl acetate and water glass (aqueous solution of alkali silicates) for the solidification of soils. Hydrolysis of the esters produces acetic acid which reacts with the alkali silicates to amorphous crosslinked silicic acid, $SiO_2 \cdot nH_2O$ (see Section 12.3.4).

*Acetone*, $(CH_3)_2CO$, is industrially produced in greater amounts than isopropanol (ca. $3 \cdot 10^6$ t/a) by direct oxidation of propene (Fig. 4-7, Route VII), dehydration of isopropanol (Fig. 4-7, Route VIII), or as a coproduct in the synthesis of phenol by the Hock process (Eq.(4-30)).

About 33 % of all industrially produced acetone is used for the synthesis of methyl methacrylate (Fig. 4-7, Route VIII). Another 17 % is reacted with phenol to bisphenol A, $HO(1,4\text{-}C_6H_4)C(CH_3)_2(1,4\text{-}C_6H_4)OH$ (Fig. 4-12, Route II).

## 4.5 $C_4$ Compounds

The $C_4$ fraction from the cat cracking of gas oil consists of ca. 50 wt% saturated $C_4$ hydrocarbons (butane, isobutane), ca. 50 wt% monounsaturated $C_4$ hydrocarbons (1-butene, 2-butenes, isobutene), and only ca. 0.5 wt% 1,3-butadiene, $CH_2{=}CH{-}CH{=}CH_2$ (Table 3-17). Much higher proportions of butadiene, the most valued $C_4$ compound, are obtained from steam cracking of naphtha (Table 3-17). Although severe steam cracking produces a smaller $C_4$ fraction, it does result in higher butadiene yields because butadiene has a conjugation energy of 14.7 kJ/mol and is therefore fairly stable at high-severity crack conditions.

The components of the $C_4$ fraction boil in a narrow range from −12°C to +4°C (Table 3-17); some also form azeotropes. These components can not be separated by simple distillation. From butadiene-rich $C_4$ cuts, one rather removes first the butadiene by extractive distillation or by complexation. Extractive distillation uses "complexing liquids" that interact physically with the double bonds of butadiene. The resulting "physical complexes" have higher boiling temperatures than butadiene itself. They remain in the sump of the distillation column whereas the more volatile butanes and butenes distill off. Complexing liquids comprise *N,N*-dimethylformamide (GPB process = Geon process butadiene of Nippon Zeon) or *N*-methylpyrrolidone (NMP process of BASF). Acetone, acetonitrile, furfurol, and dimethyl acetamide have also been used. Butadiene is obtained in 96 % yield with a polymerization purity of 99.8 %.

For butadiene-poor cuts, chemical complexation of butadiene with copper-ammonium acetate, $[Cu(NH_3)_2]OOCCH_3$, is still used (CAA process of Exxon).

Approximately 3 % of the industrially produced butadiene is obtained by endothermal dehydrogenation of hydrocarbons, either of butane at 600–700°C and 0.2–0.4 bar with Cr/Al oxide catalysts (Houdry process) or of butenes using a $Cr_2O_3$-stabilized Ca/Ni phosphate catalyst at 600–675°C and 1 bar (Dow Chemical). Every few minutes, the catalysts must be regenerated because of the deposition of coke.

After removal of butadiene, the remaining liquid still contains some butadiene which is removed by selective hydrogenation. The resulting "$C_4$ raffinate" contains ca. 7 % butane, 2 % isobutane, 26 % 1-butene, 20 % 2-butenes, and 45 % isobutene. Isobutene is isolated from the $C_4$ raffinate by reversible addition of water or methanol:

$$(4\text{-}21)\qquad (CH_3)_3C\text{–}OH \underset{-\,H_2O}{\overset{+\,H_2O}{\rightleftarrows}} H_2C\text{=}C(CH_3)_2 \underset{-\,CH_3OH}{\overset{+\,CH_3OH}{\rightleftarrows}} (CH_3)_3C\text{–}OCH_3$$

Alternatively, it can also be removed from the raffinate by oligomerization to di, tri ... isobutenes or by polymerization to poly(isobutylene), $\text{+}CH_2\text{–}C(CH_3)_2\text{+}_n$.

### 4.5.1 1,3-Butadiene

The paths from butadiene to intermediates, monomers, and polymers are shown schematically in Fig. 4-8:

I. Butadiene is polymerized directly to poly(butadiene), a rubber (BR). It is also a comonomer for styrene-butadiene (SBR) and acrylonitrile-butadiene rubbers (NBR). Copolymerization with acrylonitrile and styrene delivers a high-impact plastic (ABS).

II. Polymerization of butadiene by the 1:1 Ziegler catalyst $TiCl_4 + (C_2H_5)_2AlCl$ leads to 1,4-*trans*-poly(butadiene). However, the use of 4–5 moles of $(C_2H_5)_2AlCl$ per 1 mole of $TiCl_4$ cyclotrimerizes butadiene to 1,5,9-cyclododecatriene (CDT), predominantly to the *trans,trans,cis*-CDT. *Trans,trans,trans*-CDT, 1,5-*cyclo*-octadiene, and vinylcyclohexene are byproducts; the *cis,cis,cis*-CDT is not formed (Eq. (4-22)).

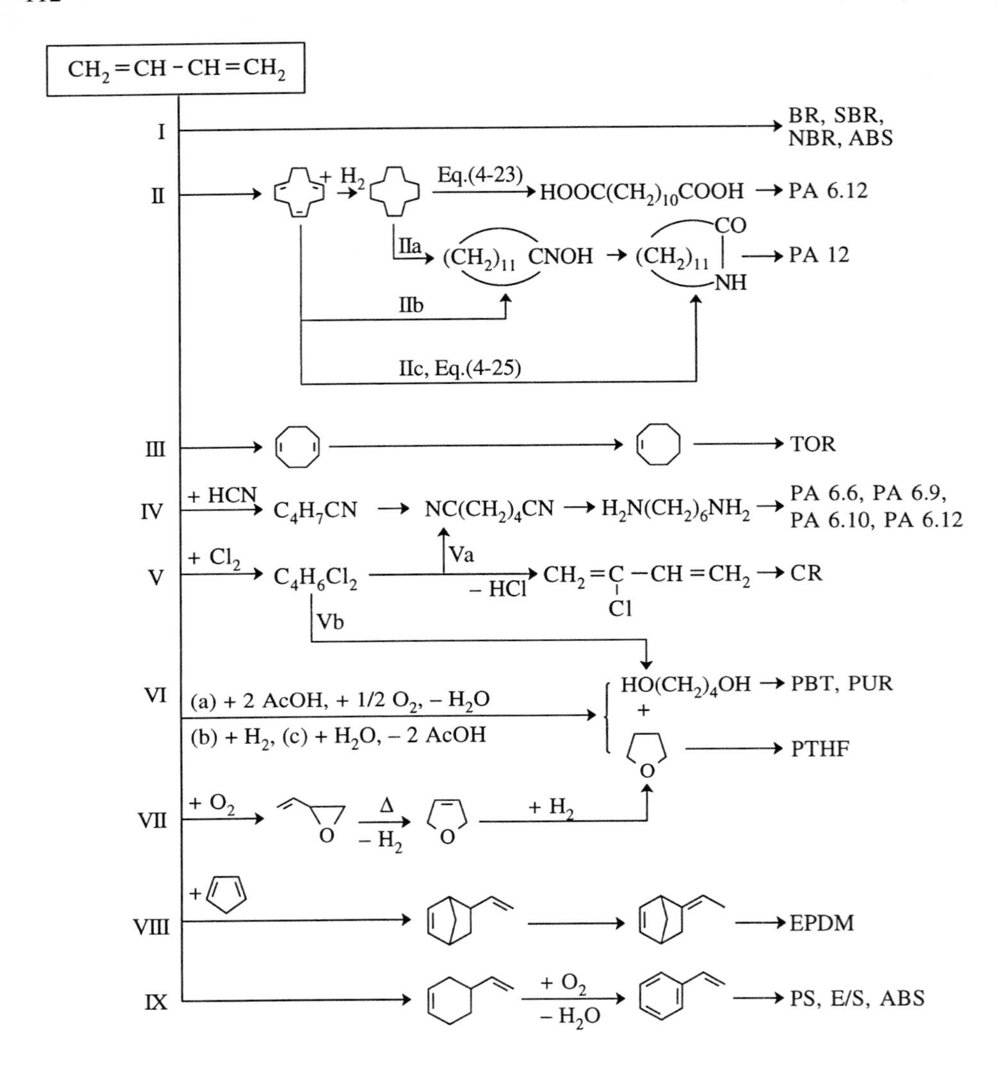


Fig. 4-8 Routes from butadiene to intermediates, monomers, and polymers.

(4-22) 3 butadiene → ttc-CDT (+ ttt-CDT + cyclooctadiene + vinylcyclohexene: oligo(butadiene)s)

ttc-CDT ttt-CDT oligo(butadiene)s

CDT (I) is hydrogenated in liquid phase at 200°C and 10–15 bar by Ni catalysts to cyclododecane, $C_{12}H_{24}$ (II, Fig. 4-23) which, in turn, is oxidized by oxygen or air with Mn or Co catalysts to an ca. 9:1 anone/anol mixture (III) of cyclododecanone and cyclododecanol. The anone/anol mixture is the feedstock for two compounds, 1,12-dodecane dicarboxylic acid (IV) and cyclododecanone (V):

(4-23)

I $\xrightarrow{+ 3 H_2}$ II $\xrightarrow{+ O_2}$ [anone + anol] III

III $\xrightarrow{+ HNO_3}$ IV (COOH, COOH)

III $\xrightarrow{- H_2}$ V

Ring-opening oxidation (Fig.(4-23)) of the anone/anol mixture (III) by $HNO_3$ delivers 1,12-dodecane dicarboxylic acid (dodecanedioic acid), $HOOC(CH_2)_{10}COOH$ (IV). This $C_{12}$ compound is polycondensed with the $C_6$ compound hexamethylenediamine to polyamide 6.12 (nylon 6.12).

The mixture (III) is converted to 100 % anol (V) by dehydrogenation with $Cu/Al_2O_3$ or Cu–Cr catalysts at ca. 240°C. Reaction of V with hydroxylamine sulfate yields cyclododecanone oxime (VI) (Eq.(4-24)). This reaction generates sulfuric acid which must be continuously neutralized by addition of ammonia in order to keep the reacting mixture at pH = 7. The oxime (VI) is then subjected to a Beckmann rearrangement in oleum (= fuming sulfuric acid = solution of $SO_3$ in concentrated $H_2SO_4$). Neutralization of VII with $NH_3$ delivers laurolactam (dodecanolactam), the monomer for poly(laurolactam) = polyamide 12:

(4-24)

V $\xrightarrow[- H_2O,\ - H_2SO_4]{+ HONH_2 \cdot H_2SO_4}$ VI (NOH) $\xrightarrow{+ H_2SO_4}$ VII (O, $NH \cdot H_2SO_4$)

The undesirable formation of ammonium sulfate by this classic route has led to several other processes. The photochemically initiated reaction of cyclododecane with nitrosyl chloride, NOCl, at 70°C directly to cyclododecane oxime (Route IIb in Fig. 4-8) and further to the lactam is already an industrial process..

Another route, IIc in Fig. 4-8, carbonylizes 1,5,9-cyclododecatriene with CO + ROH to the cyclododecanediene carboxylic ester which is converted by saponification and hydrogenation to the cyclododecane carboxylic acid (Eq.(4-25)). Reaction of this acid with nitrosyl sulfuric acid delivers directly the sulfate of laurolactam:

(4-25)

cyclododecatriene $\xrightarrow[+ ROH]{+ CO}$ COOR

$+ H_2O / + 2 H_2 \downarrow - ROH$

COOH $\xrightarrow[- CO_2]{+ NOHSO_4}$ (O, $NH \cdot H_2SO_4$)

III. Catalytic dimerization of butadiene delivers to *cis,cis*-cycloocta-1,5-diene which is partially hydrogenated to cyclooctene. Metathesis polymerization of cyclooctene (Section 6.4.3) leads to a so-called poly(octenamer) (TOR) with 80 % trans double bonds.

IV. The direct hydrocyanation of butadiene is now the preferred route to adiponitrile. It has mainly displaced the older 4-step process: chlorination of butadiene to dichlorobutenes, reaction of these with HCN or alkali cyanide to butene dinitriles, and hydrogenation of butene dinitriles to adiponitrile, $NC(CH_2)_4CN$ and then to $H_2N(CH_2)_6NH_2$.

The new route reacts butadiene with hydrogen cyanide in the presence of $Ni^0$-phosphine or $Ni^0$-phosphate complexes plus promotors to a mixture of pentenenitriles, $C_4H_7CN$. These mononitriles then add HCN in an anti-Markovnikov reaction to adiponitrile, $NC(CH_2)_4CN$. Adiponitrile is hydrogenated to 1,6-hexamethylenediamine, the monomer for many aliphatic polyamides (PA 6.6, 6.9, 6.10, 6.12).

V. The direct chlorination of butadiene to chloroprene is not very selective. Chloroprene is therefore synthesized in three steps. First, chlorine is added to gaseous butadiene in a free-radical reaction at ca. 250°C and 1–7 bar to give a mixture of dichlorobutenes: 3,4-dichloro-1-butene + *cis*-1,4-dichloro-2-butene + *trans*-1,4-dichloro-2-butene. Second, the two trans compounds are isomerized by catalytic amounts of CuCl to the 3,4-product, $CH_2{=}CH{-}CHCl{-}CH_2Cl$, which is distilled off. Third, 3,4-dichloro-1-butene is dehydrochlorinated by dilute alkali at elevated temperature to chloroprene.

In the past, the mixture of the three dichlorobutenes (see above) served as a base chemical (Route Va). On hydrocyanation with HCN or alkali cyanides, the two 1,4-dichloro-2-butenes directly deliver the corresponding 1,4-dicyano-2-butenes. The 3,4-dichloro-1-butene is converted to the 3,4-dicyano-1-butene which undergoes an allyl rearrangement by the hydrocyanation catalyst, a copper-cyano complex. Subsequent hydrogenation leads to hexamethylene-1,6-diamine.

Hydrolysis 1,4-dichloro-2-butenes by sodium formate delivers 1,4-butenediol:

$$\text{(4-26)} \qquad ClCH_2CH{=}CHCH_2Cl \xrightarrow[-\,2\,HCOOH,\,-2\,NaCl]{+\,2\,HCOONa,\,+\,2\,H_2O} HOCH_2CH{=}CHCH_2OH$$

The diol is subsequently hydrogenated to 1,4-butanediol (Route Vb), a monomer for poly(butylene terephthalate) (PBT) and polyurethanes (PUR).

VI. Acetoxylation of butadiene with acetic acid and $O_2$ at 70°C/70 bar by Pd/C catalysts and Te as a promoter delivers 1,4-diacetoxy-2-butene (DAB). The hydrogenation of DAB leads to the saturated compound which, when treated with acidic ion-exchange resins (+ $H_2O$) results in a mixture of 1,4-butanediol, tetrahydrofuran, and acetic acid.

VII. The older Route VI to 1,4-butanediol is less economical than the newer Route VII via epoxidation of butadiene with air by a silver catalyst. The resulting 3,4-epoxy-1-butene is thermally transformed to 2,5-dihydrofuran. Hydrogenation delivers tetrahydrofuran which is then hydrolyzed to 1,4-butanediol.

VIII. The Diels-Alder reaction of butadiene with cyclopentadiene leads to 5-vinylbicyclo-[2.2]-hept-2-ene which is isomerized by alkali metal catalysts to 5-ethylidenenorbornene, the termonomer in ethene-propene rubbers, EPDM.

IX. In a pilot process, butadiene is cyclodimerized to 4-vinylcyclohexene by Cu-zeolite catalysts at 100°C and 19 bar. Hydrogenation leads to styrene which is then used for polymerizations to poly(styrene) (PS) and for copolymerizations with ethene (to E/S; now discontinued) or acrylonitrile plus butadiene (to ABS).

## 4.5.2 Butane

Butane is obtained from LPGs, the light hydrocarbon fraction of refining, and the $C_4$ fraction from the cracking of gas oil or naphtha. It is difficult to separate from other hydrocarbons and it is also rather inert, but it is relatively inexpensive and therefore the new base chemical for four routes (Fig. 4-9):

I. Various processes isomerize butane to isobutane which is subsequently dehydrogenated to isobutene, the monomer for poly(isobutylene), (PIB).

II. Selective oxidation of butane (no process details known) produces 1,4-butanediol without formation of tetrahydrofuran (THF) whereas

III. another new oxidation/dehydration process (no process details known) leads to THF without 1,4-butanediol as byproduct, and

IV. a third oxidation/dehydration by vanadium oxides with various promoters delivers directly maleic anhydride for unsaturated polyesters (UP).

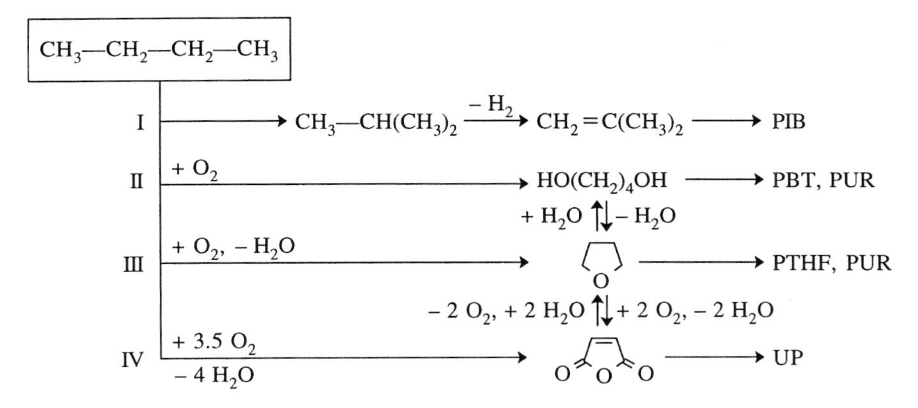


Fig. 4-9 Butane as a base chemical for intermediates, monomers, and polymers.

## 4.5.3 Butenes

Butenes are metathesized with ethenes to propene (p. 107), dehydrogenated to butadiene (p. 111), or oxidized by air to maleic anhydride with supported and modified vanadium oxide or phosphorus oxide catalysts at ca. 400°C and slight pressure.

## 4.5.4 Isobutene

Isobutene $CH_2=C(CH_3)_2$, commonly called isobutylene, is obtained from the $C_4$ raffinate (Table 3-17 and p. 111; separation of 2-butene/isobutene mixtures by distillation because of sufficiently high differences in boiling points), from dehydrogenation of isobutane (Fig. 4-9, I), and from isomerization of butenes. More than 50 % of isobutene and butenes are used for the synthesis of so-called polymer gasoline which is not a polymer but a mixture of highly branched higher hydrocarbons (isooctane, etc.) with excellent anti-knock properties.

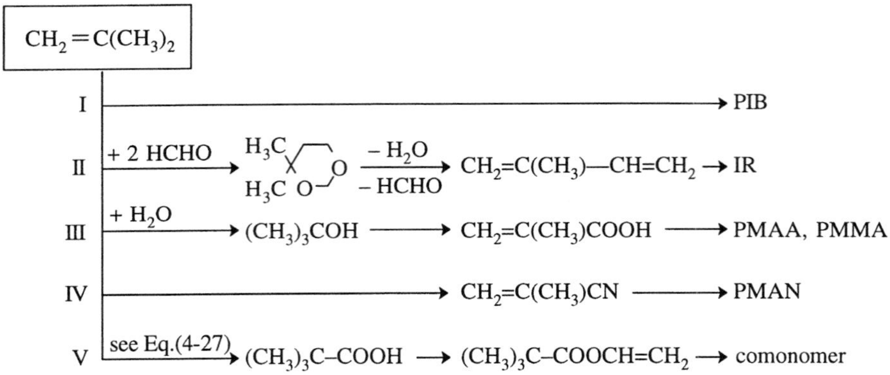

$CH_2{=}C(CH_3)_2$

I → PIB

II + 2 HCHO → 4,4-dimethyl-1,3-dioxane ($H_3C$, $H_3C$) $\xrightarrow[-\,HCHO]{-\,H_2O}$ $CH_2{=}C(CH_3){-}CH{=}CH_2$ → IR

III + $H_2O$ → $(CH_3)_3COH$ → $CH_2{=}C(CH_3)COOH$ → PMAA, PMMA

IV → $CH_2{=}C(CH_3)CN$ → PMAN

V see Eq.(4-27) → $(CH_3)_3C{-}COOH$ → $(CH_3)_3C{-}COOCH{=}CH_2$ → comonomer

Fig. 4-10 Isobutene as a base chemical for intermediates, monomers, and polymers.

Isobutene is the base chemical for a number of monomers and polymers (Fig. 4-10):

I. Polymerization of isobutene with a little isoprene leads to poly(isobutylene) (PIB).

II. The Prins reaction of formaldehyde with isobutene in a butadiene-free $C_4$ fraction at 70–95°C/20 bar is catalyzed by strong mineral acids. The resulting 4,4-dimethyl-1,3-dioxane is then thermolyzed to isoprene by phosphate catalysts at 240–400°C. Isoprene is polymerized to *cis*-1,4-poly(isoprene) (IR), a synthetic natural rubber.

III. Hydration of isobutene by 60 % $H_2SO_4$ delivers *t*-butanol. This reaction is used to retrieve isobutene from mixtures of butenes since isobutene reacts much faster with water than 1-butene and 2-butenes. Dehydration of *t*-butanol delivers a pure isobutene.

A small proportion of industrial *t*-butanol is catalytically oxidized by air with molybdenum catalysts at ca. 420°C with 94 % conversion to methacrolein which is subsequently oxidized at 300°C in the same system with 89 % conversion to methacrylic acid:

$$(4\text{-}27)\qquad H_3C{-}C(CH_3)_2{-}OH \xrightarrow[-\,H_2O]{+\,O_2} H_2C{=}C(CH_3){-}CHO \xrightarrow{+\,(1/2)\,O_2} H_2C{=}C(CH_3){-}COOH$$

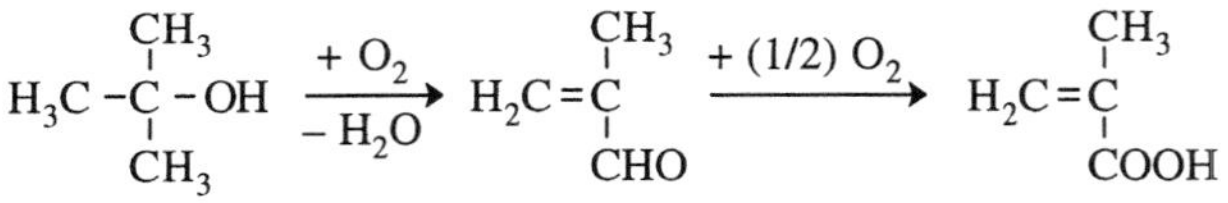

IV. Ammonoxidation of isobutene, similar to that of propene (Fig. 4-7, X), delivers methacrylonitrile, the monomer for poly(methacrylonitrile) (PMAN).

V. The Koch synthesis of carboxylic acids from isobutene, carbon monoxide, and water is catalyzed by protonic acids such as HF, $H[SbF_6]$, $H_3PO_4$, $H_2SO_4$, etc. The addition of a proton from these acids to isobutene produces a secondary carbenium cation that adds carbon monoxide, and then water to pivalic acid, $(H_3C)_3C{-}COOH$:

$$(4\text{-}28)\qquad (H_3C)_2C{=}CH_2 \underset{}{\overset{+\,H^{\oplus}}{\rightleftharpoons}} (H_3C)_2\overset{\oplus}{C}{-}CH_3 \xrightarrow{+\,CO} H_3C{-}C(CH_3)_2{-}\overset{\oplus}{C}{=}O \xrightarrow[-\,H^{\oplus}]{+\,H_2O} H_3C{-}C(CH_3)_2{-}C({=}O){-}OH$$

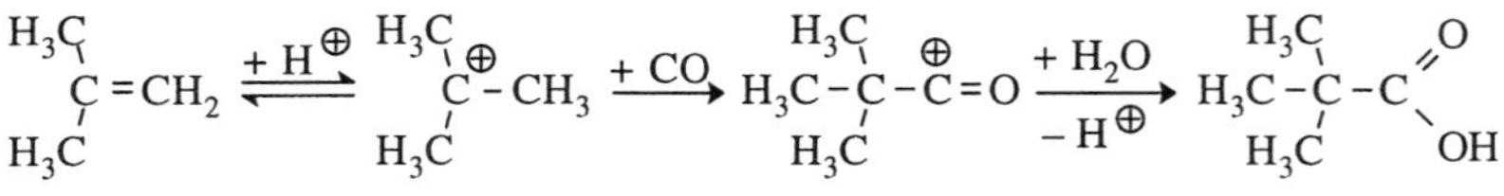

Pivalic acid is converted to vinyl pivalate by vinylation with acetylene and $Hg/H_2SO_4$ as catalyst or by trans-vinylation with acetic acid and $PdCl/_2NaCl$ as catalyst or vinyl chloride and $PdCl_2$/sodium pivalate (not a transesterification! The split is between $CH_2{=}CH{-}$ and $-O\cdots$ and not between $CH_2{=}CH{-}O{-}$ and $-CO\cdots$). The monomer is a comonomer for vinyl polymers if internal plasticization of the latter is desired.

## 4.6 $C_5$ Compounds

From the higher fractions of cracking operations (Table 3-16), only the $C_5$ fraction and crack gasoline (pyrobenzine) are used for chemical syntheses. Of the major components of $C_5$ fractions (Table 3-18), isoprene (15 wt%) and cyclopentadiene/dicyclopentadiene (14.5 wt%) serve as monomers. Neither pentane (22 wt%), isopentane (15 wt%), 1,3-pentadienes (9 wt%), nor methylbutenes (8 %) are presently chemically utilized.

### 4.6.1 Isoprene

Isoprene is obtained from the $C_5$ fraction of cracking operations by extractive distillation with acetonitrile, *N*,*N*-dimethylformamide, or *N*-methylpyrrolidone or by fractionating distillation of its azeotrope with *n*-pentane.

After removal of 2-methyl-2-butenes, $(H_3C)_2C{=}CHCH_3$, by formation of their sulfuric acid half-esters $(CH_3)_2C(CH_2CH_3)(OSO_3H)$, isopentane and isopentenes are dehydrogenated to isoprene at 600°C with a $Fe_2O_3{\cdot}Cr_2O_3{\cdot}K_2CO_3$ catalyst. Subsequent extractive distillation with acetonitrile delivers a 99.5 % pure isoprene.

Isoprene is also obtained by the Prins reaction of isobutene with formaldehyde (Fig. 4-10, II) where polyols are major byproducts. Improvements of this process include the replacement of formaldehyde by implicit formaldehyde sources such as methylal $(CH_2(OCH_3)_2)$, 1,3-dioxolane + $H_2O$, or even $CH_3OH + 0.5\ O_2$:

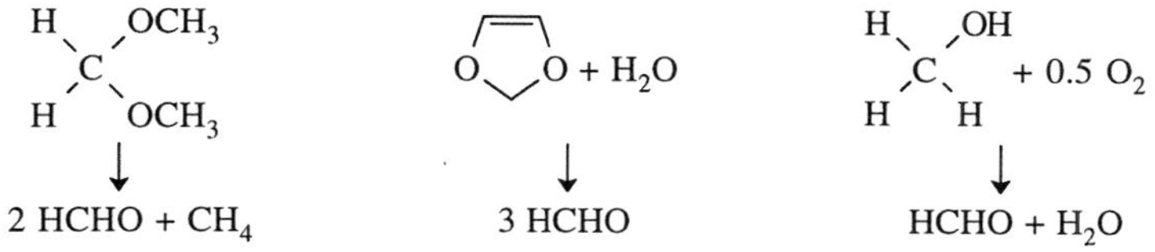

$$H_2C(OCH_3)_2 \rightarrow 2\ HCHO + CH_4 \qquad \text{1,3-dioxolane} + H_2O \rightarrow 3\ HCHO \qquad H_2C(H)OH + 0.5\ O_2 \rightarrow HCHO + H_2O$$

Isoprene has long been used in small proportions as a comonomer in isobutylene rubbers (IIR) but is now increasingly homopolymerized to a "synthetic natural rubber" (natsyn, (IR)).

### 4.6.2 Cyclopentadiene

$C_5$ fractions from steam cracking operations contain 14–25 wt% cyclopentadiene of which only a small fraction is used for the synthesis of chemicals. The bulk of the $C_5$ fraction is used to improve the octane rating of gasolines by first removing the diolefins and then hydrogenating the remaining unsaturated $C_5$ compounds.

In order to obtain cyclopentadiene, the $C_5$ fraction is heated several hours to 100°C or to higher temperatures under pressure, which causes cyclopentadiene to dimerize to dicyclopentadiene (DCP), mostly to its endo form:

(4-29) cyclopentadiene $\overset{\Delta}{\rightleftharpoons}$ dicyclopentadiene (endo, exo)

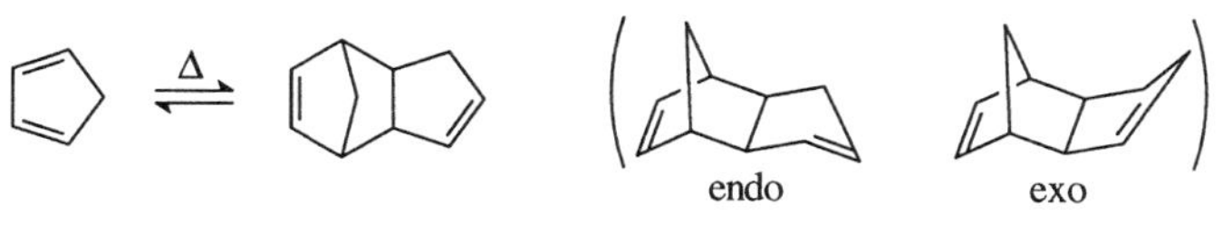

On distillation, the higher-boiling dicyclopentadiene ($T_{bp}$ = 171°C) remains in the sump of the distillation column whereas the other components of the $C_5$ fraction boil off ($T_{bp}$ = 28–69°C) (Table 3-18). Cyclopentadiene ($T_{bp}$ = 41.7°C) is stored and transported as DCP from which cyclopentadiene is reconstituted by heating to temperatures of 200°C and higher.

Cyclopentadiene is polymerized thermally to poly(dicyclopentadiene) (Section 6.4.6). It is also reacted with ethene to norbornene, the monomer for poly(norbornene) (Section 6.4.4), or with butadiene to 5-ethylidene norbornene (Fig. 4-8, VIII), a comonomer for EPDM rubbers (p. 229).

Hydrogenation of cyclopentadiene delivers cyclopentene which is polymerized to a so-called poly(pentenamer), a rubber (Section 6.4.2).

# 4.7 Aromatics

## 4.7.1 Sources and Syntheses

Aromatic hydrocarbons are key chemicals that are used world-wide in annual amounts of ca. $100 \cdot 10^6$ tons. All of them are synthesized, either from oil (> 96 wt%) or from coal (< 4 wt%). There is no ready source of aromatics in nature.

The most important group of aromatics is the so-called **BTX fraction** (benzene, toluene, xylenes) which comprises ca. 68 wt% of aromatics in the United States, ca. 70 wt% in Western Europe, and ca. 75 wt% in Japan. The second most important aromatic compound is ethyl benzene with 31 wt% (USA), 28 wt% (Western Europe), and 23 wt% (Japan), respectively. The share of naphthalene and other aromatics is less than 2 wt%.

Neither the direct isolation of aromatic compounds from coal or crude oil nor their synthesis from smaller molecules is economical. Instead, aromatic compounds are obtained from three major sources: as "coke extract" ("raw benzene") from the coking of coal (now declining), as "pyrolysis gasoline" ("pyrobenzine", "crack gasoline") from steam cracking of naphtha (mainly in Europe), and as "raffinate" from "reformate gasoline" obtained by distillation of crude oil and subsequent reforming and solvent extraction (p. 53; mainly in the United States)). Coke extract, pyrobenzine, and raffinate contain various proportions of aromatics (Table 4-2).

Table 4-2 Content of aromatic compounds in various feedstocks [1b].

| Component | Coke extract vol% | Pyrobenzine wt% | Raffinate wt% |
|---|---|---|---|
| Fraction from first running | 2 | 0 | 0 |
| Benzene | 65 | 40 | 3 |
| Toluene | 18 | 20 | 13 |
| Xylenes | 6 | 5 | 18 |
| Ethylbenzene | 2 | 3 | 5 |
| Higher aromatics | 7 | 3 | 16 |
| Non-aromatics | 0 | 29 | 45 |

**Coal as Feedstock**

On coking (p. 44), coal delivers coke as the main product (ca. 80 %) and coal tar, coke water, and coke gas as byproducts and sources of aromatic compounds. The worldwide annual production of coke has not changed much in the past decades. It is characterized by a decrease of coke production due to improved steel-making processes and a decline in steel production caused by economic conditions in the United States and Western Europe, the replacement of town gas by natural gas and crack gas in Western Europe, and the increase of steel and coke production in developing countries.

Coking of coal at 900–1200°C (p. 44) splits off low-molecular weight "substituents" from the coal polymer and simultaneously crosslinks the highly aromatic skeleton of the coal molecule (Fig. 3-6, p. 41). The crosslinked material is coke (yield: 70–75 %) which is almost pure carbon. The degradation products are light gas (ca. 20 %; mostly methane and hydrogen; used as fuel), light oil (coal naphtha; ca. 1 %), and coal tar (ca. 4 %).

Light gas contains aromatic compounds that can be obtained by two methods. Washing the gas with higher-boiling hydrocarbons delivers an oil from which the aromatics can be distilled off. The gas can also be treated with activated charcoal from which the absorbed aromatics are desorbed by water vapor to give "coke water." In order to remove nitrogen- and sulfur-containing compounds, the resulting mixture of aromatics is then either treated with sulfuric acid or catalytically hydrogenated.

Coke water contains ca. 0.3 wt% of phenolic compunds (phenol, cresols, xylenols, etc.) that can be extracted by benzene or butyl acetate.

Distillation of coal tar delivers light oil (mainly BTX and pyridine), carbol oil (mixture of phenols), naphthalene oil (naphthalene), wash oil (acenaphthalene, etc.), anthracene oil (anthracene, phenanthrene, carbazole), and pitch.

**Crude Oil as Feedstock**

The distillation of both paraffin-rich and naphthenic crude oils delivers gasolines that have low octane numbers. They thus have to be reformed, i.e., catalytically modified (see p. 52–53). **Reforming** reactions include isomerization of alkanes to isoalkanes, cyclizing dehydrogenation of alkanes to cycloalkanes, and isomerization of alkyl cyclopentanes to substituted cyclohexanes with subsequent dehydrogenation to aromatics. The distillation of the resulting *reformate gasoline* then leads to a fractionate that is rich in toluene and xylenes (Table 4-2).

**Steam cracking** of petroleum naphtha for the production of olefins delivers a benzene-rich *pyrolysis benzine* (Table 4-2) that needs to be hydrogenated in order to remove polymerizable monoolefins and diolefins as well as S, N, and O compounds.

After the components of pyrobenzine and raffinate are separated, the BTX fraction of these feedstocks is obtained by three methods:

(A) Addition of polar components such as $CH_3OH$ and $(H_3C)_2O$ to feedstocks. The azeotropes of these components with the non-aromatics of the feedstock are distilled off; the aromatics remain.

(B) Extractive distillation of feedstocks with thermally stable compounds such as dichlorobenzenes, polyglycols, *N*-methyl pyrrolidone, sulfolane (= tetrahydrothiophene dioxide), etc. Aromatics and added compounds remain in the sump, all other hydrocarbons are distilled off from the 1-phase mixture.

(C) Liquid-liquid extraction (2 liquid phases) of feedstocks with solvents such as tetraethylene glycol, sulfolane, *N*-methyl pyrrolidone, and many others. The extract contains the solvent and the aromatics. It is separated by either direct distillation or by back-extraction with a light hydrocarbon such as pentane that extracts the aromatics.

### 4.7.2 Benzene

Benzene, $C_6H_6$, was formerly isolated from coal tar (hence also called **coal tar naphtha**) but is now obtained to more than 98 % from the refining of crude oil, mainly by direct isolation. About 15 % is obtained from toluene by disproportionation ($C_7H_8 \rightarrow C_6H_6 + C_6H_4(C_2H_5)_2$) or by hydrodealkylation ($C_7H_8 + H_2 \rightarrow C_6H_6 + CH_4$). In United States commerce, a liquid containing 90–100% benzene is called **benzol(e)** which is also the German name for benzene. It was used in the 1960s for blending with gasoline.

Benzene is the base chemical for intermediates, monomers, and polymers (Fig. 4-11):

I. Alkylation of benzene in the liquid phase by less than stoichiometric amounts of ethene with $AlCl_3$ as "catalyst" and $C_2H_5Cl$ as promoter leads to ethyl benzene. Ethyl chloride reacts with benzene to ethyl benzene and HCl which coordinates with $AlCl_3$. The true catalyst is therefore $H[AlCl_4]\cdot n\ C_6H_5C_2H_5$ (see also Chapter 8.4 in Volume I).

The excess of benzene over ethene reduces further alkylations of the desired ethyl benzene to the undesired multiethyl benzenes. The classic process worked at ca. 90°C and slight pressure; it used ca. 1 kg of $AlCl_3$ per 100 kg of ethyl benzene. A newer process operates at higher temperatures of 140–200°C. It needs only 0.25 kg of $AlCl_3$ per 100 kg of $C_6H_5C_2H_5$.

Gas-phase alkylations at solid catalysts, higher temperatures, and pressures do not need corrosion-resistant plants and expensive alkaline washes for removal of the spent catalyst. An example is the Mobil-Badger process that operates at ca. 440°C and 14–28 bar with a modified zeolite as catalyst. Such plants now have nameplate capacities of up to 530 000 t/a.

Ethyl benzene is catalytically dehydrogenated to styrene by iron oxide catalysts at temperatures of 550–620°C (world production of styrene is ca. $20\cdot10^6$ t/a (2003)). Subsequent cooling condenses first tars and then styrene + ethyl benzene which, after addition of a polymerization inhibitor, are separated by distillation in vacuum (boiling temperatures at normal pressure are 136°C for ethyl benzene and 145°C for styrene).

Another process (Ia) oxidizes ethylbenzene to its hydroperoxide that is reduced to the carbinol which is then dehydrated to styrene.

Styrene is used as a monomer for poly(styrene) (PS); as a comonomer for styrene-acrylonitrile (SAN), and acrylonitrile-butadiene-styrene plastics (ABS); for styrene-butadiene rubbers (SBR); and as a "crosslinking agent" for unsaturated polyester resins (UP). It was also used for the now discontinued E/S Interpolymers®.

II. Most maleic anhydride (MA) is now produced by oxidation of butane (Section 4.5.2) or butene (Section 4.5.3), but some MA is still obtained by direct oxidation of benzene at ca. 350°C by activated $V_2O_5$ catalysts.

III. Benzene is oxychlorinated by HCl/air at ca. 240°C (Raschig–Hooker process) to chlorobenzene which is then hydrolyzed to phenol at 400–450°C by $Ca_3(PO_4)_2/SiO_2$. The latter process is highly corrosive and energy intensive; it is no longer economical.

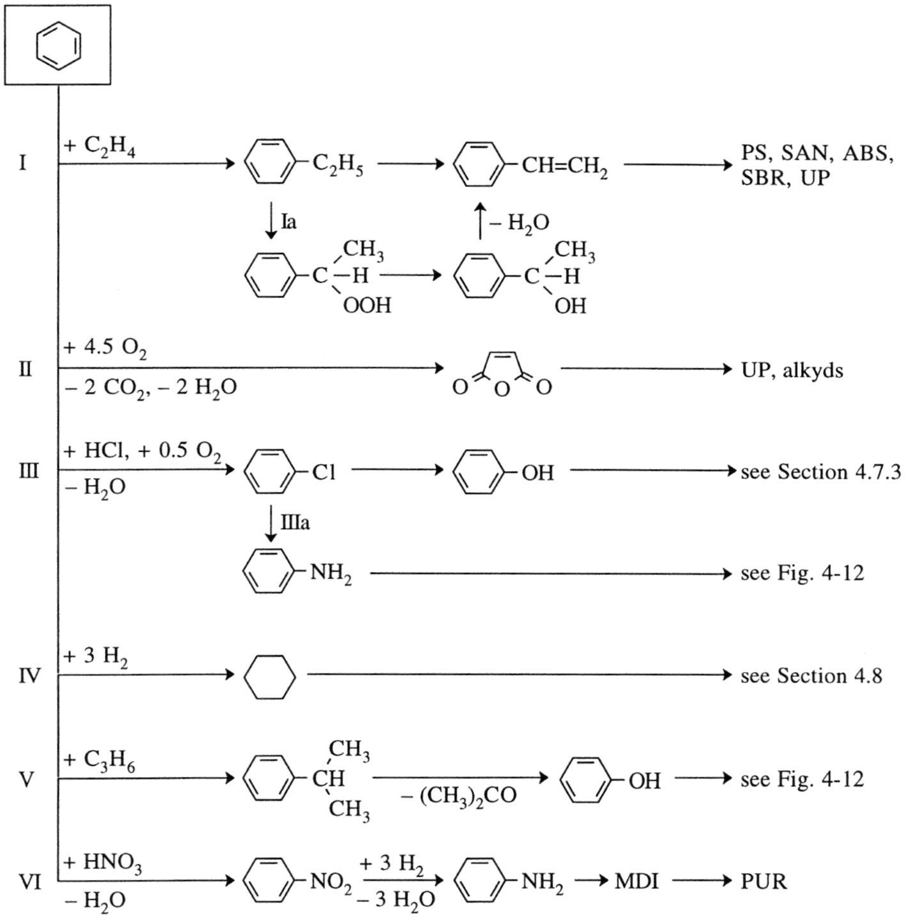

Fig. 4-11 Benzene as base chemical for intermediates, monomers, and polymers.

Two older phenol syntheses are also no longer used industrially. (1) Chlorination of benzene to chlorobenzene in the liquid phase with subsequent hydrolysis of chlorobenzene by aqueous NaOH (no longer economical). (2) Sulfonation of benzene by oleum (= solution of $SO_3$ in concentrated sulfuric acid) to benzene sulfonic acid, neutralization of the sulfonic acid, reaction of the sodium salt in molten NaOH to sodium phenolate and sodium sulfite, and reaction of $C_6H_5ONa$ with $SO_3$ and $H_2O$ to $C_6H_5OH$ + $NaHSO_3$ (large amounts of many byproducts that cannot be sold or used profitably).

Chlorobenzene serves also as a base chemical in one of the syntheses of aniline. In this process, chlorobenzene is ammonolyzed by aqueous $NH_3$ in the presence of CuCl + $NH_4Cl$ (Nieuwland catalyst) at ca. 190°C and 60–75 bar. One of the main uses of aniline is for the synthesis of MDI, a monomer for polyurethanes (PUR) (see also Route VI).

IV. Benzene is hydrogenated to cyclohexane, a base chemical for many intermediates and monomers (Section 4.8). Modern hydrogenations are performed in the gas phase or at medium pressures in the liquid phase. The older high-pressure hydrogenation processes with sulfur-resistant catalysts for sulfur-containing benzenes are now obsolete.

In medium-pressure processes, very pure benzene is hydrogenated to cyclohexane by Raney nickel in the liquid phase at ca. 210°C and medium pressures of 30 bar, using exact reaction conditions (residence time, temperature constancy, heat discharge) in order to prevent the equilibration of cyclohexane to methylcyclopentane. Gas-phase processes employ noble-metal catalysts at temperatures up to 600°C.

V. Cumene (= isopropylbenzene) is obtained exclusively from the alkylation of benzene by propene (Fig. 4-11, Route V) in amounts of ca. $8 \cdot 10^6$ t/a. About 90 % of cumene is obtained by gas-phase processes such as the UOP process at 200–250°C and 20–40 bar with $H_3PO_4/SiO_2 + BF_3$ plus water vapor (UOP = Universal Oil Products Co.). These processes use propene-propane mixtures that are poor in ethene and other olefins since these would also alkylate benzene. Water vapor fixes phosphoric acid on the carrier ($SiO_2$) via hydrate formation; it also serves as a heat sink for the exothermic reaction.

Liquid-phase processes provide another 10 % of cumene. As catalysts, they use either $H_2SO_4$ or $AlCl_3$ at ca. 35°C or HF at ca. 60°C, both at low pressures of up to 7 bar.

Practically all cumene is used to produce phenol by the Hock process in which cumene is oxidized by air to the hydroperoxide, either with redox catalysts in the liquid phase at 120°C or in emulsion at 90–130°C and pressure. The hydroperoxide is then split into phenol and acetone by 1 % sulfuric acid in acetone at 60°C or by 40 % $H_2SO_4$ at 50°C:

$$C_6H_5\text{-}CH(CH_3)_2 \xrightarrow{+O_2} C_6H_5\text{-}C(CH_3)_2OOH \xrightarrow{[+H^{\oplus}]} C_6H_5\text{-}OH + O{=}C(CH_3)_2 \qquad (4\text{-}30)$$

The Hock process is the most economical process for the synthesis of phenol; it also delivers 0.62 t acetone per ton of phenol. The very exothermic splitting reaction leads to a number of other byproducts such as $\alpha$-methylstyrene, $CH_2{=}C(CH_3)(C_6H_5)$, which is used as a comonomer in styrene polymers. More often it is hydrogenated to cumene.

VI. The classic Mitscherlich nitration of benzene by nitrating acid (= mixture of nitric acid and sulfuric acid) leads to nitrobenzene which is catalytically hydrogenated to aniline, $C_6H_5NH_2$, either in fixed beds with nickel sulfide catalysts at 300–475°C or in fluidized beds with Cu, Cr, Ba, and Zn oxides on $SiO_2$ at 270–290°C and 1–5 bar. The traditional reduction of $C_6H_5NO_2$ by iron shavings and HCl is no longer economical.

About 80 % of the world production of aniline is used for the synthesis of 4,4'-di-(phenylmethane diisocyanate), OCN($p$-$C_6H_4$)$CH_2$($p$-$C_6H_4$)NCO (= **m**ethane **d**iphenyl d**i**isocyanate, MDI), the most important diisocyanate for the synthesis of polyurethanes (PUR) (see Section 10.5.3).

## 4.7.3 Phenol

Phenol is the trivial name of hydroxybenzene, $C_6H_5OH$. It is also known as carbolic acid, phenyl hydroxide, phenylic acid, phenic acid, and oxybenzene.

The annual world production of phenol exceeds $5.2 \cdot 10^6$ tons. More than 99 % is produced synthetically; coal is now a very minor source of phenol. The cumene-based Hock process (see above) provides more than 95 % of all synthetically produced phenol (Fig. 4-11, Route V, and Eq.(4-30)).

Other less economical phenol syntheses are based on toluene or benzene:

(4-31) $C_6H_5CH_3 \rightarrow C_6H_5COOH \rightarrow C_6H_5OH$
(oxidative decarboxylation of benzoic acid produces large amounts of $CO_2$);

(4-32) $C_6H_6 \rightarrow C_6H_5Cl \rightarrow C_6H_5OH$
(requires corrosion-resistant equipment and produces chlorinated waste);

(4-33) $C_6H_6 \rightarrow C_6H_5SO_3H \rightarrow C_6H_5SO_3Na \rightarrow C_6H_5ONa \rightarrow C_6H_5OH$
(obsolete because of large amounts of hard-to-sell $Na_2SO_3$ and $Na_2SO_4$).

A new developmental process (2003) uses a Pd–chalcogen catalyst on silica at 150–200°C to produce phenyl acetate from benzene, acetic acid, and oxygen:

(4-34) $CH_3COOH + C_6H_6 + (1/2)\ O_2 \rightarrow CH_3COOC_6H_5 + H_2O$

The phenyl acetate is then hydrolyzed to phenol and acetic acid at temperatures of less than 100°C by an ion-exchange resin.

Another new process (2005) reacts benzene with air as oxidizing agent and CO as reducing agent in aqueous $CH_3COOH$ at ca. 90°C with molybdovanadophosphoric acid as catalyst. The single-pass conversion to phenol is 25–30 % with 90 % selectivity.

Phenol is the base chemical for a number of intermediates and monomers (Fig. 4-12):

I. Approximately one-third of the phenol production is used directly for the syntheses of phenol-formaldehyde resins (PF) and epoxy resins (EP).

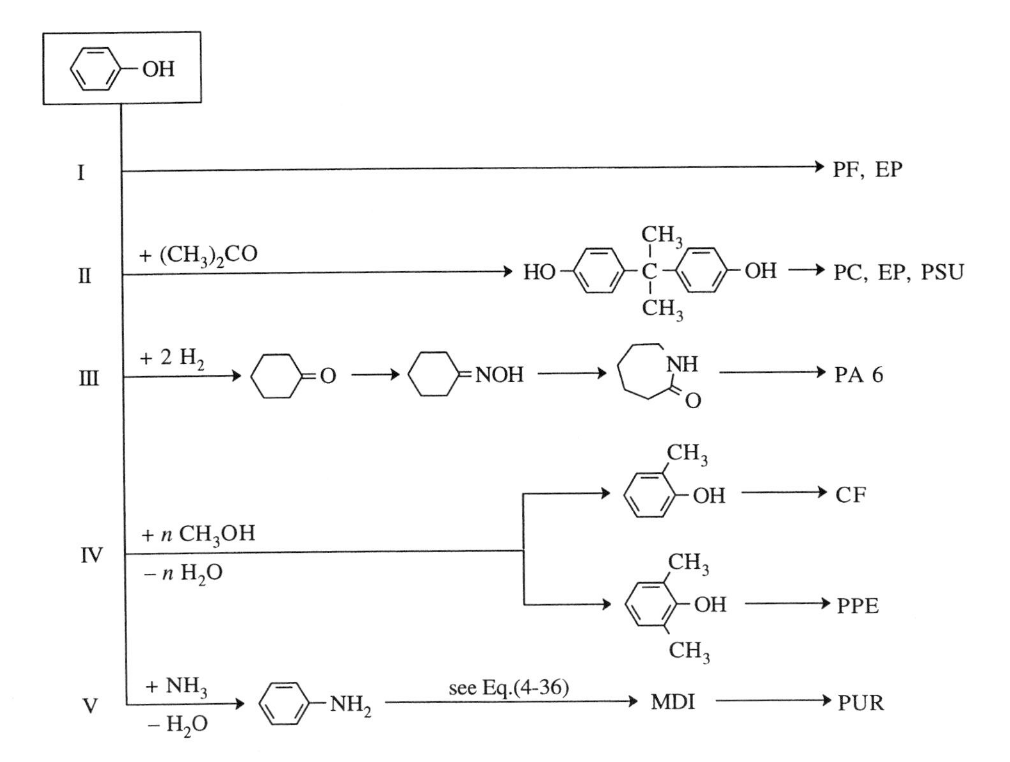


Fig. 4-12 Phenol as a base chemical for intermediates, monomers, and polymers.

II. Another one-third of phenol is condensed with acetone to 2,2-bis(4-hydroxyphenyl)propane (= bisphenol A), using either sulfuric acid, dry HCl, or acidic ion exchange resins as catalysts. Bisphenol A is the most important monomer for polycarbonates (PC). It is also a monomer for epoxides (EP) and polysulfones (PSU).

III. About 15–20 % of phenol is hydrogenated to cyclohexanone which is the major base chemical for the synthesis of ε-caprolactam (see Fig. 4-14)

Cyclohexanone was formerly synthesized from phenol by a 2-step process: first hydrogenation to cyclohexanol by nickel catalysts at 140–160°C and 15 bar and then dehydrogenation of cyclohexanol to cyclohexanone by Zn or Cu catalysts at 400–450°C and normal pressure.

A newer 1-step gas-phase process employs a palladium catalyst on, e.g., $CaO–Al_2O_3$ at 140–170°C and 1–2 bar. In this process, phenol is activated by adsorption of its keto form on the carrier followed by successive hydrogenations:

(4-35) OH ⇌ [ O ↔ O ] $\xrightarrow{+H_2}$ O $\xrightarrow{+H_2}$ O

IV. The methylation of phenol by methanol with $Al_2O_3$ as catalyst leads predominantly to a mixture of *o*-cresol and 2,6-xylenol whose ratio can be varied widely by variation of experimental conditions such as 300–450°C in the gas phase at normal pressure or 300–360°C at 40-70 bar in the liquid phase.

*o*-Cresol is a monomer for cresol-formaldehyde resins (CF). 2,6-Xylenol is the monomer for poly(2,6-dimethyl-1,4-oxyphenylene) (PPE), commonly called "poly(phenylene oxide)" (PPO®).

V. Aniline is mainly obtained from nitrobenzene (Fig. 4-11, Route VI) and to a smaller extent also from chlorobenzene (Fig. 4-11, Route IIIa). Aniline is also synthesized by ammonolysis of phenol at 425°C and 200 bar with catalysts such as $Al_2O_3$–$SiO_2$, zeolites, or mixtures of oxides of Mg, B, Al, and Ti, often with other cocatalysts. Condensation of aniline with formaldehyde delivers 4,4'diaminodiphenylmethane which is then phosgenated to 4,4'-diphenylmethanediisocyanate (= **m**ethane**d**iphenyld**i**isocyanate, MDI) (for the mechanism, see Section 10.6):

(4-36) 2 ⌬–$NH_2$ $\xrightarrow[-H_2O]{+CH_2O}$ $H_2N$–⌬–$CH_2$–⌬–$NH_2$

$+ 2\ COCl_2$ ↓ $- 4\ HCl$

O=C=N–⌬–$CH_2$–⌬–N=C=O

## 4.7.4 Toluene

The components of BTX fractions can be separated by simple distillation (boiling points: benzene 80°C, toluene 111°C, xylenes 138–144°C) if the BTX does not contain appreciable amounts of non-aromatic hydrocarbons since some of the cycloaliphatics

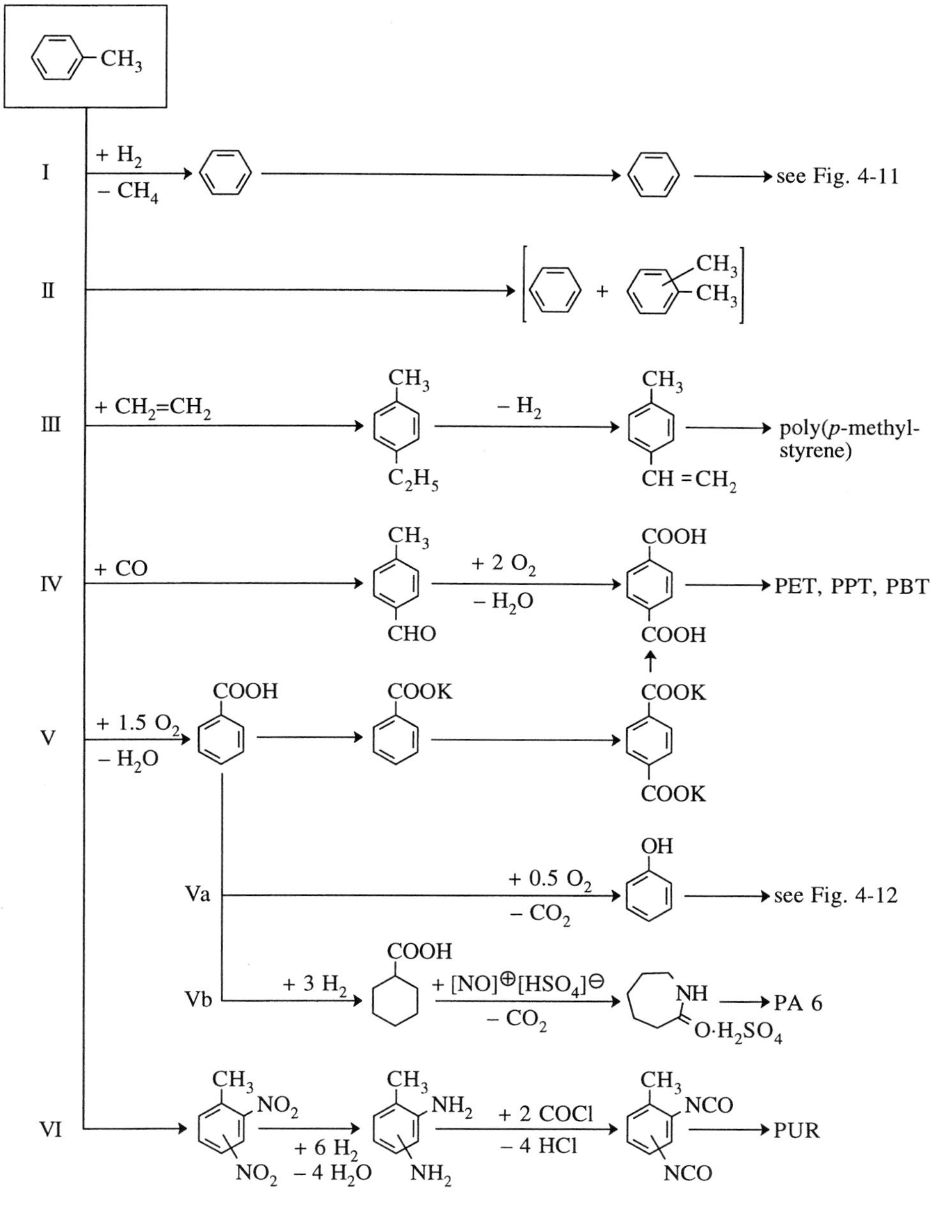

Fig. 4-13 Toluene as base chemical for intermediates, monomers, and polymers.

have very similar boiling points. BTX with larger amounts of non-aromatic hydrocarbons are separated by the methods A–C mentioned on pages 119–120.

Toluene is the base chemical for a number of polymer-related chemicals (Fig. 4-13):

I. Petrochemistry produces more toluene than is needed directly or for the synthesis of intermediates. In order to overcome the benzene shortage, toluene is therefore converted to benzene by hydrodealkylation with a circulating $H_2/CH_4$ mixture either thermally at 580–800°C and 30–100 bar or catalytically on carriers at lower pressures and temperatures, for example, with $Cr_2O_3/Al_2O_3$ at 500–650°C or with $Rh/Al_2O_3$ at 400-480°C. The circulating mixture becomes enriched in methane. On cooling, a part of the methane is separated. The remaining mixture is replenished with hydrogen which is obtained from the methane produced by the reaction $CH_4 + 2\ H_2O \rightarrow CO_2 + 4\ H_2$.

II. The disproportionation (dismutation) of toluene by $BF_3/HF$ or $AlCl_3/HCl$ in the presence of $H_2$ at 80–120°C and 35–70 bar leads to benzene and a mixture of xylenes. The proportion of *p*-xylene can be boosted to 95 % with special zeolite catalysts.

III. Ethylation of toluene by ethene and a modified zeolite at 425°C and slight pressure delivers *p*-ethyl toluene which is dehydrogenated to *p*-methyl styrene which is used as a monomer and comonomer.

IV. Terephthalic acid is usually obtained from *p*-xylene (p. 127). However, it can also be synthesized from toluene as demonstrated in a pilot plant (Fig. 4-13, IV). Reaction of toluene with CO at 30–40°C in the presence of $H[BF_4]$ delivers *p*-toluylaldehyde which is then oxidized to terephthalic acid.

V. Toluene is oxidized by air to benzoic acid, for example, with Co salts at 120°C and 3 bar. The potassium salt of benzoic acid is disproportionated at ca. 430°C by cadmium benzoate in a $CO_2$ atmosphere and the resulting dipotassium terephthalate converted to terephthalic acid, a monomer for poly(ethylene terephthalate) (PET, PES), poly(propylene terephthalate) (PPT), and poly(butylene terephthalate) (PBT).

Va. The oxidative decarboxylation of benzoic acid in the melt or in high-boiling solvents at 220–250°C by a mixture of air and water vapor in the presence of copper salts yields phenol (for its reactions, see Fig. 4-12).

Vb. Liquid benzoic acid is hydrogenated to cyclohexane carboxylic acid at 170°C and 10–17 bar using palladium on activated carbon as catalyst. The subsequent reaction of cyclohexane carboxylic acid with nitrosylsulfuric acid in oleum produces directly ε-caprolactam via a mixed anhydride and several intermediates.

VI. The nitration of toluene by nitrating acid results in a mixture of nitrotoluenes, usually in the ratio 63:34:3 = *o*:*p*:*m*. A mixture of the ortho and para isomers is subsequently nitrated further to an 80:20 mixture of 2,4- and 2,6-dinitrotoluene. This mixture is hydrogenated, preferentially with Raney-nickel or palladium catalysts at elevated temperatures and pressures of ca. 50 bar.

The mixture of 2,4 and 2,6-diaminotoluenes is then phosgenated at ca. 180°C to toluene diisocyanate (TDI), a mixture of the 2,4 and 2,6 isomers. TDI is the second most important diisocyanate after MDI for the synthesis of polyurethanes.

### 4.7.5 Higher Aromatics

The petrochemical $C_8$-aromatics fraction consists of ethylbenzene and the three xylenes (*o*, *m*, *p*). These four compounds cannot be separated economically by fractional distillation (see Table 4-3). On distillation, *o*-xylene remains in the sump whereas the mixture of ethyl benzene/*p*-xylene/*m*-xylene goes over the top. Distillation of this mixture in a distillation column with many plates removes ethyl benzene. The remaining mixture of *p*- and *m*-xylenes is subsequently separated by fractional crystallization. Alternatively, *p*-xylene can also be removed by adsorption onto solids such as zeolites.

Table 4-3 Boiling temperatures, $T_b$, and melting temperatures, $T_M$, of $C_8$ aromatics.

| Compound | $T_b$ | $T_M$ |
|---|---|---|
| Ethyl benzene | 136.2 | –95.0 |
| *p*-Xylene | 138.3 | –13.3 |
| *m*-Xylene | 139.1 | –47.9 |
| *o*-Xylene | 144.4 | –15.2 |

The three xylenes are then oxidized using $V_2O_5$ as catalyst: *o*-xylene to phthalic anhydride, a monomer for unsaturated polyesters (UP) and alkyd resins, *m*-xylene to isophthalic acid for unsaturated polyesters, and *p*-xylene to terephthalic acid for aromatic polyesters (PET, PPT, PBT; see Section 7.7.4 for the monomer synthesis).

Heavy oils of coal (p. 44) and the higher boiling fractions of reformate or pyrolysis gasoline (p. 50) contain 1,2,4-trimethylbenzene (pseudocumene) and 1,2,4,5-tetramethylbenzene (durol). On oxidation, they deliver trimellitic acid anhydride and pyromellitic acid anhydride, respectivly. Both are monomers for polyimides.

Naphthalene is obtained from the heavy distillates of coal (p. 44 ff.) and from higher–boiling fractions of crude oil. Its oxidation delivers phthalic anhydride.

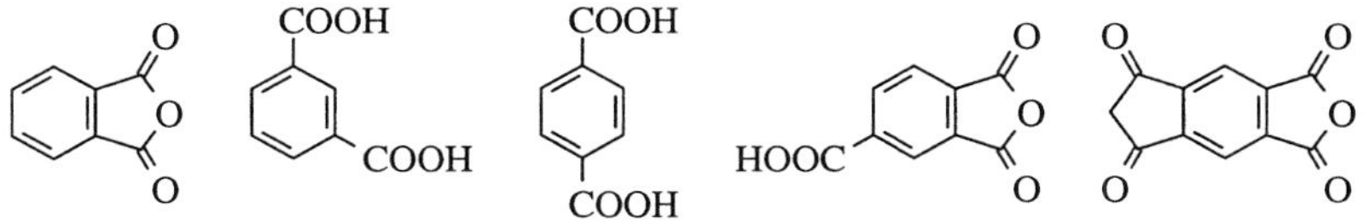

phthalic anhydride | isophthalic acid | terephthalic acid | trimellitic acid anhydride | pyromellitic acid anhydride

## 4.8 Cycloaliphatics

About 85 % of the ca. $5 \cdot 10^6$ t/a world production of cyclohexane stems from the hydrogenation of benzene (Fig. 4-11, Route IV). The remaining 15 % is obtained from raw gasoline, a mixture of aliphatic hydrocarbons and cycloaliphatics such as cyclohexane and methylcyclopentane. The fractionating distillation of raw gasoline delivers a cyclohexane of only 85 % purity which can be boosted to ca. 98 % if the methylcyclopentane is simultaneously isomerized to cyclohexane.

More than 96 % of the produced cyclohexane is used to synthesize adipic acid, hexamethylenediamine, hexamethylene-1,6-diisocyanate, and ε-caprolactam.

### 4.8.1 Adipic Acid

*Adipic acid* is obtained in two steps. In the first step, cyclohexane is oxidized by air at 125–160°C and 8–15 bar with, e.g., cobalt naphthenate. This free-radical reaction proceeds via cyclohexane peroxide as intermediate and leads to a mixture of cyclohexanol and cyclohexanone, called anol–anone or **KA oil** (*k*etone-*a*lcohol oil). The reaction is stopped at a cyclohexane conversion of only 12 % in order to prevent further oxidative degradation to glutaric acid, succinic acid, etc. Unreacted cyclohexane is distilled off and the acids are washed out.

The resulting anol-anone mixture is oxidized to adipic acid, either at 50–80°C by nitric acid with ammonium metavanadate/copper nitrate as catalyst or at 85°C by air with Cu-Mn acetate in acetic acid as catalyst. Adipic acid, $HOOC(CH_2)_4COOH$, is mainly used as a monomer for polyamide 6.6 (70 % of production in the U.S. and Western Europe).

### 4.8.2 1,6-Hexamethylenediamine

Adipic acid is also one of the sources for the second monomer for polyamide 6.6, i.e., 1,6-hexamethylenediamine. Reaction of adipic acid in the gas phase at 300–350°C with a great excess of ammonia in the presence of a boron phosphate catalyst leads to the ammonium salt. Newer processes work in the melt at 200–300°C or in solution with adiponitrile as solvent. The resulting diammonium adipate, $NH_4OOC(CH_2)_4COONH_4$, is reduced to adipodiamide, and further to adipodinitrile. Hydrogenation then results in 1,6-hexamethylenediamine, the other monomer for polyamides 6.6, 6.10, and 6.12.

Reaction of 1,6-hexamethylenediamine with phosgene leads to *hexamethylene-1,6-diisocyanate* (HDI, formerly HMDI), $OCN(CH_2)_6NCO$, the third important isocyanate for the synthesis of polyurethanes.

### 4.8.3 ε-Caprolactam

ε-Caprolactam is industrially obtained from different processes which are summarized in Fig. 4-14: ca. 50 % of the world production of ca. $3.5 \cdot 10^6$ t/a are produced from oxidation of cyclohexane, ca. 45 % from phenol, and the remaining 5 % from either photonitrozation of cyclohexane or oxidation of toluene.

The common route from phenol involves its hydrogenation to a mixture of cyclohexanol and cyclohexanone (Fig. 4-12, Route III). From this mixture, cyclohexanone is distilled off. The remaining cyclohexanol is catalytically dehydrogenated to cyclohexanone by Zn or Cu catalysts at ca. 430°C. A newer route leads directly to cyclohexanone (Fig. 4-12, Route III).

Cyclohexanone is oximated at 85°C by salts of hydroxylamine, mostly by the sulfate. The liberated sulfuric acid makes the system more acidic which requires the successive addition of $NH_3$ in order to maintain equilibrium at pH = 7. The process thus produces ammonium sulfate.

The synthesis of hydroxylamine sulfate, $NH_2OH \cdot H_2SO_4$ for the oximation also generates ammonium sulfate:

(4-37)
$$NH_4NO_2 + NH_3 + 2\,SO_2 + H_2O \rightarrow HON(SO_3NH_4]_2$$
$$HON(SO_3NH_4]_2 + 2\,H_2O \rightarrow NH_2OH \cdot H_2SO_4 + (NH_4)_2SO_4$$

Ammonium sulfate is also produced by the Beckmann rearrangement of the oxime with oleum or sulfuric acid:

(4-38) Cyclohexanone oxime $\xrightarrow{\text{oleum}}$ caprolactam $\cdot$ (1/2) $H_2SO_4$ $\xrightarrow[-(1/2)\,(NH_4)_2SO_4]{+NH_3}$ caprolactam

The oxime can also be obtained directly from cyclohexane by photonitrozation with nitrosyl chloride, NOCl, which is synthesized from sulfuric acid, nitrogen monoxide, nitrogen dioxide, and hydrochloric acid:

(4-39) $$2\,H_2SO_4 + NO + NO_2 \rightarrow 2\,NOHSO_4 + H_2O$$

$$NOHSO_4 + HCl \rightarrow NOCl + H_2SO_4$$

The Snia Viscosa process avoids the oxime as intermediate. Toluene is oxidized here to benzoic acid (Fig. 4-13, Route Vb) which is subsequently catalytically hydrogenated (Pd/activated carbon, 170°C, 15 bar) to cyclohexane carboxylic acid, $C_6H_{11}COOH$. Reaction of this acid with nitrosyl sulfuric acid, $O{=}N{-}OSO_3H$ in oleum at temperatures up to 80°C leads directly to the oxime, probably via several intermediates.

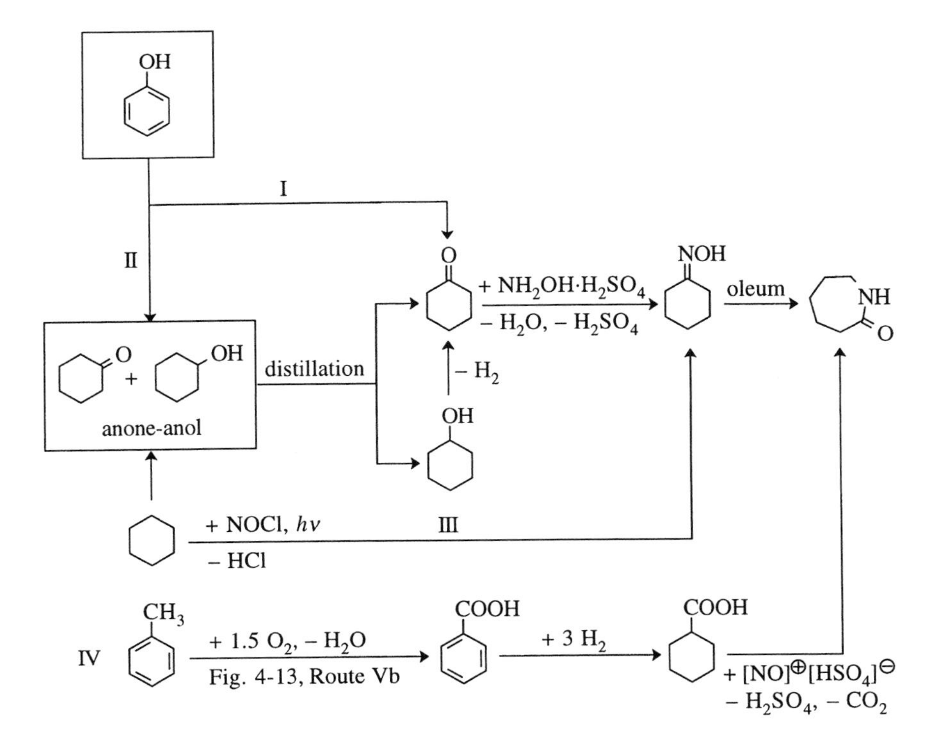


Fig. 4-14 Summary of industrial routes to ε-caprolactam.

# A 4 Appendix: Characterization of Reactions

Chemical reactions are characterized by dimensionless quantities such as conversion, yield, and selectivity as well as more complex quantities such as productivity, throughput, and activity. These quantities are reported in terms of amount-of-substances n in mol; masses m in kg; volumes $V$ in L; amount-of-substance concentrations $C$ ("molar concentrations") in mol/L or mol/kg; mass concentrations $c$ in kg/L, times $t$ in s, min, or h; and pressures $p$ in Pa, bar, or atm (in the US literature also in lbf/sq in (pound-foot per square inch) and wrongly in lb/(sq in)). For dimensionless quantities, percentages such as mol%, wt%, and vol% are also used.

An example is the characterization of the reaction, E + A ⇄ P + B, of an **educt** E (also called **substrate**) with an auxiliary educt A to a desired **product** P and a **byproduct** B. Compounds E, A, and P are also called **reactants**.

**Conversion** $u$ (more exact: molar conversion) refers to the fate of the educt E after a reaction time $t$. If $n_{E,0}$ moles of E were initially present and $n$ E moles remain unreacted after a time $t$, then the conversion of E is given by

(A 4-1) $$u = \frac{n_{E,0} - n_E}{n_{E,0}}$$

Multiplication of the dimensionless quantity $u$ by 100 delivers the so-called mole percent, $u_n$:

(A 4-2) $$u_n = 10^2 \left( \frac{n_{E,0} - n_E}{n_{E,0}} \right)$$

Industry often prefers to report mass-related educt conversions in wt% or vol%, i.e.,

(A 4-3) $$u_w = 10^2(m_{E,0} - m_E)/m_{E,0} \qquad \text{or} \qquad u_V = 10^2(V_{E,0} - V_E)/V_{E,0}$$

In polycondensations, one refers to the conversion of a functional (reactive) group of an educt (i.e., a monomer) instead of the educt itself. This quantity is then called the **extent of reaction,** $p$, of this group (see Volume I).

In reactions with two or more educts, extents of side reactions may differ. Two educts (or more) may not be present initially in stoichiometric amounts for technical, economical, or safety reasons. In these cases, conversions can be calculated for each of the educts.

**Yield** $Y$ refers to the quantity of product with respect to the quantity of initially present educt. In order to calculate yields, **stoichiometric factors** $f$ have to be known for the presumed equilibrium reaction. They are positive for products and negative for educts. For example, in the reaction of hydrogen and nitrogen to ammonia, $3\ H_2 + N_2 \rightleftarrows 2\ NH_3$, stoichiometric factors are $f_{H_2} = -3, f_{N_2} = -1, f_{NH_3} = 2$. The yield with respect to amounts (the "**molar yield**") of the product is defined as

(A 4-4) $$y = \left( \frac{f_E}{f_P} \right) \left( \frac{n_{P,0} - n_P}{n_{E,0}} \right)$$

Again, yields may be expressed not only in amounts-of-substance but also in masses or volumes and their corresponding percentages, i.e., mol%, wt%, and vol%. Since yields can be calculated for several educts and/or products of a reaction, products P and educts E must always be defined, i.e., percent of product P with respect to educt E.

In polycondensations, care has to be taken in the calculation of yields since the educt may be defined as either the monomer M itself or its monomeric unit $U_M$ and the product may be taken either as the polymer P or its repeating unit $U_P$. For a discussion of the variation of different yields of polycondensation reactions as a function of the extent of reaction, see Volume I, p. 440 ff.

**Selectivity** $\varsigma$ is defined as the ratio of yield and conversion:

$$\varsigma = \frac{y}{u} = \left(\frac{f_E}{f_P}\right)\left(\frac{(n_{P,0} - n_P)}{(n_{E,0} - n_E)}\right) \qquad \text{(A 4-5)}$$

Selectivities are indicative of the presence and extent of competing reactions which may be either side reactions (parallel reactions) or secondary reactions to byproducts P' and secondary products P", respectively:

E → P ; E → P' (side reaction)

E → P → P" (secondary reaction)

They differ from reactions, E → P + B, that inevitably deliver coproducts. Examples are decomposition or elimination reactions such as esterifications, RCOOH + HOR' → RCOOR' + $H_2O$, which produce the coproduct $H_2O$. Formation of a coproduct does not affect the selectivity for P whereas the formations of P' and P" do. A "byproduct" in the meaning of organic chemistry may be a coproduct or a competing product.

A reaction E + E' → P + P' + P", involving several educts E and E' and generating several products P, P', and P" has therefore two selectivities, $\varsigma$ and $\varsigma'$, for each of the three products. The selectivity of a reaction with respect to a product P, P', or P" may be expressed in fractions or percentages of the respective educts E or E', for example, in mol%, wt%, or vol%.

Industrial reactions are often characterized by the conversion of educts per time, volume, or catalyst:

The **throughput** is usually expressed in mass $m_E$ of the converted educt per time $t$ and volume $V$ of the reaction volume:

$$\text{throughput} = \frac{m_E}{t\,V} \qquad \text{(A 4-6)}$$

**Catalyst yield** indicates the mass $m_P$ of the product P that has been obtained by the mass $m_{cat}$ of the catalyst which may be the mass of the total catalyst complex or the mass of the effective catalyzing metal atom, for example, the zirconium atom in the catalyst $Cp_2ZrCl_2$ where Cp is the cyclopentadiene unit:

$$\text{catalyst yield} = \frac{m_P}{m_{cat}} \qquad \text{(A 4-7)}$$

Since catalyst yields depend on the reaction time $t$, a more useful quantity is the **catalyst productivity** which refers to either the amount-of-substance, $n_{cat}$, of the catalyst (for known chemical structures; i.e., homogeneous catalyses), Eq.(A 4-8), or to its mass $m_{cat}$ (for unknown chemical structures, i.e., heterogeneous catalyses), Eq.(A 4-9):

(A 4-8) catalyst productivity = $m_P / (n_{cat}\, t)$ homogeneous catalysis

(A 4-9) catalyst productivity = $m_P / (m_{cat}\, t)$ heterogeneous catalysis

**Catalyst activity** relates the mass $m_P$ of the generated product to the amount-of-substance $n_{cat}$ of the catalyst (or its metal atom), time $t$, and the concentration of the educt which may be its molar concentration [E], its mass concentration $c_E$, or its partial pressure $p_E$:

(A 4-10) $$\text{catalyst activity} = \frac{m_P}{n_{cat}\, t\,[E]} \quad \text{or} \quad \frac{m_P}{n_{cat}\, t\, c_E} \quad \text{or} \quad \frac{m_P}{n_{cat}\, t\, p_E}$$

# Literature to Chapter 4


ENCYCLOPEDIAS
-, Ullmann's Encyclopedia of Industrial Chemistry, Wiley-VCH, Weinheim, 7th ed. 2005
Kirk-Othmer, Encyclopedia of Chemical Technology, Wiley, Hoboken (NJ), 5th ed. 2004 ff.
–, Industrial Organic Chemicals. An Ullmann's Encyclopedia, Wiley-VCH, Weinheim 1999, 8 vols.

MONOGRAPHS
F.A.Lowenheim, M.K.Moran, "Faith, Keyes, and Clark's Industrial Chemicals", Wiley-Interscience, New York, 4th ed. 1975
P.Wiseman, An Introduction to Industrial Organic Chemistry, Appli.Sci.Publ., London 1976
A.L.Waddams, Chemicals from Petroleum, John Murray, London 1978
H.A.Wittcoff, B.G.Reuben, Industrial Organic Chemicals in Perspective, Wiley-Interscience, New York 1980; Part I: Raw Materials and Manufacture; Part II: Technology, Formulation, and Use
P.H.Spitz, Petrochemicals. The Rise of an Industry, Wiley, New York 1988
H.Ulrich, Raw Materials for Industrial Polymers, Hanser, Munich 1988
H.Ulrich, Introduction to Industrial Polymers, Hanser, Munich, 2nd ed. 1993
M.M.Green, H.A.Wittcoff, Organic Chemistry Principles and Industrial Practice, Wiley-VCH, Weinheim 2002
K.Weissermel, H.-J.Arpe, Industrial Organic Chemistry, Wiley-VCH, Weinheim, 4th ed. 2003
H.A.Wittcoff, B.G.Reuben, J.S.Plotkin, Industrial Organic Chemicals, Wiley, Hoboken (NJ), 2nd ed. 2004
F. Diederich, P.I.Stang, R.R.Tykwinski, Eds., Acetylene Chemistry, Wiley-VCH, Weinheim 2004


# Reference to Chapter 4


[1] K.Weissermel, H.-J.Arpe, Industrielle Organische Chemie, Wiley-VCH, Weinheim, 5th Edition (1998), (a) Table 3-1, expanded. (b) Tables 12-2 and 12-4

# 5 Polymerization Processes

## 5.1 Types of Polymerization

### 5.1.1 Monomers for Polymers

A few natural polymers are used directly in industry, for example, natural rubber and starch. Others polymer are converted chemically to semi-synthetic polymers, for example, cellulose acetate. Most industrial polymers are produced synthetically from monomers, however.

**Monomers** are low-molecular weight chemical compounds consisting of **monomer molecules** that are converted by **polymerization** to **polymers** containing **polymer molecules**. Molecules become potential monomers if they possess explicit or implicit chemical functionalities of 2 or more *at the desired reaction conditions.* This demand is fulfilled by three groups of monomers:

I. Open-chain monomers with functional groups such as COOH, OH, $NH_2$, etc., at the end of the chains. These groups and the molecules containing them are often symbolized by capital letters A, B, C, etc. An **AA** or **$A_2$ monomer** thus contains two identical functional groups; an example is adipic acid with 2 carboxylic endgroups. An **AB monomer**, on the other hand, has two different functional groups A and B; an example is lactic acid. An **$A_3$ monomer** has three identical groups, for example, glycerol, and an **$AB_2$ monomer** one group of type A and two groups of type B, for example, aspartic acid. Note that "identical" in the formal sense does not mean "equal chemical reactivity": the primary and secondary OH groups of glycerol have inherently different reactivities and so have the two COOH groups of aspartic acid.

| $H_2NCH_2COOH$ | $H_2N(CH_2)_6NH_2$ | $H_2NCH_2CH_2N(CH_2CH_2NH_2)CH_2CH_2NH_2$ | $HOCH_2CH(OH)CH_2OH$ | $H_2NCH(CH_2COOH)COOH$ |
|---|---|---|---|---|
| glycine | hexamethylene diamine | tris(2-aminoethyl)amine | glycerol | aspartic acid |
| AB | AA | $A_3$ | $A_3$ | $AB_2$ |

**AB polymerizations** are intermolecular reactions between functional groups of AB molecules. They are relatively rare because functional groups of many potential candidates often react uncontrolled at ambient temperatures.

Much more common are **AA+BB polymerizations**, i.e., intermolecular reactions of A groups of AA molecules with B groups of BB molecules. In both AB and AA+BB polymerizations, functional groups A and B are usually selected in such a way that like groups do not react with their own type at the chosen reaction conditions: neither shall A react with A nor B with B, only A with B.

AB and AA/BB polymerizations lead to **linear molecules** which are defined as molecules without *newly* formed branches. **$AB_2$ polymerizations,** on the other hand, do give branched molecules, called **hyperbranched** (see Volume I), whereas **$A_2B_2$**, **$A_2 + B_3$**, **$A_3 + B_3$**, etc. polymerizations result in branched molecules at low monomer conversions but, depending on the molar ratio of $A_2/B_2$, crosslinked ones at high degrees of reaction.

All these polymerizations are often externally catalyzed by added catalysts that can be recovered after isolation of the desired polymer. Many of these reactions, but not all, are also equilibrium reactions.

II. Cyclic monomer molecules with internal functional groups such as heteroatoms or groups of heteroatoms and double bonds. Examples are

| tetrahydrofuran | 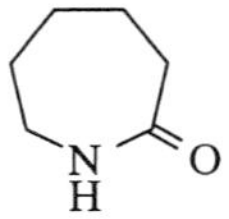 ε-caprolactam | 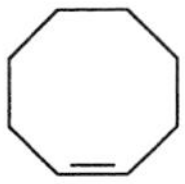 cyclooctene | 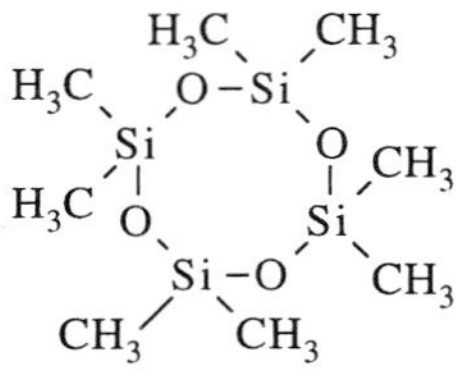 octamethyltetrasilicone |
|---|---|---|---|

Polymerizations of cyclic monomers are usually initiated, not catalyzed. They lead to linear polymer molecules in which initiator residues R* become part of polymer molecules, for example, in $R^* + C_4H_8O \rightarrow R(CH_2)_4O^*$. Initiators or their fragments are not recovered after the polymerization. However, there are some polymerizations of cyclic molecules that proceed by ring extension to larger rings.

III. Monomer molecules with multiple bonds, usually double bonds between two carbon atoms or between a carbon atom and a heteroatom such as oxygen. Examples are

| $H_2C{=}CH_2$ | $H_2C{=}CH(CH_3)$ | $H_2C{=}O$ | $H_2C{=}C(CH_3){-}CH{=}CH_2$ |
|---|---|---|---|
| Ethene | Propene | Formaldehyde | Isoprene |

Polymerization of groups with triple bonds such as –C≡C– or –C≡N are less common.

Polymerizations of unsaturated monomer molecules are usually initiated. They proceed by opening of the multiple bonds. An example is the opening of the double bond of ethene to two single bonds using free radicals, $R^{\bullet} + CH_2{=}CH_2 \rightarrow R{-}CH_2{-}CH_2^{\bullet}$.

## 5.1.2 Classification of Polymerizations

Polymers are synthesized either from monomers (see above) or so-called prepolymers by polymerization or from other polymers by so-called polymer transformations (polymer-analogous reactions). **Polymerization** is the umbrella term for all reactions leading from monomer molecules to polymer molecules. They are subdivided according to the

- source of polymers (biological, synthetic),
- chemical structure of monomers (vinyl, diene, ring-opening, etc.),
- chemical structure of polymers (linear, branched, crosslinked, ring-forming, etc.),
- composition of monomeric units relative to the composition of monomer molecules (polyaddition, polycondensation),
- type of connection of monomer molecules (stepwise or kinetic chain),
- type of initiation reaction (thermal, catalytic, photochemical, etc.),
- type of growing species (macroradical, macroanion, etc.),
- polymerization mechanism (equilibrium, irreversible, living, stepwise, etc.),
- type of incorporation of the monomeric unit (addition, insertion),
- reaction medium (melt, solution, emulsion, gas, etc.),
- physical state of the reaction system (homogeneous, heterogeneous), etc.

Polymerizations are most commonly subdivided according to: (a) the type of growth of polymer molecules (stepwise versus kinetic chain); and (b) the chemical composition of monomeric units relative to monomer molecules (with or without leaving molecules). This leads to a total of four types of polymerization (Table 5-1): 2 types of step-growth polymerizations (polycondensation and polyaddition) and 2 types of chain-growth polymerizations (polyelimination and chain polymerization). Alternatively, there are 2 types of polymerizations with formation of leaving molecules (polycondensation and polyelimination) and 2 types of polymerizations without leaving molecules (polyaddition and chain polymerization).

In chain-growth polymerizations, growing polymer chains $P_i$ (usually as $RM_iM^*$ where R is an initiator fragment, M a monomeric unit, and M* the activated terminal unit (* = radical, anion, cation)) add monomer molecules M, whereas in step-growth polymerizations, polymer molecules react with each other. In either case, leaving molecules L (usually small) may be released or not.

Historically, polymerizations were subdivided in 1929 by the American W.H.Carothers into C polymerizations (now called polycondensations) and A polymerizations (now called chain polymerizations). An example of a C polymerization is shown in Eq.(5-4) and an example of an A polymerization in Eq.(5-1). In 1937, the German Otto Bayer discovered a new type of polymerization that combined features of a C polymerization (reaction between polymer molecules) with that of an A polymerization (no release of leaving molecules (for an example, see Eq.(5-3)). He called this type of polymerization in German "Polyaddition" in order to distinguish it from the German "Polymerisation" for an A polymerization. But A polymerizations are commonly known in English as "addition polymerizations" and the confusion was complete. IUPAC suggested therefore the terms polycondensation, polyaddition, and chain polymerization for these three types of polymerization (see Table 5-1).

Table 5-1 Types and names of polymerizations. P = Polymer molecule, M = monomer molecule, L = leaving molecule, e.g., water in –COOH + HO– → –COO– + $H_2O$. IUPAC: new nomenclature proposal. "Chain" refers to kinetic chains, not to chain-like molecules. See also Volume I.

| | Step-growth polymerization | Chain-growth polymerization |
|---|---|---|
| *With formation of leaving molecules* | | |
| | $P_i + P_j \longrightarrow P_{i+j} + L$ | $P_i + M \longrightarrow P_{i+1} + L$ |
| Historic name | C polymerization | - |
| Conventional names | condensation polymerization<br>step polymerization | - |
| New IUPAC name | polycondensation | condensative chain polymerization |
| This book (see text) | polycondensation | polyelimination |
| *Without formation of leaving molecules* | | |
| | $P_i + P_j \longrightarrow P_{i+j}$ | $P_i + M \longrightarrow P_{i+1}$ |
| Historic name | - | A polymerization |
| Conventional name | adduct formation (rare) | addition polymerization |
| New IUPAC name | polyaddition | chain polymerization |
| This book | polyaddition | chain polymerization |

However, there is a fourth type of polymerization that combines features of chain polymerizations (kinetic chain reaction) with that of polycondensations (release of leaving molecules). This type of polymerization has no common name. IUPAC suggested therefore the term "condensative chain polymerization" in order to distinguish it from "chain polymerization" without release of leaving molecules. This is, of course, a linguistic blunder since *all* subgroups of a general term must carry adjectives: "chain polymerization" as a general type has two subtypes (with or without leaving molecules), hence both have to be distinguished, for example, as "additive chain polymerization" and "condensative chain polymerization". Since this would lead to further confusion, the author will use "polyelimination" instead of "condensative chain polymerization."

Chain polymerizations and polyeliminations are chain reactions in which monomer molecules are reacted with growing polymer molecules. In **chain polymerizations,** monomer molecules are either added to the active center at the end of a growing polymer chain (there is no term for this subgroup; systematically, it could be called "end-adding chain polymerization") or inserted between the active center and the polymer chain, commonly known as **coordination polymerization** (because the incoming monomer molecule coordinates with the active center before it is inserted).

An example of the end-adding of monomer molecules is the free-radical polymerization of ethene by initiator radicals $R^{\bullet}$ which leads to growing poly(ethylene) chains with the initiator fragment R– at one end of the polymer chain and the active radical $-CH_2CH_2^{\bullet}$ at the other:

$$(5\text{-}1) \qquad R(CH_2CH_2)_{n-1}CH_2\dot{C}H_2 + CH_2{=}CH_2 \longrightarrow R(CH_2CH_2)_{n-1}CH_2CH_2CH_2\dot{C}H_2$$

Similar chain reactions may be initiated by anions (for example, the anionic chain polymerization of styrene, $CH_2{=}CH(C_6H_5)$ or by cations (for example, the cationic chain polymerization of isobutene, $CH_2{=}C(CH_3)_2$). One speaks of **living polymerizations** if active chain ends remain active until all monomer is exhausted.

In **coordination polymerizations,** the monomer is inserted into the bond between the polymer chain and the "catalyst" (hence also called **insertion polymerizations**), for example, in the polymerization of ethene by Ziegler-type catalysts [cat], consisting of, e.g., $Al(C_2H_5)_3 + TiCl_3$:

$$(5\text{-}2) \qquad R(CH_2CH_2)_{n-1}[cat] + CH_2{=}CH_2 \longrightarrow R(CH_2CH_2)_{n-1}CH_2CH_2[cat]$$

**Polyeliminations** (IUPAC: condensative chain polymerizations) are either end-adding or inserting-chain polymerizations. They are not very common for synthetic polymers; an example is the the polymerization of *N*-carboxy anhydrides (Leuchs anhydrides) of $\alpha$-amino acids (see Section 11.2.1).

Polyeliminations do occur in nature; they are usually called "polycondensations" in biochemistry since they involve leaving molecules. An example is the biological polymerization of glucose to poly(glucose). The true monomer is not glucose G, though, but nucleoside diphosphate glucose, NDP-G, or, more exactly, the lipid of NDP-G. The polymerization reaction consists of additions of NDP-G to the non-reducing ends $\sim G_n$–OR of previously formed poly(glucose) molecules during which glucose residues are inserted in $\sim G_n$–OR bonds with the help of an enzyme EP and NDP is released:

$$(5\text{-}3) \qquad \sim\!G_n\text{–O–R} + \text{G–NDP} \xrightarrow{EP} \sim\!G_{n+1}\text{–O–R} + \text{NDP}$$

Step-growth polymerizations, on the other hand, are polymerizations in which polymer molecules (defined as those with degrees of polymerization of 2 and higher) react not only with monomer molecules but also with other polymer molecules. In **polycondensations** (= condensation polymerizations), leaving molecules are formed similarly to polyeliminations. However, polymer molecules are not produced by successive addition of monomer molecules to growing polymer chains but rather by intermolecular reactions between all types of reactants, i.e., monomer molecules ($X \equiv 1$) and polymer molecules ($X \geq 2$). An example is the polycondensation of 11-aminoundecanoic acid, an AB reaction, with release of water molecules:

$$(5\text{-}4) \qquad n\, H_2N(CH_2)_{10}COOH \longrightarrow H[NH(CH_2)_{10}CO]_nOH + (n-1)\, H_2O$$

**Polyadditions** proceed similarly to polycondensations albeit without formation of leaving molecules. An example is the polyaddition of diisocyanates and diols to polyurethanes:

$$(5\text{-}5) \qquad n\, OCN-Z-NCO + n\, HO-Z'-OH \longrightarrow \left[ CO-NH-Z-NH-CO-O-Z'-O \right]_n$$

These four types of polymerizations differ considerably in the required process conditions. Polycondensations and polyeliminations produce leaving molecules which will influence not only the yield (as in low-molecular weight reactions) but also the achievable molecular weights if these polymerizations lead to equilibria (Section 5.1.4). Cross-linking polycondensations and polyeliminations may also trap leaving molecules and oligomers in the resulting network.

The four types of polymerizations also differ in the time dependence of monomer conversions and resulting molecular weights. Reaction rates and thus monomer conversions depend on the types of elementary reactions present; they are thus specific for the particular type of polymerization. The dependence of molecular weight on conversion can often be generalized, however (Fig. 5-1).

In **living polymerizations**, all initiator molecules start the polymerization at the same time and all resulting active polymer chains remain active even after all monomer molecules are consumed. The ratio of initial molar concentration of monomer molecules, $[M]_0$, to initial concentration of active initiator species, $[i]_0 = f_i[I]_0$, from initiator molecules I with functionality $f_i$ determines the number-average degree of polymerization, $\overline{X}_n$, that is obtained at the (fractional) extent of reaction, $p$, of functional groups:

$$(5\text{-}6) \qquad \overline{X}_n = p[M]_0/(f_I[I]_0)$$

The number-average degree of polymerization thus increases linearly with the extent of reaction, $p \equiv (n_{A,0} - n_A)/n_{A,0}$, where $n_{A,0}$ and $n_A$ are the amounts of functional groups A at times $t = 0$ and $t > 0$ (Fig. 5-1). Such living polymerizations can be obtained fairly easily with several anionic chain initiators and some coordination catalysts but only with difficulty or not at all with cations. In a strict sense, they are not possible at all for free-radical polymerizations but there are experimental techniques that lead to linear time dependencies of $\overline{X}_n$ albeit at present not to industrially feasible ones.

Detrimental to many industrial applications are the required high ratios $[M]_0/[I]_0$ for high degrees of polymerization. This, in turn, translates to small initiator concentrations

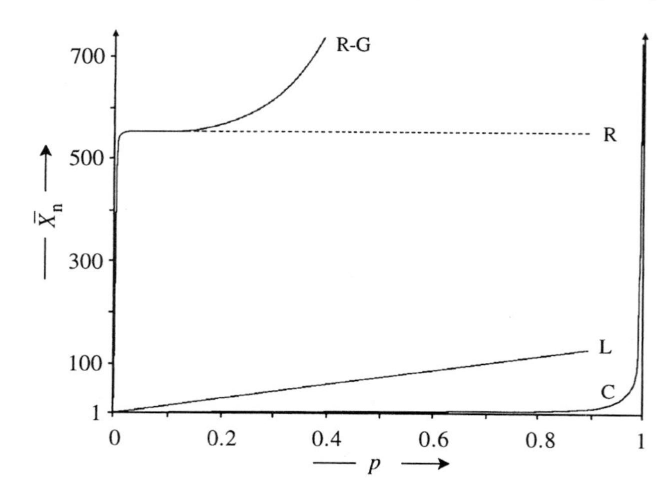


Fig. 5-1 Number-average degree of polymerization, $\overline{X}_n$, as a function of the extent of reaction, $p$, for
C: bifunctional equilibrium AA + BB polycondensations or polyadditions (system-independent).
L: living polymerization. Example: anionic polymerization of styrene by the monofunctional initiator butyl lithium at an initial molar ratio of $[M]_0/[I]_0 = 141:1$.
R: free-radical polymerization. Example: polymerization of styrene in bulk with azobisisobutyronitrile AIBN as initiator; same initial molar ratio of 141:1 as "L.". Each AIBN molecule generates two radicals from which only a fraction initiates polymer chains (see Volume I).
R-G: free-radical polymerization with gel effect (schematic).

and to low polymerization rates. Of equal if not greater importance is the fact that small initiator concentrations can be easily overwhelmed by monomer impurities, which makes the control of polymerization reactions difficult or even impossible. Polymerization-grade monomers are usually "only" 99 % to 99.99 % pure (see Chapter 4); higher purities require costly additional work-up procedures. Living polymerizations are therefore used industrially only for the synthesis of polymers with relatively low degrees of polymerizations, for example, blocks of block polymers.

Another characteristic of living polymerizations is the high molecular uniformity of the produced polymer molecules, i.e., narrow molecular weight distributions. This again is advantageous for the synthesis of block copolymers since it leads to well-defined morphologies such as spherical domains of small A blocks in a continuous matrix of large B segments (see Volume IV). It is not so advantageous for many melt-processed plastics where one often prefers polymers with relatively broad molecular weight distributions: low-molecular weight tails lead to a reasonable fluidity in the molten state whereas high-molecular weight tails provide the plastics with strength due to the entanglement of long chains.

In conventional **free-radical polymerizations**, polymer chains are not started simultaneously as in living polymerizations but successively because of the successive dissociation of initiator molecules to two free radicals. Growing macroradicals are furthermore terminated by mutual deactivation of two radicals (see Volume I, for example: (re)combination ($\sim M^{\bullet} + {}^{\bullet}M\sim \rightarrow \sim M{-}M\sim$) or disproportionation ($\sim M^{\bullet} + {}^{\bullet}M\sim \rightarrow \sim M + M\sim$)) or converted to other radicals by radical-transfer reactions to other molecules HA such as $\sim M^{\bullet} + HA \rightarrow \sim MH + {}^{\bullet}A$. The new radicals $A^{\bullet}$ can also start new chains as in vinyl polymerizations or act as radical sinks as in allyl polymerizations.

Free-radical polymerizations consist therefore of many simultaneous reactions. The interplay of free-radical generation by initiator dissociation, radical conversion by chain transfer, and radical disappearance by termination reactions leads to a steady state that is often established in seconds (Volume I, Section 10.3.3). As a result, number-average degrees of polymerization are already high at small extents of reaction and stay constant for a considerable range of monomer conversion (Fig. 5-1, R). At higher monomer conversions, viscosities increase markedly. Since the diffusion of macroradicals is impeded, this leads to a diffusion control of termination reactions in neat polymerizations whereas the diffusion of small monomer molecules to macroradicals is not affected. The resulting gel-effect causes degrees of polymerization to increase strongly (Fig. 5-1, R-G) and may even lead to runaway polymerizations. Free-radical polymerizations are therefore almost never run to very high monomer conversions but are often stopped by addition of terminating agents at monomer conversions as low as 60 %.

The situation is quite different for **equilibrium polycondensations** and **polyadditions** which require exact stoichiometries of reacting groups for high-molecular weight linear polymers. Such a stoichiometry is automatically given for AB reactions. During polymerization, dimer molecules AbaB (II) are formed with reacted groups "b" and "a" and unreacted endgroups A and B. The dimers then react with either dimer molecules AbaB (II) to tetramer molecules $A(ba)_3B$ (IV) or with monomer molecules AB (I) to trimer molecules $A(ba)_2B$ (III). Simultaneously new dimer molecules II are formed from monomers I + I. Next is the simultaneous formation of octamers (from IV + IV), heptamers (from IV and III), hexamers (IV + II, III + III), pentamers (IV + I, III + II), tetramers (III + I, II + II), trimers (II + I, IV – I), and dimers ( I + I, IV – II, III – I), etc.

All functional endgroups have the same probability to react if their reactivity is independent of the size of the molecule (*Principle of Equal Chemical Reactivity*; Volume I). However, a reaction of endgroups at larger molecules increases the average degree of polymerization much more than the reaction of two small molecules. Since larger molecules are formed only at an advanced stage of polymerization, degrees of polymerization are low at small extents of reaction but increase avalanche-like at large ones. The same is true for stoichiometric AA + BB equilibrium reactions (Fig. 5-1, C).

For this reason, truly high-molecular weight linear polymers are much more difficult to synthesize by polycondensation and polyaddition than by chain polymerizations. For example, number-average degrees of polymerizations of industrial linear condensation polymers such as polyamide 6.6 rarely exceed 100–200 whereas those of industrial polymers by chain polymerization such as poly(ethylene) are at least 2000–3000.

Monomers for polycondensations and polyadditions usually carry very polar groups whereas monomers for chain polymerizations are mostly not very polar. Since the former have relatively low molecular weights and the latter relatively high ones, intermolecular group interactions *per molecule* are about the same for both types of polymers. For example, melting temperatures of oligomers of the hexamethylene adipamide type (nylon 6.6), $H[NH(CH_2)_6NH{-}CO(CH_2)_4CO]_nOH$, reach their high–polymer melting temperature of ca. 255°C already at $n$ = 10 (i.e., 160 chain atoms) whereas alkanes $H[CH_2]_nH$ with $n$ = 390 ($T_M$ = 132°C) are still a ways off from high-molecular weight $T_M$ = 144°C of unbranched poly(methylene)s. As a result, many mechanical properties of polymers by polycondensation and polyaddition become independent of molar masses at far lower molecular weights than those of polymers by chain polymerization.

### 5.1.3 Thermodynamic Requirements

Polymerizations proceed only if the Gibbs polymerization energies of their propagation reactions are negative (Volume I). Since the Gibbs polymerization energy

(5-7) $\Delta G_p^o = \Delta H_p^o - T\,\Delta S_p^o = -RT \ln K$

depends on both the polymerization enthalpy, $\Delta H_p^o$, and the polymerization entropy, $\Delta S_p^o$, four different types of thermodynamically controlled polymerizations are possible (see Volume I, p. 199 ff.):

| | | | |
|---|---|---|---|
| I: | $\Delta H_p^o$ negative or zero; | $\Delta S_p^o$ negative: | polymerization below ceiling temperature |
| II: | $\Delta H_p^o$ negative or zero; | $\Delta S_p^o$ positive: | polymerization at all temperatures |
| III: | $\Delta H_p^o$ positive; | $\Delta S_p^o$ negative: | no polymerization at all temperatures |
| IV: | $\Delta H_p^o$ positive; | $\Delta S_p^o$ positive: | polymerization above floor temperature |

Case I is the most common. Here, the entropy term $-T\Delta S_p^o$ becomes more positive with increasing temperature until the enthalpy and entropy terms compensate each other at a **ceiling temperature** $T_c$ at which the Gibbs polymerization energy becomes zero. No *high-molecular weight* polymers are formed at temperatures $T > T_c$. This does not rule out the formation of oligomers at $T_c$ since equilibrium polymerizations are consecutive equilibria between molecules of different degrees of polymerization. Case I is found for nearly all polycondensations and polyadditions but is not so common for chain polymerizations, an example of which is the anionic polymerization of $\alpha$-methyl styrene.

Conversely, a **floor temperature** $T_f$ exists for Case IV polymerizations. No *high-molecular* weight polymers are formed below $T_f$. An example of this rare case is the polymerization of *cyclo*-octasulfur, $S_8$.

Case II is possible but cannot be tested since all polymers decompose if exposed to high temperatures for sufficiently long times. Case III can also not be tested since a negative can never be proven.

Polymerization enthalpies and polymerization entropies are affected by the physical states of monomers and polymers, the physical interactions between components in a reactor, and the pressure. As noted before, most polymerizations have negative polymerization enthalpies (Case I) and are thus exothermic. For adiabatic conditions, the resulting heat of polymerization can therefore lead to enormous increases in temperature, for example, to 1800 K if gaseous ethene is completely polymerized to crystalline poly(ethylene)! Industrial reactors must therefore allow rapid heat transfer since otherwise inhomogeneous charges would result or even explosions of reactors.

### 5.1.4 Elementary Reactions

The four basic types of polymerizations (chain polymerization, polycondensation, polyaddition, polyelimination) differ not only in the type of propagation reaction (see Volume I for details) but also in the number and types of other elementary reactions (Table 5-2). Some of these other reactions are essential for the type of polymerization, for example, initiation of free-radical polymerizations. Some of the other elementary reactions are often undesirable side reactions, for example, most chain transfer reactions.

Table 5-2 Schematic representation of elementary reactions of polycondensations (PC), polyadditions (PA), and chain polymerizations. Living chain polymerizations may either be of the end-adding (EA; here anionic) or the insertion type (I). Examples of non-living polymerizations include those of the insertion type (I) and the free-radical polymerizations of the vinyl type (RV) and the acrylic type (RA). * Activated species (radical •, anion ⊖, cation ⊕, etc.), L = leaving molecule, ( ) not always present.

| Elementary reaction Name | Process | PC | PA | Chain polymerizations living A | I | non-living I | RV | RA |
|---|---|---|---|---|---|---|---|---|
| Isomerization | $M_A \rightarrow M_B$ | – | – | – | (+) | (+) | – | – |
| Catalysis | – | + | + | – | – | – | – | – |
| Initiator dissociation | $I_2 \rightarrow 2\, I^\bullet$ | – | – | – | – | – | + | + |
| Initiator dissociation | $CA \rightarrow C^\oplus + A^\ominus$ | – | – | + | – | – | – | – |
| Initiation, $R^* = I^\bullet, A^\ominus$ ... | $R^* + M \rightarrow RM_1{}^*$ | – | – | + | + | + | + | + |
| Initiation reaction | $RM_1{}^* + M \rightarrow RM_2{}^*$ | – | – | + | + | + | + | + |
| Propagation by | | | | | | | | |
| a. Monomer addition | $RM_n{}^* + M \rightarrow RM_{n+1}{}^*$ | – | – | + | + | + | + | + |
| b. Polymer addition | $M_n + M_m \rightarrow M_{n+m}$ | – | + | – | – | – | – | – |
| c. Polymer addition | $M_n + M_m \rightarrow M_{n+m} + L$ | + | – | – | – | – | – | – |
| Transfer to | | | | | | | | |
| a. Monomer | $RM_n{}^* + M \rightarrow RM_n + M^*$ | – | – | (+) | (+) | (+) | + | + |
| b. Other molecules | $RM_n{}^* + A \rightarrow RM_n + A^*$ | – | – | (+) | (+) | (+) | + | + |
| Termination by | | | | | | | | |
| a. Combination | $RM_n{}^* + RM_m{}^* \rightarrow M_{n+m}$ | – | – | – | – | – | + | + |
| b. Disproportionation | $RM_n{}^* + RM_m{}^* \rightarrow RM_n + RM_m$ | – | – | – | – | (+) | + | + |
| c. Initiator/catalyst | $RM_n{}^* + R^* \rightarrow RM_nR$ | (+) | (+) | – | – | – | + | (+) |
| d. Monomer | $RM_n{}^* + M \rightarrow RM_n + M$ | – | – | – | – | – | (+) | + |
| Exchange reactions | $M_n + M_m \rightarrow M_p + M_q$ | + | (+) | – | – | – | – | – |
| Inhibition | – | – | – | (+) | + | (+) | + | + |

In **polycondensations** and **polyadditions**, each propagation step is catalyzed, usually by an added catalyst but sometimes also by one of the monomers themselves. The catalyst molecule and the reacting functional group form an intermediary species that reacts with another functional group. After the reaction, the catalyst molecule is split off; it is not consumed. However, at high monomer conversions (and correspondingly low concentrations of remaining functional A groups), catalyst groups compete with B groups for reaction with A groups (for example, $H_2SO_4$ as catalyst with ~COOH in esterifications). Monofunctional catalyst molecules will then seal the chain ends of the growing polymer molecules and the degree of polymerization of the polymer becomes smaller as predicted by the equations for stoichiometric reactions. Industrially, addition of a monofunctional component is used to regulate the degree of polymerization (Volume I).

Free-radical, anionic, and cationic **chain polymerizations** need initiators that provide the initiating species, for example, $C_6H_5COO^\bullet$ radicals from the thermal homolysis of dibenzoyl peroxide, $C_6H_5COO–OOCC_6H_5$, or $C_4H_9{}^\ominus$ anions from the dissociation of butyl lithium, $C_4H_9Li$. In the initiation reaction, the initiating species R*adds a monomer molecule M and becomes a monomer radical (if $R^* = I^\bullet$), monomer anion (if $R^* = I^\ominus$), etc. In subsequent propagation reactions, these species then add more monomer molecules but the initiating species always remains connected to one end of the propagating polymer molecules.

Table 5-3 Rate constants of elementary reactions of polycondensations (PC), polyadditions (PA), and anionic (A), cationic (C), free-radical (R), and insertion (I) chain polymerizations. Subgroups a, c, d, e, and f correspond to those of Table 5-2.

| Reaction | Unit | PC, PA | A | I | C | R |
|---|---|---|---|---|---|---|
| Dissociation of initiator | $s^{-1}$ | | | | | $10^{-1}$–$10^{7}$ |
| Initiation | L mol $s^{-1}$ | | | | 1-10 | |
| Propagation by | | | | | | |
| a. Monomer addition | L mol $s^{-1}$ | $10^{-6}$–$10^{-2}$ | $10^{-2}$–$10^{6}$ | $10^{-3}$–$10^{5}$ | $10^{-5}$–$10^{9}$ | 10 –$10^{5}$ |
| c. Polymer addition | L mol $s^{-1}$ | $10^{-6}$–$10^{-2}$ | | | | |
| Transfer to | | | | | | |
| a. Monomer | L mol $s^{-1}$ | | | $10^{-2}$ | | $10^{-5}$–$10^{-3}$ |
| c. Solvent | L mol $s^{-1}$ | | | | | $10^{-6}$–$10^{2}$ |
| d. Initiator | L mol $s^{-1}$ | | | $10^{-6}$ –1 | | $10^{-5}$–10 |
| e. Polymer | L mol $s^{-1}$ | | | | | $10^{-5}$–$10^{-3}$ |
| f. Inhibitor | L mol $s^{-1}$ | | | | | 10 –$10^{3}$ |
| Termination of radicals | | | | | | |
| a. Combination, disproportionation | L mol $s^{-1}$ | | | | | $10^{5}$ –$10^{9}$ |

In living polymerizations, chain ends $IM_n^*$ remain active until all monomer molecules are exhausted; examples are certain anionic polymerizations with $IM_n^{\ominus}$. In common free-radical polymerizations, propagating chain ends $IM_n^{\bullet}$ may react with each other, with initiator radicals, or with other molecules to dead chains (see Volume I).

Certain ionic chain polymerizations lead to isomerizations of monomeric units during the propagation steps, whereas certain coordination catalysts isomerize monomer molecules before they are incorporated into the growing chains. Depending on the system, either all or only some monomeric units do not resemble monomer molecules anymore. An example is the complete polymerization of $CH_2{=}CHCONH_2$ to ~$CH_2CH_2CONH$~ by strong anions instead of ~$CH_2CH(CONH_2)$~ as in free-radical polymerizations.

Chain polymerizations, especially free-radical ones, may also lead to transfer reactions by monomer, polymer, initiator, and solvent molecules as well as specially added inhibitor molecules or transfer agents (Volume I). Transfer reactions by so-called **stabilizers** are used industrially to inhibit the polymerization of stored monomers and increase the shelf-life of monomers. These inhibitors are usually not removed before polymerization; their action is rather overruled by increased concentrations of initiators. Transfer reactions by so-called **regulators** serve to control molecular weights and/or polymerization rates.

Rate constants of elementary polymerization reactions vary widely (Table 5-3), depending on the type of polymerization and the type of monomers, initiators, and transfer agents on one hand and processing conditions such as temperature and pressure on the other. The problem is compounded by the fact that industrial polymerizations are often not discontinuous as in laboratory procedures but continuous. Many industrial polymerizations are furthermore neither isothermal nor isochronic. An example is the so-called hydrolytic polymerization of lactams (Fig. 5-2), a polymerization in the presence of water which assures good heat transfer and a homogeneous melt. The modeling of industrial polymerizations can thus become very complicated.

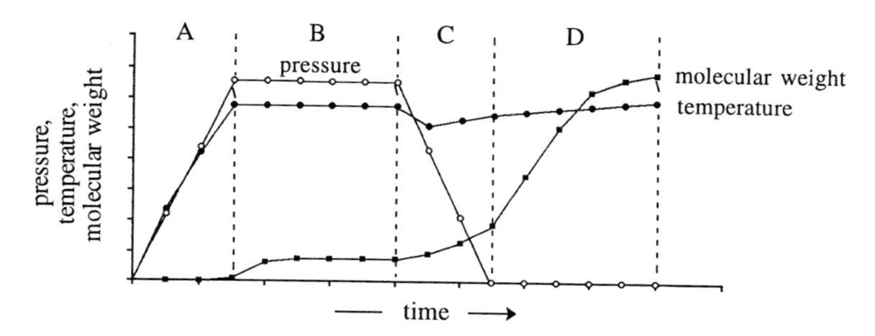


Fig. 5-2 Molecular weight build-up during hydrolytic laurolactam polymerization [1]. During the heating-up period A, some lactam molecules are converted to the corresponding amino acid molecules which initiate the polymerization in period B. During this period of constant temperature and pressure, a homogeneous precondensate is formed which minimizes the loss of monomer. Molecular weights do not grow appreciably, though. Release of water vapor reduces the pressure (C) and slightly the temperature. Stable conditions are obtained after complete release of pressure (D). The growing molecular weight increases the melt viscosity considerably which causes problems with stirring and may lead to so-called hot spots. At the end of period D, the reactor is discharged and the melt extruded.

### 5.1.5 Polycondensations and Polyadditions

#### Linear Equilibrium Reactions

Polycondensations and polyadditions are generally subdivided into bifunctional and multifunctional reactions. *Bifunctional reactions* require bifunctional monomer and polymer molecules; they lead to linear molecules. High molecular weights result if the stoichiometry is exact, functional group conversion is high, and side reactions are absent.

AB reactions automatically fulfill the first requirement since both types of functional groups are united in monomer molecules (cf. Eq.(5-4)). The functional groups should not be too reactive, though, since uncontrolled polymerizations may result on storage. It is for this reason that only a few AB reactions are used industrially. Examples are the polycondensations of 11-aminoundecanoic acid $H_2N(CH_2)_{10}COOH$ with release of $H_2O$ to the polyamide 11, $H[NH(CH_2)_{10}CO]_nOH$; of phenylsulfochloride $C_6H_5SO_2Cl$ with release of HCl to the polysulfone $H[C_6H_4SO_2]_nCl$; and of diisocyanates OCN–R–NCO with release of $CO_2$ to polycarbodiimides $\text{+}R–N=C=N\text{+}_n$.

In AA/BB reactions, the stoichiometry of monomers must be carefully controlled since only low molecular weights are obtained if the amounts of AA and BB molecules are mismatched. If, for example, 1 mole of BB molecules reacts completely with 2 moles of AA molecules, stoichiometry dictates that the *average* composition of the resulting product will be $A[a–b–b–a]_iA$ with $i = 1$. The number-average degree of polymerization with respect to *monomeric units* is 3 whereas the number-average degree of polymerization with respect to *repeating units* is just 1, as one can see from the reaction of dicarboxylic acids and diols, HOOCR'COOH + 2 HOROH → H[ORO–OCR'CO]OROH. The product does *not* consist exclusively of molecules $H[ORO–OCR'CO]_iOROH$ with $i = 1$, however, but of molecules with all possible numbers of $0 \leq i < \infty$. Note that the definition of "equilibrium" means that group conversion can never reach *exactly* 100 %!

Note that the degree of polymerization of products of such polymerizations refers to the number of monomeric units per molecule and not to the number of repeating units. The *average* degree of polymerization furthermore applies to the mixture of reactants ($1 \leq X$) and not to the mixture of polymer molecules ($2 \leq X$). Reactants thus include unreacted monomer molecules.

In stoichiometric polymerizations with incomplete monomer conversion, number-average degrees of polymerization are controlled by the *extent of reaction* of A groups, $p_A \equiv (n_{A,0} - n_A)/n_{A,0}$, where $n_{A,0}$ = initial amount of A groups and $n_A$ = amount of remaining A groups at reaction time $t$. The number-average degree of polymerization of reactants is defined as the ratio of the amount $n_{U,R}$ of monomeric units in reactants and the amount $n_R$ of reactant molecules. In AB reactions, the amount of monomeric units equals the amount of initially present A groups, $n_{U,R} = n_{A,0}$, and the amount of reactant molecules equals the amount of remaining A groups, $n_R = n_A$, since every monomer molecule and every polymer molecule just carries one A group. Thus,

$$(5\text{-}8) \qquad \overline{X}_n \equiv n_{U,R}/n_R = n_{A,0}/n_A = 1/(1 - p_A)$$

Eq.(5-8) applies to all AB and stoichiometric AA/BB polymerizations if intramolecular ring formation can be ruled out.

Industrially useful flexible-chain polymers by equilibrium polycondensation and polyaddition often require number-average degrees of polymerization of at least 100–200, i.e., extents of reaction of at least 0.99–0.995. These yields are much higher than the ones that are customary in organic-chemical reactions. Polymers from such polymerizations still contain 0.005–0.01% monomer which is toxicologically unacceptable for many polymer applications. In these cases, the remaining monomer has to be removed during work-up, often to concentrations of less than 1 ppm = 0.0001 %.

AA/BB reactions are used for many polymer syntheses. Examples of polycondensations are preparations of aliphatic and aromatic polyamides, aromatic and unsaturated aliphatic polyesters, and aromatic polysulfones. Examples of polyadditions are the formations of polyisocyanates, polyureas, and polythioesters.

**Multifunctional Equilibrium Reactions**

Polycondensations and polyadditions of multifunctional monomers such as $A_2B_2$, $A_3 + B_2$, etc., lead to (hyper)branched molecules and, at higher group conversions, also to crosslinked ones. Polymerizations of $AB_2$, $AB_3$, etc., monomers result in hyperbranched polymers but not in crosslinked ones (Volume I).

The onset of crosslinking is usually characterized by a sudden gelation of the reacting system. At this "gel point", a part of the reactant molecules unite to an "infinitely" large, insoluble molecule (or several of those?) that extends from reactor wall to reactor wall. Embedded in this gel are soluble branched and unbranched molecules including unreacted monomer molecules. The fraction of these soluble molecules decreases with increasing extent of reaction.

The onset of crosslinking occurs at a critical extent of reaction of A groups, $p_{A,crit}$, that is controlled by the functionality $f_{br}$ of branching molecules with A groups, the amount fraction (mole fraction) $x_A^\Delta$ of A groups in branching molecules as related to all A groups, and the ratio $r_o = p_B/p_A \leq 1$ of extents of reaction of all B and A groups. The gel point is calculated as (see Volume I, Eq.(13-118))

$$(5\text{-}9) \qquad p_{A,crit,theor} = \left( \frac{(f_{br} - 1)^{-1}}{r_o[x_A^\Delta + (f_{br} - 1)^{-1}(1 - x_A^\Delta)]} \right)^{1/2}$$

An example is the polycondensation of 2 mol $A_3$ + 3 mol $B_2$ with $f_{br} = 3$, $x_A^{\Delta} = 1$, and $r_o = (3{\cdot}2)/(2{\cdot}3) = 1$. The gel point of this system is thus calculated as $p_{A,crit} = 0.707$.

These theoretical gel points are generally lower than the experimental ones because crosslinking reactions are usually accompanied by intramolecular cyclizations that waste functional groups. The theoretical gel points are therefore often safe markers for the synthesis of prepolymers for thermosetting reactions. These prepolymers (often called B stage) are later simultaneously crosslinked ("cured") and shaped. Examples are the formation of alkyd, phenol, and amino resins, the curing of epoxy resins, and the synthesis of crosslinked polyurethanes.

Low conversions of multifunctional monomer molecules may even lead predominantly to intramolecular formation of small ring structures if the functional groups are adjacent to each other. At higher extents of group reaction, crosslinking sets in. Examples are the syntheses of polyimides and polybenzimidazoles.

**Kinetically Controlled Reactions**

Equilibrium constants of thermodynamically controlled polycondensations and polyadditions are usually small, for example, $K \approx 10$ for the formation of aromatic polyesters from terephthalic acid and ethylene glycol and $K \approx 100$–$300$ for the synthesis of polyamides from aliphatic diamines and aliphatic dicarboxylic acids. Hence, they lead to small extents of reaction and thus to low molecular weight products. In industrial polycondensations, leaving molecules (byproducts) are therefore continually removed which shifts the equilibria toward polymer formation. An example is the polyesterification of a diol and a dicarboxylic acid with water as leaving molecule.

Kinetically controlled polycondensations and polyadditions do not need such measures; they are sometimes not even sensitive to stoichiometric imbalances. Examples are Schotten-Baumann reactions, "activated" polycondensations, and/or heterogeneous polymerizations that are controlled physically by diffusion processes.

Industrially utilized Schotten-Baumann reactions comprise the formation of polycarbonate A from phosgene, $COCl_2$, and the disodium salt of so-called bisphenol A, $NaO(p\text{-}C_6H_4)C(CH_3)_2(p\text{-}C_6H_4)ONa$; the polycondensation of isophthalic acid dichloride and *m*-phenylene diamine to an aromatic polyamide; and the reaction of sulfone dichlorides and sulfone diphenolates. These fast reactions are kinetically controlled because of their slow back reactions and the absence of exchange reactions between chain segments, especially, if they proceed heterogeneously because of the precipitation of newly formed polymer molecules.

### 5.1.6 Ionic Chain Polymerizations

Ionic polymerizations proceed by consecutive addition of monomer molecules to growing macroions, $RM_n^* + M \rightarrow RM_{n+1}^*$, where * may be an anion $^{\ominus}$ or a cation $^{\oplus}$. In contrast to polycondensations and polyadditions, catalysis of each step is not necessary. The initiation reaction $R^* + M \rightarrow RM^*$ is not spontaneous, however. Instead, ionic polymerizations are *initiated* by suitable chemical compounds that furnish the initiating ions $R^{\ominus}$ or $R^{\oplus}$, respectively.

Correspondingly, anionic polymerizations with growing macroanions $\sim M^{\ominus}$ are distinguished from cationic polymerizations with growing macrocations $\sim M^{\oplus}$. The rates and stereospecificities of these polymerizations are strongly influenced by the counterions of the growing macroions. Only in very rare cases are macroions truly "free", i.e., totally dissociated from their counterions. They rather form solvent-separated and/or contact ion pairs, especially in anionic polymerizations (Volume I).

*Anionically polymerizable* molecules have electron-accepting (i.e., withdrawing) substituents or rings. The first group comprises dienes, styrenes, acrylic compounds, and some aldehydes and ketones, the second group lactams, lactones, and oxirane.

Anionic polymerizations are initiated by bases or Lewis bases, for example, alkali metals, alcoholates, amines, phosphines, organo-lithium compounds, or sodium naphthalene. These initiators dissociate spontaneously in polar solvents such as ethers like tetrahydrofuran or ethylene glycol dimethylether. These dissociations and the subsequent initiation reactions require little activation energy. Hence, anionic polymerizations often proceed with high speeds even at −100°C.

As a consequence of their spontaneous dissociation, initiating anions attain their effective concentrations "instantaneously". Termination reactions of growing chains are relatively rare; they are usually caused by impurities. Anionic polymerizations are therefore often "living" and the resulting degrees of polymerization are controlled by the ratio of the amount concentrations of monomer and initiator and the monomer conversion (cf. Eq.(5-6)). High molecular weights therefore require low initiator concentrations which are difficult to control.

All monomer molecules have the same statistical chance to add to growing chains if all initiating species are present at the start of polymerizations, all initiating reactions are fast, termination and chain transfer reactions are absent, and diffusion of molecules is not impeded. The number of monomeric units per macromolecule then follows a Poisson distribution (Volume I), a very narrow distribution that leads to "practically molecularly homogeneous" polymers. This molecular feature is sometimes desired, for example, for some biomaterials. However, it may be detrimental to melt processing (see p. 138, 2nd paragraph). For high molecular weights, living polymerizations also need high monomer/initiator ratios (Fig. 5-1) which requires rigorous exclusion of impurities.

These features of anionic polymerizations therefore limit their industrial applications to those where monomers are impossible or difficult to polymerize by other methods (e.g., formaldehyde to poly(oxymethylene)), high polymerization rates are desired ($\varepsilon$-caprolactam to polyamide 6 by the RIM process), a certain stereoregularity is required (butadiene to *cis*-1,4-poly(butadiene)), or a certain molecular architecture is impossible or difficult to obtain by other methods (thermoplastic block elastomers of the styrene-butadiene-styrene type).

*Cationically polymerizable* are olefin derivatives with electron-rich (i.e., donating) substituents, monomers with heteroatoms in double bonds, and certain heterocycles. Initiators comprise Brønsted acids (perchloric acid, etc.), Lewis acids ($AlCl_3$, $TiCl_4$, etc.) with so-called "co-initiators" (e.g., water), and carbenium salts (acetyl perchlorate, tropylium hexachloroantimonate, etc.). Examples of initiating species are the monomer cation $H{-}CH_2{-}C^{\oplus}(CH_3)_2$ with $ClO_4^{\ominus}$ as counterion from isobutene, $CH_2{=}C(CH_3)_2$, and perchloric acid, $HClO_4$, and the monomer cation $C_2H_5{-}CH_2{-}C^{\oplus}(CH_3)_2$ with $[R_2AlCl_2]^{\ominus}$ as counterion from isobutene and the Lewis acid $R_2AlCl$ plus the co-initiator $C_2H_5Cl$.

Growing macrocations are thermodynamically and kinetically unstable. They try to stabilize themselves by addition of nucleophilic species, which leads to desirable high polymerization rates for monomer additions on the one hand and to undesirable transfer and termination reactions with other species (monomer molecules, counteranions, solvents, etc.) on the other. Industrially, only a few cationic polymerizations leading to *high*-molecular weight polymers are performed; examples include those of isobutene, vinyl ethers, trioxane copolymers, ethylene imine, and tetrahydrofuran. However, there are many cationic polymerizations to low-molecular weight resins such as the "hydrocarbon resins" from $C_5$ and $C_9$ feedstreams and "petroleum resins" from $C_5$–$C_6$ and $C_8$–$C_{10}$ streams of petroleum processing.

### 5.1.7 Insertion Polymerizations

Polyinsertions are polymerizations in which monomer molecules are inserted during propagation between the catalyst moiety and the attached polymer chain (Eq.(5-2)). Insertions are preceded by a coordination of the monomer molecule with the polymer-catalyst complex, hence the also-used term "coordination polymerization."

In insertion polymerizations, bifunctional monomer molecules are *bifunctionally inserted* whereas in conventional chain polymerization they are *monofunctionally added.* This has two consequences. First, participating components must fit each other. Examples are Ziegler-Natta polymerizations of olefins and dienes with transition metal catalysts and metathesis polymerizations of cycloolefins with Group 8 metal compounds. Second, bifunctional insertions are much more stereospecific than monofunctional ones because for steric and electronic reasons both "ends" of the incoming monomer molecule must be fixed in defined positions relative to the catalyst and the growing chain.

The direction of monomer insertion depends on the catalyst moiety [Mt]. For example, propene $CH_2=CH(CH_3)$ is $\alpha$-inserted to ~$CH(CH_3)$–$CH_2$–[Mt] in its isospecific polymerization by $TiCl_4/AlR_3$ but $\beta$-inserted to ~$CH_2$–$CH(CH_3)$–[Mt] in its syndiospecific polymerization by $VCl_4/(C_2H_5)_2AlCl$/anisole.

Polyinsertions are industrially interesting for two reasons. First, they usually do not have transfer reactions that lead to branched polymers. An example is the polymerization of ethene by Ziegler or metallocene catalysts to high-density poly(ethylene)s, HDPE, that are far less branched than low-density poly(ethylene)s, LDPE, by free-radical polymerizations. If desired, short-chains can be introduced by copolymerization, for example, in so-called linear low-density poly(ethylene)s, LLDPE, by copolymerization of ethene with a few percent of higher $\alpha$-olefins.

Second, insertion polymerizations allow the synthesis of highly stereoregular polymers. Examples of such major industrial processes are Ziegler–Natta polymerizations of propene to isotactic poly(propylene) and 1-butene to isotactic poly(1-butene), the terpolymerization of ethene, propene, and a non-conjugated diene, and the polymerizations of butadiene and isoprene to their cis polymers.

Similar catalysts are used in *metathesis polymerizations*. These polymerizations are exchange and disproportionation reactions of double bonds, mainly of olefins and cycloolefins. Industrial metathesis polymerizations include those of cyclooctene, norbornene, and dicyclopentadiene.

### 5.1.8 Free-Radical Polymerizations

Free-radical polymerizations are initiated and propagated by free radicals, usually carbon radicals. In contrast to ionic polymerizations, they are relatively insensitive to impurities and usually simple to control. They furthermore deliver high-molecular weight polymers at low monomer conversions (Fig. 5-1). Free-radical polymerizations are thus the polymerizations of choice if monomer structures allow it and resulting polymer properties are not detrimental (branching, molecular-weight distributions, etc.).

Most free-radical polymerizations require initiators that provide initiating radicals by thermal homolysis, redox reactions, photochemical reactions, or electrolysis. Examples of industrially used initiators are shown in Table 5-4. Styrene is the only monomer that is polymerized thermally without added free-radical initiator.

Table 5-4 Free-radical initiators for the polymerization of monomers (PM) and the crosslinking (XL) of polymers. $T_S$ = storage temperature, $T_{10}$ = 10-hour lifetime. Examples: polymerizations of ethene (E), styrene (S), and/or vinyl chloride (V) in bulk (b), emulsion (e), or suspension (s) and crosslinking of poly(ethylene)s (PE) or unsaturated polyesters (UP). In chemical formulas of initiators, – indicates primary scissions and / secondary ones. In formulas for substituents, ··· is the bond

| Initiator | | Substituent R or R' | $T_S$/°C | $T_{10}$/°C | PM | XL |
|---|---|---|---|---|---|---|
| *Peroxides* RO–OR' | | | | | | |
| BPO | Dibenzoyl | $C_6H_5CO\cdots$ | | 73 | S(s) | UP |
| Dicup | Dicumyl | $C_6H_5(CH_3)_2C\cdots$ | | 117 | | PE |
| | Dilauroyl | $C_{11}H_{23}CO\cdots$ | 25 | 62 | E,V(e) | |
| ACSP | Acetylcyclohexanesulfonyl | $CH_3CO\cdots$, $C_6H_{11}SO_2\cdots$ | –12 | 40 | V(b) | |
| *Hydroperoxides* RO–OH | | | | | | |
| CHP | Cumene | $C_6H_5C(CH_3)_2\cdots$ | | 155 | | |
| *Dihydroperoxides* HO-ORO-OH | | | | | | |
| MEKP | Methylethylketone | $\cdots C(CH_3)(C_2H_5)\cdots$, $\cdots C(CH_3)(C_2H_5)O–OC(CH_3)(C_2H_5)\cdots$ | | | | UP |
| *Peroxydicarbonates* RO/C(O)O–OC(O)/OR | | | | | | |
| BCP | Bis(4-*t*-butylcyclohexyl) | $(CH_3)C(p\text{-}C_6H_{10})\cdots$ | 20 | | E,V | |
| NPP | Dipropyl | $C_3H_7\cdots$ | | | E | |
| IPP | Diisopropyl | $(CH_3)_2CH\cdots$ | –10 | 45 | E,V | |
| EHP | Bis(2-ethylhexyl) | $C_4H_9CH(C_2H_5)CH_2\cdots$ | –15 | | E,V | |
| | Dicyclohexyl | $C_6H_{11}\cdots$ | 10 | 50 | V(b) | |
| | Dimyristyl | $H(CH_2)_{14}\cdots$ | 20 | 50 | V(b) | |
| *Ethane derivates* R–R | | | | | | |
| | Benzpinacol | $(C_6H_5)_2(OH)C\cdots$ | | | S(b) | |
| | Bis[(methyl)(ethyl)(phenyl)]- | $(CH_3)(C_2H_5)(C_6H_5)C\cdots$ | | | S(b) | |
| *Azo compounds* R–N=N–R | | | | | | |
| AIBN | *N,N*-Azobisisobutyronitrile | $(CH_3)_2(CN)C\cdots$ | 25 | 64 | V | |
| *Other* | | | | | | |
| | Vinylsilane triacetate | $CH_2{=}CHSi(OOCCH_3)_3$ | | | S(b) | |
| | Dipotassium persulfate | $KO(SO_2)O–O(SO_2)OK$ | | | V(e) | |
| | Dihydrogen peroxide | HO–OH | | | V(e) | |

Table 5-5 Activation energies $E^{\ddagger}$ and half-lifes $t_{50}$ of the decomposition of free-radical initiators in various solvents. For abbreviations, see Table 5-4.

| Initiator | Solvent | $E_d^{\ddagger}$ in kJ/mol | $t_{50}$/h at 40°C | 70°C | 110°C |
|---|---|---|---|---|---|
| IPP | Dibutyl phthalate | 115.0 | 21 | 0.32 | 0.0044 |
| AIBN | Dibutyl phthalate | 122.2 | 303 | 5.0 | 0.057 |
| | Benzene | 125.5 | 354 | 6.1 | 0.076 |
| | Styrene | 127.6 | 414 | 5.7 | 0.054 |
| BPO | Acetone | 111.3 | 443 | 10.6 | 0.180 |
| | Benzene | 133.9 | 2 130 | 23.7 | 0.177 |
| | Styrene | 132.8 | 3 525 | 29.2 | 0.231 |
| | Poly(styrene) | 146.9 | 11 730 | 84.6 | 0.392 |
| Dicup | Benzene | 170 | 3 000 000 | 11 200 | 27 |
| CHP | Benzene | 100 | 4 000 000 | 60 000 | 760 |
| $K_2S_2O_8$ | 0.1 mol NaOH/L $H_2O$ | 140 | 1 850 | 11.9 | - |

Most monomers for free-radical polymerizations have carbon-carbon double bonds. Mono-unsaturated monomers produce linear or slightly branched polymers that are used as thermoplastics, fibers, coatings, packaging materials, etc. Examples are the free-radical polymerizations of ethene, styrene, vinyl chloride, methyl methacrylate, acrylonitrile, and tetrafluoroethylene. Crosslinked polymers result from the polymerization of multifunctional monomers. An example is the copolymerization of styrene with divinyl benzenes for the preparation of ion-exchange resins. Polymers from diallyl and triallyl monomers are the only thermosets from free-radical polymerizations.

**Initiation**

Industrially used initiators are mainly peroxy compounds such as peroxides, hydroperoxides, dihydroperoxides, and peroxydicarbonates (Table 5-4). Ethane derivatives and vinyl silane triacetate are used as high-temperature initiators. Polymerizations in emulsion use water-soluble initiators such as dipotassium persulfate, $K_2S_2O_8$, or redox systems like $Fe^{2+}/H_2O_2$ which require low thermal activation energies and thus allow polymerizations at ambient temperatures. Photochemical free-radical initiators are used for lithography and the hardening (curing) of lacquers and electrolytical polymerizations for the coating of metals by polymers.

The decomposition of initiators is scientifically characterized by rate constants $k_d$ of dissociation, activation energies $E^{\ddagger}$, and pre-exponential factors $A^{\ddagger}$:

$$[\mathrm{I}] = [\mathrm{I}]_o \exp(-k_d t) \quad ; \quad k_d = A^{\ddagger} \exp(-E^{\ddagger}/RT) \tag{5-10}$$

These molecular parameters vary widely (Table 5-5). Industry characterizes initiators by their 50 % decomposition: in Europe by the half-life $t_{50} = 0.693\ k_d$ for various temperatures (Table 5-5) and in the United States by the temperature $T_{10}$ at which 50 % is decomposed after 10 hours (Table 5-4). These manufacturer-supplied data do not disclose the applied solvent or dispersion agent, obviously because of the wide-spread belief that free-radical reactions are not affected by the environment (but they are (Table 5-5)). The data therefore do not necessarily apply to polymerizing systems.

Free-radical initiators should be highly polymerization active: they should deliver high yields of highly active radicals and lead to large initiation rates. Polymerizations by peroxy compounds are therefore often "activated" or "accelerated" by additional chemical compounds that induce their decomposition. Such activators are especially required for the cold-hardening of polymers or prepolymers at room temperature, for example, the copolymerization of unsaturated polyesters with styrene or methyl methacrylate. Dibenzoyl peroxide as initator is activated here by *N,N*-dimethylaniline:

$$(5\text{-}11)\qquad C_6H_5N(CH_3)_2 + C_6H_5COO-OOCC_6H_5 \xrightarrow{-\,C_6H_5COO^\bullet} [C_6H_5N(CH_3)_2]^{\oplus}{}^{\ominus}[OOCC_6H_5] \longrightarrow C_6H_5\dot{N}(CH_3)_2 + HOOCC_6H_5$$

For cold-hardening, initiators should also lead to polymers with only very small residual contents of monomer (< 1 ppm, depending on toxicity) which can be achieved by the use of two different initiators: a fast-decomposing one for the initial polymerization and a slowly decomposing one for the final monomer conversion. Dual initiators may therefore circumvent the expensive removal of residual monomer by vacuum or steam. Certain initiators may also lead to the formation of crusts at reactor walls.

On storage, initiators should not decompose prematurely. Shelf-lives are usually the larger, the better the crystallizability since the resulting spatial ordering reduces the mobility of molecules, molecule segments, and radicals. Because of the danger of explosions, peroxy compounds can only be stored in small quantities. Industrial free-radical initiators are therefore often delivered in tankers as concentrated solutions or dispersions. Examples are 40–70 % solutions of bis(2-ethylhexyl)peroxydicarbonate in isoparaffins or aromatics; 40 % aqueous dispersions of bis(4-*t*-butylcyclohexyl)peroxydicarbonate; and 80 % aqueous dispersions of bis(*o*-methylbenzoyl)peroxide.

#### Propagation and Termination

Most free-radical polymerizations are true kinetic chain reactions. Initiator radicals $I^\bullet$ from the homolysis of initiator molecules, for example, from the thermal dissociation of dibenzoyl peroxide, $C_6H_5COO\text{–}OOCC_6H_5 \rightarrow 2\ C_6H_5COO^\bullet \rightarrow 2\ C_6H_5^\bullet + 2\ CO_2$, add to monomer molecules such as $CH_2{=}CHR$ (Table 5-2). The resulting monomer radical I–M$^\bullet$, e.g., $C_6H_5COOCH_2\text{–}^\bullet CHR$, adds additional monomer molecules in a kinetic chain reaction. The kinetic chain is terminated by **(re)combination** of two polymer radicals to one dead chain (example of Eq.(5-12)) or by **disproportionation** to two dead chains (example of Eq.(5-13)):

$$(5\text{-}12)\qquad C_6H_5COO(CH_2CHR)_n^\bullet + {}^\bullet CHRCH_2(CHRCH_2)_mOOCC_6H_5 \rightarrow C_6H_5COO(CH_2CHR)_n(CHRCH_2)_{m+1}OOCC_6H_5$$

$$(5\text{-}13)\qquad C_6H_5COO(CH_2CHR)_n^\bullet + {}^\bullet CHRCH_2(CHRCH_2)_mOOCC_6H_5 \rightarrow C_6H_5COO(CH_2CHR)_{n-1}CH_2CH_2R + CHR{=}CH(CHRCH_2)_mOOCC_6H_5$$

In most free-radical polymerizations, a very short initial period (order of seconds) is followed by a **steady state** with respect to the formation and disappearance of radicals in which the initiator decomposition delivers as many initiator radicals (and thus also monomer radicals and macroradicals) as polymer radicals $P^{\bullet}$ disappear by termination reactions. The change of the radical concentration in the steady state with time is given by the difference of rates of initiation and termination by two polymer radicals:

$$(5\text{-}14) \qquad d[P^{\bullet}]/dt = R_i - R_{t(pp)} = k_i[I^{\bullet}][M] - k_{t(pp)}[P^{\bullet}]^2 = 0$$

Elementary calculations (Volume I) show that the rate of polymerization, $R_p$, in the steady state should be directly proportional to the actual monomer concentration, [M], and to the square root of the actual initiator concentration, [I]:

$$(5\text{-}15) \qquad R_p = -d[M]/dt = k_p[P^{\bullet}][M] = k_p(2\,fk_d/k_{t(pp)})^{1/2}[M][I]^{1/2}$$

$k_d$, $k_i$, $k_p$, and $k_{t(pp)}$ are the rate constants of initiator dissociation (index d), initiation reaction (i), propagation (p), and termination by mutual deactivation of two macroradicals (t(pp)). $f$ indicates the fractional yield of effective initiator radicals ($0 \leq f \leq 1$).

For the steady state, Eq.(5-15) predicts a linear decrease of the propagation rate $R_p$ with decreasing monomer concentration, i.e., increasing fractional monomer conversion, which is indeed found (Fig. 5-3). However, the steady state is maintained only for a relatively short initial period, ca. 10 % monomer conversion in the example of Fig. 5-3. The reason is the so-called **gel effect** (= Trommsdorff effect or Norrish–Trommsdorff effect), an increase of polymerization rate with increasing monomer conversion that accompanies the conversion of neat monomer to a viscous polymer solution. The increased viscosity impedes the diffusion of macroradicals to each other; the rate constant $k_{t(pp)}$ of mutual deactivation decreases accordingly. The diffusion of small monomer molecules to macroradicals is not affected, however, and the rate constant $k_p$ of propagation remains constant. The initiator dissociation is also not affected; hence, there are now more radicals generated per unit time than macroradicals disappear and the polymerization rate increases accordingly.

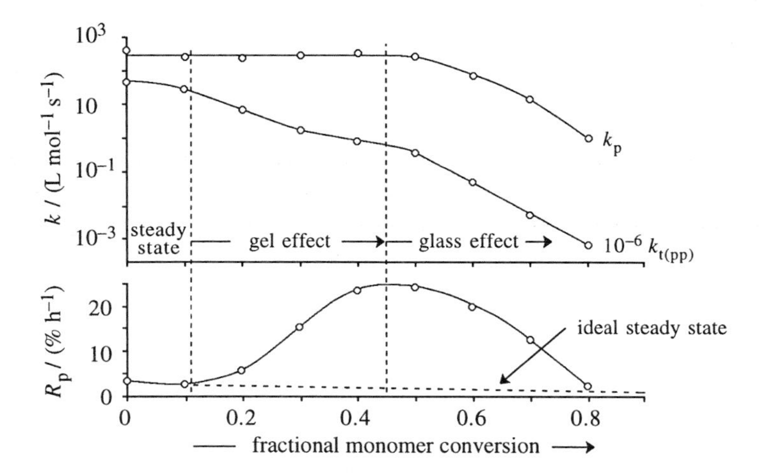


Fig. 5-3 Dependence of polymerization rates $R_p$ (in percent per hour), rate constants $k_{t(pp)}$ of termination by mutual deactivation of two macroradicals, and rate constants of propagation for the free-radical polymerization of methyl methacrylate in bulk at 22.5°C [2].

The polymerizing system finally becomes extremely viscous ("glass-like"). Monomer molecules can no longer diffuse easily to growing macroradicals, which causes rate constants of propagation, $k_p$, to drop (Fig. 5-3). This **glass effect** decreases propagation rates. Finally, the polymerization stops before all monomer is polymerized.

Because of these effects, kinetics of industrial polymerizations must be modeled separately for each monomer–initiator–temperature–reactor system. Kinetic equations for idealized mechanisms (low monomer conversions, isothermal reactions, etc. (Volume I)) can rarely be applied directly.

Polymer molecular weights and molecular weight distributions are also controlled by micro and macrokinetics. The **kinetic chain length**, $\nu_{PP}$, is defined as the number of monomer molecules that are polymerized by one initiator radical. For terminations by disproportionation and steady-state conditions, it equals the number-average degree of polymerization, $\overline{X}_n$, and is given by the ratio of rate of propagation, $R_p$, to either the rate $R_{t(pp)}$ of termination or the rate of initiation, $R_i$ (Volume I, Eq.(10-45)) (for (re)combination, $\nu_{PP} = 2\ \overline{X}_n$):

$$(5\text{-}16) \qquad \nu_{pp} = \frac{R_p}{R_{t(pp)}} = \frac{R_p}{R_i} = \frac{R_p}{2\,fk_d[\mathrm{I}]} = \frac{k_p}{(2\,fk_d k_{t(pp)})^{1/2}} \cdot \frac{[\mathrm{M}]}{[\mathrm{I}]^{1/2}} = \overline{X}$$

In the course of polymerization, number-average degrees of polymerization should thus decrease with decreasing [M] (higher monomer conversions). Experimentally, $\overline{X}_n$ often stays constant, however (Volume I, Fig. 10-11). This behavior seems to be caused by a decrease of termination rate constants (Fig. 5-3) as well as chain transfer reactions by initiator radicals that lead to branched polymer molecules. After this period, number-average degrees of polymerization increase because of the gel effect.

### 5.1.9 Copolymerizations

Copolymerizations are joint polymerizations of two or more types of monomers that lead to polymer molecules with more than one type of monomeric units per chain. Correspondingly, one distinguishes between bi, ter, quater, quinter, ... polymerizations of two, three, four, five, ... types of monomers. In the older literature, copolymers were called **interpolymers**, a term that has recently been used for ethene-styrene copolymers.

In the most simple case of a bipolymerization, the two possible active chain ends ~a* and ~b* react irreversibly with monomer molecules A or B, which leads to four different propagation rates $R_{iJ}$ (i = a, b; J = A, B) and four different propagation rate constants $k_{aA}$, $k_{aB}$, $k_{bA}$, and $k_{bB}$. This **terminal model** neglects a possible influence of penultimate units ~aa*, ba*, ~ab*, and ~bb*.

It is expedient to define **copolymerization parameters** $r_a \equiv k_{aA}/k_{aB}$ and $r_b \equiv k_{bB}/k_{bA}$ as ratios of propagation rate constants of co-additions to homo-additions. Comonomer molecules are exclusively added if $r = 0$ and preferentially if $r < 1$, the own monomer molecules preferentially if $r > 1$ and exclusively if $r = \infty$. Both types of monomers are added with equal probabilities at $r = 1$. The relative monomer consumption is given by

$$(5\text{-}17) \qquad \frac{-\mathrm{d}[\mathrm{A}]/\mathrm{d}t}{-\mathrm{d}[\mathrm{B}]/\mathrm{d}t} = \frac{R_{aA} + R_{bA}}{R_{bB} + R_{aB}} = \left(\frac{k_{bA}[\mathrm{b}^*] + k_{aA}[\mathrm{a}^*]}{k_{bB}[\mathrm{b}^*] + k_{aB}[\mathrm{a}^*]}\right) \cdot \frac{[\mathrm{A}]}{[\mathrm{B}]} = \frac{\mathrm{d}[\mathrm{A}]}{\mathrm{d}[\mathrm{B}]}$$

In cross-reactions ~a* + B → ~ab* and ~b* + A → ~ba*, chain ends ~a* are replaced by ends ~b* and vice versa. Their concentrations [a*] and [b*] remain constant with time if cross-reaction rates equal each other, $R_{aB} = R_{Ba}$. Eq.(5-17) thus becomes

$$\text{(5-18)} \qquad \frac{d[A]}{d[B]} = \frac{1 + r_a([A]/[B])}{1 + r_b([B]/[A])} \qquad \textbf{(Mayo–Lewis equation)}$$

Eq.(5-18) applies only to small *conversion intervals*, i.e., from 0 to 1 % of monomer conversion, from 50 % to 51 %, etc.. In small intervals, the relative change d[A]/d[B] of monomer concentrations equals approximately the relative composition $x_a/x_b$ of copolymers, i.e., the *instantaneous* one for the conversion interval (1 % in both examples) and not the total one (1 % in the first example, 51 % in the second).

Eq.(5-18) does not apply to greater conversion intervals which leads to a drift of the copolymer composition if $r_a \neq r_b$. Polymer compositions do not change with monomer conversion if $r_a = r_b = 1$ (ideal copolymerization); $r_a = r_b = 0$ (alternating copolymerization); d[A]/d[B] = [A]/[B] (azeotropic condition); or if the faster polymerizing monomer is constantly replenished so that its concentration stays constant.

Different types of initiators lead to different copolymerization behaviors. An example is the bipolymerization of styrene and methyl methacrylate (Fig. 5-4). The free-radical polymerization is almost azeotropic. Ionic bipolymerizations show widely different copolymerization parameters: cationic ones lead to long sequences of styrene units, anionic ones to long sequences of methyl methacrylate units. The Ziegler–Natta type of insertion polymerization by $(C_2H_5)_3Al_2Cl_3$ as catalyst is almost alternating.

Ionic copolymerizations and also coordination polymerizations usually show widely different copolymerization parameters. This leads to strong shifts of average copolymer compositions with monomer conversion and to long homosequences of monomeric units. Industrial ionic copolymerizations are therefore relatively rare (see Volume I, Table 12-1). Much more common are free-radical copolymerizations.

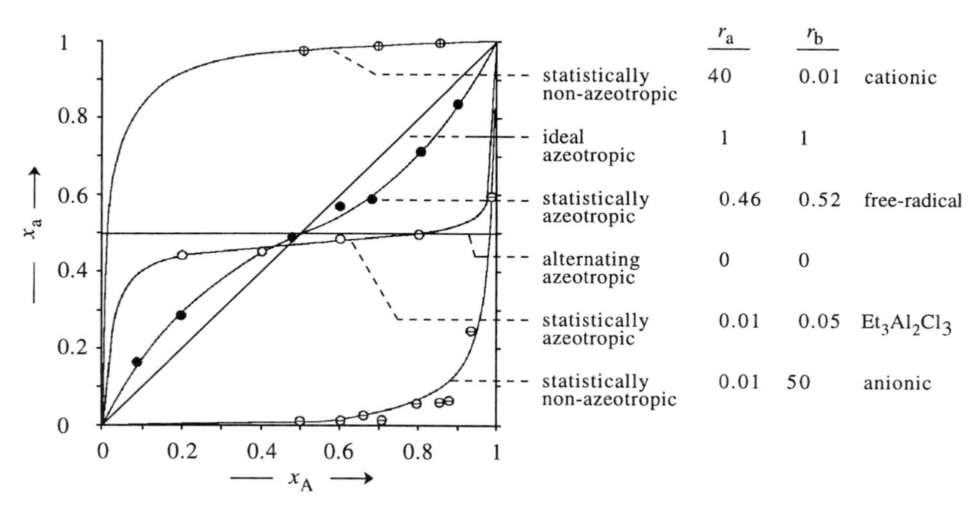


Fig. 5-4 Copolymerization diagram: Mole fraction $x_a$ of styrenic units as a function of the mole fraction $x_A$ of styrene monomer in cationic (⊕), free-radical (●), coordination (○), and anionic (⊖) copolymerizations of styrene A and methyl methacrylate B [3].

### 5.1.10 Polymer Transformations

Chemical transformations of polymers to other polymers are usually costly since they require reactions in solution and thus often expensive solvents and product and solvent work-up. Such transformations are therefore only used if (a) base polymers are inexpensive and (b) target polymers cannot be obtained (or only with difficulty) by direct polymerization of monomers. Esterifications of cellulose as base polymer are examples of the first group, saponification of poly(vinyl acetate) to poly(vinyl alcohol) is an example of the second one.

Polymer transformations are classic polymer-analog reactions. With respect to group reactions, they are identical to their low-molecular weight analogs. They differ from the latter, however, in that side reactions do not lead to easy removable byproducts but to "wrong" substituents in polymer chains and often to unsatisfactory polymer properties. For this reason, industrial polymer transformations are usually restricted to only five types: hydrogenations, transesterifications–esterifications–saponifications, etherifications, chlorinations–sulfochlorinations, and cyclizations.

The first four of these reactions usually allow group transformations between 0 % and 100 %. The partial saponification of poly(vinyl acetate), $\text{+}CH_2\text{–}CH(OOCCH_3\text{+}_n$, thus leads to pseudo-copolymers with more or less randomly distributed vinyl acetate units, $\text{–}CH_2\text{–}CH(OOCCH_3)\text{–}$, and vinyl alcohol units, $\text{–}CH_2\text{–}CH(OH)\text{–}$. The main base polymer of this group is cellulose which is transformed by acetic acid to cellulose acetates for fibers and cigarette filters; by nitric acid to cellulose nitrates for gun cotton, thermoplastics, fibers, and films; and by oxiranes to cellulose ethers for thickeners.

In irreversible formations of rings between two adjacent substituents, on the other hand, the maximum amount of group transformation is restricted for statistical reasons (see Volume I). An example is the formation of acetal groups by irreversible reaction of butyraldehyde with adjacent OH groups of poly(vinyl alcohol) which cannot exceed $1/e^2$ (= 86.5 %) of the OH groups.

Polymer transformations in the wider sense include the syntheses of graft and block polymers and the post-crosslinking of polymers. Block polymers result from either the polymerization of monomers by reactive chain ends of a polymer molecule, for example, $\sim a_n{}^* + m\ B \rightarrow \sim a_n b_m{}^*$, or by coupling of two different polymer chains, for example, $\sim a_n{}^* + {}^*b_m\sim \rightarrow \sim a_n b_m\sim$. Examples are the syntheses of diblock polymers as compatibilizers, and of triblock polymers as thermoplastic elastomers, using anionic chain ends. Segmented polymers can also be obtained by condensation type reactions of chain ends of preformed polymers.

Random grafting of monomers from or to polymer chains delivers graft copolymers. Grafting is usually by free-radical chain polymerization that is initiated by radicals on preformed polymer chains. An example is the grafting of styrene/acrylonitrile on saturated acrylic rubbers.

Another type of polymer transformation is post-polymerization crosslinking. Examples are the so-called vulcanization of unsaturated rubbers by sulfur or saturated rubbers by peroxides, the hardening of unsaturated polyester resins by styrene or methyl methacrylate, and the tanning of collagen to leather. These reactions differ from those of thermosetting resins which are not polymers but oligomers (= prepolymers) with many reactive groups.

## 5.2 Industrial Polymer Productions

### 5.2.1 Overview

Polymerizations and polymer transformations can be performed in different media: in bulk (as neat polymerization), gas phase, suspension, emulsion, or solution, by precipitation, and in rare cases, also in the solid state. The choice of the reaction medium depends not only on the monomer and resulting polymer but also on the processing conditions (continuous or discontinuous, type of reactor, removal of heat of polymerization, etc.), the work-up of the resulting polymers, and the requirements of the customers (powders, pellets, bales, etc.) (Table 5-6).

Table 5-6 Industrial polymerizations in bulk (B), emulsion (E), solution (L), or suspension (S), by precipitation (P) or in the gas phase (G), by polycondensation (PC), polyaddition (PA), or chain polymerization (CP) [free-radical (r), anionic (a), cationic (c), insertion (i), metallocene (m)].

Application as adhesive (A), absorbent (Ab), coating (C), thermoset (D), elastomer (E), fiber (F), flocculant (Fl), thermoplastic (T). + Major process, (+) minor process.

| Polymer | Polymerization Type | Polymerization Growth | B | Polymerization medium S | L | P | G | E | Application |
|---|---|---|---|---|---|---|---|---|---|
| Poly(ethylene), high density | CP | i, m | + | + | | + | + | | T,F |
| , low density | CP | r | + | + | (+) | (+) | | (+) | T |
| Poly(propylene) | CP | i, m | + | + | + | | + | | T,F |
| Poly(styrene) | CP | r, m | + | (+) | (+) | | | (+) | T |
| Poly(methyl methacrylate) | CP | r | + | + | + | | | + | T |
| ABS polymers | CP | r | | | (+) | | | + | T |
| EPDM rubber | CP | z | | | + | | | | E |
| Poly(chloroprene) | CP | r | | | | | | + | E |
| Butyl rubber | CP | c | | | + | | | | E |
| Poly(isoprene) | CP | a | | | + | | | | E |
| Poly(oxymethylene) | CP | a,c | + | (+) | | | | | T |
| Poly(vinyl acetate) | CP | r | (+) | (+) | (+) | | | + | C,A |
| Poly(vinyl chloride) | CP | r | (+) | + | (+) | | + | + | T,C |
| Poly(vinyl fluoride) | CP | r | + | | | | | | T,C |
| Poly(vinylidene fluoride) | CP | r | | + | | | | + | T,C |
| Poly(tetrafluoroethylene) | CP | r | | + | | | | + | T,C |
| Poly(acrylic acid) | CP | r | | + | | | | | Ab,Fl |
| Poly(acrylic ester)s | CP | r | | | | | | + | A,C,E |
| Poly(acrylamide) | CP | r | | | + | | | | Fl, C |
| Poly(acrylonitrile) | CP | r | | | | + | | | F |
| Polyamide 6 | CP | a | + | | + | | | | F,T |
| Poly(diallyl phthalate) | CP | r | + | | | | | | D |
| Polyamide 6.6 | PC | | + | (+) | | | | | F,T |
| Polyaramids | PC | | | | + | | | | F,T |
| Phenolic resins | PC | | + | | (+) | | | | D,F |
| Poly(ethylene terephthalate) | PC | | + | | | | | | F,T |
| Unsaturated polyesters | PC | | + | | | | | | D |
| Polycarbonates | PC | | | | + | | | (+) | T |
| Polyimides | PC | | | | + | | | | D,T |
| Amino resins | PC | | + | | | | | | D |
| Polyurethanes | PA | | + | | + | | | | E,D |

Industrial polymerizations differ considerably from those performed in the laboratory. They usually proceed to much higher monomer conversions, which in turn leads to additional phenomena such as glass and gel effects (Figs. 5-1 and 5-3), initiator exhaustion, catalyst poisoning, etc.

These effects affect not only the course of polymerization but also the chemical structure, molecular weight, and molecular weight distribution of the resulting polymer. For these reasons, industrial polymerizations are often conducted differently from those in the laboratory: instead of discontinuous isothermal polymerizations, one can add the monomer continuously, use two different initiators at different times, work non-isothermally or with pressure (Fig. 5-2), etc.

Microkinetic effects such as initiation, propagation, termination, and transfer reactions are furthermore accompanied by macrokinetic ones. The increased viscosity at higher monomer conversions affects not only the chemical kinetics but also the homogeneity of the reaction system and the transfer of heat. Effects of impurities, on the other hand, are mostly not as important as in laboratory experiments since impurities often stem from reactor walls and industrial reactors have a much higher ratio of volume to surface than laboratory vessels.

Reactors and agitators affect not only rates of polymerization, space-time yields, and residence times but also polymer structures and molecular weight distributions (Section 5.3). Decisive here is the coupling of concentration, temperature, and reaction gradients.

### 5.2.2 Two-Phase Systems

Polymerizations in bulk or solvents proceed in 1-phase systems but all others do not. Polymerizations in 2-phase systems are subdivided into

- Suspension polymerizations, defined as polymerizations of droplets of water-insoluble monomers in water (Section 5.2.4);
- Emulsion polymerizations which start in micelles filled with monomer molecules that are sparingly soluble in water; they then proceed in latex particles (Section 5.2.5);
- Inverse emulsion polymerizations in which water-soluble monomers are dispersed in hydrophobic organic solvents (Section 5.2.5);
- Precipitation polymerizations where precipitated polymers settle out of the system (Section 5.2.7);
- Slurry polymerizations where precipitated polymer molecules and remaining monomer molecules form a pasty mass.

These types of heterophasic polymerizations differ characteristically in many details (Table 5-7) such as loci of polymerizations, types of particle stabilization, polymerization rates, presence of branching reactions, molecular weight distributions, etc.

The terms "suspension polymerization," "emulsion polymerization," and "dispersion polymerization" as used by polymer science and industry differ from the terms suspension, emulsion, and dispersion as defined by colloid science (Volume IV).

In colloid science, "dispersion" is a generic term that *includes* "emulsion" and "suspension" (Volume IV). A dispersion is defined as a system consisting of one continuous phase and at least one discontinuous phase. Continuous phases are always fluid (liquids or gases) and discontinuous phases always condensed (liquids or solids). Dispersions are subdivided into *aerosols* (solids or liquids dispersed in gases), *emulsions* (liquids dispersed in liquids; either as oil-in-water or as water-in-oil), and *suspensions* (solids in liquids).

Table 5-6 Characteristics of heterogeneous polymerizations [4].

| Property | Emul-sion | Disper-sion | Suspen-sion | Precip-itation |
|---|---|---|---|---|
| Separate monomer phase | + | - | + | - |
| Dissolution of initiator in diluent | + | + | - | + |
| Formation of particles in continuous phase | + | + | - | + |
| Stabilization of particles | + | + | + | - |
| Control of the number of particles by concentration of stabilizer | + | + | + | - |
| Control of the rate of polymerization by number of particles | + | - | - | - |

However, "dispersion polymerization" refers to a polymerizing water-in-oil *emulsion* whereas "emulsion polymerization" and "suspension polymerization" are both special cases of polymerizing oil-in-water emulsions. "Emulsion polymerization" and "suspension polymerization" both lead either to suspensions of polymer particles (if the glass temperature of the resulting dispersed polymer particles is higher than the temperature of the surrounding liquid; example: polymerization of styrene) or to emulsions (if the glass temperature of the resulting dispersed polymer particles is lower than the temperature of the surrounding liquid; example: copolymerization of butadiene/styrene to rubbers).

### 5.2.3 Bulk Polymerizations

Polymerizations of neat monomers are generally conducted at elevated temperatures; they thus require thermally stable monomers and polymers. For polycondensations, they are called **melt polycondensations** since monomers generally need to be melted for polymerization. For chain polymerizations, they are referred to as **polymerizations in bulk** since monomers are usually liquid at room temperature. In industry, the term **bulk polymerization** includes not only truly **neat polymerizations** but also often those with 5–25 % solvent as a polymerization aid or transfer agent.

*Melt polycondensations* are usually conducted at temperatures between 120°C and 180°C with or without catalysts under a blanket of inert gas ($N_2$, $CO_2$, $SO_2$). An example is the synthesis of poly(hexamethylene adipamide) (nylon 6.6). In direct polycondensation, the melt of adipic acid and hexamethylenediamine leads to a precondensate that is further condensed in the solid state to the desired degree of polymerization of ca. 200. Adipic acid and hexamethylenediamine are often combined to the nylon salt (AH salt), $[NH_2(CH_2)_6NH_3]^{\oplus}[OOC(CH_2)_4COOH]^{\ominus}$, which is condensed as a slurry in methanol.

*Chain polymerizations* in bulk are usually free-radical processes. Very pure polymers result if >99.99% pure monomers and no other chemicals are used, except sometimes free-radical initiators. An example is the thermal bulk polymerization of styrene without initiator that leads to "crystal poly(styrene)", a glass-clear *amorphous* polymer. Another example is the free-radical-initiated bulk polymerization of methyl methacrylate to optically pure poly(methyl methacrylate)s, sold as Plexiglas® or Lucite®.

Free-radical bulk polymerizations lead to high heats of polymerization per unit volume which are not easy to remove from large batch reactors. Hot spots may develop locally, depending on the type of reactor and the agitator, especially after onset of the gel effect. These effects may lead to bimodal molecular weight distributions of polymers (Fig. 5-5), branched polymer molecules, reactor runaways, explosions, and degradation and/or discoloration of polymers.

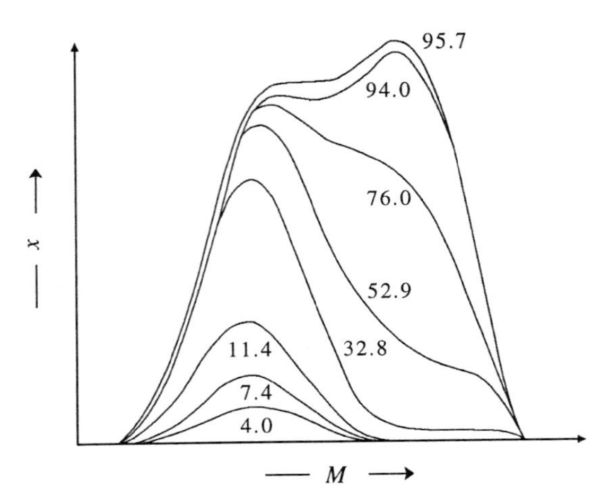


Fig. 5-5 Differential molecular weight distributions of polymers by size-exclusion chromatography from the bulk polymerization of methyl methacrylate by 0.5 wt% *N*,*N*-azobisisobutyronitrile at 70°C [5]. Numbers indicate the percentage of monomer conversion.

Industrial chain polymerizations in bulk are therefore often terminated at monomer conversions of 40–60 %. The residual monomer is distilled off. Alternatively, polymerizations can be conducted in two steps. The polymerization first proceeds in large batch reactors to medium monomer conversions. The melt is then further polymerized in thin layers at walls, through capillaries, or in thin jets of liquid in free fall.

Better heat control and narrower molecular weight distributions are obtained by continuous bulk polymerizations albeit with more complex and expensive equipment.

### 5.2.4 Suspension Polymerizations

Suspension polymerizations are "water-cooled" chain polymerizations in which water-insoluble monomers are dispersed in water as droplets of 0.001–1 mm diameter with the help of dispersants (see below). Agitation breaks up the droplets to smaller entities which are stabilized by the dispersant and polymerized by monomer-soluble ("oil-soluble") free-radical initiators. Nonporous polymer "beads" of usually 50–1000 μm diameter are formed if the polymer is soluble in the monomer. Suspension polymerizations are therefore also called "(suspension) bead polymerizations." Monomer-insoluble polymers give porous and often irregular particles that are composed of many smaller primary entities, hence the designation "suspension powder polymerization."

Suspension polymerizations are usually performed in tank reactors because the simultaneous requirements of good agitation and high monomer conversions are not easily achieved by continuous-flow processes.

The emulsions of "suspended monomers" are inherently unstable. Coalescence of the monomer droplets is prevented by addition of suspending agents such as protecting colloids, ionic surfactants, or Pickering emulsifiers (see p. 159).

*Protecting colloids* are water-soluble organic polymers such as poly(vinyl alcohol). They increase the viscosity of the system and decrease the difference in densities of monomer and water, both of which lead to less frequent collisions of monomer droplets.

Protecting colloids also increase the interfacial tension between monomer droplets and water. *Ionic detergents* charge monomer droplets which reduces the number of collisions and agglomerations; they are used in concentrations below the critical one.

*Pickering emulsifiers* are finely dispersed, water-insoluble, inorganic chemical compounds such as barium sulfate. After polymerization, they are easily separated from polymer beads; for this reason, they are often preferred over protecting colloids and ionic detergents. A certain fraction of Pickering emulsifiers is buried in the beads, however. This fraction decreases with decreasing bead diameter.

Suspension polymerizations require monomers with very low solubilities in water. A somewhat water-soluble monomer will lead to some polymerization outside the monomer droplets. The size distribution of the resulting polymer beads will then have a long tail of small particles. This "polymer dust" is undesirable since it may lead to inhomogeneous melts on polymer processing.

Copolymerizations of monomers with differing solubilities in water lead to beads that have a composition gradient between the interior and the exterior. The same effect is observed if polymerization rates differ appreciably.

Dispersions of monomers for suspension polymerizations are not thermodynamically stable since they have positive Gibbs energies of formation. As metastable systems, they would coalescence with time. Thus, they need to be kinetically stabilized by stabilizers (effect of interfacial tension) and agitation (speed and relative size of agitator), both of which control the size of the resulting polymer beads. For low concentrations of the dispersed phase and small changes in viscosity, the size is described by the Weber number

(5-19) $We = (\rho\, v^2\, d_a^3) / (\gamma g)$

where $\rho$ = total density of the 2-phase system, $v$ = rotational speed of the agitator, $d_a$ = diameter of the agitator, $\gamma$ = interfacial tension, and $g$ = acceleration due to gravity. The diameter of beads decreases strongly with increasing Weber number and then passes through a minimum (Fig. 5-6). The diameters of beads from industrial suspension polymerizations are usually 50–400 μm but sizes of up to 10 000 μm are possible.

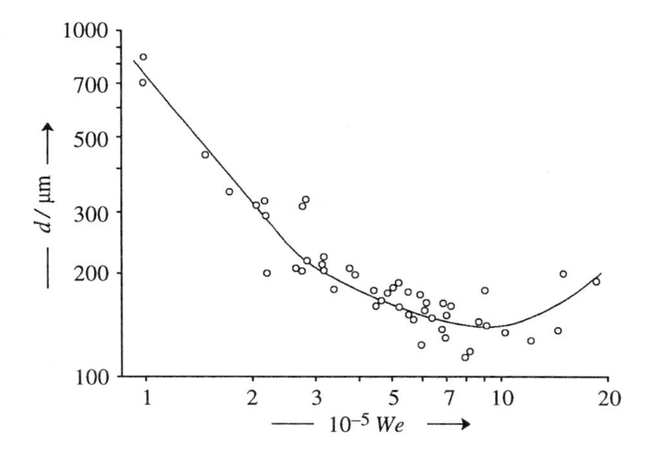


Fig. 5-6 Bead diameter $d$ as a function of the Weber number, *We*, for the suspension polymerization of vinyl chloride in water [6]. By permission of the Society of Plastics Engineers, Brookfield, CT.

Sizes of monomer droplets and emerging polymer beads are large enough to accommodate between 100 and 10 000 radicals. The polymerization thus resembles that of bulk polymerizations albeit with much better heat control because of the rapid transfer of heat of polymerization from the droplets/beads to the surrounding water. Similar to bulk polymerizations, gel and glass effects occur.

Monomers for suspension polymerizations need to lead to polymers that have glass temperatures $T_G$ higher than polymerization temperatures $T_P$. Polymers with $T_G < T_P$ deform easily and tend to agglomerate, etc.

Suspension polymerizations are easy to control (low viscosity, good heat control). They deliver polymer particles directly so that post-polymerization pelletizing is unnecessary. Disadvantageous is the small volume/time yield, the expensive removal and purification of water, and the always incomplete removal of suspending agents which can lead to undesirable product properties such as cloudy polymers and accelerated aging.

Special problems arise for polymers that are insoluble in their monomers, for example, poly(vinyl chloride). The resulting polymer particles are porous and have broad size distributions which leads to low bulk densities. Such suspension polymers also do not gel with plasticizers as nicely as polymers from bulk polymerizations. Another problem is a grafting of suspending agents on polymer molecules. This leads not only to rough surfaces but also to low bulk densities (in injection molding, higher pressures are needed for compacting) and problems with the take-up of plasticizers.

## 5.2.5 Emulsion Polymerizations

**Phenomena**

Emulsion polymerizations are heterophasic free-radical polymerizations of monomers that are sparingly soluble in solvents; totally insoluble monomers do not work, e.g., dodecyl methacrylate in water. Monomers are dispersed in these solvents by surfactants which are present in concentrations that exceed their critical micelle concentration, CMC. This feature distinguishes emulsion polymerizations from suspension polymerizations where surfactant concentrations are always smaller than CMC.

The dispersed monomers form droplets with diameters that depend on the polymerization system. One distinguishes opaque macroemulsions with droplet diameters of ca. 1–10 μm from miniemulsions with droplets of ca. 0.1–0.2 μm and transparent microemulsions with 0.005–0.1 μm. After polymerization, polymer particles are obtained with diameters between 0.01 μm and 1 μm and with special recipes also up to 100 μm. These polymer particles remain dispersed in the solvent; they are therefore also called polymer dispersions or polymer latices. Polymers from these polymerizations are known as emulsion polymers or polymer colloids.

The first recipes for emulsion polymerizations were found empirically during experiments that tried to obtain synthetic latices similar to the latex of natural rubber. These experiments showed that water-soluble initiators such as $K_2S_2O_8$ were much more effective for dispersed water-"insoluble" monomers than monomer-soluble ones such as dibenzoyl peroxide. The polymerization must therefore occur in the aqueous phase and not in the monomer droplets. This conclusion was confirmed by experiments in which the monomer was layered on the aqueous phase or in which the aqueous phase and the monomer were separated and only connected via a common gas space.

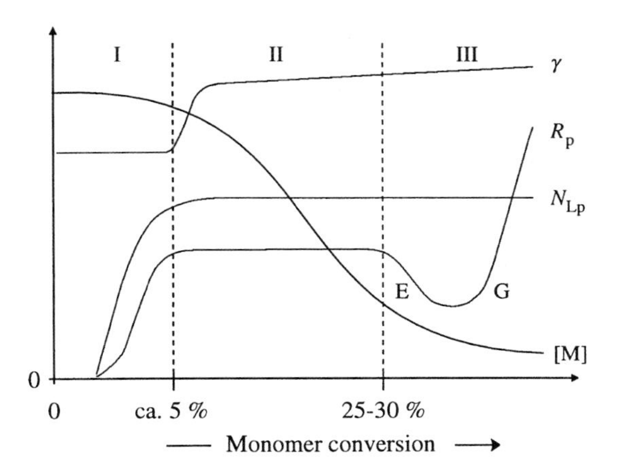


Fig. 5-7 Schematic representation of the dependence of monomer concentration, [M], surface tension $\gamma$, rate of polymerization, $R_p$, and number of latex particles, $N_{Lp}$, on monomer conversion in the *ab initio* emulsion polymerization of water-insoluble monomers.

In so-called *ab initio* polymerizations (Fig. 5-7), monomer concentrations decrease first slowly with increasing extent of polymerization (Period I), then fast (Period II), and finally slowly again (Period III). However, the polymerization rate $R_p$ is not proportional to the monomer concentration as required by Eq.(5-13). It rather increases fast in Period I, becomes constant in Period II, and decreases again in Period III before it sometimes increases again. The surface tension $\gamma$, on the other hand, is constant during Period I, increases fast after the end of Period I and then only slowly in Periods II and III. The number of latex particles, $N_{Lp}$, grows strongly during period I but then stays constant during the rest of polymerization.

Emulsion polymerizations obviously do not follow the classical schemes for homogeneous polymerizations. The key for the observed phenomena is the variation of surface tension. Surfactant molecules are amphiphilic molecules with an apolar tail and a hydrophilic head group. In water, they self-assemble spontaneously to so-called micelles in which the apolar tails associate in the interior whereas the polar head groups face the surrounding water. This behavior can be modeled by an all-or-nothing mechanism (closed association), $N\,S \rightleftarrows S_N$, in which $N$ non-associated surfactant molecules $S_1$ are *in equilibrium* with their N-mers $S_N$, called micelles. The presence of N-mers becomes only noticeable above a certain critical micelle concentration, CMC, above which the concentration $c_N$ of N-mers grows strongly with increasing total concentration $c$ of surfactant while the concentration $c_1$ of unimers increases only slightly (Fig. 5-8).

Equilibria also exist between surfactant molecules in the interior and on the surface of surfactant solutions. Above the CMC, the surface tension becomes practically independent of the total surfactant concentration (Fig. 5-8) because the water-air surface is almost completely covered by a laminar layer of surfactant molecules that have their hydrophilic heads in the water and their hydrophobic tails in the air.

The CMC by surface tension coincides approximately with the CMC as measured by molecular weight determinations. These two CMCs are not identical, however, because the arrangement of surfactant molecules at the liquid–air interface is planar whereas the micelles in the interior of liquids may adopt various shapes.

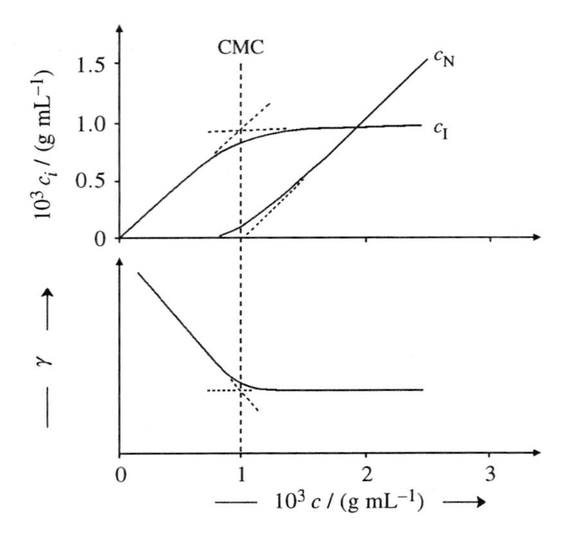


Fig. 5-8 Top: Weight concentrations $c_i$ (= $c_1$ of unimers or $c_N$ of N-mers) as a function of the total concentration $c$ for a closed association of molecules with molecular weight $M_1 = 200$, degree of association $N = 21$, and equilibrium constant of $K = [S_N]/[S_I]^N = 1 \cdot 10^{45}$ (L/mol)$^{20}$. Bottom: schematic representation of the concentration-dependence of surface tensions $\gamma$ of a surfactant.

In dilute solution, surfactant molecules associate in equilibrium to spherical micelles composed of ca. 20–100 surfactant molecules (Fig. 5-9). At higher surfactant concentrations, cylindrical (rod-like) micelles appear. At still higher concentrations, both spherical and cylindrical micelles may be composed of double layers of surfactant molecules.

The hydrophobic interiors of spherical micelles have diameters of ca. 4–10 nm. They can accommodate water-insoluble monomer molecules which swells the micelles, for example, to 5.5 nm from 4.3 nm for styrene molecules in potassium oleate micelles in water. Conventional monomer emulsions also contain much larger monomer droplets of ca. 1000–10 000 nm diameter that are non-equilibrium structures.

**Types of Emulsion Polymerizations**

Emulsion polymerizations are controlled by the so-called surface tension driving force and the concentration of the emulsifier. The "driving force" can be positive or negative and the emulsifier concentration larger or smaller than the so-called "stability threshold." There are thus four types of emulsion polymerizations (Fig. 5-10): macroemulsion and microemulsion polymerizations on one hand and so-called direct and inverse ones on the other.

The "surface tension driving force" $\Delta\gamma$ is not a force (in newton N) but a surface tension (in N/m or J/m$^2$) and thus an energy per unit area. It is defined as $\Delta\gamma = \gamma_{M-L} - \gamma_{A-H}$ where $\gamma_{M-L}$ is the energy/area for the contact between the monomer phase M and the lipophilic segments L of the surfactant molecules and $\gamma_{A-H}$ the energy/area for the contact between the aqueous phase A and the hydrophilic parts of the surfactant molecules.

Positive driving forces lead to oil-in-water dispersions and to direct emulsion polymerizations (for historic reasons usually without the adjective "direct"). Negative driving forces result in so-called inverse (i.e., water-in-oil) emulsion polymerizations.

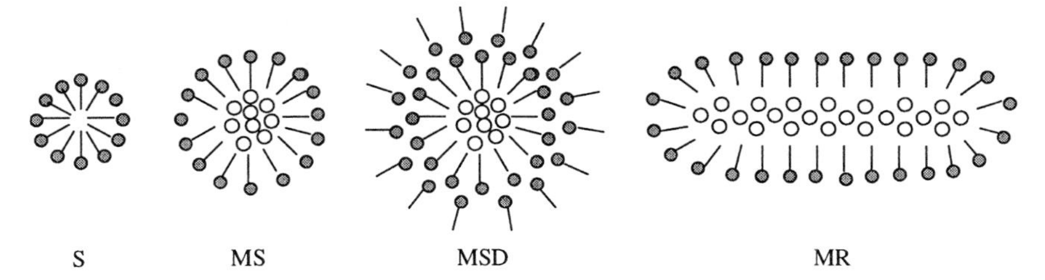


Fig. 5-9 Schematic representation of cross-sections of spherical surfactant micelles (S), spherical micelles with incorporated monomer molecules (MS), monomer-containing spherical micelles with a double layer (MSD), and monomer-loaded rod-like micelles (MR), all in water as the continuous phase. ● Polar head group of surfactant molecule, — hydrophobic tail of surfactant molecule, ○ monomer molecule.

Emulsion polymerizations may be kinetically or thermodynamically stable. Suspension polymerizations and macroemulsion polymerizations are not thermodynamically stable since both types consist of dispersions of large monomer droplets. The stability of these droplets can only be maintained by agitation, regardless of whether the surfactant concentration $c$ is lower than the CMC in the surfactant-solvent system (suspension polymerization) or somewhat higher (macroemulsion polymerization).

However, the requirement of $c$ > CMC for emulsion polymerizations gives rise to additional monomer-filled surfactant micelles that do not exist in suspension polymerizations. Polymerization in emulsion also leads to smaller polymer particles than polymerization in suspension: diameters of particles are 0.01–1 μm diameter by emulsion polymerization versus 50–1000 μm by suspension polymerization.

At much higher surfactant concentrations, surfactant micelles become stabilized by double layers of surfactant molecules (Fig. 5-9). They may also not only be spherical but rod-like. Higher surfactant concentrations disperse more monomer molecules in micelles than in monomer droplets and the emulsion becomes thermodynamically stable. There are therefore 4 types of emulsion polymerization: kinetically stable versus thermodynamically stable and oil-in-water versus water-in-oil (Fig. 5-10).

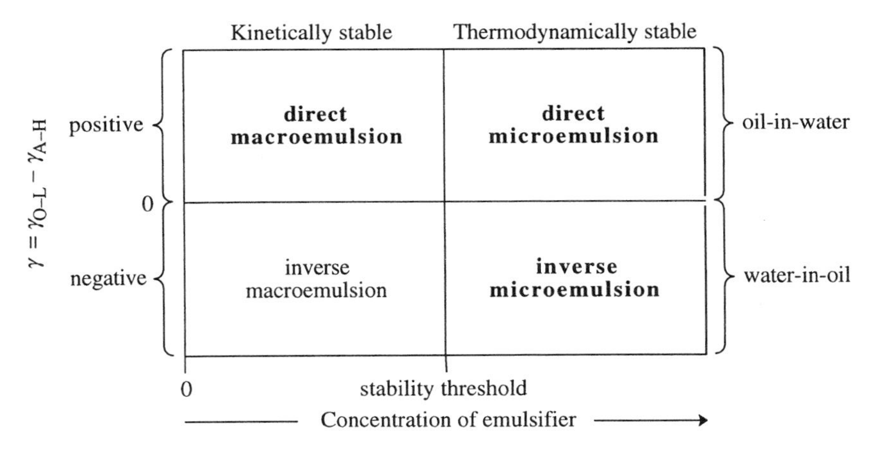


Fig. 5-10 The four types of emulsion polymerizations [7]. Bold: experimentally observed types. See text for explanation.

**Direct macroemulsion polymerizations** are usually only called **emulsion polymerizations** since they are the oldest and most common ones. They employ hydrophobic monomers that are dispersed by surfactants in water as the continuous phase. The polymerization is initiated by water-soluble initiators. Radicals from these initiators travel to the monomer-filled micelles where the polymerization is initiated. Monomer molecules from monomer droplets diffuse to the polymerizing micelles which grow to become latex particles and later polymer particles. The polymerization in the particles is stopped by the entry of another radical.

Practically all industrial emulsion polymerizations are of this type. Because of their importance, these emulsion polymerizations will be discussed in greater detail in the following subsections.

A subcategory of direct macroemulsion polymerization is **miniemulsion polymerization**. The size of the resulting polymer particles here is reduced by cosurfactants with long hydrocarbon chains such as hexadecane or cetyl alcohol that penetrate the surface of the monomer droplets less than monomer molecules (Fig. 5-11) and subject the system to high shear fields. The curvature of the surface of monomer droplets is reduced and therefore also their size. Since sizes of monomer-swollen micelles, latex particles, and monomer droplets become comparable, they all compete for radical entry, both for initiation and termination. As a consequence, the resulting polymer particles are much smaller than in (macro)emulsion polymerization, hence the term "miniemulsion."

**Direct microemulsion polymerizations** are thermodynamically stabilized by mixtures of classical surfactants and weak amphiphilic cosurfactants, for example, short-chain alcohols. Because of much higher total surfactant concentrations, microemulsions consist of much greater concentrations of monomer-filled micelles and far lower concentrations of monomer droplets than macroemulsions. Additional monomer can only be supplied to the polymerization-active micelles by either diffusion from or collision with an inactive micelle. Termination of growing chains is again by entry of a second radical. As a result, the chance of active micelles growing to larger polymer particles is restricted: the size of particles from microemulsion polymerizations is one to two decades smaller than that from macroemulsion polymerizations.

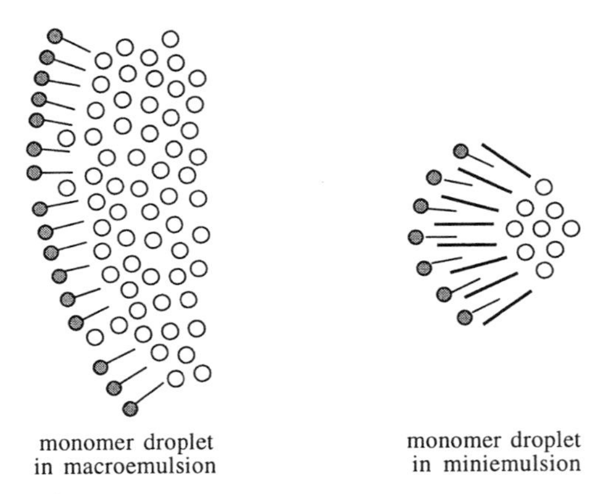


Fig. 5-11 Schematic representation of sections of monomer droplets in macroemulsions and miniemulsions. ●— surfactant molecule with hydrophilic head ● and hydrophobic tail —, ○ monomer molecule, —— additive with long hydrocarbon chain.

**Inverse microemulsion polymerizations** are polymerizations of thermodynamically stabilized water-in-oil emulsions. An example is the polymerization of a concentrated aqueous solution of acrylamide in paraffin oil with sorbitol monooleate as emulsifier and redox systems as initiators. Industrial polymerizations apparently do not exist.

**Inverse macroemulsion polymerizations** have as yet not been proven to exist.

**Processes in Direct Macroemulsion Polymerizations**

The basic processes in classical emulsion polymerization (= direct macroemulsion polymerizations of sparingly soluble monomers) were first depicted correctly in 1947 by W.D.Harkins and later refined by many workers (Fig. 5-12).

The polymerization system consists initially of a continuous aqueous phase and a dispersed oil-phase of emulsified monomer droplets. The aqueous phase contains dissolved free-radical initiator molecules, monomer-filled surfactant micelles, and, since no monomer is completely insoluble, also a very low concentration of monomer molecules (for example, the solubility of styrene in water is ca. 0.07 g/L). The hydrophilic radicals I• from the dissociation of initiator molecules pick up 1–3 dissolved monomer molecules and become amphiphilic oligoradicals $I\text{–}M_i$• that may, in principle, either travel to monomer-swollen micelles or to dispersed monomer droplets, and, in later stages of the polymerization, also to polymerizing micelles and latex particles, respectively.

In Period I (p. 161) of the emulsion polymerization of, e.g., styrene, the system contains ca. $10^{21}$ monomer-swollen micelles per liter of ca. 9 nm diameter and ca. $10^{12}$ monomer droplets of ca. 5000 nm diameter. The total surface area of all micelles (ca. $2.5 \cdot 10^{23}$ nm$^2$) is thus far greater than the total surface area of all monomer droplets (ca. $8 \cdot 10^{19}$ nm/L). The oligomer radicals will therefore preferentially attack the monomer molecules in the micelles and not those in the monomer droplets. Furthermore, the low concentration of dissolved monomer molecules and the high diffusion rate of oligomeric initiator radicals also rule out a polymerization of dissolved monomer molecules.

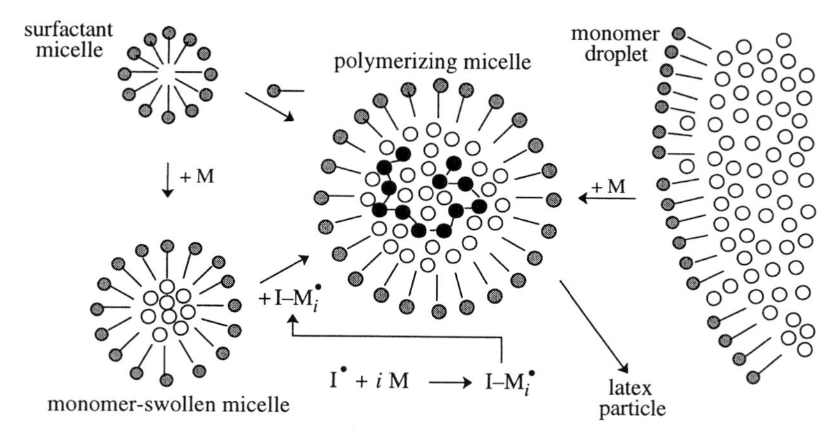


Fig. 5-12 Simplified representation of an emulsion polymerization. ●— surfactant molecules with hydrophilic head ● and hydrophobic tail —, O monomer molecule M, ●-●-●-● growing polymer chain, I = initiator residue. Note that in direct macroemulsion polymerization, conventional monomer molecules O can enter the emulsifier "wall" of monomer droplets, which increases the curvature and leads to larger monomer particles. See also Fig. 5-11.

The constancy of surface tension in Period I (Fig. 5-7) indicates a constant number of micelles. With time, more and more micelles contain a growing radical. The increasing radical concentration should thus lead to a greater rate of polymerization. However, this is counteracted by a decrease of the average concentration of monomer molecules in micelles which should finally completely stop the polymerization.

This, however, does not happen. The conversion of monomer molecules to monomeric units disturbs the solubilization equilibrium in micelles, which causes a diffusion of monomer molecules from thermodynamically unstable monomer droplets to polymerizing micelles. Each small monomer-loaded micelle grows and becomes a larger latex particle. The macroradical in such a particle adds further monomer molecules which causes its molecular weight to increase substantially. The growth of a macroradical stops when a second radical enters. This radical terminates the chain because the rather small "reaction compartment" cannot accommodate two radicals.

The conversion of monomer-swollen micelles to bigger and polymer-containing latex particles leaves the latter short of stabilizing surfactant molecules which are therefore replenished from dissolved surfactant molecules or empty surfactant micelles. At the end of Period I, this supply is exhausted. The concentration of molecularly dissolved surfactant molecules falls below the critical micelle concentration and the surface tension increases strongly (Figs. 5-7 and 5-8).

New latex particles can no longer be formed; Period II is therefore characterized by a constant number of latex particles (Fig. 5-7). However, monomer molecules continue to diffuse from monomer droplets via the aqueous phase to latex particles where they polymerize. The volume of each latex particle grows steadily with the monomer conversion $u$, i.e., the diameter $d$ of the latex particles is proportional to the cubic root of the monomer conversion $u$ (Fig. 5-13).

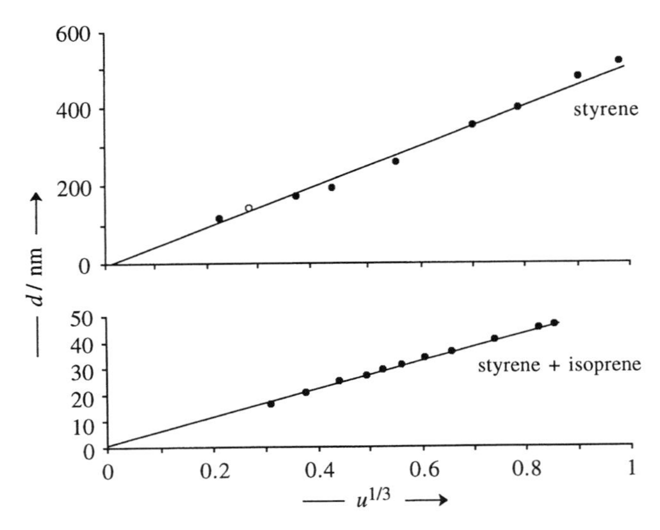


Fig. 5-13 Increase of the diameter $d$ of latex particles with the cubic root of monomer conversion $u$.

Top: 9.1 wt% styrene plus 0.18 w% dipotassium persulfate in an aqueous solution of 0.5 wt% of a sodium alkylsulfonate with $M \approx 300$ g/mol as emulsifier [8].

Bottom: 0.36 wt% of styrene/isoprene (25/75) and 2.15 wt% potassium myristinate as emulsifier plus traces of dipotassium persulfate as initiator and a mercaptan as regulator.

A latex particle can only accommodate one macroradical since the entry of another radical will immediately lead to termination because of the small size of the reaction volume. Comparison of the diameters of latex particles with the molecular dimensions of polymer molecules shows, however, that a latex particle must contain many polymer molecules. This must be caused by a succession of initiation and termination events. The time between the entry of the first (polymerizing) and the second (terminating) radical is however fairly long, so that many monomer molecules can be polymerized by one radical. Emulsion polymerizations proceed therefore with higher rates to higher molecular weights than bulk polymerizations where termination is much more frequent.

The dynamic equilibrium between the disappearance of monomer molecules by polymerization and their replenishment by diffusion leads to constant monomer concentrations in latex particles. The polymerization rate is therefore of zeroth order with respect to the monomer concentration.

At the end of Period II, all monomer from monomer droplets has been delivered to latex particles. In the subsequent Period III, polymerization steadily reduces the monomer concentration in latex particles and the polymerization rate becomes first order with respect to monomer concentration.

**Rates of Polymerization**

In 1948, Smith and Ewart were the first to categorize and quantify the processes in emulsion polymerizations which can be divided into ab initio and seeded polymerizations. *Ab initio* polymerizations start with aqueous mixtures of monomers, emulsifiers, and initiators (*de novo* systems) which subsequently lead to radicals in Period I and latex particles in Period II. For *seeded* systems, a latex with known concentration of latex particles is prepared first which is then used as seed.

In emulsion polymerizations, most monomer is polymerized in latex particles Lp (Fig. 5-7). The first part of the equation for the rate of polymerization, Eq.(5-15), is therefore replaced by

$$(5\text{-}20) \qquad R_{\mathrm{p}} = -\,\mathrm{d[M]}/\mathrm{d}t = k_{\mathrm{p}}[\mathrm{M}]_{\mathrm{Lp}}[\mathrm{R}^{\bullet}]_{\mathrm{Lp}} = k_{\mathrm{p}}[\mathrm{M}]_{\mathrm{Lp}}\{\mathrm{L}\}\cdot N_{\mathrm{R,Lp}}/N_{\mathrm{A}}$$

where the monomer concentration, $[\mathrm{M}]_{\mathrm{Lp}}$, and the radical concentration $[\mathrm{R}^{\bullet}]_{\mathrm{Lp}}$ are the molar concentrations of monomer and radicals, respectively, in all latex particles (and not in the total volume), $\{\mathrm{L}\}$ = number concentration of latex particles in the continuous phase (= number of particles per volume of aqueous phase), $N_{\mathrm{R,Lp}}$ = average number of radicals per latex particle, averaged over the distribution of latex particles, and $N_{\mathrm{A}}$ = Avogadro constant.

The rate constant of propagation, $k_{\mathrm{p}}$, is usually obtained from bulk polymerization. Without further assumptions, $k_{\mathrm{p}}$ can also be calculated from the measured propagation rate $R_{\mathrm{p}}$, the experimentally determined molar concentration $[\mathrm{M}]_{\mathrm{Lp}}$ of monomer in latex particles, and the concentration $[\mathrm{R}^{\bullet}]_{\mathrm{Lp}}$ of radicals in latex particles (Eq.(5-20)). Data from ***de novo*** (seeded) emulsion polymerizations showed that the propagation rate constant $k_{\mathrm{p}}$ remains constant to high mass fractions $w_{\mathrm{P,Lp}}$ of polymer in latex particles (Fig. 5-14). At still higher mass fractions, e.g., $w_{\mathrm{P,Lp}} > 0.7$ for the system in Fig. 5-14, $k_{\mathrm{p}}$ decreases rapidly because of diffusion control by the gel effect (see also Fig. 5-3).

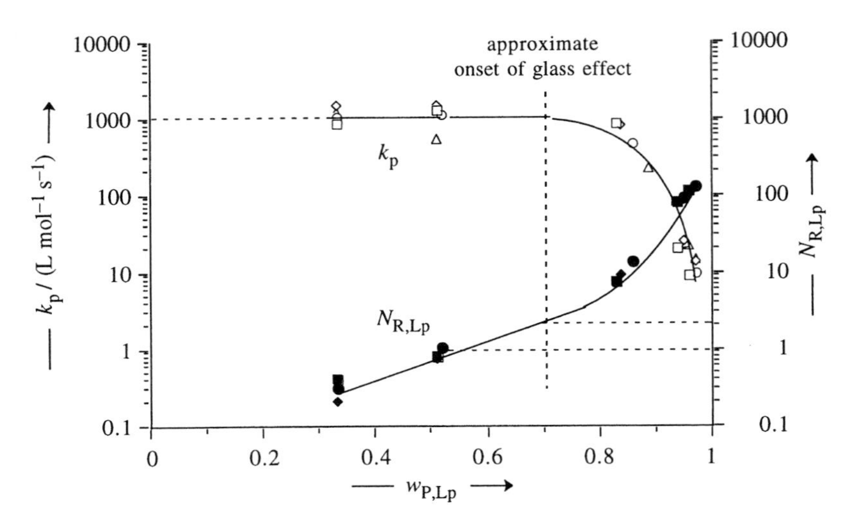


Fig. 5-14 Dependence of propagation rate constants, $k_p$, and average number of radicals per particle, $N_{R,Lp}$, on the mass fraction of polymers in latex particles in the seeded emulsion polymerization of methyl methacrylate at 50°C [9].

The molar concentration of monomer in latex particles, $[M]_{Lp}$, is assumed to be that of the equilibrium state; it is determined by swelling experiments. Use of this assumption in Eq.(5-20) presumes, however, that latex particles are the predominant loci of polymerization, which is not always true.

The number concentration of latex particles, {L}, is constant and exactly known for seeded systems.

The average number of radicals per latex particle, $N_{R,Lp} = [R^{\bullet}]_{Lp}N_A/\{L\}$, can be calculated from the number concentration of latex particles, {L}, and the molar concentration $[R^{\bullet}]_{Lp}$ of free radicals as measured by electron spin resonance (Fig. 5-14). The logarithm of the number of radicals per particle increases linearly with the mass fraction of polymer in latex particles up to the onset of the glass effect, which allows a latex particle to accommodate two or more radicals.

The situation is different for ***ab initio*** emulsion polymerizations where the concentrations of radicals and latex particles increase with time until they become constant in Period II (Fig. 5-7). This more complicated type of emulsion polymerization was first successfully modeled by Smith and Ewart for three types I, II, and III.

The **Smith–Ewart Case II** considers that the concentration $[R^{\bullet}]_{Lp}$ of radicals in latex particles can only be constant if the rate of formation of active radicals equals the rate of their disappearance. A radical that enters a latex particle will start a polymerization. However, since the reaction volume is small, an entering second radical will rather terminate the already present macroradical than start a new polymer chain. Also, termination rates are most likely higher than the entrance rates of radicals.

The time from the entry of the second radical to its deactivation is therefore much smaller than the time between the entries of the first radical and second radical. If the exit of radicals from latex particles can be neglected, then a latex particle will either harbor one radical or none at all. The time-averaged number of radicals in latex particles is thus $N_{R,Lp} = 1/2$.

This assumption is only valid for small monomer conversions. $N_{R,Lp}$ can exceed 1/2 at higher monomer conversions (glass effect) and larger sizes of latex particles (**Smith–Ewart Case III**). A rapid transport of radicals out of latex particles decreases the average number of radicals to $N_{R,Lp} < 1/2$ (**Smith–Ewart Case I**).

### Formation of Latex Particles

A part of emulsion polymers is directly used as latex (dispersion) and it is therefore of interest how the concentration and size distribution of latex particles depends on the concentrations of initiator and surfactant.

The volume $V_E$ of the emulsion is assumed to contain $N_{R,E}$ oligoradicals with a number concentration $\{R^\bullet\} = N_{R,E}/V_E$. An oligoradical enters the monomer-swollen micelle at a time $t_e$ with a rate of $R_{R,e} = d\{R^\bullet\}/dt$. The entering radical starts the polymerization whereby the micelle is converted to a latex particle. Polymerization lets the volume $V_{Lp}$ of each latex particle increase with a constant rate, i.e., $dV_{Lp}/dt = const.$ After a time interval $t - t_e$, the surface area $A_{Lp}$ of the latex particle has grown to

$$(5\text{-}21) \qquad A_{Lp} = \left[3(4\pi)^{1/2} \int_{t_e}^{t} \frac{dV_{Lp}}{dt} dt_e\right]^{2/3} = \left[3(4\pi)^{1/2} \frac{dV_{Lp}}{dt}\right]^{2/3} (t - t_e)^{2/3}$$

The total surface area $A$ of all latex particles in the total volume $V_E$ of the emulsion is

$$(5\text{-}22) \qquad A = R_{R,e} \int_0^t A_{Lp} dt_e = 0.60\, R_{R,e} \left[3(4\pi)^{1/2} \frac{dV_{Lp}}{dt}\right]^{2/3} t^{5/3}$$

Surfactant molecules with a mass concentration of $c_S = m_s/V_E$ occupy a specific area $a_S = A_s/m_S$. The area $A_s$ of all surfactant molecules equals the surface area $A_{Lp}$ of all latex particles if all micelles have disappeared and all surfactant molecules are only on the surface of the latex particles, which is the situation at the time $t_I$ at the end of Period I (Fig. 5-7). The time interval $\Delta t_I = t_I - t_0$ (where $t_0 = 0$) is therefore given by

$$(5\text{-}23) \qquad t_I = (a_s c_s)^{3/5} \{0.60\, R_{R,e} [3(4\pi)^{1/2} (dV_{Lp}/dt)]^{2/3}\}^{-3/5}$$

$$(5\text{-}24) \qquad t_I = 0.53\, (a_s c_s / R_{R,e})^{3/5} \{(dV_{Lp}/dt)^{-1}\}^{2/5}$$

At the end of Period I, the number concentration $N_{R,e}/V_E$ of the maximum number of latex particles per volume of the emulsion is therefore

$$(5\text{-}25) \qquad N_{R,e}/V_E = R_{R,e} t_I = 0.53 [R_{R,e}/(dV_{Lp}/dt)]^{2/5} (a_s c_s)^{3/5}$$

The concentration $N_{R,e}/V_E$ of the latex particles increases with the 3/5th power of the surfactant concentration $c_S$ which has been confirmed for styrene and ring-substituted styrenes. The exponent becomes smaller than 0.6 for partially soluble monomers, for example, vinyl acetate.

The factor 0.53 is an upper limit since radicals can enter not only micelles but also latex particles. Model calculations resulted in a lower limit of 0.37.

Eq.(5-25) also predicts the dependence of concentration of latex particles on surfactant and initiator concentration. According to the Smith-Ewart Case II for the beginning of Period II, each latex particle contains either one radical or no radical. The number concentration of radicals, $N_{R,E}/V_E$, is therefore proportional to the number concentration {L} of latex particles. However, the entry rate $R_{R,e}$ of radicals is proportional to their rate of formation, $R_i$, which is proportional to the initiator concentration, [I]. The mass concentration $c_s$ of surfactant molecules can furthermore be replaced by the molar concentration [S]. Eq.(5-25) becomes

(5-26) $\{L\} = const\ [I]^{0.4}[S]^{0.6}$

Discontinuous emulsion polymerizations lead to latices with narrow particle distributions. The polymerization is often stopped at monomer conversions of ca. 40 % since larger particle concentrations increase the probability of coagulation.

**Properties of Emulsion Polymers**

Emulsion polymerizations offer many advantages. Because of the smallness of latex particles, heats of polymerization are rapidly transferred to the surrounding water of which the temperature can be easily controlled. Redox initiators allow polymerizations with high rates to high molecular weights at low temperatures. Non-polymerized monomer is easy to remove by steam treatment.

A disadvantage is the difficult removal of surfactant residues. Such residues make polymers more hydrophilic, which in turn increases dielectric losses in electrical applications. In poly(vinyl chloride), one of the major emulsion polymers, certain emulsifiers also catalyze the splitting-off of hydrochloric acid. The emulsifier concentration is therefore held low. Alternatively, non-ionic or saponifiable surfactants are used.

Latices ("latexes")from emulsion polymerizations can be used directly for adhesives and coatings or for the finishing of leathers. For these applications, control of the size distribution of latex particles is crucial. For example, only initially formed latex particles will grow and no new ones will be formed if monomer and initiator are added continuously to a system that initially consisted only of water and emulsifier. The resulting latex particles are relatively small and have a narrow distribution of particle sizes.

The situation is different if a part of the water-emulsifier-monomer-initiator mixture is polymerized and then the rest of the emulsified monomer is added. The initial latex particles will become very large and the final ones relatively small, resulting in dispersions with broad distributions of particle sizes.

The concentration of polymers in latex particles is high. This increases the probability of chain transfer to polymer molecules and thus the probability of branching.

The composition of emulsion copolymers depends on the relative rate constants and the local monomer concentration. Two "insoluble" monomers lead to the same composition for emulsion and bulk copolymerizations. However, different water solubilities of comonomers will cause different copolymer compositions since the ratio of monomer concentrations in the oil phase (the growing latex particles) differs from that in the aqueous phase. The true local monomer ratio can be calculated from the initial ratio of monomer concentrations and the corresponding distribution coefficients.

### 5.2.6 Solution Polymerizations

Some chain polymerizations, polycondensations, and polyadditions are conducted in solvents. Because of the cost of solvents and solvent recovery and work-up, they are generally more expensive than polymerizations in bulk, suspension, or emulsion.

Solution polymerizations are advantageous if the polymers are subsequently processed as a polymer solution, for example, as lacquers. However, because of high solvent costs, the necessity to protect workers from vapors, and environmental concerns, solvent-based lacquers are increasingly being replaced by dispersion polymers.

Practically all industrial *ionic polymerizations* are conducted in solution. The polarities of solvent and monomer determine the extent of dissociation of initiators and growing macroions to free ions, tight and loose ion pairs, and ion associates with widely different propagation rates (see Volume I). Except for free ions (which are usually only present in very small concentrations), their temperature dependences do not follow the Arrhenius type. Generalizations about ionic polymerizations are therefore difficult.

Solvents act as diluents in *free-radical polymerizations*. They allow a better dispersion of the heat of polymerization, but they also reduce monomer concentrations, which lowers polymerization rates. Dilution also diminishes radical transfer to polymer molecules (less branching), which narrows molecular weight distributions.

Chain transfer to solvent molecules, on the other hand, may lead to very active solvent radicals and thus to higher polymerization rates. Certain solvents also induce initiator decompositions which increases polymerization rates further.

Especially strong effects are caused by polymerizations in non-solvents and thermodynamically bad solvents, i.e., those with small interactions between polymer and solvent molecules (Volume III). An example is the polymerization of methyl methacrylate in various solvents. In bad solvents, growing macroradicals are much more coiled than in good solvents which, at higher monomer conversions, leads to stronger diffusion control of terminations (gel effect). In turn, this results in higher degrees of polymerization, higher viscosities, and greater polymerization rates (Fig. 5-15).

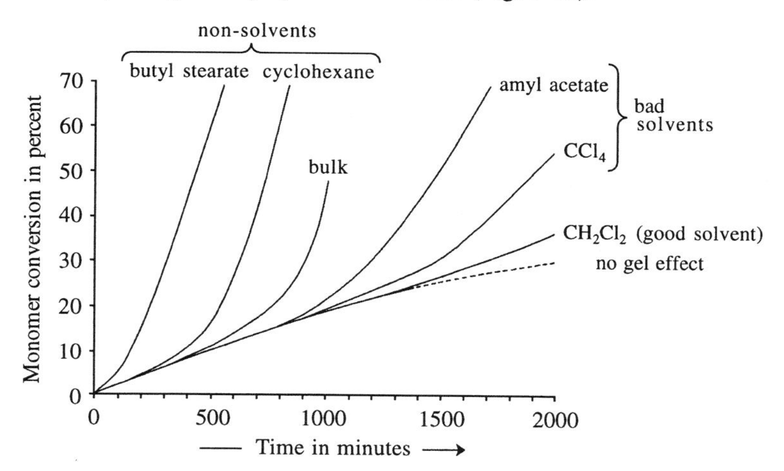


Fig. 5-15 Time-dependence of conversion of methyl methacrylate at 40°C and constant initiator concentration in various environments [10]. - - - - Ideal polymerization without gel effect.
By permission of MacMillan Magazines Ltd., London.

*Polycondensations in solution* are preferred over those in melts if monomers and/or polymers are thermally unstable. They are usually performed at monomer concentrations of ca. 20 %. Leaving molecules (water, alcohol, etc.) are usually distilled off, for example, water by azeotropic distillation with, e.g., toluene or carbon tetrachloride.

Water can also be removed by continuous evaporation in thin layers. The polymer solution enters the top of a heated packed column where the water is removed by a countercurrent of carbon dioxide. The process delivers very lightly colored products since overheating is absent. Examples of such solution polycondensations are the AA/BB polycondensations to aliphatic polyamides and aromatic polyesters.

*Interfacial polycondensation* is a special type of polycondensation. The process was formerly one of the two industrial methods for the synthesis of polycarbonate A from bisphenol A and phosgene, but is presently only used for a special copolyamide from *m*-phenylene diamine and the acid chlorides of isophthalic and terephthalic acid.

The polymerization proceeds at the interface of two immiscible liquids, usually water and an organic solvent, that each contain one of the monomers. The resulting polymer precipitates at the interface, either as a film if the liquids form two separated microphases or as a fine suspension if the two liquids form a dispersion.

The process is fast and therefore very attractive. Its disadvantages are the high costs of acid chlorides, HCl removal, and equipment (corrosion). Use of the sodium salts of bisphenols instead of bisphenols themselves leads to another problem: NaCl particles that remain in the polymer even after washing. These particles are larger than one-half the wavelength of light, hence, the scattered light renders the polymers opaque.

### 5.2.7 Precipitation Polymerizations

Some polymers are insoluble in their own monomers; examples are poly(vinyl chloride), poly(acrylonitrile), and poly(tetrafluoroethylene). Thus, they precipitate at very small monomer conversions. Such precipitations also occur if a solvent for the monomer is a precipitant for the polymer.

In the *free-radical polymerization* of acrylic acid in toluene as precipitant, polymerization rates increase strongly, pass through a maximum, and decrease (Fig. 5-16). The maximum occurs shortly before all polymer particles are formed. Subsequently, the polymerization rate decreases because the monomer diffusion to occluded radicals in the particles is impeded.

The number concentration of precipitated polymer particles, {P}, also increases strongly in the early stage of polymerization (not shown: from {P} = 0 at zero conversion to $\{P\} = 1{\cdot}10^{17}\ L^{-1}$ at 20 % monomer conversion) and then more slowly before becoming constant at ca. 40 % monomer conversion (Fig. 5-16). At this point, all initiator molecules are consumed and/or all radicals are occluded in polymer particles, which prevents termination. Since monomer can still diffuse to the radicals, the polymerization continues and the particle diameter increases linearly with monomer conversion (see also Fig. 5-13). The particles are nano-sized, however, because of the relatively high concentration of added surfactant.

Precipitation polymerizations are advantageous because polymers are directly obtained as solid particles, often even as powders. The use of polymer precipitants in solu-

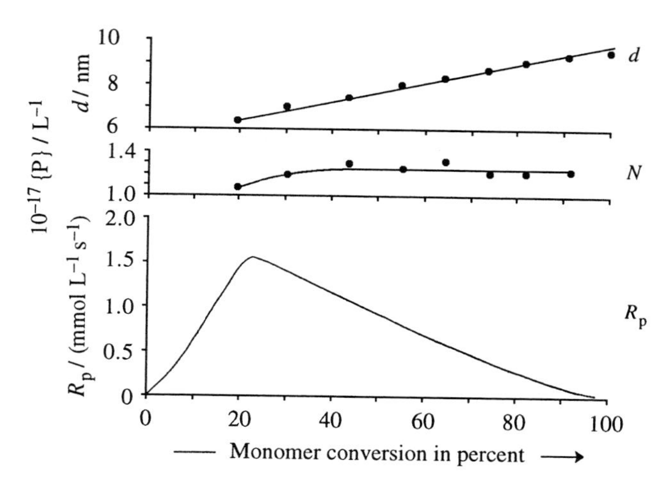


Fig. 5-16 Polymeriztion rate $R_p$ and number concentration {P} of polymer particles of diameter $d$ resulting from the free-radical polymerization of 1 mol/L acrylic acid with 3 mmol/L 2,2'-azobis(2,4-dimethylvaleronitrile) as initiator at 50°C in toluene with the addition of 0.3 mol/L water and 3 mmol/L poly(styrene)-*block*-poly(oxyethylene) as surfactant [11]. Agitation: 300 rotations per minute.

tion polymerizations is thus not uncommon. The disadvantage is of course the cost of the recovery and work-up of the liquid precipitants.

Precipitation also occurs in some solution *polycondensations* and *polyadditions*, albeit as relatively low-molecular weight polymers at medium extents of reaction. An example of such a *polycondensation in suspension* is the reaction of dicarboxylic acid diphenyl esters with diamines in aromatic hydrocarbons with release of phenol and precipitation of the aramid as a fine powder:

(5-27) $$C_6H_5OOC-Ar-COOC_6H_5 + H_2N-Z-NH_2 \longrightarrow \sim\sim OC-Ar-CO-NH-Z-NH\sim\sim + 2\,C_6H_5OH$$

This polycondensation is performed as a precondensation at temperatures of 80–100°C for amorphous polymers and at 130–160°C for semicrystalline ones. The solvent should not react with reactants, must dissolve the phenyl esters, and should not swell the resulting oligomer. The subsequent postcondensation to high molecular weights occurs in the solid state, usually in a fluidized bed, at temperatures at which particles do not stick together.

*Solid-state polycondensations* thus need to be performed at temperatures below the melting temperature of crystalline polymers and the glass temperature of amorphous ones. These polycondensations are fairly rapid and are therefore also used for many polycondensations. For example, the after-condensation of poly(hexamethylene adipamide) (melting temperature: 262°C in the dry state) requires only 2 hours at 216°C to boost the number-average molecular weight to 15 000 (industrial range) from 4000 (precondensate). Polymers from so-called **solid-stating** are more desirable than those from finishing by melt polycondensation because the lower polycondensation temperature of the former leads to less discoloration.

### 5.2.8 Gas-Phase Polymerizations

Gas-phase polymerizations use gaseous monomers. However, the resulting polymers are non-volatile and precipitate. The term "gas-phase polymerization" is thus misleading: the polymerization does not proceed in the gaseous phase but in the resulting aerosol of the precipitating polymer.

Gas-phase polymerizations are chain polymerizations that may proceed free-radically or by coordination catalysts. Because of the great dilution, each growing polymer particle contains only one active center, i.e., a macroradical or a complex between the coordination catalyst and a monomer. New monomer molecules are delivered to this center via the gas phase. The polymerization rate is thus controlled by the rate of monomer absorption by the polymer particles. These particles need to remain porous for the course of polymerization because gas molecules need to access the active centers in the interiors of the particles. Monomers must be absolutely dry since condensable impurities would act as plasticizers. The resulting polymer should not become sticky, i.e., polymerization temperatures must be below melting temperatures (if semi-crystalline) or below glass temperatures (if amorphous). These restrictions leave ethene and propene as candidates for gas-phase polymerizations which are usually performed in fluidized-bed reactors.

At higher temperatures, less monomer is absorbed. Since the monomer concentration at the reaction loci is smaller, polymerization rates will also decrease and the gross activation energy of polymerization is thus negative. The resulting polymers are very pure, especially those by free-radical polymerization. However, those by insertion polymerization may contain catalyst residues that are difficult to remove or render ineffective. A remedy for this situation is the development of highly acitve initiators.

Monomer concentrations, and thus polymerization rates, increase with increasing pressure. For gas-phase reactions, pressures of tens of megapascals have considerable effects, whereas reactions in the liquid state require pressures of hundreds or thousands of megapascals (1 MPa = 0.1 bar ≈ 9.9 physical atmospheres ≈ 10.2 technical atmospheres ≈ 145 pound-force per square inch). Elevated pressures have the following effects:

| | |
|---|---|
| $< 10^2$ MPa | densification of gases (shifts of equilibria); |
| $10^2 - 10^3$ MPa | surmounting of intermolecular forces (crystallization, viscosity); |
| $10^3 - 10^4$ MPa | change of molecular structures and electron arrangements; |
| $10^4 - 10^5$ MPa | generation or destruction of chemical bonds. |

The effect of pressure $p$ on the rate constants $k_i$ of chemical reactions at a thermodynamic temperture $T$ is described by the transition state theory as

$$(5\text{-}28) \qquad k_i = (k_B T/h) \exp [\Delta S_i^{\ddagger}/R] \exp [-(\Delta H_i^{\ddagger} + p\Delta V_i^{\ddagger})/RT]$$

$$(5\text{-}29) \qquad \partial \ln k_i/\partial p = -\Delta V_i^{\ddagger}/RT$$

where $R$ = gas constant, $k_B$ = Boltzmann constant, $h$ = Planck constant, $\Delta S_i^{\ddagger}$ = activation entropy, $\Delta H_i^{\ddagger}$ = activation enthalpy, and $\Delta V_i^{\ddagger}$ = activation volume.

The activation volume is as important for the pressure dependence of rate constants as the activation energy is for their temperature dependence. In analogy to the expression for the gross activation energy of free-radical polymerizations (Volume I, Eq.(10-42)), one obtains from the rate equation, Eq.(5-10), for the change of gross activation volume in reactions under pressure the expression

(5-30) $\Delta V^{\ddagger} = \Delta V_{p}^{\ddagger} + (1/2)\,\Delta V_{d}^{\ddagger} - (1/2)\,\Delta V_{t(pp)}^{\ddagger}$

Activation volumes of propagation reactions ($\Delta V_{p}^{\ddagger}$), initiator dissociation ($\Delta V_{d}^{\ddagger}$), and termination by mutual deactivation ($\Delta V_{t(pp)}^{\ddagger}$) can be estimated as follows.

*Initiation.* In homolytic decomposition of initiator molecules, dissociating bonds must be extended and then broken, for example, the peroxide bond –O–O– in dibenzoylperoxide, $C_6H_5COO{-}OOCC_6H_5$. The activation volume is thus positive. Its value depends strongly on the solvent (if any), i.e., from the induced decomposition by the solvent (see Volume I, Chapter 10). For example, molar activation volumes of bis(*t*-butyl)peroxide at 120°C are 5.4 mL/mol in toluene but 13.3 mL/mol in $CCl_4$.

*Propagation.* Polymerization of unsaturated compounds converts short double bonds (C=C: 0.133 nm) to longer single bonds (C–C: 0.154 nm), but the resulting volume expansion is overridden by the corresponding disappearance of intermolecular distances (ca. 0.3–0.5 nm) as monomer molecules are transformed to monomeric units. Polymerizations are therefore usually accompanied by volume contractions (Volume I, Section 6.5.1): the activation volume of the propagation reaction is negative.

An example is the activation volume, $\Delta V_{p}^{\ddagger} = -18$ mL/mol, of styrene at 60°C. Likewise, pressure also benefits propagation in ring-opening polymerizations since cyclic monomer molecules usually have greater molar volumes than their open-chain counterparts. The propagation rate constants therefore increase with increasing pressure as shown in Table 5-8 for the polymerization of ethene.

*Termination and chain transfer.* Termination by combination of two growing macroradicals generates a new bond: the *large* intermolecular distance between *two* molecules is replaced by *one shorter* intramolecular one. The activation volume is thus negative.

Termination by disproportionation and chain transfer to other molecules, on the other hand, break old bonds and form new ones; the sign of the activation volume can therefore be either negative or positive. Experiments have shown, however, that transfer reactions increase with increasing pressure. The activation volume of chain transfer must therefore be negative.

These estimates indicate negative gross activation volumes for polymerization, i.e., positive values for $(\partial \ln k_{gross})/\partial p$. Polymerization rates thus increase with increasing pressure, for example, tenfold in the polymerization of gaseous ethene if the pressure is increased to 300 MPa from normal pressure. However, the molecular weight increases by only 1.5 times.

Table 5-8 Rate constants $k_p$ of propagation and $k_{t(pp)}$ by mutual deactivation of macroradicals in the free-radical polymerization of ethene.

| State | $\frac{T}{°C}$ | $\frac{p}{MPa}$ | $\frac{k_p}{L\ mol\ s^{-1}}$ | $\frac{k_{t(pp)}}{L\ mol\ s^{-1}}$ |
|---|---|---|---|---|
| Bulk | –20 | ≈ 0.1 | 19 | 4.6 |
| Solution in benzene | 83 | < 10 | 470 | 10.5 |
| Dense gas | 130 | 176 | 5400 | 2.0 |

## 5.3 Polymerization Reactors

Industrial polymerizations are conducted in many different types of reactor. The choice of a particular reactor type depends on

- Yield of polymer: amounts and amounts per hour;
- Polymer properties: state, molecular weight, solubility, thermal stability, etc.;
- Reaction system: viscosity, heat of polymerization, etc.;
- Product properties: solution, powder, beads, pellets, bales, etc.

Industrial polymerizations differ from those on the laboratory scale by the strong influence of couplings between chemical reactions and transport processes. These couplings affect not only reaction rates but also chemical and/or physical structures and properties of polymers. Especially important are large exothermic heats of polymerization, high viscosities, low diffusion rates, and small heat conductivities.

### 5.3.1 Viscosities

Viscosities of polymerization systems depend on the constitution, configuration, and architecture of polymer molecules, the polymerization temperature, and, for solutions and dispersions, also on the polymer concentration. Viscosities of polymer solutions and melts are also dramatically affected by the molecular weight and molecular weight distribution of polymers as well as the shear rate. Since both the concentration of polymers and their molecular weights and molecular weight distributions change with the course of polymerization, very different time dependencies of viscosities are observed during polymerizations and for the various types of discontinuous and continuous reactors.

In *suspension polymerization*, soft spherical monomer droplets are converted to hard polymer spheres. Since the relative viscosity, $\eta/\eta_1$, of suspensions of spheres depends only on the volume fraction $\phi$ of spheres according to the Einstein–Batchelor equation

$$\eta/\eta_1 = 1 + 2.5\,\phi + 6.2\,\phi^2 \qquad (5\text{-}31)$$

and the number of particles does not change much during polymerization, one would expect viscosities to be practically constant during the course of suspension polymerizations. Experimentally, a small decrease of relative viscosities with monomer conversion is found (Fig. 5-17) because volume fractions become steadily smaller. The particles contract because long intermolecular distances between monomer molecules are replaced by shorter covalent bonds in monomeric units (Volume I, p. 187). This effect overrides the replacement of shorter π-bonds by longer σ-bonds in olefin and vinyl polymerizations.

In *emulsion polymerizations*, small spherical monomer-filled micelles and large monomer droplets are converted to spherical latex particles. The concentration of particles does not change, except during the initial Period I. Experimentally, viscosities are slightly higher at greater monomer conversions than predicted by Eq.(5-31) because of increased particle interactions.

*Bulk polymerizations* show more drastic increases of relative viscosities with monomer conversion. The free-radical polymerization of vinyl compounds is a relatively simple

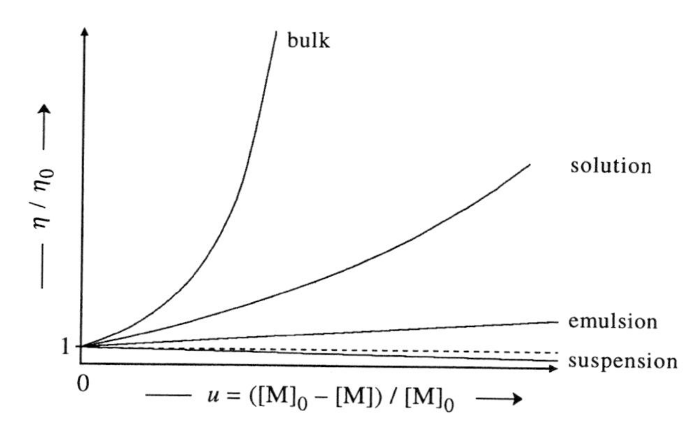


Fig. 5-17 Schematic representation of the dependence of relative viscosities on relative monomer conversion for different types of polymerizations. $\eta$ = viscosity of the polymerization system at monomer conversion $u$, $\eta_0$ = viscosity at zero monomer conversion. – – – $\eta/\eta_0 = 1$.

case because low extents of reaction already deliver high molecular weights that do not change much with monomer conversion before the gel effect sets in. Higher monomer conversions lead to higher polymer concentrations and therefore also to higher relative viscosities.

In *dilute solution polymerizations*, logarithms of relative viscosities increase with the logarithm of polymer concentration. This alone lets viscosities grow strongly with monomer conversion. At higher polymer concentrations, viscosities increase almost exponentially. Gel efects give additional boosts to molecular weights and thus to viscosities. For example, the viscosity of the polymerization system increases a million-fold in the bulk polymerization of styrene to 80 % monomer conversion.

The exact dependence of polymer solution viscosities (including its own monomer) on weight concentration $c$ and molecular weight $M$ has so far defied theoretical approaches. It is known, however, that polymers become entangled above a critical, molecular weight-dependent concentration. In general, this dependence of viscosity $\eta$ on concentration $c$ and weight-average molecular weight $\overline{M}_w$ can be expressed by

$$\eta \sim c^x \overline{M}_w^y \tag{5-32}$$

where $x < 12$, $y \approx 2–4$ below, and $y \approx 5–6$ above a critical weight-average molecular weight. Viscosities increase more slowly with monomer conversion during polymerization in solution than in bulk because of the much lower polymer concentrations.

Eq.(5-32) applies to Newtonian viscosities, i.e., at low shear rates. However, polymer solutions and polymer melts often show non-Newtonian behavior, i.e., a concentration, molecular weight, and temperature dependence of viscosity on shear rate. In most cases, this dependence is shear-thinning, leading to smaller viscosities than Newtonian ones.

The increasing viscosity impedes heat transfer and causes hot spots that need to be eliminated by suitable reactor designs and/or agitator types. They also lead to a diffusion control of elementary reactions, first of molecule-controlled ones (termination) and then to segment-controlled ones (propagation) (Fig. 5-3). At higher monomer conversions, the combination of all these effects leads to very complicated viscosity dependencies that can not be modeled as yet.

### 5.3.2 Types of Reactors

Polymerization plants range from simple vessels to very complex installations that may include not only reactors but also pumps, distillation towers, gas washers, compressors, centrifuges, control stations, etc. The heart of the polymerization plant is the polymerization reactor or a group of reactors. These reactors may be tanks, tubes, extruders, etc., or combinations thereof.

**Ideal Reactors**

Reactors for polymerizations can be divided into four idealized classes (Fig. 5-18):

- Batch reactors (BR), including stirred-tank reactors (STR);
- Continuous plug flow reactors (CPFR);
- Cascades of stirred-tank reactors (C);
- Continuous(-flow) stirred-tank reactors (CSTR).

The action of twin-screw reactors (Fig, 5-18) for highly viscous systems is intermediate between that of a continuous -low reactor and a cascade.

These four ideal types of reactors differ in the residence times of species and the variation of monomer concentration with time (Fig. 5-18). There are two borderline cases: batch reactors and continuous stirred-tank reactors.

In **ideal batch reactors**, all educt (monomer) is completely present at the beginning; neither monomer is added nor polymer removed during the course of reaction. For second-order reactions, monomer concentrations decrease rapidly and than more slowly with time (Table 5-9). However, all volume elements are present for the same length of time: there is no distribution of residence times (Fig. 5-18).

**Ideal continuous stirred-tank reactors**, on the other hand, are defined as reactors in which the concentration does not vary with time or volume element. A newly arrived volume element mixes instantaneously and completely with the volume elements that are already present. However, the various volume elements reside in the reactor for different times: there is a broad distribution of residence times with an average residence time of $\overline{\tau} = V_R/(dV/dt)$. The average residence time increases with the reactor volume and decreases with the flow velocities $dV/dt$.

Table 5-9 Characteristics of ideal reactors with volumes $V_R$ for second-order reactions with rate constants $k$. $C_0$, $C$, $C_\infty$ = amount-of-substance concentrations (in mol/L) at the beginning, at time $t$, and in equilibrium, respectively. $t_0$ = start-up time.

| | Discontinuous, batch | Continuous, plug flow | Continuous, stirred tank |
|---|---|---|---|
| Relative conversion $u/u_\infty = (C - C_0)/(C_\infty - C_0)$ | $1 - \exp(-kt)$ | $1 - \exp(-kt)$ | $kt/(1 + kt)$ |
| Output $L$ in volume per time | $\dfrac{V_R C_0 u}{t_o + t}$ | $\dfrac{V_R C_0 u}{t}$ | $\dfrac{V_R C_0 u}{t}$ |
| Effective output $L_{eff}$ in amount-of-substance per time | $\dfrac{u / u_\infty}{k(t_o + t)}$ | $\dfrac{u / u_\infty}{kt}$ | $\dfrac{1}{1 + kt}$ |

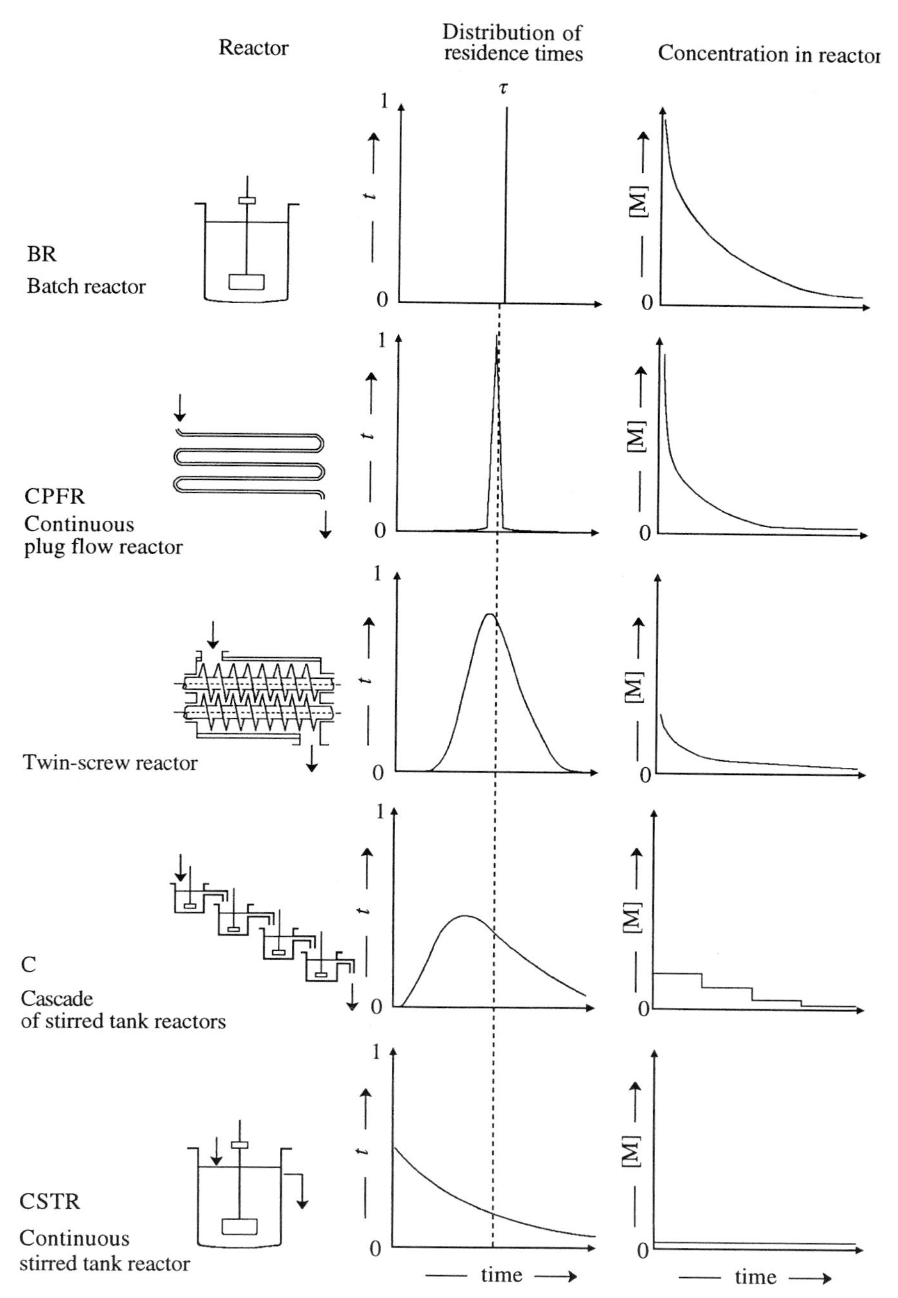


Fig. 5-18 Types of reactors and their time dependencies of distributions of residence times *t* and monomer concentrations [M] [12, 13a].

Left: Schematic representation of the various classes of reactors.

Center: Time dependence of differential distributions of residence times. The dotted line - - - represents the average residence time in continuous plug flow reactors, twin-screw reactors, cascades of stirred-tank reactors, and continuous stirred-tank reactors.

Right: Time dependence of monomer concentrations in second-order reactions.

Table 5-10 Reactors for industrial polycondensations (PC), polyadditions (PA), living polymerizations (LP), and chain polymerizations (CP) with termination by mutual deactivation in batch reactors (BR) without (O) or with additional intake during reaction but no outflow (SB = semi-batch, see Fig. 5-20), continuous plug-flow reactors (CPFR), and continuous stirred-tank reactors (CSTR), either as single reactors (S) or as cascades of reactors (C) [13b]. See also Fig. 5-19.

Emulsion and suspension chain-polymerizations behave as quasi-living polymerizations for large intervals of monomer conversions, hence CP ≈ LP.

| State | Type of polymerization | | | | | BR | | CPFR | CSTR | |
|---|---|---|---|---|---|---|---|---|---|---|
| of matter | PC | PA | CP | | LP | O | SB | | C | S |
| Melt | PC | PA | CP | | | + | + | + | + | + |
| Solution | PC | PA | CP | | | + | + | + | + | + |
| | | | | | LP | + | + | | + | + |
| Emulsion | | | CP | ≈ | LP | + | + | | + | + |
| Precipitation | | | CP | | LP | + | | + | | + |
| Solid state | PC | | | | | + | | + | | + |
| Interface | PC | | | | | + | | + | + | |
| Suspension | | | CP | ≈ | LP | + | | | | |

The distribution of residence times dictates the type of reactor that is suitable for the type of polymerization (polycondensation, living chain polymerization, chain polymerization with termination, etc.) in the desired state (melt, solution, emulsion, etc.). For example, discontinuous batch reactors can be used for almost all types of polymerization (Table 5-10), whereas semi-batch reactors are not suitable for polymerizations that lead to states with interfaces (precipitation, solid state, interfacial, suspension). Emulsion and suspension polymerizations behave largely as living polymerizations because of the very small size of their loci of reactions (see Sections 5.2.4 and 5.2.5).

### Batch Reactors (BRs)

Batch reactors come in many sizes and shapes (Fig. 5-19). They range from simple vessels (stirred-tank reactors, **STRs**) and batch reactors combined with filter press reactors or autoclaves, etc., to piston-type reactors and tumblers. In batch reactors, all reactants are completely present at the beginning of the reaction. **Semi-batch reactors** have an inlet for additions but no outlet for products during the reaction.

Volumes of the largest vessels are 200 $m^3$ for polymerizations at atmospheric pressure and 30 $m^3$ for pressures up to 10 MPa (≈ 100 bar ≈ 100 atm ≈ 1450 lbf/sq in). The larger the reactor, the more difficult it is to remove the heat of polymerization via the reactor surface since the ratio of reactor surface to reactor volume decreases linearly with the radius of the reactor. Reactors up to volumes of ca. 30 $m^3$ can therefore still be cooled by cooling jackets, whereas larger reactors need internal cooling coils. Alternatively, heat may be removed by boiling off coolants, including specially added ones.

Stirred-tank reactors can be used for practically all types of polymerization (Table 5-10): melt polycondensations, interfacial polymerizations in suspension, living polymerizations of the slurry type, and free-radical polymerizations in bulk, solution, or by precipitation. Discontinuous batch reactors are chosen for practically all suspension polymerizations. Emulsion polymerizations often suffer from variations in properties of batches; semi-batch reactors and continuous stirred-tank reactors are preferred here.

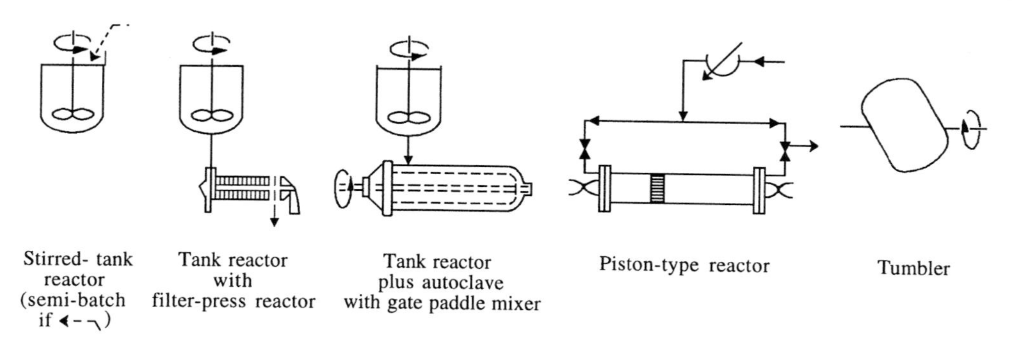


Fig. 5-19 Examples of discontinuous batch reactors [13c, 13d].

Stirring and heat transfer are difficult for highly viscous systems. The after-polycondensation of polyamide 6.6 in the solid state is therefore performed in tumblers on pellets below the melting temperature. In the polymerization of methyl methacrylate, a batch reactor is combined with a filter press reactor, whereas a combination of batch reactor and autoclave serves for the polymerization of neat vinyl chloride. Piston reactors are used for the polymerization of acrylamide to highly viscous, gel-like solutions. Two-component mixers combined with reaction injection molding (RIM) serve for the rapid polyaddition of isocyanates and polyols to polyurethanes and the activated anionic polymerization of $\varepsilon$-caprolactam.

**Continuous Stirred-Tank Reactors (CSTRs)**

Polymerizations in continuous stirred tank reactors are characterized by broad distributions of residence times and constant monomer concentrations (Fig. 5-18), whereas polymerizations in stirred-tank reactors have narrow distributions of residence times and rapidly decreasing monomer concentrations. CSTRs and STRs are thus at the opposite ends of the spectrum of reactor characteristics.

Like stirred-tank reactors or batch reactors in general, CSTRs are available in many different designs. In the simplest version, melts or solutions of reactants flow continuously through a stirred reactor (Fig. 5-20). This type of continuous-flow stirred-tank reactor is used for polymerizations in melt, solution, suspension, and emulsion (Tables 5-10 and 5-11). Copolymerizations in solution deliver copolymers with especially narrow distributions of monomeric units.

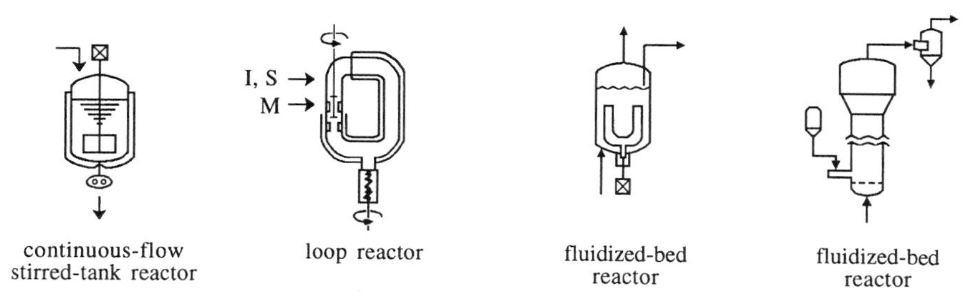


Fig. 5-20 Examples of different types of continuous-flow stirred-tank reactor [13e]. I = initiator, M = monomer, S = solvent.

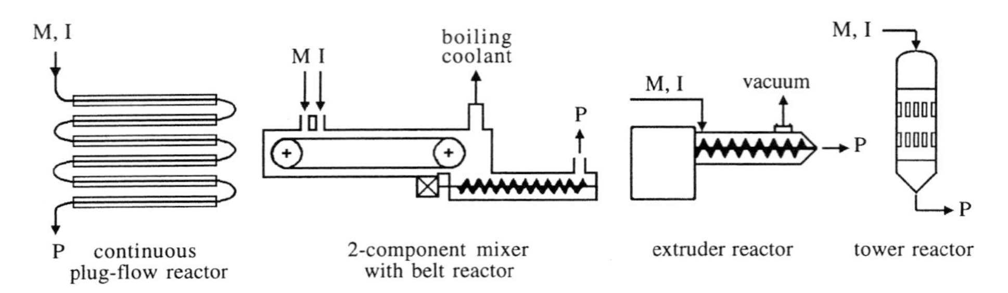


Fig. 5-21 Some continuous plug-flow reactors [13f]. I = initiator, M = monomer, P = polymer.

Loop reactors (again, many different designs), in which the reaction mixture is circulated, are special types of CSTRs since distributions of residence times approach that of CSTRs if the circulation is much larger than the run-through. Because of their favorable ratio of surface to volume, they are used for very exothermic polymerizations such as the slurry polymerization of propene that takes place in dispersions of precipitated polymer particles in their own liquid monomer, similar to a suspension polymerization.

Fluidized-bed reactors also come in many variations. They are used for so-called gas-phase polymerizations which are polymerizations in polymer particles that are dispersed in the gaseous monomer and not in a liquid as in slurries.

**Continuous Plug-Flow Reactors (CPFRs)**

Reactors with plug-flow characteristics lead to polymerizations with narrow distributions of residence times. The simplest CPFRs are horizontal tube reactors and vertical tower reactors (Fig. 5-21), both of which can operate with temperature gradients. An example of the latter is the thermal polymerization of styrene without any initiator.

Belt reactors and extruder reactors are special types of horizontal tube reactors. A belt reactor is used in the cationic polymerization of isobutene with ca. 4 % isoprene as comonomer to isobutylene rubber (IIR). A double-screw extruder delivers powder dry poly(oxymethylene)s from the precipitation polymerization of 1,3,5-trioxane.

**Cascades**

Narrow distributions of residence times can be obtained by polymerization in continuous plug-flow reactors or reactor cascades (Fig. 5-22). Most common are cascades of continuous-flow stirred-tank reactors, the second most important type of reactors after batch reactors. Such cascades are used for polymerizations in melt, solution, suspension, and emulsion (Table 5-10), not only for chain polymerizations with Ziegler-Natta catalysts or by free-radical initiators but also for polycondensations to phenol-formaldehyde and amino resins as well as to unsaturated polyesters.

Cascades are not only comprised of many STRs but may also consist of several towers, for example, in the thermal polymerization of styrene where various towers are kept at different temperatures. Also used are combinations of towers and tank reactors, for example, an STR on top of a tower reactor for the polymerization of $\varepsilon$-caprolactam in the melt. A cascade-type of reactor is the ring-disc reactor that is used for the final stage of the polycondensation of poly(ethylene terephthalate).

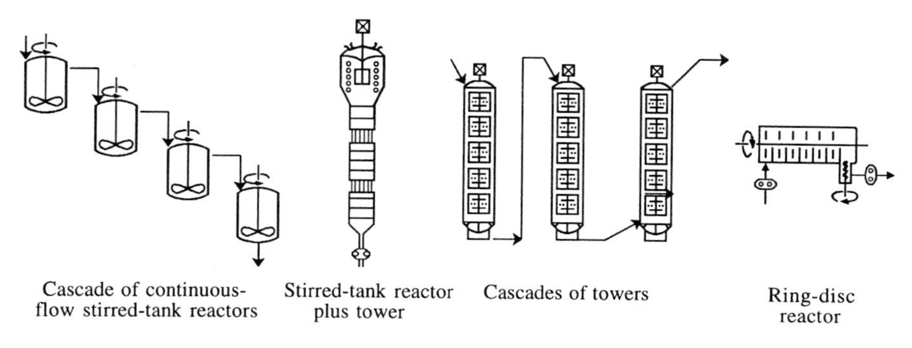


Fig. 5-22 Different types of reactor cascades [13g].

## Operation of Reactors

The four ideal types of reactors (p. 178) differ in their operations and in the widths of distribution of residence times (Fig. 5-18).

*Stirred-tank reactors* (STRs) serve for discontinuous (batch) operations in which the tank is filled with monomers, solvents, and, if necessary, auxiliary agents such as regulators (chain transfer agents, see Volume I, p. 350). Catalysts or initiators are added before, during, or after start-up which brings the content of the tank to the desired reaction temperature. Inhibitors for stabilization of monomers (if any) are not removed before the tank is filled. Their action is rather overridden by careful overdosing of initiators.

Agitation strives to assure a complete mixing in order to produce the same concentration of educts and products in each volume element. Since agitation rarely leads to complete homogeneity of temperatures and concentrations (see next Section), hot spots by local overheating and spatial gradients may appear, especially by temperature gradients near cooled or heated reactor walls (Table 5-11).

Educt concentrations decrease according to the order of reaction (Fig. 5-18, right); the concentration of reactants is therefore not stationary. The polymerization is stopped at the desired degree of monomer conversion which is usually 99–99.5 % in equilibrium polycondensations but often only 50–60 % in free-radical chain polymerizations. The temperature of the tank contents is then brought to the temperature for the work-up of the reaction mixture, for example, by cooling. The work-up includes separation of the polymer from the mixture, removal of solvent, removal of residual monomer (recovery, burning), work-up of polymers (washing, pelletizing, etc., see Section 5.4), and the like.

Table 5-11 Operational features of ideal and real basic reactor types.

| Type of reactor | | Concentration of reactants with respect to | | |
|---|---|---|---|---|
| | | space | | time |
| | | ideal | real | ideal |
| STR | | homogeneous | spatial gradient | not stationary |
| CSTR | | homogeneous | inhomogeneous | stationary |
| CPFR | axial direction | increase | redistribution | stationary |
| | radial direction | constant | concentration profile | stationary |

Tank reactors are most often the choice for many polymerizations (Table 5-12) because they are simple, inexpensive, and very flexible with respect to the desired type and grade of polymers. However, they have considerable idle times during start-up and shutdown. Relatively small differences in operations may also lead to significant variations in polymers from batch to batch which, in turn, requires considerable control efforts.

*Continuous reactors* (CSTRs, CPFRs, cascades) are mainly used for the synthesis of large-volume polymers with narrow specifications. Again, simple tank reactors are often preferred, either single ones with continuous flow (CSTR) or cascades of CSTRs. However, other types of reactors may be sometimes more advantageous. For example, high-density poly(ethylene)s with bimodal molecular weight distributions are usually obtained from polymerizations in slurry reactors in series; such HDPEs are used for film blowing, blow-molding, and pipe extrusion. About 20–40 % of the investment costs here were saved when the CSTR cascades were replaced by a single fluidized-bed reactor and a proprietary combination of catalysts.

Ideal *continuous-flow stirred-tank reactors*, like STRs, are homogeneous with respect to spatial concentrations of reactants. In real CSTRs, however, agitation is never complete, which leads to segregation (Section 5.3.3). CSTRs are advantageous for the production of large amounts of polymers with well-controlled specifications; however, they do not allow large variations of polymerization conditions and product properties. Also, changes of polymer grades or types produce transitional products that are less valuable or not valuable at all.

Table 5-11 Examples of polymerizations in various reactors. See Appendix for abbreviations.

| Type of operation and reactor | State of polymerization system<br>Melt or solution | Suspension | Emulsion | Gas phase |
|---|---|---|---|---|
| *Batch and semi-batch operations* | | | | |
| STR | LDPE, HDPE, PB, PS, PET, PMMA, PC, PF, UF, MF | PVC, PMMA, EPS | PVC, PVAC, PTFE, SBR, ABS, SAN, many other copolymers | PVC |
| inject. mold (RIM) | PUR, PA 6 | | | |
| *Continuous polymerizations with broad distributions of residence times* | | | | |
| CSTR | LDPE, PS, IR, PMMA | HDPE, PP, PAN | PVC | |
| Fluidized bed | HDPE, PP | HDPE | | HDPE, PP, LLDPE |
| Loop reactor | | HDPE, PP | | |
| *Continuous polymerizations with narrow distributions of residence times* | | | | |
| *A. CPFR* | | | | |
| Tube reactor | LDPE, EVAC, PA | | | |
| Tower reactor | HIPS, PS, PA | | | |
| Extruder reactor | POM | | | |
| Belt reactor | PIB, PUR | | | |
| *B. Cascades* | | | | |
| CSTR cascade | HDPE, HIPS, PP EPDM, BR, IR, PET, UP, PF, MF | EPDM, HIPS | SBR, ABS, SAN, CR, NBR | |
| Tower cascade | PS, PA | | | |

Ideal *continuous plug-flow reactors* (CPFRs) are characterized by an increase of concentration of reactants per volume element in axial direction; in real CPFRs, diffusion and convection change the profile. In ideal CPFRs, the concentration does not change in the radial direction; in real CPFRs, the flow generates a radial concentration profile.

### 5.3.3 Agitators

**Types**

Polymerization volumes need to be mixed thoroughly for a homogeneous distribution of educts and products and a continuous transfer of heat of polymerizations. Some reactors achieve this by their construction and operation. Examples are tumblers, piston reactors, and fluidized-bed reactors.

In stirred-tank reactors (STRs, CSTRs), mixing is achieved by agitation. The energy provided by the agitator is converted to local mixing motions near the agitator and to circulation of the reactor contents. The movements of volume elements are interrupted by deflecting blades (retarding elements, baffles, etc.) that convert tangential flows to radial and axial ones. The increased shearing of the volume elements requires more energy input by the agitator. The resulting radial, axial, and tangential flow profiles are very complex.

Deflecting elements are not necessary for suspensions because shear thinning is absent. They are obviously of no use in very viscous reaction systems.

Agitators come in very different configurations (Fig. 5-23). The choice of a particular agitator is determined not only by the size of the reactor and the viscosity of the reaction system but also by the type of polymerization and the desired polymer properties (molecular weight, molecular weight distribution, grafting, etc.). Types of polymerization, reactors, and agitators often need to be matched.

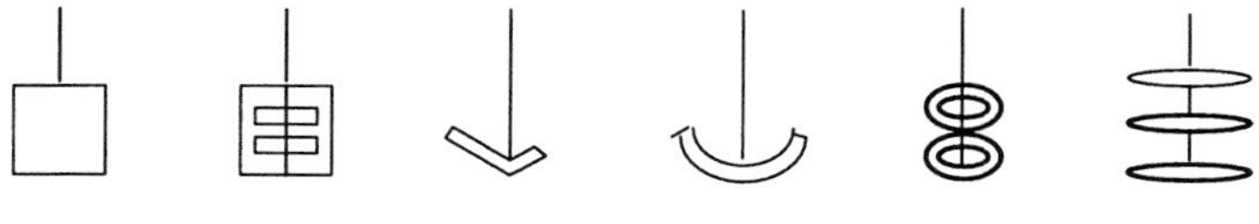

| | paddle mixer | gate paddle agitator | impeller | anchor agitator | spiral agitator | axial-flow turbine |
|---|---|---|---|---|---|---|
| Reactor volume in $m^3$ | < 10 | < 30 | < 200 | < 1 | < 70 | < 100 |
| Flow | laminar | laminar | turbulent | laminar + turbulent | laminar | turbulent |
| Direction of flow | radial | radial | radial | radial | axial | axial |
| Viscosity in Pa s | 0.5 - 5 | 0.5 - 5 | < 20 | < 20 | > 20 | < 20 |
| Polymerization | – | bulk | – | bulk | bulk | – |
| | – | solution | – | solution | solution | – |
| | – | emulsion | – | emulsion | – | – |
| | – | – | precipitation | precipitation | – | precipitation |
| | suspension | – | – | – | – | – |

Fig. 5-23 Examples of important types of agitators [14a]. Impellers are fast-rotating agitators. The view-down the shaft of the impeller of this figure is 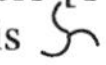 .

A certain type of agitator can only be used up to a certain reactor volume and a maximum viscosity (Fig. 5-23). The correct choice of an agitator is especially important because polymerizing systems pass through various types of flow with time: intially turbulent flows of low-viscosity systems by fast-rotating agitators are later replaced by laminar flows of high-viscosity systems.

### Power of Agitators

Agitation is energy intensive; agitators should therefore use as little energy as possible for maximum homogenization and heat transfer. A measure of the effective power of an agitator is the dimensionless Newton number $Ne$ which is calculated from the experimentally determined power $P$ (= energy/time) of agitation, density $\rho$ of the system in mass per volume, rotational velocity $\dot{N}_{ag}$ of the agitator in inverse time, and maximum diameter $d_{ag}$ of the agitation device via

$$(5\text{-}33) \qquad Ne = P / (\rho\, \dot{N}_{ag}^3\, d_{ag}^5)$$

The Newton number of an agitator varies with the type and design of the reactor (dimensional ratios, type and number of deflecting elements, etc.). Most literature data refer to agitators in reactors with equal heights $h_r$ and diameters $d_r$.

In the laminar range, Newton numbers are inversely proportional to Reynolds numbers, $Ne \sim Re^{-1}$; this range extends for many agitators to $Re \leq 10$ and for anchors and impellers to $Re \leq 100$ if they almost touch the reactor walls. In the range between laminar flow and the onset of turbulent flow, $Ne \sim Re^{-1/3}$ applies. $Ne$ becomes independent of $Re$ in the turbulent range which starts at ca. $Re \sim 1000$ for Newtonian liquids (Volume III).

Agitation should not only homogenize local concentrations but also improve by convection the heat transfer at cooling elements. The flow of heat, $Q$, is characterized by a dimensionless number, $q$, that is obtained from

$$(5\text{-}34) \qquad Q \sim C V_{int} \left[ \frac{d_{ext}}{d_{int}^6} \cdot \frac{(P/V_{int})}{Ne} \right]^{2/9} = q V_{int} \left[ \frac{(P/V_{int})^{2/9}}{d_{int}^{10/9}} \right]$$

via experimental data for $V_{int}$ = internal reactor volume, $d_{int}$ = internal reactor diameter, $d_{ext}$ = external reactor diameter, $P$ = power of agitation, and a constant $C$ as a function of the dimensionless Nusselt number $Nu$, Reynolds number $Re$, and Prandtl number $Pr$:

$$(5\text{-}35) \qquad C = Nu\, Re^{-2/3}\, Pr^{-1/3}\, f^{-1}(\eta)$$

$$(5\text{-}36) \qquad Nu = h\, d_{int} / \lambda$$

$$(5\text{-}37) \qquad Re = \rho\, d_{ag}^2\, \dot{N}_{ag} / \eta$$

$$(5\text{-}38) \qquad Pr = \eta\, c_p / \lambda$$

where $h$ = heat transfer coefficient, $\lambda$ = heat conductivity, $c_p$ = specific heat capacity, and $\eta$ = dynamic viscosity. $f^{-1}(\eta)$ is a parameter that measures the directional effect of heating and cooling, respectively.

Table 5-13 Agitator dimensions and characteristic constants in the turbulent range [14b].

| Type of agitator | $d_{int}/d_{ext}$ | $C$ | $Ne$ | $q$ |
|---|---|---|---|---|
| Anchor | 0.98 | 0.26 | 0.35 | 0.33 |
| Impeller | 0.67 | 0.31 | 0.75 | 0.3 |
| Propeller | 0.3 | 0.5 | 0.35 | 0.48 |
| Slanted blade | 0.4 | 0.61 | 1.5 | 0.45 |
| Disc | 0.3 | 0.76 | 5.0 | 0.41 |
| Blade | 0.5 | 0.8 | 9.8 | 0.42 |

Table 5-12 shows characteristic constants $C$, $Ne$, and $q$ of agitators, as calculated from Eqs.(5-35), (5-33), and (5-34), respectively. The data show that anchors and impellers are not very effective; indeed, anchors are rarely used because they do not provide good mixing from top to bottom. Propellers are good but agitators with several beams are even better ($q = 0.55$). These characteristic numbers are important for the **scale-up** from laboratory scale to production and from small reactors to larger ones.

### 5.3.4 Segregated Reactors

Well stirred *ideal* CSTRs provide molecular mixing (**micromixing**) in which all volume elements are identical with respect to temperature and concentration of reactants and products. A stirred-tank reactor with such a homogeneous content is thus called a **homogeneous continuous stirred-tank reactor** (HCSTR).

However, real CSTRs are often not completely mixed on a molecular level. In bulk polymerizations, this may be due to the high viscosity of the system which prevents fast mixing of volume elements and rapid diffusion of molecules from one volume element to another. In suspension polymerizations, the system is compartmentalized: each dispersed latex particle acts as a separate microreactor. The polymerization system of such real CSTRs thus consists of many small volume elements, each with very many molecules that differ in concentration and/or temperature. The system is segregated and the reactor is a **segregated continuous stirred-tank reactor** (SCSTR) consisting of many independent compartments.

By definition, nothing can be exchanged between volume elements in *ideal* SCSTRs where all completely segregated volume elements are ideally mixed. Residence times differ from volume element to volume element, but the whole SCSTR has the same distribution of residence times as a homogeneous continuous stirred tank reactor, HCSTR (last example in Fig. 5-18).

Segregation processes affect the residence time of volume elements and thus the monomer conversion (Fig. 5-24) as well as polymer molecular weights and molecular weight distributions (Fig. 5-25). Since molecular weight distributions are controlled by the ratio of the lifetimes of growing species to average residence times in the compartment, different types of polymerization will deliver different types of molecular weight distributions. Examples are living chain polymerizations where growing polymer chains add only monomer molecules versus polycondensations where polymer chains react not only with monomer molecules but also with oligomers and other polymer molecules.

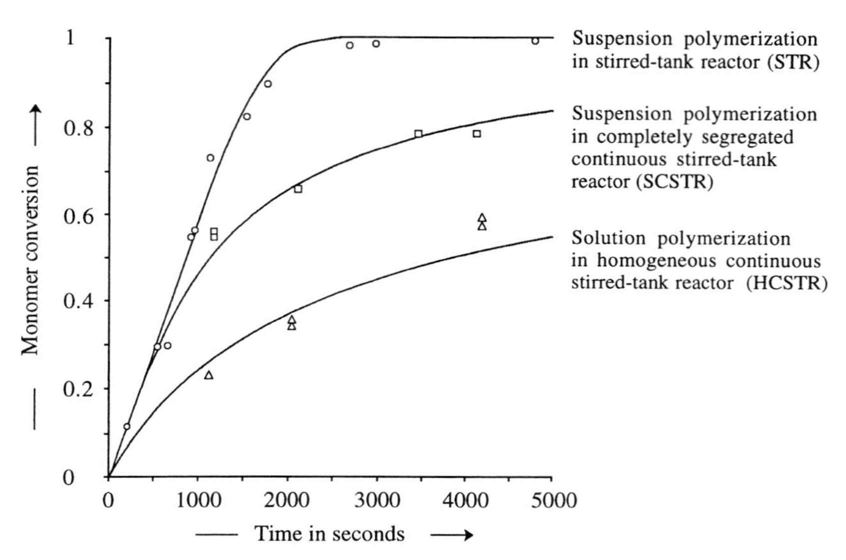


Fig. 5-24 Free-radical polymerization of vinyl acetate by dicyclohexyl peroxydicarbonate in aqueous suspension or solution at 60°C in different reactors [15a]. $[M]_0$, $[I]_0$ = initial concentrations of monomer and initiator, respectively; $k_p$ = rate constant of propagation.

| | | | |
|---|---|---|---|
| Suspension: | $[M]_0 = 10.8$ mol/L ; | $[I]_0 = 0.016\ 3$ mol/L ; | $k_p = 397$ L mol$^{-1}$ s$^{-1}$ |
| Solution: | $[M]_0 = 1.41$ mol/L ; | $[I]_0 = 0.002\ 12$ mol/L ; | $k_p = 9\ 500$ L mol$^{-1}$ s$^{-1}$ |

The *free-radical solution polymerization* of vinyl acetate in an HCSTR is initially fast (Fig. 5-24) but then becomes progressively slower because of the decrease in monomer concentration according to Eq.(5-15). The time dependence of monomer conversion in suspension polymerization of the same monomer in a (discontinuous) STR shows initially an even faster linear increase of monomer conversion with time up to ca. 90 % monomer conversion since the polymerization proceeds as "quasi-living" in a minireactor without termination. Ultimately, the monomer conversion reaches 100 %. Suspension polymerization in an SCSTR, however, behaves initially like a suspension polymerization in an batch reactor but later, after a monomer conversion of ca. 30 %, more like a solution polymerization in an HCSTR.

Calculations showed that free-radical polymerizations in HCSTRs should deliver the narrowest possible molecular weight distribution, which is the Schulz–Flory distribution with $\overline{X}_w/\overline{X}_n = 2$ (Fig. 5-25) (VAc terminates by disproportionation). The distribution is slightly broader ($\overline{X}_w/\overline{X}_n \approx 2.1$) for a batch polymerization in an STR and considerably broader ($\overline{X}_w/\overline{X}_n \approx 2.8$) for the polymer from a segregated reactor (SCSTR).

These theoretical predictions have been confirmed experimentally by free-radical polymerizations of vinyl acetate in STR, HCSTR, and SCSTR (Fig. 5-26). At low monomer conversions in STR, polymolecularity indices slightly exceed the theoretical value of $\overline{M}_w/\overline{M}_n = 2$ for chain polymerizations with termination by disproportionation, probably because of chain transfer to polymer molecules which leads to branching. The strong increase of $\overline{M}_w/\overline{M}_n$ at higher monomer conversions is caused by the gel effect which reduces the encounter of two macroradicals. This increase is only slightly affected by a polymerization of vinyl groups at the end of the chains that were created by disproportionation.

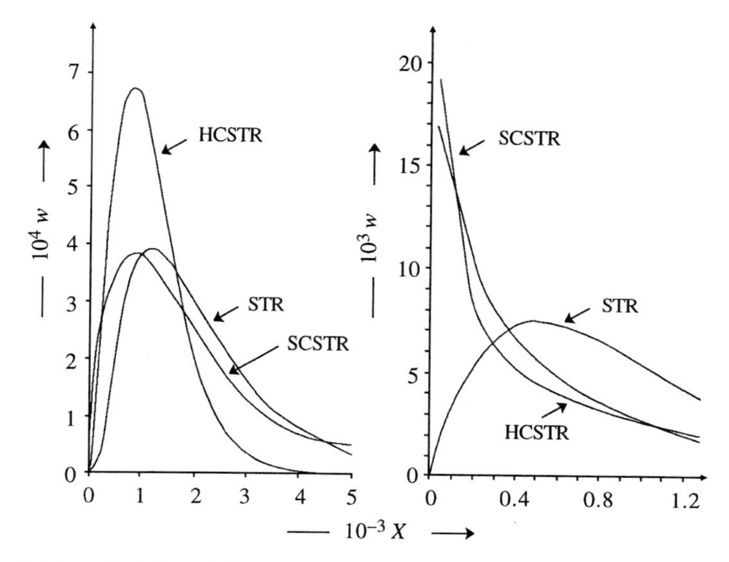


Fig. 5-25 Calculated differential mass distributions of degrees of polymerization, $X$, from polymerizations in stirred-tank reactors (STR), homogeneous continuous stirred-tank reactors (HCSTR), and segregated continuous stirred-tank reactors (SCSTR).

Left: Free-radical chain polymerization with termination by disproportionation at a fractional monomer conversion of 60 % ($\overline{X}_n$ = 1000) [16a]. With permission of the American Chemical Society, Washington (DC).

Right: Equilibrium polyaddition (or polycondensation with zero mass of leaving molecules) with extent of reaction, $p$, in STR ($p$ = 0.998), SCSTR ($p$ = 0.9887), and HCSTR ($p$ = 0.9561) [17].

By permission of the Society of Plastics Engineers, Brookfield (CT).

*Equilibrium polycondensations* and *polyadditions* to linear polymers deliver in STRs the ideal distributions of degrees of polymerization of reactants ($X \geq 1$!) (Fig. 5-25), i.e., number-average degrees of polymerization, $\overline{X}_{R,n} = 1/(1-p)$, and, if molecular weights of leaving molecules are negligible, $\overline{X}_{R,w} = (1+p)/(1-p) = 2\ \overline{X}_{R,n} - 1$.

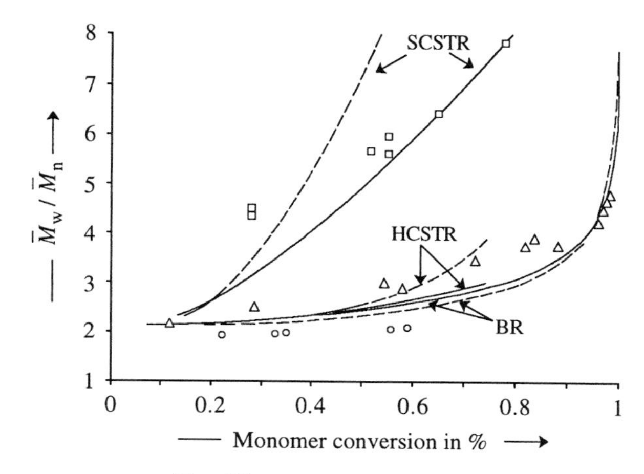


Fig. 5-26 Polymolecularity index $\overline{M}_w / \overline{M}_n$ as a function of monomer conversion in the free-radical polymerization of vinyl acetate in *t*-butanol (see Fig. 5-24) [15b]. O, Δ, □ Experimental data, —— calculated for a gel effect with polymerization of terminal vinyl groups, - - - ditto without.

The situation is different for homogeneous continuous (flow) stirred tank reactors (HCSTRs). Polymer molecules here continue to grow during the whole time that monomer and polymer molecules stay in the reactor. The broad distribution of residence times (Fig. 5-18, bottom) increases the probability that two very large molecules meet. The width of molecular weight distributions, as measured by $\overline{X}_{R,w}/\overline{X}_{R,n}$, broadens considerably with increasing monomer conversion or increasing $\overline{X}_{R,n}$ (Fig. 5-27), as the following example shows.

For bifunctional polycondensations, HCSTRs and STRs deliver the same expression for the number-average degree of polymerization: $\overline{X}_{R,n} = 1/(1-p)$, where $p$ = extent of reaction. However, the weight-average degree of polymerization is now given by $\overline{X}_{R,w} = (1+p^2)/(1-p)^2$ instead of $\overline{X}_{R,w} = (1+p)/(1-p) = 2\overline{X}_{R,n} - 1$ (Volume I, p. 446). For $p = 0.9561$ (i.e., $\overline{X}_{R,n} = 22.8$), one obtains $\overline{X}_{R,w}/\overline{X}_{R,n} = 43.6$ (HCSTR) instead of $\overline{X}_{R,w}/\overline{X}_{R,n} = 1.98$ (discontinuous batch reactor (STR)).

In completely segregated reactors, ideally mixed volume elements cannot exchange matter (p. 187). Segregation removes both the very low and very high tails of molecular weight distributions, which therefore become more narrow if obtained from polycondensations in SCSTRs instead of in HCSTRs, but are still considerably broader than those produced in STRs (Fig. 5-27). In this respect, step-growth polymerizations differ from chain polymerizations (Fig. 5-26).

According to these theoretical calculations, SCSTRs and HCSTRs should not be suited for polycondensations since polymers obtained therefrom should have very broad molecular distributions that are detrimental to most mechanical properties of these polymers. However, reactions of functional groups of monomer, oligomer, and polymer molecules are not the only reactions in many polycondensations. In polyamides, polyesters, polysiloxanes, etc., trans reactions between chain sections or between center groups and endgroups cause equilibrations that narrow the molecular-weight distributions so that they approach Schulz–Flory distributions. An example is the transamidation reaction ~Z–NH–CO–Z'~ + ~Z"–NH–CO–Z'''~ ⇄ ~Z–NH–CO–Z"~ + ~Z'–NH–CO–Z''' (Volume I, p. 574)

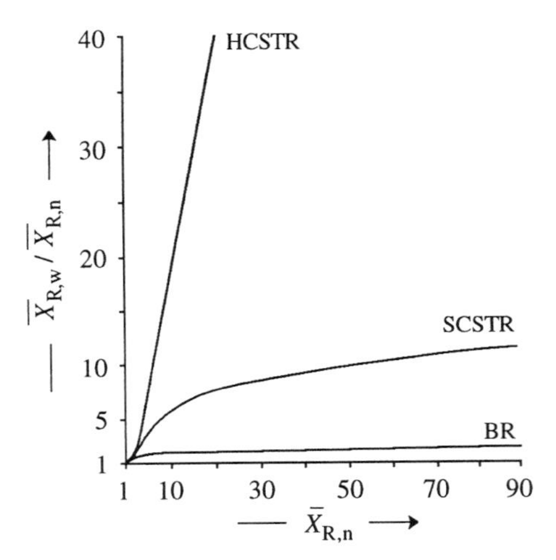


Fig. 5-27 Theoretical dependence of polymolecularity indices on the number-average degree of polymerization in polycondensations [16b], assuming absence of exchange reactions.
By permission of the American Chemical Society, Washington (DC).

## 5.4 Work-up of Polymers

Polymerizations are ideally performed by processes in such reactors that directly deliver products with the desired shipping properties, since many of the work-up procedures of conventional organic chemistry are either not possible or much too costly. For example, polymers cannot be purified by distillation or crystallization. Fractionations are avoided not only because it is an additional step but also because expensive solvents are used that later need to be removed and recycled or burnt. Extraction of initiator residues and catalysts, solvents, and residual monomers are difficult and also expensive.

Polymerizations deliver either molten, dissolved, or emulsified (dispersed) polymers or polymers as powders or beads. Powdery polymers are dried. Polymer melts and solutions are degassed in order to remove volatile components (residual monomers, solvents, leaving molecules) because of their toxicity and environmental effects or because of safety concerns.

For example, residual vinyl chloride in poly(vinyl chloride) should not exceed 1 ppm (= $10^{-4}$ %) which can be obtained by steaming the polymer with water vapor. The release of organic vapors by such operations requires special procedures. In most cases, it is far too expensive to recover monomers or solvents from these vapors; they are rather burned instead.

Residual moisture in slightly hydrophilic polymers has to be adjusted to specifications, for example, in polyamides. These specifications usually differ from country to country; they may also vary in practice. True equilibration can take long times. An example is nylon 6: ca. 30 days are needed for 2 mm-thick pellets and 120 days for 5 mm-thick ones at 20°C and 100 % relative humidity. Most shipments of such polymers are therefore not in the equilibrium state.

Polymer melts are filtered in order to remove solid particles and gels. Filters usually consist of mesh screens with very small mesh sizes, often down to 70 µm, or of sintered metal particles with pore sizes down to 5 µm. Gel particles are removed by filtration, for example, through layers of sand.

Suspension and emulsion polymers are usually filtered, washed, and dried to make them ready for shipment. Most polymers are sold as pellets of 1.5–3 mm diameter that are usually prepared by extrusion from the melt.

In polymer plants, granulates are usually transported pneumatically by air with high speeds. Collisions of granules with each other or with the tube walls breaks off or shears off protruding hair-like pieces that cluster together like angel's hair. Such lumps do not break up during processing by injection molding or extrusion where they then lead to either defective parts or machine failures. These lumps are therefore removed in the polymerization plant by cyclones.

Polymer producers usually store granulates and powders in silos which requires precautionary measures against fires and dust explosions by antistatic charging of particles. The polymers are shipped by rail or truck as free-flowing bulk products, in reusable flexible bags (up to 500 kg), in boxes, or in paper bags.

Polymers for fibers, plastics, elastomers, etc., are rarely used as such. They are rather compounded with fillers, pigments, antioxidants, biocides, plasticizers, and other adjuvants before being processed (Volume IV). For plastics, these operations are rarely performed by polymer producers but rather by special compounders.

# 5.5 Economic Aspects

## 5.5.1 Energy Consumption

Synthetic polymers became a success not only because of their properties but also because of their economics: low cost of raw materials, synthesis, processing, and application. Questioned now sometimes are their ecological costs, i.e., hidden costs of using "free" air and water, pollution of the environment (i.e., lack of bio-friendly degradation), suspected high energy consumption in the synthesis and processing, and "wasting" of oil and other fossil energy carriers.

Consumptions of energy in the production of materials and critical amounts for the pollution of the environment have been calculated by industrial companies, governments, international organizations, and non-governmental organizations, including the North Atlantic Treaty Organization (NATO), the German chemical company BASF, and the Swiss Federal Agency for Environmental Protection (BUWAL) (Table 5-13). Although the data sometimes differ considerably, they all show the same trend: the energy consumption for the synthesis of plastics on a *per-volume* basis does not differ much from that for glass, and is considerably lower than that for metals. Syntheses of synthetic polymers also cause less pollution than those of new paper and aluminum.

Table 5-13 Consumption of energy for the production of various materials according to NATO, BASF, and BUWAL and critical volumes for the pollution of air, water, and soil during the manufacture, use, and disposal of 1 kg of material as calculated by BUWAL.

Comparisons must be made per volume and not per weight because materials are always used by volume although they are sold by weight.

| Material | Density in $g/cm^3$ | Energy consumption in $kJ/cm^3$ as calculated by | | | Critical volume of | | |
|---|---|---|---|---|---|---|---|
| | | NATO | BASF | BUWAL | air $m^3$ | water $dm^3$ | soil $cm^3$ |
| Urea-formaldehyde foam | 0.012 | | 0.48 | | | | |
| Poly(styrene) foam | 0.015 | | 1.4 | | | | |
| PUR foam, soft | 0.030 | | 3.0 | | | | |
| HDPE | 0.96 | | 63 | 68 | 0.69 | 440 | 300 |
| PVC | 1.38 | 11 | 73 | 84 | 6.60 | 520 | 400 |
| PS, glass clear | 1.05 | | 84 | | | | |
| ABS | 1.06 | | 89 | 87 | 0.64 | 690 | 260 |
| Phenolic resin | ~ 1.3 | | 107 | | | | |
| PET | 1.35 | | 113 | | | | |
| PA 6 | 1.13 | | 176 | | | | |
| Lumber | ~ 0.7 | 2.0 | | | | | |
| Kraft paper, unbleached | ~ 1.6 | 40 | | 63 | 1.54 | 9400 | 340 |
| Cardboard with 20 % waste paper | ~ 1.6 | | | 89 | 1.05 | 1000 | 1000 |
| Cardboard from 100 % waste paper | ~ 1.6 | | | 18 | 0.32 | 330 | 330 |
| Aluminum, new (cast) | ~ 2.68 | < 460 | | 750 | 2.40 | 1140 | 2900 |
| Glass, new | ~ 2.5 | < 125 | | > 26 | 0.28 | 110 | 600 |
| Glass, with 43 % recyclate | ~ 2.5 | | | 19 | 0.21 | 80 | 490 |
| Steel | 7.75 | < 390 | | | | | |
| Tinplate | 7.29 | | | 210 | 0.30 | 55 | 780 |
| Cement | ~ 2.5 | 23 | | | | | |

### 5.5.2 Capacities

Production plants are characterized by their name-plate capacities and their effective capacities. The **name-plate capacity** indicates the maximum amount of material that a plant can produce per unit time over a long period. The calculated name-plate capacity considers regular (long-term) maintenance but not seasonal variations. A plant can therefore produce short-term (up to several months!) more than its theoretical capacity, i.e., more than 100 % of name-plate capacity.

**Effective capacities** of plants are usually 85–95 % of their name-plate capacities because of unplanned maintenance, changes in product mix, etc. Effective capacities do not include feedstock shortages, power and equipment failure, and other temporary fluctuations. In bad economic times, effective capacities may be far smaller than 85-95 %.

Capacity data for whole industrial branches of countries contain usually neither data for new plants under construction nor for those that were incapacitated for longer time periods because of accidents. However, they are the only data that are available internationally for different types of goods.

The internationally reported country-wide **consumption** of a good is calculated from productions plus imports minus exports. This consumption is an apparent one since it does not include storage. Western countries try to have fast turnovers of goods so that stored reserves do not exceed 10 %; strategic reserves are exceptions. One even tries to reduce warehousing to practically zero by demanding that goods are only delivered the same day they are used (just-in-time delivery).

In centrally planned economies, warehousing often exceeded 10 % since plants tried to fulfill or even overfulfill the government plan. Sales were not controlled, however, so that warehouses overflowed with difficult-to-sell goods.

The apparent per-country consumption of a polymer is often not very meaningful since consumption data do not include the amounts of polymers in imported and exported finished goods. An example is the apparent consumption of elastomers that is based on the production, import, and export of rubbers but does not consider the rubber content of tires on imported cars or the export of old tires for reclaiming.

### 5.5.3 Prices

The price of a good is based on the costs of operations, capital, and financing as well as a reasonable yield for investments, reserves, and dividends. In polymer production, operational costs are usually the biggest cost factor and here especially the costs of monomers which may reach 35–70 % of polymer prices if monomers have to be bought on the open market (Table 5-15). Hence, high costs of raw materials favor the backward integration polymer → monomer → intermediate → feedstock.

Polymer prices depend strongly on prices of raw materials and energy as well as on money markets, taxes, and costs of environmental protection and waste disposal. Price increases for raw materials, energy, and environmental protection necessitate cost reductions by technical improvements which often also lead to reductions of the work force. These improvements include larger production plants, greater volume/time yields, less difficult to sell byproducts, lower work-up costs, savings of energy costs, and the like.

Table 5-15 Comparison of prices for high-density poly(ethylene) (HDPE), isotactic poly(propylene) (PP), atactic poly(styrene) (PS), and atactic poly(vinyl chloride) (PVC) and their monomers ethene, propene, styrene, and vinyl chloride in December 1989 [18]. At that time, crude oil was 15.1 US cts/kg and benzene 46 US cts/kg. The spot price of crude oil is now ca. 75 cts/kg (April 2006).

| | Unit | PE-HD | PP | PS | PVC |
|---|---|---|---|---|---|
| Monomer | US-cts/kg | 53 | 41 | 77 | 44 |
| Polymer | US-cts/kg | 115 | 110 | 110 | 88 |
| (Monomer price/polymer price)·100 | % | 46 | 37 | 70 | 50 |

For example, the price of crude oil was practically constant between 1950 and 1972. During this time span, prices of plastics dropped considerably (Fig. 5-28) because of technical improvements (for example, larger plants, see below), whereas the wholesale price index for industrial consumer goods increased slightly (not shown).

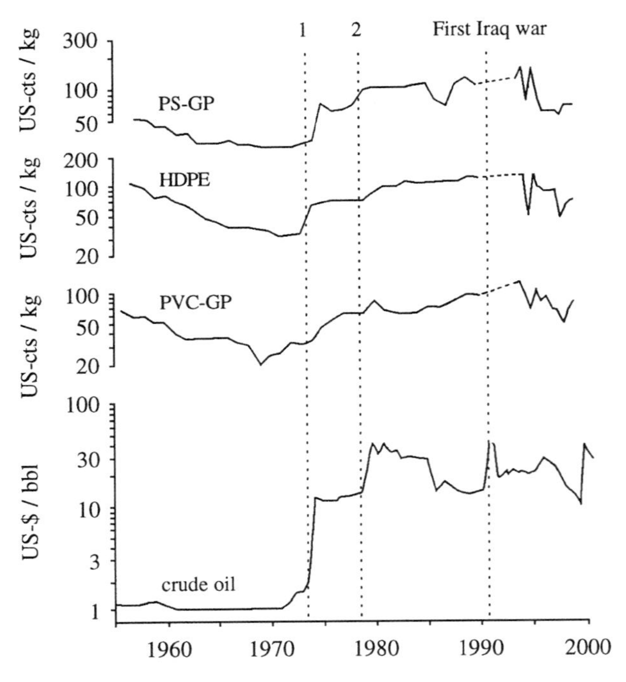


Fig. 5-28 Development of spot prices for crude oil and some standard plastics. Prices of plastics refer to general purpose grades (GP) of poly(styrene) and poly(vinyl chloride) and to high-density poly-(ethylene)s for blow forming [19]. Numbers 1 and 2 indicate the first and the second oil crises.

After nationalization of their oil industries (Fig. 3-11), OPEC members raised the price of a barrel of crude oil from 3 to 5 $ in October 1973, then from 5 to 11.50 $ in December 1973, and again from 12 to 40 $ between 1977 and 1979. The price increases were first said to be political measures in the fight against Israel, and then declared as economically motivated in order to extend the reach of oil reserves. These abrupt and extreme price increases caused price explosions of goods and economic crises that could only be overcome slowly. The resulting overproduction of oil caused a price drop to ca. 13 \$/bbl in December 1990. The subsequent invasion of Kuwait by Iraq (Fig. 3-11) let prices sore again to 33 \$/bbl; they dropped to 20 \$/bbl before the liberation of Kuwait by the United States was completed. Since then, spot prices of oil have been on a roller coaster. In the summer of 2005, they exploded again to 65 \$ per barrel. This time, the price increase was mainly driven by increased demands by China and India, transportation shortages, and speculation. Until ca. 1998, China was self-sufficient with respect to energy. It has its own oil production and vast reserves of coal. Its rapid industrialization coupled with a population growth (from $1.15{\cdot}10^9$ in 1990 to $1.29{\cdot}10^9$ in 2003 despite a one-child per couple policy!) increased its hunger for oil. Even more dramatic is the growth of population in India: from $0.84{\cdot}10^9$ in 1990 to $1.05{\cdot}10^9$ in 2003.

### 5.5.4 Economy of Scale

Economy of scale is a very important factor for cost reductions: larger plants require less operational and investment cost per unit of production. A plot of logarithms of prices of plastics versus the logarithm of their production (**experience diagram, Boston diagram**) per year shows a linear relationship between these two quantities with a slope of –0.4 (Fig. 5-29). Experience shows that the cost of plants including equipment and measuring devices usually increase by the 0.6th power. If all other costs (personnel, energy, etc.) also increase by the 0.6th power, then the cost, and by and large also the price of goods, will decrease with the 0.4th power of production volume.

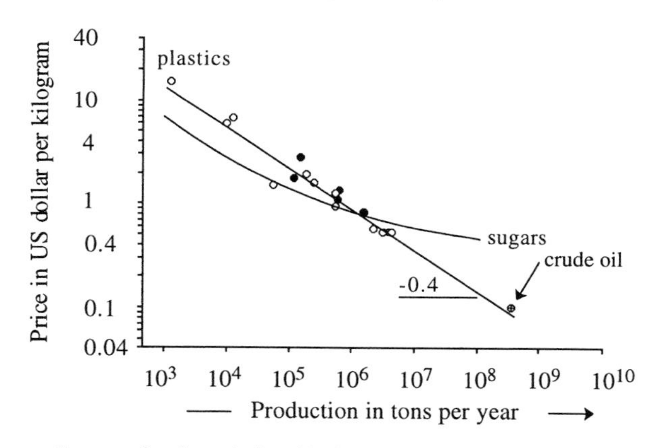


Fig. 5-29 Experience diagram for the relationship between annual US production and prices of standard and technical thermoplastics (O) and thermosets (●) in April 1990 and annual world productions and prices of sugars in 1989 (curve taken from Fig. 8-11). Note that the price of oil falls on the same experience line as that of plastics.

The strong influence of economy of scale can be especially seen for the world production of sugars (monosaccharides and disaccharides, sorbose) (Fig. 5-29). Prices for less common sugars are lower than those of plastics but follow the same economy of scale as the production of standard plastics (slope of –0.4). Prices for sugars with a greater annual world production, such as saccharose, deviate from the trendline of smaller volume sugars, however. The reason is that relatively little "raw material" (sugar beets, cane, etc.) is produced per acre, which in turn requires the raw materials to be transported to processing plants over long distances. The high costs of transportation thus limits the size of processing plants, which drives up the cost of production.

Economic data also show for comparable annual world productions that the price *per volume* of plastics is generally smaller than that of metals. Productions of plastics also show much higher annual increases. One reason is the age of the respective industries: the younger plastics industry has more room for technical innovations than the older metal industry. In general, one observes a linear relationship between the logarithm of the annual increase of annual production and the age of the industry as measured in years since the first introduction of large-scale production (Fig. 5-30).

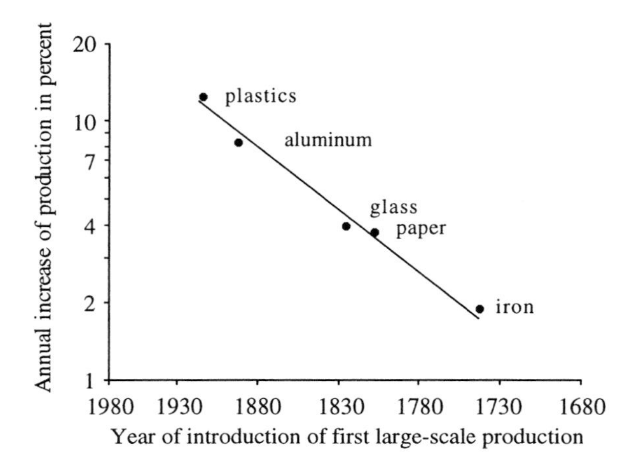


Fig. 5-30 Correlation between the logarithm of the 1975 annual growth rate of a material and the year of its first large-scale production [20]: industrial synthesis of phenolic resins (plastics), electrolysis of aluminum oxide (aluminum), mold casting of bottles (glass), paper machine (paper), and use of coke (iron). By permission of Gordon and Breach Publishers, Newark (NJ).

## Historical Notes on Emulsion Polymerization

M.Luther, C.Heuck, German patent (to IG Farben) 558 890 (1927) = U.S. patent 1 864 078
Free-radical emulsion polymerizations of sparingly water-soluble monomers by water-soluble initiators using emulsifiers.

R.P.Dinsmore, U.S.Patent 1 732 795 (1927; to Goodyear Tire and Rubber); CA **24** (1927) 266
Basic recipes for emulsion polymerizations ("thermal" polymerization of aqueous emulsions of various diene monomers using oleate salts and egg albumin as stabilizers).

H.Fikentscher, Angew.Chem. **51** (1938) 433
Investigations of the formation of latices of styrene-butadiene rubbers by the German IG Farbenindustrie show that the locus of polymerization in (direct macro)emulsion is the aqueous phase, not the oil phase (the emulsified monomer droplets).

J.Hengstenberg, I.G. Farben Report on 26 February 1938; as quoted by J.W.Breitenbach, Chemie und Technologie der Kunststoffe, Akademische Verlagsgesellschaft, Leipzig 1954
Experimental proof that the polymerization proceeds in monomer micelles.

In 1938, IG Farben and Standard Oil exchanged patents, including those on emulsion polymerization to styrene-butadiene rubbers which in World War II formed the basis of the U.S. crash program to develop so-called GR types such as GR-S = government rubber with styrene. After the end of the war, results of the many top-secret investigations were published.

W.D.Harkins, J.Am.Chem.Soc. **69** (1947) 1428
Experimental investigations of the size of surfactant micelles, loci of polymerization, etc.

W.V.Smith, R.H.Ewart, J.Phys.Chem. **16** (1948) 592
Kinetic theory of emulsion polymerization.

J.W.Vanderhoff, E.B.Bradford, H.L.Tarkowski, J.B.Shaffer, R.M.Wiley, Adv.Chem.Ser. **34** (1962) 32
Inverse emulsion polymerization.

# Literature to Chapter 5

5.0 GENERAL ASPECTS
C.A.Clausen III, G.Mattson, Principles of Industrial Chemistry, Wiley-Interscience, New York 1978

5.1 POLYMERIZATIONS (general)
G.Allen, J.C.Bevington, Eds., Comprehensive Polymer Science, Vols. 3 and 4 (Chain Polymerisations), 5 (Step Polymerisations), 6 (Polymer Reactions), Pergamon Press, Oxford 1989
P.Rempp, E.W.Merrill, Polymer Synthesis, Hüthig und Wepf, Basel, 2nd ed. 1990
N.A.Dotson, R.Galvan, R.A.Laurence, M.Tirrell, Polymerization Process Modelling, VCH, Weinheim 1996
J.-P.Vairon, N.Spassky, Industrial Cationic Polymerizations: An Overview, in K.Matyjaszewski, Ed., Cationic Polymerization, Dekker, New York 1996
M.K.Mishra, Y.Yagci, Handbook of Radical Vinyl Polymerization, Dekker, New York 1998
A.-D.Schlüter, Ed., Synthesis of Polymers, Wiley-VCH, Weinheim 1999
G.Moad, D.H.Solomon, The Chemistry of Free-Radical Polymerization, Elsevier (Pergamon Press), New York and Amsterdam 1999
F.Rodriguez, C.Cohen, C.K.Ober, L.Archer, Principles of Polymer Systems, Taylor and Francis, London 2003
G.Odian, Principles of Polymerization, Wiley-Interscience, New York, 4th ed. 2004
H.R.Kricheldorf, O.Nuyken, G.Swift, Eds., Handbook of Polymer Synthesis, Dekker, New York, 2nd ed. 2005
H.-G.Elias, Macromolecules, Vol. I: Chemical Structure und Synthesen, Wiley-VCH, Weinheim 2005

5.2 INDUSTRIAL POLYMER PRODUCTIONS (surveys)
J.L.Throne, Plastics Process Engineering, Dekker, New York 1979
J.A.Biesenberger, D.A.Sebastian, Principles of Polymerization Engineering, Wiley, New York 1983
K.H.Reichert, W.Geiseler, Eds., Polymer Reaction Engineering: Vol. 1, Hanser, Munich 1983; Vol. 2, Hüthig und Wepf, Basel 1986; Vol. 3, VCH, Weinheim 1989
N.A.Dotson, R.Galvan, R.L.Laurence, M.Tirrell, Polymerization Process Modeling, VCH, Weinheim 1995
E.S.Wilks, Ed., Industrial Polymers Handbook. Products, Processes, Applications, 4 vols., Wiley-VCH, Weinheim 2000
–, Ullmann's Chemical Processes and Process Engineering, Wiley-VCH, Weinheim, 3 vols. 2004 (updated excerpts of Ullmann's Encyclopedia of Industrial Chemistry, VCH, Weinheim, 1985-1995)
E.T.Denisov, T.G.Denisova, T.S.Pokidova, Handbook of Free-Radical Initiators, Wiley, Hoboken (NJ) 2003
T.Meyer, R.Keurentjes, Handbook of Polymer Reaction Engineering, Wiley-VCH, Weinheim 2005, 2 volumes

5.2.4 SUSPENSION POLYMERIZATIONS and 5.2.5 EMULSION POLYMERIZATIONS
F.A.Bovey, I.M.Kolthoff, A.J.Medalia, E.J.Meehan, Emulsion Polymerization, Interscience, New York 1955 (includes history of emulsion polymerization in the United States)
H.Gerrens, Kinetik der Emulsionspolymerisation, Fortschr.Hochpolym.Forschung **1** (1959) 234 (includes history of emulsion polymerization in Germany)
D.C.Blackley, Emulsion Polymerization: Theory and Practice, Halsted, New York 1975
K.E.J.Barrett, Ed., Dispersion Polymerization in Organic Media, Wiley, New York 1975
I.I.Piirma, Ed., Emulsion Polymerization, Academic Press, New York 1982
R.D.Athey, Jr., Emulsion Polymer Technology, Dekker, New York 1991
H.G.Yuan, G.Kalfas, W.H.Ray, Suspension Polymerization, J.Macromol.-Sci.-Rev.Macromol. Chem.Phys. **C 31** (1991) 215
Q.Wang, S.Fu, T.Yu, Emulsion Polymerization, Progr.Polym.Sci. **19** (1994) 703
R.G.Gilbert, Ed., Emulsion Polymerization: A Mechanistic Approach, Academic Press, London 1995
K.Matyjaszewski, Ed., Cationic Polymerization. Mechanism, Synthesis, and Applications, Dekker, New York 1996
M.Szwarc, Ionic Polymerization Fundamentals, Hanser, Munich 1996
H.L.Hsieh, R.P.Quirk, Eds., Anionic Polymerization, Dekker, New York 1996
R.M.Fitch, Polymer Colloids. A Comprehensive Introduction, Academic Press, London 1997
P.A.Lovell, M.S.El-Aasser, Eds., Emulsion Polymerization and Emulsion Polymers, Wiley, Chichester 1997
D.V.Blackley, Ed., Polymer Latices. Science and Technology, Chapman & Hall, London, 2nd ed. 1997 (3 vols.)
I.Capek, C.-S. Chern, Radical Polymrization in Direct Mini-Emulsion Systems, Adv.Polym.Sci. **155** (2001) 102

5.2.8 GAS-PHASE POLYMERIZATIONS
K.E.Waele, Addition Polymerization at High Pressure, Quart.Revs. **16** (1962) 267
Y.Ogo, Polymerizations at High Pressure, J.Macromol.Sci.-Revs.Macromol.Chem.Phys. **C 24** (1984) 1

5.3 POLYMERIZATION REACTORS
E.B.Naumann, Mixing in Polymer Reactors, Revs.Macromol.Chem. **C 10** (1974) 75
H.Gerrens, Über die Auswahl von Polymerisationsreaktoren, Chem.-Ing.Tech. **52** (1980) 477; -, How to Select Polymerization Reactors, ChemTech **12** (1982) 380, 434
H.Gerstenberg, P.Sukuhr, R.Steiner, Rührkessel-Reaktoren für Polymer-Synthesen, Chem.-Ing. Tech. **54** (1982) 541 (selection of polymerization reactors)
B.W.Brooks, Polymerization Reactors, Rev.Chem.Eng. **1** (1983) 403
J.F.MacGregor, A.Penlidis, A.E.Hamielec, Control of Polymerization Reactors: A Review, Polym.Process Eng. **2** (1984) 179
S.K.Gupta, A.Kumar, Reaction Engineering of Step Growth Polymerization, Plenum Publ., New York 1987
H.Thiele, H.D.Zettler, Auswahlkriterien für Reaktoren zum Herstellen von Polymeren, Kunststoffe **79** (1989) 687 (selection of polymerization reactors)
A.E.Hamielec, H.Tobita, Polymerization Processes, Ullmann's Encyclopedia of Industrial Chemistry, **A 21** (1992) 305
F.J.Schork, P.B.Deshpande, K.W.Leffew, Control of Polymerization Reactors, Dekker, New York 1993
C.McGreavy, Ed., Polymer Reaction Engineering, Blackie (Chapman & Hall), New York 1993
–, Ullmann's Chemical Engineering and Plant Design, Wiley-VCH, Weinheim, 2 vols. 2004 (updated excerpts of Ullmann's Encyclopedia of Industrial Chemistry, VCH, Weinheim, 1985-1995)

5.4 WORK-UP OF POLYMERS
J.A.Biesenberger, Ed., Devolatilization of Polymers, Hanser, Munich 1983

## References to Chapter 5

[1] P.Blondel, T.Briffaud, M.R.G.Werth, Macromol.Symp. **122** (1997) 243, Fig. 1
[2] P.Hayden, H.W.Melville, J.Polym.Sci. **43** (1960) 201, Fig. 1
[3] Experimental data of M.Hirooka, PhD Thesis, Kyoto 1971
[4] K.E.J.Barrett, M.W.Thompson, in K.E.J.Barrett, Ed., Dispersion Polymerization in Organic Media, Wiley, London 1975
[5] S.T.Balke, A.E.Hamielec, J.Appl.Polym.Sci. **17** (1973) 905
[6] G.R.Johnson, J.Vinyl Technol. **2** (1980) 138, Fig. 1
[7] D.Hunkeler, F.Candau, C.Pichot, A.E.Hamielec, T.Y.Xie, J.Barton, V.Vaskova, J.Guillot, M.V.Dimonie, K.Reichert, Adv.Polym.Sci. **112** (1994) 115, Fig. 1
[8] Data of D.Rahlwes, R.Casper, D.Kranz, Dechema meeting 1978 (private communication by D.Rahlwes)
[9] Data of M.J.Ballard, R.G.Gilbert, D.H.Napper, P.J.Pomery, P.W.O'Sullivan, J.H.O'Donnell, Macromolecules **19** (1986) 1303
[10] R.G.W.Norrish, R.R.Smith, Nature **150** (1942) 336
[11] S.Fengler, K.-H.Reichert, Angew.Makromol.Chem. **225** (1995) 139, Figs. 3 and 5
[12] H.Thiele, H.D.Zettler, Kunststoffe **79** (1989) 687, Fig. 4
[13] H.Gerrens, Chem.-Ing.Tech. **52** (1980) 477, (a) Fig. 1, (b) Fig. 2, (c) Fig. 4, (d) Fig. 11, (e) Fig. 13, (f) Fig. 12, (g) Fig 4, Fig. 13; see also H.Gerrens, CHEMTECH (July 1982) 434
[14] H.Gerstenberg, P.Sukuhr, R.Steiner, Chem.-Ing.Tech. **54** (1982) 541, (a) Table 2, (b) Table 3
[15] K.H.Reichert, H.U.Moritz, Makromol.Chem.-Macromol.Symp. **10/11** (1987) 571, (a) Fig. 13, (b) Fig. 14; see also W.Baade, H.U.Moritz, K.H.Reichert, J.Appl.Polym.Sci. **27** (1982) 2249
[16] Z.Tadmor, J.A.Biesenberger, Ind.Eng.Chem.Fundam. **5** (1966) 336, (a) Fig. 5, (b) Fig. 3
[17] J.A.Biesenberger, Z.Tadmor, Polym.Eng.Sci. **6** (1966) 304, Fig. 4
[18] Data from Chemical Week, end of December 1989
[19] Half-year data from Chemical Week (end of June and end of December, each year)
[20] G.Snelling, Polymer News **3/1** (1976) 35, Fig. 2

# 6 Carbon Chains

## 6.1 Carbons

The natural element carbon consists of two stable isotopes, $^{12}C$ (98.90 mol%) and $^{13}C$ (1.10 mol%), and an unstable isotope $^{14}C$ (ca. $10^{-10}$ mol%) with a half-life of 5730 years that is used for radiocarbon dating. In addition, 15 artificial isotopes are known. The shortest-lived artificial isotope, $^{8}C$, has a half-life of ca. $2 \cdot 10^{-21}$ s.

The substance carbon exists in seven polymeric allotropes (G: *allos* = other, *tropos* = direction) with, in principle, infinite molecular weights:

- 2 types of diamonds: "infinitely" large aliphatic structures (Section 6.1.1);
- 2 types of graphites: "infinitely" large aromatic structures (Section 6.1.3);
- Lonsdaleite ("meteoric diamond");
- chaoit(e) (discovered in 1968 in meteor craters): unclear structure;
- carbon VI (discovered in 1972): possibly with –C≡C–C≡C– structure elements.

In addition, there are several carbon forms that can be considered oligomeric carbon allotropes, i.e., *tecto* oligomers:

- diamondoids: small molecules with diamond-like structures (Section 6.1.2);
- fullerenes: cages consisting of rings with 5 and 6 carbon atoms (Section 6.1.4);
- nanotubes: hollow rods composed of 6-membered carbon rings (Section 6.1.5);
- nanofoams (Section 6.1.8).

Some other industrially used carbons are not pure polymeric or oligomeric allotropes but have mixed structures: carbon and graphite fibers (Section 6.1.6), glass carbon (Section 6.1.7), carbon black (Secton 6.1.9), charcoal (Section 6.1.10), activated carbon (Section 6.11), and amorphous carbon without short-range order.

### 6.1.1 Diamonds

Diamonds are natural or synthetic carbon allotropes consisting of "infinitely large" crystalline lattices in which carbon atoms are three-dimensionally bonded to each other by $sp^3$ bonds with bond lengths of 0.154 nm, thus forming three-dimensionally anellated six-membered rings. Diamonds crystallize in two modifications: the common cubic modification in which all $C_6$ rings are in the chair conformation; and the rare hexagonal modification where $C_6$ rings are in both chair and boat configuration (Fig. 6-1).

The cubic modification has a density of 3.51 g/cm$^3$, a melting temperature of 3700°C, and a boiling temperature of 4200°C. The rigid, ordered, covalent network structure leads to the lowest compressibility known to man, an extremely low thermal expansion coefficient of $1.1 \cdot 10^6$ K$^{-1}$, a heat capacity of 6.12 J/(mol K), the best thermal conductivity of 2000 W/(m K), a high modulus of elasticity of 1160 GPa, a compression strength of >110 GPa, a tensile strength at fracture >1.2 GPa, and the highest known Mohs hardness of 10 (≈ 100 GPa ), hence the name "diamond" (G: *adamas* = invincible, steel).

Diamonds are so brittle that they can be crushed in a steel mortar. They are thermodynamically metastable at room temperature and atmospheric pressure. In inert atmosphere above 1500°C, they convert to graphite.

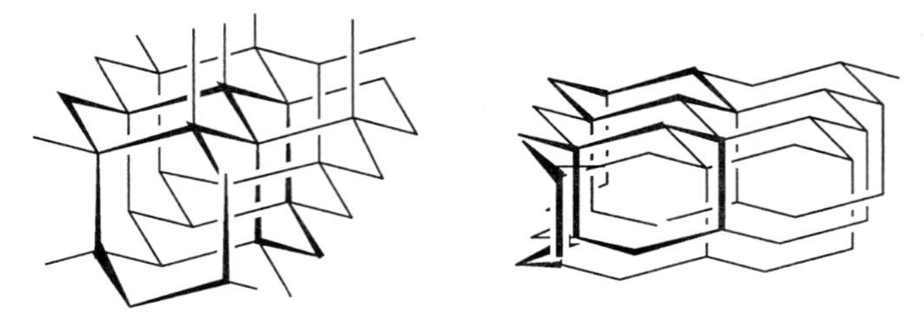

Fig. 6-1 Sections of cubic (left) and hexagonal (right) diamond structures. In each section, three six-membered rings are indicated by bold bonds in order to emphasize that in the cubic diamond structure all $C_6$ rings are exclusively in the chair formation whereas hexagonal diamond structures have both chair and boat formations.

**Natural Diamonds**

Diamonds are found in nature in streambeds and especially in the magmatic rocks of the walls of volcanic pipes. The size of diamond crystals varies between fractions of a milligram and 600 grams; the largest diamond ever found was the Cullinan (621.2 g). The color of diamonds depends on the type and proportion of impurities; it ranges from very water-clear to yellow, pink, green, or black.

Mined rocks are crushed and the diamonds mostly separated by hand into two categories: "gem" for jewelry and "industrial" for many purposes, for example, abrasives and electronics.

Of the annual world production of 26 000 kg (= $130 \cdot 10^6$ carat), less than one-third can be used as gems that are rated by the four C's: color, clarity, cut, and carat (1 ct = 0.2 g). For diamonds of less than 1 carat, points are used (1 point = 1/100 carat). The carat is an old measure. In ancient times, balances were not sensitive enough to measure small weights. Instead, the crescent-shaped seeds of the carob tree (locust bean tree) were used because they all had about the same weight of 0.2 g (G: *keration* = little horn).

Natural diamonds have the natural distribution of $^{12}C$ and $^{13}C$ isotopes (see above). They are not pure but always contain nitrogen and are therefore classified for electronic purposes according to their nitrogen content: Class Ia with less than 0.3 % N; Class Ib with less than 0.1 %; Class IIA where the nitrogen content is so small that it is very difficult to detect by ultraviolet or infrared; and Class IIb with such extremely low nitrogen contents that the diamond behaves as a p-type semiconductor (caused by uncompensated impurities).

Lonsdaleit, the black "meteoric diamond," is found in Canyon Diablo (AZ) meteorites. It is a hexagonal form of diamond with a density of 3.2–3.3 g/cm$^3$ and a Mohs hardness of only 7–8. Another hexagonal form is the dark-gray metallic Chaoit(e) from the Riesa crater in Swabia, Germany; this form has a density of 3.43 g/cm$^3$ and a Mohs hardness of 2.0.

Hexagonal diamonds can also be synthesized from graphite at temperatures above 1000°C by static pressure of 13 MPa (130 kbar).

**Synthetic Diamonds**

Synthetic diamonds are produced from graphite (Fig. 6-2) which is more stable than diamond at 30°C and normal pressure by 2.7 kJ/(mol C). At 300°C and 1500 MPa, both allotropes are in equilibrium.

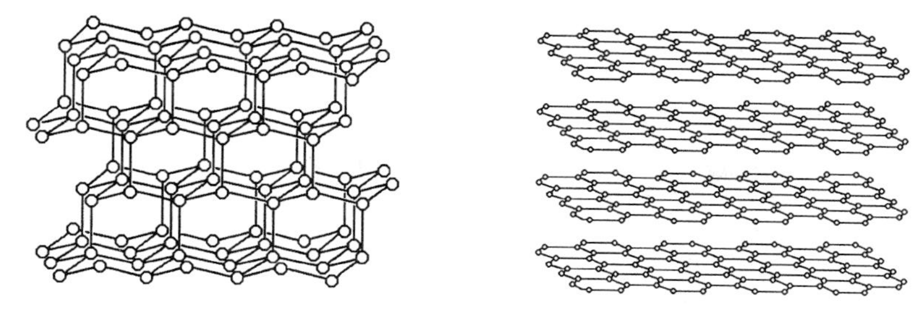

Fig. 6-2 Cubic form of diamond (left) and hexagonal form of graphite (right).

Industrial synthetic diamonds are thus produced by subjecting graphite in the presence of liquid metals (Fe, Co, Pt, etc.) to temperatures above 1400°C and pressures of ca. 8.4 GPa (≈ 83 000 atm). This **graphite process** is a thermodynamically controlled solution reaction in which graphite in molten metals as solvents is converted to the less-soluble diamond. Very small diamonds can also be produced dynamically at 1230°C by shock waves (30 GPa). All these diamonds are too flawed for gems; they are used for abrasives, drill bits, cutting tools, and the like. More than three quarters of the annual world production of diamonds are made by these processes.

Diamonds for jewelry up to 2.8 carat are produced by a process similar to the graphite process (1500°C, 5.9 GPa), albeit by seeding the molten carbon with tiny natural diamonds. These diamonds have a yellow tint because less than 0.005 % of the carbon atoms are replaced by nitrogen atoms. Another process for the synthesis of gem-type diamonds uses vapor deposition to grow single crystals (this is *not* CVD, which produces polycrystalline materials, see below). This process can control the amount and type of impurities; it delivers pink to black diamonds. Synthetic diamonds for jewelry are usually priced 30 % lower per carat than natural diamonds.

**Chemical vapor deposition** (**CVD**) decomposes methane to carbon in hydrogen at 820°C in a low-pressure (12 mbar) process using plasma glow discharges, microwaves, etc., as an energy source. The process delivers thin films composed of octahedral diamond crystallites that can be used directly as coating.

Thin platelets are also used as heat sinks in microelectronics, in glass fiber optics, and in laser technology where much heat is developed in a small space. For this purpose, isotope-pure $^{12}C$ diamonds are preferred since their heat conductivity is 1.5 times higher than that of natural diamonds or diamonds produced from graphite, 3.6 times higher than that of silver, and 8.5 times higher than that of copper.

Such isotope-pure diamonds with 99.9 % $^{12}C$ are produced from $^{12}CH_4$ as the carbon source. They cannot be obtained from isotope-pure graphite or carbon black since the conversion graphite → diamond is accompanied by an increase of the density to 3.51 g/cm$^3$ from 2.35 g/cm$^3$. This density change causes local pressure changes which promote the unwanted formation of twin crystals or polycrystalline aggregates.

Carbon-12 for $^{12}CH_4$ is obtained from byproducts of the synthesis of $^{13}C$ compounds that are used in biology and medicine as non-radioactive tracers. The carbon atoms in these byproducts are present in the natural $^{12}C/^{13}C$ ratio. The byproducts are partially oxidized to carbon monoxide; $^{12}CO$ and $^{13}CO$ are then separated by cryogenic multi-plate distillation of liquid CO. The isotope-pure $^{12}CO$ is subsequently hydrogenated to $^{12}CH_4$.

### 6.1.2 Diamondoids

Diamondoids, also called **diamond molecules** or **cage hydrocarbons**, are hydrocarbons that are based on the smallest three-dimensional carbon skeletons of cubic diamond. Their parent compound is adamantane, $C_{10}H_{16}$. It was formerly called diamantane which is now the name of the next higher homologue, $C_{14}H_{20}$, that is also known as congressane (Fig. 6-3).

Expansion of the adamantane ring system by additional 6-membered rings delivers the homologue series of **amantanes** with the composition $C_{4n+6}H_{4n+12}$ where $n \geq 1$. The names of the adamantame homologues are created by prefixing amantane with ad-, di-, tri-, tetr-, pent-, hex-, hept-, etc.

The first three representatives of this series – adamantane, diamantane, triamantane – possess only one isomer each, but tetramantane has 3, pentamantane 6, hexamantane 17, etc., regulary cata-condensed isomers. Anti- and skew-tetramantanes have two quaternary carbon atoms (carbon atoms that are completely "substituted" by other carbon atoms and not by hydrogen) while iso-tetramantane has three of these.

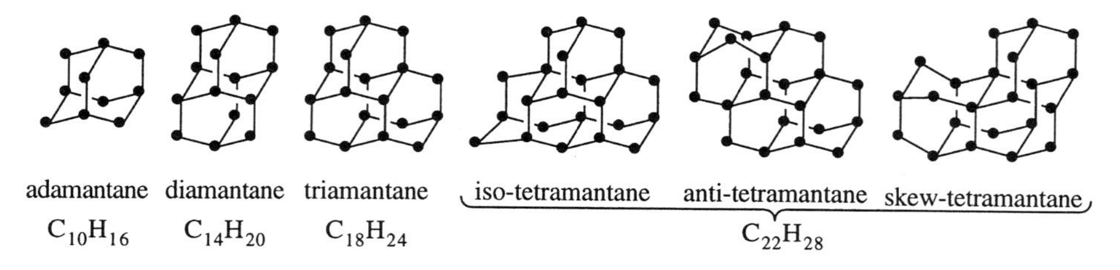


Fig. 6-3 Carbon skeletons of the first four amantanes.

In addition, there are non-isomeric amantanes, the composition of which deviates from $C_{4n+6}H_{4n+12}$. Pentamantane possesses one non-isomeric amantane with the composition $C_{25}H_{30}$, in addition to the 6 isomeric ones with the composition $C_{26}H_{32}$. Hexamantane comprises 17 regularly cata-condensed isomers with $C_{30}H_{36}$, 6 irregularly cata-condensed isomers with $C_{29}H_{34}$, and 1 peri-condensed isomer with $C_{26}H_{30}$.

Diamondoids are crystalline and have high melting temperatures (270°C for adamantane, 237°C for diamantane). Boiling temperatures are not known because amantanes sublime easily at temperatures below their melting temperatures.

Diamondoids were first isolated from material that plugged refinery equipment, for example, upon drastic changes of pressure and/or temperature on processing of natural gas and gas condensates. The reason is sublimation of diamondoids: on distillation of mixtures of aliphatic hydrocarbons, diamondoids are found in cuts with boiling points near 190°C. Diamondoids can be isolated by heating petroleum distillate fractions to 450°C which destroys less-stable linear, branched, and cyclic hydrocarbons. Purification is by shape-selective chromatography and crystallization.

Diamondoids form a continuous series of materials from lower diamondoids (< 1 nm) and higher diamondoids (~1 to 2 nm) to nanocrystalline and chemical vapor deposition diamonds (2 nm to micrometer length) and macroscopic diamonds. Lower diamondoids are envisioned as building blocks for pharmaceuticals; for example, 1-aminoadamantane is an antiviral drug. Higher diamondoids have apparently not yet found an application.

### 6.1.3 Graphite

Graphite is a steel-gray to black material that is found in nature or synthesized by man. It has a metallic luster but feels fatty and soft (Mohs hardness 1). On paper, it produces black lines which is utilized for pencils (G: *graphein* = to write).

Natural graphite is almost never pure; it can contain up to 20 wt% ash. Pure graphite consists completely of carbon. It crystallizes in a hexagonal form (α-graphite) and a rhombohedral one (β-graphite). Natural graphite contains more than 70 % of the α-form and less than 30 % of the β-form. Synthetic graphite is 100 % α-graphite, the thermo-dynamically most stable form. An amorphous graphite is also known.

Graphite consists of stacks of "infinitely" large sheets of anellated 6-membered carbon rings (L: *anellus* = small ring; *anus* = ring) (Fig. 6-4). In such a sheet, called **graphene**, each carbon atom is bonded by σ-bonds to 3 other carbon atoms. The fourth valence electron of each of the carbon atoms forms non-localized π-bonds. The resulting delocalization enables each carbon atom to maintain 8 valence electrons.

So far, graphene itself has not been synthesized. However, it has been isolated from graphite by removing graphite layer by layer with adhesive tape.

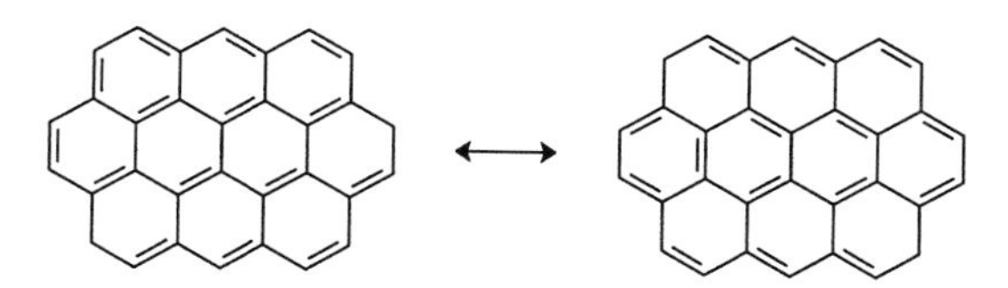

Fig. 6-4 Schematic representation of resonance structures in sections of graphene.

The bond length between $sp^2$-bonded carbon atoms is 0.1415 nm whereas the distance between graphene sheets is 0.335 nm. The latter distance corresponds approximately to the sum of van der Waals radii of carbon atoms. The weak van der Waals forces between graphene sheets explain why graphite layers are so easy to shift (use of graphite as temperature-resistant lubricant) and why the electrical conductivity perpendicular to the graphene sheets is only 1 S/cm versus 10 000 S/cm parallel to the sheets.

The arrangement of graphenes in hexagonal graphites differs from that in rhombohedral ones (Fig. 6-5). In hexagonal graphite, every second graphene B is shifted parallel to the preceeding graphene A, leading to an ...ABABAB... structure. In rhombohedral graphite, all graphenes are shifted unidirectionally to the preceding ones, which results in ABCDEF... structures. The number of intermediate structures is of course unlimited.

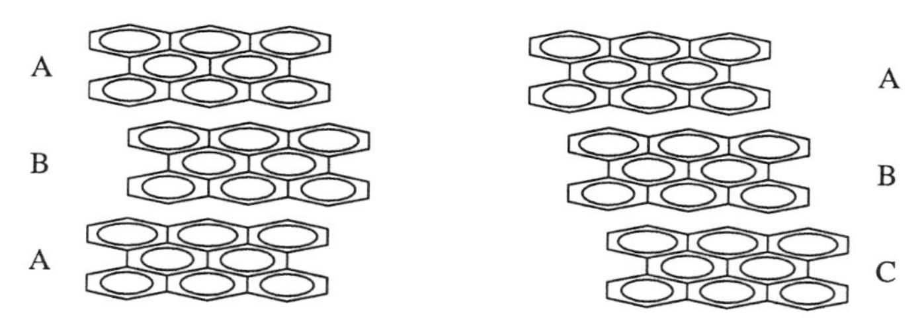


Fig. 6-5 Stacking of graphenes in hexagonal (left) and rhombohedral (right) graphite.

Table 6-1 Properties of various carbons parallel (||) and perpendicular (⊥) to the sheet direction.

| Property | | Physical unit | Graphite | Electro graphite | Glass carbon | Carbon fiber type HT |
|---|---|---|---|---|---|---|
| Density | | $g/cm^3$ | 2.27 | 1.5–1.8 | 1.5 | 1.8 |
| Thermal expansion coefficient | ‖ | $K^{-1}$ | $0.5 \cdot 10^{-6}$ | $0.6 \cdot 10^{-6}$ | $3 \cdot 10^{-6}$ | $0.6 \cdot 10^{-6}$ |
| | ⊥ | $K^{-1}$ | $27 \cdot 10^{-6}$ | $2.0 \cdot 10^{-6}$ | $3 \cdot 10^{-6}$ | |
| Heat conductivity | ‖ | $W\ m^{-1}\ K^{-1}$ | 2000 | 150 | 6 | 9 |
| | ⊥ | $W\ m^{-1}\ K^{-1}$ | 10 | 150 | 6 | |
| Electrical conductivity | ‖ | S/cm | 20 000 | 2000 | 220 | 550 |
| | ⊥ | S/cm | 2.5 | 1000 | 220 | 15 |
| Modulus of elasticity | ‖ | GPa | 965 | 10 | 35 | 240 |
| | ⊥ | GPa | 35 | 5 | 35 | |
| Tensile strength | ‖ | GPa | 96 | | | < 4.6 |
| | ⊥ | GPa | 34 | | | |

The relatively large distance between graphenes allows one to incorporate in graphite various atoms, atomic groups, or ions, for example, chlorine or potassium ions. Sudden heating causes these intercalation compounds to swell. The resulting **expandate** can be rolled to shiny flexible films or pressed to articles which can replace asbestos.

Fluorination of graphite in a fluidized bed by fluorine at 627°C leads to **poly(carbonfluoride)**, $(CF_i)_n$ with $i < 1.12$, in which the sides of graphenes contain "superstoichiometric" $CF_2$ groups. The white polymer is the most stable fluorinated carbon polymer (stability in air up to 600°C). This polymer is an excellent lubricant; It is also used as a cathode material for batteries.

The world production of graphite of ca. 500 000 t/a varies widely from year to year. Major producers are Russia and China with ca. 20 % each. Because the mining of natural graphite does not cover the demand, graphite is also synthesized. A mixture of petroleum coke (Table 3-15) and coal pitch is compression-molded to briquets that are heated to 800–1300°C. The resulting **artificial carbon** is then graphitized in electrically heated furnaces to yield the so-called **electro graphite**.

### 6.1.4 Fullerenes

Fullerenes are cage-like spheroidal oligomers of carbon atoms in anellated, unsaturated 5- and 6-membered rings (Fig. 6-6). Their name is a short form of the original name "buckminsterfullerene", a name in honor of the American architect Richard Buckminster Fuller who designed a geodesic dome consisting of 20 hexagons and 12 pentagons which is also the structure of the $C_{60}$ fullerene (IUPAC: $(C_{60}\text{–}I_h)[5,6]$fullerene).

The $C_{60}$ fullerene looks like a football (US: soccer ball) and was therefore also called "soccerene." Because it has 20 pentagons, it is occasionally adressed as icosahedron (G: *eikosi* = twenty, *hedra* = base), although its additional 12 hexagons provide it with a total of 32 lateral faces.

Fullerenes were first detected in carbon black (R.F.Crutzen, H.W.Kroto, R.E.Smalley, Nobel Prize 1996) and later obtained by evaporating graphite with a laser. They are now mostly synthesized by the Krätschmer–Huffman process in which carbon is evaporated by an electric arc in a helium or argon atmosphere. Fullerenes are extracted from the resulting materials and separated chromatographically on a graphite column.

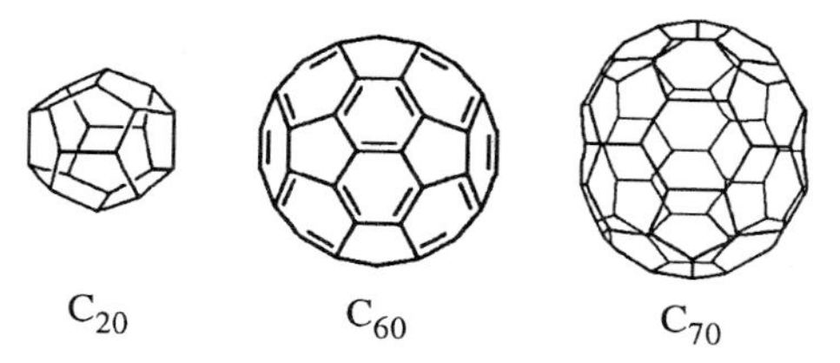


Fig. 6-6 $C_{20}$ icosahedron (not a chemical molecule), the $C_{60}$ fullerene (only top view but with double bonds), and $C_{70}$ fullerene (for clarity shown without double bonds).

The number of possible carbon atoms per fullerene is dictated by geometrical and physical rules. In tension-free fullerenes, this number should follow Euler's rule, which says that a plane consisting of hexagons can only be closed to a strain-free polyhedron if it contains exactly 12 additional pentagons. In a polyhedron of unsaturated carbon bonds, $sp^2$ bonds between carbon atoms must be bent because they are lying on the surface of a spheroid. However, the curvature cannot exceed a certain maximum value because the bond would break. According to the isolated pentagon rule (IPR), exceedingly high curvatures are avoided if all 12 pentagons are surrounded by hexagons on all sides.

This rule requires a minimum of 60 carbon atoms for the lowest fullerene. The IPR rule furthermore predicts the existence of $C_{60}$, $C_{70}$, $C_{72}$, $C_{74}$, $C_{76}$, ... fullerenes but rules out fullerenes $C_{62}$, $C_{64}$, $C_{66}$, and $C_{68}$. Experimentally, fullerenes $C_{60}$, $C_{70}$, $C_{76}$, $C_{78}$, ... have been isolated. $C_{72}$ seems to exist only as sublimate; $C_{74}$ is unknown.

The mustard-colored $C_{60}$ fullerene forms spheres of ca. 0.7 nm diameter which crystallize cubic face-centered. Higher fullerenes are not necessarily spherical: the reddish-brown $C_{70}$ fullerene resembles a rugby ball that is squeezed in the center (Fig. 6-6, $C_{70}$).

Fullerenes are somewhat soluble. Maximum concentrations at room temperature are 0.02 g/mL in 1,2,4-trichlorobenzene and 0.012 g/mL in $CS_2$. They dissolve in many other organic solvents albeit in smaller concentrations.

Fullerenes are very stable to heat, oxidation, and corrosion but have quite reactive surfaces. They have interesting electrical and optical properties, for example, a non-linear variation of transparency as a function of the intensity of incident light. Fullerenes are therefore used in electronics and optoelectronics (see Volume IV).

Real fullerenes can contain defects, for example, missing carbon atoms. Carbon atoms at these defects can be easily oxidized to >C=O groups. The oxidation is reversible at 600°C where CO and $CO_2$ are split off. Carbon atoms can also be replaced by other elements. An example is the azafullerene, $C_{48}N_{12}$, that is obtained on substrates by ultra-high vacuum magnetron sputtering. It forms thin films of 7 nm thickness which corresponds to layers of 7–10 shells. These films have tensile moduli of ca. 37 GPa and hardnesses of ca. 7 GPa.

As cage compounds, fullerenes can encapsulate atoms of other elements, for example, atoms of noble gases such as xenon in $^{129}XeC_{60}$. Higher fullerenes ($C_{72}$, $C_{74}$, etc.) can also encapsulate metal ions, examples are the **endohedral metallofullerenes** $La_2@C_{72}$ and $Eu@C_{74}$ (G: *endon* = inside; @ is the symbol for "in"). The metal ions are highly mobile in symmetric fullerenes but bound to specific sites in less symmetric ones.

Some fullerenes do not follow the IPR rule. Examples are the metallofullerene $Sc@C_{66}$, the substituted non-IPR fullerenes $C_{36}H_6$ and $C_{36}H_6O$, and the cations $C_{20}^{\oplus}$ and $C_{20}^{2\oplus}$. The latter compounds are obtained by bromination of the dodecahydran,

$C_{20}H_{20}$, which can be viewed as a fully hydrogenated and saturated $C_{20}$ fullerene derivative. Bromination of this compound leads to isomeric trienes with the average composition $C_{20}HBr_{13}$, which, on debromination in the gaseous phase, leads to $C_{20}^{\oplus}$ and $C_{20}^{2\oplus}$.

Fullerenes of different sizes can also be nested concentrically like the Russian Matuschka dolls and form a kind of hyperfullerene. The resulting carbon ions can have diameters of several hundred nanometers.

### 6.1.5 Nanotubes

Carbon atoms can also unite to form **carbon nanotubes** (**CNT**) with walls composed of anellated unsaturated 6-membered rings similar to graphene (Fig. 6-7). The seamless hollow cylinders are either open or capped at both ends. Single-wall nanotubes (**SWNT**) consist of a single carbon layer, whereas multi-wall nanotubes (**MWNT**) have walls that are two or more carbon layers thick. CNTs are synthesized by several methods:

*Arc discharge* uses a gas plasma between two carbon electrodes that are ca. 1 mm apart. It burns one electrode while depositing the CNT on the other. The eroding electrode is either carbon, e.g., graphite (for MWNT) or a composite of carbon and a catalytically acting transition metal ctalyst (for SWNT).

*Pulsed-laser vaporization* employs a continuous high-energy $CO_2$ laser to erode a carbon–metal composite at temperatures of ca. 1200°C under a high-pressure blanket of He or Ar. This process delivers only SWNTs.

*Chemical vapor deposition* (CVD) passes a precursor gas over a metal or metal oxide (e.g., Ni, Fe, Mo, Co) on a solid surface (e.g., Si, $SiO_2$, zeolite) at < 550°C. Purities of CNTs can reach 99.99 % carbon. In one variant, gases like $C_2H_2$, $CH_4$, $C_2H_4$, or CO are used as a feedstock for a plasma–enhanced deposition of SWNTs or MWNTs. In the alcohol-catalytic variant, decomposition of the alcohol delivers hydroxyl radicals that remove carbon atoms on dangling bonds at the solid carbon substrate. The atoms then cluster together and form SWNTs.

**Single-walled nanotubes** (**SWNT**s) usually have diameters of 1.2–1.4 nm and lengths of 10–1000 nm (the length record is 4 cm); they are sometimes capped by half-spheres of fullerenes. The smallest SWNT was found inside an MWNT; it had a diameter of 0.3 nm and is likely capped by half of a $C_{12}$ hexagonal prism. Shorter SWNTs are rigid, whereas longer ones bend by their own weight. They usually form bundles (ropes) of 10–100 tubes with lengths up to 2.5 mm that can be separated by ultrasonic agitation.

SWNTs exist in two ideal types (Fig. 6-7), armchair and zigzag, that differ by the arrangement of hexagons relative to the long tube axis and thus in diameter and chiral angle $\alpha$ (wrapping angle). In addition, there are many so-called chiral SWNTs with chiral angles of $0° < \alpha < 30°$ that have helical arrangements of hexagons with respect to the long axis. These structural features control electric conductivities: armchair SWNTs are metal-like conductors ($\sigma \approx 10^4$ S/cm, similar to mercury), whereas zigzag SWNTs are either semimetallic or semiconducting ($10^2$-$10^{-9}$ S/cm), depending on their diameters.

SWNTs are extremely stiff: the modulus of elasticity of a single SWNT is ca. 1 TPa whereas that of a bundle of SWNTs with a diameter of 10–20 nm is ca. 0.1 TPa. The tensile strength of ca. 30 GPa resembles that of graphite perpendicular to graphene sheets (Table 6-1), whereas the thermal conductivity of 2000 W/(m K) corresponds to that of graphite in the sheet direction. SWNTs can be extended 10–30 % without breaking.

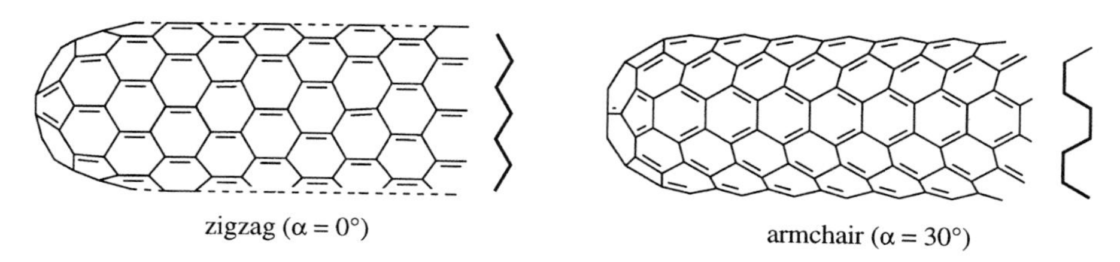


Fig. 6-7 Single-walled nanotubes (SWNT). The names refer to the arrangement of carbon bonds *at the end* of open tubes (bold patterns) and not along the tube axis where the armchair SWNT (α = 30°) has a zigzag arrangement of hexagons and the zigzag SWNT (α = 0°) an armchair one. Zigzag and armchair structures are achiral. Chiral nanotubes (0° < α < 30°) have left- and right-handed variants.

**Multi-walled nanotubes** have external diameters of 2–50 nm and internal diameters of 1–10 nm; they may be up to 2 mm long. Their modulus of elasticity is ca. 1300 GPa.

No routine processes seem to have emerged for the separation of SWNTs and MWNTs on one hand and the separation of armchair, zigzag, and chiral SWNTs on the other. The latter is especially important for electronic applications of carbon nanotubes (p. 208). SWNTs are presently used as tips in atomic and electronic force microscopy and chemical sensors for the detection of $NO_2$ or $NH_3$. They are also effective hydrogen storage materials. MWNTs are superior to graphite as the electronically conducting component in polymer composites where they dissipate the build-up of electric charge and act as a barrier against fuel diffusion.

### 6.1.6 Carbon and Graphite Fibers

Graphite is stable in inert atmosphere up to 2500°C and against oxidation up to 400°C. These properties are exploited in high-temperature resistant carbon fibers (world production: ca. 5000 t/a) that are subdivided according to their properties and carbon content into 3 types (Table 6-2):

HT fibers with high tensile strength and 92–96 % carbon, produced at 1200–1400°C;
HM fibers with high modulus and 99 % carbon, produced at 1800–2500°C;
UHM fibers with ultrahigh modulus and > 99 % carbon, produced at >2800°C.

The United Kingdom uses another classification based on the heat treatment after the thermal decomposition of feedstocks at ca. 300°C. Heat treatment up to 1000°C generates Type A fibers with low elastic moduli and tensile strengths, treatment up to 1400°C results in Type II fibers with medium elastic moduli and high tensile strengths (corresponds to HT), and treatment above 2500°C leads to Type I fibers with high elastic moduli and medium tensile strengths (corresponds to HM).

Another subdivision is according to the feedstock. Carbon fibers in the narrow sense are fibers with high carbon content that are obtained from the thermal decomposition of poly(acrylonitrile) fibers; nylon fibers and rayon are no longer used. Graphite fibers are based on pitch or tar, both materials with preformed graphitic structures. The term "graphitic fiber" is sometimes also used for proper carbon fibers that are heat-treated at high temperatures because of a supposed graphitization. It seems expedient to call all these fibers "carbon fibers" and distinguish them from each other as HT, HM, and UHM according to their carbon content and mechanical properties (Table 6-2).

Table 6-2 Properties of carbon fibers based on poly(acrylonitrile) (PAN) or pitch [1].

| Property | Physical unit | HT PAN | HM PAN | UHM PAN | HM pitch | UHM pitch |
|---|---|---|---|---|---|---|
| Carbon content | % | 93 | 99 | >99 | 99 | >99 |
| Diameter, average | μm | 6.8 | 7.2 | 8.4 | 10 | 10 |
| Density | g/cm$^3$ | 1.77 | 1.80 | 1.96 | 2.02 | 2.15 |
| Tensile modulus | GPa | 245 | 380 | 520 | 380 | 690 |
| Tensile strength at break | GPa | 3.8 | 2.5 | 1.9 | 1.9 | 2.2 |
| Elongation at break | % | 1.6 | 0.65 | 0.4 | 0.5 | 0.3 |
| Thermal conductivity | W/(m K) | 8.7 | 67 | 120 | 110 | 520 |
| Electrical conductivity | S/cm | 610 | 1050 | 1550 | 1330 | 4000 |

Present carbon fiber syntheses start with organic fibers or pitch; industrial processes do not grow carbon fibers in high-pressure arcs or by thermal decomposition of coke oven gases. Many organic fibers can serve as feedstocks (rayon, poly(vinyl alcohol), poly(acetylene), aramids, cellulose, fibers from phenolic or furan resins). but poly-(acrylonitrile) fibers (PAN) are the only ones that are used industrially (see also p. 292). The pyrolysis leads first to ladder structures that on further reaction and simultaneous stretching become crosslinked:

(6-1)

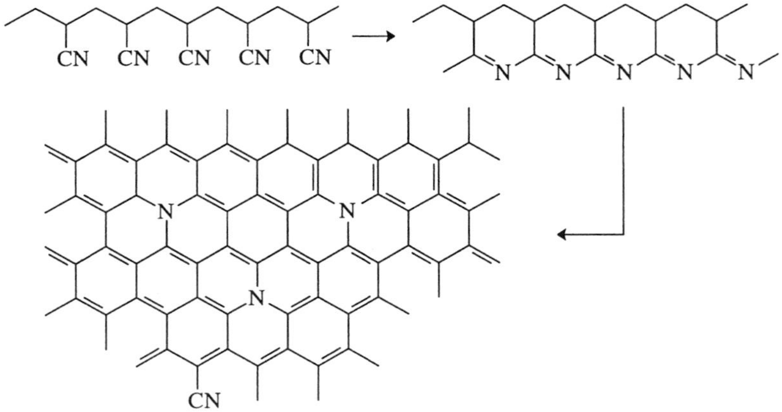

CN CN CN CN CN → N N N N N

N N N CN

The pyrolysis produces various oxidation products. Generation of volatile carbon compounds needs to be avoided because such compounds would expand the polymer and deliver porous fibers with reduced mechanical properties. Fibers should also not melt. Stretching during pyrolysis preserves the axes of the precursor fibers so that the resulting ribbon-like structures remain preferentially in the fiber direction. The parallel arrangements of these ribbons are strongly disturbed, however, because they are bent and have different sizes. Carbon fibers thus do not show a periodic three-dimensional order.

The subsequent "stretch graphitization" to HM and UHM fibers is difficult. It can be circumvented by oxidizing the stretched precursor fibers at 200-300°C. The resulting crosslinks stabilize the fiber shape, prevent shrinkage, and promote preorientation. The last step (if so desired) is the carbonization of the fibers under a hydrogen blanket for 24 hours at 200°C.

Carbon fibers serve mainly as reinforcing fibers for plastics, for example, epoxy resins, aramids, and polyimides (see Volume IV). They are also used in fabrics, both textile (car upholstery) and industrial (filter cloths).

### 6.1.7 Glass Carbon

The pyrolysis of highly crosslinked polymers such as thermosets from phenol–formaldehyde and furfural resins leads to macroscopically isotropic materials since the narrow mesh prevents the formation of well-ordered graphite sheets. These materials probably consist of coils with ribbon-like carbon structures that contain few domains with graphitic character. The material fractures like glass, hence the name "glass carbon." It is used for fuel rods in reactors and also for replacing heart valves and bones. Glass carbon is also the feedstock for one of the nanofoams.

### 6.1.8 Nanofoams

The term "carbon nanofoam" (**CNF**) is used for two very different materials. Industrial types are produced by preparing gels from organic polymers using sol-gel techniques. The gels are then air-dried and pyrolyzed to deliver porous materials, for example, a high-density CNF with $\rho$ = 0.8 g/cm$^3$ and a low-density one with $\rho$ = 0.25 g/cm$^3$. These CNFs consist of carbon particles with diameters of $d_{part}$ ≈ 100 nm that form aggregates with pores of $d_{pore}$ < 100 nm (high-density) and $d_{pore}$ > 500 nm (low-density) plus pinholes of $d_{pinhole}$ ≈ 0.7 nm diameter. The high-density CNFs have electrical conductivities of ca. 25–100 S/cm. These materials are used for electrodes.

Completely different types of carbon nanofoam result if high-power ultrafast laser pulses in the picosecond to femtosecond range ($10^{-12}$–$10^{-15}$ s) act on glass carbon. The resulting temperature of ca. 10 000°C creates single carbon atoms that polymerize to graphene-like sheets with degrees of polymerization of ca. 4000. In these sheets, some carbon hexagons are replaced by carbon heptagons and octagons which produce negative curvatures similar to the positive curvatures of fullerenes that result from carbon pentagons.

These carbon nanofoams have very low densities of ca. 0.002 g/cm$^3$. Because of many unpaired electrons (only three bonds per carbon?), they are electrical semiconductors. Most astonishing are their magnetic properties. They are ferromagnetic at room temperature (i.e., they are attracted to iron) and become magnetic themselves at –183°C.

### 6.1.9 Carbon Black

Carbon black is the general term for a series of colloidal black materials that consist essentially of non-crystalline elemental carbon particles with oxidized surfaces. These materials are produced by thermal decomposition or controlled burning of gaseous or liquid hydrocarbons by under-stoichiometric amounts of oxygen. Uncontrolled burning results in soot.

The world production of carbon black is ca. $7 \cdot 10^6$ t/a of which ca. 70 % is used for the reinforcement of elastomers in car tires and another 25 % in other rubber articles. Only 5 % is used for other purposes such as printing inks, paper, fibers, cement, and electrodes.

### Manufacture

Carbon blacks are produced in ca. 100 different types by thermo-oxidative processes (furnace blacks, gas blacks, channel blacks) or by thermal decompositions (thermal blacks, acetylene blacks).

About 98 % of all industrial carbon blacks are **(oil-)furnace blacks**. In a ceramic furnace, liquid aromatic residual hydrocarbons (heavy distillates from the cat cracking of gas oils) are sprayed continuously into a large flame that is generated from natural gas and heated air. At temperatures of 1200–1900°C, hydrocarbons are incompletely burned. Reaction gases are sprayed with cold water, carbon black is filtered off, and the remaining gases are burnt. The resulting furnace blacks are difficult to work with because of their low bulk weights; they are therefore compacted or pelletized. Furnace blacks come in numerous grades, depending on process conditions: general purpose, high and low modulus, superior-processing, reinforcing, super-abrasion, and conducting.

The furnace process dominates the carbon black production not only because it is more versatile and delivers better products in higher yields but also because its feedstock can be easily shipped so that carbon black can be produced at the sites of its main consumption, tire factories. It has not completely surplanted other processes, however, because the desired properties of other carbon blacks cannot by reproduced by the furnace process.

**Gas blacks**, on the other hand, use methane or natural gas which cannot be transported easily. Furthermore, air has free access because the system is not completely closed. Gas blacks are very finely divided carbon blacks (particle size: 10–30 nm) that are used mainly as pigments (**pigment blacks** for inkjet printers).

Related to the gas black process is the **channel black** process, the oldest thermo-oxidative process, now practically obsolete. In this process, the smoky flame from the incomplete burning of methane or natural gas impinges on iron channels from which carbon black is collected, hence the name.

**Thermal blacks** are produced by two reactors that reverse their roles in 5-minute turns. One reactor generates energy by burning natural gas in air, while the other reactor cleaves natural gas thermally. On reversal of the reactor function, carbon black is collected from the second reactor, now cooled, whereas the first reactor starts producing.

**Acetylene blacks** result from the thermal decomposition of acetylene to carbon black and hydrogen. Acetylene blacks are very pure.

### Structure and Properties

Most carbon blacks consist of spherical particles of 10–500 nm diameter (exception: acetylene blacks) that are irreversibly aggregated to larger clusters of branched strings of pearls. All carbon blacks contain small "pores" with diameters that are usually simple multiples of 0.35 nm, the distance between two graphenes. The large interior surface of carbon blacks (Table 6-3) makes them excellent absorbents.

Table 6-3 Properties of various carbon blacks (CB) [2]. n.a. = not applicable. LSO = linseed oil.

| | Physical unit | Thermal-oxidative processes Channel blacks | Gas blacks | Furnace blacks | Thermal processes Thermal blacks | Acetylene blacks |
|---|---|---|---|---|---|---|
| Particle diameter | nm | 110-120 | 10-30 | 10-80 | 120-500 | 32-42 |
| Specific surface ($N_2$ adsorption) | $m^2/g$ | 16-24 | 90-500 | 15-450 | 6-15 | ca. 65 |
| Iodine absorption | mg/g | 23-33 | n.a. | 15-450 | 6-10 | ca. 100 |
| Absorption of dibutyl phthalate | mL/100 g | 100-120 | n.a. | 40-200 | 37-43 | 150-200 |
| Oil absorption (g LSO/100 g CB) | % | 2.5-4.0 | 2.2-11.0 | 2.0-5.0 | 0.65-0.90 | 4.0-5.0 |

Carbon blacks contain many different carbon structures that range from fullerenes and nanotubes to sections of graphenes and disturbed graphite lattices. The graphene layers are parallel but are not three-dimensionally ordered.

The surface of carbon blacks is studded with considerable amounts of different chemical groups. The following amounts were found per gram of carbon black: 1.5–5.3 mmol H, 0.1–0.9 mmol OH, 0.01–0.5 mmol O (–O–, >C=O), 0.02–0.3 mmol $COO^{\ominus}$, up to 0.07 mmol COOH, and also radicals. These groups add considerably to the reinforcing action of carbon blacks on elastomers and plastics; for example, by reaction of surface radicals with olefinic double bonds of polydienes,

The various carbon blacks differ considerably in their chemical and physical properties. For example, furnace blacks have only a few oxygen groups of types I and II whereas gas blacks have many groups of types II–V.

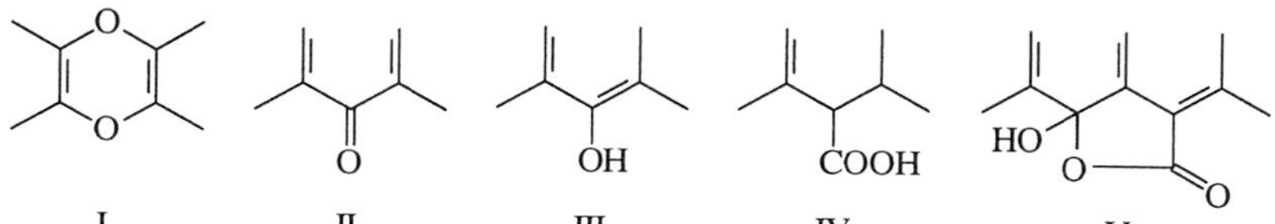

I II III IV V

Carbon blacks are subdivided according to their specific surfaces, directly by adsorption of small nitrogen molecules or large iodine molecules, and indirectly via the take-up of energy of a kneader during titration with the large molecules of dibutyl phthalate (DBP) or linseed oil (LSO) (Table 6-3). Additional classifications are according to color or application. For example, types of furnace blacks include GPF (general purpose furnace), FEF (fast extrusion furnace), HAF (high abrasion furnace), etc.

### 6.1.10 Charcoal

Charcoal is obtained from plant or animal matter by three methods: traditionally in charcoal kilns, industrially in retorts, and furthermore as a byproduct of the distillation of wood (for a discussion of these processes, see p. 74). The ineffective traditional method delivers a charcoal with 85–90 wt% carbon, 6 wt% oxygen, and 3 wt% hydrogen, as well as nitrogen, ash, and moisture. Industrial charcoals contain 95 wt% carbon, 3 wt% ash, and 2 wt% of other elements.

Industrial charcoals do not contain sulfur and are therefore used to make sulfur-free carbon steel. Other uses include refining of copper, making black gun powder, and chemical syntheses ($CS_2$, NaCN, etc.). Their high porosity (bulk density: 0.45 g/cm$^3$; true density: 1.4 g/cm$^3$; specific internal surface area: 50-80 m$^2$/g) allows applications as a gas absorbant, filter, and decolorizing medium.

### 6.1.11 Activated Carbon

Activated carbon results from the coaling of plant, animal, or mineral materials (wood, nut shells, blood, bones; coals, petrochemical products) by two different methods. In one type of process, feedstocks are distilled dry and then oxidatively activated by heating with water vapor or carbon dioxide at 700–1000°C. The other method heats feedstocks at 500–900°C with dehydration agents such as phosphoric acid or zinc dichloride, followed by washing with water. World production is ca. 400 000 t/a.

Activated carbon consists of graphite crystallites embedded in amorphous carbon. Specific surface areas of active coal are much greater than those of charcoal: 500–1500 m$^2$/g versus 50–80 m$^2$/g. Applications range from removal of odorants from gases and recovery of solvents from solvent vapors to adsorption of poisonous compounds in the stomach and colon (**medicinal coal**).

## 6.2 Poly(olefin)s

### 6.2.1 Definitions

**Olefins** are aliphatic (acyclic) or cycloaliphatic low-molecular weight hydrocarbons with one or more than one reactive carbon–carbon double bond. The designation "olefin" can be traced to the early beginnings of organic chemistry: since ethene, $CH_2{=}CH_2$, reacts with chlorine to the "oily" 1,1-dichloroethane, $CH_3CHCl_2$, it was called in French "*gaz oléfiant*", i.e., oil-forming gas. Later, the name "olefin" was applied to all hydrocarbons that contain carbon–carbon double bonds.

Hydrocarbon molecules with a single carbon–carbon double bond are also known as **alkenes** and **cycloalkenes**, respectively. **Diolefins** (= **dienes**) contain two double bonds which may be isolated (L: *insula* = island), conjugated (L: *conjugare* = to join or yoke together), or cumulated (L: *cumulare* = to heap), for example, in pentadiene:

| | |
|---|---|
| $CH_2{=}CH{-}CH_2{-}CH{=}CH_2$ | isolated double bonds |
| $CH_2{=}CH{-}CH{=}CH_2{-}CH_3$ | conjugated double bonds |
| $CH_2{=}C{=}CH{-}CH_2{-}CH_3$ | cumulated double bonds |

However, the term **poly(olefin)** refers only to polymers derived from alkenes and not to those from cycloalkenes, dienes, trienes, etc. The term is also process-based and not structure-based. Poly(olefin)s are not olefins with many double bonds since the polymerization of a mono-unsaturated alkene leads to a saturated high-molecular weight hydrocarbon. An example is the polymerization of ethene, $CH_2{=}CH_2$, to poly(ethylene),

$+CH_2-CH_2+_n$. Poly(olefin)s thus have the constitution of "**polyalkanes**", a term that is not used. **Alkanes** are saturated hydrocarbons (name from *alk(yl)* + *ane* (as in eth*ane*, prop*ane*, but*ane*, ...). Nevertheless, polymers of cycloolefins (= cycloalkenes) are often called **polyalkenes** because the double bond remains intact in polymers from ring-opening polymerizations (but not in polymers from vinyl-type polymerizations of cycloalkenes) (Section 6.4).

Polymers from dienes are known as **polydienes** if the double bonds of the monomers are conjugated or cumulated; polymers from dienes with isolated double bonds are not called polydienes. The polymerization of dienes with conjugated double bonds opens only one of the double bonds, while the other double bond is either preserved or shifted. An example is the polymerization of butadiene, $CH_2=CH-CH=CH_2$, to the monomeric units $-CH_2-CH(CH=CH_2)-$ or $-CH_2-CH=CH-CH_2-$.

The resulting polydienes are polymers with *one* double bond per monomeric unit (Section 6.3). Although they contain many double bonds per molecule, they are *not* called **polyenes** which are low-molecular weight compounds consisting of molecules with two or more double bonds (butadiene, annulenes, cyclooctatetraene, cumulenes, etc.).

### 6.2.2 Poly(methylene)

Poly(methylene), is the simplest polymeric hydrocarbon. It can be prepared by polymerization of diazomethane, $CH_2N \rightarrow +CH_2+_n + N_2$, or from the reaction of CO with $H_2$ (140°C, 50 MPa, ruthenium as catalyst). Neither reaction is industrial.

The chemical structure of diazomethane is unclear; proposed are $CH_2^{\ominus}-N^{\oplus}\equiv N$ and $CH_2=N^{\oplus}=N^{\ominus}$. Polymerization mechanisms are unknown. For gold as initiator, a polymerization via carbenes $^{\bullet}CH_2^{\bullet}$ has been discussed. The cationic polymerization of diazomethane by $BF_3/H_2O$ is assumed to involve an initiation reaction by addition of a proton to $CH_2N_2$, followed by successive eliminations of nitrogen molecules:

$$H^{\oplus} \xrightarrow{+\,CH_2N_2} CH_3N_2^{\oplus} \xrightarrow[-\,N_2]{+\,CH_2N_2} CH_3CH_2N_2^{\oplus} \quad \text{etc.} \qquad (6\text{-}2)$$

### 6.2.3 Poly(ethylene)s

The polymer $+CH_2-CH_2+_n$ from the polymerization of ethene, $CH_2=CH_2$, has the systematic name "poly(ethylene)" because of the multiple repetition of the ethylene diradical as the monomeric unit $-CH_2-CH_2-$. Its poly(monomer) name would be poly(ethene). Industry always uses "poly(ethylene)", in part, because "ethylene" is the former monomer name, and in part, because the systematic nomenclature is not widely known.

#### Natural Poly(ethylene)

**Elaterite** is a dark-brown amorphous, bitumen-like mineral that is found in the lead mines of Derbyshire (England), Perry Sound (Canada), Ukraine, and Bolivian Andes.

According to IR and NMR spectroscopy, elaterite is a highly branched amorphous poly(ethylene) that contains ca. 3 % CO groups but no double and triple bonds or aromatic structures. Elaterite (density: 0.8–1.2 g/cm$^3$) is usually soft and elastic (hence also known as **mineral rubber**) but sometimes hard and brittle. It is partially soluble in hydrocarbons such as heptane and toluene. The insoluble part is probably crosslinked via sulfur bridges (sulfur content: ca. 1.4 %).

It seems that the "elaterite" (**elastic mineral pitch, elastic bitumen**) of Colorado and Utah is not the same material. This "elaterite" is used for protective coatings, insulation, paving, and water-proofing.

Elastic bitumen is probably a decomposition product of organic matter like bitumen or pitch. The formation of mineral rubber is not understood, but it can be speculated that it is indeed a polymerization product of ethene. Nature produces ethene from L-methionine which is reacted with adenosine triphosphate (see Volume I) to S-adenosyl-L-methionine by the action of the enzyme methionine-adenosyltransferase. The enzyme ACC-synthetase splits off methylthioadenosine and the resulting aminocyclopropane carboxylic acid is converted to ethene and glycine by oxidases:

(6-3)

$H_3C-S-CH_2-CH_2-CH(NH_2)-COOH$ (L-methionine) $\xrightarrow[\text{ATP}]{\text{methionine-adenosyltransferase}}$ $H_3C-S^{\oplus}(\text{adenosine})-CH_2-CH_2-CH(NH_2)-COOH$ (S-adenosyl-L-methionine)

ACC-synthetase ↓ – methylthioadenosine

$CH_2{=}CH_2 + H_2N{-}CH_2{-}COOH \xleftarrow{\text{oxidases}}$ $c\text{-}C_3H_4(NH_2)COOH$ (aminocyclopropane carboxylic acid)

**Types of Industrial Poly(ethylene)s**

"Poly(ethylene)" is the umbrella term for all polymers from the many types of polymerization of ethene. The monomer ethene is obtained petrochemically from crude oil or natural gas (Section 4.3.2). Dehydration of alcohol is still used in developing countries on small scale. The production of ethene from coke gas is no longer economical (see Section 3.3.3).

Poly(ethylene) (**PE**) has the idealized constitution $\text{+}CH_2{-}CH_2\text{+}_n$; practically all poly(ethylene)s are more or less branched, however. Industrially, copolymers of ethene with small amounts of other monomers are also called "poly(ethylene)s" if the properties of these polymers resemble that of true poly(ethylene).

Industry subdivides homopolymers of ethene according to their density in high-density (**HDPE** (ASTM) or **PE-HD** (rest of the world, e.g., ISO or DIN)), medium-density (**MDPE**), low-density (**LDPE, PE-LD**), and very low-density types (**VLDPE**) (Table 6-4). ASTM (p. 9) has also used Roman numerals for the various types. Many LDPE types are in reality copolymers of ethene with 2–5 % vinyl acetate (VAC), although they are not identified as such ("true" copolymers have larger proportions of vinyl acetate units; see p. 230). Some European LDPEs are blends of ethene homopolymers with so-called linear low-density poly(ethylene)s (LLDPE) (see below).

Table 6-4 Classification and syntheses of poly(ethylene)s. mLLDPE = metallocene LLDPE (p. 225).

| Designation | | | Density | High-pressure syntheses | | Low-pressure syntheses | | |
|---|---|---|---|---|---|---|---|---|
| ASTM | DIN | ASTM | in g/cm$^3$ | Tube | Tank | Gas | Solution | Slurry |
| IV | | | > 0.96 | | | | | |
| III | PE-HD | HDPE | 0.94 – 0.96 | - | - | + | + | + |
| II | - | LLDPE | 0.925 – 0.94 | - | - | + | + | + |
| I | PE-LD | LDPE | 0.90 – 0.925 | + | + | - | - | - |
| - | - | mLLDPE | 0.886 – 0.935 | - | - | + | + | + |
| - | - | VLDPE | 0.863 – 0.885 | - | - | + | + | + |

So-called linear low-density poly(ethylene)s (**LLDPE, PE-LLD**) are copolymers of ethene with certain 1-olefins (1-butene, 1-hexene, or 1-octene), usually 6–8 mol% but sometimes up to 19 %. They are called "linear" because their branches are not generated by side-reactions of the (dominant) monomer (cf. the definition of "branching" in polymer science (p. 10)) but are introduced by the constitution of the comonomer. Solution polymerizations lead to random distributions of comonomeric units, but gas-phase polymerizations produce segmented copolymers because they proceed in the heterophase and are thus diffusion-controlled.

High-density poly(ethylene)s (HDPE) by coordination polymerization are relatively little branched while, because of chain-transfer reactions, low-density poly(ethylene)s (LDPE) by free-radical polymerization are much more branched. The concentration of branches increases with the residence time during polymerization: it is therefore higher for poly(ethylene)s from polymerizations in tank reactors than for those from continuous plug-flow reactors (Fig. 6-8).

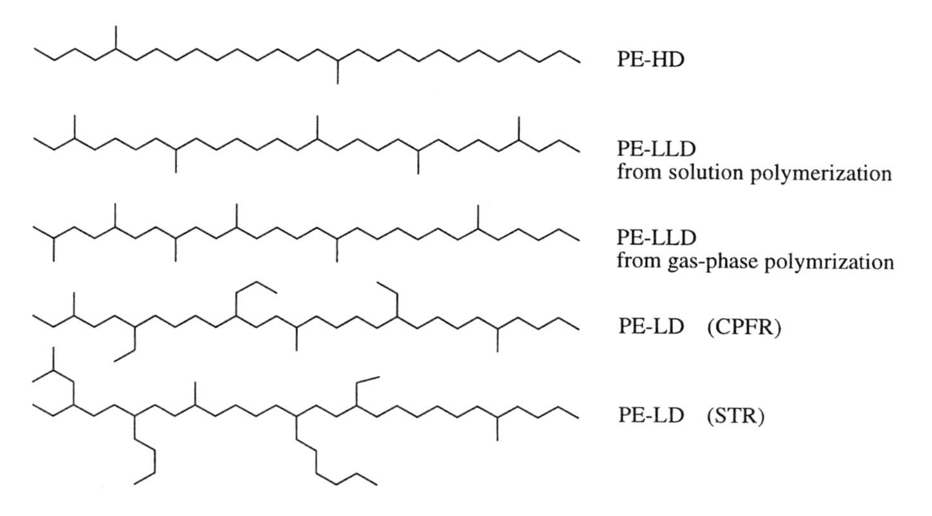


Fig. 6-8 Schematic representation of branching density and branching lengths in high-density (HD), linear low-density (LLD), and low-density (LD) poly(ethylene)s (PE). Note that side groups |, ⌈ , etc. represent *relative* lengths of side chains, not methyl, ethyl, propyl, butyl etc., *groups* (there are no methyl sidegroups in LDPE, see text). These schematic representations also show relatively more branches than are present in reality. In real PE-LDs, ca. 1–4 % of the carbon atoms are in branches. CPFR = continuous plug-flow reactor, STR = stirred-tank rector.

Conventional poly(ethylene)s have molecular weights of $M < 300\,000$. For special applications, high-density poly(ethylene)s with higher molecular weights are produced which are known as "high-molecular weight" (HMW-HDPE; $3 \cdot 10^5 < M < 4 \cdot 10^5$), "extra-high molecular weight" ($5 \cdot 10^5 < M < 1.5 \cdot 10^6$), and "ultra-high molecular weight" (UHMW-PE; $M > 3.1 \cdot 10^6$). So-called **metallocene poly(ethylene)s** are not poly(ethylene)s containing metallocenes but rather poly(ethylene)s of the LLDPE type that have been produced by metallocene catalysts (see also p. 225 and Volume I, p. 290 ff.).

Plants for poly(ethylene)s now have capacities of up to 400 000 t/a. The annual world production of all PEs is ca. $60 \cdot 10^6$ t/a (2003) wherefrom 43 % HDPE, 30 % LDPE, and 27 % LLDPE.

**Low-Density Poly(ethylene)s (LDPE, PE-LD)**

In 1933, R.O.Gibson of Imperial Chemical Industries (ICI) accidentically discovered poly(ethylene) as he tried to react ethene with benzaldehyde at 170°C and 140 MPa and found a white deposit on the inner wall of a newly developed high-pressure tank reactor. The reaction was not reproducible and it took many attempts until it was found that successful high-pressure polymerizations depended on the presence of traces of oxygen in ethene. In 1939, a plant with 200 t/a went on line. Because of their low dielectric losses, these poly(ethylene)s were used as electrical insulators in the newly developed radar which played a decisive role in World War II.

**Low-density poly(ethylene)s** are produced today by **high-pressure** polymerizations of ethene in either continuously stirred-tank reactors (CSTR) or in continuous-tube reactors (CPFR) (Fig. 6-9). Ethene is first passed over a reduced copper catalyst to remove all oxygen since traces of oxygen would prevent exact dosages of initiators and may lead to reactor run-aways or even explosions.

The polymerization is then initiated by carefully measured oxygen (ca. 0.05 %) which first probably forms ethene hydroperoxide, $CH_2{=}CH(OOH)$, that then decomposes and initiates the polymerization, for example, by $HO^\bullet$ radicals. Hydroxyl groups have been found in low-density poly(ethylene)s.

In some polymerizations, organic per compounds such as *t*-amyl perpivalate or di-2-ethylhexyl peroxydicarbonate are used. Radicals from these compounds lead to less radical chain transfer than oxygen which in turn results to less branching (see below) and higher densities of poly(ethylene)s.

The free-radical polymerization proceeds in CSTRs at > 130°C and 110–200 MPa and in CPFRs at < 350°C and < 350 MPa with throughputs of 7000 kg/(m³ h) (CSTR) and < 30 000 kg/(m³ h) (CPFR). Poly(ethylene) remains dissolved in ethene since the latter is in the supercritical state. Some swollen poly(ethylene) sticks to the reactor walls where it impedes heat transfer. It is removed by periodic brief opening of the reactor valve which causes a turbulence that dislodges the deposits and transports them out.

Monomer conversions reach 35 % in multiple-injection CPFRs, 20–25 % in single-injection CPFRs, and only 10–18 % in CSTRs. Lowering the pressure allows the separation of poly(ethylene) from oxidation products, waxes, and unreacted ethene (see Fig. 6-9).

Poly(ethylene)s from CSTRs have very broad molecular weight distributions ($\overline{M}_w/\overline{M}_n \approx 12$–16) that are mainly caused by strong long-chain branching. Molecular weight distributions are narrower for LDPEs from CPFRs: $\overline{M}_w/\overline{M}_n \approx 6$–8 for multiple feeds and 3–4 for single feeds.

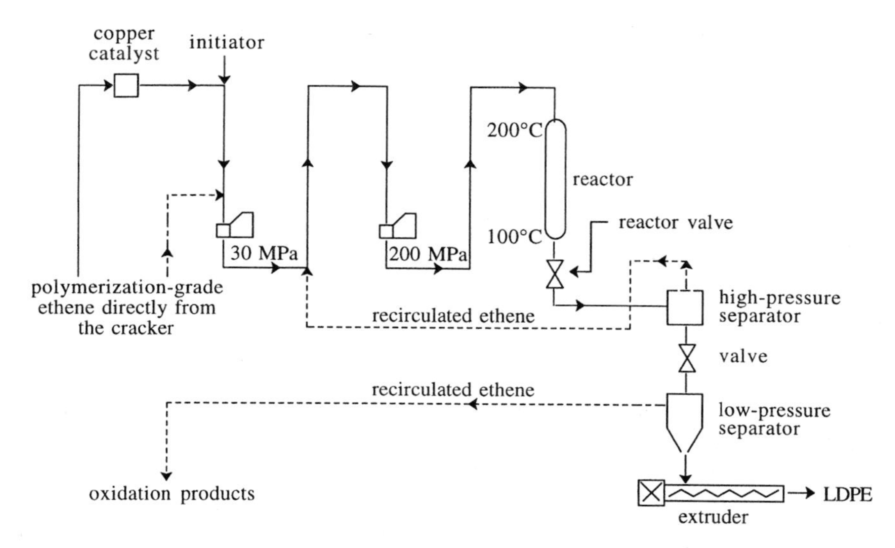


Fig. 6-9 Simplified flow scheme of the high-pressure polymerization of ethene [3]. Ethene is passed over a copper catalyst, provided with an initiator, compressed in two steps to 200 MPa, and polymerized in an STR or CPFR. The product is degassed in two steps. Unreacted ethene is recirculated, oxidation products and waxes are separated (see also p. 89), and the poly(ethylene) is extruded and pelletized. By kind permission of Chemical Publishing Co., New York.

Branching in LDPEs is caused by intramolecular and intermolecular transfer reactions of free-radical ends of growing poly(ethylene) chains, $\sim CH_2CH_2^\bullet$. *Intra*molecular transfer leads to back-biting (Eq.(6-4)). The resulting internal radicals can add ethene molecules whereas the original chain ends become side-groups; for example:

$$\text{(6-4)}\qquad \sim\!CH_2{-}CH_2{-}CH_2{-}CH_2{-}\overset{\bullet}{C}H_2 \longrightarrow \sim\!\overset{\bullet}{C}H{-}CH_2{-}CH_2{-}CH_2{-}CH_3 \xrightarrow{+\ CH_2{=}CH_2} \sim\!CH(CH_2{-}\overset{\bullet}{C}H_2){-}CH_2{-}CH_2{-}CH_2{-}CH_3$$

Intermediary six-membered rings are preferred for conformational reasons (Eq.(6-4)). Rings with seven or eight carbon atoms are also possible but rarely larger ones since the radicals are then too far away from their own chain. Correspondingly, one typically finds 9.6 butyl groups (from $C_6$ rings), 3.6 pentyl groups ($C_7$ rings), and 5.6 hexyl groups ($C_8$ rings) per 1000 carbon atoms. Propyl and methyl groups are absent. Ethyl side groups are present (ca. 1 per 1000 C atoms). They result from addition of an ethene molecule and subsequent transfer of a hydrogen atom (see also Table 6-5).

*Inter*molecular transfer reactions by polymer or initiator radicals generate internal carbon radicals. These radicals add ethene molecules and lead to long-chain branching:

$$\text{(6-5)}\qquad \sim\!CH_2{-}CH_2\!\sim\ \xrightarrow[-\ RH]{+\ R^\bullet}\ \sim\!CH_2{-}\overset{\bullet}{C}H\!\sim\ \xrightarrow{+\ C_2H_4}\ \sim\!CH_2{-}CH(CH_2{-}\overset{\bullet}{C}H_2)\!\sim$$

Table 6-5 Effect of reactor type, pressure $p$, and temperature $T$ on the number of groups per 1000 carbon atoms in low-density poly(ethylene)s.

| Reactor | $p$/MPa | $T$/°C | Number of groups per 1000 carbon atoms $-CH_3$ | $-CH=CH_2$ | $>C=CH_2$ |
|---|---|---|---|---|---|
| Stirred-tank reactor | 300 | 250 | 10 | 0.03 | 0.05 |
| | 160 | 250 | 14 | 0.04 | 0.17 |
| | 80 | 250 | 35 | 0.04 | 0.5 |
| | 80 | 130 | 15 | < 0.015 | 0.08 |
| Continuous stirred-tank reactor | 150 | 225 | 15 | | 0.018 |

The extent of branching and the concentration of endgroups depends on the polymerization temperature and pressure (and thus on the concentration of ethene), as well as on the residence time in the reactor and the type of operation (continuous versus discontinuous) (Table 6-5). At a constant temperature of 250°C, the proportions of vinylidene groups, $>C=CH_2$, and methyl groups in main and side chains, $-CH_3$, decrease with increasing pressure, whereas the proportion of vinyl groups, $-CH=CH_2$, stays practically constant. STRs produce more long-chain branching than CPFRs (Fig. 6-8).

The lower the temperature, and the higher the pressure, the less branches are formed. Conventional LDPEs from polymerizations at 130–275°C and 140–350 MPa have densities of $\rho$ = 0.912–0.935 g/cm$^3$ but those from 50–80°C and 500–700 MPa can reach $\rho$ = 0.955 g/cm$^3$. The latter polymers have densities like HDPEs (Table 6-4).

The designations LDPE, HDPE, etc., thus do not necessarily refer to actual densities but rather to the type of polymerization. There are obviously many grades that can be obtained from any given process. Some companies offer up to 60 grades of poly(ethylene)s, including filled ones; Table 6-7 shows examples.

The various grades have different processibilities (as measured by their melt flow indices; Table 6-6), different properties (Table 6-7), and different applications (as indicated by their densities; Tables 6-6 and 6-7). Melt flow indices, MFI, are indicative of fluidities; the higher the MFI, the smaller is the molecular weight (see p. 14). Ca. 75 % of all LDPEs are used for monolayered and multilayered films.

Table 6-6 Use of LDPE grades [4]. $\rho$ = density at 23°C, MFI = melt flow index at 190°C with a standard load of 2.16 kg. The melt flow index is a measure of the inverse melt viscosity, $\eta$, which in turn increases with increasing molecular weight. Note that the MFI is *not* inversely proportional to the molecular weight $M$ since MFIs are shear-dependent, $\eta \propto M$ obeys different power laws in different molecular weight regions, etc. (see Volumes III and IV). MWD = molecular weight distribution.

| $\rho$/(g cm$^{-3}$) | MFI | Processing | Use |
|---|---|---|---|
| 0.919–0.922 | 0.1 – 0.3 | film blowing | shrink–films 70–300 μm, heavy–duty bags |
| 0.919–0.926 | 1 – 5 | film blowing, extrusion | standard films 20– 70 μm |
| 0.920–0.926 | 1.5 – 8 | extrusion | thin films 20– 50 μm |
| 0.920–0.923 | 3 – 9 | extrusion | paper coating (broad MWD) |
| 0.920–0.925 | 1 – 15 | injection molding | thick–walled molded parts |
| 0.920–0.925 | 20 – 40 | injection molding | thin–walled molded parts |

Table 6-7 Six of the 17 offered unfilled injection-molding grades of poly(ethylene)s from Neste [5].
All grades: relative permittivity $\varepsilon_r = 2.3$; volume resistivity $10^{15}$ Ω cm; surface resistivity $10^{14}$ Ω; water absorption 0.01 % (23°C, saturation); moisture absorption 0.01 % (23°C, 50 % relative humidity). NB = no break. Note that 3416 is called "LD", not "HD".

| Physical property | Condition | Physical unit | LD NCPE 1515 | LLD NCPE 8030 | LLD NCPE 8644 | LD NCPE 3416 | HD NCPE 7003 | HD NCPE 7007 |
|---|---|---|---|---|---|---|---|---|
| Density | | g/cm³ | 0.915 | 0.919 | 0.935 | 0.958 | 0.958 | 0.964 |
| Heat deflection temperature (B) | | °C | 41 | 46 | | | 84 | 80 |
| Heat deflection temperature (A) | | °C | | | 39 | 58 | 50 | 54 |
| Vicat temperature (A) | | °C | 75 | 86 | 110 | | 130 | 130 |
| Vicat temperature (B) | | °C | | 48 | 65 | 75 | 85 | 80 |
| Young's modulus | 1 mm/min | MPa | 150 | 830 | 650 | 1350 | 1350 | 1600 |
| Yield strength | 50 mm/min | MPa | 8 | 12 | 17 | 30 | 30 | 31 |
| Strain at yield | 50 mm/min | % | | 12 | 12 | 8.1 | 8.2 | 7.1 |
| Strain at break | 50 mm/min | % | > 50 | > 50 | > 50 | > 50 | > 50 | > 50 |
| Impact strength | 23°C or –30°C | kJ/m² | NB | NB | NB | NB | NB | NB |
| Notched impact strength | 23°C | kJ/m² | NB | NB | 6.5 | NB | 5.2 | 5.6 |
| Notched impact strength | –30°C | kJ/m² | NB | 11 | 5.9 | 5 | 5.4 | 4.6 |
| Melt volume index | 2.16 kg, 190°C | mL/10 min | 18 | 36 | 9.5 | 0.24 | 3.6 | 8.3 |
| Melt density | 225°C | g/mL | | 0.787 | 0.794 | | 0.814 | 0.819 |
| Melt thermal conductivity | | W m⁻¹ K⁻¹ | | 0.287 | 0.303 | | 0.326 | 0.353 |
| Melt specific heat capacity | | J kg⁻¹ K⁻¹ | | 3070 | 3150 | | 3350 | 3420 |

Densities are direct measures of degrees of crystallinity (Volume III) which in turn control many other mechanical properties. Both microhardnesses $h_m$ and upper yield stresses $\sigma_Y$ increase linearly with the degree of crystallinity (Fig. 6-10). The same is true for the modulus of elasticity since it depends linearly on the microhardness.

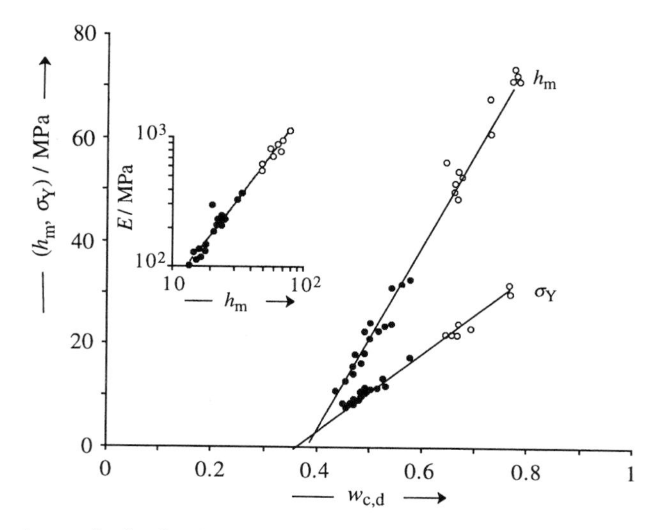


Fig. 6-10 Dependence of microhardness, $h_m$, and upper yield stress, $\sigma_Y$, of HDPEs (○) and LDPEs and LLDPEs (●) on the density-based degree of crystallinity, $w_{c,d}$ [6]. Insert: Dependence of moduli of elasticity, $E$, on microhardness.

## High-Density Poly(ethylene)s (HDPE, PE-HD)

**High-density poly(ethylene)s** (HDPE, PE-HD)) and also some low-density and very low-density poly(ethylene)s (LLDPE, VLDPE) are produced by transition metal catalysts at **low pressure** (Table 6-8). The processes leading to HDPEs were discovered almost simultaneously in 1953–1954 by Karl Ziegler and coworkers at the Max Planck Institute for Coal Research, Germany, and by workers at Standard Oil of Indiana and Phillips Petroleum, United States (for the history, see Volume I, pages 270 and 305). All processes lead to far less branching than LDPE processes. Present processes are suspension, solution, or gas-phase polymerizations. The Standard Oil process (solution polymerization of ethene in xylene) is no longer used.

In the propagation steps of all these polymerizations, ethene is inserted between the complexed transition metal (e.g., titanium) and the growing poly(ethylene) chain (Volume I, p. 274 ff.), for example:

$$\text{(6-6)} \qquad \sim\!\sim(CH_2CH_2)_n[Ti] + CH_2{=}CH_2 \longrightarrow \sim\!\sim(CH_2CH_2)_nCH_2CH_2[Ti]$$

Molecular weights are often controlled by terminating the chain with hydrogen:

$$\text{(6-7)} \qquad \sim\!\sim(CH_2CH_2)_n[Ti] + H_2 \longrightarrow \sim\!\sim(CH_2CH_2)_nH + H[Ti]$$

The classic Ziegler catalyst consisted of $TiCl_4$ and $(C_2H_5)_2AlCl$. This system produced various catalytically active species which in turn led to very broad molecular weight distributions. First-generation *industrial* Ziegler processes used the more effective system $\beta$-$TiCl_3$ + $(C_2H_5)_2AlCl$ + $AlCl_3$, which was later replaced by the second-generation system $\beta$-$TiCl_3$ + $(C_2H_5)_2AlCl$ + electron donor. The third-generation system, $TiCl_4$ + $(C_2H_5)_3Al$ + $MgCl_2$ + electron donor, replaced the expensive $TiCl_3$ by the less expensive $TiCl_4$ (which is reduced to $\gamma$-$TiCl_3$ by $(C_2H_5)_3Al$)) and used $\delta$-$MgCl_2$ as carrier which forms complexes with $TiCl_4$ (see Volume I, p. 277 ff.).

Table 6-8 Industrial low-pressure polymerizations to high-density poly(ethylene)s, HDPE. 0.1 MPa = 1 bar ≈ 1 atm. * 1-Butene (B), 1-hexene (H), 1-octene (O), propene (P); ** Ziegler (Z), Phillips, metallocene (single-site (SS); see p. 225 and Volume I). pr = partially reduced.

| | Suspension Ziegler type | Suspension Phillips type | Solution | Gas-phase many |
|---|---|---|---|---|
| Medium | hexane | butane, $C_6H_{12}$ | hexane | gaseous ethene |
| Comonomer (if any *) | B, P | H | Z: P; SS: B,H,O | B, H, O |
| Catalyst | $TiCl_4/R_3Al$ | $Cr_2O_3$ (pr) | Z, SS | all ** |
| Carrier | $MgCl_2$ | $Al_2O_3$ | – | $SiO_2$, $MgCl_2$ |
| Pressure, MPa | <1.5 | 2.8–5.0 | 5–15 | 2–2.5 |
| Temperature, °C | 75–85 | <110 | 75–300 | 90–120 |
| Type of reactor | CSTR, CSTR–C | CPFR (loop) | CSTR | fluidized bed |
| Residence time, h | 1–2 | 1.5 | 0.03–0.15 | 3–5 |
| Control of molecular weight | $H_2$ | temperature | | $H_2$ |
| Catalyst yield, g PE/g catalyst | 3000 | 3000–10 000 | | 9000 |
| Time–space yield, kg/($m^3$ h) | 100–150 | 250 | 5000 | 150 |
| PE types (density in g/$cm^3$) | >0.94 | 0.92–0.97 | 0.86–0.96 | 0.92–0.96 |

The **Standard Oil process** was a solution polymerization (Table 6-8) at temperatures above the melting temperature of poly(ethylene). It had the advantage that the resulting poly(ethylene) remained dissolved and did not occlude the catalyst particles, so that the catalyst stayed active throughout the course of polymerization. Disadvantages were the cost of the solvent and the difficult work-up of polymer and solvent. The process also worked at costly medium pressures. The Standard Oil process was thus replaced by the Ziegler and Phillips processes that work at lower pressures (Table 6-8). In both processes, ethene is continually fed to a suspension of the catalyst in liquid hydrocarbons.

The **Ziegler process** works at 75–85°C and slightly increased pressures of < 1.5 MPa with residence times of 1–2 h and throughputs of 100–150 kg/($m^3$ h). Ethene is compressed, washed with acetone, and fed to the CSTR or a cascade (Fig. 6-11). Separately, catalyst components are mixed with hexane and transferrred to the reactor. The polymerization leads to a suspension (**slurry**) of polymer-coated catalyst particles in hexane; the viscosity depends on content and morphology of solids. The polymerization rate decreases steadily since the catalyst becomes more and more occluded. The process also delivers volatile inert low-molecular weight hydrocarbons which are removed by venting.

The slurry is then transferred to another stirred–tank reactor where catalyst residues (which catalyze the aging of polymers) are removed by adding ethanol which converts the titanium components to soluble compounds. Subsequent treatment with dilute hydrochloric acid transforms these Ti compounds to $TiO_2$ which is seperately reworked to $TiCl_4$. The remaining mixture is neutralized by alkali and washed with water.

The dispersion of poly(ethylene) in hexane is then separated by centrifugation. The polymer is washed with a solution of detergents and alkali in water. After neutralizing with citric acid, the washing liquid is removed by centrifugation. The polymer is then dried, extruded, and pelletized. The liquids are worked up and hexane is recirculated.

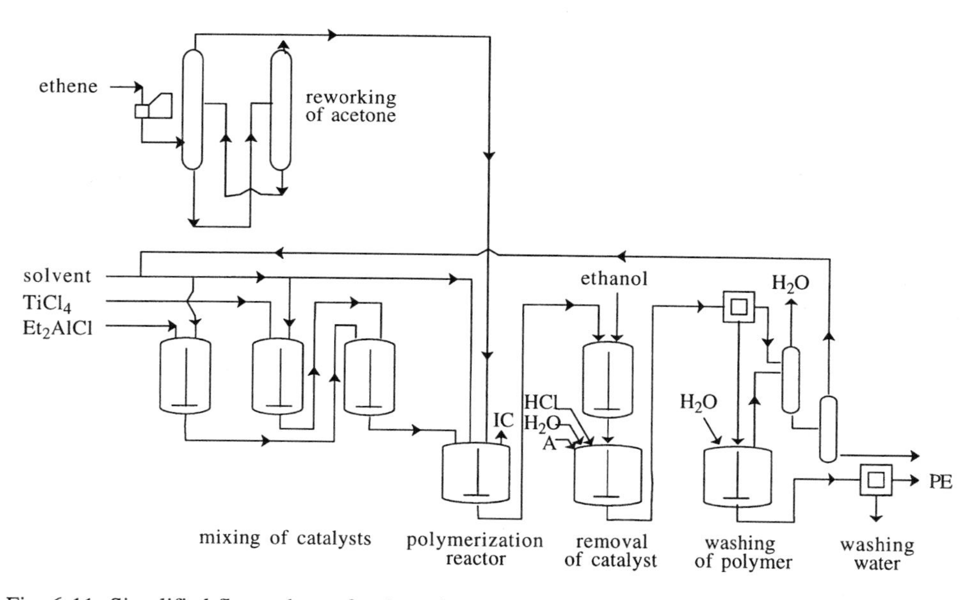


Fig. 6-11 Simplified flow scheme for the polymerization of ethene by the Ziegler process [3]. A = alcohol, IC = inert hydrocarbons. By kind permission of Chemical Publishing Co., New York.

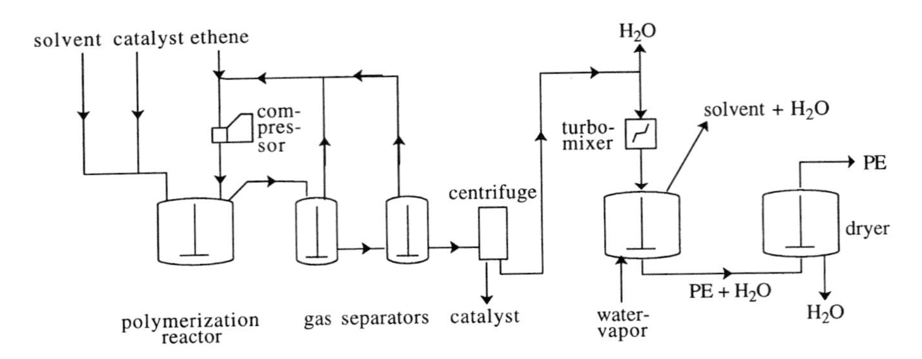


Fig. 6-12 Simplified flow scheme for the suspension polymerization of ethene by the Phillips process [3]. Newer modifications of this process use double-loop reactors instead of stirred–batch reactors.

The **Phillips process** uses somewhat higher temperatures and pressures than the Ziegler process (Table 6-8) and can proceed as either suspension or solution polymerization (Fig. 6-12). In both variants, catalyst (0.06 % $CrO_3$ on highly porous $SiO_2$), solvent or suspension agent, and compressed ethene are fed to the reactor. Molecular weights are controlled by chain transfer agents, usually hydrogen, and branching (and thus density via change of crystallinity) by small proportions of 1-butene as comonomer.

In solution polymerizations, the resulting poly(ethylene) remains dissolved in cyclohexane as a ca. 10 % solution. The catalyst is removed by heated centrifugation and the polymer solution is cooled. The precipitated polymer is filtered off and washed. Residual monomer is sometimes removed by steaming.

In suspension polymerizations, the resulting slurry of poly(ethylene) in butane is degassed in gas separators; the unpolymerized ethene is returned to the reactor. The slurry is mixed with water, agitated by a turbomixer at 55°C, and then washed by water vapor at 120°C. Solvent and water are removed over the top and the polymer is dried.

A fourth process for the preparation of HDPE is the gas-phase process (Fig. 6-13) which is discussed in the next section.

The Phillips process uses less-complicated plants than the Ziegler process, and has therefore become the process of choice for the synthesis of high-density poly(ethylene)s. Newer plants have name-plate capacities of up to 400 000 t/a. Main applications are for films, followed by injection-molded articles (packaging, household goods, and extruded materials (tubes, sheeting, sheets).

## Linear Low-Density Poly(ethylene)s (LLDPE, PE-LLD)

Copolymerizations of ethene with 5–12 % α-olefins are performed in the gas phase (1-butene) or in solution or suspension (1-hexene, 1-octene). Cracking of oil provides sufficient 1-butene but not enough 1-hexene and 1-octene, which are therefore obtained by oligomerizing ethene by $AlEt_3$ (p. 100) or new Cr catalysts or from the Sasol process (p. 45). 1-Octene gives polymers with higher tear strengths but is more expensive.

Polymerizations are also performed in solution or suspension similar to the Ziegler or Phillips processes for HDPEs (Table 6-8), or in the gas phase (for example, Unipol process) using the same catalysts and similar plants (see Table 6-8).

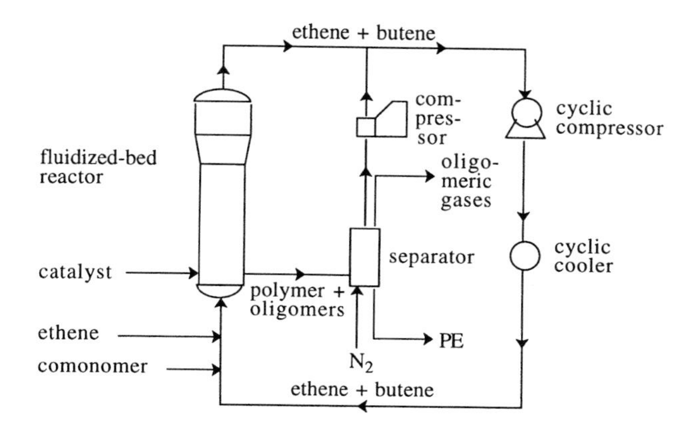


Fig. 6-13 Flow scheme of the gas-phase process (Unipol process) for LLDPE [7].

*Solution polymerizations* give LLDPEs with the best product properties. They can also be relatively easily modified for various grades but have high additional costs for solvents and solvent work-up.

*Suspension polymerizations* have the advantage of delivering powdered LLDPEs; pelletizing is not required. However, polymers for films are difficult to produce.

Very simple plants are used by **gas-phase processes** for the synthesis of HDPE or LLDPE in horizontal or vertical fluidized-bed reactors (Fig. 6-13). The resulting mixture is separated under nitrogen to polymer, oligomers, and unreacted ethene + comonomer. The latter is compressed and recirculated.

The costs of these plants are only 30 % of those for high-pressure or low-pressure polymerizations. There are no costs for solvents and their distillation; catalyst removal is unnecessary. With Ziegler or Phillips catalysts, it can only be used for 1-butene as comonomer. Metallocene catalysts (see below) also allow the copolymerization of ethene with 1-hexene; 1-octene is not reactive enough. The process uses only ca. 25 % of the energy that is required for LDPE processes. Name-plate capacities range to 225 000 t/a.

LLDPEs can also be obtained from ethene alone if it is simultaneously oligomerized and polymerized, for example, by compatible catalysts based on $\alpha$-$TiCl_3$, $MgH_2$, plus $Cp_2TiCl_2$ (SHOP process = *S*hell *h*igher *o*lefin *p*rocess).

## Very Low-Density Poly(ethylene)s (VLDPE, PE-LLD)

Copolymerization of ethene with ca. 24 % of higher olefins, usually 1-butene or 1-octene, delivers **poly(olefin) elastomers (POE)** that range in properties between LLDPE and totally amorphous elastomers. The degree of crystallinity of VLDPEs is so low that no crystalline lamellae develop (see Volume III); these polymers most likely form fringed micelles. Most VLDPEs are now made with metallocene catalysts.

## mLLDPE

Linear low-density polymers by polymerization of ethene with transition metal catalysts that are activated by methylaluminoxanes are called **metallocene-poly(ethylene)s** (mLLDPE, MPE-LLD), **single-site catalyzed poly(ethylene)s** (LLDPE(SSC)), **poly(ole-**

**fin) plastomers** (POP), or homogeneous ethylene copolymers, depending on structure and properties. These newer types of LLDPEs have the same composition as regular LLDPEs but different sequence statistics of comonomeric units and thus different properties. Transition metal catalysts usually have Ti or Zr centers that are bound to cyclopentadienyl moieties Cp or fluorenyl structures Flu (see Volume I, p. 290 ff.).

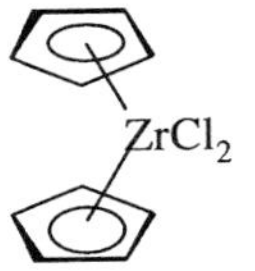

$Cp_2ZrCl_2$ 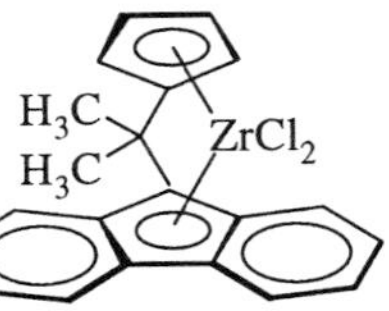 iPr[1-Flu;Cp]$ZrCl_2$ 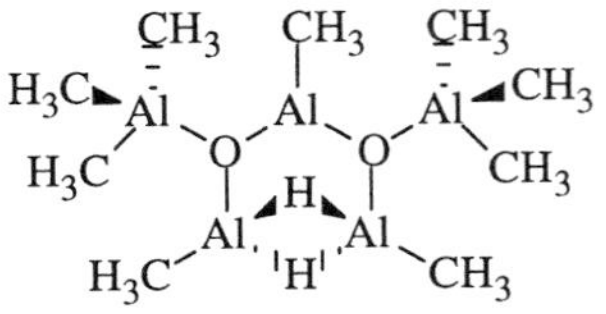 methylaluminoxane (MAO) (see also p. 236)

**Properties of Poly(ethylene)s**

Properties of poly(ethylene)s are mainly controlled by molecular weights, molecular weight distributions, and branching (via crystallinity or densities, respectively (Tables 6-9 and 6-10, see also Table 6-4)). Densities generally decrease in the order HD > LLD > mLLD > LD > VLD (Table 6-10) but there is some overlap depending on the grade. The density of 100 % crystalline, truly linear poly(ethylene) ($\rho \approx 1$ g/cm$^3$) is never reached.

Molecular weights of LDPEs are controlled by initiator concentrations and temperatures; those of HDPEs by the polymerization temperature, the catalyst system, and the concentration of chain transfer agents, e.g., hydrogen. Molecular weights of HDPEs are between 30 000 and 500 000. Ultrahigh molecular weight HDPEs from the slurry process with Ziegler catalysts range up to 4 000 000. LLDPEs mostly have molecular weights between 30 000 and 300 000.

Large differences exist in polymolecularity indices, PMI = $\overline{M}_w/\overline{M}_n$. LDPEs have very broad distributions: PMI = 8–30 for Ziegler types and 8–15 for Phillips types. For HDPEs, PMIs between 4 and 15 are reported depending on the intended application. LLDPEs have lower PMIs of 3–4. The lowest PMIs of ca. 2 are observed for metallocene LLDPEs because they have only one type of catalytically active center. mLLDPEs have a near Bernoulli-type distribution of comonomer units.

Molecular weight and molecular weight distribution control the processing. Low molecular weights lead to low melt viscosities and thus to faster processing with less energy consumption. Since they also decrease mechanical properties, applications such as film blowing therefore requires polymers with bimodal molecular weight distributions.

LLDPEs have no long-chain branches and more narrow molecular weight distributions than HDPEs. This increases entanglement densities which, in turn, leads to higher melt viscosities that require lower processing speeds. The higher entanglement density also increases the stretchability of films during film blowing which makes films less prone to damage by foreign objects. A tear during high-speed blowing can pile up hundreds of meters of film in a hurry!

The narrow molecular weight distributions of metallocene LLDPEs make them more difficult to process than conventional LLDPEs. Their processability is improved by reducing entanglements through increase of lengths of substituents (1-octene instead of 1-butene) and use of two different catalysts (which lead to bimodal weight distributions).

Table 6-9 Effect of molecular characteristics on processing and end-use properties of poly(ethylene).

| Processing/end-use property | Effect of increase of Molecular weight (increasing viscosity) | MWD | Long-chain branching (decreasing density) |
|---|---|---|---|
| Shear-thinning | increase | increase | |
| Melt strength | increase | increase | |
| Melt elasticity | increase | increase | |
| Melt fracture | increase | decrease | |
| Elongation at break | increase | | increase |
| Modulus of elasticity | | | decrease |
| Tensile strength at yield | | | decrease |
| Tensile strength at break | increase | | increase |
| Impact strength | increase | decrease | (increase) |

All poly(ethylene)s except VLDPEs are thermoplastics. They are used mainly for packaging materials (films, sheets, bottles), molded parts, tubes, cable sheetings, and, as latices, also for floor waxes (cf. Tables 6-6 and 6-9).

Table 6-10 Average properties of different types of poly(ethylene)s. * With 1-octene as comonomer. NB = no break. DSC = differential scanning calorimetry.

| Property | Physical unit | HDPE | LLDPE | mLLDPE* | LDPE |
|---|---|---|---|---|---|
| Density | $g/cm^3$ | 0.94–0.96 | 0.925–0.94 | 0.89–0.94 | 0.90–0.925 |
| Refractive index | 1 | | 1.52 | | 1.52 |
| Melting temperature (DSC) | °C | <135 | 123 | 90–125 | 105–115 |
| Heat distortion temperature (0.45 MPa) | °C | | | | 38–49 |
| Vicat temperature B | °C | 60–65 | 80–94 | 95–120 | 40 |
| Glass temperature (DSC) | °C | −123 | | | −(103–133) |
| Maximum service temperature (short term) | °C | 90–120 | 90–115 | | 80–90 |
| (long term) | °C | 70–80 | 70–95 | | 60–75 |
| Linear thermal expansion coefficient | $K^{-1}$ | $2 \cdot 10^{-4}$ | $2 \cdot 10^{-4}$ | | $1.7 \cdot 10^{-4}$ |
| Thermal conductivity (20°C) | $W\ m^{-1}\ K^{-1}$ | 0.4–0.5 | | | |
| Tensile modulus | MPa | 60–290 | 140–520 | 20–550 | 100–310 |
| Flexural modulus | MPa | | 240–800 | | 240–330 |
| Yield strength | MPa | 18–32 | 9–20 | | 6–15 |
| Tensile strength at break | MPa | 10–60 | | 30–50 | |
| Fracture elongation | % | | 100–1200 | > 700 | 100–800 |
| Impact strength (Charpy) | $kJ/m^2$ | NB | NB | | NB |
| Notched impact strength (Izod 3.1 mm) | J/m | 21–210 | 53–480 | 2500–NB | NB |
| (Charpy) | $kJ/m^2$ | 6–NB | NB | | NB |
| Hardness (Shore D) | – | | 47–58 | | 40–60 |
| Ball indentation hardness | – | ≈ 50 | | | |
| Relative permittivity | 1 | 2.4 | 2.4 | | 2.3 |
| Electrical resistance ("surface resistivity") | Ω | $10^{13}$ | $10^{13}$ | | $10^{13}$ |
| Resistivity ("volume resistance") | Ω cm | $10^{16}$ | $10^{16}$ | | $10^{16}$ |
| Electric strength | kV/mm | > 700 | > 700 | | > 600 |
| Dissipation factor (loss tangent) (50 Hz) | 1 | | < 0.0005 | | 0.0003 |
| Water absorption (96 h) | % | < 0.05 | < 0.05 | | < 0.05 |

### 6.2.4 Modified Poly(ethylene)s

Poly(ethylene)s are relatively inexpensive because of the low cost of monomer and the economy of scale of polymerization. Some of them are therefore converted chemically to other polymers.

Crosslinking of poly(ethylene)s by γ-rays or peroxides (dicumyl peroxide, di-*t*-butyl peroxide) leads to **crosslinked poly(ethylene)s (XLPE, PE-X)** with service temperatures between –118°C and +65°C, high tensile and impact strengths, high dielectric resistance, and low permeabilities to oxygen. XLPEs are used for cable and wire coatings and insulation, tubes, and tanks. Radiation in the presence of hydrophilic monomers such as acrylamide leads to grafting and surfaces that are easier to print on.

Chlorination of poly(ethylene)s in bulk (for example, fluidized bed), emulsion, or suspension leads either to fully amorphous polymers with randomly distributed Cl substituents or to partially crystalline polymers with poly(ethylene) and chlorinated poly(ethylene) segments, depending on the parent polymers and reaction conditons. **Chlorinated poly(ethylene) rubbers (CPE, PE-C)** with chlorine contents of 25–40 % are crosslinked by peroxides to **chlorinated poly(ethylene) elastomers** that have good resistance to hydrocarbon fuels and oils at elevated temperatures and are thus used for cable sheeting, seals, fume hoods, tires, hose lining, and underground electric lines. For certain rubber applications, CPEs compete with sulfochlorinated poly(ethylene) (CSM), poly(chloroprene) (CR), terpolymers from ethene, propene, and a diene comonomer (EPDM), and nitrile rubber (NBR) (Table 6-11).

Chlorinated poly(ethylene)s with chlorine contents of 45–60 % resemble poly(vinyl chloride) and are thus often called **heat-stable poly(vinyl chloride)**s. They are either used directly (for example, for hot-water tubes) or are blended with poly(vinyl chloride) in order to improve the impact strength of PVC.

Poly(ethylene)s, especially LDPEs, are also sulfochlorinated by chlorine and sulfur dioxide in hot $CCl_4$ solutions in the presence of UV light or an azo initiator to **sulfochlorinated poly(ethylene)**s (**CSM, CSR**) that contain between 25 and 42 $–CH_2CHCl–$ groups, and 1–2 $–CH_2CH(SO_2Cl)–$ groups per 100 ethylene units. Reaction of these groups with metal oxides MtO (= MgO, ZnO, PbO) leads to $MtCl_2$ and the formation of –O–Mt–O– bridges between chains. Crosslinked CSMs are very weather-resistant; hence they are used for protective coatings, cable sheetings, white-wall tires, etc.

Table 6-11 Environmental resistance of some specialty rubbers [8]. +++ (very good), ++ (good), + (sufficient), – (insufficient).

| Resistance against | CPE | CSM | CR | EPDM | NBR |
|---|---|---|---|---|---|
| Heat | +++ | ++ | + | +++/++ | + |
| Cold | ++ | ++ | +++ | ++ | ++ |
| Weather | +++ | +++ | ++ | +++/++ | – |
| Ozone | +++ | +++ | ++ | +++ | – |
| Oil and fuel | ++ | ++ | ++ | – | +++ |
| Chemicals | ++ | ++ | ++/+ | ++/+ | +++ |
| Flame | +++ | +++ | +++ | – | + |
| Color changes | ++ | +++ | + | +++ | ++ |
| Abrasion | ++ | ++ | ++ | + | + |

### 6.2.5 Ethene Copolymers

Ethene is copolymerized not only with small proportions of 1-butene, 1-hexene, or 1-octene to linear low-density poly(ethylene)s (p. 224) but also with various other monomers by free radicals or transition metal catalysts. Incorporation of these monomer units decreases sequence lengths of methylene segments and thus the tendency to crystallize.

**Copolymers with 1-Olefins or Dienes**

Copolymerization of ethene and propene in hexane by Ziegler catalysts such as $VOCl_3$ + $(C_2H_5)_3Al/(C_2H_5)AlCl_2$ delivers **EPM rubbers (E/P)** that contain 25–55 wt% propene units. Since the catalyst system loses its activity with time (see Volume I, p. 276), highly chlorinated organic compounds are added as activators, for example, perchlorocyclopentadiene.

Crosslinked EPM elastomers have excellent elasticities and high resistances to breakdown on shearing during processing. They compete with butyl rubbers (IIR) but are more easy to process than IIR. The absence of carbon–carbon double bonds provides these polymers with excellent stability against light, oxidation, and chemicals. For the same reason, EPMs cannot be crosslinked by the usual sulfur-based vulcanization systems of the rubber industry. Rather, a very special crosslinking system with peroxides had to be developed.

EPMs have therefore been widely replaced by **EPDMs**, terpolymers of ethene, propene, and a comonomer with a diene structure, that can be conventionally crosslinked by vulcanization with sulfur systems. Industrially used termonomers comprise mainly 5-ethylidene-2-norbornene, *endo*-dicyclopentadiene, and 1,4-hexadiene as well as some other comonomers (Fig. 6-14). The solution terpolymerization is by Ziegler catalysis in cascades of stirred-tank reactors at 30–60°C.

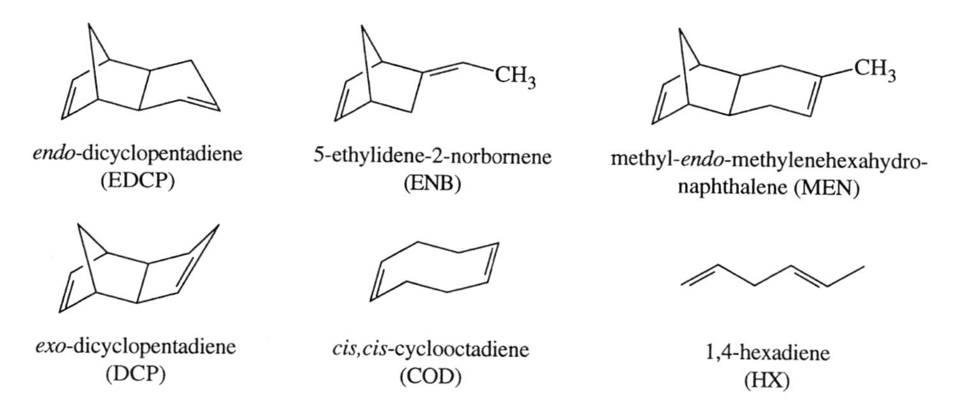


Fig. 6-14 Termonomers for EPDM rubbers.

Industrial EPDMs contain 45–80 % ethene units and only ca. 15 double bonds per 1000 carbon atoms. They are thus much more resistant to weather in general and to ozone in particular than polydienes with ca. 250 double bonds (1,4-*cis*-poly(butadiene)) and 200 double bonds (1,4-*cis*-poly(isoprene)) per 1000 carbon atoms, respectively.

ENB-containing EPDM rubbers are also sulfonated by reaction with concentrated sulfuric acid and acetic anhydride which *in situ* form acetylsulfonate, $CH_3COOSO_3H$, that then sulfonates ENB units:

(6-8) $+ CH_3COOSO_3H$ → $H$, $SO_3H$, $\oplus$ $CH_3$, $\ominus OOCCH_3$ → $SO_3H$, $CH_3$, $- CH_3COOH$

The resulting polymers are thermoplastic elastomers. Polymers with free acid groups dissolve in hydrocarbons and degrade easily to crosslinked products. Metal salts containing polymers are thermally more stable and much tougher. Their melt viscosities depend on the type of cation. Addition of zinc acetate to sulfonated EPDM polymers leads to polymers with relatively low melt viscosities that still maintain their mechanical strengths. The effect may either be caused by replacement of $SO_3H$ groups by less-ionic $Zn(SO_3)OOCCH_3$ groups or by plasticizing ion domains by zinc acetate.

Copolymerization of ethene and larger proportions of dicyclopentadiene by vanadium trisacetylacetone/$R_3Al$ as catalyst delivers copolymers with isolated double bonds that oxidize at room temperature to crosslinked films. These copolymers can be compounded with phenolic resins before conventional crosslinking.

**Poly(ethylene-*co*-norbornene)**s take up very little water and have excellent optical properties. These polymers are used as binders for toners, pharmaceutical packaging, medical equipment, and capacitors. They could not compete for CD production because of high costs and slow cycle times but are candidates for DVDs.

Other ethene copolymers include elastomers with 1-octene or norbornene as comonomers and plastics with styrene or carbon monoxide as comonomers (Section 6.13).

### Copolymers with Vinyl Monomers

Ethene is copolymerized free-radically with vinyl acetate (VAC) in bulk, solution, or emulsion. Since the copolymerization is almost ideally azeotropic ($r_{eth} = 0.88$, $r_{vac} = 1.03$), various vinyl acetate feeds can be used for different states, depending on the desired composition (Table 6-12).

**EVAC** copolymers (**EVA, E/VAC**) with more than 10 % vinyl acetate units are used as non-shrink films, those with up to 30 % VAc units as thermoplastics, and those with more than 40 % VAC units for clear sheeting. EVACs with even higher concentrations of VAC units serve as elastomers, melt or solution adhesives, or modifiers for poly(vinyl acetates). These polymers can be crosslinked by reaction with, for example, triallyl cyanurate using lauroyl peroxide as radical supplier.

Table 6-12 Processes for the copolymerization of ethene and vinyl acetate (VAC) at pressures *p*.

| Percent VAC in feed | State | *p*/MPa |
|---|---|---|
| 1 - 35 | bulk | 100 - 200 |
| 35 - 99 | solution in *t*-butanol | 10 - 40 |
| 60 - 99 | aqueous emulsion | 0.1 - 20 |

Ethene copolymers with 20–50 % vinyl acetate units are transesterified to pseudo-copolymers of ethene and vinyl acetate (**EVAL, E/VAL**). Products with small proportions of vinyl alcohol units are used as powders for coatings in fluidized beds, for example, coatings of metallic garden chairs. EVAL with higher contents of vinyl alcohol units are fairly impermeable to carbon dioxide and oxygen and are therefore used for bottles and other packaging materials.

Copolymers of ethene and chlorotrifluoroethene (**ECTFE**) are thermally stable to 200°C and do not burn. Because of their outstanding chemical inertness and good mechanical properties, they are used in medical packaging, for cable sheeting, and for chemical labware.

**Copolymers with Acrylic Monomers**

Terpolymers of ethene, methyl acrylate, and a small proportion of acrylic acid (or another monomer with a carboxylic group) are rubbers that can be crosslinked by diamines. The resulting saturated elastomers have excellent weatherabilities and are therefore used for tubes, sealants, and cushioning, especially for automobiles.

High-pressure free-radical polymerization of ethene and acrylic acid results in strongly branched **EAA** copolymers (**E/AA**) with 3–20 % acrylic acid units that are physically crosslinked via associated COOH groups. The COOH groups of these polymers furthermore adhere excellently to glass and metals whereas the methylene sequences adhere to poly(ethylene)s. EAAs are thus used as tough packaging films or as laminates with aluminum foils.

Ethene is also free-radically copolymerized with methacrylic acid to **EMA** polymers (E/MA) usually with 15 % methacrylic acid units. Partial neutralization of carboxylic acid groups by metal ions such as $Na^+$, $Mg^{2+}$, etc., delivers so-called **ionomers** in which carboxylic groups are surrounded by many metal ions and vice versa (numbers of ions depend on the coordination number and not the valency, see Volume III). The resulting ion clusters act at lower temperatures as random crosslinkers. Large crystalline regions cannot form; these polymers are therefore transparent. The ionic clusters dissociate at higher temperatures and the ionomers can therefore be processed like thermoplastics.

Physical crosslinking by ion clusters leads to tough polymers while the ionic groups are responsible for good adhesion to many materials. Ionomers are ideally suited for extrusion coatings, especially, because these are pore-free.

Ionomers comprise not only copolymers of ethene and methacrylic acid but also many other types of comonomers and even after-treated polymers (Table 6-13).

Table 6-13 Industrial ionomers that are known by their trade names.

| Trade name | Monomers | After-treatment |
|---|---|---|
| Surlyn® | $CH_2{=}CH_2 + CH_2{=}C(CH_3)COOH$ | neutralization |
| Hycar® | $CH_2{=}CH{-}CH{=}CH_2 + CH_2{=}CHCOOH$ | neutralization |
| Nafion® | $CF_2{=}CF_2 + CF_2{=}C(F){-}O{-}[(CF(CF_3){-}CF_2{-}O]_i{-}(CF_2)_2{-}SO_3H$ | neutralization |
| Flemion® | $CF_2{=}CF_2 + CF_2{=}C(F){-}[O{-}CF_2{-}CF(CF_3)]_i{-}(CF_2)_i{-}COOH$ | neutralization |
| Hypalon® | $CH_2{=}CH_2$ | chlorosulfonation |
| - | $CH_2{=}CH_2 + CH_2{=}CH(CH_3)$ + diene | chlorosulfonation |

### 6.2.6 Poly(propylene)

**History**

Poly(propylene) (**PP**), $\text{+}CH_2–CH(CH_3)\text{+}_n$, is the fastest-growing synthetic polymer because of the easy access to its monomer propene, good mechanical properties, and the now possible variation of its configuration.

Propene (historic name: propylene), $CH_2=CH(CH_3)$, is now practically exclusively obtained from petrochemical cracking (p. 107). Liquid petroleum gases (LPG) and natural gas liquids (NGL) dominate in the United States (76 %), followed by naphtha (11 %), gas oil (9 %), and refinery gases (4 %). Western Europe prefers naphtha (71 %) over LPG + NGL (17 %), gas oil (9 %) and refinery gases (3 %). Some crackers now have nominal capacities of 750 000 t/a ethene and 450 000 t/a propene.

Previously, world capacities were ca. $88 \cdot 10^6$ t/a for ethene (1997) and $43 \cdot 10^6$ t/a for propene (1995). In 2003, the world production of PP was ca. $39 \cdot 10^6$ t/a.

Half a century ago, propene could only be polymerized free-radically to branched "atactic" oligomers that formed oils. Between 1951 and 1954, however, four research groups independently obtained crystalline poly(propylene)s by polymerization of propene with either supported metal oxides (J.P.Hogan and R.L.Banks (Phillips Petroleum); A.Zletz (Standard Oil of Indiana)) or organometal catalysts (G.Natta, P.Pino, and G.Mazzanti (Montecatini); W.N.Baxter, N.G.Merkling, I.M.Robinson, and G.S.Stamatoff (DuPont)). The Hogan-Banks and Zletz discoveries were accidentally made during a search for new catalysts for the petrochemical synthesis of high-performance gasolines. Natta's group had knowledge of Ziegler catalysts for the low-pressure polymerization of ethene to high-density poly(ethylene).

All four groups claimed patent rights for the structure of isotactic polt(propylene). After a patent process of 22 years (1958-1980), a U.S. Federal Court decided that Hogan and Banks were the first to present a new composition of matter. Hence, all other parties had to pay license fees for the polymer structure but not for the various processes that lead to isotactic poly(propylene)s.

In 1982, Brintzinger's group described the synthesis of chiral metallocenes with $C_2$ symmetry. These metallocenes, when activated by aluminoxanes (first reporteded in 1980 by H.Sinn and W.Kaminsky), led to new polymerization processes and ultimately also to syndiotactic poly(propylene)s that were discovered in 1984 by J.A.Ewen's group (see Volume I). Later discoveries comprised isotactic block polymers, hemi-tactic and atactic poly(propylene)s, and many other PPs with different configurational statistics. The number of different poly(propylene) types and grades now exceeds ca. 7000.

**Isotactic Poly(propylene)s by Transition Metal Catalysts**

Classic metal oxide or transition metal-catalyzed polymerizations (Table 6-8), when applied to propene instead of ethene, lead to poly(propylene)s with moderately high isotacticities and relatively high proportions of "atactic" poly(propylene)s as byproduct (Table 6-14). Both types of polymerization processes were successively refined, as shown by the history of polymerizations by Ziegler catalysts:

Industrial *first-generation* Ziegler catalysts consisted of $TiCl_4 + (C_2H_5)_3Al$ which reacted *in situ* to form small amounts of active $\gamma$-$TiCl_3$ by exchange reactions. The low catalyst productivity was enhanced by addition of electron donors; for example, amines

Table 6-14 Ziegler-Natta polymerization of propylene. $x_{mon}$ = initial mole fraction of monomer, $p$ = pressure, $T$ = polymerization temperature. Catalyst yield = mass of poly(propylene) per mass of catalyst (Eq.(A 4-7)). Catalyst productivity (in $h^{-1}$) = catalyst yield per time (Eq.(4-9)). [a] In higher hydrocarbons; [b] gaseous in fluidized bed in propene; [c] also by simplified suspension polymerization; [d] also in bulk or in simplified suspension polymerization.

| Generation | Polymerization conditions<br>Phase | $x_{mon}$ | $p$/MPa | $T$/°C | Catalyst yield | Catalyst productivity in $h^{-1}$ | Isotacticity index in % | Percent a-PP |
|---|---|---|---|---|---|---|---|---|
| 1 | Suspension [a] | 0.20 | 1.0 | 77 | 1000 | 500 - 1 000 | 83-93 | < 15 |
| 2 | Suspension [b] | 0.39 | 1.4 | 65 | 70 000 | 3 000 - 6 000 | 94-97 | < 0.5 |
| 2 | Bulk | 0.85 | 2.2 | 54 | 90 000 | | | |
| 3 | Bulk [c] | 0.90 | 2.9 | 80 | 4 500 | 12 000 - 30 000 | 96-99 | ~ 0 |
| 4 | Gas phase [d] | | | | 12 000 | 30 000 | 96-99 | ~ 0 |

(see Volume I, p. 271 ff.). Catalyst yields and isotacticities were low. The resulting isotactic poly(propylene) precipitated whereas the also-formed atactic poly(propylene) remained dissolved in higher hydrocarbons. The catalyst was then destroyed by HCl..

The higher the isotacticity, the greater the crystallinity of poly(propylene). A convenient measure of crystallinity is density (see Volume III). However, density measurements are not sensitive enough for poly(propylene)s because of the relatively small density difference between 100 % crystalline isotactic poly(propylene) ($\rho$ = 0.938 g/cm$^3$ (monoclinic)) and 100 % amorphous poly(propylene) ($\rho \approx$ 0.850 g/cm$^3$). Isotacticities are therefore characterized by so-called **isotacticity indices**, IT, that indicate the percentage of poly(propylene) that is insoluble in boiling xylene. Despite the name, IT does not measure isotacticities (i.e., the proportion of isotactic diads, triads, etc.). It rather indicates morphologies of specimens.

In *second-generation catalysts*, $TiCl_4$ is reduced to $TiCl_3$ by $(C_2H_5)_2AlCl$. The by-product $C_2H_5AlCl_2$ is removed by extraction with ethers or thioethers because it can complex with $TiCl_3$ and thus reduce active centers. The resulting product is again treated with $TiCl_4$ to give the desired catalyst, $\delta$-$TiCl_3 \cdot Al(Et_nCl_{3-n})_xD_y$, where D is a complexing electron donor such as an ether, ester, or amine. Typical catalyst compositions are $0 \leq n \leq 2$, $x < 0.3$, and $0.009 < y < 0.11$. Catalyst yields are so high that residual catalysts need not be removed. The proportion of atactic poly(propylene) is usually negligible.

*Third-generation catalysts* employ irregular powders of $MgCl_2$ or Mg(OH)Cl as carriers and complexing agent. The polymerization processes are similar to polymerizations of ethene; most of them use slurry processes in solvents. Less frequent are slurry processes in liquid propene, gas-phase processes, and solution polymerizations.

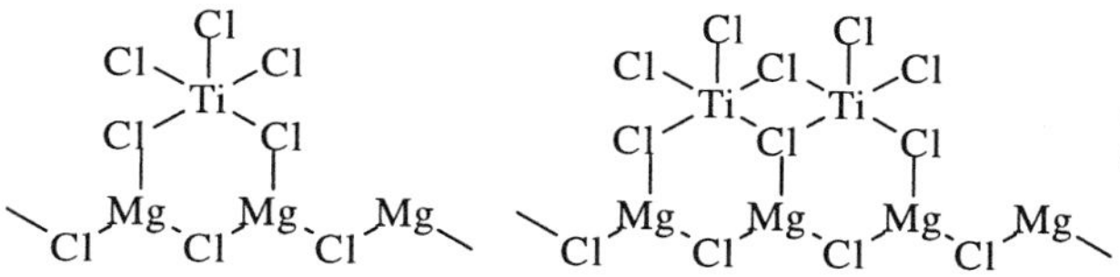

Complexation of $TiCl_3$ on $MgCl_2$ surfaces

*Fourth-generation* Ziegler catalysts employ spherical $MgCl_2$ particles of 50-80 μm diameter that result from the reaction of magnesium alkyls with, e.g., silicon tetrachloride on the surface of dried particles of silicic acid. The $MgCl_2$-coated particles are then treated with $TiCl_4$ and so-called internal electron donors (diethers, phthalic acid esters,, succinates). The resulting catalysts are then activated by $AlR_3$ (R = ethyl or isobutyl) plus silanes as external donors.

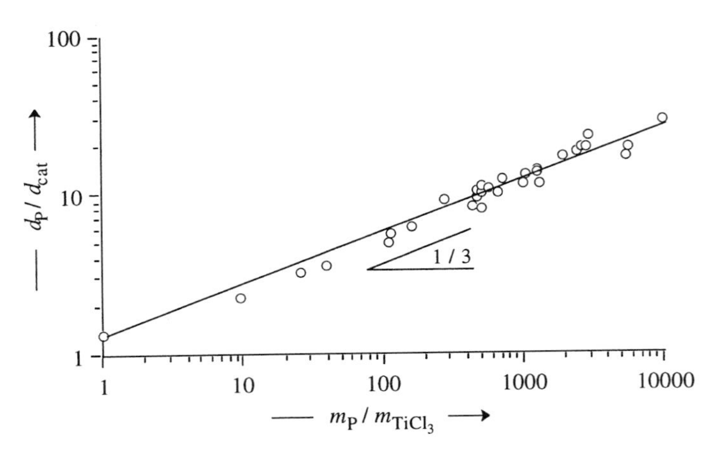


Fig. 6-15 Ratio of diameters of polymer particles and original catalyst particles, $d_P/d_{cat}$, as a function of catalyst yield, i.e., the ratio of mass of produced polymer per mass $m_{TiCl_3}$ of catalyst. Data of [9].

The resulting catalyst particles act as "microreactors" for the production of spherical poly(propylene) particles around a catalyst core. The relative diameter of polymer particles increases with the cubic root of the catalyst yield (Fig. 6-15). The polymer spheres are only slightly crystalline. On extrusion or injection molding, they thus need less energy for melting than highly crystalline poly(propylene)s. Melts crystallize rapidly, however, because the polymers are highly isotactic, which in turn leads to fast cycle times.

Isotactic poly(propylene) (**IPP, iPP, it-PP**) is now obtained by slurry polymerization of liquid propene in either cascades of loop reactors or continuous stirred tank reactors and also by gas-phase polymerization in cascades of horizontal multichamber reactors..

Highly isotactic poly(propylene) forms three crystalline modifications. Most common is the thermodynamically stable monoclinic α-modification in which the polymer chains are present as $3_1$ helices with equal proportions of $TG^+$ and $TG^-$ chains (Volume III). The compact chain conformation leads to higher melting temperatures than that of high-density poly(ethylene) (176°C versus 135°C) which crystallizes in the all-trans conformation (...TTT...). The relatively high melting temperature, low densities of 0.85–0.92 g/cm$^3$, and good mechanical properties have allowed iPP to replace metals in some applications. A disadvantage is the high glass temperature of 0°C or lower which is responsible for the relatively low toughness (Table 6-15).

Toughness is increased by either copolymerization of propene with small proportions of ethene to so-called **reactor copolymers** (**RCP**) or by blending iPP with ethene-propene copolymers, preferably by *in situ* processes (see below). Many industrial "isotactic poly(propylene)s" are in reality such copolymers or blends although they are not identified as such. Examples are random copolymers of propene with less than 8 % ethene and random terpolymers of propene with less than 12 % ethene and 1-butene.

So-called **propylene block-copolymers** are obtained as *in situ* blends that are generated in reactor trains in which the homopolymerization of propene is followed in the next reactor by a copolymerization of propene with ethene. The fast chain transfer leads to a blend of it-poly(propylene), poly(propylene-*co*-ethylene), and poly(ethylene), i.e., to a heterophasic mixture of homopolymers and copolymers and not, as the designation "propylene block-copolymer" suggests, to a block copolymer.

Table 6-15 Average properties of isotactic polymers of propene (it-PP), 1-butene (it-PB), and 4-methyl-1-pentene (it-P4MP) as compared to high-density poly(ethylene) (HDPE) at 23°C.

| Property | Physical unit | HDPE | it-PP | it-PB | it-P4MP |
|---|---|---|---|---|---|
| Density | $g/cm^3$ | 0.94–0.96 | 0.91–0.94 | 0.86–0.91 | 0.83–0.84 |
| Refractive index | 1 | | | 1.51 | 1.46 |
| Melting temperature (DSC) | °C | < 135 | 176 | 126 | 230–240 |
| Heat distortion temperature (1.8 MPa) | °C | | | 54–60 | 41–48 |
| Vicat temperature B | °C | 60–65 | 130 | 108–113 | 179 |
| Glass temperature (DSC) | °C | −123 | −(13–35) | −(33–38) | 29–40 |
| Maximum service temperature (short) | °C | 90–120 | 140 | | |
| (long) | °C | 70–80 | 100 | | |
| Linear thermal expansion coefficient | $K^{-1}$ | $2 \cdot 10^{-4}$ | $1.1 \cdot 10^{-4}$ | $1.3 \cdot 10^{-4}$ | $1.2 \cdot 10^{-4}$ |
| Thermal conductivity (20°C) | $W\ m^{-1}\ K^{-1}$ | 0.4–0.5 | 0.12–0.22 | 0.22 | 0.17 |
| Tensile modulus | MPa | 60–290 | | 210–260 | 1100–2000 |
| Flexural modulus | MPa | | 1400 | 310–370 | 770–1800 |
| Yield strength | MPa | 20 | 30 | 12–17 | 23–28 |
| Tensile strength at break | MPa | 10–60 | 30 | 31–37 | 17–28 |
| Tensile elongation at break | % | | | 280–380 | 10–50 |
| Impact strength (Izod) | J/m | no break | 27 | | |
| (Charpy) | $kJ/m^2$ | | 20 | | |
| Notched impact strength (Izod, 3.1 mm) | J/m | 21–210 | | no break | 16–64 |
| (Charpy) | $kJ/m^2$ | 6–no break | 4 | | |
| Hardness (Shore D) | – | | 74 | 55–65 | |
| Rockwell hardness (L) | | | | 80–90 | 67–74 |
| Ball indentation hardness | – | ≈ 50 | | | |
| Relative permittivity | 1 | 2.4 | 2.2–2.3 | 2.52 | 2.12 |
| Electrical resistance (surface resistivity) | Ω | $10^{13}$ | $10^{13}$ | | |
| Resistivity (volume resistance) | Ω cm | $10^{16}$ | $10^{16}$ | $10^{16}$ | $10^{16}$ |
| Electric strength | kV/mm | > 700 | 2400 | | |
| Dissipation factor (50 Hz) | 1 | 0.0003 | 0.0005 | 0.0005 | |
| Water absorption (96 h) | % | < 0.05 | | < 0.03 | 0.01 |

**Impact copolymers** of propene (**ICP**) are blends of iPP with 15–25 % EPDM that are made *in situ* in the same reactor train. Higher EPDM contents lead to **thermoplastic poly(olefin) elastomers (TPO)** which contain 60–65 EPDM and an iPP with a high melt index (i.e., low molecular weight) or less EPDM and an iPP with low melt index.

**Elastomeric poly(propylene)s** result from the homopolymerization of ethene with combinations of catalysts, for example, $V(O)Cl_2(OCH_2Si(CH_3)_3–AlR_3$, $TiCl_4$ on $MgCl_2$ as carrier, $ZrR_4$ on $Al_2O_3$ as carrier, or zirconocenes plus methyl aluminoxanes (see p. 236). In these polymers, isotactic blocks alternate with atactic blocks, which is presumably caused by fluctuations in the nature of the catalyst during polymerization.

Isotactic poly(propylene)s by either Ziegler or metallocene polymerizations (see next subsection) are used for thermoplastic parts, films, fibers and ropes, and elastic materials (see also Volume IV). Films for packaging and electrical applications are available in standard and biaxially oriented grades (BOPP) and fibers as monofilaments and multifilament yarns for carpets, thermal clothing, sportswear, etc. Applications of plastics include containers, bottles, pipes, tubes, appliance parts, fish nets, artificial grass, and chemical-resistant parts.

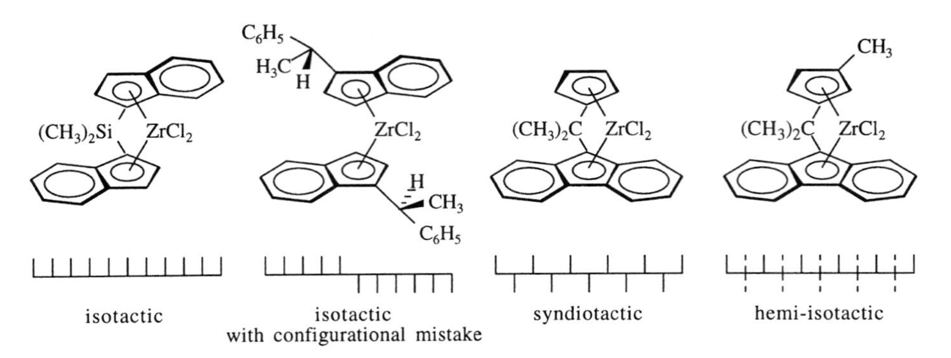


Fig. 6-16 Some metallocene catalysts and the generated poly(propylene)s (in Fischer projection).

## Metallocene Poly(propylene)s

Metallocene-poly(propylene)s (**mPP**) result from the polymerization of propene with metallocene compounds as catalysts and methyl aluminoxanes (MAO) as cocatalysts. Metallocenes are compounds in which a transition metal atom is sandwiched between two cyclopentadienyl moieties (Fig. 6-16). The structure of MAOs is still unclear (Fig. 6-17).

All active centers are identical in metallocene catalysts; they are therefore also called **single-site catalysts**. Ziegler catalysts, on the other hand, have different active centers; they are **multiple-site catalysts** (see Volume I).

Polymerization by single-site catalysts can be homogeneous or carrier-bound on spherical silica gel, zeolites, or even poly(styrene) particles. Heterogeneous systems are less expensive because they need only ca. 100 equivalents of MAO per metallocene,ß whereas homogeneous systems require 1000–20 000 equivalents. Heterogeneous systems also allow good control of polymer morphology. Such systems can replace Ziegler catalysts without remodeling existing plants (drop-in technology).

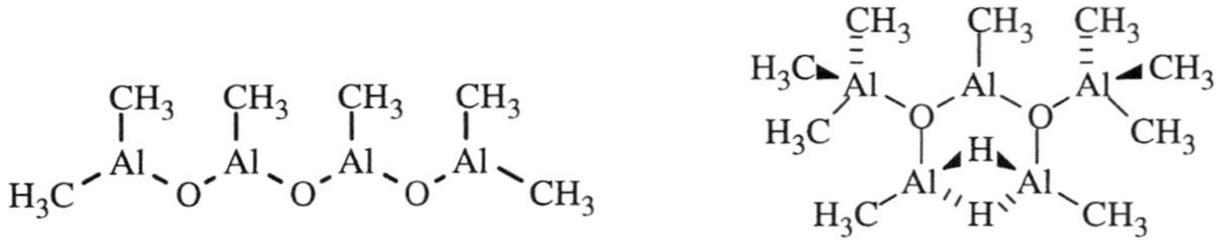

Fig. 6-17 Possible structures of methyl aluminoxanes.

## Atactic Poly(propylene)s

So-called atactic poly(propylene)s (**APP**, **aPP**) were orginally worthless byproducts of propene polymerizations by first-generation Ziegler catalysts (p. 232), which were either burned or buried in dumps. aPPs are not atactic but non-crystallizing highly branched low-molecular weight polymers with many head-to-head structures.

It was found later that their properties (Table 6-16) make them useful for improving bitumen, as melt adhesives, for carpet coatings, for laminating papers, and many other purposes. These uses created such great demands (United States > 60 000 t/a) that aPPs are now obtained by direct polymerization since they are no longer byproducts of Ziegler polymerizations. High-molecular weight aPPs are now being tested for possible applications as elastomers or blends with IPPs.

Unbranched atactic poly(propylene)s with ideally distributed it and st diads result from hydrogenating poly(2-methyl-1,3-pentadiene), $+\!\!CH_2\text{–}C(CH_3)\text{=}CH\text{–}CH(CH_3)\!\!+_n$, that is obtained by polymerizing the diene $CH_2\text{=}C(CH_3)\text{–}CH\text{=}CH\text{–}CH_3$. These polymers have only academic interest.

### Syndiotactic Poly(propylene)s

Early low-temperature polymerizations of propene by vanadium-based Ziegler catalysts resulted in predominantly syndiotactic poly(propylene)s (**SPP, sPP, st-PP**). An example is the polymerization of propene by $VCl_4/Et_2AlCl$ at –78°C to a polymer with 88 % syndiotactic diads and 76 % syndiotactic triads.

Syndiotactic poly(propylene)s with higher syndiotacticities are obtained by polymerizations with zirconium and hafnium-based metallocene catalysts (see Fig. 6-16). These SPPs are highly transparent and heat-stable. The density of their rhombic modification ($\rho$ = 0.898 g/cm$^3$) is less than that of the monoclinic modification of IPP ($\rho$ = 0.938 g/cm$^3$). The melting temperature is also lower ($T_M$ = 160°C versus 176°C), whereas glass temperatures are practically identical ($T_G \approx$ –10°C).

Table 6-16 Properties of an atactic poly(propylene) as a byproduct of a Ziegler polymerization ($M_r$ = 29 000), two atactic poly(propylene)s from metallocene polymerizations ($M_r$ = 200 000, 490 000), an elastomeric poly(propylene) (= stereoblock-poly(propylene)), and a syndiotactic poly(propylene with $x_{ssss}$ = 96.5 % syndiotactic pentads. a) Ideal tacticity and crystallinity (100 %).

| Property | Physical unit | at-PP 29 000 | at-PP 200 000 | at-PP 490 000 | e-PP | st-PP | it-PP |
|---|---|---|---|---|---|---|---|
| Density | g/cm$^3$ | 0.863 | 0.861 | 0.855 | | 0.88 | 0.92 |
| $\overline{M}_w / \overline{M}_n$ | 1 | ≈ 6 | ≈ 3.3 | ≈ 2.3 | | | |
| Melting temperature (DSC) | °C | - | - | - | 50–66 | 148 | 176 |
| (ideal) a) | °C | | | | | 217 | 212 |
| Glass temperature (DSC) | °C | –5 | | –13 | | | |
| Tensile modulus | MPa | 10 | 8 | 5 | 23–28 | 210 | 690 |
| Flexural modulus | MPa | 10 | 8 | 5 | | 760 | |
| Tensile strength (fracture) | MPa | 1 | 1 | 2 | 5–15 | 180 | 34 |
| Elongation at yield | % | | | | | 11 | |
| Elongation at fracture | % | 110 | 1400 | 2000 | > 1000 | 180 | |
| Hardness (Shore A) | - | 67 | 50 | 55 | 77–96 | | |

### Copolymer with Carbon Monoxide

Ziegler–Natta copolymerizations of propene and carbon monoxide by a modified palladium catalyst do not lead directly to the alternating copolymer II (see Section 6.13). Rather, poly[*spiro*-2,5-(3-methyltetrahydrofuran)] (I) is obtained first and then transformed to the true copolymer, poly(1-oxo-2-methyltrimethylene) (II), by heating:

(6-9) $4\,CH_2{=}CH(CH_3) + 4\,CO \longrightarrow$ I $\longrightarrow$ II

### 6.2.7 Poly(1-butene)

1-Butene, $CH_2=CH(CH_2CH_3)$, is a byproduct of the cracking of gas oil (p.110) and one of the components in $C_4$ raffinates. Formerly, it was obtained from propene by the Phillips triolefin process, Eq.(4-19).

Polymerization of 1-butene by first-generation Ziegler catalysts leads to a mixture of isotactic and atactic poly(1-butene)s of which the latter needed to be extracted. 1-Butene is now polymerized with higher-generation Ziegler catalysts (world production (2003): ca. $3{\cdot}10^4$ t/a). A non-industrial route starts with mixtures of *cis*- and *trans*-2-butenes that isomerize to 1-butene with certain Ziegler-type catalysts before they are polymerized.

Isotactic poly(1-butene) is commonly known as **poly(butylene) (PB)**. The semi-crystalline polymer crystallizes in three crystal modifications composed of different types of helices: rhombohedral ($3_1$), tetragonal ($11_3$), and orthorhombic ($4_1$). The $3_1$ helix is wider than the $3_1$ helix of it-PP because of the larger substituent ($CH_2CH_3$ versus $CH_3$), which in turn leads to a lower melting temperature (140°C versus 176°C).

Because of its high fracture strength and good resistance to stress corrosion, isotactic poly(1-butene) is used for pipes and packaging films (see Tables 6-15 and 6-17). Atactic poly(1-butene) is now obtained by direct polymerization; it is used as a melt adhesive. Syndiotactic poly(1-butene) results from hydrogenation of 1,2-poly(butadiene).

Table 6-17 Properties of poly(1-olefin)s: poly(1-butene) (PB), poly(1-hexene) (PHE), poly(4-methyl-1-pentene) (P4MP), and poly(isobutene) (PIB).

| Property | Physical unit | it-PB | it-PHE | it-P4MP | PIB |
|---|---|---|---|---|---|
| Density | $g/cm^3$ | 0.91 | 0.83 | 0.83 | 0.92 |
| Refractive index | 1 | | | 1.463 | |
| Melting temperature (DSC) | °C | 140 | −55 | 240 | 2 |
| Heat distortion temperature (1.82 MPa) | °C | 57 | | 41 | |
| Vicat temperature B | °C | 110 | | 179 | |
| Glass temperature (DSC) | °C | −20 | | 40 | −70 |
| Linear thermal expansion coefficient | $K^{-1}$ | $1.3{\cdot}10^{-4}$ | | $1.2{\cdot}10^{-4}$ | $1.2{\cdot}10^{-4}$ |
| Thermal conductivity (20°C) | $W\ m^{-1}\ K^{-1}$ | 0.22 | | 0.17 | 0.50 |
| Tensile modulus | MPa | 260 | | 1800 | 0.5-50 |
| Flexural modulus | MPa | 350 | | < 1800 | 2000 |
| Yield strength | MPa | 17 | | 25 | |
| Tensile strength at break | MPa | 34 | | 18 | 0.5-50 |
| Flexural strength | MPa | 14–16 | | 25–42 | |
| Elongation at yield | % | 24 | | | |
| Elongation at fracture | % | 340 | | 10–50 | 1000 |
| Notched impact strength (Izod, 3.1 mm) | J/m | no break | | 100–200 | |
| Hardness (Shore D) | - | 60 | | | |
| Hardness (Rockwell) | - | D60 | | L70 | |
| Relative permittivity | 1 | 2.53 | | 2.12 | 2.3 |
| Resistivity (volume resistance, 2 min) | Ω cm | $> 10^{-16}$ | | $> 10^{-16}$ | $10^{-15}$ |
| Electric strength | kV/mm | 18–40 | | 54 | |
| Dissipation factor (100 Hz) | 1 | 0.0005 | | 0.00006 | 0.004 |
| Water absorption (24 h) | % | < 0.03 | | | |
| (equilibrium) | % | | | 0.01 | |

### 6.2.8 Poly(4-methyl-1-pentene)

4-Methyl-1-pentene, $CH_2=CH(CH_2CH(CH_3)_2)$, is obtained by dimerization of propene (p. 107, II). Polymerization of this monomer by Ziegler catalysts delivers isotactic polymers (**PMP, P4MP**) with a very low density of 0.83 g/cm$^3$. The polymers are glass-clear despite X-ray crystallinities of 40–65 % because the crystalline and amorphous regions have practically the same refractive indices. Clarities can be further improved by incorporation of small proportions of other 1-olefins. However, injection molding of such polymers may lead to voids and thus to hazy parts.

Despite a glass temperature of 40°C, polymers can be sterilized and used continuously up to 170°C because of a continuous service temperature of 179°C (melting temperature: 240°C). The thermal expansion coefficient of P4MP (Table 6-17) is similar to that of water; P4MP can thus be used for graduated labware for aqueous solutions. Films of P4MP serve as packaging materials for fresh fruits and vegetables because of their high permeabilities for oxygen and carbon dioxide.

### 6.2.9 Higher Poly(1-olefin)s

Homopolymers of higher 1-olefins, $\text{+}CH_2\text{–}CH((CH_2)_iH)\text{+}_n$, have no applications because of insufficient properties. For example, melting temperatures decrease with increasing number $i$ of methylene groups in side-chains, pass through a minimum at –55°C at $i = 4$ and then increase only slowly: 17°C ($i = 5$), 5°C ($i = 6$), 19°C ($i = 7$), 34°C ($i = 8$), 36°C ($i = 9$), 45°C ($i = 10$), etc., until 80°C ($i = 16$).

However, several copolymers are utilized. The metathesis of ethene + isoprene delivers a mixture of 4- and 5-methyl-1,4-hexadiene that is copolymerized with less than 85 % 1-hexene by Ziegler catalysts such as $\alpha$-$TiCl_3$ + $(C_2H_5)_3Al$. The resulting **poly[(1-hexene)-*co*-[(4/5)-methyl-1,4-hexadiene]s** are specialty rubbers with long fatigue lifes. They are used for diaphragms of membrane pumps in artificial hearts.

The free-radical polymerization of 1-octadecene, $CH_2=CH((CH_2)_{16}H$, and maleic anhydride delivers an alternating copolymer. This so-called **polyanhydride resin** is useful as an adherent, thickener, adhesive, and curing agent.

### 6.2.10 Poly(isobutylene)s

Isobutene (= isobutylene), $CH_2=C(CH_3)_2$, is now predominantly obtained from crack gases (p. 110, 111, 115). Dehydration of *t*-butanol is no longer used.

Isobutene is industrially polymerized to various polymers: poly(butene)s, poly(isobutylene)s, and butyl rubbers, all by cationic polymerization with $BF_3/H_2O$. **Poly(butene)s** is the industrial term for viscous liquids consisting of low-molecular weight random copolymers from the polymerization of isobutene/butene fractions of the $C_4$ stream of refineries (p. 110–111). These polymers carry unsaturated endgroups that are often functionalized.

**Poly(isobutylene)s (= poly(isobutene)s, PIB)** are homopolymers of isobutene. The industrial process liquefies a mixture of isobutene plus a little diisobutene = 2,4,4-trimethyl-1-pentene, $CH_2{=}C(CH_3){-}CH_2{-}C(CH_3)_3$, adds the same amount of liquid ethene, and sprays the mixture at –80°C onto a cold moving conveyor belt. Addition of the catalyst leads to instant polymerization. The diisobutene acts as a chain transfer agent that regulates molecular weight, usually between 50 000 and 500 000. Ethene does not polymerize; instead, its vaporization removes the heat of polymerization.

PIB crystallizes only under tension. Because of its low crystallinity, low glass temperature of –70°C, and weak intermolecular forces, it is a rubber. Low-molecular weight PIBs are used as adhesives or viscosity improvers and high-molecular weight ones as rubber additives or for air-tight inner tubes of tires. Modified poly(isobutylene)s are used as protective covers for buildings and machinery, for example, carbon black or graphite-filled copolymers of isobutene and ca. 10 % styrene.

The copolymerization of isobutene and allyl chloride leads to allyl-telechelic PIBs, $CH_2{=}CH{-}CH_2{+}CH_2{-}C(CH_3)_2{+}_nCH_2{-}CH{=}CH_2$. These oligomers are used as sealants because of their crosslinkable endgroups,

**Butyl rubbers (IIR)** are copolymers that are obtained in a slurry process by copolymerization of isobutene and 0.5–2.5 wt% isoprene in boiling ethene at –90°C with $AlCl_3/CH_3Cl$ as initiator. The isoprene units, $-CH_2{-}C(CH_3){=}CH{-}CH_2-$, of these rubbers provide carbon–carbon double bonds for conventional rubber vulcanizations (cross-linking) by sulfur. The resulting elastomers have good weatherability because of very small concentrations of remaining double bonds. IIRs are not very permeable for air and are therefore used for inner tubes of tires on bicycles and large vehicles.

Polymerization in the presence of a polymeric branching agent produces **star-branched butyls** with bimodal distributions that have improved processabilities.

Vulcanization of IIRs is relatively slow since there are so few reactive double bonds per chain. Faster-vulcanizing IIRs are obtained by post-halogenation with bromine or chlorine in which a halogen cation, for example, $Br^{\oplus}$, is added to the double bond. The resulting methylcarbenium ion splits off a proton which generates an allylic carbon double bond:

$$\sim CH_2{-}\overset{\overset{CH_3}{|}}{C}{=}CH{-}CH_2\sim \xrightarrow{+Br_2,\ -Br^{\ominus}} \sim CH_2{-}\overset{\overset{CH_3}{|}}{\underset{\oplus}{C}}{-}\underset{\underset{Br}{|}}{CH}{-}CH_2\sim \longleftrightarrow \sim CH_2{-}\overset{\overset{CH_3}{|}}{C}{-}CH{-}CH_2\sim\ (\text{bridged } Br^{\oplus}) \xrightarrow{-H^{\oplus}} \sim CH_2{-}\overset{\overset{CH_2}{\|}}{C}{-}\underset{\underset{Br}{|}}{CH}{-}CH_2\sim \qquad (6\text{-}10)$$

Chlorinated butyl rubbers are used as the inner liners of tubeless tires. Such liners may also be brominated terpolymers of isobutene and small proportions of isoprene and α-methylstyrene. The total world capacity for halogenated and brominated IIRs is now more than 400 000 t/a.

Dehydrohalogenation of halogenated butyl rubbers leads to polymers with conjugated double bonds, $-CH_2{-}C({=}CH_2){-}CH{=}CH-$ and $-CH{=}C(CH_3){-}CH{=}CH-$. These **conjugated-diene butyls (CDB)** are useful for coatings.

Isobutene is also copolymerized with maleic anhydride by free-radical polymerization with redox initiators. The resulting poly(isobutenyl succinic anhydride) is used as a curing agent for epoxy resins as well as a rust inhibitor.

# 6.3 Poly(diene)s

## 6.3.1 Overview

Poly(diene)s are polymers of conjugated 1,3-dienes, $CH_2$=CR–CH=$CH_2$, such as butadiene (R = H), isoprene (R = $CH_3$), and chloroprene (R = Cl). The polymerization proceeds via both double bonds to 1,4-diene units, –$CH_2$–CR=CH–$CH_2$–, or via one double bond to 1,2-diene units, –$CH_2$–CR(CH=$CH_2$)–, or 3,4-diene units, $CH_2$–CH(CR=$CH_2$)–, or combinations thereof.

The monomeric units can also exist in different configurations: 1,4-diene units in either cis-1,4- or trans-1,4- units as well as 1,2- or 3,4-diene units (so-called vinyl structures) in either isotactic or syndiotactic diads. All constitutional and configurational units may furthermore be randomly distributed or in blocks.

$$\begin{array}{cccc} \text{~~}CH_2(R)C{=}C(H)CH_2\text{~~} & \text{~~}CH_2(R)C{=}C(H)CH_2\text{~~} & \text{~~}CH_2{-}CR(\text{~~}){-}CH{=}CH_2 & \text{~~}CH_2{-}CH(\text{~~}){-}CR{=}CH_2 \end{array}$$

cis-1,4 trans-1,4 1,2 (it, st) 3,4 (R ≠ H) (it, st)

The proportions of 1,4, 1,2, and 3,4 units as well as the proportions of geometric (cis, trans) and tactic (isotactic, syndiotactic, heterotactic, etc.) configurations depend on the diene itself as well as the polymerization conditions (initiator, solvent, temperature) (Table 6-18). Poly(diene)s with nearly 100 % cis-1,4 structures and poly(diene)s with random proportions of cis-1,4, trans-1,4, 1,2, and 3,4 structures are rubbers. Examples include natural rubber and the most important synthetic rubbers from butadiene, butadiene + styrene, and chloroprene. Poly(diene)s with nearly 100 % trans-1,4, 1,2, or 3,4 structures are thermoplastics.

Table 6-18 Structure of various industrial poly(diene)s. na = not applicable, ? = not published.

| Monomer | Polymerization Initiator | Medium | Structure in % cis-1,4 | trans-1,4 | 1,2 | 3,4 |
|---|---|---|---|---|---|---|
| Butadiene | Na | neat | 10 | 25 | 65 | na |
| | $C_4H_8Li$ | hexane | 38 | 53 | 9 | na |
| | $C_2H_5Li$ | tetrahydrofuran | 0 | 9 | 91 | na |
| | $C_2H_5Li$ | THF/benzene | 13 | 13 | 74 | na |
| | $C_2H_5Li$ | benzene/triethylamine | 23 | 40 | 37 | na |
| | $C_2H_5Li$ | toluene | 44 | 47 | 9 | na |
| | $TiCl_3/R_3Al$ | not disclosed | 95 | 3 | 2 | na |
| | alfin | solution | 20 | 80 | 0 | na |
| Butadiene/styrene | free radical | emulsion, 5°C | 13 | 70 | 17 | na |
| | free radical | emulsion, 70°C | 21 | 59 | 20 | na |
| | $C_4H_8Li$ | diglyme | | | 90 | na |
| | $C_4H_8Li$ | triethylamine | | | 20 | na |
| Isoprene | $C_4H_8Li$ | solution | 93 | 0 | 0 | 7 |
| Chloroprene | free radical | neat | 11 | 86 | 2 | 1 |

Industrial names of poly(diene)s deviate from the usual custom because they are neither poly(monomer) names (example: vinyl chloride → poly(vinyl chloride)) nor constitutional names (example: vinyl chloride → poly(1-chloroethylene)), but rather names of monomers (where numbers refer to the position of double bonds) prefixed with "poly" and the position of the linked carbon atoms. The polymerization of 1,3-butadiene via both double bonds leads to 1,4-poly(butadiene)s whereas the polymerization via one double bond results in 1,2-poly(butadiene)s (see above).

The structure of poly(diene)s is that of poly(alkene)s. The group of poly(alkene)s includes so-called poly(alkenamer)s that result from metathesis polymerizations of cyclo-olefins (Section 6.4). This group of polymers is related by structure to polymers from the polymerization of acetylene compounds (Section 6.5).

### 6.3.2 Poly(butadiene)s

Butadiene (= divinyl, vinyl ethylene, erythrene, pyrrolylene) is obtained as a byproduct from the $C_4$ cut of the cracking of naphtha, gas oil, or butane, as a byproduct from refining to motor fuels, or from oxidative dehydrogenation of butane or butene (p. 110 ff., Table 3-17). Processes based on ethanol (p. 105) or acetylene are no longer economical. The world nameplate capacity of butadiene-based rubbers is ca. $9 \cdot 10^6$ t/a.

**Anionic Polymerizations**

The oldest industrial anionic polymerization of butadiene by dispersions of sodium (G: Natrium) in hydrocarbons led to *bu*tadiene-*na*trium polymerization products (**Buna**) with high contents of 1,2-structures. Their insufficient elastomeric properties caused Germany in 1939 to replace them by copolymers of butadiene with either styrene (Buna S) or acrylonitrile (Buna N), so-called letter Bunas (see below).

The interest in anionic polymerizations of butadiene was rekindled after petrochemicals became widely available. Butadiene is present in concentrations of 30–65 % in $C_4$ cuts of naphtha cracking. The direct polymerization of these cuts by butyl lithium delivers **butadiene rubbers (BR)** with average contents of 40 % cis, 50 % trans, and 10 % vinyl structures. This elastic polymer can be processed like all diene rubbers, i.e., by heavy, energy-consuming machinery (roll-mills, calenders, extruders).

For many rubber applications, less-viscous grades are preferred. **BR liquid rubbers** with various contents of 1,2-structures are obtained by polymerization of butadiene by high concentrations of initiators. BRs with 10–20 % vinyl units are obtained by lithium alkyls in hydrocarbons, with 30–70 % vinyl units by sodium alkyls in hydrocarbons, and with up to 90 % vinyl structures at low temperatures by sodium in hydrocarbons plus some tetrahydrofuran.

The high initiator costs for the syntheses of BR liquid rubbers can be reduced if these polymerizations are conducted in the presence of chain transfer agents, so-called **telegenes**, such as amines or 1,2-butadiene (= methyl allene, $CH_2=C=CH-CH_2$). These agents react with the active site of the growing chain, leaving dead molecules behind, and become themselves new initiator molecules. In this way, many polymer chains are started by one initiator molecule (see Volume I).

Liquid BR rubbers are also obtained by using dianions as initiators and terminating the polymer chains by carbon dioxide to give poly(butadiene)s with molecular weights of ca. 10 000. The COOH endgroups of these polymers allow their crosslinking with polyisocyanates (see Volume IV).

Addition of Lewis bases as randomizers to BuLi-initiated anionic polymerizations leads to **vinyl butadiene rubbers (VBR)** with various 1,2-vinyl contents: 90% with diethylene glycol methyl ether (diglyme), 60 % with tetrahydrofuran, 40 % with diethyl ether, 70 % with tetramethylenediamine, and 20 % with triethylamine. These 1,4/1,2-poly(butadiene)s have properties similar to those of SBRs, the free-radically polymerized copolymers of butadiene and styrene (see below). In many applications, VBRs have therefore replaced SBRs since the latter became too expensive because of the high price of styrene.

## Alfin Polymerizations

Alfin polymerizations originally used initiators from an *al*coholate and an ole*fin*; for example, sodium isopropylate and allyl sodium. The most economical route prepares this initiator from isopropanol, sodium, and butyl chloride:

(6-11) $$2\,Na + (CH_3)_2CHOH + C_4H_9Cl \longrightarrow (CH_3)_2CHONa + (1/2)\,H_2 + C_4H_9Na + NaCl$$

Addition of propene converts the resulting butyl sodium to allyl sodium, which forms a catalyst with sodium isopropylate:

(6-12)
$$CH_2{=}CH{-}CH_3 \xrightarrow[-\,C_4H_{10}]{+\,C_4H_9Na} CH_2{=}CH{-}CH_2Na \xrightarrow{+\,(CH_3)_2CHONa} \begin{array}{l} H_2\overset{\ominus}{C}\cdots\overset{\oplus}{Na} \qquad CH_3 \\ HC^{\oplus} \qquad {}^{\ominus}O-C-CH \\ H_2\overset{\ominus}{C}\cdots\overset{\oplus}{Na} \qquad CH_3 \end{array}$$

Alfin polymerizations produce extremely high-molecular weight poly(butadiene)s with 65–85 % trans-1,4 structures. The molecular weights are regulated by addition of chain transfer agents such as cyclohexene or 1,4-dihydronaphthalene. Industrial types are copolymers of butadiene with 3–10 % isoprene or 5–15 % styrene.

## Free-Radical Polymerizations with Styrene

Elastomeric butadiene-styrene copolymers were first developed in the 1930s as **Buna S** in Germany and during World War II in the United States as **GR-S** (*g*overnment *r*ubber with *s*tyrene) (for the history, see Volume IV). They are now known as *s*tyrene-*b*utadiene *r*ubbers (**SBR**).

The classic free-radical emulsion polymerization was first conducted at 70°C but is now performed at 50°C to so-called "hot rubber" or, predominantly, at 5°C to "cold rubber." Polymerization at low temperatures results in polymers with more trans structures (Table 6-19). Trans-rich butadiene rubbers are preferred over cis-rich ones since the latter tend to cyclize more than the former, which leads to undesirable viscosity increases on kneading or roll-milling.

Table 6-19 Constitution of poly(butadiene)s from free-radical emulsion polymerizations.

| Polymerization temperature in °C | Percent monomeric units in polymer cis-1,4 | trans-1,4 | 1,2 | Molecular weights $\overline{M}_w$ | $\overline{M}_n$ |
|---|---|---|---|---|---|
| 70 | 20.8 | 59.4 | 19.8 | | |
| 50 | 19.0 | 62.7 | 18.8 | | 30 000 – 100 000 |
| 5 | 13.0 | 69.9 | 16.5 | < 500 000 | 110 000 – 260 000 |
| –33 | 5.4 | 78.9 | 15.6 | | |

Both hot and cold copolymerizations use the same monomer ratios of $w_{bu}/w_{sty}$ = 71:29 (butadiene/styrene), the same water to monomer ratio of $w_w/w_{mon}$ = 190:100, the same percentage (ca. 1.7 wt%) of fatty acids or rosin acids (p. 78) as emulsifiers ("soaps"), and the same polymerization time of 12 hours.

Hot polymerizations employ dipotassium persulfate as a free-radical initiator and mercaptans such as *n*-dodecanethiol as chain-transfer agents for the regulation of molecular weights. Polymerizations are led to monomer conversions of ca. 72 % since higher monomer conversions would lead to crosslinking via 1,2 double bonds.

Cold polymerizations use redox systems as initiators; for example, *p*-menthane hydroperoxide + iron(II) sulfate (ferrous sulfate) + sodium formaldehyde sulfoxylate and also the tetrasodium salt of ethylene diamine tetraacetic acid (EDTA) as chelating agent. Since cold polymerizations lead to higher molecular weights than hot ones, they are terminated at lower monomer conversions of 60–65 %.

There are many variations of this basic recipe. Standard grades usually have lower molecular weights. Because of their lower viscosity, they do not require mastication, the break-down of rubber chains by shearing in roll mills. Oil-extended rubbers have higher molecular weights which provide them with better mechanical properties; the added mineral oil reduces their viscosities. Both hot and cold types contain ca. 20–40 % styrenic units; they are mainly used for tire treads (Volume IV). SBRs with ca. 65 % styrenic units are processing aids.

Free-radical copolymerization of butadiene and styrene in bulk are known as "solution polymerizations" since residual monomers serve as the solvent for the polymer. Styrene-butadiene rubbers polymerized in solution (i.e., in "liquid", hence **L-SBR**) contain 15–25 % styrenic units. Corresponding vinyl-types (**L-VSBR**) have 15–55 % 1,2-units.

The relatively low molecular weights do not provide raw SBRs with green strength (= dimensional and tear resistance of unvulcanized rubber); this is improved by working in carbon black as a reinforcing agent. Nor do raw SBRs have sufficient inherent tack (see Volume IV).

### Free-Radical Copolymerizations with Acrylonitrile

Copolymerization of butadiene and acrylonitrile in emulsion produces oil-resistant **nitrile rubbers** (**NBR**: *n*itrile-*b*utadiene *r*ubber; formerly **Buna N** or **GR-N**). Added regulators (= chain transfer agents for the control of molecular weight) are consumed rapidly. They are therefore added stepwise because too broad molecular weight distributions would result otherwise. NBR grades contain between 18 % and 50 % acrylonitrile units; the azeotrope is at 37 mol% acrylonitrile.

**Carboxylated NBRs** are obtained by emulsion terpolymerization of acrylonitrile and butadiene with small proportions of acrylic acid. These NBRs have high strengths.

Selective hydrogenation of carbon–carbon double bonds of NBRs leads to **hydrogenated nitrile-butadiene rubbers (HNBR)** with improved resistance to heat (up to 150°C in hot air or ozone) and swelling agents (sulfur-containing crude oils, hydraulic fluids, break fluids, etc.). HNBRs are crosslinked by peroxides or high-energy radiation.

### Ziegler–Natta Polymerizations

The heterogeneous polymerization of butadiene by $VCl_3/(C_2H_5)_2AlCl$ in hydrocarbons leads to ***trans*-1,4-poly(butadiene)s** with very high contents of trans-1,4 units (cis: 0 %; 1,2: ≤ 1 %). The polymerization becomes homogeneous if $VCl_3$ is replaced by $VCl_3 \cdot 3$ THF. These polymerizations are living: the molecular weight increases linearly with increasing monomer conversion. The resulting high molecular weights at high monomer conversions are undesirable for the polymerization process because they generate high viscosities which in turn make agitation difficult and prevent homogeneous polymerizations. Polymer applications require high molecular weights, however, and for that reason, polymerizations are started with high catalyst concentrations. The resulting lower molecular weight molecules are then coupled by added alkyl or acyl halides; for example, $SOCl_2$, to higher molecular weight ones ("molecular jump reaction").

***cis*-1,4-Poly(butadiene)s** with cis contents of 99 % are obtained by polymerization of butadiene with cobalt compounds + $R_2AlCl$ or nickel compounds + $R_3Al$ + $BF_3$-etherate. The resulting low-molecular weight **poly(butadiene) oils** have only a few branches. They dry (i.e., polymerize by air) as fast as tung oil and faster than linseed oil. Their reaction with 20 % maleic anhydride leads to air-drying alkyd resins.

Emulsions of modified poly(butadiene) oils serve to solidify erosion-prone soils. The low-viscosity emulsion can easily penetrate the upper layer of soil where the poly(butadiene) oxidizes and glues together the upper soil particles. The ability of the soil to take up water is preserved, however, since no continuous film is formed.

Polymerization of butadiene by $CoHal_2$/ligands/$R_3Al/H_2O$ results in **syndiotactic 1,2-poly(butadiene)s** with high molecular weights. These thermoplastic elastomers form films with extremely high tear resistance and good permeability for gases; they are used to package fish or fresh fruits. The allylic side groups $>CH–CH=CH_2$ of these polymers crosslink under the action of air and light while simultaneous photodegradation enables the polymer to crumble into pieces after some time. For properties, see Table 6-20.

Table 6-20 Properties of poly(butadiene)s. $\rho$ = Density, $c$ = repeat distance in chain direction, $N_{mon}$ = number of monomeric units, $T_M$ = melting temperature, $T_G$ = glass temperature.

| Structure | $\frac{\rho}{\text{g cm}^{-3}}$ | Crystal system | $\frac{c}{\text{nm}}$ | $\frac{N_{mon}}{\text{unit cell}}$ | $\frac{T_M}{°C}$ | $\frac{T_G}{°C}$ |
|---|---|---|---|---|---|---|
| 1,4-trans (I) | 1.03 | pseudo-hexagonal | 0.490 | 1 | 100 | −108 |
| 1,4-trans (II) | 0.908 | monoclinic | 0.465 | 4 | 148 | |
| 1,4-cis | 1.011 | monoclinic | 0.860 | 4 | 1 | −105 |
| it-1,2 | 0.96 | rhombohedral | 0.65 | 18 | 120 | 20 |
| st-1,2 | 0.964 | orthorhombic | 0.514 | 4 | 154 | 20 |

### 6.3.3 Poly(isoprene)s

**Natural Poly(isoprene)s**

Polyprenes are oligomers and polymers of isoprene (= 2-methyl-1,3-butadiene), $CH_2{=}C(CH_3){-}CH{=}CH_2$, that are found in nature in thousands of plants; only a few of them are utilized, however. The two basic natural polymer types are the *cis*-1,4 and *trans*-1,4-poly(isoprene)s; there are no natural 1,2- or 3,4-poly(isoprene)s.

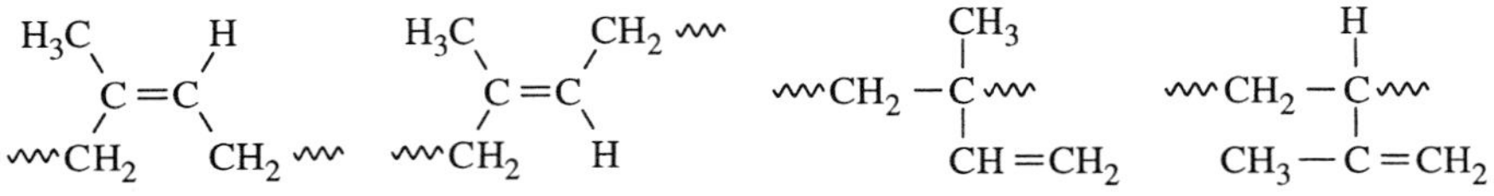

| 1,4-*cis* | 1,4-*trans* | 1,2 | 3,4 |
|---|---|---|---|
| $H_3C(\sim\!CH_2)C{=}C(H)(CH_2\!\sim)$ | $H_3C(\sim\!CH_2)C{=}C(CH_2\!\sim)(H)$ | $\sim CH_2{-}C(CH_3)(CH{=}CH_2)\sim$ | $\sim CH_2{-}C(H)(C(CH_3){=}CH_2)\sim$ |

The 1,4-cis types are rubbers ($T_M$ = 2°C, $T_G$ = –73°C) whereas the 1,4-trans varieties are thermoplastics ($T_M$ = 146°C, $T_G$ = 38°C). None of the naturally occurring molecules is constitutionally homogeneous. The *substance* "**natural rubber**" with the ideal structure of a 1,4-*cis*-poly(isoprene) consists of ca. 95 % 1,4-cis units and ca. 3 % 3,4-units plus aldehyde and epoxide structures, even if the rubber latex is harvested with exclusion of oxygen. The *material* "raw rubber", such as the so-called **crepe**, contains 89–93 wt% natural rubber molecules, 2–3 wt% acetone-extractables, 2–4 wt% proteins, 2–4 wt% moisture, and 0.1–0.5 wt% ash. Practically all natural rubber of commerce is harvested from the latex of the tree *Hevea brasiliensis* that is now grown in rubber plantations of Thailand, Indonesia, Malaysia, Sri Lanka, and Vietnam (see Volume IV). The annual world production of natural rubber increased steadily since the end of World War II; it is now ca. 7 000 000 t/a. For the history and production, see Volume IV.

The two natural 1,4-trans types **gutta percha** and **balata** also contain epoxide structures. Gutta percha is mainly obtained from the latices of the trees *Palaquium gutta* and *P. oblongifolia* of Indonesia and Southern India, whereas the similar balata is harvested from the tree *Mimusops balata* of Brazil and Venezuela. Both gutta percha and balata are used for belts and in dentistry, balata also in chewing gums (**gum chicle**).

**Chicle** (Nahuatl: *chictli*) from the plant *Achras sapota* of Central America is a mixture of 1,4-*trans*-poly(isoprene)s and triterpenes which are molecules containing six isoprene units. It is the base of some chewing gums. However, most chewing gums consist of poly(vinyl acetate), poly(vinyl ethyl ether), poly(isobutylene), poly(ethylene), or SBR.

Plants do not synthesize polyprenes from isoprene but by enzymatic polymerization of isopentenyl pyrophosphate with dimethylallyl pyrophosphate as initiator (Volume I). Like many biological polymerizations, the reaction is a polyelimination (condensating chain polymerization). The first step is

$$(6\text{-}13)\qquad CH_3{-}\underset{CH_3}{\underset{|}{C}}{=}CH{-}CH_2{-}O{-}P_2O_6^{3\ominus} + CH_2{=}\underset{CH_3}{\underset{|}{C}}{-}CH_2{-}CH_2{-}O{-}P_2O_6^{3\ominus} \longrightarrow$$

$$CH_3{-}\underset{CH_3}{\underset{|}{C}}{=}CH{-}CH_2{-}CH_2{-}\underset{CH_3}{\underset{|}{C}}{=}CH{-}CH{-}O{-}P_2O_6^{3\ominus} + H^{\oplus} + P_2O_7^{4\ominus}$$

It is unclear why some plants produce cis polymers and others trans polymers.

**Synthetic Poly(isoprene)s**

Isoprene is obtained mainly and increasingly from the $C_5$ cut of naphtha cracking (Table 3-16, Section 4.6.1) from which it is recovered by extractive distillation. The $C_5$ isoalkanes and isoalkenes are furthermore dehydrogenated to isoprene, albeit only after removal of 2-methylbutenes by reaction with sulfuric acid. Direct dehydrogenation of the $C_5$ isoalkane/isoalkene mixture would convert 2-methylbutenes to 1-methyl-1,3-butadiene (piperylene), $CH_3–CH=CH–CH=CH_2$, which is difficult to separate from isoprene by distillation and would adversely affect the polymerization of isoprene.

More than 50 chemical syntheses of isoprene are known; most are not economical (isoprene from propene: p. 116, II). Japan and Eastern Europe produce isoprene from isobutene and formaldehyde (p. 96, X; p. 116, II). The intermediary 4,4-dimethyl-1,3-dioxane here is catalytically cleaved to $CH_2=C(CH_3)–CH=CH_2 + H_2O + HCHO$. By-products of this reaction are polyols which are useful for the synthesis of polyurethanes (see Section 10.5). A small proportion of isoprene is still obtained from acetylene (p. 105, IX). Other industrial syntheses of isoprene start with propene (p. 107, III) or acetylene + acetone (p. 105, IX).

In principle, all four basic types of poly(isoprene)s (1,4-cis; 1,4-trans; 1,2; 3,4) can now be synthesized (Table 6-21). In practice, only the 1,4-cis polymer (**IR**) is important. IR is now produced in the United States, Japan, and Russia. Its world production (exclusive of Eastern Europe) peaked in 1980 at 220 000 t/a, then declined to 129 000 t/a in 1985, and is now approximately constant at ca. 150 000 t/a. There is also a production of a high 1,4-trans polymer in Japan.

Poly(isoprene) types with 90–92 % cis-1,4 structures were first synthesized by polymerizing isoprene with lithium or lithium alkyls (Table 6-21). These rubbers neither crystallized in the unstretched state nor sufficiently enough in the stretched one which led to insufficient green strengths (see p. 244). Lithium types are now replaced with titanium types that employ Ziegler catalysts.

Use of alkanes as solvents leads to polymers with gel contents of 20–35 %. The gels are probably caused by occasional 3,4-structures on the surface of catalysts; their fraction is independent of initiator concentration and monomer conversion. Only small proportions of gel are found in polymers from polymerizations in aromatic solvents because these solvents form complexes with Ziegler catalysts. These gels affect the processing but not the end-use properties of crosslinked polymers.

Table 6-21 Chemical structure of poly(isoprene)s from polymerizations at 25–30°C.

| Initiator | Solvent | % cis-1,4 | % trans-1,4 | % 3,4 | % 1,2 |
|---|---|---|---|---|---|
| $TiCl_4/Al(C_2H_5)_3$ | not disclosed | 98.5 | 1 | 0.5 | |
| Li, LiR | alkanes | 90 | 5 | 5 | |
| Na | alkanes | | 43 | 51 | 6 |
| K | alkanes | | 52 | 40 | 8 |
| Rb | alkanes | 5 | 48 | 39 | 8 |
| Cs | alkanes | 4 | 51 | 37 | 8 |
| $VCl_3/Al(C_2H_5)_3$ | alkanes | | 99 | | |
| $AlCl_3$ | ethyl bromide | | 93 | | |
| $BF_3$ | pentane | | 90 | | |
| Redox | water | | 95 | | |

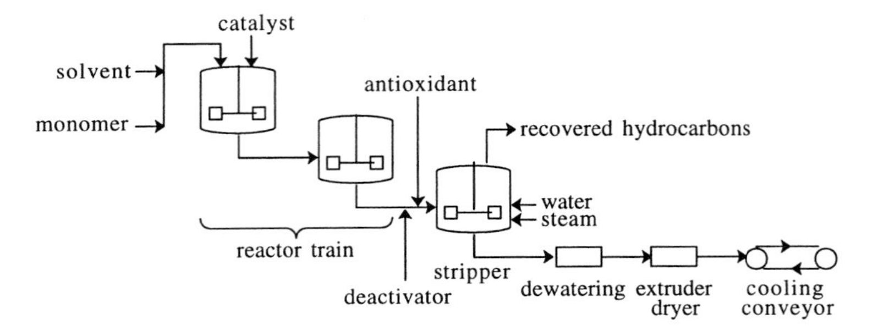


Fig. 6-18 Simplified flow scheme of the Ziegler–Natta polymerization of isoprene [10].

Figure 6-18 shows a flow scheme for the industrial polymerization of isoprene by Ziegler catalysts. Isoprene devoid of catalyst poisons (air, moisture, chemical impurities) enters a series of reactors where the catalyst is added. After the targeted monomer conversion, the polymerization is stopped by addition of a deactivator (shortstop). After addition of an antioxidant, alkanes and unreacted monomer are stripped by steam and water from the produced "polymer cement." The resulting slurry of rubber crumbs is dewatered, further dried during extrusion, and cooled on a conveyor belt. The rubber is then baled for shipment.

Table 6-22 compares the properties of vulcanized (= crosslinked) and carbon black-reinforced "synthetic natural rubber" (**natsyn**) with those of general-purpose grades of other rubbers. IRs have similar but not identical properties as natural rubber (Table 6-22). For example, natural rubber has a good inherent tack (= autohesion) and little tire heat build-up under loaded dynamic conditions (important for truck tires). SBR, on the other hand, is much more abrasion-resistant (important for tires of passenger cars).

Table 6-22 Characteristic properties of reinforced general-purpose elastomers. BR = poly(butadiene) (lithium type), *CR = non-reinforced poly(chloroprene), EPDM = ethylene-propylene-diene rubber, E-SBR = styrene-butadiene rubber (emulsion type with 40 % styrene units), IIR = poly(isobutylene) rubber, IR = synthetic 1,4-*cis*-poly(isoprene), NR = natural rubber, TPR = 1,5-*trans*-poly(pentenamer). Note that the use temperatures are higher than the melting temperatures because elastomers are crosslinked rubbers and not semicrystalline thermoplastics.

| Physical property | Physical unit | Reinforced elastomers from | | | | | | | |
|---|---|---|---|---|---|---|---|---|---|
| | | NR | IR | BR | TPR | E-SBR | IIR | EPDM | *CR |
| Density | g/cm$^3$ | 0.93 | 0.93 | 0.94 | | 0.94 | 0.93 | 0.86 | 1.23 |
| Melting temperature | °C | 28 | 2 | 1 | 20 | | −1.5 | | ≈ 60 |
| Glass temperature | °C | −73 | −72 | −95 | −97 | −50 | −66 | −55 | −42 |
| Use temperature, lower | °C | −60 | | −90 | | −40 | −30 | −35 | |
| , upper | °C | 120 | | 90 | | 140 | 190 | 180 | |
| Tensile strength at break | MPa | 32 | 26 | 14 | 18 | 29 | 22 | 13 | 20 |
| Elongation at break | % | 780 | 620 | 510 | 360 | 650 | 620 | 320 | 800 |
| 300 % modulus | MPa | 5.0 | 3.2 | 7.3 | 13 | 9.3 | 7.2 | 8.6 | 2 |
| Rebound elasticity | % | 40 | 40 | 65 | | 40 | 2 | 45 | |
| Hardness | Shore A | 50 | 55 | 60 | 64 | 60 | 55 | 65 | |
| Abrasion resistance | - | good | good | excellent | | very good | moderate | good | |

**Derivatives of Poly(isoprene)s**

Cis double bonds of natural rubber are isomerized to trans double bonds by sulfur dioxide from the heating of butadiene sulfone. This **isomerized natural rubber** crystallizes, however, and has therefore been replaced for most applications by non-crystallizing oil-extended SBRs.

Heating of natural rubber with proton donors at temperatures above 250°C cyclizes and degrades natural rubber to molecular weights of 2000–10 000. Examples are sulfuric acid, sulfonic acids, and chlorostannic acid ($H_2SnCl_6$). Other cyclizing agents are sulfonyl chloride ($RSO_2Cl$), metal halides, boron trifluoride, and especially phenol which is incorporated in part as an ether endgroup or substituted phenol (see also Eq.(6-15)); it also acts as an oxygen catcher. **Cyclized rubber** (*r*ubber *u*nits *i*somerized, **RUI**) contains mono-, di-, or triyclic structures that are separated from each other by methylene groups or non-cyclic isoprene units, depending on conversion and reaction conditions:

(6-14)

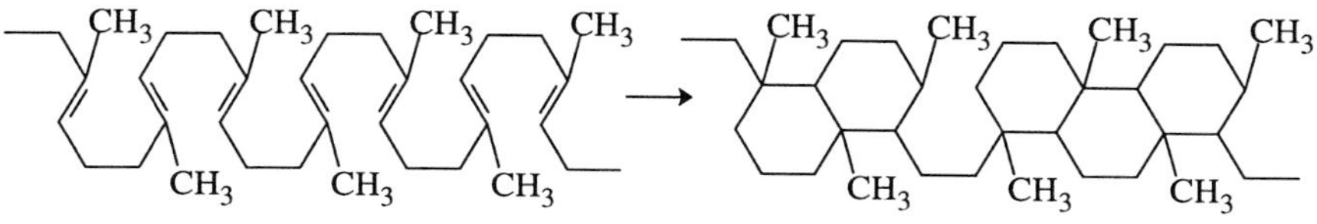

Cyclized rubber (also called isomerized rubber, but see first paragraph) has 50–90 % less double bonds than natural rubber of the same molecular weight. Depending on the pre-treatment, its properties resemble those of balata, gutta percha, or vulcanized rubber. Cyclized rubber is used as a binder for printing inks, lacquers, adhesives, etc.

Cyclic structures are also generated by reacting natural rubber with phenols and formaldehyde to **resin-modified natural rubber** which is used as a processing aid or reinforcing agent:

(6-15) $\xrightarrow[-\,CH_3OH]{+\,HCHO}$

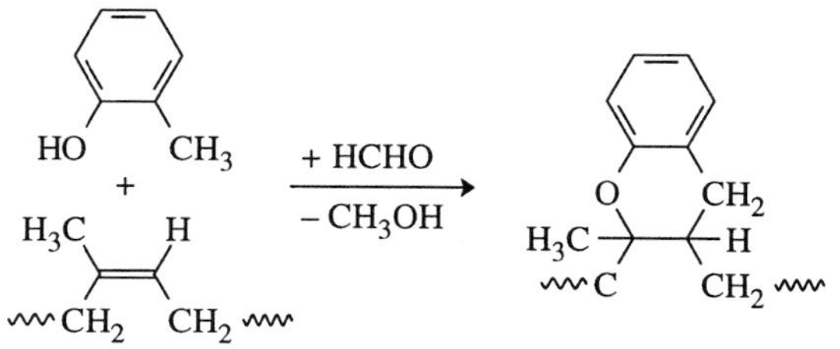

In the presence of $H_2SnCl_6$, hydrochloric acid adds to dissolved natural rubber to give **rubber hydrochloride** with $-CH_2-C(CH_3)Cl-CH_2-CH_2-$ units and a total chlorine content of ca. 30 wt% (Fig. 6-19). Because of its resistance to water, oil, and acids, it is used as protective coatings and transparent films for food packaging. However, stabilization against UV damage is required.

On heating, rubber hydrochloride splits off HCl and isomerizes the polymer to **iso rubber**, $\text{+}CH_2-C(=CH_2)-CH_2-CH_2\text{+}_n$, (Fig. 6-19).

Chlorination of natural rubber leads to **chlorinated rubber** with up to 65 wt% chlorine which indicates that chlorine adds not only to double bonds (maximum: 51 wt%) but also substitutes the polymer (Fig. 6-19). Spectroscopy shows the presence of cyclohexane structures. The polymers are adhesives for diene rubber/metal connections.

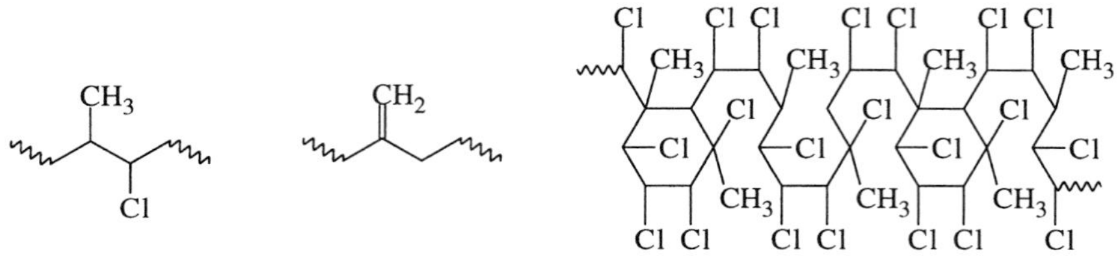

rubber hydrochloride iso rubber chlorinated rubber

Fig. 6-19 Structures of some modified isoprene rubbers.

Bromoalkanes ($CBrCl_3$, $CBr_4$, etc.) add to poly(isoprene)s in the presence of peroxides. The technical product is called **brominated isobutylene-isoprene rubber** (**BIIR**; **bromobutyl rubber**, **brominated butyl rubber**) with ca. 1.2 wt% bromine. BIIRs are elastic, vulcanizable, and flameproof. They are used as latex foams on fabrics or carpets.

A number of other modified natural rubbers and polyprenes are presently being evaluated. Examples are epoxidized natural rubber (**ENR**); reaction products with maleic anhydride; addition compounds with –N=O, –N=N–, >C=O, >C=S, or >C=C< (ene derivatives); addition compounds of $RS^•$, RSH, carbenes or nitrenes, etc.

### 6.3.4 Poly(2,3-dimethylbutadiene)

Poly(2,3-dimethylbutadiene), $+CH_2–C(CH_3)=C(CH_3)–CH_2+_n$, was first synthesized in 1912. During World War I, this **methyl rubber** was used as a substitute for natural rubber by the Central Powers (Austria-Hungary, Bulgaria, Germany, Turkey). An **H type** was obtained by enclosing the monomer in metal drums and keeping them in the sun for 1-1/2 to 3 months. The process produced a popcorn-like, white, crystalline mass that became rubber-like on grinding. It could only be used for *h*ard rubber, hence **H type**.

The **W type** for soft rubber (German: *W*eichgummi) resulted from polymerizing the monomer in 4000 L-tanks for 4–6 months at 75–85°C under pressure. Polymerization with 0.5 % sodium as initiator in the presence of carbon dioxide required "only" 2–3 weeks for a **B type** (German: beschleunigt = accelerated). After the end of World War I, production of methyl rubber ceased because of insufficient properties and high prices (see also Volume IV).

### 6.3.5 Poly(chloroprene)

Chloroprene = 2-chloro-1,3-butadiene, $CH_2=CCl–CH=CH_2$, is obtained by chlorinating butane, butene, or butadiene and subsequent dehydrochlorination (p. 114, V). The acetylene-based synthesis (p. 104, I) is no longer economical. Chloroprene monomer is extraordinarily reactive, especially with oxygen, and must therefore be stabilized by, e.g., *p*-*t*-butylpyrocatechol (TBC), phenothiazine, or nitrogen oxide as polymerization inhibitors and stored at low temperatures of 4–6°C.

Chloroprene was first industrially polymerized for ten days in the dark with oxygen from air as initiator (DuPont). Presently, it is polymerized exclusively free-radically in emulsion at 40–50°C. Since this polymerization is strongly inhibited by traces of oxygen, IG Farben tried in 1937 to remove the last traces of oxygen by adding reducing agents such as sodium dithionite, $Na_2S_2O_4$. Against all expectations, polymerizations were accelerated which led to the discovery of redox initiation.

Emulsion polymerizations of chloroprene are about 700 times faster than those of isoprene under the same conditions: complete polymerization is attained within one hour if saturated fatty acids are used as emulsifiers. Heats of polymerization must therefore be removed rapidly by strong external cooling or by use of flow tubes as reactors.

Heat flow can also be reduced by decreasing the rate of polymerization. For so-called sulfur-modified grades, this is achieved by the addition of sulfur which acts as a rate-reducing comonomer:

(6-16)

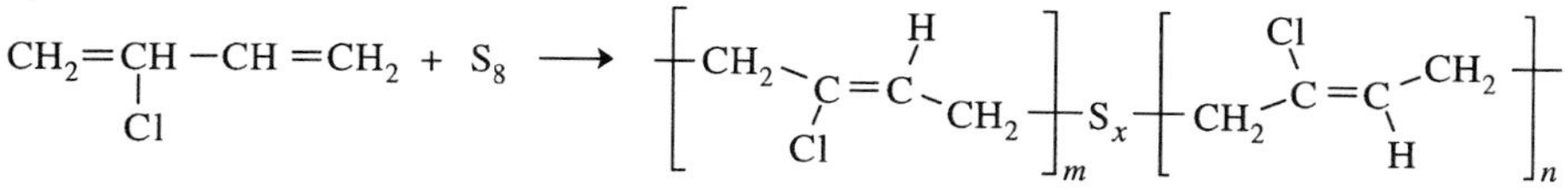

$$CH_2{=}C(Cl){-}CH{=}CH_2 + S_8 \longrightarrow \left[\left[{-}CH_2{-}C(Cl){=}CH{-}CH_2{-}\right]_m{-}S_x{-}\left[{-}CH_2{-}C(Cl){=}CH{-}CH_2{-}\right]\right]_n$$

The resulting gel-like crosslinked poly(chloroprene) is then treated with tetraethyl thiuram disulfide (TETD), $(C_2H_5)_2NC(=S)–S_2–C(=S)N(C_2H_5)_2$, which reacts with $–S_x–$ links of poly(chloroprene) by exchanging $–S_x–$ bonds and $–S_2–$ bonds. Poly(chloroprene) chains are thus split to $\sim CH_2–CCl=CH–CH_2–S_x–SC(=S)N(C_2H_5)_2$ chain ends. TETD was chosen because its surplus acts as a vulcanization accelerator (Volume IV).

Newer mercaptan grades are general purpose (GP) grades that are obtained by regulating polymerization rates and molecular weights through the addition of thiols as chain-transfer agents. These grades comprise many different configurational units (Fig. 6-20). Because they contain ca. 90 % trans-1,4 units, they crystallize rapidly on cooling or stretching. Vulcanized poly(chloroprene)s thus have very high tensile strengths without added reinforcing agents (Table 6-22). The rate of crystallization is lower for slow-crystallizing grades which have copolymeric units of 2,3-dichloro-1,3-butadiene.

Poly(chloroprene)s **CR** (formerly **GR-M**) are produced world-wide in amounts of ca. 300 000 t/a. Because of their polarities, they are also resistant against oils and solvents.

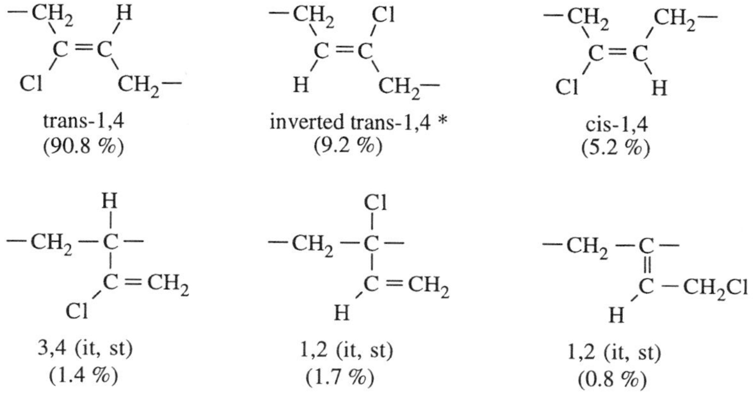

| $-CH_2(Cl)C{=}CH(CH_2-)$ | $-CH_2(H)C{=}C(Cl)CH_2-$ | $-CH_2(Cl)C{=}C(H)CH_2-$ |
|---|---|---|
| trans-1,4 (90.8 %) | inverted trans-1,4 * (9.2 %) | cis-1,4 (5.2 %) |
| $-CH_2-CH(-C(Cl){=}CH_2)-$ | $-CH_2-CCl(-CH{=}CH_2)-$ | $-CH_2-C(={CH-CH_2Cl})-$ |
| 3,4 (it, st) (1.4 %) | 1,2 (it, st) (1.7 %) | 1,2 (it, st) (0.8 %) |

Fig. 6-20 Percentage of trans-1,4 and isomerized structures (cis-1,4; 3,4; 1,2) in poly(chloroprene) from the emulsion polymerization of chloroprene at 40°C [11]. *) 4,1-structure.

# 6.4 Poly(cycloolefin)s and Poly(cyclodiolefin)s

## 6.4.1 Overview

Poly(cycloolefin)s result from the polymerization of cycloolefins and cyclodiolefins as well as from their copolymerization with other monomers. Industrially used monomers are cyclopentene, cyclopentadiene, norbornene, dicyclopentadiene, and cyclooctene, all byproducts of petrochemistry (see also Table 3-18).

Polymers and copolymers of these monomers have been known for more than 50 years, but polymers and copolymers of norbornene and cyclooctene appeared on the market only 30 years ago and cycloolefin copolymers only ca. 15 years ago. This gap between the discovery of polymers and effective industrial polymerizations occured because the catalysts were not only too expensive but also delivered polymers with insufficient properties in low time/space yields.

Cycloolefins polymerize either via double bonds with preservation of ring structures (vinyl type) to saturated polymers or by ring-opening (ring-opening type) to unsaturated ones. Vinyl polymerizations result in poly(cycloalkylene)s (= poly(alkene)s), whereas ring-opening polymerizations lead to poly(alkenylene)s, for example, in the polymerization of cyclopentene:

(6-17)

poly(cyclopentene) ⟵ vinyl polymerization — cyclopentene — ring-opening polymerization ⟶ poly(1,5-pentenylene)

The description of poly(1,5-pentenylene) as $\neq CH-CH_2-CH_2-CH_2-CH \neq_n$ reflects the ring-opening by metathesis more accurately than $+CH=CH-CH_2-CH_2-CH_2+_n$.

Vinyl polymerizations to poly(cycloalkylene)s may proceed via trans routes to threo polymers or via cis routes to erythro polymers (Volume I, Section 4.2.5; note that the erythro/threo nomenclature refers to constitutionally different adjacent stereogenic centers). Both cis and trans vinyl polymerizations may deliver diisotactic, disyndiotactic, or atactic polymers. Ring-opening polymerizations, on the other hand, result in polymers with cis-tactic and/or trans-tactic repeating units.

Industrially, poly(alkenylene)s are called **poly(alkeneamers)**. This designation is a remnant of a short-lived IUPAC nomenclature, now obsolete. According to this proposal, the systematic name of a polymer consisted of the systematic name of the monomeric unit, the prefix "poly", and the suffix "mer." The polymer by ring-opening of cyclopentene was thus called "poly(pentenamer)"; it is still the industrial designation. The systematic IUPAC name of this polymer was later changed to poly(1-pentenylene). According to the nomenclature of Chemical Abstracts Service, the systematic name of this polymer is now poly(1-pentene-1,5-diyl).

### 6.4.2 Poly(pentenamer)

Cracking of naphtha and heavy gas oils delivers a $C_5$ fraction (Table 3-18) from which cyclopentene is removed by extractive distillation together with cyclopentadiene and its dimer dicyclopentadiene (see also p. 117). Dicyclopentadiene dissociates thermally to cyclopentadiene which is then hydrogenated to additional cyclopentene.

Cyclopentene is polymerized industrially by $WCl_6/(C_2H_5)_3Al/C_2H_5OH$ to *trans*-poly-(pentenamer), $\text{+CH=CH–CH}_2\text{–CH}_2\text{–CH}_2\text{+}_n$ with ca. 89 % trans structures. The polymerization is a special case of a metathesis polymerization (see Volume I, p. 298). Such polymerizations are controlled by entropy; there is therefore practically no heat of polymerization. The polymer contains predominantly acyclic macromolecules and also some large rings.

The trans polymer ("1,5-trans-pentadiene rubber", **TPR**) is a general-purpose rubber that resembles both natural rubber and *cis*-1,4-poly(butadiene). Like NR, it has good autohesion (inherent tack). It also crystallizes on stretching; its high tensile strength (Table 6-23) thus increases with deformation. This effect is neither observed for *cis*-1,4-poly(butadiene) nor for styrene-butadiene rubbers.

Like *cis*-1,4-poly(butadiene), TPR can be highly filled with mineral oil and carbon black, is very resistant against thermal and mechanical degradation, and has high abrasion resistance. TPR can be blended in all ratios with other general-purpose rubbers.

*cis*-Poly(pentenamer)s are not industrial products. They can be obtained by polymerization of cyclopentene with $MoCl_5/R_2AlCl$.

### 6.4.3 Poly(octenamer)

Cyclooctene is obtained from the hydrogenation of 1,5-cyclooctadiene, the dimerization product of butadiene. It can be polymerized with similar metathesis catalysts as cyclopentene, either in solution by $WCl_6/RAlCl_2$ with or without $C_2H_5OH$ or in bulk by $WF_6/RAlCl_2$. The cis content of the polymer is regulated via the ratio of tungsten and aluminum in the catalyst. Chain transfer to 1-hexene adjusts molecular weights.

The commercial product is a 1,8-*trans*-poly(*o*ctenamer) *r*ubber (**TOR**) with 80 % trans units. Poly(octenamer)s with high cis contents have been developed but are not produced, probably because polymers with high cis contents crystallize rapidly and are therefore difficult to process on roll mills. Both trans and cis polymers are general-purpose rubbers that are used in blends with other rubbers.

### 6.4.4 Norbornene Polymers

Norbornene (systematic name: bicyclo[2,2,1]-2-heptene) is synthesized by the Diels–Alder reaction of ethene with either cyclopentadiene or dicyclopentadiene (p. 114):

(6-18) 2 [cyclopentadiene] + 2 $CH_2{=}CH_2$ ⟶ 2 [norbornene] ⟵ 2 $CH_2{=}CH_2$ + [dicyclopentadiene]

cyclopentadiene norbornene dicyclopentadiene

*R*ing-*o*pening *m*etathesis *p*olymerization (ROMP) of norbornene by $WCl_6/R_3Al/I_2$ or ruthenium or iridium catalysts delivers poly(norbornene):

(6-19)

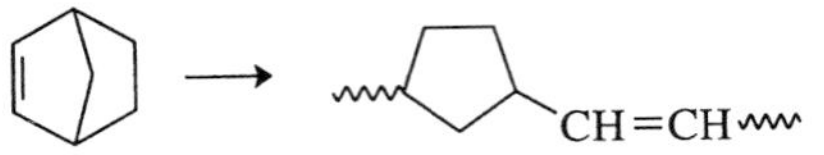

$$\longrightarrow \text{~~~}\!\!-\!\text{CH=CH~~~}$$

Poly(norbornene) (= poly(1,3-cyclopentylene vinylene)) (**PN**) has cis and trans double bonds and molecular weights of more than 2 000 000. It is a white, free-flowing powder with a powder density of ca. 0.30 g/cm$^3$. The raw polymer is a partially crystalline thermoplastic with a melting temperature of 180°C and a glass temperature of 46°C.

Poly(norbornene) is not offered as a thermoplastic, however, but as an oil-filled poly(norbornene) rubber (**PNR**). It was the first rubber that was developed specifically for the powder technology of rubber processing (see Volume IV). Porous PN powders can absorb up to ca. 400 parts of mineral oil per 100 parts of rubber. Typical PNRs consist of 100 parts of polymer, 220 parts of mineral oil, and 100 parts of carbon black, i.e., they contain only 24 wt% poly(norbornene). Poly(norbornene) rubbers can be conventionally vulcanized by sulfur (for properties see Table 6-23).

Vulcanized PNRs are used for compact, soft but strong sealants in the automobile and construction industries. Unvulcanized, highly filled PNRs serve as casting and insulation materials. PNs are also useful in removing oil slicks.

Polymerization of norbornene by zirconium-based Ziegler catalysts delivers insoluble polymers. Nickel catalysts lead to a "vinyl polymerization" via the carbon–carbon double bonds (Eq.(6-20)) to heptane-soluble polymers; industrial uses are not known.

(6-20)

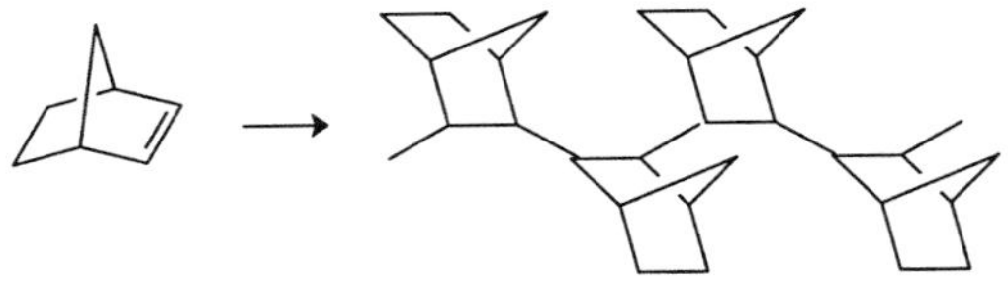

$$\longrightarrow$$

### 6.4.5 Cycloolefin Copolymers

Cycloolefin copolymers (**COC**) are copolymers of ethene and cycloolefins, usually with norbornene but sometimes with tetracyclododecene. The copolymerization of these monomers by Ziegler catalysts or zirconium-based metallocene catalysts + MAO proceeds via opening of the double bond of norbornene; the ring system remains intact:

(6-21)

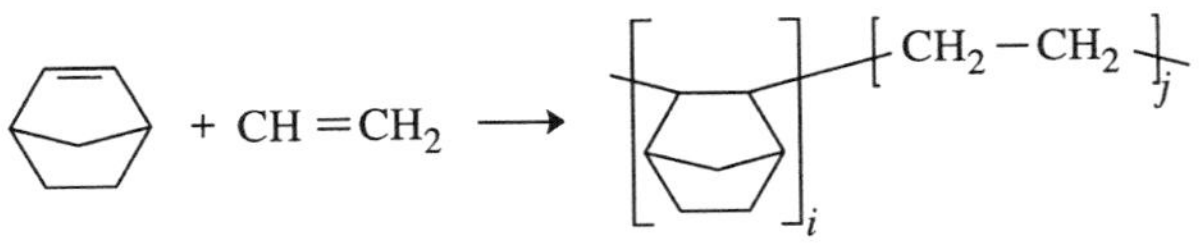

$$+ \ CH=CH_2 \longrightarrow [\ \ ]_i [CH_2-CH_2]_j$$

Commercial polymers contain ca. 50 % randomly distributed norbornene units. The size and distribution of these units prevent crystallizations: thin films pass 92 % of visible light. Since COCs also have high refractive indices ($n$ = 1.52) and low optical anisotropies, they are excellent materials for optical parts, for example, lenses.

Table 6-23 Properties of poly(cycloolefin)s. COC = Cycloolefin copolymer (copolymer of norbornene and ethene), PN = poly(norbornene), PNR = poly(norbornene) rubber, TPR = Poly(pentenamer) rubber (80 % trans), TOR = poly(octenamer) rubber. Physical properties of PNR, TOR, and TPR apply to oil-extended, carbon black-filled, vulcanized elastomers, not to raw rubbers.

| Property | Physical unit | Elastomers TPR | TOR | PNR | Thermoplastics PN | COC |
|---|---|---|---|---|---|---|
| Density | $g/cm^3$ | | 0.91 | 0.977 | 0.977 | 1.02 |
| Refractive index | 1 | | | 1.534 | 1.534 | 1.53 |
| Melting temperature (DSC) | °C | 44 | 55 | | 180 | |
| Glass temperature (DSC) | °C | –93 | –75 | –38 | 46 | 150 |
| Heat distortion temperature (0.46 MPa) | °C | | | | | 150 |
| Tensile modulus | MPa | | | | | 3100 |
| 100 % modulus (at 100 % elongation) | MPa | | | 0.4 | | |
| 300 % modulus (at 300 % elongation) | MPa | 13 | | 3.5 | | |
| Tensile strength at rupture or break | MPa | 17 | | 10 | | 66 |
| Elongation at rupture or break | % | 350 | | 500 | | 4 |
| Impact strength (Charpy) | $kJ/m^2$ | | | | | 15 |
| Notched impact strength (Charpy) | $kJ/m^2$ | | | | | 2 |
| Relative permittivity | 1 | | | | | 2.35 |
| Volume resistivity | Ω cm | | | | | > $10^{16}$ |
| Water absorption (24 h) | % | | | | | $< 0.01$ |

COCs also have low densities (80 % of that of polycarbonate A), low flexural creep moduli, good heat deflection, and low water absorptions (Table 6-23), which make them potential competitors of polycarbonates for CDs and DVDs. Their higher price and greater cycle time is detrimental, however.

COCs have very low dielectric loss factors of ca. $5 \cdot 10^{-5}$ at 1 kHz and high electric strengths which makes them useful as condensor films. Other applications comprise medical syringes, shrink films, coatings, and binders for toners.

Metatheses of norbornene by tungsten, molybdenum, or ruthenium catalysts proceed by ring-opening to **cycloolefin polymers (COP)**, $\text{+}(1,3\text{-}C_5H_8)\text{–}CH{=}CH\text{+}_n$, which are then hydrogenated to $\text{+}(1,3\text{-}C_5H_8)\text{–}CH_2\text{–}CH_2\text{+}_n$. These catalyst systems allow the copolymerization of cyclic olefins and functional monomers.

### 6.4.6 Poly(dicyclopentadiene)

Dicyclopentadiene (**DCPD**) polymerizes thermally at 250–280°C in aromatic solvents to poly(cyclopentadiene)s of undisclosed structure. The polymers are sometimes stabilized by hydrogenation. They are used as melt adhesives.

Dicyclopentadiene is also polymerized with simultaneous shaping by reaction injection molding (RIM). The catalyst system is created by mixing two solutions at elevated temperature (dissociation of DCPD to cyclopentadiene): one solution consisting of DCPD + $(C_2H_5)_2AlCl$ + Lewis base + antioxidants and the other of DCPD + $WCl_6$ + $WOCl_4$ + nonyl phenol + acetyl acetone. Nonyl phenol reacts with tungsten compounds to compounds with better solubilities. Acetyl acetone serves to increase the shelf-life of the solution. Both solutions also contain fillers.

The cycle time of this RIM process is ca. 1 minute. The resulting polymers are highly crosslinked (approximately 1 crosslink per 5 monomeric units):

(6-22)

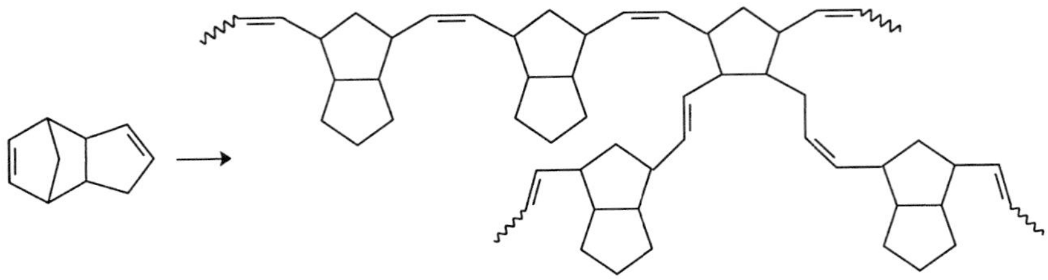

Another cyclopentadiene polymer is obtained from so-called bisdienes that result from the reaction of the sodium compound of cyclopentadiene with dichloro compounds, Cl–Z–Cl. The resulting bisdienes $C_5H_5$–Z–$C_5H_5$ oligomerize at elevated temperatures (see Eq.(6-23)). The oligomers serve as crosslinking agents for unsaturated polyesters at temperatures above 150°C.

(6-23)

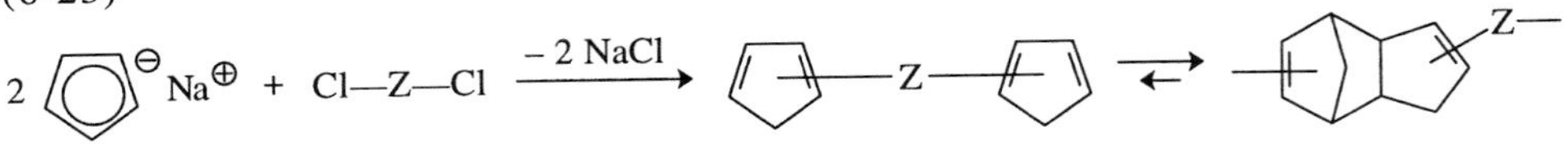

2 $C_5H_5^{\ominus}Na^{\oplus}$ + Cl—Z—Cl $\xrightarrow{-2\,NaCl}$ $C_5H_5$—Z—$C_5H_5$ ⇄ —[ ]—Z—

## 6.5 Acetylene Polymers

### 6.5.1 Poly(acetylene)s

Poly(acetylene)s, $\text{+CH=CH+}_n$, are polymers of acetylene, CH≡CH, that exist in three types: cis-transoidal, trans-cisoidal, and trans-transoidal:

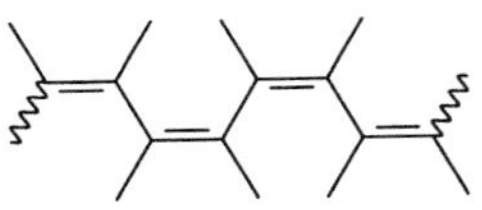

cis-transoidal 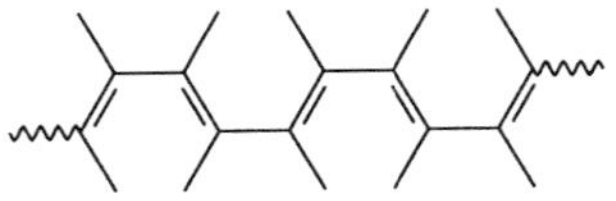 trans-cisoidal 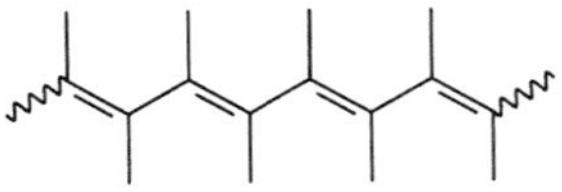 trans-transoidal

Poly(acetylene)s are produced by the Shirikawa process where acetylene is passed over a concentrated solution of $[(C_2H_5)_3Al]_4/Ti(Bu)_4$. The polymerization at –78°C leads to cis-transoidal polymers with 60–70 % cis contents, that at 45–85°C to trans-cisoidal ones, and that at 150°C and above to trans-transoidal ones. Heating at temperatures above 200°C converts cis polymers to trans polymers. All poly(acetylene)s are always accompanied by cyclo trimers.

All three types of poly(acetylene)s are highly crystalline, unmeltable, and insoluble in all known solvents. The insolubility is probably caused by spontaneous crosslinking reactions, either from Diels–Alder cycloadditions (Eq.(6-24)) or from crosslinking polymerizations by fragments of Ziegler catalysts (Eq.(6-25)).

(6-24)

(6-25) $$\begin{matrix} \sim CH=CH \sim \\ + \\ \sim CH=CH \sim \end{matrix} \xrightarrow[-\,(C_2H_5)_2AlH]{+\,(C_2H_5)_3Al,\ +\,HX} \begin{matrix} \sim CH(C_2H_5)-CH \sim \\ | \\ \sim CH(X)-CH \sim \end{matrix} \quad ;\ X = OH,\ Cl\ etc.$$

The *properties* of the cis-types (cis-transoidal and trans-cisoidal) correspond to those of poly(acetylene)s, whereas those of trans-transoidal ones resemble those of poly(cumulene)s, $\text{+}\!\!=\!C\!=\!C\!=\!\!\text{+}_n$. Films from *cis*-poly(acetylene)s have a copper-like shine whereas thin fibers are deep blue. Films from trans-poly(acetylene) are silvery-black with a metallic gloss in reflecting light whereas thin fibers appear deep red in transmitted light.

Poly(acetylene)s can be doped with $AsF_5$, $BF_3$, $I_2$, etc. This kind of doping refers to oxidizing (p-doping) or reducing (n-doping) reactions, and not to the creation of lattice defects or interstitial voids as in inorganic semiconductors. On doping with $AsF_5$, for example, the electrical conductivity of *trans*-poly(acetylene) increases tremendously to 1200 S/cm from $10^{-9}$ S/cm. Poly(acetylene)s are unsuited as semiconductors or conductors, though, because they are not stable against atmospheric gases.

Vaporization of graphite by lasers delivers fullerenes (Section 6.1.4) but vaporization in the presence of $CF_3$ or CN radicals $R^{\bullet}$ leads to true poly(acetylene)s, $R\text{+}C\equiv C\text{+}_nR$, with triple bonds in the chain and $CF_3$ or CN endgroups. Polymers with degrees of polymerization of 300-500 dissolve easily in most organic solvents.

### 6.5.2 Poly(diacetylene)s

**Poly(diacetylene)s**, $\text{+}C\equiv C\text{–}CR\text{=}CR'\text{+}_n$, with alternating double and triple bonds result from lattice-controlled polymerizations of solid diacetylenes, R–C≡C–C≡C–R', by ultraviolet light, high-energy radiation, heat, or pressure. The deep colors (yellow, red, or blue) of the polymers depend on the constitution of substituents R and R', respectively, and the planarity of chain conformations. On exposure to air and light, colors change slowly; this property is exploited for freshness codes on food packages as indicators of the expiration date.

## 6.6 Aromatic Poly(hydrocarbon)s

### 6.6.1 Poly(phenylene)s

Aromatic poly(hydrocarbons) are hydrocarbon polymers with aromatic rings in the main chain. The simplest polymers of this group are the poly(phenylene)s, also known as **polyphenyls**, **oligophenyls**, or **polybenzenes**. Poly(phenylene)s consist of phenylene groups that are interconnected in ortho, meta, or/and para positions. These polymers may also incorporate units of higher aromatic compounds such as naphthalene or anthracene.

Poly(phenylene)s comprise two major groups: linear poly(1,4-phenylene)s as basic materials for electrically conducting polymers and highly branched poly(phenylene)s for heat-stable thermosets.

### Branched Poly(phenylene)s

Benzene can be converted to poly(benzene)s in many ways, for example, by reaction of 1,4-dichlorobenzene and sodium with release of NaCl or by polymerization of 1,3-cyclohexadiene to 1,4-poly(cyclo-2,3-hexene) and subsequent aromatization to 1,4-poly(benzene) = poly(1,4-phenylene). These routes are only of academic interest.

For a while, benzene was industrially polymerized at mild temperatures by initiators such as $FeCl_3$, $AlCl_3/CuCl_2$, etc. The resulting brown to black, insoluble materials were then molded by high pressures to parts for high-temperature applications.

Soluble and melt-processable poly(phenylene)s result from oxidative-cationic polymerizations of terphenylene or terphenylene + biphenyl in the presence of benzene-1,3-disulfochloride. The resulting prepolymers (=oligomers) dissolve in chloroform or chlorobenzene. They are used either directly or with dispersed fillers as impregnating varnishes for laminates. Hardening (= crosslinking) to insoluble, unmeltable materials is by $BF_3$-diethyl ether, toluene sulfonic acid, or sulfuryl chloride.

Reactive branched oligophenylenes with acetylene endgroups result from the cyclotrimerization of *p*-diethinyl benzene, $HC{\equiv}C{-}C_6H_4{-}C{\equiv}CH$, or its copolytrimerization with phenyl acetylene, $C_6H_5{-}C{\equiv}CH$. The polymerization is stopped at a so-called B-stage of 86 % conversion since higher monomer conversions would lead to gelation (crosslinking). The resulting **H resins** can be loaded with 90 % fillers and crosslinked via acetylene groups by $TiCl_4/(C_2H_5)_2AlCl$, nickel acetylacetonate, or similar compounds. The structure of the crosslinks is unknown; they are probably not benzene rings that would have formed by trimerization of acetylene groups.

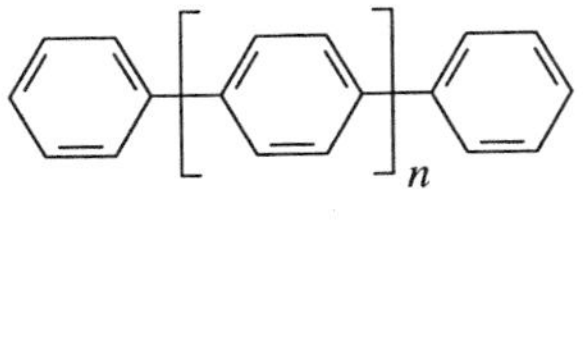

PPP 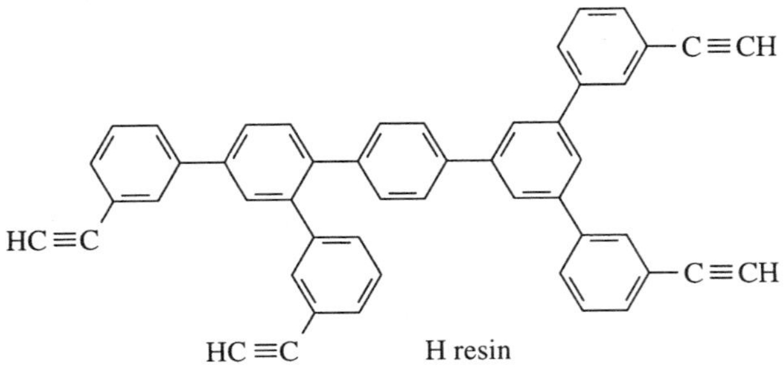 H resin

### Linear Poly(phenylene)s

Enzymatic oxidation of benzene by an enzyme ($E_1$) of the bacterium *Pseudomonas putida* leads to 5,6-dihydroxy-1,3-cyclohexadiene (I) if the successive oxidation to 1,2-dihydroxybenzene (catechol) (II) by an enzyme $E_2$ of the same bacterium can be suppressed by genetic manipulation. Genetically engineered *P. putida* produce only I, which is then excreted by the cell.

(6-26)

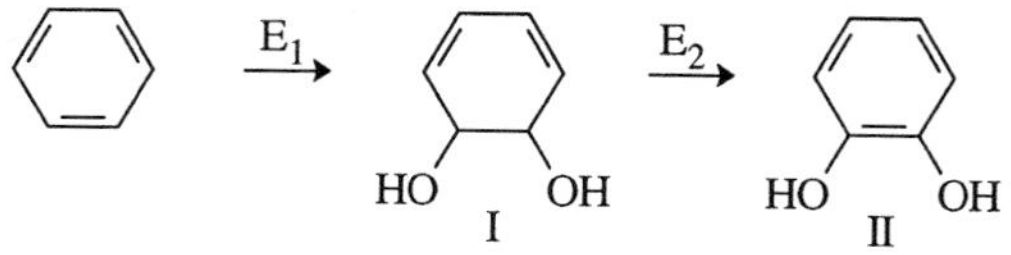

The dihydroxy compounds I can be esterified by standard procedures to the diester III. Free-radical polymerization of III delivers the soluble polymer IV which can be formed into films or spun into fibers. Heating these materials results in **poly(1,4-phenylene) = poly(*p*-phenylene) (PPP)** (V):

(6-27)

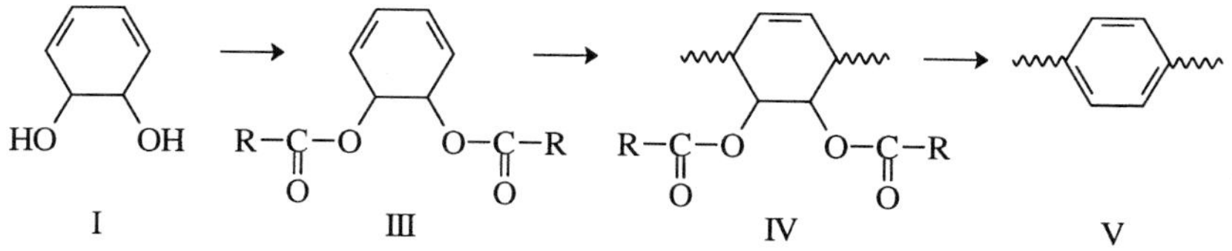

A totally synthetic route to PPP couples 1,4-dichloro benzoic acid esters. The resulting polymer is saponified and subsequently decarboxylated:

(6-28)

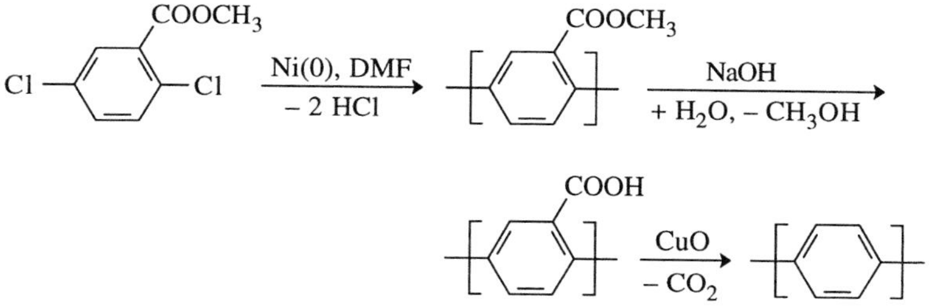

All these PPPs are oligomers; high-molecular weight linear poly(1,4-phenylene) (PPP) is not known. A truly high-molecular weight PPP should have a melting temperature of ca. 1200°C according to estimates based on the variation of melting temperatures of linear PPP oligomers with known degrees of polymerization.

Benzoyl and 4-phenoxybenzoyl substituted poly(1,4-phenylene)s with estimated degrees of polymerization of 100 were obtained by undisclosed processes. These soluble polymers carry their substituents R ($C_6H_5$–CO–, $C_6H_5O$–(1,4-$C_6H_4$)–CO–) in various positions and are said to have tensile moduli of 7–17 GPa.

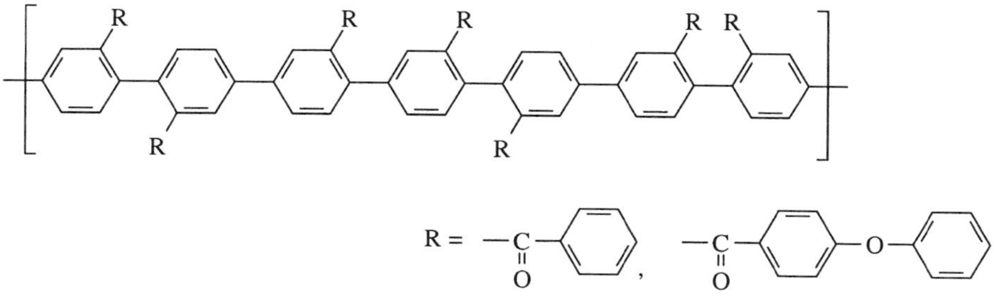

### 6.6.2 Poly(*p*-xylylene)s

Poly(*p*-xylylene) (**PPX**) is the conventional name of poly(1,4-phenylene ethylene) = poly(1,4-phenylene-1,2-ethanediyl), the polymer of the unstable *p*-xylylene (*p*-quinodimethane) (**PX**). This monomer is an intermediate in the pyrolysis of *p*-xylene (X). On quenching the reaction products with liquid xylene, PX converts to its cyclic dimer, [2.2]-*p*-cyclophane = di-*p*-xylylene (**DPX**).

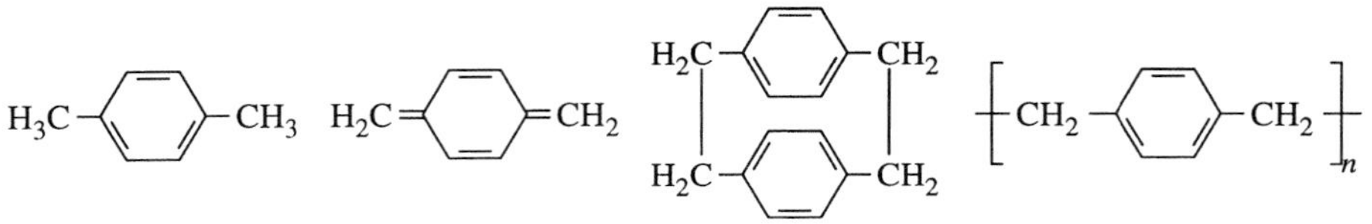

$H_3C-C_6H_4-CH_3$ $H_2C=C_6H_4=CH_2$ DPX $[-CH_2-C_6H_4-CH_2-]_n$

X PX DPX PPX

The technical route to PPX heats DPX stepwise, first to 200°C at 133 Pa and then to 680°C at ca. 67 Pa, which causes DPX to dissociate quantitatively to PX. On cooling to 25°C at 13 Pa, PX deposits on surfaces that are cooled to –70°C, polymerizes to PPX, and forms a solid layer (**Gorham process**):

(6-29)

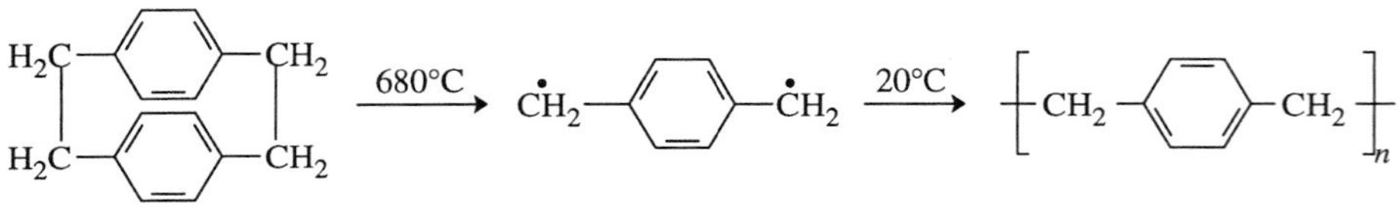

$$\text{DPX} \xrightarrow{680°C} \dot{C}H_2-C_6H_4-\dot{C}H_2 \xrightarrow{20°C} [-CH_2-C_6H_4-CH_2-]_n$$

The polymerization proceeds via living biradicals. Immediately after polymerization, electron spin resonance measurements indicate the presence of unpaired electrons. The concentration of unpaired electrons corresponds to a molecular weight of ca. 10 000 if one assumes 2 free radicals per PPX molecule (Eq.(6-29)). Exposure of PPX to air results in a loss of unpaired electrons by branching and crosslinking reactions.

DPX is known industrially as DPX-N. Also available are di-*p*-xylenes with 1 chlorine atom (DPX-C) and 2 chlorine atoms (DPX-D) per molecule and thus also PPX-N, PPX-C, and PPX-D. The number of chlorine atoms per DPX molecule and PPX monomeric unit, respectively, indicates *average* numbers. Since chlorine substituents are randomly distributed, each individual benzene moiety can thus carry between 0 and 4 of them.

Poly(*p*-xylylene)s are prepared exclusively as films on surfaces by the Gorham process. These films are produced by a gas-phase process and not from solution as other surface coatings. Hence, there is no effect of surface tension on film formation.

The Gorham process allows one to produce pore-free coatings of constant thickness even on complicated three-dimensional parts. The resulting PPX-N, PPX-C, and PPX-D polymers have high melting temperatures, good continuous service temperatures, and good mechanical and electrical properties (Table 6-24). They are also not very permeable to gases and water vapor. PPX polymers are therefore mainly used for intermediate layers in electric condensors and also for printed circuits, thin membranes, and medical implants.

Table 6-24 Properties of poly(*p*-xylylene)s.

| Property | Physical unit | PPX-N | PPX-C | PPX-D |
|---|---|---|---|---|
| Density (23°C) | $g/cm^3$ | 1.11 | 1.29 | 1.42 |
| Refractive index | 1 | 1.661 | 1.639 | 1.669 |
| Melting temperature (DSC) | °C | 420 | 290 | 380 |
| Glass temperature (DSC) | °C | 13 | | |
| Continuous service temperature (in air) | °C | 100 | 130 | |
| Linear thermal expansion coefficient | $K^{-1}$ | $6.9 \cdot 10^{-5}$ | $3.5 \cdot 10^{-5}$ | |
| Specific heat capacity (25°C) | $J\ K^{-1}\ g^{-1}$ | 1.3 | 1.0 | |
| Thermal conductivity (20°C) | $kW\ m^{-1}\ K^{-1}$ | 12 | 8.2 | |
| Tensile modulus, 23°C | MPa | 2400 | 3200 | 2800 |
| 200°C | MPa | 170 | 170 | 170 |
| Flexural modulus | MPa | 2450 | 2800 | 2800 |
| Tensile strength at yield | MPa | 42 | 55 | 60 |
| Tensile strength at fracture | MPa | 45 | 70 | 75 |
| Fracture elongation | % | < 30 | 200 | < 10 |
| Hardness (Rockwell) | - | 85 | 80 | |
| Relative permittivity (23°C), 60 Hz | 1 | 2.65 | 3.15 | 2.84 |
| 1 MHz | 1 | 2.65 | 2.95 | 2.80 |
| Electrical resistance, 50 % RH | Ω | $1 \cdot 10^{13}$ | $1 \cdot 10^{14}$ | $5 \cdot 10^{16}$ |
| Resistivity, 50 % RH | Ω cm | $1.4 \cdot 10^{17}$ | $8.8 \cdot 10^{16}$ | |
| Electric strength | kV/mm | 260–280 | 145–220 | 200–215 |
| Dissipation factor, 60 Hz | 1 | 0.0002 | 0.020 | 0.004 |
| 1 MHz | 1 | 0.0006 | 0.013 | 0.002 |
| Permeability ($10^{14}$ *P*) $N_2$ | $cm^2\ s^{-1}\ Pa^{-1}$ | 0.35 | 0.020 | 0.20 |
| $O_2$ | $cm^2\ s^{-1}\ Pa^{-1}$ | 1.76 | 0.32 | 1.44 |
| $Cl_2$ | $cm^2\ s^{-1}\ Pa^{-1}$ | 3.32 | 0.016 | 0.025 |
| $CO_2$ | $cm^2\ s^{-1}\ Pa^{-1}$ | 9.64 | 0.35 | 0.58 |
| $H_2S$ | $cm^2\ s^{-1}\ Pa^{-1}$ | 35.7 | 0.58 | 0.065 |
| $SO_2$ | $cm^2\ s^{-1}\ Pa^{-1}$ | 85.1 | 0.49 | 0.21 |
| $H_2O$ | $cm^2\ s^{-1}\ Pa^{-1}$ | 428 | 150 | |

### 6.6.3 Phenolic Resins

Phenolic resins = **phenol-formaldehyde resins** (**PF**) are usually condensation products of phenol and formaldehyde and sometimes also of other aldehydes. The formation of a resin from phenol and formaldehyde was described by Adolf von Bayer as early as 1872. Laccain®, a resin of phenol, formaldehyde, and tartaric acid as a condensation promoter was offered commercially in 1902 as a substitute for the expensive shellac (p. 78). However, molded parts were difficult to produce. The commerical breakthrough came in 1907 with the "heat-and-pressure" patent of Leo H. Baekeland.

Acid catalysis of the reaction of phenol with less than stoichiometric amounts of formaldehyde delivers so-called **novolacs** (= new (shel)lac; L: *novus* = new; Italian: *lacca* = lacquer (see p. 78)).

In contrast to this, base catalysis of the reaction of phenol with a surplus of formaldehyde results first in soluble **resols** (**A stage**), then in still soluble and meltable higher molecular weight **resitols** (**B stage**), and finally in insoluble **resites** (**C stage**).

**Acid Catalysis**

Protonation of formaldehyde, $CH_2O$, leads to the methylol cation $^{\oplus}CH_2OH$ which reacts slowly with phenol to *p*- or *o*-methylol phenols, for example,

(6-30) [phenol $\xrightarrow{+\ ^{\oplus}CH_2OH}$ o-($CH_2OH$, H) cyclohexadienyl cation $\xrightarrow{-\ H^{\oplus}}$ o-methylol phenol]

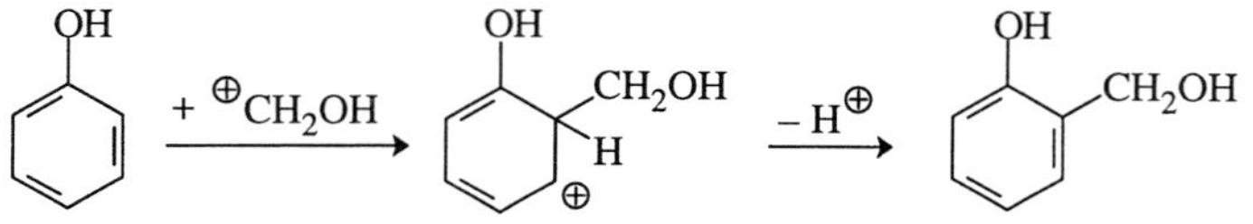

Methylol phenols cannot be isolated in the presence of protons. They react rather rapidly to methylene compounds, Eq.(6-31), and also to open-chain formals, Eq.(6-32):

(6-31) [o-$CH_2O^{\oplus}H_2$ phenol $\underset{+\ H_2O}{\overset{-\ H_2O}{\rightleftharpoons}}$ o-$CH_2^{\oplus}$ phenol] $\xrightarrow{+\ C_6H_5OH}$ [σ-complex] $\xrightarrow{-\ H^{\oplus}}$ [o,o'-methylene bisphenol]

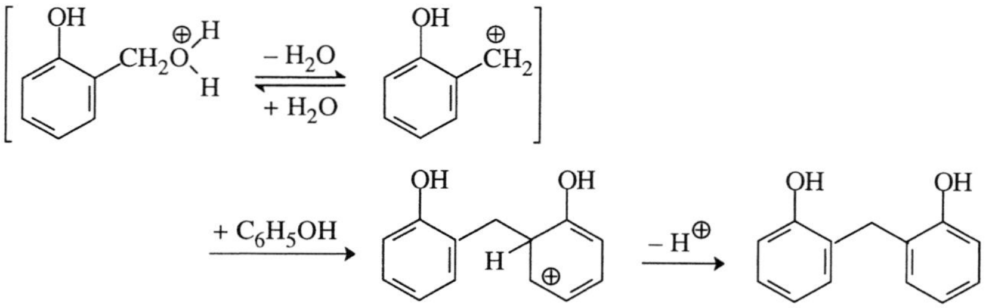

(6-32) 2 [o-methylol phenol] $\underset{-\ H_2O}{\xrightarrow{+\ HCHO}}$ [bis(o-hydroxybenzyloxy)methane]

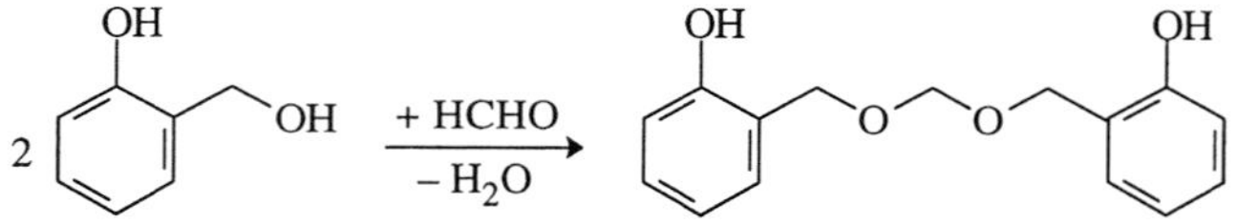

The reaction products of Eqs.(6-31) and (6-32) proceed further to **novolacs** which are *o,o*', *o,p*, and *o'p* branched oligomers with average molecular weights of ca. 1000.

For applications, novolacs should have mainly *o,o'* links (as shown for a section in Eq.(6-33)) and not be *o,p* or *o',p* substituted. The para substitution of phenols to methylols and subsequently to novolacs can be reduced if lower proton concentrations are employed in the novolac synthesis because intermediate *o*-methylol phenols are here stabilized by short-lived hydrogen bridges. Stabilities and yields of these compounds are increased further if chelating bivalent metals are added.

Novolacs, shown here as an *o,o*' section, are then crosslinked with hardeners such as hexamethylene tetramine (hexa, urotropin), Eq.(6-33). This hardening reaction is faster in the para than in the ortho position, which is the reason why one wants to have *o,o'*-substituted novolacs and not *o,p* and *o',p* ones (see preceding paragraph).

(6-33) [~~$CH_2$–phenol–$CH_2$~~ (section of a novolac) $\xrightarrow{+\ C_6H_{12}N_4}$ two novolac phenol units linked para via $CH_2$—NH—$CH_2$]

section of
a novolac

**Base Catalysis**

In base-catalyzed reactions, phenolate anions are nucleophilically added to formaldehyde, for example, in the ortho position, in a fast reaction to relatively stable methylols:

(6-34) $C_6H_5O^{\ominus} \longleftrightarrow$ (H, $\ominus$) $\xrightarrow{+ \text{HCHO}}$ ($O^{\ominus}$, $CH_2OH$) $\longrightarrow$ (OH, $CH_2O^{\ominus}$)

In basic environments, ortho and para positions have about the same reactivity. The reaction then proceeds slowly to A, B, and C stages by etherification of the methylol groups and, at elevated temperatures, also by formation of methylene bridges.

The reaction is better controlled if it is stopped at the B stage shortly before the conversion reaches the gel point (see Volume I). Addition of proton acids protonates methylol groups because they are more basic than hydroxyl groups. The resulting oxonium ions split off water and form benzylcarbonium ions which then crosslink molecules with at least two nucleophilic groups HY.

Y can be O-alkyl, S-alkyl, NH-alkyl, or a CH-acidic group. The acid-catalyzed hardening of resols and resitols leads mainly to methylene bridges because the activation energy for the formation of a methylene group is only one half of that of the reaction to an ether group. So-called uncatalyzed hardenings are crosslinkings wihout added acid or base that proceed like acid- or base-catalyzed hardenings and not via quinone methides as postulated in the past. In the absence of oxygen, significant proportions of quinone methides form only at temperatures above 600°C.

**Properties**

In the presence of oxygen, quinone methide endgroups are formed at lower temperatures:

(6-35) $\sim\!\!\sim C_6H_4{-}CH_2{-}C_6H_4{-}OH \xrightarrow[-\,H_2O]{+\,(1/2)\,O_2} \sim\!\!\sim C_6H_4{-}CH{=}C_6H_4{=}O$

Phenolic resins become yellow if quinone methides are present. The yellowing can be prevented if phenol groups are blocked, for example, by esterification.

The yellow color of most phenolic resins does not result from the presence of quinone methides, however. Rather, it is caused by a side reaction during the curing of phenolic resins by hexa which leads to the formation of secondary amines. The resulting imine structures, $-CH_2-NH-CH_2-$, dehydrogenate to azomethines, $-CH_2-N{=}CH-$, which cause the yellow color, especially, if imines are present in endgroups, $-CH_2-NH-CH_3$:

(6-36) $\sim\!\!\sim C_6H_4{-}CH_2{-}NH{-}CH_2{-}C_6H_4 \sim\!\!\sim \xrightarrow[-\,H_2]{}$

$\sim\!\!\sim C_6H_4{-}CH_2{-}N{=}CH{-}C_6H_4 \sim\!\!\sim$

Hardening of unfilled **resols** (**A stage**) in molds at elevated temperatures leads to translucent materials which are used, for example, for handles of knives. Acidic hardening of resols by phosphoric acid or aromatic sulfonic acids in the presence of benzyl alcohol results in acid-proof cements. Hard foams are obtained by hardening with benzene sulfonic acid in the presence of sodium hydrocarbonate as the $CO_2$ source.

**Resitols** (**B stage**) are used as adhesives, either as such or as resitol-impregnated papers, and in laminates with paper, wood, or fabrics. In World War II, the British fighter-bomber Mosquito was made from resitol-glued plywood.

Phenolic resins have good thermal, electrical, and flame resistance. They are hard, rigid, strong, and dimensionally stable. Phenolic resins are used not only as filled molding compounds for parts, fixtures, tubes, etc., but also as tanning agents, shell mold binders, and vulcanizing agents.

Phenolic resins can be modified in many ways, for example, for lacquers. **Novolacs** dissolve only in polar solvents such as alcohol, acetone, and lower esters. Such spirit varnishes have only limited use; they are also relatively brittle after drying. So-called **plastified** and **elastified phenolic resins** are either partially etherified (e.g., by *t*-butanol) or esterified (e.g., by fatty acids) or both etherified and esterified, e.g., by adipic acid and trimethylol propane). Plastified phenolic resins have improved elasticities and are soluble in aromatic solvents. Since they are also compatible with vinyl polymers and fatty acids, they can be used as baking varnish.

Alcohol-soluble novolacs and plastified phenolic resins do not dissolve in drying oils, such as linseed oil, however. More soluble are the so-called **modified phenolic resins** that contain glycerol esters of abietic acid (see p. 78). These resins dry better than copal-linseed resins.

**Elastified phenolic resins** contain either additional or different phenolic compounds. With bisphenol A, $HO(1,4\text{-}C_6H_4)\text{–}C(CH_3)_2\text{–}(1,4\text{-}C_6H_4)OH$, as the phenol component, one obtains not only better solubilities in drying oils but also a reduced tendency to yellow in light and air because of a smaller proportion of easily attacked methylene groups. Such resins can be combined with drying oils for use as baking varnishes. Even better elastifications are obtained with bisphenols or polyphenols that contain elastic segments. Non-saponifiable baking varnishes result from the $ZnCl_2$-moderated condensation of phenol and highly chlorinated $C_{15}$–$C_{30}$ paraffins and subsequent reaction with formaldehyde to resols.

Cocondensation of phenol and formaldehyde with phenols that carry sulfonic acid, carboxylic acid, or amino groups leads to ion-exchange resins. Space rockets have a coating of phenolic resins that carbonizes by the heat of friction and becomes an excellent heat shield.

Melt spinning of novolacs delivers fibers that are subsequently hardened with formaldehyde gas or solutions. These yellowish fibers are used for flame-retardant clothing and blankets. White phenolic fibers are obtained if endgroups are acetylated (see above).

Phenolic resins are also offered as free-flowing powders or stable aqueous dispersions. These types are obtained by condensation of phenol and formaldehyde in aqueous alkali or alkaline earth hydroxide solutions to predetermined molecular weights. Protective colloids, such as polysaccharide gums, are added as soon as the two-phase systems forms. Further reactions include condensation to oligomeric rings.

The world production of phenolic resins is now ca. $3 \cdot 10^6$ t/a (2003).

### 6.6.4 Poly(armethylene)s

Condensation of aralkyl halides or aralkyl ethers such as $\alpha,\alpha'$-dimethoxy-$p$-phenol with phenols or other aromatic, heterocyclic, or organometallic compounds in the presence of Friedel–Crafts catalysts delivers poly(arylene)s that are known as **phenol aralkyl resins**. The structure of these resins resembles that of phenol-formaldehyde resins:

(6-37) (n + 2) [phenol, OH] + (n + 1) $CH_3OCH_2-C_6H_4-CH_2OCH_3$ $\xrightarrow[-(2n+2)\ CH_3OH]{}$ $HO-C_6H_4+CH_2-C_6H_4-CH_2-C_6H_3(OH)+_n CH_2-C_6H_4-CH_2-C_6H_4-OH$

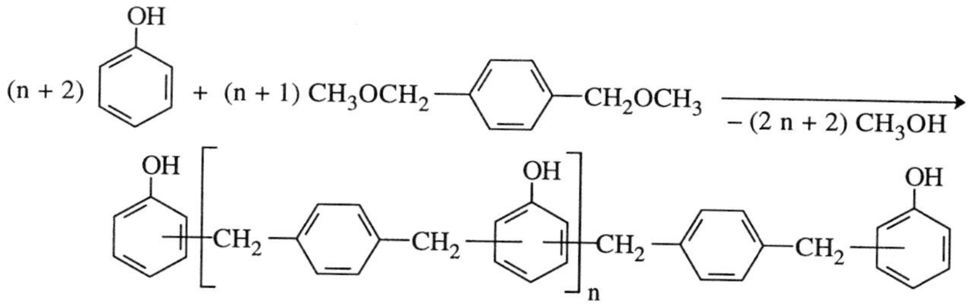

The resins have average degrees of polymerization of $n$ = 1.6. Prepolymers are offered as powders or as solutions in 2-ethoxyethanol with 50–60 % solids. Commercial solutions always contain hardeners (= crosslinking agents) such as hexamethylenetetramine or polyepoxides. They are used for composites with glass or carbon fibers.

### 6.6.5 Poly(benzocyclobutene)s

Benzocyclobutene (cyclobutabenzene, cardene; BCB) is the conventional name of bicyclo[4.2.0]octa-1,3,5-triene (= 1,2-Dihydrobenzocyclobutadiene), a chemical compound with a *saturated* butane ring (II). The true benzocyclobutene (I) of Chemical Abstracts Service has an unsaturated butane ring.

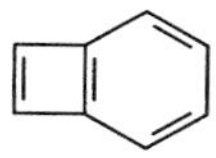

benzocyclobutene (CAS) I

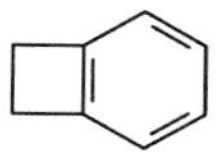

benzocyclobutene (conventional) II

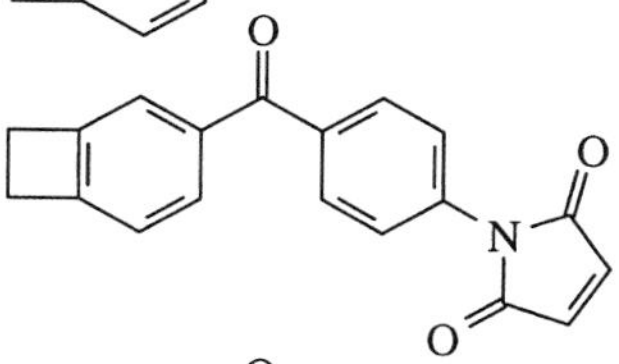

4-maleimidephenyl-4-benzocyclobutenylketone III

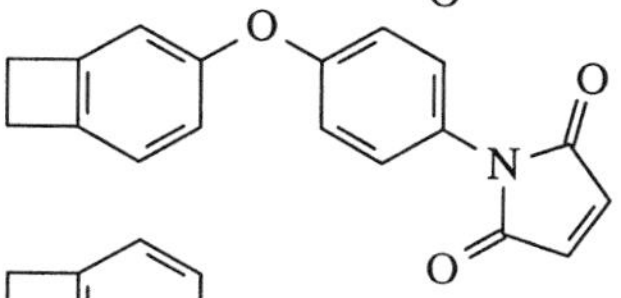

4-maleimidephenyl-4-benzocyclobutenylether IV

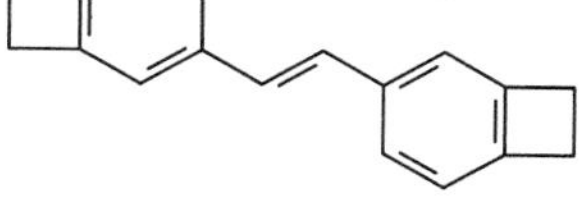

1,4-bis(4-benzocyclobutenyl)-2-butene V

On heating, conventional benzocyclobutene is thermally activated to *o*-quinodimethane (OQDM) = *o*-xylylene (OX):

(6-38) BCB (II) ⇌ OQDM, OX

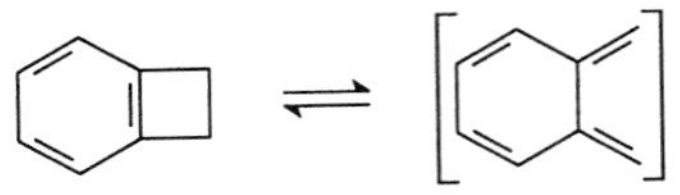

OQDM is unstable and adds easily to many carbon–carbon double bonds, forming Diels–Alder products, for example, with isoprene or maleic anhydride.

Benzocyclobutanes II–IV polymerize without catalyst at 200–250°C (II) or at lower temperatures (III–V) without forming volatile products. The polymerization mechanism is unknown. Benzocyclobutene II leads to 1,2:5,6-dibenzooctane and apparently also to linear oligomers, $+CH_2(1,2\text{-}C_6H_4)CH_2+_n$. III–V deliver crosslinked products with good properties (Table 6-25). None of these polymers has been commercialized.

Table 6-25 Properties of poly(benzocyclobutene)s [12].

| Property | Physical unit | Homopolymer of III | IV | V |
|---|---|---|---|---|
| Melting temperature of monomer | °C | 152 | 116 | |
| Density | g/cm$^3$ | 1.30 | 1.2 | |
| Glass temperature (DSC) | °C | 317 | 260 | > 350 |
| Linear thermal expansion coefficient, $T < T_G$ | K$^{-1}$ | $4.3 \cdot 10^{-5}$ | $6.0 \cdot 10^{-5}$ | |
| $T > T_G$ | K$^{-1}$ | $19.3 \cdot 10^{-5}$ | $25.0 \cdot 10^{-5}$ | |
| Flexural modulus | MPa | 3240 | 3500 | 5150 |
| Flexural strength | MPa | 207 | 180 | |
| Elongation at break | % | 6 | | |
| Critical stress intensity factor, $K_{IC}$ | MPa/m$^{1/2}$ | 1.59 | 2.31 | |
| Surface fracture energy, $G_{IC}$ | J/m$^2$ | 780 | 1530 | |
| Relative permittivity | 1 | 3.15 | | 2.7 |
| Dielectric loss factor (1 MHz) | 1 | 0.0026 | | 0.0004 |
| Water absorption (equilibrium) | % | 4.2 | | 0.87 |

## 6.7 Other Poly(hydrocarbon)s

### 6.7.1 Coumarone-Indene Resins

The tar fraction with a boiling range of 160–185°C (naphtha) is rich in cycloparaffins. It also contains up to 60 % polymerizable compounds such as 20–30 % coumarone (= benzo[b]furan) and considerable proportions of indene.

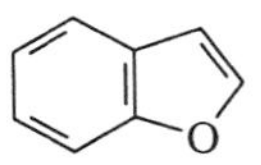

benzo[b]furan 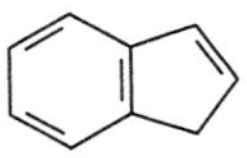 indene 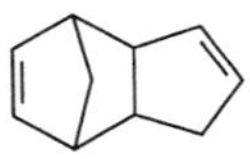 dicyclopentadiene

Benzofuran and indene have very similar boiling temperatures (174°C versus 182°C). They are therefore not separated but polymerized together by $H_2SO_4$, $H_3PO_4$, or $AlCl_3$ in solution in naphtha to resins with molecular weights between 1000 and 3000. The polymerization proceeds predominantly via carbon–carbon double bonds of the five-membered rings. Naphtha is then evaporated.

The resulting yellow to black thermosetting polymers are one of the oldest known plastics. The resins were formerly called **coumarone resins** and are also known as **poly-coumarone resins** or **indene resins**. They are used as adhesives, binders for papers, paints, waterproofing compounds, and the like. The polymers discolor in light and by air which can be prevented by post-hydrogenation. World production is ca. 100 000 t/a.

## 6.7.2 Resin Oils

Cracking of crude gasoline or gas oil in tube reactors delivers so-called resin oils consisting of a $C_8$–$C_{10}$ hydrocarbon fraction that contains both inert hydrocarbons (xylenes, naphthalenes, etc.) and polymerizable compounds (styrene, $\alpha$-methyl styrene, vinyl toluenes, indene, methyl indenes, dicyclopentadiene, etc.). Resin oils are directly polymerized by Friedel–Crafts catalysts. The yield of the resulting resins is more than 100 % of the polymerizable components since this process also alkylates some inert hydrocarbons. Copolymerization of resin oils with drying oils delivers easy-drying lacquers with good gloss and hardness.

## 6.7.3 Pinene Resins

$\alpha$-Pinene and $\beta$-pinene are components of turpentine (p. 77), a yellowish oil from pines. $\beta$-Pinene polymerizes thermally under a nitrogen blanket to crystalline polymers:

(6-39)

Cationic polymerization of $\beta$-pinene with 20 % isobutene delivers impact thermoplastics and with more than 90 % isobutene, vulcanizable rubbers.

On cationic polymerization, $\alpha$-pinene is first isomerized to dipentene (D,L-limonene) which is the true monomer:

(6-40)

### 6.7.4 Polymers from Unsaturated Natural Oils

Unsaturated natural oils are used directly as the base of oil paints (Section 3.10). They also serve as base chemicals for linoxyn and factice (**vulcanized oil**).

**Linoxyn** results from the polymerization of *lin*seed oil at 60°C in the presence of *oxygen*. Its mixture with colophony or copal resins is then homogenized at 150°C to the tough, gel-like linoleum cement which is mixed with fillers, pigments, and an oleoresin binder (p. 76), rolled on jute, burlap, or canvas fabrics, and hardened by oxygen to **linoleum** (from *linen* and *oleum* since the first linoleum was produced from linen and oils).

**Factice** (L: *factitius* = artificial) is obtained by vulcanization of fatty oils (linseed oil, ricinus oil, soybean oil, rapeseed oil; see p. 79). *Brown factice* results from heating the oil with sulfur for 6–8 hours at 130–200°C. It is a soft, crumb-like, elastic material with sulfur contents of 5–20 %. For *white factice*, rapeseed oils are treated with disulfur dichloride at room temperature. White factice contains up to 25 % sulfur; it is not elastic. *Black factice* has mineral bitumen added. Factices are used as low-cost extenders for rubber articles, for example, to improve the dimensional stability of calender films.

## 6.8 Vinyl Polymers

### 6.8.1 Overview

Vinyl polymers (**polyvinyls**) are polymers of the type $\text{+CH}_2\text{–CHY+}_n$ where Y is an aromatic substituent, a halogen, or a group that is connected to the carbon chain of these polymers via an oxygen, sulfur, or nitrogen atom. These polymers are obtained either by polymerization of vinyl monomers, $CH_2{=}CHY$, or by chemical transformation of another polyvinyl compound.

Not all chemical compounds with ethenyl groups $CH_2{=}CH–$ are called vinyl compounds. Compounds $CH_2{=}CHR$ with an aliphatic group R are referred to as 1-olefins (formerly: $\alpha$-olefins). The group $CH_2{=}CH–CH_2–$ is known as "allyl" if it is not connected to a carbon atom. Acrylic compounds have carbon–carbon double bonds in structures that are derived from acrylic acid, either directly such as $CH_2{=}CH–CO–$ or indirectly, such as $CH_2{=}CH–C{\equiv}N$.

This section is concerned with aromatic vinyl compounds, $CH_2{=}CHAr$, and their derivatives as well as important O-vinyl, N-vinyl, and S-vinyl compounds. Vinyl chloride and its polymers are discussed with other chlorinated and brominated compounds in Section 6.9, since the terms "vinyl", "vinyl monomer", "vinyl polymer", and "polyvinyls" refer exclusively to vinyl chloride-based compounds in the plastics industry. Fluorinated carbon–carbon chain polymers are treated in Section 6.10, acrylics in Section 6.11, and allylics in Section 6.12. S-vinyl compounds have only academic interest.

**Vinylidene monomers** have the general structure $CH_2{=}CY_2$ where Y can be a halogen, the nitrile group, etc., but neither an aliphatic, aromatic, or cycloaliphatic group nor –O–, –S–, or –N– residues. **Vinylene** refers to the groups –CH=CH–.

The term "vinyl" came into the chemical literature probably as follows. Ethanol was formerly called "spirit of wine." Thermal decomposition of ethanol delivered "gas of wine" (ethene). Compounds of ethene were thus called vinyl compounds (L: *vinum* = wine).

According to the theory of "radicals" (not 1-electron species!) by Kolbe (1854), methane $CH_4$ is the hydride of the methyl radical $CH_3$ (thus $CH_3H$), ethane $C_2H_5H$ the hydride of the ethyl radical $C_2H_5$, and ethene $CH_2=CHH$ the hydride of the vinyl radical $CH_2=CH$ (note that "radical" was used in the sense of our present term "group"). The theory had to be abandoned after the disubstituted ethylene glycol $HOCH_2CH_2OH$ was discovered. The term "vinyl" remained, however, for compounds such as vinyl chloride ($CH_2=CHCl$), vinyl alcohol ($CH_2=CHOH$), vinyl acetate ($CH_2=CHOOCCH_3$), etc.

### 6.8.2 Poly(styrene)s

Vinyl benzene, $CH_2=CH(C_6H_5)$, is usually called styrene since it was first isolated from styrax (storax), a gray-brown soft resin from the wood of the tree *Liquidambar orientalis* (Volume I, p. 9). The resin from styrax tree, *Styrax tonkinensis*, is not called styrax but gum benzoin or benzoin. In chemistry, benzoin is $C_6H_5CHOH–CO–C_6H_5$.

Some 85 % of the world production of styrene is now obtained from catalytic dehydrogenation of ethyl benzene (p. 120, I) which is produced by alkylation of benzene by ethene (p. 103, XIV) (now also by ethane). Approximately 15 % of styrene results from dehydration of 1-phenyl ethanol (= $\alpha$-methyl benzyl alcohol, methyl phenyl carbinol, styrolyl alcohol), $CH_3CH(C_6H_5)OH$. This compound is *the* "byproduct" if propene is oxidized to propylene oxide by ethylbenzene hydroperoxide (2.5 kg of styrene per 1 kg of propylene oxide!). A small amount of styrene stems from the oxidation of ethyl benzene to acetophenone, reduction of acetophenone to the carbinol, and further dehydration. A pilot production starts with butadiene as the base chemical (p. 114, IX).

Styrene is polymerizable by free radicals, anions, cations, and Ziegler or metallocene catalysts. Industrial homopolymerizations to general-purpose, atactic poly(styrene)s are dominated by thermal polymerizations in bulk or solution. Homopolymerizations consume ca. 40 % of styrene, copolymerizations ca. 60 %. Both homopolymers and copolymers are also components of polymer blends. More than 40 % of poly(styrene)s are used for packaging, including many foamed products. **Expanded poly(styrene)**) also serves for insulation of buildings. All styrene polymers (**PS**) are referred to as **styrenics**. The world production of styrenics, including expanded PS, is ca. $16 \cdot 10^6$ t/a (2003).

**Atactic Homopolymers**

The thermal (= initiator-free) polymerization of styrene was known in the mid-1800s (see Volume I, p. 9). In 1925, the Naugatuck Chemical Company tried the first industrial production but gave up because of technical difficulties. BASF in Germany started a batch-type thermal polymerization process in 1931 (PS Type I) and a continuous one in 1936 (Type III). In 1938, the Dow Chemical Company used 10-gallon cans as a form of batch reactor for thermal polymerizations. All of these processes led to atactic polymers.

Type III poly(styrene)s were obtained by thermal polymerization in a tower reactor under a nitrogen blanket (Fig. 6-21). Since the polymerization is highly exothermic (ca. 70 kJ/mol at 100°C), temperature control is of utmost importance. This is achieved by prepolymerization at 80°C to a styrenic solution of 30–35 % poly(styrene) and the use of zigzag heat transfer tubes in the subsequent polymerization tower. The liquid enters the tower reactor at 100°C and passes downward through increasingly hotter zones. It is finally withdrawn as a melt at the bottom of the reactor where it enters an extruder. The extruded and solidified poly(styrene) is then pelletized.

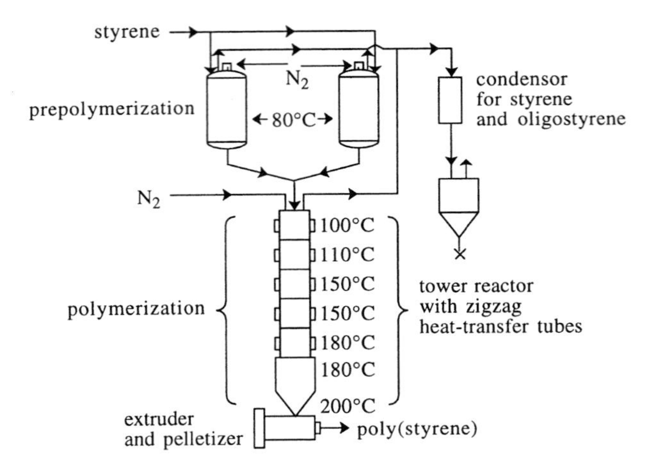


Fig. 6-21 Continuous thermal polymerization of styrene in the bulk by the tower process [13].

Thermal polymerizations start with the dimerization of styrene molecules to Diels-Alder products with axial (I) or equatorial phenyl groups (see Volume I, p. 313). The vastly more reactive axial type I then reacts with another styrene molecule II to radicals III and III that start the chain polymerization. Type III polymers had broad molecular weight distributions because of the temperature gradient, appreciable contents of oligomers V, VI, VII etc., due to reactions of I–IV, as well as considerable residual monomer contents.

$$\text{I} + \text{II} \;(CH{=}CH_2) \longrightarrow \text{III} + \text{IV} \;(H\dot{C}{-}CH_3)$$

(6-41) V VI VII

Newer industrial thermal polymerizations are performed in homogeneous continuous stirred-tank reactors (HCSTR, p. 187). In order to prevent dangerously high monomer contents in the case of a reactor run-away, one polymerizes first a small amount of monomer to the desired conversion and then adds monomer at a rate that corresponds to the polymerization rate. Once the reactor is filled to ca. 2/3 of the reactor volume, one begins to release the unpolymerized styrene and starts the continuous operation with constant addition of monomer at the top and continuous removal of polymer at the bottom. Polymerization temperatures are 130–140°C since this allows the removal of heat of polymerization by boiling off styrene ($T_{bp}$ = 146°C). In this way, one can produce ca. 5000 kg per hour in a reactor of 80 $m^3$. Alternatively, reactor cascades are used.

HCSTRs deliver much narrower molecular weight distributions, i.e., ideally Schulz–Flory distributions with $\overline{M}_w/\overline{M}_n$ = 1.5–2.0 (depending on the ratio of termination by combination and disproportionation, see Volume I). In reality, polymolecularities are 2.2–2.5 because of after-polymerizations.

General-purpose poly(styrene)s from neat thermal polymerizations have molecular weights of 100 000-400 000, since at lower molecular weight they are too brittle and at higher molecular weight too difficult to process. Molecular weights are adjusted by using chain transfer agents, for example, 5–10 % ethyl benzene as "solvent" (chain-transfer constant: $C_{ax} = 70 \cdot 10^{-6}$) and/or small proportions of dodecyl mercaptan ($C_{tr}$ = 15), terpinols, or dimeric α-methyl styrene. These agents not only lower molecular weights but also prevent clogging of reactors by gels and crosslinked polymer particles.

Styrene polymerizations are also conducted in continuous reactor cascades and by initiation with free-radical initiators such as *t*-butyl peracetate, *t*-butyl perbenzoate, 1,1-di(*t*-butylperoxy)cyclohexane, or 1,1-di(*t*-butylperoxy)-3,3,5-trimethylcyclohexane, often as mixtures of two initiators with different half-lifes. The use of (very costly!) initiators lowers the formation of unwanted Diels–Alder-type oligomers (see Eq.(6-41)).

Solvent and residual monomer are removed by heating the polymer solution to 210–250°C. The mixture is devolatilized in a flash tank under vacuum, using a flash extruder, or letting it run as a thin film on a heated vessel wall. The content of residual styrene is ca. 200–800 ppm. Lower residual monomer concentrations are obtainable by steam stripping, where the water vapor reduces the partial pressure of styrene.

There have also been attempts to remove residual monomer by so-called scavengers. Examples are substituted benzocyclobutenes (see II on p. 265) which, at $T > 200$°C, form biradicals that add to the double bond of styrene to form 3-phenyltetraline.

Poly(styrene)s from thermal polymerizations are very pure and crystal clear. They are therefore called **crystal poly(styrene)s (PS, at-PS, aPS, APS)** although they are atactic, *non-crystalline* (amorphous) polymers with a glass temperature of 100°C.

*Suspension* and *emulsion* polymerizations are not used for standard poly(styrene). These discontinuous processes use simple kettles and are easy to control, but polymers suffer from contamination by stabilizers and emulsifiers and need expensive work-up operations (washing, drying, waste-water treatment, etc.). Polymerization in suspension in presence of a blowing agent is the method of choice for expanded poly(styrene)s.

A small amount of styrene is also polymerized *cationically* to low-molecular weight polymers that are used as tackifiers for adhesives and printing inks. *Anionic* polymerizations serve for the synthesis of block polymers (see below).

Poly(styrene)s by free-radical polymerization are offered in many different grades: easy flow, medium flow, high heat, food-contact grade, low molecular weight, etc. General-purpose poly(styrene)s have the lowest density and the highest stiffness of all transparent polymers. They are easy to process and have fairly high tensile moduli but are relatively brittle and tend to stress crack. In contrast to other polymers, atactic poly(styrene)s are rarely offered as reinforced grades since neither hardness nor brittleness or stress cracking are changed appreciably upon addition of reinforcing agents.

General-purpose poly(styrene)s are not always pure poly(styrene)s. For better processing, commercial at-PS sometimes contains up to 4.5 wt% of plasticizer, usually white mineral oil, which allows easy flow and high injection molding cycles but also reduces the heat distortion temperature (Table 6-26). For improved resilience of, e.g., drinking cups, small proportions of styrene-butadiene-styrene triblock copolymers (**SBC**) are added. So-called **high-impact poly(styrene)s (HIPS)** are obtained by polymerizing a solution of a butadiene rubber in styrene; they are intimate blends of poly(styrene) and butadiene rubber (see Volume IV).

Table 6-26 Properties of atactic poly(styrene)s (at-PS), syndiotactic poly(styrene)s (st-PS), and poly-(*p*-methyl styrene) (PMS). HH = highly heat-resistant grade; EF = easy-flow grade. * Izod.

| Property | Physical unit | at-PS HH | at-PS EF | st-PS | PMS |
|---|---|---|---|---|---|
| Density | $g/cm^3$ | 1.05 | 1.05 | 1.04 | 1.05 |
| Refractive index | 1 | 1.59 | 1.59 | | |
| Melting temperature (DSC) | °C | - | - | 270 | - |
| Heat distortion temperature (1.82 MPa) | °C | 103 | 77 | 96 | 97 |
| Vicat temperature B | °C | 99 | 78 | 108 | 115 |
| Glass temperature (DSC) | °C | 100 | 80 | 100 | 90 |
| Linear thermal expansion coefficient | $K^{-1}$ | $7 \cdot 10^{-5}$ | $7 \cdot 10^{-5}$ | | |
| Specific heat capacity | $J\ K^{-1}\ g^{-1}$ | 1.3 | 1.3 | | |
| Thermal conductivity (20°C) | $W\ m^{-1}\ K^{-1}$ | 0.18 | 0.18 | | |
| Tensile modulus | MPa | 3350 | 3200 | | 2410 |
| Flexural modulus | MPa | | | 3000 | 2900 |
| Shear modulus | MPa | 1250 | 1200 | 1000 | |
| Tensile strength at fracture | MPa | 65 | 45 | 35 | 48 |
| Flexural strength | MPa | 100 | 90 | 75 | 79 |
| Elongation at fracture | % | 4 | 3 | 20 | |
| Impact strength (Charpy) | $kJ/m^2$ | 5–20 | 5–13 | | |
| Notched impact strength (Izod, 3.1 mm) | J/m | | | | 16 |
| (Charpy) | $kJ/m^2$ | 2.5 | 2.0 * | 2.0 | |
| Ball indentation hardness (60 s) | MPa | 1150 | 1100 | | |
| Relative permittivity | 1 | 2.5 | | 2.6 | |
| Electrical resistance (= surface resistivity) | Ω | $10^{15}$ | | | |
| Resistivity (= volume resistance) (2 min) | Ω cm | $10^{18}$ | | | |
| Electric strength | kV/mm | 200 | | | |
| Dissipation factor (50 Hz) | 1 | 0.0001 | | | |
| Water absorption (96 h) | % | 0.1 | 0.1 | 0.04 | |

**Stereoregular Poly(styrene)s**

Stereoregular poly(styrene)s are obtained by Ziegler–Natta or metallocene polymerizations. *Isotactic poly(styrene)s* are brittle and difficult to process because of their high melting temperature of 230°C. They are industrially unimportant.

Commercial *syndiotactic poly(styrene)s* (**SPS**, **sPS**, **st-PS**) with more than 99 % syndiotactic diads result from styrene polymerization by a Ti metallocene catalyst plus methylaluminoxane or borate. Similar to atactic poly(styrene)s, they have low density, easy processability, and low moisture uptake. However, their heat deflection temperature is considerably higher since they are crystalline and have a high melting temperature (Table 6-26). Parts made from sPS distort only a little since crystalline and amorphous regions have practically the same density. Mechanical and thermal properties of sPS resemble those of polyamide 6.6. sPS does not need to be conditioned like polyamides, however, because of its low moisture uptake.

Neat sPS is more brittle than aPS, however, which disqualifies it as an engineering plastic. All presently offered sPS grades are therefore reinforced by 30 wt% glass fiber (GF) (Questra®). According to one study, GF-sPS is tougher than GF reinforced poly-(butylene terephthalate), but according to another, it is not.

**Styrene Copolymers**

Monomeric styrene serves as a comonomer for many high-volume copolymers, for example, the world production of ABS + ASA + SAN was ca. $5 \cdot 10^6$ t/a in 2000. Major styrene copolymers include

| | |
|---|---|
| ABS | graft copolymers of styrene/acrylonitrile on butadiene rubbers; |
| ACS | graft copolymers of styrene/acrylonitrile on chlorinated poly(ethylene) rubbers; |
| AES | graft copolymers of styrene on EPDM rubbers; |
| ASA | graft copolymers of styrene and acrylonitrile on acrylic rubbers; |
| ESI | pseudo-random bipolymers of 70–80 wt% styrene and ethene ("interpolymers"); |
| SAN | random bipolymers of styrene and acrylonitrile; |
| SB | diblock polymer of styrene and ca. 25 wt% butadiene; |
| SBC | diblock polymer of styrene (65–85 wt%) and 1,3-butadiene; |
| SBS | triblock polymer of styrene-butadiene-styrene; |
| SEBS | selectively hydrogenated styrene-butadiene-styrene triblock polymer; |
| SEPS | selectively hydrogenated styrene-isoprene-styrene triblock polymer; |
| SIS | triblock polymer of styrene-isoprene-styrene; |
| SMA | alternating bipolymer of styrene and maleic anhydride; |
| SMI | alternating bipolymers of > 40 wt% styrene and *N*-phenyl maleimide; |
| SMMA | random bipolymers of styrene and methacrylic acid; |
| TPO | thermoplastic olefin elastomer (= hydrogenated SIS). |

They are produced by free-radical copolymerization (SAN, SMA, SMI, SMMA), graft copolymerization (ABS, ACS, AES, ASA), anionic block polymerization (SB, SBC, SBS, SIS), metallocene copolymerization (ESI), or post-hydrogenation (SEBS, SEPS, TPO).

Free-radical copolymerizations to SAN, SMA, SMI, and SMMA are similar to that of standard poly(styrene). SANs are usually offered with compositions that correspond to the azeotrope (24 % acrylonitrile); these polymers have higher heat distortion temperatures than standard poly(styrene) (see p. 297). SANs with other compositions require HCSTRs in order to maintain a constant monomer composition, for example, SANs with 60–70 % acrylonitrile units as barrier polymers for gas-impermeable packaging.

HCSTRs are also used to produce SMAs with other than 1:1 ratios of monomeric units since discontinuous S/MA copolymerizations lead to alternating polymers.

All graft copolymers ABS, ACS, AES, and ASA are turbid multiphasic polymers with high impact strengths (for the relationship between synthesis and properties, see Volume IV). SB and SBC diblock polymers are also biphasic but clear because of the small size of the dispersed poly(butadiene) phases; these polymers are thus called **glass-clear high-impact poly(styrene)s** (glass-clear **HIPS**).

Triblock polymers SBS, SIS, SEBS, and SEPS are thermoplastic elastomers (see Volume IV for their properties). SBS and SIS are triblock polymers but not triblock *co*-polymers because they are synthesized by either sequential polymerization of their monomers or by coupling of preformed polymer blocks (see Volume I, Section 16.2.2). The hydrogenation product carries the acronym SEBS since selective hydrogenation of the butadiene center block $-CH_2-CH=CH-CH_2-$ of SBS leads to ethylene (E) and butanediyl units (B); similarly, hydrogenation of isoprene units $-CH_2-C(CH_3)=CH-CH_2-$ of SIS results in ethylene (E) and propylene units (P), hence SEPS. Thermoplastic olefin elastomers (TPO) are either SEPS or polymer blends (see Volume IV).

The free-radical copolymerization of styrene with a small percentage of divinyl benzenes (usually as mixture of *o*-, *m*-, and *p*-isomers) delivers crosslinked polymers. These polymers from bulk polymerizations can be worked only by machining. They are used in electrical engineering applications but have little importance. Suspension polymerizations deliver beads that are sulfonated to become ion-exchange resins.

*p*-Divinyl benzene as well as *m*-diisopropenyl benzene, *m*-$C_6H_4(C(CH_3)=CH_2)_2$, can be anionically polymerized to soluble polymers if group conversions are not too high because of differences in reactivities of the first and second double bond.

### 6.8.3 Substituted Poly(styrene)s

Styrene has five monomethyl derivatives, all of which are obtained by dehydrogenation of their saturated counterparts: α-methyl styrene from cumene (p. 122, V), β-methyl styrene from propyl benzene, and *o*-, *m*-, and *p*-styrenes from the corresponding methyl ethyl benzenes (see p. 126, III).

High-molecular weight products are difficult to obtain from free-radical polymerization of β-methyl styrene for kinetic reasons (steric hindrance at the β-CH) and of α-methyl styrene for both kinetic (allyl compound!) and thermodynamic reasons (ceiling temperature in bulk: $T_c$ = 60°C). Low-molecular weight poly(α-methyl styrene)s are used as modifiers and processing aids for poly(vinyl chloride), ABS polymers, thermoplastic elastomers, etc.

*p*-Methyl styrene is industrially interesting since it is ultimately obtained from toluene (via *p*-ethyl toluene, p. 126, III). Toluene is less expensive than benzene because the latter is needed for a series of important chemical intermediates as well as for lead-free antiknock gasolines. Since benzene is in short supply, toluene is therefore dealkylated to benzene (in the United States: 65 % of toluene!).

Both *p*-methyl styrene and vinyl toluene (= mixture of 33 % *p*- and 67 % *m*-methyl styrene) are polymerized free-radically. Poly(*p*-methyl styrene)s (**PMS**) are thermoplastics with greater hardnesses and higher heat distortion temperatures than standard poly(styrene)s (Table 6-26). Also commercial are bipolymers of *p*-methyl styrene and acrylonitrile as well as terpolymers of *p*-methyl styrene, acrylonitrile, and butadiene. Poly(vinyl toluene)s serve as modifiers for unsaturated polyester resins (Section 7.7).

Also offered are homopolymers and copolymers of halogenated styrenes. Chlorostyrene often serves as a comonomer to improve the flame resistance of styrene polymers. Poly(2,4,6-tribromostyrene) is employed as a flame retardant in several polymers. Highly fluorinated poly(styrene)s such as poly(2-trifluoromethyl styrene) and poly(4-perfluoroisopropyl styrene) have high glass temperatures and excellent optical properties, for example, high refractive indices and increased Abbé numbers (Volume IV).

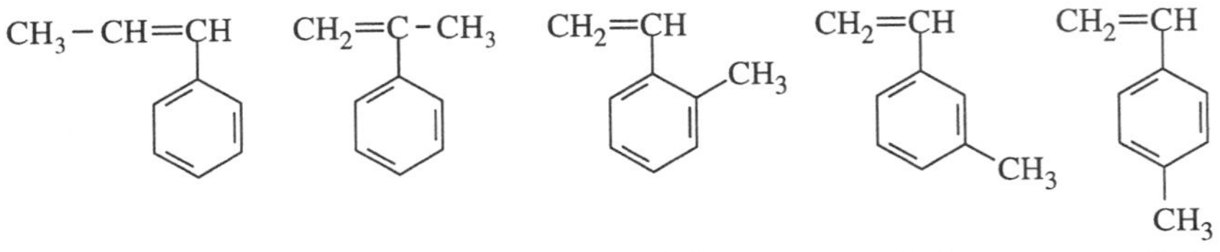

$CH_3-CH=CH-C_6H_5$ β-methyl

$CH_2=C(CH_3)-C_6H_5$ α-methyl

$CH_2=CH-C_6H_4-CH_3$ *o*-methyl

$CH_2=CH-C_6H_4-CH_3$ *m*-methyl

$CH_2=CH-C_6H_4-CH_3$ *p*-methyl

### 6.8.4 Poly(vinyl acetate)

In Western countries, vinyl acetate (**VAC**), $CH_2=CH(OOCCH_3)$, is obtained from gas-phase acetoxylation of ethene (p. 101, VI). In China and India, vinyl acetate is synthesized from acetylene and acetic acid (p. 105, V) as well as from acetaldehyde and acetanhydride (p. 106). In 1991, vinyl acetate-based polymers included ca. $1.4 \cdot 10^6$ t/a poly(vinyl acetate) (**PVAC**, "PVA"), ca. $1.3 \cdot 10^6$ t/a vinyl acetate copolymers, and ca. $0.9 \cdot 10^6$ t/a poly(vinyl alcohol) (**PVAL**, "PVA", PVOH).

All polymerization-grade vinyl esters need to be very pure because impurities from monomer syntheses either retard the polymerization (vinyl acetylene, crotonaldehyde = 2-butanal) or act as regulators for molecular weights (acetaldehyde, acetic acid, acetone, aromatics). Monomeric vinyl acetate always contains 2–50 ppm inhibitor (hydroquinone, *t*-butyl benzcatechol, etc.) that is not removed before the polymerization but whose action is overridden by a higher concentration of the free-radical initiator (AIBN, diacyl peroxides, redox initiators, etc.).

Most vinyl acetate is polymerized discontinuously in emulsion under nitrogen in reactors of 5–30 $m^3$, since the resulting dispersions can be used directly as binders in paints, as adhesives, as spray-dried additives to concrete, sizes for glass fibers, paper coatings, and the like. Emulsifiers are usually anionic; the resulting dispersions are therefore negatively charged. Positively charged dispersions are obtained if nitrogen-containing derivatives of ethoxylated poly(propylene oxide)s are used as emulsifiers. The solids content of dispersions is usually 40 %. Finely dispersed, stable dispersions with more than 50 % solids are difficult to obtain, but are possible if vinyl acetate is copolymerized with a small proportion of a hydrophilic monomer.

Polymer particles from emulsion polymerizations with anionic emulsifiers usually have diameters of 0.1–0.3 μm, those with protecting colloids (cellulose ethers, partially saponified poly(vinyl acetate)s), 1–10 μm. Suspension polymerizations with protecting colloids (poly(vinyl alcohol), poly(*N*-vinyl pyrrolidone), poly(acrylamide), etc.) deliver beads with diameters of 200–3000 μm. Both suspension and emulsion polymerizations require low polymerization temperatures because the low glass temperature of poly(vinyl acetate) ($T_G$ = 28°C) would lead to aggregation of beads and latices, respectively, at higher temperatures.

Bulk polymerizations proceed mainly continuously in tube reactors at the boiling temperature of vinyl acetate (72.5°C at atmospheric pressure (≈ 0.1 MPa)). The resulting polymers are considerably branched mainly because of strong chain transfer to the ester group (75–95 % of all branching) but also to the main chain. The polymerization proceeds to ca. 60 % monomer conversion. Neat vinyl acetate can also be polymerized in tower reactors. This process produces polymers with only moderate degrees of polymerization since viscosities (and thus flow temperatures) increase strongly with molecular weight; higher initiator concentrations are therefore needed for lower molecular weights.

Vinyl acetate is also copolymerized. Copolymers of VAC with vinyl stearate or vinyl pivalate are more resistant to moisture since the bulky, hydrophobic alkyl groups reduce the hydrolysis of PVAC. Copolymers of VAC with vinyl chloride serve as impact modifiers for poly(vinyl chloride). Copolymers of ethene and VAC are used as adhesives and binders in paint (70–95 wt% VAC), for cable insulation (40–50 wt% VAC), as coatings (30–40 wt% VAC), as sealants (15–30 wt%), and as packaging materials for wastes and deep-frozen foods (< 10 wt%).

### 6.8.5 Poly(vinyl alcohol)

Vinyl alcohol, $CH_2=CHOH$, from the hydrolysis of precursors is stable for several hours at –10°C. As an enol, it is also present in small amounts in the tautomeric equilibrium with acetaldehyde, $CH_3CHO$. Polymerization of this mixture in polar solvents by alkali alcoholates as initiators delivers poly(vinyl alcohol), $\text{+}CH_2\text{–}CH(OH)\text{+}_n$, since the steady shift of the equilibrium always furnishes new enol until all acetaldehyde is exhausted.

Hydrolysis of ketene methylvinyl acetal in acetone also leads to vinyl alcohol, which can be copolymerized with monomers such as maleic anhydride by photo-initiation with AIBN. Such copolymerizations also proceed with methylvinyl alcohol (1-propenol), $CH_3CH=CH\text{–}OH$, which is generated by isomerizing allyl alcohol, $CH_2=CHCH_2OH$.

$$\text{(6-42)} \qquad CH_2=C\begin{matrix} \diagup OCH_3 \\ \diagdown OCH=CH_2 \end{matrix} \xrightarrow{+\,H_2O} CH_3COOCH_3 + HO-CH=CH_2$$

Industry obtains poly(vinyl alcohol) (**PVAL**, formerly also **PVA** or **PVOH**) from either discontinuous or continuous transesterification of poly(vinyl acetate) by methanol or butanol that are catalyzed by the corresponding sodium alcoholates. The byproducts methyl acetate and especially butyl acetate are valuable solvents:

$$\text{(6-43)} \qquad \sim CH_2-\underset{\underset{\displaystyle O-\underset{\underset{\displaystyle O}{\|}}{C}-CH_3}{|}}{CH}\sim \;+\; ROH \longrightarrow \;\sim CH_2-\underset{\underset{\displaystyle OH}{|}}{CH}\sim \;+\; CH_3COOR$$

At group conversions of 45–85 %, a highly viscous gel-like phase is formed because of the intermolecular association of intermediate poly(vinyl acetate-*co*-vinyl alcohol) molecules. The extent of self-association depends on the type of transesterification: acidic ones lead to a more random distribution of the remaining vinyl acetate units whereas alkaline ones result in a more block-like structure.

In order to avoid the gel phase, it has been suggested to work continuously in very dilute solution, to emulsify PVAC in hydrocarbons, or to employ kneaders. Difficulties could also be circumvented if PVAC is replaced by poly(vinyl formate) which easily saponifies in hot water. The resulting formic acid is corrosive, however, which necessitates expensive equipment. Furthermore, the monomer vinyl formate hydrolyzes easily and is therefore difficult to produce.

PVAL is generally produced in three types with 1–2 %, 3–7 %, or 10–15 % vinyl acetate units. Because of its solubility in water, it is used for many purposes: in mixtures with starch as a size for polyamide or rayon fibers; as an adhesive; as an emulsifier or protecting colloid in polymerizations; as a component of printing inks, toothpastes, or cosmetics; as a working material for fuel-resistant hoses; and as a binder in paper-making. Mixtures of PVAL and dichromates serve as copy layers for off-set printing with positive copies (see Volume IV).

Spinning aqueous PVAL solutions into a bath of sodium sulfate in water, followed by drawing, annealing, and crosslinking with formaldehyde delivers a fiber (Volume IV). Pseudo-copolymers of ethene and vinyl alcohol (from transesterification of vinyl acetate units) serve as oxygen barriers in coextruded films (Volume IV).

## 6.8.6 Poly(vinyl acetal)s

Poly(vinyl acetal)s result from reacting aldehydes RCHO with PVAL or, for syntheses of poly(vinyl formal)s, directly from the acetalization of PVAC with HCHO. Aldehydes are bifunctional compounds and therefore form intramolecular rings between adjacent hydroxy groups:

(6-44) OH OH OH OH OH $\xrightarrow{+\ \text{RCHO}}$ O O (R, H) OH O O (R, H)

Since acetalizations proceed at random and not sequential, some OH groups will not find adjacent partners for *irreversible* reactions. For molecules with head-to-tail arrangements of monomeric units, theory predicts that a fraction of $f = \exp(-2) = 0.135$ hydroxy groups remain unreacted (Volume I, p. 585). Experimentally, $f$ is somewhat higher because PVAL contains some head-to-head and tail-to-tail structures.

**Poly(vinyl formal)s (PVFM)** are obtained by adding sulfuric acid to a suspension of poly(vinyl acetate) in water, acetic acid, and formaldehyde. Neutralization and subsequent addition of water precipitates poly(vinyl formal). Mixtures of resols (Section 6.6.3) with poly(vinyl formal)s containing 84 % formal groups, 6 % hydroxy groups, and 10 % acetate groups are used as wire coatings, adhesives, films, etc.

**Poly(vinyl butyral)s (PVB)** are synthesized by adding butyraldehyde and a catalyst to a suspension of poly(vinyl alcohol) in ethanol and precipitating the polymer by water. PVBs containing 80 % butyral, 18 % hydroxy, and 2 % acetate groups are plasticized by 30 % dibutyl sebacate and used as interlayers in safety glass because PVBs are tough, highly flexible, and optically exceptionally clear. PVBs also serve as adhesives, paints, and lacquers.

## 6.8.7 Poly(vinyl ether)s

Vinyl ethers, $CH_2{=}CHOR$, are obtained by addition of alcohols to acetylene (p. 105, VIII). Their syntheses from ethene and alcohols in the presence of oxygen as well as vinyl exchange with vinyl acetate are possible, but apparently not industrial processes.

Vinyl ethers are polymerized cationically by Lewis acids such as $BF_3 \cdot 2\ H_2O$. Industrial polymerizations are either continuous in liquid propane similar to the polymerization of isobutene (p. 239) or discontinuous in 1,4-dioxane at 5°C. In the latter process, a portion of the monomer is prepolymerized; after the onset of polymerization, more monomer is added so that polymerizations can proceed at 100°C under reflux.

All poly(vinyl ether)s, $\text{+}CH_2\text{–}CHOR\text{+}_n$, are viscous gums or rubbery solids that are lightproof and difficult to saponify. **Poly(vinyl methyl ether)s (PVME**; R = $CH_3$) are plasticizers for coatings. **Poly(vinyl ethyl ether)s (PVE**; R = $C_2H_5$) are plasticizers for cellulose nitrate and natural resin-based lacquers; solutions of PVE form the base of pressure-sensitive adhesives. **Poly(vinyl isobutyl ether)s (PVI**; R = $i\text{-}C_4H_9$) are also used for pressure-sensitive adhesives as well as tackifiers. **Poly(vinyl octadecyl ether)s** (R = $C_{18}H_{37}$) serve for waxes and polishes.

Alternating copolymers of **methyl vinyl ether** and **maleic anhydride** are used as protective colloids, in the textile industry, as binders, etc. The esterified polymer is the effective component of most hair sprays. Copolymers of vinyl methyl ether and vinyl chloride (**PVM**) are binders for anti-corrosive paints.

The free-radical copolymerization of divinyl ethers (**DIVE**) and maleic anhydride (**MA**) delivers a **pyran copolymer** that is known as **DIVEMA**. The copolymerization consists of alternate intermolecular and intramolecular (cyclopolymerizing) steps, leading to a copolymer with a 1:2 ratio of maleic anhydride and divinyl ether units where the latter are mainly present as six-membered rings and occasionally as five-membered ones (Eq.(6-45)). Hydrolyzed and neutralized DIVEMA is biologically active: it battles tumors, induces the formation of interferon, and has also antibacterial, fungicidal, and antiarthritic properties.

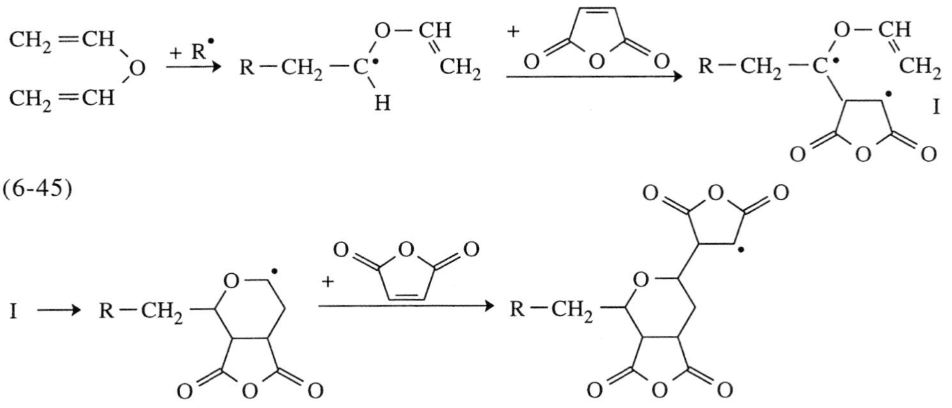

$$(CH_2{=}CH)_2O \xrightarrow{+R^\bullet} R{-}CH_2{-}\dot{C}H{-}O{-}CH{=}CH_2 \xrightarrow{+\text{MA}} \text{I} \qquad (6\text{-}45)$$

$$\text{I} \longrightarrow R{-}CH_2{-}(\text{pyran-fused anhydride radical}) \xrightarrow{+\text{MA}} R{-}CH_2{-}(\text{pyran-fused anhydride})\text{-succinic anhydride radical}$$

## 6.8.8 Poly(*N*-vinyl compound)s

Vinyl amine, $CH_2{=}CHNH_2$, the isomer of cyclic ethylene imine (I), is not known. Poly(vinyl amine) (II) is therefore prepared from other poly(*N*-vinyl compounds) by polymer-analog reactions such as hydrolysis of poly(*N*-vinyl acetamide) (III) or poly(*N*-vinyl-*t*-butyl carbamate) (IV) or by hydrazinolysis of poly(*N*-vinyl succinimide) (V) or poly(*N*-vinyl phthalimide) (VI).

Polymers III–VI are obtained by free-radical polymerization of the corresponding monomers. Poly(vinyl amine) II has limited use as a precursor for polymeric dyestuffs, while all other polymers III–V are only of academic interest. Poly(*N*-vinyl acetamide) (III) is soluble in water and organic solvents and might therefore be a possible thickener or film-former.

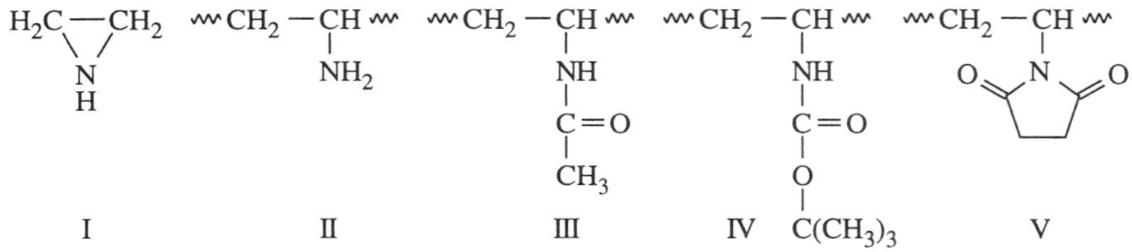

| I | II | III | IV | V |
|---|---|---|---|---|
| $H_2C{-}CH_2$ (ring, N–H) | $\sim CH_2{-}CH(NH_2)\sim$ | $\sim CH_2{-}CH(NH{-}C(=O){-}CH_3)\sim$ | $\sim CH_2{-}CH(NH{-}C(=O){-}O{-}C(CH_3)_3)\sim$ | $\sim CH_2{-}CH(N\text{-succinimide})\sim$ |

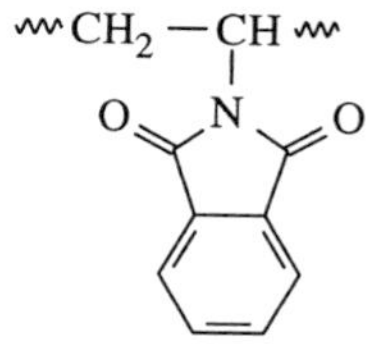

VI 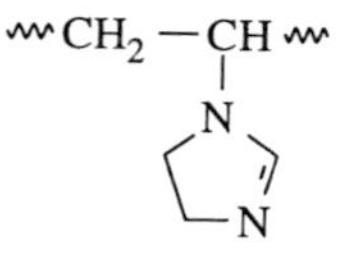 VII 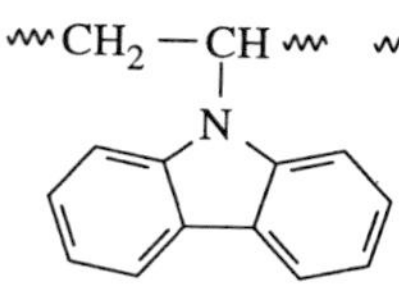 VIII 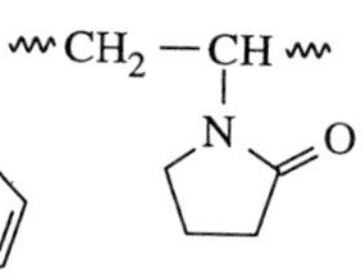 IX 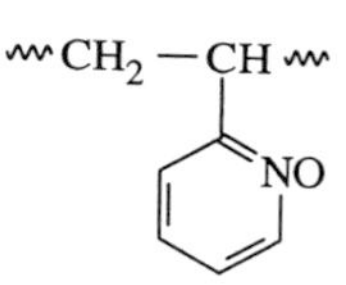 X

Poly(*N*-vinyl imidazole) (VII) is produced in small amounts, whereas poly(*N*-vinylcarbazole) (VIII) and poly(*N*-vinyl pyrrolidone) (IX) are significant industrial polymers (see below). Vinyl pyridines are comonomers in some elastomers; homopolymers are not produced. Poly(2-vinyl pyridine-1-oxide) (X) seems to protect the lung against silicosis.

### Poly(*N*-vinyl carbazole)

Carbazole from anthracene oil (p. 119) is vinylated by acetylene in the presence of ZnO/KOH as catalyst at 160–180°C and a pressure of 2 MPa. *N*-Vinyl carbazole is polymerized by non-oxidizing free-radical initiators in bulk or suspension.

Poly(*N*-vinyl carbazole) (**PVK, PVCZ**) (VIII) has excellent temperature stability (glass temperature of 227°C, heat distortion temperature of ca. 160°C). Its extreme brittleness can be overcome by copolymerization with a little isoprene. These polymers have been used as insulators in electrical parts for high-frequency applications. PVCZ is used in electrical appliances as a replacement for mica and as an impregnant for paper capacitors. When doped with electron donors such as chloranil (= 2,3,5,6-tetrachloro-1,4-benzoquinone) and tetracycanoethylene, PVCZ is used as a toner in photocopiers instead of diarsenotriselenide because of its good photo conductivity and its high electrical resistance in the dark.

### Poly(*N*-vinyl pyrrolidone)

*N*-Vinyl pyrrolidone (**NVP**), a water-soluble monomer, is obtained by vinylation of pyrrolidone (= 2-pyrrolidinone, γ-butyrolactam) with acetylene and the γ-butyrolactam by reaction of γ-butyrolactone with ammonia.

NVP is polymerized free-radically by $H_2O_2/NH_3$ at 35–65°C in alkaline aqueous solution at a pH of ca. 9 (maximum of polymerization rate) because the monomer dissociates in acidic solution into pyrrolidone and acetaldehyde. A solution polymerization is also possible by organic peroxides in organic solvents, albeit at higher temperatures of 90–150°C. Suspension polymerizations proceed in heptane or benzene at 65°C with AIBN as free-radical initiator and a graft copolymer of NVP on a $C_{16}$ olefin as suspending agent.

Poly(*N*-vinylpyrrolidone) (**PVP**) (IX) dissolves in water or polar organic solvents. It is used as a protecting colloid in suspension polymerizations, as a binder in pharmaceuticals (PVP-iodine is a common desinfectant), as a film-former in cosmetics, as a stabilizer in beverages (for example, in American beers), and, as a pseudo-copolymer with vinyl alcohol, as a hair-setting gel. During World War II, its isotonic sodium chloride solution served as an artificial blood serum.

# 6.9 Chlorine-containing Polymers

## 6.9.1 Poly(vinyl chloride)

Chlorine was formerly the undesired byproduct from the synthesis of caustic soda (= aqueous solution of NaOH) by electrolysis of sodium chloride solutions. Caustic soda was needed for the production of soda pulp (p. 72) and many other uses. Vinyl chloride and poly(vinyl chloride) were therefore good outlets for chlorine. Today, chlorine is usually the target chemical and caustic soda an often undesired byproduct.

Vinyl chloride monomer (**VC, VCM**), $CH_2$=CHCl, is obtained mainly from the intermediates of the chlorination or oxychlorination of ethene (p. 100, XII and XIII). Addition of HCl to acetylene (p. 104, II) is no longer viable in Western countries. Direct chlorination of ethane (p. 98) is not a major process.

The world production of vinyl chloride uses ca. 50 % of the world production of chlorine. With a world production of ca. $27 \cdot 10^6$ t/a (2003), poly(vinyl chloride) (including copolymers) is now the third largest synthetic polymer behind poly(ethylene) and poly(propylene), but before poly(styrene).

Approximately 95 % of vinyl chloride is used for the preparation of its homopolymer, poly(vinyl chloride) (**PVC**). 2/3 of all PVCs are so-called **rigid PVCs** (= non-plasticized PVCs; **PVC-U**), 1/3 are plasticized ones (**PVC-P**). In industrial lingo, all poly(vinyl chloride)s (and no other vinyl polymers!) are called "**vinyls**".

### Homopolymerization

Vinyl chloride is a gas (bp −13.4°C) that forms explosive mixtures with air if the VCM content is 4–22 %. Since oxygen reacts with VCM to polymeric peroxides that affect polymerization rates and lead to weak links in the polymer (see below), polymerization-grade VCM must be >99.9 % pure and stabilized by phenolic compounds.

Vinyl chloride is industrially polymerized in bulk (gas phase), emulsion, suspension, or microsuspension. The continuous *bulk polymerization* is a two-step precipitation polymerization (Fig. 6-22) that is initiated by oil-soluble free-radical initiators such as acetylcyclohexane sulfonylperoxide, $CH_3$–CO–O–O–$SO_2$–$C_6H_{11}$, or *bis*(2-ethylhexyl)-peroxy dicarbonate, $(C_2H_5)(C_4H_9)CH–CH_2–O–CO–O)_2$.

The resulting PVC is insoluble in VCM; at a monomer conversion of ca. 30 %, granules are formed that imbibe the remaining vinyl chloride. In order to avoid heat build-up during this precipitation polymerization, the heat of polymerization is removed by boiling off monomer. Polymerizations are conducted in two steps:

Granules of prepolymerized PVC are first loaded with VCM that is prepolymerized to a conversion of $u$ = 12 % (Fig. 6-22). Direct polymerizations of VCM are possible if the monomer conversion $u$ is at least 7 % but not more than 10 %: the resulting granules cannot be transferred intact from the prepolymerization reactor to the main reactor if $u$ < 7 %, and the system becomes too viscous if $u$ > 10 %.

After transfer to the post reactor(s), the monomer and initiator are added. The post reactor was a rotating horizontal tube reactor with steel beads as mixing elements, then a horizontal screw reactor, and later a cascade of vertical screw reactors. The structure of the resulting PVC granules depends critically on the impeller speed in the prepolymerization unit; it affects the take-up of plasticizer and the gelation of plasticized PVC.

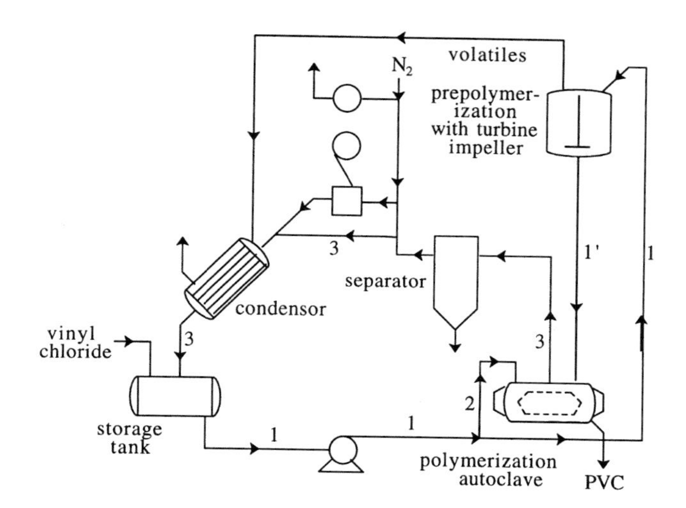


Fig. 6-22 Early bulk polymerization of vinyl chloride. The monomer is pumped from a storage tank into a prepolymerization vessel (path 1) and from there into the polymerization autoclave (path 1') where it is replenished with new monomer (path 2). Unreacted monomer is withdrawn at the top (path 3), removed from polymer dust in a separator, condensed, and transferred to the storage tank.
By permission of Chemical Engineering, New York [14].

Approximately 10 % of all PVCs stem from polymerizations in bulk, 15 % from those in emulsion (**E-PVC**), and 75 % from those in suspension (**S-PVC**). *Suspension polymerizations* employ very many different free-radical initiators, for example, dicetylperoxycarbonate, $(C_{16}H_{33}OCOO)_2$, and dilauroyl peroxide $(C_{11}H_{23}COO)_2$, besides the initiators used in bulk polymerizations. Protecting colloids are various cellulose ethers poly(vinyl alcohol)s with more than 70 mol% OH groups, often as mixtures. The systems also contains pH buffers, co-suspending agents (salts of fatty acids, non-ionic surfactants, etc.) that increase the porosity of beads, regulators (thioglycolates, aldehydes) to regulate molecular weights, and compounds that prevent wall deposits (fatty acids, etc.).

The polymerization proceeds discontinuously in stirred-tank reactors ($V = 200$ m$^3$) at 50–70°C and 0.8–1.2 MPa until it is terminated after 5–10 h at ca. 75–90 % monomer conversion. The remaining monomer is removed by steaming the resulting PVC beads with water vapor; the polymer beads are filtered or centrifuged off and dried.

*Emulsion polymerizations* are either continuous or discontinuous, usually half-continuous, by adding monomer and emulsifier to a seed latex in such a way that no new particles are formed. Preferred for plastisols (Volume IV) are bimodal distributions of particles that are obtained by mixing different latices or by control of the seeding process. Particle diameters are usually 1 μm (unimodal) or 1 μm plus 0.1–0.5 μm (bimodal).

Emulsion polymerizations of vinyl chloride do not follow standard kinetics (Section 5.2.4). Polymerization rates are not proportional to the number of latex particles but are practically independent of it. They also do not depend on the 0.4th power of the initiator concentration but on $[I]^{0.5-0.7}$ and they are not proportional to the 0.6th power of the emulsifier concentration but vary with the type and concentration of the emulsifier. These effects seem to be caused by the rapid desorption and readsorption of radicals that are formed by chain-transfer reactions.

**Structure**

Free-radical polymerizations of VCM are strongly affected by chain transfer to the monomer and formation of ethenyl radicals:

(6-46) $$\sim CH_2-\dot{C}HCl + CH_2{=}CHCl \longrightarrow \sim CH_2-CHCl_2 + CH_2{=}\dot{C}H$$

The rate of this transfer reaction is much greater than the rate of mutual deactivation of two polymer radicals. As a consequence, the degree of polymerization is therefore practically independent of the initiator concentration. Degrees of polymerization must therefore be controlled by varying the polymerization temperature.

Monomer radicals from chain-transfer reactions to monomer molecules, Eq.(6-46), start new polymer chains that have unsaturated endgroups, $CH_2{=}CH-$. Chain transfer to polymer chains also creates other unsaturated groups such as $-CH{=}CH-CH_2Cl$ and $-CH{=}CH-CHCl-$. These unsaturated groups are present in concentrations of 0.005–0.7 groups per 100 monomeric units. They are in part responsible for the zip-like dehydrochlorinations that proceed on exposure of PVC to light, oxygen, and elevated temperatures, and lead to sequences of conjugated double bonds. Increases in the length of the resulting polyene structure changes the color of PVC from white to yellow, brown, and black. The polymer crosslinks and the properties of PVC deterioriate drastically.

Head-to-head addition on propagation leads to radicals $\sim CH_2-CHCl-CHCl-{}^{\bullet}CH_2$ that are transformed to $\sim CH_2-CHCl-{}^{\bullet}CH-CH_2Cl$ and further to $\sim CH_2-{}^{\bullet}CH-CHCl-CH_2Cl$. Addition of vinyl chloride delivers $\sim CH_2-CH(CH_2CH_2Cl)-CHCl-CH_2Cl$ and thus short-chain branches. Poly(vinyl chloride)s also contain tertiary bound chlorine atoms.

Additional "wrong" structures are caused by traces of oxygen in the polymerization system. Oxygen can add to growing polymer radicals and form peroxide radicals such as $\sim CH_2-CHCl-CH_2-CHCl-O-O^{\bullet}$. These radicals decompose to $\sim CH_2-{}^{\bullet}CHCl$, HCHO, and HCOCl and the latter further to CO and HCl. Carbon monoxide copolymerizes with vinyl chloride to $\sim CH_2-CHCl-{}^{\bullet}C{=}O$ which isomerizes to $\sim CH_2-{}^{\bullet}CH-COCl$. Subsequent addition of vinyl chloride results in incorporated acryloyl chloride units.

Poly(vinyl chloride) must therefore be stabilized against dehydrochlorination and other adverse reactions (Volume IV). An alternative strategy for the improvement of heat stability of PVC is copolymerization of vinyl chloride with small proportions of other monomers. These comonomer units interrupt the sequence of vinyl chloride units and thus prevent the formation of polyene sequences.

Another peculiarity is the structure of suspension PVC. This polymerization leads to so-called primary particles of 0.5–1.5 μm diameter that aggreagate to PVC granules of 100–150 μm diameter. The granules are coated with a layer of protecting colloids and behave in flow processes as rheological units. They may also stick to the reactor wall where they cause the formation of hard particles in a new batch. The processing of such "contaminated" batches does not totally destroy the hard particles, which then leads to inhomogenieties in films.

**Properties**

PVC from polymerization at 55°C contains ca. 55 % syndiotactic units. S-PVC is slightly crystalline (< 10 %); these crystallites melt at 185–210°C.

Table 6-27 Properties of rigid PVC (as S-PVC), chlorinated PVC with 64 wt% chlorine) (PVC-C), poly(vinylidene chloride) (PVDC; no comonomer!), and plasticized PVC (PVC-P with 25 wt% diethylhexyl phthalate). [a] Amorphous; the density of crystalline PVDC is 1.96 g/cm³.

| Property | Physical unit | S-PVC | PVC-C | PVDC | PVC-P |
|---|---|---|---|---|---|
| Density | g/cm³ | 1.39 | 1.55 | 1.775 [a] | 1.32 |
| Melting temperature (DSC) | °C | 185–210 | | 200 | |
| Vicat temperature B | °C | 70-80 | 100 | | |
| Glass temperature (DSC) | °C | 80 | 110 | −18 | |
| Linear thermal expansion coeff. (100°C) | $K^{-1}$ | $(7–8)\cdot 10^{-5}$ | $6\cdot 10^{-5}$ | | |
| Specific heat capacity | $J\ K^{-1}\ g^{-1}$ | 0.85 | 0.85 | | |
| Thermal conductivity (20°C) | $W\ m^{-1}\ K^{-1}$ | 0.17 | 0.14 | | |
| Tensile modulus | MPa | 3000 | 3400 | | |
| Tensile strength at fracture | MPa | 50–60 | 75 | | 28 |
| Elongation at break | % | 10–50 | 10 | | 250 |
| Impact strength (Izod) | kJ/m | no break | 20 | | |
| Notched impact strength (Izod) | kJ/m | 2-5 | 2 | | |
| Hardness (Shore D) | - | 84 | | | |
| Relative permittivity (1 kHz) | 1 | 3.4 | | | |
| Resistivity (volume resistance) (2 min) | Ω cm | $> 10^{15}$ | $> 10^{15}$ | | $9.6\cdot 10^{14}$ |
| Electric strength | kV/mm | 10–40 | 12 | | |
| Dissipation factor (1 kHz) | 1 | 0.081 | | | |

Properties of PVCs are listed in Table 6-27. Rigid PVC is used for tubes, profiles, and films. Its poor weatherability can be improved so much by the addition of 0.15–0.2 % cadmium stabilizers that window frames made from PVC last more than 25 years. In Europe, such frames usually consist of graft copolymers of acrylates or vinyl acetate plus ethene on PVC. Cable sheetings are stabilized by lead compounds which increase the continuous service temperature and prevent electrical breakthroughs. PVC floor coverings contain neither cadmium nor lead stabilizers.

The flowability of PVC improves on addition of poly(methyl methacrylate) or by copolymerization of VCM with small proportions of vinyl acetate. Increased heat distortion temperatures (but decreased processibilities and thermal stabilities) are obtained by chlorinating PVC, the addition of compatible copolymers from α-methyl styrene/acrylonitrile or acrylonitrile/butadiene/styrene (ABS), or crosslinking PVC that contains small proportions of allyl acrylate units or by after-treatment of conventional PVC with multifunctional amines. Impact strengths of PVCs increase on the addition of modifiers (ABS, NBR, etc.) or by copolymerization of VCM with acrylic or fumaric esters.

About one-third of commercial PVCs is plasticized by adding low-molecular weight plasticizers ("monomer plasticizers"). Most common is di(2-ethylhexyl)phthalate (DEHP = dioctylphthalate (DOP)), the toxicity of which is neither acute nor chronic. Another common plasticizer is tricresol phthalate (see Volume IV).

Plasticized PVC is mostly delivered to processors as so-called **plastisol**, i.e., as a slurry of 40–70 % PVC granules in plasticizers. Plastisols also contain stabilizers, pigments, etc.; they gel at 180°C to plasticized PVC. **Organosols** (organisols) are plastisols that also contain volatile organic solvents. Plastisols and organosols are used to manufacture films and coatings as well as molded articles.

Burning of PVC waste produces hydrochloric acid; incinerators for plastics thus have alkaline scrubbing towers. PVC has been accused of releasing large amounts of dioxins on burning, but studies on household waste enriched with ca. five times the normal proportion of PVC did not indicate an increased amount of dioxins as compared to normal household wastes (ca. 200 μg dioxins per ton of waste).

**Derivatives**

PVC dissolves in only a few common solvents, for example, cyclohexanone or tetrahydrofuran. Acetone-soluble PVCs are obtained by after-chlorination in solution at 60–100°C or in the gel state at ca. 50°C. These **CPVC**s (**PVC-C**) have chlorine contents of 64–68 wt% instead of the usual 56.6 wt%. Both processes chlorinate predominantly $CH_2$. After-chlorinated PVC thus contains approximately equal amounts of –CHCl–CHCl– and $-CH_2-CHCl-$ groups and only a few $-CH_2-CCl_2-$ groups. After-chlorination in the gel state leads to PVC-C with a higher glass temperature than that from solution chlorination, probably because of a higher content of blocks of the same units. After-chlorinated PVC serves as an adhesive, binders for lacquers, or industrial fibers.

**Copolymers**

Many PVC grades contain small proportions of comonomers for better processability or improved properties (see above). Major proportions of comonomers are found in copolymers with vinyl acetate, propene, 1,3-butadiene, or butyl acrylate (graft copolymer).

Free-radical copolymerization of vinyl chloride and 3–20 % vinyl acetate in acetone or 1,4-dioxane as solvent or hexane as precipitant delivers soluble polymers (**PVCA**) for lacquers. Copolymers of VCM and 15 % VAC were used for hi-fi records.

A constant feed of propene is required for copolymerization of 90–97 % vinyl chloride and 3–10 % propene in order to prevent compositional drifts. Propene units interrupt the sequence of vinyl chloride units which in turn prevents zip-like dehydrochlorination (see above). Such copolymers are used for refrigerators, TV sets, vacuum cleaners, etc.

In the copolymerization of vinyl chloride and butadiene, ca. 40 % of butadiene units are cyclized in the 1,6-position. This copolymerization is not a cyclopolymerization, however, because the cyclohexane rings are side groups and not part of the main chain:

(6-47)

Cl Cl Cl → Cl Cl Cl

## 6.9.2 Poly(vinylidene chloride)

Vinylidene chloride, $CH_2=CCl_2$, is obtained by pyrolysis of 1,1,2-trichloroethane (p. 102, XIII). Free-radical homopolymerization of this monomer delivers a polymer (Saran A) with a melting temperature of ca. 200°C (Table 6-27) that decomposes at the required high processing temperatures.

Commercial poly(vinylidene chloride)s (**PVDC**) are therefore always copolymers: with 15–20 % vinyl chloride for packaging and shrink films (Saran B) or with vinyl chloride and acrylonitrile, alkyl acrylates, or methacrylates for coatings, e.g., with 2 % acrylonitrile and 13 % vinyl chloride. PVDC is known by many trade names; Saran is a free name in the United States but still a protected trademark elsewhere.

PVDCs have extraordinarily low permeabilities for gases and water vapor (Volume IV) and are therefore excellent packaging materials. The main disadvantage is the insufficient thermal stability on processing.

## 6.10 Fluorine-containing Polymers

### 6.10.1 Poly(tetrafluoroethylene)

The first fluorine-containing polymer was poly(tetrafluoroethylene) (**PTFE**) of DuPont. Its former trade name **Teflon** is now a free name that is used for a series of fluorinated polymers (Table 6-28) with a total world production of $1.1 \cdot 10^5$ t/a (2004).

**Polymerization**

Tetrafluoroethylene, $CF_2{=}CF_2$, is obtained in several steps from chloroform and hydrogen fluoride (p. 93, III). The monomer is a gas (bp –76.3°C) and is therefore polymerized at 10–80°C in aqueous suspension or emulsion under pressure (1–3 MPa) in order to remove the very high heat of polymerization. Local overheating must be avoided by good agitation and meticulous control of the polymerization, since the monomer can disproportionate explosively to C + $CF_4$ with the same heat development as black gunpowder ($\Delta H$ = –316 kJ/mol).

*Suspension polymerizations* employ little or no dispersants; dispersion is by strong agitation. Water acts predominantly as the heat-transfer agent. Initiators are redox systems or persulfates. Some of the Poly(tetrafluorethylene) suspension grades include $-CF_2-CF(OCF_2CF_2CF_3)-$ units for better processability.

The polymerization proceeds as a precipitation polymerization and produces porous granules with diameters of 0.2–1 mm. The granules are subsequently ground to smaller, more narrowly distributed powders since large-sized voids would lead to voids on processing. The powders are compacted and sintered to semi-finished goods by ram-extrusion at 380–400°C.

*Emulsion polymerization* employs perfluorinated organic peroxides as initiators and ammonium perfluorooctanoate as emulsifier. Hydrogen-containing initiators and emulsifiers cannot be used because they act as strong chain-transfer agents. Some grades contain hexafluoropropylene, chlorotrifluoroethylene, or perfluoropropyl vinyl ether units.

Addition of electrolytes precipitates PTFE as a fine, free-flowing powder consisting of practically pore-free spheres of ca. 0.2 μm diameter. After drying, they are converted to a paste by addition of ca. 20–25 wt% kerosene, compacted in an extruder, dried, and sintered at 360–380°C. Alternatively, one can concentrate the dispersion to 60–65 % and then apply latex processes such as spray-drying or dipping.

PTFEs are insoluble in all solvents. Molecular weights are therefore determined via initiator endgroups, for example, by saponification of the endgroups from $K_2S_2O_8$ as initiator (Eq.(6-48)). According to these data, molecular weights of PTFEs range between 1 and 10 million.

$$\text{(6-48)}\quad \sim(CF_2)_2CF_2-O-SO_2OK \xrightarrow[-\,HKSO_4]{+\,H_2O} \sim(CF_2)_2CF_2-OH \xrightarrow[-\,2\,HF]{+\,H_2O} \sim(CF_2)_3-COOH \longrightarrow \sim CF_2-CF{=}CF_2 + HF + CO_2$$

**Processing and Properties**

Poly(tetrafluoroethylene) crystallizes in a helix conformation and not in zigzag chains because the van der Waals radius of fluorine atoms (0.13 nm) is somewhat greater than that of hydrogen atoms (0.12 nm). The degree of crystallinity of PTFE is 92–98 %. Virgin PTFE melts at 342°C and molten and recrystallized PTFE at ca. 327°C. Other first-order transition temperatures are 30°C (helix-helix transition) and 19°C (triclinic-hexagonal transition). The glass temperature is 127°C (Table 6-28). PTFEs are therefore thermally stable up to continuous service temperatures of ca. 260°C. They start to decompose at 400°C; the main decomposition product is toxic perfluoroisobutylene.

Table 6-28 Properties of unfilled poly(tetrafluoroethylene) (PTFE), poly(chlorotrifluoroethylene) (PCTFE), poly(vinyl fluoride) (PVF), and poly(vinylidene fluoride) (PVDF). NB = no break.

| Property | Physical unit | PTFE | PCTFE | PVF | PVDF |
|---|---|---|---|---|---|
| Density | $g/cm^3$ | 2.28 | 2.11 | 1.38–1.57 | 1.75–1.80 |
| Melting temperature (DSC) | °C | 327 | 220 | 185–210 | 155–192 |
| Heat distortion temprature (0.46 MPa) | °C | 121 | 126 | | 140–168 |
| Glass temperature (DSC) | °C | 127 | 45 | 64 | –35 |
| Linear thermal expansion coefficient | $K^{-1}$ | $12\cdot10^{-5}$ | $6\cdot10^{-5}$ | $9\cdot10^{-5}$ | $(7–15)\cdot10^{-5}$ |
| Specific heat capacity | $J\ K^{-1}\ g^{-1}$ | 1.0 | 0.9 | | 1.3–1.4 |
| Thermal conductivity (20°C) | $W\ m^{-1}\ K^{-1}$ | 0.24 | 0.22 | | 0.18 |
| Tensile modulus | MPa | 410 | 1050–2100 | 1800 | 1000–2300 |
| Flexural modulus | MPa | 350–630 | | | 1100–2300 |
| Tensile strength at yield | MPa | 10 | | 23 | 38–52 |
| Tensile strength at fracture | MPa | 7–28 | 32–40 | 50–125 | 42–59 |
| Flexural strength | MPa | NB | 52–90 | | |
| Compressive strength | MPa | 12 | 32–52 | | 55–90 |
| Elongation at fracture | % | 100–600 | 130–175 | 115–250 | 50–300 |
| Notched impact strength (Izod, 3.1 mm) | J/m | 160 | 140 | | 75–230 |
| Hardness (Shore D) | – | 50–65 | 76–80 | | 77–80 |
| Relative permittivity (1 kHz) | 1 | 2.1 | 2.5 | 8.5 | 8–9 |
| Electrical resistance (surface resistivity) | Ω | $> 10^{18}$ | $> 1.2\cdot10^{18}$ | | |
| Resistivity (volume resistance) (2 min) | Ω cm | $10^{17}$ | $10^{16}$ | $3\cdot10^{13}$ | $2\cdot10^{14}$ |
| Electric strength | kV/mm | 60–80 | | 135 | |
| Dissipation factor (50 Hz) | 1 | 0.0002 | 0.025 | 0.008 | 0.049 |
| Water absorption (24 h) | % | < 0.01 | 0 | 0.5 | 0.04 |

Homopolymers are resistant against chemicals, especially oxygen, because PTFE does not contain C–H bonds and its C–F bonds have smaller covalent radii than C–H bonds (0.072 nm versus 0.077 nm). Unfortunately, it flows under load (cold flow).

PTFEs have very low critical surface tensions (Volume IV). They are not wetted by oils or water and serve therefore as coatings for frying pans where they compete with poly(phenylene sulfide).

Melts are extremely viscous ($10^{10}$ Pa s at 380°C). High-molecular weight PTFEs cannot be melt-processed but mixtures of high- and low-molecular weights can.

Films of PTFE can be glued if their surfaces are pretreated with sodium. The reaction $>CF_2 + Na \rightarrow >^{\bullet}CF + NaF$ produces free radicals that can react with the adhesive. Industry has also developed hot melt adhesives based on perfluorinated polymers.

## 6.10.2 Copolymers of Tetrafluoroethylene

PTFE has excellent use properties but is difficult to process. Industry has therefore developed many fluorine polymers that are more easy to process. A favorite strategy is to copolymerize tetrafluoroethylene (TFE) with small or great proportions of other monomers (Table 6-29) that deliver monomeric units which disturb crystallization. These copolymerizations usually proceed in suspension or emulsion by free-radical initiation. Industrially used comonomers include

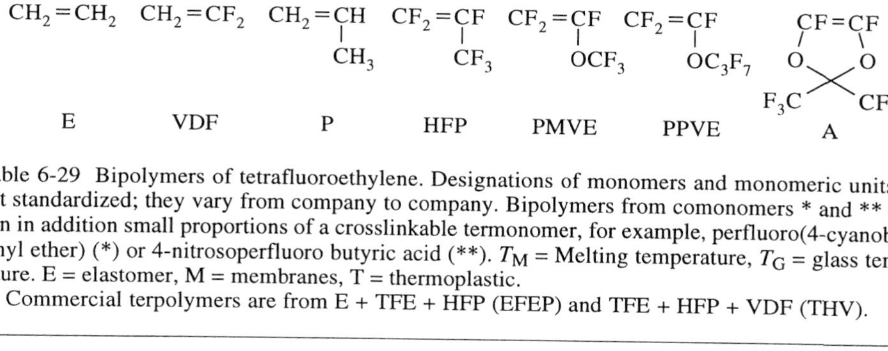

| $CH_2{=}CH_2$ | $CH_2{=}CF_2$ | $CH_2{=}CH(CH_3)$ | $CF_2{=}CF(CF_3)$ | $CF_2{=}CF(OCF_3)$ | $CF_2{=}CF(OC_3F_7)$ | $CF{=}CF$ ring with two O bridged to $C(CF_3)_2$ |
|---|---|---|---|---|---|---|
| E | VDF | P | HFP | PMVE | PPVE | A |

Table 6-29 Bipolymers of tetrafluoroethylene. Designations of monomers and monomeric units are not standardized; they vary from company to company. Bipolymers from comonomers * and ** contain in addition small proportions of a crosslinkable termonomer, for example, perfluoro(4-cyanobutyl vinyl ether) (*) or 4-nitrosoperfluoro butyric acid (**). $T_M$ = Melting temperature, $T_G$ = glass temperature. E = elastomer, M = membranes, T = thermoplastic.

Commercial terpolymers are from E + TFE + HFP (EFEP) and TFE + HFP + VDF (THV).

| Copolymer Type | Comonomer name | Comonomeric units mol% | Sequence | $T$ in °C $T_M$ | $T_G$ | Application |
|---|---|---|---|---|---|---|
| PTFE | - | 0 | - | 325 | −150 | T |
| ETFE | Ethene (E) | 50 | alternating | 270 | | T |
| | Ethene (E) | 45 | | 275 | 110 | T |
| PT | Propene (P) | 45 | | | −2 | E |
| FEP | Hexafluoropropene (HFP) | | | 265 | | T |
| PFA | Perfluoropropyl vinyl ether (PPVE) | | | 305 | | T |
| AF | 2,2-Bis(trifluoromethyl)-4,5-difluoro-1,3-dioxole | 18 | | | 160 | T |
| | | 37 | | | 240 | T |
| MFA | Perfluoro(methyl vinyl ether) (PMVE) * | < 50 | random | 285 | −12 | E |
| CNR | Trifluoronitroso methane ** | 50 | alternating | | −50 | E |
| XR | Sulfonylfluoride vinyl ether | 9–16 | | | | M |
| CF | ω-Carbalkoxy-perfluoralkoxy vinyl ether | 10–15 | | | 12 | M |

Copolymerization of TFE and ethene (E) delivers thermoplastic ETFE whereas propene (P) as comonomer leads to elastomeric PT. All thermoplastic copolymers have end-use properties similar to PTFE but can be processed as melts.

A special class of thermoplastic TFE copolymers contains comonomeric units with acid groups such as $-SO_2H$ (XR) and –COOH (CF) which result from the saponification of sulfonyl fluoride groups $-SO_2F$ or carbalkoxy groups –COOR of the original monomeric units. These copolymers form chemically resistant and temperature-stable permselective membranes.

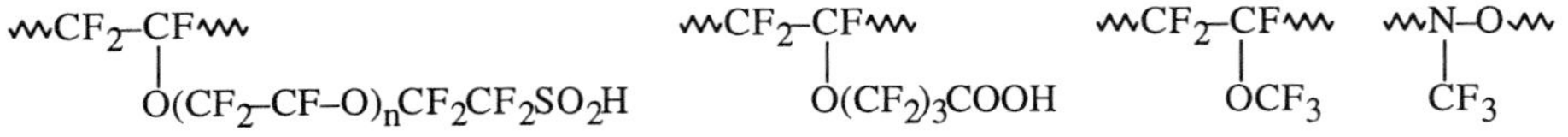

$$\sim CF_2{-}\underset{O(CF_2{-}CF{-}O)_nCF_2CF_2SO_2H}{\underset{|}{CF}}\sim \qquad \sim CF_2{-}\underset{O(CF_2)_3COOH}{\underset{|}{CF}}\sim \qquad \sim CF_2{-}\underset{OCF_3}{\underset{|}{CF}}\sim \qquad \sim \underset{CF_3}{\underset{|}{N}}{-}O\sim$$

XR CF FM CNR

TFE copolymers with large proportions of bulky comonomeric units such as FM or CNR are elastomers. **Carboxynitroso polymers** (**CNR**, **AFMU**) are crosslinked by reaction of their termonomeric 4-nitrosoperfluoro butyric acid units with diamines. Terpolymers of TFE, perfluoro(methyl vinyl ether), and perfluoro(4-cyanobutyl vinyl ether) are crosslinked by cyclotrimerization of the cyano groups of the latter monomeric units to *s*-triazine rings. Fluoroelastomers can be used in a wide temperature range, for example, as sealants in jet planes and as containers for rocket fuels.

### 6.10.3 Poly(chlorotrifluoroethylene)

Poly(chlorotrifluoroethylene) (**PCTFE**), $\text{+}CFCl{-}CF_2\text{+}_n$, was developed as a more easily processable competitor to poly(tetrafluoroethylene). Chlorotrifluoroethylene is obtained from hexachloroethane via 1,1,2-trifluoro-1,2,2-trichloroethane. Polymerization is in suspension by the redox system $K_2S_2O_8/NaHSO_3/AgNO_3$ or in bulk by bis(trichloroacetyl peroxide) as initiator.

PCTFE melts at a lower temperature (220°C) than PTFE (327°C), and can therefore be processed at 250–300°C with the usual plastics machinery. However, processing machines (extruders, etc.) need to be made of non-corroding steel since PCTFE starts to decompose at these processing temperatures. For properties of PCTFE, see Table 6-28.

The alternating copolymer of ethene and chlorotrifluoroethylene (**ECTFE**) has a higher melting temperature than PCTFE (240°C versus 220°C). It resembles both poly(vinylidene fluoride) (Section 6.10.5) and the alternating copolymer of tetrafluoroethylene and ethylene (Table 6-29). Melts of ECTFE can be spun to fibers.

### 6.10.4 Poly(vinyl fluoride)

Vinyl fluoride, $CH_2{=}CHF$, is obtained from ethene via vinyl chloride (p. 102, XIIIc) or from acetylene (p. 104, III). Because of its low boiling temperature of –72°C, it is polymerized in bulk at a pressure of 30 MPa. **PVF** is a partially crystalline polymer with physical properties (Table 6-28) that resemble those of poly(ethylene) (Table 6-10). Melting temperature (200°C) and weatherability are much higher, however. PVF is used as a film and in painting systems.

### 6.10.5 Poly(vinylidene fluoride)

Vinylidene fluoride (VDF), $CH_2=CF_2$, is synthesized by pyrolysis of 1,1-difluoro-1-chloroethane which, in turn, is obtained by reaction of hydrogen fluoride with acetylene, vinylidene chloride, or 1,1,1-trichloroethane. Its low boiling temperature (–84°C) requires polymerization under pressure, either in suspension or in emulsion. The free-radical polymerization delivers poly(vinylidene fluoride)s (**PVDF**) with considerable proportions of head-to-head arrangements of monomeric units (Volume I, p. 45).

PVDF is a thermoplastic with excellent weatherability and resistance to chemicals that crystallizes in four modifications. Its physical properties resemble more poly(ethylene) than poly(vinylidene chloride) (Table 6-28). Its orthorhombic β-modification is piezoelectric: deformation of films generates electric currents and vice versa.

Free-radical suspension polymerization of vinylidene fluoride and hexafluoroisobutene, $CH_2=C(CF_3)_2$, delivers an alternating copolymer that has not only the same melting temperature as poly(tetrafluoroethylene) but also practically the same mechanical properties, except lower fracture and impact strengths. Poly(vinylidene fluoride-*alt*-hexafluoroisobutylene) can be melt-processed.

Specialty rubbers are obtained from copolymerizations of vinylidene fluoride with either of the following monomers: chlorotrifluoroethylene; 1-hydropentafluoropropylene; 1-hydropentafluoropropylene + tetrafluoroethylene; hexafluoroisobutylene; hexafluoropropylene; hexafluoropropylene + tetrafluoroethylene (with or without crosslinkable quatermonomer); or hexafluoropropylene + tetrafluoroethylene + perfluoromethyl vinyl ether + crosslinkable quintermonomer.

Crosslinkable monomeric units allow vulcanizations by peroxides instead of diamines or bisphenols. The former crosslinked polymers do not contain hydrolyzable crosslinks and are therefore considerably more resistant against hot water and water vapor. Crosslinking by peroxides also does not release water as do those by diamines and bisphenols. This allows vulcanization to void-free products without or under only small pressure.

## 6.11 Acrylic Polymers

Acrylic compounds are derivatives of acrylic acid, $CH_2=CH–COOH$, which has a pungent smell (L: *acer* = sharp). Methacrylic compounds are similarly based on methacrylic acid, $CH_2=C(CH_3)–COOH$, and its derivatives. In the plastics industry, **acrylics** (acrylic resins, acrylate resins) are polymers of (meth)acrylic acid and its derivatives, especially poly(methyl methacrylate). In the fiber industry, "acrylics" refers to fibers that contain at least 85 % acrylonitrile units $–CH_2–CH(CN)–$.

### 6.11.1 Poly(acrylic acid)

Acrylic acid (propenoic acid), $CH_2=CHCOOH$, has been and is being prepared by many different processes. No longer used is the Reppe process based on $C_2H_2$ + CO + $H_2O$ (p. 92, 98), the thermolysis of propiolactone (from formaldehyde + ketene; p. 96, IX), and the ethylene cyanohydrin process based on dehydration and hydrolysis of hydroxypropionitrile (from ethylene oxide + HCN). The hydrolysis of acrylonitrile to acrylic acid is only used in Japan (p. 102, Xa). Presently, the main industrial process is the successive oxidation of propene → acrolein → acrylic acid (p. 108, IVa). A new process oxycarbonylates ethene (p. 102, XI).

Acrylic acid and its alkali salts are polymerized in aqueous solution by $K_2S_2O_8$ as initiator. Industrially, poly(acrylic acid) is also obtained by hydrolysis of its esters.

Poly(acrylic acid) (**PAA**) is a good flocculant for the treatment of waste water. It is also used to distribute pigments in water-based paints. As a polyfunctional compound, PAA serves to seal the surface of leathers by crosslinking collagen.

Partially neutralized, crosslinked poly(acrylic acid) is a superabsorbent that can absorb 1000 times its weight of water. Its leading application is for baby diapers (world production now approaching $10^6$ t/a; adds 2 % to landfills!). The leading commercial process copolymerizes acrylic acid and a crosslinking agent in solution, adds a crosslinked polyamide and starch, and dries the soft gel on a conveyor belt. The dried material is milled to an amorphous powder whose particles are crosslinked by heating with ethylene carbonate. For comfort, crosslinking is adjusted in such a way that the urine uptake of such diapers is limited to 30 wt%.

The highly viscous aqueous solutions of salts of poly(acrylic acid) serve as sizes, thickeners for latices or cosmetics, drilling agents in oil production, and flocculants. For the latter two applications, copolymers of acrylic acid and acrylamide are preferred. These copolymers also serve as soil improvers. On addition of alum (= potassium aluminum sulfate, $KAl(SO_4)_2 \cdot 12H_2O$), low-molecular weight copolymers improve the dry strength of paper.

Ethene-acrylic acid copolymers (**EAA**) serve as adhesives for metals, paper, and other materials. Applications range from tubes for toothpaste to wires and cables.

Oxidative polymerization of aqueous solutions of acrolein, $CH_2{=}CH{-}CHO$, by $H_2O_2$ as initiator/oxidizing agent delivers so-called **poly(aldehydocarboxylic acid)s** that contain 60-85 mol% acrylic acid units and 15-40 mol% acryl aldehyde units. These anionic polyelectrolytes complex metal ions and serve as hardness stabilizers in cooling water.

## 6.11.2 Poly(acrylate)s

Esterification of acrylic acid by alcohols delivers acryl esters, $CH_2{=}CHCOOR$. Acrylic esters have low glass temperatures (8°C (R = $CH_3$), –22°C ($C_2H_5$), –58°C ($C_6H_{13}$)) and are therefore polymerized in emulsion by free-radical initiators. The resulting emulsions are used as obtained for sizes, car paints, floor polishes, etc.

Poly(pentabromobenzyl acrylate), $\text{+}CH_2{-}CH(COOCH_2C_6Br_5)\text{+}_n$, serves as compatible flame retardant for glass fiber-reinforced polyamide 6 and poly(butyl terephthalate).

Monomeric and oligomeric acrylic esters with functional endgroups serve as solvent-free, low VOC coating compounds (VOC = volatile organic compound). Functional endgroups R comprise, for example, trimethylolpropane triacrylate (I), ethoxylated trimethylolpropane triacrylates (II), acrylated epoxides (III), and acrylated urethanes (Fig. 6-23). These compounds are mixed with suppliers of photochemically generated radicals, e.g., benzophenone derivatives, applied, and then cured-in-place (= hardened = crosslinked) by UV radiation, electron beams (EBC = electron beam curing), or thermally by added free-radical initiators (see Volume IV).

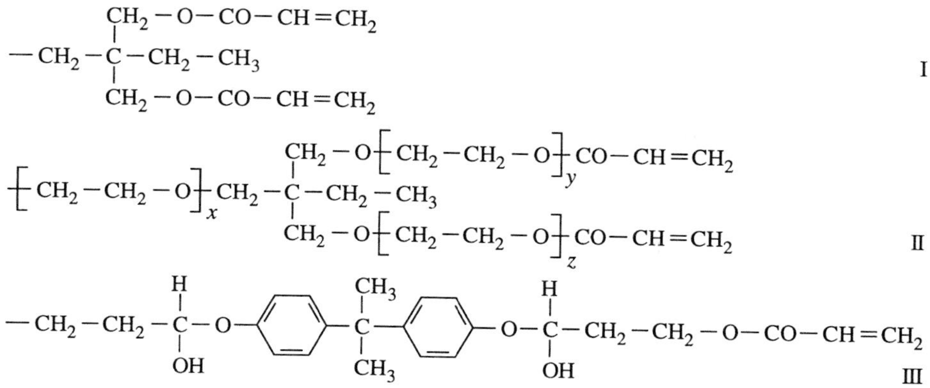

$$\begin{array}{l} \quad\quad CH_2-O-CO-CH=CH_2 \\ -CH_2-\overset{|}{\underset{|}{C}}-CH_2-CH_3 \\ \quad\quad CH_2-O-CO-CH=CH_2 \end{array} \qquad \text{I}$$

$$\begin{array}{l} \quad\quad\quad\quad\quad\quad\quad\quad CH_2-O\left[CH_2-CH_2-O\right]_y CO-CH=CH_2 \\ \left[CH_2-CH_2-O\right]_x CH_2-\overset{|}{\underset{|}{C}}-CH_2-CH_3 \\ \quad\quad\quad\quad\quad\quad\quad\quad CH_2-O\left[CH_2-CH_2-O\right]_z CO-CH=CH_2 \end{array} \qquad \text{II}$$

$$-CH_2-CH_2-\overset{H}{\underset{OH}{C}}-O-C_6H_4-\overset{CH_3}{\underset{CH_3}{C}}-C_6H_4-O-\overset{H}{\underset{OH}{C}}-CH_2-CH_2-O-CO-CH=CH_2 \qquad \text{III}$$

Fig. 6-23 Functional endgroups for low VOC acrylic coating compounds.

Copolymerization of acrylic esters with 5–15 % acrylonitrile delivers **acrylate rubbers (ANM)** that do not contain carbon–carbon double bonds and are therefore more resistant to oxidation and heat than acrylonitrile-butadiene rubbers. They are used as sealants and membranes for high-sulfur oils, for example, in cars. These rubbers are cold- vulcanized by diamines and not hot by steam since their acrylic ester side groups are prone to hydrolysis.

Other saturated rubbers are terpolymers of ethyl acrylate, (m)ethoxyethyl acrylate, and small proportions of vinyl chloroacetate (**ACR**). **Acrylic rubbers** (**AR**) are a series of terpolymers of ethene, acrylic acid, and various acrylic esters, i.e., methyl acrylate (**EMAC**, **ACM**), ethyl acrylate (**EEA**), or butyl acrylate (**EBAC**). The latter rubbers are crosslinked by hexamethylenediamine carbamate. Especially oil-resistant rubbers are based on poly(1,1-dihydroperfluorobutyl acrylate) and poly(3-perfluoromethoxy-1,1-dihydroperfluoropropyl acrylate).

Emulsion polymers of butyl acrylate (**PBA**) deliver after grafting by styrene plus acrylonitrile so-called **ASA** polymers which consist of PBA particles that are dispersed in a SAN matrix.

**Vinyl ester resins** (= **vinyl resins, modified vinyl resins**) do not contain vinyl groups. They are rather macromonomers from the reaction of an epoxide resin with acrylic acid (R = H) or methacrylic acid (R = $CH_3$):

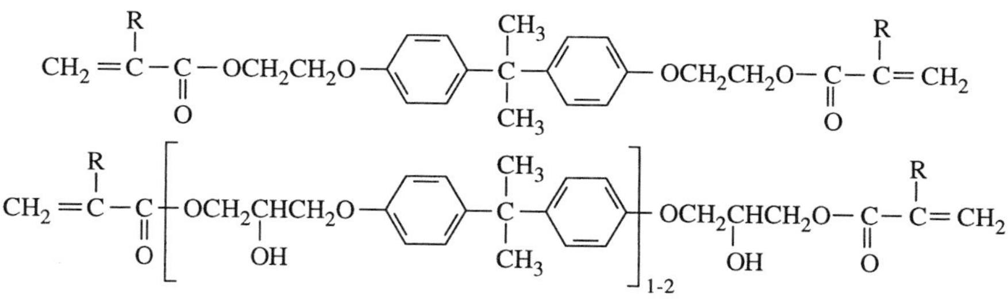

$$CH_2=\overset{R}{\overset{|}{C}}-\underset{\underset{O}{\|}}{C}-OCH_2CH_2O-C_6H_4-\overset{CH_3}{\underset{CH_3}{C}}-C_6H_4-OCH_2CH_2O-\underset{\underset{O}{\|}}{C}-\overset{R}{\overset{|}{C}}=CH_2$$

$$CH_2=\overset{R}{\overset{|}{C}}-\underset{\underset{O}{\|}}{C}\left[-OCH_2\underset{OH}{\underset{|}{C}}HCH_2O-C_6H_4-\overset{CH_3}{\underset{CH_3}{C}}-C_6H_4-\right]_{1-2}OCH_2\underset{OH}{\underset{|}{C}}HCH_2O-\underset{\underset{O}{\|}}{C}-\overset{R}{\overset{|}{C}}=CH_2$$

Vinyl ester resins are sold as solutions in styrene (S) or vinyl toluene (VT). They are "hardened" like unsaturated polyester resins (Section 7.7), i.e., by free-radical polymerization of the solvents S or VT with the vinyl ester resin as crosslinking agent.

## 6.11.3 Poly(acrolein)

Acrolein (propenal, acryl aldehyde, allyl aldehyde), $CH_2$=CH–CHO, is obtained by oxidation of propene. Dehydration of glycerol and reaction of acetaldehyde with formaldehyde are no longer industrial processes.

The free-radical polymerization of acrolein leads to polymers with many different monomeric units, for example:

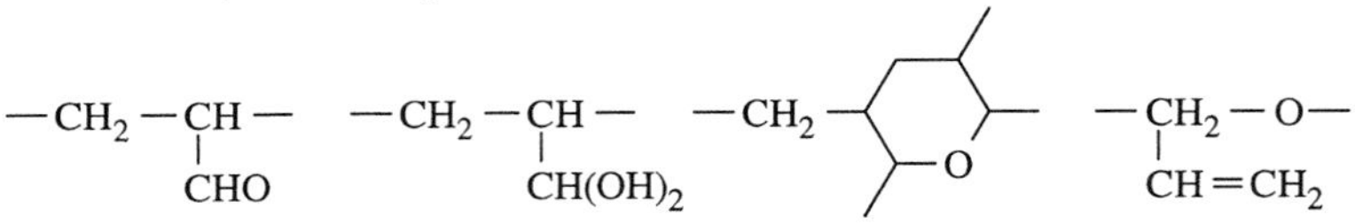

$$-CH_2-\underset{\substack{|\\ CHO}}{CH}- \quad -CH_2-\underset{\substack{|\\ CH(OH)_2}}{CH}- \quad -CH_2-\text{(tetrahydropyran ring)}- \quad -\underset{\substack{|\\ CH=CH_2}}{CH_2}-O-$$

The homopolymer has a relatively high glass temperature of 85°C, probably caused by ring structures. Ionic polymerizations proceed via the vinyl group, the aldehyde group, or both, depending on initiator and reaction conditions.

Homopolymers are not industrially produced. Oxidative copolymerization of acrolein and acrylic acid leads to rust removers. 40 % aqueous solutions of condensation products of acrolein and formaldehyde serve as biocides, for example, in circulating cooling waters.

## 6.11.4 Poly(acrylonitrile)

Acrylonitrile (vinyl cyanide, propene nitrile), $CH_2$=CHCN, is obtained by ammonoxidation of propene with $NH_3$ and $O_2$ (p. 108, X). Use of propane instead of propene reduces the selectivity to ca. 30 %.

Syntheses of acrylonitrile via hydroxypropionitriles as intermediates are no longer used, i.e., from acetaldehyde and hydrogen cyanide via $\alpha$-hydroxypropionitrile (2-hydroxypropionitrile, lactonitrile, cyanoethanol), $CH_3CH(OH)CN$, or from ethylene oxide and hydrogen cyanide via $\beta$-hydroxypropionitrile (3-hydroxypropionitrile, ethylene cyanohydrin, hydracrylic acid nitrile), $HOCH_2CH_2CN$.

The world capacity of acrylonitrile is $4.3 \cdot 10^6$ t/a. (1996). Approximately 60 % of acrylonitrile is used for acrylic fibers, ca. 20 % for acrylonitrile-butadiene-styrene and styrene-acrylonitrile thermoplastics, ca. 4 % for nitrile rubbers, ca. 8 % for adiponitrile, and the rest for the synthesis of acrylamide (Section 6.11.5), 1,4-tetramethylenediamine (Section 10.2.2), and others.

Monomeric acrylonitrile (**AN**) dissolves in water. Its polymerization to poly(acrylonitrile) I in acidic solution starts as a solution polymerization and ends as a precipitation polymerization. Polymerization in alkaline solution delivers yellowish polymers, probably because some nitrile groups polymerize to ladder structures II that are then oxidized by oxygen to nitrones III:

(6-49)

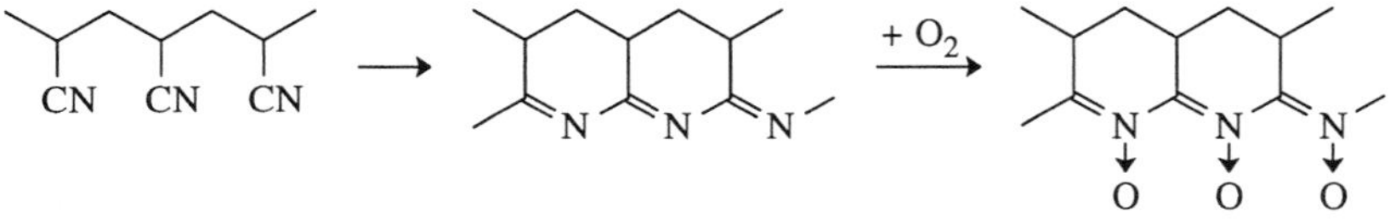

CN CN CN → N N N $\xrightarrow{+O_2}$ N→O N→O N→O

I II III

Commercial poly(acrylonitrile) is an atactic polymer with ca. 47 % syndiotactic diads and a degree of crystallization of ca. 30 %. The crystallinity of this atactic polymer is probably caused by cocrystallization of longer isotactic and syndiotactic sequences.

Poly(acrylonitrile) dissolves only in very polar solvents such as *N,N*-dimethylformamide, *γ*-butyrolactone, dimethyl sulfoxide, ethylene carbonate, and azeotropic nitric acid. Spinning from these solvents by dry or wet processes delivers fibers with good resistance to light and weather (see Volume IV). However, poly(acrylonitrile) fibers should not be ironed at temperatures above 150°C because of a possible cyclization of I → II.

Poly(acrylonitrile) fibers are difficult to dye but this can be remedied by copolymerization with ca. 4 % comonomers which can be basic (2-vinyl pyridine, *N*-vinyl pyrrolidone, ethylene imine), acidic (acrylic acid, methallylsulfonic acid, itaconic acid), or plasticizing (vinyl acetate, (meth)acrylic ester)s, vinyl chloride, vinylidene chloride).

By definition, **acrylic fibers** contain at least 85 % acrylonitrile units. **Modacrylic fibers** consist of 35–84 % AN units; comonomeric units are usually vinyl chloride or vinylidene chloride. The official ASTM symbol for acrylic fibers is **PAN** which is a registered trademark in Europe, however. The IUPAC symbol is therefore **PAC**.

Poly(acrylonitrile) fibers are precursors for carbon and graphite fibers (Section 6.1.6 and Volume IV). Stretched fibers are first heated to 200–300°C in an oxidizing atmosphere, which causes the molecules to crosslink and stabilizes the fiber shape. Further heating to 1000–2000°C in vacuum or under an argon blanket leads to ring systems II that are subsequently graphitized at 3000°C.

### 6.11.5 Poly(acrylamide)

Acrylamide, $CH_2{=}CHCONH_2$, is now mainly obtained by hydration of acrylonitrile using a copper-containing catalyst. In an obsolete method, acrylonitrile was reacted with $H_2SO_4{\cdot}H_2O$; acrylamide and its sulfate was separated by neutralization with lime (= calcium carbonate). In Japan, acrylonitrile is also hydrated biochemically in yields of 99 % by immobilized cells of *Pseudomonas chloraphis* B 23 or *Rhodococcus* sp. N 774.

Acrylamide dissolves in water and polymerizes in aqueous acidic solution by free radicals to poly(acrylamide) with monomeric units I, molecular weights between $10^3$ and $5{\cdot}10^7$, and a glass temperature of 195°C. At temperatures above 140°C, poly(acrylamide) releases ammonia, either intramolecularly to units II or intermolecularly to units III. In strong alkali solution, amide groups are hydrolyzed to carboxylic groups.

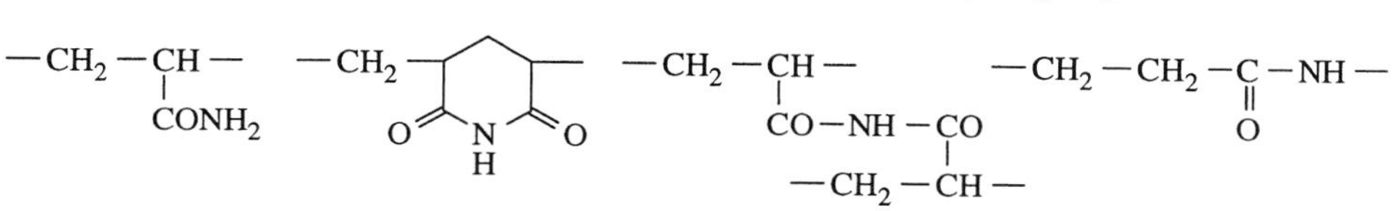

I: $-CH_2-CH(CONH_2)-$

II: $-CH_2-$ (six-membered glutarimide ring: C=O, N–H, C=O) $-$

III: $-CH_2-CH(CO-NH-CO)-$ linked to $-CH_2-CH-$

IV: $-CH_2-CH_2-C(=O)-NH-$

Anionic polymerizations by strong bases in organic solvents do not lead to poly(acrylamide) but to poly(*β*-alanine) (nylon 3, PA 3) with monomeric units IV.

In Western countries, 50–60 % of poly(acrylamide) is used for water treatment, ca. 20–40 % for paper treatment, ca. 10 % for mining operations and tertiary oil recovery, and ca. 5 % for other uses (tanning, etc.). Crosslinked poly(acrylamide)s are used as carriers for enzymes and as the matrix for electrophoresis. Reaction of poly(acrylamide)s with aldehydes delivers methylol compounds for textile applications.

The use of poly(acrylamide)s for water treatment and other applications has raised concerns because acrylamide is a potential human carcinogen with neurotoxic effects (it is carcinogenic in animals). It has therefore been suggested to replace acrylamide for certain applications by *N*-vinyl formamide (NVF), $CH_2$=CH–NH–CHO. This compound is obtained from the thermal decomposition of cyanoethylformamide (from $CH_3CHO$ + HCN + $HCONH_2$) to NVF + HCN or from the reaction of hydroxyethylformamide with formamide.

A sulfonic acid group-containing acrylamide derivative for homo and copolymerizations is obtained from acrylonitrile, isobutene, and sulfuric acid in a one-step Ritter-type reaction at temperatures between –40°C and + 60°C:

$$CH_2{=}CH(CN) + (CH_3)_2C{=}CH_2 + H_2SO_4 \longrightarrow CH_2{=}CH{-}CO{-}NH{-}C(CH_3)_2{-}CH_2{-}SO_3H \qquad (6\text{-}50)$$

## 6.11.6 Poly($\alpha$-cyanoacrylate)s

Poly($\alpha$-cyanomethyl acrylate), $+\!\!\!-CH_2{-}C(CN)(COOCH_3)\!-\!\!\!+_n$, was developed in World War II as a glass-clear plastic for gun sights but turned out to be much too sticky. Since $\alpha$-cyanoacrylic esters, $CH_2$=C(CN)(COOR), polymerize already in the presence of very weak bases such as water, they are used as fast one-component adhesives ("superglue"), usually as mixtures of monomers, thickeners, plasticizers, and stabilizers (e.g., $SO_2$).

Polymers of higher homologues (R = butyl, hexyl, heptyl) are well wetted by blood. They are therefore used not only to stop bleeding by the formation of a polymer film on the surface of tissue but also by surgeons to glue cuts and by morticians to fix lips and eye lids. Since the monomers attack neighboring cells, they can only be used as glues for those living tissues where necrosis (cell death; G: *nekros* = corpse) can be tolerated; examples are surgeries of liver and kidney but not of the heart. The polymer film decomposes biologically within 2–3 months. Formaldehyde from the decomposition of the polymer is an antiseptic that either reacts in the body with $NH_3$ to urea or is oxidized to $CO_2$ and $H_2O$.

## 6.11.7 Poly(methyl methacrylate)

The classic industrial route to methyl methacrylate (**MMA**) reacts acetone cyanohydrin, $(CH_3)_2C(OH)(CN)$, from the addition of HCN to acetone, with concentrated sulfuric acid to an amide sulfate which is then esterified to MMA by methanol (p. 97). This process produces 2.5 kg of unwanted $NH_4[HSO_4]$ per 1 kg of MMA (see p. 97). Since poly(methyl methacrylate) (**PMMA**) is a major polymer (world production ca. $1{\cdot}10^6$ t/a (2004)), industry has developed many alternative MMA syntheses.

One possibility is to replace route I→II→III of the cyanohydrin process, Eq.(4-14), by route I→IV→V→III, but this involves an additional step and also the conversion of the byproduct $HCONH_2$ to HCN. The *t*-butanol route oxidizes $(CH_3)_3COH$ (from isobutene and $O_2$) to a mixture of methacrolein, $CH_2=C(CH_3)CHO$, and methacrylic acid, $CH_2=C(CH_3)COOH$ (p. 95, V; p. 101, VII). The same mixture results from the reaction of propioaldehyde (from the oxo reaction of olefins) with formaldehyde (p. 116, III). Other new processes synthesize methacrylic acid by addition of $CH_3OH$ + CO to methylacetylene, $CH_3C{\equiv}CH$, or react ethene, methanol, and CO at 100°C and 1 MPa (Pd catalyst) to methyl propionate, $CH_3CH_2COOCH_3$, that is then converted by HCHO at 325°C and 0.1 MPa to methyl methacrylate and water.

**Polymerization in Bulk**

Methyl methacrylate is polymerized industrially by free-radical initiators mainly in bulk and also in solution, emulsion, or suspension. The choice of the medium depends on the desired application which dictates the processing method and, in turn, the maximum molecular weight. Continuous bulk polymerizations are used to synthesize PMMAs for injection molding ($\overline{M}_w \approx 120\,000$) or extrusion ($\overline{M}_w \approx 180\,000$). PMMAs with $\overline{M}_w > 300\,000$ can only be obtained by continuous or discontinuous casting polymerization and processed by machining or deep drawing. Suspension polymerization delivers beads for molding powders and ion-exchange resins (see below). PMMAs as binders in lacquers and for adhesives are usually prepared discontinuously in solution or emulsion.

*Continuous* bulk polymerization of MMA is difficult to control because of the high heat of polymerization during the conversion of the liquid to the condensed state ($\Delta H_{lc}$ = –56 kJ/mol), the low ceiling temperature ($T_{lc}$ = 102°C (Volume I, p. 213)), and the onset of the glass effect at relatively low monomer conversions (see Volume I, p. 337):

For better heat control, MMA is prepolymerized at 90°C to a viscosity of ca. 1 Pa s; this not only drives out dissolved oxygen but also reduces contraction during afterpolymerization. Higher viscosities of the prepolymer lead to formation of voids.

Most commercial PMMAs contain small proportions of methyl, ethyl, or butyl acrylate comonomer units that stop the unzipping of PMMA chains (Volume I, p. 594). The stability of PMMA chains is increased further by regulators for the polymerization process (Volume I, p. 350) which, by reaction with macroradicals, reduce the proportion of head-to-head bonds from combination of macroradicals and double bonds from disproportionation reactions of macroradicals. The incorporation of small proportions of acrylic ester units also decreases the glass temperature and thus the processability.

The glass effect is overridden by keeping the temperature during the later stages of polymerization always higher than the glass temperature.

High-molecular weight PMMA sheets are obtained by bulk polymerization in which a prepolymer of ca. 20 % PMMA in MMA plus initiator is filled into a chamber between two mirror-quality glass panes (Fig. 6-24). In order to balance the ca. 20 % contraction on polymerization, spacers consist of compressible elastomers. They are removed once the mass is sufficiently, but not completely, solidified. The panes then follow the shrinking slab. Heat of polymerization is removed by air cooling. The polymerization is slow: at 50°C, days are required for a 90 % monomer conversion to a 5-cm thick sheet. Residual monomer is polymerized just above the glass temperature of PMMA (ca. 115°C).

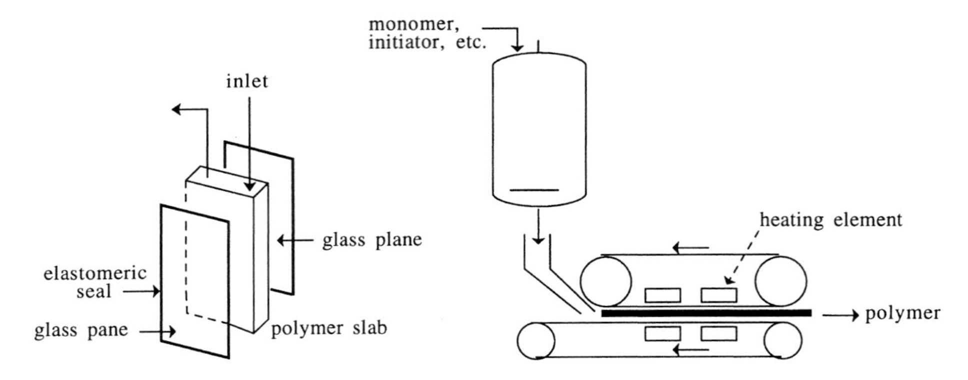


Fig. 6-24 Discontinuous (left) and continuous (right) bulk polymerization of MMA to slabs [15].

Polymerization is considerably faster by the continuous band process which uses steel conveyors (Fig. 6-24, right). Polymers from this process have lower molecular weights, though, and therefore lower strengths. Surface properties are also not as good.

PMMA has moderate strength, good weatherability, and excellent optical properties (Table 6-30). Because of its high transparency (92 % for slabs from the discontinuous process), is is called "organic glass"; famous trademarks are Plexiglas® and Lucite®. It is used for lamps, neon signs, TV screens, etc. For aircraft windows, a crosslinked copolymer of MMA and glycol dimethacrylate is used.

The excellent optical properties of PMMA are also used in fiber optics and for hard contact lenses. These lenses are not permeable for gases, though, so that the human eye cannot breathe. Gas-permeable hard contact lenses consist, for example, of copolymers from methyl methacrylate and methacrylic comonomers that carry siloxane groups in the side chains.

**Solution and Emulsion Polymerizations**

Solution and emulsion polymerizations of methacrylate monomers are used for the synthesis of air-drying lacquers such as copolymers of MMA and lauryl methacrylate as well as heat-drying ones like copolymers of MMA and glycidyl or glycol esters of methacrylic acid. Incorporation of functional groups (–COOH, $–CONH_2$, etc.) adjusts poly(methyl methacrylate) for specific uses (metals, wood, paper, etc.). Water-soluble copolymers consist of methyl methacrylate and methacrylic acid units that are neutralized by ammonia.

Methyl methacrylate and methacrylic esters of higher alcohols (for example, lauryl alcohol) are copolymerized in mineral oil. They serve as viscosity improvers for mineral oils (Volume IV, Section 14.4.3).

**Suspension Polymerization**

Suspension polymers are used as injection molding and extrusion grades. Methacrylic polymers for dental plates and tooth fillings are special types that are filled with finely milled glass or silica. An example is the polymer from isopropylidene *bis*[2(3)-hydroxy-3(2)-4-phenoxypropylene methacrylate].

Table 6-30 Properties of poly(methyl methacrylate) (PMMA), styrene-acrylonitrile copolymers (SAN) with 75/24 (w/w) = 62/38 (mol S/mol AN), poly(methacrylonitrile) (PMAN), and alternating ethene-carbon monoxide copolymers (PK).

| Property | Physical unit | PMMA | SAN | PMAN | PK |
|---|---|---|---|---|---|
| Density | $g/cm^3$ | 1.18 | 1.08 | 1.13 | 1.22-1.24 |
| Refractive index | 1 | 1.491 | 1.57 | | |
| Light transmission | % | 92 | | | |
| Melting temperature (DSC) | °C | | | | 245 |
| Heat distortion temperature (1.82 MPa) | °C | 100 | 90 | 100 | |
| Glass temperature (DSC) | °C | 113 | 106 | 12 | 25 |
| Continuous service temperture | °C | 100 | | | |
| Linear thermal expansion coefficient | $K^{-1}$ | $7 \cdot 10^{-5}$ | | | |
| Specific heat capacity (100°C) | $J\ K^{-1}\ g^{-1}$ | 1.5 | 30 | | |
| Thermal conductivity (20°C) | $W\ m^{-1}\ K^{-1}$ | 0.19 | 0.26 | | |
| Tensile modulus | MPa | 3300 | 3600 | | |
| Flexural modulus | MPa | 3100 | | 4000 | 1000-1700 |
| Tensile strength at break | MPa | 68-75 | 68 | 69 | 45-70 |
| Flexural strength | MPa | 110-130 | 125 | 90 | |
| Elongation at break | % | 3-4.5 | 2.9 | | > 300 |
| Notched impact strength (Izod, 3.1 mm) | J/m | 16-27 | 8-20 | 21 | 80-150 |
| (Charpy) | $kJ/m^2$ | 2 | 3 | | |
| Hardness (Rockwell) | - | M97 | | M95 | |
| Relative permittivity (1 kHz) | 1 | 3.3 | 5.5 | 3.83 | |
| Resistivity (volume resistance) (2 min) | Ω cm | $> 10^{15}$ | | $1.1 \cdot 10^{16}$ | |
| Electric strength | kV/mm | 30 | | | |
| Dissipation factor (1 kHz) | 1 | 0.05 | 0.085 | 0.038 | |
| Water absorption (24 h) | % | 0.4 | 0.2 | 0.24 | |

Copolymers of styrene and methyl methacrylate have increased impact strengths and only slightly reduced optical properties. In another high-impact methacrylate type, modifier particles with a core shell structure are embedded in a PMMA matrix (see Volume IV). These particles consist of a PMMA core in a shell of crosslinked poly(butyl methacrylate) (PBMA). Optical clarity is maintained if the refractive index of the shell is adjusted by incorporation of styrene units in the PBMA. A second shell of PMMA serves to anchor the modifier particles in the PMMA matrix.

### 6.11.8 Poly(2-hydroxyethyl methacrylate)

Crosslinked, hydrophilic gels are obtained by free-radical copolymerization of glycol methacrylate (= 2-hydroxyethyl methacrylate (**HEMA**)), $CH_2{=}C(CH_3)COOCH_2CH_2OH$, and glycol dimethacrylate, $CH_2{=}C(CH_3)COO(CH_2)_2OOCC(CH_3){=}CH_2$. These gels serve as soft contact lenses because they can absorb up to 37 % of lachrymal fluid (L: *lacrima* = tear). The pore diameters of these lenses are so small (0.8–3.5 nm) that bacteria (diameters ca. 200 nm) cannot enter. Nevertheless, lenses must be cleaned periodically in order to prevent deposition of proteins which are good breeding grounds for bacteria.

### 6.11.9 Poly(methacrylimide)

Poly(methacrylimide)s (**PMI**) are synthesized by adding an $NH_3$-releasing blowing agent such as $NH_4[HCO_3]$ to poly(methacrylic acid-*co*-methacrylonitrile) and heating to temperatures above the glass temperature of the copolymer:

(6-51)
$$\text{–CH}_2\text{–C(CH}_3\text{)(COOH)–CH}_2\text{–C(CH}_3\text{)(CN)–} \longrightarrow \text{–CH}_2\text{–C(CH}_3\text{)(–C(=O)–NH–C(=O)–)C(CH}_3\text{)–CH}_2\text{–} \xleftarrow[-2\,H_2O]{+\,NH_3} \text{–CH}_2\text{–C(CH}_3\text{)(COOH)–CH}_2\text{–C(CH}_3\text{)(COOH)–}$$

Imidation occurs simultaneously with foaming by the liberated gases. The resulting hard foams have open cell structures, good heat stabilities ($T_G \approx 200°C$), and high tensile and compressive strengths. Their hydrophilic nature causes them to take up considerable amounts of water.

### 6.11.10 Poly(methacrylonitrile)

Atactic poly(methacrylonitrile) (**PMAN**), $\text{+CH}_2\text{–C(CH}_3\text{)CN+}_n$, results from the free-radical polymerization of methacrylonitrile in bulk, emulsion, or solution. Polymers serve for films and coatings, in elastomers, and as photoresists. Some properties are shown in Table 6-30.

## 6.12 Allylic Polymers

Allylic molecules contain the $CH_2$=CH–$CH_2$– group. The term "allyl" stems from garlic (*Allium sativa*), the characteristic smell of which is caused by the antibioticum allicin, $CH_2$=CH–$CH_2$-S(O)–S–$CH_2$–CH=$CH_2$, an enzymatic decomposition product of the garlic component alliin, $CH_2$=CH–$CH_2$–S(O)–$CH_2$–CH($NH_2$)–COOH.

Monoallyl compounds $CH_2$=CH–$CH_2$Y with Y = OH, $OOCCH_3$, Cl, etc., polymerize by free-radical initiators only to low-molecular weight polymers because of termination by the monomer (Volume I, p. 324). All industrial allyl polymers are therefore derived from diallyl and triallyl compounds; they are thermosets, i.e., crosslinked. In industrial lingo, thermosets from diallyl and triallyl esters are sometimes called "polyesters" because they have ester groups as substituents. True polyesters, however, have ester groups –CO–O– in the main chain (Sections 7.7–7.8).

The key intermediate for the synthesis of diallyl and triallyl monomers is allyl alcohol, $CH_2$=CH–$CH_2$OH, which is obtained by

- hydrolysis of allyl chloride (from chlorination (p. 107, V) or oxychlorination of propene);
- isomerization of propylene oxide (p. 108, VI),
- selective hydrogenation of acrolein (from oxidation of propene (p. 107, IV);
- hydrolysis of allyl acetate (from propene, acetic acid, and $O_2$).

$CH_2{=}CHCH_2O{-}OC\backslash$
$\quad Z$
$CH_2{=}CHCH_2O{-}OC/$

| Z | name of monomer | |
|---|---|---|
| — | diallyl oxalate | |
| $-(CH_2)_2-$ | diallyl succinate | |
| $-(CH_2)_8-$ | diallyl sebacate | |
| (o-phenylene) | diallyl phthalate | DAP |
| (m-phenylene) | diallyl isophthalate | DAIP |
| $-CH{=}CH-$ (cis) | diallyl maleate | DAM |
| $-CH{=}CH-$ (trans) | diallyl fumarate | DAF |
| (hexachloronorbornene; Cl, Cl, Cl, Cl, Cl, Cl) | diallyl chlorendate | DAC * |
| $-OCH_2CH_2-O-CH_2CH_2O-$ | diallyl diglycol carbonate | DADC * |

| | | |
|---|---|---|
| $CH_2{=}CHCH_2{-}N^{\oplus}(CH_3)_2{-}CH_2CH{=}CH_2\ Cl^{\ominus}$ | *N,N*-diallyl dimethyl ammonium chloride | DADM * |
| triazine ring with three $OCH_2CH{=}CH_2$ groups ($CH_2{=}CHCH_2O-$) | triallyl cyanurate | TAC * |
| triazinetrione ring with three N–$CH_2CH{=}CH_2$ groups | triallyl isocyanurate | TAIC * |
| $CH_2{-}CHCH_2O{-}OC{-}C(CH_3){=}CH_2$ | allyl methacrylate | AMA |

Fig. 6-25 Some industrial allyl monomers. DAF and DAM as well as AMA (= allyl methacrylate), $CH_2{=}C(CH_3)COOCH_2CH{=}CH_2$, are known as allyl vinyl monomers. * See text below.

Very many diallyl and triallyl compounds are used as monomers (Fig. 6-25). Triallyl cyanurate and all diallylic monomers, with exception of those marked by *, are synthesized by esterification of allyl alcohol, $CH_2{=}CHCH_2OH$, with the corresponding acid. DAC is esterified by chlorendic acid anhydride from the reaction of maleic anhydride and hexachlorocyclopentadiene. Cyanuric acid = 1,3,5-triazine-2,4,6-triol, $C_3N_3(OH)_3$, for TAC is produced by heating urea, $H_2N{-}CO{-}NH_2$, to 200–300°C. DADC results from the reaction of allyl alcohol with diglycol chloroformate, $O(CH_2CH_2OCOCl)_2$.

Diallyl phthalates are polymerized free-radically to a pregelation stage ($M < 25\,000$) and then simultaneously shaped and hardened because shrinkage of the prepolymer is small compared to that of the monomer itself (1 % versus 12 %). Hardened resins of diallyl phthalate and diallyl isophthalate have electrical resistivities between those of china and Teflon; they are used for electrical devices in the computer, communication, and aerospace industries. Allyl monomers are also used as crosslinking comonomers, for example, in prepregs of UP resins (TAC, TAIC). (Prepregs are glass fiber mats that are impregnated by crosslinkable resins and ready for molding and simultaneous shaping.)

DADC polymers are optically clear compounds with about the same light transmission as poly(methyl methacrylate) but with a 30–40 % higher scratch resistance. A major use is in sunglasses, where they compete with polycarbonates.

Resin CR 39 is a mixture of DADC and TAC that polymerizes to optically clear (92 % light transmission) polymers for sheets, rods and lenses. Thin sheets of the polymer are very sensitive to ionic radiation and serve as detectors for cosmic rays, fast neutrons, α-particles, and ions of elements with atomic numbers of 10 and higher

DADM (from epichlorohydrin and dimethylamine) is used as a 60–70 % aqueous solution for water treatment. It cyclopolymerizes by free-radical initiators.

## 6.13 Aliphatic Polyketones

Aliphatic polyketones, $\text{+Z–CO+}_n$, possess alternating aliphatic units –Z– and carbonyl groups –CO–. Industrial polyketones are copolymers of ethene and/or propene and carbon monoxide. Colymerization of these monomers by palladium(II)/nickel catalysts delivers high-molecular weight alternating copolymers, for example, $\text{+}CH_2CH_2\text{–CO+}_n$, (see also p. 237). Ethene-propene-CO terpolymers are tough, semicrystalline high-molecular weight polymers with $T_M$ = 200°C and $T_G$ = 15°C. Free-radical copolymerization results in low-molecular weight ($M < 8000$) polymers with less than 50 % carbon monoxide units. The high-molecular weight types are engineering thermoplastics (for properties see Table 6-30). Their production of ca. 20 000 t/a ceased in 2000.

# Literature to Chapter 6

6.1 CARBONS

P.L.Walker, Jr., Ed., Chemistry and Physics of Carbon, Dekker, New York, Vol. **1** ff. (1965 ff.)

H.Marsh, Introduction to Carbon Science, Butterworth, London 1989

H.O.Pierson, Handbook of Carbon, Graphite, Diamonds, and Fullerenes. Properties, Processes and Applications, Noyes, Park Ridge (NJ) 1993

P.A.Thrower, L.R.Radovic, Eds., Chemistry and Physics of Carbon, Dekker, New York 1999

6.1.1 DIAMONDS

R.A.Gael, The Diamond Dictionary, Gemnolog.Inst.America, Santa Monica (CA), 2nd ed. 1977

J.E.Field, Ed., The Properties of Diamond, Academic Press, New York 1979

C.H.Yaverbaum, Ed., Synthetic Gems Production Techniques, Noyes, Park Ridge (NJ) 1980

G.Davies, Diamond, Hilger, Bristol 1984
-, Diamond Films, Techn.Insights, Fort Lee (NJ) 1987
J.Wilks, E.Wilks, Properties and Applications of Diamond, Butterworth-Heinemann, Oxford 1991
J.E.Field, Ed., The Properties of Natural and Synthetic Diamond, Academic Press, San Diego (CA) 1992
R.F.Davis, Ed., Diamond Films and Coatings, Noyes Data Corp., Park Ridge (NJ) 1993
K.E..Spear, J.P.Dismukes, Eds., Synthetic Diamond–Emerging CVD Science and Technology, Wiley-Interscience, New York 1994
M.A.Prelas, G.Popovici, L.K.Bigelow, Handbook of Industrial Diamonds and Diamond Films, Dekker, New York 1998
B.Dischler, C.Wild, Low-Pressure Synthetic Diamond, Springer, Berlin 1999

6.1.2 DIAMONDOIDS

G.A.Mansoori, http://tigger.uic.edu/–mansoori/Diamondoids.html, accessed 2006-02.22

6.1.3 GRAPHITE

B.T.Kelly, Physics of Graphite, Appl.Sci.Publ., Barking (Essex) 1981

6.1.4 FULLERENES

W.E.Billups, M.A.Ciufolini, Eds., Buckminsterfullerenes, VCH, Weinheim 1993
A.Hirsch, The Chemistry of the Fullerenes, Thieme, Stuttgart 1994
H.Schuster, Ed., Von Fuller bis zu Fullerenen, Vieweg, Braunschweig 1996
M.S.Dresselhaus, G.Dresselhaus, P.C.Eklund, Science of Fullerenes and Carbon Nanotubes, Academic Press, Orlando (FL) 1996
S.Yoshimura, R.P.H.Chang, Supercarbon. Synthesis, Properties and Applications, Springer, Berlin 1998
A.Hirsch, Fullerenes and Related Structures, Springer, Berlin 1999
J.Shinar, Z.V.Vardeny, Z.H.Kafafi, Optical and Electronic Properties of Fullerenes and Fullerene-Based Materials, Dekker, New York 1999
K.M.Kadish, R.S.Ruoff, Fullerenes: Chemistry, Physics, and Technology, Wiley, New York 2000
P.C.Eklund, A.M.Rao, Eds., Fullerene Polymers and Fullerene Polymer Composites (Springer Series in Materials Science, Volume **38**), Springer, Berlin 2002
A.Hirsch, M.Brettreich, Fullerenes, Wiley-VCH, Weinheim 2005

6.1.5 NANOTUBES

T.W.Ebbesen, Carbon Nanotubes: Preparation and Properties, CRC Press, Boca Raton (FL) 1996
P.J.F.Harris, Carbon Nanotubes and Related Structures–New Materials for the Twenty-First Century Cambridge University Press, Cambridge, UK, 2002
S.M.Bachilo, M.S.Strano, C.Kittrell, R.A.Hauge, R.E.Smalley, R.B.Weisman, Structure-Assigned Optical Spectra of Single-Walled Nanotubes, Science **298** (2002) 2361 (includes definition of zigzag and armchair types)
S.Reich, C.Thomsen, J.Maultzsch, Carbon Nanotubes. Basic Concepts and Physical Properties, Wiley-VCH, Weinheim 2004
N.N.Mallikarjuna, S.K.Manohar, T.M.Aminabhavi, Versatile Carbon Nanotubes: Synthesis, Purification, and Their Applications, Polymer News **30** (2005) 6
M.Meyyappan, Ed., Carbon Nanotubes. Science and Application, CRC Press, Boca Raton (FL) 2005
M.J.O'Connell, Carbon Nanotubes: Properties and Applications, CRC Press, Boca Raton (FL) 2006

6.1.6 CARBON AND GRAPHITE FIBERS

D.J.O'Neil, Precursors for Carbon and Graphite Fibers, Int.J.Polym.Mater. **7** (1979) 203
G.Henrici-Olivé, S.Olivé, The Chemistry of Carbon Fiber Formation from Polyacrylonitrile, Adv.Polym.Sci. **51** (1983) 1
The Plastics and Rubber Institute, Carbon Fibers: Technology, Uses and Prospects, Noyes, Park Ridge (NJ) 1986
L.H.Peebles, Jr., Carbon Fibers: Formation, Structure, and Properties, CRC Press, Boca Raton (FL) 1995
J.-B.Donnet, Ed., Carbon Fibers, Dekker, New York, 3rd ed. 1998
M.Inagaki, New Carbons. Control of Structure and Function, Elsevier, Kidlington (UK) 2000
P.Morgan, Carbon Fibers and Their Composites, CRC Press, Boca Raton (FL) 2005

6.1.7 GLASS CARBON
G.M.Jenkins, K.Kawamura, Polymeric Carbons: Carbon Fibre, Glass and Char, Cambridge Univ. Press, London 1976
M.Inagaki, New Carbons. Control of Structure and Function, Elsevier, Kidlington (UK) 2000

6.1.8 NANOFOAMS
http://en.wikipedia.org/wikiCarbon_nanofoam

6.1.9 CARBON BLACK
J.-B.Donnet, Carbon Black, Dekker, New York 1993

6.1.11 ACTIVATED CARBON
R.C.Bansal, J.-B.Donnet, F.Stoeckli, Active Carbon, Dekker, New York 1988
H.Marsh, Ed., Activated Carbon Compendium, Elsevier, Amsterdam 2001

6.2 POLY(OLEFIN)S
F.M.McMillan, The Chain Straighteners: Fruitful Innovation. The Discovery of Linear and Stereoregular Polymers, MacMillan, London 1981
H.R.Sailors, J.P.Hogan, A History of Polyolefins, Polym.News **7** (1981) 152
R.B.Seymour, T.Cheng, Eds., History of Polyolefins, Reidel, Hingham (MA) 1985
L.F.Albright, Processes for Major Addition-Type Plastics and Their Monomers, Krieger, Melbourne (FL), 2nd ed. 1985 (ethene, propene, vinyl chloride, styrene)
D.B.Sicilia, A Most Invented Invention, Amer.Heritage of Invention and Technology **6**/1 (1990) 45
L.A.M.Utracki, Polyolefin Alloys and Blends, Macromol.Symp. **188** (1997) 335
B.A.Krentsel, Y.V.Kissin, V.J.Kleiner, L.L.Stotskaya, Polymers and Copolymers of Higher α-Olefins, Hanser, Munich 1997
I.Scheirs, W.Kaminsky, Eds., Metallocene-Based Polyolefins, Wiley, Chichester 1999 (2 vols.)
C.Vasile, R.B.Seymour, Eds., Handbook of Polyolefins, Dekker, New York, 2nd ed. 2000
A.J.Peacock, Handbook of Polyethylene. Structure, Properties, and Application, Dekker, New York 2000
L.L.Böhm, Die Ethylenepolymerisation mit Ziegler-Katalysatoren, Angew.Chem. **115** (2003) 5162; -, The Ethylene Polymerization with Ziegler Catalysts, Angew.Chem.Internat.Ed.Engl. **42** (2003) 5010
A.-C.Albertsson, Ed., Long-Term Properties of Polyolefins, Springer, Berlin 2004
J.L.White, D.D.Choi, Polyolefins: Processing, Structure Development, and Properties, Hanser, Munich 2005

6.2.3a POLY(ETHYLENE) (natural)
J.Heidberg, K.Krejci-Graf, Elaterite - A Fossil Polyethylene, Naturwiss. **56** (1969) 513

6.2.3b POLY(ETHYLENE) (industrial)
K.Ziegler, Folgen und Werdegang einer Erfindung, Angew.Chem. **76** (1964) 545 (Nobel Lecture) (no English translation)
P.Ehrlich, G.A.Mortimer, Fundamentals of the Free-Radical Polymerization of Ethylene, Adv.Polym.Sci. **7** (1970) 386
F.P.Baldwin, G.Ver Strate, Polyolefin Elastomers Based on Ethylene and Propylene, Adv.Polym. Sci. **7** (1970) 386
S.Cesca, The Chemistry of Unsaturated Ethylene-Propylene Based Terpolymers, Macromol.Rev. **10** (1975) 1
J.P.Hogan, Catalysis of the Phillips Petroleum Company Polyethylene Process, in E.B.Leach, Ed., Applied Industrial Catalysis, Academic Press, New York, Vol. **1** (1983)
T.E.Nowlin, Low Pressure Manufacture of Polyethylene, Progr.Polym.Sci. **11** (1985) 29
A.J.Peacock, Handbook of Polyethylene. Structures, Properties, and Applications, Dekker, New York 2000

6.2.5 ETHENE COPOLYMERS (see also 6.2.3)
S.Schick, Ed., Ionomers. Characterization, Theory, and Applications, CRC Press, Boca Raton (FL) 1996
A.Eisenberg, J.S.Kim, Introduction to Ionomers, Wiley, New York 1998

6.2.6 POLY(PROPYLENE)S
T.O.J.Kresser, Polypropylene, Reinhold, New York 1960
G.Natta, Von der stereospezifischen Polymerisation zur asymmetrischen autokatalytischen Synthese von Makromolekülen, Angew.Chem. **76** (1964) 553 (Nobel Lecture)
H.P.Frank, Polypropylene, Gordon and Breach, New York 1968
E.G.Hancock, Ed., Propylene and Its Industrial Derivatives, Halsted, New York 1973
S. van der Ven, Polypropylene and Other Polyolefins. Polymerization and Characterization, Elsevier, Amsterdam 1990
D.B.Sicilia, A Most Invented Invention, Amer.Heritage of Invention and Technology **6**/1 (1990) 45
J.Karger-Kocsis, Ed., Polypropylene, Chapman & Hall, London 1995 (Vol. I: Structure and Morphology, Vol. II: Copolymers and Blends, Vol. III: Composites)
E.P.Moore, Jr., Ed., Polypropylene Handbook. Polymerization, Characterization, Properties, Processing, Applications, Hanser, Munich 1996
E.P.Moore, Jr., The Rebirth of Polypropylene: Supported Catalysts, Hanser, Munich 1999
H.G.Karian, Handbook of Polypropylene and Polypropylene Composites, Dekker, New York 1999
J.Karger-Kocsis, Polypropylene, Kluwer Academic, Dordrecht (NL) 1999
N.Pasquini, Ed., Polypropylene Handbook, Hanser, Munich, 2nd ed. 2005

6.2.7 POLY(1-BUTENE)
I.D.Rubin, Poly(1-butene), Gordon and Breach, New York 1968
B.A.Krentsel, Y.V.Kissin, V.I.Kleiner, L.L.Stoskaya, Eds., Polymers and Copolymers of Higher $\alpha$-Olefins, Hanser, Munich 1997

6.2.8 POLY(4-METHYL-1-PENTENE)
K.J.Clark, R.P.Palmer, Transparent Polymers from 4-Methylpentene-1, Soc.Chem.Ind., Monograph No. 20, London 1966, p.82

6.2.10 POLY(ISOBUTYLENE)S
H.Güterbock, Polyisobutylen und Isobutylen-Mischpolymerisate, Springer, Berlin 1955
J.P.Kennedy, I.Kirschenbaum, Isobutylene, in E.C.Leonhard, Eds., Vinyl and Diene Monomers, Vol. 2, Wiley, New York 1971

6.3 POLY(DIENE)S, Overview
G.S.Whitby, Synthetic Rubber, Wiley, New York 1954
S.Boström, Kautschuk-Handbuch, Berliner Union, Stuttgart, 6 vols. 1958-1962
P.W.Allen, Natural Rubber and the Synthetics, Crosby Lockwood, London 1972
W.M.Saltman, Ed., The Stereo Rubbers, Wiley, New York 1977
Vol. 2: T.Koyama, A.Steinbüchel, Eds., Polyisoprenoids (2001); = Vol. 2 of A.Steinbüchel, Ed., Biopolymers, Wiley, New York 2001-2003 (12 volumes)
J.E.Mark, B.Erman, F.E.Eirich, Eds., The Science and Technology of Rubber, Elsevier-Academic Press, Burlington (MA), 3rd ed. 2005

6.3.2 POLY(BUTADIENE)S
F.A.Howard, Buna Rubber: The Birth of an Industry, Van Nostrand, New York 1947
W.Breuers, H.Luttropp, Buna, Verlag Technik, Berlin 1954
W.Hofmann, Nitrilkautschuk, Berliner Union, Stuttgart 1965
C.Heuck, Ein Beitrag zur Geschichte der Kautschuk-Synthese: Buna-Kautschuk IG (1925-1945), Chem.-Ztg. **94** (1970) 147
H.Logemann, G.Pampus, Buna S - seine grosstechnische Herstellung und seine Weiterentwicklung - ein geschichtlicher Überblick, Kautsch.Gummi-Kunststoffe **23** (1973) 479
V.Herbst, A.Bisio, Synthetic Rubber: A Project that Had to Succeed, Greenwood, New York 1985

6.3.3 POLY(ISOPRENE)S
W.König, Cyclokautschuklacke, Colomb, Stuttgart 1966
E.Schoenberg, H.A.Marsh, S.J.Walters, W.M.Saltman, Polyisoprenes, Rubber Chem.Technol. **52** (1979) 526
A.D.Roberts, Ed., Natural Rubber Science and Technology, Oxford Univ.Press, Oxford 1988
Y.Tanaka, Structure and Biosynthesis of Natural Polyisoprenes, Progr.Polym.Sci. **14** (1989) 339
R.Friedel, Crazy about Rubber, American Heritage of Invention and Technology **5**/3 (1990) 44

6.3.5 POLY(CHLOROPRENE)
R.M.Murray, D.C.Thompson, The Neoprenes, DuPont, Wilmington (DE) 1963
P.R.Johnson, Polychloroprene Rubber, Rubber Chem.Technol. 49 (1976) 650

6.4 POLY(CYCLOOLEFIN)S AND POLY(CYCLODIOLEFIN)S
J.K.Stille, Diels-Alder Polymerisation, Fortschr.Hochpolym.-Forschg.-Adv.Polym.Sci. **3** (1961/1964) 48
A.Renner, F.Widmer, Vernetzung durch Diels-Alder-Polyaddition, Chimia **22** (1968) 219
W.J.Bailey, Diels-Alder Polymerization, Kin.Mech.Polym. **3** (1972) 333

6.5 ACETYLENE POLYMERS
S.Cesca, A.Priola, M.Bruzzone, Synthesis and Modification of Polymers Containing a System of Conjugated Double Bonds, Adv.Polym.Sci. **32** (1979) 1
C.I.Simionescu, V.Percec, Progress in Polyacetylene Chemistry, Progr.Polym.Sci. **8** (1982) 133
J.W.Chien, Polyacetylene. Chemistry, Physics, and Materials Science, Academic Press, New York 1984
- various authors, Polydiacetylenes, Adv.Polym.Sci. **63** (1984)
A.M.Saxman, R.Liepins, M.Aldissi, Polyacetylene: Its Synthesis, Doping and Structure, Progr.Polym.Sci. **11** (1985) 57

6.6 AROMATIC POLY(HYDROCARBON)S
H.F.Mark, S.M.Atlas, Aromatic Polymers, Int.Rev.Sci., Org.Chem.Ser. Two **3** (1976) 299
V.V.Korshak, A.L.Rusanov, Novel Trends in Ladder Polyheteroarylenes, J.Macromol.Sci.-Rev. Macromol.Chem. **C 21** (1981-1982) 275

6.6.1 POLY(PHENYLENE)S
G.K.Noren, J.K.Stille, Polyphenylenes, J.Polym.Sci. **D** (Macromol.Revs.) **5** (1971) 385
J.G.Speight, P.Kovacic, F.W.Koch, Synthesis and Properties of Polyphenyls and Polyphenylenes, J.Macromol.Sci. **C 5** (1971) 295
D.R.Wilson, H.Jathavedam, N.W.Thomas, in J.C.Salamone, J.S.Riffle, Contemporary Topics in Polymer Science, Vol. 7, Advances in New Materials, Plenum, New York 1992, S. 181

6.6.2 POLY(*p*-XYLYLENE)S
M.Szwarc, Poly-para-xylylene: Its Chemistry and Application in Coating Technology, Polym.Sci.Eng. **16** (1976) 473
L.Baldauf, C.Hamann, L.Libera, Parylene-Polymere I. Synthese, Eigenschaften, Bedeutung. Plaste Kautsch. **25**/2 (1978) 61

6.6.3 PHENOLIC RESINS
T.S.Carswell, Phenoplasts, Interscience, New York 1947
K.Hultzsch, Chemie der Phenolharze, Springer, Berlin 1950
R.W.Martin, The Chemistry of Phenolic Resins, Wiley, New York 1956
N.J.L.Megson, Phenolic Resin Chemistry, Butterworths, London 1958
D.F.Gould, Phenolic Resins, Reinhold, New York 1959
A.A.K.Whitehouse, E.G.K.Pritchett, G.Barnet, Phenolic Resins, Iliffe, London 1967
A.Knop, L.A.Pilato, Phenolic Resins, Springer, Berlin 1985
G.W.Becker, D.Braun, W.Woebcken, Eds., Duroplaste (= Kunststoff-Handbuch, Vol. 10), Hanser, Munich 1988
A.Gardziella, L.A.Pilato, A.Knop, Phenolic Resins, Springer, Berlin 2000

6.6.5 POLY(BENZOCYCLOBUTENE)S
R.A.Kirchhoff, K.J.Bruza, Benzocyclobutenes in Polymer Synthesis, Progr.Polym.Sci. **18** (1993) 85
R.A.Kirchhoff, K.J.Bruza, Polymers from Benzocyclobutenes, Adv.Polym.Sci. **117** (1994) 1

6.7 OTHER POLY(HYDROCARBON)S
E.Hicks, Shellac, Chem.Publ.Co., New York 1961
B.Keszler, J.P.Kennedy, Synthesis of High Molecular Weight Poly($\beta$–Pinene), Adv.Polym.Sci. **100** (1992) 1
R.Mildenberg, M.Zander, G.Collin, Hydrocarbon Resins, Wiley-VCH, Weinheim 1997

6.8.2 POLY(STYRENE)S
R.H.Boundy, R.F.Boyer, Styrene, Reinhold, New York 1952
H.Ohlinger, Polystyrol, Springer, Berlin 1955
C.H.Basdekis, ABS Plastics, Reinhold, New York 1964
H.-L.von Cube, K.E.Pohl, Die Technologie des schäumbaren Polystyrols, Hüthig, Heidelberg 1965
R.Vieweg, G.Daumiller, Kunststoff-Handbuch, Vol. V, Polystyrol, Hanser, Munich 1969
A.Echte, F.Haaf, J.Hambrecht, Fünf Jahrzehnte Polystyrol - Chemie und Physik einer Pioniersubstanz im Überblick, Angew.Chem. **93** (1981) 372
C.A.Brighton, G.Pritchard, G.A.Skinner, Styrene Polymers: Technology and Environmental Aspects, Appl.Sci.Publ., London 1979
G.W.Becker, D.Braun, H.Gausepohl, R.Gellert, Eds., Polystyrol (Kunststoff-Handbuch, Vol. 4), Hanser, Munich, 2nd ed. 1996
R.Po', N.Cardi, Synthesis of Syndiotactic Polystyrene: Reaction Mechanism and Catalysis, Progr.Polym.Sci. **21** (1996) 47
J.Scheir, D.Priddy, Eds., Modern Styrene Polymers, Wiley, Hoboken (NJ) 2003

6.8.3 SUBSTITUTED POLY(STYRENE)S
W.W.Kaeding, L.B.Young, L.Brewster, A.G.Prapas, Para-methylstyrene, CHEMTECH **12** (1982) 556
M.Camps, M.Chatzopoulos, J.-P.Monthéard, Chloromethylstyrene: Synthesis, Polymerization, Transformation, Applications, J.Macromol.-Sci.-Rev.Macromol.Chem.Phys. **C 22** (1982-1983) 343
B.Bömer, H.Hagemann, Polytrifluormethylstyrole, eine Polymerklasse mit aussergewöhnlichen optischen Eigenschaften, Angew.Makromol.Chem. **109-110** (1982) 285

6.8.4 POLY(VINYL ACETATE)
-, Polyvinyl Acetate, Chem.Soc.Monograph No.30, London 1969
G.Matthews, Ed., Vinyl and Allied Polymers, Bd. 2, Iliffe, London 1972
C.A.Finch, Ed., Polyvinyl Acetate - Properties and Applications, Wiley, New York 1973
H.Yildirim Erbil, Vinyl Acetate Emulsion Polymerization and Copolymerization with Acrylic Monomers, CRC Press, Boca Raton (FL) 2000

6.8.5 POLY(VINYL ALCOHOL)
F.Kainer, Polyvinylalkohole, Enke, Stuttgart 1949
J.G.Pritchard, Poly(vinyl alcohol) - Basic Properties and Uses, Gordon and Breach, New York 1970
I.Sakurada, Polyvinyl Alcohol Fibers, Dekker, New York 1986
C.A.Finch, Ed., Polyvinyl Alcohol. Developments, Wiley, New York 1992

6.8.7 POLY(VINYL ETHER)S
N.D.Field, D.H.Lorenz, Vinyl Ethers, in E.C.Leonard, Ed., Vinyl and Diene Monomers, Vol. 1, Wiley, New York 1970
G.B.Butler, Synthesis and Antitumor Activity of "Pyran Copolymer", J.Macromol.Sci.-Rev. Macromol.Chem.Phys. **C 22** (1982/1983) 89

6.8.8 POLY(*N*-VINYL COMPOUND)S
W.Reppe, Polyvinylpyrrolidon, Verlag Chemie, Weinheim 1954
W.Klöpffer, Polyvinylcarbazol, Kunststoffe **61** (1971) 533
R.C.Penwell, B.N.Ganguly, T.W.Smith, Poly(*N*-vinyl carbazole): A Selective Review of the Polymerization, Structure, Properties, and Electrical Characteristics, Macromol.Rev. **13** (1978) 63
J.M.Pearson, M.Stolka, Poly(*N*-Vinyl carbazole), Gordon and Breach, New York 1981

6.9.1 POLY(VINYL CHLORIDE)
H.Kainer, Polyvinylchlorid und Vinylchlorid-Mischpolymerisate, Springer, Berlin 1965
M.Kaufman, The History of PVC - The Chemistry and Industrial Production of Polyvinylchloride, MacLaren, London 1969
J.V.Koleske, L.H.Wartman, Poly(vinylchloride), Gordon and Breach, New York 1969
W.S.Penn, PVC Technology, MacLaren, London, 3rd ed. 1972
R.H.Burgess, Ed., Manufacture and Processing of PVC, Hanser, München 1981
G.Butters, Ed., Particulate Nature of PVC, Elsevier, New York 1982

W.V.Titow, Ed., PVC Technology, Elsevier Appl.Sci., New York, 4th ed. 1984
E.D.Owen, Ed., Degradation and Stabilization of PVC, Elsevier Appl.Sci., New York 1984
M.K.Naqvi, Structure and Stability of Polyvinyl Chloride, J.Macromol.Sci.-Rev.Macromol. Chem.Phys. **C 25** (1985) 119
G.W.Becker, D.Braun, Eds., Kunststoff-Handbuch, Vol. 2, H.Felger, Ed., Polyvinylchlorid, Hanser, Munich 1986 (2 parts)
J.Wypych, Polyvinyl Chloride Stabilization, Elsevier, Amsterdam 1986
L.I.Nass, C.A.Heiberger, Eds., Encyclopedia of PVC, Dekker, New York, 2nd ed. 1988-1992 (4 vols.)
C.A.Daniels, J.W.Summers, C.E.Wilkes, Eds., PVC Handbook, Hanser, Munich 2005

6.9.2 POLY(VINYLIDENE CHLORIDE)

R.A.Wessling, Polyvinylidene Chloride, Gordon and Breach, New York 1975

6.10 POLY(FLUOROHYDROCARBON)S

M.A.Rudner, Fluorocarbons, Reinhold, New York 1958
L.A.Wall, Ed., Fluoropolymers, Wiley, New York 1972
R.G.Arnold, A.L.Barney, D.C.Thompson, Fluoroelastomers, Rubber Chem.Technol. **46** (1973) 619
R.E.Banks, Ed., Preparation, Properties and Industrial Applications of Organofluoro Compounds, Wiley, New York 1982
L.D.Albin, Current Trends in Fluoroelastomer Development, Rubber Chem.Technol. **52** (1982) 902
D.P.Carlson, W.Schmiegel, Fluoropolymers, Organic, Ullmann's Encyclopedia of Industrial Chemistry **A 11** (1986) 393
A.E.Feiring, J.F.Imbalzano, D.L.Kerbow, Developments in Commercial Fluoroplastics, Elsevier Science, Amsterdam, 1992
R.E.Banks, B.E.Smart, J.C.Tatlow, Eds., Organofluorine Chemistry, Plenum, New York 1995
J.Sheirs, Eds., Modern Fluoropolymers, Wiley, New York 1997
J.G.Drobny, Technology of Fluoropolymers, CRC Press, Boca Raton (FL) 2001

6.11 ACRYLIC POLYMERS

M.B.Horn, Acrylic Resins, Reinhold, New York 1960
H.Rauch-Puntigam, Th.Völker, Acryl- und Methacrylverbindungen (= Vol. 9 of K.A.Wolf, Ed., Chemie, Physik und Technologie der Kunststoffe in Einzeldarstellungen), Springer, Berlin 1967

6.11.3 POLY(ACROLEIN)

R.C.Schulz, Polymerization of Acrolein, in G.E.Ham, Ed., Vinyl Polymerization, Vol. 1, Dekker, New York 1967
C.W.Smith, Ed., Acrolein, Hüthig, Heidelberg 1975

6.11.4 POLY(ACRYLONITRILE)

R.H.Beevers, The Physical Properties of Polyacrylonitrile and Its Copolymers, Macromol.Rev. **3** (1968) 113
M.A.Dalin, I.K.Kolchin, B.R.Serebyakov, Acrylonitrile, Technomic, Stamford (CT) 1971

6.11.5 POLY(ACRYLAMIDE)

N.M.Bikales, Acrylamide and Related Amides, in E.C.Leonard, Ed., Vinyl and Diene Monomers, Vol. 1, Wiley, New York 1970
W.-M.Kulicke, R.Kniewske, J.Klein, Preparation, Characterization, Solution Properties and Rheological Behaviour of Polyacrylamide, Progr.Polym.Sci. **8** (1982) 373

6.11.6 POLY($\alpha$-CYANOACRYLATE)S

H.Lee, Ed., Cyanoacrylate Resins - The Instant Adhesives. A Monograph of Their Applications and Technology, Pasadena Technol.Press, Pasadena (CA) 1981

6.11.7 POLY(METHYL METHACRYLATE)

R.Vieweg, F.Esser, Ed., Kunststoff-Handbuch, Vol. IX, Polymethacrylate, Hanser, Munich 1975
W.Reidt et al., Methacrylat-Reaktionsharze, expert Verlag, Sindelfingen 1986
W.Wunderlich, Polymethacrylate - Werk- und Wirkstoffe mit breiter Anwendung, Angew.Makromol. Chem. **244** (1997) 135

6.12 ALLYLIC POLYMERS

H.Raech, Allylic Resins and Monomers, Reinhold, New York 1965
H.Schildknecht, Allyl Compounds and Their Polymers, Wiley-Interscience, New York 1973
C.Wandrey, J.Hernández-Barajas, D.Hunkeler, Diallyldimethylammonium Chloride and Its Polymers, Adv.Polym.Sci. **145** (1999) 123

6.13 ALIPHATIC POLYKETONES

G.E.Ash, Alternating Olefin/Carbon Monoxide Polymers: A New Family of Thermoplastics, J.Mater.Educ. **16**/1 (1994); Int.J.Polym.Mater. **30** (1995) 1

# References to Chapter 6

[1] J.P.Riggs, Carbon Fibers, in J.Kroschwitz, Concise Encyclopedia of Polymer Science and Engineering, Wiley-Interscience, New York 1990
[2] –, Was ist Russ?, company literature of Degussa AG, Frankfurt/Main (no year given) ("What is Carbon Black")
[3] D.C.Miles, J.H.Briston, Polymer Technology, Chemical Publ., New York 1965
[4] K.Lederer, in H.Batzer, Ed., Polymere Werkstoffe, Thieme Publ., Stuttgart and New York 1984, Volume III, p. 27
[5] H.-G.Elias, An Introduction to Plastics, Wiley-VCH, Weinheim, 2nd edition, Table 12-5; data from the CAMPUS collection of Neste
[6] V.Lorenzo, J.M.Pereña, J.M.G.Fatou, Angew.Makromol.Chem. **172** (1989) 25, Figs. 1, 5, and 6
[7] I.D.Burdett, CHEMTECH (October 1992) 616, Fig. 1
[8] H.Domininghaus, Plastics for Engineers, Hanser, Munich 1993
[9] S. van der Ven, Polypropylene and other Polyolefins, Elsevier, Amsterdam 1990
[10] www.iisrp.com, based on Figure 2
[11] From a compilation by R.P.Quirk, D.L.Gomochak Pickel, in J.E.Mark, B.Erman, F.R.Eirich, Eds., Science and Technology of Rubber, Elsevier, Amsterdam, 3rd ed. (2005), p. 54
[12] R.A.Kirchhoff, K.J.Bruza, Benzocyclobutenes in Polymer Synthesis, Progr.Polym.Sci. **18** (1993) 85; Polymers from Benzocyclobutenes, Adv.Polym.Sci. **117** (1994) 1
[13] H.Hopff, Die Technik der Polymerisation, Kunststoffe **49** (1959) 495, Fig. 1; C.C.Winding, G.D.Hiatt, Polymeric Materials, McGraw-Hill, New York 1961, Fig. 8-14
[14] A.Krause, Chem.Engng. **72** (20 December 1965) 72, Fig. 6
[15] W.Wunderlich, Polymethylmethacrylat, in Winnacker-Küchler, Eds., Chemische Technologie, Hanser, Munich, 4th ed. (1982), vol. **6**, p. 413 and 414

# 7 Carbon–Oxygen Chains

## 7.1 Overview

Carbon–oxygen chains are chains with at least one oxygen group –O– as chain atom per repeating unit. They may be open-chained or part of a ring system if the oxygen-containing part of the ring can be considered the shortest path in the chain. Chains with oxygen groups outside the shortest path are not considered as carbon–oxygen chains.

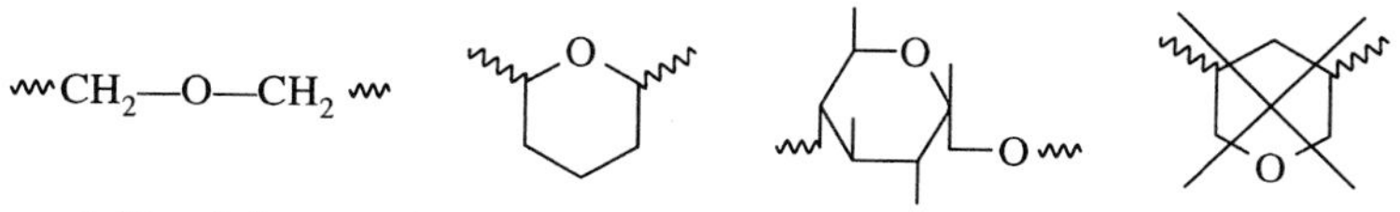

~$CH_2$—O—$CH_2$~

linear (left) and ring-type (center and right) carbon-oxygen chains   not a C–O chain

Repeating units of such carbon–oxygen chains consist of one or more oxygen units –O– and one or more carbon units –Z–, –Z'–, etc., which may be substituted or not. Examples of –Z–, –Z'–, etc., are $-CH_2-$, $-CH(CH_3)-$, $-C_6H_4-$, –C(O)–, etc. Although carbon–oxygen chains by this definition, poly(saccharides)s are discussed separately (see Chapter 8). Examples of simple carbon–oxygen chains are

~O–CHR~ polyacetals

~O–Z~ polyethers

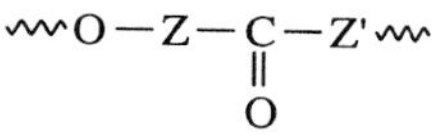

~O–Z–C(=O)–Z'~ polyetherketones

~O–C(=O)–Z~ polyesters

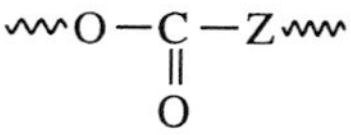

~O–C(=O)–O–Z~ polycarbonates

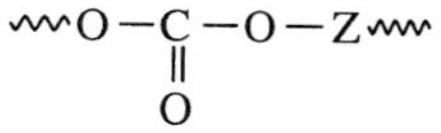

~C(=O)–O–C(=O)–Z~ polyanhydrides

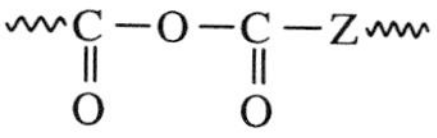

## 7.2 Polyacetals and Polyketals

### 7.2.1 Monomers

Polyacetals consist of chains in which oxygen atoms alternate with unsubstituted or substituted carbon atoms. Polyacetals are obtained by polymerization of >C=O groups of aldehydes or by ring-opening polymerization of cyclic trimers or tetramers of aldehydes such as

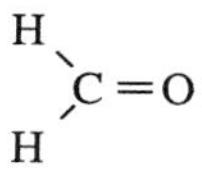

formaldehyde, 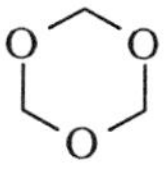 *s*-trioxane, 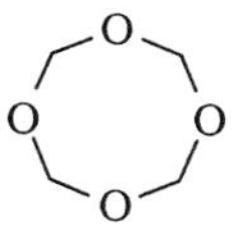 *s*-tetroxocane, 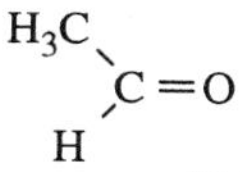 acetaldehyde, 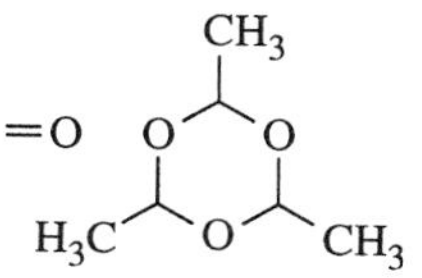 paraldehyde, 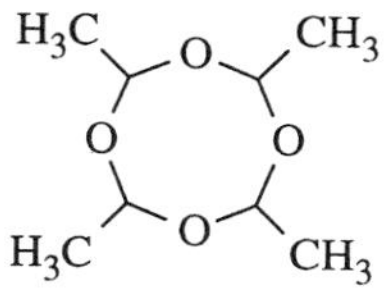 metaldehyde, 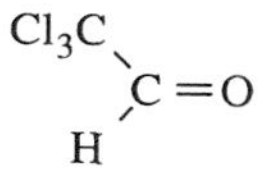 chloral

Table 7-1 Names of cyclic (c) and open-chain (o) oligomers of formaldehyde (F) and acetaldehyde (A). Subscripted numbers indicate the number of aldehyde units per molecule. *s* = symmetric.

| Type | Systematic name | Semi-systematic name | Trivial names | |
|---|---|---|---|---|
| $cF_3$ | 1,3,5-trioxane | *s*-trioxane | trioxymethylene | metaformaldehyde |
| $cF_4$ | 1,3,5,7-tetroxocane | (*not* tetroxane) | tetroxymethylene | |
| $oF_{6\text{-}10}$ | oligo(oxymethylene) | | | paraformaldehyde |
| $cA_3$ | 2,4,6-trimethyl-1,3,5-trioxane | | | paraldehyde |
| $cA_4$ | 2,4,6,8-tetramethyl-1,3,5,7-tetroxocane | | | metaldehyde |
| $cA_{4\text{-}6}$ | *cyclo*-oligo(oxy-(1-methyl)ethylene) | | | metaldehyde |

Many of these compounds are known by various historic and semi-systematic similar-sounding names (Table 7-1) that may easily lead to mix-ups (metaformaldehyde versus metaldehyde, paraformaldehyde versus paraldehyde). "Metaldehyde" may be both the cyclic tetramer of acetaldehyde or all cyclic oligomers of acetaldehyde with 4–6 monomeric units. The cyclic tetramer of formaldehyde, 1,3,5,7-tetroxocane, is often called "tetroxane", but the true tetroxane is a six-membered ring with 4 oxygen atoms in positions 1, 2, 4, and 5, i.e., a cyclic diperoxide.

Trivial names are also commonly used for halogenated acetaldehydes, for example, chloral for tri*chlor*oacet*al*dehyde, $CCl_3CHO$, and fluoral for the corresponding trifluoroacetaldehyde, $CF_3CHO$. Fluoraldehyde® is not a simple fluoroaldehyde but a dye for amino acids that is used in chromatography.

## 7.2.2 Poly(oxymethylene)s

Poly(oxymethylene) (**POM**; **poly(methylene oxide)**) consists ideally of monomeric units $-CH_2O-$ which leads to chains with acetal structures $-O-CH_2-O-$. Polymers containing exclusively or predominantly $-CH_2-O-$ units are therefore industrially referred to as **polyacetals**, **acetal plastics**, or just **acetals**.

POM is obtained by polymerization of formaldehyde or 1,3,5-trioxane (= *s*-trioxane. The formaldehyde polymer is called **poly(formaldehyde)** or **polyacetal homopolymer**. The trioxane polymer usually contains a small proportion of comonomeric units. **Poly(trioxane)** is therefore also known as **polyacetal copolymer**. The world production of polyacetals is 560 000 t/a (2003) (22 % homopolymers, 78 % copolymers).

### Monomers

Formaldehyde (methanal), HCHO, is the aldehyde of formic acid (L: *formica* = ant, because ants exude pungent formic acid when disturbed). It is presently obtained almost exclusively from methanol, either by dehydrogenation ($CH_3OH \rightleftarrows HCHO + H_2$), oxydehydrogenation ($CH_3OH + (1/2)\ O_2 \rightarrow HCHO + H_2O$), or oxidation (ditto) (see p. 94–95); formerly also by free-radical oxidation of propane and butane (USA) or by oxidation of dimethyl ether (Japan). The partial oxidation of methane to formaldehyde has been attempted without success: it requires such high temperatures that any HCHO formed disintegrates immediately.

The hot gases from the (oxy)dehydrogenation/oxidation reaction are cooled fast to 150°C and washed with and absorbed in water. Pure formaldehyde is not a commercial product. It is rather sold as solution, paraformaldehyde, or trioxane:

- 35–55 % aqueous solutions are usually stabilized with 1–2 % methanol. The solutions do not contain formaldehyde itself but its hydrate, i.e., methylene glycol, $HOCH_2OH$, and its oligomeric dehydration products, $H(OCH_2)_nOH$. Aqueous solutions with up ca. 30 % HCHO are clear; higher concentrations lead to the precipitation of **paraformaldehyde**, $H(OCH_2)_mOH$, where $m > n$.
- Paraformaldehyde is obtained by evaporation of aqueous formaldehyde solutions. It depolymerizes to formaldehyde at 180-200°C.
- Trioxane is produced by heating aqueous formaldehyde solutions with 2 % sulfuric acid, followed by chloroform extraction. For polymerization, trioxane is purified by fractional distillation and recrystallization from methylene chloride or petrol ether.

**Poly(formaldehyde)**

Formaldehyde can be polymerized to either poly(oxymethylene), $+OCH_2+_n$, or to poly(hydroxymethylene, $+CH(OH)+_n$.

**Poly(hydroxymethylene)s** are low-molecular weight sugars. They are usually only obtained in small yields; only TlOH as catalyst delivers 90 % monomer conversions. Cannizzaro reactions restrict such polymerizations to hexoses and smaller compounds.

Polymerization to poly(formaldehyde)s delivers **acetal homopolymers**. Such polymerizations may be cationic, anionic, insertion chain polymerizations or polycondensations. Industrial polymerizations are presumably anionic.

*Polycondensations* of formaldehyde, i.e., its hydrate, occur spontaneously in water or alcoholic solutions that contain some water (see above). The driving force is crystallization since the resulting oligo(oxymethylene)s are insoluble in water or aqueous ethanol. Because of the absence of >C=O groups, chain polymerizations are not possible.

*Cationic polymerizations* with, e.g., proton acids proceed via carbenium ions:

$$(7\text{-}1) \qquad H^{\oplus} + O{=}CH_2 \longrightarrow HO{-}\overset{\oplus}{C}H_2 \xrightarrow{+\,HCHO} HO{-}CH_2{-}O{-}\overset{\oplus}{C}H_2 \text{ etc.}$$

*Anionic suspension polymerization* is the method of choice for industrial polymers. Initiators are amines, ammonium salts, alcoholates, phosphines, etc., which react with traces of water to form the true initiator. An example is the reaction of tertiary amines, $R_3N$, with water to $[R_3NH]^{\oplus}[OH]^{\ominus}$. Reaction of hydroxy anions with formaldehyde delivers alkoxy anions that initiate and then propagate the polymerization:

$$(7\text{-}2) \qquad HO^{\ominus} + H_2C{=}O \longrightarrow HO{-}CH_2{-}O^{\ominus} \xrightarrow{+\,HCHO} HO{-}CH_2{-}O{-}CH_2{-}O^{\ominus} \text{ etc.}$$

Transfer of growing macroanions to water molecules terminates polymer chains:

$$(7\text{-}3) \qquad HO{+}CH_2{-}O{+}_nCH_2{-}O^{\ominus} + H_2O \longrightarrow HO{+}CH_2{-}O{+}_nCH_2{-}OH + OH^{\ominus}$$

The termination reaction, Eq.(7-3), regenerates hydroxy anions $HO^{\ominus}$ which start new polymer chains. The kinetic chain is maintained.

In industrial polymerizations, highly purified gaseous formaldehyde is fed to the initiator solution in cyclohexane at room temperature below atmospheric pressure, which allows the heat of polymerization of ca. 63 kJ/mol to be removed by refluxing the monomer. The polymer precipitates as a fine powder.

Polymer chains have half-acetal endgroups (Eq.(7-2)) that are capped by reaction with acetanhydride; this reaction causes a loss of ca. 10 % formaldehyde. The resulting ester endgroups, ~O—$CH_2$–O–$OCCH_3$, prevent depolymerization of polymer chains by unzipping to formaldehyde from chain ends but not after chain scissions (Volume I, Section 15.3.3). The latter occurs on processing since the thermodynamic ceiling temperature $T_c$(lc') of formaldehyde is only 120°C (Volume I, p. 213). As a consequence, poly(formaldehyde)s need always to be stabilized (see below).

### Acetal Copolymers

Acetal copolymers are produced by cationic copolymerization of neat *s*-trioxane and a small proportion (0.1–0.15 mol%) of a cyclic ether such as ethylene oxide, 1,3-dioxolane, diethylene glycol formal, or 1,3-dioxepan by borontrifluoride hydrate, $H[BF_3OH]$, or its etherate, $C_4H_9[BF_3{\cdot}OC_4H_9]$, in the presence of traces of water. Rigorous exclusion of water prevents a polymerization by $BF_3$ (such as $BF_3 + O< \rightleftarrows F_3B^{\ominus}{-}^{\oplus}O<$). The real initiator is $H^{\oplus}$ from traces of water, i.e., from $H[BF_3OH]$, which adds to the trioxane ring that then opens because the open-chain carbenium ion is resonance stabilized. Furthermore, formaldehyde is split from the end of the trioxymethylene segment:

$$\text{(7-4)}\quad \text{trioxane} \underset{-H^{\oplus}}{\overset{+H^{\oplus}}{\rightleftharpoons}} H{-}\overset{\oplus}{O}(\text{trioxane ring}) \longrightarrow HOCH_2OCH_2O\overset{\oplus}{C}H_2 \rightleftharpoons HOCH_2OCH_2\overset{\oplus}{O}{=}CH_2$$

$$+\,HCHO \uparrow\downarrow -\,HCHO$$

$$HOCH_2O\overset{\oplus}{C}H_2 \rightleftharpoons HOCH_2\overset{\oplus}{O}{=}CH_2$$

The chain grows via oxonium structures by addition of trioxane (see Eq.(7-5)), tetraoxocane ("tetroxane", see Table 7-1) formed by backbiting reactions (see Eq.(7-7)), or formaldehyde (Eq.(7-6); from its equilibrium with trioxane, see Eq.(7-4)):

$$\text{(7-5)}\quad HOCH_2O\overset{\oplus}{C}H_2 + \text{trioxane} \longrightarrow HOCH_2OCH_2{-}\overset{\oplus}{O}(\text{trioxane ring}) \longrightarrow HO(CH_2O)_4O\overset{\oplus}{C}H_2 \text{ etc.}$$

$$\text{(7-6)}\quad HOCH_2O\overset{\oplus}{C}H_2 + O{=}CH_2 \longrightarrow HOCH_2OCH_2O\overset{\oplus}{C}H_2 \text{ etc.}$$

$$H(OCH_2)_nOCH_2OCH_2OCH_2OCH_2O\overset{\oplus}{C}H_2$$

$$\downarrow$$

$$\text{(7-7)}\quad H(OCH_2)_nOCH_2{-}O(\text{ring},\ {}^{\oplus}CH_2) \rightleftharpoons H(OCH_2)_nO\overset{\oplus}{C}H_2 + \text{tetraoxocane}$$

Comonomer units are mainly incorporated in blocks but are later redistributed randomly by transacetalization reactions (see below).

Because chain lengths are short at the initial stage of polymerization, formation of formaldehyde (Eq.(7-4)) and tetraoxocane (Eq.(7-7)) is relatively more important than at later stages. Since these reactions compete with chain propagation, relatively fewer polymer molecules are formed and the polymerization shows an induction period. Fast polymerizations occur only after concentrations of formaldehyde and tetraoxocane pass through maxima.

Induction periods are larger for ethylene oxide as comonomer than for 1,3-dioxolane because ethylene oxide can react with trioxane to many other polymerizable cyclic compounds in addition to its incorporation as comonomeric unit into polymer chains:

1,3,5,7,10-pentaoxacyclododecane $\xleftarrow{+\,2\ \text{ethylene oxide}}$ 1,3,5-trioxane $\xrightarrow{+\ \text{ethylene oxide}}$ 1,3,5,7-tetraoxacyclononane $\xrightarrow{-\,HCHO}$ 1,3,6-trioxepane $\xrightarrow{-\,HCHO}$ 1,3-dioxolane

Besides their role as initiators, water molecules in the polymerization system also act as chain-transfer agents that terminate the polymer chain and regenerate protons:

(7-8) $$H(OCH_2)_nOC\overset{\oplus}{H}_2 + H_2O \longrightarrow H(OCH_2)_nOCH_2OH + H^{\oplus}$$

Another strong chain-transfer agent is dimethylformal that is added to the polymerization system as a stabilizer. Each newly formed cation $^{\oplus}CH_2OCH_3$ can start up to 40 new polymer chains.

(7-9) $$\sim\!(OCH_2)_nOC\overset{\oplus}{H}_2 + CH_3OCH_2OCH_3 \longrightarrow \sim\!(OCH_2)_nOCH_2OCH_2\overset{\oplus}{\underset{CH_3}{O}}CH_2OCH_3$$

$$\downarrow$$

$$\sim\!(OCH_2)_nOCH_2OCH_3 + [\overset{\oplus}{C}H_2OCH_3 \longleftrightarrow CH_2{=}\overset{\oplus}{O}CH_3]$$

Cationic trioxane polymerizations are also accompanied by transacetalizations (Eq.(7-10)). These equilibrium reactions lead to more random distributions of comonomeric units within chains as well as to Schulz–Flory distributions of degrees of polymerization while maintaining the number-average molecular weight (see Volume I, p. 574).

(7-10) $$\begin{matrix}\sim\!(CH_2O)_mO \\ | \\ \sim\!(OCH_2)_pOCH_2\end{matrix} + \begin{matrix}CH_2O(CH_2O)_n\sim \\ | \\ OCH_2(OCH_2)_q\sim\end{matrix} \rightleftharpoons \begin{matrix}\sim\!(CH_2O)_mCH_2OCH_2O(CH_2O)_n\sim \\ + \\ \sim\!(OCH_2)_pOCH_2OCH_2(OCH_2)_q\sim\end{matrix}$$

A new industrial process polymerizes formaldehyde instead of trioxane. The copolymerization of HCHO with a cyclic formal as a comonomer proceeds in the gas phase by $BF_3$ or its complex with diethyl ether.

Diblock copolymers are obtained by polymerization of formaldehyde in the presence of a functionalized polymer $R{-}Z_m{-}OH$.

**Stabilization of Acetal Polymers**

Incorporated ethylene oxide units stabilize acetal copolymers against unintended chain scission by heat and alkali. Unstable hemiacetal endgroups of copolymers are removed by treating the copolymers with alkali at elevated temperatures. This causes zip-like eliminations of hemiacetal groups at the end of polymer chains. The reaction stops at an acetal group, i.e., a terminal ethylene oxide unit:

$$\text{(7-11)} \quad \sim O{-}CH_2{-}O{-}CH_2{-}CH_2{-}O{-}(CH_2{-}O)_n{-}CH_2{-}O{-}CH_2OH \xrightarrow{-(n+2)\,HCHO} \sim O{-}CH_2{-}O{-}CH_2{-}CH_2{-}OH$$

POMs so stabilized are thermally stable because the ceiling temperature of poly(oxyethylene) units is much higher than that of poly(oxymethylene) units. They are also resistant against alkali because they contain only alkali-resistant acetal groups and no longer alkali-sensitive hemiacetal units. Similar stabilizing action is provided by comonomeric units $-O-CH_2-O-(CH_2)_i-$ from dioxolane (= 1,3-dioxacyclopentane; $i = 2$) and dioxepane (= 1,3-dioxacycloheptane = butanediol formal; $i = 4$).

Carbon black provides the best stabilization of POMs against degradation by ultraviolet light. Organic UV stabilizers are less efficient.

**Properties**

Polyacetals are translucent whitish engineering plastics (glass temperature –75°C) that are resistant to weak acids and bases, aliphatics, aromatics, halogenated hydrocarbons, and detergents (no stress corrosion). Their water uptake is small (Table 7-2) and their dimensional stability is correspondingly high. They dissolve in hexafluoroacetone hydrate at room temperature and in *m*-cresol at elevated temperatures.

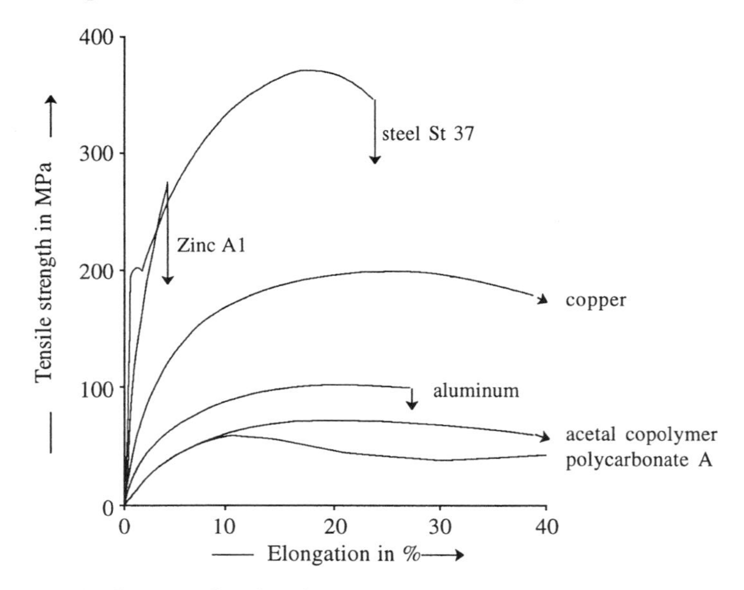


Fig. 7-1 Stress-strain diagrams of engineering materials [1a]. By permission of Hanser Publ., Munich.

Table 7-2 Properties of polyacetals. * Onset of decomposition; nb = no break.

| Property | Physical unit | Acetal homopolymer | Acetal copolymer | Poly-(chloral) |
|---|---|---|---|---|
| Density | $g/cm^3$ | 1.42 | 1.41 | 1.9 |
| Melting temperature (DSC) | °C | 175 | 165 | >260 |
| Heat distortion temp. (1.82 MPa), short time | °C | 124 | 120 | |
| long time | °C | 110 | 110 | |
| Vicat temperature B | °C | 173 | 163 | 200* |
| Linear thermal expansion coefficient | $K^{-1}$ | $90 \cdot 10^{-6}$ | $110 \cdot 10^{-6}$ | |
| Thermal conductivity (20°C) | $W\ m^{-1}\ K^{-1}$ | 0.8 | 1.1 | |
| Tensile modulus | MPa | 2800 | 3200 | ~3000 |
| Flexural modulus | MPa | | 3300 | 2200 |
| Tensile strength at yield | MPa | 65–70 | 67–72 | 38 |
| Tensile strength at fracture | MPa | 69 | 71 | ~45 |
| Flexural strength | MPa | 120 | 90 | |
| Compressive strength | MPa | | | 10 |
| Elongation at yield | % | | | 5 |
| Elongation at fracture | % | 25–70 | 25–70 | |
| Impact strength (Izod, 3.1 mm) | J/m | 69–122 | 53–80 | 60–80 |
| (Charpy) | $kJ/m^2$ | | nb | nb |
| Hardness (Rockwell) | – | 94 | 80 | R10. M50 |
| Ball indentation hardness (30 s) | – | 145 | 160 | |
| Relative permittivity (1 MHz) | 1 | 3.7 | 3.7 | 2.8 |
| Electrical resistance (surface resistivity) | Ω | $10^{13}$ | $10^{13}$ | |
| Resistivity (volume resistance) (2 min) | Ω cm | $10^{15}$ | $10^{15}$ | $4 \cdot 10^{15}$ |
| Electric strength | kV/mm | 700 | 700 | |
| Dissipation factor (100 Hz) | 1 | 0.0055 | 0.001 | 0.003 |
| Water absorption (96 h) | % | 0.7 | | |

Homopolymers are ca. 70 % crystalline, copolymers ca. 55 %, as determined by X-ray crystallography. The high crystallinity and the dense packing of polymer chains provide metal-like properties. In contrast to many other engineering polymers, polyacetals do not show significant yielding (= maxima in stress–strain curves) (Fig. 7-1). Because of these properties, they have replaced injection-molded zinc as well as brass, aluminum, and steel in mountings, car parts, pumps, screws, etc.

### 7.2.3 Poly(acetaldehyde)

Acetaldehyde, $CH_3CHO$, can only be polymerized at very low temperatures because of its low ceiling temprature of –60°C. Anionic polymerizations lead to highly syndiotactic, crystalline polymers, $\text{+O–CH(CH}_3\text{)+}_n$, whereas cationic polymerizations deliver "atactic" rubber-like ones. Copolymers result from the reaction of divinyl ether and diols (Eq.(7-12)). These polymers are not produced industrially because they are easily oxidized (tertiary H atoms!).

$$\text{(7-12)} \quad CH_2{=}CH{-}O{-}CH{=}CH_2 + HO{-}Z{-}OH \longrightarrow \sim\!\!\sim CH(CH_3){-}O{-}CH(CH_3){-}O{-}Z{-}O \sim\!\!\sim$$

### 7.2.4 Poly(fluoral)

Poly(fluoral), $\text{+O–CH(CF}_3\text{)+}_n$, depolymerizes to its monomer at temperatures above ca. 380°C without any indication of a melting or glass temperature. It is extraordinarily resistant to 10 % caustic soda solution or boiling fumic nitric acid. However, its very difficult processability prevents commercial uses.

### 7.2.5 Poly(chloral)

Chloral, $CCl_3CHO$, polymerizes anionically or cationically to $\text{+O–CH(CCl}_3\text{)+}_n$. Its polymerization is initiated by mixing chloral with phosphines or lithium *t*-butoxide above the ceiling temperature of $T_c$ = 58°C and then cooling it to temperatures far below $T_c$. Tertiary amines produce only poly(chloral)s of low thermal stability. Anionic copolymerization of chloral with an excess of isocyanates leads to alternating copolymers as does the cationic copolymerization with 1,3,5-trioxane.

Poly(chloral) is predominantly isotactic; it crystallizes in a $4_1$ helix conformation. The polymer is very resistant to chemicals but depolymerizes completely at 200°C to its (combustable!) monomer.

Poly(chloral) does not dissolve in any solvent. It can therefore only be processed by machining or by monomer casting, i.e., polymerization in a mold by cooling of a monomer/initiator mixture below the ceiling temperature. Because of these properties, it is not a commercial product, although it has fairly good mechanical properties (Table 7-2).

### 7.2.6 Poly(phthaldialdehyde)

Phthaldialdehyde cyclopolymerizes either anionically or cationically:

(7-13) [n phthaldialdehyde → poly(phthaldialdehyde) repeat unit with ring O and chain O, subscript n]

Poly(phthaldialdehyde)s are used as photo resists since they deliver positive patterns by "self-development." In this process, some C–O bonds are first broken catalytically by photochemically generated acids. The polymer chains then unzip to the monomer (i.e., they "self-develop") because of the low ceiling temperature of –40°C.

### 7.2.7 Poly(diphenolformal)s

Polycondensation of methylene dichloride with diphenols such as 1,3-dihydroxybenzene (resorcinol), 1,4-dihydroxybenzene (hydroquinone), 4,4'-dimethylmethylidene bisphenol (dioxydiphenylpropane, dian, bisphenol A), or 4,4'-thiodiphenol and 4,4'-sulfonyldiphenol (both waste products of chemical wood pulp (p. 72)) in the presence of NaOH delivers formal polymers, for example, with bisphenol A:

(7-14) $n\ \text{HO}-C_6H_4-C(CH_3)_2-C_6H_4-\text{OH} + n\ \text{Cl}-CH_2-\text{Cl} \xrightarrow{-2\,n\,\text{HCl}} \left[-C_6H_4-C(CH_3)_2-C_6H_4-O-CH_2-O-\right]_n$

The phenolic chain ends are capped by reaction with 4-*t*-butyl phenol. Byproducts of this polymerization (in yields up to 50 %!) are cyclic oligomers with 2–40 repeating units per ring. Formation of these rings can be surpressed if some *N*-methyl pyrrolidone is added to the monomers.

The poly(4,4'-isopropylidene diphenylformal) shown in Eq.(7-14) has a flexural strength of 83 MPa, a tensile strength of 49 %, and a fracture elongation of 80 %. Its low glass temperature of 94°C prevents it from being a good amorphous engineering plastic.

### 7.2.8 Poly(1,3-dioxolane)

1,3-Dioxolane = dihydro-1,3-dioxol (bp 78°C) is synthesized by the reaction of ethylene glycol with aqueous formaldehyde. Below its ceiling temperature, it polymerizes cationically with $HClO_4$, $H[BF_3OH]$, or $BF_3{\cdot}O(C_2H_5)_2$ to linear and/or cyclic polymers with repeating units $-O-CH_2-O-CH_2-CH_2-$. Ceiling temperatures are 144°C (liquid → crystalline polymer), 87°C (gas → amorphous polymer), and –8°C (4.95 mol/L solution in benzene → crystalline polymer).

The formation of open chains or rings depends on a critical monomer concentration, $[M]_{crit}$ (see Volume I, Section 7.5.1). Below 25°C, practically only polymeric rings are formed for $[M]_0 < [M]_{crit}$ but predominantly linear chains for $[M]_0 > [M]_{crit}$:

(7-15)

$$\text{cyclo-}\left[O-CH_2-O-CH_2-CH_2\right]_{n \geq 0} \xleftarrow{[M] < [M]_{crit}} n\ \text{(1,3-dioxolane)} \xrightarrow{[M] > [M]_{crit}} \left[O-CH_2-O-CH_2-CH_2\right]_n$$

Poly(1,3-dioxolane) (**PDXL**) is a semi-crystalline polymer with degrees of crystallinity of 55–80 %. Densities ($\rho$/(g cm$^{-3}$)) are ca. 1.00 in the amorphous state and 1.325 (triclinic), 1.414 (orthorhombic), and 1.331 (hexagonal) in the crystalline states. Melting temperatures are 93°C (triclinic), 55°C (orthorhombic), and 60°C (hexagonal). Conventional polymers melt between 52°C and 93°C. The glass temperature was determined as –63°C. Industrial production and use of PDXL is unknown.

4-Methylene-1,3-dioxolanes contain two polymerizable groups: the formaldehyde group and a vinylidene group. They can therefore polymerize via the C=C double bond, by ring-opening, and by elimination of ketones to poly(4-methylene-1,3-dioxolane)s with very different types and proportions of chain units:

(7-16)

vinyl polymerization ring opening elimination

## 7.2.9 Poly(acetone)

Chain polymerization of acetone, $(CH_3)_2C{=}O$, to polymers $\text{+O–C(CH}_3)_2\text{+}_n$ with a ketal structure seems to be possible below the ceiling temperature of –80°C for polymerization of liquid monomer to crystalline polymer using Mg atoms, $Al(C_2H_5)_3 + TiCl_3$, or anionic initiators. The unstabilized polymer depolymerizes at $T > T_c$ to acetone.

Acids condense acetone to phorone: 3 $(CH_3)_2CO \rightarrow (CH_3)_2C{=}CH{-}O{-}CH{=}C(CH_3)_2 + 2\ H_2O$. Under these conditions, phorone condenses further to hyperbranched oligomers. Isophorone is not formed.

# 7.3 Polyethers

Polyethers are polymers with repeating units –O–Z–, where Z is an aromatic, a cycloaliphatic, or an aliphatic group with at least 2 carbon atoms in the backbone of the chain unit. The ether moiety, –O–, can also be a part of a ring system (Section 7.3.8).

## 7.3.1 Poly(oxyethylene)s

Polymers with repeating units $–O–CH_2–CH_2–$ carry the systematic name **poly(oxyethylene)**. Industrial polymers are called **poly(ethylene glycol) (PEG)** if the molecular weight is $M \leq 40\ 000$ and **poly(ethylene oxide) (PEOX; PEO)** if $M \geq 100\ 000$.

Both PEG and PEOX are polymerization products of ethylene oxide (**EO**; oxirane) which is exclusively synthesized by oxidizing ethene with oxygen (no longer with air) at 250–300°C and 10–20 bar, using a silver catalyst (p. 102, X). The selectivity of this process is less than 75 %; ca. 80 % of the cost is tied to the price of ethene. Dehydrochlorination of ethylene chlorohydrin, $HOCH_2CH_2Cl$, is too expensive (price of chlorine; $CaCl_2$ as inexpensive byproduct) and no longer used industrially.

Only ca. 15 % of the total EO production of ca. $10{\cdot}10^6$ t/a is utilized for poly(oxyethylene)s, polyols for polyurethanes (p. 493), and other direct polymer uses. Approximately 60 % is converted to ethylene glycol which, in turn, serves mainly for the synthesis of polyesters (Sections 7.5–7.7). The remaining ethylene oxide is converted to non-ionic surfactants, ethanol amines, and glycol ethers.

All poly(oxyethylene)s, PEG and PEOX, are crystalline polymers that dissolve in water and, except for those with very high molecular weights, also in most common organic solvents but not in alkanes, diethyl ether, or carbon disulfide.

**Poly(ethylene glycol)s** (PEG) with molecular weights below ca. 40 000 are obtained by polymerization of ethylene oxide with sodium methylate or alkali hydroxides. Since industrial systems always contain some water, the resulting polymers are poly(ethylene glycol)s, $H{+}OCH_2CH_2{+}_nOH$, or their alkali alcoholates.

Poly(ethylene glycol)s are somewhat hygroscopic, colorless, odorless, non-toxic waxes that serve for very many purposes from adjuvants in rubbers and plastics to pharmaceuticals. They are cherished in cosmetics because their melting temperatures can be adjusted to different climes and body temperatures by simple mixing of PEGs of different chain lengths, $n$. For example, melting temperatures are 2.1°C ($n$ = 6), 7.7°C ($n$ = 7), 30.0°C ($n$ = 9), and 50.0°C ($n$ = 45).

**Poly(ethylene oxide)s** (PEOX) result from the polymerization of dry ethylene oxide with alkaline earth oxides or carbonates as catalysts. These polymers have molecular weights between 100 000 and 3 000 000. In very dilute solution, they act as drag reducers (Volume IV, Section 14.4.2). More concentrated solutions are strongly viscoelastic which makes them useful as thickeners. They also serve as emulsifiers, detergents, protecting colloids, and packaging films.

### 7.3.2 Poly(propylene oxide)

Approximately 50 % of propylene oxide (**PO**) is obtained by dehydrochlorination of propylene chlorohydrin, $CH_3$–CHOH–$CH_2Cl$, (synthesis similar to epichlorohydrin; p. 107, V) and the other 50 % by a 2-step oxidation of propene with hydroperoxides ($t$-butyl hydroperoxide, $\alpha$-methylbenzyl hydroperoxide) or peroxycarboxylic acids (p. 108, VI). In this process, hydroperoxides are converted to alcohols and peroxy acids to acids. Direct oxidation of propene by oxygen delivers mainly acrolein and not propylene oxide. The world production of propylene oxide is now ca. $4 \cdot 10^6$ t/a.

Propylene oxides exist in two antipodes. The (+) compound with the [$R$] configuration is the D monomer but is called the L monomer in older publications. The (–) compound has the [$S$] configuration (L-propylene oxide; formerly D-propylene oxide). Each antipode leads to the corresponding stereoregular poly(propylene oxide). The polymerization of the racemate, on the other hand, delivers "atactic" products that have many head-to-head arrangements.

Anionic polymerizations cleave the oxirane ring mainly in the β-position, cationic ones predominantly in the α-position:

CH3
H2C—CH
O
β cleavage α cleavage

anionic: $\beta > \alpha$
cationic: $\alpha > \beta$

Low-molecular weight atactic poly(propylene oxide)s with molecular weights of 250–4000 are **poly(propylene glycol)s (PPG)**, $HO{+}CH_2{-}CH(CH_3){-}O{+}_nH$. Their dimers, trimers, and tetramers are also known as **polyols**; this group of polymers also comprises low-molecular weight copolymers of propylene oxide and ethylene oxide (EO) (p. 493). Higher molecular weight copolymers of PO and EO are segmented copolymers that associate in water because of their hydrophobic/hydrophilic nature; they are used as surfactants. Copolymers of PO with non-conjugated dienes deliver oil-resistant and low-temperature stable elastomers that can be cured by sulfur. Copolymers of PO with allyl glycidyl ether or other unsaturated epoxides are resistant to ozone and low temperatures but not very resistant to oils.

High-molecular weight poly(propylene oxide)s (**PPOX**) do not find industrial uses. Atactic poly(propylene oxide) is an amorphous polymer with a glass temperature of ca. –70°. Isotactic PPOX crystallizes orthorhombically with a melting temperature of 75°C.

### 7.3.3 Poly(epichlorohydrin)

Epichlorohydrin (= α-epichlorohydrin, 1-chloro-2,3-epoxypropane, EPI) is obtained from allyl chloride either in a two-step process by reaction with HOCl and then with $Ca(OH)_2$ (p. 107, Va) or in a one-step process by oxidation with perpropionic acid (p. 108, Vb). A Japanese route chlorinates allyl alcohol to 2,3-dichloro-1-hydroxypropane which is then reacted with $Ca(OH)_2$ to epichlorohydrin.

Homopolymerization of EPI by $(C_2H_5)_3Al/H_2O$/acetyl acetone (= 2,4-pentanedione) delivers poly(epichlorohydrin), $\text{+}OCH_2CH(CH_2Cl)\text{+}_n$, a linear, somewhat regioirregular, and probably atactic polymer (**CO** (ASTM), **PECH**). Copolymers of EPI include those with ethylene oxide (EO), allyl glycidol ether (AGE), propylene oxide (PO), or EO + AGE. PECH, $EPI_{68}$-$EO_{32}$ (**ECO**), $EPI_{95}$-$AGE_5$, and $EPI_{63\text{-}70}$-$(EO\text{-}AGE)_{37\text{-}30}$ are usually crosslinked by 2-mercaptoimidazoline via chloromethyl groups of EPI monomeric units. EPI-PO copolymers are vulcanized by sulfur systems. These elastomers are resistant to ozone, oil, and low temperatures.

Glass temperatures of poly(epichlorohydrin) are recorded as –15°C to –22°C. The polymer has a 100 % modulus of ca. 5 MPa and a 200 % modulus of ca. 12 MPa, a tensile strength at fracture of 17 MPa, and an elongation at rupture of ca. 280 %. After 70 hours, compression sets are 26 % at 100°C and 57 % at 150°C.

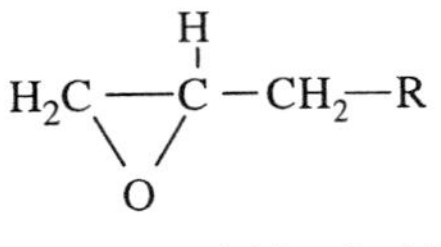

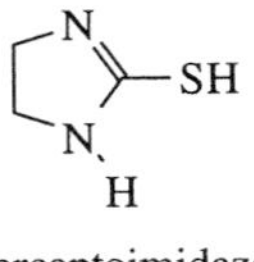

$H_2C$—C(H)—$CH_2$—R (epoxide ring via O) 

N—SH, N—H (imidazoline ring)

R: Cl — α-epichlorohydrin, 1-chloro-2,3-epoxypropane
OH — glycidol, 2,3-epoxy-1-propanol
O—$CH_2$—CH=$CH_2$ — allyl glycidol ether, allylglycidyl ether

2-mercaptoimidazoline, NA-22®

### 7.3.4 Poly[1,2-di(chloromethyl)ethylene oxide]

1,2-Di(chloromethyl)ethylene oxide = 1,3-dichloro-2,3-epoxybutane is an epichlorohydrin-related monomer that is obtained by chlorination of butadiene and subsequent oxidation of the resulting $ClCH_2CH{=}CHCH_2Cl$. The monomer can be polymerized with dialkyl magnesium + $H_2O$ or dialkyl zinc + $H_2O$. Depending on the catalyst, chains with various tacticities are formed, for example, the diisotactic polymer:

(7-17) $4n$ (R)(H)C–C(R)(H) epoxide (O) $\longrightarrow$ $[\text{–O–CHR–CHR–O–CHR–CHR–O–CHR–CHR–O–CHR–CHR–}]_n$ $\quad R = CH_2Cl$

The resulting polymers are engineering thermoplastics; they can also be spun to wool-like fibers. Industrial uses are not known, however. Similar polymers can be obtained from 3,3-bis(chloromethyl)oxacyclobutane:

(7-18) $$n\ ClCH_2-\underset{H_2C-O}{C}-CH_2 \longrightarrow \left[ O-CH_2-\underset{CH_2Cl}{\overset{CH_2Cl}{C}}-CH_2 \right]_n$$

## 7.3.5 Phenoxy Resins

Phenoxy resins are obtained from bisphenols and epichlorohydrin in a 2-step reaction. The first step employs stoichiometric amounts of sodium hydroxide and an excess of epichlorohydrin. The resulting phenolate anion attacks the oxirane ring which opens and attaches to the phenolate by forming a secondary hydroxyl group:

(7-19)

$$\sim C_6H_4-O^{\ominus} + \underset{\diagdown O \diagup}{H_2C-CH}-CH_2Cl \longrightarrow \sim C_6H_4-O-CH_2-\underset{O^{\ominus}}{CH}-CH_2Cl$$

The resulting anions can either be dehydrochlorinated by sodium ions

(7-20) $$\sim C_6H_4-O-CH_2-\underset{O^{\ominus}}{CH}-CH_2Cl \xrightarrow[-\ NaCl]{+\ Na^{\oplus}} \sim C_6H_4-O-CH_2-\underset{\diagdown O \diagup}{HC-CH_2}$$

or react with epichlorohydrin to crosslinked polymers. Such crosslinking is prevented by a surplus of epichlorohydrin which also restricts the polymerization to the oligomer stage. Formation of high-molecular weight polymers requires exact stoichiometry, however. Because of this, any remaining epichlorohydrin and sodium hydroxide are removed after the first step and replaced by the calculated amount of bisphenol and catalytic amounts of NaOH. The oxirane ring opens and attaches to the phenolate:

(7-21)

$$\sim C_6H_4-O-CH_2-\underset{\diagdown O \diagup}{HC-CH_2} + \sim C_6H_4-O^{\ominus} \longrightarrow \sim C_6H_4-O-CH_2-\underset{O^{\ominus}}{CH}-CH_2-O-C_6H_4\sim$$

The resulting alkoxide anions then react with phenolic hydroxy groups which restores phenolate anions.

Low-molecular weight phenoxy resins are produced in the bulk, whereas polymers for coatings are synthesized in butanone and those for injection moldings in water-soluble solvents. The latter polymers are then precipitated by water.

Phenoxy resins are excellent primers because of their many hydroxy groups per molecule. In the automobile industry, metal sheets are first treated by phenoxy resins, then by special epoxy resins, and finally by an acrylic resin as lacquer. Phenoxy resins have only limited use for injection molding because of their low glass temperature of 80°C.

### 7.3.6 Epoxy Resins

**Epoxy resins** (= **epoxies, EP**) are oxirane groups containing aliphatic, cycloaliphatic, or aromatic oligomers that are crosslinked to thermosets by so-called curing. They are synthesized either by reaction of epichlorohydrin with a compound containing an active H atom (such as phenol) (Eqs.(7-19) and (7-20)) or by direct epoxidation of olefins with peracids. More than 85 % of epoxy resins are produced from bisphenol A and epichlorohydrin.

Other epoxy resins are based on epoxidized phenol-formaldehyde or cresol-formaldehyde resins, hydantoin, hexahydrophthalic acid, bisamine A tetraglycidyl ether, epoxidized oils from unsaturated fatty acids, etc. Commercial epoxy resins are always formulated, i.e., they contain hardeners, plasticizers, diluents, fillers, etc.

#### Syntheses

*Liquid epoxy resins* are industrially produced by reaction of bisphenol A and an excess of epichlorohydrin at 60°C. Released HCl is continuously neutralized by powdery NaOH. The exothermic reaction leads to a mixture of products, for example,

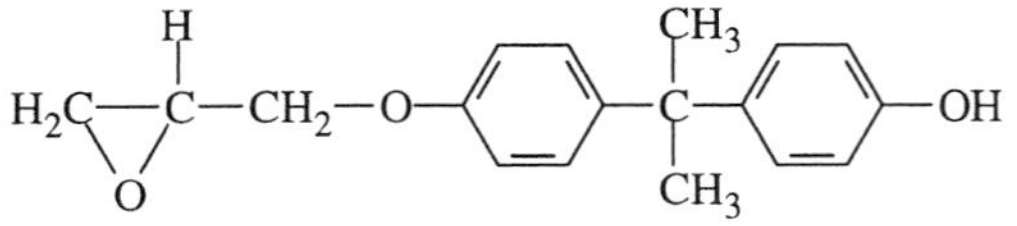

$$\underset{\diagdown O \diagup}{H_2C{-}CH}{-}CH_2{-}O{-}C_6H_4{-}C(CH_3)_2{-}C_6H_4{-}OH \qquad \text{I}$$

$$\underset{\diagdown O \diagup}{H_2C{-}CH}{-}CH_2{-}O{-}C_6H_4{-}C(CH_3)_2{-}C_6H_4{-}O{-}CH_2{-}\underset{\diagdown O \diagup}{CH{-}CH_2} \qquad \text{II}$$

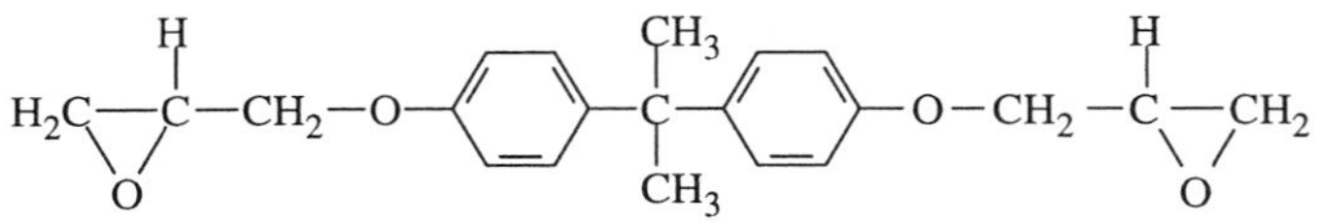

A subsequent slow temperature raise to 120°C results in further condensation. NaCl and any remaining NaOH are removed by washing and volatile components by applying vacuum at 140°C. The resulting epoxy resin consists of low-molecular weight **di**glycidyl **e**thers of **b**isphenol **A** (**DGEBA**) with the idealized structure

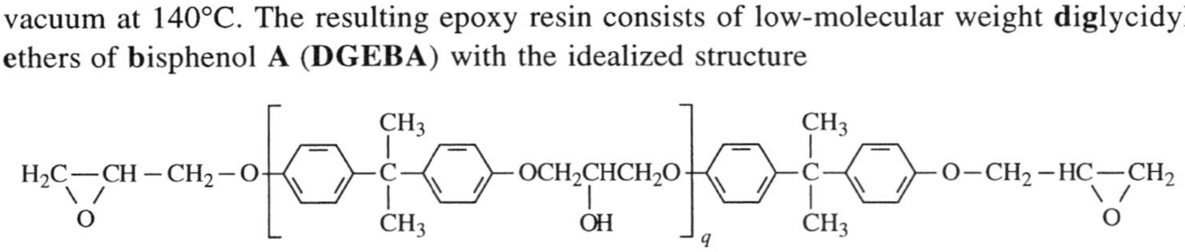

$$\underset{\diagdown O \diagup}{H_2C{-}CH}{-}CH_2{-}O{-}\left[C_6H_4{-}C(CH_3)_2{-}C_6H_4{-}OCH_2\underset{OH}{CH}CH_2O\right]_q{-}C_6H_4{-}C(CH_3)_2{-}C_6H_4{-}O{-}CH_2{-}\underset{\diagdown O \diagup}{HC{-}CH_2}$$

Table 7-3 Effect of the molar ratio of epichlorohydrin (EPH) and bisphenol A (BPA) on the degree of polymerization $q$, number-average molecular weight, $\overline{M}_n$, and softening temperature, $T_S$, of oligomers [2].

| mol EPH/mol BPA | $q$ | $\overline{M}_n$/(g mol$^{-1}$) | $T_S$/°C |
|---|---|---|---|
| 1.57 : 1 | 2 | 900 | 65 – 75 |
| 1.22 : 1 | 3.7 | 1 400 | 95 – 105 |
| 1.15 : 1 | 8.8 | 2 900 | 125 – 135 |
| 1.11 : 1 | 12.0 | 3 750 | 145 – 155 |

Molecular weights are low: the distributions are 87, 11, and 1.5 % for $q = 0$, 1, and 2! DGEBAs with $0.1 < q < 0.6$ are liquids, those with $2 < q < 30$, solids (Table 7-3).

*Low-molecular weight solid epoxies* are produced by the **Taffy process** in which epichlorohydrin is added at 45°C to a rapidly agitated mixture of bisphenol A (BPA) and the stoichiometric amount of 10 % NaOH. The highly exothermic reaction makes the temperature climb to 95°C. It results in a tough DGEBA with q ≈ 3.7 ($\overline{M}_n$ ≈ 1400 g/mol) that is thoroughly washed and dried.

*Higher molecular weight solid epoxies* with $q > 4$ are synthesized by the **advancement** or **fusion process** where liquid DGEBA is extended with bisphenol A by the catalytic action of, e.g., triethanolamine, quaternary ammonium salts, alkali hydroxides, alkali carbonates, etc.

### Curing

Curing (= hardening) of epoxy resins is through a crosslinking reaction by either di- or trifunctional agents or catalytically acting compounds. *Warm hardening* is performed at temperatures of 80–100°C by polycondensing agents such as liquid acid anhydrides (NMA, DDSA, HHPA, alkendic anhydride) or aromatic diamines (MPDA, MDA, DDS), or by chain-polymerization catalysts such as piperidine, boron trifluoride/ethylamine, or benzyl dimethylamine:

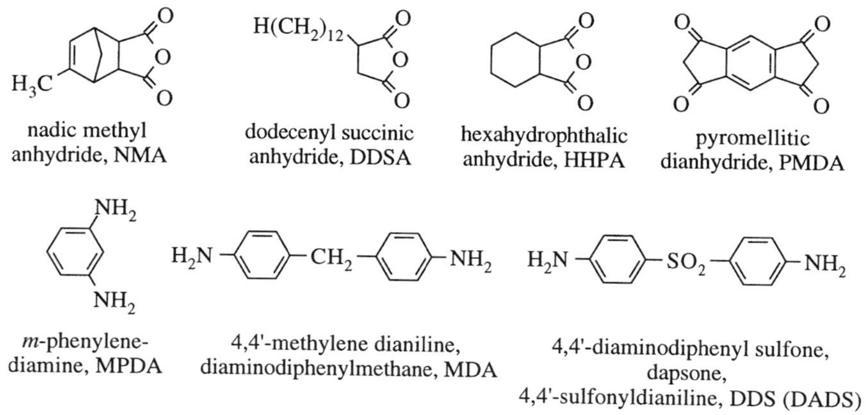

Fig. 7-2 Some common anhydride (top) and diamine (bottom) curing agents.

Warm hardening of epoxy resins by anhydrides leads to polyester and polyetherester structures:

(7-22)

$$\sim\!\mathrm{CH}(\mathrm{OH})\!\sim \; + \; \text{anhydride} \xrightarrow{-\mathrm{H}^{\oplus}} \sim\!\mathrm{CH}(\mathrm{O{-}C(=O){-}C{-}C{-}C(=O){-}O^{\ominus}})\!\sim$$

$$\text{epoxide (O)} \xrightarrow{+\mathrm{H}^{\oplus}} \mathrm{HO}^{\oplus} \longleftrightarrow \mathrm{OH}, \oplus$$

$$-\mathrm{H}^{\oplus} \downarrow + \sim\!\mathrm{CH}(\mathrm{OH})\!\sim \qquad \downarrow + \mathrm{RCOO}^{\ominus}$$

$$\mathrm{HO} \;\; \mathrm{O{-}CH}\!\sim \qquad \mathrm{HO} \;\; \mathrm{O{-}C(=O){-}R}$$

The chemical structure of the anhydride and the type of the accelerator used (often a tertiary amine) influence not only the curing time but also the end-use properties of the resulting thermoset (Table 7-4).

*Cold hardening* at room temperature by di- and triamines such as *m*-phenylenediamine (MPDA), 4,4'-methylene diamine (MDA), 4,4'-diaminodiphenylsulfone (DDS), etc., leads to $\beta$-hydroxypropylamine structures as shown in Eq.(7-23) for the action of just one amine group of the curing agent, ®–N(R)H, where ® is the rest of the molecule:

(7-23)

$$\text{®}(\mathrm{R})\mathrm{N{-}H} + \mathrm{H_2C{-}CH{-}CH_2}\!\sim \text{(epoxide, O)} \longrightarrow \text{®}(\mathrm{R})\overset{\oplus}{\mathrm{N}}(\mathrm{H}){-}\mathrm{CH_2{-}CH(O^{\ominus}){-}CH_2}\!\sim$$

$$\downarrow$$

$$\text{®}(\mathrm{R})\mathrm{N{-}CH_2{-}CH(OH){-}CH_2}\!\sim$$

Table 7-4 Properties of some anhydride-cured epoxy resins [3]. PA = phthalic anhydride; for the other abbreviations, see preceding page. phr = parts per hundred parts of resin.

| Property | Physical unit | Type and concentration of curing agent | | | | |
|---|---|---|---|---|---|---|
| | | DDSA 130 phr | PA 75 phr | HHPA 80 phr | NMA 80 phr | PMDA/PA 17/23 phr |
| Heat distortion temperature | °C | 66–70 | 110–152 | 110–130 | 150–175 | 225 |
| Flexural modulus | MPa | 2620 | 2770 | 2760 | 3030 | 2900 |
| Flexural strength | MPa | 93 | 110 | 130 | 97 | 75 |
| Compressive strength | MPa | 73 | 152 | 116 | 126 | 317 |
| Tensile strength | MPa | 56 | 81 | 79 | 69 | 25 |
| Ultimate elongation | % | 4.5 | 4.8 | 7.4 | 2.5 | 0.9 |

The degree of crosslinking and thus the glass temperature increases with increasing degree of curing. Polymer segments become less and less mobile and the accessibility of remaining functional groups decreases but does not stop completely: cured epoxy resins show an after-cure which may extend for many days.

Epoxy resins are often not used as such but in combination with reactive diluents (i.e., monoepoxides, low-viscosity epoxy resins), non-reactive diluents = plasticizers (styrene, hydrocarbon oils, dibutyl phthalate, etc.), flexibilizing agents (polyesters, silicones, nitrile rubber, phenolic modifiers, coal tar, etc.), fillers (silica, talc, etc.), and reinforcing fibers (glass, graphite, aramids (= polyaramids), etc.) (see also Volume IV). Table 7-5 shows how flexibilizers and fillers affect properties. It must also be noted that epoxy resins wet surfaces exceptionally well because of their polarity.

Conventional epoxy resins are all based on bisphenol A; they are mainly used in interior applications. Much more weather-resistant and usable for outdoor applications are cycloaliphatic epoxides (names are common but not systematic), for example,

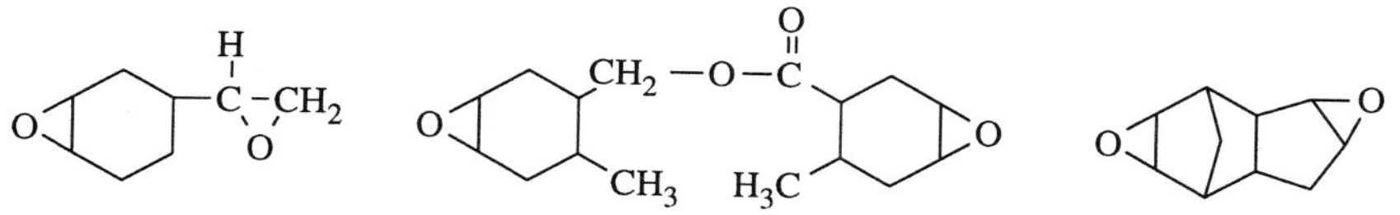

vinylcyclohexanedioxide 3,4-epoxy-2-methylcyclohexylmethyl dicyclopentadienedioxide

Cycloaliphatic epoxy resins are not only cured with anhydrides such as poly(azelaic anhydride) or amines but also with poly(*p*-vinyl phenol) that is formed *in situ* at 100–200°C from added *p*-vinyl phenol, $CH_2=CH(p\text{-}C_6H_4OH)$, without any initiator.

Cured epoxies from bisphenol A are fairly heat-resistant (Tables 7-4 and 7-5), aliphatic and cycloaliphatic ones less so, both because of the lack of aromatic structures and the former also because of the higher flexibility of aliphatic chain segments.

Table 7-5 Properties of cured unfilled (u), flexibilized (f), or silica-filled epoxy molding resins based on bisphenol A (BPA) or a non-specified cycloaliphatic compound (CA).

| Property | Physical unit | CA u | BPA u | BPA f | BPA g |
|---|---|---|---|---|---|
| Density | $g/cm^3$ | 1.16–1.21 | 1.2–1.3 | 0.96–1.35 | 1.6–2.0 |
| Heat distortion temperature (1.82 MPa) | °C | 97–237 | 47–287 | 23–117 | 67–287 |
| Thermal conductivity (20°C) | $W\ m^{-1}\ K^{-1}$ | | 0.19 | | 0.42–0.84 |
| Tensile modulus | GPa | | 3–5 | | |
| Compression modulus | GPa | 3.4 | | 0.007–2.4 | |
| Tensile strength at fracture | MPa | 55–83 | 28–90 | 14–70 | 48–90 |
| Flexural strength | MPa | | 55–170 | | |
| Compressive strength | MPa | 100–140 | 100–170 | 7–97 | 100–240 |
| Elongation at fracture | % | 2–10 | 3–6 | 20–85 | 1–3 |
| Impact strength (Izod, 3.1 mm) | J/m | | 10–50 | 120–170 | 16–24 |
| Relative permittivity | 1 | | | | 4.6 |
| Resistivity (volume resistance) (2 min) | Ω cm | | | $10^{16}$ | |
| Electric strength | kV/mm | | | | 14.2 |
| Water absorption (24 h) | % | | < 1.0 | < 0.5 | < 0.1 |

More than 50 % of epoxy resins are used for coatings, the rest for printed wiring boards, adhesives, flooring and paving, tool casting and molding, etc. Glass fiber-reinforced epoxies can be used as enginering plastics but suffer from higher resin prices and longer curing times compared to the competing unsaturated polyester resins.

### 7.3.7 Poly(tetrahydrofuran)

Tetrahydrofuran (**THF**) (= tetramethylene oxide, diethylene oxide, 1,4-epoxybutane, oxolane) is obtained from agricultural waste (p. 81), dehydration of 1,4-butanediol (p. 96, VIIIa; p. 115, II), butadiene via 1,4-diacetoxy-2-butene (p. 114, VI), oxidation/dehydration of butane (p. 115, III), and hydration/deoxygenation of maleic anhydride (p. 115, IV). THF polymerizes cationically at temperatures below 83°C (ceiling temperature, see Volume I, p. 202) via growing tertiary oxonium ions to **poly(tetrahydrofuran)** (**PTHF**), also called **poly(tetramethylene oxide)**:

$$\text{(7-24)} \quad \text{THF} \xrightarrow{+\,R^{\oplus}X^{\ominus}} \text{THF-}O^{\oplus}\!-R\;X^{\ominus} \xrightarrow{+\,\text{THF}} \text{THF-}O^{\oplus}\!-CH_2CH_2CH_2CH_2OR\;X^{\ominus}$$

Low molecular weight PTHFs are viscous oils, whereas high molecular weight ones are crystalline solids ($T_M$ = 59°C, $T_G$ = –84°C). They dissolve in benzene, tetrahydrofuran, chloroform, and ethanol but not in aliphatic hydrocarbons, methanol, and water. High-molecular weight PTHFs have tensile moduli of 97 MPa, tensile strengths at fracture of 29 MPa, and fracture elongations of 820 %. PTHFs with two hydroxy endgroups are known as **poly(tetramethylene ether) glycols** (**PTMEG**, **PTMG**). They are used for the soft segments in elastic polyurethane fibers or polyetherester elastomers.

### 7.3.8 Furan Resins

Heating furfural (from agricultural waste, see p. 81), furfuryl alcohol (Eq.(7-25)), or furfural plus a ketone at 100°C in the presence of acids delivers a dark liquid polymeric resin consisting of two types of monomeric units:

$$\text{(7-25)} \quad \text{furan-}CH_2OH \xrightarrow{-\,H_2O} \underbrace{-\text{furan-}CH_2-}_{\text{main unit}} \;/\; \underbrace{-\text{furan-}CH_2-O-CH_2\text{-furan-}}_{\text{secondary unit}}$$

The liquid resin is neutralized, dehydrated, and mixed with large proportions of urea-formaldehyde or phenol-formaldehyde resins for use as a binder in foundry sand cores for iron castings. Addition of weak acids furnishes products with long shelf lifes that harden at 100–200°C. Strong acids cure at room temperature, probably by crosslinking via the double bonds.

Furan plastics (furan polymers), based on furan resins and usually reinforced, are used for pipes, fittings, asphalt pavement coatings, adhesives, etc.

### 7.3.9 Poly(phenylene oxide)s

Oxidative coupling of 2,6-disubstituted phenols leads to **substituted poly(oxyphenylene)s (poly(phenylene oxide)s)**. Poly(oxy-2,6-dimethyl-1,4-phenylene), *scientifically* abbreviated as **PPO**), a **poly(2,6-xylenol)** developed by General Electric, is no longer sold as homopolymer in the United States but apparently still in Poland and Russia.

The *commercial* abbreviation PPO® (a registered trademark) was originally used for PPO itself and then for a blend of PPO and poly(styrene). PPO® is now a blend of PPO and rubber-modified poly(styrene) (**modified PPO, PPOm**, for example, Noryl®). Another commercial blend consists of PPOm and polyamides (for example, Luranyl®). All of these products are often sold as **polyphenylene ethers (PPE)**. However, "polyphenylene ether" sometimes means only copolymers with 90–95 % 2,6-xylenol and 5–10 % 2,3,6-trimethylphenol units. Both the parent polymer poly(oxy-1,4-phenylene) and the unmodified poly(oxy-2,6-dimethyl-1,4-phenylene) are not commercial polymers. The world production of modified PPO is ca. 300 000 t/a (2003).

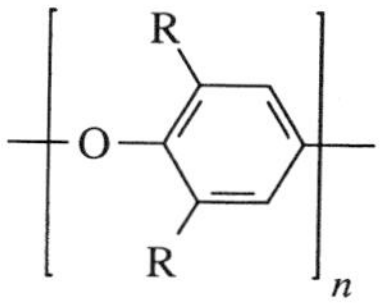

| | | |
|---|---|---|
| R = H | poly(oxy-1,4-phenylene), | POP |
| R = $CH_3$ | poly(oxy-2,6-dimethyl-1,4-phenylene), | PPO |
| R = $C_6H_5$ | poly(oxy-2,6-diphenyl-1,4-phenylene) | |
| R = Br | poly(oxy-2,6-dibromo-1,4-phenylene) | |

**Synthesis**

**Poly(oxy-1,4-phenylene)** = poly(*p*-phenylene oxide) is obtained by polymerization of 4-chlorophenol, Cl(*p*-$C_6H_4$)OH, or the corresponding bromine compound in the presence of CuCl. **Poly(oxy-2,6-dimethyl-1,4-phenylene)**, the "scientific PPO", results from the copper/amine complex catalyzed oxidative coupling of 2,6-dimethylphenol in aromatic solvents at 35–40°C (Eq.(7-26)). According to one source, "diamines" are used. Other sources point out that primary and secondary amines do not work because they are oxidized to azo compounds and (probably) also to hydrazo compounds.

(7-26) (ArOH) $\xrightarrow[-\ HO^{\bullet}]{+\ (1/2)\ O_2}$ [ I ⟷ II ] (ArO•) $\xrightarrow{+\ ArO^{\bullet}}$ … $\underset{-\ H^{\oplus}}{\overset{+\ H^{\oplus}}{\rightleftarrows}}$ … $\xrightarrow{-\ H^{\oplus}}$ ArO–(C₆H₂R₂)–OH

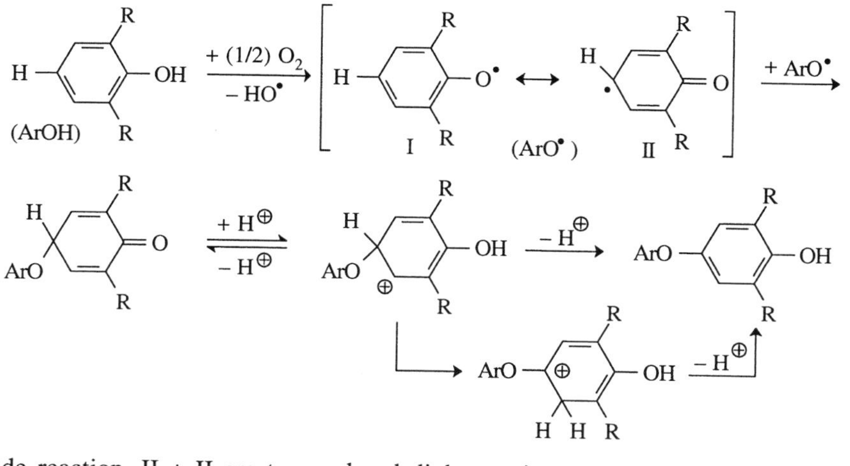

In a side reaction, II + II are to a colored diphenoquinone which is in equilibrium with its dihydroxy compound. Incorporation of the latter causes the beige color of PPO.

These reactions are polyeliminations that proceed probably via quinone mechanisms consisting of couplings of molecules ArOH (Eq.(7-26)). This view is supported by the fact that couplings occur only if the para position to the hydroxyl group is occupied by H (as in Eq.(7-26)), $C(CH_3)_3$, or $CH_2OH$, but not if $CH_3$, $C_2H_5$, or $C_6H_5$ are present. Couplings also do not happen if substituents R are large or strongly negative ($NO_2$, $CH_3O$). Polymerizations are terminated by introducing nitrogen instead of oxygen.

Modified PPO is produced by post-polymerization grafting of, e.g., styrene, similar to the ABS technology (see Volume IV). Poly(oxy-2,6-dimethyl-1,4-phenylene) and poly-(styrene) are compatible over the whole range of compositions.

**Properties**

**Poly(oxy-1,4-phenylene)** is a crystalline polymer that is soluble in *N*-methylpyrrolidone, hexamethylphosphoric triamide, diphenyl ether, tetralene, naphthalene, or benzophenone. It is not a commercial product.

Substituted **poly(phenylene ether)s** play a role in reprography, the reproduction (copying) by electromagnetic radiation, which employs light-induced crosslinkings of quinone azides, in the simplest case:

(7-27) $$O{=}C_6H_4{=}N_2 \xrightarrow[-N_2]{h\nu} O{=}C_6H_4{:} \longrightarrow [\cdot O{-}C_6H_4{\cdot}] \longrightarrow -[O{-}C_6H_4]-$$

**Poly(oxy-2,6-dimethyl-1,4-phenylene) = poly(2,6-dimethyl-1,4-phenylene oxide)** (PPO) ($\overline{M}_w$ = 40 000; $\overline{M}_n$ = 18 000) crystallizes as a $4_1$ helix, but can also be obtained as an amorphous polymer if its melt is cooled by 12 K/h. The PPOs of the 1960s were about 40 % crystalline (calorimetry; melting temperature ≈ 262°C) and probably slightly branched. They also contained hydroperoxide groups from oxidations during polymer formation. The polymer was soluble in benzene, toluene, and halogenated hydrocarbons as well as in $CH_2Cl_2$ above its theta temperature of 69.1°C. It had very good end-use properties (Table 7-6) but degraded readily in air at temperatures above 110°C. For this reason, and because of its high melt viscosity, it was difficult to process above its glass temperature and subsequently replaced by its 1:1 blend with styrene (see above).

Modified polymers (PPO®, PPOm) are hard and tough but not transparent. Their improved processability is counterbalanced by less favorable thermal properties. PPOm is available in various grades, not only with poly(styrene) (PPO/S) and polyamides (PPO/S/PA) as components but also with, for example, poly(1,4-butylene terephthalate).

**Poly(oxy-2,6-diphenyl-1,4-phenylene)** is also synthesized by oxidative coupling. The polymer has a glass temperature of 235°C and a melting temperature of 480°C. It is stable in air up to 175°C and can be dry spun from organic solvents. Fibers become highly crystalline upon drawing. Short fibers are processed to papers that serve for insulation of cables for high-voltage applications.

**Poly(2,6-dibromo-1,4-phenylene oxide)** is obtained by free-radical-initiated elimination of HBr from the 4-position of 2,4,6-tribromophenol in the presence of small proportions of an oxidant, probably $OH^{\ominus}/K_3[Fe(CN)_6]$. The polymer serves as a flame retardant for polymers that need to be processed at high temperatures, for example, thermoplastic polyesters or glass fiber-reinforced polyamides.

Table 7-6 Properties of poly(oxy-1,4-phenylene) (POP), poly(oxy-2,6-dimethyl-1,4-phenylene) (PPO), a 50:50 blend of PPO and grafted poly(styrene) S (= PPO®), and a blend of PPO/S and a polyamide (PPO/S/PA). nb = no break.

| Property | Physical unit | POP | PPO | PPO/S Noryl® N 300 | PPO/S/PA Luranyl® KR 2401 |
|---|---|---|---|---|---|
| Density, melt | g/cm$^3$ | | 0.958 | | |
| Density, amorphous polymer | g/cm$^3$ | 1.270 | 1.06 | 1.06 | 1.07 |
| Density, 100 % crystalline | g/cm$^3$ | 1.407 | | | |
| Melting temperature (DSC) | °C | 262 | 267 | | |
| Heat distortion temperature (1.82 MPa) | °C | | 174 | 149 | 115 |
| Glass temperature (DSC) | °C | | 205 | 140 | |
| Linear thermal expansion coefficient | | | | | |
| amorphous, $T < T_G$ | K$^{-1}$ | $6.2 \cdot 10^{-5}$ | $5.2 \cdot 10^{-5}$ | $6 \cdot 10^{-5}$ | $6 \cdot 10^{-5}$ |
| crystalline, $0.7\ T_M < T < 0.95\ T_M$ | K$^{-1}$ | $9.3 \cdot 10^{-6}$ | | | |
| Thermal conductivity (20°C) | W m$^{-1}$ K$^{-1}$ | | | | 0.18 |
| Tensile modulus | MPa | | 2690 | | 2500 |
| Flexural modulus | MPa | | 2590 | 2410 | |
| Tensile strength at yield | MPa | | | | 52 |
| Tensile strength at fracture | MPa | | 80 | 76 | 45 |
| Flexural strength | MPa | | 114 | 104 | |
| Elongation at yield | % | | | | 4 |
| Elongation at fracture | % | | 20–40 | 20 | 28 |
| Impact strength (Izod, 3.1 mm) | J/m | | | | nb |
| Notched impact strength (Izod, 3.1 mm) | J/m | | 64 | 530 | |
| (Charpy) | kJ/m$^2$ | | | | 11 |
| Hardness (Rockwell) | - | | M78 | R119 | |
| Relative permittivity ("dielectric constant") | 1 | 4.8 | 2.6 | 2.6 | |
| Electrical resistance (surface resistivity) | Ω | | $> 10^{15}$ | | $10^{14}$ |
| Resistivity (volume resistance) (2 min) | Ω cm | | $> 10^{15}$ | | $10^{15}$ |
| Electrical strength | kV/mm | | 20 | 20 | 80 |
| Dissipation factor (100 Hz) | 1 | 0.0005 | 0.00035 | | |
| Water absorption (24 h) | % | | | 0.06 | < 0.1 |

## 7.4 Polyketones

Polyketones are defined as polymers with intact ketone groups, –CO–, in the chain (see also Section 6.13) and not as polymers with polymerized ketone groups.

### 7.4.1 Poly(carbon suboxide)

The simplest polyketones are obtained by polymerization of carbon suboxide (propadiene-1,3-dione), O=C=C=C=O. The colorless poisonous gas polymerizes exothermally in the presence of impurities to an amorphous red-black material ("red coal") that contains pyranone structures (pyranones are ketones of pyran, see next page).

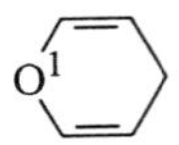

4H-pyran 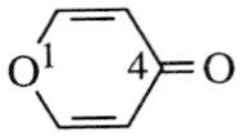 4-pyranone (pyrone)

Carbon suboxide can be polymerized by anions $R^{\ominus}$, cations $R^{\oplus}$, or radicals $R^{\bullet}$ in solvents such as $CCl_4$ to different units that combine to polypyrone = polypyranone structures. The polymers have not found industrial applications.

(7-28)

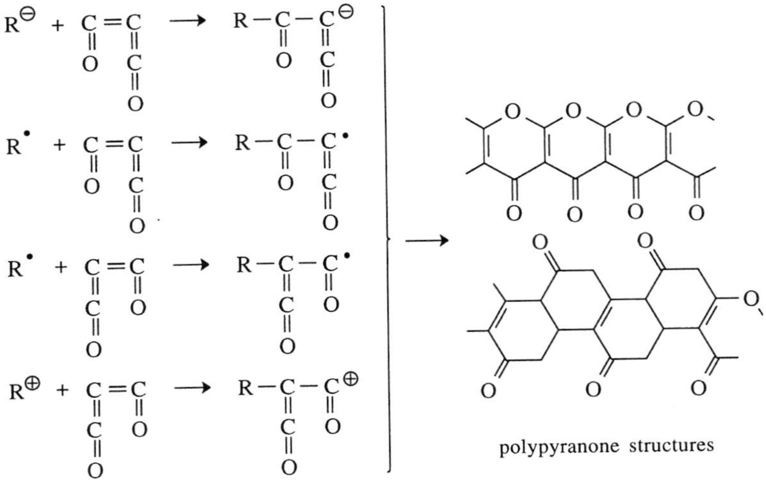

polypyranone structures

## 7.4.2 Polyaryletherketones

**Polyetherketones** contain in their chains ether and keto groups; all industrial representatives also contain 1,4-phenylene units (**polyaryletherketones, PAEK**). The alternative designation as **polyarylene ethers (PAE)** also comprises polyethersulfones.

The simplest **polyetherketone (PEK)** results from the Friedel–Crafts acylation of the acid chloride of diphenyl ether in $CH_2Cl_2$ at 20°C, an electrophilic AB reaction,

(7-29) $\xrightarrow{-n\,HCl}$

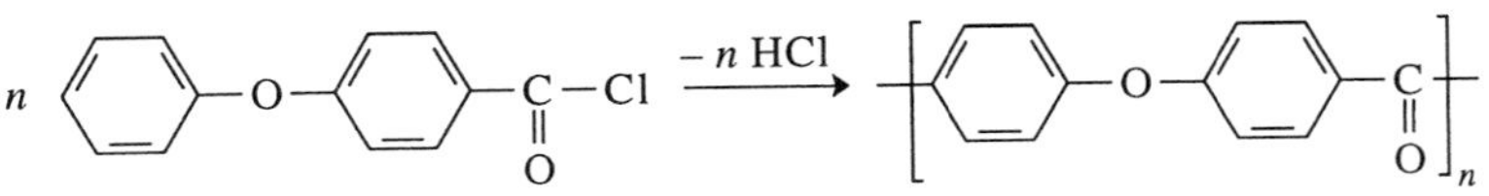

or from the corresponding AA/BB reaction, for example, of diphenyl ether, $C_6H_5OC_6H_5$, and isophthalic acid dichloride, $ClOC(i\text{-}C_6H_4)COCl$, using $BF_3$ in liquid hydrofluoric acid as catalyst, i.e., an aromatic electrophilic substitution.

These syntheses, and also the self-polycondensation of $KO(p\text{-}C_6H_4)\text{–}CO\text{–}(p\text{-}C_6H_4)F$ in diphenylsulfone, $(C_6H_5)_2SO_2$, to PEK at high temperatures, deliver only low-molecular weight products since initially formed crystalline oligomers are insoluble, precipitate, and cease to react further to high-molecular weight polymers.

However, syntheses to high-molecular weight PEK are possible by an AA/BB reaction of the dipotassium compound of 4,4'-dihydroxybenzophenone with 4,4'-difluorobenzophenone:

(7-30)

$$n\ KO{-}C_6H_4{-}CO{-}C_6H_4{-}OK + n\ F{-}C_6H_4{-}CO{-}C_6H_4{-}F \xrightarrow{-2n\ KF} {+}O{-}C_6H_4{-}CO{-}C_6H_4{-}O{-}C_6H_4{-}CO{-}C_6H_4{+}_n$$

Polyetheretherketone (**PEEK**) is synthesized by the nucleophilic aromatic substitution reaction of 4,4'-difluorobenzoquinone and the dipotassium salt of hydroquinone (= 1,4-dihydroxybenzene) in diphenylsulfone, $(C_6H_5)_2SO_2$, at ca. 330°C:

(7-31)

$$n\ KO{-}C_6H_4{-}OK + F{-}C_6H_4{-}CO{-}C_6H_4{-}F \xrightarrow{-2n\ KF} {+}O{-}C_6H_4{-}O{-}C_6H_4{-}CO{-}C_6H_4{+}_n$$

The very many types of industrially synthesized polyaryletherketones are distinguished by their sequence of ether groups (symbol E) and ketone groups (symbol K) in the repeating units:

| | |
|---|---|
| $-O{-}C_6H_4{-}CO{-}C_6H_4{-}$ | PEK |
| $-O{-}C_6H_4{-}CO{-}C_6H_4{-}CO{-}C_6H_4{-}$ | PEKK |
| $-O{-}C_6H_4{-}CO{-}C_6H_4{-}O{-}C_6H_4{-}CO{-}C_6H_4{-}CO{-}C_6H_4{-}$ | PEKEKK |
| $-O{-}C_6H_4{-}O{-}C_6H_4{-}CO{-}C_6H_4{-}$ | PEEK |
| $-O{-}C_6H_4{-}O{-}C_6H_4{-}CO{-}C_6H_4{-}CO{-}C_6H_4{-}$ | PEEKK |

Polyaryletherketones are semicrystalline polymers with high melting temperatures but without extraordinarily high melt viscosities. After pre-drying at 370–400°C, they can be processed with the usual machinery. The polymers have high moduli, strengths, and impact strengths (Table 7-7). They are inert against most organic solvents but are attacked by halogenated hydrocarbons. PEEK, for example, dissolves in hydrofluoric acid, trifluoromethane sulfonic acid, and benzophenone. Annual productions of these high-performance engineering plastics are ca. 500 t/a to 1500 t/a per type.

Table 7-7 Properties of some polyaryletherketones.

| Property | Physical unit | PEK Kadel® | PEEK Victrex® | PEEKK Hostatek® | PEKEKK Ultrapek® |
|---|---|---|---|---|---|
| Density, 100 % amorphous | $g/cm^3$ | 1.272 | 1.264 | | |
| commercial grade | $g/cm^3$ | | 1.320 | 1.30 | 1.32 |
| 100 % crystalline | $g/cm^3$ | 1.430 | 1.401 | | |
| Melting temperature (DSC) | °C | 364 | 334 | 363 | 377 |
| Heat distortion temperature (ISO A) | °C | | 140 | 103 | 170 |
| Glass temperature (DSC) | °C | 153 | 143 | 167 | 175 |
| Continuous service temperature | °C | | 250 | 220 | 260 |
| Linear thermal expansion coefficient | $K^{-1}$ | | $4.7 \cdot 10^{-6}$ | | $4.2 \cdot 10^{-6}$ |
| Thermal conductivity (20°C) | $W\ m^{-1}\ K^{-1}$ | | 0.25 | | 0.24 |
| Tensile modulus | MPa | 3190 | 3650 | 4000 | 4700 |
| Tensile strength at yield | MPa | | 92 | 100 | |
| Tensile strength at fracture | MPa | 104 | 92 | 90 | 118 |
| Elongation at yield | % | | 4.9 | 5.5 | |
| Elongation at fracture | % | | 50 | 28 | 13 |
| Impact strength (Izod, 3.1 mm) | J/m | | no break | | |
| (Charpy) | $kJ/m^2$ | | no break | no break | |
| Notched impact strength (Izod, 3.1 mm) | J/m | 59 | 83 | | 80 |
| (Charpy) | $kJ/m^2$ | | 8.2 | 8 | 10 |
| Hardness (Rockwell) | - | | M99 | | |
| Relative permittivity (50 Hz) | 1 | | 3.2 | | 3.4 |
| Electrical resistance (surface resistivity) | Ω | | | $10^{15}$ | |
| Resistivity (volume resistance) | Ω cm | | $5 \cdot 10^{16}$ | | $>10^{16}$ |
| Dissipation fctor (50 Hz) | 1 | | 0.003 | | 0.002 |
| Water absorption (24 h) | % | | 0.5 | | 0.2 |

## 7.5 Aliphatic AB Polyesters

Aliphatic AB polyesters $\text{+O–Z–CO+}_n$ with one ester group –COO– and an aliphatic moiety Z per repeating unit can be synthesized by many methods: self-condensation of $\alpha,\omega$-hydroxy acids, ring-opening polymerization of lactones, and polyeliminations of O-carboxyanhydrides of $\alpha$- or $\beta$-hydroxycarboxylic acids. Some polyesters are also produced by nature or biotechnology.

Chain units Z may consist of one or more unsubstituted or substituted skeletal carbon atoms such as $–CH_2–$, $–CH_2–CH(CH_3)–$, $–(CH_2)_i–$, etc. Polymers with one, two, three, ... skeletal carbon atoms in Z are called **poly($\alpha$-ester)s, poly($\beta$-ester)s, poly($\gamma$-ester)s**, ...

Aliphatic AB polyesters always result from homopolymerizations, which distinguishes them from aliphatic AA/BB polyesters with repeating units –O–Z'–O–CO–Z"–CO– that are usually obtained from joint polymerizations of two types of monomers. In AA/BB polyesters, Z' is a unit consisting of one or more unsubstituted or substituted skeletal carbon atoms, whereas the number of skeletal carbon atoms in Z" may number zero, one, two, ... AA/BB polyesters are usually synthesized by polycondensation of dicarboxylic acids or their diacid chlorides with diols or their alkali alcoholates.

### 7.5.1 Poly(α-hydroxyacetic acid)

The AB polycondensation of α-hydroxyacetic acid (glycolic acid), $HOCH_2COOH$, does not lead to high-molecular weight polymers. Poly(α-hydroxyacetic acid)s (= **poly(glycolic acid)s (PGA)**) are therefore produced by anionic polymerization of the cyclic dimer of α-hydroxyacetic acid, i.e., glycolide (I), (Eq.(7-32)). The solid-state polycondensation of the sodium salt of chloroacetic acid as well as the polyelimination of $CO_2$ from the *O*-carboxyanhydride of glycolic acid (II) are both laboratory procedures.

(7-32) $(1/2)\,n$ I $\longrightarrow$ $\mathrm{[\!-O-CH_2-C(=O)-\!]_n}$ $\xleftarrow{-n\,CO_2}$ $n$ II

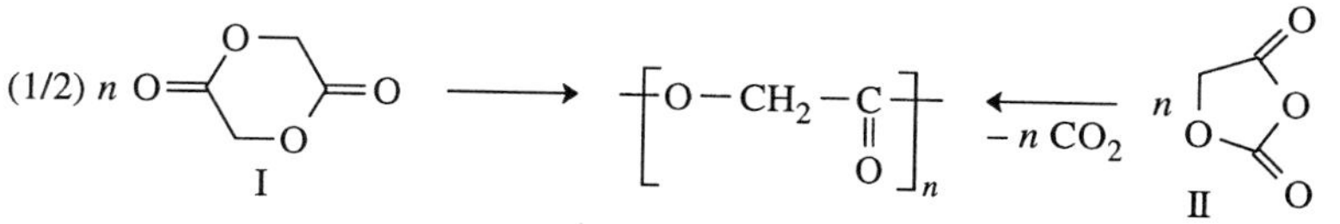

Poly(glycolide) is used for surgical threads since it is neither inflammatory nor encapsulated by the body but rather resorbed by biological degradation.

### 7.5.2 Poly(lactide)s

Poly(α-ester)s with lactide units $-O-CH(CH_3)-CO-$ are prepared similar to poly(glycolide)s by the self-polycondensation of lactic acid, $HO-CH(CH_3)-COOH$, its lactone, or its *O*-carboxyanhydride. Lactic acid is now almost exclusively obtained by fermentation of saccharides, mainly dextrose (D-glucose) from the hydrolysis of starch from corn. Only one plant worldwide (2003) is still using a fully synthetic method in which acetaldehyde and hydrogen cyanide are reacted in a cyanohydrin synthesis and the resulting cyanohydrin hydrolyzed to D,L-lactic acid:

$$\text{(7-33)}\quad CH_3-\overset{\delta+}{C}H=O^{\delta-} \xrightarrow{+\,CN^\ominus} CH_3-C(CN)(H)-O^\ominus \xrightarrow{+\,H^\oplus} CH_3-C(CN)(H)-OH \xrightarrow[-\,NH_3]{+\,H_2O} CH_3-C(COOH)(H)-OH$$

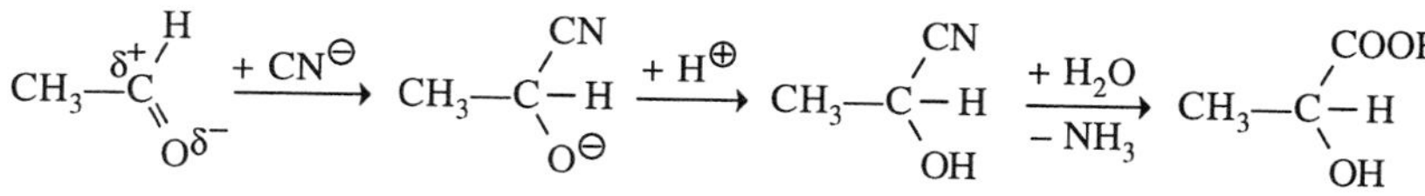

Poly(lactide) (**PLA**), $\mathrm{+OCH(CH_3)CO+_n}$, has a stereogenic center ("asymmetric carbon atom") and can thus form several stereoisomers: two isotactic polymers (PLLA, PDLA), a syndiotactic polymer with alternating D and L units, configurational block copolymers, configurational copolymers with randomly distributed D and L units, etc. The proportion of D and L units as well as the lengths of stereoregular sequences determine whether the polymer will be amorphous, semi-crystalline, or crystalline.

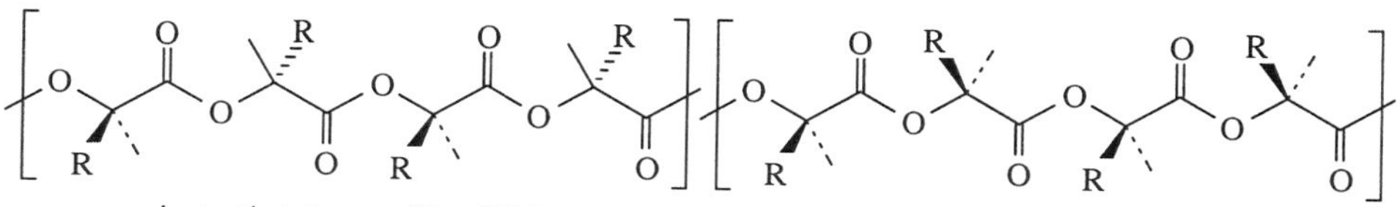

isotactic tetramer (R = $CH_3$)    syndiotactic tetramer (R = $CH_3$)

The first poly(lactide)s were produced by polymerization of the lactide, the cyclic dimer of D,L-lactic acid; they are used to encapsulate pharmaceuticals. A copolymer, poly-

Table 7-8 Properties of poly(glycolide) (PGA), poly(lactide) (PLA), poly(3-hydroxybutyrate) (P(3HB), and a poly[(β-D-hydroxybutyrate)-co-(3-hydroxyvalerate)] (P(3HB-3HV) at 23°C. Refractive indices parallel ($n_{\parallel}$) and perpendicular ($n_{\perp}$) to the fiber direction of highly oriented fibers.

| Property | Physical unit | PGA | PLA L | PLA D,L | P(3HB) 100 | P(3HB-3HV) 80/20 |
|---|---|---|---|---|---|---|
| Density, amorphous | g/cm³ | 1.45 | 1.248 | 1.25 | 1.177 | |
| commercial product | g/cm³ | 1.5–1.64 | | | 1.250 | |
| 100 % crystalline | g/cm³ | 1.707 | 1.290 | | 1.262 | |
| Refractive index $n_{\parallel}$ | 1 | 1.556 | | | | |
| $n_{\perp}$ | 1 | 1.466 | | | | |
| Melting temperature (DSC) | °C | 230 | 188 | 170 | 176 | 145 |
| Glass temperature (DSC) | °C | 36 | 64 | 57 | –1 | –1 |
| Specific heat capacity | J K⁻¹ g⁻¹ | 2.08 | 0.54 | | | |
| Tensile modulus, film | GPa | | 1.2–3.0 | 1.9–2.4 | 3.5 | 1.2 |
| fiber | GPa | | < 9.2 | | | |
| Flexural modulus | GPa | | 2.8 | | 4.0 | 0.8 |
| Tensile strength (fracture), film | MPa | | 28–50 | 29–35 | 40 | 20 |
| fiber | MPa | 340–390 | < 870 | | | |
| Flexural strength | MPa | | 132 | | | |
| Elongation at yield, film | % | | 1.8–3.7 | 3.5–4.0 | | |
| Elongation at fracture, film | % | 15–35 | 2.0–6.0 | 5.0–6.0 | 8 | 50 |
| fiber | % | | 25 | 50–60 | | |
| Notched impact strength (Izod) | J/m | | | | 50 | 200 |
| Water absorption (24 h) | % | 28 | | 0.4–0.6 | 0,7 | |

[(lactide)$_8$-*stat*-(glycolide)$_{92}$], serves for biodegradable surgical threads. Copolymers of L-lactic acid and ε-caprolactone are applied as orthopedic materials, for example, for the repair of bones and tissue engineering.

Newer poly(lactide)s (production ca. 130 000 t/a) rely on corn starch as the feedstock. Starch is hydrolyzed to dextrose (D-glucose) which is fermented to lactic acid. Evaporation of water at 180–200°C converts lactic acid to the lactide which is purified by distillation (bp = 255°C at normal pressure) and then polymerized.

Poly(lactide)s are hydrophobic polymers that dissolve in acetone, 1,4-dioxane, benzene, chloroform, and other organic solvents. They decompose thermally at temperatures above 235°C and degrade biologically.

PLA is processed by injection molding, blow forming, extrusion, and fiber spinning. It has fairly high tensile strengths and good barrier properties and competes with PET and PS for packaging and films. PLA fibers compete with PET but are more easy to dye.

### 7.5.3 Poly(hydroxypropionic acid)s

Polymerization of β-propiolactone (2-oxetanone; I) delivers **poly(propiolactone)** (II), whereas hydrogen transfer on heating acrylic acid (III) at temperatures above 120°C leads to **poly(β-hydroxypropionic acid) = poly(3-hydroxypropionic acid)** which is identical with poly(propiolactone), except for the endgroups (Eq.(7-34)). The polymer is not used industrially, probably because β-propiolactone is very carcinogenic.

$$(7\text{-}34) \qquad n\ \underset{\text{I}}{\text{(oxetan-2-one)}} \longrightarrow \underset{\text{II}}{\left[ -O-CH_2-CH_2-\overset{O}{\overset{\|}{C}}- \right]_n} \longleftarrow \underset{\text{III}}{n\ CH_2=CH-\overset{O}{\overset{\|}{C}}-OH}$$

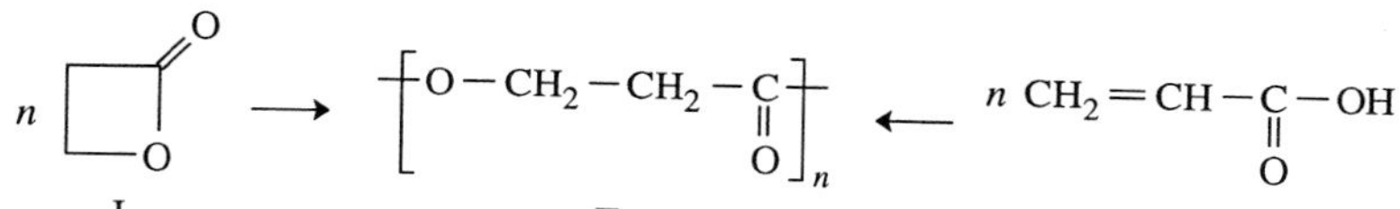

### 7.5.4 Poly(hydroxyalkanoic acid)s

Poly(hydroxyalkanoic acid)s = poly(hydroxyalkanoate) (**PHA**) = poly(hydroxy fatty acid)s (**PHF**) are biologically produced linear polyesters that are derived from $\alpha,\omega$-hydroxy fatty acids with the general composition HO–Z–COOH, where Z is an unsubstituted or substituted, saturated or unsaturated, aliphatic moiety with at least two skeletal carbon atoms. Unsubstituted parent compounds $HO(CH_2)_iCOOH$ ($i$ = 2–4) are produced, for example, by the bacterium *Alcaligenes eutrophus* (Ae):

| | | | |
|---|---|---|---|
| $HO-CH_2-CH_2-COOH$ | 3-hydroxypropionic acid | 3HP | (Ae) |
| $HO-CH_2-CH_2-CH_2-COOH$ | 4-hydroxybutyric acid | 4HB | (Ae) |
| $HO-CH_2-CH_2-CH_2-CH_2-COOH$ | 5-hydroxyvaler(ian)ic acid | 5HV | (Ae) |

More than 40 different substituted PHAs are generated in nature by more than 100 different microorganisms (see also below). Only one PHA with a $C_5$ chain is known (5HV) and only two $C_4$ chains (4HB and the 2-methyl substituted 4HV, see below). However, there are very many PHAs that derive from 3-hydroxypropionic acid: saturated and unsaturated ones; aliphatics, cycloaliphatics, and aromatics; hydroxylated or halogenated ones, etc. for example:

| | | | |
|---|---|---|---|
| $HOOC-CH_2-CH(CH_3)-CH_2-OH$ | 4-hydroxyvaler(ian)ic acid | 4HV | (Ae) |
| $HOOC-CH=C(CH_3)-OH$ | 3-hydroxy-2-butenoic acid<br>3-hydroxycrotonic acid | 3HB:2en | (*Nocarda*) |
| $HOOC-CH_2-CH(CH_3)-OH$ | 3-hydroxybutyric acid | 3HB | (Bm) * |
| $HOOC-CH_2-CH(CH_2-CH_3)-OH$ | 3-hydroxyvaler(ian)ic acid | 3HV | (Ae) |
| $HOOC-CH_2-CH(CH=CH_2)-OH$ | 3-hydroxy-4-pentenoic acid | 3HV:4en | (Rr) |
| $HOOC-CH_2-CH(CH_2-CH(CH_3)_2)-OH$ | 3-hydroxy-5-methylhexanoic acid | 3HHx5Me | (Po) |
| $HOOC-CH_2-CH((CH_2)_7-CH_2Br)-OH$ | 3-hydroxy-11-bromoundecanoic acid | 3HVUD11Br | (Po) |
| $HOOC-CH_2-CH((CH_2-CH=CH)_2-(CH_2)_5H)-OH$ | 3-hydroxy-5,8-tetradecenoic acid | 3HTD:5,8dien | (Po) |

* Polymer now available by stereospecific copolymerization of propylene oxide and carbon monoxide.

Instead of names for PHFs, literature often uses symbols that are derived from conventional (and not systematic) organic-chemical names of compounds. For example, the symbol for 5-hydroxypentanoic acid is not 5HP but 5HV (from 5-hydroxyvaler(ian)ic acid). A prefixed number indicates the position of the hydroxyl group. Position numbers and symbols for substitutents are denoted by suffixes, for example, the 5-position of $CH_3$ in 3-hydroxy-5-methylhexanoic acid by x5Me in 3HHx5Me.

Aliphatic PHF are synthesized by the bacteria *Alcaligenes eutrophus* (Ae), *Bacillus megaterium* (Bm), *Pseudomonas oleovorans* (Po), *Rhodospirillum rubrum* (Rr), and *Nocarda* whereas 2-hydroxysuccinic acid (malic acid), HOOC–$CH_2$–CH(COOH)–OH is also produced by fungi. An aromatic PHF, $\text{+}CO\text{-}CH_2\text{-}CH(CH_2CH_2C_6H_5)\text{-}O\text{+}_n$, is synthesized by the bacterium *Pseudomonas oleovorans* from 5-phenylvaler(ian)ic acid, $C_6H_5(CH_2)_4COOH$ (no OH!), as the single carbon source. PHFs are furthermore formed by plants, for example, *Arabidopsis thaliana* (Thale cress, mouse-ear cress).

Bacteria, fungi, and plants produce PHFs instead of new cells only if the carbon source (sugars, alkanes, etc.) is present in surplus and one important element is missing (N, S, P, Mg, Fe, K, etc.). Several enzymes participate in the syntheses. In the best known Ae-PHF synthesis of 3HB, carbon sources are first transformed to the acetyl group in activated acetic acid (acetyl-CoA), the S-acetic acid ester of coenzyme A, which is also the key chemical in the biosynthesis of poly(isoprene) (Volume I, p. 560):

adenosine-5'-diphosphate-3'-phosphate unit

pantothein unit

$CH_3\text{-}C(=O)\text{-}S\text{-}CH_2\text{-}CH_2\text{-}N(H)\text{-}C(=O)\text{-}CH_2\text{-}CH_2\text{-}N(H)\text{-}C(=O)\text{-}CH(OH)\text{-}C(CH_3)_2\text{-}CH_2\text{-}O\text{-}P(=O)(O^{\ominus})\text{-}O\text{-}P(=O)(O^{\ominus})\text{-}CH_2\text{-}$[ribose with 3'-$O\text{-}PO_3^{2\ominus}$, 2'-OH, and adenine ($NH_2$)]

acetyl | cysteamine unit | pantothenic acid unit

Two molecules of acetyl-CoA are then dimerized to acetoacetyl-CoA by the enzyme β-ketothiolase in a kind of Claisen condensation (Eq.(7-35)). The $CH_3CO$ group of this molecule is subsequently reduced to a $HOCH(CH_3)$– group by an NADPH-dependent acetoacetyl-CoA reductase. The resulting D-(–)-3-hydroxybutyryl-CoA is the true monomer which is converted by the enzyme P(3HB)-synthase to poly(3-hydroxybutyric acid) in a polyelimination in which the coenzyme A unit is split off as HSCoA:

$$\text{(7-35)} \qquad 2\, H_3C\text{—}C(=O)\text{—SCoA} \xrightarrow[\text{– HSCoA}]{\beta\text{-ketothiolase}} O{=}C(CH_3)\text{—}CH_2\text{—}C(=O)\text{—SCoA}$$

$$\xrightarrow[\text{– NADP}^{\oplus}]{\text{acetoacetyl-CoA reductase} \quad + \text{NADPH} + \text{H}^{\oplus}} HO\text{—}CH(CH_3)\text{—}CH_2\text{—}C(=O)\text{—SCoA}$$

$$\xrightarrow[\text{– HSCoA}]{\text{PHF synthase}} \text{+}O\text{—}CH(CH_3)\text{—}CH_2\text{—}C(=O)\text{+}$$

$NADP^{\oplus}$ = Nicotinamide-adenine-dinucleotidephosphate; $NADP^{\oplus} + 2\,H \rightleftarrows NADPH + H^{\oplus}$.

*In vivo*, PHFs are stored in bacterial cells, mainly as amorphous hydrophobic granules of ca. 500 nm diameter that serve as carbon and food reserves, similar to the role of starch in plants and glycogen in animals. Chemical compositions of PHFs depend on both the carbon source and the type of bacterium. Molecular weights may be up to 3 million; they decrease with increasing fermentation time and by hydrolysis during the work-up of the reaction mixture. 3HP, 4HB, and 5HV have no center of chirality. The repeating units of all other PHFs are D(–) enantiomers; polymers are 100 % isotactic.

**Poly(β-*D*-hydroxybutyrate)** (P(3HB)), $+O–CH(CH_3)–CH_2–CO+_n$, can be produced synthetically and biotechnologically. Synthetic routes polymerize [*S*]-*β*-butyrolactone by $Zn(C_2H_5)_2$ to a 100% isotactic polymer ($x_i$ = 1), whereas polymerization of the [*R*] monomer by 1-ethoxy-3-chlorotetrabutyldistannoxane delivers only $x_i$ = 0.94 and that of the [*R*,*S*] monomer only $x_i$ = 0.30. None of these polymerizations has reached full production stage, apparently mainly for cost reasons.

Biotechnologically produced PHAs comprise the homopolymeric P(3HB) and the copolymeric $P[(3HB)_{80}$-*co*-$(3HV)_{20}]$. The polymers are produced by two different processes. Biopol® processes employ glucose for P(3HB) and glucose + propionic acid for P(3HB-*co*-3HV) as carbon sources and the enzymes of the bacterium *Alcaligenes eutrophus* as catalysts. The other process uses saccharose or starch syrup as the carbon source and the nitrogen-fixing bacterium *Alcaligenes latus* as the catalyst.

The Biopol® process proceeds in fermenters of ca. 200 $m^3$ volume. In the first stage, fermentation is advanced up to 60 hours to a certain density of bacterial cells; it is controlled by the phosphate concentration. In the second stage, carbon sources are fed to the fermenter over a 60-hour period until the broth contains ca. 75 % polymer.

In copolymerizations, propionic acid as comonomer must be carefully added because it acts as a cell poison in concentrations as low as 0.1 %. However, successive small additions allow one to produce copolymers with up to 30 mol% propionic acid units. Use of valeric acid can even lead to 3HB-3HV copolymers with up to 90 mol% 3-hydroxyvaleric units.

The solids content of the slurry is ca. 10 %. Heating the slurry precipitates the cells. The wet cells are then treated with hydrolases and detergents. The resulting insoluble polymers have degrees of polymerization of ca. 20 000 and very narrow molecular weight distributions.

PHAs are biologically degradable plastics (for properties, see Table 7-8). P(3HB) and $P[(3HB)_i$-*co*-$(3HV)_j]$ with $i$ = 80–93 % and $j$ = 7–20 % are available as flocculants and granules, P(3HB) also as films, sheets, staple fibers, and non-woven fabrics. They are used for orthopedic and surgical devices, personal hygiene packaging materials, and slow-release systems for drugs, herbicides, etc. They did not succeed as expected, however, for various reasons.

P(3HB) has only a relatively narrow processing window between the melting temperature (175°C) and the beginning of thermal decomposition (205°C). It is also fairly brittle, which is one of the reasons that led to the development of the more easy to process copolymer $P[(3HB)_{80}$–*co*–$(3HV)_{20}]$. However, this copolymer requires the relatively expensive propionic acid, which lessens its competitiveness even more. There is also the question whether a biological degradation by bacteria and/or the atmosphere to ecologically harmless but basically wasted compounds is preferrable over less costly and more easy-to-process synthetic plastics that can be reused after recycling.

### 7.5.5 Poly(pivalolactone)

The polycondensation of hydroxypivalic acid, HO–$CH_2$–$C(CH_3)_2$–COOH, does not result in sufficiently high-molecular weight polymers. The corresponding polymer is therefore obtained by polymerization of pivalolactone with tributylphosphine as initiator that leads to living zwitterions which add more monomer molecules:

$$(7\text{-}36)\qquad (C_4H_9)_3P + H_3C\text{–}\underset{\text{(ring: }C(CH_3)\text{–}C{=}O\text{, }CH_2\text{–}O)}{C(CH_3)} \longrightarrow (C_4H_9)_3\overset{\oplus}{P}\text{–}CH_2\text{–}C(CH_3)_2\text{–}CO\text{–}O^{\ominus}\ \text{etc.}$$

Unstrained tertiary amines act similarly, whereas strained ones are not only initiators but also comonomers:

$$(7\text{-}37)\qquad \sim\!\!\sim C(=O)\text{–}O^{\ominus} + \text{N(quinuclidine)} \longrightarrow \sim\!\!\sim C(=O)\text{–}O\text{–}CH_2\text{–}CH_2\text{–}\overset{\oplus}{N}\text{(ring)}$$

Higher temperatures lead to transfer reactions that produce endgroups which stabilize the polymer against unzipping:

$$(7\text{-}38)\qquad 2\ (C_4H_9)_3\overset{\oplus}{P}\sim\!\!\sim COO^{\ominus} \longrightarrow \begin{cases} (C_4H_9)_3\overset{\oplus}{P}\sim\!\!\sim COOC_4H_9 + (C_4H_9)_2P\sim\!\!\sim COO^{\ominus} \\ (C_4H_9)_3\overset{\oplus}{P}\sim\!\!\sim COC_4H_9 + (C_4H_9)_2\underset{O}{\underset{\|}{P}}\sim\!\!\sim COO^{\ominus} \end{cases}$$

Poly(pivalolactone) decomposes above its melting temperature of 245°C to pivalolactone and further to isobutene and carbon dioxide. For processing to molded articles, films, or fibers, the polymers must be heated rapidly and the polymer must also contain nucleation agents for rapid crystallization. Process difficulties and transformations of the various crystal modifications are probably the reason why poly(pivalolactone)s are not industrial polymers, despite their good mechanical properties.

### 7.5.6 Poly($\beta$-malic acid ester)s

Poly($\beta$-malic acid ester)s III result from the cationic or anionic polymerization of the corresponding lactone esters II, which in turn can be obtained from hydroxysuccinic acid esters I or from bromosuccinic acid, HOOC-CHBr$CH_2$-COOH. The lactone acid (II with R = H) may also be polymerized directly to poly($\beta$-malic acid) (III with R = H):

$$(7\text{-}39)\qquad n\ HO\text{–}\underset{\underset{\text{I}}{COOR}}{CH}\text{–}CH_2\text{–}COOH \longrightarrow n\ \underset{\text{II}}{\boxed{O\text{–}CO}}\ (\text{ring with ROOC}) \longrightarrow \underset{\text{III}}{\left[\text{O–}\underset{COOR}{CH}\text{–}CH_2\text{–CO}\right]_n}$$

Cations attack the $CH_2$ group of the ring and anions the CH(COOR) group. Both polymerizations lead to isotactic polymers albeit with different mirror planes. Poly(pivalolactone)s are used as carriers for pharmaceuticals.

### 7.5.7 Poly(ε-caprolactone)

Free-radical polymerization of lactones leads to high yields but only to low molecular weights because of pronounced radical-transfer reactions. High molecular weights are obtained by cationic or anionic polymerization, for example, with zinc octanoate in the presence of an initiator with an active H atom. Higher lactones such as ε-caprolactone open via acyl cleavage whereas an alkyl opening was proposed for β-propiolactone (Section 7.5.3).

(7-40)

$$2\ CH_3C(=O)^{\oplus} + 2\ (\varepsilon\text{-caprolactone}) \longrightarrow (CH_3COO\text{-substituted cyclic oxonium ion})^{\oplus} + CH_3C(=O){-}O(CH_2)_5C(=O)^{\oplus}$$

$$2\ R^{\ominus} + 2\ (\varepsilon\text{-caprolactone}) \longrightarrow (R,\ O^{\ominus}\text{-substituted ring}) + R{-}C(=O)(CH_2)_5O^{\ominus}$$

Poly(ε-caprolactone) (**PCL**) is a semicrystalline polyester ($T_M$ = 58°C, $T_G$ = –72°C) that dissolves in benzene, chloroform, and *N*,*N*-dimethylacetamide. Because of its low glass temperature and good miscibility with many other polymers, it is used as a polymeric plasticizer and as an additive for the improvement of dyeability and impact strength of poly(olefin)s. The diol of poly[(ε-caprolactone)-*co*-ethylene] serves as a flexible extender in polyurethanes.

### 7.5.8 Other Poly(ω-hydroxyalkanoate)s

Nature produces not only polyesters of phosphoric acid (= nucleic and teichoic acids) (Volume I, p. 516) and polymers of ω-hydroxycarboxylic acids (Section 7.5.4) but also many other aliphatic AB and AA/BB polyesters.

Earth-dwelling bees line their nests with copolyesters of 18-hydroxyoctadecanoic acid, $HO(CH_2)_{17}COOH$, and 20-hydroxyeicosanoic acid = 20-hydroxyarachi(di)c acid, $HO(CH_2)_{19}COOH$ which they produce from the corresponding lactones.

**Cork** (L: *cortex* = tree bark) is the spongy bark of at least 25 and up to 150-year-old cork oaks (*Quercus suber*) that are harvested every 8–10 years. It contains ca. 50 % air and consists mainly of hydrophobic **suberin**, a high-molecular weight polyester with ester and lactone units from hydroxycarboxylic acids and dicarboxylic acids, such as

| | |
|---|---|
| $HOOC–(CH_2)_7–CHOH–CHOH–(CH_2)_7–COOH$ | phloionic acid |
| $HOOC–(CH_2)_7–CHOH–CHOH–(CH_2)_7–CH_2OH$ | phloinolic acid |
| $HOOC–(CH_2)_{20}–COOH$ | phello(ge)nic acid |

**Cutin** consists of $C_{18}$ fatty acids with 2–3 hydroxyl groups per molecule that are crosslinked via ester or peroxide bridges. The wax-like cutin is the essential component of the cuticle of plants (L: *cutis* = skin, *cuticula* = little skin), the outer protective layer of the epidermis (G: *derma* = skin; *epi* = after, on), the upper layer of the skin of plants, and, with a different composition, the upper layer of the skin of animals (see also Section 11.4.3 and Volume IV).

# 7.6 Aliphatic $AA_i/BB_j$ Polyesters

Aliphatic $AA_i/BB_j$ polyesters result from the joint polycondensation of compounds $A_i$ with $i \geq 2$ hydroxyl groups per molecule and compounds $B_j$ with $j \geq 2$ carboxyl groups per molecule. Hydroxyl and/or carboxyl compounds may also be replaced by acid anhydrides or lactones.

## 7.6.1 Poly(alkylene carbonate)s

Poly(alkylene carbonate)s have the general structure $-[Z-O-CO-O]_n-$ where Z is an aliphatic group. **Poly(ethylene carbonate)** (II) does not result from the ring-opening polymerization of succinic anhydride (I) since this polymerization with cationic or anionic initiators releases $CO_2$ and delivers poly(oxyethylene) (III). Polymerization (V) of I by metal alkoxides such as VI leads to a polymer IV with alternating ethylene carbonate and ethylene oxide units.

(7-41)

$$n\ \text{I} \not\longrightarrow -[CH_2-CH_2-O-C(=O)-O]_n- \quad \text{II}$$

$$n\ \text{I} \xrightarrow[-n\,CO_2]{R^{\oplus},\ R^{\ominus}} -[CH_2-CH_2-O]_n- \quad \text{III}$$

$$n\ \text{I} \xrightarrow[-(n/2)\,CO_2]{\text{V}} -[CH_2-CH_2-O-C(=O)-O-CH_2-CH_2-O]_{n/2}- \quad \text{IV}$$

VI: cyclic $-O-Sn[(CH_2)_4H]_2-O-$

Poly(ethylene carbonate) (II) can be obtained by copolymerization of oxiranes, such as ethylene oxide or propylene oxide, and carbon dioxide under pressure, using $(C_2H_5)_2Zn$ or similar compounds as catalysts (Eq.(7-42)). Poly(ethylene carbonate)s are clear elastomers with low glass temperatures (Table 7-9) that are biologically degradable.

$$(7\text{-}42)\qquad n\ \text{(ethylene oxide)} + n\,CO_2 \longrightarrow -[CH_2-CH_2-O-C(=O)-O]_n- \quad \text{II}$$

**Poly(1,3-trimethylene carbonate)**, $-[(CH_2)_3-O-CO-O]_n-$, results from the polycondensation of 1,3-propanediol, $HO(CH_2)_3OH$, and diethylcarbonate, $C_2H_5O-CO-OC_2H_5$.

Table 7-9 Properties of poly(alkylene carbonate)s and poly($\varepsilon$-caprolactone). $T_M$ = melting temperature, $T_G$ = glass temperature, $E$ = modulus of elasticity, $\sigma_B$ = tensile strength at break, and $\varepsilon_B$ = elongation at break.

| Polymer | $T_M$/°C | $T_G$/°C | $E$/MPa | $\sigma_B$/MPa | $\varepsilon_B$/% |
|---|---|---|---|---|---|
| Poly(ethylene carbonate) | | 5 | 205 | 580 | |
| Poly(1,2-propylene carbonate) | | 30 | 295 | | |
| Poly(1,3-trimethylene carbonate) | 36 | −15 | 295 | 50 | 160 |
| Poly($\varepsilon$-caprolactone) | 59 | −60 | 2770 | 246 | 750 |

### 7.6.2 Other Poly(alkylene alkanoate)s

Polycondensation of ethylene glycol, $HOCH_2CH_2OH$, with dicarboxylic acids, especially adipic acid, $HOOC(CH_2)_4COOH$, or sebacic acid, $HOOC(CH_2)_8COOH$, delivers aliphatic polyesters with molecular weights of several thousands. Because of their low melting and glass temperatures (**poly(ethylene adipate)**: $T_M$ = 55°C, $T_G$ = –50°C; **poly(ethylene sebacate)**: $T_M$ = 79°C, $T_G$ = –30°C), they serve as the soft segments in elastomers and elastic fibers, secondary polymeric plasticizers, non-fatty bases of ointments, and because of their water-repellance, also in leather impregnations.

The synthesis of adipic acid by oxidation of cyclohexane (p. 127) is accompanied by a Dieckmann condensation as a side reaction which converts adipic acid to cyclopentanone that in turn is oxidized to glutaric acid, $HOOC(CH_2)_3COOH$. This acid is polycondensed with unspecified glycols to **poly(alkylene glutarate)s**, highly viscous liquids that are used as plasticizers for poly(vinyl chloride), adhesives, or synthetic rubbers.

### 7.6.3 Alkyd Resins

**Alkyd resins (AK)** are (hyper)branched polyesters from multifunctional alcohols (glycerol, trimethylol propane, pentaerythritol, sorbitol, etc.), dicarboxylic acids (phthalic, succinic, maleic, fumaric, adipic, azelaic) or their anhydrides (phthalic acid anhydride), and fatty acids (from oils of linseeds, castor beans, soy beans, or coconuts; Section 3.10). Glycerol can be partly substituted by ethylene glycol. Other modifications include addition of urea-formaldehyde or melamine-formaldehyde resins or ethyl cellulose.

The name *alkyd resin* is derived from their monomers: *alc*ohol + ac*id*. Alkyd resins based on phthalic acid are also called **phthalate resins** and those based on phthalic acid and glycerol, **glyptal resins**.

Alkyd resins are produced by the fatty acid method or by alcoholysis. In the *fatty acid method*, polyols, polybasic acids, and fatty acids are directly polycondensed to group conversions just short of the gel point (Volume I, Section 13.9). The *alcoholysis method* first reacts a glyceride oil (such as a triglyceride of a fatty acid RCOOH) and a polyol (such as glycerol) at 225-250°C to a monoglyceride:

$$\begin{array}{c} CH_2OOCR \\ | \\ CHOOCR \\ | \\ CH_2OOCR \end{array} + 2 \begin{array}{c} CH_2OH \\ | \\ CHOH \\ | \\ CH_2OH \end{array} \rightleftharpoons 3 \begin{array}{c} CH_2OOCR \\ | \\ CHOH \\ | \\ CH_2OH \end{array} \qquad (7\text{-}43)$$

The monoglyceride is then reacted with a dibasic acid (for example, phthalic acid) to a soluble homogeneous resin with conversions just short of the gel point.

Alkyd resins based on half-drying or drying oils (Section 3.10) act as "binders" in paints that also contain pigments, solvents, and small proportions of so-called driers. Driers are chemical compounds such as cobalt ethylhexanoate that catalyze the cross-linking oxidation of carbon–carbon double bonds of the oils by oxygen from air (these driers are called "siccatives" (L: *siccus* = dry) if applied to solutions). Fatty acid-containing alkyd resins with 56-70 % oxidizing oils are called **long-oil alkyds** in order to distinguish them from **short-oil alkyds** with 30-45 % non-oxidizing oils.

Air-drying alkyd resins are relatively brittle. Less brittle and more light-colored are oil-free alkyd resins that do not contain fatty acids but many unreacted OH and COOH groups. Alkyd resins can furthermore be modified by addition of other monomers, for example, styrene, acrylates, silicones, isocyanates, epoxides, etc. All these alkyd resins or their solutions in water or organic solvents serve as varnishes, baking varnishes (with high solid contents), coating and patching compounds, cements, etc. So-called **alkyd plastics** are not true alkyd resin-based materials but rather diallyl phthalate-based materials that contain inorganic fillers.

Alkyd resins are a special class of hyperbranched polyesters (see Volume I, pp. 54 and 480) of which other classes are presently being explored for industrial uses.

### 7.6.4 Unsaturated Polyesters

Unsaturated polyesters are polycondensates from maleic (I) and/or phthalic anhydride (II), isophthalic (III) and/or terephthalic acid (IV), adipic acid ($HOOC(CH_2)_4COOH$), or chlorendic acid (= *h*exachloro-*e*ndomethylene *t*etrahydrophthalic acid = HET acid® (V) on one hand and ethylene glycol ($HOCH_2CH_2OH$ (VI)), 1,2-propylene glycol (VII), 1,4-butanediol ($HO(CH_2)_4OH$), diethylene glycol ($HOCH_2CH_2OCH_2CH_2OH$), neopentylglycol (VIII), and/or oxyethylated bisphenols (IX) on the other.

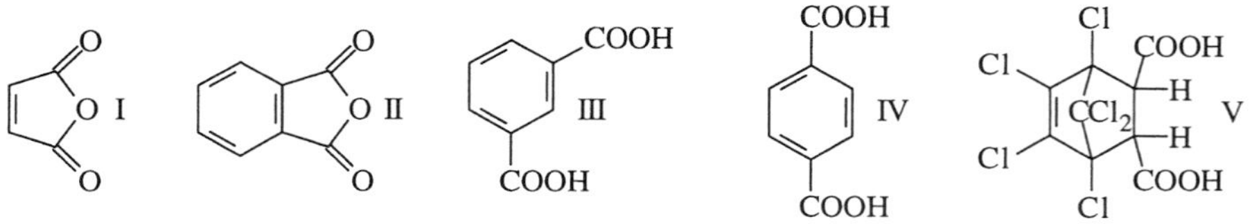

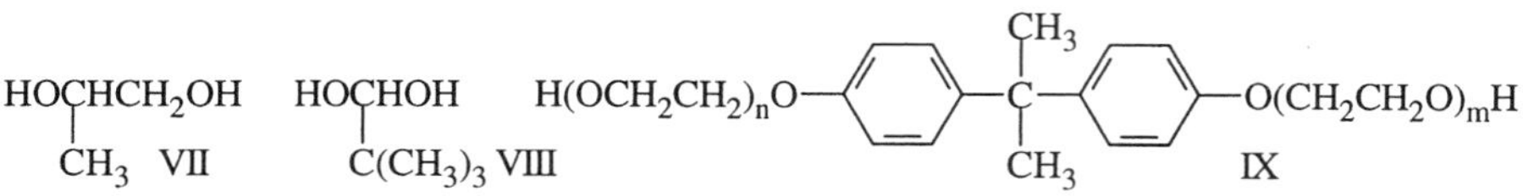

I II III IV V

$HOCH(CH_3)CH_2OH$ VII  $HOCH(C(CH_3)_3)OH$ VIII  $H(OCH_2CH_2)_nO{-}C_6H_4{-}C(CH_3)_2{-}C_6H_4{-}O(CH_2CH_2O)_mH$ IX

In polycondensations of maleic anydride (I) with, e.g., ethylene glycol (VI), most of the maleic anhydride is converted to the industrially more desirable fumaric acid units (X):

$$n\ HOCH_2CH_2OH + n\ \text{(maleic anhydride)} \longrightarrow \left[ OCH_2CH_2O{-}C(=O){-}CH{=}CH{-}C(=O) \right]_n \qquad (7\text{-}44)$$

VI I X

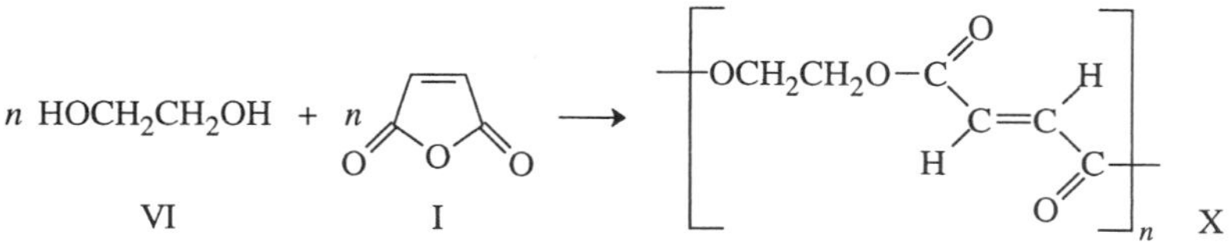

Up to 15 mol% of the carbon–carbon double bonds of the original maleic anhydride units also form ether bonds from the OH groups of ethylene glycol. The polycondensation reaction, Eq.(7-44), thus does not proceed stoichiometrically.

In industry, **unsaturated polyesters** or **unsaturated polyester resins** (**UP**) are *not* the polymer molecules X but the mixtures of these polymers with 45–55 % styrene, methyl methacrylate, or other monomers as "crosslinking agents" (the true crosslinkers are, of course, the unsaturated polyester molecules themselves). UPs are also known as **thermosetting polyesters**. The mixtures are then hardened (i.e., crosslinked) by free-radical copolymerization of the double bonds of the unsaturated polyester chains with the double bonds of the added monomers, using a free-radical initiator.

Variation of acids, glycols, and vinyl or acrylic monomers allows one to fine-tune the properties of hardened resins for the intended use. For example, copolymerization of unsaturated polyester molecules with more electronegative monomers such as styrene or vinyl acetate leads to more "alternating copolymers", i.e., shorter crosslinks, and thus harder thermosets. More electropositive comonomers such as methyl methacrylate deliver longer crosslinks between polyester chains and thus softer thermosets. Specialty products also contain vinyl toluene, $\alpha$-methyl styrene, or diallyl phthalate units.

Unsaturated polyester resins are used as adhesives and for coatings, appliances, boat hulls, chemical tanks, bath tubs, etc. For many of these purposes, they are reinforced by glass fiber mats (see Volume IV). World production is ca. $2.5 \cdot 10^6$ t/a (2000).

# 7.7 Aromatic Polyesters

## 7.7.1 Polycarbonates

Polycarbonates (**PC**) are polyesters from diols and derivatives of carbonic acid such as diphenyl carbonate, $C_6H_5O{-}CO{-}OC_6H_5$, or phosgene, $COCl_2$. Industrial aromatic polycarbonates employ as diols predominantly bisphenols (BP) such as bisphenol A for standard types and bisphenol I, isatine biscresol (IBK), and others for special types. Crosslinked aromatic polycarbonates contain trisphenols (e.g., THPE) and tetrakisphenols. World consumption of all aromatic polycarbonates is ca. $2 \cdot 10^6$ t/a (2003).

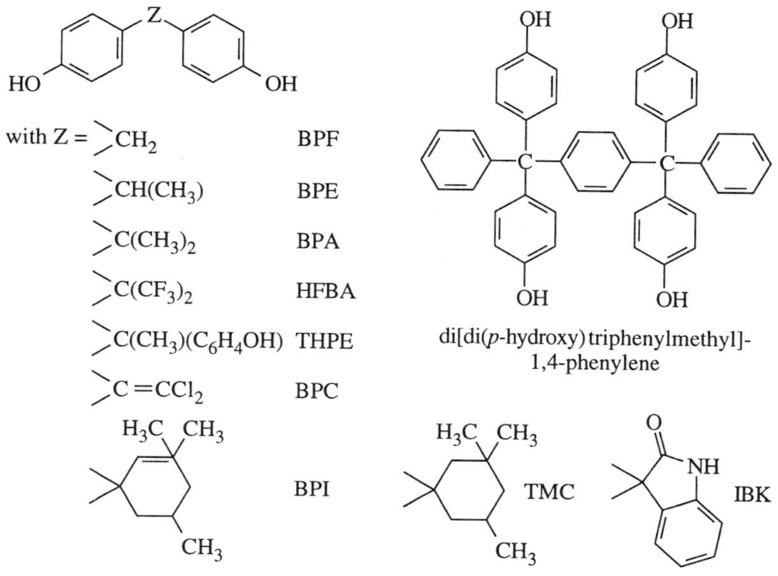

HO–C₆H₄–Z–C₆H₄–OH

with Z = $>CH_2$ BPF

$>CH(CH_3)$ BPE

$>C(CH_3)_2$ BPA

$>C(CF_3)_2$ HFBA

$>C(CH_3)(C_6H_4OH)$ THPE

$>C{=}CCl_2$ BPC

$H_3C$ $CH_3$ … $CH_3$ BPI

$H_3C$ $CH_3$ … $CH_3$ TMC

O NH IBK

OH OH OH OH

di[di(*p*-hydroxy)triphenylmethyl]-1,4-phenylene

Most bisphenols result from the condensation of two phenol molecules with one ketone molecule, 2 $C_6H_5OH$ + >CO → $HOC_6H_4$—C(<)–$C_6H_4OH$ + $H_2O$,: bisphenol A if the ketone is *a*cetone, $(CH_3)_2CO$; bisphenol I if *i*sophorone; bisphenol TMC if 3,3,5-*t*ri*m*ethyl*c*yclohexanone, etc. (see previous page). Bisphenol F is based on *f*ormaldehyde.

**Syntheses**

Polycarbonates are synthesized by either interfacial polymerization of a bisphenol and phosgene or by transesterification of a bisphenol and diphenyl carbonate. Solution polycondensation of bisphenol A and phosgene in pyridine and chlorohydrocarbons as solvents is no longer viable because of the high cost of solvent recovery and work-up.

Most polycarbonates are now obtained at room temperature by *interfacial polymerization* of phosgene with, for example, bisphenol A. The system consists of an emulsion with an organic phase (a chlorohydrocarbon such as $CH_2Cl_2$ or $C_6H_5Cl$) and an alkaline aqueous phase containing the bisphenol (present as the sodium salt) and a tertiary amine or a quaternary ammonium salt as a phase-transfer agent. Introduced $COCl_2$ reacts fast with the phenolate ions to $ClOC–OC_6H_4–C(CH_3)_2–C_6H_4O–COCl$, the di(chlorocarbonic acid ester) of bisphenol A, which is only slightly soluble in water and reacts only slowly with phenolate ions of the aqueous phase. As a consequence, it moves to the organic phase where it forms the ammonium salt (~OCOCl + $NR_3$ → ~$OCONR_3^{\oplus}$ $Cl^{\ominus}$). This polar intermediate travels to the interface between organic solvent and water where it reacts with phenolate ions, etc. The gross reaction of, e.g., bisphenol A is thus

$$n\,\text{NaO}-C_6H_4-C(CH_3)_2-C_6H_4-\text{ONa} \xrightarrow[-\,2\,n\,\text{NaCl}]{+\,n\,\text{COCl}_2} \left[-O-C_6H_4-C(CH_3)_2-C_6H_4-O-\underset{\substack{\|\\O}}{C}-\right]_n \quad (7\text{-}45)$$

Phosgene must be used in excess because 10–20 % of it is saponified. After polymerization, the phases are separated. The organic phase is washed with water to remove electrolytes; the solvent is separated by spray-drying or extrusion.

*Interfacial polycondensation* is less expensive than the older *ester exchange* (bisphenol in 10-fold amount of pyridine plus phosgene). It also leads to higher molecular weights of 40 000-150 000 and delivers uniform products if solvents for the resulting polycarbonates are present (see above). Molecular weights are regulated and chain ends are capped by reaction with monofunctional phenols, since uncapped chain ends are prone to photo-Fries rearrangements (Volume I, p. 375). Sodium chloride must also be removed, for example, by steam extrusion; otherwise, hazy polymers result.

In *melt transesterifications*, bisphenols are reacted with a slight excess of diphenyl carbonate (DPC; from $C_6H_5OH$ + $COCl_2$), using basic transesterification catalysts:

$$n\,\text{HO}-C_6H_4-C(CH_3)_2-C_6H_4-\text{OH} \xrightarrow[\downarrow -\,2\,n\,C_6H_5OH]{+\,(n+1)\,C_6H_5O-CO-OC_6H_5} C_6H_5O-\underset{\substack{\|\\O}}{C}\left[-O-C_6H_4-C(CH_3)_2-C_6H_4-O-\underset{\substack{\|\\O}}{C}-\right]_n OC_6H_5 \quad (7\text{-}46)$$

Acid catalysts lead to high polymerization rates but also generate branches by the Kolbe reaction:

$$\text{(7-47)} \quad \sim\!\!\sim C_6H_4{-}C(CH_3)_2{-}C_6H_4{-}O{-}\underset{\substack{\|\\ O}}{C}{-}O\sim\!\!\sim \longrightarrow \sim\!\!\sim C_6H_4{-}C(CH_3)_2{-}C_6H_3(COOH){-}O\sim\!\!\sim$$

Industrially, alkali is used as a catalyst but alkali also promotes at $T > 150°C$ the decomposition of phenolic chain ends to *p*-isopropenyl structures and phenol:

$$\text{(7-48)} \quad \sim\!\!\sim C_6H_4{-}C(CH_3)_2{-}C_6H_4{-}OH \xrightarrow[{[OH]^{\ominus}}]{> 150°C} \sim\!\!\sim C_6H_4{-}C(=CH_2){-}CH_3 + C_6H_5{-}OH$$

The industrial process therefore consists of two steps. First, a non-volatile oligomer with protecting phenolester endgroups is produced by pre-polymerization at 180-200°C. Second, the temperature is slowly raised to 300°C for transesterification. The high reaction temperature is necessary because of the very high viscosity of the melt; even so, molecular weights (ca. 30 000) are lower than those from interfacial polymerization. The high viscosity of the melt also necessitates special reactors, for example, extruder reactors (see p. 182).

Transesterifications do not require solvents and their costly recovery; polymers are also more easy to work up. A new transesterification process (2002) is even less costly since it does not require expensive phosgene for the synthesis of diphenyl carbonate (DPC). DPC is here obtained by a 3-step process: ethylene oxide + $CO_2$ → ethylene carbonate in 99 % yield; ethylene carbonate + methanol → dimethyl carbonate + ethylene glycol in 99 % yield; dimethyl carbonate + phenol → diphenyl carbonate + methanol. The overall consumption of chemicals is thus:

$$\text{(7-49)} \quad n\, H_2C\underset{O}{-}CH_2 + n\, CO_2 + 2n\, C_6H_5OH \longrightarrow n\, C_6H_5O{-}\underset{\substack{\|\\ O}}{C}{-}OC_6H_5 + n\, HOCH_2CH_2OH$$

The new process controls the molecular weight by letting the molten reaction mixture flow by gravity through a non-agitated reactor at temperatures above 200°C. Excess phenol is recovered from the top of the reactor.

**Properties**

Polycarbonates are transparent, glossy, amorphous engineering thermoplastics that have very low water absorptions, moderate heat stabilities, and excellent dimensional stabilities and impact strengths (Table 7-10). They can be easily molded (injection, blow, rotational), extruded, and processed by thermoforming and vacuum forming (see Volume IV). PCs are used for injection-molded articles (for example, CDs), prescription lenses, insulating films, and fracture- and bullet-proof panels (for example, windows; larger panels are usually made from poly(methyl methacrylate), however). Flame retardancy is reduced by blending PCs with polyphosphonate (Section 12.5.4).

Bisphenol A polycarbonate (**PC-A**) is resistant to water, dilute acids and alkali, ethanol and higher alcohols. It dissolves in tetrahydrofuran, *p*-dioxane, cyclohexanone, chloroform and other chlorinated hydrocarbons, *N,N*-dimethylformamide, benzene, toluene, and xylenes. Methanol and strong bases and acids induce stress cracking.

Fracture strengths become constant at molecular weights above 30 000 because chains are then completely entangled (Volume III). Polymers with such high molecular weights have high melt viscosities which are detrimental to injection molding of CDs where one needs high flowability for short cycle times, i.e., molecular weights of ca. 10 000. Polymers with lower molecular weights have less favorable mechanical properties, however. This reduction in properties can in part be compensated by using PCs with bulky end-groups.

Polycarbonate fibers are blended with cellulose for easy-care garments that can be laundered hot. Monofilament PC fibers are used for temporary stitches in tailoring because these fibers can be removed by dry-cleaning solvents.

High-impact polycarbonates result from partial replacement of BPA by TMC (p. 343). The dihydroisophorone group impedes the rotations of chain segments around the chain axis and increases the glass temperature from 150°C (100 % BPA-PC) to ca. 239°C (100 % poly(TMC)). Glass temperatures of such copolymers vary as a function of composition between 150°C and 239°C. At temperatures above $T_G$, continuous chair ⇄ boat transitions of the cyclohexane moiety generate higher chain flexibilities and thus reduced melt viscosities.

BPA is also copolymerized with bisphenols that increase the resistance against heat distortion (I), improve flame resistance (II), or increase notched impact strengths (III). Polycarbonates with chain ends capped by benzocyclobutane units (IV) can be processed like thermoplastics but crosslink thermally by ring-opening of the cyclobutane ring (see also Section 6.6.5). These polymers are tough engineering plastics at low cross-linking densities and high-performance thermosets at high ones.

**Poly(ester-*co*-carbonate)s** result from the copolycondensation of phosgene with a mixture of diphenols and dicarboxylic acids. Depending on composition and the way the monomers are added, random or block copolymers are obtained. Poly(ester-*co*-carbonate)s with up to 10 % of flexibilizing dodecanedioic acid, $HOOC(CH_2)_{10}COOH$ serve for thin-walled articles. **Polyphthalate carbonates** are copolymers of bisphenol A and terephthalic acid. **Poly(ether-*co*-carbonate)s** with up to 20 wt% poly(ethylene glycol) units are used for dialysis membranes.

**Polycarbonate blends** are mixtures of bisphenol A polycarbonates with PE, ABS, PET, poly(butylene terephthalate) (PBT), poly(oxy-2,6-dimethyl-1,4-phenylene) (PPO, PPE), poly(methyl methacrylate) (PMMA), or propylene copolymers.

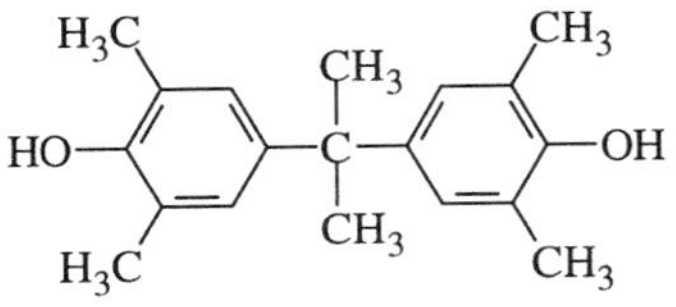

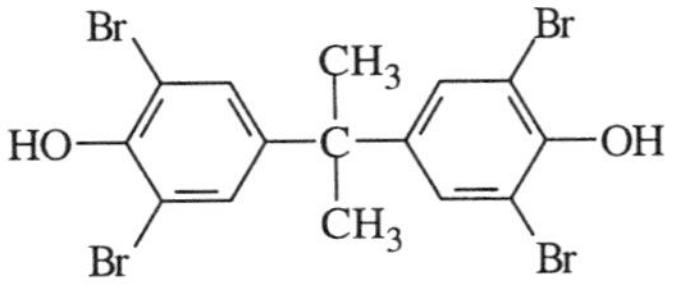

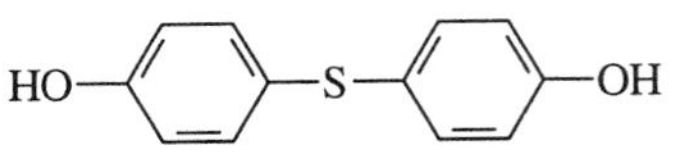

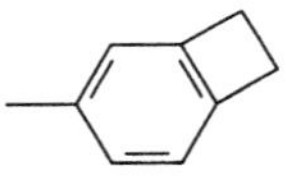

Table 7-10 Properties of injection-molded semicrystalline aromatic polyester resins. nb = no break

| Property | Physical unit | PC-A | POB | PET | PTT | PBT | PEN |
|---|---|---|---|---|---|---|---|
| Density | $g/cm^3$ | 1.22 | | 1.38 | 1.35 | 1.34 | 1.35 |
| Refractive index | 1 | 1.586 | | | | | |
| Melting temperature (DSC) | °C | | 550 | 255 | 233 | 227 | 266 |
| Heat distortion temperature (1.82 MPa) | °C | 142 | | 80 | 59 | 54 | |
| Vicat temperature B | °C | 157 | | 115 | | | |
| Glass temperature | °C | 150 | 161 | 70 | 42 | 60 | 121 |
| Continuous service temperature | °C | | 450 | | | | |
| Linear thermal expansion coefficient | $10^5$ $K^{-1}$ | 6.5 | | 7 | | 7 | 4.4 |
| Specific heat capacity | $J\ K^{-1}\ g^{-1}$ | | | 1.2 | | 1.35 | |
| Thermal conductivity (20°C) | $W\ m^{-1}\ K^{-1}$ | 0.21 | | 0.29 | | 0.21 | |
| Tensile modulus | MPa | 2400 | 6900 | 2800 | | 2600 | 2440 |
| Flexural modulus | MPa | 2340 | | 3100 | 2800 | 2300 | 2500 |
| Tensile strength at yield | MPa | 64 | | 81 | | 52 | |
| Tensile strength at fracture | MPa | 69 | | 73 | 68 | 57 | 83 |
| Flexural strength | MPa | 93 | 76 | | | | 108 |
| Elongation at yield | % | 6 | | 4 | | 4 | |
| Elongation at fracture | % | 115 | | 70 | | 120 | 49 |
| Impact strength (Izod, 3.1 mm) | J/m | | | 37 | 48 | 53 | |
| (Charpy) | $kJ/m^2$ | nb | | nb | | nb | |
| Notched impact strength (Izod, 3.1 mm) | J/m | 850 | | 90 | | | |
| (Charpy) | $kJ/m^2$ | 28 | | 3 | | 3.5 | |
| Hardness (Rockwell) | - | R122 | | R105 | | R120 | |
| Relative permittivity (50 Hz) | 1 | 3.0 | | 3.0 | 3.0 | 3.1 | |
| Electrical resistance (surface resistivity) | Ω | $>10^{15}$ | | $6 \cdot 10^{14}$ | | $5 \cdot 10^{13}$ | |
| Resistivity (volume resistance) (2 min) | Ω cm | $>10^{16}$ | $1 \cdot 10^{15}$ | $1 \cdot 10^{15}$ | $1 \cdot 10^{16}$ | $1 \cdot 10^{16}$ | |
| Electric strength | kV/mm | $>80$ | 22 | 21 | | 16 | |
| Dissipation factor (1 MHz) | 1 | 0.008 | | 0.02 | 0.015 | 0.02 | |
| Limiting oxygen index | 1 | 25 | | 20 | | | |
| Water absorption (24 h, 50 % RH) | % | 0.15 | | 0.1 | | 0.09 | |

### 7.7.2 Poly(*p*-hydroxybenzoate)s

*p*-Hydroxybenzoic acid, HO(*p*-$C_6H_4$)COOH, is easily available from potassium phenolate and carbon dioxide by the Kolbe–Schmitt reaction. The acid decarboxylates at temperatures above ca. 200°C and cannot be directly polymerized by melt polycondensation. Instead, its polymer is obtained by polycondensation of the phenyl ester in terphenyl as heat-transfer agent. Both the homopolymer and the copolymers with 6-hydroxy-2-naphthoic acid or terephthalic acid/ethylene glycol are known in the plastics industry simply as **polyoxybenzoates** or "**aromatic polyesters**" (**ARP**), notwithstanding other aromatic polyesters such as poly(ethylene terephthalate), etc.

Poly(*p*-hydroxybenzoate) (**POB**; PHB is now predominantly used for poly(3-hydroxybutyrate) has a melting temperature of at least 550°C and is extraordinarily thermally stable (for properties, see Table 7-10). It is insoluble in all known solvents and can only be processed by machining or by hammering, sintering, or plasma spraying. POBs filled with aluminum or bronze powders are used for bearings, joints, etc.

### 7.7.3 Liquid Crystalline Polyesters

**Liquid crystalline polymers** (**LCP**) are polymers that show liquid crystalline behavior in bulk or solution, i.e., states with ordered regions that flow like liquids. **Thermotropic liquid crystalline polymers** show this behavior as neat substances, whereas **lyotropic liquid crystalline polymers** act this way in solution above a certain critical concentration (Volume III). The lyotropic liquid state is utilized in polymer technology for the formation of oriented fibers and films; the polymers used are all aromatic polyamides, polyhydrazides, polybenzoxazoles and related compounds (see Chapter 10). Thermotropic liquid crystalline polymers serve as **self-reinforcing plastics**; industrial thermotropic LCPs are all based on 1,4-hydroxybenzoic acid (Table 7-11).

Several copolymers are offered. As is often custom in industry, trade names are no guarantee of the same composition. Several Vectra® grades such as Vectra® 950 consist of 1,4-oxybenzoyl (I) and 2,6-oxynaphthoyl units (V) in the molar ratio 73:27, but other Vectra® grades may contain other units such as 1,4-terephthaloyl and 1,4-biphenol. Vectra® B 950 is a terpolymeric polyester amide from 60 wt% 6-hydroxy-2--naphthoic acid, 20 wt% terephthalic acid, and 20 wt% aminophenol; it is totally devoid of 1,4-oxybenzoyl units. Vectran® fibers also contain additional comonomeric units.

Table 7-11 Composition (in mol%) of commercial copolymers of *p*-hydroxybenzoic acid (I) and other monomers (II–VI). + Present in unknown proportion, (+) present in very small proportion.

| Monomeric unit | | X7G | Xydar® | Vectra® grades | Econol® HITBP |
|---|---|---|---|---|---|
| I | –O–(1,4-$C_6H_4$)–CO– | 60 | 67 | + | + |
| II | –O–$CH_2$–$CH_2$–O– | 20 | | | |
| III | –CO–(1,4-$C_6H_4$)–CO– | 20 | + | + | + |
| IV | –O–(1,4-$C_6H_4$)–(1,4-$C_6H_4$)–O– | | + | + | + |
| V | –O–(2,6-$C_{12}H_6$)–CO– | | | + | |
| VI | –CO–(1,3-$C_6H_4$)–CO– | | | | (+) |

Thermotropic liquid crystalline aromatic polyesters are synthesized either in bulk (both melt and solid phase) or in non-aqueous suspension. In all three processes, monomers HO–Z–COOH (I, V), HO–Z'–OH (II, IV), and/or HOOC–Z"–COOH (III, VI) are not directly polymerized because direct esterification of phenols (themselves weak acids) requires rather drastic conditions. The monomer mixture of, for example, 1,4-hydroxybenzoic acid and 2,6-hydroxynaphthoic acid is rather reacted first with an excess of acetic anhydride for several hours at temperatures above 150°C under a nitrogen blanket at reflux, either in bulk or in the suspending liquid. The reaction converts phenol groups to phenol ester groups which can be polycondensed at milder conditions than the phenol groups themselves.

The phenol esters then transesterify with carboxyl groups to oligomers, schematically shown in Eq.(7-50) whereby acetic acid is released. At a monomer conversion of 70 %, only oligomers are formed ( $\overline{X}_w \approx 5.7$, $\overline{X}_n \approx 3.3$ of reactants (including monomer molecules!) for stoichiometric reactions in batch reactors, see Volume I, p. 446):

(7-50) $n\ CH_3COO-C_6H_4-COOH + n\ CH_3COO-C_{10}H_6-COOH \xrightarrow{-2n\ CH_3COOH} [-O-C_6H_4-CO-O-C_{10}H_6-CO-]_n$

Ia Va

The resulting monomer/oligomer mixture contains molecules with phenolic ester endgroups, $CH_3CO–O–Ar\sim$ and carboxyl endgroups $–Ar–COOH$ (not shown in Eq.(7-50)). It is heated first rapidly to 240°C and then more slowly to 300°C, whereupon a polymer with a medium degree of polymerization is formed. The released acetic acid and acetic anhydride are distilled off and recycled to acetic anhydride.

The fate of the resulting "intermediate polymer system" then depends on which of the three types of polymerization was chosen:

In *suspension polymerizations*, the intermediate polymer precipitates out of the still present suspending liquid as powder which can be filtered off. The "intermediate polymer" itself must be postpolymerized by heating just below its melting point which (a) increases the molecular weight to the desired degree, (b) removes the acid/anhydride, and (c) drives out remaining suspending liquid.

In the *solid-state process*, the "intermediate polymer" is melt-extruded and pelletized. The pellets are postpolymerized like the powders from suspension polymerizations but this takes more time because pellet particles have greater volumes than powder particles which requires more time for heat transport, homogenization, and removal of volatiles.

*Melt polymerizations* also need postpolymerizations of intermediate polymers; this is not done by solid-stating but rather directly in the melt, which for this purpose is heated far above the melting point of the ultimate polymer, sometimes to 370°C. The melt viscosity of the ultimate polymer is much higher than that of the intermediate polymer from the solid-state process. The driving-out of the melt of the ultimate polymer takes much more time (up to several hours) and this may lead to complications.

At monomer conversions greater than ca. 70 % and high temperatures, side reactions occur that change the composition and sequence statistics of comonomeric units. For example, a part of the end units $\sim O(1,4-C_6H_4)COOH$ of X is decarboxylated to Xa whereas end units $\sim O(2,6-C_{10}H_6)COOH$ remain intact (Eq.(7-51), p. 350). Some endgroups $CH_3COO–Ar–$ (XIa) in *both* units $CH_3COO(1,4-C_6H_4)–$ and $CH_3COO(2,6-C_{10}H_6)–$ are furthermore pyrolyzed to HO–Ar– groups (XIb). The phenol ester (Xa) from the decarboxylation then reacts with phenol endgroups from the pyrolysis (XIb) or from polymer molecules. This phenolysis reaction splits off phenol.

Ultimately, one monomeric unit (XII) replaces two monomeric units (Xa + XIa) which changes both the copolymer composition and the sequence statistics of the monomeric units. Phenol from this side reaction acts as a plasticizer; it also produces unwanted coloration of the polymer.

LCPs are self-reinforcing because cooling their melts leads to the formation of liquid-crystalline regions that act like embedded fibrils and make the polymer stiffer and stronger and thus excellent engineering plastics. Their processing requires extra care, though, because a property gradient may develop on cooling the melt.

(7-51)

~COO–$C_6H_4$–COOH (X) $\xrightarrow[\text{decarboxylation}]{-CO_2}$ ~COO–$C_6H_5$ (Xa)

$CH_3COO$–$C_6H_4$–COO~ (XIa) $\xrightarrow[\text{pyrolysis}]{-CH_2=C=O}$ HO–$C_6H_4$–COO~ (XIb)

Xa + XIb $\xrightarrow[\text{phenolysis}]{-HO-C_6H_5}$ ~COO–$C_6H_4$–COO~ XII

A copolyester, X7G, with 60 mol% *p*-hydroxybenzoate units and 40 % terephthaloyl-glycol units was the first self-reinforcing thermoplastic since it is a liquid-crystalline polymer. It is no longer produced, though, because the same properties can be obtained from far less expensive glass fiber-reinforced unsaturated polyesters.

### 7.7.4 Poly(ethylene oxybenzoate)

Reaction of *p*-hydroxybenzoic acid with ethylene oxide delivers *p*-$\beta$-hydroxyethoxy-benzoic acid, $HOCH_2CH_2O(p\text{-}C_6H_4)COOH$. The methyl ester of this compound is poly-condensed in vacuum at ca. 250°C to $H\!\!+\!\!OCH_2CH_2O(p\text{-}C_6H_4)CO\!\!+\!\!_n$. Fibers from this polymer (**PEOB**) have a silk-like touch and are extraordinarily wrinkle-resistant (see Volume IV, Section 6.3.3).

### 7.7.5 Poly(ethylene terephthalate)

Poly(ethylene terephthalate), $\!+\!OCH_2CH_2OOC(p\text{-}C_6H_4)CO\!+\!_n$, consists of alternating ethylene glycol and terephthalic acid units. Its poly(monomer) name is **poly(ethylene glycol terephthalate)** and its systematic name, **poly(oxyethyleneoxyterephthaloyl)**. In the plastics industry, it is known as **thermoplastic polyester** in order to distinguish it from the thermosetting "unsaturated polyester" (Section 7.6.4). The textile industry calls it **polyester fiber**. Recommended abbreviations are **PET** for plastics, **PETE** or **1** for re-cyclable plastics, and **PES** for polyester fibers. World consumption: $29 \cdot 10^6$ t/a (2000).

**Chemical Processes**

Monomers for PET and PES are ethylene glycol (EG) and terephthalic acid (TPA) or its methyl ester (dimethyl terephthalate, DMT). Ethylene glycol is now predominantly obtained from ethene via ethylene oxide (p. 102, X) and to a smaller extent from methanol via dimethyl oxalate (p. 94, VI). Acetoxylation of ethene (p. 101, VI) and hydrating carbonylation of formaldehyde (p. 95, IV) are no longer viable syntheses for intermediates on the path to ethylene glycol.

Terephthalic acid is now predominantly produced by oxidation of *p*-xylene, as dis-cussed in greater detail below. It can also be synthesized from toluene via either benzoic acid (p. 126, V) or *p*-tol(u)ylaldehyde (p. 126, IV). The history of terephthalic acid is directly and intimately connected with the history of poly(ethylene terephthalate) fibers and plastics and therefore deserves a little more discussion.

At the time of the first poly(ethylene terephthalate) patent in 1941, terephthalic acid was a laboratory curiosity; for example, it is not mentioned in the long list of dicarboxylic acids in Carothers' umbrella patents for polyesters and polyamides in the mid-1930s. Early oxidation of *p*-xylene by air in dilute nitric acid delivered a TPA that contained colored impurities. This TPA was difficult to purify by distillation (sublimation above its melting temperature of at least 300°C) or recrystallization (soluble only in solvents such as pyridine, dimethylsulfoxide, *N*,*N*-dimethylformamide). It was therefore converted to its dimethyl ester (**DMT**) as the monomer for the polycondensation with ethylene glycol.

However, this DMT did not have an acceptable purity either and needed to be repeatedly recrystallized and distilled before polycondensation. The next route to DMT (1953) oxidized *p*-xylene by air, but this led only to *p*-toluic acid, $CH_3(1,4\text{-}C_6H_4)COOH$, and not to TPA. *p*-Toluic acid was therefore esterified with methanol and this ester then oxidized to the TPA monomethyl ester, $HOOC\text{–}(1,4\text{-}C_6H_4)\text{–}COOCH_3$ which was finally esterified to DMT. In this process, both oxidations and both esterifications could be performed in the same reactor.

*Transesterification* of DMT and EG is a 2-step process in which DMT is condensed with a 1.7-fold molar excess of EG, using acid catalysts. Upon slow temperature increase to 245°C, this transesterification/methanolysis delivers a mixture of oligomers that consists mainly of *bis*(*h*ydroxy*e*thylene) *t*erephthalate (**BHET**):

$$\text{(7-52)} \qquad CH_3OOC\text{–}C_6H_4\text{–}COOCH_3 + 2\,HOCH_2CH_2OH$$

$$\longrightarrow HOCH_2CH_2O\text{–}OC\text{–}C_6H_4\text{–}CO\text{–}OCH_2CH_2OH + 2\,CH_3OH$$

In the second step, vacuum is applied in order to remove ethylene glycol which is liberated by two transesterifications/glycolyses, 2 BHET $\rightleftarrows$ $(PET)_2$ + EG, and $i\ (PET)_j \rightleftarrows (PET_{ij} + (i-1)$ EG ($i \geq 2$). These reactions are acid-catalyzed, i.e., free –COOH groups act as both reagents and catalysts (see Volume I, p. 450 ff.). Industrially, they are accelerated by antimony compounds. Less than 30 % of PET is synthesized by this route.

*Direct esterification* of TPA became viable in 1955 after it was found that a bromide-controlled air oxidation of *p*-xylene at 200°C in acetic acid as solvent directly delivers terephthalic acid. The first step to PET is here an esterification/hydrolysis, TPA + 2 EG $\rightleftarrows$ BHET + 2 $H_2O$ which, in the second step, is followed by the two *melt* transesterification/glycolyses (see previous paragraph). For textile fiber grades, polymers are then worked up. For bottle grades or technical fiber grades, the two melt transesterifications/-glycolyses are followed by a *solid-state* transesterification/glycolysis at 235–240°C (melting temperature of PET: 255°C) in vacuum or in an inert gas atmosphere, i.e., $(PET)_n + (PET)_m \rightleftarrows (PET)_{n+m}$ + EG. More than 70 % of PET is now produced by the transesterification route.

Poly(ethylene terephthalate)s for use as textile or industrial fibers (2/3 of production), bottles, packaging films, and tapes differ in their molecular weights (Table 7-12) which are usually reported as intrinsic viscosities ($[\eta]$ in mL/g or IV in dL/g).

Table 7-12 Molecular properties and production (2000) of types of poly(ethylene terephthalate) [4a].

| Grade | Molecular weight | $[\eta]$/(mL $g^{-1}$) range | typical | Crystal-linity | World capacity in $10^6$ t/a | Typical plant size in t/d |
|---|---|---|---|---|---|---|
| Fiber, textile | 15 000 - 20 000 | 55 - 67 | 64 | medium | staple 9.1<br>filament 11.1 | 100 - 200<br>100 - 300 |
| Fiber, technical | 40 000 - 50 000 | >100 | | | yarns 1.2 | 20 - 40 |
| Bottle | 24 000 - 36 000 | 75 - 100 | 80 | low | 7.9 | 200 - 600 |
| Packaging film, tapes | | | 64 | low | 6.0 | |
| Engineering grade | | | | high | | |

**Reaction Engineering**

Until 1963, PET was mostly synthesized discontinuously from DMT and EG. The *transesterification batch process* requires one esterification reactor for the esterification-methanolysis for the ethylene glycol capped oligomer preparation, DMT + 2 EG ⇄ BHET + 2 $CH_3OH$, and at least one polycondensation reactor for the two transesterifications/glycolyses since esterification and transesterification need different residence times.

The same problem arises in the *direct esterification batch process* with an additional difficulty since terephthalic acid (TPA) does not dissolve well in either EG or BHET. TPA is therefore mixed with the prepolymer to a slurry that is reacted at 235–265°C under a nitrogen blanket and a slight overpressure. In the subsequent polycondensation at 270-295°C, the equilibrium is shifted by removing ethylene glycol in vacuum. Since the viscosity raises from 0.8 Pa s to 400 Pa s (at 280°C), anchor stirrers (Fig. 5-23) are used.

*Continuous direct esterifications* face the same problems, especially for the melt condensation stage where special agitation devices have been developed. Examples are disc-ring reactors (see Fig. 7-3) and twin-screw reactors (see Fig. 5-18).

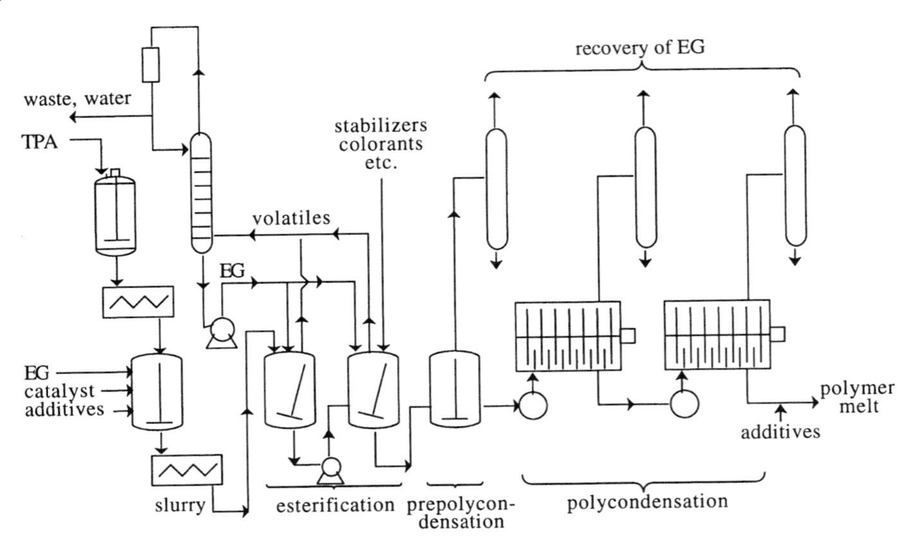


Fig. 7-3 Simplified representation of a plant for continuous direct polycondensation of terephthalic acid (TPA) and ethylene glycol (EG) with five stirred-tank reactors in series [4b].

Plants for the continuous synthesis of PET first prepare a slurry of terephthalic acid (TPA) with a 5–15 % molar excess of ethylene glycol (EG). The slurry plus EG and a catalyst then enter a reactor train consisting of 1–2 esterification reactors, 1–2 prepolycondensation reactors, and 1–2 polycondensation reactors, so-called finishers (Fig. 7-3). The plant also includes a vacuum system and a distillation unit for the removal and/or recovery of glycol, byproducts, etc. A typical fiber-grade process proceeds as follows:
– The esterification at 255–265°C and a pressure of 120 000-180 000 Pa (i.e., slightly above atmospheric pressure (ca. 100 000 Pa) has a residence time of 180–360 min; it delivers a prepolymer with a degree of polymerization of ca. 4–6.
– At the prepolycondensation stage, the temperature is raised to 265–275°C and the pressure lowered to 2500–3000 Pa. After 50–70 minutes, the prepolymer attains a degree of polymerization of ca. 15-20.
– The finisher at 275–295°C and a pressure of 50–150 Pa produces after 90–150 minutes a polymer with a degree of polymerization of ca. 100 ($[\eta]$ = 100 mL/g). The resulting fiber-grade PET is then supplied with $TiO_2$ as a delustering agent before the melt is either directly spun to fibers or extruded and then granulated. For packaging films, $SiO_2$ or china clay are added as deblocking agents, i.e., to prevent sticking.
– The vacuum/distillation system removes volatile monomers and low-molecular weight condensation products such as ethylene glycol, diethylene glycol, terephthalic acid, hydroxyethylene terephthalate, oligomers, etc., as well as byproducts CO, $CO_2$, methane, ethene, acetaldehyde, benzene, and *p*-dioxane from thermal degradations, back-biting reactions, oxidations, etc.

**Chemical Structure**

Continuous melt polycondensations usually deliver polymers that are better resistant to thermal and thermo-oxidative degradation than PETs from batch processes, probably because of a lower content of $-OCH_2CH_2OCH_2CH_2O-$ units which stem from either the direct incorporation of diethylene glycol into the chain or from the etherification of ethylene glycol endgroups and their subsequent transesterification.

Endgroups comprise not only –OH and –COOH but also $\sim COO-CH=CH_2$ from the thermal cleavage of ester groups:

(7-53)

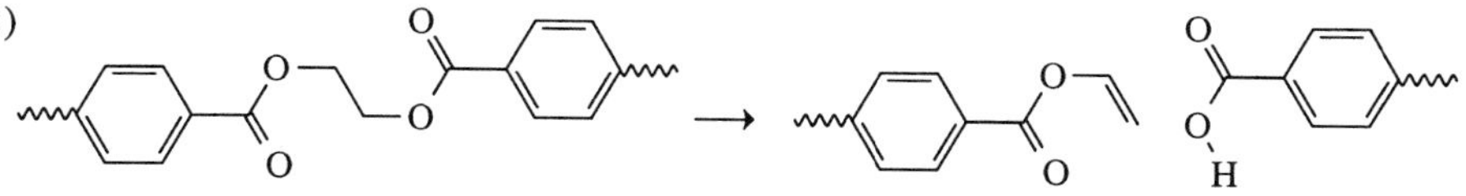

PET contains ca. 2–3 % of short-chain oligomers which can be reduced by solid-state after-polycondensation to ca. 1.4 %. The main oligomeric compound is *cyclo*-tri(ethylene terephthalate). This impurity may crystallize on the fiber during melt spinning and then deposit on rolls and bobbins, where it may cause fibers to break at the spinning speeds of 1500 to 7000 m/min.

PET yellows, which is a severe problem for fibers and bottles. Yellowing is caused by thermal and oxidative degradations that lead to conjugated double bonds. The formation and nature of these structures is not well understood; polyenes from aldol condensations of acetaldehyde, quinoids from decarboxylations, etc., have been discussed

**Properties**

PET fibers have excellent wash and wear properties; they can also be produced with more silk-like or more cotton-like characteristics (Volume IV). Their hydrophobicity causes soiling but this can be remedied by introducing hydrophilic groups, either by grafting of acrylic acid or by partial hydrolysis of fiber surfaces which leads to free carboxyl groups. Dyeability is improved by reducing crystallization through incorporation of adipic acid or polar monomeric units.

PET films are used for packaging and in electronics. They are less permeable to water vapor than polyamides but more than poly(olefin)s. PETs for packaging films are often copolymers with less than 5 mol% diethylene glycol or poly(ethylene glycol) which reduces crystallinity and improves clarity.

After addition of nucleating agents, PET can be injection molded, albeit with little tolerance from exact processing conditions. The polymer has excellent mechanical properties (Table 7-10) which are highly dependent on molding conditions since these affect crystallinity. Engineering plastics are often reinforced by glass fibers, clay, etc.

Bottles are produced by extrusion blowing. The required transparency is achieved by reducing crystallinity to less than 15 % by suppressing crystallization through rigorous exclusion of nucleation and incorporation of a few percent of comonomeric units from diethylene glycol, isophthalic acid, or cyclohexane-1,4-dimethylol. This, in turn, decreases melt strength which therefore has to be boosted by increasing the molecular weight (Table 7-12). Processability is improved by addition of, e.g., polyamides.

**PETG** is a copolymer of terephthalic acid, ethylene glycol, and ca. 35 mol% cyclohexane-1,4-dimethylol in its natural 30/70 ratio of cis/trans isomers. The amorphous polymer is used for impact-resistant, high-clarity, injection-molded articles.

### 7.7.6 Poly(1,4-bismethylenecyclohexane terephthalate)

Hydrogenation of the benzene ring of dimethyl terephthalate with a palladium catalyst leads to dimethyl 1,4-cyclohexanedicarboxylate, $CH_3OOC(1,4\text{-}C_6H_{10})COOCH_3$, which is then hydrogenated with $CuCrO_2$ as catalyst to cyclohexane-1,4-dimethylol (CHDM), $HOCH_2(1,4\text{-}C_6H_{10})CH_2OH$. CHDM exists in a cis and trans form:

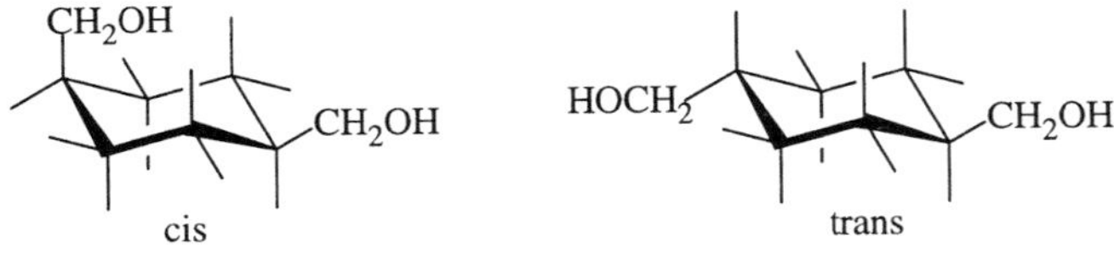

$CH_2OH$ $CH_2OH$ — cis

$HOCH_2$ $CH_2OH$ — trans

The industrial polymer (**PCT, PCDT**) consists of cis and trans isomeric units of CHDM in the ratio 30:70. The polymer was originally developed for use as a textile fiber in order to circumvent the Calico Printer/DuPont patents for poly(ethylene terephthalate) (see p. 351). As a fiber, it is more easily dyeable with more brilliant colors than PET. PCT is now finding more use as a reinforced thermoplastic because it has a considerably higher melting temperature (290°C versus 255°C) and glass temperature (90°C versus 70°C) than PET. The higher cost of PCT is somewhat offset by its ca. 12 % lower density compared to PET.

### 7.7.7 Poly(trimethylene terephthalate)

Poly(trimethylene terephthalate) (**PTT**), also called **poly(propylene terephthalate)**, is a relatively new polymer since no industrially viable methods were available for trimethylene glycol (= 1,3-propanediol, **PDO**), $HO(CH_2)_3OH$. At present, three intermediates compete for the most economical synthesis of 1,3-propane diol: oxirane (hydroformylation, p. 102, Xa), acrolein (hydration followed by hydrogenation, p. 107, IV), and glucose (from corn or cane sugar) which is converted by natural yeasts to glycerol that is subsequently transformed to 1,3-propanediol by genetically modified bacteria.

Poly(trimethylene terephthalate) is obtained by a continuous direct esterification process similar to that of PET, albeit at lower temperatures (maximum 250–270°C), using a train of five reactors: two agitated esterification reactors with removal of excess PDO, followed by two precondensation reactors, one agitated, the next a disc-ring type (see Fig. 7-3), and a disc-ring polycondensation reactor.

The polycondensation to PTT produces different byproducts than that of PET: acrolein and allyl alcohol instead of acetaldehyde, the cyclic dimer ($T_M$ = 254°C) of PTT instead of a cyclic trimer, and no cyclic ether such as oxetane (trimethylene oxide) or 1,5-trioxacyclooctane. Incorporated in PTT are dipropylene ether glycol units from the dimerization of propylene glycol and allylic endgroups, $CH_2{=}CHCH_2O–$, from a reaction similar to Eq.(7-53).

PTT has an equilibrium melting temperature of 248°C (DSC: 233°C) and a glass temperature of >45°C (depends on crystallinity. It crystallizes more slowly than PET but faster than PBT (Section 7.7.8) with which it competes as an engineering plastic. The main application of PTT is as fiber, especially for carpets and upholstery, because of a very favorable combination of properties. Their tensile elastic recovery is better than that of PBT and PET; it competes with that of nylon 66. Because of its lower glass temperature, PTT can by dyed with disperse dyes at lower temperatures than PET. The absence of basic groups for dyeing with reactive dyes also means that PTT fibers are not prone to stains which are mostly acidic. The additional $CH_2$ group makes the chains more flexible, which gives carpets a softer feel and textiles some extra softness.

### 7.7.8 Poly(butylene terephthalate)

Poly(butylene terephthalate) (**PBT**) is the polyester from terephthalic acid and 1,4-butylene glycol (1,4-butanediol, tetramethylene glycol). It is also called **poly(tetramethylene terephthalate)** (**PTMT**). World production is ca. $4 \cdot 10^5$ t/a.

1,4-Butanediol is obtained by the old process of ethynylation of acetylene with formaldehyde (p. 96, VIII and p. 105, VI); the newer routes from butadiene via chlorination (p. 114, V), acetoxylation (p. 114, VI), or epoxidation (p. 114, VII); and the newest route from butane (p. 115, II). For the synthesis of terephthalic acid, see p. 350.

#### Conventional Poly(butylene terephthalate)

PBT synthesis resembles that of PET but uses only one catalyst (e.g., a tetraalkyl titanate) instead of two. The catalyst is not deactivated after polymer synthesis but stays in the polymer where it promotes additional polycondensation on processing.

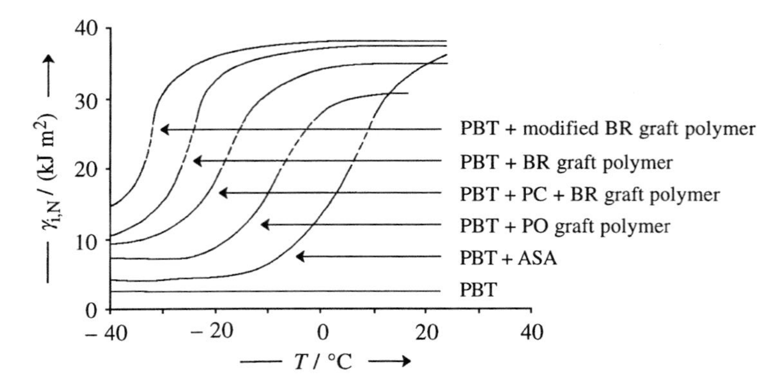


Fig. 7-4 Temperature-dependence of notched impact strength of an unmodified poly(butylene terephthalate (PBT) and its blends with acrylic ester-styrene-acrylonitrile terpolymers (ASA), butadiene rubber (BR), polycarbonate A (PC), and poly(olefin)s (PO). Dotted lines indicate transitions from brittle to ductile behavior. With kind permission by Hanser Publishers, Munich [1b].

Properties of PBT resemble that of PET (Table 7-10). However, PBT can be processed at considerably lower temperatures than PET because of its lower melting temperature (227°C versus 255°C). This advantage is somewhat offset by a lower glass temperature (60°C versus 70°C) and poorer mechanical properties (Table 7-10). In contrast to PET, PBT does not need nucleating agents for crystallization; it is always semicrystalline.

Impact strengths are improved by blending PBT with other polymers (Fig. 7-4) which may lead to grafting (see Volume IV). Addition of ASA decreases crystallinity which lowers shrinkage and reduces internal stresses. Other commercial blends include those of PBT with poly(ethylene terephthalate) or polycarbonate A.

Most PBT grades are glass fiber-reinforced which slightly lowers the melting temperature, considerably lowers the expansion coefficient and the mold shrinkage, and drastically decreases the fracture elongation (Table 7-13). Heat distortion temperatures rise notably, especially at higher loads. Moduli and flexural and tensile strengths are greatly increased. As an engineering plastic, glass fiber-reinforced PBT competes with nylon 6 and PET.

Table 7-13 Properties of some glassfiber (GF)-reinforced engineering thermoplastics [5]. * Dry state.

| Property | Physical unit | PBT unreinforced | PBT +30% GF | PA 6* +30% GF | PET +30% GF |
|---|---|---|---|---|---|
| Density | g cm$^{-3}$ | 1.31 | 1.53 | 1.36 | 1.60 |
| Melting temperature | °C | 230 | 220 | 223 | 245 |
| Heat distortion temperature, (1.06 MPa) | °C | 154 | 216 | 215 | 249 |
| (1.82 MPa) | °C | 54 | 207 | 200 | 227 |
| Linear thermal expansion coefficient, 25°C | $10^5$ K$^{-1}$ | 8.1 | 2.5 | | 1.8 |
| Mold shrinkage in flow direction, average | % | 1.9 | 0.65 | 0.3 | 0.15 |
| Flexural modulus | MPa | 2350 | 7600 | 8300 | 8300 |
| Flexural strength | MPa | 83 | 190 | 280 | 235 |
| Tensile strength at yield | MPa | 52 | 120 | 180 | 165 |
| Tensile elongation at break | % | 300 | 3 | 3 | 3 |
| Notched impact strength | J/m | 53 | 85 | 117 | 91 |

**Cyclic Butylene Terephthalate**

Cyclic butylene terephthalate (**CBT**) is a solid mixture of cyclic oligomers ($i \geq 2$) that can be obtained in high yields by equilibrium-controlled thermal chain scission of linear poly(butylene terephthalate) or by kinetically controlled pseudo-dilute condensation of terephthaloyl chloride and 1,4-butanediol.

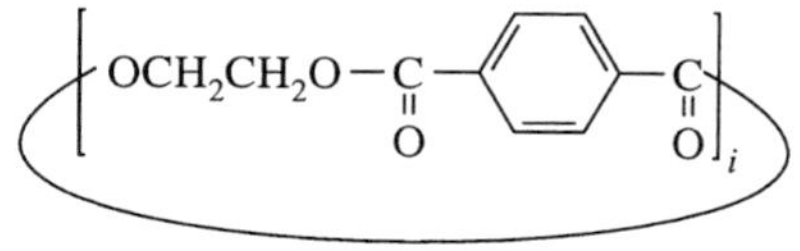

cyclic butylene terephthalate 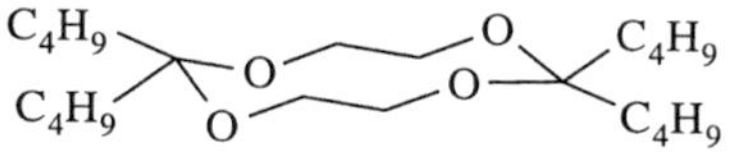 cyclic butyl stannoxane 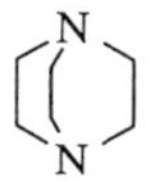 DABCO

Equilibrium-controlled polycondensations deliver cyclics only in very small yields, as the following considerations show. Conventional stoichiometric AA/BB polycondensations are equilibrium reactions in which equilibria are continually shifted by removal of byproducts, i.e., water (AB) in the reaction of dicarboxylic acids AXA and diols BYB. These reactions have four sets of competing and consecutive equilibria:

- to linear molecules by $i$ AXA + $i$ BYB $\rightleftarrows$ A(XY)$_i$B + (2$i$ – 1) AB (plus 2 sets of reactions to linear molecules with odd numbers of monomeric units) and
- to cyclic molecules by $i$ AXA + $i$ BYB $\rightleftarrows$ [XY]$_i$ + 2 $i$ AB.

On condensation in a closed system to a *given degree of polymerization* of repeating units, the total number of new molecules, $N$, does not change in linear reactions but increases in cyclic ones, leading for the latter to a gain in translational and external rotational entropy in very dilute systems. This entropic gain is somewhat decreased by the greater rigidity of small rings compared to equal-sized linear chains (loss of internal rotational and vibrational entropy). The incremental entropic gain on formation of cyclics is rapidly lost at higher degrees of polymerization, however (at $i$ = 1: 2 versus 3 molecules; at $i$ = 5: 10 versus 11, etc.). It decreases furthermore with increasing total concentration because translations and external rotations become more restricted.

For these reasons, ring formation can only dominate in thermodynamically controlled reactions if initial monomer concentrations are very small. At a certain higher initial concentration, the molar concentration of ring units in all rings reaches the value given by the equilibrium constant of the reaction. Above this value, the concentration of all rings stays constant (see Volume I, p. 214). The excess monomer concentration is then used for the formation of linear polymer molecules, which is the reason why thermodynamically controlled *neat* polycondensations proceed mainly intermolecularly and deliver only a few percent of cyclic oligomers (p. 352 and 354). One may argue that *complete* reaction of functional groups in *exact* stoichiometry should lead to a 100 % yield of rings because otherwise only one molecule with infinite molecular weight would exist. However, this reasoning ignores the definition and existence of thermodynamic equilibria where concentrations of participants can never become zero.

Larger concentrations of oligomers are obtainable by equilibrium depolymerization of linear polyesters, which are not reversals of polycondensations but catalyzed elimination reactions of the type A(XY)$_i$B $\rightleftarrows$ [XY]$_j$ + A(XY)$_{i-j}$B. Yields of cyclics increase with dilution. For example, after 1 hour in boiling $o$-$C_6H_4Cl_2$, yields of cyclics (CBT) from poly(butylene terephthalate) (PBT) were 42 % for $[PBT]_0$ = 0.3 mol/L but 96 % for $[PBT]_0$ = 0.05 mol/L. The CBTs consisted mainly of dimers and trimers (Table 7-14).

Table 7-14 Yields $Y$ in percent of total cyclics [6, 7] and melting temperatures $T_M$ [8] of cyclic butylene terephthalates (CBT) and cyclic ethylene terephthalates (CET) with degrees of polymerization of repeating units, $X$, by depolymerization (dp) of PBT at 179°C in $o$-dichlorobenzene with cyclic butyl stannoxane as catalyst [6] or by pseudo-dilution Schotten–Baumann reaction (sb) of terephthaloyl dichloride (TC) and 1,4-butanediol (BD) or ethylene glycol (EG), respectively, at 25°C [7]. Melting temperatures indicate the range of data that have been reported by various authors for allegedly authentic oligomers.

| $X$ | $Y_{CBT}$ / % dp | $Y_{CBT}$ / % sb | $Y_{CET}$ / % sb | $T_M$ / °C CET | $T_M$ / °C CBT |
|---|---|---|---|---|---|
| 2 | 32 | 51 | 39 | 224 - 229 | 196 - 199 |
| 3 | 34 | 26 | 21 | 314 - 327 | 168 - 171 |
| 4 | 17 | 11 | 14 | 225 - 331 | 243 - 251 |
| 5 | 11 | 8 | 10 | 247 - 264 | 207 |
| 6 | 2 | 2 | 7 | 304 - 306 | |
| 7 | 4 | 2 | 3 | 238 - 240 | |

High concentrations of cyclic polyesters can also be obtained from kinetically controlled reactions of terephthaloyl dichloride (TPDC) with diols, for example, 1,4-butanediol (BD). For example, the simultaneous drop-wise addition of 1 mol/L TPDC in $CH_2Cl_2$ and 0.8 mol neat BD to 1.7 mol $Et_3N$ + 40 mmol DABCO® (triethylene diamine) in 3 L $CHCl_3$ delivered a yield of 82 % cyclic oligomers. The distribution of oligomers deviated of course from that of equilibrium depolymerization (Table 7-14).

CBT and CET can be melt-polymerized by ring-opening polymerization to their polyesters. These polymerizations, for example, by titanates as catalysts proceed rapidly to high molecular weights (e.g., in 6 min to $M \approx 100\ 000$), probably to a mixture of cyclic and linear polyesters (see Volume I, p. 471, for similar polymerizations to polycarbonates)). Polymerization of CBT is industrially more attractive than that of CET because of its lower melting range of 140–190°C. Melt viscosities are very low, only 0.03 Pa s at 190°C (water at 20°C: $\eta \approx 0.001$ Pa s), which makes melt processing easy.

### 7.7.9 Thermoplastic Polyetherester Elastomers

Transesterification of dimethyl terephthalate, $CH_3OOC(p\text{-}C_6H_4)COOCH_3$, with poly(tetramethylene glycol) (PTMG) = poly(tetrahydrofuran) (PTHF), $HO[(CH_2)_4O]_kH$, and 1,4-butanediol, $HO(CH_2)_4OH$, delivers multiblock copolymers with "hard" blocks of oxybutyleneoxyterephthaloyl (4GT) and "soft" blocks of poly(tetrahydrofuran)-terephthaloyl (PTHFT). The butylene segments ($i, j$ = 4) of both blocks may also be replaced by ethylene segments ($i, j$ = 2)

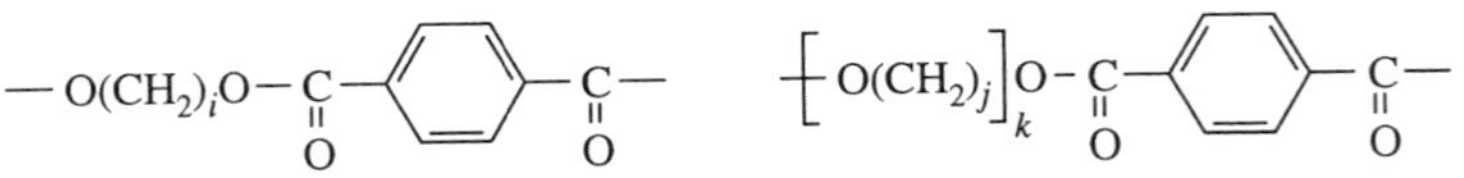

$$-O(CH_2)_iO-\underset{\underset{O}{\|}}{C}-\langle C_6H_4\rangle-\underset{\underset{O}{\|}}{C}- \qquad \left[O(CH_2)_j\right]_k O-\underset{\underset{O}{\|}}{C}-\langle C_6H_4\rangle-\underset{\underset{O}{\|}}{C}-$$

hard segments ($i$ = 2, 4) soft segments ($j$ = 2, 4; $k$ = 8-50)

"Hard" and "soft" refer neither to hardness (resistance to pressure) nor to mechanical strength (resistance against tension) but to the relative magnitude of the transition temperatures: "hard segments" have high melting temperatures and "soft segments" have low glass temperatures. These polymers are thermoplastic elastomers in which crystalline clusters of hard segments act as physical crosslinks in a matrix of soft (rubbery) segments. At elevated temperatures, clusters melt and the polymer can be processed like a thermoplastic. On cooling, hard segments cluster together.

Of the many possible soft polyether blocks, only PTHFT blocks are suitable because they provide full hardness immediately after cooling. Thermoplastic polyetheresters with exclusive poly(oxyethylene)-terephthaloyl segments crystallize much more slow and need approximately 1 day for full hardness.

### 7.7.10 Poly(alkylene naphthalate)s

Naphthalene-2,6-dicarboxylic acid (naphthalenedicarboxylate, NDC) is obtained by oxidation of 2,6-dimethylnaphthalene in 95 % acetic acid, with Co/Mn acetate as the catalyst, and the cocatalyst $NH_4Br$ + $CBr_4$ as a constantly renewable radical source. The same system also serves for the production of fiber-grade terephthalic acid from *p*-xylene (Amoco).

**Poly(ethylene-2,6-naphthalate) (PEN)** is produced by melt polycondensation in batch reactors similar to PET: an ester exchange to oligomers is followed by polycondensation, summarily shown in Eq.(7-53). The melt from the polycondensation reactor is extruded and the exiting strand cooled and cut. World production is ca. $4 \cdot 10^4$ t/a.

(7-54) $HOCH_2CH_2OH$ + HOOC–$C_{10}H_6$–COOH $\xrightarrow{-2\,H_2O}$ —$OCH_2CH_2O$–C(=O)–$C_{10}H_6$–C(=O)—

The resulting amorphous pellets are dried at 190°C. Injection-molded parts or thermoformed sheets are amorphous (APEN).

PEN is mainly extruded to sheets at 200–300°C; these films are highly ductile because of the low crystallization rates. For biaxially oriented films, the sheet is first drawn forward and then sideways at temperatures slightly above the glass temperature of ca. 120°C. Subsequent drawing of the oriented film between rollers at ca. 215°C induces crystallization.

One present main application is for the Advantix® film system which requires strong thin films with good optical properties in the visible region. These conditions cannot be fulfilled by conventional photographic films (cellulose triacetate, poly(ethylene terephthalate), polycarbonate). Other applications of biaxially oriented films are for packaging materials because of its clarity and low gas permeation (less than PET).

**Poly(1,4-butylene-2,6-naphthalate)** has a greater resistance to heat and aging than poly(butylene terephthalate), which makes it interesting for applications in electronics where miniaturization allows higher operating temperatures.

### 7.7.11 Polyarylates

Polyarylates (**PAR**) are copolyesters of diphenols and aromatic dicarboxylic acids. A typical polyarylate consists, for example, of terephthalic acid, isophthalic acid or *p*-hydroxybenzoic acid, and bisphenol A (below left) units in molar ratios of 1:1:2. There are also reports of other monomers such as substituted 1,2-bis(phenoxy)ethane-4,4'dicarboxylic acids, substituted hydroquinones, and sterically hindered bisphenols (center and right). Sequence distributions (random, blockiness) are usually not known.

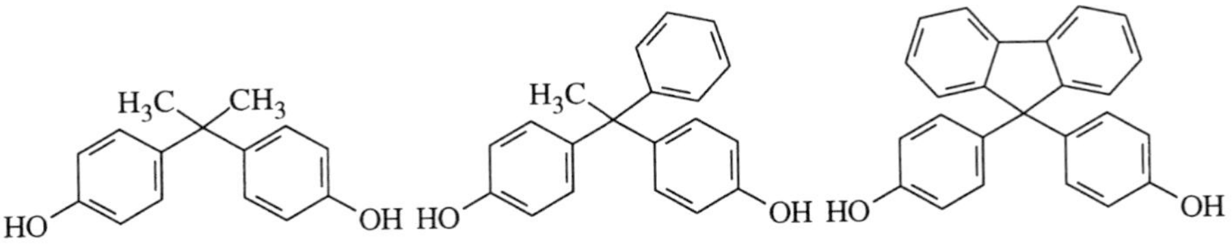

Some polyarylates seem to be produced by melt polycondensation of diacids and bisphenols, others by interfacial polymerization of diacid chlorides in chlorocarbons and aqueous solutions of alkali salts of bisphenols with quaternary ammonium salts as catalysts.

Polyarylates are transparent to opaque, amber-colored engineering plastics with good mechanical strengths, excellent elastic recovery, good impact resistance, and high thermal stability (continuous service temperature up to 240°C (unreinforced) and 343°C (reinforced)). They are used as components of engines, kitchenware, power tools, etc.

## 7.8 Polyorthoesters

Orthoformic acid, $HC(OH)_3$, and orthocarbonic acid, $HOC(OH)_3$, do not exist as free acids but only in the form of their open or cyclic esters, i.e., orthoformates and orthocarbonates, respectively. In polyorthoesters, at least one –C–O– bond is part of a polymer chain whereas the other –C–O– bonds may be part of substituents.

Polyorthoesters are not identical with the **orthopolyesters** of industry. The latter polymers are polyesters that are based on *ortho*-phthalic acid. They were so named in order to distinguish them from **isopolyesters**, i.e., polyesters based on *iso*-phthalic acid.

Poly(orthocarbonic ester)s can be obtained by transesterification of orthocarbonic esters with di- and polyfunctional alcohols. Because of the functionality of the monomeric acid, such esterifications lead to branched and later crosslinked polymers.

Linear polyorthoesters with *one* –C–O– chain bond per orthoester group result from the reaction of 1,1,4,4-tetramethoxybutadiene with diols, Eq.(7-55). At higher monomer conversions, side-reactions lead to branching and crosslinking.

(7-55)

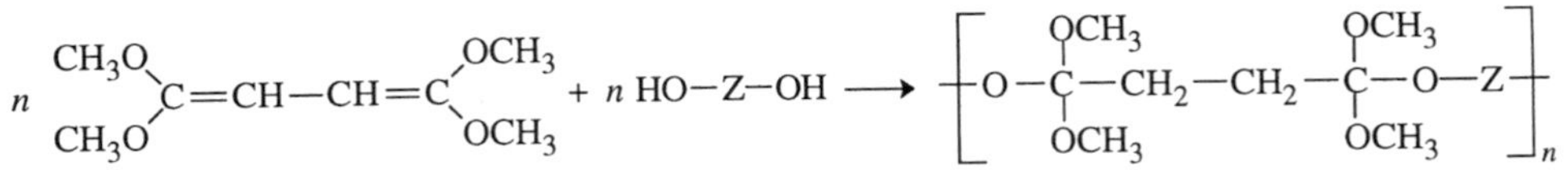

$$n\,(CH_3O)_2C{=}CH{-}CH{=}C(OCH_3)_2 + n\,HO{-}Z{-}OH \longrightarrow \left[ O{-}C(OCH_3)_2{-}CH_2{-}CH_2{-}C(OCH_3)_2{-}O{-}Z \right]_n$$

Polyorthoesters with *two* ortho –C–O– bonds per orthoester group are produced by reaction of diols and diketene acetals:

(7-56)

$$n\ CH_2{=}\underset{OR}{C}{-}O{-}Z{-}O{-}\underset{OR}{C}{=}CH_2 + n\ HO{-}Z'{-}OH \rightarrow \left[{-}O{-}\underset{OR}{\overset{CH_3}{C}}{-}O{-}Z{-}O{-}\underset{OR}{\overset{CH_3}{C}}{-}O{-}Z'{-}\right]_n$$

or by reaction of cyclic orthoformic acid esters with diols:

(7-57)

$$n\ C_2H_5O\ \ OC_2H_5\ (\text{cyclic orthoester}) + n\ HO{-}Z{-}OH \xrightarrow[-2\,n\,C_2H_5OH]{} \left[{-}O\ \ O{-}Z{-}\right]_n$$

The constitutional unit –C(CH$_3$)(OR)–O–Z– may also be part of a spiro system in which all *three* –C–O– bonds are part of the chain structure:

(7-58)

$$n\ CH_3{-}CH{=}(\text{spiro diketene acetal}){=}CH{-}CH_3 + n\ HO{-}Z{-}OH \rightarrow \left[CH_3CH_2\ \ O\ \ O\ \ CH_2CH_3\ \ {-}O\ \ O\ \ O\ \ O{-}Z{-}\right]_n$$

Polyorthoesters degrade biologically; for example, the polymer of Eq.(7-57) by hydrolysis to the lactone and further to the $\omega$-hydroxy acid:

(7-59)

$$\left[{-}O\ \ O{-}Z{-}\right] \xrightarrow[-\ HO-Z-OH]{+\ H_2O} (\text{lactone}) \xrightarrow{+\ H_2O} HO(CH_2)_3COOH$$

Polyorthoesters erode hydrolytically from their surfaces; hence, they do not break down catastrophically. Their slow degradation makes them useful for compounding with pharmaceuticals that should be gradually released.

## 7.9 Polyanhydrides

Polyanhydrides are reaction products of dicarboxylic acids and acid anhydrides:

(7-60)

$$n\ HO{-}\underset{O}{\underset{\|}{C}}{-}Z{-}\underset{O}{\underset{\|}{C}}{-}OH \xrightarrow[-\,n\,H_2O]{+\,n\,(CH_3CO)_2O}$$

$$n\ H_3C{-}\underset{O}{\underset{\|}{C}}{-}O{-}\underset{O}{\underset{\|}{C}}{-}Z{-}\underset{O}{\underset{\|}{C}}{-}O{-}\underset{O}{\underset{\|}{C}}{-}CH_3 \xrightarrow[-n\,(CH_3CO)_2O]{} \left[{-}\underset{O}{\underset{\|}{C}}{-}Z{-}\underset{O}{\underset{\|}{C}}{-}O{-}\right]_n$$

Anhydride copolymers of sebacic acid and 1,3-bis(*p*-carboxyphenoxy) propane with Z = (1,4-C$_6$H$_4$)O(CH$_2$)$_3$O(1,4-C$_6$H$_4$) are crystalline polymers that, like polyorthoesters, erode from the surface. Pharmaceuticals that are dispersed or dissolved in polyanhydrides are thus released in constant doses per unit time.

Examples are platelets of polyanhydride copolymers of sebacic acid and the dimer of erucic acid (p. 79):

$$\left[\begin{array}{l} CO-(CH_2)_7-CH-(CH_2)_8-CH_3 \\ CH_3-(CH_2)_8-CH-(CH_2)_7-CO-O \end{array}\right]$$

Platelets of this polymer with a diameter of 14 mm and a thickness of 2.7 mm erode under physiological conditions with rates of 0.3 mg/h. Hydrophobic pharmaceuticals are released at a rate of 1-3 % per day, hydrophilic ones, at 3-6 % per day.

## Literature to Chapter 7

7.1 OVERVIEW

H.-G.Elias, Neue polymere Werkstoffe 1969-1974, Hanser, Munich 1975; –, New Commercial Polymers, Gordon and Breach, New York 1977

H.-G.Elias, F.Vohwinkel, Neue polymere Werkstoffe, 2. Folge, Hanser, Munich 1983; –, New Commercial Polymers 2, Gordon and Breach, New York 1986

L.Bottenbruch, Ed., Engineering Thermoplastics. Polycarbonates, Polyacetals, Polyesters, Cellulose Esters, Hanser Gardner, Cincinnati 1996

7.2 POLYACETALS AND POLYKETALS

J.Furukawa, T.Saegusa, Polymerization of Aldehydes and Oxides, Wiley, New York 1963

M.Sittig, Polyacetal Resins, Gulf Publ., Houston (TX), 3rd ed. 1964

O.Vogl, Polyaldehydes, Dekker, New York 1967

S.J.Barker, M.B.Price, Polyacetals, Iliffe, London 1970

H.Tani, Stereospecific Polymerization of Aldehydes and Epoxides, Adv.Polym.Sci. **11** (1973) 57

O.Vogl, Kinetics of Aldehyde Polymerizations, J.Macromol.Sci.Rev. **C 12** (1975) 109

K.Neeld, O.Vogl, Fluoroaldehyde Polymers, J.Polym.Sci.-Macromol.Rev. **16** (1981) 1

J.Masamoto, Modern Polyacetals, Progr.Polym.Sci. **18** (1993) 1

A.L.Rusanov, Condensation Polymers Based on Chloral and Its Derivatives, Progr.Polym.Sci. **19** (1994) 589

7.3 POLYETHERS

J.Furukawa, T.Saegusa, Polymerization of Aldehydes and Oxides, Wiley, New York 1963

A.F.Gurgiolo, Poly(alkylene oxides), Rev.Macromol.Chem. **1** (1966) 39

F.E.Bailey, Jr., J.V.Koleske, Poly(ethylene oxide), Academic Press, New York 1976

P.Dreyfuss, Poly(tetrahydrofuran), Gordon and Breach, New York 1982

G.W.Gokel, S.H.Korzeniowski, Macrocyclic Polyether Syntheses, Springer, Berlin 1983

F.E.Bailey, Jr., J.V.Koleske, Alkylene Oxides and Their Polymers, Dekker, New York 1991

J.M.Harris, Ed., Poly(ethylene glycol) Chemistry: Biotechnical and Biomedical Applications, Plenum Press, New York 1992

P.M.Hergenrother, J.W.Lonnell, J.W.Labadie, J.L.Hedrick, Poly(Arylene Ether)s Containing Heterocyclic Units, Adv.Polym.Sci. **117** (1994) 67

V.M.Nace, Ed., Nonionic Surfactants. Polyoxyalkylene Block Copolymers, Dekker, New York 1996

7.3.6 EPOXY RESINS

A.M.Paquin, Epoxydverbindungen und Epoxydharze, Springer, Berlin 1958

I.Skeist, Epoxy Resins, Reinhold, New York 1958

H.Lee, K.Neville, Handbook of Epoxy Resins, McGraw-Hill, New York 1967

P.F.Bruins, Ed., Epoxy Resins Technology, Interscience, New York 1968
W.G.Potter, Epoxide Resins, Iliffe, London 1970
H.S.Eleuterio, Polymerization of Perfluoro Epoxides, J.Polym.Sci. [A-1] **6** (1972) 1027
C.A.May, Y.Tanaka, Eds., Epoxy Resins, Chemistry and Technology, Dekker, New York 1973
J.I.DiStasio, Epoxy Resin Technology, Noyes, Park Ridge (NJ) 1982
C.A.May, Ed., Epoxy Resins. Chemistry and Technology, Dekker, New York, 2nd ed. 1988
B.Ellis, Ed., The Chemistry and Technology of Epoxy Resins, Blackie, Glasgow 1993

7.3.8 FURAN RESINS
C.R.Schmitt, Polyfurfuryl Alcohol Resins, Polym.-Plast.Technol.Eng. **3** (1974) 121
A.Gandini, The Behavior of Furan Derivatives in Polymerization Reactions, Adv.Polym.Sci. **25** (1977) 47

7.3.9 POLY(PHENYLENE OXIDE)S
A.S.Hay, Aromatic Polyethers, Adv.Polym.Sci. **4** (1967) 496
A.S.Hay, Polymerization by Oxidative Coupling - A Historical Review, Polym.Eng.Sci. **16** (1976) 1
R.C.Cotter, Engineering Plastics. A Handbook of Polyarylethers, Gordon and Breach, New York 1995

7.4 POLYKETONES
M.J.Mullins, E.P.Woo, The Synthesis and Properties of Poly(aromatic ketones), J.Macromol.Sci.-Rev.Macromol.Chem.Phys. **C 27** (1987) 313

7.5–7.6 ALIPHATIC and UNSATURATED POLYESTERS
J.Bjorksten, H.Tovey, B.Harker, J.Henning, Polyesters and Their Applications, Reinhold, New York 1959
H.B.Boenig, Unsaturated Polyesters, Elsevier, Amsterdam 1964
V.V.Korshak, S.V.Vinogradova, Polyesters, Pergamon Press, Oxford 1965
I.Goodman, J.A.Rhys, Polyesters, Vol. I, Saturated Polyesters, Iliffe, London 1965
B.Parkyn, F.Lamb, B.V.Clifton, Polyesters, Vol. II, Unsaturated Polyesters, Iliffe, London 1967
P.F.Bruins, Unsaturated Polyester Technology, Gordon and Breach, New York 1976
S.Inoue, High Polymers from $CO_2$, CHEMTECH **6** (1976) 588
K.Holmberg, High Solids Alkyd Resins, Dekker, New York 1987
Y.Doi, Microbial Polyesters, VCH, Weinheim 1990
D.Byrom, Ed., Polyhydroxyalkanoic Acid, MacMillan, Basingstoke, UK, 1991
D.P.Mobley, Ed., Plastics from Microbes, Hanser, Munich 1995
S.L.Needleman, A.I.Laskin, Eds., Biodegradable Polyesters, Academic Press, San Diego 1996
C.Scholz, R.Grass, G.Leatham, Polymers from Renewable Resources. Polyesters of Biomedical and Environmental Importance, Oxford University Press 2000 (ACS Symposium Series)
W.Babel, A.Steinbüchel, Biopolyesters, Springer, Berlin 2001
Y.Doi, A.Steinbüchel, Eds., Polyesters I - Biological Syntheses and Biotechnological Products (2001); –, Polyesters II - Properties and Chemical Structures (2001); –, Polyesters III - Applications and Commercial Products (2002); = (Vols. 3a, 3b, and 4 of A.Steinbüchel, Ed., Biopolymers, Wiley, New York 2001–2003 (12 volumes))
A.-C.Albertson, I.K.Varma, Aliphatic Polyesters: Synthesis, Properties and Applications, Adv.Polym.Sci. **157** (2002) 1

7.7 AROMATIC POLYESTERS
H.Schnell, Chemistry and Physics of Polycarbonates, Interscience, New York 1964
I.Goodman, J.A.Rhys, Polyesters, Vol. I, Saturated Polyesters, Iliffe, London 1965
H.Ludwig, Polyester-Fasern, Akademie-Verlag, Berlin, 2nd ed.1975; –, Polyester Fibers, Wiley, New York 1971
R.Vieweg, L.Leonhard, Ed., Kunststoff-Handbuch, Vol. VIII, Polyester, Hanser, Munich 1973
H.Ludewig, Polyester Fibers, Chemistry and Technology, Wiley, New York 1979
B.M.Walker, Handbook of Thermoplastic Elastomers, Van Nostrand Reinhold, New York 1979
N.R.Legge, G.Holden, H.Schroeder, Thermoplastic Elastomers, Research and Development, Hanser, Munich 1987
H.H.Yang, Aromatic High-Strength Fibers, Wiley, New York 1989

L.Bottenbruch, Ed., Polycarbonate, Polyacetale, Polyester, Celluloseester, Hanser, Munich 1992 (= G.W.Becker, D.Braun, Eds., Kunststoff-Handbuch **3**/1)
D.G.LeGrand, J.T.Bendler, Handbook of Polycarbonate Science and Technology, Dekker, New York 2000
S.Fakirov, Handbook of Thermoplastic Polyesters, Wiley, New York 2002
J.Scheirs, T.E.Long, Eds., Modern Polyesters, Wiley, Chichester, UK, 2003

7.9 POLYANHYDRIDES
S.Einmahl, K.Schwach-Abdellaoui, J.Heller, R.Gurny, Poly(Ortho Esters): Recent Developments for Biomedical Applications, Chimia **55** (2001) 218

# References to Chapter 7

[1] H.Domininghaus, Plastics for Engineers, Hanser, Munich 1993, (a) Figs. 207 and 208, (b) Fig. 416
[2] L.V.McAdams, J.A.Gannon, Concise Encyclopedia of Polymer Science and Engineering (J.Kroschwitz, Ed.), Wiley, New York 1990, p. 345
[3] P.F.Bruins, Ed., Epoxy Resins Technology, Interscience, New York 1968
[4] (a) Data of Th.Rieckmann, S.Völker, in J.Scheirs, T.E.Long, Eds., Modern Polyesters, Wiley, Chichester, UK, 2003, p. 31 ff. (amended); (b) p. 97, company literature of Zimmer AG
[5] Data of R.R.Gallucci, B.R.Patel, in J.Scheirs, T.E.Long, Eds., Modern Polyesters, Wiley, Chichester, UK, 2003, p. 295
[6] D.J.Brunelle, in J.Scheirs, T.E.Long, Eds., Modern Polyesters, Wiley, Chichester, UK, 2003, p. 129
[7] D.J.Brunelle, J.E.Bradt, J.Serth-Guzzo, T.Takekoshi, T.L.Evans, E.J.Pearce, P.R.Wilson, Macromolecules **31** (1998) 4782, Table 2
[8] G.Wick, H.Zeitler, Angew.Makromol.Chem. **112** (1983) 59, Table 4

# 8 Polysaccharides

## 8.1 Saccharides

### 8.1.1 Overview

Polysaccharides (G: *sakcharon* = sugar; Sanskrit: *sarkara* = sand, gravel) are homopolymers or copolymers of sugars that occur in animals and plants; synthetic polysaccharides presently are not of interest in industry. They are subdivided according to their chemical structure (type and connection of sugar units) or their action in living matter. **Structural polysaccharides** are fibrous or sheet-like linear sugar polymers that make up the physical structure of an organism. Examples are cellulose and chitin. **Reserve polysaccharides** are branched sugar polymers with spheroidal shapes that allow them to be easily stored in cells as reserve food. Examples are amylose and glycogen. A third group consists of physically crosslinked, gel-forming polysaccharides that serve either as structural components or as lubricants; examples are mucopolysccharides and plant gums. Polysaccharides may also be metabolites of bacteria and fungi.

Degrees of polymerization of polysaccharides from plants and animals vary with the species, the organ, the age of the individual, and the preparation. Many "native" polysaccharides are not "pure" but contain a few percent of possibly covalent-bound peptide groups. These peptides originated in polysaccharide biosynthesis; industrially, they are usually removed.

Polysaccharides are used by man in enormous quantities, either directly (lumber, foodstuffs, etc.) or after mechanical processing (cotton, paper, etc.), physical isolation (starches, pectins, etc.), chemical isolation (rayon, etc.), chemical transformation to other polymers with preservation of the main chain (cellulose acetate, chitosan, etc.), chemical transformation to other polymers with transformation of the main chain (pullulan, etc.), or as raw material for the manufacture of other products (corn syrup, ethanol, etc.).

Natural polysaccharides and their derivatives serve as foodstuffs, fibers, thickeners, working materials, explosives, blood plasma expanders, etc. Statistical data for the world consumption of most polysaccharides are practically non-existent. Even data for a single country are only rough estimates (Table 8-1). However, the combination of all the scattered data shows the enormous economic importance of polysaccharides: in 2004, ca. $2000 \cdot 10^6$ t were used worldwide compared to "only" $250 \cdot 10^6$ t of *all* synthetic polymers.

### 8.1.2 Simple Sugars

Sugars are monooxo-polyhydroxy compounds that are also called **saccharides** after their main representative, saccharose, the common household sugar (cane or beet sugar) with the chemical composition $C_{12}H_{22}O_{11}$ (G: *sakcharon* = sugar). The historical name for members of this family of chemical compounds is **carbohydrates** because most of them have the composition $C_m(H_2O)_n$. The names of simple sugars with the composition $C_nH_{2n}O_n$ are formed by the numeral giving the number of carbon atoms per molecule and the ending "**ose**": tetroses ($n$ = 4), pentoses ($n$ = 5), hexoses ($n$ = 6), etc. Practically all sugar units of industrially used polysaccharides are hexoses or pentoses.

Table 8-1 Annual world production and annual US consumption (= production + import – export) of polysaccharides and/or their raw materials or products and some low-molecular weight sugars [1–3].

a) Includes newsprint. b) CMC = sodium carboxymethyl cellulose. c) HAC = hydroxyalkyl celluloses (ethyl and propyl). d) Highly industrialized countries now import far less industrial plant fibers from Third-World countries: between 1963 and 1984, US imports of sisal and henequen from Brazil and Kenya decreased to 1500 t/a from 92 500 t/a, that of jute from Bangladesh and Thailand to 9500 t/a from 75 500 t/a. e) Unclear whether "sorghum" refers to the plant *Sorghum bicolor* (mainly for food), milo (for starch), *S. saccharatum* (for syrup), or to sorghum syrup, also called "sorghum."

Consumption = (production + import) – (export + storage).

| Polysaccharide | Source | World production in t/a | | US consumption in t/a | |
|---|---|---|---|---|---|
| | | 1983 | 2004 | 1983 | 2004 |
| **Wood and wood products (wood fuel and sawnwood in $m^3/a$)** | | | | | |
| Wood fuel | trees, shrubs | 600 000 000 | 1 772 000 000 | 58 000 000 | 43 700 000 |
| Sawnwood | trees | | 415 600 000 | 110 000 000 | 127 000 000 |
| Paper and board a) | wood | 200 000 000 | 354 400 000 | 86 000 000 | 92 200 000 |
| Wood charcoal | wood | | 41 500 000 | | 980 000 |
| **Poly(β-1,4-D-glucose) (cellulose and cellulose derivatives)** | | | | | |
| Cotton lint | cotton bush | 18 000 000 | 21 000 000 | 1 300 000 | 1 605 000 |
| Rayon, acetate | wood | 3 000 000 | | ≈ 500 000 | |
| CMC b) | wood | | | 29 000 | |
| HAC c) | wood | | | 24 000 | |
| Methyl cellulose | wood | | | 12 000 | |
| Jute d) | jute shrub | 3 500 000 | 2 861 000 | 15 400 | 1 440 |
| Jute-like fibers | various | | 386 000 | | |
| Kenaf | *Hibiscus cannabinus* | 1 060 000 | | | |
| Coir | coconut | 285 000 | 185 000 | | |
| Flax, fiber, tow | *Linum* | 640 000 | 790 000 | | |
| Sisal d) | sisal agave | 350 000 | 314 000 | 2 530 | –58 |
| Ramie | *Boehmeria nivea* | 100 000 | 249 000 | | |
| Kapok fiber | silk-cotton tree | | 123 000 | | |
| Hemp fiber, tow | *Cannabis sativa* | 120 000 | 66 500 | 43 | |
| Abaca | *Musa textilis* | 99 000 | 102 000 | | 2 000 |
| Agave | *Agave* | | 59 000 | | 1 000 |
| **Poly(α-1,4-glucose)s** | | | | | |
| Starch | different plants | | | 37 000 000 | |
| Starch syrup | different plants | | | 1 200 000 | |
| Tapioca | *Manihot* roots | 33 000 | | | |
| Dextran | saccharose | 62 600 | | | |
| **Poly(galactose)s** | | | | | |
| Gum arabic | sick acacias | 70 000 | | 14 000 | |
| Agar | red algae | 10 000 | | 350 | |
| Pectin | higher plants | 10 000 | | 3 200 | |
| Carrageen | red algae | 9 000 | | 4 100 | |
| Karaya gum | *Karaya/Sterculia* trees | 4 500 | | 3 600 | |
| Tragacanth | *Astragalus* shrubs | 1 500 | | 500 | |
| Gum Ghatti | *Anogeissus latifolia* | 1 200 | | 900 | |
| Furcellerane | *Furcellaria fastigiata* | 1 200 | | | |
| **Poly(mannose)s** | | | | | |
| Guar gum | Guar seeds | 90 000 | 160 000 | 30 000 | 50 000 |
| Carob fruit | carob tree | | 185 000 | 8 000 | |
| Algin | brown algae | 13 000 | | 5 800 | |
| **Low-molecular weight sugars** | | | | | |
| Sugar, refined | cane, beets | | 133 600 000 | | 9 680 000 |
| Sorghum e) | sorghum plants | | 57 900 000 | | |

Sugars exist as **aldehyde sugars (aldoses,** I) with oxo groups at the $C^1$ carbon and **keto sugars (ketoses,** II) with oxo groups at the $C^2$ carbon. An example are hexoses:

$$\underset{\text{OH}}{\overset{6}{\text{H—CH}}}\text{—}\underset{\text{OH}}{\overset{5^*}{\text{CH}}}\text{—}\underset{\text{OH}}{\overset{4^*}{\text{CH}}}\text{—}\underset{\text{OH}}{\overset{3^*}{\text{CH}}}\text{—}\underset{\text{OH}}{\overset{2^*}{\text{CH}}}\text{—}\underset{\text{O}}{\overset{1}{\text{CH}}} \quad \text{I} \qquad \underset{\text{OH}}{\overset{6}{\text{H—CH}}}\text{—}\underset{\text{OH}}{\overset{5^*}{\text{CH}}}\text{—}\underset{\text{OH}}{\overset{4^*}{\text{CH}}}\text{—}\underset{\text{OH}}{\overset{3^*}{\text{CH}}}\text{—}\underset{\text{O}}{\overset{2}{\text{C}}}\text{—}\underset{\text{OH}}{\overset{1}{\text{CH}}}\text{—H} \quad \text{II}$$

Aldoses have $j = n - 2$ stereogenic centers (asymmetric carbon atoms) and ketoses $j = n - 3$ (marked by * in the formulas above). There are therefore $2^2$ aldotetroses, $2^3$ aldopentoses, $2^4$ aldohexoses, etc., and $2^2$ ketopentoses, $2^3$ ketohexoses, etc., each in D and L configuration (Fig. 8-1).

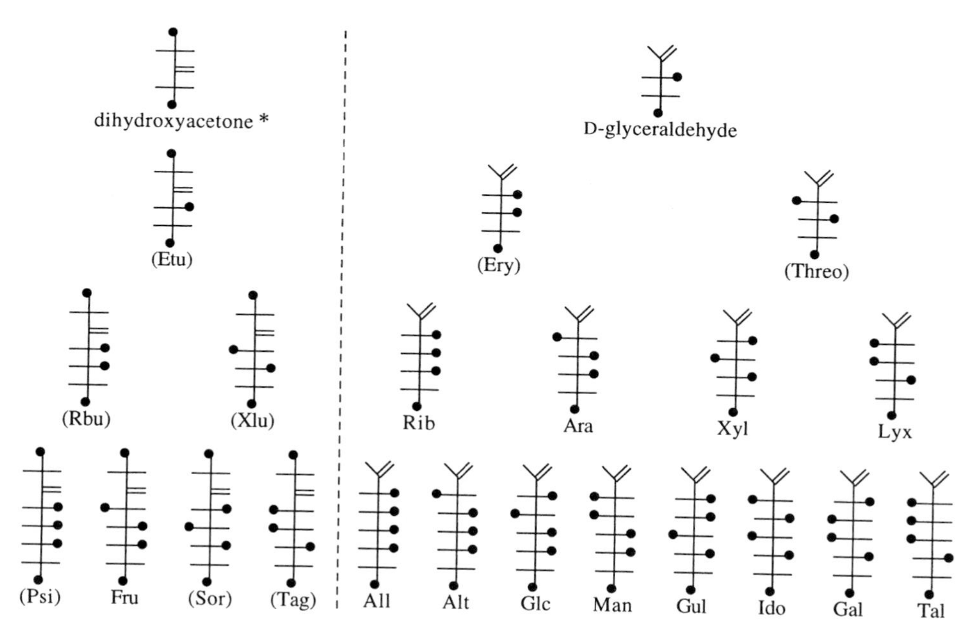


Fig. 8-1 Ketoses (left) and aldoses (right). Fischer projections of D configurations of tetroses (2nd row), pentoses (3rd row), and hexoses (4th row). OH groups are symbolized by ●, C=O groups by =. * No stereoisomers. Official abbreviations and acronyms of names of sugars use a three-letter code where the first letter is now capitalized (in parentheses: unofficial abbreviations and acronyms):

All allose
Alt altrose
Ara arabinose
(Ery) erythrose
(Etu) erythrulose
Fru fructose
Gal galactose
Glc glucose
Gul gulose
Ido idose
Lyx lyxose
Man mannose
(Psi) psicose (allulose)
(Rbu) ribulose (araboketose)
Rib ribose
(Sor) sorbose
(Tag) tagatose
Tal talose
(Threo) threose
(Xlu) xylulose
Xyl xylose

*Trivial names* of sugars are usually derived from their sources: glucose from grapes (G: *gleukos* = sweet new wine); galactose from milk (G: *gala* = milk); fructose from fruits (L: *fructus* = fruit); sorbose from the fruits of European mountain ash (*Sorbus aucuparia*); mannose from the sap of the manna ash tree *Fraxinus ornus* (see p. 419); xylose from wood (G: *xylos* = wood), arabinose from Gum arabic. Other names are arbitrary alterations, for example, ribose from arabinose.

Replacement of the CHO group of aldoses, $C_nH_{2n}O_n$, by a $CH_2OH$ group generates sugars of the composition $C_nH_{2n+2}O_n$. These sugars carry the same trivial names as aldoses but with the suffix "itol" instead of "ose." The four aldopentoses thus lead to three pentitols: ribitol = ribite = adonite = adonitol), arabitol = lyxitol, and xylitol, and the eight aldohexoses to six hexitols: allitol, altritol = talite, glucitol = gulite = sorbite = sorbitol), mannitol = mannite = manna sugar, iditol, and galactitol = dulcite = dulcitol, euonymit = melampyrit.

All "oses" have D and L configurations; the R,S nomenclature is not used in sugar chemistry (Fig. 8-2). L configurations are the mirror images of the D configurations shown in Fig. 8-1. Configurations of aldoses refer to that carbon atom which is adjacent to the CHOH group that is farthest away from the C=O group. By convention, D-aldoses have the same configuration as D-glyceraldehyde (Fig. 8-2). Ketoses are derived from dihydroxyacetone; configurations are here based on that of erythrulose (Etu).

Some "oses" have **epimers** which are sugars that differ only in the configuration at one carbon atom (G: *epi* = upon, over, at, after). Examples are D-mannose (epimer of D-glucose with respect to $C^2$) and L-galactose (epimer of L-glucose with respect to $C^4$).

**Anomers** are cyclic epimers with half-acetal structures that are formed by intramolecular reaction of an OH group and the >C=O of aldoses (Fig. 8-3) or ketoses (Fig. 8-4). The resulting six-membered rings are derivatives of pyran (oxacyclohexane) whereas five-membered rings are derivatives of furan (oxacyclopentane). *Systematic names* of sugars thus insert "pyran" and "furan", respectively, between the trivial sugar name and the suffix "ose".

Each D-pyranose and each D-furanose can exist in two configurations that are distinguished as "$\alpha$" (anomeric OH axial, anomeric H equatorial) and "$\beta$" (anomeric OH equatorial, anomeric H axial).

In the crystalline state, a hexose exists exclusively as one of its anomers, i.e., as a cyclic compound; the same is true for pentoses. In solution, simple sugars may exist in many types of cyclic anomers in various concentrations; the concentration of open sugar structures (if any) is usually small.

$\beta$-D-Glucopyranose, for example, can form an intramolecular hydrogen bridge between $C^6H_2OH$ and the equatorial anomeric OH group at $C^1$ (Fig. 8-3) but $\alpha$-D-glucopyranose cannot do so for steric reasons. D-Glucose thus crystallizes as $\beta$-D-glucopyranose, its thermodynamically most stable form. Dissolution of D-glucose in water severs the hydrogen bridge $-C^6H_2OH \cdots HO^1C(H)<$ and an equilibrium between $\alpha$ and $\beta$ forms (38/62 at 40°C) is established via the open form (**mutarotation** of optical activity).

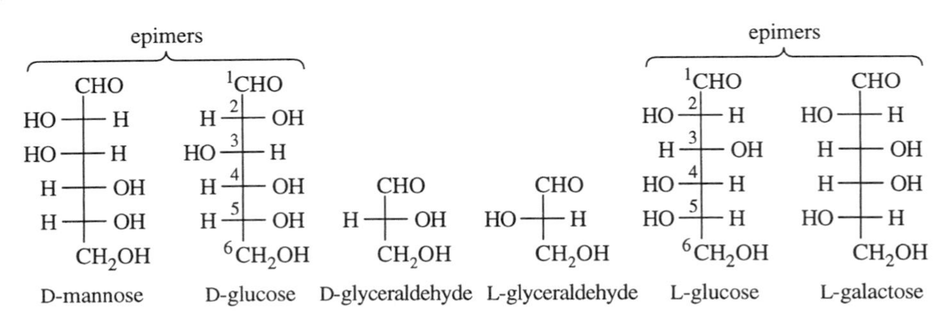


Fig. 8-2 Definition of L and D for some sugars and their epimers.

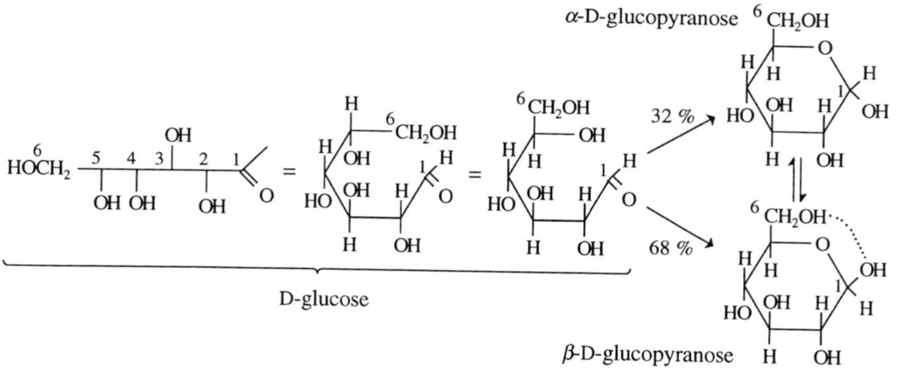

Fig. 8-3 Open-chain D-glucose and its two pyranosidic anomers. ····· Hydrogen bridge.

Intramerar hydrogen bonds lower the asymmetry of the molecule. β anomers dominate therefore in aldoses with D-configured $C^2$ atoms (allose, glucose, gulose, galactose, see Fig. 8-1). In aldoses with L-configured $C^2$ atoms either the α anomer is favored (mannose, talose) or α and β are present in about equal proportions (altrose, idose).

D-Glucose does not form furanoses but D-fructose does. In aqueous solution ($D_2O$) at 25°C, all four D-fructose forms are present (Fig. 8-4). Again, β structures are thermodynamically favored for both pyranose and furanose forms.

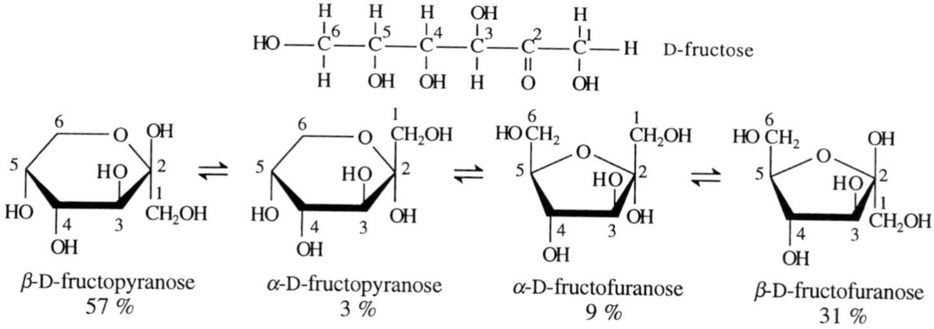

Fig. 8-4 The ketohexose D-fructose and its anomers (in Haworth projection). For clarity, H as a substituent is not shown. Percentages indicate presence in $D_2O$ solution.

In principle, hexoses may exist in chair or boat conformation, but in reality, chair forms dominate. For illustration, Fig. 8-5 shows the chair forms of the four possible glucoses. In polysaccharides, conformations of monomeric sugar units are fixed.

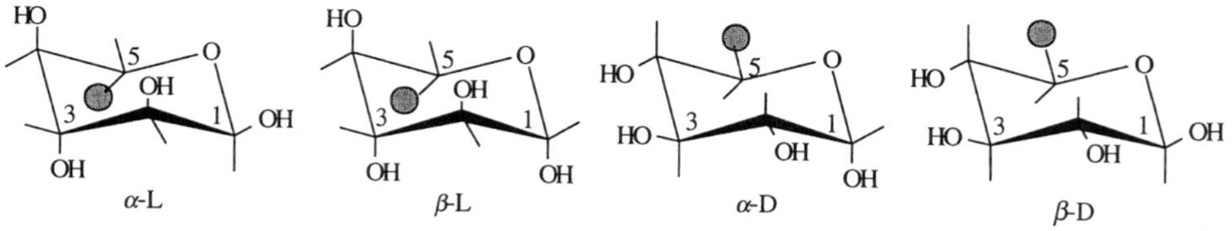

Fig. 8-5 Chair conformations of the four theoretically possible glucoses. ● indicates $CH_2OH$.

The $C^1$ atom of aldoses is in an aldehyde group in open sugars and in a hemiacetal group in rings. Acetalization of the anomeric OH group of this carbon atom removes the reducing end and freezes a certain conformation: the resulting alkyl glucopyranoses no longer mutarotate. The bond between the sugar moiety and the substituent is called **glycosidic**. Oligosaccharides and polysaccharides consist therefore of sugar units that are connected via glycosidic bonds. **Glycosides** are correspondingly all organic chemical compounds that on hydrolysis produce sugars or sugar derivatives and a non-sugar compound (**aglycon** (**genin**)). An example is the methyl glycoside of glucose, 1-*O*-methyl-*β*-D-glucopyranoside (Fig. 8-6).

*Glyco* is a variant of *gluco* (G: *gleukos* = sweet, see p. 367). Older literature therefore sometimes uses gluco instead of glyco; for example, aglucone instead of aglycone (see above) or glucoprotein instead of glycoprotein (see Sections 8.4 and 11.1.1). The suffix glyco is used for sugars (see below) and for derivatives of glycine, $H_2NCH_2COOH$, such as glycylglycine, ($H_2NCH_2CO–NHCH_2COOH$), and glycocholic acid, a peptide-like conjugate of glycine and cholic acid, $C_{24}H_{40}O_5$.

Hydroxyl groups of sugars may also be substituted, for example, by etherification, esterification, or anhydride formation. OH groups may also be replaced by other groups, for example by H in deoxy sugars and by $NH_2$ in aminodeoxy sugars.

Oxidation of $CH_2OH$ to COOH leads to **uronic acids**, for example, glucuronic acid (see below). Other examples are the monosaccharide galacturonic acid and the polysaccharide hyaluronic acid (see Section 8.4.1). Names of these acids are formed by replacing the "ose" of sugars by "uronic acid." Uronic acids are aldehyde acids whereas the oxidation of aldehyde groups of aldoses to carboxylic groups leads to **aldonic acids**. These acids form $\gamma$ and $\delta$ lactones on dehydration.

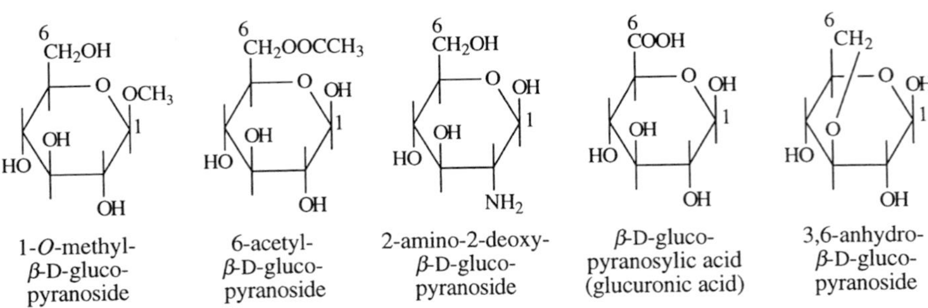


Fig. 8-6 Chemical structure of some *β*-D-glucose derivatives.

## 8.1.3 Oligosaccharides and Polysaccharides

Simple sugars and their derivatives of the last section are called **monosaccharides**. Their dimers of two identical or non-identical simple sugars are **disaccharides**, their trimers, **trisaccharides**, etc. **Oligosaccharides** consist of a "few" sugar units, i.e., 2 and more but usually less than ca. 20 and **polysaccharides** of "many" sugar units.

The term "saccharides" for this group of chemical compounds is derived from **saccharose**, the most common sugar (cane sugar, beet sugar, sucrose). Saccharose is a disaccharide from *β*-D-fructose and *α*-D-glucose, i.e., a *β*-D-fructofuranosyl-*α*-D-glucopyranoside. Since here a glycosidic bond exists between two sugar units, systematic nomenclature calls this molecule a **glycan**. However, it is not customary to talk about di-, tri-, ... oligo, and polyglycans.

Homopolymers from a single type of low-molecular weight sugar are called **homoglycans** and those with two or more types, **heteroglycans**. Terms like coglycan or copolyglycan, in analogy to "copolymer", are not used. In food chemistry, one rather talks about "complex polysaccharides" or "complex carbohydrates" if polysaccharides with two or different types of sugar units are meant.

"Homoglycan" does not imply that all sugar units are interconnected in the same way. Because hexose and pentose molecules carry several hydroxyl groups, different glycosidic bonds are possible between the 1 position and the 3, 4, or 6 positions which are indicated by arrows. Examples are polymers of the pentose xylose (Fig. 8-7).

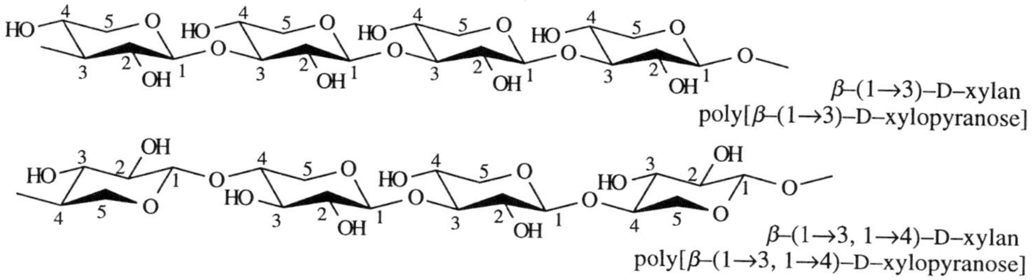

β-(1→3)-D-xylan
poly[β-(1→3)-D-xylopyranose]

β-(1→3, 1→4)-D-xylan
poly[β-(1→3, 1→4)-D-xylopyranose]

Fig. 8-7 Sections of poly(xyloses) from the seaweed *Palmariales*. For explanation of →, see p.372.

Many polysaccharide chains are regularly (in defined intervals) or irregularly substituted by sugar units of a different type. An example is a poly(D-mannose) that is substituted by D-galactose units (Fig. 8-8).

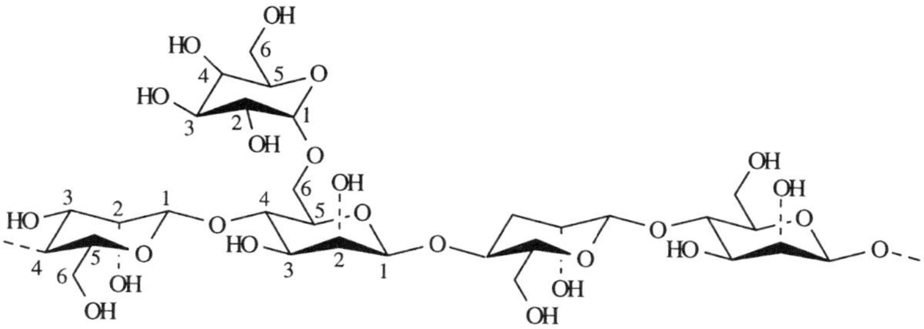

Fig. 8-8 Structure of a poly(D-mannose) that is substituted by D-galactose units.

## 8.1.4 Nomenclature

Chemical structures of polysaccharides are described by various nomenclatures. For simplicity, trivial names are used if common. Examples are cellulose, chitin, guar, and xanthan.

If trivial names are uncommon or one wants to draw attention to the constituent sugar units, poly(monomer) names are employed, either with or without designation of the configuration of the sugar. Examples are poly(glucose) or poly(D-glucose).

Another nomenclature considers the glycosidic α- and β-acetalic bonds between the anomeric hydroxy group of one monomeric unit and the non-anomeric hydroxy group of the next one. If the type of this glycosidic bond is not known, one simply characterizes the polymer by the name of its monosaccharide using the prefixes "anhydro" and "poly." An example is poly(anhydro-D-glucose).

Bonds between two adjacent sugar units are indicated by the position numbers of the participating carbon atoms of the ring. Since one considers intermerar bonds between two sugar units and not intramerar ones of the same unit, one connects these position numbers by an arrow. An example is poly[(1→4')-anhydro-D-glucose].

In principle, one has also to indicate whether the sugar unit is in the pyranose or furanose form, for example, poly[(1→4')-anhydro-D-glucopyranose]. This designation is often omitted since sugar units of natural polysaccharides are usually aldehyde sugars in the pyranose form.

In addition, the type of anomerism should be indicated, i.e., $\alpha$ or $\beta$ in the old nomenclature and "ax" (axial) or "eq" (equatorial) in the newer one (Fig. 8-9).

| | $\alpha$-D-Glucose | $\beta$-D-Glucose |
|---|---|---|
| anomeric OH group | axial | equatorial |
| anomeric H | equatorial | axial |

Fig. 8-9 Definition of axial and equatorial positions.

Cellulose (Section 8.3) therefore has systematic names such as poly[$\beta$-(1→4')-anhydro-D-glucopyranose] or poly[$\beta$-(1eq-4eq)-anhydro-D-glucopyranose]. The length and clumsiness of such names are good reasons to avoid them and use trivial names. Shortened names like the ones in Figs. 8-6 and 8-7, on the other hand, usually do not immediately convey all structural details which may leave the non-specialist somewhat at a loss. It is for this reason that modern literature often prefers the glycan nomenclature which does away with the prefix "poly" (since it is inherent in "glycan") and replaces the suffixes "pyranose" and "furanose" by "an" (see Fig. 8-7). The polymer of Fig. 8-8 is correspondingly a D-galacto-D-mannan.

Microbial polysaccharides (**exopolysaccharides**) often have very complex chemical structures and are therefore characterized by a special short-hand notation. These polysaccharides are either homopolymers or periodic copolymers of up to seven different types of sugar units; they are also usually regularly branched. Examples are the *Klebsialla* types K-4 and K-17 where the repeating units are composed of axially (α) or equatorially (eq) 1,2, 1,3, 1,4, or 1,6 connected glucose (glc), glucuronic acid (glcA), mannose (man), and rhamnose (rha) units.

$$\overset{3}{—}\,\text{glc}\,\underset{\alpha}{\overset{1\quad 2}{——}}\,\text{glcA}\,\underset{\alpha}{\overset{1\quad 3}{——}}\,\text{man}\,\underset{\alpha}{\overset{1\quad 3}{——}}\,\text{glc}\,\underset{\beta}{\overset{1}{——}}$$

K–4

$$\overset{4}{—}\,\text{glcA}\,\underset{\beta}{\overset{1\quad 3}{——}}\,\text{rha}\,\underset{\alpha}{\overset{1\quad 4}{——}}\,\text{glc}\,\underset{\alpha}{\overset{1\quad 2}{——}}\,\underset{\text{rha}}{\underset{1\,|}{\underset{\alpha}{\underset{3\,|}{\text{rha}}}}}\,\underset{\beta}{\overset{1}{——}}$$

K–17

The newer IUPAC system is slightly different: it capitalizes abbreviations and acronyms and writes the type of anomerism and connecting positions on one line (Fig. 8-10). The newest CFG system uses graphic symbols and colors to symbolize sugar units.

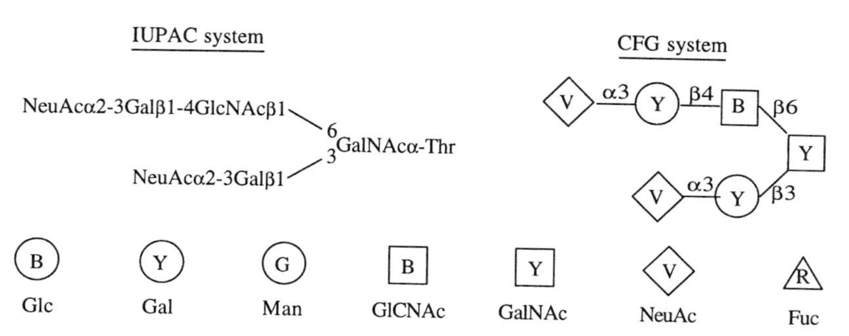


Fig. 8-10 IUPAC and CFG representation of a glycan [4]. CFG uses colored symbols, here symbolized by inscribed letters: B = blue, G = green, R = red, V = violet, Y = yellow. IUPAC = International Union of Pure and Applied Chemistry, CFG = Consortium for Functional Glycomics. Fuc = fucose, Gal = galactose, GalNAc = *N*-acetylgalactosamine, Glc = glucose, GlcNAc = *N*-acetylglucosamine, Man = mannose, NeuAc = *N*-acetylneuraminic acid, Thr = threonine.

The linkage is still indicated by α or β but the first numeral of a connecting bond is dropped since sugars are always connected via the half-acetal position (carbon 1) whereas the other connection point may vary.

Polysaccharides can have many different structures: linear, branched, and crosslinked; true homopolymers and those with some "incorrect" sugar units; copolymers with regio and block structures; and polysaccharides with peptide, protein, lipid, or nucleic acid components (**conjugated polysaccharides**) where the latter may be bound to the former either covalently or non-covalently. **Proteoglycans** are polysaccharides with protein structures as side groups which, in turn, may by substituted by oligosaccharides. **Glycoproteins** are proteins with oligosaccharide side chains (see Section 8.4.1).

**Heteroglycans** contain two or more types of sugar units. Glycans with distributions of molecular weights are called **polymolecular**, similar to synthetic polymers. "**Polydisperse glycans**", on the other hand, consist of mixtures of constitutionally similar glycans that differ in smaller constitutional features such as proportion and lengths of side-chains, degrees of acetylation, etc. The use of "polydisperse" thus differs from that in polymer science (Volume I, p. 44). "**Polydiverse glycans**" are mixtures of polysaccharide molecules with different constitutions of monomeric units or repeating units.

### 8.1.5 Biosynthesis

The biosynthesis of polysaccharides is not a reversal of their hydrolysis since the direct "polycondensation" of monosaccharides in water has rather a positive Gibbs energy of polymerization. The monomer molecules are correspondingly not the hydrated monomeric units, i.e., the corresponding low-molecular weight sugar molecules.

Instead, sugar molecules (S) are first converted to sugar-6-phosphates (S-6-P) by reaction of the sugar with adenosine triphosphate (ATP) and adenosine diphosphate. S-6-P is then converted enzymatically to S-1-P (sugar-1-phosphate) which reacts with a nucleoside triphosphate by release of $H_4P_2O_7$ to a nucleoside diphosphate-sugar (NDP-S) which is then converted to the true monomer, a phosphorylated carrier lipid, NDP-S/lipid (see Volume I, p. 352 ff.).

Different sugars and nucleosides are used for the various polysaccharides. Other nucleosides serve to connect sugar units and proteins and oligopeptides, respectively. For example, cellulose chains are started from a protein with uridine diphosphate glucose but the propagation steps involve guanosine diphosphate glucose. The different structure of nucleoside diphophate sugars for start and propagation explains why homo-polysaccharides often carry different sugar units at their chain ends. Glucosamine glycan chains, for example, are started from a protein to which xylose-galactose-galactose-glucuronic acid units are attached.

### 8.1.6 Economic Importance

Each year, nature produces and destroys some $3 \cdot 10^{11}$ t of carbon compounds wherefrom 95 % is in plants and 5 % is in animals. 95 % of the annual biomass consists of mono, oligo and polysaccharides wherefrom ca. 40 % is cellulose. Biomass could therefore be an important renewable source of raw materials. Only ca. 3 % is used, however: some mono, di, and trisaccharides as foodstuffs and some polysaccharides as foods, construction materials, fibers, plastics, thickeners, etc. A small percentage of biomass is also used as energy carriers (wood, biodiesel, ethanol) and raw materials for intermediates where plants grow rapidly (e.g., sugar cane in Brazil) and/or production is subsidized.

Low-molecular weight sugars used by man are usually either the natural products themselves or hydrolysis products of natural oligo or polysaccharides; enzymatically or fermentatively obtained sugars are not common. Polysaccharides for human use are also generally the natural products, albeit often derivatized. A few polysaccharides are obtained synthetically by enzymatic or fermentative polymerization of mono and disaccharides. The polymerization of other saccharide monomers, such as vinyl saccharides, is presently of academic interest only.

The reliance on natural products and their derivatives is in part caused by cost factors and in part by the intrinsic properties of saccharides and polysaccharides. Saccharose, the sugar produced worldwide in the largest amounts (ca. $10^8$ t/a), was sold in 1989 for similar prices as vinyl chloride and ethene (Fig. 8-11). It can be clearly seen that world market prices for commodity sugars are considerably higher than those of plastics of the same volume, even if they are not subsidized. Although agricultural products are usually less capital intensive than industrial products, their costs will never be able to compete with industrial products on a per volume basis for commodity products. The reason is the relatively low production per acreage: 1 hectare of land (10 000 $m^2 \approx 2.47$ acres) produces annually ca. 0.03 t wool, 0.5 t natural rubber, 0.5–0.7 t line fiber, 1 t cotton, 4 t corn starch, 5–9 t beet sugar (US), or 8–13 t cane sugar (Brazil). As a result, agricultural products must be transported over relatively large distances to processing facilities. The high transportation costs for products from farther outlaying areas limit the size of sugar refineries which can therefore not exploit the economy of scale. The same effect is observed for the inverse situation: transport of an industrial byproduct to agricultural users. An example is the fertilizer ammonium sulfate which is the byproduct of the production of ε-caprolactam (p. 465).

With increasing production, costs and prices of agricultural products thus approach a limiting value whereas those of industrial products can enjoy the economy of scale.

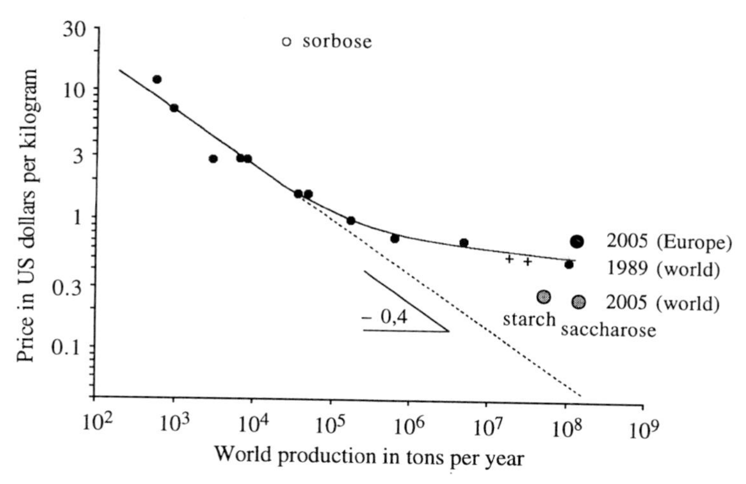


Fig. 8-11 Price of D-monosaccharides and D-disaccharides (•) and L-sorbose (o) as a function of annual world productions in 1989 [5] and corresponding data (+) for ethene and vinylchloride. The slope of the experience line for small-production sugars (–0.4) corresponds to that of synthetic polymers (solid and dotted line) (Fig. 5-29). For comparison, world production and world market prices of starch and saccharose ( ● ) and European subsidized beet sugar price (●) are shown for 2005.

## 8.2 Poly($\alpha$-glucose)s

### 8.2.1 Overview

In principle, D-glucose units can be interconnected to linear poly(glucopyranoside)s in 8 different ways: each of the four possible bonds (1→2, 1→3, 1→4, 1→6) can lead to one of two anomers (α and β). In nature, 1→3 and 1→4 bonds dominate (Table 8-2). 1→6 bonds are found in bacteria and frequently in branch units.

Linear *homoglycans* of D-glucose may have their glycosidic bonds exclusively in α(1-4) (**amylose**), α(1-6) (**pseudo-nigeran**), β(1-6) (**pustolan**), β(1-4) (**cellulose**), β(1-3) (**callose**), or β(1-2) positions (***Agrobacterium tumefaciens***) (Table 8-2). Anomeric bonds may also alternate, for example, α(1-4) and α(1-6) in **pullulan** or α(1-3) and α(1-4) in **nigeran**, or they may be randomly distributed along the chain, for example, β(1-3) and β(1-4) in **lichenin**. Repeating units may also consist of two or more sugar units in the same sequence, albeit apparently not in poly(D-glucose)s.

By definition, branch units of branched homoglycans have the same sugar units as the units of the main chain, although their bond to the main chain is necessarily different (**amylopectin, glycogen**). Replacement of OH groups of D-glucose leads to *heteroglycans*, for example, by *N*-acetyl-2-amino groups in **chitin**.

Polysaccharides produced by lower organisms usually have many functions. **Curdlan** from agrobacteria is both a structural and a reserve polysaccharide. Polysaccharides from higher organisms are more specialized. **Cellulose** (Section 8.3), for example, serves only as a structural polysaccharide; it cannot by digested by humans but happily by cows. **Amylose** (Section 8.2.3), on the other hand, is a reserve polysaccharide without any structural functions.

Table 8-2 Polyglycans with chains of glucose units Glc, GlcA = glucuronic acid, AcN2 = *N*-acetyl-2-amino, Su6 = 6-sulfate, Su2 = 2-sulfate. A = alternating, B = branched, L = linear, LR = linear regio-periodic, PS = polysaccharide. α = ax = axial, br = branched, β = eq = equatorial. [a] Scleroglucan.

| Polymer | Unit | Bond | Type | Source (* bacterium) |
|---|---|---|---|---|
| Dextran | D-Glc | α(1-6)-*br*-α(1-3) | B | bacteria |
| Amylose | D-Glc | α(1-4) | L | many plants |
| Amylopectin | D-Glc | α(1-4)-*br*-α(1-6) | B | many plants |
| Glycogen | D-Glc | α(1-4)-*br*-α(1-6) | B | animals, microorganisms |
| Pullulan | D-Glc | α(1-4)-*alt*-α(1-6) | A | yeast, *Aureobasidium pullulans* |
| Pseudonigeran | D-Glc | α(1-3) | L | fungi, *Streptococcus m.* |
| Nigeran | D-Glc | α(1-3)-*alt*-α(1-4) | LR | fungi |
| Pneumococcus-PS | D-Glc | α(1-3)-*alt*-α(1-6) | L | *Pneumococcus* |
| Pustulan | D-Glc | β(1-6) | L | lichen, yeasts |
| Cellulose | D-Glc | β(1-4) | L | plants (cell walls) |
| Xanthan | D-Glc | β(1-4)-*br*-(Man-Glc-Man) | B | *Xanthomonas campestris** |
| Chitin | D-Glc(NAc2) | β(1-4) | L | arthropods (rind) |
| Hyaluronic acid | D-Glc(NAc2) | β(1-4)-GlcA-β(1-3) | A | tears, synovial fluid |
| Chondroitin-6-sulf. | D-Gal(NAc2,Su6) | β(1-4)-GlcA-β(1-3) | A | matrix of cartilage |
| Chondroitin-4-sulf. | D-Gal(NAc2,Su4) | β(1-4)-GlcA-β(1-3) | A | matrix of cartilage |
| Callose | D-Glc | β(1-3) | L | higher plants |
| Curdlan | D-Glc | β(1-3) | L | some bacteria |
| Pachyman | D-Glc | β(1-3) | L | some fungi, protozoae |
| Laminaran | D-Glc | β(1-3) | (B) | some brown algae |
| - | D-Glc | β(1-3)-*br*-β(1-6) | B | many yeasts |
| Schizophyllan [a] | D-Glc | β(1-3)-*br*-β(1-6) | B | *Schizophyllum commune* |
| Lichenin | D-Glc | β(1-3)-*co*-β(1-4) | ? | lichen, oats, barley |
| Isolichenin | D-Glc | α(1-3)-*co*-α(1-4) | LR | lichen |
| - | D-Glc | β(1-2) | L ? | *Agrobacterium tumefaciens* |

## 8.2.2 Starch

### Occurence

Starch (Middle English: *sterchen* = to stiffen with starch; from Germanic: *stärken* = to make strong) is a fine, white, amorphous, tasteless powder that is produced as a reserve polysaccharide by many plants. Chemically, starch is a mixture of amylose, a linear molecule (Section 8.2.3) and amylopectin, a branched molecule (Section 8.2.4), both with main chains of α(1→4) connected D-glucose units (Table 8-1) (L: *amylum* = starch; G: *amylon* = (powder) not (*a-*) ground in a mill (*myle*)).

Starches are stored in plants as granules of 2–900 μm diameter (Table 8-3) in

- bulbs or roots: potato, sweet potato, maranta, manioc;
- seeds: corn, wheat, rye, barley, oats, millet, sorghum, spelt, amaranth;
- fruits: pulse (peas, beans, etc.), chestnut, acorn, banana;
- pith (medulla): sago palm.

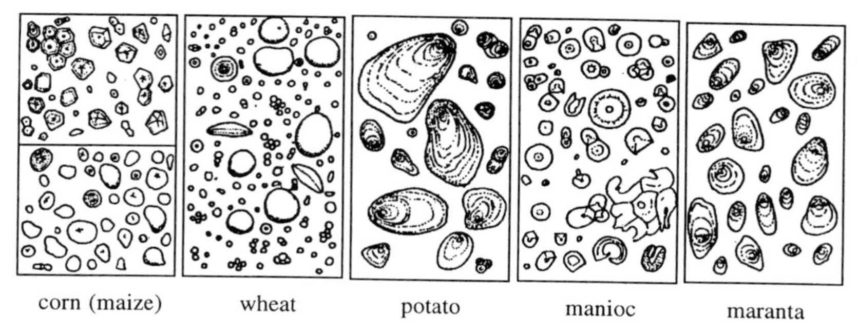


Fig. 8-12 Starch granules (200-fold magnification) [6]. Reproduced with kind permission by Georg Thieme Verlag, Stuttgart, Germany.

**Sweet potato** (*Ipomoea batate*) is a tropical vine with bulby roots that is cultivated in Africa, China, and the Southern states of the United States.

Roots of **maranta** contain ca. 84 % digestible starch which is called arrowroot after one of the species of maranta. "**Arrowroot**" is now also used for edible starches from roots of other tropical plants such as *Cannacea* (African or Australian arrowroot) or *Tacca pimatifolia* (Tahitian arrowroot). Maranta roots were used by American Indians to remove poison from arrow wounds, hence the name.

**Manioc** (manioca, cassava, kassava, yucca) is a spurge that is cultivated in Africa, India, East Asia, and the tropical areas of the United States (world production: $\approx 150 \cdot 10^6$ t/a (1999), wherefrom ca. 10 % for starch). The beady starch from cassava roots is called **tapioca**.

**Spelt** is an old hardy wheat that is now being cultivated again in Europe.

**Sorghum** is a type of millet that is cultivated in Africa, Asia, and the United States. Its subspecies **milo** is mainly processed to starch and its subspecies *Sorghum saccharatum* to sorghum syrup.

The **sago palm** (*Metroxylon sago*) delivers **sago. True sago** is produced from tapioca starch and **potato sago** from potatoes. **Pearl sago** for food is obtained by forcing a sago dough through a sieve. **Bullet sago** is a lower grade of pearl sago that is used as a size.

**Corn** (US) is called by its original name **maize** in International English where "corn" designates all kinds of hard grains. American "corn" is the short form of "Indian corn" (*Zea mays*).

Starches from various sources differ in granule size (Fig. 8-12), the ratio of amylose to amylopectin (Table 8-3), and the type and concentration of byproducts (proteins, lipids, phosphate). The amylose/amylopectin ratio controls the crystallinity of starches and thus their swelling and solubility in water. Starches do not dissolve in water *per se*; they form aqueous dispersions. The gelatinization temperature is the temperature at which dispersed starch granules start to swell and the system becomes heterogeneous..

Table 8-3 Contents $w_{st}$ of starches in bulbs and seeds; diameters $d$ and gelatinization temperatures $T_{gel}$ of granules; viscosity $\eta$ and shear stability of paste; and solubilities $w_L$ (at 95°C) and amylose contents $w_{am}$ of starches. Waxy starches do not contain wax; their name was derived from the waxy appearance of corn or rice seeds.

| Source | Type | $w_{st}$/% | $d$/μm | $T_{gel}$/°C | $\eta$ | Stability | $w_L$/% | $w_{am}$/% |
|---|---|---|---|---|---|---|---|---|
| Corn | amylocorn | | | > 100 | low | stable | 50–85 | |
| | normal | 60–70 | 5–26 | 62–70 | low | stable | 25 | 27 |
| | hybrid waxy | | | 63–72 | high | unstable | <2 | |
| Potato | normal | 12–20 | 15–100 | 58–62 | very high | unstable | 17–23 | |
| Wheat | normal | 53 | 25–45 | 53–64 | | | 41 | 19 |
| Manioc | tapioca | 20–30 | 5–35 | 52–64 | high | unstable | 48 | 17–23 |
| Rice | normal | 70–75 | 3–9 | | | | 19 | |

**Biosynthesis**

All green plants and algae produce starch intracellularly in chloroplasts (chlorophyl-containing plastids = specialized cytoplasmic structures) from adenosyldiphosphate glucose as monomer and the enzyme transglucosylase as catalyst (see Volume I, p. 552).

*In-vitro* syntheses can utilize instead uridinyldiphosphate glucose and the enzyme phosphorylase, which is contained in the sap of squeezed potatoes. Such enzyme preparations produce only low-molecular weight amyloses since they also contain a hydrolyzing enzyme. Heating the sap destroys the thermolabile hydrolase and amyloses with degrees of polymerization of ca. 10 000 are obtained (Section 8.2.3). *In vivo*, amylopectin (Section 8.2.4) is probably produced by a branching enzyme.

Starch is stored by the plant in plastids. During biosynthesis, short amylopectin side chains form rigid double helices that combine laterally to crystalline lamellae (Fig. 8-13). The resulting physical network prevents the crystallization of the much longer amylose chains from which only short segments can be incorporated as double helices in crystallites. The remaining segments of amylose molecules stay in the amorphous state where they and the main chains of amylopectin molecules form amorphous interlayers similar to the ones in semi-crystalline poly(ethylene (Volume III).

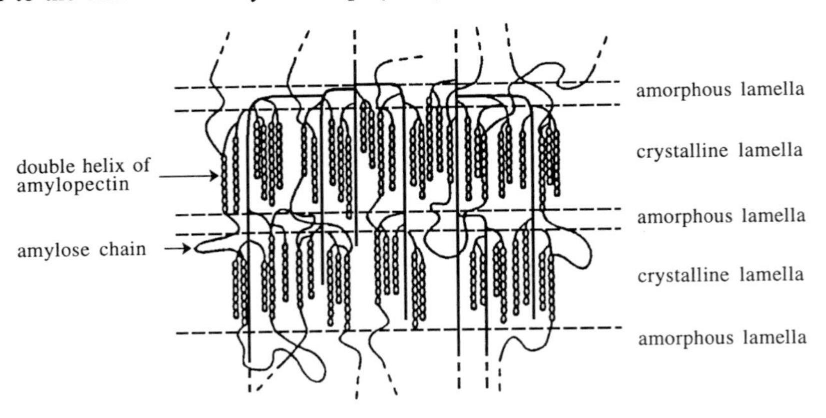


Fig. 8-13 Schematic representation of the structure of starch granules [7a].

**Industrial Isolation**

Starch is obtained in the United States predominantly from corn (maize), in Europe mainly from maize and potatoes, and in the tropics almost exclusively from manioc (tapioca roots). Yields of starch from corn are ca. 60 %, from wheat ca. 50 %, and from potaoes ca. 20 %. World production of starches climbed rapidly from $37 \cdot 10^6$ t/a (1995) to $53 \cdot 10^6$ t/a (2002) of which 62 % was from corn, 30 % from manioc and wheat, 6 % from potatoes, and 2 % from other sources (sorghum, sweet potato, etc.).

The U.S. industrial production of starch from corn first removes cobs, sand, dust, chaff, and other foreign material. Grains are subsequently softened at 40°C by water that contains ca. 0.1 % $SO_2$ and coarsely ground ("wet milling process"). The resulting slurry is fed to a hydrocyclone where the germ fraction is separated from the underflow that contains the endosperm and fiber. The germ fraction is used for the production of corn oil whereas the twice remilled underflow furnishes starch, gluten, and fiber. After removal of fiber, the resulting suspension of starch and gluten (mill starch) is concentrated by centrifugation. Starch granules are filtered off. The combined gluten and fiber are used as animal feed. Also used for corn is a dry-milling process with dry degermination.

For tapioca, dried and shredded manioc roots are milled and pressed through sieves. The granules (Tupi Indian: *tipioca* = residue) are roasted in order to destroy the toxic cyanogenic glycoside.

### Technical Use

The use of the various starches depends on the amylopectin/amylose ratio, the number and structure of amylopectin side chains, and the contents of phosphate, proteins, fats, and lipids. Starches are used directly or as modified starch as thickeners, gelation agents, etc., in foodstuffs and for many industrial applications from indispensable paper additives to uses as textile sizes and plastics additives and in building materials (gypsum, concrete), ore processing, pharmaceuticals, and cosmetics.

The ratio of food/non-food applications is estimated as 55:45 (world) and 47:53 (European Union) (2000). Approximately 21 % of starches are used in paper-making, 6 % for card boards, 6 % in chemical processes, and 20 % for other technical uses (1988), for example, in the exploration and production of crude oil. Starches also serve as raw materials for the production of ethanol (p. 105) or glucose for lactic acid (p. 333) as well as that of glucose syrup.

The application of *food-grade* starches depends on their amylose contents because amylose retrogrades easily in water, i.e., it becomes again insoluble (see p. 383 ff.). For this reason, corn starch (ca. 27 % amylose) cannot be used for deep-frozen ready-to-eat meals since they would develop an unappealing texture. Ready-to-eat meals thus contain either starches that are more rich in amylopectin or modified starches. Retrogradation also seems to be responsible for the staling of bread (Section 8.2.4) since helical and non-helical structures cause different taste sensations.

*Technical uses* are very diversified and can therefore not be discussed in detail; some of them are mentioned in Volume IV (textiles, paper, etc.). It is noteworthy, however, that plasticized starch can even be injection-molded to biodegradable articles in a very narrow range of plasticizer content, temperature, and pressure. This **thermoplastic starch** is obtained by swelling starch in a plasticizer such as glycerol and subsequent homogenizing by kneading. The resulting product becomes pseudo-plastic at temperatures above 160°C. The molded material is semi-crystalline with glass temperatures between –60°C and +150°C, depending on plasticizer content. It becomes rubber-elastic between –45°C and +65°C and has similar properties as plasticized poly(vinyl chloride).

### Modified Starches

Only amorphous regions are affected if starch is hydrolyzed below the gelatinization temperature. This **acid-modified starch** is less viscous and has a lower gel strength.

Oxidation of starch by NaOCl produces aldehyde and carboxyl groups in amorphous regions (Eq.(8-1)). The resulting **chlorite-oxidized oxyamylose** (COAM) has a lower gelatinization temperature and is less thickening than starch; it is also antiviral and only slightly toxic.

(8-1) $\sim\!\!\sim$[glucose unit with $CH_2OH$, OH, OH]$-O\sim\!\!\sim$ $\xrightarrow[-H_2O]{+O_2}$ $\sim\!\!\sim CH(CHO)-CH(CH_2OH)-O-CH(COOH)-O\sim\!\!\sim$

The many hydroxyl groups per monomeric unit allow starch to be modified in many different ways. Similar to cellulose, it can be reacted with ethylene oxide to **hydroxy-ethylated starch** (**HES**) or with propylene oxide to hydroxypropylated starch (**HPS**) with degrees of substitution of 0.05-0.10. Such starches are used to improve the strength and stiffness of papers; paper coated with HES or HPS can be printed with more intensive colors. These modified starches also serve as sizes in the textile industry.

**Cationic starches** contain tertiary or quaternary amino groups. For their synthesis, hydroxyl groups of dispersed starches are reacted with reactive groups (epoxy, β-halogenated alkyl) of a tertiary amine. The resulting tertiary amino groups are then quaternized. Cationic starches are mainly used in paper making.

**Starch phosphates** (**SP**) and **starch acetates** (**SAC**) result from the esterification of starch with appropriate acids or acid anhydrides, respectively. Phosphates of corn starch and potatoes with degrees of substitution usually below 0.15 have similar properties. Monophosphates are suitable for ready-to-eat meals because of their good freezing and thawing properties. Starch acetates have higher degrees of substution of 0.5-3. The low-substituted types are used in foods, the highly substitutes ones for packaging films.

Free-radical grafting of acrylonitrile onto starch and subsequent alkaline hydrolysis delivers graft polymers with carboxyl and carboxyamide groups. The acidic form of these polyelectrolytes dissolves easily in water. On neutralization, a thick elastic mass results that consists of tightly packed gel particles. These highly concentrated dispersions can be cast to self-supporting films that retain their shapes over a large range of temperatures and pH values.

Strong stirring of these dispersions reduces their viscosity by a factor of 1000. The resulting true solutions can be cast to films that become insoluble by aging at high humidities, heating, or irradiation by $^{60}Co$. Films absorb water up to 2000 times their weight and have therefore been called "super slurpers." Superabsorbents are now mainly based an acrylic acid, however (p. 290).

### Starch as Raw Material

Complete acid hydrolysis of starch delivers D-glucose (= dextrose) which is mainly used as such. 1-Methyl-α-D-glucose serves as a polyol in the synthesis of polyurethanes or as a modifying agent in the synthesis of phenol, amino, and alkyd resins.

Partial hydrolysis of starch by acids, heating, or enzymes delivers **dextrins** (**starch gum**; see Section 8.2.6). A more complete hydrolysis by acids or the enzyme α-amylase leads to **glucose syrup** (**starch syrup**), an ca. 80 % aqueous solution of dextrins, maltose, and at least 20 % dextrose that is used in confectionery, tanning, etc. The sweetness increases if part of the dextrose is converted to fructose by the enzyme glucose isomerase, resulting in **high-fructose syrup**. Glucose syrup is also converted to lactic acid, the monomer for polylactide (p. 333), and enzymatically to **ethanol**.

Table 8-4 Amylose content and apparent (at $c = 0.1$ %!) weight-average molecular weights of starches and amyloses and amylopectins thereof in aqueous solution [7b].

| Source | Amylose content in % | Weight-average molecular weight Starch | Amylose | Amylopectin |
|---|---|---|---|---|
| Waxy maize | 0 | 77 000 000 | | 77 000 000 |
| Wheat | 20 | 64 000 000 | 8 500 000 | 78 000 000 |
| Maize (corn) | 22 | 88 000 000 | 2 100 000 | 122 000 000 |
| Potato | 24 | 51 000 000 | 20 000 000 | 61 000 000 |
| Barley | 30 | 49 000 000 | 12 500 000 | 65 000 000 |
| Smooth pea | 42 | 33 500 000 | 5 500 000 | 54 000 000 |
| Wrinkled pea | 63 | 30 500 000 | 2 600 000 | 78 000 000 |
| Amylomaize | 76 | 16 700 000 | 2 500 000 | 69 000 000 |

**Molecular Weights**

Native starch consists of suprastructures that only convert to "molecular" solutions if aqueous starch dispersions are autoclaved at 135-160°C. These solutions consist in part of self-associates of amylose and amylopectin molecules since successive extractions of starch granules delivers molecular weights that are up to 50 times smaller. It is unclear whether amyloses and amylopectins of native starches are molecularly uniform and polymolecularities are produced by association; whether native amyloses and amylopectins do indeed have molecular weight distributions; or whether the observed distributions are artefacts of the isolation of starch from plants.

Observed weight-average molecular weights depend on the origin and age of plants, the processing, and the state of solution. Except for corn starch, molecular weights of amylopectins do not vary much with origin; they are ca. $(70 \pm 15) \cdot 10^6$ (Table 8-4). Amyloses always have lower molecular weights than amylopectins; the average molecular weight of starch thus decreases with increasing amylose content.

## 8.2.3 Amylose

Treating starch with hot water separates "soluble" amylose and "insoluble" amylopectin. Alternatively, amylose can either be extracted from starch with liquid ammonia or obtained from aqueous starch solutions after precipitation of amylopectin by butanol.

Amyloses are presently obtained from starches since no plant produces pure amylose; genetic engineering of plants for the production of pure amylose seems to have been unsuccessful so far. A new biocatalytic process (2005) allows amylose to be synthesized from sucrose, however. Sucrose is first treated with the enzyme sucrose phosphorylase to yield glucose-1-phosphate. The enzyme glucan phosphorylase then converts glucose-1-phosphate to amylose. Ultimately, 1 kg of sucrose yields 450 g of amylose; the direct conversion of glucose (from sucrose hydrolysis) boosts the yield to 900 g.

Amyloses are used by the food industry for instantaneously soluble products, edible sausage casings, puddings, thickeners, etc. In the pharmaceutical industry, they serve for encapsulations; in medicine as binders, sponges, or bandages; in the paper industry as wet strength improvers; and in the textile industry as sizes.

amylose
poly[α-(1→4)-anhydro-D-glucopyranose]

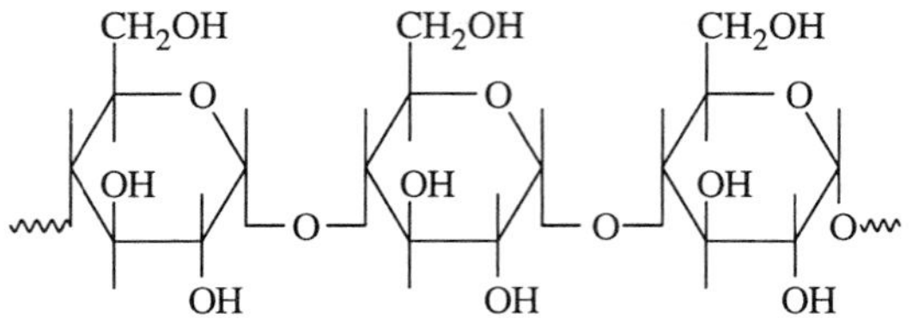

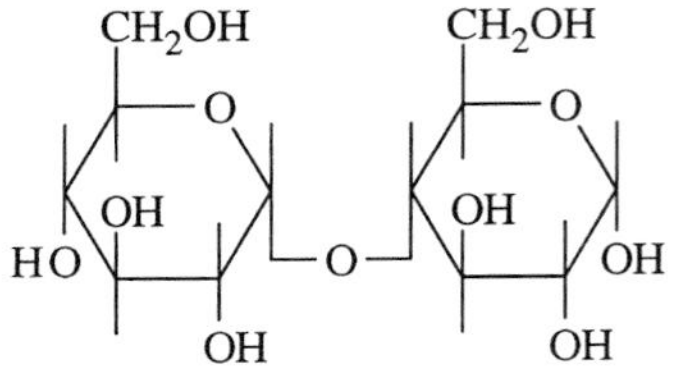

maltose

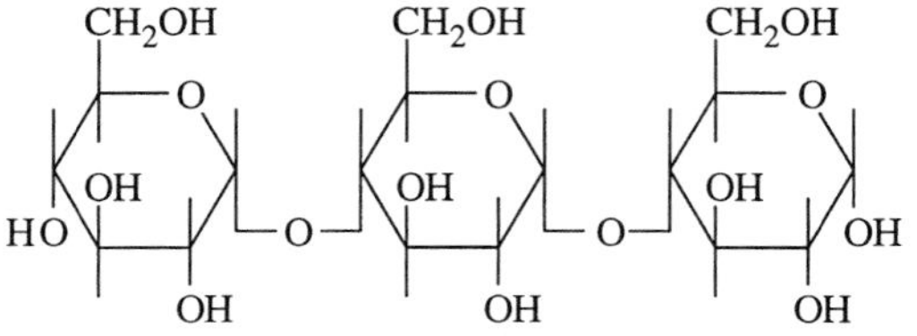

maltotriose

Fig. 8-14 Structure of $\alpha$-(1→4)-D-glucoses.

Acid hydrolysis of amylose delivers almost exclusively D-glucose; amylose is thus a linear poly[$\alpha$-(1→4)-anhydro-D-glucopyranose] (Fig. 8-14). Enzymatic cleavage leads to maltose. Permethylation followed by hydrolysis indicates only minimal branching.

Anomeric OH groups of amylose are axially oriented, which promotes the formation of a helix (Fig. 8-15). Such a helix can be stabilized by the inclusion of iodine, lipids, or phosphatides. Iodine exists in these clathrates as long one-dimensional chains with $I_5^{\ominus}$ groups, which tints the clathrate dark blue. Simple helices can also stabilize themselves by forming double helices through "retrogradation" following "gelatinization."

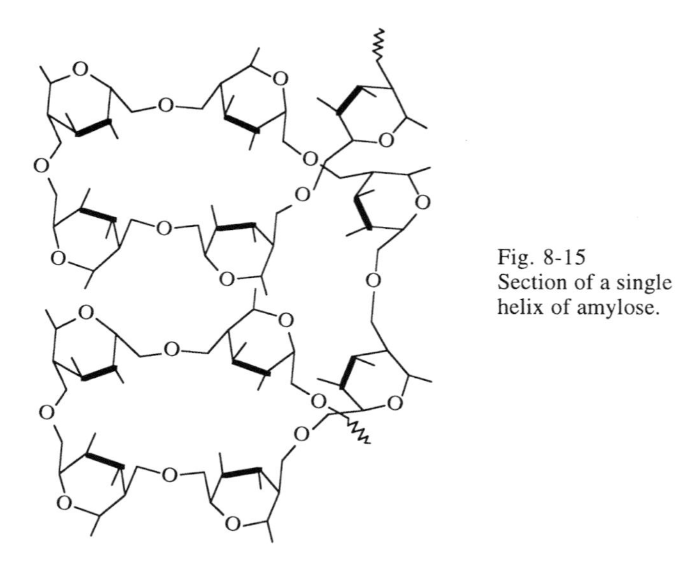

Fig. 8-15
Section of a single helix of amylose.

Helix formation explains the peculiar solution behavior of amylose. In starch, amylose molecules are embedded in a physical network of amylopectin molecules (Fig. 8-13) which prevents the formation of complete amylose helices. Dissolution of starch in hot water detaches amylose molecules from the amylopectin scaffold and the amylose molecules enter the water in their "native" form as random coils composed of disordered segments and short helical sections. In *dilute solution*, amylose molecules try to stabilize themselves by growing short helical segments to longer ones. The resulting more or less perfect helical structures self-associate to double helices which are insoluble in water: thus, amylose crystallizes slowly and becomes insoluble. This **retrogradation** is most pronounced for degrees of polymerization of ca. 80, since longer chains do not crystallize well for kinetic reasons. Dried amylose no longer dissolves in water.

In *concentrated solution*, the formation of longer double helices is impeded by the neighboring segments of other amylose molecules. Shorter double-helical segments are still able to self-associate intermolecularly and intramolecularly with other segments. The resulting partial crystallization generates a physical network and the amylose gelates.

### 8.2.4 Amylopectin

The main chains of amylopectin molecules have the same constitution as amylose chains: D-glucose units connected by α-(1→4) bonds. However, permethylation and subsequent degradation of amylopectin delivers 4–5 % tetramethylglucose compared to ca. 0.3 % tetramethylglucose from amylose. Since tetramethylglucose units can only be present as terminal units and the molecular weights of amylopectins are high (Table 8-4), amylopectin molecules must be branched via 1→6 positions (Fig. 8-16). The degree of polymerization of amylopectin side chains is ca. 18–27.

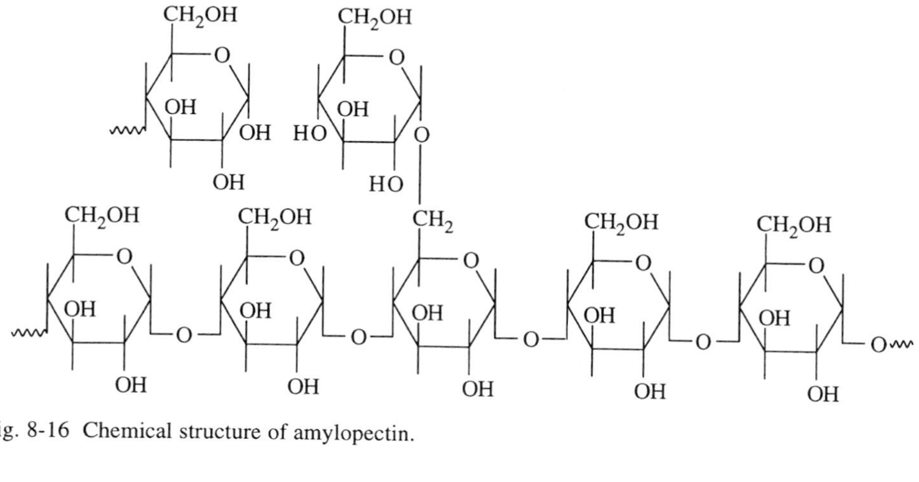

Fig. 8-16 Chemical structure of amylopectin.

For steric reasons, helix formation of amylopectin side chains is severely impeded while that of amylose main chains is almost impossible: amylopectin and amylopectin-rich starches practically do not retrograde. Added amylose molecules can form short double helices with side chains of amylopectin molecules, however, which promotes retrogradation.

Crystalline regions of amylopectin consist of double helices that are not "molten" by hot water: amylopectin does not dissolve in hot water. However, amylopectin can be colloidally dispersed ("dissolved") in cold water by strong stirring. Drying of these dispersions produces amorphous powders that can be dispersed again in cold water.

Heating of aqueous slurries of amylopectin to 160°C under pressure dissolves semi-crystalline regions; after cooling, amorphous amylopectin remains. Under the same conditions, amylose dissolves since it is amorphous in the native state. However, the amylose–water system is thermodynamically unstable, and on cooling, random coils of amylose molecules start to self-organize to helices which are insoluble and cluster together: amylose precipitates and slowly crystallizes. This phenomenon is called **retrogradation** (L = *retrogradus* = a step back; *retro* = backward, *gradus* = a step)

Retrogradation is responsible for the staling of bread, which leads to a coarser texture, a stronger crumbling, and an unpleasant smell. The taste changes because taste buds now face compact helix structures instead of open coils.

At higher temperatures, the helix formation of amylose is relatively fast. Furthermore, chain segments can also form double helices with segments of other molecules, which leads to highly swollen physical networks. At low temperatures, on the other hand, intramolecular helix formation is slow and segments are not very mobile. Consequently, there must be a temperature at which well-ordered helical segments are formed without being incorporated into intermolecular structures. For doughs, this temperature is just above the freezing temperature of water, i.e., at normal refrigerator temperatures. Storing bread in the refrigerator thus promotes staling. It is better to either deep-freeze bread or store it at room temperature.

Staling is retarded if the bread contains gluten, the sticky protein of cereal seeds. Amide groups of these proteins form hydrogen bonds with hydroxyl groups of glucose units, which blocks the helix formation. Whole-grain breads that contain plenty of gluten thus stale more slowly than white breads made from finely ground flours. Staling can be prevented by adding poly(oxyethylene)-20-sorbitanmonostearate to flours. This compound forms hydrogen bonds with OH groups of glucose, which stops helix formation.

Doughs of flour and water or milk are leavened by baking powder, yeast, or sour dough. Yeasts and sour doughs contain enzymes that convert a fraction of the starch to glucose and further to carbon dioxide and ethanol. The latter forms hydrogen bonds with the OH groups of glucose units and prevents helix formation. Baking does not remove all alcohol, but the alcohol slowly diffuses out on storing bread. Alcohol evaporation (and thus helix formation and staling) can be delayed by wrapping breads tightly.

Amylopectin is now produced economically on a large scale by genetically engineered plants which produce starches that are totally devoid of amylose (see Table 8-4).

### 8.2.5 Glycogen

Glycogen is the main reserve polysaccharide of animals, where it is found in the liver (up to 10 %) and in many muscles. It is also present in some yeasts and fungi. In glycogen-containing cells, it is stored as so-called β-particles of 20–40 nm diameter and in the liver as α-particles of 100-200 nm which are rosette-like aggregates of β-particles. It can be extracted from tissue by dimethylsulfoxide or aqueous trichloroacetic acid.

Glycogen consists of α-(1→4)-D-glucose main chains that are much more branched than amylopectin; its branches-upon-branches, make it a spheroidal molecule. Molecular weights range from $1.6 \cdot 10^6$ for muscle glycogen to $300 \cdot 10^6$ for liver glycogen. It is not clear whether these numbers refer to molecules or their associates.

### 8.2.6 Dextrins

Degradation products, $(C_6H_{10}H_5)_i$ $x$ $H_2O$, of starch, amylose, amylopectin, or glycogen are called **dextrins** (L: *dexter* = right; aqueous solutions turn the plane of light to the right). An old term for dextrin is **amylin** but this is also the name of a peptide.

Dextrins are mainly prepared from corn (US) or potatoes (Europe) as white to yellow powders that dissolve easily in water but very sparingly in alcohol. Molecular weights vary between ca. 2000 and 30 000, depending on preparation. Dextrins are used as components of adhesives, binders, sizes, etc. They are subdivided into

- **acid dextrins** (British gums) from the partial hydrolysis of starch by dilute acids;
- **roast dextrins** from heat degradation, for example, baking;
- **cyclodextrins** from the enzymatic degradation by $\beta$-amylase. These dextrins are rich in 1,6-glycosidic bonds which cannot be attacked by the enzyme.
- **Starch gums** are sticky syrups consisting of dextrins and a little water.

**Cyclodextrins** (**Schardinger dextrins**) are unbranched oligomeric cycloamyloses from the degradation of amylose, amylopectin, or glycogen by *Bacillus macerans* or *B. circulans*. Their D-glucose units are interconnected in the $\alpha$-(1→4) position to rings with degrees of polymerization of 6, 7, or 8 that are called $\alpha$-cyclodextrin (= cyclohexaamylose), $\beta$-cyclodextrin (= cycloheptaamylose), and $\gamma$-cyclodextrin (= cyclooctaamylose). Depending on the preparation, these oligomers are present in various proportions.

The physical structure of these rings is predetermined by their chemical structure. In solution and the solid state, cyclodextrins adopt the shape of conical cylinders with a height of 0.8 nm and ca. 0.7 nm-deep "holes" (Fig. 8-17). Diameters are ca. 0.45 nm (cyclohexaamylose), ca. 0.70 nm (cycloheptaamylose), and ca. 0.85 nm (cyclooctaamylose). The interior of the cylinders is lipophilic because of intermerar hydrogen bridges, ~OH···O···HO~. The holes can thus accommodate oil-soluble guest molecules, which makes cyclodextrins useful for applications in cosmetics.

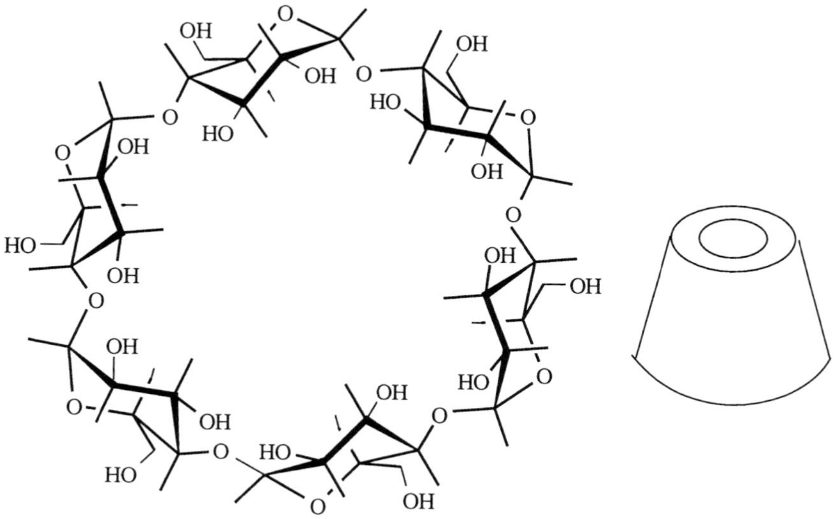

Fig. 8-17 Chemical (left) and physical (right) structure of cyclohexaamylose ($\alpha$-cyclodextrin).

### 8.2.7 Dextrans

Dextrans are poly[α-(1→6)-D-glucose]s that are branched via α-(1→3) linkages (Fig. 8-18). Their formation was first observed in sugar processing plants as annoying slimes which were produced by bacteria and prevented the crystallization of sugar.

Dextran is produced extracellularly from saccharose by the enzyme dextran saccharase. The propagation step consists of the elimination of fructose from a primarily formed enzyme-dextran-glucose-saccharose complex (see Volume I, p. 556). The reaction does not need a primer as does amylose synthesis. Molecular weights are already high at low saccharose conversions. Some acceptors increase polymerization rates (glucose, maltose, isomaltose), while others lower it (fructose, glycerol, saccharose). Simultaneously, low molecular weights are formed.

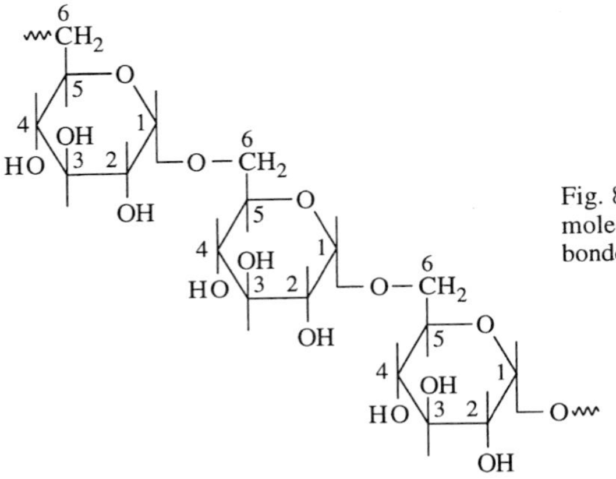

Fig. 8-18 Main chain of dextran molecules consisting of α-(1→6)-bonded D-glucose units

Practically all dextrans of American and European commerce are produced by dextran saccharase of the bacterium *Leuconostoc mesenteroides* NRRL B-512(F), either in the presence or absence of bacteria lcells. This dextran contains 95 % α-(1→6) bonds and 5 % α-(1→3) bonds. Some 80 % of the branches are only one glucose unit long, and the remaining 20 % are long-chain branches. Number average molecular weights of native dextrans are ca. 200 000. *Apparent* weight-average molecular weights can reach more than 500 million because of self-association in aqueous solutions.

Although Japan and Russia use other bacterial strains, resulting dextrans have apparently the same structures as those from B-512. *Leuconostoc mesenteroides* NRRL B-742, on the other hand, leads to a more strongly branched dextran with 50 % α-(1→6) bonds and 50 % α-(1→3) bonds. Different strains of *L. mesenteroides* and *L. dextranicum* produce up to 50 % α-(1→3), up to 35 % α-(1→2), and up to 13 % α-(1→4). *Streptococcus muteus* even makes 95 % α-(1→3) and only 5 % α-(1→6) bonds.

High-molecular weight dextrans from these polymerizations are precipitated from their aqueous solutions by methanol or acetone, and purified. For use as blood plasma expanders, their molecular weight is reduced by acid hydrolysis and adjusted to molecular weights of 40 000 to 60 000 by subsequent precipitation fractionation. As the kidney threshold is ca. 60 000; dextrans with $M$ > 60 000 cannot be processed by the kidneys but are instead stored in the body.

Crosslinked dextrans serve as columns in gel chromatography. Derivatized dextrans are known and used in great numbers. Dextran sulfates serve as blood anticoagulants and for the treatment of stomach ulcers. Diethylaminoethyl dextrans are used to treat foot-and-mouth disease; cationic dextrans are employed in paper production, etc.

### 8.2.8 Poly(dextrose)

Melt polycondensation of a mixture of 89 wt% dextrose (= D-glucose), 10 wt% sorbitol (= D-glucitol), and 1 wt% citric acid produces a randomly branched poly(dextrose) with mainly α-(1→6) bonds and molecular weights up to 22 000. The polymer is used as a water-soluble filler and texturizing agent in reduced-energy foodstuffs.

### 8.2.9 Poly[α–(1→3)-D-glucose]

Poly[α-(1→3)-D-glucopyranose] (pseudonigeran) has a band-like macroconformation that resembles that of cellulose (Section 8.3). It is synthesized by fungi and not by higher plants (as cellulose), and therefore should be more widespread. However, it is not very common since for selectivity reasons, its synthesis seems to be restricted to fungi, streptococci, and other lower organisms.

The chains of solid pseudonigeran run antiparallel, which is much more difficult to achieve by organisms than aligning them parallel by simultaneous biosynthesis and crystallization, as in native cellulose. This may also be the reason why some lower organisms connect D-glucose units alternatingly, for example, α-(1→3)-*alt*-α-(1→4) in nigeran and α-(1→3)-*alt*-α-(1→6) in pneumococcus polysaccharide (see Table 8-2).

### 8.2.10 Pullulan

Treatment of starch by the yeast-like fungus *Pullularia pullulans* (now *Aureobasidium pullulans*) generates linear polymers consisting primarily of α-(1→6)-connected D-maltotriose units (p. 382; Fig. 8-19). In the language of synthetic macromolecular chemistry, this "pullulan" is a homopolymer with respect to its monomeric unit, D-glucose, but a periodic copolymer with respect to the interconnections of its units, i.e., $\{[(\alpha\text{-}(1\rightarrow4))\text{-}per\text{-}(\alpha\text{-}(1\rightarrow4)\text{-}per\text{-}(\alpha\text{-}(1\rightarrow6)]\}_n$. Some maltotriose units are also connected in the α-(1→3) position. The polymer also contains maltotetrose units.

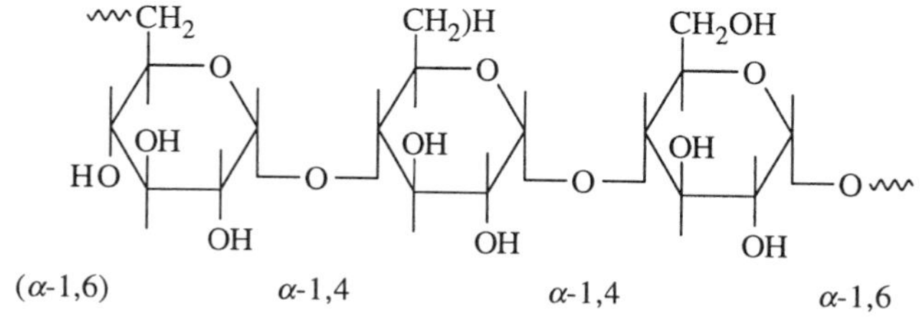

(α-1,6) α-1,4 α-1,4 α-1,6

Fig. 8-19 Repeating unit of pullulan.

Pullulans with molecular weights between 10 000 and 400 000 are very soluble in water. The addition of a little water to pullulan generates a dough that can be molded into biodegradable sheets, films, and articles. Packaging films are edible, transparent, and not very permeable to oxygen. Pullulan is also used to coat and laminate plywood.

# 8.3 Cellulose

## 8.3.1 Chemical Structure

The word "cellulose" has different meanings in different branches of science. In 1847, the botanist Payen called the main components of cell walls of plants "celluloses." Botanists continue to use the word cellulose in this sense, regardless of whether the plant is a flowering plant, a fern, or an algae (see Section 8.7!). The "celluloses" of fiber technologists are materials that can be obtained from a small group of higher plants by certain basic processes. To crystallographers, celluloses are crystalline materials with certain defined unit cells. The "cellulose" of chemists, on the other hand, is a high-molecular weight substance composed of D-glucose units that are interconnected in the β-(1→4) position, i.e., a poly[β-(1→4)-D-glucopyranose] with cellobiose as the repeating unit (Fig. 8-20). Cellulose is the most important β-anomer of poly(glucose)s.

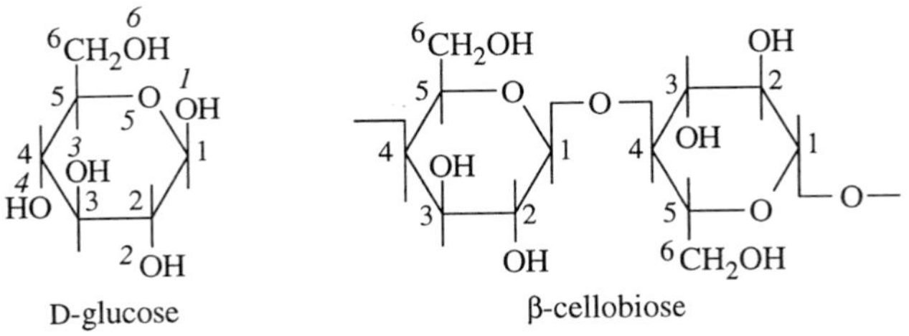

D-glucose β-cellobiose

Fig. 8-20 Left: D-glucose with position numbers of carbon atoms (upright) and oxygen atoms (*italics*). Right: β-cellobiose, the repeating unit of cellulose.

However, only a few native celluloses are pure poly[β-(1→4)-anhydro-D-glucopyranose]s, for example, the so-called α-celluloses of the algae *Valonia* and *Cladophora*. Most other celluloses contain different proportions of other sugar units; an example is cotton with 1.5 % xylose and smaller proportions of mannose, galactose, and arabinose units.

On the other hand, α-cellulose of the red alge *Rhodumenia palmata* consists of 50 % glucose and 50 % xylose units, but delivers the same X-ray diagram as the α-cellulose from *Valonia*. The α-cellulose from *R. palmata* must therefore consist of a crystalline core of poly[β-(1→4)-anhydro-D-glucopyranose] that is surrounded by a non-crystalline hull of xylose units. Native celluloses furthermore always contain COOH groups: ca. 1 COOH per 500–1000 glucose units in cotton and 1 COOH per 100–3000 glucose units in chemical woodpulp (p. 72).

The terms α-, β-, and γ-cellulose have different meanings in different fields. α sometimes refers to a cellulose of highest purity, β to a less pure one, etc., for example, the

Table 8-5 Number-average degree of polymerization, $\overline{X}_n$, of various celluloses.

| Source | $\overline{X}_n$ | Source | $\overline{X}_n$ |
|---|---|---|---|
| *Valonia* (algae) | < 44 000 | Mechanical wood pulp | < 10 000 |
| Cotton, anaerobically harvested | < 18 000 | Chemical wood pulp | < 2 000 |
| Cotton, conventionally harvested | 7 000 | Viscose yarn | < 450 |
| Cotton, cleaned | < 1 500 | Acetate yarn | 250 |
| Cotton, linters | 6 500 | Cellophane® | 300 |
| Flax | 8 000 | Bacterial cellulose | 2 700 |
| Ramie | 6 500 | Acetobacter cellulose | 600 |

cellulose from *Valonia*. In botany, α-celluloses are microfibrils of 10–30 nm diameter that are obtained from cell walls after removal of non-cellulosic components by boiling in water, chlorination, and subsequent treatment with liquid caustic potash (= aqueous solution of KOH). In the pulping business, α-cellulose is the high-molecular weight fraction ($X > 200$) that is insoluble in 17.5 % NaOH or 24 % KOH in water; β-cellulose the fraction that is soluble in 17.5 % aqueous NaOH but precipitates on addition of methanol; and γ-cellulose the remaining soluble fraction. The β- and γ-celluloses of wood pulps are partially oxidized celluloses of low degrees of polymerization.

Native celluloses have high molecular weights and narrow molecular weight distributions (Table 8-5). Cotton cellulose has a degree of polymerization of $\overline{X}_n \approx 18\,000$ if seed pods are opened under nitrogen in the dark. Processing lowers $\overline{X}_n$, whether by conventional harvesting of fibers, pulping of wood (p. 71 ff.), or viscose preparation (Section 8.3.7). On hydrolysis of cotton cellulose, the number-average degrees of polymerization first approach a plateau at $\overline{X}_n \approx 360 \pm 40$ before decreasing further. The plateau value indicates the presence of a substructure, since 360 just amounts to 2 % of the native degree of polymerization of 18 000 and cotton contains ca. 2 % of sugar units that are more easy to hydrolyze than glycosidic bonds, i.e., "weak links."

Strong hydrolytic degradation of cellulose leads to so-called **hydrocelluloses** or **microcrystalline celluloses** with degrees of polymerization of 100–200. Rapid agitation of aqueous dispersions of these celluloses delivers creamy masses that are used as non-digestible thickeners by the food industry. The creamy consistency is caused by the physical crosslinking of cellulose crystallites.

### 8.3.2 Physical Structure

Cellulose exists in various crystalline modifications (Fig. 8-21) that differ in the relative direction of cellulose chains, lattice constants $a$, $b$, and $c$, and unit cell angles $\alpha$, $\beta$, and $\gamma$ (Table 8-6). In modification I, cellulose chains are extended and parallel to each other (Fig. 8-22) as shown by X-ray analysis of oligomers and the cutting of ramie (from China grass). In the latter experiments, parallel elementary fibrils were cut perpendicular to the fiber axis by a microtome in distances that corresponded to the known degree of polymerization. On average, each cellulose chain was cut just once. The number average degree of polymerization dropped to one-half of its original value, which is to be expected for straight chains, regardless of the arrangement of chain segments.

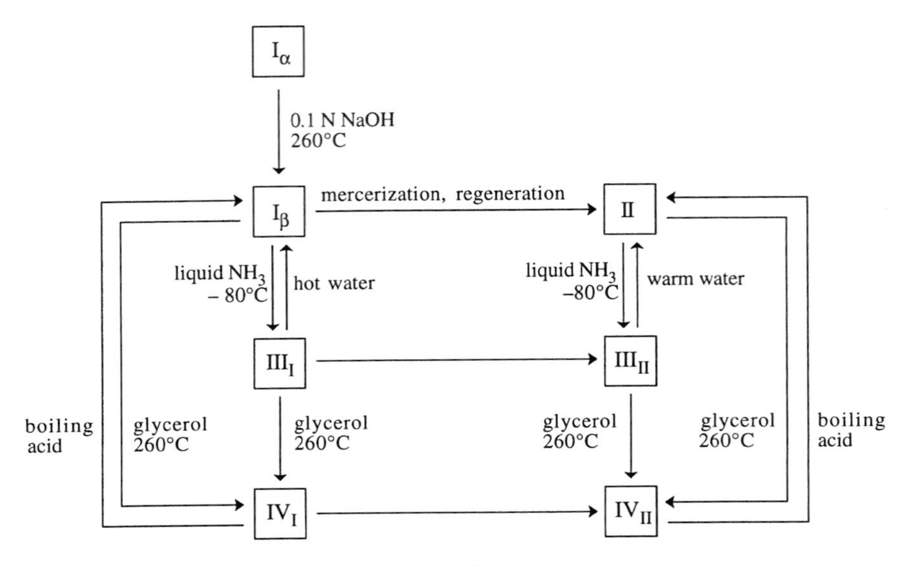


Fig. 8-21 Transformation of various cellulose modifications.

For folded chains, one would expect that some chains are hit very often near their chain folds, but others not at all if the fold lengths are smaller than the contour lengths of the chains. In this case, original molecular weight distributions would not change much on cutting, but a small secondary distribution at low molecular weights should appear. For straight (extended) chains, on the other hand, molecular weight distributions are shifted to smaller values of molecular weights as found experimentally: hence, cellulose chains of ramie cellulose are extended and not chain-folded.

Cellulose I is not physically homogeneous but consists of two crystalline allomorphs that also differ in their infrared spectra. According to X-ray and electron diffraction, cellulose $I_\alpha$ is triclinic whereas cellulose $I_\beta$ is monoclinic. $I_\alpha$ is metastable; it is easily converted to $I_\beta$.

Table 8-6 Crystal modifications of cellulose (see text for cellulose I).

| Modification | Origin | | Lattice constants | | | Angle |
|---|---|---|---|---|---|---|
| | natural | synthetic | $a$/nm | $b$/nm | $c$/nm | $\gamma$/ ° |
| Cellulose $I_\alpha$ | *Cladosphora sp.* | - | 0.674 | 0.593 | 1.036 | 117 |
| Cellulose $I_\beta$ (native c.) | ramie, cotton | from III with water under pressure | 0.801 | 0.817 | 1.036 | 97.3 |
| Cellulose II (hydrate c.; regenerated c.) | *Helicystis* algae | dissolution and re-precipitation of I; mercerized fibers | 0.801 | 0.899 | 1.036 | 116.6 |
| Cellulose III (ammonia c.) | - | controlled decomposition of ammonia cellulose (from II with $NH_3$) | 0.774 | 1.03 | 0.99 | 122 |
| Cellulose IV (high temp. c.) | coltsfoot | heating of III in glycerol to 260°C | 0.811 | 1.03 | 0.791 | 90 |

Celluloses from algae and bacteria are rich in $\mathbf{I}_\alpha$, for example, 76 % in *Cladosphora sp.* and 64 % in *Valonia ventricoa.* $I_\beta$ is predominantly found in ramie, cotton, and woody plants, and to 100 % as tunicine, the structural cellulose of tunicates (chordate marine animals with a tough outer covering (tunic)).

Physical structures of celluloses $I_\beta$ and II are very well established by fiber X-ray diffraction data, computer modeling and energy minimizing, and, for cellulose II, also by comparison with synchrotron diffraction data of the hemihydrate of *β*-D-cellotetrose.

In **cellulose** $\mathbf{I}_\beta$, successive chains are arranged parallel to the *c* axis and *parallel* to each other. The second, fourth, sixth, ... *β*-D-glucose unit is turned 180° to the preceding one (Fig. 8-22). Each chain is stabilized by intramolecular, intermerar hydrogen bridges O*3*···O*5* and O*6*···O*2* between one glucose unit and the next.

In *a–c* planes, the chains are interconnected by intermolecular hydrogen bridges between the O*3* of one chain and the neighboring O*6* of the next chain. This results in layers of cellulose molecules that are separated from each other in the *b* direction by ca. 0.4 nm (not shown in Fig. 8-22). Layers are *not* interconnected by hydrogen bonds.

The situation is quite different for **cellulose II**. Chains here are also parallel to the *c* axis, but alternating chains are *antiparallel* to each other, resulting in an ...ABABAB... structure. In both types of chains, C6–O6 bonds have the same trans–gauche orientation according to data from cellotetrose hemihydrate. However, torsional angles for the orientation of glycosidic bonds differ by 11°.

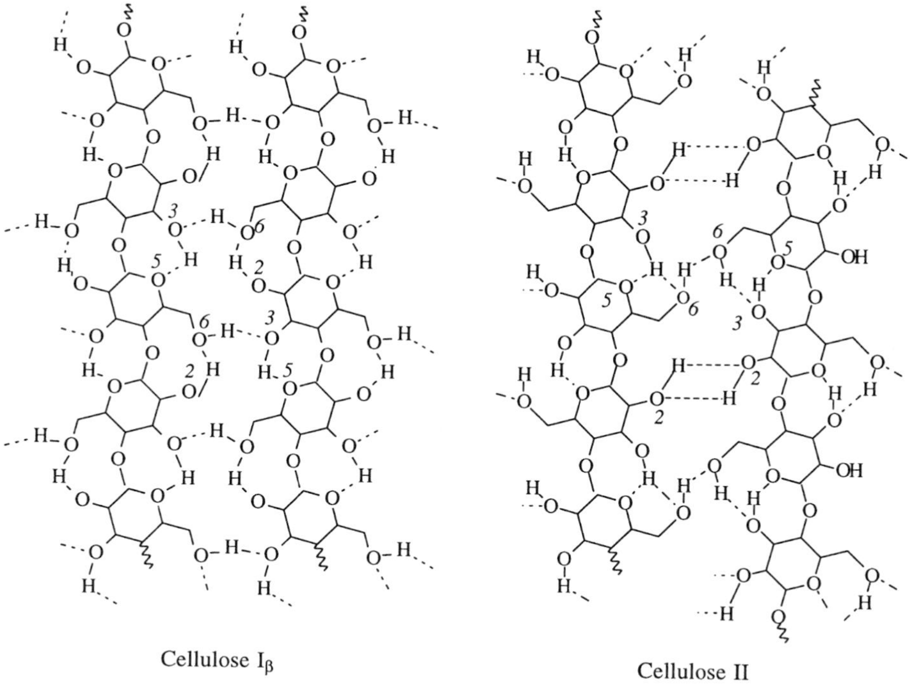

Cellulose $I_\beta$

Cellulose II

Fig. 8-22 Sections of the structures of cellulose $I_\beta$ (native cellulose, [8]) and cellulose II (cellotetrose as model, [9]). Chains run in the *c* direction ↕ and are parallel to each other in the *a* direction ↔ . Hydrogen bonds are indicated by - - -. *Numbers* refer to position numbers of *oxygen* atoms (not carbon atoms (cf. Fig. 8-20)).

These probably most reliable data for cellotetrose contradict older proposals for cellulose II which suggested A chains with trans–gauche orientation of $C^6$-$O^6$ bonds and B chains with gauche–trans orientation of these bonds. The newer measurements indicate three-center hydrogen bonds with $O^3H$ as donor and $O^5$ and $O^6$ of the next glucose residue as acceptors. Intermolecular hydrogen bridges exist *within* a layer between $O^2A$ and $O^2B$ and between $O^6A$ and $O^6B$ (Fig. 8-22) and, different from cellulose $I_β$, also *between* layers at $O^2A \cdots O^6A$ and $O^2B \cdots O^6B$, respectively (not shown in Fig. 8-22).

**Celluloses III** and **IV** have mixed structures. Their Raman spectra indicate linear combinations of the spectra of celluloses I and II.

### 8.3.3 Physical Properties

Celluloses from different sources have different crystallinities. According to X-ray measurements, cotton is ca. 70 % crystalline and rayon ca. 40 %.

Different methods measure different degrees of order and thus different crystallinities, however. For cotton, acid hydrolysis indicates 85 % crystallinity, X-ray 70 %, and density or deuterium exchange ca. 60 %. Melting temperatures must be higher than 400°C since cellulose self-ignites at ca. 400°C.

Intramolecular and intermolecular hydrogen bonds are responsible for the high glass temperature of dry cellulose, $T_G$ = 220–245°C. Water severs these bonds and the glass temperature drops to ca. 20°C: thus, clammy cotton and rayon shirts crease easily and moist fabrics can be nicely ironed.

High crystallinity and many hydrogen bonds prevent the dissolution of celluloses in water. Cellulose does dissolve in many aqueous solutions of metal complexes such as $[Cu(NH_3)_4]^{2+}$ (cuoxam), $[Cu(NH_2CH_2CH_2NH_2)_2]^{2+}$ (cuen), or the 3:1 complex of $[(C_4H_3O_6)Fe]_3Na$ and HOOC–CHOH–CHOH–COOH. Cuoxam solutions, for example, can be strongly diluted by water before cellulose precipitates. Cellulose also dissolves in highly concentrated aqueous solutions of diethyldibenzylammonium hydroxide.

Cellulose is soluble in chloral (forms hemiacetals), hydrazine, methylamine + dimethylsulfoxide, and many ionic liquids such as 3-methylimidazolium chloride. The most important organic solvent is *N*-methylmorpholine-*N*-oxide (Section 8.3.7).

### 8.3.4 Plant Fibers

#### Occurrence

Plant fibers, sometimes called **vegetable fibers**, are fibers rich in cellulosic components. They are usually divided into seed, bast, leaf, and miscellaneous fibers.

The mechanical separation of cellulose fibers from seed hairs, stems, and leaves is a thousands-of-years-old technology that is still practiced today. Far younger, only ca. 150 years old, are non-mechanical separation processes that allow one to obtain celluloses from hard and soft woods and from stems of annual plants. Many of these processes do not deliver pure celluloses since the proportions of pectins, lignins, fats, waxes, and hemicelluloses (which are not celluloses (Section 8.9)) can be considerable (Table 8-7). Some of these materials are removed by further processing but some are not.

Table 8-7 Composition of some dry natural cellulose fibers and woods.

| Fiber | Composition in wt% of | | | | |
|---|---|---|---|---|---|
| | Cellulose | Hemicelluloses | Pectins | Lignins | Extract |
| Cotton | 92.9 | 2.6 | 0.9 | 2.6 | 0.4 |
| Ramie | 76.2 | 14.6 | 0.7 | 2.1 | 6.4 |
| Hemp | 74.4 | 17.9 | 0.9 | 3.7 | 3.1 |
| Sisal | 73.1 | 13.3 | 11.0 | 0.9 | 1.6 |
| Jute | 71.5 | 13.3 | 11.0 | 0.9 | 1.6 |
| Flax | 71.2 | 18.6 | 2.2 | 2.0 | 6.0 |
| Hard woods | ≈ 45 | ≈ 30 | | ≈ 20 | ≈ 5 |
| Soft woods | ≈ 42 | ≈ 27 | | ≈ 28 | ≈ 3 |
| Bagasse | ≈ 40 | ≈ 30 | ≈ 20 | | ≈ 10 |

**Cotton** (Arabic: *qutn*) is the seed hair of the fruits (bolls) of a 1–2.5 m high, bushy, perennial subtropical plant (genus *Malva*, order *Gossypium*) that is grown as an annual for commercial purposes. Early cotton cultures existed in India (ca. 1500 B.C.) and Peru (ca. 500 B.C.). The largest producers of cotton (2001) are China (24 %), United States (21 %), India (13 %), and Pakistan (8.6 %) (Volume IV).

Cotton plants need 3–4 months from planting to flowering and, after a flowering time of only 10 hours, another 2–3 months for ripening. Bolls are picked by hand or with machines; they consist of the seed (ca. 2/3 by weight) and seed hairs (ca. 1/3 by weight). Seeds and seed hairs are separated by saw gins (upland cotton) or roller gins (extra-long staple) ("gin" is an abbreviation of "engine"); these are rolls with many saw blades that reach through grids into the chamber with the seeds. On the other side of the grid is a brush roll that removes the seed hairs from the brush roll. The seed hairs are blown away, collected, and pressed into bales. Bale sizes are standardized albeit differently (ISO, US); bale weights range from 85 kg (China Bale I) to 327 kg (Egypt). The typical weight of a US bale is 225 kg. Seed hairs (**lints**) are used for fiber spinning. Egyptian cotton fibers from *G. barbadense* are 33–36 mm long (ELS = extra-long staple), US cotton from *G. hirsutum,* 26–30 mm. The world production is ca. $23 \cdot 10^6$ t/a (2004).

The seeds are shorn again by a kind of electric razor to deliver **linters** which are shorter, more coarse, and often colored fibers with hardly any lumen. The longer first-cut linters (< 15 mm) are used for mattresses and upholstery, the shorter second-cut linters (2–3 mm) as chemical feedstock for the production of rayon, nitrocellulose, cellulose esters, specialty papers, filters, fillers, etc.

The remaining seeds are pressed to give cotton seed oil (ca. 15–20 % of kernels). The remaining cake is used as animal fodder.

**Flax** is not a seed hair but a fiber from the bast layer of the stems of the flax plant *Linum usitatissimum* that is mainly planted in Northern countries but also in China, Egypt, Turkey, and Japan. The plants are uprooted shortly before the seeds ripen, dried in the field for 1–2 days, and then bundled and dried in ventilated piles for another 10–14 days. Leaves and branches are removed mechanically and the seeds by threshing. The seeds are pressed to deliver linseed oil.

Fibers are extracted from the stems by natural or chemical retting. Natural retting was formerly done by leaving stems for 3–5 weeks in the field. The best commercial natural retting method uses cascade tanks with circulating luke-warm water for 3–14 days; the water is renewed every 2 days. Retting removes lignins oxidatively, pectins by bacterial action, sugars and minerals by leaching, etc. For chemical degumming, the straw is treated with either very dilute sulfuric acid or a dilute aqueous emulsion of naphtha.

In order to remove woody parts, the dried stalks are smashed by fluted rollers and scutched clean by rotating bladed wheels. During this process, some flax fibers are broken. The resulting "scutched tow", which includes pieces of stalk and shive, is used for ropes and coarse fabrics.

The remaining long fiber bundles are fed to a kind of combing machine which separates the individual fibers called **line**. Thus, 100 kg of flax plants deliver 12–16 kg of line fibers which are spun to **linen**.

**Hemp** (*Cannabis sativa var indica*) grows in temperate climates; its structure, harvesting, and work-up is similar to that of flax. Its fibers have high strength and have therefore always been used for ropes. It is now utilized as a reinforcing fiber in plastics for insulating materials. For fabrics, it has been mainly replaced by jute and for ropes by sisal hemp or manila hemp (both *not* hemps!) or nylon, respectively. In the United States, its cultivation is forbidden since it delivers marijuana from its dried flower clusters and leaves and hashish from the resin of its flower sprouts.

**Jute** (Sanskrit: *juta* = plait, braid) is the name of plants of the genus *Corchorus* and of fibers from their stems. Fibers are rich in lignin and are used for mats, bags, and as a reinforcing agent for linoleum. The water-loving jute is cultivated along the great rivers in the tropics and subtropics, e.g., Ganges, Amazon, Yangtze, and Irrawaddy.

**Kenaf** (genus *Hibiscus*) delivers jute-like bast fibers but needs less water for cultivation. It is grown in the United States, Europe, Mexico, South America, China, and Japan.

**Sunn** is a plant (*Crotalaria juncea*) as well as a fiber (**Bombay hemp**, **Madras hemp**) that is used for cordage. It is not a hemp.

**Abaca** (*Musa textilis*) is a hard leaf fiber from East India that is related to the banana plant and provides a leaf fiber also called **Manila hemp**. The plant does not grow near Manila and does not deliver a soft fiber-like hemp.

**Banana fiber** comes from the stems of banana plants (*M. cavendishi*, *M. sapientum*).

**Sisal hemp** and **New Zealand hemp** are also not hemps but leaf fibers, the former from *Agave sisalava* (Mexico, India) and the latter from swamp lily (*Phormium texax)*.

**Henequen** is a coarse reddish fiber from the leaves of *Agave fourcroydes*.

**Coir** (Malayalam: *kayar* = cord) is the fiber from the husk of the coconut.

**Ramie** (**Rhea, China grass**) is a broad-leaved woody Asian plant with two subspecies: white ramie (*Boehmeria nivea*) and green ramie (*B. tenacissima*). Raw fibers contain 25–35 % xylans and arabinans which are removed by cooking the fiber with alkali. The plant is cultivated in China, Japan, Thailand, India, Malaysia, Mexico, and Florida. Ramie fibers are thicker than those of flax. They are used for clothing (usually combined with cotton or viscose) and for tear-proof paper (Japanese paper).

**Esparto grass** grows in North Africa. It is used for fine papers and mats.

**Kapok** is the silky air-filled fiber from the fruit of the silk-cotton tree (*Ceiba pentandra*). It cannot be spun to fibers but is used as a down-like filler for upholstery and lifebelts as well as for thermal and acoustical insulation.

**Bagasse** (L: *bacca* = berry; UK: **megass**) is the residue of sugar cane (world production of bagasse: ca. $25 \cdot 10^6$ t/a). Most bagasse is burned for energy, but a part is converted to cellulose for paper and cardboard or to chemicals (furfural).

**Structure of Natural Fibers**

The terms "fiber", "microfiber", "filament", "fibril", "microfibril", "strand", "thread", etc., have different meanings in common language, botany, food science, and industry, especially in the textile industry in general and in the synthetic fiber industry in particular. The "fiber" of the food industry is a misnomer for food that is not digested in the stomach but by bacteria in the intestines ("soluble fiber") or not ("insoluble fiber").

In everyday language, **fiber** (L: *fibra*) is a thin elongated structure whose length is much greater than its diameter, regardless of its length or internal substructure. **Synthetic fibers** leave the spinneret as "endless" fibers with no supermolecular structures (crystalline structures are not considered superstructures for the purpose of this discussion). Such endless structures are now called **filaments** in the synthetic fiber industry, which distinguishes **monofilaments** from spinning heads with one hole from **multifilaments** produced by spin heads with several holes. Filaments cut to shorter lengths are called **staple fibers**. Synthetic **microfibers** are filaments of small diameter (see Volume IV).

"Fiber" as used in commerce for **natural fibers** refers to the elongated structures that emerge from a living organism such as silk threads (4000 m long; see Section 11.4.5), flax fibers (40–150 cm), wool fibers (3–35 cm), or cotton fibers (1.5–6 cm). These fibers always have substructures or hierarchies thereof which carry different names.

The long **plant fibers** of commerce (ramie, flax, etc.; Table 8-8) consist of bundles of interlocking structural elements that are usually called **ultimate cells** (or **ultimates**) by botanists. These "ultimate fibers" are the "fibers" of the non-botanical scientific literature. The bundling, structure, and composition of ultimate fibers control the use properties of fibers; cotton is an example.

Cotton ultimates have a fairly complicated physical structure (Fig. 8-23). They consist of bundles of ultimate fibers with cylindrical to oval and smooth to irregular cross sections (Table 8-8). The ultimates are surrounded by the cuticle which is a chemically resistant outer layer composed of lignin, pectin, fat, and waxes. Below the cuticle is the net-like primary cell wall that contains only 8 % cellulose and a little pectin but a large proportion of hemicelluloses. The primary cell wall is separated from the secondary cell wall by a pectinic middle lamella called winding that acts like a glue. The secondary cell wall consists of highly oriented macrofibrils that are arranged in several layers with opposite spiral directions which provides the fiber with the highest possible strength. The center of the fiber is hollow (L: *lumen*: light, eye, opening) (see also the photograph in Volume IV) which allows cotton fibers to take up large quantities of water.

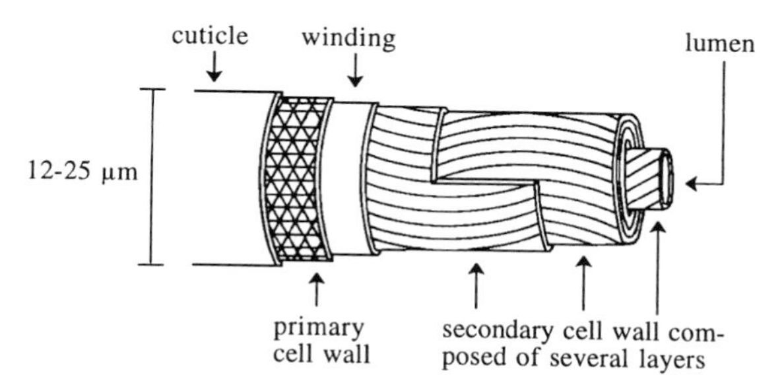


Fig. 8-23 Structure of an ultimate fiber of cotton (schematic). The secondary cell wall consists of several layers (only three shown) of macrofibrils with different orientation of fibril axes. Line fiber, jute, ramie, etc., are similarly structured but not identical (different bundling, cross-sections, etc.).

Table 8-8 Lengths $L_F$ of commercial fibers and $L_{UC}$ of ultimate cells (ultimate fibers, ultimates) and diameters $d_F$ of fibers and $d_{UC}$ of ultimate cells. Amended data of [10]. a) *Valonia*.

| Fiber | $L_F$ / mm | $L_{UC}$ / mm | $d_F$ / mm | $d_{UC}$ / mm | Cross-section of ultimate cell |
|---|---|---|---|---|---|
| Ramie | 400 - 1500 | 40 - 250 | 0.06 - 0.90 | 0.016 - 0.126 | hexagonal / oval |
| Flax | 400 - 1500 | 4 - 77 | 0.04 - 0.62 | 0.005 - 0.076 | polygonal |
| Sisal | 400 - 1000 | 0.8 - 8 | 0.1 - 0.46 | 0.007 - 0.047 | cylindrical |
| Coir | | 0.3 - 1 | 0.1 - 0.45 | 0.012 - 0.024 | |
| Abaca | 600 - 2000 | 3 - 12 | 0.01 - 0.28 | 0.006 - 0.046 | round / oval |
| Jute | 1500 - 3600 | 0.8 - 6 | 0.03 - 0.14 | 0.015 - 0.025 | polygonal / oval |
| Kenaf | | 1.5 - 11 | | 0.012 - 0.036 | cylindrical |
| Bagasse | | 1 - 4 | 0.01 - 0.04 | | |
| Banana | 800 - 2800 | 0.9 - 5.5 | 0.011 - 0.034 | 0.018 - 0.030 | cylindrical |
| Hemp | 1000 - 3000 | 5 - 55 | 0.015 - 0.028 | 0.010 - 0.051 | polygonal |
| Sunn | | 2 - 14 | | 0.008 - 0.061 | irregular |
| Cotton | 15 - 56 | 15 - 56 | 0.012 - 0.025 | 0.012 - 0.025 | round / oval |
| Algae cellulose a) | | | 0.010 - 0.020 | | |
| Microbial cellulose | | | 0.007 | | |

The various plant fibers differ in lengths and diameters as well as in the bundling, packing, lengths, and diameters of their ultimate fibers (Table 8-8). They also vary in the chemical make-up of the cuticle, the primary cell wall, the interstitial matter between ultimate cells, etc., i.e., the chemical composition of the plant fiber (Table 8-7).

Ultimate cotton fibers of 12–25 μm diameter (Table 8-8) consist of macrofibrils of 0.06–0.90 μm diameter. Macrofibrils contain microfibrils of 0.01–0.05 μm diameter and 30–80 μm length which in turn consist of elementary fibrils of 0.0035 μm diameter and 10–20 μm length. Lignin serves as a kind of cement between macrofibrils and microfibrils.

Between elementary fibrils are smaller gaps of ca. 1 nm width that can accommodate water, iodine, or zinc chloride but not dyestuffs. Interdispersed are also vacuoles (intermicellar spaces). For these reasons, macroscopic densities of cellulose fibers are smaller than densities by X-ray measurements (1.50–1.55 g/cm$^3$ versus 1.59 g/cm$^3$).

Cellulosic fibers are dominated by their highly ordered, cellulose-rich secondary cell walls which provide them with the stiffness that makes the fibers useful for textiles and other purposes. Cotton fibers (= ultimates) can be used directly for spinning, whereas longer plant fibers (ramie, hemp, etc.) first have to be broken up to smaller fragments which are the equivalents of the staple fibers of the cotton, wool, and chemical fiber industries.

Fibers with small diameters such as cotton ($d_F$ = 0.012–0.025 mm) serve for textiles, whereas those with larger fiber diameters are preferred for ropes and cables, for example, sisal ($d_F$ = 0.1–0.46 mm). Ramie can be relatively thick ($d_F$ = 0.06–0.90 mm) and is therefore sometimes used to stiffen cotton fabrics.

Reorientation of the fiber structure delivers additional useful products. Examples are mercerized cotton, papers, parchment paper, and vulcanized fiber (all Volume IV). Other properties are exhibited by regenerated celluloses (Section 8.3.7). Cellulose also serves for a number of derivatives such as cellulose ethers and cellulose esters (Section 8.3.9).

### 8.3.5 Microbial Cellulose

Until recently, celluloses were only obtained from plants. A new industrial process produces microbial celluloses by the action of *Acetobacter xylinum*, a Gram-negative, aerobic, rod-like bacterium on glucose, saccharose, molasses, or other carbon sources. The same bacterium is used in the Philippines to obtain *Nata de Coco*, a fermented food.

Microbial cellulose (bacterial cellulose, Cellulon®) is produced in 180 000 L fermenting reactors from glucose, starch syrup, and various salts by a special strain of *A. xylinum* which has been engineered by mutation and selection. Wild strains cannot be used because they stop synthesizing cellulose if suspensions are stirred too vigorously while oxygen is introduced. The engineered strain also yields fewer byproducts.

*A. xylinum* bacteria produce cellulose chains in particles of ca. 12 nm diameter that contain a multi-enzyme complex. After extrusion from the particles, chains associate via hydrogen bonds to elementary fibrils that contain ca. 63 % $I_\alpha$ cellulose. Since synthesis-extrusion compartments are arranged regularly on cell surfaces, elementary fibrils cluster to microfibrils that organize further to macrofibrils consisting of flat ribbons.

The synthesis is stopped at 5–6 % solids, bacteria are killed by hot lye, and cellulose is continually filtered off and drained. Yields are ca. 0.2 g of cellulose per 1 g of glucose.

Microbial celluloses have very fine ultimate fibers of ca. 0.1 μm diameter whereas those from coniferous woods have diameters of ca. 200 μm. Microbial celluloses therefore have ca. 200 times greater specific surfaces. Similar to non-woven textiles, fibrils are strongly intertwined and crosslinked. Single fibrils do not exist.

Microbial celluloses are used to coat printing papers because they are less expensive than starches and latices. They also serve as flotation aids for ore concentrates and in hydraulic liquids for the production of crude oils.

### 8.3.6 Synthetic Cellulose

*β*-D-Cellobiosylfluoride has been polymerized *in vitro* by the enzyme cellulase in an aqueous $CH_3CN$ buffer (5:1) to cellulose II on a laboratory scale:

(8-2) β-D-cellobiosyl fluoride $\xrightarrow[-\ HF]{\text{Cellulase}}$ cellulose chain

### 8.3.7 Regenerated Celluloses

Regeneration of celluloses by cuoxam, viscose, or organosolv processes converts linters (Section 8.3.4) or pulps (Section 3.7.8) to molecularly soluble intermediates from which celluloses are reconstituted. These processes allow the manufacture of fibers (rayon) and films (cellophane) from inexpensive raw materials. They suffer from environmental problems and/or less than optimal fiber properties.

The world production of regenerated celluloses is ca. 4 000 000 t/a. About 75 % of this is used for the manufacture of fibers (viscose, acetate) and 25 % for cellophane, cellulose ethers, special papers, and others. "Regenerated cellulose fiber" is not synonymous with "regenerated fiber" since the latter also comprises those from non-cellulosic materials such as alginate fiber (from algae) or ardein fiber (from proteins of peanuts).

### Cuoxam Process

In the cuoxam process (Bemberg process) for cupro silk and cellophane, linters are dissolved in ammoniacal copper solution, a solution of $[Cu(NH_3)_4](OH)_2 \cdot 3\ H_2O$ in water (Schweizer's reagent; commonly mispelled as Schweitzer's reagent. Schweizer was a professor at what is now ETH Zurich). Cupro silk and cellophane are produced by slightly different cuoxam processes.

**Cupro silk** (**cuprammonium rayon, cuprammonium cellulose, cupro, cupra**) is manufactured by mixing solid cellulose in kneaders with a solution of 40 % copper sulfate in 25 % aqueous ammonia to which 8 % NaOH is added. The resulting solution is subjected to air that partially oxidizes and degrades cellulose molecules which adjusts the viscosity for fiber spinning. Filtration and deaeration under vacuum delivers a clear solution that is stable in the dark and under nitrogen. The process itself is simpler than the viscose process but is still more expensive because auxiliary compounds can only be partially recovered (recoveries: $CuSO_4$ 95 %; $NH_3$ 80 %).

The solution is then spun continuously into water. The emerging coagulated filaments are stretched by flowing through several hot baths where they are first treated with hot sulfuric acid to remove the copper and then by water to remove the acid. The resulting yarn is lubricated by oil and dried.

The manufacture of **cellophane** (a free name in the United States but a registered trademark in many other countries) requires higher cellulose concentrations than the cupro silk process since freshly formed films would otherwise contain too much solvent and tear easily. Increasing the copper sulfate concentration is detrimental since it would deliver too much sodium sulfate which reduces the solubility of cellulose in cuoxam. Copper sulfate is therefore replaced by a basic copper sulfate or copper hydroxide.

### Viscose Process

Cellulose (cell–OH, usually sulfite pulp (p. 73)) is converted to alkali cellulose (I) and then to cellulose xanthate (II). Solutions of II are extruded through orefices (fibers) or slits (cellophane) into a sulfuric acid bath which regenerates cellulose:

$$\text{Cell—OH} \xrightarrow[-H_2O]{+NaOH} \underset{I}{\text{Cell—ONa}} \xrightarrow{+CS_2} \underset{II}{\text{Cell—O—}\underset{|\atop SNa}{C}\text{=S}} \xrightarrow[-CS_2;\ -NaHSO_4]{+H_2SO_4} \text{Cell—OH} \qquad (8\text{-}3)$$

The viscose process consists of various steps (Fig. 8-24). Most alkali celluloses are now obtained by steeping, a process in which pulp in 18–20 % caustic soda solution is beaten to a homogeneous mash at 40–55°C. The excess lye is removed by pressing the slurry through sieves until it contains ca. 33 % solids; this also removes dissolved hemicelluloses and short-chain celluloses.

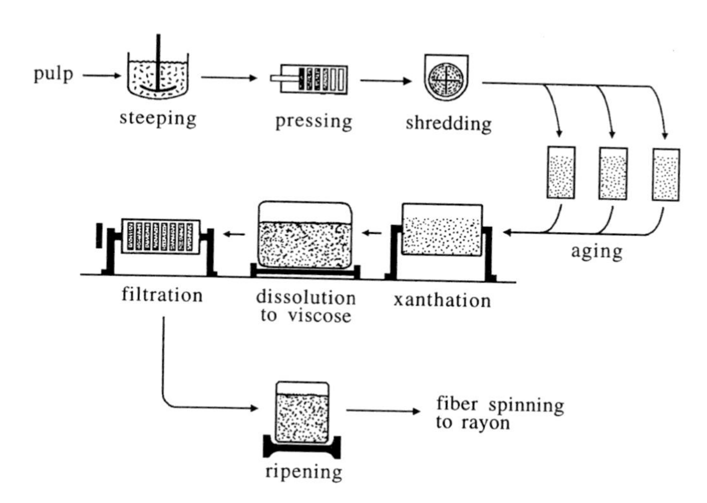


Fig. 8-24 Production of rayon by the viscose process. For each ton of cellulose, 2 tons of lignin sulfate and large volumes of waste water are produced. Worldwide, ca. 600 000 t/a $CS_2$ are consumed.

The compressed mass is then opened by shredding to "crumbs"; this "fluffs" the fibers so that they can be homogeneously degraded by hydrolysis and oxygen from the air in the subsequent aging process. For normal viscoses, degrees of polymerization of 300–350 are targeted because this would produce managable viscosities at the high concentrations that are required for fiber spinning.

The subsequent xanthation of the alkali cellulose uses $CS_2$ in the amount that corresponds to ca. 1/3 of that for complete reaction. The resulting sodium cellulose xanthate (II) contains ca. 0.5–0.6 xanthate groups per glucose unit. The products are reddish-orange because the reaction of $CS_2$ with NaOH leads to the colored byproduct sodium trithiocarbonate, $Na_2CS_3$. Pure sodium cellulose xanthogenate is colorless.

Dissolution of II in 3 % aqueous sodium lye delivers a raw **viscose** that contains ca. 8–10 % cellulose and ca. 6 % alkali. Repeated filtration removes solid particles and successive degassing eliminates small air bubbles, both of which would lead during fiber spinning to either clogging of spinnerets or fiber cleavages.

Freshly prepared viscose cannot be spun to fibers or extruded to cellophane. For 10–100 hours at 15-20°C, it is therefore post-ripened during which time the xanthate continuously loses $CS_2$ which either rexanthates OH groups or reacts with NaOH to $Na_2CS_3$ and $Na_2S$. Redistributions of xanthate groups occur for the following reason:

Xanthation is a heterogeneous reaction on alkali cellulose molecules that form intra- and intermolecular hydrogen bonds. Because of these bonds, OH groups at $C^2$ react faster than the OH at $C^6$ (see also Fig. 8-22), although the primary OH is usually more reactive than the secondary OHs. During ripening, xanthate groups at $C^2$ and $C^3$ are preferentially split off whereas OH at $C^6$ is rexanthated. Ripening thus simultaneously homogenizes the distribution of xanthate groups and decreases molecular weights.

As a result, the viscosity of viscose first decreases, passes through a minimum, and then increases; at long times, coagulation may occur. The viscosity decrease is caused by the redistribution of xanthate groups as well as the resulting dissolution of supramolecular regions. The subsequent viscosity increase is produced by the reformation of hydrogen bonds, especially intermolecular ones.

For fiber spinning, viscose is pressed through a spinneret with hole diameters of 40-100 μm into a solution of $Na_2SO_4$ in $H_2SO_4$ which sometimes contains $ZnSO_4$ and, in special cases, also $(NH_4)_2SO_4$. The xanthate coagulates to fibers and decomposes to cellulose, $CS_2$, and $Na_2SO_4$. Sulfuric acid also converts dissolved $Na_2CS_3$ and $Na_2S_x$ of viscose to $CS_2$ and elementary sulfur which partially deposits on fibers. The spinning process also stretches the fibers.

The fibers are then deacidified and desulfurized, partially bleached, washed, lubricated for better processing, and dried. The resulting fibers are called **rayon** (F: *rayon* = light beam, because of the shinyness of the fibers). Rayons are produced as endless filaments and as staple fibers, both in many variants. So-called **modal fibers** have increased wet strengths. The **polynosic** variant achieves this by lower acid and higher $CS_2$ concentrations, whereas for the **high-modul type**, amines and polyglycols are added to viscose and $ZnSO_4$ to the precipitation bath (F: *polynosic* = *poly*mère *no*n *s*ynthét*iq*ue).

**Organosolv Process**

The newer organosolv process (Lyocell process) dissolves pulp in an aqueous solution of *N*-methylmorpholine-*N*-oxide (NMMO). This solvent is obtained by oxidizing *N*-methylmorpholine (from dehydration of $(HOCH_2CH_2)_2NCH_3$) under a $CO_2$ blanket:

(8-4) $$\text{O(CH}_2\text{CH}_2)_2\text{NCH}_3 \xrightarrow[-\,H_2O]{+\,H_2O_2} \text{O(CH}_2\text{CH}_2)_2\text{N(CH}_3)\rightarrow\text{O}$$

The solution is filtered and spun into a water bath. The process requires far fewer steps than the viscose process and is also more environmentally friendly since less water and air have to be purified and 99 % of NMMO can be recovered. Fibers also have better properties since celluloses are less degraded. These **reconstituted fibers** are called **lyocell**. Approximately 1/3 each are used as textile fibers, treated fibers, and nonwovens.

### 8.3.8 Mercerization

Mercerization is the treatment of celluloses with 8–22 % aqueous sodium hydroxide. The process was invented in 1844 by the Englishman John Mercer. It is applied to cotton yarn and fabrics and to modal and polynosic fibers where mercerization shrinks the fibers and increases the strength, luster, dye uptake, and hand.

The process converts the crystal lattice of cellulose I to that of sodium cellulose I which becomes cellulose II on neutralization. The concentration of sodium cellulose I is always higher than that of cellulose II (Fig. 8-25), probably because not all segments of the linear macroconformation of cellulose I are converted to that of sodium cellulose I which is twisted at the $C^1$–$C^4$ bond according to $^{13}$C NMR and wide-angle X-ray measurements. On neutralization, some segments of sodium cellulose are converted back to cellulose I instead of cellulose II.

During mercerization, yarns and fabrics are under tension in order to maintain their dimensional characteristics. Mercerization without applied stress shrinks the fibers: yarns become elastic and fabrics become stretch articles.

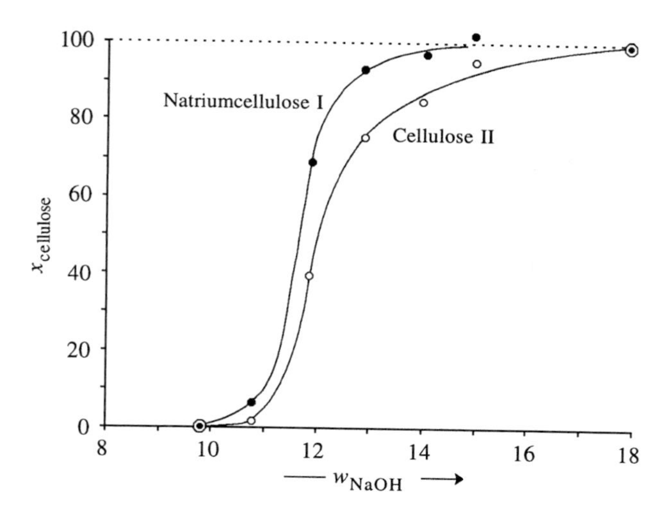


Fig. 8-25 Mol fractions $x_{\text{cellulose}}$ of sodium cellulose I and cellulose II as functions of the weight fraction $w_{\text{NaOH}}$ for the conversions cellulose I → sodium cellulose I and sodium cellulose I → cellulose II [11]. With kind permission of Gordon and Breach, Newark (NJ).

### 8.3.9 Cellulose Derivatives

Cellulose derivatives are celluloses in which hydrogens of one, two, or all three hydroxyl groups of glucose units are replaced by substitutents R, R', and/or R". All glucose *units* need not have the same degree of substitution. The degree of substitution of the *compound* may thus not be a whole number. For example, cellulose 2-1/2 acetate contains *on average* 2.5 acetate groups per glucose unit while the number of acetate groups of any glucose unit itself may vary between 0 and 3. Examples of cellulose derivatives without further reaction of the new substitutents are

CH₂OR, OR', OR", O

glucose unit of a cellulose

| R, R', and/or R" | Name of derivative |
|---|---|
| $COCH_3$ | cellulose acetate |
| $COCH_2CH_3$ | cellulose propionate |
| $COCH_2CH_2CH_3$ | cellulose butyrate |
| $NO_2$ | cellulose nitrate |
| $CH_3$ | methyl cellulose |
| $CH_2CH_2OH$ | *β*-hydroxyethyl cellulose |
| $CH_2COOH$ | carboxymethyl cellulose |
| $CH_2C_6H_5$ | benzyl cellulose |
| $CH_2CH_2CN$ | cyanoethyl cellulose |
| C(S)SNa | cellulose xanthate |

#### Reactions of Cellulose

Cellulose derivatives are synthesized by reaction of swollen cellulose or alkali cellulose with the corresponding reagent. Reactions are thus initially heterogeneous and may remain so if the resulting cellulose derivatives are insoluble. As a result, cellulose derivatives may have broad distributions of compositions of monomeric units.

Furthermore, some of the initially introduced groups may react still further. An example is the reaction of an OH group of a glucose unit, cell–OH, with ethylene oxide, $C_2H_4O$, according to cell–OH + $C_2H_4O$ → cell–$OCH_2CH_2OH$ and its subsequent reaction cell–$OCH_2CH_2OH$ + $C_2H_4O$ → cell–$O(CH_2CH_2O)_2H$. Thus, one has to distinguish between the degree of reaction and the degree of substitution. The **degree of reaction** (DR) describes the average number of reagent molecules that reacted with one glucose unit; this quantity is sometimes also called the **molar degree of substitution** (MS). The **degree of substitution** (DS), on the other hand, indicates the average number of substituted OH groups per glucose unit.

Monofunctional reagents always lead to DR = MS = DS whereas multifunctional reagents usually produce DR = MS > DS. An example is the reaction of six propylene oxide molecules with one cellobiose unit (DR = 3) which generates DS = 3 for the left glucose unit but DS = 2 for the right one:

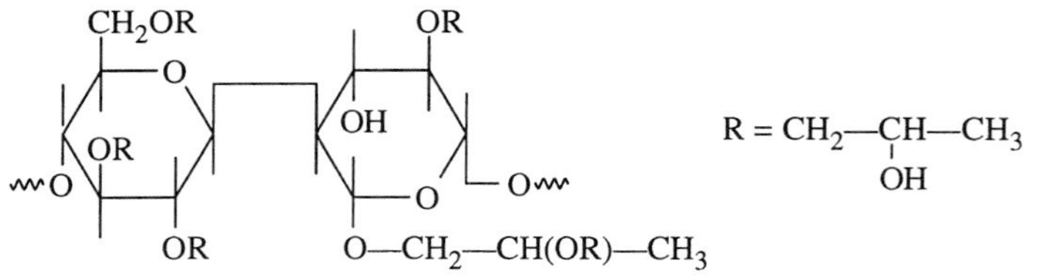

For *equal* reaction probabilities of OH groups at $C^2$, $C^3$, and $C^6$, mole fractions $x_i$ of unsubstituted ($i$ = 0), single ($i$ = 1), double ($i$ = 2), and triple ($i$ = 3) substituted glucose units can be calculated by the **Spurlin equation** from the number $N$ of functional groups per glucose unit and the degree of substitution, $DS = x_1 + 2\,x_2 + 3\,x_3$,

$$x_i = \left(\frac{N!}{3!(N-3)!}\right)\left(\frac{DS}{3}\right)^N\left(1-\frac{DS}{3}\right)^{3-N} = \binom{3}{N}\left(\frac{DS}{3}\right)^N\left(1-\frac{DS}{3}\right)^{3-N} \qquad (8\text{-}5)$$

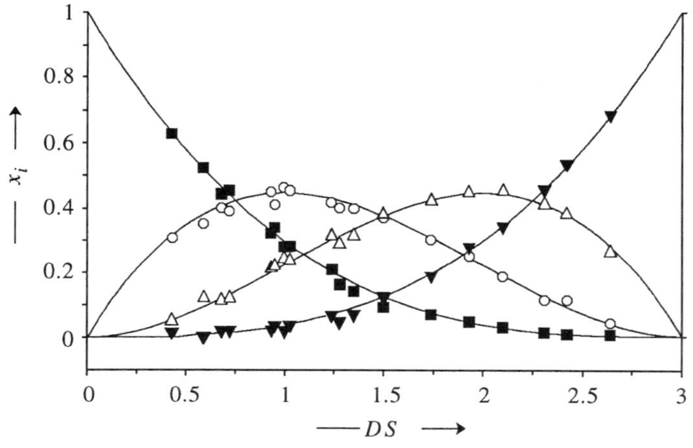

Fig. 8-26 Molar fraction $x_i$ of unsubstituted glucose units ($i$ = 0; ■), mono-*O*-carboxymethylcellulose units ($i$ = 1; O), di-*O*-carboxymethyl cellulose units ($i$ = 2; Δ), and tri-*O*-carboxymethyl cellulose units ($i$ = 3; ▼) after hydrolysis of the polymers from the reaction of alkali cellulose with chloroacetic acid [12a]. Experimental data ■, O, ▼) by high-performance liquid chromatography; calculated lines for 1, 2, and 3 correspond to the Spurlin equation.

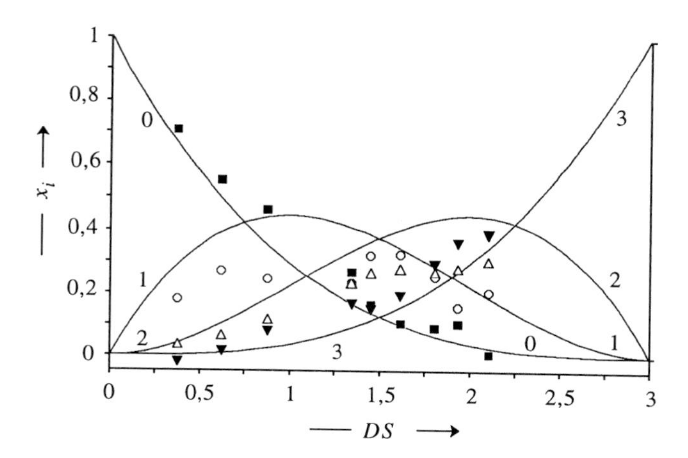


Fig. 8-27 Molar fraction $x_i$ of glucose units with 0, 1, 2, or 3 substituted OH groups per unit after hydrolysis of polymers from the reaction of cellulose with dichloroacetic acid in *N*,*N*-dimethylacetamide + LiCl [12b]. Lines correspond to the Spurlin equation. Compare with Fig. 8-26.

The Spurlin equation is well obeyed for the distributions of *O*-carboxymethyl groups in products from the reaction of chloroacetic acid with alkali cellulose that is swollen in aqueous solutions of NaOH (Fig. 8-26). It is not fulfilled for molecular solutions of cellulose in *N*,*N*-dimethylacetamide + LiCl that are reacted with granular NaOH (Fig. 8-27), i.e., *O*-carboxymethyl groups are no longer distributed at random.

**Cellulose Nitrates**

Cellulose nitrates (**CN**) are synthesized by reacting celluloses with concentrated nitrating acid ($HNO_3$ + $H_2SO_4$ in the ratio 2:1 to 1:2). The composition of nitrating acid controls the degree of substitution which is DS = 2.7-2.9 for guncotton, 2.5-2.6 for photo films, 2.25-2.6 for nitro lacquers, 2.25-2.4 for celluloid, and ca. 2 for pyroxylin. The polymers contain nitrate groups, ~CO–$NO_2$, but not nitro groups, ~$CR_2NO_2$; they are *not* "nitrocelluloses." Linters are the cellulose source for photo films, nitro lacquers, and celluloid (Grade A), whereas wood pulps can only be used for guncotton (Grade B) because they contain carbonyl and carboxyl groups which makes them sensitive to light.

Cellulose nitrates are white odor-free materials that start to decompose at ca. 90°C and self-ignite in air at ca. 160°C. After ignition, they burn smokeless to $CO_2$, CO, $H_2O$, $N_2$, and $H_2$ even if oxygen is absent. Remaining traces of $H_2SO_4$ lead slowly to the evolution of $NO_2$ and $HNO_3$.

Cellulose nitrates and products thereof are sometimes called **pyroxylins** (G: *pyr* = fire; *xylon* = wood). Patent leathers are fabrics that are coated with pyroxylins. **Collodion** (G: *kollodes* = sticky) is a solution of pyroxylin in a mixture of ether and ethanol. Cellulose nitrates with 11 % N for explosives are therefore also known as **collodion wools**.

**Chardonnet silk** (**nitrosilk, collodion silk**) was the first industrial artificial silk. Its fibers and fabrics are extraordinarily flammable and are therefore no longer used.

**Guncotton** (N > 13.5 %) is sold with a water content of 15–25 %. It is commonly not used as such but as a mixture with nitro compounds and other nitrate compounds.

Cellulose nitrates for **celluloid** contain 40–50 % water after nitration; this is partially removed by centrifugation or pressing with alcohol. The resulting product contains 30–45 % moisture which is an 80:20 mixture of ethanol and water. It is mixed with 20–30 % camphor and, after addition of ethanol, gelled in kneaders. Subsequent rolling reduces the ethanol content to 12–18 %. The rolled sheets are then fused to solid blocks by pressure at 80–90°C. Machining of the blocks produces semi-finished products. Celluloid is easy to process and extraordinarily easy to dye, which was the main reason for its use for early films. Disadvantages are its flammability and and its wage-intensive manufacture.

**Cellulose Acetates**

**Cellulose acetates** (**acetyl celluloses, CA**) are prepared from linters or from chemical pulps with low proportions of hemicelluloses. In contrast to cellulose nitrates, no *direct* partial acetylation of cellulose *molecules* is possible since partially acetylated molecules remain undissolved until a very high degree of acetylation. Therefore, one obtains mixtures of unacetylated and completely acetylated molecules. Partially acetylated molecules are produced by partial saponification of primary cellulose triacetates with a degree of acetylation of more than 92 %. Hence, triacetates are called **primary acetates** and 2-1/2 acetates also **secondary acetates**.

Commercial productions of triacetate use three different methods: "acetic acid" (with acetic anhydride and sulfuric acid as catalyst), "methylene chloride" (a solvent for cellulose triacetate, with acetic anhydride and sulfuric acid), and "heterogeneous" for dry celluloses (mixture of benzene, acetic anhydride, and perchloric acid).

**Cellulose triacetate** is spun to fibers from its solution in methylene chloride. **Triacetate fibers** (**CTA**) are crease- and weather-resistant. Some triacetate is also oxidatively degraded and then spun from methylene chloride or chloroform solution to fibers for cable sheating.

Partial saponification of triacetate delivers various products, mainly **cellulose 2-1/2 acetate** (commonly called **cellulose acetate**; **CA**) for **acetate rayon** (**acetate silk**), cigarette filters (presently the main application), and a smaller part with DS = 2.2–2.8 similar to celluloid for injection-molded articles, photofilms, or films and sheets for properties, see Table 8-9). Complete saponification of acetate rayon delivers very fine, highly oriented cellulose fibers.

**Cellulose propionate** (**CP**) is obtained b y reaction of celluloses with mixtures of propionic acid and propionic anhydride with concentrated sulfuric acid as catalyst. Continuous service temperatures range from –40°C to +115°C, i.e., a larger range than that of cellulose acetate. It also takes up less water and has better resistance to light.

Esterification of cellulose acetate by butyric acid delivers **cellulose acetobutyrate** (**cellulose acetate butyrate, butyrate plastic**; **CAB**) with 17–48 wt% butyryl groups and 6–29 wt% acetyl groups (**acetyl butyryl cellulose**). CABs are used for car parts and pipes in the petroleum industry. Corrosion-resistant packaging is obtained by dipping goods into molten CAB. **Cellulose acetate propionate** (**CAP**) with 0.6-2.5 wt% acetyl groups and 42.5–46 wt% propionyl groups has slightly better mechanical properties than CAB.

**Cellulose acetophthalate** from the reaction of cellulose acetate with phthalic anhydride dissolves in the intestines but not in the stomach. It is therefore used for packaging those pharmaceuticals that should be resorbed by the intestines.

Some insoluble esters and ethers of cellulose serve as ion-exchange resins. Examples of anion-exchange resins are aminoethyl, diethylaminoethyl, guanidinoethyl, triethylaminomethyl, and *p*-aminobenzyl celluloses. Cationic ion-exchange resins comprise carboxymethyl and sulfoethyl celluloses as well as cellulose phosphates.

Table 8-9 Some physical properties of cellulose ester plastics at 23°C unless noted otherwise. Note that impact strengths are measured as energy per cross-sectional area (1 ft lbf/in$^2$ = 2106 J/m$^2$). Notched impact strengths are reported in the United States as fracture energy per width of notch (1 ft lbf/in = 53.5 J/m) but in Europe as fracture energy per width of notch and thickness of specimen, i.e., in J/m$^2$. nb = no break.

| Property | Physical unit | Celluloid | CA | CAB | CAP |
|---|---|---|---|---|---|
| Density | g/cm$^3$ | 1.35 | 1.22–1.34 | 1.15–1.22 | 1.17–1.24 |
| *Thermal properties* | | | | | |
| Melting temperature | °C | – | 235–255 | 155–200 | 190 |
| Softening temperature (Vicat B) | °C | 70–75 | 60–97 | | |
| Heat distortion temperature (B) | °C | | 55–120 | 60–110 | 70–120 |
| (C) | °C | | 44–113 | 50–100 | 50–110 |
| Continuous service temperature | °C | | –20–95 | | |
| Linear expansion coefficient | $10^{-5}$ K$^{-1}$ | 10 | 11–16 | 11–16 | 11–16 |
| Specific heat capacity | J K$^{-1}$ g$^{-1}$ | | 1.2–1.6 | 1.2–1.6 | 1.2–1.6 |
| Thermal conductivity | W m$^{-1}$ K$^{-1}$ | 0.25 | 0.16–0.33 | 0.16–0.33 | |
| *Mechanical properties* | | | | | |
| Tensile modulus | MPa | 1800 | | | |
| Flexural modulus | MPa | | 620–1800 | 48–1400 | 690–1900 |
| Compression strength at yield | MPa | 60 | 13–64 | 8–52 | 21–69 |
| Flexural strength at yield | MPa | 60–65 | 14–110 | 10–64 | 21–76 |
| Tensile strength at yield | MPa | 40–60 | 14–48 | 10–48 | 10–48 |
| Tensile strength at break | MPa | | 13–59 | 14–52 | 14–52 |
| Elongation at break | % | 30–50 | 50–6 | 74–38 | 60–35 |
| Impact strength | kJ m$^{-2}$ | nb | | | |
| Notched impact strength, (Europe) | kJ m$^{-2}$ | 20–30 | | | |
| (US, Izod, 23°C) | J m$^{-1}$ | | 7–133 | 10–150 | 13–180 |
| (US, Izod, –40°C) | J m$^{-1}$ | | 2–14 | 7–24 | 2–19 |
| Hardness, Rockwell R | – | | 40–120 | 30–120 | 20–120 |
| *Electrical properties* | | | | | |
| Relative permittivity. 50 Hz | 1 | 7.0–7.5 | | | |
| 1 kHz | 1 | 6.0–7.0 | | | |
| 1 MHz | 1 | 6.0–6.5 | | | |
| Dielectric loss factor | | | | | |
| 50 Hz | 1 | 0.09–0.12 | | | |
| 1 kHz | 1 | 0.02–0.03 | | | |
| 1 MHz | 1 | 0.06–0.09 | | | |
| Electric strength (0.125 cm) | V mm$^{-1}$ | | 250–3670 | 250–400 | 300–450 |
| Resistivity | Ω cm | $10^{10}$–$10^{13}$ | $10^{10}$–$10^{12}$ | $10^{12}$–$10^{15}$ | |
| *Optical properties* | | | | | |
| Refractive index (25°C) | 1 | 1.50 | 1.46–1.50 | 1.46–1.49 | 1.46–1.48 |
| Light transmission | % | | 88 | 88 | 88 |
| *Other* | | | | | |
| Water absorption, 24 h | % | | 1.9–7 | 0.9–2.2 | 1.0–3.0 |
| Weight loss, 72 h at 82°C | % | | 0.4–12 | 0.1–4.0 | 0.1–2.0 |

**Cellulose Ethers**

Cellulose ethers are not prepared from celluloses but from alkali celluloses since the latter have an expanded crystal lattice which allows better access to hydroxyl groups. Industry distinguishes between processes with and without alkali consumption,

Cellulose methyl and ethyl ethers are prepared by reaction of alkali cellulose with the corresponding alkyl chlorides; this generates alkali chlorides from which the alkali cannot be recovered. The resulting polymers are washed with water, centrifuged to a water content of 55–60 wt%, and homogenized and densified in screw presses. Commercial **cellulose methyl ethers** (**methyl celluloses, MC**) have degrees of substitution of DS = 1.5–2.0, **cellulose ethyl ethers** (**ethyl celluloses, EC, AT celluloses**), those of DS = 2.1–2.6. Methyl celluloses serve as thickening and sizing agents, adhesives, and protective colloids in water-borne coatings. Ethyl cellulose is used as hot-melt adhesive and coating for paper, textiles, and cables, molding compounds, binders in ceramics, in printing inks, for microencapsulation of pharmaceuticals, and many other applications.

Alkali hydroxide is recovered in the reaction of alkali cellulose with ethylene oxide or propylene oxide. The primarily formed alkylene hydroxide groups can add additional oxide molecules. The resulting **hydroxyethylcelluloses** (**HEC**) and **hydroxypropylcelluloses** (**HPC**) have degrees of reaction of DR ≈ 4. The polymers dissolve in water at temperatures below ca. 38°C. As 2–3 % aqueous gels, they are used as thickeners and stabilizers for whipped cream, and also as binders for ceramics, suspension agents in emulsion polymerizations, etc. They can be thermoplastically processed, e.g., to packaging films.

Reaction of alkali cellulose with the sodium salt of chloroacetic acid delivers the sodium salt of **carboxymethylcellulose** (**CMC, cellulose glycolate**) with DS = 0.4–1.4 that is known as **cellulose gum** in the food industry. Technical grades contain ca. 95–98 wt% CMC (US) or 60–95 wt% (Europe); purified grades always contain more than 95 wt%. CMCs serve as thickeners in foods, emulsifying agents, protective colloids, for paper coatings, drilling agents, sizings, etc.

Use of two etherification agents results in mixed cellulose ethers. Examples are **ethylmethylcellulose** (**EMC**), **hydroxypropylmethylcelluloses** (**HPMC**), **hydroxybutylmethylcelluloses** (**HBMC**), and methylcelluloses, all sold as Methocel®. Reaction of hydroxyethylcellulose and sodium chloroacetate leads to the mixed ether **sodium carboxymethylhydroxyethylcellulose** (**CMHEC**).

Cellulose ethers are produced worldwide in many varieties with ca. 300 000 t/a. Most important are carboxymethylcelluloses, followed by methylcelluloses, methylhydroxyalkylcelluloses, and hydroxyethylcelluloses.

### 8.3.10 Chitin and Chitosan

Chitin is a cellulose derivative in which the two hydroxyl groups of the $C^2$ groups of β-cellobiose repeating units are replaced by *N*-acetylamino groups. This polymer is thus a poly[β-(1→4)-*N*-acetyl-2-amino-2-desoxy-D-glucopyranose].

Chitin is the structural polysaccharide of arthropods, i.e., insects (bugs, spiders, etc.), crustaceans (lobsters, shrimps, barnacles, etc.), molluscs (oysters, krill, etc.), where it is always associated with calcium carbonate and/or proteins with which it forms a very hard composite (G: *chiton* = tunic). It is also found in cell walls of algae, yeasts, fungi, etc.

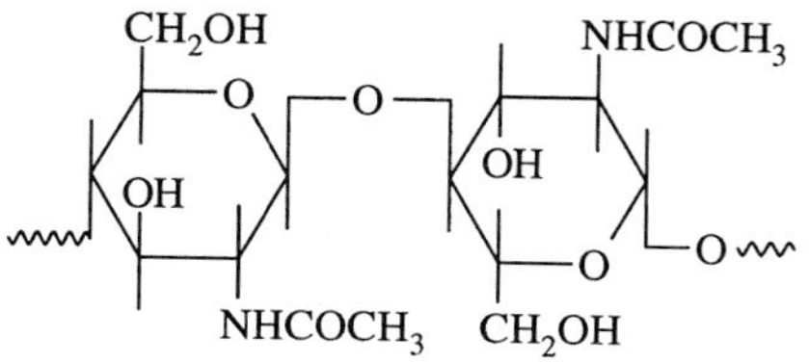

**Chitin** is isolated from lobster and crawfish shells by dissolving calcium carbonate with cold 5 % hydrochloric acid. After filtering and washing, the resulting powders are treated with boiling 4 % soda lye or proteolytic enzymes in order to remove proteins. The bleached chitin is insoluble in water, dilute acids and bases, and organic solvents. It dissolves and hydrolyzes in formic acid and concentrated mineral acids.

Treating chitin with 40 % soda lye at elevated temperatures delivers **chitosan**, a 60 % deacetylated chitin, which dissolves in dilute acids. It forms biologically degradable films for food packaging and is used as an additive for the improvement of the wet strength of paper, as an ion-exchange resin, for the protection of wounds, and in hair care products.

Aqueous solutions of ***N,O*-carboxymethylchitosan** are used to produce films on fruits and vegetables which allows one to prolong the storage time.

### 8.3.11 Xanthan

Xanthan is a poly(disaccharide) with a cellulosic main chain of β-(1→4)-D-glucopyranosyl repeating units. Every second glucose unit of the main chain carries a trisaccharide chain, β-D-mannopyranosyl-α-(1→4)-D-glucopyranosyl-β-(1→2)-dimannopyranoside-6-*O*-acetate which is connected to the main chain via β-(1→3) bonds. The end of the side chain is a ketal of a metal salt of pyruvic acid with the $C^4$ and $C^6$ hydroxyls.

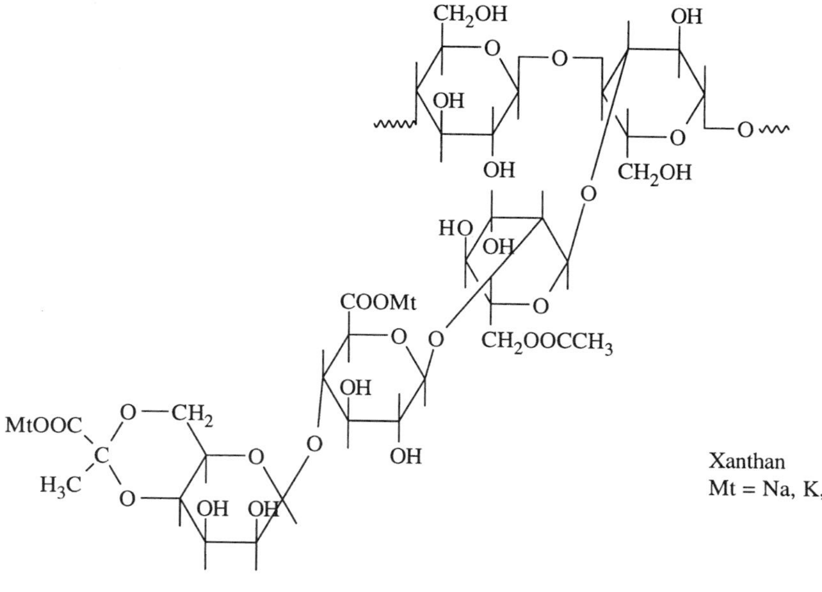

Xanthan
Mt = Na, K, Ca

Xanthan is produced under aerobic conditions by the action of the bacterium *Xanthomonas campestris* NRRL B-1459 on aqueous solutions of glucose that contain a nitrogen source, dipotassium sulfate, and trace elements. Its molecular weights are extraordinarily high, ca. 5 000 000.

The crowding and interaction of the regularly arranged side chains force xanthan molecules to adopt the macroconformation of a fairly stiff helix with a length of 600 nm and a diameter of 5–6 nm. At concentrations above ca. 2.5 g/L, salt-free xanthan solutions become birefringent because of the formation of lyotropic liquid crystals.

The structure of this helix conveys interesting rheological properties to aqueous xanthan solutions, which show pronounced shear thinning but little thixotropy. Viscosities vary little with pH (2–12) and temperature (< 90°C). Small xanthan concentrations of $(5–10)\cdot 10^{-3}$ % reduce turbulence considerably, which leads to applications as drag reducers (Volume IV). Salts in concentrations smaller than 0.01 % reduce viscosities of aqueous xanthan solutions but salt concentrations greater than 1 % increase them. Bivalent cations precipitate xanthan at pH > 9 but trivalent ones do so at a much lower pH.

Xanthan (gum) is used for the production of secondary and tertiary crude oils, as a carrier for agricultural chemicals, gelation agent for explosives, thickener for cosmetics, etc. It does not metabolize and is therefore used as a low-calorie additive in puddings, salad sauces, dry milk, fruit drinks, etc.

## 8.4 Mucopolysaccharides

Mucopolysaccharides are usually part of proteoglycans (Fig. 8-28). They are found in the skin, connective tissue, cartilage, sweat, mucus (L: *mucus* = phlebs), and in cell walls of bacteria. The name "mucopolysaccharide" for this groups of polysaccharides is outdated but is still used widely instead of the recommended "**glycosaminoglycan**."

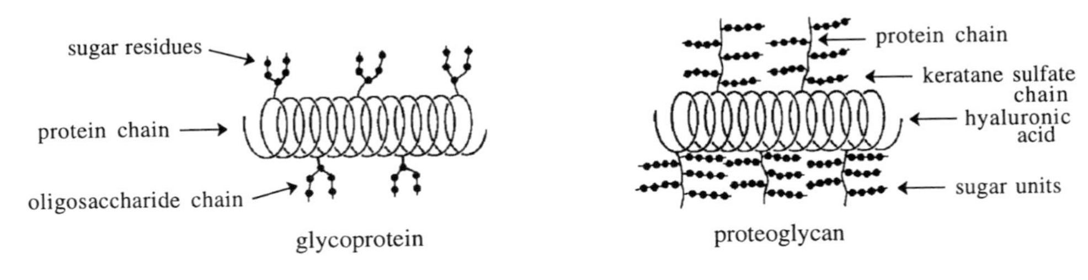


Fig. 8-28 Examples of a glycoprotein and a proteoglycan (Volume I, Section 14.4). Keratane sulfate may be replaced by chondroitin sulfate.

Most mucopolysaccharides have disaccharides as repeating units, mainly Gal-Glc but also Glc-Glc, Glc-Ido, or Gal-Ido (Table 8-10). There are also mucopolysaccharides with tetrasaccharide repeating units.

One of the two sugar units of the repeating disaccharide units is either an *N*-acetylated or an *N*-sulfated amino sugar. The same unit or the second sugar unit is a uronic acid. Uronic acids are aldoses in which the $CH_2OH$ group is replaced by a COOH group (see also Fig. 8-6).

Table 8-10 Structure of mucopolysaccharides. All sugar units are D-configured.

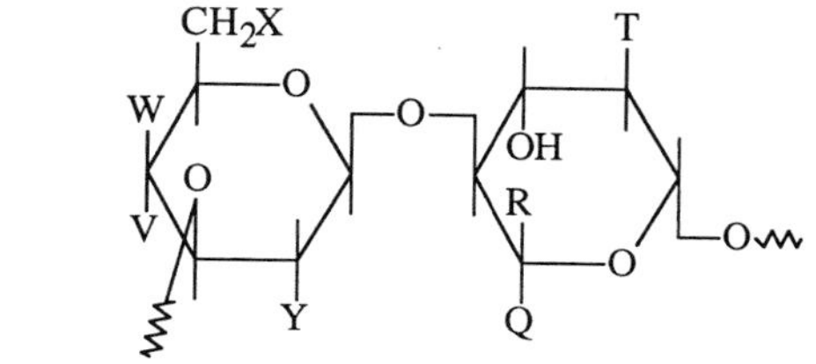

(1→3) A 1→4 B 1→3

| Name | A | B | Q | R | T | V | W | X | Y |
|---|---|---|---|---|---|---|---|---|---|
| Hyaluronic acid | β-Glc | β-Glc | COOH | H | OH | OH | H | OH | $NHCOCH_3$ |
| Heparin | β-Glc | β-Glc | COOH | H | OH | $OSO_3H$ | H | OH | $NHSO_3H$ |
| | β-Glc | α-Ido | H | COOH | OH | $OSO_3H$ | H | OH | $NHSO_3H$ |
| Chondroitin | β-Gal | β-Glc | COOH | H | OH | H | OH | OH | $NHCOCH_3$ |
| Chondroitin-4-sulfate | β-Gal | β-Glc | COOH | H | OH | H | $OSO_3H$ | OH | $NHCOCH_3$ |
| Chondroitin-6-sulfate | β-Gal | β-Glc | COOH | H | OH | H | OH | $OSO_3H$ | $NHCOCH_3$ |
| Keratane sulfate | β-Gal | β-Glc | $CH_2OSO_3H$ | H | $NHCOCH_3$ | H | OH | $OSO_3H$ | OH |
| Dermatane sulfate | β-Gal | α-Ido | H | COOH | OH | H | $OSO_3H$ | OH | $NHCOCH_3$ |

In mucopolysaccharides, uronic acid units are glycosidically bound to the $C^3$ of the other sugar unit of the repeating unit. ~OH groups are frequently replaced by $\sim OSO_3H$ and $\sim NHCOCH_3$ groups by $\sim NHSO_3H$. Mucopolysaccharides are therefore acidic.

### 8.4.1 Hyaluronic Acid

The repeating unit of hyaluronic acid is a disaccharide composed of D-glucuronic acid and *N*-acetylglucosamine units that are interconnected by *β*-(1→4) bonds (deviation from Fig. 8-10). It is found in all connective tissues (except the cornea), the vitreous body of the eye (G: *hyalos* = glass), and the synovia (synovial fluid) of joints where it binds more water than any other biochemical compound: 200 times its weight. Hyaluronic acid from combs of roosters is now (2005 ff.) injected into knees of patients with damaged menisci where it brings relief for ca. six months, although its half-life in the knee is only ca. 12 hours. Apparently, it stimulates its own production.

### 8.4.2 Chondroitin Sulfates

Chondroitin sulfates are poly(disaccharides) in which one of the saccharide units is based on galactose and the other one on either glucose or idose (see Fig. 8-1). The various chondroitin sulfates also differ in their degrees of polymerization and sulfation.

Like hyaluronic acid, chondroitin-4-sulfate (**chondroitin sulfate A**) and chondroitin-6-sulfate (**chondroitin sulfate C**) are alternating copolymers albeit of D-glucuronic acid and *N*-acetyl-D-galactosamine sulfate (Table 8-10). The two chondroitin sulfates and hyaluronic acid are the main components of connective tissue and the stroma (tissue framework). Chondroitin sulfates are bound to proteins: salt-like in connective tissue but covalently in cartilage (G: *chondros* = cartilage) where a connecting tetrasaccharide is bound to chondroitin sulfate at one end and to a serine group of the protein at the other.

~chondroitin sulfate–glucuronic acid–galactose–galactose–xylose–serine<

Chondroitin sulfates A and C participate in cell division and the development of neurons. They are responsible for the sulfate exchange in the body and for the calcification of bones. Their concentrations decrease with the age of humans and animals. Chondroitin sulfate and glucosamine are now taken orally by elderly persons in the hope that their intake will help to renew cartilage and connective tissue and reduce arthritis (not proven clinically). It is not clear which chondroitin sulfate is actually sold since different sources (lobster shells, etc.) certainly deliver different compounds.

The disaccharidic repeating unit of **chondroitin sulfate B (dermatane sulfate; β-heparin)** consists of *N*-acetyl-D-galactosamine-4-sulfate and L-iduronic acid. Dermatane sulfate forms the intercellular matrix of the skin (G: *derma* = skin). Unlike heparin, it does not prevent blood clotting.

### 8.4.3 Heparin

Heparin is not a poly(disaccharide) like hyaluronic acid and chondroitin sulfates A and C but a heteroglycan with D-glucuronic acid, *N*-acetyl-D-glucosamine sulfate, D-iduronic acid, and *N*-acetyl-D-glucosamine sulfate (32 possible disaccharide units!). It is highly charged because of many anionic groups. **Heparin sulfate** resembles heparin but contains only one sulfate group per tetrasaccharide.

Heparin is found in the liver (G: *hepar*), heart, lung, and mucous membrane of the intestines. Heparin sulfate is present in the human aorta. Its sulfate groups are mainly (but not exclusively) responsible for its fast action as a blood anticoagulant, especially in surgery. It needs to be injected subcutaneously into fatty tissue since it cannot be taken orally. As its action lasts less than 24 hours (usually ca. 6 hours), low-molecular weight coumarin derivatives are used for longer-lasting action.

**Heparinoids** are compounds that act like heparin. Examples are the protein hirudin (from leeches), dextran sulfate, and poly(vinyl sulfonic acid), $-[CH_2-CH(SO_3H)]-_n$.

### 8.4.4 Keratane Sulfate

Keratane sulfate is much more sulfated than other mucopolysaccharides (Table 8-10). It is combined with collagen in the proteoglycans of cartilage and found in ossein (cartilage of bones; L: *os* = bone) and cornea (L: *cornu* = horn; G: *keras* = horn).

### 8.4.5 Murein

Murein is the structural material of the cell walls of bacteria (L: *murus* = wall). Its repeating unit is a disaccharide composed of alternating 2-*N*-acetylglucosamine and 2-*N*-acetylmuramic acid units. The latter are 2-*N*-acetylglucosamine units that carry a lactic acid group in the 3-position which is substituted by oligopeptides, usually tetrapeptides from L-lysine, L- and D-alanine, and D-glutamic acid as well as diamino acids. Since murein resembles proteoglycans, it is also called a **peptidoglycan**.

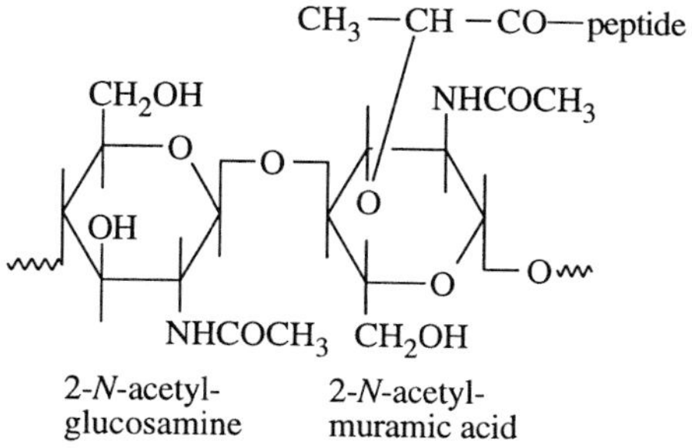

2-*N*-acetyl-glucosamine　　2-*N*-acetyl-muramic acid

murein

The oligopeptide units form intermolecular crosslinks between their lysine, glutamic acid, and diamino acid units, which generates a giant bag-like molecule (***murein sacculus***) that encloses the whole bacterium. The composition of oligopeptide units vary from bacterium to bacterium. In Gram-positive bacteria, murein carries teichonic acid units; in Gram-negative bacteria, lipids. The murein bags of Gram-negative bacteria are single-walled. The walls of Gram-positive bacteria consists of multiple layers since the teichonic acid units (Section 12.5.3) of their mureins can form intermolecular bonds.

## 8.5 Poly[β-(1→3)-D-glucose]s

Poly(β-(1→3)-D-glucose)s are found in many lower organisms (Table 8-2). **Callose** ($X \approx 100$) is not branched; it is found in cells of roots, pollen, etc., of higher plants, brown algae, and fungi. The unbranched **curdlan** is present in agrobacteria; it is used as a food additive. **Laminarin** contains some additional D-mannitol units. It is the reserve polysaccharide of some brown algae and is used to some extent as a surgical powder.

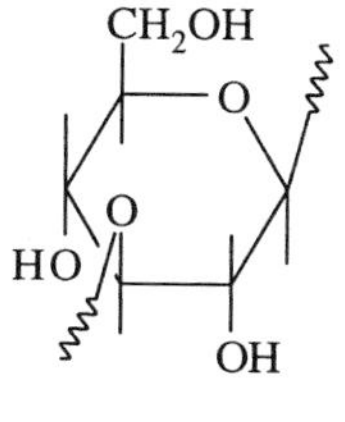

β-(1→3)-D-glucose unit

In **schizophyllan** (from the fungus *Schizophyllum commune*) and **scleroglycan** (from *Sclerotium rolfsii*), the $CH_2OH$ group of every third glucose unit is substituted by a D-glucose unit via a β-(6→1) bond (Fig. 8-29). This substitution and the interaction between these side groups stiffens the chain (persistence length: 200 nm (see Volume III)) and bestows high viscosities on dilute aqueous solutions. At higher concentrations, schizophyllan molecules self-associate to physical networks with triple helices as crosslinks and form thermoreversible gels. These rheological properties make schizophyllan useful as a drilling fluid in the production of crude oil.

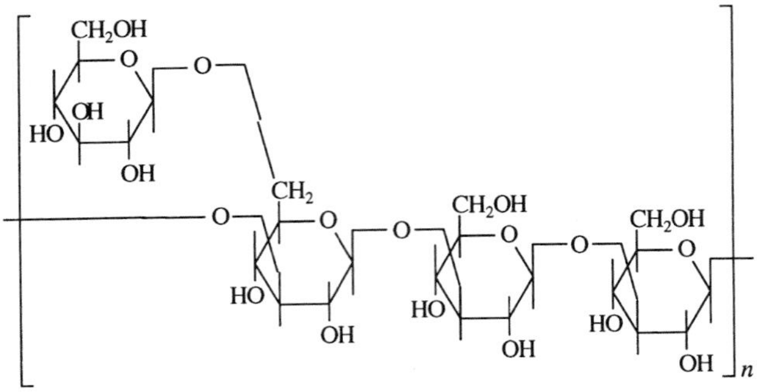

Schizophyllan
Scleroglucan

Fig. 8-29 Structure of schizophellan and scleroglucan. The main chain is a poly($\beta$-(1→3)-D-glucopyrane). Schizophyllan and scleroglucan differ only in the distribution of the $\beta$-(1→6)-bonded D-glucose side groups. A commercial scleroglucan with $X \approx 800$ is sold as Biopolymer CS®.

**Lichenin** (starch from lichen or moss) has linear molecules with $\beta$-1,3- and $\beta$-1,4-bonds in the ratio 1:2. In **isolichenin,** these bonds are in the α-position.

## 8.6 Gellan Group

Polysaccharides of the gellan group are produced by microbes: gellan by *Pseudomonas* (*Auromonas*) *elodea* and wellan by *Alcaligenes spp.*

The repeating unit of these polysaccharides is a tetrasaccharide with the sequence glucose–glucuronic acid–glucose–rhamnose (Fig. 8-30). Gellan itself is a linear polymer whereas wellan and rhamsan have sugar side groups (Table 8-11). Wellan and gellan carry *O*-acetyl groups: 1 group per 2 monosaccharide side groups in wellan and 1 group per disaccharide side group in rhamsan (here at the secondary OH).

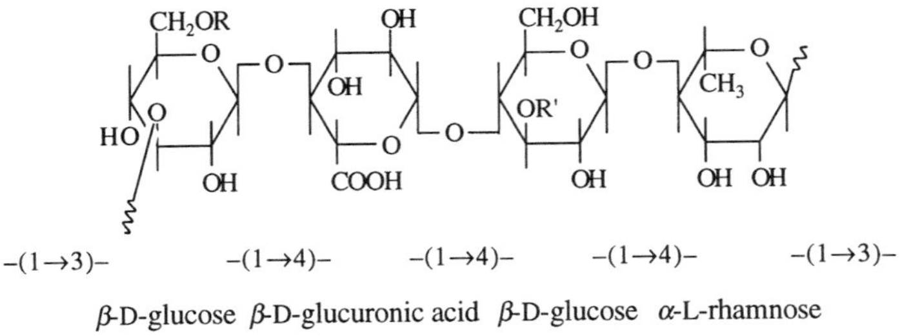

–(1→3)– –(1→4)– –(1→4)– –(1→4)– –(1→3)–

$\beta$-D-glucose $\beta$-D-glucuronic acid $\beta$-D-glucose $\alpha$-L-rhamnose

Fig. 8-30 Tetrasaccharide unit of gellan. For the structure of wellan and rhamsan see Table 8-11.

In the solid state, alkali salts of gellan form double helices that are packed parallel in the antiparallel direction with a slight shift of every second double helix. The double helices are stiffened by intracatenary hydrogen bridges between glucuronic acid and glucose units. The physical structure is stabilized by the bound cations.

Polymers of the gellan group dissolve in water but self-associate and form gels at relatively low concentrations. For this thickening action, gellan and wellan are used as additives in the food industry.

 

Table 8-11 Chemical structure of substituents R, R', and R" of gellan, wellan, and rhamsan (see tetrasaccharide unit in Fig. 8-30). * Bond to $-OCH_3$ or –O–.

| | R | R' | R" |
|---|---|---|---|
| Gellan | H | H | – |
| Wellan | H | HO, O*, R", OH OH | 2 : 1 ratio of $CH_3$ (α-L-rhamnose group) and $CH_2OH$ (α-L-mannose group) |
| Rhamsan | $CH_2OH$, O, OH, HO, OH, O, $CH_2$, O, OH, HO, *, OH<br>β-D-Glucose-(1→6)-α-D-glucose-(1→6)- | H | – |

# 8.7 Poly(galactose)s

Nature produces many types of polysaccharides but man uses mainly only glucose polymers (Section 8.3) and, to a much smaller extent, those derived from galactose (this section) and mannose (Section 8.8). Galactose-based polysaccharides are either extracts (e.g., from algae), exudates (e.g., from trees), seed or tree resins, or fermentation products (Table 8-12).

Table 8-12 Polyglycans based on galactose (gal). Su = Sulfate groups, *alt* = alternating, B = branched, L = linear, α = ax = axial, β = eq = equatorial. ? = not known with certainty.

| Polymer | Monomeric units | Bonds | Type | Occurrence |
|---|---|---|---|---|
| Fucoidan | L-gal (6-deoxy) (Su) | 1α-2α | B? | brown algae |
| | D-gal | 1β-3β | L | *Rosa glauca* |
| Arabinogalactans II | D-gal | (1β-3β)-*co*-(1β-6β) | B | needle trees |
| Arabinogalactans I | D-gal | 1β-4β | L | |
| | D-gal | (1β-4β)-*co*-(1β-6β) | B | |
| | β,L-, β,D-gal | (1-4)-*co*-(1-6) | B | snails |
| Carrageen | α,β-D (Su) | (1α-4α)-*alt*-(1β-3β) | L | red algae |
| Agarose | α-L, β-D | (1α-4α)L-*alt*-(1β-3β)D | L | |
| Pectin | α-D | 1-4 | L | |

Only a few of these polysaccharides are homopolymers. Many of them are branched, carry ionic groups, or are block or segmented copolymers. Most of them are soluble in water where they self-associate and form viscous solutions. For this reason, they are used as thickeners for food, emulsifiers, and protective colloids (cosmetics, adhesives, flocculants, film formers).

Industrially, these polysaccharides are known as gums (see also Section 3.9). In the past, "gum" was another term for "natural resin", a term that comprised not only these polysaccharides but also proteins and terpenes. Today, "gum" is the umbrella term for all industrially useful, water-soluble polysaccharides and their derivatives that form viscous solutions or dispersions. In botany, "gum" is the mucilage from the exudation of trees whereas "mucilage" (L: *mucus* = slime) is the term for carbohydrates that are stored in plants as reserve food.

## 8.7.1 Agar

**Agar** (**agar**-agar, **gum agar**, **gelose**) is obtained from seaweeds and marine algae, for example, certain red algae of the Pacific and Indian oceans and from kelp (brown algae) of the coast of California and Mexico. The world production is ca. 40 000 t/a.

Production procedures depend on the species, the desired purity, and the country. Algae are generally harvested by hand, even by divers in deep-sea diving suits. A typical procedure includes washing of algae, cooking in water for 2 hours and then in dilute sulfuric acid for 14 hours at 80°C, bleaching with sulfite, and filtering off the liquids. The remaining gel (Malayan: *agar* = gel) is cooled, cut in pieces, frozen, and rethawed, which disintegrates the cell walls. Soluble fractions are extracted by cold water. The remainder is again frozen, thawed, and washed with cold water in which agar is insoluble. Because of this elaborate procedure, agar is more expensive than similar gums from land plants. However, it has certain properties that make it indispensible for certain applications. For this reason, it is now also produced biosynthetically.

Agar dissolves in hot water but forms gels in cold water. It consists of at least two polysaccharides: agarose (up to 70 %) and agaropectin (up to 30 %) (see next section), furthermore cellulose (10 %) and xylan (ca. 3 %). Agar is used as such or for the preparation of agarose, which is the component responsible for gelation.

Agar is used directly as a thickening agent in foods, for example, in marmalades, sweets, American cheeses, yogurts, canned tuna, sorbets, etc. Other applications comprise photographic emulsions, and dental impressions. It has been mainly replaced by agarose as a carrier for immunodiffusion, immunoelectrophoresis, gel electrophoresis, and gel chromatography. Asian agar is such an important culture medium for bacteria (2000 t/a) that it is classified in the United States as a strategic reserve.

## 8.7.2 Agarose and Agaropectin

**Agarose** is a linear polysaccharide consisting of **agarobiose** as the repeating unit, a disaccharide from $\beta$-(1→3)-bound D-galactose units and $\alpha$-(1→4)-bound 3,6-anhydro-L-galactopyranose units. Galactose units are in part methylated in the 6-position. Contrary to older reports, no sulfate groups are present in the 6-position. However, agarose is said to contain some $\alpha$-L-arabinose units.

Chains of agarose form double helices with segments contributed by different molecules. The resulting network swells in water and forms gels that "melt" at 60–97°C. Gelation occurs at concentrations as small as 0.04 %. Lower gelation temperatures are produced by a commercially available partially hydroxyethylated agarose (replacement of –OH by $-OCH_2CH_2OH$).

**Agaropectin** is similar to agarose but some saccharide units are replaced by 4,6-*O*-(1-carboxyethylidene)-D-galactopyranose units or by sulfated or methylated sugar units.

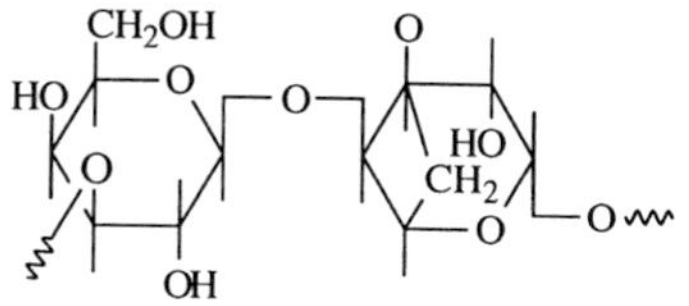

agarobiose repeating unit of agarose

## 8.7.3 Carrageen

**Carrageen (carageen, carragheen)** is the umbrella term for a group of polysaccharides of seaweeds that contain several types of cations (Na, K, $NH_4$, Ca, Mg). It is found in the red algae *Chondrus crispus* and *Gigartina stellata* of the Atlantic Ocean, especially as **Irish moss** near the Irish coastal town of Carragheen (name). "Carrageen" is sometimes used only for dried algae and "carrageenane" for extracts of algae. "Carrageenates" are semi-synthetic products that contain only one type of cation.

Like agar, carrageen is collected by hand, mainly in Ireland, Northern France, and the United States. So far, mechanical harvesting has not been successful.

Carrageens are linear polysaccharides that consist of alternating units of 1,3-bonded *β*-D-galactose units and 1,4-bonded *α*-D-galactose units (Table 8-13). In some units, one hydroxyl group (or several of them) is replaced by the half-ester of sulfuric acid. There are at least 12 types of carrageens that differ in the presence, number, and position of sulfate groups as well as in the presence or absence of anhydrogalactose units.

Table 8-13 Structures of important carrageens. Also known are α, β, γ, ξ, π, χ, and ω types.

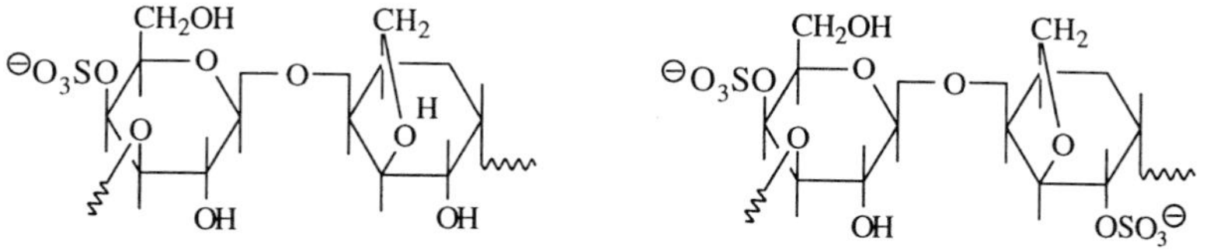

kappa　　　iota

| Name | | 1,3-bonded | 1,4-bonded |
|---|---|---|---|
| μ-Carrageen | (mu) | D-galactose-4-sulfate | D-galactose-6-sulfate |
| ι-Carrageen | (iota) | D-galactose-4-sulfate | D-galactose-2,6-disulfate |
| κ-Carrageen | (kappa) | D-galactose-4-sulfate | 3,6-anhydro-D-galactose |
| ν-Carrageen | (nu) | D-galactose-4-sulfate | 3,6-anhydro-D-galactose-2-sulfate |
| λ-Carrageen | (lambda) | D-galactose-4-sulfate (70 %) + D-galactose-2-sulfate (30 %) | 3,6-anhydro-D-galactose-2-sulfate |

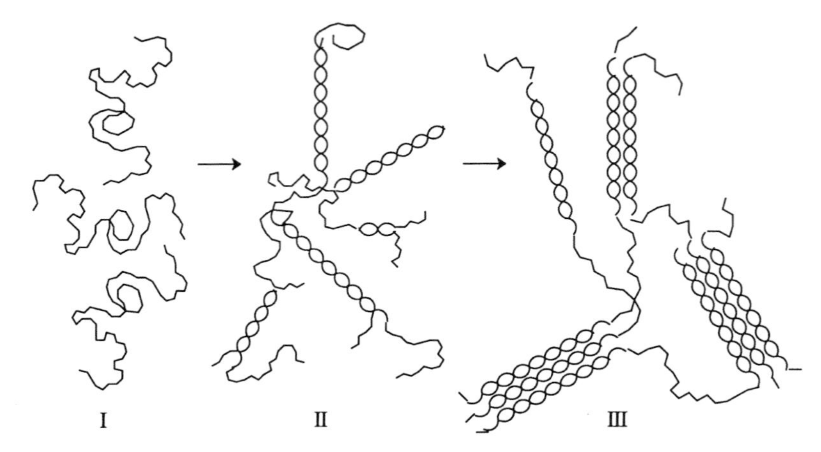


Fig. 8-31 On cooling dilute aqueous solutions of κ- and λ-carrageen, random coils (I) associate in part intermolecularly to double helices (II) which cluster to aggregates (III) and form an elastic network.

μ-Carrageen is the biological precursor of κ-carrageen, and ν-carrageen that of ι-carrageen. Industrially used are κ-, ι-, and λ-carrageens. On cooling their aqueous solutions, κ- and ι-carrageen deliver elastic, thermoreversible gels by forming associated partial double-helix structures (Fig. 8-31), especially in the presence of $K^+$ and $Ca^{2+}$. Hot solutions of sodium carrageens remain highly viscous on cooling.

The formation of these helical segments (and thus the gelation of solutions) is not impeded by sulfate ester groups in the $C_4$ position at 1,3-bonded sugar units, but very much so by such groups in the $C^2$ position at 1,4-bonded galactose units. For this reason, cold aqueous solutions of κ-carrageen form gels but those of λ-carrageen do not. ι-Carrageen both dissolves and gels in cold water.

The food industry consumes more than 80 % of carrageens as gelling agents, stabilizers in ice cream, indigestible but very filling dietary components, etc. The rest of the carrageens are used in pharmacy and cosmetics.

### 8.7.4 Furcelleran and Funoran

**Furcelleran** (Danish agar) is obtained from the red algae *Furcellaria fastigiata.* Its exact chemical structure is unknown; ca. half of it is a less-sulfated κ-carrageen. Like carrageen, it is used by the food industry.

**Funoran** is a sulfated poly(galactose) that is used in Japan as a size and adhesive.

### 8.7.5 Pectin

Pectins are gelling polysaccharides (G: *pektos* = coagulated) that are present in the cell fluids, cell walls, and primary cell membranes of fruits, leaves, and roots of all higher plants. Lemons and oranges contain up to 30 % pectins. Commercially important are also the pectins of sugar beets. Even young cotton contains up to 5 % pectins which is however reduced to 0.8 % upon ripening.

Pectins serve as a kind of cement for cell walls, probably regulate ion permeabilities, and are possibly employed in the metabolism of reserve polysaccharides. Commercially, they are used as gelling agents or thickeners.

The main chain of citrus pectins consists solely of poly[$\alpha$-D-galacturonic acid]. All other pectins contain up to 20 % neutral sugar units such as L-rhamnose, L-arabinose, D-glucose, D-galactose, or D-xylose. L-Rhamnose units are always 1,2-bonded in main chains. All other neutral sugar units are in short side chains, mainly at rhamnose units but also at D-galactose units.

About 20-80 % of carboxylic groups of galacturonic acid units are esterified by methanol. One thus distinguishes between highly esterified (> 50 %) and less-esterified (< 50 %) pectins. Degrees of esterification and molecular weights (20 000–40 000) vary with the origin and preparation of pectins. Hydroxyl groups of citrus pectins are not acetylated but those of sugar beets are, though to a small extent.

The intended use of a pectin controls its production. Pectins for gelling are extracted from plants with slightly acidic water and precipitated from solutions by alcohol. Pectins for thickening are extracted by alkaline solutions.

Gelling agents may or may not contain calcium ions. For calcium types, pectic acids should contain only a few ester groups so that many carboxyl groups are available for a coordination with $Ca^{2+}$. Since one $Ca^{2+}$ ion coordinates with carboxyl groups from many chains, physical crosslinks are formed that lead to gelling of dilute solutions.

Crosslinking of calcium-free types is promoted by high degrees of esterification. Such pectins contain many hydrophobic groups per molecule that try to cluster in hydrophobic domains. Since chains are stiffened by ionized carboxyl groups, such clusters can only be formed intermolecularly. This is the reason why the stiffest gels are produced by pectins with degrees of esterification of ca. 50%. Gelling is advanced by addition of acids (decreases dissociation of COOH groups) or sugars or glycerol (promotes dehydration of OH and COOH groups).

### 8.7.6 Tragacanth

**Tragacanth (Gum tragacanth)** is a plant exudate of thorny shrubs, also called tragacanth, of the genus *Astralagus* from which it is extracted by incision of twigs or trunks. The excreted sap congeals after a few days.

Tragacanth consists of a mixture of up to 40 % water-soluble, slightly acidic tragacanthin ($M \approx 10^4$) and less than 60 % water-insoluble (but swellable) bassorin ($M \approx 10^5$). The polysaccharide chains contain L-arabinose, D-galactose, L-fucose, and D-xylose, as well as some D-galacturonic acid.

Aqueous solutions of tragacanth have extraordinarily high viscosities at very low concentrations (Fig. 8-32) which makes it a very much sought after hydrocolloid. It can be added to food without volume restrictions, for example, as a thickener in ice cream. The food industry uses it as a protective colloid and emulsion stabilizer. It also serves as a binder in pharmaceuticals, base for hand creams and lotions in cosmetics, size in the textile industry, thickener for printing inks, etc. The very expensive gum is often replaced by sodium alginates (Section 8.8.3), Guar gum (Section 8.8.1), or carob meal. It is also not always available and therefore (and because of its high price) often adulterated.

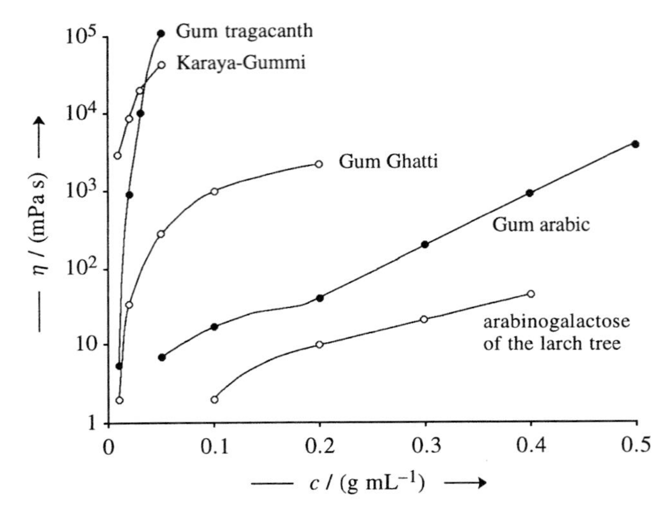


Fig. 8-32 Concentration-dependence of viscosities $\eta$ of cold aqueous solutions of various gums [13]. At 20°C, the viscosity of water is ca. 1 mPA s.

## 8.7.7 Gum Ghatti

**Gum Ghatti (Indian gum)** is not a poly(galactose) but is mentioned here because it is also used as a hydrocolloid. G. Ghatti is obtained in India and Sri Lanka from exudates of the tree *Anogeissus latifolia*.

G. ghatti is a branched polysaccharide that contains L-arabinose, D-galactose, D-mannose, D-xylose, and D-glucuronic acid in the ratio 10:6:2:1:2. The main chain consists of 1,6-connected $\beta$-D-galactopyranosyl units. Side chains are connected to the main chain via L-arabinofuranose units. G. ghatti is used as emulsifier in the food and pharmaceutical industries and as thickener in other industries.

## 8.7.8 Gum Arabic

**Gum arabic (Sudan gum, acacia gum)** is the dried exudate of sick acacias; healthy trees do not produce it. The annual yield is ca. 250 gram per tree. About three quarters of the annual world production of ca. 70 000 t are gathered in Sudan; the rest in Western Africa. Arabia does not produce G. arabic. It is called G. arabic because it was formerly imported from Arabia.

G. arabic is a mixture of calcium, magnesium, and potassium salts of arabic acid, a polysaccharide (a mixture of polymers) consisting of L-arabinose, D-galactose, L-rhamnose, and D-glucuronic acid units in the ratio 3:3:1:1. It may be called a poly(galactose) or a poly(arabinose). Galactose units are connected via 1,3 bonds. Some main chain units are substituted in the $C^6$ position.

G. arabic is used mainly as a thickener in the food and cosmetics industry and also as an ingredient in textile colorants, inks, textile sizes, and as a binder in matches. It is practically no longer used as an adhesive and for the packaging of pharmaceuticals.

## 8.8 Poly(mannose)s

Polysaccharides with mannose units in the main chain are also known as **mannans**. Mannose is an epimer of glucose. The name is derived from the Aramaic *manna* (= gift), the "sweet bread" from heaven miraculously provided during the flight of the Israelites from Egypt (Exodus 16:14-36). Manna was probably the secretion of the scale insect *Trabutina mannipar* that feeds on the fruits of tamarisks. The yellow, honey-like substance solidifies in the cool of the desert night but softens when the sun is shining.

A similar sweet sap is found as honeydew on the leaves of many trees (cherry, plum, maple, linden, oak, etc.). It comes in part from aphids and in part from the transpiration of leaves. Present-day commercial manna is the dried sap of the manna ash tree of Italy (*Fraxinus ornus*), an olive tree. The sap contains 40–60 % mannitol (manna sugar). Madagascar manna delivers dulcitol.

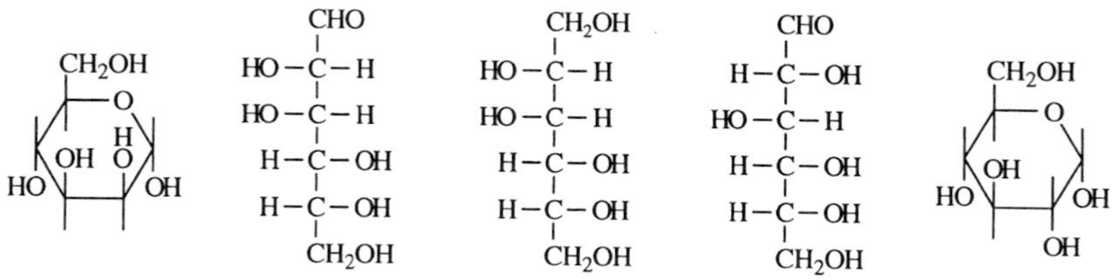

D-mannose | D-mannose | mannitol | D-glucose | D-glucose

### 8.8.1 Guar

Guar (*Cyanmopsis tetragonolaba*) is planted in India, Pakistan, and the Southwestern United States. The seeds contain guar gum, a polysaccharide with a poly[$\beta$-(1→4)-D-mannopyranose] main chain and D-galactose side groups that are bonded via α-(1→6) links to every second mannopyranosyl unit. Production of guar is ca. 150 000 t/a.

Industrial products have been reacted in part with ethylene oxide or propylene oxide to undisclosed degrees of reaction and/or substitution. H atoms marked by an * in the formula below are thus replaced by $(CH_2CH_2O)_nH$ or $(CH_2CH(CH_3)O)_nH$ with $n \geq 1$.

Hydroxyalkylation reduces or even eliminates the tendency to flocculate. In water, hydroxyalkylated guars dissolve faster than non-hydroxyalkylated ones. Solutions gel on addition of multivalent ions.

Guar and its derivatives are used in mining as flotation and flocculation aids, as a thickening agent, in paper manufacturing, as food additive, etc.

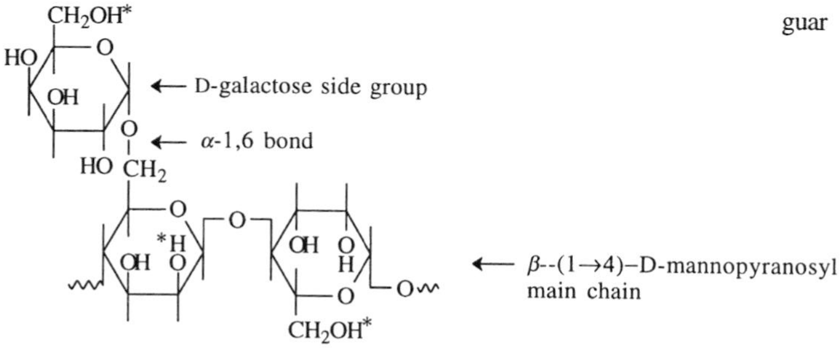

guar

← D-galactose side group

← $\alpha$-1,6 bond

← $\beta$--(1→4)–D-mannopyranosyl main chain

### 8.8.2 Carob

Carob (**locust bean gum**) is the polysaccharide from the unripe seeds (St. John's bread) of the evergreen carob (tree) (*Ceratonia siliqua*) (see also p. 202). The main chain of carob consists of D-mannose units. Ca. 25 % of these units carry a single galactose unit. Galactomannopyranosyl units are probably arranged in block-type segments.

Carob is a brownish flour that dissolves in water at tempratures above 82°C. The solutions do not gel on cooling. The polysaccharide is undigestible but swells in the intestines, binds acids, and absorbs poisons. Ground carob has a chocolate-like taste and is used as a thickener in ice cream, "chocolate" coating of nuts and raisins, etc.

### 8.8.3 Algins

Algins are polysaccharides of the cell walls of brown algae (L: *alga* = see grass) where they are responsible for ion exchange. The parent compound of algins is **alginic acid** which consists of 31–65 % mannuronic acid and 35–69 % guluronic acid. Alginic acid is a linear multiblock polymer consisting of three types of blocks: (1) 13–40 % blocks of $\beta$-(1→4)-D-mannuronic acid (ManA), (2) 18–60 % blocks of $\alpha$-(1→4)-L-guluronic acid (GulA), and (3) 27–42 % blocks of alternating mannuronic and guluronic acid units (ManA-alt-GulA).

Extraction of algae by soda solution delivers the sodium salt of alginic acid. The resulting **alginates** have lower molecular weights (20 000–60 000) than native alginic acids (ca. 200 000). Alginates are produced in England, France, Norway, Japan, Southern California, and Australia.

1-2 % aqueous solutions of sodium alginate are used as suspension and emulsion agents. Ammonium alginate is a thickener in ice creams. Fibers from water-soluble alginates are used for throw-away parachutes of commando units. Fibers from water-insoluble calcium alginate serve as fire-resistant fabrics and for resorbable surgical threads.

Reaction of alginic acid with propylene oxide delivers non-toxic propylene glycol alginate which does not gel at higher concentrations. Typical use is as a stabilizer for puddings, ice cream, orange juice, foam stabilizer of beer (US), inks, cosmetics, etc. In Germany, propylene glycol alginate is not allowed as a food additive.

## 8.9 Hemicelluloses

"Hemicellulose" is the international name of polysaccharides that are associated with celluloses and lignins. Nature produces them in enormous amounts ,but man has so far little used them as such, as raw materials for chemicals, or as a source of energy.

"Hemicellulose" referred originally to polysaccharides that could be extracted from cellulosic materials and were thus thought to be precursors of cellulose. Despite the name, hemicelluloses are neither low-molecular weight nor partial celluloses (G: *hemi* = half). They are not composed of $\beta$-1,4-glucose segments; glucose units are only a small proportion of hemicellulose molecules. Instead, hemicelluloses consist of many different hexoses (glucose, galactose, mannose, etc.) and pentoses (xylose, arabinose, etc.).

Because of the multitude of sugar types (and not because of being polymers), they are also called **polyoses** or, in some languages, also **wood polyoses**. The latter term is not justified either because hemicelluloses are not necessarily associated with wood but are also found in grasses, stems of plants, peels of potatoes and sugar beets, etc., where they are present in amounts of 15–45 % (Table 3-24). In these plant parts, hemicelluloses are essential components of cell walls.

Hemicelluloses are isolated from plants by removing lignins and then celluloses. Lignin is degraded by chlorine, chlorine dioxide, or peracetic acid. Cellulose is extracted by a mixture of dimethylsulfoxide and water, which allows the preservation of ester bonds of hemicelluloses. Hemicelluloses are partially degraded by hot water, lime water, and 10 % soda lye) (caustic soda solution) and totally destroyed by sulfite and sulfate processes.

Hemicelluloses dissolve in water. They serve to thicken foodstuffs, prevent staling of bread, and stabilize foam of beers. Derivatives of hemicelluloses are used as printing agents, in photographic processes, and in cosmetics. Small amounts of hemicelluloses are also feedstocks for the production of furfural, methanol, glycols, and formic acid.

Hemicelluloses are today subdivided into xylans, mannans, β-glucans with mixed linkages, and xyloglucans. The latter two groups are classified as gums if they can be extracted by alkali. An example is the xylan from the seaweed *Palmariales* (Fig. 8-7).

### 8.9.1 Hemicelluloses of Hard Woods

Hemicelluloses of hard woods consist of mixtures of xylanes (major component, see below) and glucomannanes (minor component). In contrast to hemicelluloses of soft woods, they do not contain galactose units

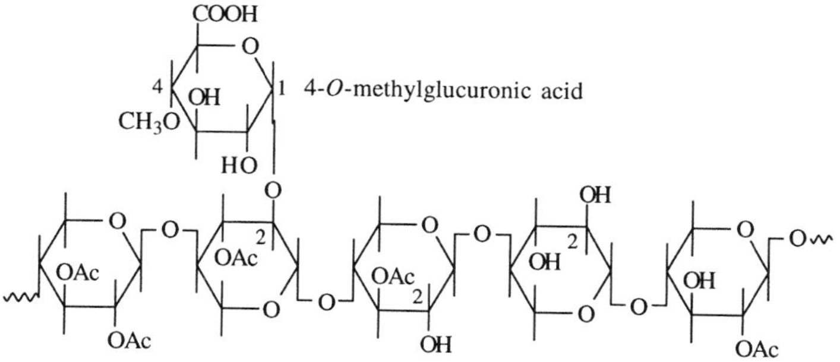

4-*O*-methylglucuronic acid

Partially acetylated poly[*β*-(1→4)-xylopyranose]; Ac = $CH_3CO$

Xylanes are poly[*β*-(1→4)-xylopyranose]s with degrees of polymerization of ca. 200 in which some xylose units of the main chain are replaced by rhamnose and galacturonic acid units. Approximately 70 % of the $C^2$ and $C^3$ hydroxyl groups of xylose units are acetylated; one unit may carry 0, 1, or 2 acetyl groups. Approximately 10 % of xylose units are furthermore substituted by 4-*O*-methylglucuronic acid which is bound via the 1 position to the $C^2$ of a xylose unit of the main chain.

Hemicelluloses of hard woods contain also 2–5 % of glucomannan, a polymer with a main chain of *β*-(1→4)-connected D-glucose and D-mannose units (see top of the next page), apparently in random sequence. The glucose/mannose ratio is usually 1:2, but in some species it is 1:1.

poly[$\beta$-(1→ 4)-glucopyranose-*co*-mannopyranose]

### 8.9.2 Hemicelluloses of Soft Woods

Hemicelluloses of soft woods have more complicated structures than those of hard woods. Although the main components are also poly[$\beta$-(1→4)-xylopyranose]s, these chains have lower degrees of polymerization of ca. 150 and are not acetylated. The content of 4-*O*-methylglucuronic acid units increases to 70 % from 10 %. About 11 % of the xylose units are substituted by arabinose units via 1→3 bonds.

4-*O*-methylglucuronic acid

poly[$\beta$-(1→4)-xylopyranose] chain

$\alpha$-L-arabinose

Hemicelluloses of soft woods contain at least two different galactomannans. The major fraction is a water-soluble copolymer of $\beta$-(1→4) connected, apparently randomly distributed, mannopyranosyl, glucopyranosyl, and galactopyranosyl units in the ratio 3:3:1 that are probably all acetylated in the native state. The degree of acetylation drops to 6–9 % after isolation and purification.

$\alpha$-D-galactopyranose

partially acetylated poly[$\beta$-(1→4)-D-galactose-*co*-D-glucose-*co*-D-mannose]

According to some reports, the major component carries acetyl groups at the $C^2$ and $C^3$ positions of mannose units, according to others, only at the $C^3$ position but at both mannose and glucose units. Sugar units are also substituted by (1→6)-connected α-D-galactopyranose groups.

The minor component has a similar structure as the major component albeit with mannose, glucose, and galactose units in the ratio 3:1:0.1. Because of the small fraction of galactose, it is often simply called a glucomannan.

### 8.9.3 Hemicelluloses of Grasses

The composition of xylanes of grasses varies widely. For example, the xylan of esparto grass is a linear poly[β-(1→4')-D-xylopyranose] with a degree of polymerization of ca. 70. Hemicelluloses of other grasses are branched as shown below.

Poly[β-(1→4')-D-xylopyranose]s of the other grasses contain less glucuronic units than the hemicelluloses of hard woods but considerably more α-L-arabinose units, for example, 1 arabinose per 7 xylose in the knots of wheat stems. Arabinose units are bound to both $C^3$ and $C^2$ of main chain xylose units.

In the hemicellulose of barley straw, ca. 6–7 % of arabinose units are esterified by ferulic acid, $HO(CH_3O)C_6H_3–CH=CH–COOH$, and ca. 3 % by *p*-coumar(in)ic acid (= 4-hydroxycinnamic acid, $HOC_6H_4–CH=CH–COOH$.

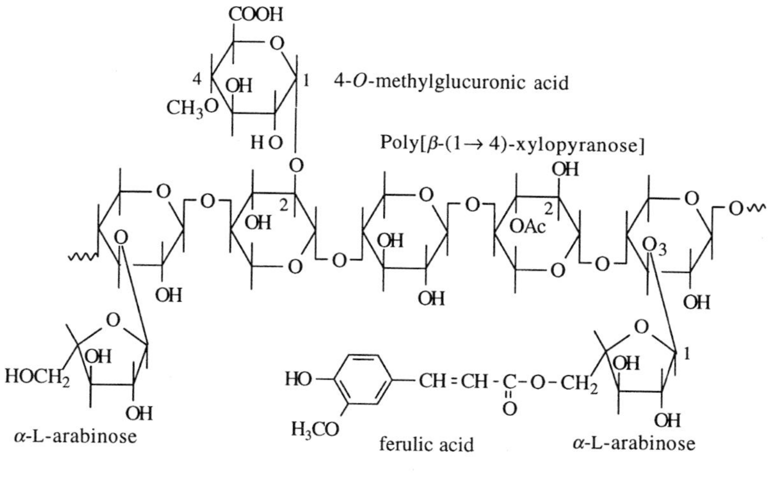

COOH
4-*O*-methylglucuronic acid
Poly[β-(1→ 4)-xylopyranose]
α-L-arabinose
ferulic acid
α-L-arabinose

## 8.10 Poly(fructose)s

Poly(fructose)s are polymers with fructose units in the main chain. They are also called **poly(fructosane)s, fructosanes**, or **fructanes**. The subgroup of **phleans** has 2,6-connected sugar groups, the subgroup of **inulin**, 1,2-glycosidic ones.

Poly(fructose)s are reserve poly(saccharide)s in roots, leaves, and seeds of plants. They are also produced by certain bacteria, for example, levans of the phlean group from saccharose as substrate.

fructose unit

glucose endgroup

inulin

saccharose unit

**Inulin** is a linear polysaccharide consisting of $\beta$-1,2-connected furanosidic fructose units that carries a glucose endgroup. It was first obtained by heating the rhizome (rootstock) of the yellow-flowered, 1.5 m-high scabwort (*Inula helenium* L.) with water and precipitating the dissolved polymer ($X \approx 30$) by ethanol. Commercial inulin with $X \approx$ 10–12 is obtained from chicory roots (*Cichorium intybus*) that contain ca. 15-20 % inulin and 5–10 % **oligofructose**, a low molecular weight inulin with $X \approx 4$. A newer **HP inulin** has a degree of polymerization of ca. 25 ($M \approx 4050$).

Inulin is found in ca. 22 000 plants of the *Compositae* and *Asteracae* families such as the tubers of Jerusalem artichoke (*Helianthus tuberosus*), hull leaves of the true (globe) artichoke (*Cynara scolymus*), sunflowers, chrysanthemums, asters, etc.

Inulin gels in water. The bland to mildly sweet inulin fills like starch and has long been used in dietary foods. Inulin is called a soluble fiber by the food industry because it is not absorbed in the stomach but digested in the large intestine by lactic acid bacteria (p. 395). Ingestion of large quantities may lead to gas and bloating. Inulin is increasingly used in other foods where it serves as a fat replacement similar to gelatin, carrageen, or pectin. Roasted chicory roots have served as coffee "ersatz" (substitute) for more than 220 years. Inulin is also hydrolyzed to fructose.

# Literature to Chapter 8

8.1a INTRODUCTION: Oligosaccharides and Carbohydrates (general)

P.M.Collins, Ed., Carbohydrates, Chapman & Hall, London 1987
H.S.El Khadem, Carbohydrate Chemistry, Academic Press, San Diego, CA 1988
J.F.Kennedy, Carbohydrate Chemistry, Oxford Science Publ., New York 1988
A.Lipták, P.Fügedi, Z.Szurmai, J.Harangi, CRC Handbook of Oligosaccharides, CRC, Boca Raton, 3 Volumes: I Disaccharides (1990); II Trisaccharides (1990); III: Higher Oligosaccharides (1991)
F.W.Lichtenthaler, Ed., Carbohydrates as Organic Raw Materials, VCH, Weinheim 1991 (mono and disaccharides, no polysaccharides)
D.L.Kaplan, Ed., Biopolymers from Renewable Sources, Springer, Heidelberg 1998
G.Zaikov, Ed., Chemistry of Polysaccharides, Brill/VSP, Leiden, Netherlands, 2005

8.1b INTRODUCTION: Polysaccharides

-, Nomenclature of Polysaccharides, J.Org.Chem. **28** (1963) 281
B.Bernfeld, Ed., Biogenesis of Natural Compounds, Pergamon Press, London 1967
A.G.Walter, J.Blackwell, Biopolymers, Academic Press, New York 1973 (no synthesis)

G.D.Fasman, Ed., CRC Handbook of Biochemistry and Molecular Biology, Section C: Lipids, Carbohydrates and Steroids, CRC Press, Boca Raton (FL), 3rd ed. 1976
E.A.MacGregor, C.T.Greenwood, Polymers in Nature, Wiley, Chichester 1980
C.R.Cantor, P.R.Schimmel, Biophysical Chemistry, Freeman, San Francisco (CA) 1980 (3 vols.)
G.O.Aspinall, Ed., The Polysaccharides, Academic Press, New York, Vol. 1 (1982), Vol. 2 (1983)
V.Creszenzi, I.C.M.Dea, S.S.Stivala, New Developments in Industrial Polysaccharides, Gordon and Breach, New York 1985
P.M.Collins, Ed., Carbohydrates, Chapman & Hall, London 1987
M.Yalpani, Ed., Industrial Polysaccharides, Elsevier Sci.Publ., New York 1987
E.Dickinson, Food Polymers, Gels, and Colloids, CRC Press, Boca Raton (FL) 1991
D.L.Kaplan, Ed., Biopolymers from Renewable Resources, Springer, Berlin 1998
R.Gross, C.Scholz, G.Leatham, Polymers from Renewable Resources, Carbohydrates and Agroproteins, Oxford Univ.Press, Oxford 2000 (ACS Symp. Series)
A.Steinbüchel, E.J.Vandamme, S.De Baets, Eds., Polysaccharides I - Polysaccharides from Prokaryotes (2002); A.Steinbüchel, S.De Baets, E.J.C.Vandamme, Eds., Polysaccharides II - Polysaccharides from Eukaryotes (2002). These two volumes are Volumes 5 and 6 of A.Steinbüchel, Ed., Biopolymers, Wiley, New York 2001-2003 (12 volumes)
A.Steinbüchel, S.K.Rhee, Eds., Polysaccharides and Polyamides in the Food Industry, Wiley-VCH, Weinheim 2005 (2 volumes)
K.J.Yarema, Handbook of Carbohydrate Engineering, CRC Press, Boca Raton (FL) 2006

8.2.2 STARCH
J.A.Radley, Starch and Its Derivatives, Chapman & Hall, London, 4th ed. 1968
W.Banks, C.T.Greenwood, Starch and Its Components, Edinburgh University Press, Edinburgh 1975
J.A.Radley, Ed., Examination and Analysis of Starch and Its Components, Appl.Sci.Publ., Barking, Essex 1976
J.A.Radley, Ed., Starch Production Technology, Appl.Sci.Publ., Barking, Essex 1976
J.A.Radley, Ed., Industrial Uses of Starch and Its Components, Appl.Sci.Publ., Barking, Essex 1976
J.C.Johnson, Industrial Starch Technology, Noyes Publ., Park Ridge (NJ) 1979
R.L.Whistler, J.N.BeMiller, E.F.Paschall, Starch: Chemistry and Technology, Academic Press, New York, 2nd ed. 1984
O.B.Wurzburg, Modified Starches: Properties and Uses, CRC Press, Boca Raton (FL) 1986
T.Galliard, Ed., Starch: Properties and Potential, Wiley, Chichester 1987

8.2.6 DEXTRINS
M.L.Bender, M.Komiyama, Cyclodextrin Chemistry, Springer, Berlin 1978
J.Szejtli, Cyclodextrins and Their Inclusion Complexes, Akademiai Kiado, Budapest 1982; -, Cyclodextrin Technology, Kluwer Academic, Dordrecht 1988

8.2.7 DEXTRANS
A.Grønwall, Dextran and Its Use in Colloidal Infusion Solutions, Almquist & Wiksell, Stockholm 1957
A.N. de Belder, Dextran, Pharmacia, Uppsala, 2nd ed. 1990

8.2.10 PULLULAN
S.Yuen, Development of Pullulan: Its Characteristics and Applications, in R.D.Deanin, Ed., New Industrial Polymers, ACS Symp.Ser. **4** (1974) 172

8.3 CELLULOSE (including derivatives)
F.D.Miles, Cellulose Nitrate, Interscience, London 1953
E.Ott, H.M.Spurlin, Eds., Cellulose and Cellulose Derivatives, Interscience, New York, 2nd ed. 1956, volumes 1–3; ditto, N.M.Bikales, L.Segal., Eds., volumes 4–5 (1971)
V.E.Yarsley, W.Flavell, P.S.Adamson, N.G.Perkins, Cellulosic Plastics, Iliffe, London 1964
R.D.Preston, The Physical Biology of Plant Cell Walls, Chapman & Hall, London 1974
A.Frey-Wyssling, The Plant Cell Wall, Gebr.Bornträger, Berlin 1976
R.M.Rowell, R.A.Young, Ed., Modified Celluloses, Academic Press, Orlando, FL 1978
S.M.Hudson, J.A.Cuculo, The Solubility of Unmodified Cellulose: A Critique of the Literature, J.Macromol.Sci.-Rev.Macromol.Sci. **C 18** (1980) 1

A.Hebeish, J.T.Guthrie, The Chemistry and Technology of Cellulosic Copolymers, Springer, Berlin 1981 (graft copolymers)
R.M.Brown, Jr., Ed., Cellulose and Other Natural Systems: Biogenesis, Structure, and Degradation, Plenum, New York 1982
D.T.Clark, A.H.K.Fowler, P.J.Stephenson, Application of Modern Analytical Techniques to the Investigation of Cellulose Nitrates, J.Macromol.Sci.-Rev.Macromol.Chem.Phys. **C 23** (1983) 217
T.P.Nevell, S.H.Zeronian, Ed., Cellulose Chemistry and Its Applications, Wiley-Horwood, Chichester 1985
R.A.Young, R.M.Rowell, Cellulose. Structure, Modifications and Hydrolysis, Wiley-Interscience, New York 1986
L.-T.Fan, M.M.Gharpuray, Y.-H.Lee, Cellulose Hydrolysis, Springer, Berlin 1987
J.F.Kennedy, G.O.Phillips, P.A.Williams, Cellulose: Structural and Functional Aspects, Prentice Hall, Englewood Cliffs (NJ) 1989
Th.Nevell, P.Zeronian, S.Haig, Eds., Cellulose Chemistry. Its Application, Horwood Publ., Chichester 1989
L.V.Backinowski, M.A.Chlenov, Cellulose. Biosynthesis and Structure, Springer, Berlin 1990
L.A.Tarchevsky, G.N.Marchenko, Cellulose: Biosynthesis and Structure, Springer, Berlin 1991
H.Krässig, Ed., Cellulose. Structure, Accessibility, and Reactivity, Gordon and Breach, Yverdon, Switzerland, 1993
J.F.Kennedy, G.O.Phillips, P.A.Williams, Eds., Cellulosics: Chemical, Biochemical and Material Aspects, Ellis Horwood, Chichester 1993
E.Doelker, Cellulose Derivatives, Adv.Polym.Sci. **107** (1993) 199
R.D.Gilbert, Ed., Cellulosic Polymers. Blends and Composites, Hanser, Munich 1994
J.F.Kennedy, G.O.Phillips, P.A.Williams, L.Piculell, Eds., Cellulose and Cellulose Derivatives: Physicochemical Aspects and Industrial Applications, Woodhead Publ.Ltd., Cambridge 1995
J.C.Roberts, The Chemistry of Paper, Royal Soc.Chem., London 1996
D.Klemm, B.Philip, T.Heinze, Comprehensive Cellulose Chemistry, VCH-Wiley, Weinheim 1998 (mainly viscose and rayon)
D.-N.S.Hon, N.Shiraishi, Ed., Wood and Cellulosic Chemistry, Dekker, New York, 2nd ed. 2000
P.Rustemeyer, Ed., Cellulose Acetates: Properties and Applications, Wiley-VCH, Weinheim 2004

8.3.10 CHITIN AND CHITOSAN

R.A.Muzarelli, Chitin, Pergamon, Oxford 1976
W.D.Comper, Heparin (and Related Polysaccharides), Gordon and Breach, New York 1981
J.P.Zikakis, Ed., Chitin, Chitosan, and Related Enzymes, Academic Press, New York 1984
R.A.Muzarelli, C.Jeuniaux, G.W.Gooday, Eds., Chitin in Nature and Technology, Plenum, New York 1986
A.F.Roberts, Chitin Chemistry, MacMillan Press, London 1986
E.R.Pariser, D.P.Lombard, Chitin Sourcebook, Wiley, New York 1989
G.Skjåk-Bræk, T.Anthonsen, P.Sandford, Eds., Chitin, Elsevier Applied Science, London 1989
P.Jollès, R.A.A.Mazzarelli, Eds., Chitin and Chitinases, Birkhäuer, Basel 1999
E.Khor, Chitin: Fulfilling a Biomaterials Promise, Elsevier, Amsterdam 2002

8.4 MUCOPOLYSACCHARIDES

J.S.Brimacombe, J.M.Webber, Mucopolysaccharides, Elsevier, Amsterdam 1964
R.W.Jeanloz, E.A.Balasz, Eds., The Amino Sugars, Academic Press, New York 1965 ff. (4 vols.)
R.A.Bradshaw, S.Wessler, Eds., Heparin: Structure, Function and Clinical Implications, Plenum, New York 1975
N.M.McDuffie, Ed., Heparin. Structure, Cellular Functions and Clinical Applications, Academic Press, New York 1979
W.D.Comper, Heparin (and Related Polysaccharides), Gordon and Breach, New York 1981
J.P.Zikakis, Chitin, Chitosan and Related Enzymes, Academic Press, New York 1984

8.5-8.8 OTHER POLYSACCHARIDES

M.Glicksman, Gum Technology in the Food Industry, Academic Press, New York 1969
N.Sharon, Complex Carbohydrates: Their Chemistry, Biosynthesis, and Functions, Addison-Wesley, Reading, MA 1975
R.L.Davidson, Ed., Handbook of Water-Soluble Gums and Resins, McGraw-Hill, New York 1980

R.L.Whistler, J.N.BeMiller, Eds., Industrial Gums: Polysaccharides and Their Derivatives, Academic Press, San Diego (CA), 3rd ed. 1993
A.N. de Belder, Industrial Gums, Academic Press, New York 1993
R.Lapasin, S.Priel, Rheology of Industrial Polysaccharides, Blackie, London 1995
J.Visser, A.G.J.Voragen, Eds., Pectin and Pectinases, Elsevier Science, Amsterdam 1996

8.9 HEMICELLULOSES
R.L.Whistler, E.L.Richards, Hemicelluloses, in W.Pigman, D.Horton, Eds., The Carbohydrates, Chemistry and Biochemistry, Academic Press, New York, Vol. **11B** (1970) 447
R.L.Whistler, R.N.Shah, Recent Developments in the Industrial Use of Hemicelluloses, in R.M.Rowell, R.A.Young, Eds., Modified Cellulosics, Academic Press, New York 1977, 341
G.T.Maloney, Chemicals from Pulp and Wood Waste, Noyes Data, Park Ridge (NJ) 1978
K.C.B.Wilkie, The Hemicelluloses of Grasses and Cereals, Adv.Carbohydr.Chem.Biochem. **36** (1979) 215
C.-S.Gong, L.F.Chen, M.C.Flickinger, G.T.Tsao, Conversion of Hemicellulose Carbohydrates, in A.Fiechter, Ed., Adv.Biochem.Eng. Bioenergy, Springer, Berlin 1981, p.93
K.C.B.Wilkie, Hemicellulose, CHEMTECH (May 1983) 306
J.W.Rowe, Ed., Natural Products Extraneous to the Lignocellulosic Cell Wall of Woody Plants, Springer, Berlin 1987
D.N.-S.Hon, Ed., Chemical Modification of Lignocellulosic Materials, Dekker, New York 1995
A.Ebringerová, Z.Hromádková, T.Heinze, Hemicelluloses, Adv.Polym.Sci. **186** (2005) 1

## References to Chapter 8

[1] 1983 world and US data from scattered data in the literature
[2] 2004 world and some US data from the United Nations Food and Agricultural Organization (http//faostat.fao.org 2005, accessed 12 February 2006)
[3] US Statistical Abstracts 2006 (www.census.gov/statab, accessed 12 February 2006)
[4] –, Chem.Eng.News (8 August 2005) 44
[5] F.Lichtenthaler, Nachr.Chem.Techn.Lab. **38** (1990) 860, Table 1
[6] J.Falbe, M.Regitz, Eds., Römpp Chemie Lexikon, Georg Thieme Verlag, Stuttgart-New York, 10th ed. (1999), Vol. **5**, p. 4210, Fig. 3
[7] T.Aberle, W.Burchard, G.Galinsky, R.Hanselmann, R.W.Klingler, E.Michel, Macromol. Symp. **120** (1997) 47, (a) Fig. 1, (b) Table 1
[8] L.M.J.Kroon-Batenburg, J.Kroon, M.G.Nordholt, Polym.Commun. **27** (1986) 290, Fig. 1
[9] K.Gessler, N.Krauss, Th.Steiner, Ch.Betzel, C.Sandmann, W.Saenger, Science **266** (1994) 1027, Figs. 2A and 2B; –, Nachr.Chem.Tech.Lab. **43**/3 (1995) Fig. 1B
[10] Amended data collected by S.K.Batra, in M.Lewin, E.M.Pearce, Handbook of Fiber Chemistry, Dekker, New York, 2nd ed. (1998), p. 526-527
[11] B.Philipp, H.-P.Fink, Polymer News **24** (1999) 122, Fig. 2
[12] T.Heinze, Macromol.Chem.Phys. **199** (1998) 2341, (a) Fig. 13, (b) Fig. 14
[13] J.K.Baird, in Kirk-Othmer, Encycl.Chem.Technol., 3rd ed., **12** (1980) 45, Table 10

# 9 Carbon–Sulfur Chains

Carbon–sulfur chains with sulfur atoms or groups in the main chain are similar to the corresponding carbon–oxygen chains but far less industrially important. Sulfur can be present as a main chain unit such as –S–, $-S_i-$, –SO–, $-SO_2-$, –O–CO–S–, –NH–CS–, etc., or as part of a main chain ring or ring system, such as a thiophene unit.

## 9.1 Aliphatic Polysulfides with Monosulfur Groups

Aliphatic polymers with monosulfur groups in the main chain have the general constitution $+\!\!\!\!-\text{Z–S}\!-\!\!\!\!+_n$ where Z is a bivalent aliphatic or cycloaliphatic group.

### 9.1.1 Polythioacetals

Polymers with the simple chain structure $+\!\!\!\!-\text{S–CHR}\!-\!\!\!\!+_n$ with R = H, $CH_3$, etc., are not polysulfides but polythioacetals. **Poly(thioformaldehyde)**, $+\!\!\!\!-\text{S-CH}_2\!-\!\!\!\!+_n$, is obtained by polymerization of thioformaldehyde, HCHS, or its cyclic trimer, trithiane. Neither poly(thioformaldehyde) nor other polythioacetals (R = $CH_3$, $C_2H_5$, etc.) or polythioketals, with the repeating unit –S–CRR'– are industrially important because such polymers degrade easily to their monomers.

### 9.1.2 Polysulfides

Polysulfide is the name for polymers with the repeating unit $-S_i-Z-$ if Z contains at least two carbon atoms as chain atoms in aliphatic, cycloaliphatic, or aromatic groups and either $i = 1$ (monosulfide) or $i \geq 2$ (polysulfide; "poly" refers here to the number of sulfur atoms per repeating unit and not to the number of repeating units per molecule).

Aliphatic polysulfides with two or more carbon chain atoms per repeating unit are obtained by (a) ring-opening polymerization of cyclic sulfides; (b) polycondensation of sulfides, $R_2S$, or dithioles, $Z(SH)_2$, with appropriate partners; or (c) polyaddition of dithiols with divinyl compounds. Alternatively, (d) preformed thioether groups in a monomer may react with functional groups of another monomer.

Ring-opening polymerization of ethylene sulfide (thiuram) with an initiator from the reaction of diethyl zinc with water delivers a high-molecular weight **poly(ethylene sulfide)** with a melting temperature of ca. 210°C which dissolves in *o*-dichlorobenzene, dimethylsulfoxide, or nitrobenzene at temperatures above 140°C:

$$\text{(9-1)} \qquad \underset{\text{S}}{\triangledown} \longrightarrow \left[\!-\text{CH}_2\text{CH}_2\text{S}-\!\right]_n$$

Polycondensation of 1,2-dibromoethane and dipotassium sulfide delivers only low-molecular weight products:

(9-2) $$n\,BrCH_2CH_2Br + n\,K_2S \longrightarrow -[CH_2CH_2S]_n- + 2\,n\,KBr$$

Similar to Eq.(9-1), high-molecular weight **poly(propylene sulfide)** results from the ring-opening polymerization of propylene sulfide. However, the anionic polymerization with lithium alkyls differs from that of cycloaliphatic ethers where an alkyl anion attacks a carbon atom. Instead, the ethyl lithium-initiated polymerization of propylene sulfide leads first to lithium ethane thiolate which then initiates the polymerization of propylene sulfide, $C_3H_6S$:

(9-3) $$C_2H_5^{\ominus} \xrightarrow[-\,C_3H_6]{+\,C_3H_6S} C_2H_5S^{\ominus} \xrightarrow{+\,C_3H_6S} C_2H_5S{-}CH(CH_3){-}CH_2{-}S^{\ominus}$$

Anionic as well as cationic polymerizations with, e.g., $BF_3$-ether, lead to amorphous, atactic polymers ($T_G \approx -48°C$). These polymers have properties similar to styrene-butadiene rubbers and, like these, can be filled with carbon black and vulcanized with sulfur. They have not found industrial applications although they are weather- and solvent-resistant. Functionalized oligomers may serve as adhesives or sealing compounds.

Zinc-containing coordination catalysts lead to crystalline isotactic polymers ($T_M \approx 52°C$). Use of optically active coordination catalysts results in optically active polymers.

Aliphatic polysulfides of the type ~S–R–S–R'~ are obtained by the polymerization of dithiols and divinyl compounds:

(9-4)
$$HS{-}Z{-}SH + CH_2{=}CH{-}Z'{-}CH_2{=}CH_2 \longrightarrow -[S{-}Z{-}S{-}CH_2{-}CH_2{-}Z'{-}CH_2{-}CH_2]_n-$$

This polyaddition is a free-radical reaction that uses peroxides, electron beams, or UV light for initiation. Industry uses multifunctional monomers for the preparation of cross-linked coatings.

Longer carbon sections between sulfur chain atoms result from the polymerization of thiacyclobutanes where rings are directly attacked (compare Eq.(9-3)) and the polymerization is propagated by carbanions:

(9-5) $$C_2H_5Li + \text{(2-methylthiacyclobutane)} \longrightarrow C_2H_5{-}S{-}CH(CH_3){-}CH_2{-}CH_2^{\ominus}Li^{\oplus}$$

A spiro compound from pentaerythritol, $C(CH_2OH)_4$, and chloroacetaldehyde can by polycondensed with disodium sulfide:

(9-6) $$ClCH_2{-}\text{(spiro-bis-dioxane)}{-}CH_2Cl \xrightarrow[-\,2\,NaCl]{+\,Na_2S} -[CH_2{-}\text{(spiro-bis-dioxane)}{-}CH_2{-}S]_n-$$

Polymers are stabilized by capping chain ends with ethylene chlorohydrin. The polymer is processed to tough films at 200-260°C.

Instead of introducing sulfur by polymerization, one can of course also employ monomers that contain sulfur in non-functional groups. An example is thiodiethanol (thioglycol, bis(2-hydroxyethyl)sulfide)), $HOCH_2CH_2–S–CH_2CH_2OH$, which is obtained by reaction of ethylene oxide and hydrogen sulfide. Acid-catalyzed polyondensation of this monomer in vacuum delivers **poly(thiodiethanol)**, $HO\text{+}CH_2CH_2SCH_2CH_2O\text{+}_nH$. Industrial polycondensations are conducted in the presence of another glycol (in order to suppress the crystallization) and a third monomer (to introduce unsaturated side groups). The resulting elastomer is crosslinked conventionally by sulfur + zinc mercaptobenzthiazole or by a zinc-free system consisting of sulfur and polyamines.

Polymeric acrylates, butadienes, urethanes, etc., that carry two or more sulfhydryl endgroups, –SH, are industrially called **polymercaptanes**. "Poly" refers here to the "many" endgroups and not to the many monomeric units. According to IUPAC, "mercapto" should only be used instead of "thiol" if the molecule contains substituents with higher seniority.

## 9.2 Aliphatic Polysulfides with Polysulfur Groups

Industrially important aliphatic polysulfides, $\text{+}S_i–Z\text{+}_n$, with $i$ = 2, 3, ... result from the polycondensation of $\alpha,\omega$-dichloro compounds, Cl–Z–Cl, with sodium polysulfides, $Na_2S_i$, where $i$ = 1–5. The average number $i$ of sulfur atoms per repeating unit is known as the **polysulfide rank**.

Polymers of this type are known since 1926 as **thioplastics, polysulfide rubbers**, or **Thiokols®** (**T**, **TR**, **TM**). The first representative of this group was obtained by reaction of 1,2-dichloroethane with sodium polysulfide to a polymer with a polysulfide rank of ca. 4. Most important is now bis(2-chloroethyl)formal, $(ClCH_2CH_2O)_2CH_2$, which leads to a polysulfide rank of 2. Also used is a mixture of bis(2-chloroethyl)formal and 1,2-dichloroethane (leads to polysulfide rank 2.2). 1,2,3-Trichloropropane is often employed as a branching agent. Bis(2-chloroethyl)ether is no longer utilized.

$Cl(CH_2)_2S(CH_2)_2Cl$ cannot be employed as monomer since it is a skin poison. It was used in World War I as a poison gas (mustard gas).

Reaction components need not be used in stoichiometric proportions since a surplus of $Na_2S_i$ leads to NaS– or $NaS_i$– endgroups which either disproportionate (Eq.(9-7)) or are oxidized (Eq.(9-8)). These reactions increase molecular weights beyond those calculated from stoichiometry.

$$\text{(9-7)} \qquad \sim\sim Z—S_i—Na + Na—S_i—Na \rightarrow \sim\sim Z—S_i—Z\sim\sim + Na_2S_i$$

$$\text{(9-8)} \qquad 2 \sim\sim Z—S—Na + 1/2\ O_2 + H_2O \rightarrow \sim\sim Z—S—S—Z\sim\sim + 2\ NaOH$$

Polysulfides are easily oxidized, a property which is exploited for industrial crosslinkings of high-molecular weight polysulfide rubbers with lead dioxide, organic peroxides, or $p$-quinone dioxime. The reaction is very exothermic and therefore not suitable for low-molecular weight products with many SH endgroups.

Crosslinking converts –SH endgroups to –S– center groups:

$$\text{(9-9)} \qquad 4 \sim\!\sim Z—SH \xrightarrow[-2\ H_2O]{+PbO_2} \sim\!\sim Z—S—Pb—S—Z\sim\!\sim \; + \; \sim\!\sim Z—S—S—Z\sim\!\sim$$

(the Pb-containing product is converted to the disulfide by $+ PbO_2, - 2\ PbO$)

The less-poisonous $MnO_2$ acts similarly. It has a longer shelf life and delivers more lightproof products but needs an activator such as tetramethylthiurame disulfide, $(CH_3)_2NC(S)S_2C(S)N(CH_3)_2$, or 1,3-diphenylguanidine, $(C_6H_5NH)_2C{=}NH$:

(9-10) $$2 \sim\sim Z{-}SH + MnO_2 \rightarrow \sim\sim Z{-}S{-}S{-}Z\sim\sim + MnO + H_2O$$

Other curing agents are $ZnCrO_4$, $Na_2Cr_2O_7$, $CaO_2$, $NaBO_2$, $H_2O_2$, and coumene hydroperoxide. Crosslinking is also possible by reaction with diisocyanates or epoxides (addition reactions) or with phenol-formaldehyde resins (condensation reactions).

Polysulfide rubbers also contain plasticizers (e.g., phthalates), fillers (e.g., $CaCO_3$), pigments ($TiO_2$, carbon black), thixotropy-causing agents (e.g., montmorillonite; see Section 12.3.8), coupling agents, and other adjuvants (see Volume IV). Such additives may be present in formulated products in amounts up to 70 wt%.

Properties of crosslinked polymers depend foremost on polysulfide rank (Table 9-1). In polymers with high polysulfide rank, sulfur is present as oligomeric sulfur units, Higher polysulfide ranks can be lowered by exchange reactions with sodium polysulfide, for example, according to $\sim S_4{-}R\sim + Na_2S_4 \rightleftarrows \sim S_{3.1}{-}R\sim + Na_2S_{4.9}$.

Solid polysulfides are resistant against solvents, oxygen, and ozone and are therefore used as sealants (insulating glass, constructions), in blends with epoxides for adhesives, and as coating compounds in street construction. Mixtures of polysulfides and certain oxidizing compounds burn with high intensity and gas evolution which makes them useful as rocket propellants.

Table 9-1 Effect of polysulfide rank on the consistency of aliphatic polysulfides, $\text{+}(CH_2)_n{-}S_i\text{+}_n$.

| Polysulfide rank $i$ | Number $n$ of methylene groups per repeating unit | | |
|---|---|---|---|
| | 1 | 2 | 3 |
| 1 | powdery | powdery | powdery |
| 2 | solid, malleable | horn-like, cold stretchable | |
| 3 | | rubber-like | |
| 4 | rubber-like | rubber-like, crystallizes on standing | rubber-like |

# 9.3 Poly(phenylene sulfide)

## 9.3.1 Syntheses

Poly(*p*-phenylene sulfide) (**PPS**, poly(1,4-thiophenylene)), $\text{+}S{-}(p\text{-}C_6H_4)\text{+}_n$, is industrially synthesized from 1,4-dichlorobenzene, $Cl(p\text{-}C_6H_4)Cl$, and disodium sulfide, $Na_2S$, in *N*-methyl-2-pyrrolidone as solvent at 260°C and 1.1 MPa (Phillips process):

(9-11) $$n\ Na_2S + n\ Cl{-}C_6H_4{-}Cl \longrightarrow \left[ S{-}C_6H_4 \right]_n + 2n\ NaCl$$

The Macallum synthesis of PPS from $Cl(p\text{-}C_6H_4)Cl$, S, and $Na_2CO_3$ failed industrially.

Eq.(9-11) leads at low monomer conversions to molecular weights that are higher than predicted for equilibrium conditions, probably because of branching reactions. The reaction also generates some –S–S– bonds that disproportionate to $S_8$ molecules at temperatures above ca. 300°C. Eq.(9-11) can therefore not be a nucleophilic substitution. It is believed to be a one-electron process with radical cations as reactive intermediates.

Syntheses of PPS from disodium sulfide and 1,4-dichlorobenzene lead only to medium molecular weights of 15 000-20 000 ($X$ = 150-200). Higher molecular weights are obtained by curing (see below). It is possible, however, to obtain high molecular weights directly without curing if alkali carboxylates are employed as modifiers. Formation of branched products can be suppressed if water is added to the last reaction step.

The reaction proceeds only in *N*-methyl-2-pyrrolidone and not in any other solvent, probably because this solvent has the appropriate oxidation potential for the formation of $NaS^{\bullet}$ radicals from $NaS^{\ominus}$ anions:

(9-12) $NaS^{\ominus}$ + N-methyl-2-pyrrolidone ⇌ $NaS^{\bullet}$ + radical anion ($O^{\ominus}$, N–$CH_3$)

The radicals then add to dichlorobenzene:

(9-13)

Cl–C6H4–Cl —(+ $NaS^{\bullet}$)→ Cl–(radical adduct with SNa, Cl) —(– $Cl^{\ominus}$)→ Cl–(radical cation ⊕)–SNa

Cl–(radical cation ⊕)–SNa —(– $Na^{\oplus}$)→ Cl–C6H4–S• —(+ $C_6H_4Cl_2$)→ Cl–C6H4–S–(radical adduct, Cl)–Cl

Cl–C6H4–S• ⇌ Cl–C6H4–S–S–C6H4–Cl

↓ – $Cl^{\ominus}$: Cl–C6H4–$\overset{\oplus}{S}$=(radical)–Cl

The mesomeric form of the newly formed radical cation reacts further according to

(9-14) Cl–C6H4–$\overset{\oplus}{S}$=(radical)–Cl ⟷ Cl–C6H4–S–(radical cation ⊕)–Cl

↓ + $NaS^{\ominus}$: Cl–C6H4–S=(radical, SNa, Cl)

—($Cl^{\ominus}$)→ Cl–C6H4–S–C6H4–SNa

↓ – $Na^{\bullet}$: Cl–C6H4–S–C6H4–$S^{\bullet}$

The primary radical cation Cl($p$-$C_6H_4$)–S–($p$-$C_6H_4$)$^{\bullet\oplus}$ $Cl^{\ominus}$ then reacts with $NaS^{\ominus}$ to 4,4'-di(chlorophenyl)sulfide:

(9-15) $$\text{Cl}-\text{C}_6\text{H}_4-\text{S}-\text{C}_6\text{H}_4^{\bullet\oplus}-\text{Cl} \xrightarrow[-\ \text{NaS}^{\bullet}]{+\ \text{NaS}^{\ominus}} \text{Cl}-\text{C}_6\text{H}_4-\text{S}-\text{C}_6\text{H}_4-\text{Cl}$$

and the growth cycle starts again. Radical cations are very reactive, which leads to high molecular weights at low monomer conversions. However, increasing resonance stabilization of larger molecules causes reactivities of polymeric radical cations to approach a limiting value. At $X > 20$, the formation of polymer molecules is determined only by the number concentration of reactants and the reaction then simply follows the laws of equilibrium polycondensation. For properties, see Table 9-2.

PPS with $X = 150–200$ can be employed for coatings but the melt viscosity is too low for injection molding and extrusion. PPS is thus oxidized ("cured") with hot air just below the melting temperature which increases $X$. The polymer becomes dark brown.

A poly(phenylene sulfide) with thiophenylene and dithiophenylene units is obtained if 1,4-diiodophenylene and sulfur are reacted in the melt:

(9-16) $$\text{I}-\text{C}_6\text{H}_4-\text{I} \xrightarrow[-\ \text{I}_2]{+\ \text{S}_i} -\text{S}-\text{C}_6\text{H}_4- \quad + \quad -\text{S}-\text{S}-\text{C}_6\text{H}_4-$$

1,4-Diiodophenylene is regenerated by reacting the byproduct iodine with benzene and oxygen according to $I_2 + C_6H_6 + O_2 \rightarrow I(p\text{-}C_6H_4)I + H_2O$.

Table 9-2 Properties of unfilled and 40 wt% glass fiber-filled high-molecular weight poly(phenylene sulfide)s ($X$ = 160-320). Data for 25°C unless noted otherwise. a) 2 minutes.

| Property | Physical unit | Types of poly(phenylene sulfide)s linear, unhardened, unfilled | branched, unhardened, unfilled | branched, hardened, unfilled | branched, hardened, filled |
|---|---|---|---|---|---|
| Density | $g/cm^3$ | 1.425 | 1.35 | | 1.6 |
| Melting temperature (DSC) | °C | 315 | 285 | | |
| Glass temperature (DSC) | °C | 85 | | | |
| Linear thermal expnsion coeff. | $K^{-1}$ | $4.9 \cdot 10^{-5}$ | $4.9 \cdot 10^{-5}$ | | $4 \cdot 10^{-5}$ |
| Specific heat capacity | $J\ K^{-1}\ g^{-1}$ | | | 1.09 | 1.05 |
| Heat conductivity (20°C) | $W\ m^{-1}\ K^{-1}$ | | | 0.29 | 0.29 |
| Flexural modulus | MPa | 3 700 | | 3 900 | 12 000 |
| Flexural strength | MPa | 130 | | 96 | 160 |
| Compression strength | MPa | | | 110 | 145 |
| Tensile strength (fracture) | MPa | 86 | 66 | 65 | 122 |
| Yield strength | MPa | 80 | | | |
| Extension at break | % | 3–21 | | 1.6–2 | 0.5–1.7 |
| Notched impact strength | J/m | 26 | | 16 | 69 |
| Impact strength | J/m | 900 | | 100 | 180 |
| Hardness (Rockwell) | - | | | R120 | R123 |
| Relative permittivity | 1 | | | | 3.8 |
| Electric strength | kV/mm | | | | 17.7 |
| Dissipation factor | 1 | | | | 3.8 |
| Resistivity (volume resistance) a) | Ω cm | | | | $4.5 \cdot 10^{16}$ |
| Oxygen index | 1 | | | 44 | 46.5 |

### 9.3.2 Properties

Uncured PPS has a molecular weight between 15 000 and 25 000. It dissolves in biphenyl, terphenyl, or 1-chloronaphthalene at temperatures above ca. 200°C.

Pure PPS is white but industrial grades have a yellowish tint which is caused by traces of $FeCl_3$ that stem from the reactor steel. On heating in air, PPS discolors to brown and becomes insoluble.

The theoretical density of PPS is 1.440 g/cm$^3$. Uncured PPS has a density of 1.425 g/cm$^3$ (Table 9-2); it is ca. 65 % X-ray crystalline.

PPS is a high-modulus polymer (Table 9-2) with moderate strength; it is inflammable and stable in air up to 500°C. It is used for corrosion-resistant coatings of pumps, valves, and cooking pots and pans. The largest use is by the electrical and electronic industries (switches, regulators, packaging of electronic parts, etc.), followed by the automotive industry (pumps, valves, sealings, etc.). PPS for these applications is usually reinforced by glass or carbon fibers or filled with inorganics such as chalk or iron oxide since mineral fillers improve processing and reduce shrinkage. Unfilled PPS is used for fibers and nonwovens (films, sheets, etc.). World production (unfilled): 28 000 t/a (2003).

## 9.4 Aromatic Polysulfide Ethers

Aromatic polysulfide ethers contain aromatic, aliphatic, ether, and thioether groups in the main chain. An example is a polymer with four methylene groups per repeating unit. The monomer synthesis from phenol and thiophene leads to cations,

(9-17) $$\text{C}_4\text{H}_8\text{S} + \text{C}_6\text{H}_5\text{–OH} \xrightarrow[\text{HCl},\ -\text{Cl}^\ominus]{+\text{Cl}_2} \text{C}_4\text{H}_8\text{S}^\oplus\text{–C}_6\text{H}_4\text{–OH}$$

that are converted by sodium methylate or anion-exchange resins to zwitterions which polymerize in a "death charge" polymerization to a polysulfide ether:

(9-18) $$\text{C}_4\text{H}_8\text{S}^\oplus\text{–C}_6\text{H}_4\text{–O}^\ominus \longrightarrow \left[\text{(CH}_2)_4\text{–S–C}_6\text{H}_4\text{–O}\right]_n$$

The monomer for a polymer with five methylene groups per repeating unit is synthesized via a ring-closure reaction, however:

(9-19) $$\text{Br(CH}_2)_5\text{Br} + \text{CH}_3\text{–S–C}_6\text{H}_4\text{–OH} \xrightarrow[-\text{CH}_3\text{Br},\ -\text{Br}^\ominus]{} \text{C}_5\text{H}_{10}\text{S}^\oplus\text{–C}_6\text{H}_4\text{–OH}$$

These reactions are industrially interesting since they deliver water-resistant coatings from aqueous monomer solutions. Linear polymers from bifunctional monomers are fairly soft, but hard coatings can be obtained by copolymerization with multifunctional zwitterions. The hardness of coatings can be improved further by film formation in the presence of latices or colloidal silica. Polymers are not commercial products, however, because the monomers are highly toxic.

## 9.5 Polysulfones

The main chain of polysulfones, $-[Z-SO_2]_n-$, is composed of sulfone groups $-SO_2-$ and bifunctional aliphatic, cycloaliphatic, or aromatic groups –Z–. Systematic IUPAC names of these polymers such as poly(sulfonyl-1-alkylene), poly(sulfonyl-1-arylene), etc., are rarely used.

Industry reserves the name "polysulfone" for sulfone polymers with aromatic ether groups. These polysulfones are **aromatic polyether sulfones** with the general composition $-[O-Ar-SO_2-Ar']_n-$ where Ar and Ar' are identical or different aromatic groups (Section 9.5.3). In order to distinguish their products from that of competitors, companies use different designations such as polysulfone, polyarylsulfone, polyphenylene sulfone, polyether sulfone, polyphenylene ether sulfone, or even polyarylene ether or polyaryl ether which almost never can be correlated with their chemical structures.

### 9.5.1 Poly(alkylene sulfone)s

Polysulfones with aliphatic or cycloaliphatic groups in the main chain are obtained either by free-radical copolymerization of sulfur dioxide with olefins or cycloolefins, Eqs.(9-20) and (9-21), or by oxidation of the corresponding polysulfides, Eq.(9-22):

(9-20) $$CH_2{=}CH_2 + SO_2 \longrightarrow -[SO_2-CH_2-CH_2]_n-$$

(9-21) $$\text{cyclopentene} + SO_2 \longrightarrow -[SO_2-\text{(1,3-cyclopentylene)}]_n-$$

(9-22) $$-[S-CH_2-CH_2]_n- + O_2 \longrightarrow -[SO_2-CH_2-CH_2]_n-$$

Free-radical copolymerizations, Eqs.(9-20) and (9-21)), to high-molecular weight polymers have to be performed below the thermodynamic ceiling temperature (p. 140). Since sulfur–carbon bonds are much weaker than carbon–carbon bonds (ca. 240 kJ/mol versus ca. 345 kJ/mol), poly(alkylene sulfone)s are easily cleaved homolytically. Above the ceiling temperature (Table 9-3), the resulting macroradicals start unzipping reactions to monomers.

Aliphatic polysulfones cannot be used as thermoplastics because they degrade easily. For the very same reason, they are used as resists in the manufacture of integrated circuits (Volume IV) where the workhorse polymer is poly(1-butene sulfone). Poly(2-methyl-1-pentene sulfone) has a low $T_c$ and the fastest depolymerization.

### 9.5.2 Poly(sulfo-1,4-phenylene)

Poly(sulfo-1,4-phenylene), $-[SO_2-(1,4\text{-}C_6H_4)]_n-$, results from the oxidation of suspensions of poly(thio-1,4-phenylene) at 40–90°C:

(9-23) $$-[S-C_6H_4]_n- + O_2 \longrightarrow -[SO_2-C_6H_4]_n-$$

Table 9-3 Densities $\rho$, glass temperatures $T_G$, and melting temperatures $T_M$ of 1:1 copolymers of sulfur dioxide and olefins or cycloolefins, respectively. Ceiling temperatures $T_c$ refer to copolymerizations with a surplus of $SO_2$ from thermodynamic states I to II which may be dissolved (s), amorphous (a), or crystalline (c). * 31 % $SO_2$.

| Olefin | $\frac{T_c}{°C}$ | $\frac{\rho}{g\ cm^{-3}}$ | $\frac{T_G}{°C}$ | $\frac{T_M}{°C}$ |
|---|---|---|---|---|
| Ethene | > 135 (sc) | | | 135 * |
| Propene | 90 (sa) | 1.457 | | 280 |
| 1-Butene | 64 (ss) | 1.245 | 81–95 | 150–160 |
| 1-Hexene | 60 (ss) | 1.220 | | 76 |
| 1-Hexadecene | 69 (sa) | 0.990 | 77 | |
| 1-Octadecene | | 0.990 | | 45 |
| Isobutene | 5 (sa) | 1.406 | | 230 |
| 2-Methyl-1-pentene | –34 (ss) | | | |
| 4,4-Dimethyl-1-pentene | 14 (sc) | | | |
| 2,4,4-Trimethyl-1-pentene | < –80 | | | |
| Cyclopentene | 103 (ss) | | 82 | |
| Cyclohexene | 24 (ss) | | | 200-205 |
| Cycloheptene | 11 (ss) | | | |
| Bicyclo[2.2.1]heptene | | | 117 | 240-290 |
| 1-Butene + cyclopentene (1:1) | | | 57 | |

The light beige polymer is commonly called **poly(phenylene sulfone)** (**PPSU**), but PPSU is more commonly used for another polymer (Section 9.5.3) and sold as Ceramer (however, this is also a trademark for an olefin/maleic polymer and, as "ceramer", the generic term for hybrids of inorganic and organic materials).

Poly(sulfo-1,4-phenylene), $+SO_2—(p\text{-}C_6H_4)+_n$, has a glass temperature of ca. 360°C and, depending on the grade, very high melting temperatures of 460°C or 520°C which rules out conventional processing. It is ca. 60 % crystalline and has a high flexural modulus of 3850 MPa and a moderate flexural strength of 84 MPa (both at 23°C). The polymer is offered as a fine spherical powder with diameters of 15–60 μm for use as a reinforcing filler for poly(tetrafluoroethylene) and other engineering thermoplastics (PBT, POM, PEEK, LCP, etc.) where it increases the abrasion resistance and, in coatings, improves the adhesion to metals. PTFE composites can be compression-molded, thermo-sprayed, or extruded as a paste.

### 9.5.3 Aromatic Polyether Sulfones

Aromatic polysulfones (**PSU**) can be processed as thermoplastics if their chains contain flexibilizing ether groups (low rotational barrier, see Volume III). Such polymers with 1,4-phenylene, $–SO_2–$, and –O– units in their chains are thus more precisely called aromatic polyether sulfones. They are known under many, often confusing, names and even more, and often conflicting, abbreviations and acronyms. Not all PSUs are homopolymers; some commercial products are blends with, e.g., ABS or SAN, and many are reinforced with 30 wt% glass or carbon fibers.

Commerical aromatic polyether sulfones always contain the characteristic group $–C_6H_4–SO_2–C_6H_4–O–$. However, this group is a monomeric unit only in the first of the following four commercial products. In this book, these products are discussed as PESU, PPSU, PBSU, and PAESU; literature uses not only the same abbreviations for other types but also many other ones such as PES, PSF, or PSO. The constitution of PAESU has not been revealed but is seems to have the structure noted below.

Polyarylene ether sulfones can be produced by polysulfonations or by polyether syntheses, both as AB or AA/BB reactions.

| | | |
|---|---|---|
| PESU (PES) | $–C_6H_4–SO_2–C_6H_4–O–$ | polyether sulfone |
| PPSU (PPSF) | $–C_6H_4–SO_2–C_6H_4–O–C_6H_4–C_6H_4–O–$ | polyphenylene sulfone |
| PBSU (PSF) | $–C_6H_4–SO_2–C_6H_4–O–C_6H_4–C(CH_3)_2–C_6H_4–O–$ | polybisphenol sulfone |
| (PAESU) | $–C_6H_4–SO_2–C_6H_4–O–C_6H_4–SO_2–C_6H_4–$ | polyarylether sulfone |

**Polysulfonylations**

In Friedel–Crafts polysulfonylations, aromatically bound hydrogens undergo electrophilic substitution by sulfonylium ions. Reactions proceed at 100–250°C with catalytic amounts of Lewis acids, for example, $AlCl_3$, $FeCl_3$, $SbCl_3$, or $InCl_3$, in the melt or in solvents such as nitrobenzene, tetrachloroethylene, or sulfones:

(9-24) $$C_6H_5–O–C_6H_4–SO_2Cl \xrightarrow[-\,HCl]{} \text{[}–C_6H_4–O–C_6H_4–SO_2–\text{]}_n$$

This AB reaction leads to almost 100 % para substitution. It is not used for the synthesis of PESU, though, because the monomer is too expensive. In principle, the same repeating unit $–C_6H_4–O–C_6H_4–SO_2–$ can be obtained by an AA/BB reaction of diphenyl ethers:

(9-25) $$C_6H_5–O–C_6H_5 + ClSO_2–C_6H_4–O–C_6H_4–SO_2Cl$$
$$\xrightarrow[-\,2\,HCl]{} \text{[}–C_6H_4–O–C_6H_4–SO_2–C_6H_4–O–C_6H_4–SO_2–\text{]}_n$$

However, this reaction leads to ca. 80 % para and 20 % ortho substitution and is useless since PESU polymers become too brittle at high ortho contents. PESU is therefore synthesized by a polyether synthesis, Eq.(9-27).

The type of synthesis of Eq.(9-25) can be used, however, for the synthesis of PAESU:

$$\text{(9-26)}\quad C_6H_5\text{–O–}C_6H_5 + ClSO_2\text{–}C_6H_4\text{–}C_6H_4\text{–}SO_2Cl \xrightarrow[-2\ HCl]{} \left[\text{–}C_6H_4\text{–O–}C_6H_4\text{–}SO_2\text{–}C_6H_4\text{–}C_6H_4\text{–}SO_2\text{–}\right]_n$$

**Polyether Syntheses**

Polyether syntheses are nucleophilic substitutions of aromatically bound halogen atoms by phenoxy ions with release of metal chlorides, MtCl, that can be performed as either AB or AA/BB reactions. The AB reaction is the method of choice for PESU:

$$\text{(9-27)}\quad Cl\text{–}C_6H_4\text{–}SO_2\text{–}C_6H_4\text{–OMt} \xrightarrow[-\ MtCl]{} \left[\text{–}C_6H_4\text{–}SO_2\text{–}C_6H_4\text{–O–}\right]_n$$

The corresponding AA/BB reaction is used for the synthesis of PPSU

$$\text{(9-28)}\quad Cl\text{–}C_6H_4\text{–}SO_2\text{–}C_6H_4\text{–Cl} + MtO\text{–}C_6H_4\text{–}C_6H_4\text{–OMt} \xrightarrow[-2\ MtCl]{} \left[\text{–}C_6H_4\text{–}SO_2\text{–}C_6H_4\text{–O–}C_6H_4\text{–}C_6H_4\text{–O–}\right]_n$$

as well as that of PBSU if (MtO($p$-$C_6H_4$)–$C(CH_3)_2$–($p$-$C_6H_4$)OMt is used instead of MtO(4,4'-$C_6H_4$–$C_6H_4$)OMt. These AB and AA/BB reactions proceed at 130–250°C in solvents that dissolve both monomers and polymers, for example, mixtures of chlorobenzene and dimethylsulfoxide.

In AB polyether syntheses of PSUs, monomer molecules react more slowly than polymer molecules (Volume I, p. 458 ff.). Initially, monomer molecules are thus present in greater proportions than predicted by classical equilibrium theory. Molecular weight distributions become broader and number-average degrees of polymerization increase first only slowly and then faster with the reaction parameter $k[M]_0t$, where $k$ is the rate constant and $[M]_0$ is the initial monomer concentration (Volume I, Fig. 13-15). In AA/BB polyether syntheses, it is just the opposite.

Nucleophilic reactions, Eqs.(9-27) and (9-28), produce polymer molecules with reactive endgroups that continue to react during melt processing. These post-polymerization reactions increase molecular weights and melt viscosities. This requires not only continuous readjustment of machine parameters but also changes molecular parameters and costs energy. Reactive phenolate endgroups are therefore converted to methoxy groups by reaction with $CH_3Cl$.

**Polyarylene Sulfide Sulfones**

Polyarylene sulfide sulfones are synthesized by copolycondensation of Cl($p$-$C_6H_4$)Cl, Cl($p$-$C_6H_4$)$SO_2$($p$-$C_6H_4$)Cl, and $Na_2S$. Polymers are amorphous with a glass temperature of 215°C.

**Properties**

Aromatic polyether sulfones are slightly yellow, transparent polymers that degrade in ultraviolet light with $\lambda_0 < 320$ nm. The degradation is caused mainly by sulfone groups and less by impurities; it leads to phenol groups. The degradation can be suppressed by addition of peroxide deactivation agents and sterically hindered amines (Volume IV).

Polymers are amorphous and have high glass temperatures between 190°C and 230°C (Table 9-4). They are thermally and hydrolytically stable and have good creep resistance. Their melt viscosities are high but are little affected by shear rates, leading to little molecular orientation on injection molding. Properties are therefore practically independent of direction.

Because of their good thermal and hydrolytic properties, aromatic polyether sulfones are especially suited for applications where parts are continually and/or repeatedly subjected to hot water or water vapor. Examples are sterilizable medical devices, coatings of

Table 9-4 Properties of aromatic polyether sulfones. nb = no break.

| Property | Physical unit | PESU Victrex® 4100 G | PPSU Radel® R 5000 | PAESU Radel® A 400 | PBSU Udel® P 1700 |
|---|---|---|---|---|---|
| Density | $g/cm^3$ | 1.37 | 1.29 | 1.37 | 1.24 |
| Refractive index | 1 | 1.545 | | | 1.63 |
| Heat distortion temperature (1.82 MPa) | °C | 210 | 204 | 204 | 174 |
| Vicat temperature B | °C | 222 | | | 188 |
| Glass temperature (DSC) | °C | 230 | 220 | 217 | 192 |
| Linear thermal expansion coefficient | $K^{-1}$ | $55 \cdot 10^{-6}$ | $56 \cdot 10^{-6}$ | $49 \cdot 10^{-6}$ | $56 \cdot 10^{-6}$ |
| Heat conductivity (20°C) | $W\ m^{-1}\ K^{-1}$ | 0.18 | | | 0.26 |
| Tensile modulus | MPa | 2440 | 2140 | 2650 | 2480 |
| Flexural modulus | MPa | 2570 | 2280 | | 2690 |
| Yield strength | MPa | 64 | 72 | 83 | 70 |
| Tensile strength at break | MPa | 84 | | 91 | |
| Flexural strength | MPa | 129 | 86 | | 106 |
| Compression strength | MPa | | | | 96 |
| Elongation at yield | % | | 7 | 6.5 | 6 |
| Elongation at fracture | % | 60 | 60 | 13–40 | 75 |
| Impact strength (Izod, 3.1 mm) | J/m | | | | nb |
| (Charpy) | $kJ/m^2$ | | 340 | | nb |
| Notched impact strength (Izod, 3.1 mm) | J/m | 90 | 640 | 85 | 69 |
| (Charpy) | $kJ/m^2$ | | | | 5 |
| Hardness (Rockwell) | - | M88 | | M110 | M69 |
| Relative permittivity | 1 | 3.5 | 3.4 | 3.5 | 3.2 |
| Electrical resistance | Ω | | | | $3 \cdot 10^{16}$ |
| Resistivity (volume resistance, 2 min) | Ω cm | $1 \cdot 10^{17}$ | $8.9 \cdot 10^{14}$ | $7.7 \cdot 10^{16}$ | $5 \cdot 10^{16}$ |
| Electrical strength | kV/mm | 16.0 | 14.6 | | 20 |
| Dissipation factor (50 Hz) | 1 | | | 0.002 | 0.001 |
| Arc resistance | s | 100 | 41 | | 122 |
| Oxygen index | % | 34 | 38 | | 30 |
| Flammability | - | V-0 | V-0 | V-0 | V-2 |
| Water absorption (24 h) | % | 0.43 | 0.3 | | 0.22 |
| (equilibrium) | % | 1.1 | 1.8 | 0.62 | |

pans and pots, microwave-resistant containers, and electrical parts. These polymers are also used as membranes for separation of hot gases and, as sulfonated products, also as supports for desalination membranes from asymmetric polyamides. Their good dielectric properties have led to applications as injection-molded circuit boards.

Solubilities of these polyether sulfones in organic solvents (aniline, DMF, DMSO, etc.) decrease in the order of PBSU > PAESU > PESU > PPSU. The polymers stress crack in organic solvents but are resistant against mineral acids, alkali, and salt solutions.

# 9.6 Polymers with Sulfur in Ring Structures

## 9.6.1 Poly(thiophene)

Poly(2,5-thiophene) exists in an aromatic and a quinoid form:

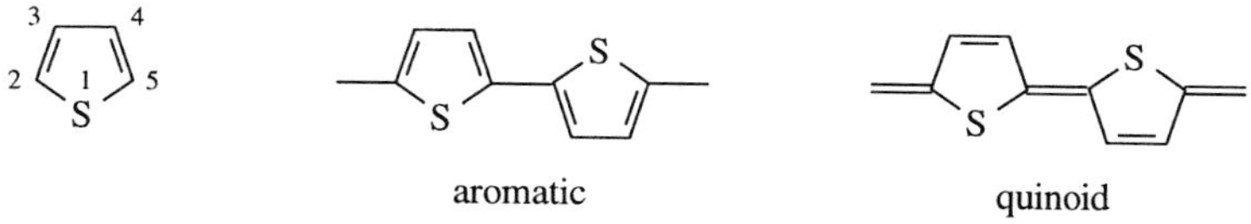

aromatic quinoid

**Syntheses**

Substituted and unsubstituted poly(thiophene)s can be synthesized in various ways. The self-condensation of 3-substituted 2,5-dichlorothiophenes needs magnesium as a chloride acceptor and nickel complexes as catalysts:

(9-29)

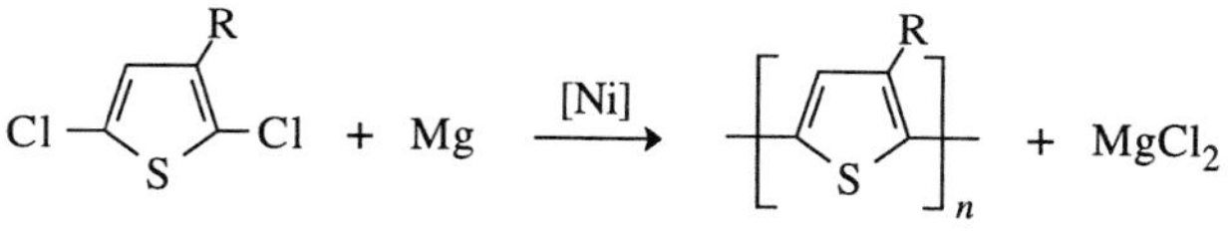

The reaction proceeds mainly by head-to-tail coupling in the 2,5-position (α,α' coupling) but the coupling in 2,4-position (α,β' coupling) is not negligible. Greater regiospecificities of ca. 90 % are obtained by a one-pot reaction, consisting of successive reaction of 2-bromo-3-dodecylthiophene with lithium diisopropylamide in tetrahydrofuran at –40°C (I), followed by the replacement of Li– by BrMg– on addition of $MgBr_2$-$O(C_2H_5)_2$ (II), and splitting off of $MgBr_2$ and condensation on addition of $NiCl_2$·(1,3-bis(diphenylphosphino)propane) (III):

(9-30)

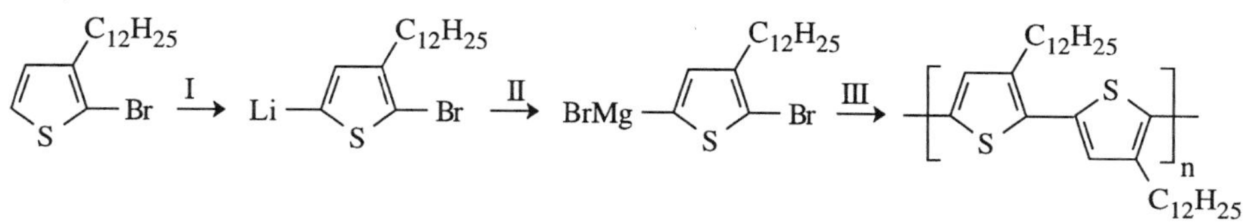

Higher molecular weights as well as greater percentages of "incorrect" units are obtained on oxidative coupling of thiophenes and bithiophenes in organic solvents, using

$FeCl_3$ as an oxidation agent. Plasma polymerization (Volume I) is possible but only for the preparation of thin films.

Thin films of poly(3-methylthiophene) are prepared by electrolytic deposition of monomers dissolved in electrolyte solutions (quaternary ammonium salts, etc.) at platinum or gold electrodes. In the first step, an electron is removed from the monomer molecule. The resulting radical cation then either combines with another monomeric, oligomeric, or polymeric radical cation or reacts with a neutral monomer molecule and releases hydrogen. Electroneutrality is maintained by the counterions of the electrolyte, for example, $[BF_4]^{\ominus}$.

**Properties**

Poly(thiophene) itself is amorphous but substituted poly(thiophene)s are slightly crystalline. The conjugated double bonds of these polymers produce a $\pi$-system that leads to electrical conductivities which depend on the constitution of the monomeric units, the type and proportion of regioregularities (in the case of monosubstituted thiophene polymers: head-to-head, head-to-tail, tail-to-head, and tail-to-tail), the macroconformation of polymer molecules, the morphology of the polymers, and the type and concentration of dopants (see Volume IV). Undoped poly(thiophene) has an electrical conductivity of $10^{-10}$ S/cm which increases to 7 S/cm by doping with iodine and even to 100 S/cm by doping with $SO_3CF_3^{\ominus}$. Poly(thiophene)s are also electrochromic and thermochromic.

Both doped and undoped poly(thiophene)s are fairly stable against oxygen and humidity. They are used for antistatic films and coatings. Possible applications are as biosensors, batteries, diodes, and transistors.

## 9.6.2 Other Cyclic Sulfur Polymers

Poly(3,4-ethylene dioxythiophene) is used for the antistatic packaging of electronic components. Very thin, almost invisible layers of this polymer make plastics electrically conductive.

Very many thiophene derivatives have been investigated as monomers for electrically conducting polymers (see Volume IV), including those with more complicated ring systems like tetrathiapentalene and tetrathiafulvalene. There are also industrially important polymers based on thiazoles such as poly(*p*-phenylene benz*bis*(1,3-thiazole) which are discussed with the corresponding oxazoles in Section 10.8.4.

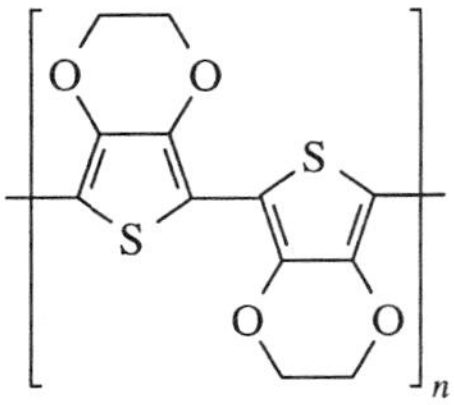

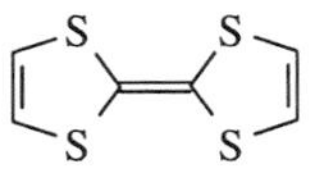

tetrathiafulvalene

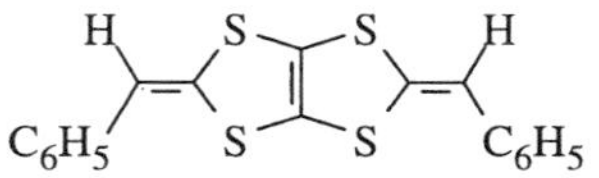

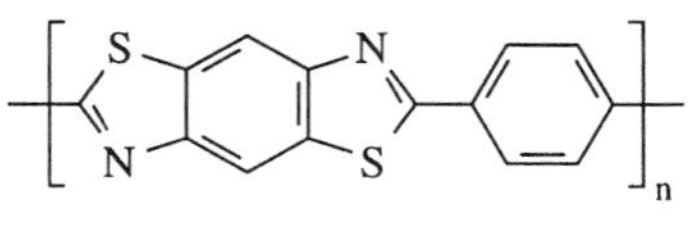

poly(3,4-ethylene dioxythiophene) tetrathiapentalene poly(*p*-phenylene benzbis(1,3-thiazole))

# Literature to Chapter 9

9.0 GENERAL SURVEYS

E.J.Goethals, Sulfur-Containing Polymers, J.Macromol.Sci. [Revs.] **C 2** (1968) 73; –, Topics in Sulfur Chem. **3** (1977) 1

9.1 ALIPHATIC POLYSULFIDES WITH MONOSULFUR GROUPS

P.Sigwalt, Stereoregular and Optically Active Polymers of Episulfides, Int.J.Sulfur Chem. **C 7** (1972) 83

W.H.Sharkey, Polymerization through the Carbon-Sulfur Bond, Adv.Polym.Sci. **17** (1975) 73

9.2 ALIPHATIC POLYSULFIDES WITH POLYSULFUR GROUPS

G.Gaylord, Polyethers, Pt. 3, Polyalkylene Sulfides and Other Polythioethers, Wiley, London 1962

E.R.Bertozzi, Chemistry and Technology of Elastomeric Polysulfide Rubbers, Rubber Chem. Technol. **41** (1968) 114

C.Placek, Polysulfide Manufacture, Noyes Data Corp., Park Ridge (NJ) 1970

W.Cooper, Polyalkylenesulphides, Brit.Polym.J. **3** (1971) 28

F.Lautenschlaeger, Alkylene Sulfide Polymerizations, J.Macromol.Sci. [Chem.] **A 6** (1972) 1089

9.3 POLY(PHENYLENE SULFIDE)

D.G.Brady, Poly(phenylene sulfide) - How, When, Why, Where, and Where Now, J.Appl.Polym.Sci. - Appl.Polym.Symp. **36** (1981) 231

L.C.Lopez, G.L.Wilkes, Poly(*p*-Phenylene Sulfide) - An Overview of an Important Engineering Thermoplastic, J.Macromol.Sci. - Revs.Macromol.Chem.Phys. **C 29** (1989) 83

9.4 AROMATIC POLYSULFIDE ETHERS

D.L.Schmidt, H.B.Smith, M.Yoshimine, M.J.Hatch, Preparation and Properties of Polymers from Aryl Cyclic Sulfonium Zwitterions, J.Polym.Sci. [Chem.] **10** (1972) 2951

9.5 POLYSULFONES

K.J.Ivin, J.B.Rose, Polysulphones, Organic and Physical Chemistry, Adv.Macromol.Sci. **1** (1976) 336

V.J.Leslie, J.B.Rose, G.O.Rudkin, J.Feltzin, Polyethersulphone - A New High Temperature Engineering Thermoplastic, in R.D.Deanin, Ed., New Industrial Polymers [ACS Symp.Ser. **4**], Amer.Chem.Soc., Washington, DC 1974, 63

9.6 POLYMERS WITH SULFUR IN RING STRUCTURES

S.Roth, One-Dimensional Metals. Physics and Materials Science, VCH, Weinheim 1995

G.Schopf, G.Kossmehl, Polythiophenes - Electrically Conductive Polymers, Adv.Polym.Sci. **129** (1997)

D.Fichou, Ed., Handbook of Oligo- and Polythiophenes, Wiley-VCH, Weinheim 1998

# 10 Carbon–Nitrogen Chains

## 10.1 Cyanide Polymers

The simplest carbon-nitrogen polymers are, in principle, obtained from monomers with polymerizable **cyano groups**, –C≡N. Such monomers are called **cyanides** in inorganic chemistry and **nitriles** in organic chemistry, the latter especially if the relationship to the carboxyl group is emphasized. The cyano group is isomeric to the **isocyanide** or **isonitrile** group, –NC, which was formerly called **carbamyl**. These polymers have interesting chemistries but are generally not very important.

### 10.1.1 Polycyanides

A linear polymer with the structure of a polymerized hydrogen cyanide, H–C≡N, is obtained by ring-opening polymerization of *s*-triazine (= 1,3,5-triazine),

(10-1) $n/3$ [s-triazine] ⟶ $\left[ \underset{\text{H}}{\text{C}} = \text{N} \right]_n$

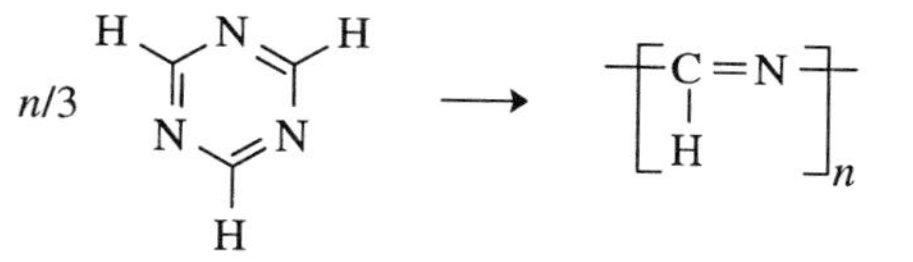

but not by direct polymerization of the extremely toxic **hydrogen cyanide** (**hydrocyanic acid**, **formonitrile, prussic acid**) itself (synthesis of HCN: p. 93). Hydrogen cyanide exists as a dimer, $(HCN)_2$, below its boiling point of 26.5°C and as a monomer H–C≡N above it. In contact with alkali (glass walls!), it converts slowly to a black-brown crosslinked polymer.

Anionic polymerization of $(HCN)_2$ delivers so-called **polymeric hydrogen cyanide** (**azulmic** acid) for which two ideal structures with fused rings and the same overall composition of repeating units, $C_4H_4N_4$, have been proposed (Fig. 10-1). Literature often quotes an average composition of $C_4H_5ON_5$ which would correspond to a low-molecular weight compound with OH and CN endgroups.

**Cyanogen** (**dicyan, oxalonitrile**), N≡C–C≡N, is obtained from many reactions, for example, from hydrogen cyanide (hydrocyanic acid) via 4 HCN + $O_2$ → 2 $(CN)_2$ + 2 $H_2O$, or 2 $Cu^{2+}$ + 4 HCN → 2 CuCN + $(CN)_2$ + 2 $H_2O$, etc. Electropolymerization of cyanogen delivers a complex **poly(cyanogen)** whose idealized structure is depicted on top of the next page. Threads can be drawn from concentrated solutions of poly(cyanogen) in tetrahydrofuran. Pyrolysis of poly(cyanogen) at 1800°C leads to electrically conducting carbon fibers (see Section 6.1.6).

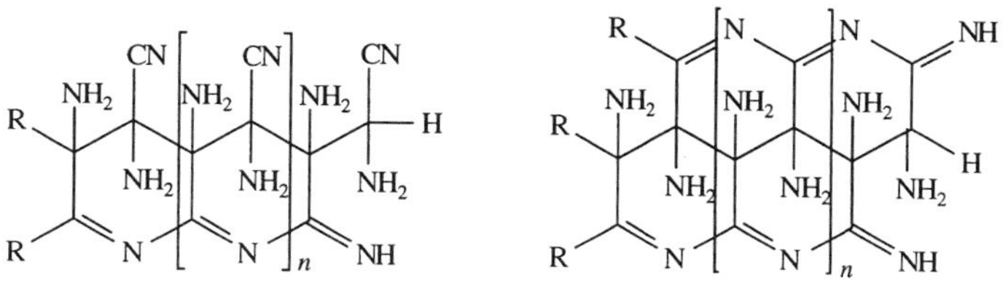

Fig. 10-1 Proposed ideal structures of azulmic acid [1]. R may be OH or CN.

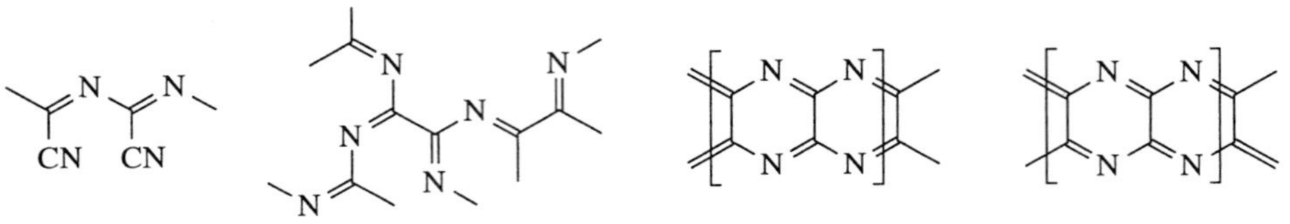

poly(cyanogen) idealized | paracyanogen real | paracyanogen ideal cis | paracyanogen ideal trans

**Paracyanogens (paracyans)** are obtained by pyrolysis of cyanogen or silver cyanide at 300–500°C. Compression of cyanogen at 4–10 GPa triggers a reversible reaction to a linear polymer with the structure of a poly(2,3-diiminosuccinonitrile), $+C_2N_2+_n$. The solid material is converted by pressure ($p > 10$ GPa) to a crosslinked black paracyanogen. In contrast to polycyanogen, this polymer cannot be converted to graphite fibers.

Paracyanogen with an ideal (hypothetical) ladder structure is predicted to be a metallic conductor. Experimental products are semiconductors ($\sigma = 3 \cdot 10^{-5}$ S/cm (25°C)).

Cyanogen has two possible isomers, isocyanogen **:**C=N–C≡N and diisocyanogen **:**C=N–N=C**:** but only the former is known. Isocyanogen polymerizes at temperatures above –80°C to **paraisocyanogen** which is not a semiconductor like its isomer, paracyanogen, but an insulator ($\sigma = 10^{-8}$ S/cm).

### 10.1.2 Polyisocyanides

Isocyanides (I) are *N*-substituted imino carbenes. The polymerization of their alkyl derivatives (for example, R = *t*-$C_4H_9$) with Ni(II) compounds, $H_2SO_4/O_2$, or $BF_3$ as initiators leads to high-molecular weight **poly(alkylisocyanide)s** (II) ($M \approx 10^5$):

(10-2) $\left[ \begin{matrix} C^{\ominus} \\ ||| \\ N^{\oplus} \\ | \\ R \end{matrix} \longleftrightarrow \begin{matrix} \bullet C \bullet \\ || \\ N \\ | \\ R \end{matrix} \right]_{I} \longrightarrow \left[ \begin{matrix} C \\ || \\ N \\ | \\ R \end{matrix} \right]_n \; II$

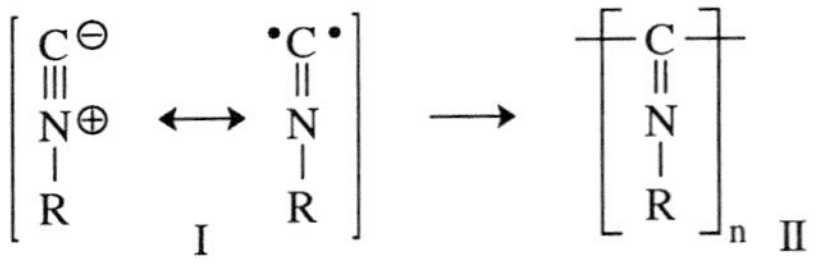

The polymerization does not proceed via the N–C multiple bond. The polymers rather have the structure of a **poly(alkyliminomethylene)**, i.e., they belong to the class of carbon chain polymers and not to the carbon–nitrogen chains of this chapter. They are treated here, however, because the monomer is related to other carbon–nitrogen compounds that do deliver carbon-nitrogen chains.

Poly(isocyanide)s with bulky alkyl substituents (R = *t*-$C_4H_9$, $CH(CH_3)C_6H_5$, etc.) seem to be predominantly stereoregular polymers whose stereoregular sequences form $4_1$ helix structures in the solid state. The cylindrical rod structure of these segments (persistence length: ca. 3 nm) is maintained in solution but the overall structure of the dissolved molecules is that of a disturbed random coil ($\alpha = 1.35$ in $[\eta] = K_v M^{\alpha}$).

Poly(isocyanate)s with optically inactive substituents form equal amounts of polymer molecules with opposite screw sense. The antipodes can be separated chromatographically in left and right-turning antipodes since the rigid helix structures survive in solution.

Heating of poly(alkylisocyanate)s (II) with small proportions of protons leads to planar **poly(azaethenylene)s** (III). The polymers are also obtained by cationic polymerization of nitriles (IV) such as benzonitrile (R = $C_6H_5$) or acetonitrile (R = $CH_3$):

$$(10\text{-}3)\quad \left[\begin{array}{c}\text{C}\\ \| \\ \text{NR}\end{array}\right]_n \xrightarrow{+\,H^{\oplus},\ \Delta} \left[\begin{array}{c}\text{C=N}\\ | \\ \text{R}\end{array}\right]_n \longleftarrow \begin{array}{c}\text{C}\equiv\text{N}\\ | \\ \text{R}\end{array}$$

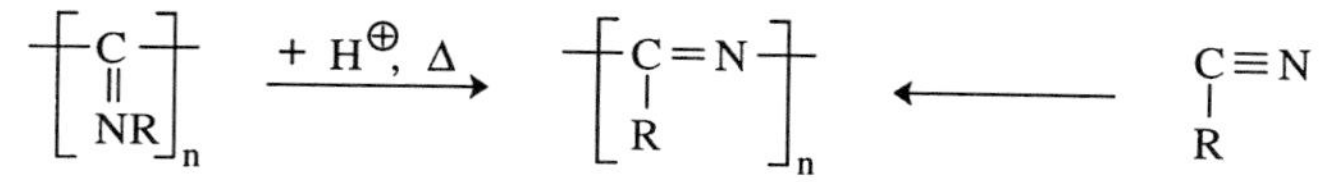

II III IV

Hydrogenation of poly(azaethenylene)s (III) results in the formation of **poly(alkylmethyleneimine)s**, $\text{+NH–CHR+}_n$.

Polymerization of substituted aromatic isocyanides leads to polyazines (Section 10.9) which are quite different types of polymers.

### 10.1.3 Polydinitriles and Polytetranitriles

Dinitriles, NC–CH=CH–CN, such as fumaronitrile and maleic dinitrile polymerize easily anionically via the nitrile groups. The polymerization yields only molecular weights of 1000 and less because oligomers are increasingly more resonance stabilized at higher degrees of polymerization. The homopolymerization of tetracyanoethylene has been unsuccessful but its copolymers with *N*-vinyl carbazole or Schiff bases are known.

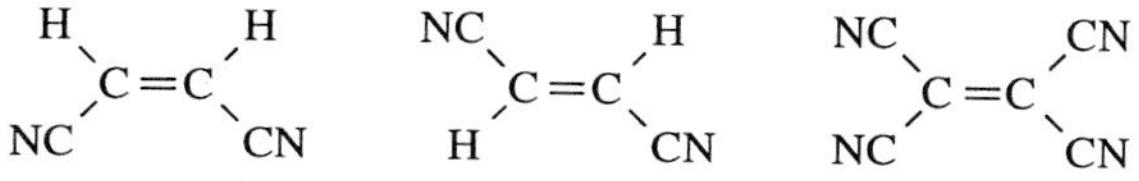

(H)(NC)C=C(H)(CN) (NC)(H)C=C(H)(CN) (NC)(NC)C=C(CN)(CN)

maleic dinitrile fumaronitrile tetracyanoethylene

### 10.1.4 Polycyanoacetylenes

Dicyanoacetylene, N≡C–C≡C–C≡N, polymerizes via the carbon–carbon triple bond with anionic initiators such as $CH_3ONa$, sodium naphthalene or $(C_6H_5)_3CNa$ and with transition metal catalysts such as $MoCl_5$ or $TiCl_4/Et_3Al$. The polymers have molecular weights of several thousands, are soluble, and show in the undoped state similar electrical conductivities ($\sigma \approx 10^{-6}$ S/cm) as *cis*-poly(acetylene) ($\sigma \approx 10^{-5}$ S/cm).

Cationic polymerization of dicyanoacetylene with $AsF_5$ at –78°C delivers black polymers with molecular weights of ca. 20 000 that do not contain carbon triple bonds. The undoped polymers have electrical conductivities of ca. 0.45 S/cm.

Dicyanodiacetylene, N≡C–C≡C–C≡C–C≡N, results from the $Cu_2Cl_2$-induced coupling of dicyanoacetylene. The higher oligomers $C_iN_2$ ($i$ = 8, 10, 12, 14, 16) result from the reaction of cyanogen, N≡C–C≡N, with vaporized carbon.

## 10.2 Imine and Amine Polymers

### 10.2.1 Nomenclature

Compounds of the type R'R"C=NH are called imines in organic chemistry. Azomethines (Schiff bases) such as R'R"C=NR'" are therefore considered substituted imines (anils if R" = $C_6H_5$). However, "imines" are also those organic compounds in which the

NH group connects identical substituents with constitutional seniority. An example is 4,4'-iminodibenzoic acid, HOOC–$C_6H_4$–NH–$C_6H_4$–COOH. Dialkylamines RR'NH, on the other hand, belong to the class of secondary amines.

Polymer chemistry uses mostly generic names of polymers, i.e., poly(monomer) names. The polymer of *N*-methyl vinyl amine, $CH_2$=CH–NH($CH_3$), is therefore called poly(*N*-methyl vinyl amine), $\text{-[}CH_2\text{–CH(}NH_2\text{)]}_n\text{-}$, whereas poly(ethylene imine) is the name of polymers of the three-membered ring molecule ethylene imine, $C_2H_5N$. The generic name of the polymer of methyl isocyanide, $CH_3$–NC, Eq.(10-2), would thus be poly(methyl isocyanide) but this name does not let one recognize the polymer constitution, $\text{-[C(=}NCH_3\text{)]}_n\text{-}$, which is that of a poly[(*N*-methylimine)methylene].

The systematic IUPAC nomenclature is practically only used for archival purposes such as Chemical Abstracts but rarely in the scientific literature and scarcely in industry. The systematic nomenclature would use poly(iminoethylene) instead of poly(ethylene imine) and poly((1-methylamino)ethylene) instead of poly(*N*-methyl vinyl amine).

## 10.2.2 Poly(ethyleneimine)s

**Linear Poly(ethyleneimine)**

Poly(alkyleneimine)s are polymers with two or more carbon atoms between nitrogen chain atoms. Linear poly(ethyleneimine), $\text{-[NH-}CH_2\text{-}CH_2\text{]}_n\text{-}$, results from the isomerizing polymerization of 4,5-dihydrooxazole (formerly: 2-oxazoline; for the nomenclature, see Section 10.8.5) and subsequent saponification of the polymer:

$$\text{(10-4)} \quad n\ \text{(4,5-dihydrooxazole ring: N3, C4, C5, O1, C2)} \longrightarrow \left[ \begin{matrix} NCH_2CH_2 \\ | \\ CHO \end{matrix} \right]_n \xrightarrow[-n\ HCOONa]{+n\ NaOH} \text{-[}NHCH_2CH_2\text{]}_n\text{-}$$

Rings such as 4,5-dihydrooxazole usually do not polymerize, but this cyclic monomer does because the iminoether group –N=CH–O– isomerizes to a group >N–CHO which has a greater resonance energy.

Linear (unbranched) poly(ethylene imine)s are crystalline ($T_M$ = 58.5°C), very hygroscopic, and soluble in hot water. They are not industrial products because of the high cost of the monomer, although they have interesting properties. Since the nitrogen chain unit, –NH–, carries two unequal substituents (H and an electron pair), linear poly(ethyleneimine) exists in two ideal macroconformations, isotactic and syndiotactic. In the isotactic species, successive N-substituents are on *opposite* sides in trans conformations of chains but on the same side in cis conformations (Fischer projection on a plane) (Fig. 10-2). In syndiotactic polymers, it is just the opposite.

The three possible segment types, –$CH_2$–NH–$CH_2$–, –$CH_2$–$CH_2$–NH– (iso) and –$CH_2$–$CH_2$–NH– (syndio), are always in the sequence trans-trans (TT) of microconformations if electrostatic forces such as hydrogen bonds are present. Such secondary bonds exist in the sesquihydrates and dihydrates of linear poly(ethylene imine)s which crystallize as zig-zag chains. Absence of electrostatic forces leads to microconformations TT for –$CH_2$–NH–$CH_2$–, $TG^-$ for –$CH_2$–$CH_2$–NH– (iso), and $TG^-$ for –$CH_2$–$CH_2$–NH– (syndio). Dry linear poly(ethyleneimine) forms $5_1$ helices in the crystalline state which aggregate to double helices.

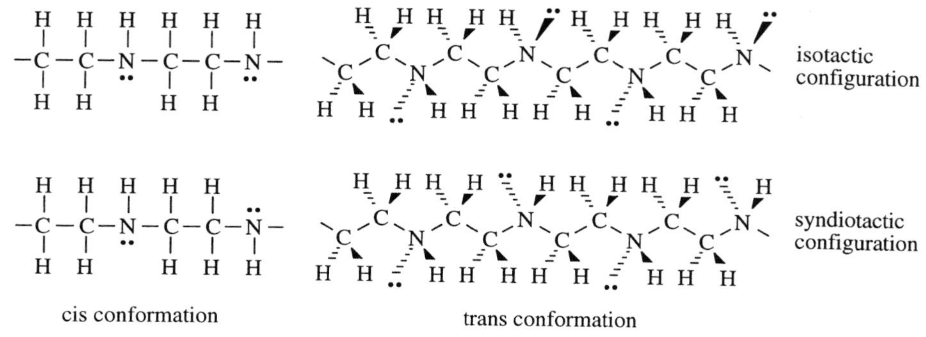

isotactic configuration

syndiotactic configuration

cis conformation

trans conformation

Fig. 10-2 Configuration and conformation of linear poly(ethyleneimine)s.

Since the barriers between the three energy minima at 0°, 120°, and 240° are low, chains are expected to have high kinetic flexibilities. In aqueous solutions, however, chains are hydrated and stiffened by the resulting electrostatic effects. Theory expects chains of linear poly(ethyleneimine) to be stiffer than those of poly(ethylene oxide), ca. two times if isotactic and ca. four times if syndiotactic, although the chains still form random coils. Since hydration decreases with increasing temperature, coils should become more compact if the temperature is raised.

### Linear *N*-Substituted Poly(ethyleneimine)s

2-Alkyl-2-oxazolines (= 2-alkyl-4,5-dihydrooxazoles; see p. 514) polymerize to poly(2-acyl-2-oxazoline)s, $\text{+N(COR)-CH}_2\text{-CH}_2\text{+}_n$, because the iminoether group isomerizes to a tertiary amide (see Section 10.8.5). These polymers are used as adhesives (for paper, glass, aluminum, polyamides, Cellophane®), as sizes, dispersing agents, etc.

### Branched Unsubstituted Poly(ethyleneimine)s

Ethyleneimine (**aziridine**) is produced industrially by two processes. The Wencker process reacts ethanolamine, $H_2NCH_2CH_2OH$, (from ethyleneoxide and ammonia) with sulfuric acid to $\beta$-aminoethylsulfuric acid, $H_2NCH_2CH_2OSO_3H$. On heating with NaOH under pressure, this acid dissociates to aziridine, sulfuric acid (as sodium salt), and water. In the Dow process, 1,2-dichloroethane is reacted with $NH_3$ and CaO to aziridine, $CaCl_2$, and $H_2O$.

Cationic polymerization of aziridine with proton acids or alkylating agents delivers poly(ethyleneimine)s (= poly(iminoethylene)s; **PEI** (also used for polyether imide!)):

$$\text{(10-5)}\qquad H^{\oplus} + HN\triangleleft \longrightarrow H_2\overset{\oplus}{N}\triangleleft \xrightarrow{+n\,HN\triangleleft} H(NHCH_2CH_2)_n\overset{\oplus}{N}\triangleleft \quad \text{etc.}$$

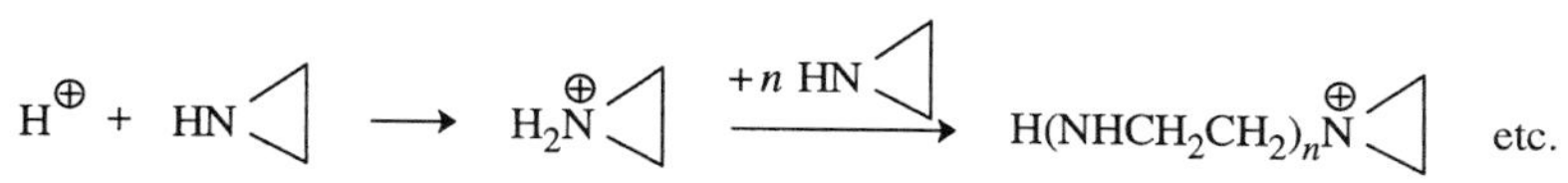

The propagation is accompanied by chain-transfer reactions to –NH– groups. The resulting polymers are thus hyperbranched via tertiary nitrogen atoms; they also carry many primary amino endgroups. In commercial products, the ratio of primary:secondary:tertiary amino groups is ca. 1:2:1. Molecular weights range from 500 to $10^5$.

These hyperbranched poly(ethyleneimine)s dissolve in cold water. They are used to improve the wet strength of paper and as binders for pigments or polyester cord in tires. Quaternized poly(ethylenimine) serves as a flocculant for polyanions in water treatment.

**Substituted Polyethyleneimines**

N-Substituted ethyleneimines polymerize cationically similar to the unsubstituted parent compound, i.e., with chain transfer and chain termination. An exception is *N-t*-butylethylenimine which polymerizes cationically without termination and transfer.

C-substituted ethyleneimines such as 2-methylaziridine (= "propyleneimine") were never commercial products (the monomer is a carcinogen for animals).

Poly(alkyleneimine)s with longer alkylene units between nitrogen chain units result from N-alkylation of tertiary diamines (Menshutkin reaction). These polymers are strong polyelectrolytes; they are known as **ionenes**:

$$n\,(CH_3)_2N-(CH_2)_i-N(CH_3)_2 + n\,Br-(CH_2)_j-Br \longrightarrow \left[ -N^{\oplus}(CH_3)_2-(CH_2)_i-N^{\oplus}(CH_3)_2-(CH_2)_j- \right]_n \; 2n\,Br^{\ominus} \qquad (10\text{-}6)$$

## 10.2.3 Poly(formaldazine)

According to IUPAC, **aldazines** are 2:1 condensation products of **ald**ehydes, R–CHO, and hydr**azine**, $H_2N–NH_2$. The simplest aldazine, formaldazine ($CH_2=N–N=CH_2$), is the nitrogen analog of butadiene. Like butadiene, it polymerizes anionically. Unlike butadiene, it polymerizes cationically but not with free radicals. Like poly(butadiene), its monomeric units may be present as cis-1,4, trans-1,4, it-1,2,or st-1,2 structures.

## 10.2.4 Poly(aniline)

Oxidation of hydrochloric acid solutions of aniline, $C_6H_5NH_2$, or other aniline salts with dichromates or other oxidizing agents in the presence of oxygen-transfer agents such as iron, copper, or vanadium salts produces deep blue to black powders that are known as **aniline black** or **poly(aniline)**, a pigment dye of the azine dye group. Aniline black consists of polymers with chain-like but not strictly linearly connected phenazine groups (for azine polymers, see Section 10.9). These polymers also contain several other groups with nitrogen–carbon, carbon–carbon, and nitrogen–nitrogen bonds and probably also branch sites. Aniline blacks are usually directly produced on cotton or silk where they dye the fabric in extraordinarily beautiful black colors of great light fastness.

phenazine
(dibenzopyrazine)

Linear poly(aniline)s (**PANI**) are produced industrially by oxidizing aniline with ammonium persulfate, $(NH_4)_2S_2O_8$, in an excess of acid (Versicon®). Alternatively, one can oxidize with $FeCl_3$ or polymerize electrochemically. A biochemical method uses $H_2O_2$ and horseradish peroxidase as catalyst. During polymerization, the expensive peroxidase is destroyed because the polymerization requires low pH.

Depending on reaction conditions and post reactions, **linear poly(aniline)** can be obtained in several forms, all of which are electrical insulators:

- fully reduced yellow **leukoemeraldine** (G: *leukos* = white, clear (!));
- 50 % oxidized blue **emeraldine (base)** with benzoid and quinoid structures;
- 100 % oxidized purple **pernigraniline** with alternating benzoid and quinoid structures.

Partial oxidation of leukoemeraldine and protonation of emeraldine base leads to the electrically conducting, green **emeraldine salt** (hence the name). Total oxidation results in blue-purple **pernigraniline salt** (for structures, see next page).

In the synthesis of PANI, aniline (I) is oxidized to a radical cation (II) which dimerizes with its mesomeric form (III) to a dication (IV). Deprotonation delivers neutral dianiline (V) which is oxidized further to a mono(radical cation) (VI) and finally to a di(radical cation) (VII). Coupling of VII and II eliminates two protons. The resulting trication (IX) is mesomeric with X which already contains the repeating unit of pernigraniline.

(10-7)

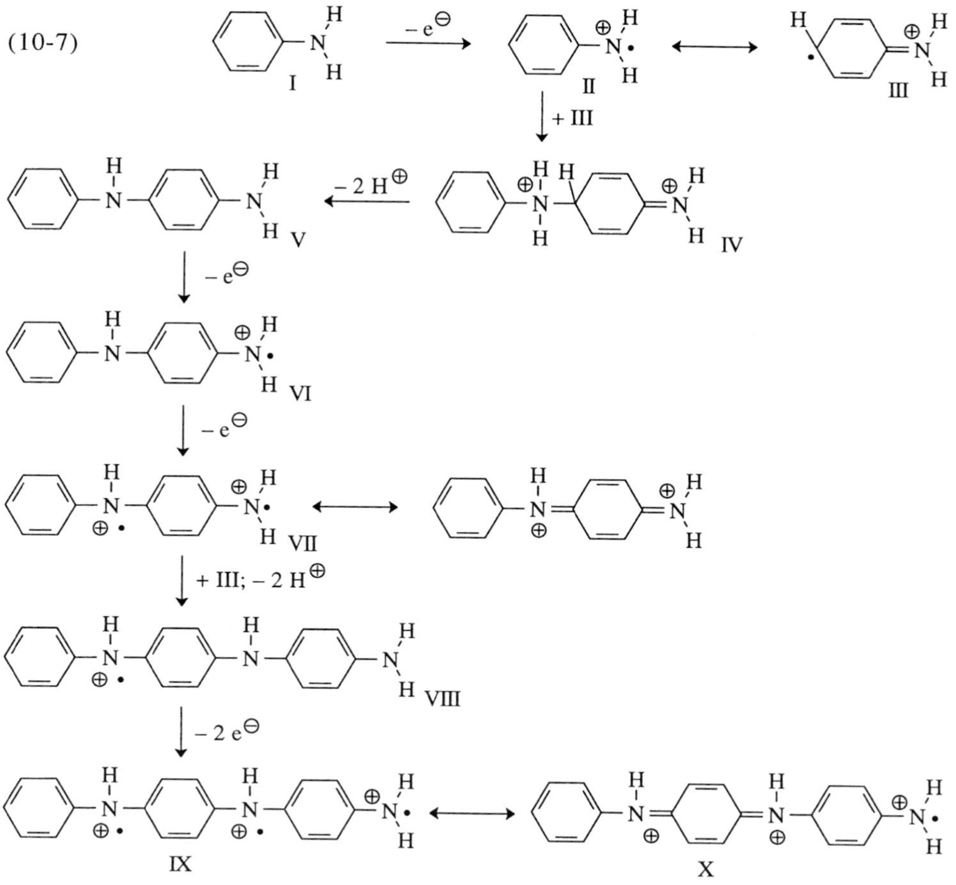

I → (− e⊖) → II ⟷ III

II ↓ + III

IV → (− 2 H⊕) → V

V ↓ − e⊖ → VI

VI ↓ − e⊖ → VII

VII ↓ + III; − 2 H⊕ → VIII

VIII ↓ − 2 e⊖ → IX ⟷ X

The synthesis leads directly to the pernigraniline salt which is however immediately reduced to emeraldine salt by excess aniline. In all steps, anions $A^{\ominus}$ of excess acid stabilize the produced polycations as salts. The acid furthermore solubilizes the aniline in its aqueous solution and provides anions as dopants for polymers.

Poly(aniline)s can be transformed to other poly(aniline)s by changing the state of oxidation or the pH of the solution, as shown in Eq.(10-8) for a dianiline unit. The repeating unit of leukoemeralidine consists of one unit of IX, that of emeraldine of one unit of IX and one unit X, and that of pernigraniline of two units X (see also Fig. 10-3).

(10-8)

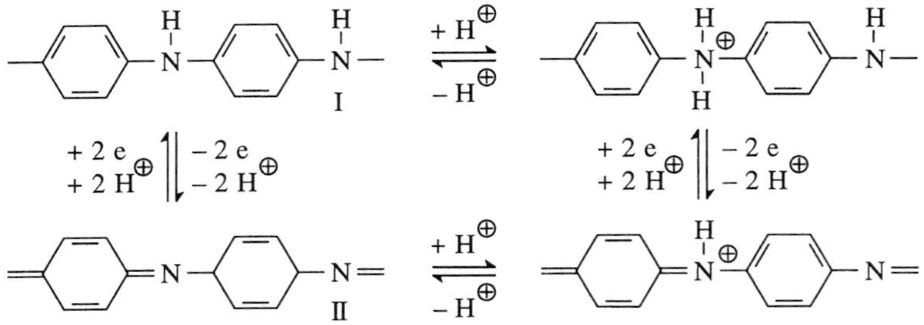

$$-C_6H_4-NH-C_6H_4-NH- \ (I) \underset{-H^{\oplus}}{\overset{+H^{\oplus}}{\rightleftharpoons}} -C_6H_4-NH_2^{\oplus}-C_6H_4-NH-$$

$$+2\,e,\ +2\,H^{\oplus} \uparrow\downarrow -2\,e,\ -2\,H^{\oplus}$$

$$=C_6H_4=N-C_6H_4-N= \ (II) \underset{-H^{\oplus}}{\overset{+H^{\oplus}}{\rightleftharpoons}} =C_6H_4=NH^{\oplus}-C_6H_4-N=$$

Chemical and physical properties of poly(aniline)s are strongly affected by their state of oxidation, their charges, the type of their counteranions, $A^{\ominus}$, and their crystallinity (see Fig. 10-3). For example, leukoemeraldine has a completely non-conjugated polymer chain whereas the chain of pernigraniline is completely conjugated. Polymer structures and properties can also be modified by introducing various substitutents.

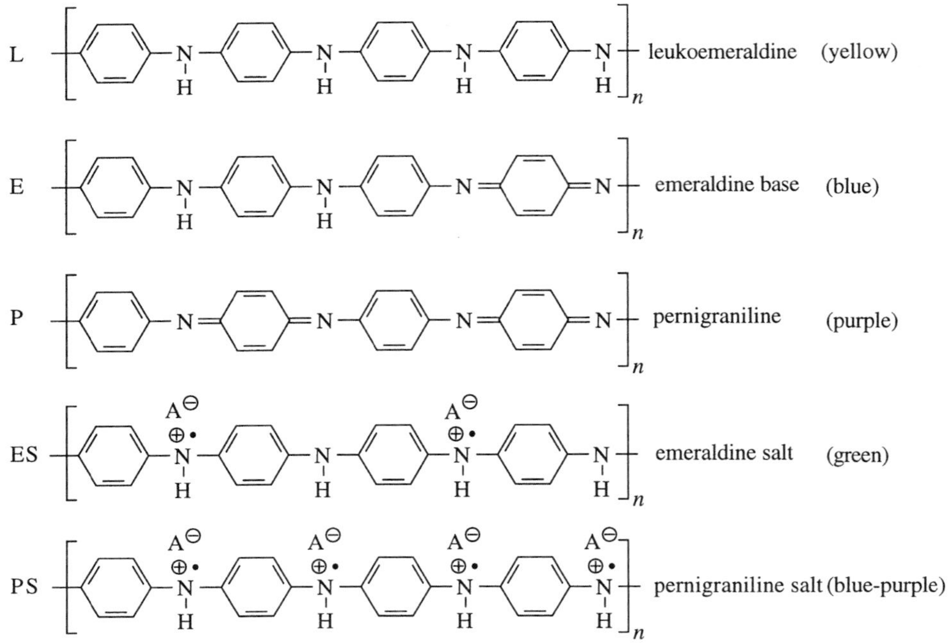

L $\left[-C_6H_4-NH-C_6H_4-NH-C_6H_4-NH-C_6H_4-NH-\right]_n$ leukoemeraldine (yellow)

E $\left[-C_6H_4-NH-C_6H_4-NH-C_6H_4-N=C_6H_4=N-\right]_n$ emeraldine base (blue)

P $\left[-C_6H_4-N=C_6H_4=N-C_6H_4-N=C_6H_4=N-\right]_n$ pernigraniline (purple)

ES $\left[-C_6H_4-\overset{A^{\ominus}}{\overset{\oplus\bullet}{N}}H-C_6H_4-NH-C_6H_4-\overset{A^{\ominus}}{\overset{\oplus\bullet}{N}}H-C_6H_4-NH-\right]_n$ emeraldine salt (green)

PS $\left[-C_6H_4-\overset{A^{\ominus}}{\overset{\oplus\bullet}{N}}H-C_6H_4-\overset{A^{\ominus}}{\overset{\oplus\bullet}{N}}H-C_6H_4-\overset{A^{\ominus}}{\overset{\oplus\bullet}{N}}H-C_6H_4-\overset{A^{\ominus}}{\overset{\oplus\bullet}{N}}H-\right]_n$ pernigraniline salt (blue-purple)

Fig. 10-3 Basic forms of poly(aniline).

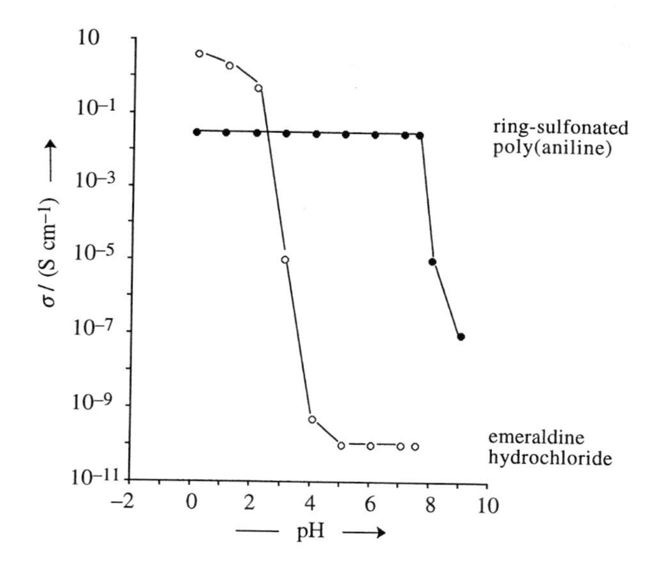


Fig. 10-4 pH-dependence of electrical conductivity $\sigma$ of emeraldine hydrochloride (○) and a poly-(aniline) with sulfonated phenylene groups (●) [1].

For example, the electrical conductivity of emeraldine hydrochloride varies almost 11 decades, from ca. $10^{-10}$ S/cm in the neutral range to ca. 40 S/cm in the strongly acidic range (Fig. 10-4). Emeraldine salts have the largest electrical conductivities; for example, $\sigma$ = 200 S/cm for a fiber that was spun from *m*-cresol and doped by (±)-camphor sulfonic acid. Such fibers have a textile modulus of elasticity of $E_{tex}$ = 7.3 gf/den (corresponds to $E$ = 1.1 GPa at $\rho \approx 1.7$ g/cm$^3$) and a tenacity of $\sigma_{tex}$ = 0.2 gf/den (corresponds to a strength at break of $\sigma_B$ = 30 MPa). Glass temperatures are ca. 200°C.

Appropriately doped poly(aniline)s maintain their electrical conductivities for many years. They serve for various electronic applications: light-emitting diodes, rechargeable batteries, biosensors, organic semiconductors, and also for other applications such as membranes for gas separations.

### 10.2.5 Amine Dendrimers

Amine dendrimers are dendritically branched polymers (Fig. 10-5) with tertiary amine groups as branching points.

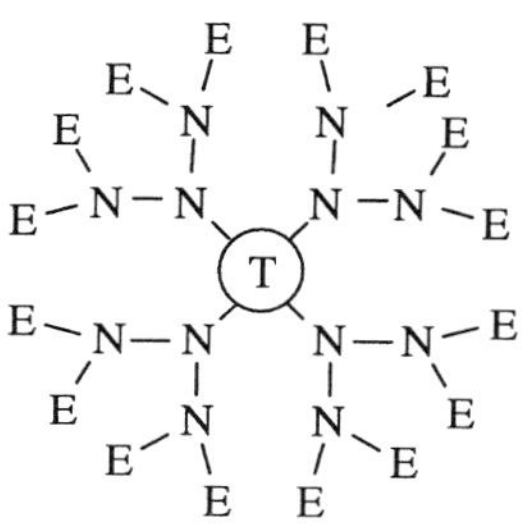

```
        E   E
    E\  /   \  /E
  E   N     N   E
   \  |     |  /
E─N─N       N─N─E
       \   /
        (T)
       /   \
E─N─N       N─N─E
   /  |     |  \
  E   N     N   E
    E/ \   / \E
        E   E
```

Fig. 10-5 Schematic representation of an amine dendrimer with a tetravalent core T, two generations of branching units, and endgroups E. Lines between T and N, N and N, and N and E do *not* indicate bonds but segments (see Table 10-1 for examples).

Table 10-1 Amine dendrimers. *) Formerly known as POPAM = *poly*or*tho**p*ropyl*am*ide amine, then called polypropylene imine (PPI). However, propylene imine is the common name for 2-methylaziridine = 2-methylethyleneimine and not for azetidine (= 1,3-propyleneimine). Ring-opening polymerization of azetidine would give $-NH-CH_2CH_2CH_2-$ units.

| Type | Core T | Branching unit | Endgroup E |
|---|---|---|---|
| PPI *) | $>N(CH_2)_4N<$ | $-CH_2CH_2CH_2N<$ | $-(CH_2)_2CN$, $-(CH_2)_3NH_2$ |
| PAMAM | $>N(CH_2)_2N<$ | $-CH_2CH_2C(=O)NHCH_2CH_2N<$ | $-(CH_2)_2COOCH_3$, $-(CH_2)_2CONH(CH_2)_2NH_2$ |
| PAMAMOS | $>N(CH_2)_2N<$ | $-CH_2CH_2C(=O)NHCH_2CH_2N<$ | $-(CH_2)_2COO(CH_2)_3Si(OCH_3)_3$, $-(CH_2)_2COO(CH_2)_3Si(CH_3)(OCH_3)_2$ |

Only one type of amine dendrimer is presently commercially produced: PAMAM polymers with amido-amine units, $-(CH_2)_2CONH(CH_2)_2N<$, as branching units. A special subclass of the PAMAM polymers are PAMAMOS polymers that carry silicon groups as endgroups (Table 10-1). The commercial production of PPI polymers with *t*-amino-1,3-propylene units, $-(CH_2)_3N<$, has ceased.

**Poly(1,3-trimethylenimine) Dendrimers (PPI)**

For the synthesis of PPI dendrimers, 1,4-diaminobutane as the core is reacted with acrylonitrile in a double Michael addition. Subsequent hydrogenation with Raney cobalt delivers a tetrafunctional core unit with four 1,3-propylene amine branches as the first generation of branching units:

$$H_2N(CH_2)_4NH_2 + 4\,CH_2{=}CH-C{\equiv}N \longrightarrow (N{\equiv}C-CH_2CH_2)_2N(CH_2)_4N(CH_2CH_2-C{\equiv}N)_2$$

$$\downarrow + 8\,H_2$$

$$(H_2NCH_2CH_2CH_2)_2N(CH_2)_4N(CH_2CH_2CH_2NH_2)_2 \qquad (10\text{-}9)$$

Repeated Michael additions of acrylonitrile produce the second, third, ... generations of branching units.

PPI molecules with nitrile endgroups adopt in solution the shape of slightly swollen spheroids with specific volumes that do not markedly exceed the Einstein value of $[\eta] = 5/(2\,\rho)$ where $\rho$ is the density of a solid sphere. Glass temperatures are very low.

**Polyamidoamine Dendrimers (PAMAM)**

PAMAM dendrimers use 1,2-diaminoethane as the core molecule. In the first step, a double Michael addition of methyl acrylate is followed by amidation of the resulting carbomethoxy groups with a great excess of ethylene diamine:

$$H_2N(CH_2)_2NH_2 \xrightarrow{+\,4\,CH_2{=}CHCOOCH_3} (CH_3OOCCH_2CH_2)_2N(CH_2)_4N(CH_2CH_2COOCH_3)_2$$

$$\xrightarrow[-\,4\,CH_3OH]{+\,4\,H_2N(CH_2)_2NH_2} (H_2N(CH_2)_2NHCOCH_2CH_2)_2N(CH_2)_4N(CH_2CH_2CONH(CH_2)_2NH_2)_2 \qquad (10\text{-}10)$$

The next generations are synthesized analogously: Michael addition of methyl acrylate followed by amidation with ethylene diamine. Commercial PAMAM and PAMAMOS dendrimers are available with 0–7 generations of branching units. In laboratory preparations, a 10th generation PAMAM molecule was obtained ($M$ = 934 720). Surface crowding of endgroups prevented the synthesis of higher generations.

In theory, PAMAM and PPI dendrimers are molecularly homogeneous. In practice, constitutional mistakes cannot be totally avoided.

At the same molecular weight, dissolved PAMAM molecules occupy a larger space than PPI molecules (Table 10-2) as indicated by intrinsic viscosities that measure specific volumes, certainly because PAMAMs have larger branch cells than PPIs and possibly because of electronic repulsion (measurements in citric acid solution). Molecules are characterized by relatively low densities in the interior and fairly high densities on the "surface". The "hollow" interior and the many reactive surface groups allow amine dendrimers to encapsulate metals; act as molecular absorbants, carriers for pharmaceuticals, etc.

Table 10-2 Properties of PAMAM (with $NH_2$ endgroups) and PPI dendrimers (with benzyl ether endgroups) [2]. $N_{gen}$ = number of generations, $N_{end}$ = number of endgroups per molecule, $M$ = molecular weight, $R_G$ = radius of gyration by small-angle neutron scattering, $R_H$ = hydrodynamic radius (from dilute solution viscometry, $T_G$ = glass temperature. $R_G$, $R_H$: measurements in 0.1 mol/L aqueous citric acid (PAMAM) or $D_2O$ (PPI).

| Type | | PAMAM | | | | PPI | | | |
|---|---|---|---|---|---|---|---|---|---|
| $N_{gen}$ | $N_{end}$ | $M$ | $R_G$/nm | $R_H$/nm | $T_G$/°C | $M$ | $R_G$/nm | $R_H$/nm | $T_G$/°C |
| 0 | 4 | 517 | | 0.77 | −11 | 317 | 0.44 | 0.65 | −107 |
| 1 | 8 | 1 430 | | 1.01 | −3 | 773 | 0.69 | 0.92 | −97 |
| 2 | 16 | 3 256 | | 1.44 | 0 | 1 687 | 0.93 | 1.21 | −90 |
| 3 | 32 | 6 909 | 1.65 | 1.75 | 11 | 3 514 | 1.16 | 1.54 | −87 |
| 4 | 64 | 14 215 | 1.97 | 2.5 | 14 | 7 168 | 1.39 | 1.98 | −84 |
| 5 | 128 | 28 826 | 2.43 | 2.72 | 14 | | | | |
| 6 | 256 | 58 048 | 3.03 | 3.37 | 16 | | | | |
| 7 | 512 | 116 493 | 3.58 | 4.05 | | | | | |
| 8 | 1024 | 233 383 | | 4.85 | | | | | |
| 9 | 2048 | 467 162 | 5.7 | 4.92 | | | | | |
| 10 | 4096 | 934 720 | | 6.75 | | | | | |

### 10.2.6 Benzoquinone Polyimines

Polycondensation of diamines with *p*-benzoquinone proceeds with formation of 1,4-dihydroxybenzene to polymers that contain both benzoquinone and imine groups per repeating unit:

(10-11)

$$(n+1)\ H_2N-Z-NH_2 + 3\ n\ \text{(}p\text{-benzoquinone, O=C}_6\text{H}_4\text{=O)} \longrightarrow$$

$$H_2N-Z-NH\!\left[\text{-C}_6\text{H}_2\text{(=O)}_2\text{-}NH-Z-NH\right]_n\!H + 2\ n\ \text{HO-C}_6\text{H}_4\text{-OH}$$

These polymers resemble marine cement which is a 3,4-dihydroxyphenyl-L-alanine rich protein that is produced by molluscs. Like marine cement, benzoquinone polyimines are completely water-resistant and non-wettable. They adhere well on metals, wood, concrete, asphalt, and electronic materials where they act as moisture-repelling coatings. Such coatings can also be cured thermally or chemically.

## 10.3 Polyamides

### 10.3.1 Overview

Polyamides are polymers with amide groups –NH–CO– in the main chain. They are usually subdivided in either aliphatic and aromatic types or in AB polymers of the type $\text{+NH–Z–CO+}_n$ and AA/BB polymers of the type $\text{+NH–Z}^1\text{–NH–CO–Z}^2\text{–CO+}_n$ where bifunctional groups Z, $Z^1$, and $Z^2$ can be aliphatic, cycloaliphatic, or aromatic. Branched polyamides are presently niche products.

The first industrially usable polyamide, poly(hexamethylene adipamide) with the constitution $\text{+NH(CH}_2)_6\text{NHCO(CH}_2)_4\text{CO+}_n$, was patented in 1937 by Wallace Hume Carothers for the DuPont Company. Fibers from this polyamide were produced in 1939 as nylon fibers (now known as nylon 6.6 or 66). "Nylon" is no longer a protected trademark but a general name for all *aliphatic* polyamides. The US Federal Trade Commission (FTC) restricts the use of "nylon" to polyamide *fibers* where more than 85 % of the amide linkages are attached directly to aliphatic groups (see Volume IV).

Carothers called aliphatic polyamides with molecular weights greater than 10 000 "superpolyamides." During more recent years, this term was occasionally used for aromatic polyamides. The latter polymers are now known as **aramids**. Again, the FTC has ruled that "aramids" are fiber-forming aromatic polyamides in which more than 85 % of amide groups are attached to aromatic groups.

In 1930/31, Carothers also condensed 1-aminocaproic acid but considered the resulting polymer useless. Only one year after Carothers' basic 1937 patent, Paul Schlack of IG Farbenindustrie patented the ring-opening polymerization of ε-caprolactam to a polymer with the same chemical structure, poly(ε-caprolactam), $\text{+NH(CH}_2)_5\text{CO+}_n$. Fibers from this polymer were marketed as Perlon®, also known as nylon 6 (PA 6).

Nylon 6.6 fibers were originally used for ladies' stockings where they replaced expensive silk and not so expensive but ugly cotton. This use probably led to the term "nylon" (there are many guesses about the origin of the word including silly and jingoistic ones). Since, when damaged, the new nylon stockings did not run as silk stockings did, the proposed trade name was first "No-run" which later mutated to "Nuron", "Niron", "Nilon", and finally "Nylon." The origin of the trade name "Perlon" is also unknown; "perlon" is probably a contraction of "su*per*ny*lon*."

The types of monomeric units of aliphatic polyamides are usually indicated by the number of their carbon chain atoms whereas those of aromatic polyamides are characterized by letters, for example, T for terephthalic acid. Poly(ε-caprolactam) is therefore polyamide 6 (nylon 6). For AA/BB polymers, the first number refers to the number of carbon chain atoms in diamine units, the second to the number of carbon chain atoms in the dicarboxylic acid unit. Poly(hexamethylene adipamide) is thus polyamide 6.6 or nylon 66 (pronounced "nylon six-six" and not "nylon sixty six"). ASTM and IUPAC separate numbers by dots but ISO and DIN do not. The latter practice may cause some people to question the morals of those who produce PA 666, the copolymer of ε-caprolactam, hexamethylene diamine, and adipic acid (see Revelations *13*:18).

PA 6.6 and PA 6 are the classical polyamides. Since DuPont and IG Farben exchanged patents and divided markets, PA 6.6 still dominates in the United States and England (where DuPont had an agreement with ICI) whereas PA 6 is the main polyamide in Germany and Japan. At present, many other aliphatic polyamides have been developed by the more than one hundred companies that produce them.

## 10.3.2 Linear Aliphatic AA/BB Polyamides

### Types

At present, six homopolymeric aliphatic AA/BB types are produced:

| | | |
|---|---|---|
| PA 4.6 | $\text{+NH(CH}_2)_4\text{NH–OC(CH}_2)_4\text{CO+}_n$ | poly(tetramethylene adipamide) |
| PA 6.6 | $\text{+NH(CH}_2)_6\text{NH–OC(CH}_2)_4\text{CO+}_n$ | poly(hexamethylene adipamide) |
| PA 6.9 | $\text{+NH(CH}_2)_6\text{NH–OC(CH}_2)_7\text{CO+}_n$ | poly(hexamethylene azelaamide) |
| PA 6.10 | $\text{+NH(CH}_2)_6\text{NH–OC(CH}_2)_8\text{CO+}_n$ | poly(hexamethylene sebacamide) |
| PA 6.12 | $\text{+NH(CH}_2)_6\text{NH–OC(CH}_2)_{10}\text{CO+}_n$ | poly(hexamethylene dodecanoamide) |
| PA 10.10 | $\text{+NH(CH}_2)_{10}\text{NH–OC(CH}_2)_8\text{CO+}_n$, | poly(decamethylene sebacamide) |

Poly(tetramethylene oxalamide) (PA 4.2) is made in a pilot plant:

| | | |
|---|---|---|
| PA 4.2 | $\text{+NH(CH}_2)_4\text{NH–OCCO+}_n$ | poly(tetramethylene oxalamide) |

whereas nylon 12.12 was discontinued and nylon 13.13 never made it to the market:

| | | |
|---|---|---|
| PA 12.12 | $\text{+NH(CH}_2)_{12}\text{NH–OC(CH}_2)_{10}\text{CO+}_n$ | poly(dodecamethylene dodecanoamide) |
| PA 13.13 | $\text{+NH(CH}_2)_{13}\text{NH–OC(CH}_2)_{11}\text{CO+}_n$ | poly(tridecanomethylene decosenamide) |

In addition, there are a number of copolymers of the aliphatic AA/BB type (e.g., 6.6 + 6.12), the aliphatic AB type (e.g., 6.6 + 6), or with aromatics (e.g., 6.6 + 6.T + 6.I).

In industry, one either uses poly(monomer) names, short designations (i.e., numbers and letters), or tradenames (for example: Zytel® (DuPont), Ultramid® (BASF), Grilon® (Ems), etc.). Tradenames may refer to the whole spectrum of polyamides of a company or to just one type, for example (DSM): Stanyl® (PA 4.6) but Akulon® (PA 6.6, PA 6).

The scientific literature also uses short forms and generic names but in addition "systematic names" that were introduced by the *International Union of Pure and Applied Chemistry* (IUPAC) and/or *Chemical Abstracts Service* (CAS) and *CAS Registry Numbers.* Systematic names are derived from constitutional formulae that arrange chain units according to their seniorities. Not only do these seniorities not conform to the familiar generic formulae but they also use different designations which have changed several times during the last thirty years. Examples are the systematic names for polyamide 6.6 (nylon 6.6) = poly(hexamethylene adipamide), $\text{+NH(CH}_2)_6\text{NH–OC(CH}_2)_4\text{CO+}_n$:

| | |
|---|---|
| poly[imino(1,4-dioxo-1,4-butandiyl)imino-1,6-hexanediyl] | CAS, outdated |
| poly[imino(1,6-dioxohexamethylene)iminohexamethylene] | IUPAC (1975) |
| poly[iminoadipoyliminohexamethylene) | IUPAC, alternative (1975) |
| poly(iminoadipoyliminohexane-1,6-diyl] | CAS + IUPAC (1996) |
| poly[imino(1,6-dioxohexamethylene)iminohexane-1,6-diyl] | CAS + IUPAC, alternative (1996) |

**Monomers**

Production and use of aliphatic AA/BB polyamides is controlled not only by product properties but also by the availability and production costs of monomers. Lower dicarboxylic acids are based on petrochemical products whereas higher ones stem predominantly from natural sources:

*Oxalic acid*, HOOC–COOH, (for PA 4.2), is obtained as the disodium salt by dimerizing sodium formate, HCOONa, (from formic acid, p. 96) in a melt of NaOH. The reaction releases hydrogen.

*Adipic acid*, $HOOC(CH_2)_4COOH$, (for PA 4.6 and PA 6.6), results from the oxidation of cyclohexane (p. 127). A new process proposes to oxidize benzene with $N_2O$ (byproduct of the oxidation of KA oil, p. 128) to phenol which is then hydrogenated to cyclohexanone. The ketone is oxidized with $HNO_3$ to adipic acid which produces $N_2O$ for the benzene oxidation.

*Azelaic acid,* $HOOC(CH_2)_7COOH$, (for PA 6.9), is obtained from the oleic acid of rice (bran) oil by ozonolysis (Eq.(3-15)).

*Sebacic acid*, $HOOC(CH_2)_8COOH$, (for PA 6.10), results from the alkali cleavage of ricinoleic acid (Eq.(3-12)), the glycerol ester of which is the main component of castor oil (main producers: North Africa, India, Brazil). Because of the unstable supply of castor oil, sebacic acid is increasingly replaced by dodecanoic acid.

*Dodecanoic acid,* $HOOC(CH_2)_{10}COOH$, (for PA 6.12), is made by oxidation of cyclododecatriene (p. 112, II) from the cyclotrimerization of 1,4-butadiene.

*Brassidic acid* (tridecanoic acid), $HOOC(CH_2)_{11}COOH$, (for PA 13.13), is produced as the monomethyl ester by ozonolysis of the methyl ester of *cis*-erucic acid (*cis*-13-docosenic acid), $H(CH_2)_8CH{=}CH(CH_2)_{11}COOCH_3$, (p. 80), which is the main component of the fatty esters of the plant *Crambe abyssinica*. Production of PA 13.13 as well as PA 13 (see below) is not economical because the press cake from pressing of crambe seeds cannot be sold as fodder for cows and chickens (transfer of taste to milk and eggs).

Diamines for all presently marketed aliphatic polyamides are petrochemical products:

*Tetramethylenediamine* (1,4-diaminobutane), $H_2N(CH_2)_4NH_2$, (for PA 4.2), is produced by hydrogenation of succinonitrile, $NC(CH_2)_2CN$, which is obtained by addition of HCN to acrylonitrile, $CH_2{=}CH{-}CN$.

*Hexamethylendiamine* (1,6-diaminohexane), $H_2N(CH_2)_6NH_2$, (for PA 6.6, 6.9. 6.10, and 6.12) is now predominantly obtained hy hydrogenation of adiponitrile from the direct addition of HCN to butadiene (p. 114, IV). Adiponitrile can also be obtained by reductive (cathodic) dimerization of acrylonitrile according to $H_2C{=}CHCN + 2\ e^{\ominus} + 2\ H^{\oplus} \rightarrow NC(CH_2)_4CN$. Another route dehydrates the ammonium salt of adipic acid (p. 128). The old 4-step process, butadiene → dichlorobutenes → butene dinitriles → adiponitrile → hexamethylenediamine, is no longer economic.

*1,10-Decanodiamine*, $H_2N(CH_2)_{10}NH_2$, (for PA 10.10) results from the hydrogenation of $NC(CH_2)_8CN$, the dinitrile of sebacic acid, $HOOC(CH_2)_8COOH$, (see above).

*1,12-Dodecanodiamine*, $H_2N(CH_2)_{12}NH_2$, (for PA 12.12) is produced by hydrogenation of the dinitrile $NC(CH_2)_{10}CN$ that is obtained from dodecanedioic acid (IV in Eq.(4-23)) and thus by a 6-step process from butadiene.

**Industrial Syntheses**

Aliphatic polyamides of the AA/BB type are produced by polycondensation of diamines, $H_2N{-}Z{-}NH_2$, with dicarboxylic acids, HOOC–Z'–COOH, or dicarboxylic acid esters, ROOC–Z'–COOR, where Z and Z' are bivalent unbranched aliphatic groups with at least two carbon chain atoms and R is H or an alkyl group such as $C_2H_5$, for example:

(10-12) $$H_2N{-}Z{-}NH_2 + HOOC{-}Z'{-}COOH \longrightarrow \text{+}NH{-}Z{-}NH{-}CO{-}Z'{-}CO\text{+}_n + 2\ H_2O$$

Such polycondensations require exact stoichiometries of the two monomers because otherwise high molecular weights would not be obtained (p. 144 and Volume I, Section 13.2). Industrial aliphatic polyamides require a number-average degree of polymerization of reactants of $\overline{X}_n = 200$ which corresponds to an extent of reaction of $p = 0.995$ (99.5 % monomer conversion) for exact stoichiometries. A 1 mol% excess of one monomer reduces the the degree of polymerization to 100 from 200.

The required stoichiometry is obtained by using not the monomers themselves but their **nylon salts**, $[H_3N{-}Z{-}NH_3]^{2\oplus}\ [OOC{-}Z'{-}COO]^{\ominus}$. The nylon salt of *a*dipic acid and 1,6-*h*examethylenediamine (**AH salt**) ($T_M$ = 198-203°C) can be used directly for melt polycondensation after recrystallization. The amidation equilibrium of AA/BB polycondensations to aliphatic polyamides are so favorable (equilibrium constants in the range 100–1000) that polycondensations can be performed in the presence of water which is used as an internal heat transfer agent.

In the discontinuous polycondensation of AH salt to PA 6.6 (Fig. 10-6), a 60–80 % aqueous slurry of the salt with some acetic acid as regulator (see Volume I) is subjected for 1–2 h to a precondensation at 275–280°C and 13–17 bar, i.e., the vapor pressure of steam, which prevents the evaporation of water. The vapor pressure is then continuously lowered, which reduces the water content and promotes the polycondensation. After a monomer conversion of 80–90 % (i.e., $\overline{X}_{R,n}$ = 5–10), the polycondensation is continued in the melt above the melting temperature of the polymer ($T_M$ = 264°C) under vacuum. Alternatively, the last step may be **solid-stating**, i.e., the polycondensation in the solid state at 150–200°C.

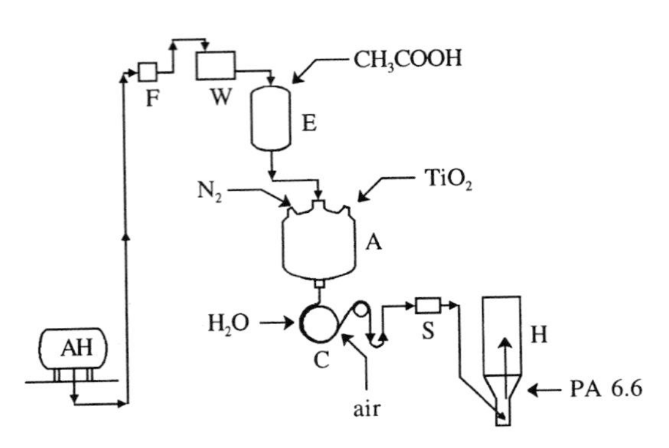


Fig. 10-6 Early discontinuous polycondensation of AH salt to PA 6.6 [3]. AH salt from a storage tank is pumped through a filter (F) into a weigh tank (W). Aqueous acetic acid (catalyst and regulator) is added. The resulting slurry passes through an evaporator (E) and enters a heated autoclave (A) where it is polycondensed under a nitrogen blanket after addition of $TiO_2$ (matt finish for fibers). After polycondensation, the exiting polymer melt enters a casting machine (C) where it is cooled, first with water and then with air. The solidified polymer is diced (S) and blended in a homogenizer (H) with polymer from other batches.

Activity coefficients of reactants deviate from unity since the AH salt is ionized in water. Use of molar concentrations instead of activities in calculations of equilibrium constants $K$ = [–CONH][$H_2O$]/([–COOH][–$NH_2$]) results in dependencies of "equilibrium constants" on water concentrations and thus also in systematic variations of apparent enthalpies of reaction (Table 10-3).

Polycondensations can also be performed continuously in tube reactors (Fig. 10-7) where residence times of melts are ca. 1 hour at 290°C and 27 bar. The melt flows laminarly and continuously along the tube walls whereas the core of the tube consists of escaping water vapor. The resulting polymer melt is pumped continuously into a spinneret for fiber spinning.

On processing melts by extrusion, fiber spinning, injection molding, etc., reactive endgroups –COOH and –$NH_2$ of these aliphatic polyamides can undergo post-condensations. Such post-condensations would lead to molecular weights and melt viscosities that increase with time which, in turn, would necessitate constant readjustments of machine power. In order to avoid this, endgroups –$NH_2$ and –COOH are reacted at the end of polycondensation proper with acetic acid, $CH_3COOH$, acetic anhydride, $(CH_3CO)_2O$, or ketene, $CH_2$=C=O, which cap the endgroups by formation of –CO–$OOCCH_3$ and –NH–$COCH_3$. These capping agents are known industrially as regulators, stabilizers, or chain terminators.

Table 10-3 Industrial polycondensation of aqueous AH salt [4a].

| wt% AH salt before reaction | Range in °C of equilibrium | wt% of water in equilibrium | Apparent enthalpy of reaction in kJ/mol |
|---|---|---|---|
| 80 | 200 – 220 | 29.5 – 29.6 | + 13 |
| 90 | 200 – 230 | 20.9 – 21.2 | + 4 |
| 100 | 210 – 250 | 12.4 – 12.8 | – 12.5 |
| AH precondensate | 250 – 270 | 4.6 – 4.9 | – 27 |

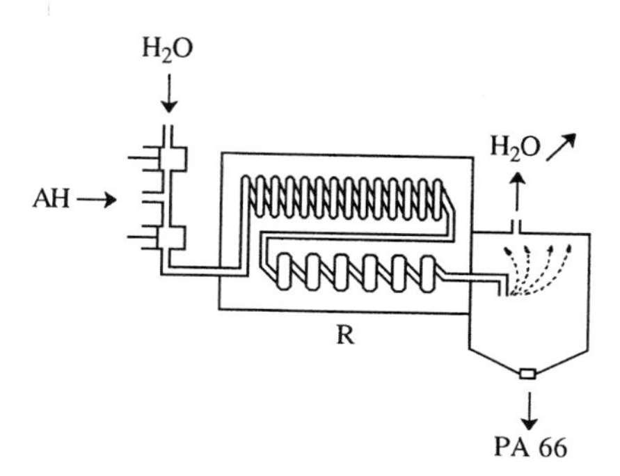


Fig. 10-7 Continuous polycondensation of AH salt [5]. A 50 wt% aqueous solution of AH salt is diluted with water and pumped into a tube reactor (R) that is heated to 290°C. Water vapor is vented at the end of the reactor and the polymer melt at the bottom of the reactor is pumped into the spinneret.
With kind permission of the American Chemical Society, Washington (DC).

Polymerizations to other aliphatic AA/BB monomers proceed similarly. Differences exist with respect to solubilities in water since higher nylon salts are far less soluble in water: 5 wt% of the 10.10 salt and < 10 wt% of the 12.12 salt versus 63–69 wt% for the 6.6, 6.10, and 6.12 salts at 90°C.

Only one aliphatic AA/BB polyamide, PA 4.2, is produced by the reaction of a diester and a diamine. The polycondensation proceeds under pressure at 140°C in a slurry of 1,4-tetraethylenediamine and diethyl oxalate in phenol/1,2,4-trichlorobenzene followed by a postcondensation in the melt in an extruder or in a fluidized bed.

**Properties**

Molecular weights and molecular weight distributions of aliphatic polyamides from AA/BB polycondensations follow practically, but not completely, the predictions of the simple theory of bifunctional equilibrium reactions. In contrast to theory, however, polymers contain small proportions of cyclic oligomers, for example, ca. 1–2 wt% in PA 6.6 (here mainly the dimer with a 14-membered ring), which act as plasticizers. An additional plasticizing effect is caused by water molecules which are attracted by hydrophilic amide groups. Moisture contents of aliphatic PAs thus increase with decreasing length of aliphatic units (Table 10-4). As a consequence, polyamides always need to be "conditioned" for shipping and processing, usually at 50 % relative humidity.

Moisture contents vary strongly with relative humidity, temperature, time, and thickness of the specimen (Fig. 10-8). Equilibrium is approached after days. Since absorbed water acts as a plasticizer, moduli of elasticity and strengths decrease with increasing moisture content whereas extensions and impact strengths increase (Table 10-5).

Commercial AA/BB polyamides are ca. 40–60 % X-ray crystalline. Crystalline forms are mainly monoclinic and triclinic (PA 4.6 and 6.6) or triclinic (PA 6.10) and usually exist in several crystal modifications. Melting temperatures increase with increasing proportion of amide groups (Table 10-4) because of the increased ability to form intermolecular hydrogen bridges (4.6 > 6.6 > 6.10 > 6.12). Polyamides with odd numbers of carbon atoms in their monomeric units have lower melting temperatures than those with comparable even numbers because they are less able to form regular hydrogen bonds.

The higher the percentage of intermolecular hydrogen bonds per chain atom, the higher are heat distortion temperatures, moduli, and mechanical strengths (PA 4.6 and 6.6 versus 6.10 and 6.12). For engineering purposes, mechanical properties can be further improved by reinforcement with glass fibers, usually 30 wt% (see Volume IV). PA 6.10 and 6.12 have higher dimensional stabilities because of lower moisture uptake.

Mechanical properties, and, in part, thermal ones are also affected by various other factors, especially for use as fibers (Volume IV). Data of Tables 10-4 and 10-5 are thus only rough indicators of properties of polyamides as engineering plastics. Each polymer type is usually offered in many grades that differ from company to company with respect to molecular weight, stabilizers, nucleation agents, etc. Reported physical data are affected by processing (for example, molding versus injection molding), testing conditions (type of test bar, conditioning, etc.), and specifications (ISO, ASTM, CAMPUS®).

Table 10-4 Average thermal and mechanical properties of unfilled aliphatic AA/BB polyamides at 23°C (unless noted otherwise). RH = relative humidity.

| Property | | Physical unit | PA 4.6 | PA 6.6 | PA 6.9 | PA 6.10 | PA 6.12 |
|---|---|---|---|---|---|---|---|
| Density | | g/cm³ | 1.18 | 1.14 | 1.08 | 1.08 | 1.07 |
| Melting temperature (DSC), dry | | °C | 295 | 262 | 210 | 227 | 218 |
| Heat distortion temperature (1.82 MPa) | | °C | 160 | 95 | 75 | 55 | 55 |
| Vicat temperature B | | °C | 287 | 250 | | 160 | 160 |
| Glass temperature (DSC, dry) | | °C | 80 | 48 | | 46 | 45 |
| (DSC, 50 % RH) | | °C | 35 | 15 | | 10 | 20 |
| (DSC, 100 % RH) | | °C | –37 | –32 | | | |
| Linear thermal expansion coefficient | | $10^{-5}$ $K^{-1}$ | 8.4 | 8.1 | 8.1 | 10 | 9 |
| Specific heat capacity | | J $K^{-1}$ $g^{-1}$ | 2.1 | 1.7 | 1.7 | 1.8 | 1.7 |
| Thermal conductivity (20°C) | | W $m^{-1}$ $K^{-1}$ | 0.29 | 0.23 | | 0.23 | |
| Tensile modulus | dry | MPa | 3000 | 3200 | 1900 | 2400 | 2100 |
| | 50 % RH | MPa | 1000 | 1600 | | 1500 | |
| Flexural modulus | dry | MPa | 3100 | 2830 | 2300 | 1970 | 2030 |
| | 50 % RH | MPa | 1000 | 1210 | 1070 | 1100 | 1240 |
| Yield strength | dry | MPa | 79 | 83 | 70 | 59 | 61 |
| | 50 % RH | MPa | 40 | 59 | 45 | 49 | 51 |
| Tensile strength | dry | MPa | 99 | 83 | | 59 | 61 |
| | 50 % RH | MPa | 65 | 77 | | 49 | 61 |
| Flexural strength | dry | MPa | 150 | 117 | | | 76 |
| | 50 % RH | MPa | 50 | 42 | | | |
| Extension at yield | dry | % | | 5 | 10 | 10 | 7 |
| | 50 % RH | % | | 25 | 10 | 30 | 40 |
| Extension at break | dry | % | | 60 | 50 | 100 | 15 |
| | 50 % RH | % | | > 300 | 115 | 220 | 34 |
| Impact strength (Izod) | dry | J/m | 96 | 53 | 53 | 53 | 53 |
| | 50 % RH | J/m | 400 | 107 | 85 | 85 | 75 |
| Notched impact strength (Izod) | dry | J/m | 96 | 53 | | 69 | 53 |
| | 50 % RH | J/m | 400 | 112 | | | 75 |
| Rockwell hardness (Shore) | dry | - | R 123 | R 119 | M 61 | R 110 | R 114 |
| | 50 % RH | - | R 107 | R 108 | | | |
| Water absorption, 24 h | 50 % RH | % | 3.8 | 2.8 | 1.8 | 1.4 | 1.0 |
| | 100 % RH | % | 15 | 8.5 | 4.7 | 3.3 | 2.7 |

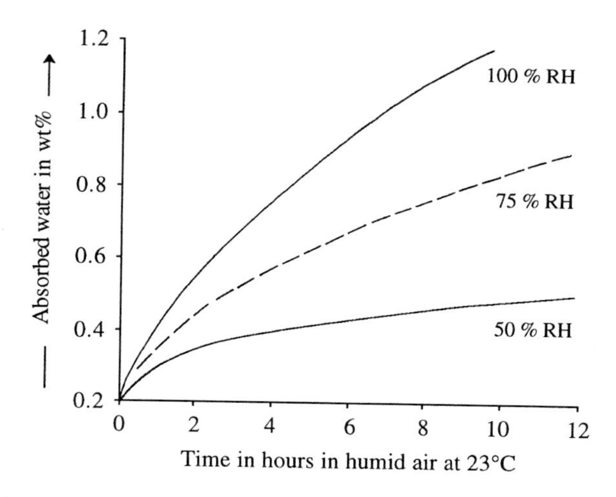


Fig. 10-8 Absorption of moisture by PA 6.6 pellets at various relative humidities [6].

Aliphatic polyamides are relatively good insulators at low temperatures and relative humidities (Table 10-5). Dry polyamides have relative permittivities of ca. 3.5–4 that increase somewhat with decreasing frequency but are practically independent of constitution. Permittivities and dissipation factors increase drastically at higher relative humidities, especially at lower frequencies and for polyamides with higher proportions of amide groups.

Table 10-5 Electrical properties of unfilled polyamides 4.6, 6.6, 6.10, and 6.12 at 23°C and 24 h at relative humidities (RH) of 0 %, 50 %, and 100 %. Relative permittivity = "dielectric constant."

| Property | | Physical unit | 0 % RH 4.6 | 6.6 | 6.10 | 6.12 | 50 % RH 4.6 | 6.6 | 6.12 | 100 % RH 6.6 | 6.12 |
|---|---|---|---|---|---|---|---|---|---|---|---|
| Relative permittivity | 50 Hz | - | 3.9 | 3.9 | 3.9 | 4.0 | 22 | 7.0 | 6.0 | 31 | 12 |
| | 1 kHz | - | 3.8 | 3.8 | 3.6 | 4.0 | 11 | 6.5 | 5.3 | 29 | |
| | 1 MHz | - | 3.6 | 3.5 | 3.3 | 3.5 | 4.5 | 4.1 | 3.3 | 18 | |
| Dissipation factor | 50 Hz | - | 0.01 | 0.02 | 0.04 | 0.02 | 0.87 | 0.11 | 0.08 | 0.50 | 0.25 |
| | 1 kHz | - | 0.01 | 0.02 | 0.04 | 0.02 | 0.35 | 0.10 | | 0.23 | |
| | 1 MHz | - | 0.03 | 0.03 | 0.03 | 0.02 | 0.12 | 0.08 | | 0.28 | |
| Resistivity (volume resistance) | | Ω cm | $10^{15}$ | $10^{15}$ | $10^{15}$ | $10^{15}$ | $10^{9}$ | $10^{13}$ | $10^{13}$ | $10^{9}$ | $10^{11}$ |

## Applications

The worldwide nameplate capacity of all polyamides (aliphatic + aromatic, AB and AA/BB) is ca. $6 \cdot 10^6$ t/a, the production ca. $4 \cdot 10^6$ t/a (2003). Approximately 95 % of all polyamides are PA 6 and PA 6.6 most of which (75–90 %) are used for fibers, depending on the country. During the last decade, the production of polyamides increased but their proportion to all synthetic polymers decreased; it is now ca. 2 %.

The silk-like properties of polyamide fibers led in 1939 to the production of silk stockings ("nylons"), and their good tear strength in World War II to their use for parachutes. Textiles from PA 6, 6.6, and 6.10 serve especially in sport and leisure wear (see Volume IV). Their good wear resistance has led to applications as carpet and technical fibers (e.g., tire cord), bristles, ropes, etc. About 10-25 % of all polyamides are used as engineering plastics for tubes, gears, sheets, conveyor belts, automotive fuel tanks, etc.

### 10.3.3 Aliphatic AB Polyamides

**Survey**

In contrast to AA/BB polyamides, AB polyamides, $\text{+NH-Z-CO+}_n$, are synthesized by many different methods (Table 10-6). Syntheses comprise direct polycondensation of aliphatic $\alpha,\omega$-amino acids (to PA 9 and PA 11):

(10-13) $$n\ H_2N\text{–}Z\text{–}COOH \longrightarrow \text{+NH–Z–CO+}_n + n\ H_2O$$

polycondensation of esters of aliphatic $\alpha,\omega$-amino acids (to PA 7),

(10-14) $$n\ H_2N\text{–}Z\text{–}COOR \longrightarrow \text{+NH–Z–CO+}_n + n\ ROH$$

polycondensation of aromatic acid chlorides (to poly($p$-benzamide), PPBA),

(10-15) $$n\ H_2N\text{–}Z\text{–}COCl \longrightarrow \text{+NH–Z–CO+}_n + n\ HCl$$

hydrolytic polymerization (to PA 6, PA 8, PA 12) and anionic polymerization (to PA 6 and dimethyl substituted PA 3) of lactams,

(10-16) $$n\ \left[\text{cyclic: }(CH_2)_i\text{–NH–C(=O)}\right] \longrightarrow \left[\text{NH}-(CH_2)_i-\underset{\underset{O}{\|}}{C}\right]_n$$

anionic hydrogen-transfer polymerization (acrylamide to poly($\beta$-alanine) (see Section 11.3),

(10-17) $$n\ CH_2\text{=CH-CO–}NH_2 \longrightarrow \text{+}CH_2\text{–}CH_2\text{–CO–NH+}_n$$

ring-opening polymerization of imides to poly(amino acid)s with mixed α and β structures (see Section 11.2),

(10-18) $$\text{(succinimide ring, N-substituted)} \longrightarrow \text{—}\underset{CH_2COOH}{\text{CH}}\text{—CO—NH—}\ /\ \text{—}\underset{COOH}{\text{CH}}\text{—}CH_2\text{—CO—NH—}$$

and polyelimination of $N$-carboxyanhydrides (Leuchs-anhydrides) of $\alpha$-amino acids:

(10-19) $$n\ \text{(R-substituted N-carboxyanhydride: R, HN, O, O, O)} \longrightarrow \left[\text{NH}-\text{CHR}-\underset{\underset{O}{\|}}{C}\right]_n$$

Table 10-6 Common polymerizations to AB polyamides, $-[NH-Z-CO]_n-$, by polyelimination of *N*-carboxyanhydrides (NCA), anionic (A) or hydrolytic (H) chain polymerization of lactams, and/or polycondensation of $H_2N-Z-COX$ with X = OR (amino esters, OH (amino acids), or Cl (acid chlorides).
*) see Section 11.2. R' = various substitutents, e.g., H, $CH_3$, benzyl, COOH, etc.

| Chemical structure Type | –Z– | Common name | Polymerization of NCA, imide, etc. | Lactam A | Lactam H | –COX with X = OR | OH | Cl |
|---|---|---|---|---|---|---|---|---|
| 2 | $-CHR'-$ | Poly(α-amino acid)s *) | + | - | - | - | - | - |
| 3 | $-CH_2CHR'-$ | Poly(β-amino acid)s *) | + | - | - | - | - | - |
| 4 | $-(CH_2)_3-$ | Poly(γ-butyrolactam) | - | + | - | - | - | - |
| 6 | $-(CH_2)_5-$ | Poly(ε-caprolactam) | - | + | + | - | - | - |
| 7 | $-(CH_2)_6-$ | Poly(enanthlactam) | - | - | - | + | - | - |
| 8 | $-(CH_2)_7-$ | Poly(capryllactam) | - | - | + | - | - | - |
| 9 | $-(CH_2)_8-$ | Poly(ω-aminopelargonic acid) | - | - | - | - | + | - |
| 11 | $-(CH_2)_{10}-$ | Poly(11-aminoundecanoic acid) | - | - | - | - | + | - |
| 12 | $-(CH_2)_{11}-$ | Poly(laurolactam) | - | - | + | - | + | - |
| PPBA | $-p-C_6H_4-$ | Poly(*p*-benzamide) | - | - | - | - | - | + |

**Monomers**

Linear aliphatic AB polyamides of the type $-[NH(CH_2)_iCO]_n-$ with $i \geq 4$ are obtained by anionic or hydrolytic ring-opening polymerization of lactams or by polycondensation of amino acids or esters (Table 10-6).

The monomer for PA 4 is *2-pyrrolidone* (α-pyrrolidone, 2-pyrrolidinone, γ-butyrolactam) which is obtained industrially from γ-butyrolactone (from dehydrogenation of 1,4-butanediol or reduction of maleic anhydride) and ammonia via the oxime. A proposed synthesis reacts acrylonitrile with HCN to succinonitrile. Partial hydrogenation to $NCCH_2CH_2NH_2$ and subsequent hydrolytic cyclization delivers 2-pyrrolidone.

*ε-Caprolactam,* the monomer for PA 6, is obtained by many different types of syntheses (p. 128 ff.). Industrially most important are the routes from phenol or cyclohexane that both lead to the oxime which is converted to the lactam by Beckmann rearrangement. All processes via the oxime produce large amounts of $(NH_4)_2SO_4$ (up to 5 kg per 1 kg of ε-caprolactam) which is an inexpensive, inferior fertilizer that acidifies the soil. As a byproduct of a lactam plant, ammonium sulfate can only be sold at a profit if routes to farmers are short which therefore may limit the size of a lactam plant.

Newer processes such as the toluene-based Snia–Viscosa process thus avoid the formation of the oxime (Fig. 4-14). In addition, a recently proposed caprolactam synthesis no longer needs expensive aromatics such as toluene or benzene as feedstocks. It rather starts with less expensive butadiene that is reacted with HCN in several steps to ε-caprolactam:

(10-20) $$\xrightarrow{+\,2\,HCN} \; (CN, CN) \; \xrightarrow{+\,2\,H_2} \; (CH_2NH_2, CN) \; \xrightarrow[-\,NH_3]{+\,H_2O} \; (O, NH)$$

In another proposed strategy, HCN is replaced by the less expensive CO:

(10-21) [reaction scheme: butadiene + CO, + $CH_3OH$ → methyl pentenoate ($COOCH_3$); + CO, + $H_2$ → ($COOCH_3$, $CH_2$—CHO); + $NH_3$, + $H_2$, − $H_2O$ → ($COOCH_3$, $CH_2NH_2$); − $CH_3OH$ → ε-caprolactam (C=O, NH)]

*Laurolactam* (dodecanolactam) is produced like ε-caprolactam, i.e., mainly via oximation of cyclododecanol (Fig. 4-14, Routes I and II). Both the photochemical process III and the nitrosyl sulfuric acid process IV of Fig. 4-14 are also viable. A new Ube process oxidizes cyclohexanone in the presence of $NH_3$. The resulting peroxydicyclohexylamine is hydrolytically cleaved to ω-cyanoundecanoic acid (with cyclohexanone and ε-caprolactam as byproducts) which is then hydrogenated to ω-aminododecanoic acid:

(10-22) 2 cyclohexanone (=O) $\xrightarrow[-\,2\,H_2O]{+\,H_2O_2,\;+\,NH_3}$ peroxydicyclohexylamine (H–N, O–O) $\xrightarrow{\Delta}$ $NC-(CH_2)_{10}-COOH$

**Hydrolytic Polymerization**

The discontinuous hydrolytic polymerization is the most important process for fiber-grade polyamide 6 (Fig. 10-9). Overall, it is an acetic acid mediated ring-opening polymerization of a highly concentrated (80–90 %) solution of ε-caprolactam in water which consists of three types of reactions: ring-opening (Eq.(10-23), polycondensation (Eq.(10-24), and polyaddition (Eq.(10-25).

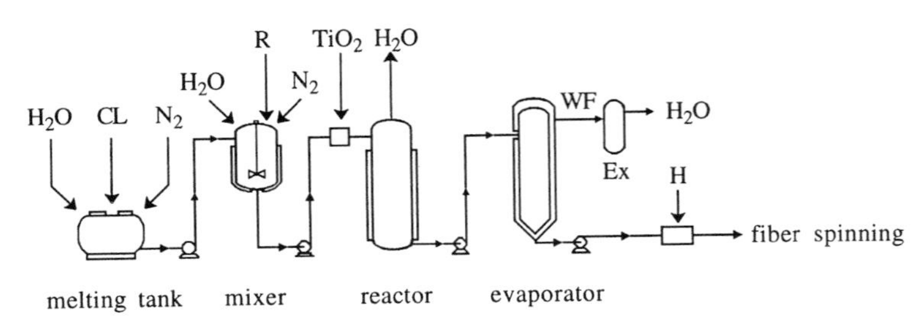


Fig. 10-9 Hydrolytic polymerization of ε-caprolactam (CL) to fiber-grade polyamide 6. The monomer is melted with water vapor in a melting tank which generates a small concentration of ω-amino acid. The melt is mixed with water to a 80–90 % aqueous solution of CL in a heated mixer. A small proportion of acetic acid (ca. 0.2–0.5 %) is added as a regulator (R). After addition of $TiO_2$, a matting agent, the mixture is hydrolytically polymerized at ca. 260°C which generates water vapor. The melt of the resulting polyamide 6 is pumped into an evaporator where water-soluble fractions (WF) are removed from the top. By a vacuum system, these fractions are then separated to water (vapor) and extractables Ex (monomers, oligomers); the latter are returned to the mixer (not shown). The remaining PA 6 is provided with a heat stabilizer (H) and spun to fibers.

The added water initially hydrolyzes a small proportion of ε-caprolactam to ε-aminocaproic acid (6-aminocaproic acid), $H_2N(CH_2)_5COOH$. The amino and carboxy groups of this acid then initiate the ring-opening polymerization of ε-caprolactam, Eq.(10-23).

Table 10-7 Equilibrium constants and enthalpies of reactions in the hydrolytic polymerization of ε-caprolactam with small proportions of water (after [4b]).

| Reaction | Physical unit | Equilibrium constants at $T$ = 220°C | 240°C | 260°C | 280°C | Reaction enthalpy in kJ/mol |
|---|---|---|---|---|---|---|
| Ring opening | mol/kg | 2500 | 2700 | 3000 | 3200 | + 9 |
| Polycondensation | - | 770 | 590 | 460 | 370 | − 27 |
| Polyaddition | mol/kg | 1.9 | 1.6 | 1.4 | 1.2 | − 18 |

Ring-opening is responsible for both initiation ($i$ = 1) and polymerization ($i \geq 2$):

(10-23) $H_2N(CH_2)_5COOH + n$ [ε-caprolactam ring, N–H, C=O] $\rightleftharpoons H[NH(CH_2)_5CO]_{n+1}OH$

The equilibrium constant of ring opening is ca. 1 decade greater than that of polycondensation, (Eq.(10-24), and about 3 decades greater than that of polyaddition (Eq.(10-25) (Table 10-7):

(10-24) $H[NH(CH_2)_5CO]_pOH + H[NH(CH_2)_5CO]_qOH \rightleftharpoons H[NH(CH_2)_5CO]_{p+q}OH + H_2O$

(10-25) [ε-caprolactam ring, N–H, C=O] $+ H[NH(CH_2)_5CO]_pOH \rightleftharpoons H[NH(CH_2)_5CO]_{p+1}OH$

The simultaneous reactions (10-23)–(10-25) give rise to a complicated reaction pattern (Fig. 10-10). The initial autocatalysis is followed by an equilibration that is constantly shifted by removal of water and monomer (see Fig. 10-9).

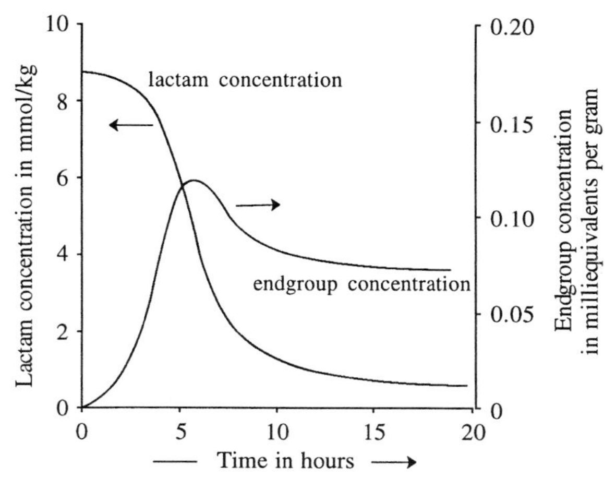


Fig. 10-10 Time dependence of monomer and endgroup concentration during the polymerization of ε-caprolactam with an initial concentration of 1.06 wt% water at 221.5°C [7].

Hydrolytic polymerizations are performed in many variants. For example, in the so-called VK process (German: *v*ereinfacht-*k*ontininuierlich (= simplified continuous)), the polymerization is initiated with 6-aminocaproic acid or AH salt (nylon 6.6 salt, p. 459). The melt of ε-caprolactam + 0.3–0.5 wt% water are continually added on top of a heated reactor. It polymerizes while flowing down. The long residence times of 15–30 hours can be shortened by a prepolymerization with 2–8 wt% water.

The polymerization leads not only to linear polymer molecules but also to cyclic oligomers with $i \geq 2$ monomeric units. At 250°C, their proportion is ca. 5 wt% ($2 \leq i \leq 6$). Ends of molecules carry not only carboxyl and amino groups but also some semicyclic amidine endgroups which also initiate the polymerization, albeit with about a decade lower rate than amine salts.

In air, amino endgroups slowly form pyrrole groups which leads with time to yellowing of polyamide 6 fibers. In addition, PA 6 fibers are not necessarily homopolymers. Most of them contain a few percent of ethylene diamine groups, $-NHCH_2CH_2NH-$, which increases the amine equivalent and thus the dyability of fibers.

Hydrolytic polymerization is also used for the synthesis of PA 12 which has a far lower monomer and oligomer content and tendency to depolymerization than PA 6, since 13-membered rings are more difficult to form than 7-membered ones. The polymerization to PA 8 apparently never made it beyond the pilot stage.

$$-\!\!\left[\!-C(=O)-(CH_2)_5-NH\!-\right]_n\!-C(=NH^{\oplus})\text{(ring)} \qquad -\!\!\left[\!-C(=O)-(CH_2)_5-NH\!-\right]_n\!-C(=O)-CH_2-\text{(pyrrole, NH)}$$

semicyclic amidine endgroup                pyrrole endgroup

## Anionic Lactam Polymerization

4- to 7-membered lactam rings have been polymerized anionically by strong bases to PA 3 to PA 6. An example is the polymerization of ε-caprolactam (I) to **poly(ε-caprolactam) (PA 6)** by sodium or alkaline earth hydroxides in the presence of coinitiators such as *N*-acetylcaprolactam (II) which is formed from the lactam by reaction with added acetanhydride or ketene. The bases abstract protons from I and form lactam anions (III) which react very rapidly with I to an *N*-substituted lactam anion (IV). Proton exchange of IV with I results in molecules (V) with lactam and amide endgroups. The regenerated lactam anions (III) start new chains.

$$\text{III (lactam } N^{\ominus}) + \text{II (lactam } N-C(=O)-CH_3) \longrightarrow \text{IV (lactam } N-C(=O)-(CH_2)_5-N^{\ominus}-C(=O)-CH_3)$$

$$\text{IV} + \text{I (lactam NH)} - \text{III (lactam } N^{\ominus}) \longrightarrow \text{V (lactam } N-C(=O)-(CH_2)_5-NH-C(=O)-CH_3)$$

(10-26)

Coinitiators (II) produce very rapid initiation reactions. Without them, lactam anions have to be acylated by lactam molecules which is a very slow reaction.

Anionic lactam polymerizations are very fast. They are used in industry for casting large molded parts from ε-caprolactam. An alternative method, **reaction injection molding (RIM)**, apparently never caught on. The nylon RIM process injects two liquids simultaneously into a mold: a caprolactam solution of caprolactam-$MgBr_2$ with adipoyl-*bis*-caprolactam as coinitiator and a liquid elastifying block component, e.g., a polyether-based polyol such as poly(propylene glycol). The rapid RIM polymerization generates block copolymers with caprolactam and polyether blocks.

Anionic lactam polymerizations exhibit characteristics of living polymerizations with an initially homogeneous distribution of inititator molecules. However, the initial polymolecularity of $\overline{M}_w/\overline{M}_n = 1.2–1.3$ broadens with time while the number-average molecular weight remains constant. The observed increase of melt viscosities (which depends on weight-average molecular weights) can therefore not be caused by an additional polymerization. It rather results from transamidation by acid-catalyzed aminolysis:

(10-27)
$$\begin{matrix} \sim M_k-CO-NH-M_m\sim \\ + \\ \sim M_n-NH_2 \end{matrix} \rightleftharpoons \begin{matrix} \sim M_k-CO \\ | \\ \sim M_n-NH \end{matrix} + H_2N-M_m\sim$$

In industrial polymerizations, amide endgroups are transformed to acetamide endgroups by the added regulators ketene or acetanhydride. Here, no transamidation is observed which indicates the absence of direct transamidation between amide groups within the chains.

**Polyamide 4** (poly(α-pyrrolidone), poly(2-pyrrolidinone), poly(γ-butyrolactam)) is obtained by anionic polymerization of pyrrolidone with alkali pryrrolidone as initiator and acyl compounds or carbon dioxide as coinitiators, similar to Eq.(10-26). Polymerizations initiated by acyl compounds lead to broader molecular weight distributions than those by carbon dioxide, probably because the former cause a trans-initiation, for example, in the case of *N*-acetyl pyrrolidone:

(10-28)
$$\begin{matrix} CH_3-CO \\ | \\ NH\sim \end{matrix} + {}^{\ominus}N\text{(pyrrolidone ring, C=O)} \longrightarrow CH_3-CO-N\text{(pyrrolidone ring, C=O)} + {}^{\ominus}NH\sim$$

Trans-initiations regenerate coinitiator molecules which assures that new polymer molecules are formed even at the lowest initiator concentrations. In the $CO_2$–initiated polymerizations, on the other hand, $-CH_3$ end-units are replaced by $-O^{\ominus}$. In the transition state, negative charges $-O^{\ominus}$ and ${}^{\ominus}N<$ repel each other which prevents trans-initiations. Neither PA 4 nor PA 5 = poly(piperidone) = poly(valerolactam) are commercial products. For PA 3 and its derivatives, see Section 11.3.

### Cationic Lactam Polymerizations

Lactams can also be polymerized by strong protonic acids at temperatures above 200°C. Such polymerizations lead only to small monomer conversions and low degrees of polymerization, probably because of the formation of amidine endgroups that are unable to add further lactam molecules:

$$\text{(10-29)} \qquad \sim\!\sim(CH_2)_5-NH-\underset{\underset{O}{\|}}{C}-(CH_2)_5-\overset{\oplus}{N}H_3 \xrightarrow[-H_2O]{} \sim\!\sim(CH_2)_5-NH-\overset{H\ \oplus}{\text{(azepine ring with } C=N\text{)}}$$

**Polycondensations**

As a rule, direct polycondensations of $\omega$-amino acids,

$$\text{(10-30)} \qquad H_2N-Z-COOH \longrightarrow \left[ NH-Z-CO \right]_n + H_2O$$

require high reaction temperatures in order to overcome the resonance stabilization of COOH groups. However, higher temperatures lead to side reactions which restrict obtainable degrees of polymerization: cyclodimerizations of $\alpha$-amino acids, release of ammonia and formation of acrylic acid from $\beta$-amino acid, and intramolecular cyclizations to lactams from $\gamma$- and $\delta$-amino acids. Predominance of polycondensation on heating is found only for $\delta$-aminocaproic acid, $H_2N(CH_2)_5COOH$, and higher $\omega$-amino acids.

In the former Soviet Union, PA 9 was synthesized by polycondensation of $\vartheta$-aminopelargonic acid but the production was discontinued because it was not economical.

PA 11 is based on castor oil as feed stock; its economical production thus depends heavily on the harvest of castor beans (see p. 79). In Russia, it is obtained from products of the telomerization of ethene with carbon tetrachloride (see Volume I, p. 350).

In principle, PA 12 is available by direct polycondensation of the $\omega$-amino acid since $\omega$-aminododecanoic acid, $H_2N(CH_2)_{11}COOH$, is directly obtainable by hydrogenation of $\omega$-cyanoundecanoic acid, $NC(CH_2)_{10}COOH$, (see Eq.(10-22)). The pilot production to poly(aminododecanoic acid) has been recently enlarged.

Amino esters polycondense more easily than amino acids because the ester group is less resonance stabilized than the acid group. However, their polycondensation is more costly because the released alcohol has to be recovered. For this reason, poly(7-aminoheptanoic acid) = poly($\omega$-aminoenanthic acid) = PA 7 is not economical.

Syntheses of PA 2 and its derivatives are discussed in Section 11.2.2. PA 3 is used in-house as a stabilizer for poly(oxymethylene). The pilot production of PA 4 has been terminated, apparently because of its thermal instability. Even below its melting temperature of ca. 260°C, it degrades by a back-biting mechanism to its monomer.

**Properties and Applications**

Polyamide 6 has the same proportions of amide and methylene groups as PA 6.6, and thus has similar properties (compare Tables 10-8 and 10-4). Like PA 6.6, PA 6 is predominantly used for fibers (for fiber properties, see Volume IV). A smaller proportion serves as an engineering plastic where it competes with PA 4.6, PA 6.6, PA 6.10, PA 6.12, PA 10.10, PA 11, and PA 12. PA 6 and PA 6.6 are more heat resistant and have greater strengths and flexural moduli, whereas the reduced proportion of amide groups provides PA 11 and 12 with improved dimensional stability and impact strength.

Table 10-8 Average thermal and mechanical properties of unfilled AB polyamides at 23°C unless noted otherwise. RH = relative humidity.

| Property | | Physical unit | PA 3 | PA 4 | PA 6 | PA 11 | PA 12 |
|---|---|---|---|---|---|---|---|
| Density | | $g/cm^3$ | 1.32 | 1.25 | 1.13 | 1.03 | 1.01 |
| Melting temperature (DSC), dry | | °C | | 260 | 220 | 190 | 180 |
| Heat distortion temperature (1.82 MPa) | | °C | | | 62 | 55 | 52 |
| Glass temperature (DSC, dry) | | °C | 123 | | 56 | 43 | 42 |
| (DSC, 50 % RH) | | °C | | | 3 | | |
| (DSC, 100 % RH) | | °C | | | –22 | | 42 |
| Linear thermal expansion coefficient | | $10^{-5}$ $K^{-1}$ | | | 8.5 | 9 | 13 |
| Specific heat capacity | | J $K^{-1}$ $g^{-1}$ | | | 1.7 | 1.26 | 1.26 |
| Thermal conductivity (20°C) | | W $m^{-1}$ $K^{-1}$ | | | 0.23 | 0.19 | 0.24 |
| Tensile modulus | dry | MPa | | | 3300 | 1350 | 1450 |
| | 50 % RH | MPa | | | 1300 | 1100 | 1100 |
| stretched fiber | 50 % RH | MPa | < 12 000 | | 3200 | | |
| Flexural modulus | dry | MPa | | | 2720 | 1170 | 1410 |
| | 50 % RH | MPa | | | 970 | 1030 | 1030 |
| Tensile strength at yield | dry | MPa | | | 81 | 36 | 45 |
| | 50 % RH | MPa | | | 45 | | 38 |
| Tensile strength at break | dry | MPa | | | 81 | 57 | 49 |
| | 50 % RH | MPa | | | 69 | 54 | 47 |
| stretched fiber | 50 % RH | MPa | < 360 | | 970 | | |
| Flexural strength | dry | MPa | | | 113 | | 56 |
| | 50 % RH | MPa | | | 40 | | 40 |
| Elongation at yield | dry | % | | | 9 | 22 | 10 |
| | 50 % RH | % | | | | | 20 |
| stretched fiber | 50 % RH | % | 3 | | 33 | | |
| Elongation at break | dry | % | | 36 | 200 | 120 | 250 |
| | 50 % RH | % | | | 300 | 300 | 250 |
| stretched fiber | 50 % RH | % | < 20 | | | | |
| Impact strength (Izod) | dry | J/m | | | 58 | 40 | 58 |
| | 50 % RH | J/m | | | 215 | | 64 |
| (Charpy) | 50 % RH | $kJ/m^2$ | | | | | 6 |
| Rockwell hardness (Shore) | dry | - | | | R117 | R108 | R108 |
| | 50 % RH | - | | | R99 | R108 | |
| Water absorption | 50 % RH | % | 7 | < 10 | 3.0 | 0.8 | 0.7 |
| | 100 % RH | % | | | 9.5 | 2.4 | 2.0 |

Polyamides 3, 6, 11, 12, 6.6, 6.9, 6.10, and 6.12 dissolve at room temperature in formic acid, dichloroacetic acid, and trifluoroacetic acid. At higher temperatures, PA 3 dissolves in chloral hydrate and PA 6 also in *m*-cresole and dimethylsulfoxide.

3,3-Dimethyl-substituted PA 3 (for the synthesis, see Section 11.3.2) can be spun from solutions of calcium thiocyanate in methanol. Fibers are highly crystalline, even without stretching. Because of their high melting temperature of 270°C (crystal modification II) and resistance against oxidation, they serve as sewing threads for fast-running industrial sewing machines. On interruption, fibers of lower-melting polymers would melt on contact with the very hot needles, which would bring production to a very costly standstill as threading up is tedious and very time-consuming.

## 10.3.4 Branched AA/BB Polyamides

**Versamid®**

Randomly branched aliphatic polyamides result from the polycondensation of dibasic fatty acids ($C_{13}$, $C_{19}$, $C_{21}$, $C_{36}$) or "polymerized" vegetable oils and difunctional or multifunctional amines. Industry uses mainly the so-called **dimer acid** which results from the heating of two unsaturated $C_{18}$ fatty acids (see p. 79) with clay. The dimer acid is a mixture of acyclic, monocyclic, and bicyclic molecules with 2 carboxylic groups and 36 carbon atoms The similar **trimer acid** from three unsaturated $C_{18}$ fatty acids contains 3 carboxylic groups and 54 carbon atoms.

These branched fatty-acid polyamides are called Versamid®. They have low to medium molecular weights, melting temperatures between room temperature and 185°C, and good solubilities in various solvents.

Versamids® based on ethylene diamine and dimer acid are used as adhesives which can be stored in the cold but deliver "hard" polymers on short heating. Diethylene triamine, on the other hand, leads to "soft" polymers which can be combined with epoxy, phenolic, or colophonic resins. Versamids® are also used for coatings and in printing inks.

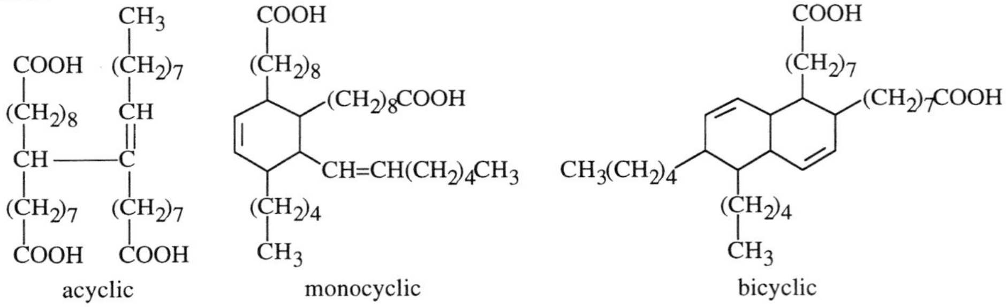

acyclic monocyclic bicyclic

**Hyperbranched Polyamides**

Polycondensation of acid anhydrides such as

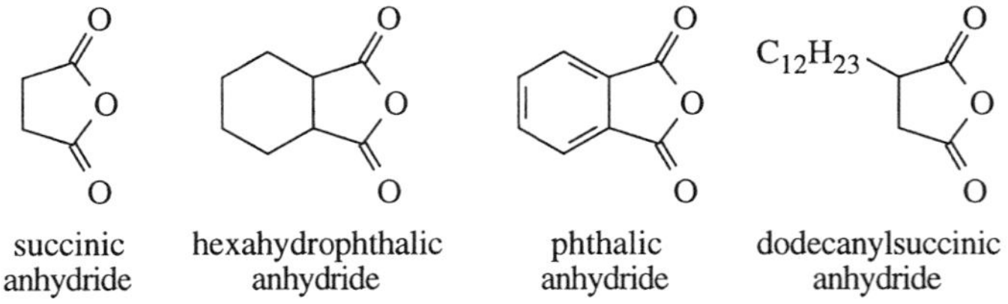

succinic anhydride hexahydrophthalic anhydride phthalic anhydride dodecanylsuccinic anhydride

with diisopropanol amine, $[CH_3CH(OH)CH_2]_2NH$, proceeds *via* reactive oxazolinium intermediates to hyperbranched amide-ester polymers which are called dendritic Hybran® polymers (Eq.(10-31)).

The reaction delivers mainly type III units and a few type IV ones. Commercial products have number-average molecular weights between 670 and 2700 which corresponds to degrees of polymerization of 2.9–9.4 for type III units. Polymolecularity indices are ca. 3–4. Hydroxyl groups average 5–13 per molecule. Modification of monomers leads to oligomers with various endgroups such as esters, tertiary amines, trialkoxysilanes, etc.

(10-31)

Polymers do not dissolve in water but in aromatic or chlorinated hydrocarbons, esters, alcohols, ketones, and tetrahydrofuran. Glass temperatures increase with degrees of polymerization from 10°C to 40°C for succinic acid polymers, from 30°C to 70°C for hexahydrophthalic acid polymers, and from 60°C to 100°C for phthalic acid polymers.

Hybran® polymers have been proposed as cores of star-like condensation polymers, branching units in elastomers and epoxy resins, dispersants for lubricants and fillers, and also as adhesives, antifreeze in diesel oils, additives in toners, and for use in cosmetics.

## 10.3.5 Aramids

In polymer chemistry, aromatic polyamides (**polyaramides, aramides**) are polymers with amide groups that are connected directly or via $CH_2$ groups to aromatic moieties of the main chain. The US Federal Trade Commission defines **aramids** (without "e") as those long-chain synthetic polyamides in which at least 85 % of amide groups are bound directly to two aromatic rings. According to ISO, up to 50 % of amide groups may be replaced by imine groups in aramids.

**AB Types**

Poly(*p*-benzamide) can be synthesized by the Schotten–Baumann reaction of the dissolved acid chloride which is obtained by reacting *p*-aminobenzoic acid, PABA, with thionyl chloride:



(10-32)

Melt polycondensations are not possible since poly(*p*-benzamide) decomposes below its melting temperature of 550°C. Alternative polycondensations are that of *p*-benzoic acid isothiocyanate, Eq.(10-33), and that of PABA by $SiCl_4$ in pyridine, Eq.(10-34):

(10-33)

$$H_2N{-}C_6H_4{-}\underset{\displaystyle O}{\overset{\|}{C}}{-}OH \xrightarrow[-\ H_2S]{+\ CS_2} SCN{-}C_6H_4{-}\underset{\displaystyle O}{\overset{\|}{C}}{-}OH \xrightarrow[-\ COS]{} \left[{-}HN{-}C_6H_4{-}\underset{\displaystyle O}{\overset{\|}{C}}{-}\right]$$

$$(10\text{-}34) \qquad 2\,n\,H_2N{-}C_6H_4{-}\underset{\displaystyle O}{\overset{\|}{C}}{-}OH \xrightarrow[-\ n\ SiO_2,\ -\ 4\ n\ HCl]{+\ n\ SiCl_4} \left[{-}HN{-}C_6H_4{-}\underset{\displaystyle O}{\overset{\|}{C}}{-}\right]_{2n}$$

The name of the polymer is commonly abbreviated as **PBA.** but this is the official ASTM abbreviation for poly(butyl acrylate) and should be avoided. Better are the also used abbreviations **PAB** or **PPBA**.

PPBA is a fairly stiff polymer. In 96 % sulfuric acid as solvent, the exponent $\alpha$ of the dependence of intrinsic viscosity on the mass-average molar mass, $[\eta] = K_v \overline{M}_w^{\alpha}$, is $\alpha$ = 1.7 for $\overline{M}_w$ < 12 000 g/mol (Fig. 10-11). The persistence length is ca. 50 nm (see Volume III). At molar masses above ca. 12 000 g/mol, molecules become more coil-like and the exponent drops to 1.07 which is still higher then the limiting value of $\alpha$ = 0.764 for random coils of perturbed chains (Volume III, Section 12.3.7).

Modification I of PPBA crystallizes orthorhombically with molecules in the macroconformation (trans–cis)$_n$. At 214°C, modification I changes to modification II which converts at 544°C to thermotropic nematic liquid crystals.

In tetramethylurea, *N*,*N*-dimethylacetamide, or *N*-methylpyrrolidone, PPBA also forms lyotropic liquid crystals from which films can be cast or fibers spun. PPBA fibers have very high tensile moduli and tensile strengths (Table 10-9) which can be further increased by using spinning solutions of higher concentrations, enlarging spin-stretch factors $A_0/A_2$, or annealing of fibers. Because of the high temperature stability, fibers are substitutes for asbestos.

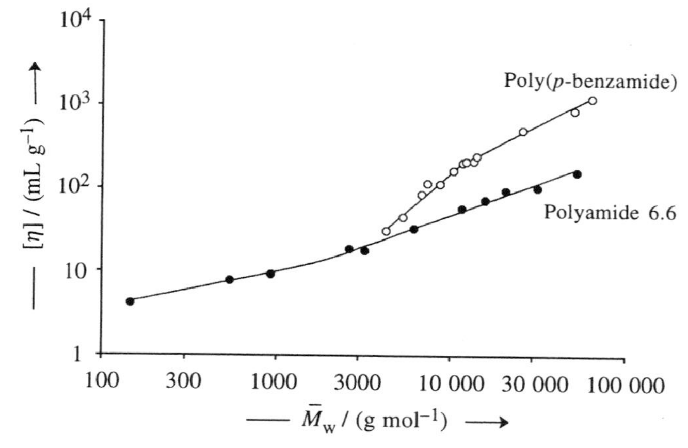


Fig. 10-11 Dependence of intrinsic viscosity in concentrated sulfuric acid at 25°C on weight-average molecular weight of PPBA [9] and PA 6 [8].

Table 10-9 Tensile moduli, $E$, tensile strengths at break, $\sigma_B$, and elongations at break, $\varepsilon_B$, of PPBA fibers spun from anisotropic solutions [8] or melts. $A_0/A_2$ = Ratio of cross-sectional area of spinneret capillary/cross-sectional area of fiber at solidification.

| Process | $E$/GPa | $\sigma_B$/GPa | $\varepsilon_B$/% |
|---|---|---|---|
| Solution spinning, 4.6 wt% PPBA in tetramethylurea + 6.5 % LiCl | 23.8 | 0.85 | 10.9 |
| Solution spinning, 5.8 wt% PPBA in tetramethylurea + 6.5 % LiCl | 43 | 1.11 | 9.7 |
| Solution spinning, 6.7 wt% PPBA in tetramethylurea + 6.5 % LiCl | 55 | 1.27 | 0.86 |
| Melt spinning with $A_0/A_2$ = 3.2 | 65 | 1.05 | 3.1 |
| Melt spinning with $A_0/A_2$ = 3.2, short annealing at 525°C | 137 | 2.2 | 1.9 |

PPBA was introduced by DuPont as Fiber B but its production was terminated after some years, probably because of high monomer costs, the short shelf-life of the dope, and/or difficult dissolution of the polymers. In the Western World, it was replaced by poly($p$-phenylene terephthalamide) which initially was also called Fiber B. PPBA is reportedly still produced as Terlon in Russia.

**AA/BB Types**

Aramids of the AA/BB type have melting temperatures that are higher than their decomposition temperatures. They too are produced in solution by a Schotten–Baumann reaction of aromatic dicarboxylic acid chlorides and aromatic diamines (as diaminodihydrochlorides since –Ar–$NH_2$ is resonance-stabilized):

$$\text{(10-35)} \qquad \text{ClOC–Ar–COCl} + \text{H}_2\text{N–Ar'–NH}_2 \rightarrow \text{+CO–Ar–CO–NH–Ar'–NH+}_n + 2\ \text{HCl}$$

An example is the polycondensation of $p$-phenylene diamine with terephthaloyl chloride to **poly($p$-phenylene terephthalamide), PPTA**, which is performed in a 2:1 mixture of $N$-methylpyrrolidone and hexamethylphosphoric triamide (+ ca. 3 % LiCl) (DuPont) and in $N$-methylpyrrolidone + $CaCl_2$ (Akzo). Addition of water precipitates the polymer.

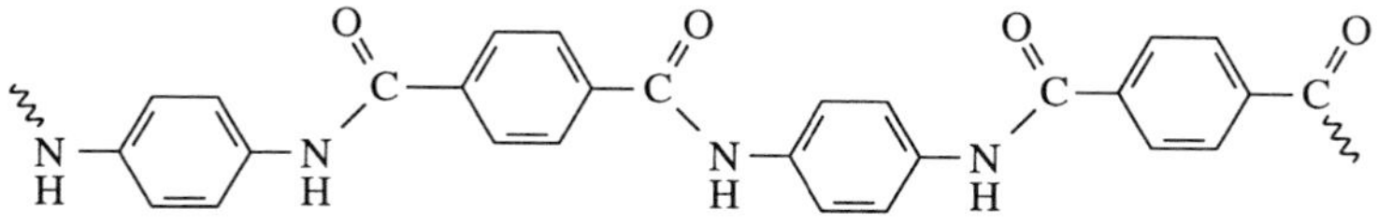

poly($p$-phenylene terephthalamide) (Kevlar®, Arenka®, Vniivlon®)

Fibers are spun at 80°C from solutions of 20 wt% PPTA in concentrated sulfuric acid into a precipitating bath. The concentration is higher than the critical concentration for the transition of dilute isotropic to more concentrated anisotropic solutions. These lyotropic solutions consist of domains in which polymer segments of PPTA-$H_2SO_4$ complexes are predominantly aligned parallel (see Volume III, Section 8.2) which lowers solution viscosities drastically. Since the high orientation of segments persists after precipitation, fibers thus have high strengths even without stretching (Table 10-10).

PPTA combines a low density with high modulus, low extension, high tensile strength, low electrical conductivity, and excellent dimensional stability. It has good chemical sta-

Table 10-10 Properties of aramids. Kevlar® 49 and Arenka® are highly drawn fibers (HM = high modulus). $\rho$ = density, $T_G$ = glass temperature, $E$ = tensile modulus in the fiber direction, $\sigma_B$ = tensile strength in fiber direction, $\varepsilon_B$ = elongation at break, $\Delta_w$ = water absorption at 21°C and 65 % relative humidity, LOI = limiting oxygen index (= lowest percent oxygen at which a polymer will burn).

| Trade name | $\rho$/(g cm$^{-3}$) | $T_G$/°C | $E$/GPa | $\sigma_B$/GPa | $\varepsilon_B$/% | $\Delta_w$/% | LOI/% |
|---|---|---|---|---|---|---|---|
| Poly(*p*-benzamide) | | | | | | | |
| Terlon | 1.4 | | 83 | 0.69 | 23 | 4.8 | 27.5 |
| Poly(*p*-phenylene terephthalamide) | | 345 | | | | | |
| Kevlar | 1.44 | | 50 | 2.9 | 4.5 | 1.5 | 31 |
| Kevlar 49 | 1.45 | | 138 | 2.8 | 2.1 | | |
| Arenka | 1.44 | | 61 | 2.7 | 3.7 | 6 | 29 |
| Arenka HM | 1.45 | | 125 | 2.7 | 2.3 | 3.5 | 29 |
| Poly(*m*-phenylene isophthalamide) | | 280 | | | | | |
| Nomex | 1.38 | | 15 | 0.67 | 23 | 4.8 | 27.5 |
| Conex | 1.37 | | 12 | 0.60 | 43 | | 28 |
| Phenylon | 1.37 | | 14 | 0.60 | 23 | | 27.5 |
| Poly[(*p*-phenylene-*co*-3,4'-diphenylether)$_{1:1}$terephthalamide] | | | | | | | |
| HM-50 | 1.40 | 242 | 69 | 3.7 | 5.5 | | |

bility and flame resistance and is self-extinguishing. PPTA is used as a fiber mainly for tire cords and also as highly drawn high-modulus fibers (Kevlar 49, Arenka HM) for cut- and bullet-proof, heat-stable body armor, in sports equipment, and as a reinforcing fiber for compositions in aviation and space applications.

The difficult processability of PPTA is improved by incorporating 4,4'-diaminodiphenyl ether groups in the chains whose angled structure interrupts the "rigid" PPTA chain. The high-modulus fiber **HM-50** (Technora®) is a copolymer from terephthaloyl chloride and a 50:50 mixture of *p*-phenylene diamine and 3,4'-diaminodiphenyl ether. The polymer is spun to fibers from 6 wt% isotropic solutions of monomers in *N*-methylpyrrolidone + $CaCl_2$, followed by drawing at 460–500°C.

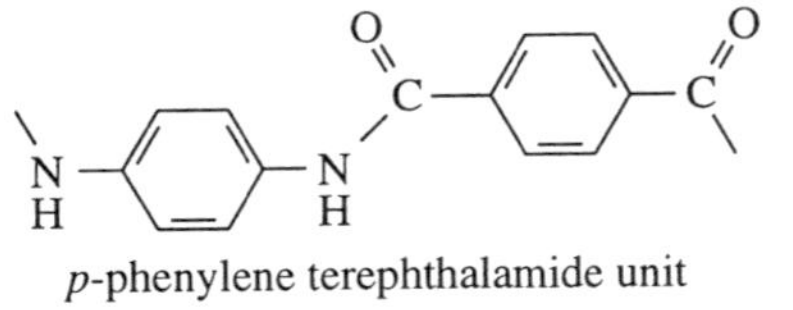

*p*-phenylene terephthalamide unit

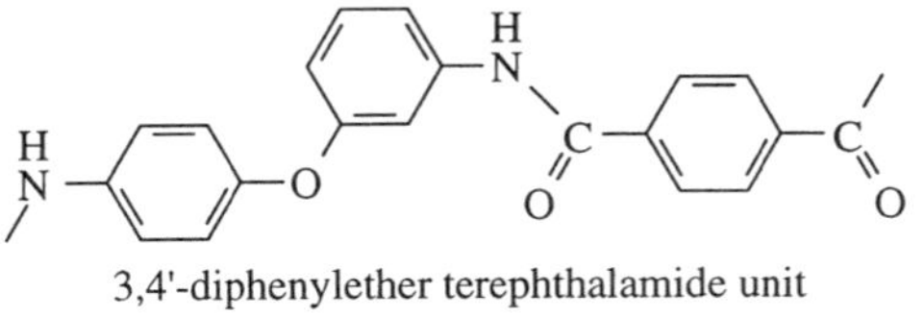

3,4'-diphenylether terephthalamide unit

**Poly(*m*-phenylene isophthalamide), PMPI**, is synthesized from *m*-phenylene diamine and isophthalic acid in dimethylacetamide as solvent with trimethylamine hydrochloride as catalyst and NaOH as HCl acceptor. After polymerization, the solution is neutralized by addition of $Ca(OH)_2$ and dry-spun to fibers or processed to films. PMPI does not form liquid crystals. The heat-resistant fibers and films are clear but yellowish.

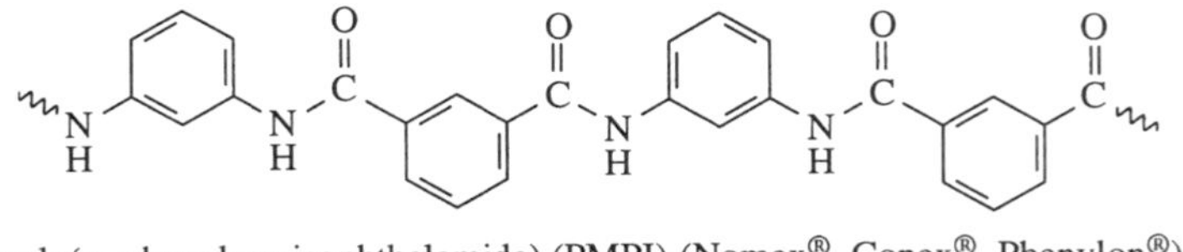

poly(*m*-phenylene isophthalamide) (PMPI) (Nomex®, Conex®, Phenylon®)

Somewhat exotic is poly(quinazolinedione isophthalamide) (ATF-2000) which uses a heterocyclic diamine that is prepared in several steps from 1-amino-4-nitrobenzoic acid (I). The resulting 3-(*p*-aminophenyl)-7-amino-2,4(1H,3H)-quinazolindione (II) is dissolved in *N*-methylpyrrolidone or dimethylacetamide and reacted with isophthaloyl chloride to ATF-2000:

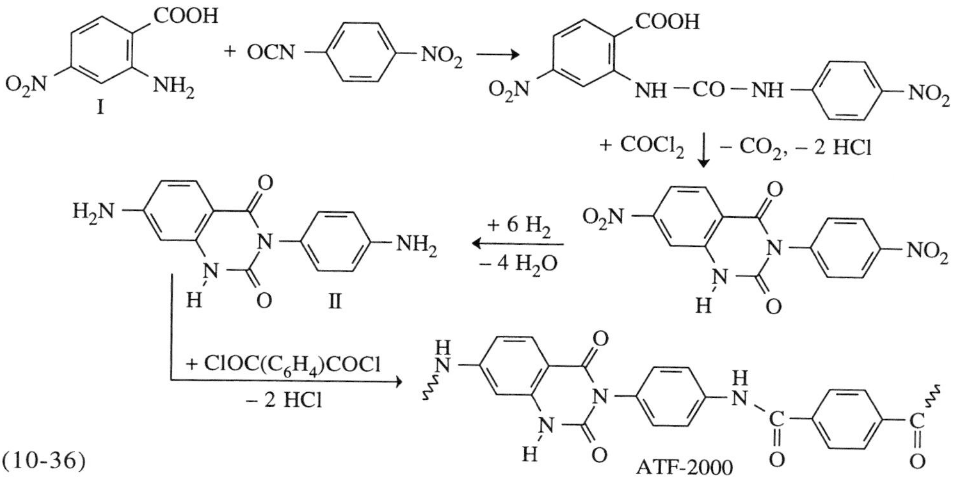


(10-36)

The viscous solution is dry or wet spun to hygroscopic fibers with excellent mechanical properties ($E \approx 9.8$ GPa, $\sigma_B \approx 700$ MPa, $\varepsilon_B = 14$ %). The temperature- and flame-resistant fibers (LOI = 38 %) serve for protective clothing and filters for hot gases.

### 10.3.6 Partly Aromatic Polyamides

The polyamide from 1,6-hexamethylenediamine and terephthalic acid (**PA 6.T**) is industrially useless since its melting temperature (371°C) is higher than its decomposition temperature (360°C). In order to maintain the improved thermal and mechanical properties that are bestowed by the aromatic group, either the hexamethylenediamine or the terephthalic acid moiety have to be replaced in part by other units.

Higher diamines are too costly as a replacement. Industry has therefore generally replaced the terephthalic acid moiety in part by the relatively inexpensive isophthalic acid or by other dicarboxylic acids. The resulting copolyamides are known as **partly aromatic polyamides** and sometimes also as **polyphthalamides** where the latter name refers to isophthalic acid; phthalic acid itself is not used as a monomer.

Partly aromatic polyamides can be subdivided into

- *semicrystalline* polyamides with hexamethylenediamine + terephthalic acid as majority components, and
- *amorphous* polyamides where terephthalic acid (if present) is usually the minority component and diamines have uncommon structures.

Amorphous polyamides of this type are usually transparent. The first commercial amorphous polyamide was polyamide TMDT = polyamide 6-3.T.

**Polyamide 6-3.T (PA TMDT)**

This polyamide is a copolymer of terephthalic acid (T) and a 50:50 mixture of 2,2,4- and 2,4,4-*tri*methyl-1,6-hexamethylene*dia*mine (TMDA; also known as 6-3 because of the 6-carbon hexamethylene unit with 3 methyl substituents):

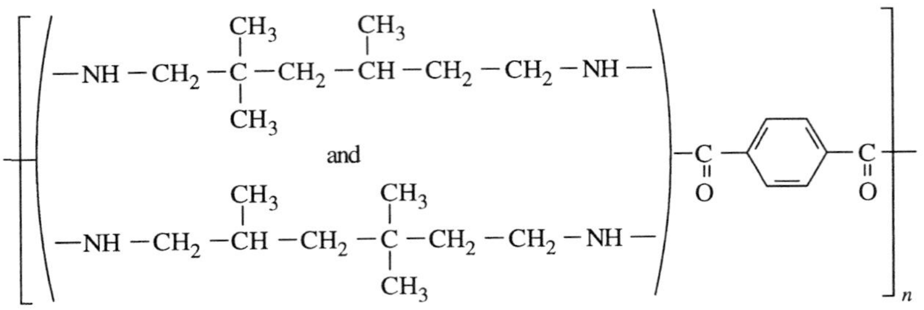

$$\left[\left(\begin{array}{c} -NH-CH_2-C(CH_3)_2-CH_2-CH(CH_3)-CH_2-CH_2-NH- \\ \text{and} \\ -NH-CH_2-CH(CH_3)-CH_2-C(CH_3)_2-CH_2-CH_2-NH- \end{array}\right)-CO-C_6H_4-CO-\right]_n \qquad \text{PA 6-3.T}$$

Branched diamines are uncommon monomers. They owe their existence to a search for the use of acetone, which is the major byproduct of the synthesis of phenol by the Hock process, Eq.(4-30). Of the many possible reactions, the trimerization of acetone to isophorone seemed most promising because isophorone (I) can be easily hydrogenated to its cyclohexanol derivative (II). Oxidation of II delivers the mixture of the isomeric 2,2,4- and 2,4,4-dicarboxylic acids (III):

$$(10\text{-}37) \qquad 3\ (CH_3)_2CO \xrightarrow{-2\ H_2O} \text{I} \xrightarrow{+2\ H_2} \text{II} \xrightarrow{+2\ O_2,\ -H_2O} HOOC-\overset{2}{C}(CH_3)_2-\overset{3}{C}H_2-\overset{4}{C}H(CH_3)-\overset{5}{C}H_2-COOH \quad \text{III}$$

(I: isophorone, 3,5,5-trimethylcyclohex-2-enone; II: 3,3,5-trimethylcyclohexanol; III here only as 2,2,4-isomer)

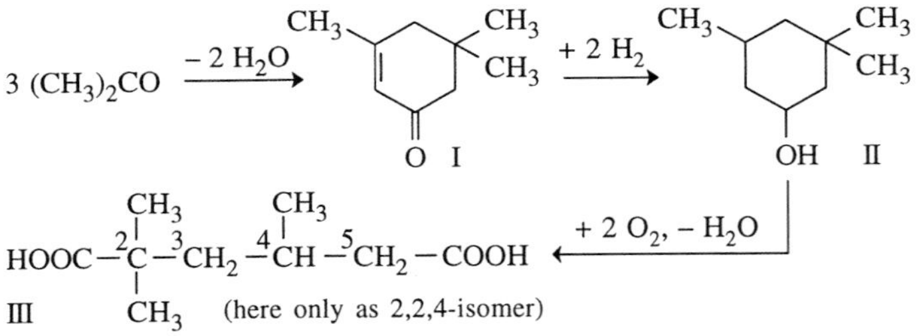

The dicarboxylic acids (III), HOOC–Z–COOH, are then converted to the diamides, $H_2NOC$–Z–$CONH_2$, and further to the dinitriles, NC–Z–CN, and finally to the diamines, $H_2N$–$CH_2$–Z–$CH_2$–$NH_2$ (1:1 mixture of 2,2,4- and 2,4,4-isomers). The diamines are polycondensed with terephthalic acid to polyamide TMDT = PA 6-3.T (Trogamid T®).

This polyamide has two different diamine units which each can be incorporated in head-to-tail, tail-to-tail, or head-to-head positions. The resulting irregular structure prevents crystallization. As a consequence, the amorphous polymer is glass-clear even in thick sheets. However, it turns white if it is exposed to 90°C water for a week.

The water uptake of PA TMDT is relatively high (7.5 %) but the glass temperature is reduced only to 135°C (wet) from 145°C (dry). Plasticizing by water increases fracture elongations but decreases yield strength. In contrast to competing polycarbonate A, no stress cracking is observed. See also Tables 10-11 and 10-12.

**Other Transparent Polyamides**

Subsequently, many other transparent polyamides were developed by other chemical companies. The structures of all these polyamides follow the same principle: prevention

Table 10-11 Monomers for commercial amorphous (transparent) polyamides. Numbers indicate composition in mol%. ● Composition was not revealed.

| Industry code | Lactams | | Diamines | | | | | Dicarboxylic acids | | | Other |
|---|---|---|---|---|---|---|---|---|---|---|---|
| | 6 | 12 | PACM | IPD | TMD | 6 | MACM | 6 | I | T | X |
| *Commercial amorphous polyamides* | | | | | | | | | | | |
| 6I | 50 | | | | | | | | 50 | | |
| 6I/6 | ● | | | | | ● | | | ● | | |
| 6I/6T | | | | | | 50 | | | 35 | 15 | |
| 6I/6T/MACMI | | | | | | ● | ● | | ● | ● | |
| 6I/6T/MACMI/MACMT | | | | | | ● | ● | | ● | ● | |
| 6I/6T/PACMI/PACMT | | | ● | | | ● | | | ● | ● | |
| 6/PACMT | ● | | ● | | | | | | | ● | |
| 12/MACMI | | ● | | | | | ● | | ● | | |
| 12/MACMT | | ● | | | | | ● | | | ● | |
| TMDT or 6-3T | | | | | 50 | | | | | 50 | |
| *Pilot amorphous polyamides* | | | | | | | | | | | |
| 6/TMDT/6T | ● | | | | ● | ● | | | | ● | |
| 66/TMDI/TMDT | | | | | ● | ● | | ● | ● | ● | |
| 66/6I/MACMI/MACMT | | | | | | ● | ● | ● | ● | ● | |
| 6I/6T/6IPD | | | | ● | | ● | | | ● | ● | |
| MACMI/MACMX | | | | | | | ● | | ● | | ● |
| XT/MACMT | | | | | | | ● | | | ● | ● |

of crystallinity by replacing linear structures (aliphatic zig-zag chains, para-substituted aromatics, 1,4-cyclohexane units, and the like) by angled ones (1,3-substituted ring systems, cyclohexane units, etc.), use of substituted monomers, and/or more than two types of monomer. According to DSC, some of the polymers have block character. However, most of these polymers never made it beyond the pilot stage. The surviving ones are usually niche products because of expensive monomers.

Monomers for the various commercially available transparent polyamides include two types of lactam, five types of amines, three types of dicarboxylic acids, and one type each of an unknown diamine and an unknown dicarboxylic acid (Table 10-11). Very often, technical and/or company literature does not reveal compositions.

Besides standard monomers (1,6-hexamethylenediamine; adipic, terephthalic, and isophthalic acid; lactams 6 and 12) and TMDA (6-3; see p. 478) the following diamines are used:

MACM 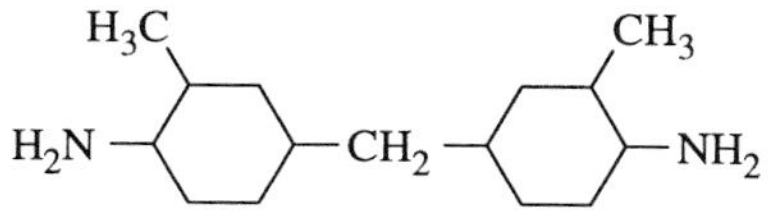 3,3'-dimethyl-4,4'-diaminodicyclohexylmethane

PACM 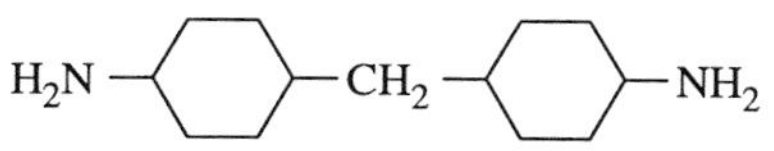 bis(*p*-aminocyclohexyl)methane

IPD 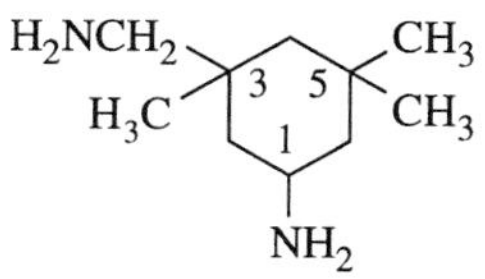 1-amino-3-aminomethyl-3,5,5-trimethylcyclohexane (isophorondiamine)

Transparent polyamides with several other diamines and dicarboxylic acids never made it to the market:

| | | |
|---|---|---|
| AMNB | $H_2NCH_2$, $CH_2NH_2$; $CH_2NH_2$, $H_2NCH_2$ | 1,4- and 1,3-bis(aminomethyl)norbornane |
| TCD | $CH_2NH_2$, $CH_2NH_2$, $(CH_2)_3$ | 2,2-bis(aminocyclohexyl)propane |
| MC | $CH_2NH_2$, $H_2NCH_2$ | 1,3-bis(aminomethyl)cyclohexane |
| 5BI | $(CH_3)_3C$, COOH, COOH | 5-*t*-butylisophthalic acid |
| | HOOC, COOH | 1,4-cyclohexanedicarboxylic acid |
| N | COOH, HOOC | 2,6-naphthalenedicarboxylic acid |
| | $CH_3$, $CH_3$, HOOC, $CH_3$, COOH, $C_6H_5$ | 1,1,3-trimethyl-3-phenylindane-4,5-dicarboxylic acid |
| MDI | OCN, $CH_2$, NCO | diphenylmethane-4,4'-diisocyanate |

**Semicrystalline Aromatic Polyamides**

Semicrystalline aromatic polyamides are all based on terephthalic acid and 1,6-hexamethylene diamine, usually at 50 mol% (Table 10-12). These polymers dissolve only in hexafluoroisopropanol except PA 6T/6I/66 which is also soluble in hot phenol or hot concentrated $H_2SO_4$. Correspondingly, they have excellent resistance against chemicals.

Their aromatic components provide them with higher melting, glass, and heat distortion temperatures than polyamide 6.6 (Table 10-13). Their excellent mechanical and electrical properties make them useful for temperature-resistant parts such as radiators for cars, printed circuit boards, and the like. Many of these polymers are offered as glass fiber-reinforced products.

Table 10-12 Monomers for commercial semicrystalline polyamides. Numbers indicate composition in mol%.

| Industry code | Lactam | Diamine | Dicarboxylic acids | | |
|---|---|---|---|---|---|
| | 6 | 6 | 6 | I | T |
| 6T/6 | 30 | 35 | | | 35 |
| 6T/6I | | 50 | | 15 | 35 |
| 6T/66 | | 50 | 25 | | 25 |
| 6T/6I/66 | | 50 | 5 | 12.5 | 32.5 |

## 10.3.7 Other Polyamides

**Poly(*m*-xylylene adipamide)**

The polymer MXD6 (Reny®, IXEF®) from *m*-xylylenediamine and adipic acid

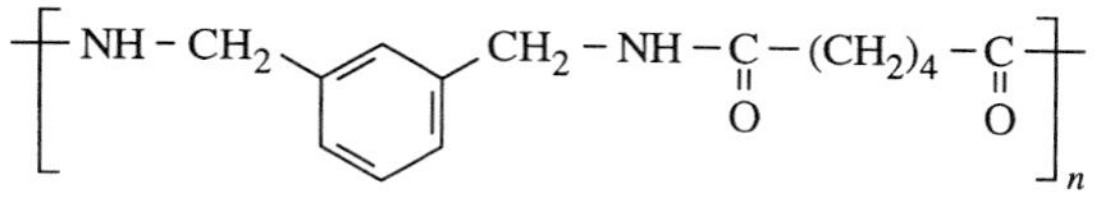

$$\left[ -NH-CH_2-(1,3\text{-}C_6H_4)-CH_2-NH-\underset{\substack{\| \\ O}}{C}-(CH_2)_4-\underset{\substack{\| \\ O}}{C}- \right]_n$$

poly(*m*-xylylene adipamide) (MXD6)

is usually considered an aliphatic polyamide because its amine groups are not directly bound to the 1,3-phenylene group. The polymer has better mechanical properties then polyamides 6.6 and 6 (see Table 10-13) but it is still relatively inexpensive.

The polymer is used for monofilaments (brushes, filter cloth) and, as glass fiber-reinforced grades, for automotive parts (air filter housings), industrial equipment parts (gears, rolls), sports equipment, and electric/electronic parts. The low permeablity for oxygen even at high humidities allows its use as blow-molded bottles and extruded films for food packaging, in part as laminates with poly(olefin)s, poly(ethylene terephthalate), or polyamide 6.

**Poly[bis(4-aminocyclohexane)methylene dodecanamide]**

Monomers for this polymer are dodecanedioic acid (1,12-dodecane dicarboxylic acid), $HOOC(CH_2)_{10}COOH$, synthesized from butadiene (p. 112, Route II) and bis(4-aminocyclohexyl)methane, $H_2N(C_6H_{10})CH_2(C_6H_{10})NH_2$, which is obtained from the condensation of aniline with formaldehyde and subsequent hydrogenation of 4,4'-diaminodiphenylmethane (= 4,4'-methylenedianiline), $H_2N(p\text{-}C_6H_4)CH_2(p\text{-}C_6H_4)NH_2$.

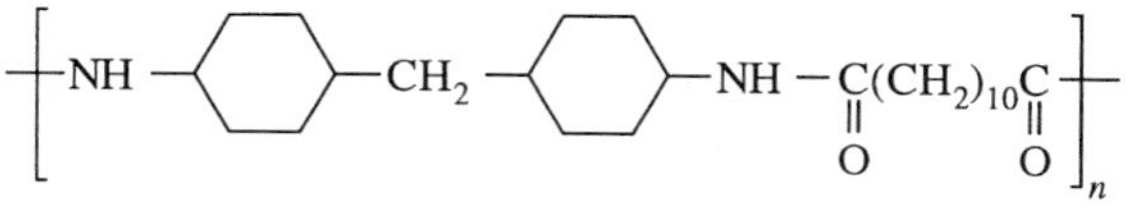

$$\left[ -NH-(C_6H_{10})-CH_2-(C_6H_{10})-NH-\underset{\substack{\| \\ O}}{C}(CH_2)_{10}\underset{\substack{\| \\ O}}{C}- \right]_n$$

poly[bis(4-aminocyclohexane)-methylene dodecaneamide] (Qiana® fiber)

Some 70 % of the cyclohexane moieties of this polymer are in the trans configuration. Melt spinning at 275°C delivers silk-like fibers (Qiana®) that are wrinkle-resistant and can be dyed with brilliant colors. Fabrics can by dry-cleaned because the polymer does not dissolve in chlorinated solvents.

Table 10-13 Average thermal and mechanical properties of unfilled semicrystalline (sc) or amorphous (a) aromatic polyamides at 23°C as compared to PA 6.6.

| Property | | Physical unit | PA 6.6 (sc) | 6.T/6.I/6.6 (sc) | 6.T/6 (sc) | MXD.6 (sc) | 6-3.T (a) |
|---|---|---|---|---|---|---|---|
| Density | | g/cm$^3$ | 1.14 | 1.17 | 1.18 | 1.22 | 1.12 |
| Melting temperature (DSC), dry | | °C | 262 | 310 | 298 | 243 | - |
| Heat distortion temperature (1.82 MPa) | | °C | 90 | 120 | | 96 | 120 |
| Vicat temperature B | | °C | 200 | 130 | 120 | | 146 |
| Glass temperature (DSC, dry) | | °C | 48 | 127 | 113 | 85 | 155 |
| (DSC, 50 % RH) | | °C | 15 | | 40 | 52 | |
| (DSC, 100 % RH) | | °C | –32 | | | 15 | |
| Linear thermal expansion coefficient | | $10^{-5}$ K$^{-1}$ | 8.1 | | | 5.1 | 6.0 |
| Specific heat capacity | | J K$^{-1}$ g$^{-1}$ | 1.7 | | | | 1.45 |
| Thermal conductivity (20°C) | | W m$^{-1}$ K$^{-1}$ | 0.23 | 0.24 | | 0.38 | 0.21 |
| Tensile modulus | dry | MPa | 3200 | 3200 | 3200 | 4700 | 2800 |
| | 50 % RH | MPa | 1600 | | 3600 | | |
| Flexural modulus | dry | MPa | 2830 | 3650 | 3500 | 4400 | 2660 |
| Yield strength | dry | MPa | 83 | 105 | 100 | | 95 |
| | 50 % RH | MPa | 59 | | 110 | | 70 |
| Fracture strength | dry | MPa | 83 | 110 | 120 | 99 | |
| | 50 % RH | MPa | 77 | | 104 | 75 | |
| Flexural strength | dry | MPa | 117 | 310 | | 160 | 135 |
| Elongation at yield | dry | % | 5 | 3 | 4.5 | | 8 |
| | 50 % RH | % | 25 | | | | |
| Elongation at fracture | dry | % | 60 | | > 10 | 2 | 70 |
| | 50 % RH | % | > 300 | | | > 10 | 180 |
| Impact strength, (Izod) | dry | J/m | 53 | | | 20 | kB |
| | 50 % RH | J/m | 107 | | | | |
| Notched impact strength, (Izod) | dry | J/m | 53 | 110 | 35 | | 360 |
| | 50 % RH | J/m | 112 | 100 | 40 | 20 | 13 |
| Rockwell hardness (Shore) | dry | - | R 119 | | | R108 | M93 |
| | 50 % RH | - | R 108 | | | | |
| Water absorption, 24 h | 50 % RH | % | 2.5 | 1.5 | 1.8 | 1.9 | 1.8 |
| | 100 % RH | % | 8.5 | 5.8 | 7.5 | 5.8 | 7.5 |
| Relative permittivity | dry, 1 kHz | - | 3.8 | | 4.0 | | 3.5 |
| | 1 MHz | - | 3.5 | | | 3.9 | 3.1 |
| | 50 % RH, 1 kHz | - | 6.5 | | | | 3.9 |
| | 1 MHz | - | 4.1 | | | | 3.4 |
| Dissipation factor | dry, 1 MHz | - | 0.03 | | | | 0.02 |
| | 50 % RH, 1 MHz | - | 0.08 | | | | 0.03 |
| Resistivity | dry | Ω cm | $10^{15}$ | | | $10^{16}$ | $10^{15}$ |
| | 50 % RH | Ω cm | $10^{13}$ | | | | $10^{15}$ |
| | 100 % RH | Ω cm | $10^{9}$ | | | | |

## 10.4 Polyureas and Amino Resins

### 10.4.1 Polyureas

Polyureas contain the characteristic ureylene group –NH–CO–NH–. As derivatives of urea, $H_2N–CO–NH_2$, they are known as **polyureylenes** and, since urea is the monoamide of the hypothetical carbamic acid, $H_2N–CO–OH$, also as **polycarbamides**.

In principle, polyureas can be synthesized from the reaction of "poly"amines (i.e., di, tri, ...) with $CO_2$ (or its derivatives such as $CO(OR)_2$, $COCl_2$, COS), with $H_2N–CO–NH_2$ (or its derivatives such as $H_2N–CO–OR$), or with OCN–R –CO as well as the reaction of "poly"isocyanates with water. In practice, only a few types of reactions are important.

Polyurea fibers are obtained industrially from the melt polycondensation of 1,9-nonamethylenediamine with urea in the melt at 140–160°C and subsequent post-condensation in vacuum at ca. 250°C:

$$(10\text{-}38) \quad H_2N-CO-NH_2 + H_2N-Z-NH_2 \rightleftharpoons \text{+}\!\!\left[NH-CO-NH-Z\right]\!\!\text{+} + 2\,NH_3$$

The polymer is spun to a fiber ($T_M$ = 240°C) that is better resistant against alkali than poly(ethylene terephthalate) and can by dyed with acidic dyes.

Most important by volume is the reaction of "poly"isocyanates with "poly"amines, ~NCO + $H_2N$~ → ~NH–CO–NH~. This reaction (including copolymerizations leading to additional urethane, amide, imide, etc., groups) is especially important for the production of polyurea foams which constitute ca. 75 % of the worldwide polyurea production of more than $6 \cdot 10^6$ t/a.

Block copolyureas are produced by the RIM process (reaction-injection molding) in which a stream of MDI-based "poly"isocyanate and a stream of amine-functionalized polyether is simultaneously injected into a mold. Addition of water produces additional urethane-urea structures. Polyurethaneurea and polyurea elastomers find application in the automobile industry.

Polyureas are also produced by reaction of diamines with $CO_2$ (or derivatives) and of diisocyanates with water. The reaction of mixtures of diamines with urea plays a minor role. Applications of polyureas and polycoureas include not only foams, elastomers, and moldings but also films, membranes, coatings, lubricants, and adhesives.

### 10.4.2 Amino Resins

**Monomers**

Amino resins (**amino plastics**) are condensation products of NH group-containing compounds with nucleophilic components and carbonyl group-containing compounds. The three components are interconnected by a kind of Mannich reaction which is called α-ureidoalkylation since urea is the most important NH component:

$$(10\text{-}39) \quad \sim\!Z-H \;+\; \underset{R'}{\overset{R}{C}}\!=\!O \;+\; H-N\!\!< \;\longrightarrow\; \sim\!Z-\underset{R'}{\overset{R}{C}}-N\!\!< \;+\; H_2O$$

nucleophilic component | carbonyl compound | NH component | amino resin

The most important *NH components* are urea for **urea resins** (= urea-formaldehyde resins, **UF**) and melamine (= 2,4,6-triamino-1,3,5-triazine) for **melamine resins** (= melamine-formaldehyde resins, **MF**). Other monomers for amino resins comprise substituted and cyclic ureas, thioureas, guanidines, cyanoamides, acidic amides, etc.

Urea is obtained from ammonia and carbon dioxide (p. 482, last line) and melamine by trimerization of molten urea under pressure or pressure-less in flow reactors:

$$(10\text{-}40)\quad 6\,H_2N-\underset{\underset{O}{\parallel}}{C}-NH_2 \longrightarrow \text{melamine } (C_3N_3(NH_2)_3) + 6\,NH_3 + 3\,CO_2$$

Formaldehyde (synthesis: p. 95) was originally the only *carbonyl component*. Newer amino acid syntheses also employ higher aldehydes and ketones, but the usefulness of these compounds is restricted by aldolizations, Cannizzaro reactions, enamine formations, and steric hindrance.

*Nucleophilic components* are all H-acidic compounds that have an unpaired electron pair at the condensation site. This group comprises halogenated hydrocarbons; OH-acidic compounds such as alcohols, carboxylic acids, and hemi-acetals; NH-acidic compounds like carboxylic amides, ureas, guanidines, melamines, urethanes, and primary and secondary amines; and SH-acidic compounds such as mercaptans. Also usable are all compounds that convert to carbanions by losing protons (CH-acidic compounds) or that transform to tautomeric forms by prototropy such as enolizable ketones. The latter group comprises compounds that are activated by COOH, $NO_2$, CN, etc., groups as well as some substituted aromatic compounds such as aniline.

**Synthesis**

The primary step in the synthesis of amino resins is the acid- or base-catalyzed reaction between a carbonyl component and an NH component:

$$(10\text{-}41)\quad H_2N-\underset{\underset{O}{\parallel}}{C}-NH_2 \xrightarrow[-H_2O]{+OH^{\ominus}} H_2N-\underset{\underset{O}{\parallel}}{C}-\overset{\ominus}{N}H \xrightarrow{+CH_2O} H_2N-\underset{\underset{O}{\parallel}}{C}-NH-CH_2-O^{\ominus}$$

$$(10\text{-}42)\quad H_2N-\underset{\underset{O}{\parallel}}{C}-NH-CH_2-O^{\ominus} \xrightarrow{+H^{\oplus}} \underset{I}{H_2N-\underset{\underset{O}{\parallel}}{C}-NH-CH_2-OH} \rightleftharpoons \text{(cyclic H-bonded form: } H_2N-C(=O\cdots H-O-CH_2-NH-)\text{)}$$

The resulting *N*-methylol urea (I) is stabilized by an intramolecular hydrogen bond. In base-catalyzed systems, the reaction stops here. In acid-catalyzed ones, the methylol compound converts to a resonance-stabilized carbonium/immonium ion:

$$(10\text{-}43)\quad \underset{I}{H_2N-\underset{\underset{O}{\parallel}}{C}-NH-CH_2-OH} \xrightarrow[-H_2O]{+OH^{\oplus}} \left[H_2N-\underset{\underset{O}{\parallel}}{C}-NH-\overset{\oplus}{C}H_2 \longleftrightarrow H_2N-\underset{\underset{O}{\parallel}}{C}-\overset{\oplus}{N}H=CH_2\right]$$

The resulting $\alpha$-ureidoalkyl(carbonium/immonium) ions then react with suitable nucleophilic partners in an electrophilic substitution reaction which extends the chain. Such a partner may be urea itself:

(10-44)

$$H_2N-\underset{\underset{O}{\|}}{C}-NH-\overset{\oplus}{C}H_2 + H_2N-\underset{\underset{O}{\|}}{C}-NH_2 \xrightarrow{-H^{\oplus}} H_2N-\underset{\underset{O}{\|}}{C}-NH-CH_2-NH-\underset{\underset{O}{\|}}{C}-NH_2$$

It seems that this reaction leads to linear polymers, i.e., one methylol group per $NH_2$ group. The reaction produces a colloidal dispersion from which the particles agglomerate and precipitatem which leads to a "hardening" of the resin by hydrogen bonding. The hardened resins dissolve in formaldehyde or sulfuric acid; the dissolved compounds have the same chemical structures as the non-hardened resins.

The rate of methylolization of NH compounds by formaldehyde is of first order with respect to the NH compound, formaldehyde, and catalyst. Since a termolecular reaction is unlikely, an associate must be formed first, for example, between formaldehyde and catalyst. The rate-determining step is then the reaction between the associate and the NH compound. Catalysts such as $HCO_3^{\ominus}$, $H_2PO_4^{\ominus}$, or $HPO_4^{2\ominus}$ give rise to faster reactions than $CH_3COO^{\ominus}$ or $HR_3N^{\oplus}$ because compounds of the first group are bifunctional catalysts in the sense of accepting protons as acids and donating them as bases.

α-Ureidoalkylations are accompanied by trans-ureidoalkylations in which an H-acidic component is replaced nucleophilically by another nucleophilic compound:

(10-45)
$$\rangle N-CH_2-OH + H-Z\sim\sim \longrightarrow \rangle N-CH_2-Z\sim\sim + H_2O$$

Such trans-ureidoalkylations are important not only in the synthesis of amino resins but also in the hardening of novolacs by polymethylene ureas and in the wash-and-wear finishing of cotton (Volume IV).

Amino resin formations from melamine or aniline and formaldehyde proceed analogously. With melamine, 2 molecules of formaldehyde react with one $NH_2$ group and not one as with urea. With aniline in an acidic environment, the aromatic ring acts as a nucleophilic partner which leads to crosslinking because of three possible ring substitutions (2 ortho, 1 para) and the bifunctional amino group.

**Industrial Resins**

Urea resins comprise ca. 86 % and melamine resins ca. 14 % of the total world production of amino resins of ca. $10.7 \cdot 10^6$ t/a (2003).

Urea resins are inexpensive, water-soluble materials that are predominantly used as glues for wood products such as plywood, fiber boards, chip boards, etc. (p. 69) and in the United States also for shingles. Smaller percentages serve as molding compounds, in lacquers, as foams, for wrinkle-proofing of cotton, or as fertilizers. Urea resins by polycondensation in 5–30 % aqueous solutions delivers spherical particles with large interior surfaces that serve as fillers and pigments of paper.

Use of *n*-butanol or *i*-butanol as an additional nucleophilic component leads to amino resins that are soluble in organic solvents and serve as lacquers. Alcohols with shorter chains produce insufficiently soluble polymers whereas those with longer chains do not etherify fast enough. Such resins are usually delivered as a 50 % solution in butanol or butanol/xylene. Partial etherification by methanol leads to water-soluble lacquer resins.

Alcohol-modified urea resins are difficult to fill with pigments. Since their baking-finishes are also relatively brittle and not very elastic, such resins are often "plastified" by combining them with cellulose nitrate and plasticizers for air-drying lacquers or with alkyd resins for baking varnishes; in the latter case, mainly by mechanical mixing and not by in-situ polymerization. Urea resins and alkyd resins thus react with each other only on hardening.

For molding, urea resins are sometimes combined with fillers (wood flour, cellulose, etc.). Amino resins are more colorless and less light-sensitive than phenolic resins but more sensitive to humidity and heat. Urea resins can be employed up to 90°C, melamine resins up to 120°C.

Foams of urea resins compete with expanded poly(styrene) for the insulation of buildings. This use, as well as that of urea resins in wood products, has raised health concerns since such resins slowly emit formaldehyde gas.

Urea resins serve as a fertilizer by releasing ammonia; they do not acidify soils as ammonium sulfate does. Mixtures of urea resins and, for example, poly($\varepsilon$-caprolactone) are pressed into containers for flowers, seedlings, etc., that slowly decompose in the soil.

### 10.4.3 Polyhydrazides

Terephthaloyl chloride and $p$-aminobenzhydrazide react to form polyhydrazides

$$n\ \mathrm{ClOC{-}C_6H_4{-}COCl} + n\ \mathrm{H_2N{-}C_6H_4{-}CO{-}NH{-}NH_2}$$

$$\xrightarrow{-2n\ \mathrm{HCl}} \left[\mathrm{CO{-}C_6H_4{-}CO{-}HN{-}C_6H_4{-}CO{-}NH{-}NH}\right]_n \qquad (10\text{-}46)$$

in which the aminobenzhydrazine unit may be incorporated as shown or in reverse, giving rise to "partially ordered" polyamidehydrazides. These polymers are spun to fibers with high tensile moduli for use as tire cords or reinforcing fibers for plastics. They are sensitive to pH and tend to yellow due to keto-enol tautomerization.

## 10.5 Polycyanates

### 10.5.1 Introduction

Compounds consisting just of one atom each of carbon, hydrogen, nitrogen, and oxygen are acids: cyanic acid and the tautomeric isocyanic acid and fulminic acid and the tautomeric isofulminic acid:

$$\underset{\text{cyanic acid}}{\mathrm{H{-}O{-}C{\equiv}N}} \rightleftharpoons \underset{\text{isocyanic acid}}{\mathrm{O{=}C{=}N{-}H} \longleftrightarrow \mathrm{O{-}\overset{\ominus}{C}{\equiv}\overset{\oplus}{N}{-}H}}$$

$$\underset{\text{fulminic acid}}{\mathrm{H{-}O{-}\overset{\oplus}{N}{\equiv}\overset{\ominus}{C}}} \rightleftharpoons \underset{\text{isofulminic acid}}{\mathrm{H{-}\overset{\ominus}{C}{=}\overset{\oplus}{N}{=}O} \longleftrightarrow \mathrm{H{-}C{\equiv}\overset{\oplus}{N}{-}\overset{\ominus}{O}}}$$

Salts and esters of these acids are called cyanates (from cyanic acid), isocyanates (from isocyanic acid), and fulminates (from fulminic acid). Inorganic salts of isofulminic acid are known as isofulminates whereas organic esters of isofulminic acid are called nitrile oxides. On heating, nitrile oxides convert to isocyanates.

In principle, all four acids and their esters can dimerize, trimerize, and polymerize as well as add other chemical compounds. However, no polymers of fulminic acid or isofulminic acid are known, probably because these monomers are not stable. But there are many cyanic acid-based polymers (Section 10.5.2) and extraordinarily many polymers based on isocyanates, either by chain polymerization (Section 10.5.5) or by polyaddition (Section 10.6).

### 10.5.2 Polyamide 1 and Cyamelide

Cyanic acid, H–O–C≡N, melts at –86°C. At temperatures below 0°C, it is stable for weeks. At temperatures above 0°C, it converts to a white product that constists of a mixture of cyanuric acid and cyamelide. Cyanuric acid is in tautomeric equilibrium with isocyanuric acid. Cyamelide has long been assumed to be a trimer of cyanic acid with the structure shown below. As a suspected isomer of cyanuric acid and because of its insolubility in all known solvents, it is also called "insoluble cyanuric acid." In fact, it is a crosslinked polymer which may have some cyamelide units as crosslinking sites but certainly some of the units shown in Eq.(10-47) as chain units.

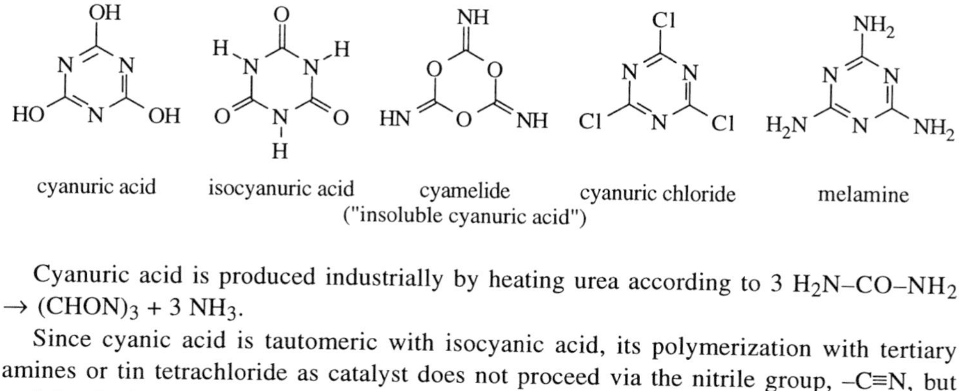

cyanuric acid | isocyanuric acid | cyamelide ("insoluble cyanuric acid") | cyanuric chloride | melamine

Cyanuric acid is produced industrially by heating urea according to 3 $H_2N–CO–NH_2 \rightarrow (CHON)_3 + 3\ NH_3$.

Since cyanic acid is tautomeric with isocyanic acid, its polymerization with tertiary amines or tin tetrachloride as catalyst does not proceed via the nitrile group, –C≡N, but mainly via the –N=C< group and somewhat via the >C=O group to poly(cyanic acid):

$$\text{(10-47)}\qquad HN{=}C{=}O \longrightarrow -NH-\underset{\underset{O}{\|}}{C}- \;+\; \left(-\underset{\underset{NH}{\|}}{C}-O-\right)$$

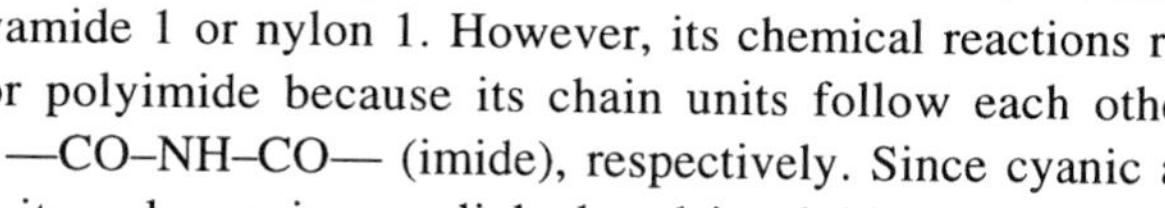

Because of amide groups as main monomeric units, resulting polymers are often called polyamide 1 or nylon 1. However, its chemical reactions rather resemble that of a polyurea or polyimide because its chain units follow each other as —NH–CO–NH— (urea) and —CO–NH–CO— (imide), respectively. Since cyanic acid is potentially tetrafunctional, its polymer is crosslinked and insoluble; it is identical with the "insoluble cyanuric acid" (cyamelide).

### 10.5.3 Polyisocyanates

Substituted poly(isocyanate)s, $\text{+NR–CO+}_n$, are obtained by polymerization of isocyanates, R–N=C=O, with, e.g., KCN as initiator. Polymer molecules adopt helical conformations. They serve as model compounds in the study of solution behavior of "rod-like" molecules but are not of industrial interest.

### 10.5.4 Cyanate Ester Resins

In industry, the term "cyanate ester resin" refers to both the prepolymer and the hardened (= crosslinked) polymers that result from cyclotrimerization of aromatic dicyanate esters, N≡C–O–Ar–O–C≡N (Eq.(10-48)). The resulting thermosets are also called **poly(cyanurate)s, poly(cyanate)s**, or **triazine resins**.

Monomers, N≡C–O–Ar–O–C≡N, are synthesized by reacting bisphenols, HO–Ar–OH, with either cyanic acid, HO–C≡N, with release of $H_2O$, or with gaseous cyanuric chloride, Cl–C≡N, with release of HCl. Industrially used are various bisphenols as well as trisphenols based on dicyclopentadiene:

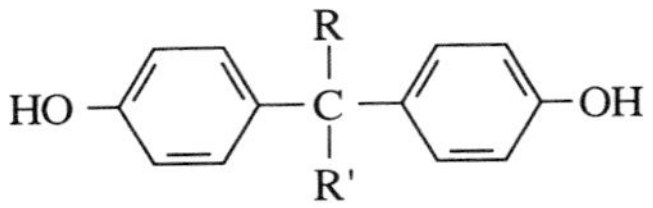

| R | R' | |
|---|---|---|
| $CH_3$ | $CH_3$ | bisphenol A |
| $CH_3$ | H | bisphenol E |
| $CF_3$ | $CF_3$ | hexafluorobisphenol A |

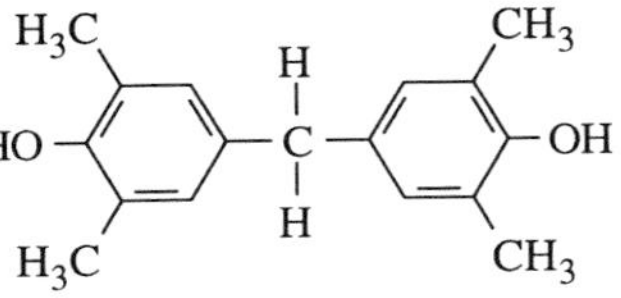

tetramethylbisphenol F

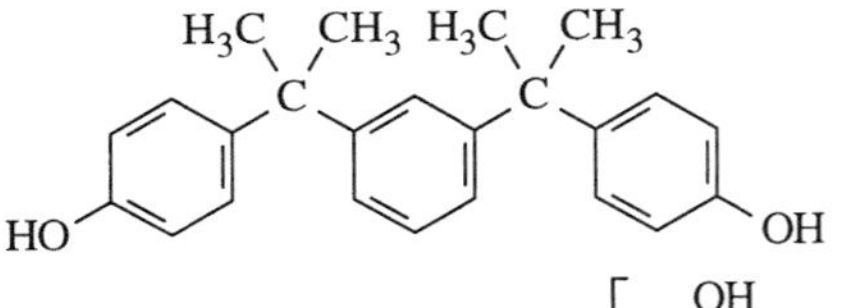

bisphenol M

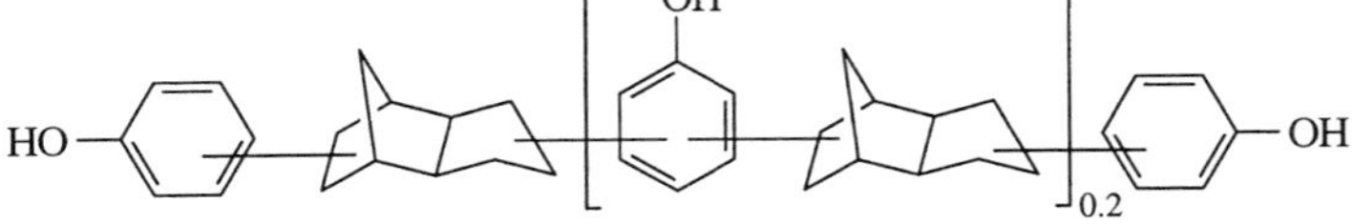

Very pure cyanate esters do not react. However, they usually contain phenolic impurities from the monomer synthesis that react according to Eq.(10-48). The neat reaction without added catalysts leads exclusively to *s*-triazine rings and thus to crosslinked polymers that are used mainly for printed circuits and load-bearing composites:

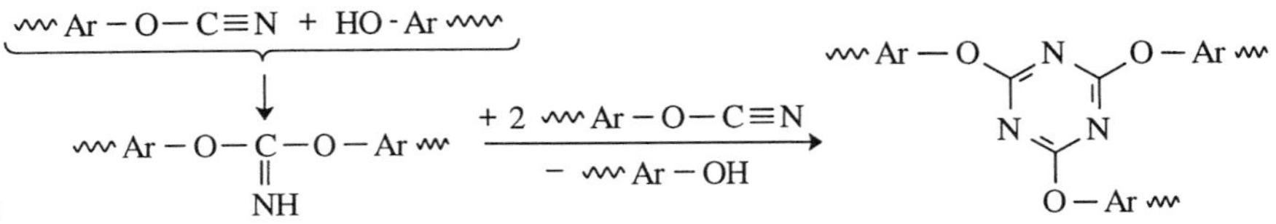

$$\sim\text{Ar}-\text{O}-\text{C}\equiv\text{N} + \text{HO-Ar}\sim \;\rightarrow\; \sim\text{Ar}-\text{O}-\underset{\text{NH}}{\underset{\|}{\text{C}}}-\text{O}-\text{Ar}\sim \xrightarrow[-\;\sim\text{Ar}-\text{OH}]{+\,2\;\sim\text{Ar}-\text{O}-\text{C}\equiv\text{N}} \text{triazine ring with three }\text{O}-\text{Ar}\sim \qquad (10\text{-}48)$$

Reaction in bulk or solution with *t*-$C_4H_9OK$ or in bulk with triethylenediamine as catalysts deliver predominantly biologically degradable polyiminocarbonate structures besides *s*-triazine units, for example:

$$HO-C_6H_4-CRR'-C_6H_4-OH + N{\equiv}C-O-Ar-O-C{\equiv}N$$

$$(10\text{-}49) \qquad \longrightarrow \left[ O-C_6H_4-CRR'-C_6H_4-O-\underset{\underset{NH}{\|}}{C}-O-Ar-O-\underset{\underset{NH}{\|}}{C} \right]$$

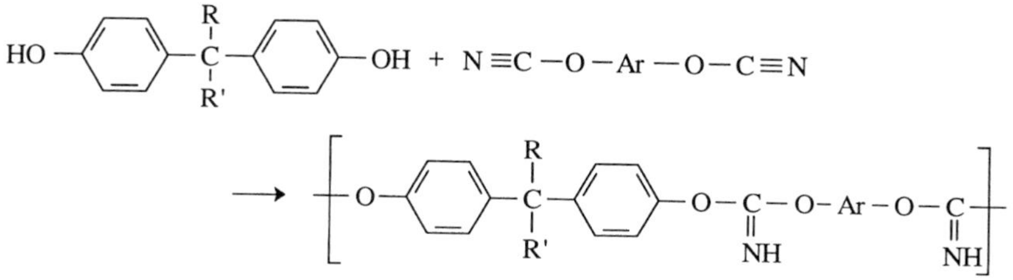

### 10.5.5 Polyisocyanurates

Polyisocyanurates (**PIR**) result from cyclotrimerizations of diisocyanates and polydi-isocyanates ("poly" in the sense of "more than two", not "polymer") (Eq. 10-50)):

(10-50) $O{=}C{=}N-Ar-N{=}C{=}O \longrightarrow$ [isocyanurate ring with three Ar–N=C=O substituents] etc.

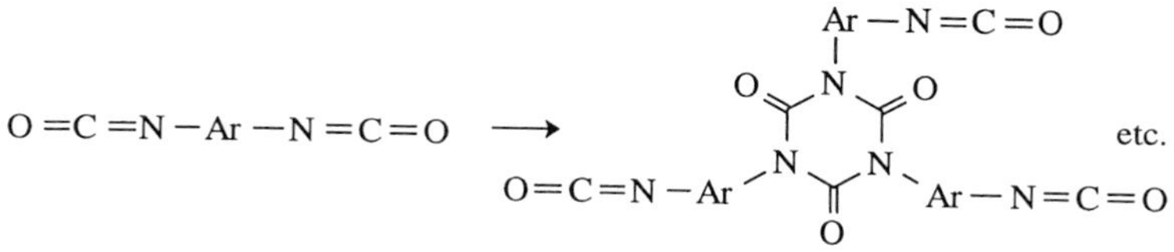

Industry uses mainly polymeric methylene-diphenylene 4,4'-diisocyanate (PMDI; see p. 497). The reaction is catalyzed by phenolates, tertiary amines, or tin compounds.

The reaction develops much heat which is used for the formation of rigid foams by evaporation of added chlorofluoroalkanes. However, PIR based on PMDI alone are very brittle. Industrial PIR foams contain therefore always flexibilizing urethane units which are generated by added polyols (see p. 493). Isocyanurate structures also serve as cross-linking agents in lacquers and adhesives.

## 10.6 Polyurethanes

### 10.6.1. Overview

Polyurethanes are polymers with the characteristic chain group –NH–CO–O–. "Ure-than(e)" was originally the trivial name for what is now called ethyl urethane (ethyl carb-amate), $H_2NCOOC_2H_5$, from the reaction of urea, $H_2N–CO–NH_2$, and ethanol $C_2H_5OH$. Since urea was first isolated from urine, ethyl carbamates were called "urethan(e)" (from (*urina* (L) + éthanol (F).

Polyurethanes were discovered in 1937 by Otto Bayer (IG Farbenindustrie). They are almost exclusively prepared by reaction of di- or triisocyanates with di- or polyhydroxy compounds. The supply of these isocyanates and their many possible reactions have since led to the use of isocyanates for many other types of polymers.

Isocyanate groups can react in many ways. Monoisocyanates polymerize to polyisocyanates (Section 10.5.3). Diisocyanates dimerize to polyuretdiones, trimerize to polyisocyanurates (Section 10.5.5), and form polycarbodiimides by release of $CO_2$ (Section 10.7.4) that react with other isocyanate groups, ~Z–NCO, to polyuretdioneimines.

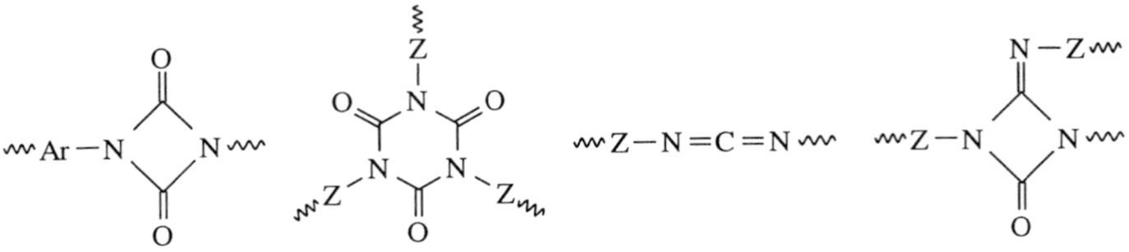

$$\sim Z-N=C=N\sim$$

polyuretdione | polyisocyanurate | polycarbodiimide | polyuretdioneimine

More important are additions of isocyanate groups to other chemical compounds, especially to hydroxyl groups with the formation of polyurethanes (this Section):

$$(10\text{-}51) \qquad \sim N=C=O + HO\sim \longrightarrow \sim NH-CO-O\sim$$

Isocyanate groups also add to ~SH, ~$NH_2$, ~NRH, ~$PH_2$, and ~SiH. Suitably substituted aromatic amines may lead to heterocyclic groups in polymers, for example, in poly-quinazolinedione)s (p. 477). Heterocyclic groups can also result from 1,3-dipolar reactions, for example,

$$(10\text{-}52) \qquad R-N=C=O + \overset{\oplus}{Q}-Z-\overset{\ominus}{Q'} \longrightarrow \text{ring: } R-N-C(=O)-Q'-Z-Q-(N)$$

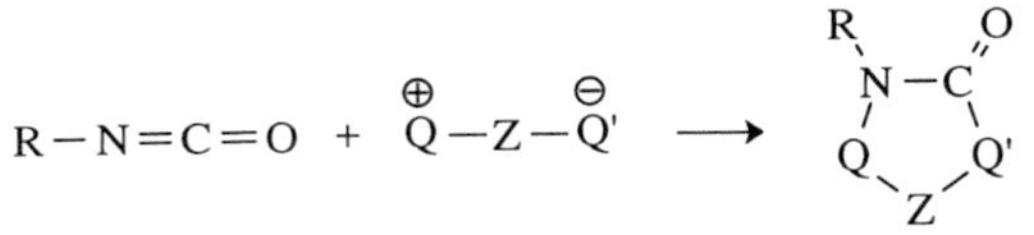

An example is the addition of isocyanate groups to epoxides which leads to oxazolidones. Correspondingly, reaction of diisocyanates and diglycidyl ethers delivers poly(2-oxazolidone)s (Section 10.8.6).

Some reactions of isocyanates proceed with release of carbon dioxide. Examples are reactions with carboxylic groups to amides (Table 10-14) and with acid anhydride groups to imides. Addition of isocyanate groups to oxamide esters produces polyparabanic acids and alcohols (Section 10.8.10).

## 10.6.2 Syntheses

The addition of diisocyanates to chemical compounds of the general type H–Z–H is the most important industrial reaction of isocyanates. Reactions with diols and diamines are polyadditions whereas those with dicarboxylic acids are polycondensations. Depending on the molar equivalency of functional groups or an excess of isocyanate groups, different chemical structures are produced (Table 10-14).

Catalysts for these reactions vary with the type of monomers and the type of processing. The standard catalyst for the reaction of aliphatic isocyanates with hydroxy compounds (including water) is dibutyltin dilaurate, which also catalyzes the hydrolysis of

Table 10-14 Isocyanate additions.

| Addition of –NCO to | Resulting structures from equivalency of groups | | surplus of –NCO groups | |
|---|---|---|---|---|
| HO~~~ | ~~~NH–CO–O~~~ | (urethane) | ~~~N–CO–O~~~<br>\|<br>CO–NH~~~ | (allophanate) |
| $H_2N$~~~ | ~~~NH–CO–NH~~~ | (urea) | ~~~N–CO–NH~~~<br>\|<br>CO–NH~~~ | (biuret) |
| HOOC~~~ | ~~~NH–CO~~~ + $CO_2$ | (amide) | ~~~N–CO~~~<br>\|<br>CO–NH~~~ | (acyl urea) |

ester groups. Zirconium catalysts are even more active. Catalysts for foaming include dibuyltin compounds, stannous octoate, triethylenediamine, and 1,4-diaza-bicyclo[2.2.2]-octane (DABCO). Blowing catalysts are tertiary polyamines such as pentamethyl diethylene triamine, tetramethyl ethylenediamine, 1,4-dimethylpiperazine, and 1,2-dimethylimidazole. They catalyze the reaction of isocyanate groups with water to primary amine groups + $CO_2$. Trimerization catalysts are usually carboxylic salts of strong bases such as potassium octoate and quaternary ammine carboxylates.

Proposed mechanisms include a general base catalysis

$$\text{(10-53)} \quad \text{ROH} + \text{:B} \rightleftharpoons \text{ROH}\cdots\text{B} \xrightarrow{+\text{R'NCO}} \text{R}-\text{NH}-\text{CO}-\text{O}-\text{R'} + \text{:B}$$

and an activation of isocyanate groups:

$$\text{(10-54)} \quad \text{R'NCO} + \text{:B} \rightleftharpoons \text{R}-\text{N}-\underset{\substack{|\\ \text{B}}}{\ddot{\text{C}}}-\text{O} \xrightarrow{+\text{ROH}} \text{R}-\text{NH}-\text{CO}-\text{O}-\text{R'} + \text{:B}$$

For a given isocyanate, reaction rates of reaction partners decrease in the order $-NH_2$ (to amide), > –OH (to urethane) > –NH–CO–NH< to biuret > –NH–COO– (to allophanate).

Addition of isocyanate groups to hydroxyl groups is an equilibrium reaction. The stability of the resulting urethanes is generally inversely proportional to their rate of formation: urethanes from aliphatic isocyanates are more stable than those from aromatic ones, and those from secondary alkohols more stable than those from primary ones. However, secondary alcohols may lead to olefin scission as a side reaction:

$$\text{(10-55)} \quad \sim\!\text{C}_6\text{H}_4-\text{NH}-\text{CO}-\text{O}-\text{CHR}-\text{CH}_2\text{R'} \longrightarrow \sim\!\text{C}_6\text{H}_4-\text{NH}_2 + \text{CHR}{=}\text{CHR'} + \text{CO}_2$$

The limited stability of many isocyanate-addition products is utilized for the synthesis of capped isocyanates which allows their physiologically harmless manipulation at room temperature (isocyanates themselves are very poisonous and very irritating). Capped isocyanates dissociate at elevated temperatures where the released isocyanate groups react with targeted compounds. Capping is by reaction of isocyanates with phenols, ethyl acetacetate, malic esters, acetoneoxime, or caprolactam. Uretdiones are capped isocyanates: no capping agent need be removed after application.

### 10.6.3 Monomers

**Diisocyanates**

**M**ethylene-**d**iphenylene 4,4'-**di**isocyanate (**MDI**; diphenylmethane 4,4'-diisocyanate, methylene(bisphenylene isocyanate) and **t**oluene-2,4-**di**isocyanate (**TDI**; *m*-tolylene diisocyanate) are standard diisocyanates which are obtained from diphenylmethane and toluene, respectively, by nitration to dinitro compounds and subsequent hydrogenation to the corresponding diamines.

tolylene-2,4-diisocyanate tolylene-2,6-diisocyanate

TDI 80:20 or TDI 65:35 (2,4:2,6)

diphenylmethane-4,4'-diisocyanate
(methanediphenyl-4,4'-diisocyanate, MDI)
4,4'-diisocyanatodiphenylmethane

Nitration of toluene delivers *o*-, *p*-, and *m*-nitrophenol which are separated by distillation and crystallization. Either pure *o*-nitrophenol or a mixture of *o*- and *p*-nitrophenol is then nitrated again to dinitrophenol. Hydrogenation to the amines and subsequent phosgenation delivers TDI which is offered in three types: pure toluene-2,4-diisocyanate and mixtures of 2,4- and 2,6-types in the ratios 80:20 and 65:35.

MDI is obtained by acid-catalyzed condensation of aniline and formaldehyde. The reaction leads to a mixture of 4,4'-diaminodiphenylmethane, its 2,2'- and 2,4'-isomers, and its homologues with higher functionalities. Separation of the mixture by distillation delivers MDI and its higher homologue, the "polymeric MDI" (PMDI; p. 497).

Two processes are employed for phosgenation. The so-called base phosgenation proceeds in two steps in order to suppress the formation of polyureas. In the first step, a solution or dispersion of the diamine, $H_2N–Z–NH_2$, is treated with a surplus of phosgene at 0–50°C. This *cold phosgenation* delivers a mixture of aromatic dicarbamic acid chloride and its hydrochloride (a deficit of phosgene would lead to polyureas and the diamine hydrochloride):

(10-56)

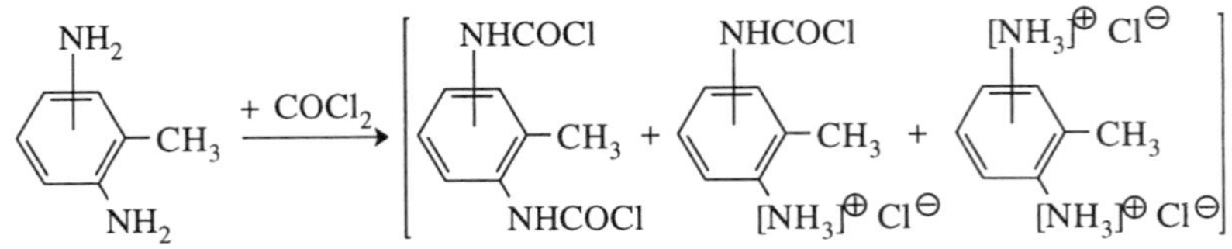

The resulting suspension of chlorides and hydrochlorides is then treated again with phosgene at 170-185°C (*hot phosgenation*) which leads to the isocyanate, for example:

(10-57) [2-(NHCOCl)-6-methylanilinium chloride structure: NHCOCl, $CH_3$, $[NH_3]^{\oplus}Cl^{\ominus}$] $\xrightarrow[-\,4\,HCl]{+\,COCl_2}$ [toluene diisocyanate structure: N=C=O, $CH_3$, N=C=O]

The second industrial process reacts hydrochlorides of diamino compounds with phosgene and separates the various isocyanates by distillation.

Since these processes waste chlorine as HCl and phosgene is both toxic and expensive, many phosgene-free syntheses have been developed. However, none of these syntheses has succeeded for TDI and MDI for technical, economic, and ecological reasons.

For example, reaction of $C_6H_5NO_2$, CO, and $C_2H_5OH$ with selenium catalysts leads to $C_6H_5{-}NH{-}CO{-}O{-}C_2H_5 + O_2$. Condensation of the urethane with $CH_2O$ splits off $H_2O$ and delivers $CH_2[C_6H_4NHCOOC_2H_5]_2$ which can be thermally cleaved to MDI. The urethane results also from the oxidative carbonylation of aniline, $C_6H_5NH_2$ with 1/2 $O_2$, CO, and $C_2H_5OH$ which produces $C_6H_5{-}NH{-}CO{-}O{-}C_2H_5 + H_2O$. MDI can also be obtained by 2 $(CH_3)_2CO + CH_2[(p\text{-}C_6H_4)NH_2]_2 \rightarrow CH_2[(p\text{-}C_6H_4)NCO]_2 + 4\ CH_4$.

However, phosgene-free syntheses are economical for two other diisocyanates, phenylene-1,4-diisocyanate (PHDI), $OCN(1,4\text{-}C_6H_4)NCO$, and cyclohexane-1,4-diisocyanate (CHDI), $OCN(1,4\text{-}C_6H_{10})NCO$. In these sequences of reactions, 1,4-dicarboxylic acids, for example, terephthalic acid, are converted to their acid amides which are then oxidized by bromine to the diacyl nitrene which rearranges to the diisocyanate:

(10-58) $H_2NOC{-}C_6H_4{-}CONH_2 \xrightarrow[-\,2\,HBr]{+\,Br_2} [NOC{-}C_6H_4{-}CON] \longrightarrow OCN{-}C_6H_4{-}NCO$

The conversion of acid amides to amines via isocyanates is called "Hofmann reaction" whereas "Hofmann degradation" is the thermal cleavage of quaternary ammonium hydroxides. In the original German, the "Hofmann reaction" is called "Hofmann-Abbau" (literally: "Hofmann degradation"!) whereas the "Hofmann degradation" is called "Hofmann-Elimination."

Industry also uses many other diisocyanates that are specific for applications as elastomers, foams, coatings, etc., and these are discussed in Section 10.6.4.

### Polyols

Diisocyanates are reacted with a plethora of compounds with two or more hydroxyl groups per molecule which are known industrially as **polyols**. The term "poly" may or may not refer to the number of OH groups per molecule and/or the repetition of groups, since "polyol" may refer to both low-molecular weight molecules such as glycerol and high-molecular weight ones such as poly(tetramethylene glycol) and also to functionalities between 2 and 8.

More than 90 % of polyols are **polyether-polyols** with two or more OH groups per molecule which are synthesized by polymerization of cyclic ethers, ethylene oxide, propylene oxide, and tetrahydrofuran with water, ethylene glycol, or propylene glycol. Polyether-polyols of functionality 2 comprise

| | | |
|---|---|---|
| PEG | poly(ethylene glycol) | $H[OCH_2CH_2]_nOH$<br>$H[OCH_2CH_2]_nOCH_2CH_2O[CH_2CH_2O]_nH$ |
| PPG | poly(propylene glycol) | $H[OCH(CH_3)CH_2]_nOH$<br>$H[OCH(CH_3)CH_2]_nOCH(CH_3)CH_2O[CH(CH_3)CH_2O]_mH$ |
| - | poly[(propylene oxide)-*co*-(ethylene oxide)] | $H[(OCH(CH_3)CH_2)_n$-*co*-$(OCH_2CH_2)_m]OH$ |
| PTMG | poly(tetramethylene glycol) = poly(tetrahydrofuran) | $H[OCH_2CH_2CH_2CH_2]_nOH$ |
| - | Poly[(tetrahydrofuran)-*co*-(ethylene oxide)] | $H[(OCH_2CH_2CH_2CH_2)_n$–*co*-$(OCH_2CH_2)_m]OH$ |

Some poly(propylene oxide)s are capped with ethylene glycol units, $–CH_2CH_2OH$. In addition, various **polyester-polyols**, $HO(C_4H_8OOC–C_4H_8–COOC_4H_8)_nOH$, as well as poly($\varepsilon$-caprolactone)s with hydroxyl endgroups are used for flexible ("soft") polyurethanes. "Rigid" polyurethanes utilize low-molecular weight diols such as 1,4-butanediol (BDOL) and hydroxyethylhydroquinone (HEHQ).

Polyols with higher functionalities are obtained by reaction of propylene oxide with core molecules of functionalities $f \geq 3$:

$f = 3$ glycerol or trimethylol propane,
$f = 4$ pentaerythritol, ethylene diamine, or phenolic resins,
$f = 5$ diethylene triamine,
$f = 6$ sorbitol,
$f = 8$ saccharose,
$f > 8$ starch.

**Modified polyether-polyols** are polymeric particles with grafted hydroxyl group-containing compounds. The three main types are

(1) so-called **polymer-polyols** with, e,g., acrylonitrile/styrene copolymers,
(2) **PHD-polyols** based on polyurea dispersions (D: "**P**oly**h**arnstoff-**D**ispersionen" = polyurea dispersions),
(3) **PIPA**-polyols based on polyurethane dispersions (**p**oly**i**socyanate-**p**oly**a**ddition).

Polyols with functionalities of 3–8 and molecular weights of 400–1200 are used for rigid foams, plastic materials, and coatings which have short segments between cross-linking sites. Polyols with functionalities of 2–3 and molecular weights of 1000–6500 serve for flexible foams and elastomers.

Polyurethanes with isocyanate endgroups can be reacted with chain extenders to poly–mers with higher molecular weights. Such extenders are low-molecular weight alcohols such as 1,4-butanediol or hydroxyethylhydroquinone on one hand and diamines such as methylene-bis(orthochloroaniline) (MOCA), methylenedianiline, and ethylene diamine on the other hand.

### 10.6.4 Applications

Most polymers (plastics, rubbers, fiber-forming polymers, etc.) are synthesized by chemical companies and sold as such to compounders (if plastics) and/or processors for conversion to articles, films, elastomers, fibers, etc. In contrast, polyurethanes (**PUR**) are directly produced from isocyanates and polyols by processors which appreciate the speed of the polymerization and the easy modification of chemical structures. An additional attraction is the good stability of polyurethanes against alkaline or acidic saponification. Polyurethanes are prepared in two steps (elastomers) or one step (foams, etc.).

These properties have led to a multitude of applications which range from hard plastics to soft elastomers. Worldwide consumption of polyurethanes is ca. $8.6 \cdot 10^6$ t/a (2000) wherefrom ca. 40 % are used as soft cellular plastics (mattreces, car seats, packaging, etc.), 30 % as rigid foams (insulation), and the rest for elastomers, thermoplastic PUR, lacquers, etc. About 1/3 of PUR production is used by the construction industry and about 1/6 each in the automobile industry and for refrigerators and cooling aggregates.

**Elastomers**

Polyurethanes were originally developed in Germany as possible substitutes for natural rubber. However, the first elastomeric materials from aliphatic diisocyanates and hydroxyl group-capped polyesters were so full of pores and holes that the testing department suggested their use as Swiss cheese imitations (for the connoisseur: Emmentaler). But necessity is the mother of invention and it took only four years after the basic PUR patent until the first flexible and rigid expanded polyurethanes appeared in 1941.

All modern PUR elastomers consist of two types of segments. Flexible segments are usually polyether-polyols and in part also polyester-polyols, while rigid segments consist mainly of MDI. Also used are hydrogenated MDI ($H_{12}$MDI) and *p*-phenylenediisocyanate (PPDI), 1,5-naphthalenediisocyanate (NDI), and 2,4'-methylenediphenylisocyanate (MDPI).

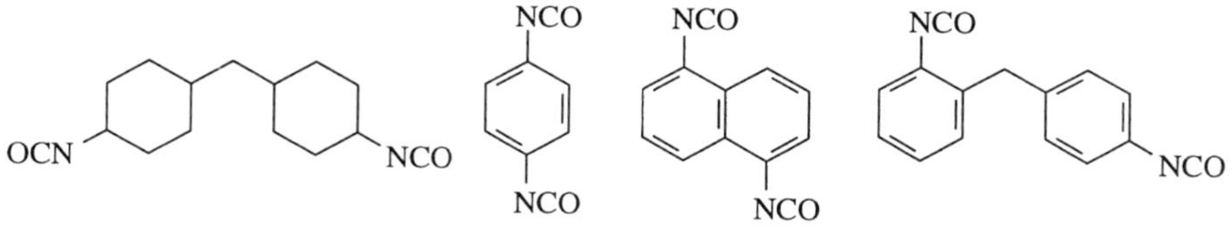

OCN–C₆H₁₀–CH₂–C₆H₁₀–NCO    $H_{12}$MDI

PPDI    NDI    MDPI

PUR elastomers are produced in two steps. In the first step, polyols are mixed with excess diisocyanate (ratio 2:3.5) which leads to "extended diisocyanates," i.e., copolymers with isocyanate endgroups. In the second step, these prepolymers are crosslinked by three means:

a. Less than stoichiometric amounts of aromatic diamines lead to chain extension by incorporation of urea groups which are subsequently crosslinked through excess isocyanate groups.

b. Less than stoichiometric amounts of glycols extend the chain via urethane groups. Reaction of urethane groups with excess isocyanate groups delivers allophanate groups (Table 10-14) which are cleaved at ca. 150°C: the polymers are crosslinked at room temperature but can be processed like thermoplastics at elevated temperatures.

3. Careful control of equivalency (with weak excess of isocyanate groups because of side reactions) delivers "linear" (i.e., weakly crosslinked) polymers that serve as **elastic fibers** (USA: Spandex; Europe: Elastan (formerly: Elasthan)). The U.S. Federal Trade Commission defines spandex as "a manufactured fiber in which the fiber-forming sub-

stance is a long-chain synthetic polymer composed of at least 85 % of a segmental polyurethane." In Lycra® fibers of DuPont, soft blocks (flexible segments) contain macroglycol units whereas hard blocks (rigid segments) contain the –NH–NH– group from hydrazine, $H_2N–NH_2$. Other elastic fibers contain different chain extenders (E ≠ 0).

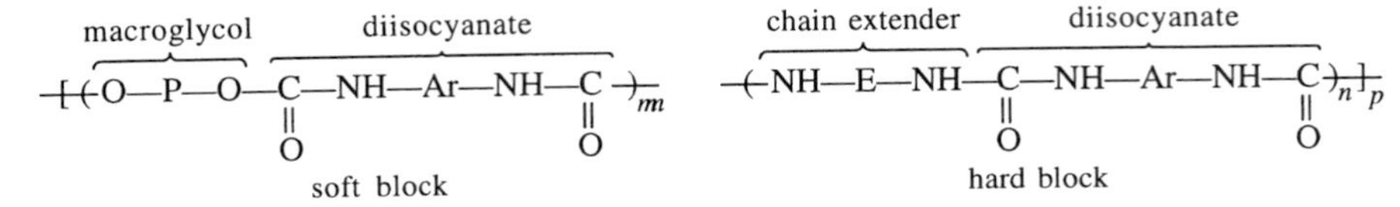

$$\left[\left(\underbrace{O{-}P{-}O}_{\text{macroglycol}}{-}\underbrace{\overset{O}{\underset{}{C}}{-}NH{-}Ar{-}NH{-}C(=O)}_{\text{diisocyanate}}\right)_m\left(\underbrace{NH{-}E{-}NH}_{\text{chain extender}}{-}\underbrace{C(=O){-}NH{-}Ar{-}NH{-}C(=O)}_{\text{diisocyanate}}\right)_n\right]_p$$

soft block hard block

An off-shoot of traditional PUR elastomer fabrication is PUR reaction injection molding (**RIM**) which uses a "liquid MDI" consisting of a mixture of MDI ($T_M$ = 38°C) with a small proportion of a trifunctional cyclic adduct. This adduct is an uretoneimine which results from a carbodiimine as intermediate:

(10-59)

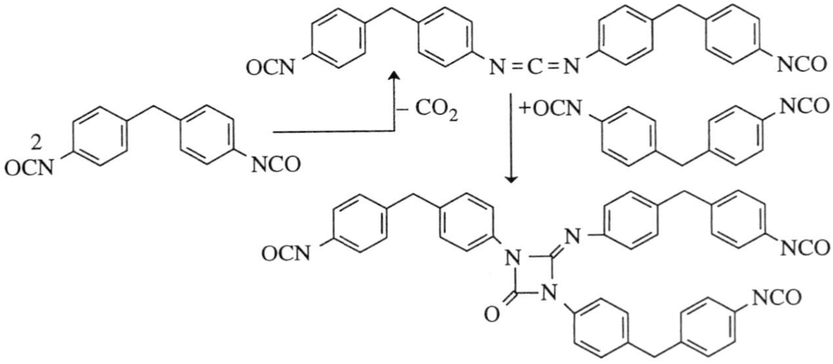

2 OCN–C₆H₄–CH₂–C₆H₄–NCO → (– $CO_2$) OCN–C₆H₄–CH₂–C₆H₄–N=C=N–C₆H₄–CH₂–C₆H₄–NCO

+OCN–C₆H₄–CH₂–C₆H₄–NCO ↓

uretoneimine (four-membered ring N–C(=N–Ar)–N–C(=O)) bearing three OCN/NCO-terminated MDI units

The carbodiimine is both a crosslinking agent and a stabilizer against degradation by acids because it reacts with carboxylic groups to acyl ureas which are stable against acids.

Conventional RIM-PUR is brittle and has a low impact strength. Both properties can be considerably improved, however, if the polymers are provided with "rigid" urea groups by addition of diamines as chain-extension agents. Even higher proportions of urea groups (up to 70 %) are obtained if aminated polyols are used ("polyetheramines"), i.e., polyethers with amino endgroups.

### Foams

In contrast to elastomers, foams are prepared in one step (see Volume IV). *Soft foams* from a mixture of toluylenediisocyanate (TDI) and polyether-polyols or polymer-polyols as soft segments are generated by $CO_2$ as blowing agent which is produced by added water through a sequence of reactions: ~NCO + $H_2O$ → ~NHCOOH → ~$NH_2$ + $CO_2$, The amino group reacts with isocyanate groups to urea groups: ~$NH_2$ + OCN~ → ~NHCONH~. Sometimes, coblowing agents are used ($CH_2Cl_2$, $CH_3CCl_3$, $(CH_3)_2CO$).

Polymer formation and foaming by $CO_2$ are simultaneous. The dosage of water must be carefully controlled because the foam would collapse if $CO_2$ were to escape prematurely (incomplete network formation) and would tear the already formed network if it were to develops too late. Such foams are produced at rates of up to 250 kg/min.

*Rigid foams* are prepared physically by expansion of compressed gases or evaporation of volatile liquids ($CFCl_3$, hydrochlorofluorocarbons, hydrofluorocarbons; see Volume IV). They are prepared from multifunctional isocyanates and low-molecular weight diols which generate networks with short segments between crosslinks.

The preferred isocyanate is "polymeric MDI" which is a mixture of di-, tri-, and multifunctional isocyanates which contains also some urea and biuret groups:

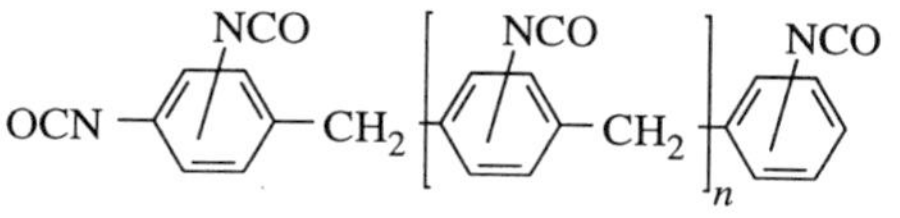

OCN–$C_6H_3$(NCO)–$CH_2$–[$C_6H_3$(NCO)–$CH_2$]$_n$–$C_6H_4$(NCO)

polymeric MDI (PMDI)

| | |
|---|---|
| $n = 0$ | (30-70) % |
| $n = 1$ | (15-40) % |
| $n > 1$ | (15-30) % |

## Coatings and Paints

Polyurethanes are very well suited for coatings and paints because of their chemical stabilities, fast hardening, good flexibilities, and low abrasion. Because of their lightfastness, aliphatic and cycloaliphatic diisocyanates are used: $H_{12}MDI$ (see p. 495), 1,6-hexamethylenediisocyanate (HDI; used as biuret), 1,4-cyclohexanediisocyanate (CDI), isophorondiisocyanate (IPDI), and *m*-xylylenediisocyanate (XDI).

HDI: $OCN-(CH_2)_6-NCO$

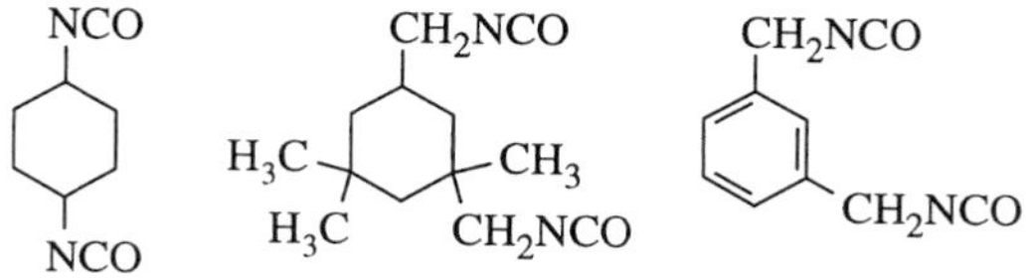

CDI: cyclohexane ring with NCO (1,4)

IPDI: cyclohexane ring with $CH_2NCO$, $H_3C$, $H_3C$, $CH_3$, $CH_2NCO$

XDI: benzene ring with $CH_2NCO$, $CH_2NCO$

Coatings can also be prepared from aqueous PUR dispersions. All known processes use isocyanate-prepolymers from TDI, HDI, and/or IPDI and liquid polyols, sometimes also glycols. Prepolymers are generated by reaction of isocyanates and glycols in the presence of 5–70 % organic solvents such as acetone, *N*-methylpyrrolidone, toluene, etc.

The prepolymers are then dispersed in water together with capped amines; examples are ketazines, $(CH_3)_2C{=}N{-}N{=}C(CH_3)_2$, aldazines, R–CH=N–N=CH–R', and hydrazones, $(CH_3)_2C{=}N{-}NH_2$. The capped amines are hydrolyzed by water to hydrazine, $H_2N{-}NH_2$, and acetone, $(CH_3)_2CO$, and aldehydes, R'CHO and R"CHO, respectively. The hydrazine then reacts with the isocyanate prepolymers by chain extension.

PUR lacquers are produced by reaction of triisocyanates with three- or multi-functional polyols in suitable solvents. The exothermic reaction proceeds at room temperature which means that the shelf-life depends on the size of the container (heat build-up due to bad heat transfer). Storage-proof laquers are obtained from capped isocyanates, for example, baking varnishes from phenol-capped isocyanates.

## Adhesives

Polyurethane are good adhesives. Their action results from the combination of several processes: removal of water films deposited on surfaces by generating urea structures (p. 496, 7th line from below), formation of hydrogen bridges and/or chemical bonds between OH groups on surfaces (silanol groups of glass, hydroxyl groups of celluloses, hydroxides of metals, etc.) and components of PUR adhsives, etc.

**Reprography**

Diisocyanates are intermediates in a type of reprography, the reproduction of printed materials such as graphics. Stable aromatic carboxylic acid azide esters decompose on exposure to light and form isocyanates that react with OH groups of poly(vinyl alcohol) to crosslinked polyurethanes (Eq.(10-60)). The printing plate is produced by dissolving uncrosslinked poly(vinyl alcohol).

(10-60) $N_3OC-C_6H_4-CON_3 \xrightarrow[-2\,N_2]{h\nu} OCN-C_6H_4-NCO$

$\xrightarrow{+\,2\,\sim CH_2-CH(OH)\sim} \sim CH_2-CH(\sim)-O-CO-NH-C_6H_4-NH-CO-O-CH(\sim)-CH_2\sim$

## 10.7 Polyimides

Imide groups, –CO–NH–CO–, result formally by replacing OH groups of two carboxylic groups by a common NH group. All industrially used polyimides (**PI**) contain imide groups as parts of rings in chains. Imide groups can be formed *in situ* during polymer formation (Section 10.7.1) or be preformed in monomers (Section 10.7.2). Each type of synthesis can lead to either crosslinked or uncrosslinked polymers.

Acid components of preformed imides are predominantly based on maleic anhydride, whereas those of the in-situ-type use mainly pyromellitic acid (= 1,2,4,5-benzenetetracarboxylic acid) and trimellitic acid (= 1,2,4-benzenetricarboxylic acid) and their anhydrides or acid chlorides, respectively.

Pyromellitic acid and trimellitic acid obtained their curious names (G: *mel* = honey, *melitta* = honey bee) as derivatives of mellitic acid (benzenehexacarboxylic acid). Mellitic acid was first found in lignite deposits as a component of a honey-colored mineral, mellit, which is the aluminum salt of mellitic acid, $Al_2C_{12}O_{12}\cdot 18\ H_2O$. Pyromellitic acid was named because mellitic acid loses on heating two $CO_2$ molecules. The name "trimellitic acid" was probably formed by analogy.

Of the three possible tetracarboxylic acids of benzene, only pyromellitic acid (1,2,4,5) is important for polymers while mellophanic acid (1,2,3,4) and prenithic acid (1,2,3,5) are not. Similarly, only trimellitic acid (1,2,4) plays a role while hemimellitic acid (1,2,3) and trimesic acid (1,3,5) do not.

### 10.7.1 *In-situ* Formation

**Polyimides (PI)**

Poly(pyromellitimide) results from the 2-step polycondensation of pyromellitic dianhydride (I) with 4,4'-diaminodiphenylether (4,4'-oxydianiline) (II). In the first step, a polyamic acid (III) is formed in polar solvents such as *N,N*-dimethylformamide, *N,N*-dimethylacetamide, tetramethylurea, or dimethylsulfoxide under a nitrogen blanket with exclusion of humidity.

The reaction proceeds mainly in the para position, Eq.(10-61), and somewhat in meta. Crosslinking is avoided by restricting solids contents to 10–15 % and group conversions to 50%. The second step consists of heating the prepolymer (III) to 300°C, which leads to ring closure, elimination of water, Eq.(10-62), and crosslinking.

(10-61)

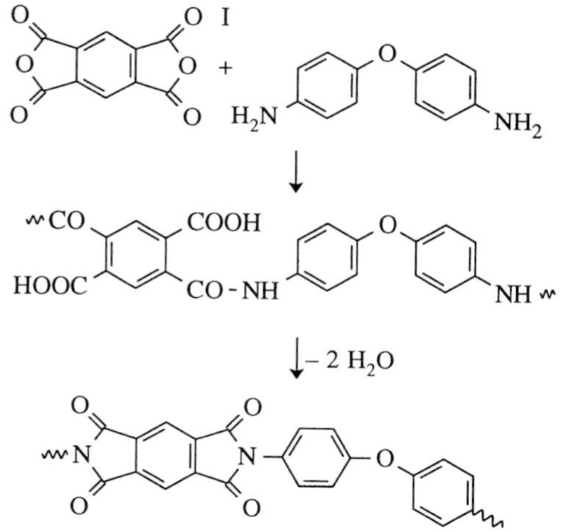

I + $H_2N$–$C_6H_4$–O–$C_6H_4$–$NH_2$ II

↓

~CO–$C_6H_2$(COOH)(HOOC)–CO-NH–$C_6H_4$–O–$C_6H_4$–NH~ III

↓ $- 2\ H_2O$

IV

PMDA-ODA

The resulting polymers (IV) have excellent heat stabilities with a glass temperature of 410°C and a heat distortion temperature of 265°C (Table 10-15). Because of crosslinking, shaping is done simultaneously with polymerization, e.g., during coating of wires or impregnation of films. Coating is from solutions of polyamic acids (III) in mixtures of *N*-methylpyrrolidone with either *N,N*-dimethylformamide or xylene. A disadvantage is the limited shelf-life of prepolymers (III) which are prone to hydrolysis. Since water is difficult to remove by heating, films are soaked with acceptors for water such as acetic anhydride and pyridine.

The same principle is used for the synthesis of many other polyimides from other acid dianhydrides (V), trimellitic acid anhydride (VI), or other diamines (VII) including cycloaliphatic moieties instead of aromatic ones.

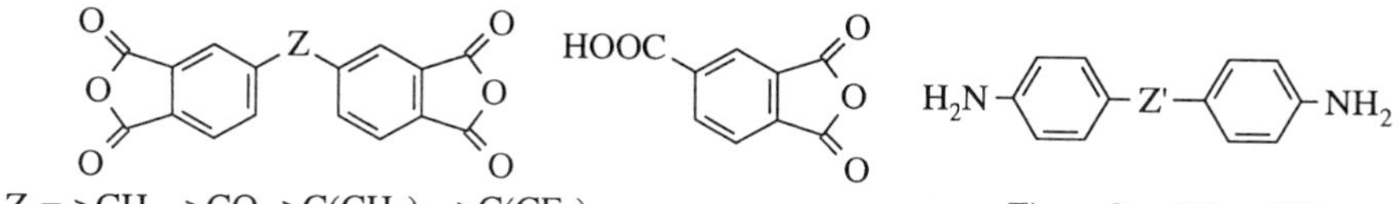

Z = >$CH_2$, >CO, >$C(CH_3)_2$, >$C(CF_3)_2$ — V

HOOC — VI

$H_2N$–$C_6H_4$–Z'–$C_6H_4$–$NH_2$; Z' = >O, >CO, >$CH_2$ — VII

Amines are strong bases and always difficult to purify. They are therefore often replaced by less basic and more easy to handle capped amine derivatives such as carbamic esters, ureas, aldimines, and ketimines.

Isocyanates can also be considered capped amines. Their use avoids the cyclization III → IV in Eq.(10-61). The reaction between acid anhydrides and isocyanate groups is catalyzed by strong bases (alkoxides, tertiary amines, etc.) and proceeds in aprotic solvents. In the first step, anhydride groups, ~CO–O–CO~, react with two isocyanate groups, OCN–$C_6H_4$~, which requires a molecule of water for the ring-opening of anhydride groups and intermediary formation of one stable amide group, ~CO–NH–$C_6H_4$~, one unstable *N*-carboxyanhydride groups, ~CO–O–CO–NH–$C_6H_4$~, and release of $CO_2$ (Eq.(10-62)). The ring is closed to the imide group by releasing one molecule of HOOC–NH–$C_6H_4$–Z–$C_6H_4$–NCO, Eq.(10-62):

Table 10-15 Thermal and mechanical properties of some unfilled polyimides at 23°C unless otherwise noted. There are very many grades of many types of polyimides from very many companies. A great number of polyimides is offered as glass fiber or carbon fiber-reinforced grades.

Repeating units of polyimides in Table 10-15:

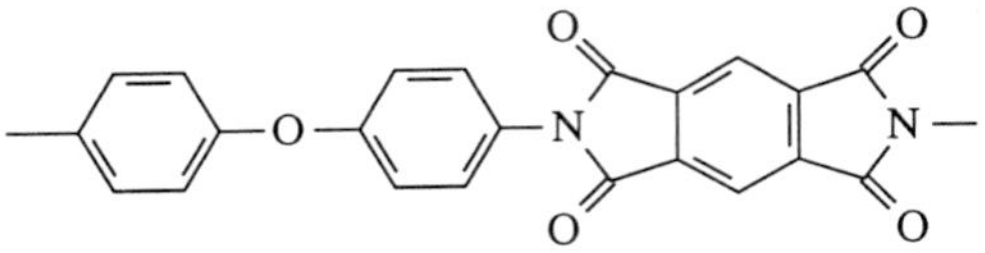

Kapton® SP1

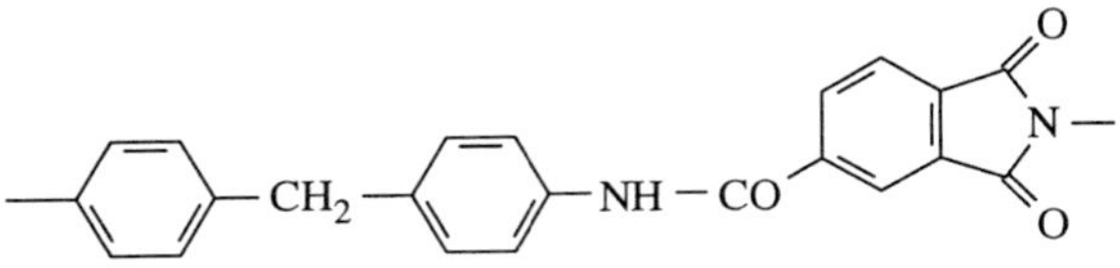

Torlon® 4000T

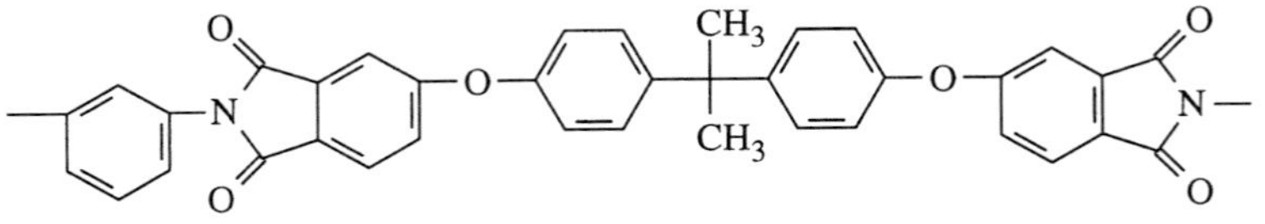

Ultem® 1000

| Property | | Physical unit | PMDA-ODA Kapton® | PAI Torlon® 2000 | PEI (DIN) Ultem® 1000 |
|---|---|---|---|---|---|
| Density | | $g/cm^3$ | 1.43 | 1.41 | 1.27 |
| Heat distortion temperature (1.82 MPa) | | °C | 260 | 282 | 200 |
| Vicat temperature B | | °C | | | 219 |
| Glass temperature (DSC) | | °C | 410 | | |
| Linear thermal expansion coefficient | | $10^{-5}$ $K^{-1}$ | 5.4 | 3.6 | 5.6 |
| Thermal conductivity (20°C) | | $W\ m^{-1}\ K^{-1}$ | 0.35 | 0.24 | 0.22 |
| Tensile modulus | | MPa | | | 3000 |
| Flexural modulus | 23°C | MPa | 3100 | 4900 | 3300 |
| | 260°C | MPa | 1700 | 3000 | |
| Tensile strength | 23°C | MPa | 86 | 93 | 105 |
| | 260°C | MPa | 41 | 62 | |
| Flexural strength | 23°C | MPa | 117 | 164 | 150 |
| | 260°C | MPa | 62 | 100 | |
| Compressive strength | 23°C | MPa | | 234 | |
| Elongation at yield | | % | | | 7 |
| Elongation at fracture | | % | 7.5 | 2.5 | 60 |
| Impact strength (Izod) | | J/m | 750 | 1088 | 1030 |
| Notched impact strength (Izod) | | J/m | 43 | 2.8 | 50 |
| Rockwell hardness | | - | 97M | 98E | 109M |
| Water absorption | 24 h | % | 0.32 | | 0.25 |
| Relative permittivity | 1 kHz | 1 | 3.6 | 3.7 | 3.15 |
| Dissipation factor, | 100 Hz | 1 | 0.0018 | 0.009 | 0.0015 |
| | 1 MHz | 1 | 0.0034 | | |
| Resistivity, dry | | Ω cm | $10^{14}$ | $3 \cdot 10^{13}$ | $1 \cdot 10^{15}$ |
| Electrical resistance | | Ω | $10^{15}$ | $1 \cdot 10^{17}$ | |
| Oxygen index (LOI) | | % | 53 | 47 | |

(10-62)

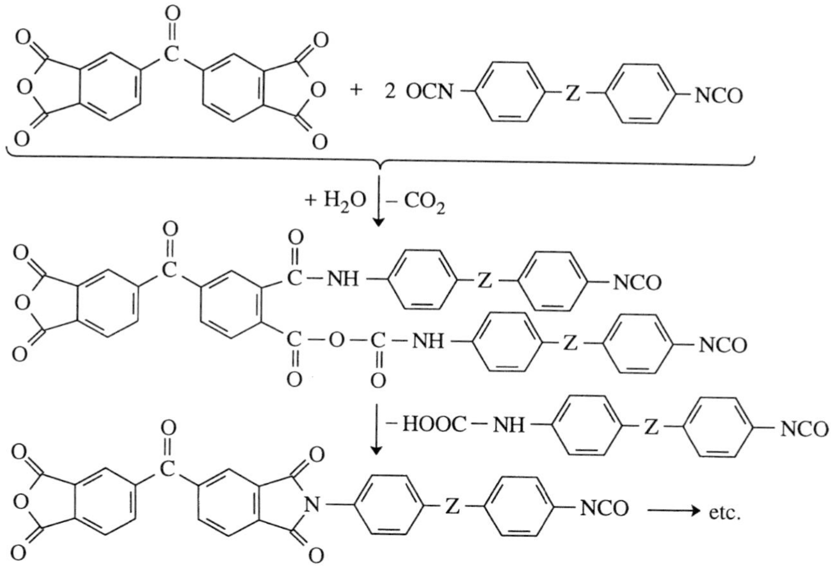

The iminocarboxylic acid loses carbon dioxide and becomes an amine, which also participates in the polymer formation:

(10-63)

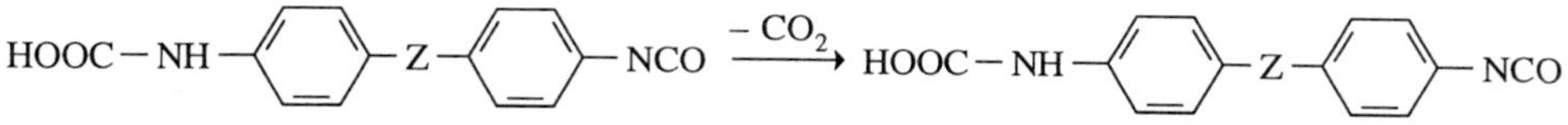

The resulting polyimides are soluble and thermoplastic. Their solutions can be either dry spun or wet spun directly to fibers.

The *in-situ* formation requires two functions, isocyanate and anhydride. These two functions can be united in one molecule which allows in principle the synthesis not only of AA/BB polyimides but also of AB ones, for example:

(10-64)

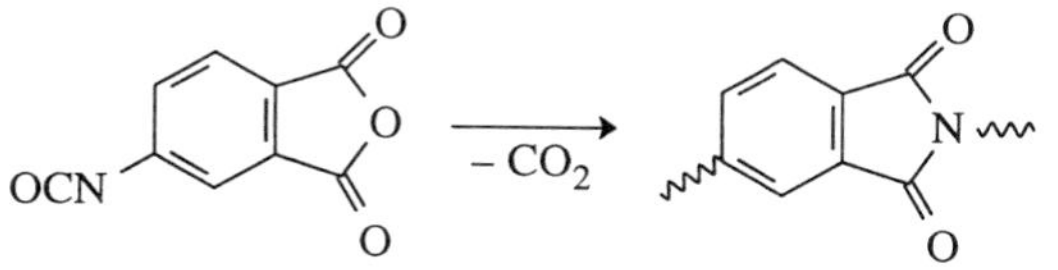

**Polyesterimides**

Regular polyimides (PI) contain only imide and arylene rings plus bridging groups between aromatic rings such as O, CO, $CH_2$, $C(CH_3)_2$, and $C(CF_3)_2$. These polyimides do not deform markedly at elevated temperatures but are difficult to prepare and process. Polyesterimides (**PEI**) and polyester amides (**PAI**) do not have these disadvantages but are less heat stable because of the presence of ester and amide units, respectively.

Polyesterimides are synthesized from diamines and ester dianhydrides. For wire coatings, ester anhydrides are obtained from the reaction of diphenylesters (I) with anhydrides such as trimellitic acid anhydride (II), pyromellitic dianhydride, or benzophenone dianhydride. Alternatively, anhydrides can also be reacted with diesters of ethylene glycol or glycerol with terephthalic acid (III), itaconic acid, phenylindane dicarboxylic acid, or benzophenone dicarboxylic acid.

$$C_6H_5-\underset{\underset{O}{\|}}{C}-O-Ar-O-\underset{\underset{O}{\|}}{C}-C_6H_5 \qquad I$$

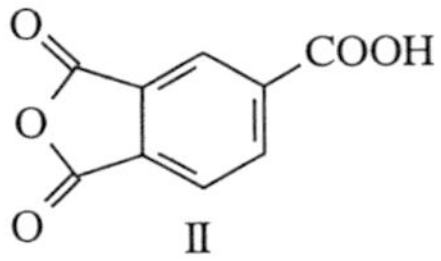

II (trimellitic anhydride, with COOH group)

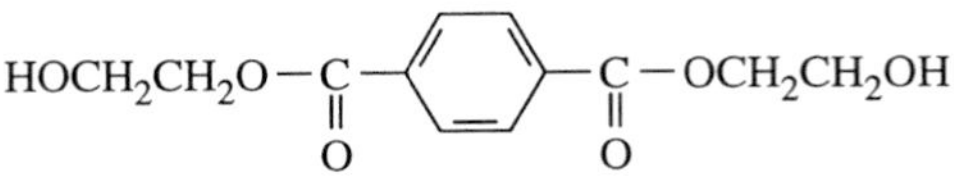

$$HOCH_2CH_2O-\underset{\underset{O}{\|}}{C}-C_6H_4-\underset{\underset{O}{\|}}{C}-OCH_2CH_2OH \qquad III$$

The resulting dianhydrides of diesters, for example, from II + III,

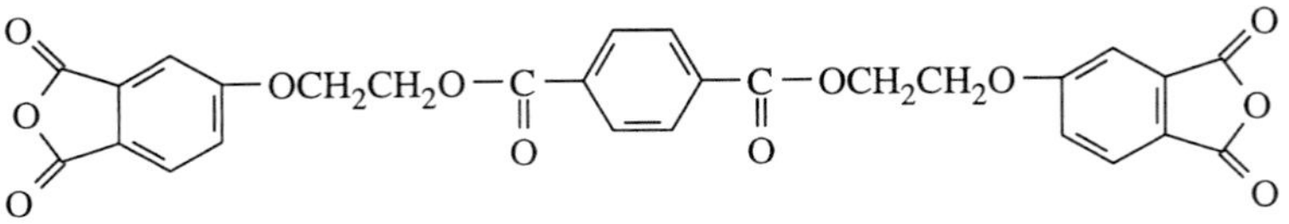

$$\text{(anhydride)}-OCH_2CH_2O-\underset{\underset{O}{\|}}{C}-C_6H_4-\underset{\underset{O}{\|}}{C}-OCH_2CH_2O-\text{(anhydride)} \qquad II + III$$

are then subjected to conventional imidization reactions with, e.g., 4,4'-diaminodiphenylmethane, *p*-phenylenediamine, *p*-aminobenzoic acid, aminoethanol, or aminoacetic acid to polyesterimides. Again, use of trifunctional or tetrafunctional compounds leads to crosslinking on heating.

**Polyamideimides**

Polyamideimides contain both imide and amide groups in their chains. They are used as heat-stable molding materials, wire and baking enamels, films, and fibers.

Polyamidimides are synthesized by either of two methods:

1. Molding compounds are obtained from phosgenation of, for example, trimellitic anhydride, and subsequent reaction of acid chlorides with diamines such as 4,4'-diaminodiphenylmethane in, for example, *N*-methylpyrrolidone as solvent at room temperature. The resulting polyamic acid is then cyclized to the polyamideimide (see Eq.(10-61)).

2. For baking varnishes and electrical insulation, one prefers the reaction of trimellitic anhydride and 4,4'-diisocyanatodiphenylmethane (MDI) (similar to Eq.(10-62)) in aprotic solvents such as *N*-methylpyrrolidone. The solutions can also be used directly for fiber spinning.

**Polyetherimides**

Polyetherimides are prepared from bis(etherphthalic acid)s or their anhydrides and aromatic di- and multi-amines with (for examples, see Table 10-15). An example is an unconventional reaction of *N*-phenyl-4-nitrophthalimide with the sodium salt of bisphenol A, where the nucleophilic attack of the bisphenolate anion removes the $NO_2^{\ominus}$ group as $NaNO_2$ (Eq.(10-65)). The subsequent high-temperature hydrolysis with aqueous NaOH at 160–175°C converts phthalimide groups to COOH groups, while aniline is removed by steam distillation. Polycondensation of either the tetracarboxylic acid or its dianhydride with aromatic amines delivers polyetheramines:

(10-65)

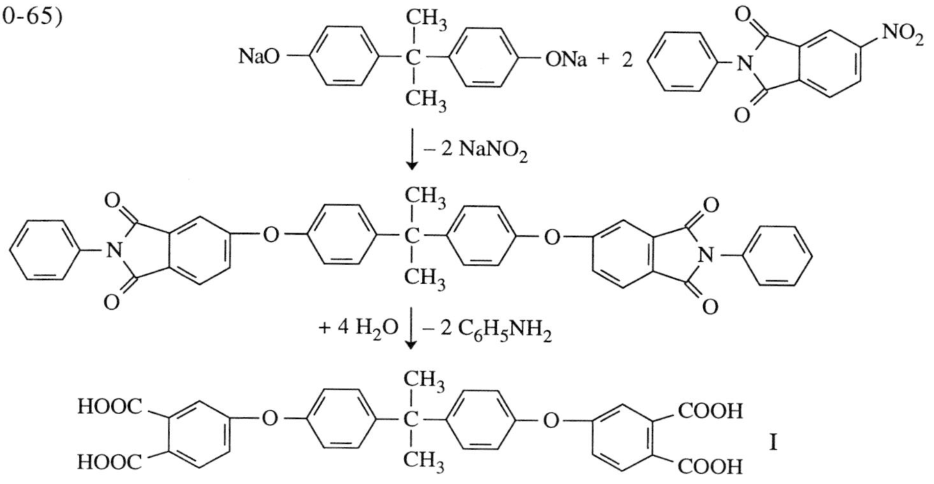

$$\downarrow -2\ NaNO_2$$

$$+4\ H_2O \downarrow -2\ C_6H_5NH_2$$

I

The resulting tetracarboxylic acid is condensed directly or as dianhydride with 4,4'-diaminodiphenylmethane, *m*-phenylenediamine, or 4,4'-diaminodiphenylether in solvents such as *N*-methylpyrrolidone, *o*-dichlorobenzene, or toluene. Depending on the reaction conditions, solutions of more or less imidated polyetherimides are obtained that are used for coating or film casting. Thermoplastic polyetherimides are obtained by melt polycondensation of I with diamines (see Ultem 1000 in Table 10-16).

**PMR Polyimides**

PMR is short for *in-situ* **p**olymerization of **m**onomer **r**eactants. The process was developed by NASA and is characterized by the use of an alcoholic solution of a monomer mix containing all reaction partners (acids + amines + regulators (= chain stoppers)). The components of such solutions do not react at room temperature; the solutions have thus excellent shelf-lifes. Table 10-16 lists some acids, amines, and chain regulators that are used for PMR polyimide compositions.

Fibers intended for reinforcement are impregnated with the solution. After removal of alcoholic solvents (methanol or ethanol), the temperature is slowly raised to more than 300°C. Melting of monomers starts below 100°C and imidization at ca. 140°C with simultaneous formation of chains via polyamic acid. At ca. 200°C, synthesis of polyamic acid is complete. Prepolymers melt between 170°C and 250°C while crosslinking starts at higher temperatures *via* simultaneously formed endgroups.

Table 10-16 Monomes for PMR polyimides. For chemical structures of BTDE, HFDEm MDA, PPDA, Jeffamine, and NE, see Fig. 10-12.

| Designation | Acid | Amine | Regulator | Molar ratio |
|---|---|---|---|---|
| PMR 15 | BTDE | MDA | NE | $n$ : $n$+1 : 2 |
| PMR 15 II | HFDE | PPDA | NE | 4 : 5 : 2 |
| LARC 160 | BTDE | Jeffamine AP-22 | NE | 0.335 : 0.610 : 0.539 |

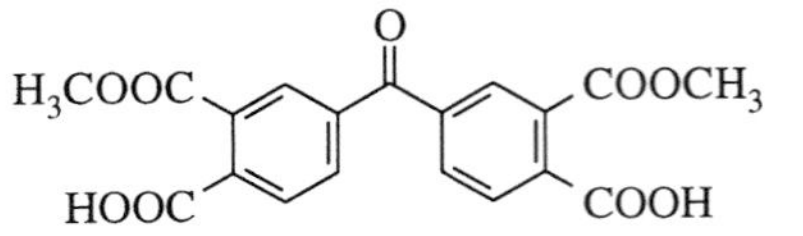

$H_3COOC$ / $HOOC$ … $COOCH_3$ / $COOH$

3,3',4,4'-benzophenonetetracarboxylic acid dimethylester (BTDE)

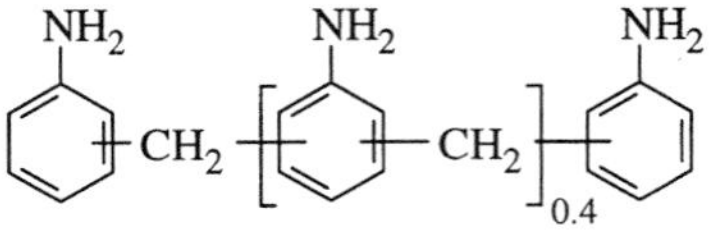

Jeffamine® AP-22

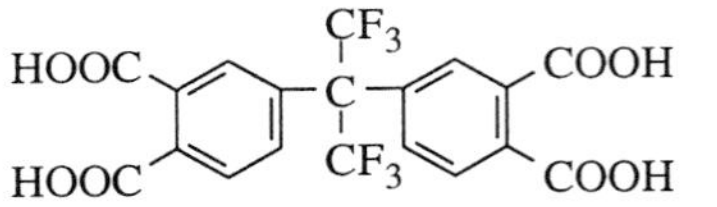

4,4'-(hexafluoropropylidene)bis(phthalic acid) (HFDE)

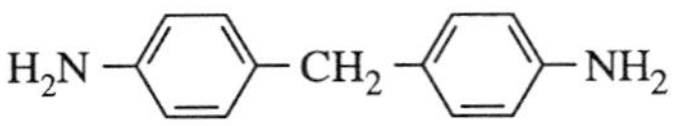

4,4'-di(aminophenyl)methane (MDA)

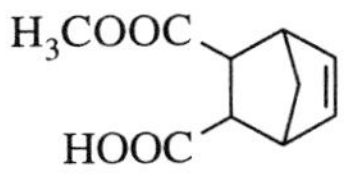

4-*endo*-methylenetetrahydrophthalic acid monomethylester (Nadic ester, NE)

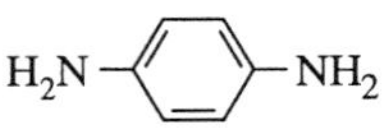

*p*-phenylenediamine (PPDA)

Fig. 10-12 Acids/esters and amines used in the PMR process (see also Table 10-16)

The process avoids the preparation and subsequent isolation of prepolymers. Prepregs (= preimpregnated fibers and fiber mats) are easy to produce from these solutions because of their excellent shelf-life; conventional prepregs from oligomers have limited shelf-lifes. Residual solvent is easy to remove since the high-boling aprotic solvents of other polyimide preparations are absent. The chance for residual solvent remaining in the resin is thus minimized.

## 10.7.2 Preformed Imide Groups

**Polybismaleimides (BMI)**

*In-situ* synthesis of polyimides produces simultaneously imide groups and polymer molecules. Crosslinking cannot be avoided; polymerization and shaping must take place simultaneously. Furthermore, byproducts and residual solvents are difficult to remove.

These problems led to the development of monomers with preformed imide groups. Since polymerization of such monomers does not lead to crosslinking, resulting polymers are called **thermoplastic polyimides**.

Suitable monomers are bismaleimides (I) from maleic anhydride and diamines such as 4,4'-di(aminophenylene)methane that are obtained in the presence of acetic anhydride and catalytic amounts of nickel acetate and triethylamine:

(10-66) $$2\ \text{(maleic anhydride)} + H_2N-Z-NH_2 \xrightarrow[-2\,H_2O]{} \text{(maleimide)}N-Z-N\text{(maleimide)} \quad \text{I}$$

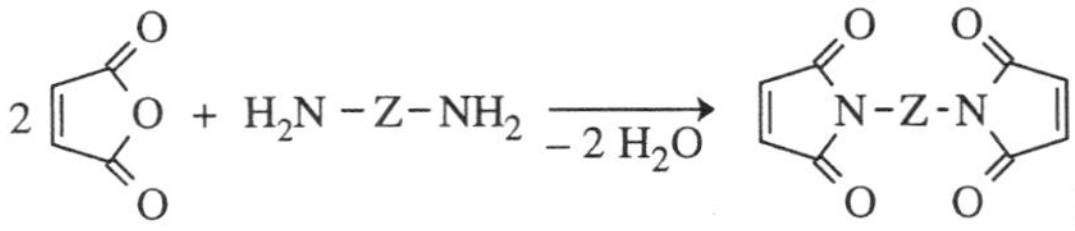

Bismaleimides are cured (crosslinked) free-radically by peroxides or azo compounds (see Volumes I and IV) or ionically by 2-methylimidazole (II), triphenylphosphine (III),

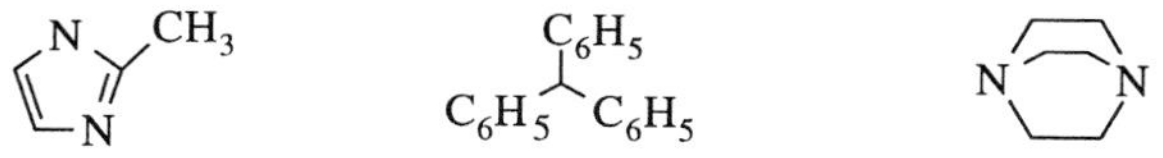

2-methylimidazol (II) triphenylphosphine (III) DABCO® (IV)

or 1,4-diazabicyclo[2.2.2]octane) (= DABCO, IV), ("thermal hardening"). The cross-linked polymers are brittle but become tougher by blending with rubbers or thermoplastics. Alternatively, "chain extenders" can be introduced, for example by a Michael addition of aromatic diamines to the carbon–carbon double bonds of bismaleimides:

(10-67)

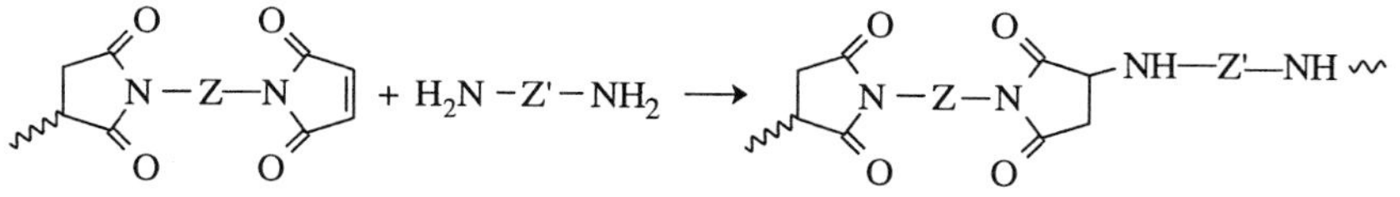

Chain extensions are also possible by Diels–Alder reactions, for example, with benzocyclobutene derivatives or divinyl benzenes (an additional ene synthesis).

Instead of AA/BB polymerizations, one can also use prepolyimides with polymerizable endgroups, for example, prepolymeric polyimides with norbornene endgroups that polymerize through ring-opening:

(10-68)

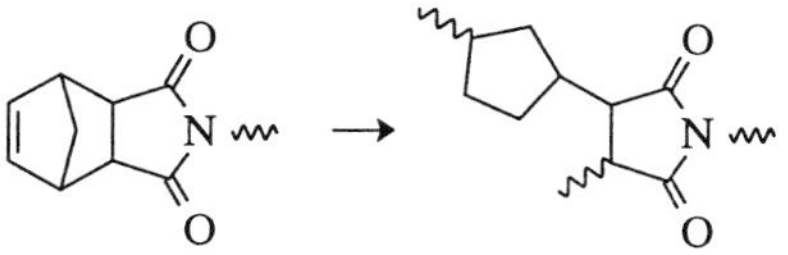

Also used are monomers and prepolymers with acetylene endgroups, but the structure of the resulting polymers does not seem to be known with certainty.

## BT Resins

Monomers for BT resins are bismaleimides such as I and dicyanates, for example, the dicyanate II that results from bisphenol A and cyanuric chloride (= 2,4,6-trichloro-1,3,5-triazine):

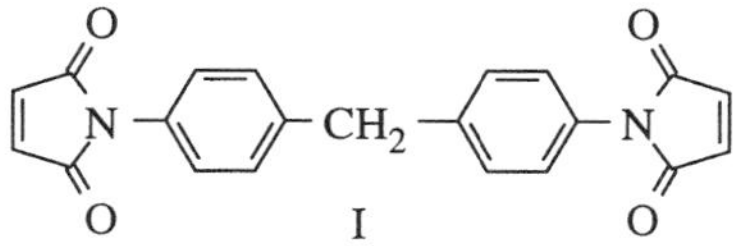

I 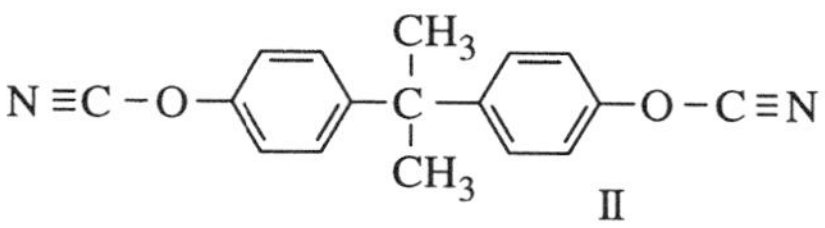 II

The monomer mixture polymerizes on heating to crosslinked polymers (Fig. 10-13). Crosslinks are generated by many reactions: (1) polymerization of C=C double bonds of maleimides; (2) trimerization of cyanate groups to *s*-triazines; (3) addition of cyanate group to C=C double bonds; (4) formation of triazine-imidazole structures by reaction of two cyanate groups with one C=C double bond; or (5) reaction of this structure with a C=C double bond, etc. The resins can further be modified by reaction with very many other monomers, oligomers, and polymers such as silicones, alkyds, polyesters, polyurethanes, phenolics, melamine-formaldehyde, diallyl phthalates, poly(vinyl butyral), poly(butadiene), etc.

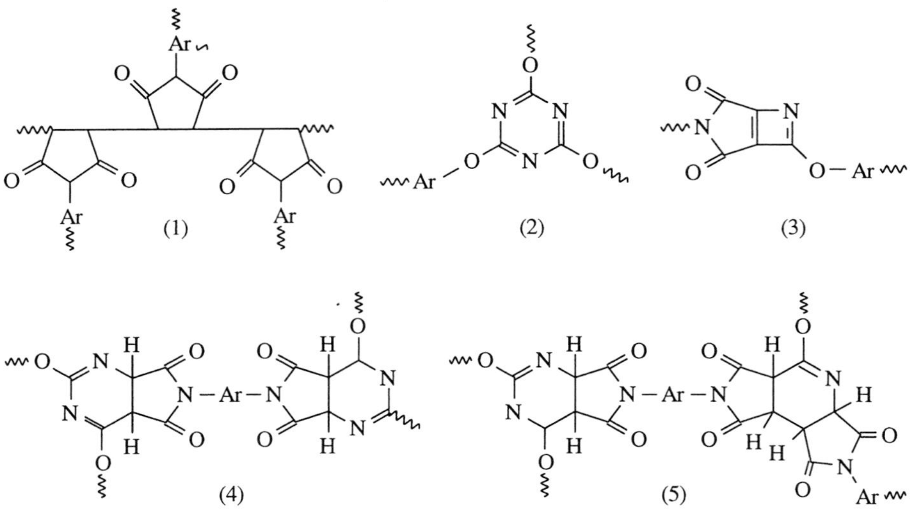

Fig. 10-13 Crosslinking sites of BT resins.

### 10.7.3 Other Polyimides

Polyimides also result from isomerizing polymerizations of lactams with amide groups that interact intramolecularly with carboxylic groups. The polymerization is irreversible and therefore unlike regular ring-opening polymerizations of lactams, which are equilibrium reactions. Monomer molecules and monomeric units differ in their constitutions, as shown for the polymerization of $\beta$-carboxymethylcaprolactam to poly[(2,6-dioxo-1,4-piperidinyl)trimethylene)]:

(10-69)

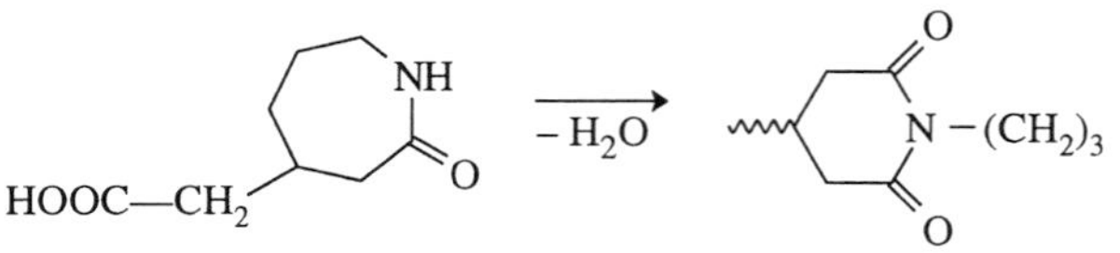

HOOC—$CH_2$ ... NH ... O $\xrightarrow{-H_2O}$ ... N$-(CH_2)_3$

### 10.7.4 Polycarbodiimides

Polyeliminations convert diisocyanates to polycarbodiimides if dihydrophospholeoxides are used as catalysts:

(10-70) $O{=}C{=}N{-}Z{-}N{=}C{=}O \longrightarrow \sim N{=}C{=}N{-}Z\sim + CO_2$

Five-membered, saturated, non-aromatic heterocyclics with one phosphorus atom per ring are called phospholanes. Dihydrophospholes (formerly: "phospholenes") have one C=C double bond while phospholes have two double bonds.

Completely reacted polycarbodiimides are light-weight, open-cell rigid foams which can be compression-molded into parts. Incompletely reacted solutions can be cast to clear films that can be cured by heating.

### 10.7.5 Polyuretdiones

Uretdiones can be viewed as modified imides. They result from dimerizations of aromatic diisocyanates, OCN–Ar–NCO, triisocyanates, etc., with acids or bases. Addition of isocyanate groups delivers crosslinked products which are utilized in the modification of polyurethanes:

(10-71)

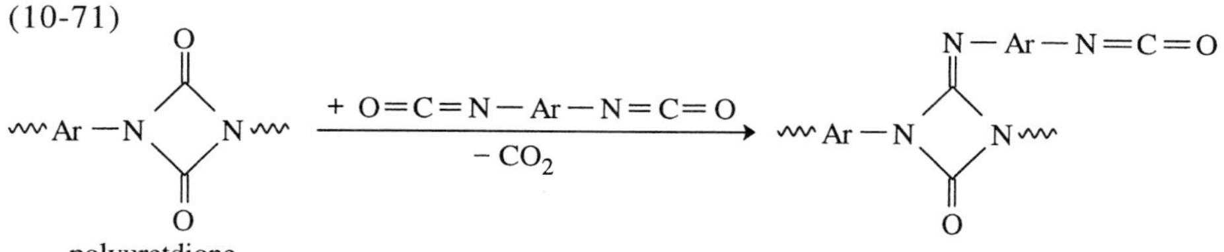

$$\text{polyuretdione} \xrightarrow[-\,CO_2]{+\,O{=}C{=}N{-}Ar{-}N{=}C{=}O} \ldots$$

polyuretdione

## 10.8 Polyazoles

### 10.8.1 Overview

Azoles are defined by IUPAC as unsaturated 5-membered rings with 1–5 ring nitrogen atoms of which at least one participates in a double bond. Azole itself has one nitrogen per ring and thus one nitrogen–carbon double bond, –N=C–. Di-, tri-, and tetrazoles contain 2, 3, and 4 ring nitrogens. Positions of imine groups, –NH–, are characterized by prefixed numbers and "H" whereas positions of double bonds are not indicated. In 2*H*-1,2,3,4-tetrazole, one of the double bonds is thus –N=N– and the other –CH=N–. Examples are

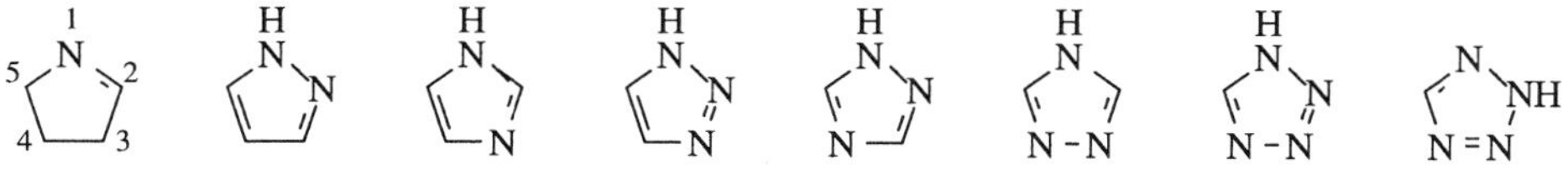

azole | pyrazole | imidazole | 1*H*-1,2,3-triazole | 1*H*-1,2,4-triazole | 1*H*-1,3,4-triazole | 1*H*-1,2,3,4-tetrazole | 2*H*-1,2,3,4-tetrazole

In the past, names of singly hydrogenated azoles carried the ending "ine" and doubly hydrogenated ones the ending "idine". The position of the remaining double bond was indicated by the number of the position at which the double bond begins, counting clockwise, sometimes also with the additional symbol Δ. Examples with outdated names in ("..."):

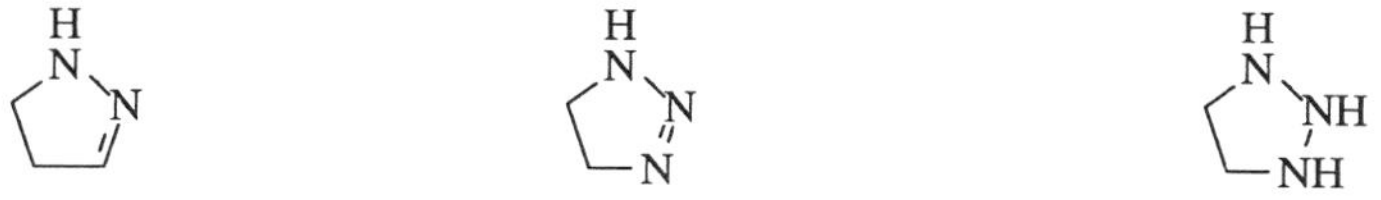

4,5-dihydro-1*H*-pyrazole ("2-pyrazole") ("$\Delta^2$-pyrazole")

4,5-dihydro-1*H*-1,2,3-triazole ("$\Delta^2$-1,2,3-triazoline")

4,5-dihydro-1*H*,2*H*,3*H*-1,2,3-triazole ("tetrahydro-1,2,3-triazole") ("1,2,3-triazolidine")

Additional hetereoatoms or hetero groups in ring such as –O– (oxa) and –SH– (thia) have seniority over nitrogen atoms or groups. Ketones of 5-membered rings are characterized by the end syllable "one." The trivial name of the ketone of oxazol is thus "oxazolone." Examples of systematic and generic names are

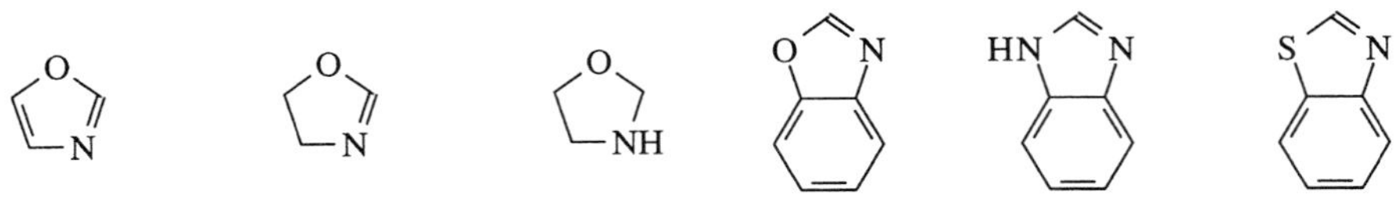

1,3-oxazole; 4,5-dihydrooxazole ("2-oxazoline"); tetrahydrooxazole ("oxazolidine"); benzoxazole; benzimidazole; benzthiazole

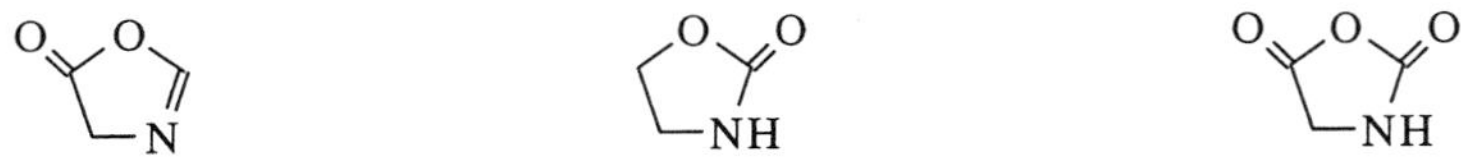

4,5-dihydrooxazole-5-one ("2-oxazolin-5-one") ("azlactone") ("oxazolinone"); 4,5-dihydro-3H-oxazole-2-one ("2-oxazolidinone") ("2-oxazolidone"); 4-hydro-3*H*-oxazole-2,5-dione (Leuchs anhydride or *N*-carboxyanhydride of glycine)

In many cases, trivial names are still used, for example, glycine *N*-carboxy anhydride for the 4-hydro-3*H*-oxazole-2,5-dione. Another example is "urazole" instead of 1,2,4-triazolidine-3,5-dione.

Polyazoles are not polymers of azole group-containing monomers. Azole groups of polymer molecules are rather created during the polymerization of suitable monomers.

### 10.8.2 Poly(pyrrole)s

The f5-membered ring of pyrrole contains one nitrogen atom and two carbon–carbon double bonds. For this reason, it is included in Section 10.8 although it is not an azole as it has no >C=N– double bond.

At room temperature, monomeric pyrrole is a colorless liquid that becomes brown and resinified on standing in air. Acids convert pyrrole to the polymeric **pyrrole red** which gave pyrrole its name (G: *pyr* = fire, *pyrros* = fiery red; L: *oleum* = oil). The electrochemical polymerization of pyrrole in a solution of acetonitrile (+ 1% water) in the presence of oxygen and the electrolyte $[(C_2H_5)_4N]^{\oplus}[BF_4]^{\ominus}$ produces at the anode a bluey-black polymer (**pyrrole black, PYR**) with the idealized structure shown in Eq.(10-72) and $[BF_4]^{\ominus}$ as counter anion:

(10-72) $2n$ pyrrole $\xrightarrow[-\,2\,H_2O]{+\,O_2}$ [polymer, α, β]$_n$

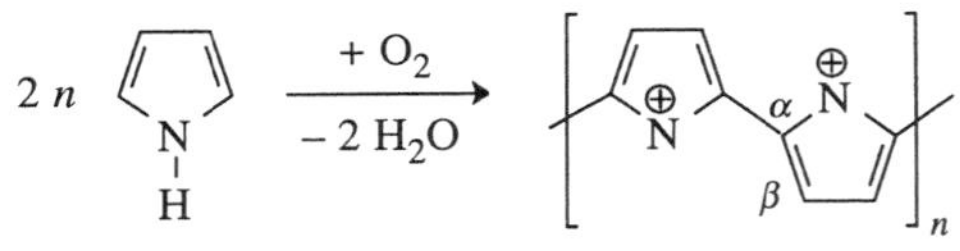

Polymerization of pyrrole in acetonitrile with $AgClO_4$ as electrolyte in the absence of water delivers a polymer with copper-bronze color.

The polymerization proceeds mainly as shown to α,α'-connected pyrrole units which, in an ideal constitution, would lead to polymer chains with planar conformations. However, pyrrole polymers contain more hydrogen than predicted by the ideal composition $(C_4H_3N)_n$. The polymerization is furthermore accompanied by isomerizations, branching, and crosslinking. The polymers also contain unpaired electrons, as shown by electron spin resonance data.

Doping of polymers with $AsF_5$, $I_2$, $BF_3$, etc. produces semiconductors. Such doped polymers can be directly generated on electrodes as flexible, semiconducting films with electrical conductivities of 50–100 S/cm and energy densities that corrspond to those of nickel-cadmium accumulators (see Volume IV). In contrast to semiconducting poly(acetylene)s, poly(*p*-phenylene)s, and poly(*p*-phenylene sulfide)s, poly(pyrrole)s are resistant to air for months and to temperature up to 250°C, which makes them useful for electronic and optoelectronic devices.

Doped poly(pyrrole) films can be reduced electrochemically in acetonitrile solutions of $[C_4H_9N]^{\oplus}[ClO_4]^{\ominus}$ to yellowish-green neutral polymers. Such polymers are rapidly oxidized by oxygen in air. Within 15 minutes, films turn black and the electrical conductivity rises to $10^{-2}$ S/cm from ca. $10^{-5}$ S/cm. Neutral polymers can be oxidized with Ag, Cu, or Ni salts to electrically conducting polymers.

## 10.8.3 Poly(benzimidazole)s

Poly(benzimidazole)s (**PBI**) are synthesized by polycondensation of dicarboxylic acids and aromatic tetramines. Industrial processes prefer the tetrahydrochloride of 3,3'-diaminobenzidine (I) and diphenyl isophthalate (II). The phenyl ester (II) is used because: (a) the free acid decarboxylates at the required high reaction temperatures of 250-400°C, (b) acid chlorides would react too fast for an orderly ring-closure, and (c) use of methyl esters would lead to a partial methylation of amino groups. Amine hydrochlorides (I) are required because free amino groups are sensitive to oxidation.

The melt polycondensation proceeds in two steps. In the first step, an intermediate (III) is formed that loses phenol to (IV):

(10-73)

$H_2N$, $H_2N$ (I), $NH_2$, $NH_2$ + $C_6H_5OOC$–(II)–$COOC_6H_5$

↓

$\sim HN$ ... $H_2N$ ... $NH_2$ ... $NH-C(OC_6H_5)(OH)$–...–$C(OC_6H_5)(OH)\sim$ III

$\xrightleftharpoons[+2\,C_6H_5OH]{-2\,C_6H_5OH}$ $\sim HN$ ... $H_2N$ ... $NH_2$ ... $NH-C(=O)$–...–$C(=O)\sim$ IV

The solidified prepolymer is pulverized and subjected to a solid-state polycondensation in the presence of 5–50 % phenol as plasticizer under a nitrogen blanket at 260–425°C. The resulting poly(benzimidazole)s (V) have molecular weights of ca. 20 000:

(10-74)

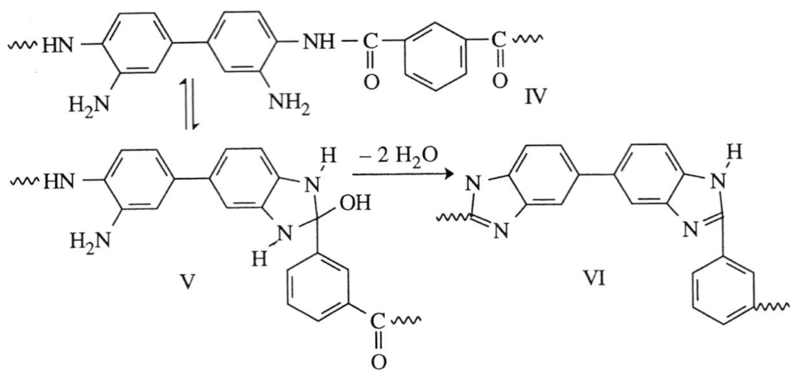

Probability predicts that complete reaction of amino groups will lead not only to linear molecules with high molecular weights but also to crosslinking between chains. Ring-closures in industrial poly(benzimidazole)s are therefore never complete and the polymer molecules always have some open-chain structures like IV and V besides VI.

Gold-colored poly(benzimidazole)s are heat-resistant and flame-retardant polymers with excellent mechanical and electrical properties (Table 10-17). Even at temperatures of 350°C, they are stable for hundreds of hours in the absence of air. Because of the labile NH, they degrade rapidly in air at temperatures above 200°C. They dissolve in *N,N*-dimethylacetamide, dimethylsulfoxide, and *N*-methylpyrrolidone but are resistant against other solvents such as 100 % acetic acid, *m*-cresol, etc.

Poly(benzimidazole)s are mainly used for specialty fibers for space and military applications. Other applications include temperature-resistant protective clothings and precursors for graphite fibers. Films and hollow fibers serve for the treatment of sea water and brackish water by reverse osmosis. Their good adhesion to metals makes them useful as heat-stable metal adhesives.

Table 10-17 Properties of unstabilized (N) and stabilized (S) fibers and untreated, annealed (A), and plasticized (P) films of poly(benzamidazole)s.

| Property | | Physical unit | Fiber N | Fiber S | Film N | Film A | Film P |
|---|---|---|---|---|---|---|---|
| Density | | g cm$^{-3}$ | 1.39 | 1.43 | 1.2 | 1.3 | 1.4 |
| Glass temperature | | °C | | | 430 | 500 | |
| Tensile modulus | | N/tex | 79 | 40 | | | |
| | | MPa | 110 000 | 57 000 | 2750 | 3790 | 2270 |
| Tensile strength | | N/tex | 2.3 | 2.3 | | | |
| | | MPa | 3 200 | 3290 | 117 | 186 | 103 |
| Elongation at break | | % | 30 | 30 | 14 | 24 | 20 |
| Relative permittivity | 100 Hz | 1 | | | 5.4 | | |
| Electrical resistance | 100 Hz | Ω | | | $10^{11}$ | | |
| Resistivity | 100 Hz | Ω cm | | | $10^{13}$ | | |
| Water absorption | 65 % RH | % | | 15 | 10 | 5 | 12 |
| Dissipation factor | 100 Hz | 1 | | | 0.013 | | |
| Electrical strength | 100 Hz | kV/m | | | 3.9 | | |
| Oxygen index (LOI) | | 1 | | 41 | | | |

There have been many attempts to improve the heat stability of PBI by incorporation of other monomers. Replacement of dicarboxylic acids by tetracarboxylic acids or their dianhydrides and polycondensation with diamines leads to more or less perfect ladder polymers. These syntheses must be conducted in solvents such as poly(phosphoric acid), zinc chloride, or eutectic mixtures of aluminum chloride and sodium chloride because of the difficult solubility of polymers.

These polymers have heat stabilities that are ca. 100 K higher than PBI and can be used up to 600°C. Examples are pyrrone, BBB, and polypyrrolone:

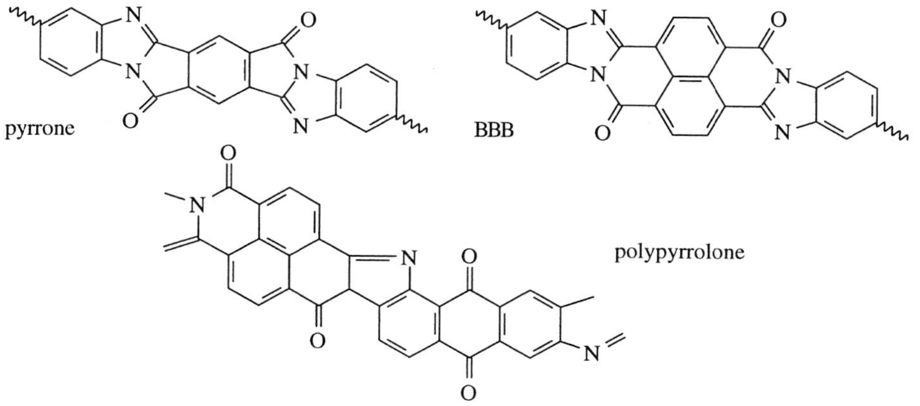

pyrrone BBB polypyrrolone

## 10.8.4 Poly(benzoxazole)s and Poly(benzthiazole)s

Poly(benzoxazole)s (**PBO**) and poly(benzthiazole)s (**PBTZ**) have similar constitutions as conventional poly(benzimidazole)s. They can be synthesized by AB or AA/BB polycondensations, either in melt or in solutions of poly(phosphoric acid) (PPA), methane sulfonic acid, or chlorosulfonic acid. Polymers also dissolve in antimony pentachloride or in $AlCl_3$ + nitroalkanes.

AB polycondensations employ amine hydrochlorides since amino groups oxidize easily. Examples are polycondensations of 3-amino-4-hydroxybenzoic acid hydrochloride (Z = O) and 3-mercapto-4-amino-benzoic acid hydrochloride (Z = S):

(10-75) $Cl[H_3N]$, HZ, COOH $\xrightarrow{-2\ H_2O,\ -\ HCl}$ N, Z

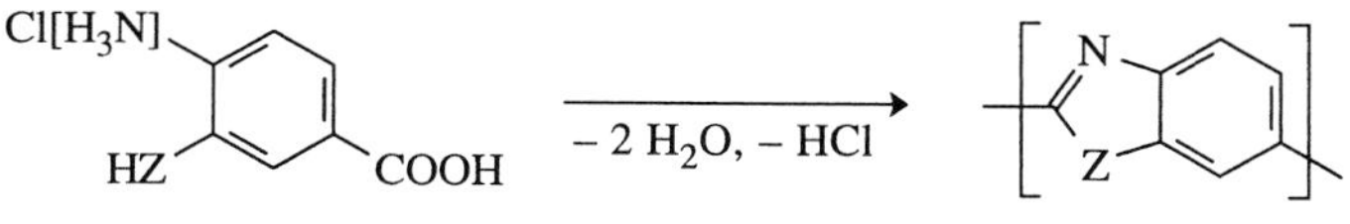

AA/BB polycondensations with terephthalic acid also use hydrochlorides, for example, 2,5-diamino-1,4-benzenedithiol dihydrochloride for "cis"-PBO and 4,6-diamino-1,3-benzenediol dihydrochloride for "trans"-PBTZ (see Eq.(10-76)).

In solution polycondensations, poly(phosphoric acid) (PPA) acts as solvent, catalyst, and dehydration agent. Liberated water is continually diluting the concentration of poly(phosphoric acid) which leads to a slowing down of the reaction and ultimately to precipitation of the polymer. The concentration of poly(phosphoric acid) is therefore maintained by continuous addition of diphosphorous pentoxide.

(10-76)

$$\text{Cl[NH}_3\text{]}(\text{HO})\text{C}_6\text{H}_2(\text{OH})\text{[NH}_3\text{]Cl} + \text{HOOC}(p\text{-C}_6\text{H}_4)\text{COOH} \xrightarrow[-4\ \text{H}_2\text{O},\ -2\ \text{HCl}]{} \text{PBO}$$

$$\text{Cl[NH}_3\text{]}(\text{HS})\text{C}_6\text{H}_2(\text{SH})\text{[NH}_3\text{]Cl} + \text{HOOC}(p\text{-C}_6\text{H}_4)\text{COOH} \xrightarrow[-4\ \text{H}_2\text{O},\ -2\ \text{HCl}]{} \text{PBTZ}$$

This PBO is called industrially the "cis" compound and the PBTZ the "trans" or "symmetric" polymer. In each of the two polymers, all repeating units are identical which contrasts with the situation for poly(benzimidazole)s. This structural regularity and the rigidity of repeating units are the reasons why PBO and PBTZ form thermotropic and lyotropic liquid crystals.

Since lyotropic solutions form only above a certain critical concentration, and since lyotropic solutions have lower viscosities than isotropic ones, solution polycondensations are conducted above the critical concentration. During solution spinning, preferential orientation of chain segments is preserved in domains, but the domains themselves become more oriented. The resulting parallelization in fiber direction leads to extraordinarily large mechanical properties, for example, elastic moduli of $E_{||}$ =320 GPa and tensile strengths of $\sigma_{||}$ = 4.2 GPa for as-spun PBTZ (Table 10-18).

Molecular orbital theory predicts a tensile modulus of $E_{||}$ = 730 GPa for the completely oriented PBTZ. Experimentally, tensile moduli of 350 GPa were measured by X-ray experiments and up to 320 GPa by stress–strain data. Heat-treated fibers showed tensile strengths of 4.2 GPa and fracture elongations of 1.7 % PBTZ.

PBTZ has higher thermal and oxidative stabilities than PBO but relatively low compressive strengths. Both PBO and PBTZ are used as high-performance fibers, films, and coatings. A commercial PBO brand is Zylon®.

### 10.8.5 Oxazoline Polymers

Oxazoline (= 2-oxazoline) is the old name for what is now called systematically 4,5-dihydrooxazole, i.e., a 5-membered cyclic compound with one oxygen group –O– and one –N=CH– group in the ring (p. 507). Polymers with such rings in the chain do not seem to be known.

2-Alkyl-2-oxazolines polymerize cationically by ring-opening. The polymerization is living if the counterion is not nucleophilic. Alkyl sulfonates such as alkyl *p*-toluenesulfonate = alkyl tosylate (R–TsO), for example, initiate a cationic polymerization because the cationic endgroup is more nucleophilic than the monomer molecule:

(10-77)

$$\text{R}\sim\text{N}^{\oplus}(\text{oxazolinium}, \text{C}_2\text{H}_5)\ \text{TsO}^{\ominus} + \text{C}_2\text{H}_5\text{-oxazoline} \longrightarrow \text{R}\sim\text{N(COC}_2\text{H}_5)\text{—CH}_2\text{—CH}_2\text{—N}^{\oplus}(\text{oxazolinium}, \text{C}_2\text{H}_5)\ \text{TsO}^{\ominus}$$

Table 10-18 Properties of PBTZ as ribbons; as untreated (U) (= as spun) fibers or as fibers that were heat-treated (T) at various temperatures; as untreated (U) (= as spun) or heat-treated (T) filaments; and as uniaxially (UA) or biaxially (BA; quasi-isotropic) films from PBTZ and PBO. All data in fiber direction at room temperature.

| Property | | Physical unit | Ribbons | Fibers | | | | Filaments | | Films | |
|---|---|---|---|---|---|---|---|---|---|---|---|
| | | | | U | T 600°C | T 650°C | T 665°C | U | T | UA | BA |
| *Poly(p-phenylene-2,6-benzobisthiazoldiyl) (PBTZ)* | | | | | | | | | | | |
| Density | | g/cm$^3$ | | 1.50 | 1.58 | | | | | 1.56 | 1.56 |
| Linear thermal expansion coefficient | | K$^{-1}$ | | | | | | | | $5 \cdot 10^{-6}$ | $10 \cdot 10^{-6}$ |
| Tensile modulus | | GPa | 40 | < 320 | | | | < 170 | < 330 | 270 | 34 |
| Torsion modulus | | GPa | | 1.2 | | | | | | | |
| Tensile strength | | GPa | 0.5 | < 4.2 | 2.7 | | | 2.3 | < 4.2 | 2.0 | 0.55 |
| Compression strength | | GPa | | < 0.4 | | | | | | | |
| Fracture elongation | | % | | < 7.1 | | | | < 7.1 | < 1.4 | 0.88 | 2.5 |
| Relative permittivity | | - | | | | | | | | 2.8 | 2.8 |
| Dissipation factor | | - | | | | | | | | 0.005 | 0.005 |
| Electrical strength | | V/mm | | | | | | | | 8900 | 8900 |
| Oxygen index | (from above) | % | | 35.7 | | | | | | | |
| | (from below) | % | | 22.6 | | | | | | | |
| *Poly(p-phenylene-2.6-benzoxazoldiyl) (PBO)* | | | | | | | | | | | |
| Density | | g/cm$^3$ | | 1.50 | 1.58 | | | | | | |
| Linear thermal expansion coefficient | | K$^{-1}$ | | $\approx 8 \cdot 10^{-6}$ | | | | | | | |
| Modulus of elasticity (X-ray data) | | GPa | | 387 | 477 | | 433 | | | | |
| Modulus of elasticity (stress–strain data) | | GPa | 7.6 | 166 | 320 | | 290 | | | | |
| Compression modulus | | GPa | | 240 | | | | | | | |
| Torsional modulus | | GPa | | 1 | | | | | | | |
| Tensile strength | | GPa | 0.3 | 4.6 | 5.0 | 3.4 | 3.0 | | | | |
| Compression strength | | GPa | | 0.5 | | | | | | | |
| Elongation at break | | % | 0.8 | 3.0 | 1.8 | 1.3 | 1.2 | | | | |
| Oxygen index | (from above) | % | | 36.1 | | | | | | | |
| | (from below) | % | | 22.8 | | | | | | | |

Alkyl halides such as RI initiate covalent polymerizations because the ends of the growing chains are unstable:

(10-78)

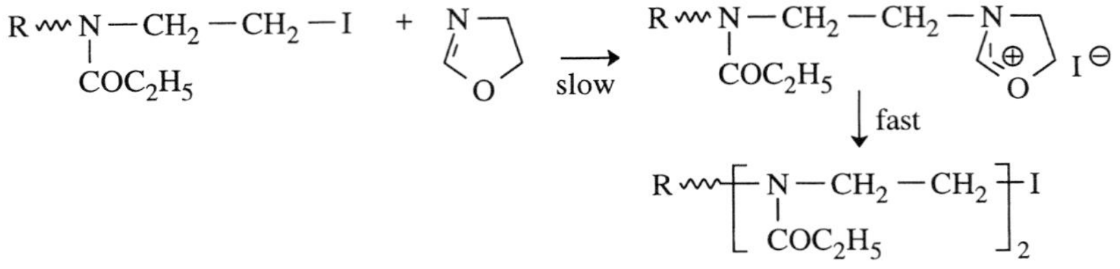

$$\mathrm{R{\sim}N(COC_2H_5)-CH_2-CH_2-I} + \text{2-ethyl-2-oxazoline} \xrightarrow{\text{slow}} \mathrm{R{\sim}N(COC_2H_5)-CH_2-CH_2-N^{\oplus}(\text{oxazolinium})\ I^{\ominus}} \xrightarrow{\text{fast}} \mathrm{R{\sim}[N(COC_2H_5)-CH_2-CH_2]_2-I}$$

The polymerization of 2-alkyl-2-oxazolines does not lead to polymers with intact oxazoline rings but rather to open-chain polymers such as the poly(*N*-propionylethyleneimine)s of Eqs.(10-77) and (10-78). The constitution of these polymers is that of a poly(*N*-acylethyleneimine).

The polymers are amorphous and dissolve in both water and organic solvents such as chloroform, methyl acetate, acetonitrile, and ethanol. The glass temperature is ca. 70°C. The polymers are used as adhesives and sizes (see Section 10.2.2).

### 10.8.6 Oxazolidone Polymers

Oxazolones are ketones that are derived from 1,3-oxazole. Two representatives, their IUPAC names (see p. 506), and their common names (in parentheses) are shown below.

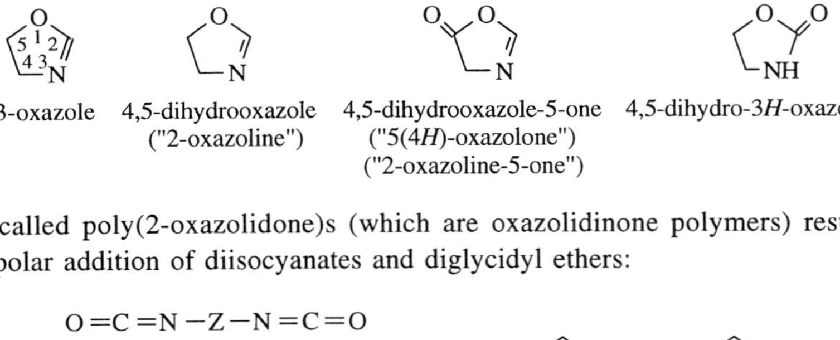

| 1,3-oxazole | 4,5-dihydrooxazole ("2-oxazoline") | 4,5-dihydrooxazole-5-one ("5(4*H*)-oxazolone") ("2-oxazoline-5-one") | 4,5-dihydro-3*H*-oxazole-2-one |
|---|---|---|---|

So-called poly(2-oxazolidone)s (which are oxazolidinone polymers) result from the 1,3-dipolar addition of diisocyanates and diglycidyl ethers:

(10-79)

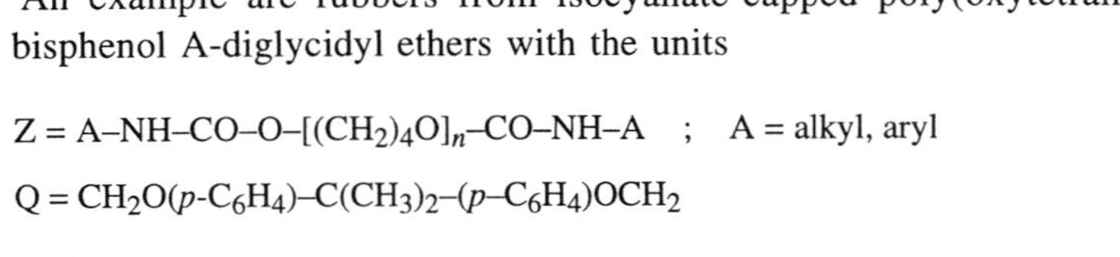

$$\mathrm{O{=}C{=}N-Z-N{=}C{=}O} + \text{diglycidyl ether (epoxide–Q–epoxide)} \longrightarrow \sim\text{oxazolidinone–N–Z–N–oxazolidinone–Q}\sim$$

Depending on the chain units Z and Q, these polymers are either thermosetting resins or rubbers. An example are rubbers from isocyanate-capped poly(oxytetramethylene glycol) and bisphenol A-diglycidyl ethers with the units

$$Z = \text{A–NH–CO–O–}[(CH_2)_4O]_n\text{–CO–NH–A} \quad ; \quad A = \text{alkyl, aryl}$$

$$Q = CH_2O(p\text{-}C_6H_4)\text{–}C(CH_3)_2\text{–}(p\text{-}C_6H_4)OCH_2$$

that compete with polyurethane elastomers.

### 10.8.7 Oxadiazole Polymers

Oxadiazoles are a group of four double-unsaturated 5-membered rings with two nitrogen atoms and one oxygen atom per ring:

1,2,3-oxadiazole 1,2,4-oxadiazole 1,2,5-oxadiazole ("furazane") 1,3,4-oxadiazole

Rings of oxadiazole polymers are not present in monomers but are formed by the polymerization reactions. For exmple, **poly(phenylene-1,3,4-oxadiazole)**s (**POD**) are obtained in one step from the polycondensation of terephthalic and/or isophthalic acid with hydrazine in poly(phosphoric acid) or oleum:

(10-80) HOOC–C$_6$H$_4$–COOH + $H_2N-NH_2$ $\xrightarrow{-3\,H_2O}$ POD

Yellow films can be cast directly from the polymerizing solutions. The transparent (but partly crystalline) films do not dissolve in organic solvents. They decompose at ca. 440°C but remain flexible in liquid nitrogen. Their mechanical properties are good, their electrical properties, excellent. A disadvantage is the relatively large uptake of water, which is 2.5 % at 25°C and 50 % relative humidity.

A similar process is used for the synthesis of high-modulus fibers of **poly[(phenylene-1,3,4-oxadiazole)-*co*-(*N*-methylhydrazide)]** in three steps, albeit without isolation of intermediary products. The three monomers terephthalic acid, dimethyl terephthalate, and hydrazine sulfate are heated in fuming sulfuric acid, $H_2SO_4/SO_3$, as solvent and dehydration agent. The reaction delivers poly(*p*-phenylene-1,3,4-oxadiazole) (I) and methyl sulfuric acid, $CH_3S_nO_{3n+1}H$, (the acid with $n = 2$ is methylpyrosulfuric acid). Methylation of some oxadiazole rings of I by this acid produces polymeric groups II, the substituents of which are subsequently hydrolyzed to give polymeric groups III:

(10-81) I $\xrightarrow{+\,CH_3S_nO_{3n+1}H}$ II ($H_3C$, $\oplus$, $HS_nO_{3n+1}^{\ominus}$) $\xrightarrow{+\,H_2O,\ -\,HS_nO_{3n+1}H}$ III (–C(=O)–N(CH$_3$)–NH–C(=O)–)

The resulting white to pale yellow copolymer can be wet spun from oleum in a bath with aqueous sulfuric acid.

The fibers were developed for use as tire cords. Their high tensile strengths are preserved to 90 % even after seven days in boiling water. Wear and tear of tire cords is smaller than that of steel, glass, and poly(*p*-phenylene terephthalamide).

### 10.8.8 Poly(terephthaloyloximidrazone)s

Poly(terephthaloyloximidrazone)s are synthesized from terephthalic acid dichloride and oxamidrazone. The polymer has the constitution of a poly[di(2,5-(1,3,4-oxadiazole)) terephthalate] (see below).

The addition of 2 molecules of hydrazine, $H_2N–NH_2$, to 1 molecule of cyanogene, N≡C–C≡N, delivers oximidrazone, $H_2N–NH–C(=NH)–C(=NH)–NH–NH_2$, which isomerizes to $H_2N–N=C(NH_2)–C(NH_2)=N–NH_2$, which is then reacted with terephthaloyl dichloride:

(10-82)

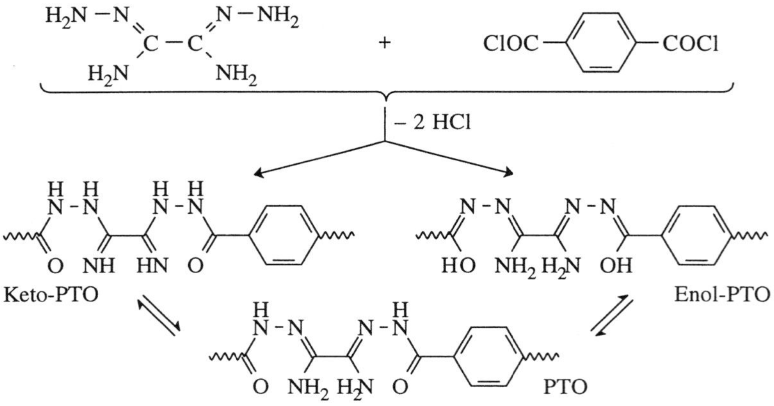

$H_2N–N=C(NH_2)–C(NH_2)=N–NH_2$ + ClOC–C₆H₄–COCl → (– 2 HCl) Keto-PTO ⇌ PTO ⇌ Enol-PTO

The resulting poly(terephthaloyloximidrazone) (PTO) can be converted to a polytriazole (PTA) by dehydration, or to a polyoxadiazole (POD) by release of ammonia. In concentrated aqueous solutions of alkali hydroxides, the yellow PTO dissolves to a solution of its ionized polymer (PTOA):

(10-83)

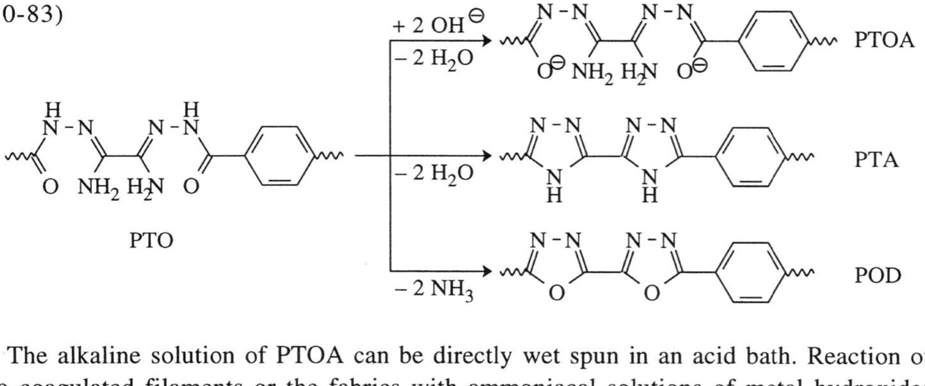

PTO → (+ 2 $OH^{\ominus}$, – 2 $H_2O$) PTOA; PTO → (– 2 $H_2O$) PTA; PTO → (– 2 $NH_3$) POD

The alkaline solution of PTOA can be directly wet spun in an acid bath. Reaction of the coagulated filaments or the fabrics with ammoniacal solutions of metal hydroxides unites one oxamidrazone group per two terephthaloyl oxamidrazone units per pseudo-cyclization to a chelate:

(10-84)

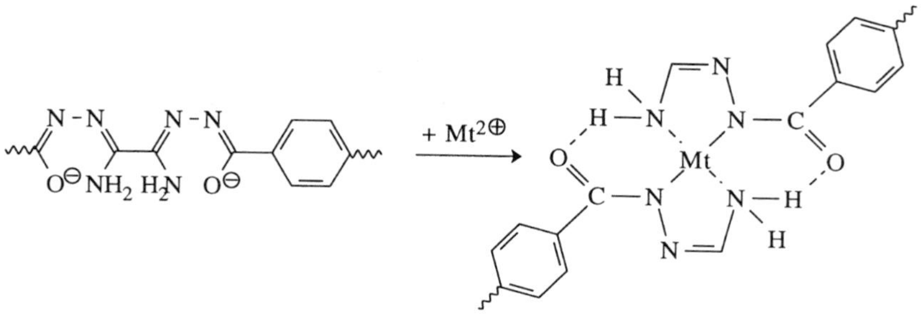

The complexation results in an insoluble coordination network which contains many different groups besides the one shown. The color of the metal chelate varies with the type of metal ion and the molar ratio metal/PTO from yellow ($Zr^{4\oplus}$/PTO = 0,35) and deep yellow ($Sn^{4\oplus}$/PTO = 1) to orange ($Zn^{2\oplus}$/PTO = 2), deep red ($Pb^{2\oplus}$/PTO = 1), olive green ($Cu^{2\oplus}$/PTO = 0,66), light brown ($Ni^{2\oplus}$/PTO = 1), and brown ($Ca^{2\oplus}$/PTO = 1) to black ($Fe^{2\oplus}$/PTO = 1). White and blue colors are not obtained.

Mechanical properties of chelated polymers correspond to that of rayon. In contrast to rayon, however, chelated polymers have extraordinarily good thermal properties. Polymers chelated by zinc, strontium, or calcium do not burn at temperatures below 1000°C. They do not shrink and melt. Burning releases only water, carbon dioxide, and ammonia. Mercury chelates are radiation-resistant but not flame-resistant.

Poly(terephthaloyloximidrazone)s have been proposed as heat- and flame-resistant fibers but never became commercial, perhaps dur to price or to restrictions with color selection.

## 10.8.9 Polyhydantoines

A number of other polymers is based on 5-membered rings with two nitrogen atoms per ring and one, two or three keto groups that are conventionally known as derivatives of urea (i.e., ethylene urea, glycolylurea, and oxalylurea) but are systematically called derivatives of imidazole. Commercial polymers contain hydantoin groups (this section) and parabanic acid groups (Section 10.8.10). Also offered are polymers with triazole rings in the chain (Section 10.8.11).

imidazole (1,3-diazole) 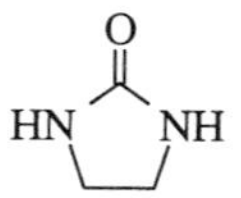 | imidazolidine-2-one (2-imidazolid(in)one) (ethylene urea) 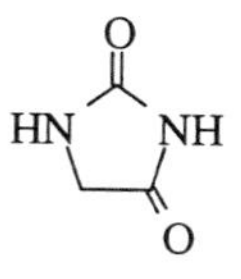 | hydantoin (glycolylurea) 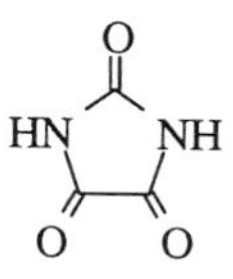 | parabanic acid (oxalylurea) (imidazoletrione) 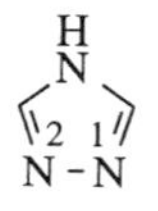 | 1,2,4-triazole

The industrial synthesis of aromatic polyhydantoines involves the condensation of aromatic diamines with the ethyl ester of dimethylchloroacetic acid to I, which is then reacted with aromatic diisocyanates in solvents such as phenol or cresol:

(10-85)

$$H_2N{-}Ar{-}NH_2 + 2\,Cl{-}C(CH_3)_2{-}C(=O)OC_2H_5 \xrightarrow{-2\,HCl} C_2H_5OOC{-}C(CH_3)_2{-}NH{-}Ar{-}NH{-}C(CH_3)_2{-}COOC_2H_5 \quad \text{I}$$

$$\text{I} + OCN{-}Z{-}NCO \rightarrow \text{urea adduct} \xrightarrow{-2\,C_2H_5OH} \text{polyhydantoine}$$

Aromatic groups Ar and Z of commercial products have not been disclosed, but Ar is probably –($p$-$C_6H_4$)–O–($p$-$C_6H_4$)– and Z probably either –($p$-$C_6H_4$)–O–($p$-$C_6H_4$)– or –($p$-$C_6H_4$)–$CH_2$–($p$-$C_6H_4$)–. Aromatic polyhydantoines are used as films for electrical insulation. They have good thermal stabilities (270°C in tensile experiments) but absorb relatively large amounts of water (4.5 % in 24 h). They are not attacked by most organic solvents or aqueous solutions of acids and bases.

Polyhydantoines with both aromatic and aliphatic units in the chain result from the polycondensation of aromatic diisocyanates with the reaction product of fumaric esters and aliphatic diamines:

(10-86)

$$2\,ROOC{-}CH{=}CH{-}COOR + H_2N{-}Z{-}NH_2 \rightarrow ROOC{-}CH_2{-}CH(COOR){-}NH{-}Z{-}NH{-}CH(COOR){-}CH_2{-}COOR \xrightarrow[-2\,ROH]{+\,OCN{-}Ar{-}NCO} \text{polyhydantoine}$$

These aromatic/aliphatic polyhydantoines serve as insulation lacquers. The viscosities of solutions (usually in methylene chloride) are high, though, so that either diluents are added or linear polymers are converted to branched ones by copolymerization with three and higher functional monomers.

## 10.8.10 Polyparabanic Acids

Chains of polyparabanic acids contain 2,4,5-triketoimidazolidine units and are thus related to those of polyhydantoines (Section 10.8.9). They can be obtained by different syntheses, for example, from oxamic acid esters and capped isocyanates:

(10-87) $\sim Z-NH-CO-CO-OR + OCN-Z^* \sim \xrightarrow[-ROH]{} \sim Z-N\langle\text{imidazolidinetrione}\rangle N-Z^* \sim$

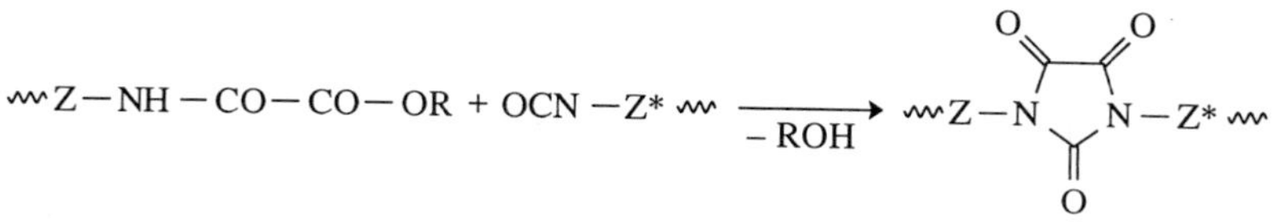

An alternative synthesis leads in three steps from isocyanates and hydrogen cyanide to cyanoformamides (I) and further to substituted ureas (II), poly(iminoimidazolidinone)s (III), and finally to polyparabanic acids (IV):

(10-88) $\sim Z-NCO \xrightarrow{+HCN} \sim Z-NH(CO-CN)$ I $\xrightarrow{+OCN\sim} \sim Z-N(CO-CN)-CO-NH\sim$ II $\longrightarrow$ III $\xrightarrow[-NH_3]{+H_2O}$ IV

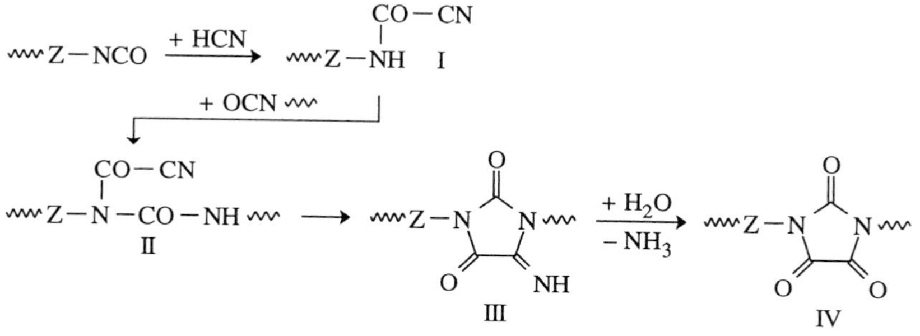

Polyparabanic acids are amorphous polymers which can be processed by film casting and compression molding to films and coatings. Solution of polyparabanic acids in, for example, *N*-methylpyrrolidone, are used as adhesives or for wire coatings.

### 10.8.11 Triazole Polymers

Polytriazoles with two triazole units and one 1,4-phenylene unit per repeating unit are obtained by the process of Eq.(10-83). Similar polymers but with only one triazole unit can be synthesized from hydrazine and terephthalic acid. The polycondensation delivers poly(phenylene hydrazide) (I) which is then cyclized to the polytriazole (II)

(10-89) I $\xrightarrow[-2\,H_2O]{+C_6H_5NH_2}$ II

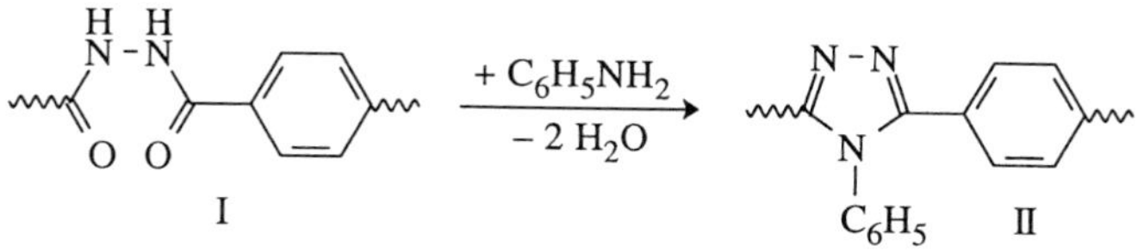

This polymer can be dry or wet spun from formic acid. It has a high glass temperature of ca. 260°C. Even at 300°C, fibers still retain 30 % of their original fracture elongation.

Polyamino triazoles are synthesized from dicarboxylic esters and hydrazine and further reaction of the primary dihydrazides with excess hydrazine:

(10-90) $H_2N-NH-C(=O)-(CH_2)_8-C(=O)-NH-NH_2 \xrightarrow[-2\,H_2O]{} \sim[\text{triazole}(N-NH_2)]-(CH_2)_8\sim$

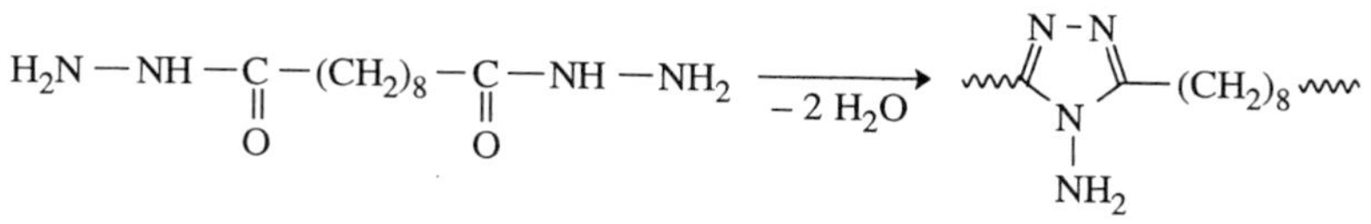

# 10.9 Polyazines

## 10.9.1 Overview

Azines are unsaturated 6-membered cyclic hydrocarbons with 1-4 nitrogen atoms per ring. Examples are pyrazine, triazine, melamine, etc. Correspondingly, polyazines are polymers with an intact azine moiety in the repeating unit. The group of polyazines comprises very many polymers but only a few are of any industrial importance.

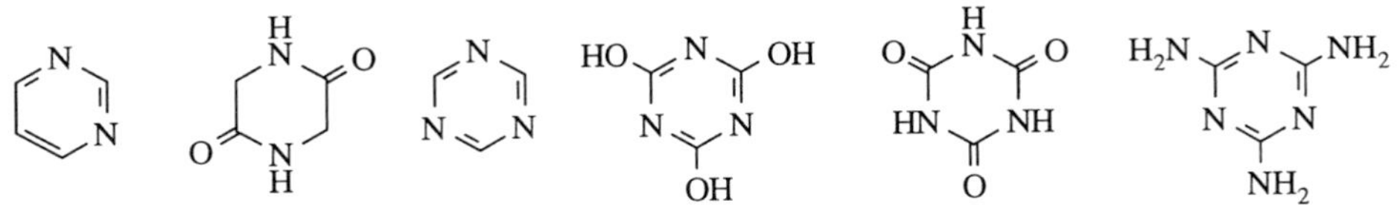

pyrazine diketopiperazine *s*-triazine cyanuric acid isocyanuric acid melamine

Azines can also be anellated with other rings (L: *anulus* = little ring, from *anus* = ring; *annus* = year!). Important for polymers are quinoline (Section 10.9.2), quinoxaline (Section 10.9.3), and quinazolindione (Section 10.9.4).

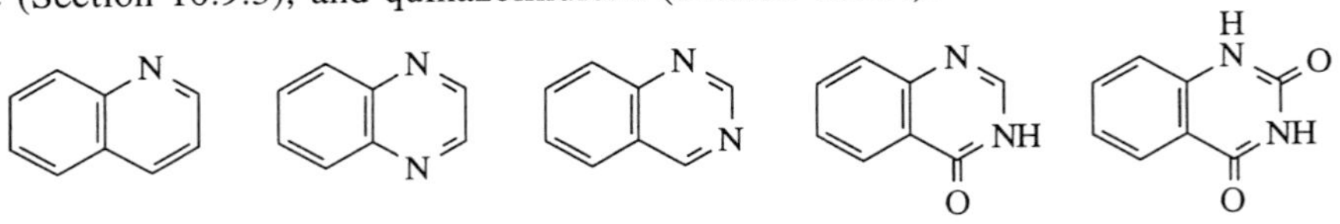

quinoline quinoxaline (benzpyrazin) (1,4-benzodiazine) quinazoline (benzopyrimidine) quinazolone quinazolinedione

## 10.9.2 Polyquinolines

Polyquinolines can be synthesized by several reactions. The highest molecular weights are apparently obtained by polycondensation of bis(aminoketone)s with diketones in *m*-cresol + di(*m*-cresyl phosphate), for example, 4,4'-diamino-3,3'-dibenzoyldiphenyl ether with 4,4'-diacetyldiphenyl ether (or 4,4'-dibenzoyldiphenyl ether):

(10-91)

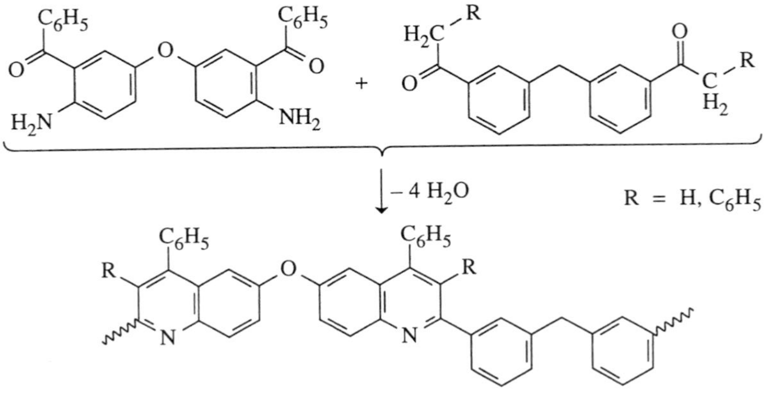

$- 4\ H_2O$ $R = H, C_6H_5$

Polyquinolines are semi-rigid to rigid polymers, depending on the bis(aminoketone) and the diketone. Semi-rigid polymers are usually amorphous; rigid polymers may be crystalline. Tensile moduli vary between ca. 2 GPa (semi-rigid) and 5 GPa (rigid); tensile strengths are ca. 100 MPa. Glass temperatures range from 250°C to 390°C.

### 10.9.3 Polyquinoxalines

A direct way to synthesize polymers with quinoxaline rings, albeit an academic one, is the transition-metal mediated polymerization of 1,2-diisocyanobenzenes such as I with, for example, *trans*(*o*-toluidine)$NiLi_2Cl$ as catalyst. The polymerization to substituted poly(quinoxalino-2,3-diyl)s is terminated by addition of methylmagnesium bromide:

(10-92)

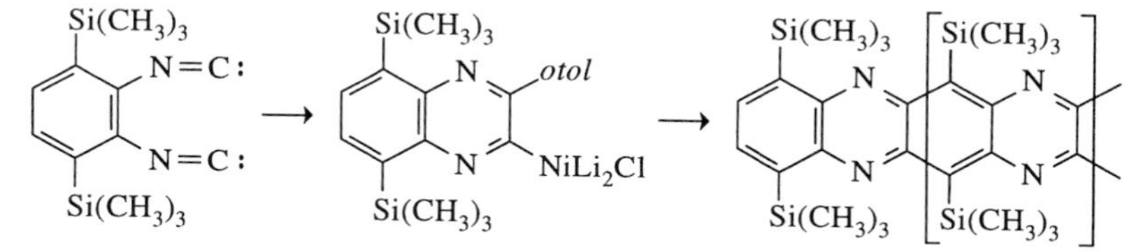

$Si(CH_3)_3$ N=C: N=C: $Si(CH_3)_3$ → $Si(CH_3)_3$ N *otol* N $NiLi_2Cl$ $Si(CH_3)_3$ → $Si(CH_3)_3$ N N $Si(CH_3)_3$ [$Si(CH_3)_3$ N N $Si(CH_3)_3$]

Polyphenyl quinoxalines are obtained by ring-closure reactions of bis(1,2-dicarboxyl) compounds with aromatic tetramines in which Z is a bifunctional aromatic or aliphatic group, Q the ether group, and R alkyl, hydroxyl, ester, alkoxy, nitrile, or halogen. The synthesis is usually performed in the melt but may also be conducted in slurries in chloroform, 1,1,2,2-tetrachloroethane, or *m*-cresol/xylene with a surplus of the dicarbonyl compound.

(10-93)

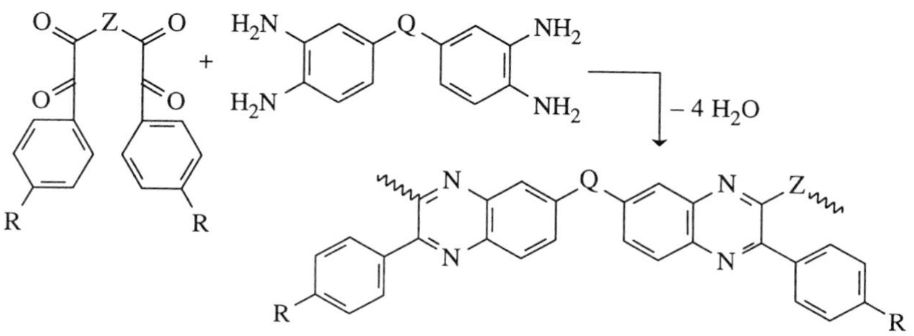

O Z O O O R R + $H_2N$ Q $NH_2$ $H_2N$ $NH_2$ → $-4\ H_2O$ N Q N Z N N R R

Films of this polymer discolor on heating but retain their transparency and most of their mechanical properties. Polyphenyl quinoxalines are used as heat-resistant adhesives and as a matrix for composites for which a post-cure at higher temperatures is required. During post-cure, polymers degrade thermally or by thermo-oxidation. Since they crosslink simultaneously, end-use properties are not affected.

### 10.9.4 Polytriazines

Like the benzene ring, the *s*-triazine ring is highly resonance stabilized and thus very resistant to heat. This attractive property was first exploited in melamine-formaldehyde resins (p. 483) by polycondensation of amino groups of preformed triazine rings. Formation of triazine rings by polymerization is utilized in the synthesis of triazine resins (= polycyanurates = polycyanates, p. 488), BT resins (p. 505), ®Triazine A resins, and ®NCNS resins.

®Triazine A resins result from the cyclotrimerization of the bisphenol A cyanuric ester, (Eq.(10-94)), which is obtained from the esterification of bisphenol A with cyanuric chloride. The resulting prepolymer ($M \approx 2000$) is sold as 70 % solution in butanone. The prepolymer is cured by zinc octoate or a mixture of zinc octoate with catechol and triethylene diamine, which crosslinks the prepolymer.

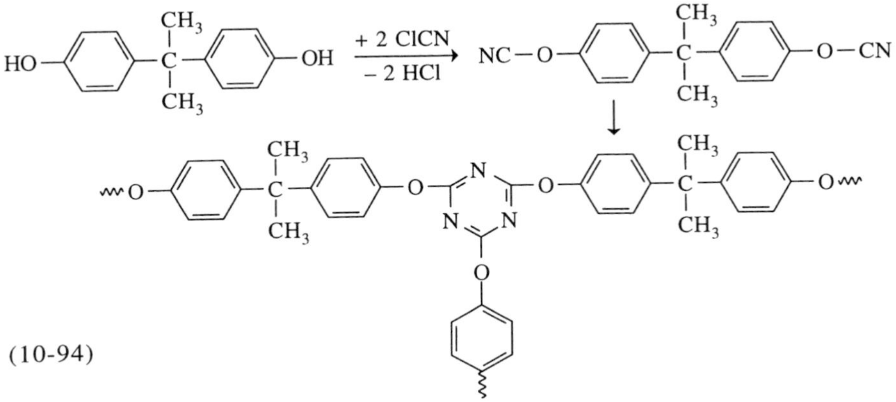

$$\text{HO}-C_6H_4-C(CH_3)_2-C_6H_4-\text{OH} \xrightarrow[-\,2\ \text{HCl}]{+\,2\ \text{ClCN}} \text{NC}-\text{O}-C_6H_4-C(CH_3)_2-C_6H_4-\text{O}-\text{CN}$$

(10-94)

Triazine rings of the melamine type are produced during the polymerization to *N-cyanosulfonamide* polymers (®NCNS polymers). Monomers for these polymers are primary or secondary aromatic biscyanoamides with Ar, Ar' = aromatic groups and R = arylsulfonyl or other electrophilic groups. The polycondensation of these monomers in the ratio 1:1 to 1:2 in lower alcohols or ketones delivers a soluble prepolymer (A stage)

(10-95)
$$\text{N}{\equiv}\text{C}-\text{N(R)}-\text{Ar'}-\text{N(R)}-\text{C}{\equiv}\text{N} + \text{N}{\equiv}\text{C}-\text{NH}-\text{Ar}-\text{NH}-\text{C}{\equiv}\text{N} \longrightarrow \left[ -\text{C(=NH)}-\text{N(R)}-\text{Ar'}-\text{N(R)}-\text{C(=NH)}-\text{N(C}{\equiv}\text{N)}-\text{Ar}-\text{N(C}{\equiv}\text{N)}- \right]$$

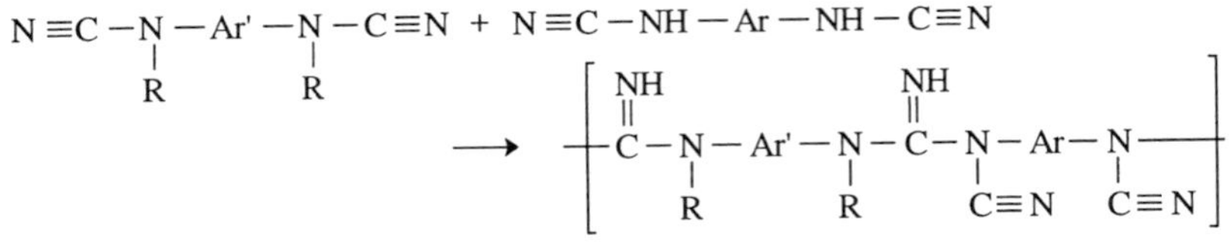

that on further heating to 100–120°C results in the commercially sold B-stage prepolymer. Curing at 150–180°C leads to the triazine resins:

(10-96)
$$\left[ -\text{C(=NH)}-\text{N(R)}-\text{Ar'}-\text{N(R)}-\text{C(=NH)}-\text{N(CN)}-\text{Ar}-\text{N(CN)}- \right] \longrightarrow \left[ -\text{N(R)}-\text{Ar'}-\text{N(R)}-(C_3N_3)-\text{NH}-\text{Ar}-\text{NH}- \right]$$

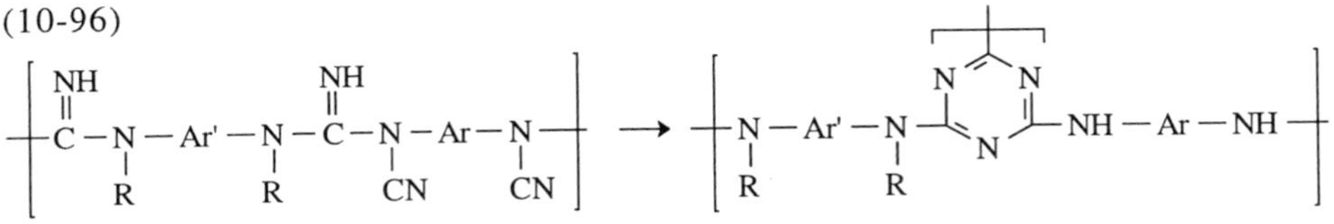

1:1 resins can be processed by injection molding or transfer molding, and 2:1 resins by compression molding. They replace epoxy resins if better dimensional and temperature stabilities as well as improved dielectric properties are required.

### 10.9.5 Heptazines

The simplest carbon-nitrogen polymer would be $[CN]_\infty$, the unknown and unnamed analog of graphite. This hypothetical compound cannot be called "carbonitride" because carbonitride is the common name of mixed crystals of carbides and nitrides of transition metals. However, there are oligomers and polymers that can be viewed as sections of $[CN]_\infty$ or, conversely, as polymerized melamine. The trimeric fusion products of *s*-triazine are known as heptazines because they consist of seven (hepta) nitrogen atoms (az). Cyameluric acid has 17 tautomers of which the trioxo compound is the most stable one since it is a triamide.

Melam, a triazine dimer, is the fusion product of *s*-triazine and 2,4-diamino-6-chloro-*s*-triazine. Melem is a triazine trimer, i.e., 2,5,8-triamino-tri-*s*-triazine, a near-planar molecule that stacks in layers. Melon is the corresponding polymer with the idealized structure shown below. Melame, melem, and melon are offered as substitutes for melamine and melamine cyanurate in resin formulations where higher temperature stabilities are required for processing and applications. Thermal decomposition temperatures are 400°C (melam), 500°C (melem), and > 500°C (melon).

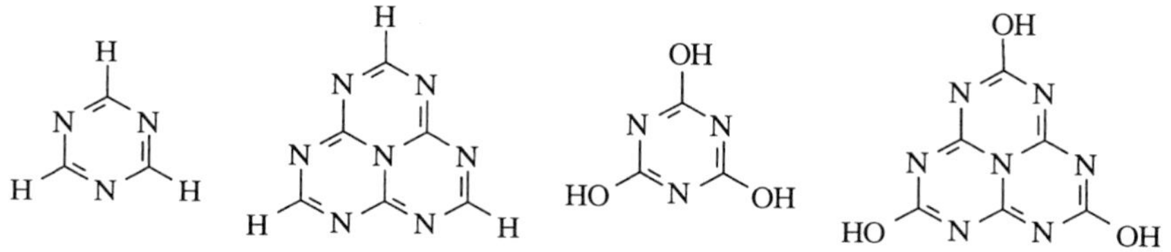

*s*-triazine

cymelurine (tri-*s*-triazine)

cyanuric acid

cyameluric acid

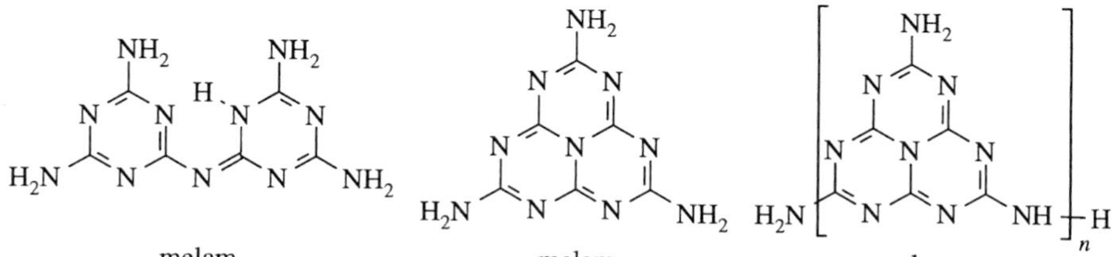

melam

melem

melon

## 10.10 Other Carbon–Nitrogen Polymers

### 10.10.1 Nitroso Rubber

Trifluoronitrosomethane and tetrafluoroethylene react at lower temperatures spontaneously to a high-molecular weight, alternating polymer that decomposes at higher temperatures to a 4-membered ring:

$$(10\text{-}97)\quad CF_3{-}N{=}O + CF_2{=}CF_2 \xrightarrow{0^\circ C} \sim\!N(CF_3){-}O{-}CF_2{-}CF_2\!\sim \xrightarrow{20^\circ C} CF_3{-}N{-}CF_2{-}CF_2{-}O \text{ (ring)}$$

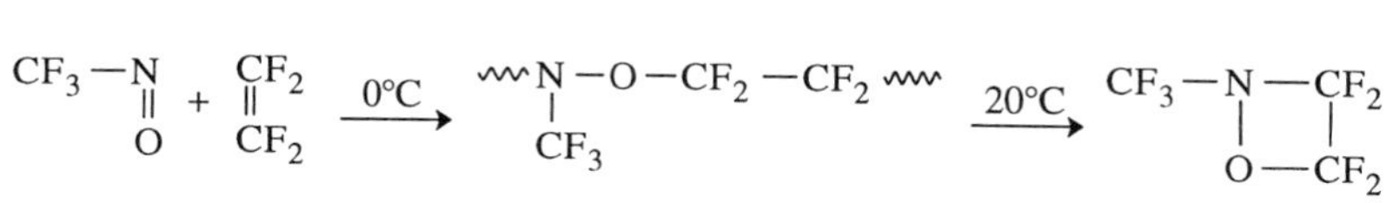

The resulting nitroso rubber has a glass temperature of –51°C. It does not burn even in pure oxygen. It is no longer produced because its tensile strength was not sufficient for most military and space applications.

### 10.10.2 Azo Polymers

Azo polymers are polymers with nitrogen–nitrogen double bonds in the main chain. Their simplest representative is poly(formaldezine) (p. 450).

Aliphatic **azo polymers** result from the polymerization of dinitroso compounds, for example, to poly(azoalkylene-*N*,*N*′-dioxide)s,

$$\text{(10-98)} \quad O{=}N{-}Z{-}N{=}O \longrightarrow \left[ {-}Z{-}\underset{\downarrow\atop O}{N}{=}\overset{O\atop\uparrow}{N}{-} \right]$$

and cycloaliphatic azo polymers from the bromine oxidation of hydroxylamines, for example, to poly(azo-1,4-cyclohexyleneisopropylidene-1,4-cyclohexylene-*N*,*N*′-dioxide):

$$\text{(10-99)} \quad HONH{-}C_6H_{10}{-}C(CH_3)_2{-}C_6H_{10}{-}NHOH \xrightarrow[-4\,HBr]{+2\,Br_2} \sim\underset{\downarrow\atop O}{N}{=}\overset{O\atop\uparrow}{N}{-}C_6H_{10}{-}C(CH_3)_2{-}C_6H_{10}\sim$$

The white polymer converts at 170°C reversibly to a yellow solid which melts at ca. 215°C to a reddish-brown liquid.

Aromatic nitroso compounds such as $ClOC(1,4\text{-}C_6H_4)NO$ dimerize to 4-membered rings which can be reacted to polyamides that are used as vulcanization agents (VANAX-PY®):

$$\text{(10-100)} \quad ClOC{-}C_6H_4{-}N\overset{O}{\underset{O}{\diamond}}N{-}C_6H_4{-}COCl + H_2N{-}Ar{-}NH_2$$

$$\xrightarrow[-2\,HCl]{} \sim C({=}O){-}C_6H_4{-}N\overset{O}{\underset{O}{\diamond}}N{-}C_6H_4{-}C({=}O){-}NH{-}C_6H_4{-}NH\sim$$

## Literature to Chapter 10

10.0 SURVEYS

H.-G.Elias, Neue polymere Werkstoffe 1969-1974, Hanser, München 1975; New Commercial Polymers, Gordon and Breach, New York 1977

K.-U.Bühler, Spezialplaste, Akademie-Verlag, Berlin 1978

H.-G.Elias, F.Vohwinkel, Neue polymere Werkstoffe, 2.Folge, Hanser, München 1983; New Commercial Polymers 2, Gordon and Breach, New York 1986

10.1 CYANIDE POLYMERS

Th.Völker, Polymere Blausäure, Angew.Chem. **72** (1960) 379 (review of azulmic acid)
D.Wöhrle, Polymere aus Nitrilen, Adv.Polym.Sci. **10** (1972) 35 (polymers from nitriles)
F.Millich, Polyisocyanides, J.Polym.Sci.-Macromol.Rev. **15** (1980) 207
R.P.Subrayan, P.G.Rasmussen, An Overview of Materials Composed of Carbon and Nitrogen, Trends Polym.Sci. **3** (1995) 165
M.Suginome, Y.Ito, Transition-Metal Mediated Polymerization of Isocyanides, Adv.Polym.Sci. **171** (2004) 77

10.2.2 POLY(ETHYLENEIMINE)S

O.C.Dermer, G.E.Ham, Ethyleneimine and Other Aziridines, Academic Press, New York 1969
M.Hauser, Alkyleneimines, in K.C.Frisch, S.L.Reegen, Eds., Ring-Opening Polymerizations, Dekker, New York 1969
G.E.Ham, Alkyleneimine Polymers, Encycl.Polym.Sci.Technol., Suppl. **1** (1976) 25
W.Drenth, R.J.M.Nolte, Poly(iminoethylenes): Rigid Rod Helical Polymers, Acc.Chem.Res. **12** (1979) 30
P.Ferruti, R.Barbucci, Linear Amino Polymers: Synthesis, Protonation and Complex Formation, Adv.Polym.Sci. **58** (1984) 55
S.Kobayashi, Ethylenimine Polymers, Progr.Polym.Sci. **15** (1990) 751

10.2.5 AMINE DENDRIMERS

E.M.M. de Brabender-van den Berg, A.Nijenhuis, M.Mure, J.Keulen, R.Reintjes, F.Vandenbooren, B.Bosman, R. de Raat, Large-Scale Production of Polypropyleneimine Dendrimers, Macromol. Symp. **77** (1994) 51
D.A.Tomalia, P.R.Dvornic, Dendritic Polymers, Divergent Synthesis (Starburst Polyamidoamine Dendrimers, in J.C.Salamone, Ed., Polymeric Materials Encyclopedia, CRC Press Boca Raton (FL) 1996, Vol. 3, p. 1814
G.R.Newkome, C.N.Moorefield, F.Vögtle, Dendrimers and Dendrons. Concepts, Syntheses, Applications, Wiley-VCH, Weinheim 2001
J.M.J.Fréchet, D.A.Tomalia, Eds., Dendrimers and Other Dendritic Polymers, Wiley, New York 2002
P.R.Dvornic, PAMAMOS: The First Commercial Silicon-Containing Dendrimers and Their Applications, J.Polym.Sci. **A** (Polym.Chem.) **44** (2006) 2755

10.3 POLYAMIDES

H.Hopff, A.Müller, F.Wenger, Die Polyamide, Springer, Heidelberg 1954
V.V.Korshak, T.M.Frunze, Synthetic Heterochain Polyamides; Akad.Wiss.USSR, Moskau 1962; Israel Program Sci.Transl., Jerusalem 1964
R.Graf, G.Lohan, K.Börner, E.Schmidt, H.Bestian, $\beta$-Lactame, Polymerisation und Verwendung als Faserrohstoff, Angew.Chem. **74** (1962) 523
K.Dachs, E.Schwarz, Pyrrolidon, Capryllactam und Laurinlactam als neue Grundstoffe für Polyamidfasern, Angew.Chem. **74** (1962) 540
C.F.Horn, B.T.Freure, H.Vineyard, H.J.Decker, Nylon 7, ein faserbildendes Polyamid, Angew.Chem. **74** (1962) 531 (nylon 7, a fiber-forming polyamide)
M.Genas, Rilsan (Polyamid 11), Synthese und Eigenschaften, Angew.Chem. **74** (1962) 535
R.Vieweg, A.Müller, Eds., Polyamide (Kunststoff-Handbuch, Bd. VI), Hanser, München 1966
M.I.Kohan, Eds., Nylon Plastics, Wiley, New York 1973
W.E.Nelson, Nylon Plastics Technology, Newnes-Butterworth, London 1976
H.K.Reimschuessel, Nylon 6, Chemistry and Mechanisms, Macromol.Rev. **12** (1977) 65
J.Sĕbenda, Recent Progress in the Polymerization of Lactams, Progr.Polym.Sci. **6** (1978) 123
R.S.Lenk, Post-Nylon Polyamides, Macromol.Rev. **13** (1978) 355
Z.Tuzar, P.Kratochvíl, M.Bohdanecký, Dilute Solution Properties of Aliphatic Polyamides, Adv.Polym.Sci. **30** (1979) 117
E.H.Pryde, Unsaturated Polyamides, J.Macromol.Sci.-Rev.Macromol.Chem. **C 17** (1979) 1
R.J.Gaymans, V.S.Venkatraman, J.Schuijer, J.Polym.Sci.-Polym.Chem.Ed. **22** (1984) 1373
R.J.Gaymans, A.G.J.Van der Ham, Nylon 4,I: An Amorphous Polyamide, Polymer **25** (1984) 1755
R.J.Gaymans, The Synthesis and Some Properties of Nylon 4,T, J.Polym.Sci.-Polym.Chem.Ed. **23** (1985) 1599
R.Puffr, V.Kubánek, Lactam-Based Polyamides, CRC Press, Boca Raton, 1990 and 1991 (2 Bde.)
M.I.Kohan, Ed., Nylon Plastics Handbook, Hanser, München 1995

E.E.Magat, In the Pursuit of Strength: The Birth of Kevlar, Eugen Magat, Chapel Hill (NC) 1996 (a publication of DuPont)
S.M.Aharoni, n-Nylons. Their Synthesis, Structure and Properties, Wiley, New York 1997
A.Steinbüchel, S.R.Fahnestock, Eds., Polyamides and Complex Proteinaceous Materials I (2002) (= Volume 7 of A.Steinbüchel, Ed., Biopolymers, Wiley, New York 2001-2003 (12 volumes)
S.R.Fahnestock, A.Steinbüchel, Eds., Polyamides and Complex Proteinaceous Materials II (2003); = Volume 8 of A.Steinbüchel, Ed., Biopolymers, Wiley, New York 2001-2003 (12 volumes)
A.Steinbüchel, S.K.Rhee, Eds. Polysaccharides and Polyamides in the Food Industry, Wiley-VCH, Weinheim 2005 (2 volumes)
E

10.4 POLYUREAS AND AMINO RSINS

J.F.Blais, Amino Resins, Reinhold, New York 1959
C.P.Vale, W.H.G.K.Taylor, Aminoplastics, Iliffe, London 1964
B.Meyer, Urea-Formaldehyde Resins, Addison-Wesley, Reading (MA) 1979

10.5 POLYCYANATES

I.Hamerton, Hrsg., Chemistry and Technology of Cyanate Ester Resins, Blackie Academic, London 1994
C.P.R.Nair, D.Mathew, K.N.Ninan, Cyanate Ester Resins. Recent Developments, Adv.Polym.Sci. **155** (2001) 1

10.6 POLYURETHANES

O.Bayer, Das Di-Isocyanat-Polyadditionsverfahren (Polyurethane), Angew.Chem. **59** (1947) 257
J.H.Saunders, K.C.Frisch, Polyurethanes, Chemistry and Technology, Interscience, New York 1961 (Vol. I), 1962 (Vol.II)
B.A.Dombrow, Polyurethanes, Reinhold, New York, 2nd ed. 1965
J.M.Buist, H.Gudgeon, Eds., Advances in Polyurethane Technology, Maclaren, London 1968
P.Wright, P.C.Cumming, Solid Polyurethane Elastomers, Maclaren, London 1969
P.F.Bruin, Polyurethane Technology, Interscience, New York 1969
N.Doyle, The Development and Use of Polyurethane Products, McGraw-Hill, New York, 2nd ed. 1971
K.C.Frisch, S.L.Reegen, Eds., Advances in Urethane Science and Technology, Technomic, Westport (CN), 7 volumes 1971-1979
M.N.Berger, Addition Polymers of Monofunctional Isocyanates, J.Macromol.Sci.[Rev.] **C 9** (1973) 269
D.J.Walsh, Newer Synthetic Routes to Isocyanates and Urethanes, Dev.Polyurethanes **1** (1978) 9
J.M.Buist, Developments in Polyurethanes, Applied Sci., London 1978
Z.W.Wicks, Jr., New Developments in the Field of Blocked Isocyanates, Progr.Org.Coatings **9** (1981) 3
D.Dieterich, Aqueous Emulsions, Dispersions and Solutions of Polyurethanes; Synthesis and Properties, Progr.Org.Coatings **9** (1981) 281
G.Woods, Flexible Polyurethane Foams - Chemistry and Technology, Appl.Sci.Publ., London 1982
C.Hepburn, Polyurethane Elastomers, Appl.Sci.Publ., Barking, Essex 1982
G.Woods, The ICI Polyurethanes Book, Wiley, New York, 2.Aufl. 1990
G.Oertel, Ed., Polyurethane (= G.W.Becker, D.Braun, Eds., Kunststoff-Handbuch, Volume 7), Hanser, Munich, 3rd ed. 1993; G.Oertel, Ed., Polyurethane Handbook, Hanser, München, 2nd ed. 1994
Z.Wirpsza, Polyurethanes. Chemistry, Technology and Applications, Ellis Horwood, New York 1993
H.Ulrich, Chemistry and Technology of Isocyanates, Wiley, Chichester 1996
M.Szycher, Szycher's Handbook of Polyurethanes, CRC Press, Boca Raton (FL) 1999
K.Uhlig, Polyurethan-Taschenbuch, Hanser, München 1998; -, Discovering Polyurethanes, Hanser, München 1999
D.Randall, S.Lee, Eds., The Polyurethane Book, Wiley-VCH, Weinheim 2003
M.Ionescu, Chemistry and Technology of Polyols for Polyurethanes, Rapra, Shropshire, United Kingdom (2005)

10.7 POLYIMIDES

H.Lee, D.Stoffey, K.Neville, New Linear Polymers, McGraw-Hill, New York 1967

M.W.Ranney, Polyimide Manufacture, Noyes Data Corp., Park Ridge (NJ) 1971
C.E.Sroog, Polyimides, Macromol.Rev. **11** (1976) 161
K.Wagner, K.Findeisen, W.Schäfer, W.Dietrich, $\alpha,\omega$-Diiocyanato-carbodiimide und -polycarbodiimide sowie ihre Derivate, Angew.Chem. **93** (1981) 855
M.I.Bessonov, M.M.Koton, V.V.Kudryavtsev, L.A.Lains, Polyimides - Thermally Stable Polymers, Plenum, New York 1987
T.Takekoshi, Polyimides, Adv.Polym.Sci. **94** (1990) 1
C.Feger, M.M.Khojastech, J.E.McGrath, Eds., Polyimides: Materials, Chemistry and Characterization, Elsevier, Amsterdam 1990
D.Wilson, H.D.Stenzenberger, P.M.Hergenrother, Eds., Polyimides, Chapman & Hall, New York 1990
M.J.M.Abadie, B.Sillion, Eds., Polyimides and Other High-Temperature Polymers, Elsevier, Amsterdam 1991
M.I.Bessonov, V.A.Zubkov, Eds., Polyamic Acids and Polyimides, CRC Press, Boca Raton (FL) 1993
C.Feger, M.M.Khojastech, M.S.Htoo, Eds., Advances in Polyimide Science and Technology, Technomic, Lancaster (PA) 1993
H.D.Stenzenberger, Addition Polyimides, Adv. Polym.Sci. **117** (1994) 165
W.Volksen, Condensation Polyimides: Synthesis, Solution Behavior, and Imidization Characteristics, Adv.Polym.Sci. **117** (1994) 111
K.Horie, T.Yamashita, Photosensitive Polyimides, Technomic Lancaster (PA) 1995
M.K.Gosh, K.L.Mittal, Eds., Polyimides. Fundamentals and Applications, Dekker, New York 1996
-, Polyimide, Adv.Polym.Sci. **140** (1999) and **141** (1999) (various authors)
W.Mittal, Ed., Polyimides and Other High Temperature Polymers, VSP/Brill, Leiden, Netherlands, Volume 3 (2005)

10.8 POLYAZOLES

J.P.Critchley, A Review of the Poly(azoles), Progr.Polym.Sci. **2** (1970) 47
V.V.Korshak, M.M.Teplyakov, Synthesis Methods and Properties of Polyazoles, J.Macromol.Sci. [Rev.] **C 5** (1971) 409
P.E.Cassidy, N.C.Fawcett, Thermally Stable Polymers: Polyoxadiazoles, Polyoxadiazole-*N*-oxides, Polythiazoles, and Polythiadiazole, J.Macromol.Sci.-Rev.Macromol.Chem. **C 17** (1979) 209
E.W.Neuse, Aromatic Polybenzimidazoles: Syntheses, Properties, and Applications, Adv.Polym.Sci. **47** (1982) 1
J.P.Critchley, G.J.Knight, W.W.Wright, Heat-Resistant Polymers, Plenum, New York 1983
G.M.Moelter, R.F.Tetreault, M.J.Hefferon, Polybenzimidazole Fiber, Polymer News **9** (1983) 134

10.9. POLYAZINES

P.M.Hergenrother, Linear Polyquinoxalines, J.Macromol.Sci. **C 6** (1971) 1
J.K.Stille, Polyquinolines, Macromolecules **14** (1981) 870
J.P.Critchley, G.J.Knight, W.W.Wright, Heat-Resistant Polymers, Plenum, New York 1983
M.Suginome, Y.Ito, Transition-Metal Mediated Polymerization of Isocyanides, Adv.Polym.Sci. **171** (2004) 77

## References to Chapter 10

[1] J.Yue, Z.H.Wang, K.R.Cromack, A.P.Epstein, A.G.MacDiarmid, J.Am.Chem.Soc. **113** (1991) 2665, Fig. 3
[2] From a compilation by P.R.Dvornic, S.Uppuluri, in J.M.J.Fréchet, D.A.Tomalia, Eds., Dendrimers and Other Dendritic Polymers, Wiley, New York 2002, Chapter 16
[3] Adopted from C.C.Winding, G.D.Hiatt, Polymeric Materials, McGraw-Hill, New York 1964, p.258
[4] P.Matthies, Polyamide, Ullmann's Enzyklopädie der technischen Chemie **19** (1979) 39, Verlag Chemie, Weinheim, (a) Table 2, (b) Table 3
[5] -, Chem.Engng.News **43**/26 (28.6.1965) 49
[6] E.I. DuPont, Bulletin E-97221, Molding Guide for Zytel® Nylon Resins (1988), p. 22
[7] P.H.Hermans, D.Heikens, P.F. van Velden, J.Polym.Sci. **30** (1958) 81, Fig. 2A
[8] J.R.Schaefgen, V.S.Voldi, F.M.Logullo, V.H.Good, L.W.Gulrich, F.L.Killian, ACS Polymer Preprints **17**/1 (1976) 69
[9] H.-G.Elias, R.Schumacher, Makromol.Chem. **76** (1964) 23

# 11 Peptides and Proteins

## 11.1 Overview

### 11.1.1 Definitions

$\alpha$-Amino acids, $H_2N–CHR–COOH$, form a great number of naturally occurring or man-made oligomers and polymers consisting of monomeric units with or without different substituents R. Monomeric units are connected by amide groups that are called **peptide bonds**, –NH–CO–. Depending on the composition and molecular weight of oligomers and polymers, one distinguishes between

- **poly($\alpha$-amino acid)s**, $\text{+NH–CHR–CO+}_n$ homooligomers or homopolymers of $\alpha$-amino acids, commonly used only for man-made compounds; there is no special name for copolymers;
- **peptides** (exact: **$\alpha$-peptides**): the term used mainly for naturally occuring or synthetically produced oligomeric to polymeric homopolymers or copolymers of $\alpha$-amino acids. This group is usually subdivided into
  **oligopeptides** with degrees of polymerization of 2 to ca. 10,
  **polypeptides** with degrees of polymerization of ca. 10 or more,
  **retropeptides** with peptide units in inverse sequence of naturally occurring peptides;
  **isopeptides**, with peptide bonds between an $\alpha$-group ($–NH_2$ or –COOH) and a $\beta$-, $\gamma$-, $\delta$-group, etc., of two $\alpha$-amino acid units and not between two $\alpha$-groups as in regular peptides (see below);
- **depsipeptides**: copolymers with alternating $\alpha$-amino acid and $\alpha$-hydroxy acid units;
- **proteins**: natural (and occasionally synthetic) copolymers from various $\alpha$-amino acids or $\alpha$-amino acids and imino acids;
- **nucleoproteins**: associates of proteins and nucleic acids (for example, nucleosomes);
- **glycoproteins**: proteins with less than ca. 4 % polysaccharide units;
- **mucoproteins**: proteins with more than ca. 4 % polysaccharide units, and
- **conjugated proteins**: compounds of proteins with so-called prosthetic groups (i.e., groups that are not based on $\alpha$-amino acids), for example:
  **chromoproteins**: proteins with bound dye molecules,
  **flavoproteins**: proteins with bound flavin derivatives,
  **hemoproteins**: proteins with iron-porphyrine compounds,
  **lipoproteins**: proteins with bound lipids,
  **metalloproteins**: proteins with bound metal atoms, etc.

Examples of structures of units of $\alpha$-peptides, isopeptides, and depsipeptides are

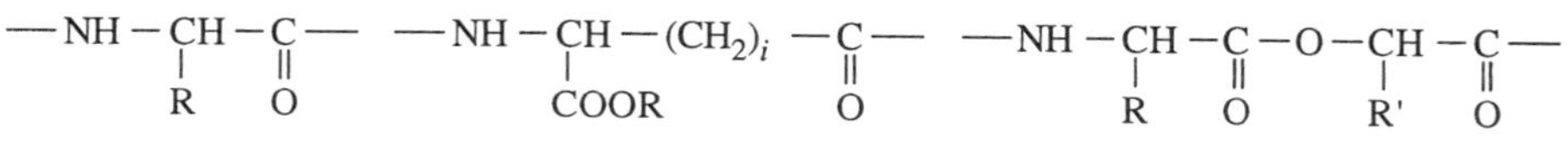

$$\text{—NH—CH(R)—C(=O)—} \qquad \text{—NH—CH(COOR)—(CH}_2)_i\text{—C(=O)—} \qquad \text{—NH—CH(R)—C(=O)—O—CH(R')—C(=O)—}$$

$\alpha$-peptide isopeptide depsipeptide

Besides $\alpha$-peptides with one –CHR– group as center group between the –NH– and the –CO– group of a monomeric unit, there are also **$\beta$-peptides** with –CHR–CHR'– centers, **$\gamma$-peptides** with –CHR–CHR'–CHR"–, **$\delta$-peptides** with –CHR–CHR'–CHR"–CHR'"– centers, etc. (Fig. 11-1). In synthetic polymer science, aliphatic homo polymers of such

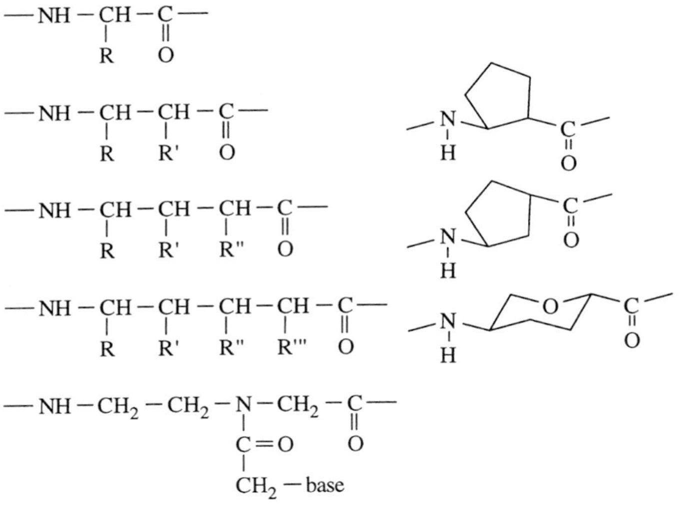

$$-NH-CH(R)-C(=O)-$$ α-peptide

$$-NH-CH(R)-CH(R')-C(=O)-$$ β-peptide

$$-NH-CH(R)-CH(R')-CH(R'')-C(=O)-$$ γ-peptide

$$-NH-CH(R)-CH(R')-CH(R'')-CH(R''')-C(=O)-$$ δ-peptide

$$-NH-CH_2-CH_2-N(C(=O)-CH_2-\text{base})-CH_2-C(=O)-$$ peptide nucleic acid

Fig. 11-1 Monomeric units of peptides and peptide nucleic acid.

β-, γ-, δ-, etc., peptides would be called substituted polyamides 2, 3, 4, etc., if R, R', R", etc., is a substituent. Note that in higher peptides, center groups of amino acid units may be part of a ring system which must contain at least 3 ring atoms for β-peptides, 4 for γ-peptides, 5 for δ-peptides, etc. Decisive for the classification as β-, γ-, δ-, etc., peptide is not the number of ring atoms but the number of carbon atoms between –NH– and –CO– peptide groups. Such rings may be sugar units, as shown for the ring atoms of a sugar unit of the δ-peptide in Fig. 11-1.

Related to higher peptides are **peptide nucleic acids** which are neither peptides nor nucleic acids. They can rather be viewed as nucleic acids in which the phosphate diester backbone of nucleic acids is replaced by a polyamide backbone, for example, the *N*-(2-aminoethylene)glycyl unit of the peptide nucleic acid of Fig. 11-1. The polyamide backbone is selected in such a way that the macroconformation of the peptide nucleic acid resembles that of the corresponding nucleic acid. Like true nucleic acids, peptide nucleic acids carry purine and/or pyrimidine bases.

In contrast to poly(α-amino acid)s, naturally occurring peptides and proteins consist of various peptide units, the sequence of which is identical in all molecules of the peptide or protein. By convention, the N-terminal amino acid unit is written to the left and the C-terminal one to the right. For example, the tripeptide Gly-Ala-Ser from the three α-amino acids glycine (Gly), $H_2N–CH_2–COOH$, alanine (Ala), $H_2N–CH(CH_3)–COOH$, and serine (Ser), $H_2N–CH(CH_2OH)–COOH$, is thus written as (see Table 11-1)

$$H[NHCH_2CO–NHCH(CH_3)CO–NHCH(CH_2OH)CO]OH \quad \text{or} \quad \text{Gly–Ala–Ser} \quad \text{or} \quad \text{G–A–S}$$

The number of peptide units in the smallest relevant sequence of these units is called **complexity**. A polymer H[gly–ala–ser]$_n$OH thus has the complexity 3 since the smallest relevant unit is the tripeptide –gly–ala–ser–.

The complexity of homopolymeric poly(α-amino acid)s is 1, and so is that of many enzyme molecules. Some other protein molecules have also low complexities, for example, silk proteins, but others have high ones, for example, wool.

### 11.1.2 α-Amino Acids

The smallest units of peptides, poly(α-amino acid)s, and proteins are the chain units –NH–CHR –CO– of α-amino acids and $-\underbrace{N-C}R'-CO-$ of **heterocyclic α-amino acids** (Table 11-1). Heterocyclic α-amino acids are usually called **imino acids,** but true imino acids have the chemical structure NH=CR–COOH since they are dehydrogenation products of α-amino acids ($H_sN$–CHR–COOH – $H_2$ → NH=CR–COOH) or ammination products of of α-keto acids (R–CO–COOH + $NH_3$ → NH=CR–COOH + $H_2O$). α-Amino acid units have a stereogenic ("asymmetric") carbon atom. They thus exist as D and L isomers (the R,S nomenclature is usually not used in protein chemistry).

In primeval times, α-amino acids and their heterocyclic cousins were probably created as racemic mixtures from methane, ammonia, and water by repeated electrical discharges (lightnings). Present biological *in-vivo* syntheses rely on the amination of α-keto acids, R–CO–COOH, by ammonia to imino acids, NH=CR–COOH, that are subsequently hydrogenated. Higher living beings incorporate in peptides and proteins exclusively L-α-amino acids; with few exceptions, these amino acids are in the [S] configuration. Peptides and proteins of lower animals contain also D-α-amino acids, for example, up to 15 % in the cell walls of bacteria. The reason for the biological selectivity in higher organisms is not fully clear, but it seems now that violation of subatomic parity is the most likely reason for the preponderance of L configurations (sse Volume I, p. 115).

Nature produces ca. 260 α-amino acids plus the heterocyclic amino acid proline but only 19 α-amino acids plus proline are genetically encoded in mammals (those with capital single letter symbols in Table 11-1) and selenocysteine and pyrrolysine in some lower organisms. Since selenocysteine units are incorporated in glycine reductase, glutathione peroxidase, and some other enzymes, selenium is an essential trace element.

The ca. 240 other biological α-amino acids are produced by post-translational reactions such as hydrogenation of glutamine to ornithine which is specific for birds. In addition, man has produced ca. 260 additional α-amino acids by synthesis.

α-Amino acids and imino acids are produced industrially by different methods, as indicated in Table 11-1:

- *extraction* (X) of hydrolyzed proteins, for example, L-α-amino acids Arg, Asn, Cys, Leu, and Tyr from casein or sugar beet residues;
- *fermentation* (F) by unregulation processes, for example, of glucose to the L-α-amino acids Arg, Gln, Glu, His, Ile, Leu, Lys, Ornithine, Phe, Pro, Thr, Trp, Tyr, or Val, or from Gly to Ser;
- *enzymatic conversion* (E) of fumaric acid to L-asparagine, L-asparagine to L-alanine, and from undisclosed substrates to Cys, Met, Phe, Trp, and citrulline which is the 2-amino-5-ureidovaleric acid, $H_2N–CO–NH–(CH_2)_3–CH(NH_2)–COOH$;
- *chemical syntheses,* either by conversion of other amino acids or by total synthesis. Ornithine is converted to arginine and glutamic acid to glutamine. Glutamic acid is produced from acrylonitrile and lysine from acrylonitrile or ε-caprolactam. Ala, Gly, Met, Phe, Ser, Thr, Trp, and Val are obtained by the Strecker synthesis: addition of $NH_3$ and HCN to aldehydes and subsequent hydrolysis of the resulting aminonitriles, R–CH($NH_2$)–CN. Racemates are resolved, L-α-amino acids separated, and D-α-amino acids racemized. Racemic methionine does not require resolution because the D compound is converted enzymatically to the L form by the human body.

Table 11-1 Constitution, trivial names, symbols, and symbols of *α*-amino acids coded in higher organisms (symbols in capital letters), *α*-amino acids coded in other organisms (selenocysteine, pyrrolysine), and some important amino acids by post-translational modification (cystine, hydroxylysine, ornithine, sarcosine, 4-hydroxyproline), all in alphabetical order [1]. Production methods [2] and annual production [2,3] in the world (roman) and in Japan (italic). Letters L and DL indicate the isomer that is obtained by the particular production process.

Symbols of amino acids: Asx or B is used if aspartic acid or asparagine cannot be distinguished; Glx or Z, ditto for glutamic acid or glutamine; Xaa or X = unknown or "other.".

Conformation of the most stable microconformation of amino acid units in poly(*α*-amino acid)s (PAS) and protein- (PR): α = α-helix, β = folded sheet, $10_3$, $3_1$ = other types of helices; [ ] after drawing. b = helix breaking, h = helix-forming, o = indifferent.

Production symbols: E = enzymatic, F = fermentation, S = synthetic, X = by extraction.

| Substituent R of $H_2N–CHR–COOH$ | Amino acid Trivial name | Symbols | | Conformation PAS | PR | Production (1987) Method | t/a | |
|---|---|---|---|---|---|---|---|---|
| $–CH_3$ | alanine | Ala | A | α [β] | h | E | *150* | L |
| | | | | | | S | *1 500* | DL |
| $–CH_2CH_2N{=}C(NH_2)_2$ | arginine | Arg | R | – | o | F,X | 1 000 | L |
| $–CH_2CONH_2$ | asparagine | Asn | N | – | b | X | 30 | L |
| $–CH_2COOH$ | aspartic acid | Asp | D | α | o | E | 4 000 | L |
| $–CH_2CH_2CH_2NHCONH_2$ | citrulline | – | – | - | - | E | *50* | L |
| $–CH_2SH$ | cysteine | Cys | C | – | o | E,X | 1 000 | L |
| $–CH_2S–SCH_2–$ | cystine | cyS–Scy | – | – | o | | | |
| $–CH_2CH_2COOH$ | glutamic acid | Glu | E | α | h | F | 340 000 | L |
| $–CH_2CH_2CONH_2$ | glutamine | Gln | Q | β | h | F | *850* | L |
| –H | glycine | Gly | G | β | b | S | 6 000 | - |
| see below | histidine | His | H | α | h | F,X | *250* | L |
| $–CH_2CH_2CHOHCH_2NH_2$ | hydroxylysine | hyl | – | – | – | – | – | |
| see below (OH in 4-pos.) | hydroxyproline | hyp | – | $3_1$ | - | X | 50 | |
| $–CH(CH_3)CH_2CH_3$ | isoleucine | Ile | I | β | (h) | F | *200* | L |
| $–CH_2CH_2CH_2CH_2NH_2$ | lysine | Lys | K | α | h | F | 70 000 | L |
| $–CH_2CH(CH_3)_2$ | leucine | Leu | L | α | h | F,X | *200* | L |
| $–CH_2CH_2SCH_3$ | methionine | Met | M | α | h | E | *150* | L |
| $–CH_2CH_2CH_2NH_2$ | ornithine | - | - | - | - | - | - | |
| $–CH_2C_6H_5$ | phenylalanine | Phe | F | α | h | S,E,F | 3 000 | L |
| see below | proline | Pro | P | $10_1$,$3_1$ | b | F | 100 | |
| see below | pyrrolysine | - | - | - | - | - | - | |
| see below | sarcosine | - | - | - | - | - | - | |
| $-CH_2SeH$ | selenocysteine | Sec | U | - | 0 | S | 100 | |
| $–CH_2OH$ | serine | Ser | S | β | o | E,F | *60* | L |
| | | | | | | S | | DL |
| $–CH(OH)CH_3$ | threonine | Thr | T | β | o | S,F | *200* | L |
| $–CH(CH_3)_2$ | valine | Val | V | β | h | S,F | *200* | L |
| see below | tryptophane | Trp | W | α | h | S,E,F | *250* | L |
| $–CH_2(p\text{-}C_6H_4)COOH$ | tyrosine | Tyr | Y | α | b | F,X | *60* | L |

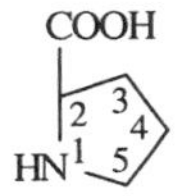

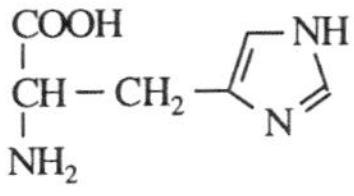

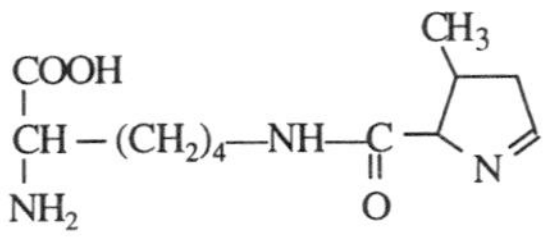

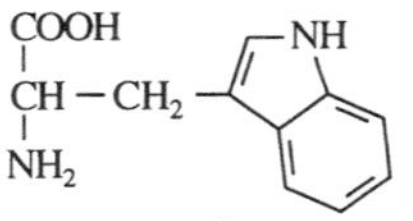

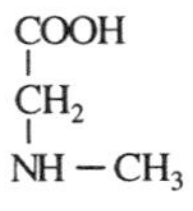

### 11.1.3 Macroconformations

α-Peptides, poly(α-amino acids), proteins, etc., all belong to the class of polyamides 2, $\text{+NH–Z–CO+}_n$. However, peptide bonds, –NH–CO–, are shorter and carbonyl groups, >C=O, larger in PA 2 types than in higher polyamides. Because of the double-bond character of α-peptide bonds, intramolecular hydrogen bonds between non-successive peptide bonds, and interactions between non-bonded successive side groups R in peptide groups, –NH–CHR–CO–, homopolymeric α-peptides and poly(α-amino acid)s of L-amino acids adopt the macroconformation of a helix in the crystalline state and in certain solvents (see Fig. 14-10 in Volume I). The helices are usually right-winding; an exception is the left-winding helix of poly(L-β-benzyl aspartate).

There are many different types of helices. In so-called α-helices of polyamides from α-L-amino acids, 3.6 peptide units are required for one full turn, whereas the larger β-peptides need only 3.0 units and γ-peptides only 2.6. The helices of α-peptides are left-winding (P) whereas those of β- and γ-helices are right-winding (M). α-Peptides also form γ-helices with 5.1 units per turn, π-helices with 4.4 units per turn, etc. In proteins, helical sections usually consist of different α-amino acid units.

In the crystalline state, other α-amino acids and all higher aliphatic amino acids exist as zigzag chains that are intermolecularly connected with adjacent chains to form pleated sheets (ses Fig. 14-10 in Volume I). In pleated sheets, adjacent chains are usually anti-parallel. Denatured proteins form random coils.

## 11.2 Poly(α-amino acid)s

### 11.2.1 Syntheses

Neither α-amino acids nor their esters can be polycondensed to high-molecular weight polymers since such polycondensations require high temperatures at which cyclodimerizations or even degradations prevail. Acid chlorides are much too reactive and ring-opening polymerizations of cyclic dimers (= diketopiperazines) do not lead to satisfactory results.

However, high-molecular weight poly(α-amino acid)s do form by base-initiated polymerization of *N*-carboxy anhydrides (NCAs, Leuchs anhydrides) of α-amino acids. Leuchs anhydrides are obtained from amino acids (I) and phosgene. In these polymerizations (Volume I, p. 245), initiating anions of primary amines, $RNH^{\ominus}$, attack NCAs nucleophilically at the $C^5$ atom resulting in carbamate anions (III) that are in equilibrium with their amine anions (IV) and carbon dioxide:

$$\text{(11-1)}\quad \underset{\text{I}}{\text{HO(O=)C–CH(R')–NH}_2} \xrightarrow[-\,2\,\text{HCl}]{+\,\text{COCl}_2} \text{II (NCA)} \xrightarrow{+\,\text{RNH}^{\ominus}}$$

$$\underset{\text{III}}{\text{R–NH–C(=O)–CH(R')–NH–COO}^{\ominus}} \underset{-\,\text{CO}_2}{\overset{+\,\text{CO}_2}{\rightleftharpoons}} \underset{\text{IV}}{\text{R–NH–C(=O)–CH(R')–NH}^{\ominus}}$$

Amine anions (IV) then trigger the polymerization of II according to the so-called amine mechanism, whereas carbamate anions (III) cause a polymerization of II according to the carbamate mechanism. Both polymerizations are polyeliminations since $CO_2$ is released.

With chiral initiators, polymerizations proceed enantioasymmetrically (Volume I, p. 179 ff.), i.e., only one of the two antipodes of racemic II is polymerized. With non-chiral initiators, polymerizations are almost completely enantiosymmetric and produce a 1:1 mixture of (almost completely) isotactic L- and D-polymers.

NCA polymerizations are living if the polymer remains dissolved. In this case, the achievable degree of polymerization is controlled by the molar ratio of NCA to initiator; it rarely exceeds 100.

Far higher degrees of polymerization of ca. 5000 are obtained by initiation with tertiary amines. In the slow, rate-determining initiation reaction, an NCA anion (V) attacks an NCA molecule which causes the NCA ring to open:

$$\text{(11-2)}\quad \text{NCA} \xrightarrow[-\,R_3\overset{\oplus}{N}H]{+\,R_3N} \text{V} \xrightarrow[k_i]{+\,\text{NCA}} \text{VI (N–CO–CHR'–NH–COO}^{\ominus}\text{)}$$

The resulting carbamate anion (VI) may react with $R_3NH^{\oplus}$, which releases $CO_2$ from V and regenerates the initiator $R_3N$. Reaction of $R_3N$ and NCA produces a new $NCA^{\ominus}$ (V).

Because of the low initiator concentration, NCA is practically only consumed by the propagation reaction between polymer (VII; from growth of VI) and NCA anions (V):

$$\text{(11-3)}\quad \text{V} + \text{VII (N–CO}\sim\text{)} \xrightarrow[-\,CO_2]{k_p\,;\,+\,H^{\oplus}} \text{VIII (N–COCHR'NH–CO}\sim\text{)}$$

Polymerizations of L-NCAs of alanine, lysine, glutamic acid, methyl glutamate, and benzyl glutamate delivers poly(α-amino acid)s in helix conformation (α structures), whereas those of NCAs of glycine, valine, serine, and cysteine lead to pleated sheats (β structures). α structures dissolve in helicogenic solvents such as *N,N*-dimethylformamide, whereas β structures are insoluble in all solvents because of their highly regular intermolecular physical crosslinks.

### 11.2.2 Polymers

Several α-amino acids are produced in large amounts (Tables 11-1 and 11-2) for the food industry: the sodium salt of glutamic acid as taste enhancer and aroma (MSG = monosodium glutamate) and the essential amino acids methionine and lysine as food supplements for man and mammals. Most other α-amino acids are niche products for the medical and pharmaceutical sector. Very few are used as monomers for polymers and if they are, then in onlysmall amounts.

Table 11-2 Change of production of major $\alpha$-amino acids with time ([2]–[4]).

| Amino acid | Configuration | Annual production in t/a | | | | |
|---|---|---|---|---|---|---|
| | | 1967 | 1977 | 1987 | 1997 | 2005 |
| Glutamic acid | L | | 250 000 | 340 000 | | |
| Methionine | DL | 15 000 | 100 000 | 100 000 | | |
| Lysine | L | 1 500 | 25 000 | 70 000 | | |
| Glycine | - | | 3 000 | 6 000 | | |
| Aspartic acid | L | | 1 000 | 4 000 | | > 20 000 |
| Phenyl alanine | L | | | 3 000 | | |

Both **poly(glycine)** = polyamide 2 (PA 2), $\text{+NH–CH}_2\text{–CO+}_n$, and **poly(L-alanine)**, $\text{+NH–CH(CH}_3\text{)–CO+}_n$, are produced in small amounts by polymerization of their Leuchs anhydrides. The polymers serve as model compounds in the study of proteins.

**Poly(L-leucine)**, $\text{+NH–CH[CH}_2\text{CH(CH}_3\text{)}_2\text{]–CO+}_n$, is also obtained from its Leuchs anhydride. Wool-like fibers are obtained by spinning poly(L-leucine) solutions from helicogenic solvents at concentrations below their critical ones. On stretching and subsequent storing, the α-helical molecules of wool-like fibers convert to more desirable silk-like fibers with pleated sheet structures. Cooking these fibers in certain solvents converts β-structures to α-structures. Fibers with β-structures cannot be obtained directly since any attempt would lead to irregular crosslinking. No production exists.

**Poly($\gamma$-D-glutamic acid)**, $\text{+NH–CH(CH}_2\text{CH}_2\text{COOH)–CO+}_n$, exists in nature in the anthrax-causing *Bacillus anthracis.* **Poly($\alpha,\gamma$-D,L-glutamic acid)** is produced extracellularly by *Bacillus licheniformi*. Similar to poly(L-leucine), **poly($\gamma$-D-glutamic acid)** can be spun to wool-like or silk-like fibers that can be dyed in brillant colors because of their free COOH groups. Again, no production exists, obviously for cost reasons.

In Japan, **poly($\gamma$-methyl-L-glutamate)** is applied as coating on synthetic leathers. Fibers are no longer produced.

**Poly($\gamma$-benzyl-L-glutamate)** serves to encapsulate pharmaceutically active hydrophobic liquids. It is also used as a stationary phase for the separation of racemates.

# 11.3 Poly($\beta$-amino acid)s

## 11.3.1 Poly($\beta$-alanine)s

The simplest poly($\beta$-amino acid), **poly($\beta$-alanine) = poly($\beta$-aminopropionic acid) (PA 3)**, $\text{+NH–CH}_2\text{–CH}_2\text{–CO+}_n$, can be obtained by two methods. The polymerization of acrylamide, $CH_2\text{=CH–CONH}_2$, by strong bases such as the potassium salt of *t*-butanol, $(CH_3)_3CO^{\ominus}K^{\oplus}$, in the presence of inorganic salts (prevents vinyl polymerization via the carbon-carbon double bond) seems to be of academic interest only (Volume I, p. 244):

$$(11\text{-}4) \qquad CH_2{=}CH{-}CONH_2 \underset{+H^{\oplus}}{\overset{-H^{\oplus}}{\rightleftharpoons}} CH_2{=}CH{-}CO\overset{\ominus}{N}H \xrightarrow{+\,CH_2=CH-CONH_2}$$

$$\rightarrow CH_2{=}CH{-}CO{-}NH{-}CH_2{-}\overset{\ominus}{C}H{-}CONH_2 \rightarrow CH_2{=}CH{-}CO{-}NH{-}CH_2{-}CH_2{-}CO\overset{\ominus}{N}H, \text{ etc.}$$

The second method seems to be the industrial method of choice. Introduction of gasous ammonia to acrylonitrile, followed by addition of water, results in $\beta$-amino-propionitrile (I) which converts to poly($\beta$-alanine) (III) via intermediate poly(imino-methylene ethylene) (polyamidine) (II):

(11-5)

$$CH_3-\underset{NH_2}{\underset{|}{CH}}-C\equiv N \ (I) \longrightarrow \left[CH_2-CH_2-\underset{NH}{\underset{\|}{C}}-NH\right] \underset{-NH_3}{\overset{+H_2O}{\rightleftharpoons}} \left[CH_2-CH_2-\underset{O}{\underset{\|}{C}}-NH\right]$$

Poly($\beta$-alanine) is used as a stabilizer for poly(oxymethylene).

Substituted linear **oligo($\beta$-alanine)s**, i.e., **$\beta$-peptides**, have been investigated as model compounds for the study of helix formation in proteins and as possible anti-bacterial compounds. Cyclic tetra($\beta$-peptide)s form tube-like supramolecules by stacking themselves on top of each other.

a cyclic tetra($\beta$-peptide)

**Poly(3,3-dimethyl-$\beta$-alanine)** results from the anionic polymerization of an intermediate (I). This intermediate is obtained from the addition of isobutene, $CH_2{=}C(CH_3)_2$, to *N*-carbonylsulfamoylchloride = chlorosulfonyl isocyanate, $ClSO_2NCO$ (from $SO_3$ and ClCN). I is saponified and the resulting monomer II polymerized anionically:

(11-6)

$$\text{I (N-}SO_2Cl\text{ 3,3-dimethylazetidin-2-one)} \xrightarrow[-\,HSO_3Cl]{+\,H_2O} \text{II (N-H 3,3-dimethylazetidin-2-one)} + \sim NH-\underset{O}{\underset{\|}{C}}-\underset{CH_3}{\overset{CH_3}{C}}-CH_2\sim$$

Fibers of this polymer are used for industrial sewing yarns (p. 471).

## 11.3.2 Poly($\beta$-aspartate)s

Aspartic acid, $H_2N–CH(CH_2COOH)–COOH$, can also be viewed as $\alpha$-aminosuccinic acid, $H_2N–CH(COOH)–CH_2–COOH$. In principle, aspartic acid may thus lead to two linear polyamides, poly($\alpha$-aspartic acid), $+\!NH–CH(CH_2COOH)–CO\!+_n$, as a PA 2 derivative, and poly($\beta$-aspartic acid), $+\!NH–CH(COOH)–CH_2–CO\!+_n$, as a PA 3 derivative. The poly(aspartic acid)s of commerce are either copolymers consisting of L-$\alpha$-amino acid and L-$\beta$-amino acid units, Eq.(11-7), or polymers with D,L-$\beta$-amino units, Eq.(11-8).

**Polymeric aspartic acids**, called **thermal aspartates**, can be obtained either by dehydration of aspartic acid (I) or by amination/dehydration of maleic anhydride (II) and reaction of succinamide structures with aqueous NaOH, as shown in Eq.(11-7):

(11-7) [Reaction scheme: aspartic acid (I) $-2\,H_2O$, or maleic anhydride (II) $+ NH_3$, $- H_2O$ → polysuccinimide; $+ NaOH$ → α and β sodium polyaspartate units]

The ratio of α and β structures in the resulting **thermal aspartates** is ca. 30:70. The water-soluble polymer is a dispersant and sequestering agent for the complexation of metal ions. It is used in water treatment, crude oil production, agriculture, and industry. In drainage pipes, it reduces lime scale and inhibits corrosion. The polymer is a "green chemical" since it can be attacked and degraded by bacteria.

## 11.4 Proteins

### 11.4.1 Survey

Proteins are copolymers of $\alpha$-amino acids and heterocyclic amino acids (imino acids). The number of naturally occurring proteins is estimated as 50 000 to 2 000 000; the number of types of possible macroconformations as ca. 5000. Macroconformations are controlled by the presence, proportion, and sequence of constituent amino acids and their intramolecular and intermolecular interactions. These complicated dependencies are explored experimentally and theoretically by **proteomics**, the study of proteins.

The smallest unit of a protein is a chain of covalently bonded $\alpha$-amino acids, a so-called **subunit** (see also Volume I, Section 14.3). A **protein monomer** (molecule) consists of one subunit or several of them which are bonded *covalently*. Molecular weights of such protein monomers rarely exceed 200 000.

Entities consisting of two or more, *non-covalently* bonded protein monomer molecules were known in the past as **quaternary structures** and are now called **protein oligomers**. In these associates of several true molecules, the sum of non-covalent interactions between protein monomer molecules is so great that the entity appears to be a **supramolecule**.

The biological synthesis of proteins is called **gene expression**. It usually consists of two steps, transcription of the genetic code into the code of the messenger ribonucleic acid (mRNA), followed by translation of the mRNA code into the signal for the incorporation of the particular amino acid (Volume I, p. 537). The codes are triplet codes which are universal for all eukaryotes and, with exception of the initial step, also for prokaryotes. The place of the protein synthesis varies with the protein. Some peptides are not synthesized by coding but rather directly enyzmatically, for example, glutathion (= $\gamma$-L-glutamyl-L-cysteinyl-glycine) of mammals and gramicidines which are pentadecapeptides of some bacteria.

Proteins were named because it was once assumed that they were the first molecules from which life sprang. They can be subdivided into two large groups:

- **spheroproteins**: proteins with spherical or ellipsoidal shapes that dissolve in dilute salt solutions. Examples are enzymes and blood and milk proteins.
- **scleroproteins**: fibrillar molecules that do not dissolve in dilute salt solutions. Examples are protein fibers, collagen, and protein gums.

Proteins are used by man in large amounts as food (meat), fibers (wool, silk), fibrous materials (leather), adhesives (bone glue), thickeners (gelatin), catalysts (enzymes), animal fodder (natural and from crude oil), and plastics (casein). The following sections deal mainly with the technical applications of proteins and not with those in the food and biomedical industries.

## 11.4.2 Enzymes

### Classification

With the exception of so-called RNA enzymes (Volume I, p. 511), all enzymes are globular proteins that act as biocatalysts. Enzymes are found in all living beings, either as intracellular enzymes that are bound to cell membranes of organelles or as extracellular enzymes that are excreted by cells.

Enzymes are characterized by a number system consisting of four numbers that are separated by points. The first number indicates the class, the second number the subclass, the third number the sub-subclass, and the fourth number the sub-sub-subclass. Depending on the enzyme action, six classes of enzymes are distinguished:

Class 1: **oxidoreductases** transfer electrons;
Class 2: **transferases** transfer chemical groups, for example, acyl groups (subclass 2.3);
Class 3: **hydrolases** transfer functional groups to water;
Class 4: **lyases** add groups to or transfer from double bonds;
Class 5: **isomerases** catalyze isomerizations;
Class 6: **ligases** (**synthetases**) form C–C, C–O, C–S, and C–N bonds by cleaving pyrophosphate bonds of adenosine triphosphate.

For example, an enzyme 1.1.3.X is an oxidoreductase (1...) which oxidizes CHOH groups (1.1...) with the help of molecular oxygen as acceptor (1.1.3).

### Structure of Enzymes

Enzymes are extraordinarily efficient biocatalysts. In contrast to many man-made catalysts, they are specific for just one substrate. They permit no side reactions and produce no byproducts. Reaction rates are higher than those of comparable non-enzymatic reactions by factors of $10^8$ to $10^{20}$. Disadvantageous is often that most enzymes function only in dilute aqueous solutions at moderate temperature and pH. However, there are certain enzymes that act also in organic solvents.

Simple enzymes ("monomers") consist of chains of 60–2000 peptide units which corresponds to molecular weights between ca. 6000 and 200 000. **Conjugated enzymes** contain not only the **apoenzyme** (the protein part) but also a non-protein (prosthetic) group which is also called the **coenzyme** or **cofactor**.

The interaction of primary, secondary, and tertiary structures (for these old terms, see Volume I) and perhaps quaternary structures leads to the "folding" of protein chains to globular entities with spherical to ellipsoidal shapes. In these structures, hydrophilic amino acid groups are mainly on the surface and hydrophobic ones usually in the interior.

All spheroproteins contain a cleft which harbors the receptor or the catalytically active center, respectively. These clefts are formed by folding of primary chains in most respiratorily acting enzymes (which assimilate $O_2$ and release $CO_2$), but by association of protein monomers (= subunits) in most regulatory enzymes (which control cycles).

Clefts can accommodate only certain substrates of which only some can position their reactive groups in such a way that they fit sterically and/or electronically the catalytically active groups of the active center. An example is the enzyme chymotrypsin with an active center composed of 2 amino groups (from 2 histidine units) and 1 hydroxyl group (from a serine unit). The catalytic action of an enzyme is optimal if substrate and active center behave like lock and key. In such cases, enzymes absorb substrates until the active center is saturated.

The binding of the substrate to the active center increases enormously the *effective* substrate concentration which, in turn, increases the reaction rate. A typical value of an equilibrium constant for the binding of a substrate to an active centers is $10^4$ L/mol, which contrasts with $10^{-8}$ L/mol for the complexation of the same group in solution. The effective substrate concentration and thus the reaction rate are thus increased by a factor of $10^4/10^{-8} = 10^{12}$!

The binding of the substrate to the enzyme decreases the entropy which is overcompensated by the change of enthalpy caused by non-covalent hydrophilic and hydrophobic interactions of the substrate with the peptide units in the cleft. The resulting bonding energy is one of many factors that decrease the activation energy of enzymatic reactions. Catalytic effects decrease the activation energy further. In nucleophilic catalyses by serine, tyrosine, histidine, cysteine, and/or lysine units, the enzyme and substrate molecules often form covalent intermediates. The total reaction thus consists of two steps, substrate → intermediate → product, which distributes and thus lowers the activation energy.

In electrophilic syntheses, the coenzyme or cofactor, respectively, is the electrophilic species and not the enzyme. In the enzyme thiamine pyrophosphatase, the quaternary nitrogen acts as stabilizer for carbanions. In acid/base reactions, active centers have both donor and acceptor groups, which also lowers activation energies.

In the simplest type of enzyme reactions, an enzyme–substrate complex ES is formed reversibly from an enzyme E and a substrate S. ES then reacts irreversibly to a product P with regeneration of the enzyme:

$$\text{E} + \text{S} \underset{k_{-1}}{\overset{k_1}{\rightleftharpoons}} \text{ES} \xrightarrow{k_2} \text{E} + \text{P} \qquad (11\text{-}8)$$

The resulting simple **Michaelis–Menten kinetics** can be modified by many factors:
- Interactions between two enzyme molecules or an enzyme molecule and a low-molecular weight effector may modify enzyme molecules either constitutionally or conformationally. These effects are called **allosteric** if they change the action of the enzyme.
- The enzyme–substrate complex may not react directly to P but to a product $M_n$. An example is the reaction of a polymerase E with a sugar M to an enzyme–substrate complex EM, which in turn reacts with the polymeric sugar $M_n$ to a complex $EM_nM$ which then reacts to $EM_{n+1}$ + a leaving molecule L (see Volume I).
- The enzyme molecule combines with an inhibitor, for example, SH groups of the enzyme with heavy metals (non-competitive inhibition). The inhibition is competitive if the inhibitor competes with the substrate for the active group of enzyme molecules.
- Enzymes may act as their own substrates and digest themselves. Such **autolyses** can be prevented by adding certain metal ions or salts. Autolyses by proteases do not happen to native proteins, only to denatured ones.

Enzeme kinetics is furthermore affected by process conditions, for example, diffusion effects if immobilized enzymes are used (see below). Industrial enzyme reactions prefer batch-type operations since continuous stirred reactors and cascades of reactors produce broad distributions of residence times (see Fig. 5-18). In continuous plug-flow reactors, pH is difficult to control.

**Immobilization of Enzymes**

Enzymes are expensive and are therefore recovered, although their small concentrations make it difficult to do so. Therefore, industrially used enzymes are increasingly immobilized in or at carriers by inclusion, microencapsulation, covalent bonding, adsorption or crosslinking (Fig. 11-2). Not only enzymes themselves, but also enzyme-containing organelles (nuclei, chloroplasts, mitochondria) and whole cells (of bacteria and plants, even animals and man) can be immobilized or microencapsulated. Such procedures avoid expensive and time-consuming isolations and purifications of enzymes. An example is the immobilization of *Escherichia coli* bacteria by poly(acrylamide) gels for the conversion ot sodium fumarate to L-aspartic acid.

Immobilization of organelles and cells has three disadvantages. First, a cell contains more than one type of enzyme. Second, the concentration of enzymes in cells is very low (usually, there are only a few enzyme molecules per cell) which lowers reaction rates considerably. Third, products cannot be easily excreted from immobilized cells.

The cell walls are therefore often swollen by treating cells with dimethylsulfoxide so that products can more easily permeate through walls so treated. The cell walls are then restored by washing out DMSO so that immobilized cells can be used again.

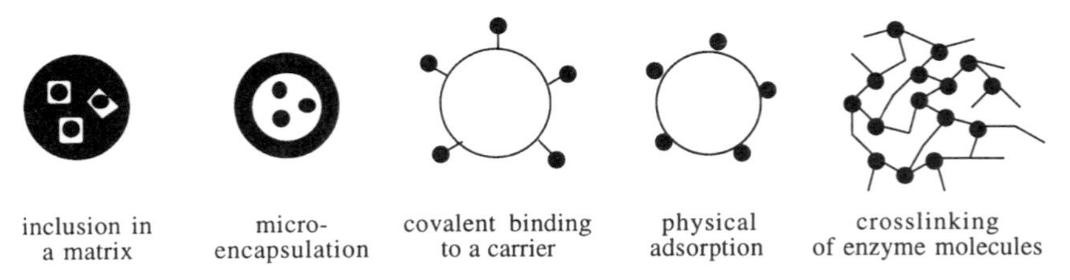


Fig. 11-2 Schematic representation of immobilized enzyme molecules (●).

Activities of immobilized cells can also be reduced if reagents for immobilization are toxic to cells and their contents. However, immobilized cells may also be more active than free cells because a partial or complete destruction of cell walls removes proteases that can hydrolyze enzymes. Enzymes can be immobilized by various methods:
- *Precipitation*. The enzyme and a water-insoluble polymer, for example, poly(styrene), are precipitated jointly from an organic solvent. The process is simple but the poly(styrene) coating of cells can not be easily penetrated by substrate and product. Cells may also be killed by the required non-physiological solvents.
- *Gelation*. Enzymes and cells may be encapsulated by a gel, for example, agar or collagen. However, such gels are mechanically unstable.
- *Ionotropic gelation*. Ionotropic gels are formed if multivalent ions diffuse into a solution of polyelectrolytes (Volume III). An example is the layering of aqueous calcium chloride solution on an aqueous solution of an alginate (or chitosan, carrageen, or carboxymethyl cellulose) that contains the enzyme. The alginate precipitates and forms a porous, ordered structure that encapsulates the enzyme. A disadvantage is the limited stability of ionic networks against ionic substrates and buffer solutions.
- *Polymerization*. The enzyme is dissolved in a monomer which is polymerized. Examples are methacryloamide or 2-hydroxyethyl methacrylate as monomers for free radical polymerizations; epoxide or diisocyanates + polyols for polyadditions; and Schotten-Baumann reactions for polycondensations.
- *Microencapsulation* encloses enzymes in capsules of 5–300 μm diameter from polyamides, cellulose acetate, ethyl cellulose, or polyesters. The thin walls of capsules allow a relatively unimpeded entry of substrates into the capsule and a fairly unrestricted exit of products.
- *Adsorption* of enzyme molecules onto the surface of a macroscopic carrier is the oldest method for the preparation of immobilized enzymes. The carrier can be any solid material with binding surface groups. Examples include gelatin, aluminum oxide, glass, activated carbon, ion-exchange resins, celluloses, or clay.
– *Crosslinking* interconnects enzyme molecules by their reaction with multifunctional agents such as diisothiocyanates, alkylation agents, aldehydes, etc. Even enzyme molecules in macrocrystals of enzymes have been crosslinked by glutaric dialdehyde. The products are rarely attacked by proteases, solvents, and changes in pH.
- *Chemical binding* of enzymes to a carrier is the most common method for immobilization. In this method, the peptide groups of enzymes are coupled by isocyanate, carbodiimide, or azide reactions to cellulose, carboxymethyl cellulose, or poly(glycidyl methacrylate), to silanol groups on the surface of glass spheres, or to surface-modified poly(ethylene) particles.

Immobilized enzymes last between days and two years. Immobilizations saves some of the cost of enzyme recovery, but not all because of the cost of immobilization itself.

**Industrial Use of Enzymes**

Enzymes have been used in **fermentation**s for thousands of years. Fermentations do not work with pure enzymes but rather with living organisms containing enzymes. However, the growth of yeasts, bacteria, fungi, and algae is not easy to control so that industry prefers to use enzymes for controlled chemical reactions.

Enzymes are preferentially isolated from microorganisms which, in contrast to plants and animals, grow fast and under controlled conditions. However, a bacterium cell contains 1000–2000 different proteins so that the desired enzymes are only present in fractions of a percent. Nevertheless, selection, mutation, and optimized growth conditions have allowed in certain cases enzymes to be obtained in concentrations of up to 10 % of the bacterium weight. Cells are then destroyed and enzymes isolated and purified by combinations of precipitation, chromatography, centrifugation, etc.

About 150 of the ca. 2200 known enzymes are produced in milligram to kilogram quantities; most of them are used in medicine, analytics, and biochemical research. Industrially used enzymes are mostly hydrolases (Table 11-3), but there is an increasing demand for enzymes that produce bioethanol from sugars. Enzymatic processes for reactions in organic solvents have not reached production stage.

The worldwide enzyme market is estimated as ca. $10^9$ $ per year. Major applications are in laundry agents (hydrolysis of proteins) for starch conversion (high-fructose syrups, bioethanol, etc.), and in cheese production. Only small concentrations of enzymes are needed for the various processes, but the economic multiplication factor is large: the $35 \cdot 10^6$ $/a US market for glucose isomerase generates a $3000 \cdot 10^6$ $ /a market for fructose syrup.

Table 11-3 Industrially used enzymes.

| Enzyme | Source | Application |
|---|---|---|
| Hydrolases | | |
| Proteases | | |
| Pancreatin | mammal pancreas | digestion aid |
| Bromelain | *Ananas comosus* | digestion aid |
| Papain | papaya | meat softening |
| Pepsin | stomach of hogs | digestion aid, milk curdling |
| Rennin | cloning, stomach of hogs | cheese production (cleavage of κ-casein) |
| Endo proteases | | laundry agent (protein hydrolysis) |
| Trypsin | | removal of hair from hides (leather) |
| Xylanase | | bleaching of pulp and paper |
| Aminoacylases | | |
| L-Aminoacylase | *Aspergillus oryzae* | separation of racemic α-amino acids |
| Penicillinacylase | *Escherichia coli* | 6-aminopenicillanic acid from penicillin G |
| Carbohydrases | | |
| Bacterial amylase | *B. subtilis* | starch liquefaction, beer brewing |
| Glucoamylase | *Aspergillus niger* | glucose from starch, starch liquefaction |
| Fungus amylase | *Aspergillus oryzae* | corn syrup |
| Invertase | *Saccharomyces cerevisiae* | invert sugar from saccharose |
| Pectinase | *Aspergillus niger* | purification of fruit juice, wine, vegetable oil |
| β-Galactosidase | | cleavage of lactose |
| Cellulase | *Asp. niger, Trichoderma viridae* | cellulose treatment ("stonewashing" of jeans) |
| Oxidoreductases | | |
| Glucose oxidase | *Aspergillus niger* | food conservation |
| Isomerases | | |
| Glucose isomerase | *Streptomyces sp.* | fructose syrup from glucose |
| Lipases | various | leather, laundry agents, fat cleavage |

### 11.4.3 Collagen

**Scleroproteins**

Scleroproteins are fibrillar proteins. Some of them support animal structures and are therefore also called **structural proteins**; examples are the keratins of hair, feathers, and nails; the collagens of skin, bones, and cartilage; and the elastins of connective tissue. Others form true fibers; examples of **fiber proteins** are wool and silk. Unlike enzymes, scleroproteins are not folded and are therefore also known as **linear proteins**. Contrary to the use of the word "linear" in synthetic macromolecular chemistry, some of these "linear proteins" are crosslinked. Scleroproteins are usually subdivided according to their macroconformations:

- In scleroproteins with pleated sheet structures (β structures), peptide chains reside in a plane, either parallel like the β-keratins of bird feathers or antiparallel like the proteins of highly crystalline silks. Scleroproteins with β structures have high tear strengths but low stretchabilities.
- Scleroproteins with α structures have helical conformations of their molecules. Examples are wool keratin, myosin (muscles), fibrinogen (blood), and collagen (bones). These scleroproteins are highly elastic and can be stretched to ca. twice their lengths.
- Denatured proteins and regenerated protein fibers exist as random coils. An example of the first group is gelatin, examples of the second group comprise arachin (peanut protein), zein (from corn), casein (from milk), and egg albumin.

In general, molecules of scleroproteins consist of many short blocks (triplets, quadruplets, quintuplets, sextuplets) that are coupled directly or via other peptide units. Their complexities are therefore much lower than those of spheroproteins.

Scleroproteins do not dissolve in the liquids of plants and animals. They are therefore produced as short chains that travel to the locus of function where they are assembled (see Volume I, Section 14.3.11).

**Structure of Collagen**

Connective tissues (skin, tendons, cartilage, intestines, blood vessels, etc.) consist mainly of collagen and elastin. Collagen provides these organs with tear strength, whereas elastin is responsible for elastic recovery at small deformations.

The biosynthesis of collagen is well known (Volume I, p. 542). The body synthesizes first pro-$\alpha_1$ and pro-$\alpha_2$ chains in the ratio 2:1. The chains consist of coupled triplets of the type Gly-X-Y where Gly = glycine, X = proline, leucine, phenylalanine, or glutamic acid, and Y = mainly hydroxyproline or arginine. Some proline and lysine units are then hydroxylated and the hydroxylysine units substituted by galactose and then glucose. The resulting pro-collagen molecule forms a triple helix consisting of two $\alpha_1$ chains and one $\alpha_2$ chain. Chain ends are connected by disulfide bridges.

The procollagen molecules travel to the place of their destination where most of the peptide units at chain ends are split off. The remaining, random coil-forming 41 peptide units per chain are known as telopeptides.

The center section of the resulting tropocollagen molecule has the shape of a rod with a length of 300 nm and diameter of 1.2 nm. These protofibrils assemble to subfibrils in which each unordered region consisting of polar sections with predominantly positively charged sidegroups of amino acid units is facing a polar region with predominantly negatively charged groups. Tropocollagen molecules are furthermore crosslinked via their sugar units. The subfibrils are assembled in collagen fibrils that are bundled to collagen fibers.

#### Use of Collagen

Chemical or physical crosslinking of collagens of skins and hides leads to **natural leathers** (Volume IV). Treatment of collagens with $Ca(OH)_2$ partially severs the crosslinks between fibrils. After washing out calcium ions and adding formaldehyde as crosslinking agent, the resulting collagen material can be extruded to **staple fibers** (for, e.g., artifical horse hair) or **sausage casings**. Endless **collagen fibers** are obtained by treating native collagen with enzymes and subsequent wet spinning.

Most collagen is converted to **gelatin(e)**. In Europe, bones are demineralized to ossein by hydrochloric acid. The subsequent extraction of ossein delivers gelatin in high yields. In the United States, gelatin is obtained directly by cooking of bones under pressure which delivers a low-molecular weight gelatin in yields of only ca. 12 %. High-quality gelatin for photographic purposes is obtained in the United States from hog skins in yields of 25–30 %. Another source of gelatin are pieces of animal skins and hides that are unusable for the preparation of leathers as well as fish skins. **Vegetable gelatin** is not a protein but a polysaccharide (algin, usually sodium alginate) from seaweeds.

Gelatin is obtained from ossein, bones, or pig skins in several steps. Cooking with acids or alkali removes impurities. A-gelatin (from *a*cids) is mainly used for foods and the B-gelatin (from *b*ases) for photographic films. After washing to remove traces of acid or alkali, gelatin is extracted by steam or hot water which deintegrates tropocollagen to the constituent $\alpha_1$ and $\alpha_2$ chains that are partially cleaved hydrolytically. The gelatin solution is concentrated and then "dried" by blowing cold air over the solution which produces a removable skin. Alternatively, gelatin may be isolated by spray-drying.

The resulting **gelatin** is used as thickener and for sausage casings in the food industry (ca. 55 %) and as an encapsulating agent in the pharmaceutical industry (25 %). Gelatin for photographic films (ca. 15 %) cannot be replaced since yet no synthetic polymer has been found that provides the same or better conditions for the ripening of silver crystals. Low-molecular weight gelatin serves as **bone glue**.

### 11.4.4 Protein Gums

Muscles and tendons contain highly elastic scleroproteins that are known as **protein gums**; examples are elastin, abductin, and resilin. Protein gums are not used industrially but are mentioned here because of their extraordinary properties.

#### Elastin

Elastin is the most important elastic protein of vertebrates, where it is the major component of the elastic fibers of connective tissues. In organs with high elasticities such as skin, tendons, blood vessels, and lungs, it is usually associated with collagen.

The biological synthesis of elastin is similar to that of collagen. Proelastin molecules consist of α-elastin (on average, 17 peptide chains of 35 amino acid units each) and β-elastin (2 chains with 17 peptide groups each), none of which contains cystine. During the conversion of proelastin to elastin, cystin is incorporated.

Peptides are arranged mainly in triplets, $ala_2lys$, and quadruplets, $ala_3lys$. Other sequences comprise ser-ala-lys, ala-pro-gly-lys, and try-gly-ala-arg. Some 90–95 % of elastin is X-ray amorphous; 1–5 % are fibrillous. Amorphous regions comprise 983 peptide units (wherefrom 4 cys), fibrillar regions, 1331 (wherefrom 48 cys).

The rebound elasticity of elastin is almost 100 %. This high elasticity is not caused by the few disulfide bridges but rather by crosslinking via desmosine and isodesmosine structures that are characteristic for elastin. In desmosin, a pyridine ring is substituted in 1-position by 2-aminocaproic acid, in 3- and 5-positions each by 2-aminobutyric acid, and in 4-position by 2-aminovaleric acid (in isodesmosin: in 2-position). These structures with 4 lysine groups each are synthesized from 4 lysine molecules, as shown schematically for desmosin:

$$HOOC-CH(NH_2)-(CH_2)_4-NH_2 \longrightarrow HOOC-CH(NH_2)-(CH_2)_2-CH(CHO)-CH(CHO)-(CH_2)_2-CH(NH_2)-COOH$$

$$\text{(11-9)} \quad \text{desmosin: pyridinium ring with } N\text{-}(CH_2)_4-CH(NH_2)-COOH \text{ (1-position)}, \; (CH_2)_3-CH(NH_2)-COOH \text{ (4-position)}, \; HOOC-CH(NH_2)-(CH_2)_2- \text{ and } -(CH_2)_2-CH(NH_2)-COOH \text{ (3- and 5-positions)}$$

The stress–strain behavior of swollen elastin follows the classic theory of rubber elasticity: on stretching, the entropy of the system decreases. This seems to be at odds with the observation that stretched elastin can take up fluorescing dyes but unstretched cannot, which points to hydration of hydrophobic regions. Such hydration would make protein chains more flexible which, in turn, should increase the entropy. However, polar water molecules cannot interact with hydrophobic groups and any water take-up near such groups must therefore lead to the formation of a highly ordered water layer. This **hydrophobic bonding** decreases the entropy of water. Because of the many water molecules involved, this is the larger effect: the entropy of the system decreases on stretching.

### Resilin and Abductin

Abductin is an elastic protein in the ligaments of molluscs (mussels, oysters, etc.) and resilin an elastic protein in arthropods (insects, spiders, crab, crayfish, etc.).

The cuticulum of all winged insects consists of 2–5 μm-thick resilin layers that are separated by 0.2 μm-thick chitin lamellae. Resilin is composed of 30–40 % glycine,

16 % aspartic and glutamic acid, 14–19 % hydroxyamino acids, and ca. 10 % alanine; it does not contain cystine. Its peptide chains are crosslinked via dimers and trimers of tyrosine, similar to desmosin from lysine.

On flying, resilin at the connection of the wings to the body is deformed if the wings reach their extremal positions. The deformation slows the movement of wings and the kinetic energy is stored as extensional energy. This energy is released on the next movement which allows for fast, steady flights.

Jumping fleas deliver energy slowly to resilin by muscle contraction. At the instant of jumping, the energy is suddenly released, which allows wide and high jumps.

The rebound elasticity of resilin is 96–97 %, and thus much higher than that of the best elastomers (ca. 70 % for vulcanized poly(butadiene) rubbers). This high value decreases heat production which would overheat the wings.

The high rebound elasticity and considerable elasticity of resilin is probably caused by a very homogeneous distribution of crosslinking sites and an absence of entanglements. The excellent properties of resilin has led to its biosynthesis. The resilin gene from the fruitfly *Drosophila melanogaster* has been inserted into the bacterium *Escherichia coli* where it produced a proresiline. Crosslinking proresilin with ruthenium catalysts and visible light via dithyrosine links produced a synthetic resilin that may be useful for artificial spinal discs.

Abductin of mussels has a much lower rebound elasticity (ca. 80 %) since, unlike insects, mussels need to move neither frequently nor rapidly.

### 11.4.5 Protein Fibers

#### Silk

Natural silks are produced by certain caterpillars and spiders. In the Antique, and in Southern Italy up to the 1920s, fine silken cloth (byssus) was also woven from sea silk, the <20 cm-long filamentous tuft ("beard") with which the up to 1 m-long bivalve mollusc *Pinna nobilis* (pen shell) attaches itself to rocks. Presently traded silks are obtained exclusively from the cocoons of the silkworm *Bombyx mori* L. This silkworm silk has the same properties as the dragline silk from the spider *Nephila clavipes* if it is extracted *directly* from silkworm glands and not obtained by the traditional commercial method.

*Byssus* (G: *bussos* = linen) was in ancient times the term for very fine cloth and was apparently used for very different textiles. In old Egypt, byssus referred to sea silk from *Pinna nobilis*; the very soft cashmere-like sea silk could only be worn by royalty (Genesis *41*:42). In the Middle East, byssus denoted very fine linen (Exodus *25*:4). English-language bibles generally translate "byssus" as "linen."

The silkworm forms a cocoon that consists 78 % of silk fibroin (raw silk fibers) which is surrounded and glued together by sericin (22 %). Raw silk is obtained by killing the larvae by steam or hot air, softening the sericin by dipping the cocoons in hot water, and catching the ends of raw fibers with rotating brushes. Between 4–10 fibers are then wound together on a reel and dried. Since the outer and inner layers of the cocoons are dirty, only ca. 900 m of the total 3000–4000 m long fiber of a cocoon can be used as silk. The rest (and also fiber of damaged cocoons) is processed by the so-called schappe spinning to less-valuable schappe silk.

The raw silk fibers are made supple by dipping in oil and then "debasted" by removing sericin with alkali-free soap. Debasting causes silk to lose 25 % of its weight. To make up for the loss, silk is made heavy again by treating it with aqueous solutions of $SnCl_4$ and $Na_2HPO_4$. These compounds react on the fiber to tin phosphate which is then converted by water glass (see Section 12.3.4) to silicates.

**Sericin** is a protein composed of serine (37 %), aspartate (26 %), glycine (17 %), and various other amino acids. It is used as culture medium for bacteria.

**Silk fibroin** is comprised of a crystalline part (60 %) and an amorphous one (40 %). The crystalline part consists of hexapeptides, ser–gly–ala–gly–ala–gly. In the silk fibroin of *Bombyx mori*, 10 of these hexapeptides (i.e., 60 peptide units) are combined with 33 peptide units of the amorphous part. The many different peptide units of the amorphous part are arranged in peptide sequences of various lengths and compositions.

In crystalline regions, peptide chains are densely packed in pleated sheet structures which provide silk fibers with high tensile strengths. The amorphous regions are responsible for the extensibility of silk. The high gloss of silk is caused by the triangular cross-section of its fibers (for textile properties of silk, see Volume IV).

## Wool

Wool and hair are the fibrous coverings of sheep, goats, llamas, rabbits, etc. The fibers of raw wool are glued together by fat, sweat, and plant material which are removed by "carbonization", a process consisting of the mechanical removal of impurities by beating, treatment with 4–7 % sulfuric acid, and drying at 100–120°C. This treatment involves several chemical processes: an N/O peptidyl rearrangement of serine units,

(11-10)

$$\text{HN(–C=O)}\sim\text{ ... OH} \longrightarrow \text{HN ... O (ring, C(OH))} \xrightarrow{-H_2O} \text{N ... O (oxazoline)} \xrightarrow[+H^{\oplus}]{+H_2O} H_3N^{\oplus}\text{ ... O–C=O}\sim \xrightarrow{+H_2O} H_3N^{\oplus}\text{ ... OH} + \text{HOOC}\sim$$

an esterification of serine with β-elimination and subsequent decomposition,

(11-11)

$$\begin{matrix}\sim NH \\ | \\ CO \\ | \\ CHCH_2OH \\ | \\ {}^{\oplus}NH_3\end{matrix} \xrightarrow[-H_2O]{+H_2SO_4} \begin{matrix}\sim NH \\ | \\ CO \\ | \\ CHCH_2OSO_3H \\ | \\ {}^{\oplus}NH_3\end{matrix} \xrightarrow[-H_2SO_4]{} \begin{matrix}\sim NH \\ | \\ CO \\ | \\ C{=}CH_2 \\ | \\ {}^{\oplus}NH_3\end{matrix} \xrightarrow{+H_2O} \begin{matrix}\sim {}^{\oplus}NH_3 \\ + \\ COOH \\ | \\ C{=}CH_2 \\ | \\ NH_2\end{matrix} \xrightarrow[-NH_3]{+H_2O} \begin{matrix}COOH \\ | \\ C{-}CH_3 \\ \| \\ O\end{matrix}$$

a sulfidation of tyrosine, and possibly also a formation of sulfamic acids (= amidosulfuric acids), ~NH–$SO_3H$, from ~$NH_2$ and $H_2SO_4$.

Wool fibers have a scale-like structure (Volume IV). They are bicomponent fibers that are composed of two side-by-side parts, paracortex and orthocortex, with different chemical compositions and unlike properties. Chemically, wool consists of ca. 200 different macromolecular compounds of which 80 % are **keratins** (G: *keras* = horn), 17 % are non-keratin proteins, 1.5 % polysaccharides, and 1.5 % lipids and inorganics.

Keratins come in three groups: sulfur-poor helix-forming proteins (ca. 20 types), cystine-rich proteins (ca. 100 types), and glycine/tyrosine-rich proteins (ca. 50 types). Most peptide units of wool proteins carry voluminous side-groups that prevent the formation of pleated sheet structures. Helical structures are crosslinked by disulfide bridges, –S–S–, and $N_{\varepsilon}$-($\gamma$-glutamyl)lysine units, $-CH_2CH_2CO-NH(CH_2)_4-$. In contrast to all other natural fibers, wool is therefore insoluble in all solvents. For the textile properties of wool, see Volume IV.

### Protein Fibers

The term "protein fiber" is usually reserved for fibers from regenerated proteins which may be plant products (e.g., ardein(e), zein) or animal materials (e.g., casein). Ardein is obtained from peanuts; its main components are the proteins arachin and conarachin. Kernels of corn contain ca. 4 % zein. Casein is a component of milk (see below).

Proteins from plants are dissolved in alkali solutions and spun into an acid bath where they are regenerated. After drawing, fibers are hardened, i.e., crosslinked, with aldehydes or aluminum sulfate. The resulting fibers are *not* "regenerated protein fibers" because there was no protein *fiber* to begin with.

About 50 years ago, ardein was used to manufacture **ardein** or **arachin** fiber and zein to produce the fiber Vicara® (ca. 2300 t/a). Zein is now used for microencapsulations, binders in printing inks, coating of tablets, paper coatings, and manufacture of laminated boards. It is also the raw material for the production of glutamic acid.

**Textur(iz)ed proteins** are not texturized fibers but protein-rich plant products such as soy that have been converted to meat-like foodstuffs by wet-spinning or extrusion.

### Casein

Casein is the most important protein of milk. It is usually separated into three fractions ($\alpha$, $\beta$, $\gamma$) which are further subdivided into subfractions. The $\alpha$- and $\beta$-fractions contain phosphate groups that are bound to serine whereas the $\gamma$-fraction is phosphate-poor. The most important subfraction of $\alpha$-casein is $\kappa$-casein which is a phosphate-poor glycoprotein.

Casein is processed to **artificial horn** (Galalith®), casein wool, glues, and binders for paints. The world production is ca. 200 000 t/a.

Artificial horn is produced from skimmed milk that is treated at 35°C with rennin (also called rennet, rennase, or chymosin), the ferment from the inner lining of the fourth stomach of suckling calves. After raising the temperature to 65°C or addition of acids, the proteins denature and coagulate to lean quark which is washed, dried, and chopped up. Ultimately, 30 kg of skimmed milk deliver ca. 1 kg dried casein. The commercial product has a yellowish-milky tint because it still contains some fat.

Because of their different colors, batches are mixed. After swelling in water, casein is dyed in the desired colors and plastified in heated presses. The resulting sheets and rods are then dipped in formaldehyde, often for days. Subsequent treatment with plasticizers (glycerol or oils) at 100°C leads to the desired artifical horn which can be cut and machined, for example, to haberdasheries. Artificial horn still has a market since it is easy to dye in many colors, which is an advantage in the rapidly changing world of fashion.

**Casein wool** is similarly produced. Like natural wool, it is sensitive to acids, alkali, and heat. Its wet strength is lower and it also deforms plastically on stretching.

Grafting of acrylonitrile on casein in the ratio of ca. 2:1 and subsequent spinning leads to a silk-like fiber that is less sensitive to light than silk. It also has better dry and wet strengths.

### 11.4.6 Synthetic Proteins

**Single-cell proteins** (**SCP**) are produced microbiologically by yeasts, bacteria, fungi, or algae from ammonia and various organic substrates such as paraffins, alcohols, sugars, starches, celluloses, etc. SCPs consist of proteins (up to 81 %), nucleic acids (5-8 % if yeasts are used, fats, carbohydrates, salts, and water. SCPs resemble fish meal; they are used as animal fodder (Volume I, p. 544).

To date, there is no industrial synthesis available for **poly(oligopeptide)s**, whether by chemical means or by gene expression. The smallest units of these polymers are pentapeptides such as GEGFP, GVGVP, GVGFP, etc., that are then coupled to hexa(pentapeptide)s, and further to, for example, hexatricosapeptides. These polymers have properties similar to collagen and may be useful as biosensors, biologically degradable matrices, etc.

## Literature to Chapter 11

11.0 GENERAL
A.G.Walter, J.Blackwell, Biopolymers, Academic Press, New York 1973 (no synthesis)
E.A.MacGregor, C.T.Greenwood, Polymers in Nature, Wiley, Chichester 1980
C.R.Cantor, P.R.Schimmel, Biophysical Chemistry, Freeman, San Francisco 1980 (3 vols.)
P.E.Nielsen, M.Egholm, Ed., Peptide Nucleic Acids. Protocols and Applications, Horizon Sci. Press, Wymondham, UK 1999
N.Sewald, H.-D.Jakubke, Peptides: Chemistry and Biology, Wiley-VCG, Weinheim 2002
A.Pandey, Ed., Concise Encyclopedia of Bioresource Technology, Haworth Press, New York 2004
H.M.Berman, K.Henrick, H.Nakamura, Announcing the Worldwide Protein Data Bank, Nature Structural Biology **10**/12 (2003) 980

11.1.2 α-AMINO ACIDS
T.Kaneka, Y.Izumi, I.Chibata, T.Itoh, Synthetic Production and Utilization of Amino Acids, Kodansha, Tokio, and Halsted, New York 1974
K.Yamada, S.Kinoshita, T.Tsunoda, K.Aida, Eds., The Microbial Production of Amino Acids, Kodansha, Tokio, and Halsted, New York 1974
K.Aida, I.Chibata, K.Nakayama, K.Takinami, H.Yamada, Eds., Biotechnology of Amino Acid Production. Elsevier, Amsterdam 1986

11.2 POLY(α-AMINO ACID)S
C.A.Bamford, A.Elliott, W.E.Hanby, Synthetic Polypeptides, Academic Press, New York 1956
M.Szwarc, The Kinetics and Mechanism of *N*-Carboxy-α-Amino-Acid Anhydride (NCA) Polymerization to Polyamino Acids, Adv.Polym.Sci. **4** (1965) 1
J.Noguchi, S.Tokura, N.Nishi, Poly-α-Amino Acid Fibres, Angew.Makromol.Chem. **22** (1972) 107
H.Block, Poly(γ-Benzyl-L-Glutamate) and Other Glutamic Acid Containing Polymers, Gordon and Breach, New York 1983
H.R.Kricheldorf, α-Aminoacid-*N*-Carboxyanhydrides and Related Heterocycles, Springer, Berlin 1988

11.4 PROTEINS (Survey)

H.Neurath, R.L.Hill, Eds., The Proteins, Academic Press, New York, 3rd ed. 1975-82 (5 vols.)
G.D.Fasman, Ed., CRC Handbook of Biochemistry and Molecular Biology, Section A: Proteins (3 vols.), CRC Press, Baton Rouge (FL), 3rd ed. 1976
S.P.Bragg, The Physical Behavior of Macromolecules with Biological Functions, Wiley, New York 1980
R.E.Dickerson, I.Geis, The Structure and Action of Proteins, W.A.Benjamin, New York 1981
T.E.Creighton, Proteins: Structures and Molecular Properties, Freeman, New York 1983
A.M.Lesk, Protein Architecture; A Practical Approach, Oxford Univ. Press, New York 1991
C.Branden, J.Tooze, Ed., Introduction to Protein Structure, Garland Publ., New York 1991
S.Tuboi, N.Taniguchi, N.Katunuma, Post-Translation Modification of Proteins, CRC Press, Boca Raton (FL) 1992
A.Steinbüchel, S.R.Fahnestock, Eds., Polyamides and Complex Proteinaceous Materials I (2002); S.R.Fahnestock, A.Steinbüchel, Eds., Polyamides and Complex Proteinaceous Materials II (2003); = Vols. 7 and 8 of A.Steinbüchel, Ed., Biopolymers, Wiley, New York 2001-2003 (12 volumes)

11.4.2A ENZYMES (general)

-, Enzyme Nomenclature: Recommendations (1984) of the Nomenclature Committee of the International Union of Biochemistry, Academic Press, Orlando (FL) 1985
P.D.Boyer, Ed., The Enzymes, Academic Press, New York, 3rd ed. 1970-1983 (15 volumes)
D.Schomburg, M.Salzmann, Eds., Enzyme Handbook, Springer, Berlin, Vol. 1 (Class 4: Lyases) (1990), Volume 2 (Class 5: Isomerases; Class 6: Ligases) (1990), Volumes 3-5 (Class 3: Hydrolases) (1991)
S.Doonana, Peptides and Proteins, Wiley, Hoboken (NJ) 2003
R.Breslow, Ed., Artificial Enzymes, Wiley-VCH, Weinheim 2005

11.4.2B ENZYMES (structure)

M.A.Dayhoff, Ed., Atlas of Protein Sequence and Structure, Natl.Biomed.Res.Found., Silver Spring (MD), 1972-1976
S.B.Needleman, Ed., Protein Sequence Determination, Springer, New York, 2nd ed. 1975
T.L.Blundell, L.N.Johnson, Protein Crystallography, Academic Press, New York 1976
S.Blackburn, Ed., Amino Acid Analysis, Dekker, New York, 2nd ed. 1978
S.Lapanje, Physicochemical Aspects of Protein Denaturation, Wiley, New York 1978
G.E.Schulz, R.H.Schirmer, Principles of Protein Structure, Springer, New York 1979
L.R.Croft, Handbook of Protein Sequence Analysis, Wiley, New York, 2nd ed. 1980
H.Bisswanger, E.Schmincke-Ott, Eds., Multifunctional Proteins, Wiley, New York 1980
G.Walton, Polypeptide and Protein Structure, Elsevier, Amsterdam 1981
C.Frieden, L.W.Nichol, Protein-Protein Interactions, Wiley-Interscience, New York 1981
C.C.Ghélis, J.Yon, Protein Folding, Academic Press, New York 1982
A.McPherson, Preparation and Analysis of Protein Crystals, Wiley, New York 1982
P.M.Harrison, Ed., Metalloproteins, VCH, Weinheim 1985 (2 parts)

11.4.2C ENZYMES (action)

M.V.Volkenstin, Enzyme Physics, Plenum, New York 1969
P.D.Boyer, Ed., The Enzymes, Academic Press, New York, 3rd ed. 1970-1982 (15 vols.)
E.Zeffren, P.L.Hall, The Study of Enzyme Mechanisms, Wiley, New York 1973
I.H.Segel, Enzyme Kinetics, Wiley, New York 1975
J.Tze-Fei Wong, Kinetics of Enzyme Mechanisms, Academic Press, London 1975
K.G.Serimgeour, Chemistry and Control of Enzyme Reactions, Academic Press, London 1977
A.Cornish-Bowden, Fundamentals of Enzyme Kinetics, Butterworths, London 1979
C.Walsh, Enzymatic Reaction Mechanisms, Freeman, San Francisco 1979
H.Bisswanger, Theorie und Methoden der Enzymkinetik, Verlag Chemie, Weinheim 1979
M.Dixon, E.C.Webb, C.J.R.Thorne, K.F.Tipton, Enzymes, Academic, New York, 3rd ed. 1980
S.Kobayashi, S.-I.Shoda, H.Uyama, Enzymatic Catalysis, in S.Kobayashi, Ed., Catalysis in Precision Polymerization, Wiley, New York 1997
A.G.Marangoni, Enzyme Kinetics, Wiley, Hoboken (NJ) 2003

11.4.2. ENZYMES (Industrial uses)
-, Immobilized Enzymes, Corning Glass Works, Corning, NY, Vol. I (1972), Vol. II (1973), Vol. III (1974)
O.R.Zaborsky, Immobilized Enzymes, CRC Press, Cleveland 1973
R.A.Messing, Ed., Immobilized Enzymes for Industrial Reactors, Academic Press, New York 1975
J.Konecny, Enzymes as Industrial Catalysts, Chimia **29** (1975) 95
K.J.Skinner, Enzyme Technology, Chem.Eng.News (18.Aug.1975) 22
H.T.Weetall, S.Suzuki, Ed., Immobilized Enzyme Technology, Plenum, New York 1975
L.B.Wingard, Jr., E.Katchalsky-Katzir, L.Goldstein, Immobilized Enzyme Principles (= Appl.Biochem.Bioeng. 1), Academic Press, New York 1976
L.Chibata, Ed., Immobilized Enzymes, Wiley, New York 1978
J.C.Johnson, Immobilized Enzymes: Preparation and Engineering, Noyes Data, Park Ridge (NJ) 1979
N.D.Pintauro, Food Processing Enzymes, Noyes Publ., Park Ridge (NJ), 1979
K.Buchholz, Ed., Characterization of Immobilized Biocatalysts, Dechema-Monograph 84 (1979)
D.I.C.Wang, C.L.Cooney, A.L.Demain, P.Dunnill, A.E.Humphrey, M.D.Lilly, Fermentation and Enzyme Technology, Wiley, New York 1979
M.G.Halpern, Ed., Industrial Enzymes from Microbial Sources, Noyes Publ., Park Ridge (NJ) 1981
L.Chibata, L.B.Wingard, Jr., Eds., Immobilized Microbial Cells, Academic Press, New York 1983 (= Appl.Biochem.Bioeng. **4** (1983))
T.Godfrey, J.Reichelt, Eds., Industrial Enzymology: The Application of Enzymes in Industry, Macmillan, The Nature Press, London 1983
W.Gerhartz, Ed., Enzymes in Industry, VCH, Weinheim 1990
W.Aele, Ed., Enzymes in Industry, Wiley-VCH, Weinheim, 2nd ed. 2003
A.S.Bommarius, B.R.Riebel, Biocatalysis. Fundamentals and Applications, Wiley-VCH, Weinheim 2004
K.Buchholz, V.Kasche, U.T.Bornscheuer, Biocatalysis and Enzyme Technology, Wiley-VCH, Weinheim 2005
L.Cao, Carrier-Bound Immobilized Enzymes, Wiley-VCH, Weinheim 2005
S.Kobashi, H.Ritter, D.Kaplan, Eds., Enzyme-Catalyzed Synthesis of Polymers, Adv.Polym.Sci. **194** (2006)

11.4.3a SCLEROPROTEINS (general)
E.D.T.Atkins, A.Keller, Eds., Structure of Fibrous Biopolymers, Butterworths, London 1975
D.A.D.Pery, L.K.Creamer, Fibrous Proteins, Academic Press, New York 1979 (2 vols.)
J.F.V.Vincent, Structural Biomaterials, Wiley, New York 1982

11.4.3b COLLAGEN
G.N.Ramachandran, A.H.Rheddi, Eds., Biochemistry of Collagen, Plenum, New York 1976
J.Woodhead-Galloway, Collagen, The Anatomy of a Protein, Arnold, London 1980
M.Nimni, Ed., Collagen: Biochemistry, Biotechnology and Molecular Biology, Vol. 3, CRC Press, Boca Raton (FL) 1988

11.4.3c GELATIN
A.Veis, Macromolecular Chemistry of Gelatin, Academic Press, New York 1964
R.H.Cox, Ed., Photographic Gelatin, Academic Press 1972
A.G.Ward, A.Courts, Eds., The Science and Technology of Gelatin, Academic Press, London 1977

11.4.4 PROTEIN GUMS
L.B.Sandberg, W.R.Gray, C.Franzblau, Eds., Elastin and Elastic Tissue, Plenum, New York 1977

11.4.5a PROTEIN FIBERS (general)
R.L.Wormell, New Fibres from Proteins, Academic Press, New York 1954
R.S.Asquith, Chemistry of Natural Fibers, Plenum, New York 1977
M.Gutcho, Textured Protein Products, Noyes Publ., Park Ridge (NJ) 1977
D.A.D.Parry, L.K.Creamer, Eds., Fibrous Proteins: Scientific, Industrial, and Medical Aspects, Academic Press, New York 1980

11.4.5b WOOL
W. von Bergen, Wool Handbook, American Wool Handbook Co., New York 1963 (2 vol.)
C.Earland, Wool, Its Chemistry and Physics, Chapman & Hall, London, 2nd ed. 1963
R.D.B.Fraser, T.P.McRae, G.E.Rogers, Keratins - Their Compositions, Structure, and Biosynthesis, Thomas, Springfield (IL) 1972

11.4.5c PROTEIN FIBERS
J.H.Collins, Casein Plastics and Allied Materials, Plastics Inst., London 1952
H.D.McKenzie, Milk Proteins, Academic Press, New York 1970
P.F.Fox, Developments in Dairy Chemistry, Appl.Sci.Publ., London 1982

11.4.6 INDUSTRIAL PROTEINS
D.W.Urry, Molecular Machines: How Motion and Other Functions of Living Organisms Can Result From Reversible Chemical Changes, Angew.Chem. **105** (1993) 859; Angew.Chem.Int.Ed.Engl. **32** (1993) 819;

## References to Chapter 11

[1] Updated Table 14-9 of Volume I of this work.
[2] S.Kinoshita, Proc. 4th European Congress on Biotechnology, Amsterdam 1987
[3] J.P.O'Brien, Trends Polym.Sci. **1**/8 (1993) 228, Tab. 1
[4] Y.Izumi, I.Chibata, T.Itoh, Angew.Chem. **90** (1978) 187, Table 1

# 12 Inorganic and Semi-Organic Polymers

## 12.1 Introduction

Carbon forms with itself (Chapter 6) or with other chemical elements such as oxygen (Chapters 7 and 8), sulfur (Chapter 9), or nitrogen (Chapters 10 and 11) polymers that are called **organic polymers** because their chains and substituents consist mainly of carbon atoms. **Inorganic polymers**, on the other hand, have chains that consist exclusively of non-carbon chain atoms; this term usually (but not always) contains polymers with carbon-containing sidegroups. **Semi-organic polymers** are those where 50 % or more of their chain atoms are non-carbon atoms. For the purpose of the following discussions it should be remembered that polymers such as poly(methylene), $-\!\!+CH_2+\!\!-_n$, are called *unsubstituted* in organic chemistry since alkanes, $H(CH_2)_nH$ are defined as parent compounds and not the diamond lattice, $\gtrdot C \lessdot$. Inorganic chemistry, on the other hand, considered polymers such as $-\!\!+SiH_2+\!\!-_n$ as substituted because the parent compound is silicone, $\gtrdot Si \lessdot$.

Polymers with metal atoms in chains or as substituents are called **metallo-organic**, **organometallic**, or **metal-containing polymers**, and, in the Russian scientific literature, also **element-organic polymers**. Most of these polymers do not belong to the group of inorganic polymers since their backbones usually consist of more than 50 % carbon atoms as the following examples of repeating units show (Mt = metal, L = ligand:

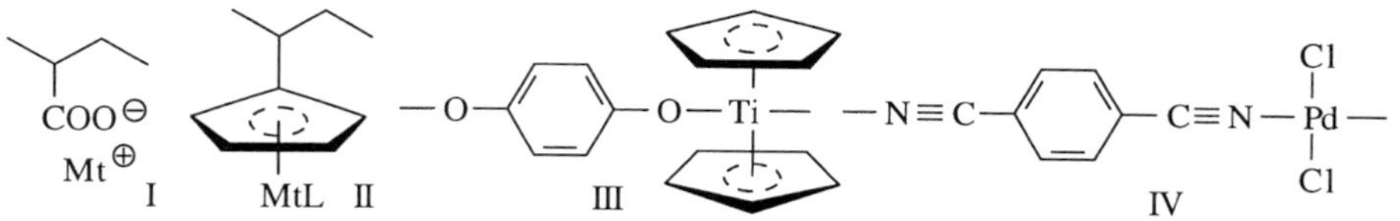

COO⊖ Mt⊕ I — MtL II — —O—C₆H₄—O—Ti— III — —N≡C—C₆H₄—C≡N—PdCl₂— IV

There are many inorganic and semi-inorganic polymers (see also Volume I, p. 37 ff.), both natural and synthetic, but relatively few of the synthetic ones are utilized industrially because of their structures, properties, or processing.

Of ca. 1200 crystal structures of inorganic compounds of *two elements*, only ca. 5 % turned out to be non-polymeric. Some 86 % were lattice polymers, 7.5 % sheet polymers, and just 1.5 % linear ones. Sheet and lattice polymers require simultaneous synthesis and shaping, usually at high temperatures, which rules out many candidates.

Some inorganic bonds possess considerably higher bond energies than carbon–carbon bonds (ca. 320 kJ/mol); for example carbon–boron (370 kJ/mol), silicon–oxygen (370 kJ/mol), boron–nitrogen (440 kJ/mol), and boron–oxygen (500 kJ/mol). Polymers with such heterochains should thus be much more thermally stable than carbon–carbon chains. However, such bonds are usually very much polarized. There are also often free electron pairs or incomplete electron shells. Reactions of such bonds therefore need only small activation energies. Hence, many polymers with inorganic heterochains oxidize or hydrolyze easily, and their good thermal stabilities can only be exploted in inert atmosphere, for example, in space.

Resistance against oxidation and/or hydrolysis can be improved by several means. Substitution of chains often protects the chains sterically or electronically. For ladder, sheet and lattice polymers, simultaneous attacks on two neighboring bonds are unlikely.

Another problem is the limited number of suitable types of syntheses. Polymer molecules are generally synthesized by four classes of reaction:

- Chain polymerizations of unsaturated organic monomers are common, but there are only few unsaturated inorganic monomers for such polymerizations.
- Ring-opening polymerizations are less common for organic polymers, but the major type of polymerizations for inorganic ones.
- Polycondensations or polyadditions of functional groups to linear inorganic or semi-organic polymers are relatively rare since most inorganic functional groups do not react monofunctionally.
- Because of side reactions and high costs, organic polymers are only interconverted to other polymers if parent polymers are inexpensive, reactions specific, and/or target polymers cannot be obtained by other means. For inorganic and semi-inorganic polymers, such polymer transformations are often the method of choice

Because of its place in the Periodic System of Elements, it is not surprising that inorganic and semi-inorganic polymers are dominated by silicon. Phosphorus-containing chains are far less common and boron- and sulfur-containing ones even less so. In all of these compounds, inorganic atoms may be incorporated in sidegroups, in backbones of linear or branched chains, or in sheet or lattice polymers.

## 12.2 Boron Polymers

### 12.2.1 Elementary Boron

**Boron** (name derived from the mineral borax (Armenian: *buraq*; Persian: *burah*)) has several allotropes. The red α form and the shiny dark β form both crystallize rhombohedrally. The α form is not polymeric since it consists of cubic-closed packed $B_{12}$ icosahedrons that are held together by weaker intermolecular bonds ($T_M$ = 2180°C; sublimation at 3650°C). The β form is polymeric: it consists also of $B_{12}$ icosahedrons but these are bonded to each other by boron–boron bonds.

Older reports about two black tetragonal forms (also called α and β) are now disputed. These forms were said to consist of layers of $B_{12}$ icosahedrons that are separated by layers of single boron atoms. The so-called amorphous boron allegedly has a "melting temperature" of 2300°C and a sublimation temperature of 2550°C.

Boron is synthesized by reduction of diboron trioxide (boron oxide, boric anhydride), $B_2O_3$, with magnesium oxide (Moisson process) or, in a more pure form, by thermal decomposition of boron hydrides such as $B_2H_6$. Boron oxide is obtained by dehydration of (ortho)boric acid, $B(OH)_3$, which is produced from borax, $Na_2B_4O_7 \cdot 10\ H_2O = Na_2O \cdot 2\ B_2O_3 \cdot 10\ H_2O$, or ulexite (boronatrocalcite), $NaCaB_5O_9 \cdot 8\ H_2O$.

**Boron fibers** result from the thermal deposition of gaseous $BCl_3$ in streaming hydrogen on tungsten or carbon fibers of ca. 10 μm diameter. The fibers with diameters of 120-140 μm have a density of 2,58 g/cm$^3$, a melting temperature of 2100°C, a tensile modul of 400 GPa, and a tensile strength of ca. 3.5 GPa. Because of their high strength and thermal resistance, they are used to reinforce engineering plastics, aluminum, and titanium (Volume IV).

### 12.2.2 Boron–Nitrogen Polymers

Boron forms with nitrogen a number of different types of chemical compounds. Parent compounds include the borane–nitrogen analogs of ethane, ethene, and benzene (R = H) and their derivatives, as well as a graphite analog:

- **borazanes** (amineboranes) $R_3N^{\oplus}-B^{\ominus}R_3$;
- **borazenes** (aminoboranes), $R_2N^{\oplus}=B^{\ominus}R_2$. The ethene analog, $H_2N^{\oplus}=B^{\ominus}H_2$, starts to polymerize slowly to $[H_2BNH_2]_n$ at –196°C;
- **borazines** (cyclotriborazanes), $B_3R_6N_3$, also called **borazol** ("inorganic benzene");
- **boron nitride**, $[BN]_n$.

Boron nitride is synthesized by heating poly(aminoborane), $(H_2BNH_2)_n$, slowly to 135–200°C where it loses hydrogen and forms the white boron nitride, $(BN)_n$:

(12-1) $$[H_2BNH_2]_n \longrightarrow [BN]_n + 2\,n\,H_2$$

Industry uses the so-called **CCPF** process (*C*hemical *C*onversion of *P*recursor *F*ibers) in which precursor fibers of diboron trioxide, $B_2O_3$, are reacted with ammonia to BN fibers with nitrogen contents of 48–52 wt% (Eq.(12-2)). The BN fibers are then continuously stretched to white boron nitride fibers ((hexagonal boron nitride (hBN); theoretical nitrogen content: 56.4 wt%) at temperatures above 1800°C. hBN exists as sheet polymer that is stable against hydrolysis and oxidation at temperatures up to 200°C. A denser cubic form is obtained by subjecting hBN to very high temperatures and pressures.

(12-2) $$B_2O_3 \xrightarrow[200°C]{+\,NH_3} (B_2O_3)_n\cdot NH_3 \xrightarrow[-\,H_2O]{>350°C} (BN)_x(B_2O_3)_y(NH_3)_z \xrightarrow[-\,H_2O]{>1800°C} \text{[hexagonal B–N sheet structure]}$$

### 12.2.3 Boron–Carbon Polymers

Fibers from boron carbide, $[B_4C]_n$, are produced by the CCPF process similar to boron nitride fibers. Instead of a diboron trioxide fiber, one uses a carbon fiber as matrix which is then reacted at 1800°C with $BCl_3 + H_2$. Both boron carbide and boron nitride fibers serve as reinforcing fibers for composites.

### 12.2.4 Carborane–Siloxane Polymers

Carborane–siloxane polymers contain *m*-carborane and siloxane groups in the chain. Their synthesis starts with decaborane, $B_{10}H_{14}$ (or pentaborane, $B_5H_9$) which is functionalized with acetylene (Fig. 12-1). The resulting *o*-carborane (= 1,2-dicarbacloso-decaborane), *o*-$B_{10}C_2H_{12}$, transforms at 475°C to *m*-carborane, *m*-$B_{10}C_2H_{12}$ and at 650-700°C to *p*-carborane. Reaction of *m*-carborane with butyl lithium, $C_4H_9Li$, to dilithium *m*-carborane, $Li_2B_{10}C_2H_{10}$, releases butane, $C_4H_{10}$:

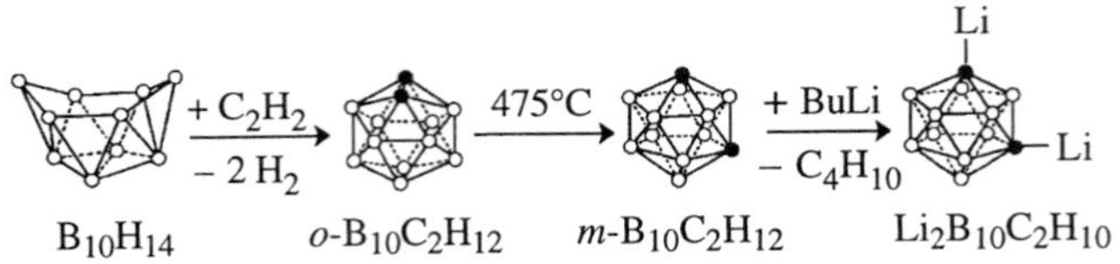

$$B_{10}H_{14} \xrightarrow[-\,2\,H_2]{+\,C_2H_2} o\text{-}B_{10}C_2H_{12} \xrightarrow{475°C} m\text{-}B_{10}C_2H_{12} \xrightarrow[-\,C_4H_{10}]{+\,BuLi} Li_2B_{10}C_2H_{10}$$

Fig. 12-1 Synthesis of dilithium *m*-carborane, $Li_2[m\text{-}B_{10}C_2H_{10}]$, from decaborane, $B_{10}H_{14}$, acetylene, and butyl lithium. ○ Boron atoms, ● carbon atoms; hydrogen atoms are not shown.

The subsequent reaction of dilithium *m*-carborane, $Li_2B_{10}C_2H_{10}$, with dichlorodisiloxanes such as $Cl–Si(CH_3)_2–O–Si(CH_3)_2–Cl$ is catalyzed by acids. The resulting dichloro compound is then polycondensed with water to carborane–siloxane polymers with a repeating unit I and molecular weights between 15 000 and 30 000:

$$(12\text{-}3)\qquad Li[m\text{-}CB_{10}H_{10}C]Li + 2\,Cl{-}Si(CH_3)_2{-}O{-}Si(CH_3)_2{-}Cl$$

$$\xrightarrow[-\,2\,LiCl]{} Cl{-}Si(CH_3)_2{-}O{-}Si(CH_3)_2{-}[m\text{-}CB_{10}H_{10}C]{-}Si(CH_3)_2{-}O{-}Si(CH_3)_2{-}Cl$$

$$\xrightarrow[-\,2\,HCl]{+\,H_2O} \sim\!Si(CH_3)_2{-}O{-}Si(CH_3)_2{-}[m\text{-}CB_{10}H_{10}C]{-}Si(CH_3)_2{-}O{-}Si(CH_3)_2{-}O\!\sim \qquad \text{I}$$

The polycondensation succeeds only with siloxane groups, $\sim Si(CH_3)_2\text{-}O\text{-}Si(CH_3)_2\text{-}Cl$, as endgroups but not with terminal silyl groups, $\sim Si(CH_3)_2Cl$.

Disiloxanes with silyl endgroups are obtained by the reaction of $mLiCB_{10}H_{10}CLi$ with $(CH_3)_2SiCl_2$. The resulting dichloro compound (II) can be converted to the dimethoxy compound (III). II and III can then be polycondensed with $FeCl_3$ as catalyst to a polymer with rubber-like properties ($T_G$ below room temperature (Table 12-2)):

$$(12\text{-}4)\qquad \underset{\text{II}}{Cl{-}Si(CH_3)_2{-}[m\text{-}CB_{10}H_{10}C]{-}Si(CH_3)_2{-}Cl} + \underset{\text{III}}{CH_3O{-}Si(CH_3)_2{-}[m\text{-}CB_{10}H_{10}C]{-}Si(CH_3)_2{-}OCH_3}$$

$$\xrightarrow[-\,2\,CH_3Cl]{} \underset{\text{IV}}{\sim\!Si(CH_3)_2{-}[m\text{-}CB_{10}H_{10}C]{-}Si(CH_3)_2{-}O{-}Si(CH_3)_2{-}[m\text{-}CB_{10}H_{10}C]{-}Si(CH_3)_2{-}O\!\sim}$$

Industrial polymers have the repeating unit V, usually with R = R' = R" = $CH_3$:

$$\text{V}\qquad \sim\!Si(R)(CH_3){-}[m\text{-}CB_{10}H_{10}C]{-}Si(R)(CH_3){-}\left[O{-}Si(R')(R'')\right]_n{-}O\!\sim \qquad \begin{array}{l}\text{SiB-1: } n = 0\\ \text{SiB-2: } n = 1\end{array}$$

Table 12-2 Melting temperatures $T_M$, glass temperatures $T_G$, relative permittivities ("dielectric constants") $\varepsilon_r$, limiting oxygen indices LOI (= flammability index), and pyrolysabilities $P$ (in argon at 800°C) of some carborane–siloxane polymers (V). The industrial production stopped in the 1970s.

| $n$ | R | R' | R" | $T_M$/°C | $T_G$/°C | $\varepsilon_r$ | LOI/% | $P$/% |
|---|---|---|---|---|---|---|---|---|
| 0 | $CH_3$ | | | 240 | 25 | | | 20 |
| 1 | $CH_3$ | $CH_3$ | $CH_3$ | 66 | – 30 | 2,27 | | 29 |
| 2 | $CH_3$ | $CH_3$ | $CH_3$ | 40 | – 68 | | | 36 |
| 3 | $CH_3$ | $CH_3$ | $CH_3$ | | – 70 | 5,92 | | 47 |
| 4 | $CH_3$ | $CH_3$ | $CH_3$ | | – 88 | | | 48 |
| 1 | $CH_3$ | $CH_3$ | $C_6H_5$ | | – 12 | | 62 | |
| 1 | $CH_3$ | $C_6H_5$ | $C_6H_5$ | | + 22 | | | |
| 1 | $CH_3$ | $CH_3$ | $(CH_2)_2CF_3$ | | – 29 | | | |
| 1 | $(CH_2)_2CF_3$ | $(CH_2)_2CF_3$ | $(CH_2)_2CF_3$ | | – 3 | | | |

These polymers are either homopolymers with units SiB-1 ($n$ = 0) or SiB-2 ($n$ = 1), copolymers of SiB-1 and SiB-2, or derivatives of homopolymers with phenyl, trifluoropropyl, and/or vinyl groups (see Table 12-2). Carborane–siloxane have excellent temperature stabilities and flame resistancies. Polymers with vinyl groups can be crosslinked and processed similar to silicone polymers; they served for gaskets, O rings, and wire coatings. Uncrosslinked polymers are also used as liquid phases in gas chromatography.

### 12.2.5 Boron–Hydrogen and Boron–Oxygen Compounds

Boranes are boron–hydrogen compounds with both boron–boron and boron–hydrogen–boron bonds. High-molecular weight boranes are not known.

Polymeric boron-oxygen compounds exist as polymeric boron oxides, $(BO)_n$, and polymeric metaborates with the unit $(B_2O_3)^{3\ominus}$ with either closed-ring or open-chain structures. Zinc borate, 2 $ZnO \cdot 3\ B_2O_3 \cdot (7/2)\ H_2O$, is used as a flame-retarding additive.

## 12.3 Silicon Polymers

Like carbon, silicon forms with itself and with other elements a series of polymeric compounds that range from linear chains to three-dimensional lattices. The crystal lattice of solid silicon resembles that of diamond. Pure silicon is an electrical insulator with an electrical conductivity of ca. $10^{-18}$ S/cm. It cannot form graphite-like sheets.

Chains of substituted silicon atoms, $\text{+}SiR_2\text{+}_n$, are known as **polysilanes** (Section 12.3.1). Chains in which silicon atoms alternate with carbon, oxygen, or nitrogen are called **polycarbosilanes,** $\text{+}SiR_2\text{–}CR_2'\text{+}_n$, (Section 12.3.2) **polysiloxanes,** $\text{+}SiR_2\text{–}O\text{+}_n$, (Section 12.3.11) and **polysilazanes,** $\text{+}SiR_2\text{–}NR'\text{+}_n$, (Section 12.3.3). Such heterosilicon chains can also combine to form ladder, double ladder, layer, and lattice polymers.

The vast class of inorganic silicon–oxygen compounds (Section 12.3.5-12.3.10)) derives from ortho-silicic acid, $Si(OH)_4$, (Section 12.3.4) and ranges from low- and high-molecular weight **silicates** to the tecto polymer **quartz,** $[SiO_2]_\infty$, (Section 12.3.12).

### 12.3.1 Silanes and Polysilylenes

**Silanes**, $H(SiH_2)_nH$, are linear silicon–hydrogen compounds with degrees of polymerization od $1 \leq n \leq 12$. At room temperature, silane, $SiH_4$, and disilane, $Si_2H_6$, are gases; tri-, tetra-, penta-, and hexasilanes are liquids; and decasilane is crystalline. Tetrasilane and higher silanes exist in various constitutional isomers. **Cyclosilanes**, $[SiH_2]_n$, are ring-shaped with up to 6 Si atoms per molecule. Silanes and cyclosilanes are unstable molecules; they oxidize explosively to $SiO_2$ in the presence of air.

The yellow, self-igniting **poly(silene)**, $[SiH_2]_\infty$, seems to be a polymer. **Poly(siline)**, $[SiH]_\infty$, exists in graphene-like sheets of hexagonally interconnected silicon atoms that are substituted by one hydrogen atoms and interconnected by single bonds. The corresponding poly(silicon monochloride), $[SiCl]_\infty$, is a white layer polymer which forms from the gaseous, carbene-like silicon dichloride (dichlorosilylene, dichlorosilandiyl), $:SiCl_2$. Alkyl-substituted poly(siline)s, $[SiR]_\infty$, are soluble. They consist of highly branched silicon networks that are separated by organic substitutents.

Substituted silanes are considerably more stable than unsubstituted (= hydrogen-substituted) ones. Linear **silicon chlorides** have been prepared up to $Si_{25}Cl_{52}$, and their cyclic counterparts up to $Si_{10}Cl_{20}$.

Even more stable are **polysilylenes** which are alkyl- or aryl-substituted **polysilanes** of the type $\text{+SiRR'+}_n$. These polymers can be obtained by very many reactions such as the reductive coupling of silicon dichlorides with sodium in toluene, a heterogeneous reaction on the surface of sodium metal, probably a chain reaction,

$$(12\text{-}5) \qquad n\,R_2SiCl_2 + 2\,n\,Na \rightarrow \text{+SiR}_2\text{+}_n + 2\,n\,NaCl \quad ; \quad R_2 = 2\,CH_3 \text{ or } CH_3 + C_6H_5$$

the dehydrogenative coupling of silanes in the presence of a transition metal catalyst,

$$(12\text{-}6) \qquad n\,R_2SiH_2 \rightarrow \text{+SiR}_2\text{+}_n + n\,H_2$$

the polymerization of a masked disilene,

$$(12\text{-}7) \qquad 2n \; \text{R}_2\text{Si–SiR}_2 \text{(bridged by cyclohexadiene ring)} \longrightarrow \left[ \text{–SiR}_2\text{–SiR}_2\text{–} \right]_{2n} + n \; \text{biphenyl}$$

and the anionic ring-opening polymerization,

$$(12\text{-}8) \qquad \text{cyclo-}[Si(H)(C_6H_5)]_4 \longrightarrow \left[ \text{–Si(C}_6\text{H}_5\text{)(H)–} \right]_{4n}$$

Dehydrogenative couplings (Eq.(12-6)) can also be performed with monoalkyl silanes instead of dialkyl ones. The resulting polymers have a reactive hydrogen substituent which can be used to introduce ino polysilylene various substituents by the addition of monounsaturated compounds, $CH_2{=}CHR'$:

$$(12\text{-}9)\qquad n\ \mathrm{RSiH_3} \xrightarrow[-n\,\mathrm{H_2}]{} \left[\!-\mathrm{Si(R)(H)}-\!\right]_n \xrightarrow{+n\,\mathrm{CH_2{=}CHR'}} \left[\!-\mathrm{Si(C_6H_5)(CH_2CH_2R')}-\!\right]_n$$

Polysilylenes as well as polycarbosilanes with three or more sequential Si atoms per repeating unit and alkyl substitutents R and R' undergo photochemical scissions in solution. Depending on conditions, either silylenes (reaction A) or silyl radicals (reaction B) or both silylenes and silyl radicals are obtained (reaction C):

(12-10)

$$\sim\mathrm{SiRR'}-\mathrm{SiRR'}-\mathrm{SiRR'}\sim$$

reaction A: $\mathrm{R'RSi:} + \sim\mathrm{SiRR'}-\mathrm{SiRR'}\sim$ or reaction B: $3\ \mathrm{R'RSi\bullet}$ or reaction C: $\mathrm{R'RSi:} + 2\ \sim\mathrm{SiRR'\bullet}$

Polysilanes strongly absorb ultraviolet light between 290 nm and 420 nm. Photochemical degradations, (Eq.(12-10)), shorten the sequence lengths of Si chain atoms and thus $\sigma$ conjugations. As a result, both maxima of light absorption as well as absorptions per Si–Si bond decrease. Light can thus penetrate thicker films, which is important for the applications of polysilanes as positive photo resists (see Volume IV). Poly(methylphenylsilane) is used in optical lithography; copolymers with methylphenylsilane and dimethylsilane units serve also as initiators for vinyl polymerizations.

### 12.3.2 Polycarbosilanes

Polycarbosilanes are defined as polymers with alternating silicon and carbon atoms as chain atoms, i.e., as polymers with repeating units –SiRR'-CR"R'"–. The term "polycarbosilane" sometimes also includes polymers with more than one carbon chain atom per repeating unit, i.e., polymers of the type $\!\!+\!\mathrm{SiRR'\text{-}(CR''R''')}_i\!+_n$ with $i \geq 2$. These polymers serve as precursors for silicon carbide fibers.

The simplest polycarbosilane is the linear **poly(silaethylene)**, $+\mathrm{SiH_2{-}CH_2}+_n$, which is obtained by ring-opening polymerization of 1,1,3,3-tetrachloro-1,3-disilacyclobutane (I) with $H_2PtCl_6$ as catalyst to the dichloro polymer (II) and subsequent reduction of II by $LiAlH_4$ to III:

(12-11)

$$n\ \mathrm{Cl_2Si{<}(CH_2)_2{>}SiCl_2} \longrightarrow \left[\!-\mathrm{SiCl_2}-\mathrm{CH_2}-\mathrm{SiCl_2}-\mathrm{CH_2}-\!\right]_n \xrightarrow[-\,\mathrm{LiCl},\ -\,\mathrm{AlCl_3}]{+\,\mathrm{LIAlH_4}} \left[\!-\mathrm{SiH_2}-\mathrm{CH_2}-\mathrm{SiH_2}-\mathrm{CH_2}-\!\right]_n$$

Use of $Cl_3SiCH_2Cl$ as a monomer for a Grignard reaction delivers a highly branched, relatively low-molecular weight polymer known as **hydridopolycarbosilane (HPCS, HBPSE)**.

Hydridopolycarbosilane, poly(silaethylene), poly(dimethylsilane), and poly(dimethylsilane-*co*-methylphenylsilane) serve as precursors for silicon carbide fibers. For example, heating to 450°C converts insoluble and difficult to process poly(dimethylsilane) (I) in part to heptane soluble **poly(silapropylene)**, also called **poly(methylsilamethylene)** (II):

$$\text{(12-12)} \qquad \left[ -\underset{CH_3}{\overset{CH_3}{\underset{|}{\overset{|}{Si}}}}- \right]_n \text{ I} \longrightarrow \left[ -\underset{H}{\overset{CH_3}{\underset{|}{\overset{|}{Si}}}}-\underset{H}{\overset{H}{\underset{|}{\overset{|}{C}}}}- \right]_n \text{ II} \longrightarrow \text{silicon carbide fibers}$$

In part, it also loses hydrogen and methane and is crosslinked to poly(carbosilane)s.

Melt spinning of poly(silapropylene) (II) delivers fibers that are dimensionally stabilized by surface oxidation. Further heating of these fibers to 1300°C under an argon blanket leads to **silicon carbide fibers** that consist mainly of crystalline β-silicon carbide, $[SiC]_n$, plus smaller proportions of amorphous $[SiC]_n$ and $SiO_2$. Silicon carbide fibers are obtained industrially not only by conversion of poly(dimethylsilane) (I) and poly(methylsilamethylene) according to Eq.(12-12) but also from dodecamethylcyclohexasilane, $Si_6(CH_3)_{12}$.

Polycondensation of mixtures of dimethyldichlorosilane, $(CH_3)_2SiCl_2$, and phenylmethyldichlorosilane, $(C_6H_5)(CH_3)SiCl_2$, with sodium results in a so-called **poly(silastyrene)**, $+Si(CH_3)_2-Si(CH_3)(C_6H_5)+_n$, with molecular weights up to 400 000 which can also be converted to silicon carbide with loss of hydrogen, methane, and benzene.

Another process for silicon carbide fibers employs a soluble polycarbosilane from the reaction of $ClCH_2Si(CH_3)_2Cl$, $CH_2{=}CHSi(CH_3)Cl_2$, and $(CH_3)_2SiCl_2$ with potassium.

Polycarbosilanes also result if $(CH_3)_4Si$ or alkylarylchlorosilanes are passed through pipes that are heated to 700°C. The resulting polymers are processed to fibers, films, and foamed materials which are converted to silicon carbide by pyrolysis at 800–2000°C.

**SiC fibers** result from the deposition of β-silicon carbide on a core of carbon fibers by passing a mixture of $Si(C_2H_5)_4$ and $H_2$ as carrier gas through an electric arc at temperatures of 1100–1300°C.

Polycarbosilanes with two $CH_2$ groups between Si chain atoms result from hydrosilylation, and those with three $CH_2$ groups from ring-opening polymerization.

Poly(carbosilane) with longer methylene sequences between silicon chain atoms are available by various reactions: with two methylene groups by hydrosilylation and with three methylene groups by ring-opening polymerization:

$$\text{(12-13)} \qquad H-\underset{R}{\overset{R}{\underset{|}{\overset{|}{Si}}}}-CH{=}CH_2 \xrightarrow{H_2PtCl_6} \sim\!\underset{R}{\overset{R}{\underset{|}{\overset{|}{Si}}}}-CH_2-CH_2\sim \quad \text{; also as AA/BB reaction}$$

$$\text{(12-14)} \qquad CH_3-\underset{CH_3}{\underset{|}{Si}}(-CH_2-CH_2-CH_2-) \longrightarrow \sim\!\underset{CH_3}{\overset{CH_3}{\underset{|}{\overset{|}{Si}}}}-CH_2-CH_2-CH_2\sim$$

### 12.3.3 Polysilazanes

Polysilazanes with Si–N chain bonds are interesting as precursors for silicon nitride ceramics, [$Si_3N_4$]. Such precursors must be easy to process; they can thus be branched but not crosslinked. Linear or lightly branched polysilazane are not easy to synthesize, though, since silicon has four valences and nitrogen three, and the choice of substituents is limited. Organic substituents would lead to undesirable carbon-containing pyrolysates at the high temperatures required for the preparation of ceramics, and hydrogen, halogen, and nitrogen compounds of silicon rearrange easily.

Aminolysis of dichlorosilane results in a mixture of ca. 70 % linear oligosilazenes

$$(12\text{-}15) \qquad (3\,n+1)\,CH_3NH_2 + n\,H_2SiCl_2 \xrightarrow{-2\,n\,[H_3CNH_3]Cl} CH_3\text{—}NH\text{—}\left[SiH_2\text{—}N(CH_3)\right]_n\text{—}H$$

and ca. 30 % cyclic ones. Aminolysis of methyldichlorosilane leads to cyclic oligomers which then polymerize with potassium hydride to preceramic materials:

$$(12\text{-}16) \qquad CH_3HSiCl_2 \xrightarrow[-\,NH_4Cl]{+\,NH_3} (CH_3HSiNH)_{3\text{-}4} \xrightarrow[-\,H_2]{KH} (CH_3HSiNH)_i(CH_3SiN)_j$$

### 12.3.4 Poly(silicic acid)s

**Silicic acid** is the umbrella term for compounds with the composition $SiO_2 \cdot n\ H_2O$. Hydrolysis of $SiCl_4$ leads first to **orthosilicic acid**, $Si(OH)_4$, a weak acid that survives for a while at pH 3.2 but condenses rapidly at other pH values, first to **orthodisilicic acid (pyrsilicic acid)**, $(HO)_3Si\text{–}O\text{–}Si(OH)_3$, and cyclic oligomers, $(O\text{–}Si(OH)_2)_{3\text{-}6}$, then to **poly(silicic acid)**s, $H_{2n+2}Si_nO_{3n+1}$, and finally to **metasilicic acid**, $H\text{–}[Si(OH)_2\text{–}O]_n\text{–}OH \approx H_2SiO_3$. Chains can combine to ladder and sheet structures (Sections 12.3.5–12.3.8) and finally to quartz (Section 12.3.12). No polymeric structures exist in aqueous solutions of silicic acids, but rather square tetramers, cubic octamers, etc.

Sodium (and potassium) salts of silicic acid with the composition $Na_2O \cdot m\ SiO_2 \cdot n\ H_2O$ as well as the aqueous solutions of these salts are known as **water glass**. Solid water glass is produced by melting quartz sand with soda or pottash at 1400–1500°C.

Depending on the molar ratio, $n_{rel} = [SiO_2]/[Na_2O]$, one distinguishes high-silicon water glass ($n_{rel}$ = 3.9–4.1), soda water glass ($n_{rel}$ = 3.3–3,5), and alkaline water glass ($n_{rel}$ = 2.0–2.2). Treating water glass with mineral acids (usually sulfuric acid) leads to **silica gel**, a highly condensed, poly(silicic acid) with sheet-like structures that are substituted with siloxane and silanol groups (L: *silex* = flint).

Water glass is almost exclusively produced as a 35 % aqueous solution in which it is present as a mixture of sodium compounds: monosilicate, $Na_4SiO_4$; disilicate, $Na_6Si_2O_7$; polymetasilicate, $(Na_2SiO_3)_n$; and poly(disilicate) $(Na_2Si_2O_5)_n$. The high viscosity of these solutions results in part from the polymers and in part from the polyelectrolyte character, i.e., the strongly hydrated $\text{–}O^{\ominus}$ ions.

The annual production of water glass exceeds 3.5 million tons. Approximately 25 % is used as additive to detergents, ca. 50 % for pigments, catalysts, zeolites, etc., and the rest as binders for sand in foundries, soil improvers, flotation agents, adhesives, etc.

### 12.3.5 Silicates

**Structures**

Silicates are naturally occurring salts and esters of orthosilicic acids and its oligomers and polymers. In some of silicates, silicon is replaced by metal atoms such as Mg, Al, Fe, Ca, Sc, etc. In other silicates, true silicate structures are combined wuth other structures, for example, spiro structures of polymeric $Mg(OH)_2$.

Neso-, soro-, and cyclosilicates are low molecular weight molecules (Fig. 12-2). **Neso-silicates** (G: *nesos* = island) are derived from orthosilicic acid, $Si(OH)_4$, whereas **soro-silicates** (G: *soros* = heap) are based on pyrosilicic acid, $H_6Si_2O_7$. **Cyclosilicates** (G: *kyklos* = circle) exist as 6-, 8-, and 12-membered rings whereas **inosilicates** (G = genetive of *is* = fiber) have linear chains, **phyllosilicates** (G: *phyllon* = leaf) sheaf-like structures, and **tectosilicates** (G: *tecton* = carpenter, builder) form lattice structures. Examples are shown in Table 12-3.

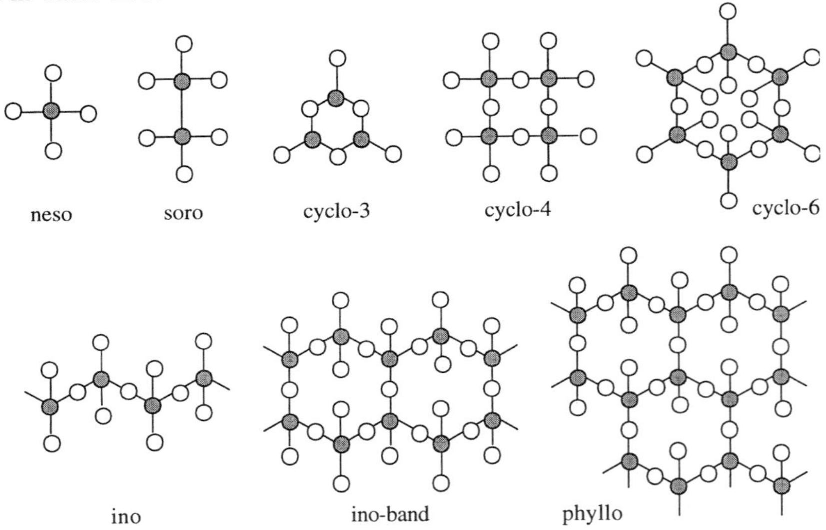


Fig. 12-2 Schematic representation of structures of low-molecular weight (top) and high-molecular weight (bottom) silicates. ● Silicon atoms, O oxygen atoms. Bond angles do not correspond to actual structures. Negative charges of oxygen atoms are not shown.

Chain silicates may be single chains (inosilicates; I) or ladder structures (ino-bands; $I_B$). Several single and double chains may also be combined to tube-like silicates (Fig. 12-3). An example is imogolite, a tube-like, soluble silicate of the gibbsite type, $(OH)_3Al_2O_3SiOH$, with lengths of up to several hundred nanometers (see Volume III) that is found in Japanese volcanic ashes. Some tube-like silicates are shown in Fig. 12-3.

The polymeric character of high-molecular weight silicates is not indicated by their conventional structural formulas (Table 12-3). Similar to the formulas for low-molecular weight silicates, one writes the silicate part of the repeating unit in square brackets that are prefixed by the counter cations. OH groups are considered part of the silicate repeating unit and thus included in the square brackets, even in those cases where they are bound to counterions and not to silicon. Examples are Mg in talc and Al in montmorillonite (Table 12-3).

Table 12-3 Examples of silicates.

| Group | Silicate type | Form | Anion | O/Si | Examples Name | Composition |
|---|---|---|---|---|---|---|
| N | neso | island | $[SiO_4]^{4-}$ | 4.00 | forsterite | $Mg_2[SiO_4]$ |
| S | soro | group | $[Si_2O_7]^{6-}$ | 3.50 | thortveitite | $(Sc,Y)_2[Si_2O_7]$ |
| | | | | | gehlenite | $Ca_3(Al.Mg)Ti_2[(Al,Si)SiO_7]$ |
| $C_3$ | cyclo | 6-ring | $[Si_3O_9]^{6-}$ | 3.00 | wollastonite | $Ca_3[Si_3O_9]$ |
| $C_4$ | cyclo | 8-ring | $[Si_4O_{12}]^{8-}$ | 3.00 | neptunite | $KNa_2Li(Fe^{II}.Mn)Ti_2[O\|Si_4O_{11}]_2$ |
| $C_6$ | cyclo | 12-ring | $[Si_6O_{18}]^{12-}$ | 3.00 | beryl | $Be_3Al_2[SiO_3]_6$ |
| I | ino | chain | $[Si_4O_{12}]^{8-}$ | 3.00 | augite | $Ca_2Mg_2[Si_4O_{12}]$ |
| $I_B$ | ino | ladder | $[Si_4O_{11}]^{6-}$ | 2.75 | chrysotile | $Mg_3[Si_4O_{11}] \cdot 3\ Mg(OH)_2{\cdot}H_2O$ |
| P | phyllo | sheet | $[Si_4O_{10}]^{4-}$ | 2.50 | muscovite | $KAl_2(OH.F)_2[AlSi_3O_{10}]$ |
| | | | | 2.50 | talc | $Mg_3[(OH)_2\|Si_4O_{10}]$ |
| | | | | 2.50 | montmorillonite | $Al_2[(OH)_2\|Si_4O_{10}]$ |
| T | tecto | lattice | $[SiO_2]$ | 2.50 | quartz | $SiO_2$ |

Atomic symbols are put in parentheses if silicon or the prevalent counterion are partially replaced by other elements; an example is the replacement of Ca by Al or Mg in gehlenite (Table 12-3). Very often so-called oxide formulas are used (see chrysotile in Table 12-3). For example, thortveitite, $(Sc)_2[Si_2O_7]$ (if present without yttrium), is written as $Sc_2O_3{\cdot}2\ SiO_2$ and the gibbsite unit as $SiO_2{\cdot}Al_2O_3{\cdot}2\ H_2O$.

About 95 % of the Earth's upper crust consists of quartz and silicates which are used as sand, gravel, stones, clays, feldspars, etc., in tremendous amounts as construction materials, fillers, ion exchangers, catalysts, etc. The properties of these materials depend predominantly on their physical structure. Band-silicates form fibrous crystals; an example is asbestos. Phyllosilicates exist as layer silicates, for example, in mica (New Latin: *micare* = to shine; from L: *mica* = grain). Other phyllosilicates such as clays can take up water between their layers and swell considerably; an example is montmorillonite.

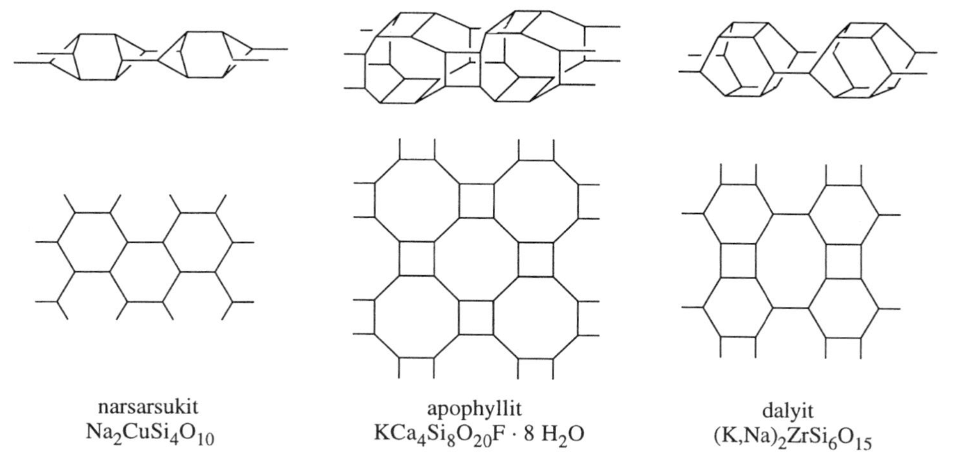


Fig. 12-3 Examples of tube-like silicates [1]. Top: three-dimensional depiction of silicate scaffolds (corners occupied by Si; connecting oxygen not shown). Bottom: tubes opened longitudinally and projected on a plane.

**Syntheses**

Silicates are formally condensation products of orthosilicate anions:

(12-17) $2\,[SiO_4]^{4-} \rightleftarrows [Si_2O_7]^{6-} + O^{2-}$, etc.

Equal molarity, $x_i$, of $SiO_2$ and oxides of bivalent metals, MtO, should lead to linear chains with "infinitely" large number-average molecular weights. $x_{SiO_2}/x_{MtO} > 1$ delivers branched silicates because of insufficient proportions of countercations. Even higher ratios of $x_{SiO_2}/x_{MtO}$ force chains to assemble in band and layer structures.

The position of the equilibrium is controlled by the counterions. In the absence of cyclization, the theory of consecutive equilibria predicts a dependence of the mole fraction, $x_{SiO_2}$, in melts on the activity, $a_{MtO}$, of metal oxides where $K$ = equilibrium constant, $b$ = 1 for linear chains, $b$ = 3 for branched ones, and $B$ = constant:

(12-18) $$\frac{1}{x_{SiO_2}} = 2 + \frac{1}{1 - a_{MtO}} + \frac{b}{1 + a_{MtO}(BK^{-1} - 1)}$$

According to the **Temkin law**, activities $a_{MtO}$ of metal oxides equal molar concentrations of $O^{2-}$ ions. These concentrations can be determined by mass spectroscopy from $CH_3$ contents after silicate anions have been capped by trimethylsilyl radicals, $(CH_3)_3Si^{\bullet}$. According to these experiments, NiO + $SiO_2$ form linear chains, whereas SnO, FeO, PbO, and CaO deliver branched silicates as indicated by $a_{MtO} = f(a_{SiO_2})$(Fig. 12-4).

Number-average degrees of polymerization, $\overline{X}_n$, of linear and branched polymers are calculated from the mole fraction, $x_{SiO_2}$, of $SiO_2$ and activities, $a_{MtO}$, of metal oxides:

(12-19) $$1/\overline{X}_n = (1 - a_{MtO})[(1/x_{SiO_2}) - 2]$$

Number-average degrees of polymerizations calculated with this equation are low for melts of $SiO_2$ and MtO. For a 6:4 ratio of MtO:$SiO_2$, they assume values of 2.1 (CaO, 1600°C; $K$ = 0,0016), 6.3 (SnO, 1100°C; $K$ = 2.55), and 11.8 (NiO, 1700°C; $K$ = 46).

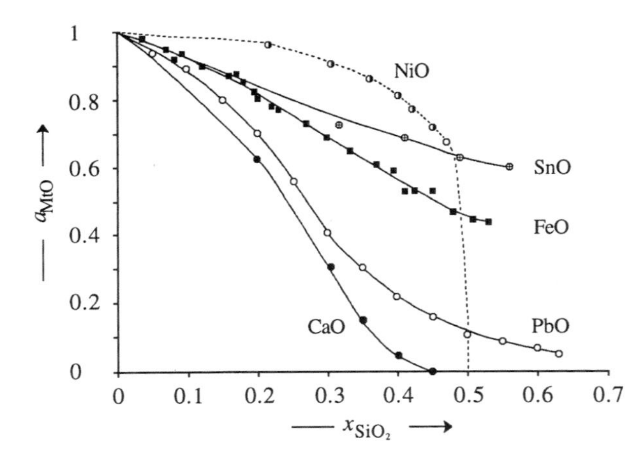


Fig. 12-4 Dependence of activity, $a_{MtO}$, of various bivalent metal oxides, MtO, on the mole fraction $x_{SiO_2}$ of $SiO_2$ in silicate melts at temperatures of 1650-1950°C (NiO-$SiO_2$), 1100°C (SnO-$SiO_2$), 1785-1960°C (FeO-$SiO_2$), 1000°C (PbO-$SiO_2$), and 1600°C (CaO-$SiO_2$) [2]. Lines calculated by Eq.(12-18) for linear (- - -) and branched (——) polymers.

### 12.3.6 Silicate Glasses

"Glass" is usually understood to be silicate glass although there are many other inorganic glasses (for example, those of $B_2O_3$, $P_2O_5$, or $As_2O_3$) and several organic ones (for example, acrylic glass = poly(methyl methacrylate)).

**Ordinary glass** (window and container glass) is a soda-lime glass made from silica sand ($SiO_2$), soda ash (calcined soda; $Na_2CO_3$), and lime (a mixture of calcium oxide and magnesium oxide) in the weight ratio 75:20:5. Its average composition is 75 wt% $SiO_2$, 13 wt% $Na_2O$, and 12 wt% CaO which corresponds to $Na_2O{\cdot}CaO{\cdot}6\ SiO_2$ and thus O:Si = 2.33 This composition puts it between phyllo- and tectosilicates.

In **Bohemian crystal glass** with the composition $K_2O{\cdot}CaO{\cdot}8\ SiO_2$, potassium replaces sodium and the silica content is increased (O:Si = 2.25). In chemically resistant laboratory glassware, such as from **Jena(er) glass**®, $Na_2O$ and CaO are substituted largely by $B_2O_3$, $Al_2O_3$ and BaO. A similar glass, **Pyrex**®, contains 81 % $SiO_2$, 13 % $B_2O_3$, 3.5 % $Na_2O$, and 2.5 % $Al_2O_3$.

A similar water- and weather-resistant borosilicate glass is **E glass** which was originally developed for *e*lectrical isolation (hence the designation). This glass is also spun to filaments and fibers with 5–13 μm diameter that are used for reinforcement of plastics. Newly developed adhesion promoters allow the use of less-expensive *a*lkali-rich **A glass** for the same purpose. **C glass** is *c*hemically resistant; **R glass** is *r*esistant against acids. **D glass** has good *d*ielectric properties, whereas **S glass** has good *s*trength and **T glass** good *t*ensile strength. For compositions, see Table 12-4.

Thermal and mechanical properties of silicate glasses are directly determined by chain structures. A sodium silicate glass with the composition $Na_2SiO_3$ (49 wt% $SiO_2$) has an O:Si ratio of ca. 3.0. It consists of linear chains and has a glass temperature of ca. 420°C. An increase of the $SiO_2$ content to ca. 70 wt% changes the O:Si ratio to ca. 2.33, which leads to polymers with short side chains. The presence of side chains interferes with the parallelization of main chains, produces an additional free volume, and reduces the glass temperature to 355°C. A further increase of the $SiO_2$ content to 92 wt% unites many –SiO– groups in band, sheet, and lattice structures; this stiffens the polymer and increases the glass temperature to 540°C.

Table 12-4 Composition in wt% of technical silicate glasses [3, 4].

| Component | A | C | D | E | ECR | R | S, T |
|---|---|---|---|---|---|---|---|
| $SiO_2$ | 70- 72 | 60-65 | 73-74 | 52-56 | 58-63 | 60 | 60-65 |
| $Al_2O_3$ | 0- 2.5 | 2-6 | 0-0.3 | 12-16 | 10-13 | 25 | 20-25 |
| CaO | 5-10 | 14 | 0.3-0.6 | 16-25 | 21-23 | 9 | 0 |
| MgO | 1-4 | 1-3 | 0 | 0-5 | 2-4 | 6 | 10 |
| $B_2O_3$ | 0-0.5 | 2-7 | 22-23 | 5-10 | 0.1 | 0 | 0-1.2 |
| F | 0 | 0 | 0 | 0-0.7 | 0.15 | 0 | 0 |
| $Na_2O$ | 12-15 | 8-10 | 1-1.3 | 0-2 | 0-1.2 | 0 | 0-1.1 |
| $K_2O$ | 0 | 0 | 1.5 | < 1 | 0.4 | 0 | 0 |
| $Fe_2O_3$ | 0 | 0.2 | trace | 0-0.8 | 0-0.4 | 0 | 0 |
| $TiO_2$ | 0 | 0 | 0 | < 1 | 2.1 | 0 | 0 |
| ZnO | 0 | 0 | 0 | 0 | 0-3.5 | 0 | 0 |

### 12.3.7 Fibrous Silicates

**Asbestos** is not a defined chemical compound but the common name of a group of fibrous silicates with high tensile strength and excellent thermal resistance. All these silicates are band silicates, i.e., silicates with ladder structures. The various types of asbestos differ in chemical composition, crystal structure, and fiber size. They crystallize usually in non-fibrous, stone-like shapes but in cartein conditions also as bundles of hundreds and thousands of fibrils of 1–2 cm length and axial ratios up to 1000.

Asbestos is commonly subdivided into serpentines and amphiboles. "**Serpentine**" refers to the often snake-like texture and color of these minerals (L: *serpes* = snake). This mineralogical term comprises both band silicates such as chrysotile (a fiber serpentine) and sheet silicates such as antigorite (a layer serpentine). The commonly given mineralogical composition $Mg_3Si_2O_5(OH)_4$ of both chrysotile and antigorite is better written as $Mg_3[Si_4O_{11}] \cdot 3\ Mg(OH)_2 \cdot H_2O$ since the former does not reveal for chrysotile that ino-bands (ladder structures) of magnesium silicate, $Mg_3[Si_4O_{11}]$, alternate with "spiro" ino-chains ("spiro" structures) of $Mg(OH)_2$ (Fig. 12-5).

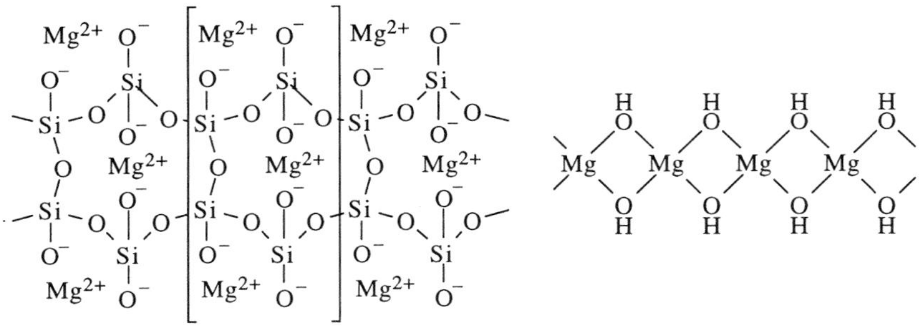

Fig. 12-5 Schematic representation of the ladder structure of magnesium silicate, $Mg_3[Si_4O_{11}]$, and the band structure of $Mg(OH)_2$. Both structures alternate in chrysotile.

The white, silken-shining hollow fibers of chrysotile melt at ca. 1100°C (G: *chrysos* = gold; *tylos* = fine hair). Chrysotile is the economically most important asbestos. It less cancerogenic than the amphiboles.

**Amphiboles** (G: *amphi bolos* = ambiguous; from *amphi* = around, on both sides; *ballein* = to throw around) are silicates with the general composition $X_2Y_5[Si_3ZO_{11}]A_2$ where X = Li, Na, K, Mg, Ca, or Fe(II); Y = Mg, Al, Ti(III), Fe, or Mn; Z = Al or Si; and A = F or OH. This group of silicates includes

| | | | |
|---|---|---|---|
| tremolite | $Ca_2Mg_5[OH\vert Si_4O_{11}]_2$ | actinolite | $CaMg_3Si_4O_{12}$ |
| crocidolite | $Na_4Mg_5[OH\vert Si_4O_{11}]_2$ | anthophyllite | $(OH)_2(Mg,Fe)_7[Si_4O_{11}]_2$ |

Writing the chemical formula of tremolite as $2\ CaMg_2[Si_4O_{11}] \cdot Mg(OH)_2$ reveals that an ino-chain of $Mg(OH)_2$ is sandwiched by two ino-bands of Ca/Mg silicate.

Because of its fireproofing properties, asbestos has been used in the past as insulating material and also as reinforcing fiber in vinyl floor coverings. In the long term, breathing in fibers of asbestos may lead to asbestosis, a clogging of lung epithelia similar to silicosis. Many buildings are presently freed of asbestos-containing materials.

### 12.3.8 Layer Silicates

Layer silicates (phyllosilicates, (Fig. 12-2)) consist of stacked layers of silicates, aluminum silicates, and magnesium silicates which are sometimes separated by layers of alkali or alkaline earth ions and/or water layers (Fig. 12-6).

The units of silicate layers have tetrahedral structures, whereas those of the associated aluminum and aluminum/magnesium structures are octahedral. Layer silicates may also contain bands of brucite units, $Mg(OH)_2$ (see Fig. 12-5), and/or layers with gibbsite units, $Al_2O_3 \cdot SiO_2 \cdot 2\ H_2O$. In imogolite, gibbsite units are bent: 10 units enclose cylindrically one orthosilicate anion that is bound to the vacant octahedral site within each gibbsite unit.

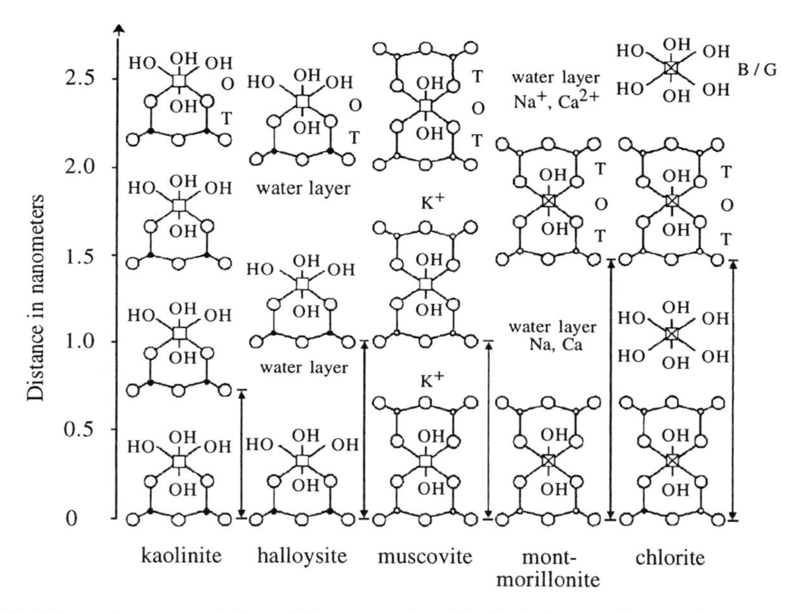


Fig.. 12-6 Schematic representation of the succession of units in layers of some clay minerals [5]. • Silicon, o silicon or aluminum, O oxygen, □ aluminum, ⊠ aluminum or magnesium; T = tetrahedral, O = octahedral, B = brucite, G = gibbsite.

For example, triclinic **kaolinite**, a major component of kaolin (see below), consists of $Al_4(OH)_8[Si_4O_{10}] = Al_2O_3 \cdot 2\,SiO_2 \cdot 2\ H_2O$. In **halloysite**, the same units are separated by water layers.

**Muscovite** with the general composition $K_2Al_4(Al_2Si_6O_{20})(OH)_4$, on the other hand, has double layers that are separated by potassium ions. The pale muscovite and the dark biotite belong to the glittering **mica** family of minerals. Muscovite is used as insulation material in plates of up to 2 m in diameter, as filler for plastics in small platelets, and as a pigment because of its pearly gloss. Biotite is commercially insignificant.

Double layers are also present in 150°C-dried montmorillonite, $Al_2(OH)_2[Si_4O_{10}]$, in which aluminum may be partially substituted by magnesium. Ionic interlayers are here composed of sodium and/or calcium ions and not of potassium ions as in halloysite. In general, montmorillonite has the composition $[Al_2O_3 \cdot 4\ SiO_2 \cdot H_2O] \cdot n\ H_2O$ because it

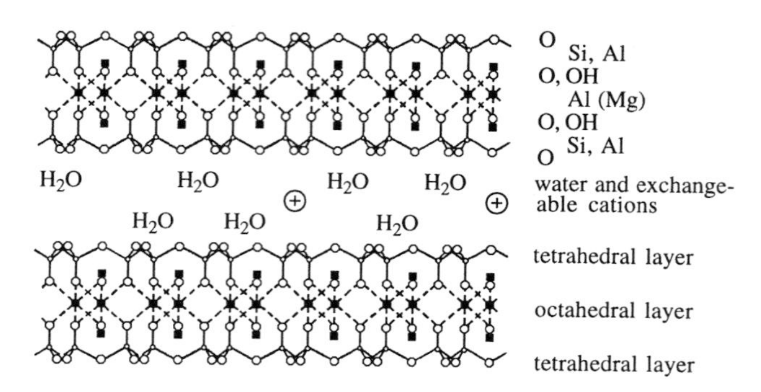


Fig. 12-7 Two of the many successive layers of montmorillonite [6]. Right: composition of a layer (top) and structures (bottom).

takes up water (Fig. 12-7). In **chlorite**, aluminum/magnesium silicate layers are not separated by Na/Ca ions but by brucite and/or gibbsite units (Fig. 12-6).

The two-dimensional, "infinitely" large polyanions of montmorillonite are electrostatically bound together by the sodium and calcium ions of the interlayers. Because of this structure, montmorillonite can take up large proportions of water as well as other cations. Montmorillonite dispersions are used as flooding agent for drillings, adsorbent, filler, coating of plant seeds, catalyst, etc. Montmorillonite as matrix can also promote the *in-vitro* polymerization of simple $\alpha$-amino acids to polypeptides.

The distance between octahedral layers is usually 1–2 nm. Swelling montmorillonite in water increases this distance considerably, which causes the attractive forces between layers to decrease to that of thermal energy, provided that only few 1,1 electrolytes are present. Hence, each layer becomes an independent kinetic unit.

Such independent layers can replicate themselves since they act as nuclei for new layers in dilute solutions of, e.g., $Na^+$, $K^+$, $Mg^{2+}$, $Al^{3+}$, $Si(OH)_4$, etc. In nature, this process is repeated during each cycle in which wet periods alternate with periods of drought. During wet periods, high–swollen montmorillonite takes up only small amounts of electrolytes but their content increases when water evaporates during dry periods.

Related to montmorillonite are several other **clays**. **Vermiculite** with the approximate composition $Mg_3(OH)_2[(Al,Si)_4O_{10}]\cdot 4\ H_2O$ loses water on heating and expands to about 50 times its original volume. Layers delaminate, covalent bonds are formed between platelets, and the material becomes rigid and brittle.

However, if the "binding" outer layers of bivalent magnesium ions are replaced by monovalent ions by washing vermiculite with aqueous solutions of $[(C_4H_9)_4N]Cl$ and NaCl, lamellae are only partially delaminated to platelets of ca. 30 nm thickness and axial ratios of ca. 200 (Fig. 12-8, I). The resulting "crystal bridges" and "anchor molecules" are broken by grinding the material in ball mills (Fig. 12-8, II) which leads to single platelets of ca. 1 nm thickness (Fig. 12-8, III), a nano material that can be used as reinforcing filler.

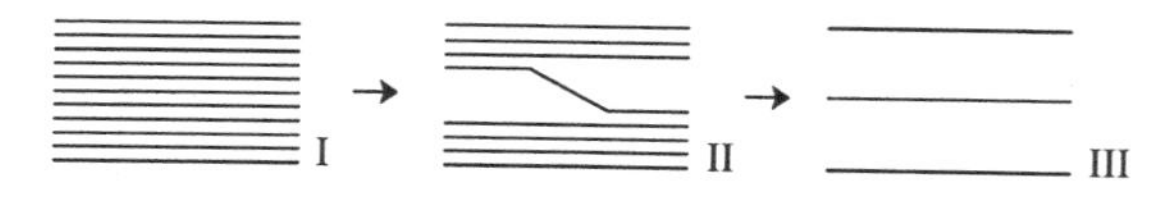


Fig. 12-8 Chemical delamination and mechanical grinding deliver single platelets.

**Kaolin** (Dinese: *kao ling* = high hill) is a mixture of layer silicates (kaolinite, dickite, nacrite), $Al_4(OH)_8[Si_4O_{10}]$, and montmorillonite, $Al_2(OH)_2[Si_4O_{10}]$, respectively, with gel-like alum earth (argillaceous earth) silicates (allophanes). Kaolin is the end product of weathering of granite and feldspar.

### 12.3.9 Cement and Concrete

**Cement** denotes in general language any material that acts as glue on hardening. "Cement" in the original narrower meaning refers to an aqueous paste of powdered calcium silicates and aluminates with varying contents of CaO, $SiO_2$, $Al_2O_3$, and $Fe_2O_3$ that hardens to a rock-like material (L: *caementum* = rough, untreated stone). The commercially most important cement is **Portland cement**, consisting of tricalcium silicate ($Ca_3SiO_5$), β-dicalcium silicate ($Ca_2SiO_4$), quicklime (CaO), and aluminum silicates of complex composition. Portland cement got its name from the Isle of Portland (a peninsula in southern England) where a stone of the same color is found.

Cement is often mixed with sand or gravel to **concrete** (L: *concrescere* = to grow together). Concrete typically consists of 16 wt% Portland cement, 7 wt% water, 33 wt% sand, and 44 wt% gravel of different sizes. The cement binds with water to cement lime, then to cement gel, and, after some hours, to cement stone.

Sand and gravel have far higher strengths than cement stone. The strength of concrete is thus controlled by cement stone which in turn depends on the weight ratio of water (w) and cement (c), $\omega = m_w/m_c$. A 100 % setting of cement requires ca. 40 wt% water (hence the term "hydraulic cement"): 25 wt% are bound chemically and the remaining 15 wt% produce gel pores of 1–10 nm diameter on "drying" (actually: hydration). Water in such small pores freezes only at temperatures of –20°C to –80°C. In contrast to this, water contents of more than 40 wt% on preparing cement produce capillary pores of 10–$10^5$ nm diameter in which the water freezes at temperatures of –10°C.

During setting, crystalline silicate and aluminosilicate fibers grow from cement particles in all directions. These fibers embed sand and gravel particles. Compression strengths increase rapidly and then more slowly: ultimate strengths are obtained after months and often years, even up to 50 years. Under pressure, concrete creeps but this is (in part) remedied by steel reinforcement. On concrete streets, dynamic stresses caused by cars produce ruts. Since one 25-ton truck causes the same damage as 6500 passenger cars trucks of 1 ton each, trucks should be taxed accordingly to cover the cost of repairs.

Concrete does not stop reacting after it reaches its ultimate strength. Already during setting, concrete shrinks and internal voids form. Water penetrates these voids where it freezes and expands causing the concrete to crack. Furthermore, carbon dioxide from air converts unreacted CaO to $CaCO_3$ and concrete loses it alkalinity and becomes neutral.

The result is a loss of protection for the steel armament since iron is attacked in an acidic environment. Water and air enter the pores and convert the iron to rust which, being less dense than iron, expands and cracks the concrete. The process is promoted by salting the streets because salt attracts water, which promotes the formation of pores and rust.

Since the volume fraction of pores in concrete may reach 25 vol%, polymers are sometimes added to fresh cement before setting. These additives may be either disper sions of dextran, polymeric sulfur, natural rubber, poly(vinyl chloride), silicones, etc., (**PPCC** = ***p****olymer-****p****ortland* ***c****ement* ***c****oncrete*).

In another approach, concrete parts are saturated with monomers such as styrene, methyl methacrylate, styrene/acrylonitrile, vinyl acetate, etc., which are then polymerized by irradiation or added AIBN (**PIC** = ***p****olymer* ***i****mpregnated* ***c****oncrete*). The latter process reduces creep and water absorption (from 7 % to 0.4 %) and increases tensile moduli (25 MPa → 37 MPa), tensile strengths (1.7 MPa → 9.5 MPa), compression strengths (35 MPa → 140 MPa), and shear strengths (5 MPa → 15 MPa).

**Polymer concrete** (**PC**) is not a polymer-modified concrete but the polymerized mixture of monomers (unsaturated polyesters, epoxy resins, methyl methacrylate) with aggregates (sand, gravel) and perhaps pigments. The binder is the organic polymer; Portland cement is not present. Polymer concrete is used for artificial marble, foundations for heavy equipment, etc., since it combines good impact strengths with tensile and compressive strengths that exceed those based on Portland cement (Table 12-5).

### 12.3.10 Ceramic Materials

**Ceramic** (G: *keramos* = clay) was originally a material that was obtained from clay ($Al_2O_3 \cdot 2\text{–}4\ SiO_2 \cdot 1\text{–}2\ H_2O$) with or without added lean materials (sand, etc.), fluxes (feldspar, iron oxide, etc.), and colorants (metal oxides) by firing (sintering) without melting. In modern terminology, ceramic is any matter composed of different inorganic materials (including carbides, nitrides, etc.) that is difficult to melt.

The **sol-gel process** to ceramics or glasses consists of transforming a liquid (the "sol") to a gel that is converted to the ceramic by firing. An example is the formation of a sodium borosilicate glass from the liquid mixture of sodium alcoholates, NaOR, boric acid esters, $B(OR)_3$, and silicic acid esters, $Si(OR)_4$. Hydrolysis delivers NaOH, $B(OH)_3$, and $Si(OH)_4$. Dehydration converts this mixture to a gel, $(Na_2O \cdot B_2O_3 \cdot SiO_2) \cdot H_2O$, which, in the language of polymer science, consists of a swollen hyperbranched and crosslinked mixed polyanhydride of boric and silicic acid. Upon firing, the gel dehydrates and becomes a rigid network (glass) with the composition $Na_2O \cdot B_2O_3 \cdot SiO_2$.

Table 12-5 Properties of cement concrete and polymer concretes based on poly(methyl methacrylate) (PMMA), unsaturated polyesters (UP; probably with styrene), and an epoxy resin (EP). Numbers in parentheses: weight ratio of polymer and aggregate. ? = Weight ratio not reported.

| Property | Physical unit | Cement concrete (1:5) | Polymer concrete PMMA (1:10) | UP (1.2:9.6) | EP ? |
|---|---|---|---|---|---|
| Thermal expansion coefficient | $10^5\ K^{-1}$ | 1.0–1.4 | 1.7 | 1.5–2.5 | 1.45 |
| Heat distortion temperature | °C | – | 106 | 45–90 | 60 |
| Modulus of elasticity | GPa | 30 | 18–38 | 20–40 | 2.8–12.7 |
| Fracture elongation | % | | 0.1 | | |
| Tensile strength | MPa | 4 | 15 | 10–20 | 45–55 |
| Flexural strength | MPa | | 29–43 | 18–16 | 50–110 |
| Compressive strength | MPa | 40 | 105–140 | 80–130 | 85–120 |
| Impact strength | $kJ/m^2$ | | 0.6–0.8 | 11 | 15 |

Sol-gel processes produce articles that are far more homogeneous and have less voids than those obtained by classic ceramic processes. Yields and properties of ceramic masses by sol-gel processes depend strongly on the chemical nature of the gel-forming polymers. Linear polymers deliver only low ceramic yields since firing produces large proportions of low-molecular weight cyclics and other volatile compounds. Cyclic and cage molecules lead to high ceramic yields since the formation of low-molecular weight volatiles requires simultaneous breaking of two (or more) neighboring bonds in one molecule, which is highly unlikely. Best ceramic yields are obtained from branched cyclics but the properties are not always optimal.

### 12.3.11 Silicones

"Silicone" is the umbrella term for both low-molecular weight and polymeric silicon compounds with silicon-oxygen bonds. The name was introduced by the English researcher F.S.Kipping, who believed that his newly synthesized compound with the composition $R_2SiO$ was a silicon–organic analog to the ketones $R_2CO$ of carbon chemistry, hence the name was derived from *silic*on ket*one*.

However, all chemical compounds with this (approximate!) composition are high molecular weight compounds, $+\!\!\!-SiR_2–O\!-\!\!\!+_n$, with –Si–O– main chains that do no contain >Si=O groups. A more appropriate name is therefore **polysiloxanes**. The chemistry of organosilicon compounds also differs in many respects from that of carbon chemistry because of the free 3d orbitals of silicon and the polarity of the Si–O bond, which is partially ionic and has partial double bond character, a consequence of the large difference in electronegativities of Si and O (1.8 *versus* 3.5 in Pauling's scale).

Like carbon and carbon–oxygen polymers, silicones can be linear, branched, and crosslinked. They can also form copolymers of silicon-containing monomers as well as with carbon compounds (Fig. 12-9). Industrially, all silicones are synthesized from low molecular weight silicon monomers; the world–wide production exceeds ca. 600 000 t/a. However, some silicones are also available by transformation of silicates.

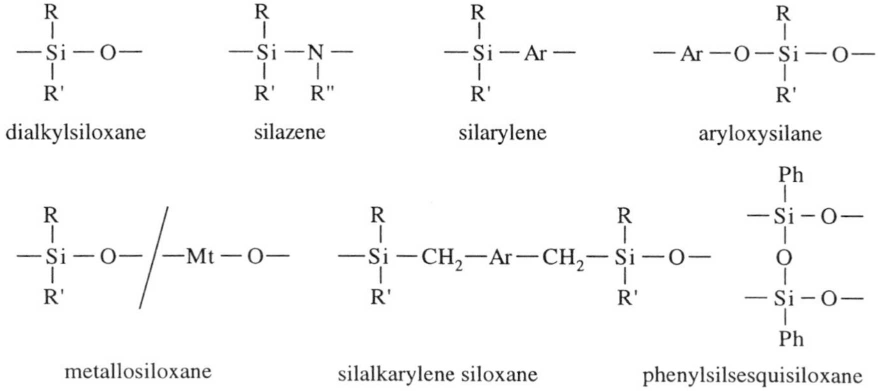

$$-\underset{R'}{\overset{R}{Si}}-O- \qquad -\underset{R'}{\overset{R}{Si}}-\underset{R''}{N}- \qquad -\underset{R'}{\overset{R}{Si}}-Ar- \qquad -Ar-O-\underset{R'}{\overset{R}{Si}}-O-$$

dialkylsiloxane  silazene  silarylene  aryloxysilane

$$-\underset{R'}{\overset{R}{Si}}-O-/-Mt-O- \qquad -\underset{R'}{\overset{R}{Si}}-CH_2-Ar-CH_2-\underset{R'}{\overset{R}{Si}}-O- \qquad \begin{matrix} Ph \\ -Si-O- \\ O \\ -Si-O- \\ Ph \end{matrix}$$

metallosiloxane  silalkarylene siloxane  phenylsilsesquisiloxane

Fig. 12-9 Siloxane and other repeating units in silicon polymers. R, R' = organic substitutents (usually alkyl), Ar = arylene, Mt = metal

### Silicate Transformations

Some silicates can be converted to silicones by reaction with hexamethyldisiloxane, isopropanol, and hydrochloric acid at 75°C. In this process, hexamethyldisiloxane is first hydrolyzed to trimethylsilanol,

$$(12\text{-}20) \qquad (CH_3)_3Si{-}O{-}Si(CH_3)_3 + H_2O \longrightarrow 2\,(CH_3)_3SiOH$$

whereas silicates (here depicted as $\text{+}O{-}Si(OMt)_2\text{+}_n$) are first transformed to poly(dihydroxysiloxane)s, $\text{+}O{-}Si(OH)_2\text{+}_n$, by losing metal hydroxides, MtOH. The poly(dihydroxysiloxane)s are then transformed by reaction with trimethylsilanol to soluble poly[di(trimethylsiloxy)siloxane]s:

$$(12\text{-}21) \qquad \sim\!Si(OMt)_2{-}O\!\sim \xrightarrow[-\,2\,MtOH]{+\,2\,H_2O} \sim\!Si(OH)_2{-}O\!\sim \xrightarrow[-\,2\,H_2O]{+\,2\,(CH_3)_3SiOH} \sim\!Si(OSi(CH_3)_3)_2{-}O\!\sim$$

It is not clear why some silicates can be transformed but not others. All investigated ino-silicates either do not react, are only partially silylated, or deliver only insoluble products. Some layer silicates such as vermiculite can be converted to fully trimethylsilylated silicones in yields of up to 18 % if metal cations are completely removed; these polymers are soluble in organic solvents. The reaction does not proceed with linear silicates or ino-bands, either. Reactivities of silicates seem to be connected to the aluminum content of silicates since silicates with high aluminum contents deliver only low-molecular weight products whereas those with low aluminum contents lead to high-molecular weight ones.

### Polymerizations

Monomers and starting compounds for silicones are obtained from silicon, [Si], and $CH_2Cl_2$ with copper catalysts which leads to dimethylsilicon dichloride (I) plus other chlorinated compounds:

$$(12\text{-}22) \qquad Si + CH_2Cl_2 \rightarrow Cl{-}Si(CH_3)_2{-}Cl \qquad (+\,(CH_3)_3SiCl,\ (CH_3)SiCl_3,\ SiCl_4)$$

Hydrolysis of I delivers the "hydrolyzate", which contains linear and cyclic molecules in an approximate ratio of 50:50. Equilibration of the hydrolyzate by exchange reactions between –Si–O– bonds produces the desired cyclics and low-molecular weight polymers. From these, industrial silicones are produced exclusively by polycondensation or ring-opening polymerization.

**Polycondensations** of dialkylsilane dichlorides with water proceed via intermediarily formed dialkylsilanols to both linear poly(dialkylsiloxane)s and oligocyclosiloxanes with $i$ = 3–20, Eq.(12-22). Tri- and tetrachlorosilanes lead to branched and crosslinked polymers, depending on monomer conversion.

$$(12\text{-}23) \qquad RR'SiCl_2 \xrightarrow[-\,2\,HCl]{+\,2\,H_2O} RR'Si(OH)_2 \xrightarrow[-\,H_2O]{} HO(SiRR'O)_nH + (RR'SiO)_{3\text{-}20}$$

**Ring-opening polymerization** of cyclosiloxanes such as octaalkylcyclotetrasiloxane, $(R_2SiO)_4$, with, for example, the dipotassium compound of tetramethyldisiloxanediol as initiator delivers high-molecular weight poly(dialkylsiloxane)s:

$$\text{KO}-\underset{\text{R}}{\overset{\text{R}}{\text{Si}}}-\text{O}-\underset{\text{R}}{\overset{\text{R}}{\text{Si}}}-\text{OK} \xrightarrow{+2\,n\,(R_2SiO)_4} \text{KO}(\underset{\text{R}}{\overset{\text{R}}{\text{Si}}}\text{O})_{4n}-\text{O}-\underset{\text{R}}{\overset{\text{R}}{\text{Si}}}-\text{O}-\underset{\text{R}}{\overset{\text{R}}{\text{Si}}}-\text{O}-(\underset{\text{R}}{\overset{\text{R}}{\text{Si}}}\text{O})_{4n}\text{K} \qquad (12\text{-}24)$$

Poly(dialkylsiloxane)s are capped at both ends by reaction with trimethylchlorosilane. Reaction (12-2) is an equilibrium reaction; the resulting product contains not only linear polymers (shown) but also cyclosiloxanes, $(R_2SiO)_i$, with $i \geq 4$.

Octamethylcyclotetrasiloxane, $((CH_3)_2SiO)_4$, also polymerizes with alkali in aqueous emulsion in the presence of, for example, benzyldimethyldodecylammonium chloride, as emulsifier, presumbly by a ring-opening polymerization to polymers with OH end-groups and subsequent polycondensation.

Ring-opening polymerizations also proceed with strong proton acids since the oxygen atoms of siloxanes are basic and can thus be protonated. The polymerization of cyclotrisiloxanes, $(R_2SiO)_3$, with trifluoromethanesulfonic acid (triflic acid), $F_3C–SO_3H$, is controlled kinetically and delivers both linear polymers and cyclics, $(R_2SiO)_i$ with $i \geq 1$, probably by simultaneous non-living chain polymerization and polycondensation. The reason for this scenario is the observation that high molecular weight polymers are already formed at small monomer conversions. The molecular weights of these polymers are proportional to polymer yields and to acid concentrations, which points to a chain polymerization. However, polycondensation cannot be excluded.

The ring-opening polymerization of higher cyclosiloxanes, $(R_2SiO)_m$, with $m \geq 4$, on the other hand, is independent of the concentration of triflic acid. The polymerization finally approaches a state in which the concentration and molecular weight of the polymer become independent of the concentration of the acid. Polymerization rates increase and yields of larger rings decrease with increasing size of the cyclic monomer. These larger rings result from intramolecular back-biting reactions of the activated chain ends (see the analogous reaction in ethylene polymerizations, Eq.(6-4)). All participating reactants are in polymerization-depolymerization equilibria; a polycondensation via activated silanol groups does not take place.

**Hydrosilation** can be utilized for polyadditions of silanes and divinyl or diallyl compounds, respectively, to polycarbosiloxanes, for example,

$$n\,\text{H}-\underset{\text{R}}{\overset{\text{R}}{\text{Si}}}-\text{O}-\underset{\text{R}}{\overset{\text{R}}{\text{Si}}}-\text{H}\ (\text{I}) + n\,\text{CH}_2{=}\text{CH}-\underset{\text{R}}{\overset{\text{R}}{\text{Si}}}-\text{O}-\underset{\text{R}}{\overset{\text{R}}{\text{Si}}}-\text{CH}{=}\text{CH}_2\ (\text{II}) \xrightarrow{\text{catalyst}} \left[-\underset{\text{R}}{\overset{\text{R}}{\text{Si}}}-\text{O}-\underset{\text{R}}{\overset{\text{R}}{\text{Si}}}-\text{CH}_2-\text{CH}_2-\right]_{2n}\ (\text{III}) \qquad (12\text{-}25)$$

Hydrosilations with $H_2PtCl_6$ as catalysts are used for room-temperature vulcanization of vinyl group-containing silicone rubbers to silicone elastomers. The same catalyst,

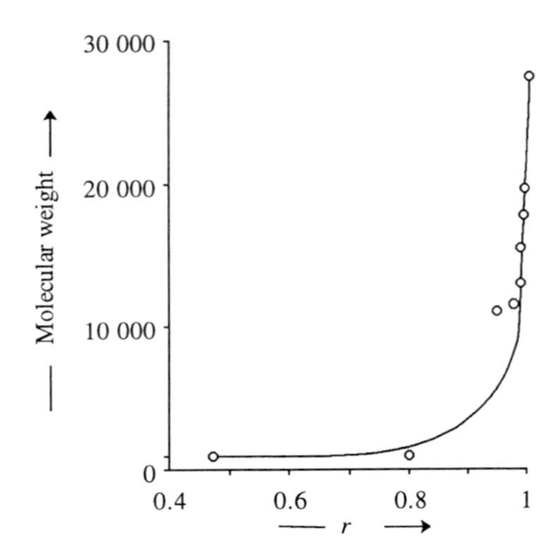


Fig. 12-10 Molecular weights of alternating copolymers (III) from the polyaddition by hydrosilylation of monomers I and II (Eq.(12-25)) as a function of the molar ratio, $r = n_I/n_{II}$, of monomers [7]. Molecular weights were calculated from measured intrinsic vicosities using the known viscosity–molecular weight relationship of poly(dimethylsiloxane)s instead of the unknown one of III.

if applied to linear polymerizations, delivers only oligomers because $H_2PtCl_6$ can be partially decomposed during the polyaddition. The decomposition product, HCl, can than sever the Si–$CH_2$ bond which leads to a degradation of the silalkylene chains, ~Si($R_2$)–$CH_2$~ + HCl → ~Si$R_2$–Cl + H–$CH_2$~.

The situation can be remedied by use of other platinum catalysts such as Karstedt's catalyst, the platimum complex with 1,3-divinyltetramethyldisiloxane (II). With such catalysts, molecular weights do indeed reach the value expected for equilibrium polyadditions (Fig. 12-10).

**Equilibration**

The easy establishment of equilibria between linear polysiloxanes and cyclosiloxanes is exploited industrially for adjustments of molecular weight distributions in "hydrolyzates" and/or proportions of cyclics in polymers. Equilibrations are not always desirable, though. In the Apollo 8 spaceship, for example, the windows fogged by cyclosiloxanes that were formed by the heat developed during the ascent but then evaporated.

Equilibration has been used for the synthesis of **ladder polymers**. Phenyltrihydroxysilane, $C_6H_5Si(OH)_3$, and phenyltrialkoxysilane, $C_6H_5Si(OR)_3$, respectively, have been equilibrated in solvents to low-molecular weight cage compounds of the composition $(C_6H_5SiO_{3/2})_i$ that precipitate from the solution The degree of oligomerization depends on the solvent: hot toluene leads to $i = 8$, acetone to $i = 10$, and tetrahydrofuran to $i = 12$ (Fig. 12-11). On heating, oligomers polymerize to poly(phenylsilsesquioxane)s.

The structure of the resulting polymer is probably controlled by the way the synthesis is conducted. Most probably, one does not obtain 100 % ladder structures (I, see also the ino-band in Fig. 12-2, p. 562) but also necklace structures (II), partial cage structures, and randomly branched units. The polymers have excellent heat stabilities (< 500°C). They are used as dielectric interlayers and anti-reflecting coatings.

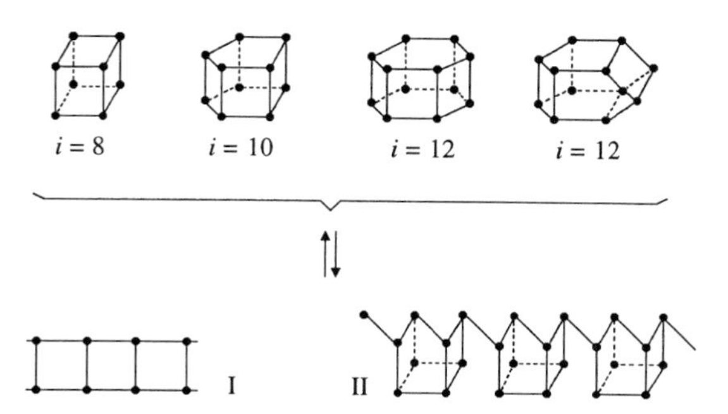


Fig. 12-11 Polymerization of phenylsilsesquioxanes, $(C_6H_5SiO_{3/2})_i$ to polymers with ladder structures (I), pearl structures (II) derived from cage structures (upper row), and random structures (not shown). Full circles indicate positions of Si atoms; oxygen atoms connecting Si atoms and phenyl groups as substituents are not shown.

## Organofunctional Polysiloxanes

Silicones may carry non-carbon substituents and various functional groups. These groups are usually separated from the –Si–O– main chain by at least one methylene group. For example, fluorinated silicones can be obtained by polycondensation according to Eq.(12-23) if R = $CF_3CH_2CH_2$ and R' = $CH_3$. Since α- and β-substituted fluorosilicones are thermally and hydrolytically unstable, only γ-substituted ones are used.

In many cases, substituents must be introduced by transformation of polymers since they would interfere with the polymerization of appropriately substituted monomers. Again, many functional groups (i.e., groups to be utilized in polymer transformations) must be separated from the main silicone chain by at least one methylene group. Examples are polysiloxanes with hydroxy groups as substituents; these are of interest as vulcanizable rubbers. The OH groups are here introduced by reaction of allyl alcohol with silane groups of polysiloxanes:

$$\text{(12-26)} \qquad \sim\!\!\sim\underset{\text{H}}{\overset{\text{CH}_3}{\text{Si}}}\text{—O}\sim\!\!\sim \; + \; \text{CH}_2\text{=CHCH}_2\text{OH} \longrightarrow \sim\!\!\sim\underset{\text{CH}_2\text{CH}_2\text{CH}_2\text{OH}}{\overset{\text{CH}_3}{\text{Si}}}\text{—O}\sim\!\!\sim$$

The direct polymerization of $CH_3Si(CH_2CH_2CH_2OH)Cl_2$ (obtained from $CH_3SiHCl_2$ and $CH_2$=$CHCH_2OH$) is not possible since the OH group of the monomer would interfere in the polycondensation (see Eq.(12-23)).

## Properties and Applications

Si–O bonds of polysiloxanes are longer than those of C–C bonds (0.164 nm *versus* 0.1.54 nm) whereas valence angles Si–O–Si are generally greater than those of valence angles of C–C–C (104° to almost 180°, depending on substituents, *versus* 111.5°). As a result, rotational barriers for the conformational transition between the lowest energy minimum and the highest energy maximum are much lower than those of carbon chains (ca. 1.3 kJ/mol *versus* 15.9 kJ/mol for, e.g. butane).

Table 12-6 Densities $\rho$ and glass ($T_G$), melting ($T_M$), and clearing temperatures ($T_{N\rightarrow I}$) of high molecular weight linear and cyclic polysiloxanes, $-[SiRR'-O]_n-$,

| R | R' | $\rho$ / g cm$^{-3}$ | $T_G$ / °C | $T_M$ / °C | $T_{N\rightarrow I}$ / °C |
|---|---|---|---|---|---|
| *Linear polysiloxanes* | | | | | |
| $CH_3$ | $CH_3$ | 0.970 | −123 | −37 | |
| $C_2H_5$ | $C_2H_5$ | 0.990 | −141 | 17 | 53 |
| $C_3H_7$ | $C_3H_7$ | 1.015 | −109 | 60 | 55 |
| $C_4H_9$ | $C_4H_9$ | | −116 | −19 | 300 |
| $C_5H_{11}$ | $C_5H_{11}$ | | − 106 | −23 | 330 |
| $C_6H_{13}$ | $C_6H_{13}$ | | −105 | 23 | 330 |
| $CH_3$ | $C_6H_5$ | 1.115 | −26 | 35 | |
| $C_6H_5$ | $C_6H_5$ | 1.22 | 45 | 268 | 540 |
| *Cyclic polysiloxanes* | | | | | |
| $CH_3$ | $CH_3$ | 0.972 | −123 | −35 | |
| $CH_3$ | $C_6H_5$ | | −28 | | |
| $CH_3$ | $CH{=}CH_2$ | 1.006 | −128 | | |

As a result, linear polysiloxanes have very low glass temperatures (Table 12-6). At room temperature, they are low- to high-viscosity liquids (depending on molecular weight) which are excellent lubricants.

In neat silicone oils, polymer molecules are present in partial helix conformation in which the inorganic backbone chain is screened by a hull of alkyl groups. Silicones are therefore not very polar: they are hydrophobic and have relatively low surface tensions. On polar substrates, polar Si–O bonds face the polar substrate (e.g., water or metals) whereas alkyl substituents are in contact with air. Poly(dimethylsiloxane)s have therefore high surface solubilities and low volume solubilities which makes them excellent anti-foaming and parting agents. Organofunctional polysiloxanes, on the other hand, have strongly negative groups and are therefore foaming agents and adhesion promoters, for example, coupling agents for fabrics of glass fibers.

The bond energy of the Si–O bond is somewhat higher than that of the C–C bond (373 kJ/mol *versus* 343 kJ/mol); polysiloxanes are therefore relatively thermally stable. However, the bond energy of Si–C bonds is considerable smaller than that of C–C bonds (243 kJ/mol versus 343 kJ/mol). Since Si–C bonds are strengthened by electron-donating methyl substituents but weakened by electron-withdrawing phenyl groups, poly(dimethylsiloxane)s are less prone to oxidation than poly(diphenylsiloxane)s.

On the other hand, siloxane backbones are more strongly polarized by methyl groups than by phenyl groups. Phenyl-substituted polysiloxanes are thus less hydrolyzable than methyl-substituted ones. Since *technical* thermal stabilities of polymers depend on both the resistance against chain scissions as well as against oxidation and hydrolysis, copolymers with methyl and phenyl substitutents are preferred over homopolymers.

Crosslinked organofunctional polysiloxanes serve as elastomers. Silicones containing ca. 0.2 % vinyl groups are crosslinked at elevated temperatures by either peroxides or hydrosilylation of vinyl groups with SiH-group containing compounds. Silicones with hydroxy groups are crosslinked at room temperature with compounds such as methyltriacetoxysilane or tetrabutyltitanate (**RTV** = room temperature vulcanization) (Volume IV). Silicon elastomers are always filled, predominantly with highly disperse silicon dioxide (see below). Crosslinking of unfilled rubbers does not lead to useful elastomers.

Block copolymers with dimethylsiloxane units are produced in many varieties, for example, with blocks of poly(styrene) or polycarbonate units. Triblock copolymers with "soft" central siloxane blocks and "hard" thermoplastic end blocks show the typical properties of thermoplastic elastomers (see Volume IV).

A curious silicone elastomer is **bouncing putty**, a children's toy, that is produced by heating poly(dimethylsiloxane)s with $B_2O_3$ as trifunctional crosslinking agent in the presence of $FeCl_3$, fillers, and plasticizers. Because of the crosslinking, balls of this material bounce back elastically. However, the crosslinks are weak so that polymer sheets break in a brittle manner if they are bent rapidly. If a sphere of this material is left laying on a surface, it slowly deforms to a sheet because of exchange equilibria.

## 12.3.12 Silicon Dioxide

**Quartz**, $[SiO_2]_\infty$, is a crystalline or amorphous tectopolymer (from West Slavic *kwarty*). Natural quartz is crystalline, either colorless and transparent (rock crystal) or colored by impurities as semi-precious stones (rose quartz, citrin, amethyst, opal, etc.), or as a microcrystalline whitish variety (quartz sand).

Quartz exists in various crystalline forms: as α-quartz at temperatures up to 573°C, as β-quartz in the temperature range 573–870°C, as tridymite from 870–1470°C, and as cristobalite 1410–1710°C. Cooling a quartz melt below its melting temperature of 1713°C results only in the high-temperature forms. Synthetic α-quartz is therefore produced hydrothermally by seeding a supersaturated solution of sodium silicate in NaOH or $Na_2CO_3$ with quartz crystallites. Synthetic quartz crystals are mainly used because of their piezoelectric properties (electrical filters, oscillators, etc.), and to some extent also because of their optical properties (prisms).

Synthetic quartz crystals can take up 0.1 % water. In contrast to α-quartz, this plasticised quartz is deformable at temperatures as low as 400°C. Quartz fibers are produced by softening and drawing of natural quartz crystals in a flame of oxyhydrogen gas.

Pyrolysis of $SiCl_4$ in an oxyhydrogen gas flame ($2\ H_2 + O_2$) delivers fumed silica (Aerosil®), a pyrogenic silicic acid with more than 99.8 % $SiO_2$:

$$(12\text{-}27) \qquad n\ SiCl_4 + 2\,n\ H_2 + n\ O_2 \longrightarrow (SiO_2)_n + 4\,n\ HCl$$

Aerosils consist of amorphous, spheroidal particles of ca. 10–20 nm diameter. Because of their low bulk density of 0.067 g/cm$^3$ and their high specific surface area of up to 800 m$^2$/g, aerosils are used as active fillers in silicon rubbers, thickeners, thixotropic agent in paints, etc.

Oil-rich but loose-packed sandstone formations are solidified directly by hydrolysis of silicon tetrachloride, Eq.(12-28). Alternative methods are the mechanical removal of loose sand or its solidification by an *in-situ* polymerization of furfuryl alcohol with $HClO_4 + FeCl_3$ (se also p. 326).

$$(12\text{-}28) \qquad n\ SiCl_4 + 2\,n\ H_2O \longrightarrow [SiO_2]_n + 4\,n\ HCl$$

## 12.4 Germanium and Tin Polymers

As may be expected from its position in the Periodic Table, **germanium** crystallizes in a diamond lattice and is thus a polymer. **Germanes**, $H(GeH_2)_iH$, are known up to the nonamer, $i = 9$). **Poly(dialkylgermane)s** and **poly(alkylphenylgermane)s** can be synthesized to high molecular weights, albeit in small yields. These polymers show strong UV absorptions because of delocalization of σ electrons along the chain. They are also thermochromic and photo-active.

**Tin**, the next highest element, forms three crystal modifications. "White tin" (β-tin) exists between 13.2°C and 161°C; it has a tetragonal lattice and a high density of 7.31 $g/cm^3$ (20°C). At 161°C, it converts to the orthorhombic γ-tin; pieces of this "brittle tin" can be easily pulverized. At 232°C, tin melts.

Below 13.2°C, "gray tin" (α-tin) is the stable modification. This tin has a diamond structure; it is thus a polymer. The transformation β → α is very slow between 13.2°C and 0°C but proceeds in hours at –50°C. Since α-tin is less dense than β-tin, articles from tin decompose to grqy powders. Since this transformation can be initiated by infection of β-tin with α-tin, it is also called **tin pest**.

## 12.5 Phosphorus Polymers

Group 15 of the Periodic Table consists of the elements N, P, As, Sb, and Bi. Elementary nitrogen has no polymeric modification but phosphorus does.

Nitrogen–hydrogen compounds ("nitranes") seem to exist as polymers, $[NH]_n$, at low temperatures (Volume I) but the more stable form is ammonium azide, $NH_4N_3$. Phosphanes are known up to $P_{10}H_{12}$, arsanes up to $As_5H_7$, and stilbanes up to $Sb_3H_5$. The also known yellow liquid diphosphine, $P_2H_4$, converts at room temperature to an amorphous yellow solid with the composition $P_2H$, probably a polymer.

### 12.5.1 Elementary Phosphorus

Phosphorus (G: *phosphoros* = light bearing, because its white form is phosphorescent in air) exists in several allotropic modifications. White phosphorus consists of discrete $P_4$ tetrahedron molecules. It melts at 44°C and dissolves in $CS_2$. Catalysts convert white phosphorus at 20°C and pressures of more than 3500 MPa to red, violet, and then black phosphorus, which has a complex graphite-like sheet structure and is insoluble in $CS_2$. Sheet structures are less regular in violet and red phosphorus, which also have lower degrees of polymerization.

### 12.5.2 Poly(phosphoric acid) and its Salts

**Poly(phosphoric acid)**, $HO{+}P(=O)OH{-}O{+}_nH \approx HPO_3$, is formally the polycondensation product of phosphoric acid = the tribasic ortho-phosphoric acid = $H_3PO_4$ = $(HO)_3P{=}O$. Poly(phosphoric acid) is obtained by dissolving diphosphorus pentoxide,

$P_2O_5$, in phosphoric acid. Due to its formation, and depending on the water content of $P_2O_5$ and $H_3PO_4$, poly(phosphoric acid) has a broad distribution of degrees of polymerization. Its properties are therefore not easy to reproduce and control if it is used as solvent or condensation agent.

So-called **metaphosphoric acids**, $[HPO_3]_i$ have the same composition as poly(phosphoric acid). They are only known as salts, $[MtPO_3]_i$, which are actually *cyclo*-oligophosphates with $i = 3, 4, 5$, etc. (for $i = 3$, see IV below). The "metaphosphoric acids" of industry are not metaphosphoric acids but in reality poly(phosphoric acid)s .

**Polyphosphates** are salts of poly(phosphoric acid). They result from the controlled dehydration of orthophosphates. Sodium dihydrogenphosphate (I) converts at < 160°C to the diphosphate (II) which polymerizes at higher temperatures to the **Maddrell salt** (III; C in Fig. 12-12) with degrees of polymerization of $X > 1000$.

(12-29)
$$\underset{\text{I}}{NaH_2PO_4} \longrightarrow \underset{\text{II}}{Na_2[H_2P_2O_7]} \longrightarrow \underset{\text{III}}{\sim\!\!\sim P(=O)(ONa){-}O\sim\!\!\sim} \longrightarrow \underset{\text{IV}}{\text{cyclo-}[P(=O)(ONa){-}O]_3}$$

Quenching the melt delivers water-soluble **Graham salt** ($X$ = 40–50), the low-molecular weight ($X$ = 15–20) form is known as **Calgon®**, used for water softening (name from *cal*cium *gone*). In Graham salt, $PO_4$ tetrahedrons are interconnected *via* two oxygen atoms each. Heating of III to 600°C converts it to the cyclic metaphosphate (IV) with a melting temperature of ca. 620°C.

On annealing at 500°C, Maddrell salt converts to one of the two types A and B of **Kurrol salt** (E and F in Fig. 12-12) which both form high-molecular weight, water-soluble fibers. All these forms are known as **fused phosphates**.

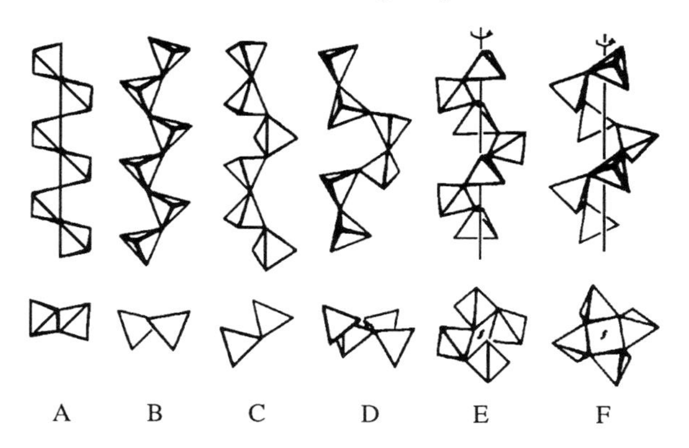


Fig. 12-12 The six chain conformations of crystalline polyphosphates [8]. Projection along (top) or perpendicular (bottom) to the chain direction.
From left to right: (A) $[RbPO_3]_n$, $[CsPO_3]_n$; (B) $[KPO_3]_n$, $[LiPO_3]_n$; (C) $[Na_2H(PO_3)_3]_n$, Maddrell salt; (D) $[Pb(PO_3)_2]_n$, $[Ca(PO_3)_2]_n$; (E) $[NaPO_3]_n$, sodium-Kurrol salt A; (F) sodium-Kurrol salt B.

Individual polymer chains are not preserved on melting and crystallization. They are rather split by trans reactions and rebuilt as new crystal forms. Such exchange equilibria have been confirmed by X-ray *via* the distribution of arsenic atoms in phosphate-arsenate copolymers. The Graham salt has a random distribution of As and P atoms whereas the Maddrell salt has a regular structure.

Similar to silicates, linear polyphosphates exist in melts in equilibrium with low-molecular weight cyclic phosphates and with branched and crosslinked polyphosphates. The relative proportion of each type of molecule depends on the temperature, the ratio Mt/P, and the water content. Endgroups consist of –OH and –OMt.

Oligo and polyphosphates complex multivalent cations so strongly that these ions can no longer be detected by addition of the usual precipitation agents. These phosphates are therefore used as water softeners and also as dispersion agents in the food industry for the preparation of cheeses and sausages.

### 12.5.3 Esters of Poly(phosphoric acid)

Nature produces two types of organic polymeric esters of phosphoric acid, teichoic acids and nucleic acids. **Teichoic acids** are found in the cell walls of bacteria. They are polyesters of phosphoric acid and substituted glycerol (I; R = H, D-alanyl, or a sugar) or phosphoric acid and ribitol (II; R' = H; R" = H or *N*-acetylglucosamine):

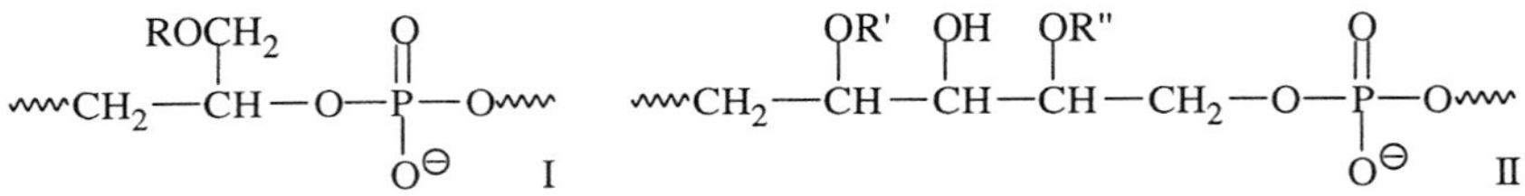

$$\sim CH_2-\underset{\displaystyle CH_2OR}{CH}-O-\overset{O}{\underset{O^{\ominus}}{\overset{\|}{P}}}-O\sim \quad \text{I}$$

$$\sim CH_2-\underset{OR'}{CH}-\underset{OH}{CH}-\underset{OR''}{CH}-CH_2-O-\overset{O}{\underset{O^{\ominus}}{\overset{\|}{P}}}-O\sim \quad \text{II}$$

**Nucleic acids** are linear polyesters of phosphoric acid and the sugars ribose or 2'-deoxyribose. The sugar units of ribonucleic acids (RNA) are substituted by the bases adenine, guanine, cytosine, or uracil, whereas those of deoxyribonucleic acids (DNA) carry adenine, guanine, cytosine, and thymine (Volume I).

Aliphatic and aromatic mono, di, and triesters of phosphoric acid, $(HO)_3P(O)$, are used in the polymer industry as plasticizers, flame retardants, and hardening agents. **Poly(estradiol phosphate)**, a polymeric phosphoric acid ester of estradiol with a molecular weight of ca. 26 000 is used as an estrogen to treat of prostate carcinoma.

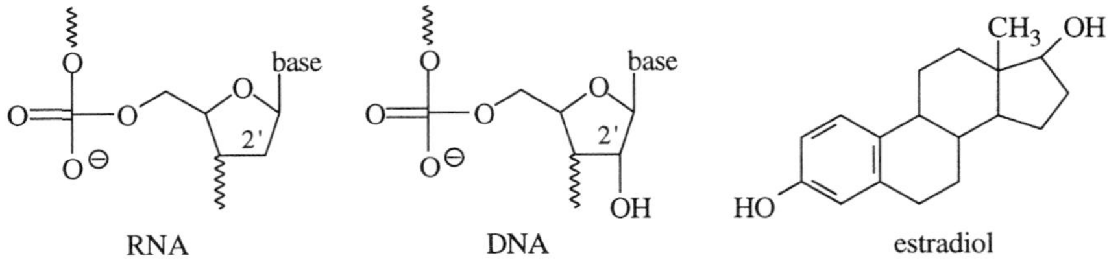

RNA DNA estradiol

### 12.5.4 Polyphosphonates

The hypothetical **phosphorous acid**, $P(OH)_3$, is only known in the form of its trialkyl and triaryl esters, the alkyl and aryl phosphites, respectively, $P(OR)_3$. However, the "phosphites" of inorganic chemistry are not phosphites but phosphonates: primary "phosphites" with the structure HP(O)(OH)(OMt) and secondary "phosphites" with the structure $HP(O)(OMt)_2$..

The unstable phosphorous acid, $P(OH)_3$, is tautomeric to the stable phosphonic acid, $HP(O)(OH)_2$, and the hypothetical phosphonous acid, $HP(OH)_2$, tautomeric to the stable phosphinic acid, $H_2P(O)(OH)$, formerly called "hypophosphorous acid." Phosphinous acid has the composition $H_2P(O)$.

The sequence of structures and names can be memorized by noting that the names of P(V) acids always end in "ic" whereas those of P(III) acids end in "ous". In each series, the two letters preceding "ic" and "ous", respectively, always have the sequence or-on-in.

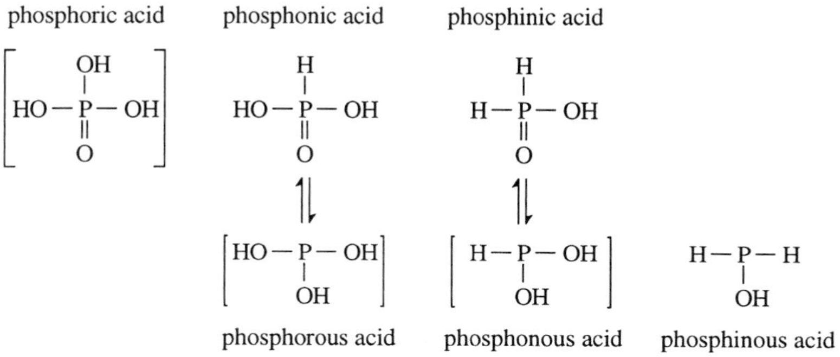

phosphoric acid: $[HO-P(=O)(OH)-OH]$ phosphonic acid: $HO-P(=O)(H)-OH$ phosphinic acid: $H-P(=O)(H)-OH$

$\upharpoonleft\downharpoonright$ $\upharpoonleft\downharpoonright$

$[HO-P(OH)-OH]$ $[H-P(OH)-OH]$ $H-P(OH)-H$

phosphorous acid phosphonous acid phosphinous acid

Polyphosphonates, $+O-Ar-O-P(=O)(R)+_n$, are produced industrially by polycondensation of phosphonic acid diphenyl esters and 4,4'-dihydroxybiphenyl with alkaline transesterification catalysts (alkali phenolates, alkaline earth hydrides, sodium carbonate, etc.):

(12-30) $C_6H_5O-P(=O)(C_6H_5)-OC_6H_5 + HO-C_6H_4-C_6H_4-OH \xrightarrow[-2\,C_6H_5OH]{} \sim P(=O)(C_6H_5)-O-C_6H_4-C_6H_4-O\sim$

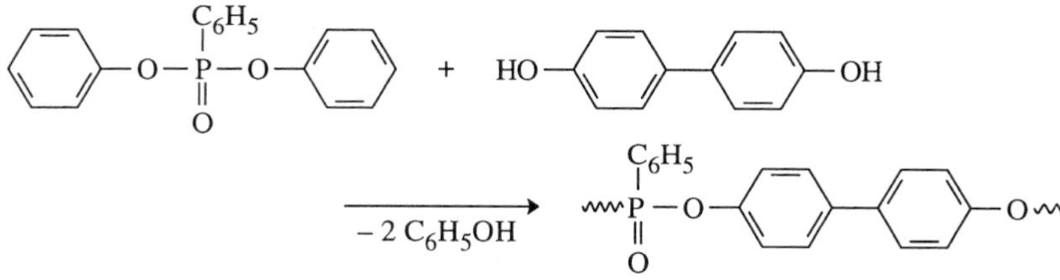

Polyphosphonates are also synthesized with other aromatic diphenols, for example, $HO(p\text{-}C_6H_4)\text{–}Z\text{-}(p\text{-}C_6H_4)OH$ (Z = S, $SO_2$). Transesterification catalysts are neutralized by equivalent amounts of acidic compounds (dialkyl sulfates, carboxy chlorides). The resulting polymers are brownish-colored by impurities from the air-oxidation of phenols. The resulting polymers are used as flame retardants since they have high thermal decomposition temperatures, for example, 395°C ($T_G$ = 120°C) for the polymer of Eq.(12-30). The thermal decomposition temperature can be increased to 465°C ($T_G$ = 146°C) if $-C_6H_4-S-C_6H_4-$ is replaced by $-C_6H_4-SO_2-C_6H_4-$.

## 12.5.5 Polyphosphazenes

Heating a chlorobenzene or tetrachloroethane solution of phosphorus pentachloride and ammonium chloride delivers white, soluble (in organic solvents) "phosphornitrilic chloride", $[NPCl_2]_i$, a mixture of cyclic and linear oligomers:

(12-32) $PCl_5 + NH_4Cl \longrightarrow (PNCl_2)_i + 4\,HCl$

Heating this crystalline solid to 230–300°C in vacuum results in an equilbrium polymerization to poly(dichlorophosphazene) = poly(phosphornitrile chloride) (**PNC**, II) as shown in Eq.(12-33) for the cyclic trimer, hexachlorocyclotriphosphazene (I):

$$\text{(12-33)}\quad (\text{NPCl}_2)_3\ (\text{I}) \underset{350°\text{C}}{\overset{250°\text{C}}{\rightleftharpoons}} \sim\!\sim P(Cl)_2\!=\!N\sim\!\sim\ (\text{II})$$

This ring-opening polymerization is probably not "thermal" but initiated by traces of cationic impurities. Endgroups are mostly $PCl_3$. Polymers from higher monomer conversions are crosslinked and show all properties of inorganic elastomers.

These polymers can also be obtained by polycondensation

$$\text{(12-34)}\quad PCl_5 + (NH_4)_2SO_4 \longrightarrow Cl_2P(=O)-N=PCl_3 \longrightarrow \sim\!\sim P(Cl)_2\!=\!N\sim\!\sim$$

and by a $PCl_5$–initiated living polymeriztion of a phosphoranimine at room temperature:

$$\text{(12-35)}\quad Cl_3P=N-Si(CH_3)_3 \longrightarrow \sim\!\sim P(Cl)_2\!=\!N\sim\!\sim + (CH_3)_3SiCl$$

PNCs hydrolyze in humid air and depolymerize at higher temperatures to hexachlorocyclotriphosphazene (I) and the octamer, octachlorocyclotetraphosphazene. Stable phosphazene chains are obtained if chlorine groups are replaced by organic substituents. However, it is not possible to polymerize hexaorganocyclotriphosphazenes, $(R_2PN)_3$, to the corresponding polymers, $+PR_2=N+_n$, by ring-opening polymerization because ceiling temperatures are low and polymerization equilibria are therefore on the side of oligomers. Organic groups are therefore introduced by post-reactions.

Commercial phosphazene polymers are obtained by alcoholysis of PNC with mixtures of fluorinated alcoholates, for example, in tetrahydrofuran with a mixture of $CF_3CH_2ONa$ and $H(CF_2)_iCH_2ONa$ or with a mixture of $C_6H_5ONa$ and R'(*p*-$C_6H_4$)ONa (aryloxyphosphazene, APM):

$$\text{(12-36)}\quad \sim\!\sim P(Cl)_2\!=\!N\sim\!\sim + 2\ NaOR \longrightarrow \sim\!\sim P(OR)_2\!=\!N\sim\!\sim + 2\ NaCl$$

The resulting fluorinated copolymers (**PNF**) resist hydrolysis. They have very low glass temperatures and can be crosslinked (vulcanized) by organic peroxides, sulfur, or high-energy radiation. The tensile strengths of these specialty elastomers are greater than those of silicone elastomers and approach those of conventional carbon-chain elastomers over a broad temperature range from –60°C to +200°C. The relatively expensive polymers are used in polar regions as seals, damping elements, and fuel pipes, especially for military applications.

Polyphosphazenes with organic substituents R ($CH_3$, $C_6H_5$) that are directly bound to phosphorous chain atoms can be obtained by direct polycondensation, albeit only with $CF_3CH_2OSi(CH_3)_3$ as leaving group (Eq.(12-37)). The resulting polymers are so stable that they can be nitrated with $HNO_3/H_2SO_4$ without degradation.

$$\text{(12-37)} \quad CF_3CH_2O-\underset{R}{\overset{R}{|}}{P}=N-Si(CH_3)_3 \longrightarrow \sim\!\!\sim \underset{R}{\overset{R}{|}}{P}=N\sim\!\!\sim \; + \; CF_3CH_2O-Si(CH_3)_3$$

Polycarbophosphazenes with carbon–nitrogen–phosphorus chains are obtainable by ring-opening polymerization, (Eq.(12-37)). Transformation of the resulting polymers by reaction of Cl substituents with $NaOC_6H_5$ introduces $-OC_6H_5$ (NaCl as leaving molecule) and with $C_6H_5NH_2$ introduces $-NH-C_6H_5$ (HCl as leaving molecule).

$$\text{(12-38)} \quad \text{cyclo-}[ClC{=}N{-}PCl_2{=}N{-}PCl_2{=}N] \longrightarrow \sim\!\!\sim C(Cl)=N-P(Cl_2)=N-P(Cl_2)=N\sim\!\!\sim$$

## 12.6 Sulfur Polymers

### 12.6.1 Elementary Sulfur

At room temperature, elementary sulfur consists mainly of the ordinary yellow sulfur modification, the orthorhombic cyclooctasulfur ($\alpha$-sulfur), $S_8$. This modification converts at 96°C to the monoclinic $S_8$-sulfur ($\beta$-sulfur) which at 119°C melts to a light yellow low-viscosity liquid which consists of ca. 95 % of $S_8$ molecules and of 5 % of $S_\pi$ sulfur. The terminology of the various sulfur modifications is somewhat confusing, but can be delineated as follows:

- $\lambda$-sulfur denotes all modifications of cyclic $S_8$ molecules, regardless of whether the modification is orthorhombic ($\alpha$-sulfur) or monoclinic ($\beta$-sulfur) or refers to cyclic $S_8$ molecules in the melt.
- $\mu$-sulfur is **polymeric sulfur** (**catena sulfur**) consisting of linear sulfur chains.
- $\pi$-sulfur comprises all other cyclic sulfur modifications ($S_6$, $S_7$, $S_9$, $S_{10}$, etc.).

The concentration of the various soluble species can be determined by high-performance liquid chromatography on extracts of samples from quenched melts, whereas the concentration of insoluble (polymeric) sulfur is obtained from the weight of the residue.

According to these measurements, the weight fraction $w_8$ of $S_8$ rings starts to decrease at 140°C (Fig. 12-13). It passes through a weak minimum at ca. 255°C, increases slightly, and then becomes constant with $w_8 \approx 0.56$ at temperatures above ca. 350°C.

The concentration of (polymeric) $S_\mu$-sulfur begins to climb at ca. 140°C, passes through a weak maximum at 255°C, decreases, and becomes constant with $w_\mu \approx 0.34$.

The concentration of all other cyclics ($\pi$-sulfur) has a small maximum at $T_\pi$ = 159°C but is practically constant below $T_\pi$ ($w_\neq \approx 0.065$) and above $T_\pi$ ($w_\neq \approx 0.10$).

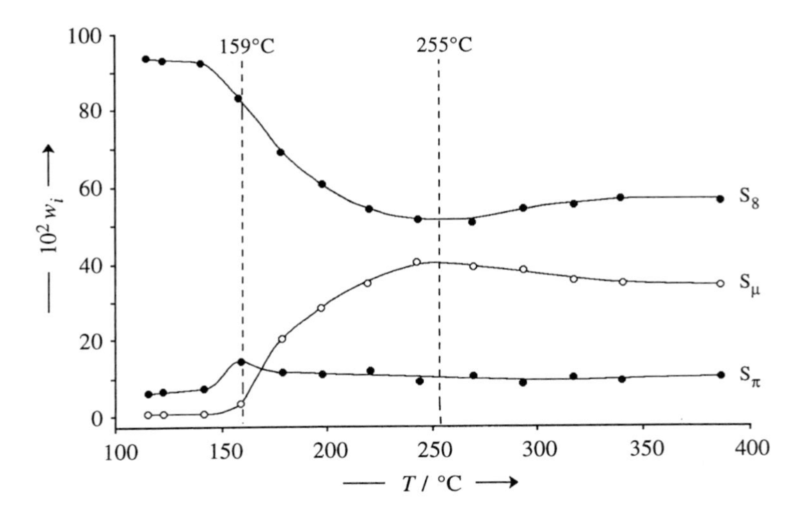


Fig. 12-13 Temperature-dependence of mass fractions $w_i$ of $S_8$-rings ($S_8$), polymeric sulfur ($S_\mu$), and other cyclic sulfur molecules ($S_\pi$) [9].

The temperature of 159°C is a transition temperature since it is marked not only by a maximum in the concentration of non-$S_8$ cyclics and the beginning of a stronger increase of the weight fraction of polymeric sulfur but also by a sudden increase of the melt viscosity (Fig. 12-14). The latter passes through a maximum at ca. 187°C and then decreases slowly. Electron spin resonance measurements show that free radicals first appear at ca. 170°C.

The concentration of these radicals can be determined by titration with iodine. Since in polymeric sulfur each free radical is an endgroup and according to X-ray measurements each sulfur atom has only two neighbors (no measurable branching), then the concentration of free radicals can be used to calculate the number-average degree of polymerization of polymeric sulfur. The values so obtained are high but not very probable for the following reasons.

Considerable proportions of $S_8$ molecules and other cyclics are still present at temperatures above 159°C (Fig. 12-13). The transition temperature of 159°C thus cannot be a floor temperature for an all-or-none process of the type $n\ S_8 \rightleftarrows {}^{\bullet}S_{8n}{}^{\bullet}$.

Electron spin resonance does not detect radicals in the temperature range 160-170°C but only above 170°C (Volume I, Fig. 2-3). This observation does not necessarily preclude a sequence $n$ c-$S_8 \rightleftarrows n\ {}^{\bullet}S_8{}^{\bullet} \rightleftarrows {}^{\bullet}S_{8n}{}^{\bullet}$ since the concentration of sulfur biradicals is small and the degree of polymerization may be high. Note that the degrees of polymerization in Fig. 12-14 are not measured but calculated..

Chain scissions of the type ${}^{\bullet}S_{m+n}{}^{\bullet} \rightleftarrows {}^{\bullet}S_m{}^{\bullet} + {}^{\bullet}S_n{}^{\bullet}$ are not very probable since the homolytic scission of the S–S bond is strongly exothermic. A possible explanation for an onset of transformations at 159°C may be ring extension reactions to larger rings, as shown in Eq.(12-39) for some relevant ring types:

(12-39) [ring $S_x$(S–S)S(S–S)] + [ring S–S–$S_y$–S–S–S] $\rightleftharpoons$ [fused ring $S_x$ … $S_y$] $\rightleftharpoons$ [ring $S_x$(S–S)S(S–S)] + [ring S–S–S–$S_y$–S–S–S]

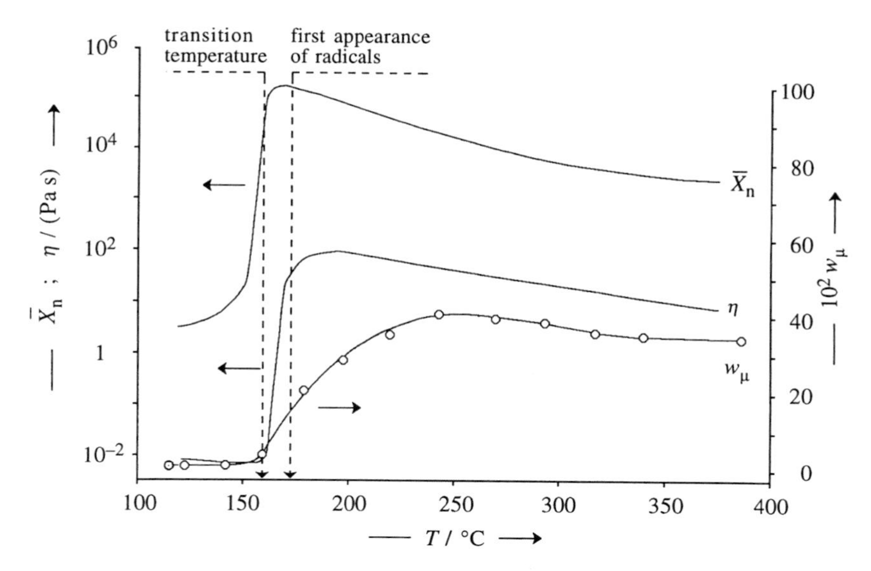


Fig. 12-14 Temperature-dependence of weight fractions $w_\mu$ of polymeric sulfur (right scale; data of Fig. 12-13), melt viscosities $\eta$ [10] (left scale), and number-average degrees of polymerization calculated for an all-or-none mechanism, $n\ S_8 \rightleftarrows [S_8]_n$ [11] (left scale).

The polymerization of $S_8$ rings is initiated by radicals from the unstable $S_{\neq 8}$ cyclics. The positive enthalpy of polymerization of $\Delta H \approx 19$ kJ/(mol $S_8$) is overcompensated by the term $-T\Delta S$ with $\Delta S = +44.7$ J/(mol K). The polymerization entropy is positive since the negative contribution from the decrease of numbers of molecules during polymerization is more than compensated by the positive contribution from the gain in conformational entropy since rigid, crown-like $S_8$ rings are converted to flexible sulfur chains. At temperatures above ca. 200°C, random chain scissions cause a decrease of melt viscosities.

Quenching the melt causes the polymerization equilibrium to freeze. The resulting **plastic sulfur** is a polymeric sulfur (μ-sulfur) that is plasticized by cyclic sulfur molecules (λ- and π-sulfur). Extraction of the cyclics causes polymeric sulfur to crystallize and to become brittle. Polymeric sulfur crystallizes as $7_2$ helices.

The equilibria are important for the various applications of elementary sulfur, especially in the vulcanization of rubbers (Volume IV) where one employs not only $S_8$ sulfur but also polymeric sulfur.

## 12.6.2 Poly(thiazyl)

Tetrasulfurtetranitride, $S_4N_4$, from $S_2Cl_2 + NH_3$, forms orange-yellow crystals that are stable in air. In these crystals, molecules have a basket-like structure (I). $S_4N_4$ melts at 178°C and boils at 185°C. Passing $S_4N_4$ vapor over silver wool at 200–300°C converts tetrasulfurtetranitride to the potentially explosive cyclic dimer (II). The colorless, paramagnetic crystals of the dimer polymerize in the solid state to a bluish-black, paramagnetic solid that later converts to a golden-shimmering diamagnetic material (III), probably *via* a series of biradicals:

(12-40) I $\longrightarrow$ 2 II $\longrightarrow$ $\left[ N{=}S^{\oplus}{-}N^{\ominus}{-}S \right]$ III

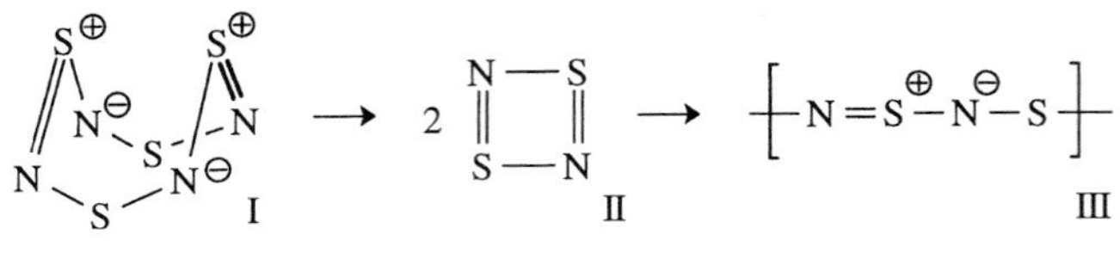

S₄N₄ vapor can also be polymerized either photochemically or by a helium plasma.

Also possible are solution polymerizations in liquid sulfur dioxide (boiling temperature: –10°C) of $S_2NAsF_6$ with $N_3^{\oplus}$ as initiator at –20°C and of $(NSCl)_3$ with $(CH_3)_3SiN_3$ at –18°C. The electrochemical polymerization of $S_5N_5Cl$ also succeeds in liquid $SO_2$. $S_iN_jCl_k$ can be polymerized to $+\!SN\!+_n$ in acetonitrile at –15°C.

The resulting **poly(thiazyl) = poly(azasulfene) = poly(sulfurnitride) = SNX polymer** forms "single crystals" that consist of bundles of highly oriented fibrils. It behaves like a metal with respect to its reflection, specific heat capacity, magnetic susceptibility, and electrical conductivity (at room temperature with respect to the *b* axis: 2000 S/cm (parallel), 40 S/cm (perpendicular); at $T < 20$ K: $10^6$ S/cm). Undoped poly(thiazyl) becomes supraconductive at 0.26 K.

Crystalline, film-like, or pasteous poly(thiazyl) is processed to electrodes. However, the polymer explodes if hit by a hammer and decomposes to a gray powder on extended exposure to air or water. Prolonged heating in air decomposes poly(thiazyl) to nitrogen and sulfur dioxide.

# 12.7 Selenium, Tellurium, and Gallium Polymers

## 12.7.1 Elements

The silvery element **selenium** (G: *selene* = moon) has two low-molecular weight modifications and one polymeric one. The low-molecular weight ones crystallize in orthorhombic and monoclinic crystal modifications, respectively; they consist probably of $Se_8$ rings, similar to sulfur.

Selenium dissolves in $CS_2$ to a dark-red solution. Low-molecular weight crystals ($T_M$ = 119°C) are formed if $CS_2$ is evaporated from these solutions at temperatures below 72°C. Evaporation above 72°C as well as quenching of a selenium melt delivers a gray, crystalline selenium modification that is more stable than $Se_8$. This trigonal modification ($T_M$ = 221°C) looks metallic, is photoconductive, and insoluble in any known solvent. It consists of $3_1$ helices with only weak interactions between chains.

**Tellurium** (L: *tellus* = earth) has only one silvery-white metallic modification that is isomorphous with the gray selenium. It is also insoluble in all known solvents.

## 12.7.2 Gallium Chalcogenides

Compounds of gallium and chalcogenes (O, S, Se, Te) are interesting for materials that are sensitive to changes of magnetic fields. For this purpose, the solid phase of these chalcogenides must be hexagonal, i.e., polymeric. The cubane, $(t\text{-}C_4H_9)_4Ga_4S_4$, of gal-

(12-41)

lium sulfide, loses at 350°C all four butyl groups and forms on solid substrates only face-centered cubic (non-polymeric) GaS phases. The corresponding selenide (I) does not lose all butyl groups, however. Rather, some Ga–Se bonds are broken, resulting in an intermediary segment (II) that polymerizes on solid surfaces as hexagonal phase with Ga–Ga and Se–Se bonds (III) (Eq.(12-41)). The corresponding telluride also delivers a hexagonal phase; however, this phase converts on heating to a monoclinic phase that consists of 5- and 6-rings.

## 12.8 Organometallic Polymers

Polymers with *metals in side-groups* can be synthesized by both polymerizations (chain polymerizations, polycondensations, polyadditions, polyeliminations) or by polymer transformations. These reactions are analogous to those of metal-free monomers and polymers. An example of a polymer transformation is the reaction of poly(*p*-chloromethylstyrene) with sodium tungstenpentacarbonyl:

(12-42) $\sim CH_2-CH \sim$ (p-$C_6H_4CH_2Cl$) $\xrightarrow[-\ NaCl]{+\ Na[W(CO)_5]}$ $\sim CH_2-CH \sim$ (p-$C_6H_4CH_2W(CO)_5$)

A special group of metal-containing polymers are **metallocene polymers**. Metallocenes are defined as complexes between metals and *bis*($\eta$-cyclopentadienyl) units, commonly called dicyclopentadienyl units. Metallocene polymers are correspondingly polymers that contain metallocene units.

It is regrettable that industry uses the term "metallocene polymer" in a different manner, namely, for polymers that have been produced by polymerization of olefinic monomers such as ethene and propene with low-molecular weight metallocene catalysts (see p. 236 ff. of this volume and especially Volume I). These "metallocene polymers" are not polymers that contain metallocene units as either side groups or in the main chain but are rather polymers made with metallocene catalysts.

True metallocene polymers are polymers that contain either in main chains or as side-groups aromatic ring systems that are bound to transition metals via π-bonds. An example of a side-group metallocene polymer is poly(ferrocenyl ethylene) from the free-radical chain polymerization of vinyl ferrocene with AIBN (dibenzoyl peroxide leads to side reactions!):

(12-43)

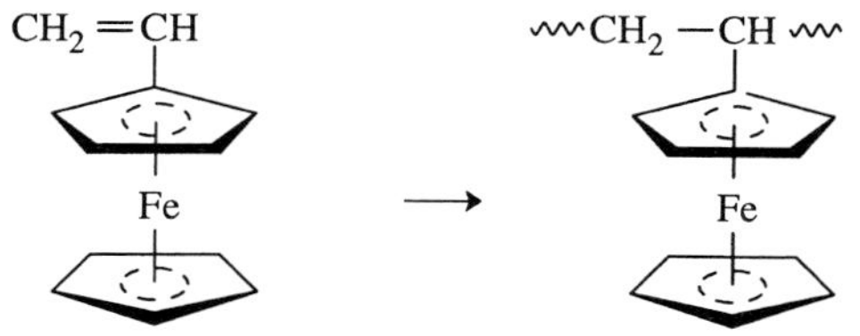

In rare cases, metallocene polymers can also be obtained by ring-opening polymerization of strained rings, for example, by polymerization of ferrocenophanes to poly-(ferrocenyl-X) where X is an ethylene or a dimethylsilane group:

(12-44)

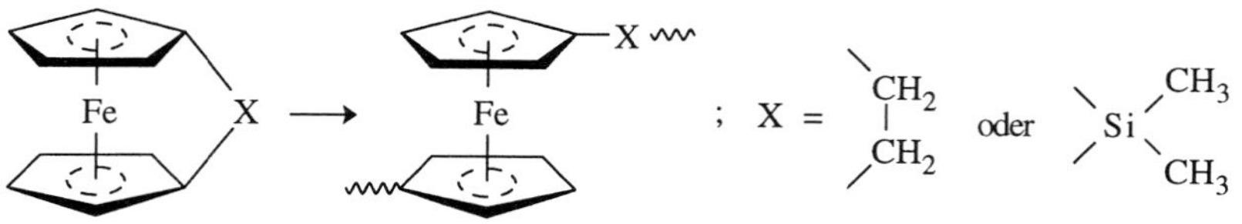

Most organometallic polymers with metal atoms in the main chain can only be produced by polycondensation. In some of these reactions, chelating metal groups are already present in monomers, for example, in the polycondensation of ethylacetoacetate derivatives of copper with glycols:

(12-45)

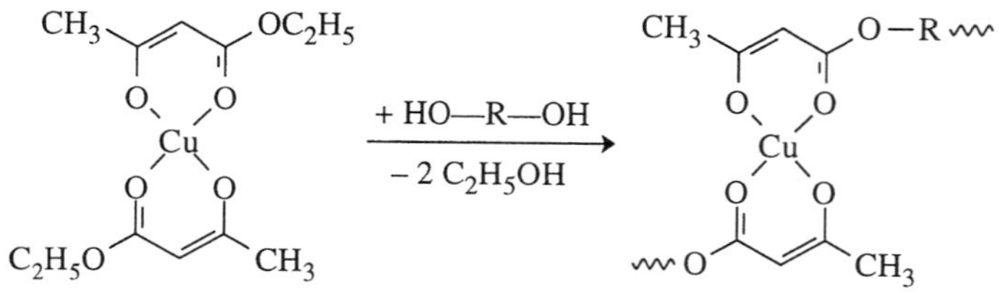

In other polycondensations, metals are participating directly in the polymer formation. An example is the formation of uncrosslinked, acetone- or butanone-soluble, **coordination polymers** from the reaction of organic acids, $ROO^-$ and $Fe^{3+}$ (or other metals of the iron group) in the presence of aldehydes:

(12-46)

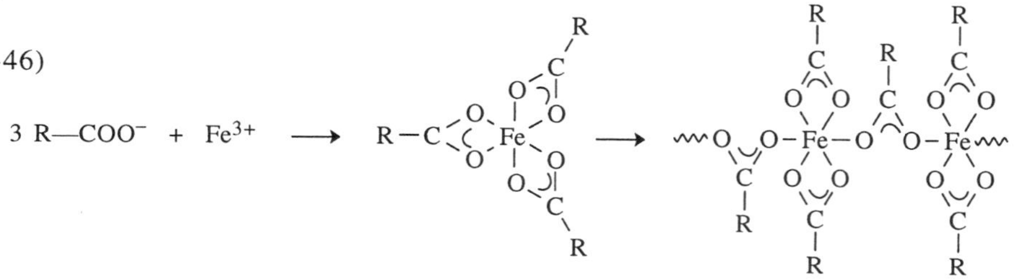

These polymers decompose slowly in water, which makes them useful as herbicides, insecticides, etc., if the acid component has the desired biological activity.

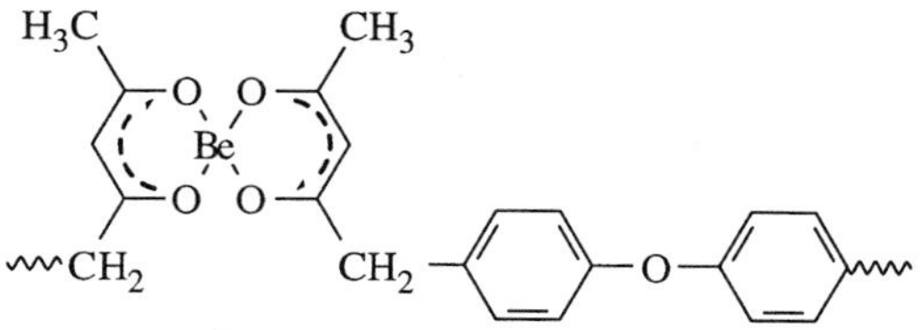

I

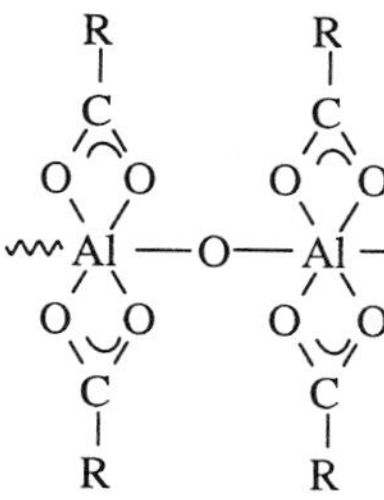

II

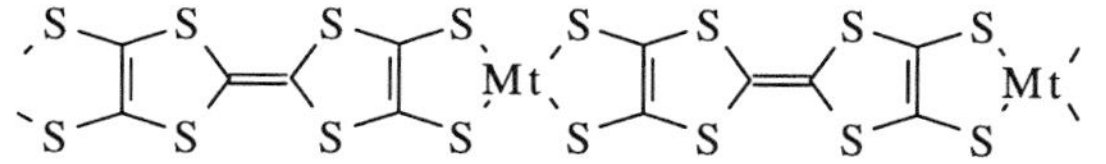

III

Mt = Ni, Cu, Pd, Fe

Other examples of coordination polymers are shown above (I, II, and III). Aluminum soaps (II) with substituents R from *na*phthenic, *palm*itic, and oleic acids gel gasoline which led to the napalm of World War II. Present-day incendiary mixtures consist of gels of 25 % poly(styrene), 25 % benzene, and 25 % gasoline without any aluminum soaps.

Another example of a polycondensation of a metal compound is the reaction of metal salts $MtX_2$ with 3,3',4,4'-tetracyano compounds, $(NC)_2C_6H_3$-Z-$C_6H_3(CN)_2$, where Z can be $(CH_2)_n$, CO, $SO_2$, or O. The resulting polymers (IV) may be useful as catalysts or photo sensitizers.

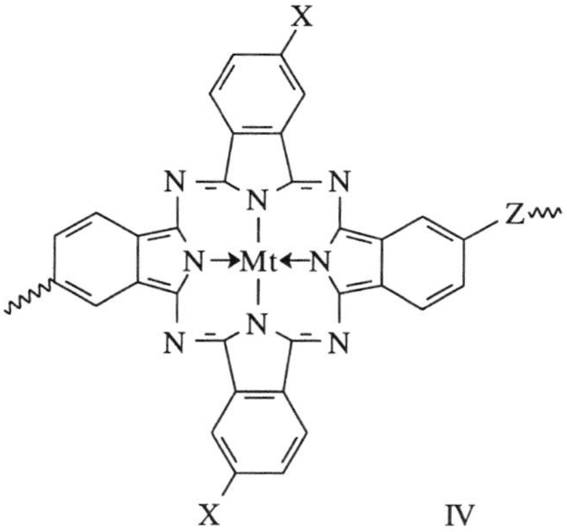

IV

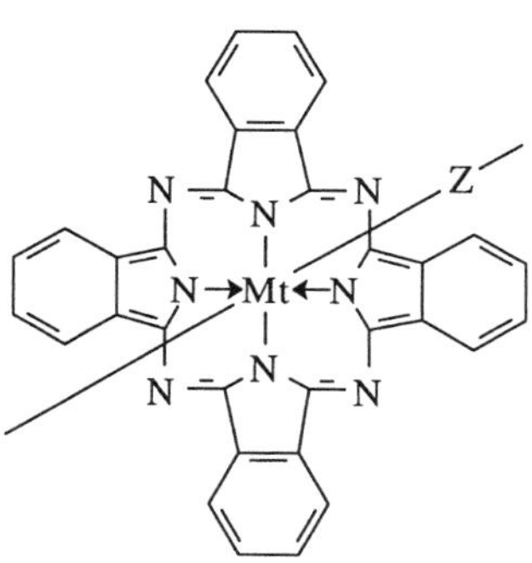

V

Coordination polymers of phthalocyanine (Pc) with the structure $\text{+}Mt(Pc)\text{–}Z\text{+}_n$ (V) can be doped with $SbF_5$ to become semiconducting polymers with electrical conductivities of up to 0.1 S/cm. In such polymers, Z may be oxygen and Mt either silicon, germanium, or tin.

Polymers of the structure V may also have the general structure $\text{+}Mt(Pc)\text{–}L\text{+}_n$ where Mt is either $Fe^{2+}$, $Fe^{3+}$, $Co^{2+}$, $Co^{3+}$, $Mn^{2+}$, $Mn^{3+}$, or $Cr^{3+}$ and L one of the following ligands:

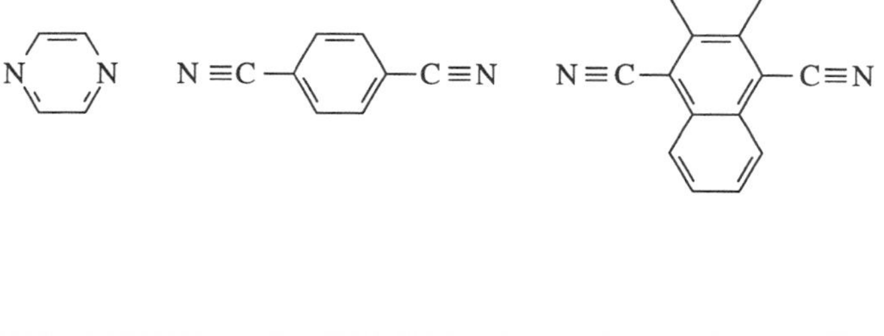

# Literature to Chapter 12

12.1 GENERAL SURVEYS
F.G.A.Stone, W.A.G.Graham, Eds., Inorganic Polymers, Academic Press, New York 1962
F.G.R.Gimblett, Inorganic Polymer Chemistry, Butterworth, London 1963
D.N.Hunter, Inorganic Polymers, Wiley, New York 1963
K.A.Andrianov, Metalorganic Polymers, Wiley-Interscience, New York 1965
H.R.Allcock, Heteroatom Ring Systems and Polymers, Academic Press, New York 1967
N.H.Ray, Inorganic Polymers, Academic Press, London 1978
A.D.Wilson, S.Crisp, Organolithic Macromolecular Materials, Appl.Sci.Publ., Barking 1978
C.E.Carraher, J.E.Sheats, C.U.Pittman, Eds., Organometallic Polymers, Academic Press, New York 1978
J.E.Harrod, R,M.Laine, Eds., Inorganic and Organometallic Oligomers and Polymers, Kluwer, Dordrecht 1990
J.E.Mark, H.R.Allcock, R.West, Inorganic Polymers: An Introduction, Prentice-Hall, Englewood Cliffs, NJ 1992 (contains only polyphosphazenes, polysiloxanes, and polysilanes)
R.M.Lane, Ed., Inorganic and Organometallic Polymers with Special Properties, Kluwer, Dordrecht 1992
A.D.Pomogailo, V.S.Savost'yanov, Synthesis and Polymerization of Metal-Containing Monomers, CRC Press, Boca Raton (FL) 1994
I.Manners, Polymers and the Periodic Table: Recent Developments in Inorganic Polymer Science, Angew.Chem.Int.Ed.Eng. **35** (1996) 1602
K.H.Büchel, H.-H.Moretto, P.Woditsch, Industrial Inorganic Chemistry, Wiley-VCH, Weinheim, 2nd ed. 2000
R.D.Archer, Inorganic and Organometallic Polymers, Wiley-VCH, Weinheim 2001
A.S.Abd-El-Aziz, C.E.Carraher, Jr., C.U.Pittman, Jr., J.E.Sheats, M.Zeldin, Macromolecules Containing Metal and Metal-Like Elements, Wiley-VCH, Weinheim
Vol. 1, A Half-Century of Metal- and Metalloid-Containing Polymers (2003); Vol. 2, Organoiron Polymers (2003); Vol. 3, Biomedical Applications (2004)
I.Manners, Synthetic Metal-Containing Polymers, Wiley-VCH, Weinheim 2004
A.S.Abd-El-Aziz, C.E.Carraher, Jr., C.U.Pittman, Jr., M.Zeldin, J.E.Sheats, Macromolecules Containing Metal and Metal-Like Elements, Wiley-VCH, Weinheim
Vol. 4, Group IVA Polymers (2005)
A.S.Abd-El-Aziz, C.E.Carraher, Jr., C.U.Pittman, Jr., M.Zeldin, Macromolecules Containing Metal and Metal-Like Elements, Wiley-VCH, Weinheim
Vol. 5, Metal-Coordination Polymers (2005); Vol. 6, Transition Metal-Containing Polymers (2005); Vol. 7, Nanoscale Interactions of Metal-Containing Polymers (2005)
T.Chandrasekhar, Inorganic and Organometallic Polymers, Springer, Berlin 2005

12.2 BORON POLYMERS
H.A.Schroeder, Polymer Chemistry of Boron Cluster Compounds, Inorg.Macromol.Rev. **1** (1970) 45
I.B.Atkinson, B.R.Currell, Boron-Nitrogen Polymers, Inorg.Macromol.Rev. **1** (1971) 203
E.N.Peters, Poly(dodecacarborane-siloxanes), J.Macromol.Sci.-Rev.Macromol.Chem. **C 17** (1979) 173
P.R.Dvornic, R.W.Lenz, High Temperature Siloxane Elastomers, Hüthig und Wepf, Basel 1990, Chapter 5

12.3 SILICON POLYMERS in general
R.G.Jones, Ed., Silicon-Containing Polymers, Royal Soc.Chem., Cambridge, UK 1996
M.A.Brook, Silicon in Organic, Organometallic and Polymer Chemistry, Wiley-VCH, Weinheim 1999

12.3.4 POLY(SILICIC ACID)S
R.Iler, The Chemistry of Silica: Solubility, Polymerization, Colloid and Surface Properties and Biochemistry, Wiley, Chichester 1979

12.3.5 SILICATES
W.Eitel, Silicate Science, Academic Press, New York 1964
F.Liebau, Structural Chemistry of Silicates, Springer, Berlin 1985

12.3.6 SILICATE GLASES
D.R.Uhlmann, N.J.Kreidel, Eds., Glass - Science and Technology, Academic Press, New York, Bd. **1** (1983) ff.
R.H.Doremus, Glass Science, Wiley, New York, 2.Aufl. 1994
M.Vogel, Glass Chemistry, Springer, Berlin 1994
J.E.Shelby, Introduction to Glass Science and Technology, Royal Soc. Chem., Cambridge, UK 1997

12.3.7 FIBROUS SILICATES
H.C.W.Skinner, M.Ross, C.Frondel, Asbestos and Other Fibrous Minerals, Oxford Univ. Press, London 1988

12.3.8 LAYER SILICATES
R.W.Grimshaw, The Chemistry and Physics of Clays, E.Benn, London, 4th ed. 1971
B.K.G.Theng, Formation and Properties of Clay-Polymer Complexes, Elsevier, New York 1979
A.Weiss, Replikation und Evolution in anorganischen Systemen, Angew. Chem. **93** (1981) 843 (= replication and evolution in inorganic systems)
A.C.D.Newman, Chemistry of Clays and Clay Materials, Wiley, New York 1987

12.3.9 CEMENT AND CONCRETE
E.Kirlikovali, Polymer/Concrete Composites - A Review, Polym.Eng.Sci. **21** (1981) 507
T.M.Aminabhavi, P.E.Cassidy, L.E.Kukacka, Use of Polymers in Concrete Technology, J.Macromol.Sci. - Rev.Macromol.Chem.Phys. **C 22** (1982-1983) 1
J.J.Beaudoin, Handbook of Fiber-Reinforced Concrete, Noyes Data, Park Ridge, NJ 1990
H.F.W.Taylor, Cement Chemistry, Academic Press, London 1990
S.Chandra, Y.Ohama, Polymers in Concrete, CRC Press, Boca Raton (FL) 1994
Y.Ohama, Handbook of Polymer-Modified Concrete and Mortars, Noyes Publ., Park Ridge, NJ 1995

12.3.10 CERAMIC MATERIALS
K.J.Wynne, R.W.Rice, Ceramics via Polymer Pyrolysis, Annu.Rev.Mater.Sci. 14 (1984) 297
L.L.Hench, D.R.Ulrich, Eds., Science of Ceramic Chemical Processing, Wiley-Interscience, New York 1986
J.D.Mackenzie, D.R.Ulrich, Eds., Ultrastructure Processing of Advanced Ceramics, Wiley, New York 1988
C.J.Brinker, G.W.Scherer, Sol-Gel Science, Academic Press, San Diego 1990
A.R.Bunsell, M.-H.Berger, Eds., Fine Ceramic Fibers, Dekker, New York 1999
J.D.Wright, N.A.J.M.Sommerdijk, Sol-Gel Materials. Chemistry and Applications, Gordon and Breach, Amsterdam 2001

12.3.11a SILICONES (History)
H.A.Liebhafsky, S.S.Liebhafsky, G.Wise, Silicones under the Monogram, Wiley, New York 1979 (history of General Electric silicones)
E.G.Rochow, Silicon and Silicones, Springer, Berlin 1987 (mainly history of General Electric silicones)
E.L.Warrick, Forty Years of Firsts - The Recollections of a Dow Corning Pioneer, McGraw-Hill 1990 (history of development of Dow Corning silicones)

12.3.11b SILICONES (Syntheses and Properties)
R.J.H.Voorhoeve, Organohalosilanes, Precursors to Silicones, Elsevier, Amsterdam 1967
S.N.Borisova, M.G.Voronkov, E.Ya.Lukevits, Organosilicon Heteropolymers and Heterocompounds, Plenum, New York 1970
A.D.Wilson, S.Crisp, Organolithic Macromolecular Materials, Appl.Sci.Publ., Barking 1978
B.A.Currell, J.R.Parsonage, Trimethylsilylation of Mineral Silicates, J.Macromol.Sci.-Chem. **A 16** (1981) 141
R.W.Dyson, Ed., Siloxane Polymers, Chapman & Hall, New York 1987
P.R.Dvornic, R.W.Lenz, High Temperature Siloxane Elastomers, Hüthig und Wepf, Basel 1990
A.Tomanek, Ed., Silicone und Technik, Hanser, München 1990; -, Silicones and Industry - A Compendium for Practical Use, Instruction and Reference, Hanser, München 1991

G.Koerner, M.Schulze, J.Weis, Silicones: Chemistry and Technology, CRC Press, Boca Raton (FL) 1992
S.J.Clarson, J.A.Semlyen, Eds., Siloxane Polymers, PTR Prentice-Hall, Englewood Cliffs (NJ) 1993
N.Auner, J.Weis, Organosilicon Chemistry II. From Molecules to Materials, VCH, Weinheim 1995
R.G.Jones, Ed., Silicon-Containing Polymers, Royal Soc.Chem., London 1996
R.M.Hill, Ed., Silicone Surfactants, Dekker, New York 1999
R.G.Jones, W.Ando, J.Chojnowski, Eds., Silicon-Containing Polymers, Kluwer, Dordrecht 2000
P.Jershow, Silicone Elastomers, Rapra, Shawbury, Shrewsbury, Shropshire, UK, 2001
P.Jutzi, U.Schubert, Eds., Silicon Chemistry. From the Atom to Extended Systems, Wiley-VCH, Weinheim 2003

12.5 PHOSPHORUS POLYMERS

E.Thilo, Zur Strukturchemie der kondensierten anorganischen Phosphate, Angew.Chem. **17** (1965) 1065 (= structural chemistry of condensed inorganic phosphates)
M.Sander, E.Steiniger, Phosphorous Containing Polymers, J.Macromol.Sci.Rev. **C 1** (1967) 1, 7, 91; **C 2** (1968) 1, 33, 57
H.R.Allcock, Phosphorus-Nitrogen Compounds. Cyclic, Linear and High Polymeric Systems, Academic Press, New York 1972
M.Schmidt, D.Freitag, L.Bottenbruch, U.Reinking, Aromatische Polyphosphonate: Thermoplastische Polymere von extremer Brandwidrigkeit, Angew.Makromol.Chem. **132** (1985) 1 (= aromatic polyphosphonates: thermoplastic polymers of extreme flame resistance)
R.C.Ropp, Inorganic Polymeric Glasses, Elsevier, Amsterdam 1992 (4 chapters on phosphate glasses, 1 chapter on silicate glass)
J.U.Otaigbe, G.H.Beall, Inorganic Phosphate Glasses as Polymers, Trends Polym.Sci. **5** (1997) 369
H.R.Allcock, Chemistry and Applications of Polyphosphazenes, Wiley-Interscience, New York 2003

12.6 SULFUR POLYMERS

A.V.Tobolsky, W.J.MacKnight, Polymeric Sulfur and Related Polymers, Interscience, New York 1965
E.J.Goethals, Sulfur-Containing Polymers, J.Macromol.Sci.-Rev.Macromol.Chem. **C 2** (1969) 74
L.Pintschovius, Polysulfur Nitride, $(SN)_x$, the First Example of a Polymeric Metal, Colloid Polym.Sci. **256** (1978) 883
M.M.Labes, P.Love, L.J.Nichols, Polysulfur Nitride - A Metallic Superconducting Polymer, Chem.Rev. **79** (1979) 1
P.Love, Some Properties and Applications of Polysulfur Nitride, Polymer News **7** (1981) 200
R.Steudel, Molekulare Zusammensetzung von flüssigem Schwefel, Z.Anorg.Allgem.Chem. **478** (1981) 139 (= molecular composition of liquid sulfur)
A.Müller, B.Krebs, Eds., Sulphur - Its Significance for Chemistry, for the Geo-, Bio- and Cosmosphere and Technology, Elsevier, Amsterdam 1984
B.Meyer, Elemental Sulfur: Chemistry and Physics, Interscience, New York 1985

12.8 ORGANOMETALLIC POLYMERS

K.A.Andrianov, Metalorganic Polymers, Interscience, New York 1965
E.W.Neuse, H.Rosenberg, Metallocene Polymers, J.Macromol.Sci.Rev. **C 4** (1970) 1
C.E.Carraher, Jr., J.E.Sheats, C.U.Pittman, Jr., Organometallic Polymers, Academic Press, New York 1978
R.S.Ward, E.Nyilas, Organometallic Polymers, Academic Press, New York 1978
R.V.Subramanian, B.K.Grag, Recent Advances in Organotin Polymers, Polym.-Plast.Technol.Eng. **11** (1978) 81
J.E.Sheets, History of Organometallic Polymers, J.Macromol.Sci.Chem. **A 15** (1981) 1173
N.Hagihara, K.Sonogashira, S.Takahashi, Linear Polymers Containing Transition Metals in the Main Chain, Adv.Polym.Sci. **41** (1981) 149
H.Szalinska, M.Pietrzak, Metal-Containing Polymers, Polym.-Plast.Technol.Eng. **19** (1982) 107
D.Wöhrle, Polymer Square Planar Metal Chelates for Science and Industry. Synthesis, Properties and Applications, Adv.Polym.Sci. **50** (1983) 45
A.D.Pomogallo, V.S.Savostyanov, Advances in the Synthesis and Polymerization of Metal-Containing Monomers, J.Macromol.Sci.-Rev.Macromol.Chem.Phys. **C 25** (1985) 375

A.D.Pomogailo, V.S.Savost'yanov, Synthesis and Polymerization of Metal-Containing Monomers, CRC Press, Boca Raton (FL) 1994
F.Ciardelli, E.Tsuchida, D.Wöhrle, Eds., Macromolecule-Metal Complexes, Springer, Heidelberg 1996
C.U.Pittman, Jr., C.E.Carraher, Jr., M.Zeldin, J.E.Sheats, B.M.Culbertson, Metal-Containing Polymeric Materials, Plenum Press, New York 1996
R.D.Archer, Inorganic and Organometallic Polymers, Wiley-VCH, Weinheim 2001
D.Wöhrle, Metal Complexes and Metals in Macromolceules, Wiley-VCH, Weinheim 2003
I.Manners, Synthetic Metal-Containing Polymers, Wiley-VCH, Weinheim 2004

# References to Chapter 12

[1] J.Hefter, M.E.Kenney, Inorg.Chem. **21** (1982) 2810, data of Figs. 4, 8, and 9
[2] C.R.Masson, J.Non-Cryst.Solids **25** (1977) 1, Figs. 1, 2, and 3
[3] R.Kleinholz, G.Heyn, R.Stolze, in R.Gächter, H.Müller, Eds., Plastics Additives Handbook, Hanser, München 1990, Chapter 10, Table 3
[4] S.Kessler, In J.Edenbaum, Ed., Plastics Additives and Modifiers Handbook, Van Nostrand Reinhold, New York 1992, Table 48-1
[5] T.Dombrowski, in Kirk-Othmer, Encyclopedia of Chemical Technology, Wiley, New York, 4th ed. 1993, Volume 6, p. 381, Fig. 1. After G.Brown, Ed., The X-ray Identification and Crystal Structure of Clay Minerals, Mineralogical Society, London 1961
[6] U.Hofmann, Angew.Chem. **80** (1968) 736; Angew.Chem.Int.Ed.Engl. **7** (1968) 681, Fig. 7
[7] P.R.Dvornic, V.V.Gerov, Macromolecules **27** (1994) 1068, Fig. 3
[8] E.Thilo, Angew.Chem. **77** (1965) 1056; Angew.Chem.Int.Ed.Engl. **4** (1965) 1061, Fig. 10
[9] R.Steudel, R.Strauss, L.Koch, Angew.Chem.Int.Ed.Engl. **24** (1985) 59, Table 1
[10] R.F.Bacon, R.Fanelli, J.Am.Chem.Soc. **65** (1963) 639, taken from Fig. 3
[11] A.V.Tobosky, A.Eisenberg, J.Am.Chem.Soc. **81** (1959) 780, Fig. 4

# 13 Appendix

## 13.1 Physical Quantities, Physical Units, and Numbers

The **International System of Units (SI system)** consists of 7 base physical quantities and 7 SI base units plus a number of derived SI units. It has replaced the formerly used CGS and MKS systems but it is not "the metric system." For details, see Volume I.

Table 13-1 Physical quantities and their SI base units.

| Base physical quantity SI symbol | SI name | SI base unit English name | American name | SI symbol |
|---|---|---|---|---|
| $l$ | length | metre | meter | m |
| $m$ | mass | kilogramme | kilogram | kg |
| $t$ | time | second | second | s |
| $I$ | electric current | ampere | ampere | A |
| $T$ | thermodynamic temperature | kelvin | kelvin | K |
| $n$ | amount of substance | mole | mole | mol |
| $I_v$ | luminous intensity | candela | candela | cd |

Table 13-2 Derived SI units and their IUPAC-IUPAP symbols. Symbols for vectorial physical quantities are printed in bold face italic type, for example, ***F*** for force. Names of units derived from a personal name are printed in lower-case letters, for example, hertz, except Celsius.

| Physical quantity Symbol | Name | SI name | SI unit SI symbol and unit(s) | | | |
|---|---|---|---|---|---|---|
| $\nu$ | frequency | hertz | Hz | $= s^{-1}$ | | |
| $P$ | power, radiant flux | watt | W | = V A | $= J\ s^{-1}$ | $= m^2\ kg\ s^{-3}$ |
| $E$ | energy, work, heat | joule | J | = N m | | $= m^2\ kg\ s^{-2}$ |
| $\boldsymbol{F}$ | force | newton | N | $= J\ m^{-1}$ | | $= m\ kg\ s^{-2}$ |
| | impact strength (US) | newton | | $J\ m^{-1}$ | | $= m\ kg\ s^{-2}$ |
| $G$ | weight | newton | N | $= J\ m^{-1}$ | | $= m\ kg\ s^{-2}$ |
| | impact strength (Europe) | - | | $J\ m^{-2}$ | | $= kg\ s^{-2}$ |
| $\gamma$ | interfacial tension | - | | $J\ m^{-2}$ | $= N\ m^{-1}$ | $= kg\ s^{-2}$ |
| $p, \sigma$ | pressure, stress | pascal | Pa | $= N\ m^{-2}$ | $= J\ m^{-3}$ | $= m^{-1}\ kg\ s^{-2}$ |
| | impulse, momentum | - | | N s | | $= m\ kg\ s^{-1}$ |
| $Q$ | electric charge | coulomb | C | = A s | | |
| $U$ | electric potential, electromotive force | volt | V | $= W\ A^{-1}$ | $= J\ C^{-1}$ | $= m^2\ kg\ s^{-3}\ A^{-1}$ |
| $R$ | electric resistance | ohm | Ω | $= V\ A^{-1}$ | | $= m^2\ kg\ s^{-3}\ A^{-2}$ |
| $G$ | electric conductance | siemens | S | $= \Omega^{-1}$ | | $= m^{-2}\ kg^{-1}\ s^3\ A^2$ |
| $C$ | electric capacitance | farad | F | $= C\ V^{-1}$ | | $= m^{-2}\ kg^{-1}\ s^4\ A^2$ |
| $t, \theta$ | Celsius temperature | degree Celsius | °C | $[\theta/°C = (T/K) - 273.15$ | | |
| $\Phi$ | magnetic flux | weber | Wb | = V s | | $= m^2\ kg\ s^{-2}\ A^{-1}$ |
| $L$ | magnetic inductance | henry | H | $= V\ s\ A^{-1}$ | | $= m^2\ kg\ s^{-2}\ A^{-2}$ |
| $B$ | magnetic flux density | tesla | T | $= Wb\ m^{-2}$ | | $= kg\ s^{-2}\ A^{-1}$ |
| $A$ | radioactivity | becquerel | Bq | $= s^{-1}$ | | |

Table 13-3 SI prefixes for SI units. ISO = International Standardization Organization. NIST = National Institute for Science and Technology (United States). * See Table 13-4.
Origin: D = Danish, G = Greek, I = Italian, L = Latin, N = Norwegian.

| Factor | Prefix | Symbol [a)] | Common name: United States | Common name: Europe (general) | Origin of prefix | |
|---|---|---|---|---|---|---|
| $10^{24}$ | yotta [b)] | Y | septillion | quadrillion * | L: | *octo* = eight [$10^{24} = (10^3)^8$] |
| $10^{21}$ | zetta [c)] | Z | sextillion | 1000 trillion * | L: | *septem* = seven [$10^{21} = (10^3)^7$] |
| $10^{18}$ | exa | E | quintillion | trillion * | G: | *hexa* = six [$10^{18} = (10^3)^6$] |
| $10^{15}$ | peta | P | quadrillion | 1000 billion * | G: | *penta* = five [$10^{15} = (10^3)^5$] |
| $10^{12}$ | tera | T | trillion | billion * | G: | *teras* = monster |
| $10^9$ | giga | G | billion | 1000 million * | G: | *gigas* = giant |
| $10^6$ | mega | M | million | million | G: | *megas* = big |
| $10^3$ | kilo | k | thousand | thousand | G: | *khilioi* = thousand |
| $10^2$ | hekto [d)] | h | hundred | hundred | G: | *hekaton* = hundred |
| $10^1$ | deka [e)] | da | ten | ten | G: | *deka* = ten |
| $10^{-1}$ | deci | d | one tenth | one tenth | L: | *decima pars* = one tenth |
| $10^{-2}$ | centi | c | one hundredth | one hundredth | L: | *pars centesima* = one hundredth |
| $10^{-3}$ | milli | m | one thousandth | one thousandth | L: | *pars millesima* = one thousandth |
| $10^{-6}$ | micro [f)] | μ | one millionth | one millionth | G: | *mikros* = small |
| $10^{-9}$ | nano | n | one billionth | one milliardth | G: | *nan(n)os* = dwarf |
| $10^{-12}$ | pico | p | one trillionth | one billionth | I: | *piccolo* = small |
| $10^{-15}$ | femto | f | one quadrillionth | one billiardth | D, N: | *femten* = fifteen |
| $10^{-18}$ | atto | a | one quintillionth | one trillionth | D, N: | *atten* = eighteen |
| $10^{-21}$ | zepto [c)] | z | one sextillionth | one trilliardth | L: | *septem* = seven [$10^{-21} = (10^{-3})^7$] |
| $10^{-24}$ | yocto [b)] | y | one septillionth | one quadrillionth | L: | *octo* = eight [$10^{-24} = [10^{-3})^8$] |

a) US finance and gas businesses use the following prefixes for dollar, cubic feet of gas, etc.:
$10^3 \equiv$ M; $10^6 \equiv$ MM $\equiv \overline{\text{M}}$; $10^9 \equiv$ B; $10^{12} \equiv$ T, $10^{15} \equiv$ Q.
In the technical literature, "M" is sometimes used for "million" (example: 2 Mt = $2 \cdot 10^6$ tons). In some US government publications, "M" (or "m") indicates "metric" (example: 2 Mt = 2 tons) in order to distinguish "metric" tons (mt, m.t., Mt) from short tons (sht, sh.t.) or long tons (lt, l.t.).
b) ISO adds the letter "y" because a prefix "o" would be misleading.
c) ISO replaced "s" by "z" in order to avoid the double use of "s" ("s" is the symbol for second).
d) ISO uses "hecto" but NIST recommends the etymologically correct "hekto."
e) ISO uses "deca" but NIST recommends the etymologically correct "deka."
f) USA: μ as the symbol for "micro" is neither known to the general public nor to newspapers and magazines; it is also not on typewriter keyboards. The prefix "μ" is therefore sometimes replaced by either "u" or the non-SI prefix "mc" (from "micro"; 1 mcg = 1 microgram = 1 μg). The non-SI prefix "ml" is then substituted for the SI prefix "m" = "milli" (1 mlg ≡ 1 milligram = 1 mg).

Table 13-4 Names of numbers in various countries. a) Reverted to the US System in 1974.

| Factor | ISO symbol | United States, United Kingdom [a)] | Europe (general) | Germany | France (before 1948) |
|---|---|---|---|---|---|
| $10^{24}$ | Y | septillion | quadrillion | Quadrillion | septillion |
| $10^{21}$ | Z | sextillion | 1000 trillion | Trilliarde | sextillion |
| $10^{18}$ | E | quintillion | trillion | Trillion | quintillion |
| $10^{15}$ | P | quadrillion | 1000 billion | Billiarde | quadrillion, quatrillion |
| $10^{12}$ | T | trillion | billion | Billion | trillion |
| $10^9$ | G | billion | 1000 million | Milliarde | milliard, billion |
| $10^6$ | M | million | million | Million | million |
| $10^3$ | k | thousand | thousand | Tausand | mille (still used) |

## 13.2 Common Physical Quantities and Units

All countries except the United States, Liberia, and Myanmar have adopted the **International System of Units** (systéme international: **SI system**) whereas governmental, commercial, and technical literature in the United States sometimes still use American and old Imperial British units although the forerunner of the SI system, the metric system, was introduced by an act of Congress in 1896. For SI base units and SI derived units, see page 595 and Volume I,

Table 13-5 Conversion of non-SI units to SI units.
≡ Identical by definition, = equal, ≈ approximately equal, ∴ equivalent to.

| Name of non-SI unit | Non-SI unit | = SI unit |
|---|---|---|
| *Length* | | |
| Nautical mile (sea mile) | 1 n | = 1852 m |
| Statute mile (land mile) | 1 mile | = 1609.344 m |
| Rod (perch, pole) | 1 rod | = 5.029 2 m |
| Fathom | 1 fathom | = 1.828 8 m |
| Yard | 1 yd | = 0.914 4 m (exact) |
| Foot | 1 ft = 1' = 12" | = 0.304 8 m (exact) |
| Inch | 1 in = 1" | = 2.54 cm (exact) |
| Mil | 1 mil | = 25.4 μm (exact) |
| Micron | 1 μ | = $10^{-6}$ m = 1 μm |
| Millimicron | 1 mμ | = $10^{-9}$ m = 1 nm |
| Ångstrøm | 1 Å | = $10^{-10}$ m = 0.1 nm |
| *Area* | | |
| Square mile | 1 sq. mile | = 2 589 988.110 $m^2$ |
| Hectare (land) | 1 ha | ≡ 10 000 $m^2$ |
| Acre | 1 acre | = 4406.856 $m^2$ |
| Square yard | 1 sq. yd. | = 0.836 127 36 $m^2$ |
| Square foot | 1 sq. ft. | = 9.203 04·$10^{-2}$ $m^2$ |
| Square inch | 1 sq. in. | = 6.451 6·$10^{-4}$ $m^2$ |
| *Volume* | | |
| Cubic yard | 1 cu. yd. | = 0.764 554 857 $m^3$ |
| Imperial barrel | 1 barrel | = 0.1636 $m^3$ |
| US barrel petroleum | 1 bbl = 42 US gal | = 0.158 987 $m^3$ |
| US barrel | 1 barrel | = 0.119 $m^3$ = 119 L |
| Bushel | 1 bu. | = 3.524·$10^{-2}$ $m^3$ = 35.24 L |
| Cubic foot | 1 cu. ft. | = 2.381 685·$10^{-2}$ $m^3$ |
| Board foot | 12·12·1 cu. in. | = 2.3597·$10^{-3}$ $m^3$ |
| Gallon (British or Imperial) | 1 gal | = 4.545 96·$10^{-3}$ $m^3$ = 4.545 96 L |
| Gallon (US dry) | 1 gal | = 4.405·$10^{-3}$ $m^3$ |
| Gallon (US liquid) | 1 gal = 4 US qt. | = 3.785 412·$10^{-3}$ $m^3$ = 3.785 412 L |
| Liter (cgs) | 1 L | = 1.000 028·$10^{-3}$ $m^3$ |
| Liter (SI) | 1 L | ≡ 1.000 000·$10^{-3}$ $m^3$ |
| Quart (US dry) | 1 qt. | = 1.101 L |
| Quart (US liquid) | 1 qt. = 2 US pints | = 0.946 335 L |
| Pint (US liquid) | 1 pt. = 2 US cups | = 0.473 168 L |
| Pint (US dry) | 1 pt | = 0.550 6 L |
| Cup (US) | 1 cup = 8 fluid oz. | = 0.236 534 L |
| Ounce (British liquid) | 1 oz. | = 0.028 413 L |
| Ounce (US fluid ounce) | 1 oz. = 2 table sp. | = 0.029 574 L |
| Cubic inch | 1 cu. in. | = 0.016 387 064 L |

Table 13-5 (continued)

| Name of physical unit | Old unit | = SI unit |
|---|---|---|
| *Mass* | | |
| Long ton (UK) | 1 l.t. = 2240 lb | = 1016.046 909 kg |
| Ton (SI) ("metric ton") | 1 t (= "m.t", "Mt") | ≡ 1000.000 000 kg |
| Short ton (US) | 1 sh.t. = 2000 lb | = 907.184 74 kg |
| Hundredweight (UK) | 1 cwt | = 50.802 3 kg |
| Short hundredweight | 1 sh. cwt | = 45.359 2 kg |
| Slug | 1 slug | = 14.593 9 kg |
| Stone | 1 stone = 14 lb | = 6.350 293 18 kg |
| Pound (international) | 1 lb | = 453.592 37 g |
| Pound (avoirdupois) (US) | 1 lb = 16 oz. | = 453.592 427 7 g |
| Pound (apothecaries' or troy, US) | 1 lb = 8 drams | = 373.242 g |
| Ounce (avoirdupois) (US) | 1 oz. | = 28.349 52 g |
| Ounce (troy) | 1 oz. | = 31.103 5 g |
| Dram (apothecaries') | 1 dram | = 3.888 g |
| Dram (avoirdupois) | 1 dram | = 1.772 g |
| Pennyweight | 1 pennyweight | = 1.555 g |
| Carat | 1 ct | = 0.2 g |
| *Time* | | |
| Year | 1 a (US: 1 yr) | ≡ 365 days (statistics only) |
| Month | 1 mo (US: 1 mon) | ≡ 30 days (statistics only) |
| Day | 1 d | ≡ 24 h = 86 400 s |
| Hour * | 1 h (US = 1 hr) | ≡ 60 min = 3600 s |
| Minute * | 1 min | ≡ 60 s (US: 60 sec) |
| *Temperature* | | |
| Degree Celsius * | y°C – 273.16°C | = x K |
| Degree Fahrenheit | (z°F – 32°F)(5/9) | = y°C |
| *Density* ($1\ kg\ m^{-3} = 1 \cdot 10^{-3}\ g\ cm^{-3}$) | | |
| Specific gravity | 1 lb/cu.in. | = $27.679\ 904\ 71\ g\ cm^{-3}$ |
| | 1 oz/cu.in. | = $1.729\ 993\ 853\ g\ cm^{-3}$ |
| | 1 lb/cu.ft. | = $1.601\ 846\ 337 \cdot 10^{-2}\ g\ cm^{-3}$ |
| | 1 lb/gal US | = $7.489\ 150\ 454 \cdot 10^{-3}\ g\ cm^{-3}$ |
| *Energy, work* (1 J ≡ 1 N m ≡ 1 W s) (see also p. 29 for other commercial energy units) | | |
| - | 1 Quad | = 1060 PJ |
| Coal unit (US) | 1 ton coal | ∴ 27.92 GJ (annually adjusted) |
| Coal unit (UK) | 1 ton coal | ∴ 24.61 GJ (annually adjusted) |
| Short ton bituminous coal | 1 T | ∴ 26.58 GJ (annually adjusted) |
| Kilowatt hour | 1 kWh | = 3.6 MJ |
| Horse-power hour | 1 hph | = 2.685 MJ |
| Cubic foot-atmosphere | 1 cu.ft.atm. | = 2.869 205 kJ |
| British thermal unit | 1 $Btu_{mean}$ | = 1.055 79 kJ |
| British thermal unit | 1 $Btu_{IT}$ | = 1.055 056 kJ |
| - | 1 cu.ft.lb(wt)/sq.in. | = 195.237 8 J |
| Liter atmosphere (cgs) | 1 L atm | = 101.325 0 J |
| Meter kilogram-force | 1 m kgf | = 9.806 65 J |
| Calorie, international | 1 $cal_{IT}$ | = 4.186 8 J |
| Calorie, thermochemical | 1 $cal_{th}$ | = 4.184 J |
| Foot pound-force | 1 ft-lbf | = 1.355 818 J |
| Foot poundal | 1 ft-pdl | = 4.215 384 J |
| Erg | 1 erg | = $1.10^{-7}$ J = 0.1 μJ |
| Electron volt | 1 eV | = $1.602\ 177\ 33 \cdot 10^{-19}$ J |

Table 13-5 (continued)

| Name of physical unit | Old unit | = SI unit |
|---|---|---|
| *Force* (1 N ≡ 1 J m$^{-1}$ ≡ 1 kg m s$^{-2}$) | | |
| - | 1 ft-lbf/in. notch | = 53.378 64 N |
| Kilogram force | 1 kgf | = 9.806 65 N |
| Pound-force | 1 lbf | = 4.448 22 N |
| Ounce-force | 1 oz.f. | = 0.2780 N |
| Poundal | 1 pdl | = 0.138 255 N |
| Gram-force | 1 gf | = 9.806 65·10$^{-3}$ N |
| Pond | 1 p | = 9.806 65·10$^{-3}$ N |
| Dyne | 1 dyn | = 1·10$^{-5}$ N |
| *Length-related force* | | |
| - | 1 kp/cm | = 980.665 N m$^{-1}$ |
| - | 1 lbf/ft | = 14.593 898 N m$^{-1}$ |
| - | 1 dyn/cm | = 1·10$^{-3}$ N m$^{-1}$ |
| *Pressure, mechanical stress* (1 MPa ≡ 1 MN m$^{-2}$ ≡ 1 N mm$^{-2}$) | | |
| Physical atmosphere | 1 atm ≡ 760 torr | = 0.101 325 MPa |
| Bar | 1 bar | = 0.1 MPa |
| Technical atmosphere | 1 at | = 0.098 065 MPa |
| - | 1 kp/cm$^2$ | = 0.098 065 MPa |
| - | 1 kgf/cm$^2$ | = 0.098 065 MPa |
| - | 1 lbf/sq.in. | = 6.894 76·10$^{-3}$ MPa |
| Pound-force per square inch | 1 psi | = 6.894 76·10$^{-3}$ MPa |
| Inch mercury (32°F) | 1 in.Hg | = 3.386 388·10$^{-3}$ MPa |
| Inch water (39.2°F) | 1 in.$H_2O$ | = 249.1 Pa |
| Torr | 1 torr | = (101 325/760) Pa ≈ 133.322 Pa |
| Millimeter mercury | 1 mm Hg | = 13.5951 · 9.806 65 Pa ≈ 133.322 Pa |
| - | 1 dyn/cm$^2$ | = 1·10$^{-5}$ MPa |
| Millimeter water | 1 mm $H_2O$ | = 9.806 65·10$^{-6}$ MPa |
| Poundal per square foot | 1 pdl/sq.ft. | = 1.488 649·10$^{-6}$ MPa |
| *Power* (1 W = 1 J s$^{-1}$) | | |
| Horsepower (boiler) | 1 hp | = 9810 W |
| Horsepower (electric) | 1 hp | = 746 W |
| Horsepower (UK) | 1 hp | = 745.700 W |
| Horsepower (metric) | 1 PS | = 735.499 W |
| British thermal unit per hour | 1 Btu/h | = 0.293 275 W |
| - | 1 cal/h | = 1.162 222·10$^{-3}$ W |
| *Thermal conductivity* | | |
| - | 1 cal/(cm s °C) | = 418.6 W m$^{-1}$ K$^{-1}$ |
| - | 1 Btu/(ft h °F) | = 1.731 956 W m$^{-1}$ K$^{-1}$ |
| - | 1 kcal/(m h °C) | = 1.162 78 W m$^{-1}$ K$^{-1}$ |
| *Heat transfer coefficient* | | |
| | 1 cal/(cm$^2$ s °C) | = 4.186 8·10$^4$ W m$^{-2}$ K$^{-1}$ |
| | 1 BTU/(ft$^2$ h °F) | = 5.682 215 W m$^{-2}$ K$^{-1}$ |
| | 1 kcal/(m$^2$ h °C) | = 1.163 W m$^{-2}$ K$^{-1}$ |
| *Length-related mass* (= linear density) | | |
| Fineness (metric) | 1 tex | = 1·10$^{-6}$ kg m$^{-1}$ |
| Fineness (denier) | 1 den | = 0.111·10$^{-6}$ kg m$^{-1}$ |
| *Fracture length* | | |
| - | 1 g/den | = 9·10$^3$ m |

Table 13-5 (continued)

| Name of physical unit | Old unit | = SI unit |
|---|---|---|
| *Textile strength* | | |
| Tenacity | 1 gf/den = 1 gpd | = 0.082 599 N $tex^{-1}$ = 0.082 599 $m^2\ s^{-2}$ = 98.06 MPa · (density in g $cm^{-3}$) |
| *Dynamic viscosity* | | |
| Poise | 1 P | = 0.1 Pa s |
| Centipoise | 1 cP | = 1 mPa s |
| *Kinematic viscosity* | | |
| Stokes | 1 St | = $1 \cdot 10^{-4}\ m^2\ s^{-1}$ |
| *Heat capacity* | | |
| Clausius | 1 Cl | = 1 $cal_{th}$/K = 4.184 J $K^{-1}$ |
| *Molar heat capacity* | | |
| Entropy unit | 1 e.u. | = 1 $cal_{th}\ K^{-1}\ mol^{-1}$ = 4.184 J $K^{-1}\ mol^{-1}$ |
| *Relative permittivity* | | |
| "Dielectricity constant" | 1 | 1 |
| *(Electrical) conductance* | | |
| Inverse ohm | 1 mho | = 1 S |
| *Electrical field strength* | | |
| - | 1 V/mil | = $3.937\ 008 \cdot 10^4$ V $m^{-1}$ |

Table 13-6 Conversion factors for crude oil production and trading. Note: the barrel (bbl) is a measure of capacity, the value of which depends on the substance stored (see p. 597).

| Conventional unit | ISO unit | Ton | Kiloliter | US barrel | US gallon |
|---|---|---|---|---|---|
| Ton (ISO = "metric") | 1000 kg | 1 | 1.165 | 7.33 | 307.86 |
| Cubic meter (= "kiloliter") | 1000 $m^3$ | 0.858 1 | 1 | 6.289 8 | 264.17 |
| US barrel petroleum | 0.158 987 $m^3$ | 0.136 4 | 0.159 | 1 | 42 |
| US liquid gallon | 3.785 412 L | 0.003 25 | 0.003 8 | 0.023 8 | 1 |

Table 13-7 Approximate conversion factors for the production and trading of natural gas (NG) and liquefied natural gas (LNG). BTU = British thermal unit, OE = oil equivalent, TOE = ton oil equivalent.

| Conventional unit | - | $10^9\ m^3$ NG | $10^9$ cu.ft. NG | $10^6$ t LNG | $10^6$ t OE | $10^6$ bbl OE | $10^{12}$ BTU |
|---|---|---|---|---|---|---|---|
| $10^9\ m^3$ | NG | 1 | 35.3 | 0.73 | 0.90 | 8.29 | 36 |
| $10^9$ cu.ft. | NG | 0.028 | 1 | 0.021 | 0.026 | 0.18 | 1.03 |
| $10^6$ t | LNG | 1.38 | 48.7 | 1 | 1.23 | 8.68 | 52.0 |
| $10^6$ | TOE | 1.111 | 39.2 | 0.805 | 1 | 7.33 | 40.4 |
| $10^6$ bbl | OE | 0.16 | 5.61 | 0.12 | 0.14 | 1 | 5.8 |
| $10^{12}$ | BTU | 0.028 | 0.98 | 0.02 | 0.025 | 0.17 | 1 |

## 13.3 Abbreviations and Acronyms

Conventional names of polymers (plastics, rubbers, fibers, etc.) and their monomers are often abbreviated in the technical and scientific literature. The recommended abbreviations and acronyms vary; ISO, IUPAC, ASTM, BS, DIN, etc., sometimes use the same codes for different plastics or give the same plastic different codes. There are also many "wild" abbreviations and acronyms that are not standardized or recommended by international or national, scientific or industrial, organizations.

Table 13-8 contains a list of abbreviations and acronyms that were encountered by the author of this book. Abbreviations and acronyms are listed in alphabetical order with numbers having precedence over letters and capital letters over lower-case letters. Hyphens are ignored. An asterik * *does not* indicate a footnote but is unfortunately an integral part of the abbreviation recommended by ASTM (example: PA = polyamide, PA* = saran-coated PA, PA** = metal-coated PA).

Care has to be taken if certain abbreviations and acronyms are used since some of them are registered trademarks in either some or all countries of the world. Examples are EVAL® (worldwide), PAN® (in Europe (DIN uses PAS)), PPO® (worldwide)), and SAN® (Japan and the United States).

For **plastics**, industry in most countries now adheres to the *ISO system* of abbreviations and acronyms [ISO 1043-1986(ε)]. This system cites the abbreviation of the main component first, usually prefixed by P for homopolymeric thermoplastics.

Abbreviations for comonomers are separated from the main component by a dash; they are not prefixed by P but may be followed by M (for *m*onomer). The copolymer of ethene (E), propene (P), and a diene monomer (DM) is thus called EPDM.

After-treatments are also indicated by letter *after* the symbol for the parent polymer; the two symbols are separated by a dash. For example, acronyms are PE for poly(ethylene) and PE-C for chlorinated poly(ethylene).

Differences in properties are also indicated by letters or combinations thereof *after* the symbol for the polymer. An example is PE-HD for high-density poly(ethylene).

*ASTM abbreviations* for plastic names are less systematic (ASTM-D 1600-94a); they usually adhere to a "natural" system. They thus abbreviate the spoken language by indicating special characteristics, modifications, etc., by letters *in front* of that of the symbol for parent polymer. Examples are HDPE = high-density poly(ethylene), CPE = chlorinated poly(ethylene), and BOPP = biaxially oriented poly(propylene).

*IUPAC* (Pure Appl. Chem. **59** (1987) 691) uses its own system that sometimes adheres to ISO, sometimes to ASTM, and sometimes to no other system.

The same *polymer* may carry different abbreviations for plastics, fibers, rubbers, and recyclables. An example is poly(ethylene terephthalate): PET (plastics), PES (fiber), PETE (recyclable plastics). Another example is poly(tetrafluoroethylene): PTFE (plastics) and PTF (PTFE as fiber raw material). A comprehensive list of abbreviations and acronyms for fibers can be found in Volume IV.

**Recyclable plastics** use numbers instead of letters for the different types.

**Rubbers** mostly have abbreviations and acronyms that end with an R or nowadays also with an M; silicone elastomers being an exception (see below). Apparently, the same symbols are used for rubbers (in this book defined as uncrosslinked polymers) and elastomers (defined as crosslinked rubbers). Some authors restrict the word "rubber" to natural rubber.

**Fibers**, especially natural ones, have their own abbreviations and acronyms (see Volume IV).

**Silicones (polysiloxanes)** are characterized by two different sets of abbreviations. ISO characterizes *polymers* using Q as general symbol for silicones. Substituents are indicated by letters: F = fluoro, M = methyl, P = phenyl, V = vinyl. FMQ is thus a silicone, $\text{+SiRR'–O+}_n$, with fluorine and methyl substituents.

The *General Electric* terminology, on the other hand, characterizes siloxanes and polysiloxanes by the functionality of their *monomers* and *monomeric units*, respectively: M indicates monofunctional, D difunctional, T trifunctional, and Q quadrifunctional siloxane units with the implicit assumption that substituents are usually methyl groups. Substitution by groups other than methyl are indicated by a prime (M', D', T', Q') but the type of group is not specified.

Since M, D, T, and Q symbolize the units $(CH_3)_3SiO_{0.5}$, $(CH_3)_2SiO_{1.0}$, $(CH_3)SiO_{1.5}$, and $SiO_2$, respectively, symbols for *siloxanes* are obtained by adding the appropriate letters: MM indicates $(CH_3)_3SiOSi(CH_3)_3$ and $MD_2M$ the linear tetramer $(CH_3)_3SiO[Si(CH_3)_2O]_2Si(CH_3)_3$ whereas the cyclic tetramer, $[(CH_3)_3SiO]_4$ has the symbol $D_4$. MM' may thus be $(CH_3)_3SiOSi(CH_3)_2(C_6H_5)$ or $(CH_3)_3SiOSi(CH_3)(C_6H_5)_2$, etc. The same symbols apply to the repeating units of *polysiloxanes*, i.e., D for $–(CH_3)_2SiO–$; D' for $–(CH_3)(C_6H_5)SiO–$, $–(CH_3)(F)SiO–$, or $–(CH_3)(CH_2{=}CH)SiO–$, etc.

Table 13-8 Common abbreviations and acronyms for polymers. The list retains older chemical monomer names, e.g., ethylene instead of ethene.

| Abbreviation | Common polymer name or preparation | Abbreviation recommended by ISO | IUPAC | ASTM |
|---|---|---|---|---|
| 1 | recyclable poly(ethylene terephthalate) | | | |
| 2 | recyclable high-density poly(ethylene) | | | |
| 3 | recyclable poly(vinyl chloride) | | | |
| 4 | recyclable low-density poly(ethylene) | | | |
| 5 | recyclable isotactic poly(propylene) | | | |
| 6 | recyclable atactic poly(styrene) | | | |
| 7 | other recyclable plastics, including multiple layer plastics | | | |
| ABA | NBR-acrylic ester copolymer | A/B/A | A/B/A | ABA |
| ABR | acrylic ester-butadiene rubber | | ABR | ABR |
| ABS | graft copolymer from NBR and styrene | ABS | ABS | ABS |
| ACM | acrylic rubber = copolymer of ethyl acrylate, butyl acrylate, and/or 2-methoxyethyl acrylate | | | |
| ACR | acryl rubber = copolymer of ethyl acrylate, vinyl chloroacetate, or chloroethyl vinyl ether | | | |
| ACS | graft copolymer of S/AN on PE-C | | | |
| AES | acrylonitrile-ethylene-styrene copolymer | | | AES |
| AES | graft copolymer of styrene on EPDM rubber | | | |
| AF | 2,2-bis(trifluoromethyl)-4,5-difluoro-1,3-dioxolane | | | |
| AFMU | nitroso rubber, see CNR | | | AFMU |
| AK | alkyd resin | | | |
| AMMA | acrylonitrile-methyl methacrylate copolymer | A/MMA | | AMMA |
| ANM | acrylate rubber = ethylene acrylate + acrylonitrile copolymer or ethylene acrylate + 2-chloroethylvinyl ether copolymer | | | |
| APB | atactic poly(1-butene) | | | |
| APEN | amorphous poly(ethylene-2,6-naphthenate) | | | |
| APP | atactic poly(propylene) | | | |
| aPP | atactic poly(propylene) | | | |
| APS | atactic poly(styrene) | | | |
| AR | acrylic rubber in general | | | |
| ARP | "aromatic polyester" = homo and copolymers of *p*-hydroxybenzoic acid | | | |
| ASA | PBA particles in SAN matrix | A/S/A | | ASA |
| AU | polyurethane with polyester segments | AU | | AU |
| AU-I | ditto, crosslinkable with isocyanates | | | |
| AU-P | ditto, crosslinkable with peroxides | | | |
| BHET | *bis*(hydroxyethylene) terephthalate | | | |
| BIIR | bromobutyl rubber (= brominated IIR) | | | |
| BIMS | brominated isobutylene *p*-methyl styrene rubber | | | |
| BMC | bulk molding compound | | | |
| BMI | polybismaleimide | | | |
| BOPP | biaxially oriented poly(propylene) film, balanced oriented poly(propylene) film | | | |
| BR | butadiene rubber | | BR | BR |
| C | cellophane (a registered trademark in Germany!) | | | C |
| C* | saran-coated cellophane (saran, see PVDC) | | | C* |

Table 13-8 (continued)

| Abbrevi-ation | Common polymer name or preparation | Abbreviation recommended by ISO | IUPAC | ASTM |
|---|---|---|---|---|
| CA | cellulose acetate (= cellulose 2 1/2-acetate) | CA | CA | CA |
| CAB | cellulose acetate butyrate | CAB | CAB | CAB |
| CAP | cellulose acetate propionate | CAP | CAP | CAP |
| CDB | conjugated-diene butyl rubber | | | |
| CE | cellulose plastics in general | | | CE |
| CF | cresol-formaldehyde resin | CF | CF | CF |
| | $\omega$-carbalkoxy-perfluoralkoxy vinyl ether | | | |
| CFM | fluoro rubber (poly(chlorotrifluoroethylene)) | | | |
| CHDM | cyclohexane-1,4-dimethylol | | | |
| CHR | elastomeric poly(epichlorohydrin) = epichlorohydrin-ethylene oxide copolymer | CHR | | ECO |
| CIIR | chlorobutyl rubber (post-chlorinated IIR) | | | CIIR |
| CM | chlorinated poly(ethylene), formerly: CPE | | | CM |
| CMC | carboxymethyl cellulose | CMC | CMC | CMC |
| CMHEC | carboxymethyl/hydroxyethyl cellulose | | | |
| CN | cellulose nitrate | CN | CN | CN |
| CNF | carbon nanofoam | | | |
| CNR | nitroso rubber from tetrafluoroethylene, trifluoro-nitrosomethane, and an unsaturated termonomer | | | AFMU |
| CNT | carbon nanotube | | | |
| CO | poly(chloromethyl oxirane) = poly(epichlorohydrin) | | CO | CO |
| COAM | chlorite-oxidized starch | | | |
| COC | cycloolefin copolymer (ring preservation polym.) | | | |
| COP | cycloolefin polymer (ring opening polymerization) | | | |
| COX | carboxylic rubber | | | |
| CP | cellulose propionate | CP | CP | CP |
| CPE | chlorinated poly(ethylene), now: CM (ASTM) | PE-C | PE-C | CPE |
| CPET | crystallizable poly(ethylene terephthalate) = PET with fast-acting nucleating agents | | | |
| CPVC | chlorinated poly(vinyl chloride) | PVC-C | PVC-C | CPVC |
| CR | poly(chloroprene) rubber | | CR | CR |
| CS | casein | CS | CS | CS |
| CSM | chlorosulfonated poly(ethylene) | | | CSM |
| CSR | chlorosulfonated poly(ethylene) (rubber) | | | |
| CT | cellulose triacetate | | | |
| CTA | cellulose triacetate | CTA | | CTA |
| DIVEMA | divinyl ether-maleic anhydride copolymer | | | |
| EAA | ethene-acrylic acid copolymer | E/AA | | EAA |
| EAM | ethene-ethyl acrylate copolymer (rubber) | | | EAM |
| EBAC | ethene-butyl acrylate-acrylic acid copolymer (rubber) | | | |
| EC | ethyl cellulose | EC | EC | EC |
| ECB | ethene copolymer-bitumen blend | | ECB | |
| ECO | elastomeric epichlorohydrine-ethylene oxide copolymer | | | |
| ECTFE | ethene-chlorotrifluoroethylene copolymer | | | |
| EEA | ethene-ethyl acrylate-acrylic acid copolymer | E/EA | E/EA | EEA |
| EFEP | terpolymer of ethene (E), tetrafluoroethylene (TFE), and hexafluoropropylene (HFP) | | | |
| EHEC | ethyl hydroxyethyl cellulose | | | |

Table 13-8 (continued)

| Abbreviation | Common polymer name or preparation | Abbreviation recommended by ISO | IUPAC | ASTM |
|---|---|---|---|---|
| ELO | epoxidized linseed oil | ELO | | |
| EMA | ethene-methacrylic acid copolymer | E/MA | | EMA |
| EMAC | ethene-methyl acrylate-acrylic acid copolymer | | | |
| EMC | ethylmethylcellulose | | | |
| ENM | hydrogenated NBR | | | |
| ENR | epoxidized natural rubber | | | |
| EP | epoxy resin | EP | EP | EP |
| E/P | ethene-propene copolymer | E/P | | |
| EPDM | ethene-propene-non-conjugated diene copolymer | EPDM | EPDM | EPDM |
| EP-HI | high-impact epoxy resin | | | |
| EPI | poly(epichlorohydrin) (or its monomer) | | | |
| EPM | ethene-propene rubber | E/P | | EPM |
| EPS | expandable poly(styrene) (uncrosslinked) | | | |
| EPT | ethene-propene-termonomer (diene) copolymer | | | EPD |
| E-PVC | PVC by emulsion polymerization | | | |
| ETFE | copolymer of ethylene and tetrafluoroethylene | | | |
| ETP | engineering thermoplastic | | | |
| ESI | ethene-styrene copolymer ("interpolymer") | | | |
| ESO | epoxidized soybean oil | ESO | | |
| ETFE | ethene-tetrafluoroethylene copolymer | E/TFE | | ETFE |
| EVA | ethene-vinyl acetate copolymer (plastic) | E/VAC | | EVA |
| EVAC | ethene-vinyl acetate copolymer (plastic) | E/VAC | | EVA |
| EVAL® | ethene-vinyl alcohol copolymer | E/VAL | | EVAL |
| EVM | ethene-vinyl acetate copolymer (rubber) | | | |
| FEP | tetrafluoroethylene-hexafluoropropylene copolymer | FEP | | FEP |
| FF | furan-formaldehyde resin | FF | | FF |
| FM | perfluoro(methyl vinyl ether) | | | |
| FMQ | methyl fluorosilicone rubber (formerly: FSI) | | | |
| FPM | vinylidene fluoride-hexafluoropropylene copolymer | | | FPM |
| FSI | methyl fluorosilicone rubber (now: FMQ) | | | |
| GP | general purpose (resin) | | | |
| GPO | allyl glycidyl ether-propylene oxide copolymer | | | GPO |
| GR-I | government (butadiene) rubber with isoprene (outdated) | | | |
| GR-M | government (butadiene) rubber with chloroprene (outdated) | | | |
| GR-N | government (butadiene) rubber with acrylonitrile (outdated) | | | |
| GR-S | government (butadiene) rubber with styrene (outdated) | | | |
| HBMC | hydroxybutylmethylcellulose | | | |
| HBP | hyperbranched polymer | | | |
| HDPE | high-density poly(ethylene) | PE-HD | HDPE | HDPE |
| HEC | hydroxyethyl cellulose | | | |
| HEMA | poly(2-hydroxyethyl methacrylate) | | | |
| HES | hydroxyethylated starch | | | |
| HIPS | high-impact poly(styrene) | PS-HI | | |
| HNBR | hydrogenated NBR (also H-NBR or NBR-H) | | | |
| HPC | hydroxypropyl cellulose | | | |

Table 13-8 (continued)

| Abbreviation | Common polymer name or preparation | Abbreviation recommended by ISO | IUPAC | ASTM |
|---|---|---|---|---|
| HPMC | hydroxypropyl methyl cellulose | | | |
| HPS | hydroxypropylated starch | | | |
| ICP | poly(propylene), impact type (= blend of IPP and E/P copolymer | | | |
| IIR | isobutene-isoprene copolymer (butyl rubber) | | IIR | IIR |
| IPN | interpenetrating network | | | |
| IPP | isotactic poly(propylene) | | | |
| iPP | isotactic poly(propylene) | | | |
| IPS | impact-modified poly(styrene) | | | IPS |
| IR | synthetic *cis*-1,4-poly(isoprene) rubber | IR | | IR |
| LCP | liquid-crystalline polymer | | | LCP |
| LDPE | low-density poly(ethylene) | PE-LD | LDPE | LDPE |
| LLDPE | linear low-density poly(ethylene) | PE-LLD | | LLDPE |
| LMDPE | linear medium-density poly(ethylene) | | | LMDPE |
| L-SBR | solution-polymerized styrene-butadiene rubber | | | |
| L-VSBR | solution-polymerized vinyl-styrene-butadiene rubber | | | |
| MABS | methyl methacrylate-acrylonitrile-butadiene-styrene copolymer | | | |
| MBS | methyl methacrylate-butadiene-styrene copolymer | | MBS | MBS |
| MC | methyl cellulose | MC | | MC |
| MCLCP | main-chain liquid-crystalline polymer | | | |
| MDPE | medium-density poly(ethylene) | PE-MD | | MDPE |
| MF | melamine-formaldehyde resin | MF | MF | MF |
| MFA | copolymer of tetrafluoroethylene and perfluoromethyl vinyl ether | | | |
| MFQ | methyl fluorosilicone rubber | MFQ | | FMQ |
| mLLDPE | metallocene-catalyzed LLDPE | | | |
| mPE | metallocene-catalyzed poly(ethylene) | | | |
| MPF | melamine-phenol-formaldehyde resin | MPF | | MPF |
| MPIA | poly(*m*-phenylene isophthalamide) | | | |
| mPP | metallocene-catalyzed poly(propylene) | | | |
| MPQ | methyl phenyl silicone rubber | MPQ | | PSI |
| MPVQ | methyl phenyl vinyl silicone rubber | MPVQ | | PVMQ |
| MQ | methyl silicone rubber with vinyl and phenyl endgroups (PVMQ) | | | |
| MVQ | methyl vinyl silicone rubber | MVQ | | |
| MVFQ | methyl vinyl fluorosilicone rubber | MVFQ | | FVMQ |
| MWTN | multi-wall carbon nanotube | | | |
| MXD | poly(*m*-xylylene adipamide) | | | |
| NADIC® | methylbicyclo[2.2.1]heptene-2,3-dicarboxylic anhydride isomers | | | |
| NBR | acrylonitrile-butadiene rubber (nitrile rubber) | NBR | NBR | NBR |
| NCR | acrylonitrile-chloroprene rubber | | NCR | NCR |
| NDC | naphthalenedicarboxylate = naphthalene-2,6-dicarboxylic acid | | | |
| NIR | acrylonitrile-isoprene-nitrosomethane-nitrosoperfluorobutyric acid copolymer (rubber) | | NIR | |

Table 13-8 (continued)

| Abbreviation | Common polymer name or preparation | Abbreviation recommended by ISO | IUPAC | ASTM |
|---|---|---|---|---|
| NR | natural rubber (isoprene rubber) | NR | NR | NR |
| OER | oil-extended rubber | | | |
| OPET | oriented poly(ethylene terephthalate) | | | |
| OPP | oriented isotactic poly(propylene) | | | |
| OPS | oriented poly(styrene) film | | | |
| OPVC | oriented poly(vinyl chloride) | | | |
| OT | polysulfide rubber | | | |
| P4MP | poly(4-methyl-1-pentene) | | | |
| PA | phthalic anhydride (monomer) | | | |
| PA | aliphatic polyamide (nylon) | PA | PA | PA |
| PA* | saran-coated polyamide | | | PA* |
| PA** | metal-coated polyamide | | | PA** |
| PA 2 | poly(glycine) | | | |
| PA 3 | poly($\beta$-alanine = poly($\beta$-aminopropionic acid) | | | |
| PA 4.2 | poly(tetramethylene oxalamide) | PA 42 | PA 4.2 | PA 4.2 |
| PA 4.6 | poly(tetramethylene adipamide) | PA 46 | PA 4.6 | PA 4.6 |
| PA 6 | poly($\varepsilon$-caprolactam) | PA 6 | PA 6 | PA 6 |
| PA 6.6 | poly(hexamethylene adipamide) | PA 66 | PA 6.6 | PA 6.6 |
| PA 6.9 | poly(hexamethylene azelamide) | PA 69 | PA 6.9 | PA 6.9 |
| PA 6.10 | poly(hexamethylene sebacamide) | PA 610 | PA 6.10 | PA 6.10 |
| PA 6.12 | poly(hexamethylene dodecanoamide) | PA 612 | PA 6.12 | PA 6.12 |
| PA 6.T | poly(hexamethylene terephthalamide) | | | |
| PA 11 | poly(11-aminoundecanoic acid) | PA 11 | PA 11 | PA 11 |
| PA 12 | poly($\omega$-laurolactam) | PA 12 | PA 12 | PA 12 |
| PAA | poly(acrylic acid) | | | PAA |
| PAB | poly(1,4-benzamide) | | | |
| PAC | poly(acrylonitrile) | | PAC | PAN [1)] |
| PADC | poly(allyl diglycol carbonate) | | | PADC |
| PAE | polyarylene ether | | | PAE |
| PAEK | polyaryletherketone | | | PAEK |
| PAESU | aromatic polyether sulfone | | | |
| PAI | polyamide-imide | PAI | | PAI |
| PAK | polyester alkyd | | | PAK |
| PAMAM | polyamidoamine dendrimers | | | |
| PAMAMOS | polyamidoamine organosilicon dendrimers | | | |
| PAMS | poly($\alpha$-methylstyrene) | | | |
| PAN [1)] | poly(acrylonitrile) (also a trademark) | | | PAN |
| PANI | linear poly(aniline) | | | |
| PAPA | poly(azeleic anhydride) | | | |
| PAR | polyarylate | | | |
| PARA | polyaryl amide | | | PARA |
| PAS | polyarylsulfone | | | PASU |
| PASU | polyarylsulfone | | | PASU |
| PAT | polyaminotriazole | | | |
| | polyarylate | | | PAT |
| PAUR | poly(ester urethane) | | | PAUR |
| PB | poly(1-butene) | PB | PB | PB |
| PBA | poly(butyl acrylate) | | | PBA |
| PBAN | poly(butadiene-*co*-acrylonitrile) | | | PBAN |

Table 13-8 (continued)

| Abbrevi-ation | Common polymer name or preparation | Abbreviation recommended by ISO | IUPAC | ASTM |
|---|---|---|---|---|
| PBI | poly(benzimidazole) | PBI | PBI | PBI |
| PBMA | poly(butyl methacrylate) | | | |
| PBN | poly(1,4-butylene-2,6-naphthalate) | | | |
| PBO | poly(2,6-benzobisoxazolediyl-1,4-phenylene) | | | |
| PBOX | poly(2,6-benzobisoxazolediyl-1,4-phenylene) | | | |
| PBR | vinyl pyridine-butadiene copolymer | | PBR | PBR |
| PBS | poly(butadiene-*co*-styrene) | | | PBS |
| PBSU | aromatic polyether sulfone | | | |
| PBT | poly(1,4-butylene terephthalate) | PBTP | | PBT |
| PBTP | poly(butylene terephthalate) | PBTP | | PBT |
| PBTZ | poly(2,6-benzobisthiazolediyl-1,4-phenylene) | | | |
| PC | polymer concrete | | | |
| | polycarbonate | PC | PC | PC |
| PC-A | polycarbonate based on bisphenol A | | | |
| PCD | polycarbodiimide | | | |
| PCDT | poly(1,4-cyclohexyldimethylol terephthalate) = PCT | | | |
| PCL | poly($\varepsilon$-caprolactone) | | | |
| PCT | poly(1,4-cyclohexylene terephthalate) = PCDT | | | |
| PCTFE | poly(chlorotrifluoroethylene) | PCTFE | PCTFE | PCTFE |
| PDAF | poly(diallyl fumarate) | | | PDAF |
| PDAIP | poly(diallyl isophthalate) | | | PDAIP |
| PDAM | poly(diallyl maleate) | | | PDAM |
| PDAP | poly(diallyl phthalate) | PDAP | PDAP | PDAP |
| PDCPD | poly(dicyclopentadiene) | | | |
| PDLA | poly(D-lactic acid) | | | |
| PDMS | poly(dimethyl siloxane) | | | |
| PDO | 1,3-propanediol | | | |
| PDXL | poly(1,3-dioxolane) | | | |
| PE | poly(ethylene) | PE | PE | PE |
| PEA | poly(ethyl acrylate) | | | |
| PEBA | polyether-*block*-amide thermoplastic elastomer | PEB | | PEBA |
| PECH | linear poly(epichlorohydrin) | | | |
| PEEK | polyetheretherketone | PEEK | | PEEK |
| PEEKK | polyetheretherketoneketone | | | |
| PEG | poly(ethylene glycol) | | | |
| PE-HD | high-density poly(ethylene) | PE-HD | HDPE | HDPE |
| PE-HMW | high molecular weight poly(ethylene) | PE-HMW | | HMWPE |
| PEI | polyether imide | PEI | | PEI |
| | polyether imide (DIN) | | | |
| | poly(ethylene imine) | | | |
| PEK | polyetherketone | | | PEK |
| PEKEKK | polyetherketoneetherketoneketone | | | |
| PEKK | polyetherketoneketone | | | |
| PE-LD | low-density poly(ethylene) | PE-LD | | LDPE |
| PE-LLD | linear low-density poly(ethylene) | PE-LLD | | LLDPE |
| PE-MD | medium-density poly(ethylene) | PE-MD | | MDPE |
| PEN | poly(ethylene 2,6-naphthalate) | | | |
| PEO | poly(ethylene oxide) | PEOX | PEO | PEO |
| PEOB | poly(ethylene oxobenzoate) | | | |
| | poly($p$-$\beta$-hydroxyethoxybenzoic acid) | | | |
| PEOX | poly(ethylene oxide) | PEOX | PEO | PEO |

Table 13-8 (continued)

| Abbrevi-ation | Common polymer name or preparation | Abbreviation recommended by | | |
|---|---|---|---|---|
| | | ISO | IUPAC | ASTM |
| PES | polyester fibers (= poly(ethylene terephthalate)), | | | |
| | aromatic polyether sulfone | PES | | PES |
| PESU | aromatic polyether sulfone | PESU | | |
| PET | poly(ethylene terephthalate) plastics | PETP | PETP | PET |
| PET* | saran-coated poly(ethylene terephthalate) | | | PET* |
| PETE | poly(ethylene terephthalate), recyclable thermoplastic | | | |
| PETG | poly(ethylene terephthalate) with add'l glycol | | | PETG |
| PETP | poly(ethylene terephthalate) | PETP | PETP | PET |
| PE-UHMW | ultrahigh-molecular weight poly(ethylene) UHMWPE | PE-UHMW | | |
| PEUR | poly(ether urethane) | | | PEUR |
| PE-X | crosslinked poly(ethylene) | | | |
| PF | phenol-formaldehyde resin | PF | PF | PF |
| PFA | perfluoro(alkoxy alkane) = tetrafluoroethylene-perfluorinated alkyl vinyl ether copolymer | PFA | PFA | PFA |
| PFF | phenol-furfural polymer | | | PFF |
| PGA | poly(glycolic acid) | | | |
| PHA | poly(hydroxy alkanoate) | | | |
| PHAS | poly($\alpha$-hydroxy acrylic acid) | | | |
| PHB | poly(*p*-hydroxy benzoate) = POB | | | |
| | poly(3-hydroxybutyrate) = poly($\beta$-hydroxybutyrate) | | | PHB |
| PHE | poly(1-hexene) | | | |
| PHEMA | poly(2-hydroxyethyl methacrylate) | | | |
| PHF | poly(hydroxy fatty acid) | | | |
| PI | polyimide | PI | | PI |
| PIB | poly(isobutylene) (*not* the rubber, see IIR) | PIB | PIB | PIB |
| PIBO | poly(isobutylene oxide) | | | |
| PIC | polymer impregnated concrete | | | |
| PIP | synthetic poly(isoprene) | | | |
| PIR | poly(isocyanurate) | PIR | | |
| PISU | polyimidesulfone | | | PISU |
| PLA | poly(lactic acid) | | | |
| PLLA | poly(L-lactic acid) | | | |
| PMAA | poly(methacrylonitrile) | | | |
| PMAN | poly(methacrylonitrile) | | | |
| PMCA | poly(methyl $\alpha$-chloroacrylate) | | | PMCA |
| PMI | poly(methacrylimide) | PMI | | |
| PMMA | poly(methyl methacrylate) | PMMA | PMMA | PMMA |
| PMMI | polypyromellith imide | | | |
| PMP | poly(4-methyl-1-pentene) | PMP | PMP | PMP |
| PMPI | poly(*m*-phenylene isophthalamide) | | | |
| PMQ | phenyl-methyl silicone rubbers | | | PMQ |
| PMS | poly(*p*-methyl styrene) | PMS | | PMS |
| PN | poly(norbornene) | | | |
| PNC | poly(phosphonitrile dichloride) | | | |
| PNF | phosphorus nitrile fluoroelastomer | | | |
| PNR | poly(norbornene) rubber = PN + mineral oil | | | |
| PO | elastomeric poly(propylene oxide) | | | PO |
| POB | poly(oxybenzoate) = poly(*p*-hydroxybenzoate) = PHB | | | POB |
| POD | poly(phenylene-1,3,4-oxadiazole) | | | |
| POE | poly(olefin) elastomer | | | |

Table 13-8 (continued)

| Abbreviation | Common polymer name or preparation | Abbreviation recommended by ISO | IUPAC | ASTM |
|---|---|---|---|---|
| POM | poly(oxymethylene), polyacetal, poly(formaldehyde) | POM | POM | POM |
| POP | poly(olefin) plastomer | | | |
| POPAM | polypropylene imine amine dendrimers, later called PPI | | | |
| POR | propylene oxide-allyl glycidyl ether copolymer | | | |
| PP | isotactic poly(propylene) | PP | PP | PP |
| PP* | metallized poly(propylene) | | | PP* |
| PPA | poly(parabanic acid) | | | |
| | polyphthalamide | | | PPA |
| PPBA | poly(*p*-benzamide) | | | |
| PPCC | polymer-portland cement concrete | | | |
| PPD-T | poly(*p*-phenylene terephthalamide) | | | |
| PPE | polyphenylene ether, commonly used for poly(oxy-2,6-dimethyl-1,4-phenylene), see also PPO + PPO® | | | PPE |
| PPG | poly(propylene glycol) | | | |
| PPI | polymeric polyisocyanate | | | |
| | polypropylene imine amine dendrimers (formerly POPAM) | | | |
| PPO | poly(propylene oxide) | | | |
| | commonly used abbreviation for poly(oxy-2,6-dimethyl-1,4-phenylene) | | | |
| PPO® | "poly(phenylene oxide)", now a commercial blend of PPO and rubber-modified PS | | | |
| PPOm | modified PPO | | | |
| PPOX | poly(propylene oxide) | PPOX | PPOX | PPOX |
| PPP | poly(*p*-phenylene) | | | |
| PPS | poly(*p*-phenylene sulfide) | | | PPS |
| PPSF | aromatic polyether sulfone | | | |
| PPSU | aromatic polyphenylene sulfone | PPSU | | PPSU |
| PPT | poly(1,3-propylene terephthalate) | | | |
| PPTA | poly(*p*-phenylene terephthalamide) | | | |
| PPX | poly(1,4-phenylene-1,2-ethanediyl) | | | |
| PQ | phenyl-substituted silicone | | | PQ |
| PS | atactic poly(styrene) | PS | PS | PS |
| PSBR | vinyl pyridine-styrene-butadiene rubber | | | PSBR |
| PSF | bisphenol A polysulfone | | | |
| PS-HI | high-impact poly(styrene) | PS-HI | | HIPS |
| PS-I | impact-modified poly(styrene) | PS-I | | IPS |
| PSU | aromatic polysulfone | PSU | | PSU |
| PTAC | poly(triallyl cyanurate) | | | PTAC |
| PTF | poly(tetrafluoroethylene) as raw material for fibers | | | |
| PTFE | poly(tetrafluoroethylene) | PTFE | PTFE | PTFE |
| PTHF | poly(tetrahydrofuran) | | | |
| PTMA | poly(tetramethylene adipate) | | | |
| PTMEG | poly(tetramethylene ether glycol) | | | |
| PTMG | poly(tetramethylene glycol) | | | |
| PTMT | poly(tetramethylene terephthalate) = PBT | | | |
| PTO | poly(terephthaloyloximidrazone) | | | |
| PTT | poly(trimethylene terephthalate) = poly(propylene terephthalate) | | | |
| PUR | polyurethane (rubber) | PUR | PUR | PUR |

Table 13-8 (continued)

| Abbreviation | Common polymer name or preparation | Abbreviation recommended by ISO | IUPAC | ASTM |
|---|---|---|---|---|
| PVA | poly(vinyl acetate) | | | |
| PVAC | poly(vinyl acetate) | PVAC | PVAC | PVAC |
| PVAL | poly(vinyl alcohol) | PVAL | PVAL | PVAL |
| PVB | poly(vinyl butyral) | PVB | PVB | PVB |
| PVC | poly(vinyl chloride) | PVC | PVC | PVC |
| PVC-C | chlorinated poly(vinyl chloride) | PVC-C | | CPVC |
| PVC-P | plasticized poly(vinyl chloride) | | | |
| PVC-U | unplasticized poly(vinyl chloride) | | | |
| PVCA | vinyl chloride-vinyl acetate copolymer | | PVCA | PVCA |
| PVCH | poly(vinyl cyclohexane) | | | |
| PVCZ | poly(*N*-vinyl carbazole) | | | |
| PVDC | poly(vinylidene chloride) (always a copolymer) | PVDC | PVDC | PVDC |
| PVDF | poly(vinylidene fluoride) | PVDF | PVDF | PVDF |
| PVE | poly(vinyl ethyl ether) | | | |
| PVF | poly(vinyl fluoride) | PVF | PVF | PVF |
| PVFM | poly(vinyl formal) | PVFM | PVFM | PVFM |
| PVI | poly(vinyl isobutyl ether) | | | |
| PVK | poly(*N*-vinyl carbazole) | PVK | PVK | PVK |
| PVM | vinyl chloride-vinyl methyl ether copolymer | | | |
| PVME | poly(vinyl methyl ether) | | | |
| PVMQ | methyl silicone rubber with vinyl and phenyl endgroups (now: MQ) | | | |
| PVOH | poly(vinyl alcohol) | PVAL | PVAL | PVAL |
| PVP | poly(*N*-vinyl pyrrolidone) | PVP | | PVP |
| PYR | pyrrole black | | | |
| Q | silicones (generic) | | | |
| RCP | reactor copolymer (from propene and some ethene) | | | |
| RF | resorcinol-formaldehyde resin | | | |
| RUI | cyclized isoprene rubber | | | |
| SAC | starch acetate | | | |
| SAN [1)] | poly(styrene-*co*-acrylonitrile) plastic | SAN | SAN | SAN |
| SB | styrene-butadiene copolymer (thermoplastic) | S/B | S/B | PBS |
| SBC | styrene-butadiene-styrene triblock copolymer | | | |
| SBR | styrene-butadiene copolymer (rubber) | SBR | SBR | SBR |
| SBS | styrene-butadiene-styrene triblock copolymer | | | |
| SCLCP | side-chain liquid-crystalline polymer | | | |
| SCP | single-cell protein | | | |
| SCR | styrene-chloroprene copolymer (rubber) | SCR | SCR | SCR |
| SEBS | styrene–ethylene/butylene–styrene polymer = selectively hydrogenated SBS | | | |
| SEPS | styrene-ethylene/propylene-styrene polymer = selectively hydrogenated SIS | | | |
| SI | silicone plastics | SI | | Si |
| Si | silicone plastics | SI | | Si |
| SIR | styrene-isoprene rubber | SIR | | SIR |
| SIS | styrene-isoprene-styrene triblock copolymer | S-I-S | | |
| SMA | styrene-maleic anhydride copolymer | S/MA | SMA | S/MA |
| SMC | sheet molding compound | | | |

Table 13-8 (continued)

| Abbreviation | Common polymer name or preparation | Abbreviation recommended by ISO | IUPAC | ASTM |
|---|---|---|---|---|
| SMI | styrene-*N*-phenylmaleimide copolymer | | | |
| SMMA | styrene-methacrylic acid copolymer | | | |
| SMS | styrene-$\alpha$-methylstyrene copolymer | S/MS | | SMS |
| SP | saturated polyester plastic | | | SP |
| | starch acetate | | | |
| SPP | syndiotactic poly(propylene) | | | |
| sPP | syndiotactic poly(propylene) | | | |
| SPS | syndiotactic poly(styrene) | | | |
| S-PVC | suspension-polymerized PVC | | | |
| SR | synthetic rubber | | | |
| SRP | styrene-rubber plastic | | | SRP |
| SWNT | single-wall carbon nanotube | | | |
| T | polysulfide rubber (Thiokol®) | | | |
| TEEE | segmented ether-ester TPE | TPE-E | | TEEE |
| TETD | tetraethyl thiuram disulfide | | | |
| TM | polysulfide rubber | | | |
| TOR | 1,8-*trans*-poly(octenamer) rubber | | | |
| TP | thermoplastic | | | |
| TPE | thermoplastic elastomer | | | |
| TPE-A | segmented ether-amide TPE | | | |
| TPE-E | segmented ether-ester TPE | TPE-E | | TEEE |
| TPE-O | thermoplastic olefin elastomer (mostly PP + EPDM) | TPE-O | | TEO |
| TPE-S | styrene-butadiene-styrene triblock TPE | TPE-S | | TES |
| TPE-U | urethane TPE with ether or ester segments | | | |
| TPEL | thermoplastic elastomer | TPE | EU | TPEL |
| TPES | thermoplastic polyester | | | TPES |
| TM | polysulfide rubber | | | |
| TPO | thermoplastic olefin elastomer (EPDM/IPP blend) | TPE-O | | |
| TPR | 1,5-*trans*-poly(pentenamer) rubber | | | |
| TPU | thermoplastic polyurethane | | | |
| TPUR | thermoplastic polyurethane | | | TPUR |
| TR | polysulfide rubber | | | |
| TSUR | thermoset polyurethane | | | TSUR |
| UF | urea-formaldehyde resin | UF | UF | UF |
| UHMWPE | ultrahigh molecular weight poly(ethylene) UHMWPE | | | |
| UP | unsaturated polyester, includes "hardening monomer" | UP | UP | UP |
| V | poly(vinyl chloride), recycle code | | | |
| VAC | vinyl acetate (monomer) | | | |
| VBR | vinyl butadiene rubber | | | |
| VCE | vinyl chloride-ethylene resin | VC/E | | VCE |
| VC/E/MA | vinyl chloride/ethylene/methyl acrylate | VC/E/MA | | |
| VCEV | vinyl chloride-ethylene-vinyl | | | VCEV |
| VC/E/VAC | vinyl chloride/ethylene/vinyl acetate | VC/E/VAC | | |
| VCMA | vinyl chloride-methyl acrylate polymer | VC/MA | | VCMA |
| VCMMA | vinyl chloride-methyl methacrylate polymer | VC/MMA | | VCMMA |
| VCOA | vinyl chloride-octyl acrylate resin | VC/OA | | VCOA |
| VCVAC | vinyl chloride-vinyl acetate resin | VC/VAC | | VCVAC |

Table 13-8 (continued)

| Abbrevi-ation | Common polymer name or preparation | Abbreviation recommended by | | |
|---|---|---|---|---|
| | | ISO | IUPAC | ASTM |
| VCVDC | vinyl chloride-vinylidene chloride polymer | VC/VDC | | VCVDC |
| VE | vinyl ester resin | | | |
| VF/HFP | vinylidene fluoride-hexafluoropropylene copolymer | VF/HFP | | |
| VLDPE | very low density poly(ethylene) | | | VLDPE |
| VMQ | methyl silicone rubber with vinyl groups | | | |
| VPE | crosslinked poly(ethylene) | | | |
| XABS | acrylonitrile-butadiene-styrene-carboxylic group | | | XABS |
| XLPE | crosslinked poly(ethylene) | | | |
| XNBR | acrylonitrile-butadiene-carboxylic group copolymer | | | XNBR |
| XPE | crosslinked poly(ethylene) | PE-X | | XLPE |
| XPS | expandable poly(styrene) (crosslinkable) | | | |
| XSBR | styrene-butadiene-carboxylic group copolymer | | | XSBR |

Table 13-9 Common abbreviations and acronyms of monomers, oligomers, solvents, initiators, etc.

| Abbreviation | Name |
|---|---|
| 3HB | 3-hydroxybutyric acid |
| 3HB:en | 3-hydroxy-2-butenoic acid = 3-hydroxycrotonic acid |
| 3HHx5Me | 3-hydroxy-5-methylhexanoic acid |
| 3HP | 3-hydroxypropionic acid |
| 3HTD:5,8dien | 2-hydroxy-5,8-tetradecenoic acid |
| 3HV | 3-hydroxyvaler(ian)ic acid |
| 3HV:en | 3-hydroxy-4-pentenoic acid |
| 3HVUD11Br | 3-hydroxy-11-bromoundecanoic acid |
| 4HB | 4-hydroxybutyric acid |
| 4HV | 4-hydroxyvaler(ian)ic acid |
| 5HV | 5-hydroxyvaler(ian)ic acid |
| 3HB | 3-hydroxybutyric acid |
| | |
| A | alanine |
| ACSP | acetylcyclohexanesulfonyl peroxide |
| ADC | allyl diglycol carbonate |
| AF | 2,2-bis(trifluoromethyl)-4,5-difluoro-1,3-dioxolane |
| AGE | allyl glycidol ether |
| AIBN | *N,N'*-azobisisobutyronitrile |
| Ala | alanine |
| Arg | arginine |
| Asn | asparagine |
| | |
| BCP | bis(4-*t*-butylcyclohexyl) peroxydicarbonate |
| BDMA | *N*-benzyl-*N,N*-dimethylamine |
| BDOL | 1,4-butanediol |
| BHET | *bis*(hydroxyethylene) terephthalate |
| BP | bisphenol in general |
| BPA | bisphenol A, based on acetone |
| BPC | bisphenol C = 1,1-di(4-hydroxyphenyl-2,2-dichloroethene) |
| BPE | bisphenol E, based on ethanal (= acetaldehyde) |
| BPF | bisphenol F, based on formaldehyde |
| BPI | bisphenol I, based on isophorone |
| BPO | dibenzoyl peroxide |
| BTDE | 3,3',4,4'-benzophenonetetracarboxylic acid |
| BTX | benzene, toluene, xylenes (as aromatic precursors) |
| | |
| C | cysteine |
| CBT | cyclic butylene terephthalate (mixture of dimer, trimer, and tetramer) |
| CF | $\omega$-carbalkoxy-perfluoralkoxy vinyl ether |
| CHDI | cyclohexane-1,4-diisocyanate |
| CHDM | cyclohexane-1,4-dimethylol |
| CHP | cumene hydroperoxide |
| CNR | trifluoronitroso methane |
| COD | *cis,cis*-cyclooctadiene |
| CSI | chlorosulfonyl isocyanate = *N*-carbonylsulfamoylchloride |
| CTFE | chlorotrifluoroethylene monomer |
| Cys | cysteine |
| | |
| DABCO® | triethylene diamine = 1,4-*diazabi*cyclo[2.2.2]*o*ctane |
| DAC | diallyl chlorendate = diallyl 1,4,5,6,7,7-hexachloro-bicyclo[2.2]hept-5-ene-2,3-dicarboxylate |

Table 13-9 (continued)

| Abbreviation | Name |
|---|---|
| DADC | diallyl diglycol carbonate |
| DADM | *N,N*-diallyl dimethyl ammonium chloride |
| DADS | 4,4'-diaminodiphenylsulfone |
| DAF | diallyl fumarate |
| DAIP | diallyl isophthalate |
| DAM | diallyl maleate |
| DAP | diallyl phthalate |
| DCP | *exo*-dicyclopentadiene |
| DCPD | dicyclopentadiene |
| DDS | 4,4'-diaminodiphenylsulfone |
| DDSA | dodecenyl succinic anhydride |
| DGEBA | diglycidyl ether of bisphenol A |
| Dicup | dicumyl peroxide |
| DIVE | divinyl ether |
| DMT | dimethyl terephthalate |
| DPC | diphenyl carbonate |
| DPX | di-*p*-xylylene = [2.2]-*p*-cyclophane |
| DVB | divinyl benzene (mixture of isomers) |
| | |
| E | glutamic acid |
| EDCP | *endo*-dicyclopentadiene |
| EG | ethylene glycol |
| EHP | bis(2-ethylhexyl) peroxydicarbonate |
| ELO | epoxidized linseed oil |
| ENB | 5-ethylidene-2-norbornene |
| EO | ethylene oxide (= oxirane) |
| EPH | epichlorohydrin |
| EPI | epichlorohydrin (or its polymer) |
| ESO | epoxidized soybean oil |
| ET | ethene |
| | |
| F | phenylalanine |
| | |
| G | glycine |
| Gln | glutamine |
| Glu | glutamic acid |
| Gly | glycine |
| | |
| H | histidine |
| $H_{12}$MDI | hydrogenated MDI |
| HDI | 1,6-hexamethylenediisocyanate |
| HEHQ | hydroyethylhydroquinone |
| HET acid® | hexachloro-endomethylene tetrahydrophthalic acid |
| HFBA | bisphenol HF-A, based on hexafluoroacetone |
| HFDE | 4,4'-(hexafluoropropylidene)bis(phthalic acid) |
| HFP | hexafluoropropylene |
| HHPA | hexahydrophthalic anhydride |
| HX | 1,4-hexadiene |
| hyl | hydroxylysine |
| hyp | hydroxyproline |
| | |
| I | isoleucine |

Table 13-9 (continued)

| Abbreviation | Name |
|---|---|
| IBK | bisphenol IBK, based on isatine biscresol |
| ILE | isoleucine |
| IPDI | isophorondiisocyanate |
| IPP | diisopropyl peroxydicarbonate |
| K | lysine |
| L | leucine |
| Leu | leucine |
| Lys | lysine |
| M | methionine |
| MA | maleic anhydride |
| MAA | methacrylic acid |
| MAO | methylaluminoxane |
| MDA | 4,4'-di(aminophenyl)methane = 4,4'-methylene dianiline = diaminodiphenylmethane |
| MDI | methylene-diphenyl-4,4'-diisocyanate |
| MDPI | 2,4'-methylenediphenylisocyanate |
| MEKP | methylethylketone dihydroperoxide |
| MEN | methyl-*endo*-methylenehexahydronaphthalene |
| Met | methionine |
| MMA | methyl methacrylate |
| MOCA | methylene-bis(orthochloroaniline) |
| MPDA | *m*-phenylene diamine |
| N | asparagine |
| NADIC® | methylbicyclo[2.2.1]heptene-2,3-dicarboxylic anhydride isomer |
| NCA | *N*-carboxyanhydride (of $\alpha$-amino acids) |
| NDC | naphthalenedicarboxylate = naphthalene-2,6-dicarboxylic acid |
| NDI | 1,5-naphthalenediisocyanate |
| NE | 4-*endo*-methylenetetrahydrophthalic acid monomethyl ester (Nadic ester) |
| NMA | nadic methyl anhydride |
| NMMO | *N*-methylmorpholine-*N*-oxide |
| NPP | dipropyl peroxydicarbonate |
| NVP | *N*-vinyl pyrrolidone |
| OQDM | *o*-quinomethane |
| OX | *o*-xylene |
| Q | glutamine |
| P | proline |
| PA | phthalic anhydride |
| PDO | 1,3-propanediol |
| PFA | perfluoropropyl vinyl ether |
| PHDI | phenylene-1,4-diisocyanate |
| Phe | phenylalanine |
| PMDA | pyromellitic dianhydride |
| PMDI | polymeric methylene-diphenylene-4,4'-diisocyanate |
| PO | propylene oxide |
| PPDA | *p*-phenylenediamine |
| PPDI | *p*-phenylenediisocyanate |
| PPVE | perfluoropropyl vinyl ether |

Table 13-9 (continued)

| Abbreviation | Name |
|---|---|
| Pro | proline |
| PVME | perfluoromethyl vinyl ether |
| PX | *p*-xylylene = *p*-quinodimethane |
| R | arginine\ |
| Sec | selenocysteine |
| Ser | serine |
| T | threonine |
| TAC | triallyl cyanurate |
| TAIC | triallyl isocyanurate |
| TDI | *m*-tolylene diisocyanate |
| TETD | tetraethyl thiuram disulfide |
| TFE | tetrafluoroethylene |
| THF | tetrahydrofuran |
| Thr | threonine |
| THV | terpolymer from tetrafluoroethylene (TFE), hexafluoropropylene (HFP), and vinylidene fluoride (VDF) |
| TMC | bisphenol TMC, based on 3,3,5-trimethylcyclohexanone |
| TPA | terephthalic acid |
| Trp | tryptophane |
| Tyr | tyrosine |
| U | selenocysteine |
| V | valine |
| VAC | vinyl acetate |
| Val | valine |
| VC | vinyl chloride |
| VCM | vinyl chloride monomer |
| W | tryptophane |
| XDI | *m*-xylylenediisocyanate |
| XR | sulfonylfluoride vinyl ether |
| Y | tyrosine |

# 14 Subject Index

Entries are listed in strict alphabetical order; they may consist of a single word, abbreviations, acronyms, or combinations thereof. For alphabetization, technical terms consisting of two nouns were considered to be one word, whether written as two words (example: acetal polymer), with a hyphen (for example, tension-thinning), or in parentheses, brackets, or braces (example: "Catalyst, def.", comes before "Catalyst efficiency"). Qualifying numbers and letters as well as hyphens, parentheses, brackets, and braces in names of chemical compounds such as 1-, 1,4-, $\alpha$-, $\beta$-, *o*-, *m*-, *p*-, L-, D-, etc., also have been disregarded for alphabetization. Terms consisting of an adverb and a noun are arranged according to the noun (example: Molecular mass → Mass, molecular).

The following abbreviations are used: abbr. = abbreviation, def. = definition; eqn. = equation; ff. = and following; PM = polymerization, ZN = Ziegler-Natta.

## *Related Titles*

Krzysztof Matyjaszewski, Yves Gnanou, Ludwik Leibler (eds.)

**Macromolecular Engineering**

**Precise Synthesis, Materials Properties, Applications**

4 Volumes
2007
ISBN-10: 3-527-31446-6
ISBN-13: 978-3-527-31446-1

Maartje F. Kemmere, Thierry Meyer (eds.)

**Supercritical Carbon Dioxide in Polymer Reaction Engineering**

2005
ISBN-10: 3-527-31092-4
ISBN-13: 978-3-527-31092-0

Thierry Meyer, Jos Keurentjes (eds.)

**Handbook of Polymer Reaction Engineering**

2 Volumes
2005
ISBN-10: 3-527-31014-2
ISBN-13: 978-3-527-31014-2

Marino Xanthos (ed.)

**Functional Fillers for Plastics**

2005
ISBN-10: 3-527-31054-1
ISBN-13: 978-3-527-31054-8

Hans-Georg Elias

**An Introduction to Plastics**

2003
ISBN-10: 3-527-29602-6
ISBN-13: 978-3-527-29602-6

Edward S. Wilks (ed.)

**Industrial Polymers Handbook**

**Products, Processes, Applications**

4 Volumes
2001
ISBN-13: 978-3-527-30260-4
ISBN-10: 3-527-30260-3

Hans-Georg Elias

**An Introduction to Polymer Science**

1997
ISBN-13: 978-3-527-28790-1
ISBN-10: 3-527-28790-6